MARINE AUTHIGENESIS: FROM GLOBAL TO MICROBIAL

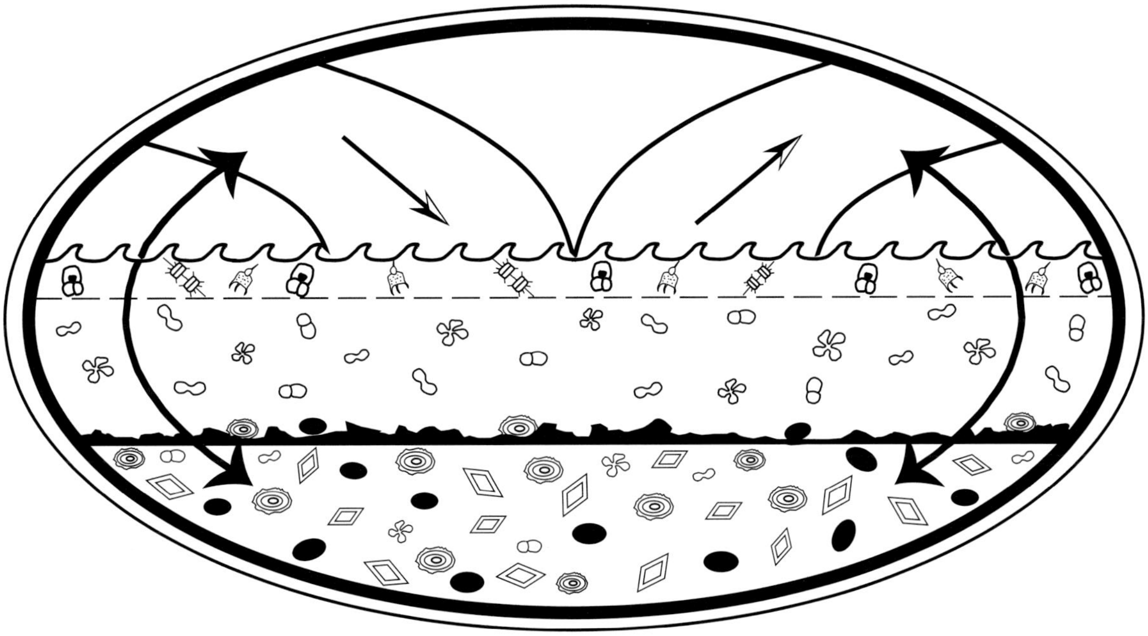

Edited by

Craig R. Glenn, Department of Geology and Geophysics, University of Hawaii

Liliane Prévôt-Lucas, Centre de Géochimie de la Surface, Strasbourg

and

Jacques Lucas, Université Louis Pasteur, Institut des Sciences de la Terre, Strasbourg

Robert W. Dalrymple, Editor of Special Publications

Special Publication No. 66

Tulsa, Oklahoma, U.S.A. *January, 2000*

ISBN: 1-56576-064-6

SEPM (Society for Sedimentary Geology)

1731 E. 71st Street

Tulsa, OK 74136-5108

1-800-865-9765

Printed in the United States

CONTENTS

Edited by
Craig R. Glenn, Liliane Prévôt-Lucas and Jacques Lucas

SECTION I:
GLOBAL MODELS AND EXPERIMENTAL STUDIES OF AUTHIGENIC MINERAL GENESIS AND CYCLING THROUGH TIME

SECTION II: PHOSPHORITES, GLAUCONITES AND ASSOCIATED FACIES ON THE MODERN SEAFLOOR

SECTION III: SEAMOUNT PHOSPHORITES AND FE-MN DEPOSITS

SECTION IV: STRATIGRAPHIC, SEQUENCE STRATIGRAPHIC AND CHRONOSTRATIGRAPHIC STUDIES OF MARINE AUTHIGENESIS

SECTION V: ORIGIN OF PHOSPHORITES, GLAUCONITES AND ASSOCIATED FACIES THROUGH TIME: CASE STUDIES

DEDICATION

Richard Sheldon

Arthur Notholt

This volume is dedicated to the late Richard (Dick) Sheldon and the late Arthur Notholt, two people who had a major impact on our knowledge of phosphate deposits.

Their methods were quite different: Dick Sheldon was the innovative thinker with ideas that challenged orthodoxy whereas Arthur Notholt was the painstaking collector of critical information. They provided the basis for a new level of understanding of phosphate deposits that was brought together (along with co-author Dave Davidson) in their major Cambridge University Press volume *Phosphate Rock Resources*. There were parallels in their careers in that both spent most of their working lives in Geological Surveys; both worked extensively in developing countries and, sadly, both died prematurely.

Dick Sheldon was born in Tulsa, Oklahoma, in 1923. He had a B.S. Degree in Geology from Yale University and a Ph.D. in Geology from Stanford University. Dick was formerly Chief Geologist of the U.S. Geological Survey in Reston, Virginia, and was widely recognized as an expert on world phosphate resources. He joined the USGS in 1947 to study phosphate deposits of the western United States. He spent most of his career with the USGS, except for a brief stint as an independent and oil company geologist. Also, from 1964 to 1965, he was a Visiting Professor of Geology at Yale University. During his career with the USGS, Sheldon served in a variety of research and managerial positions. His research publications included reports on phosphates, oil, and iron ore, and he undertook assignments in Turkey, Chile, Peru, Iran, the Middle East, and several countries in Southeast Asia, in addition to domestic geologic studies.

My first meeting with Dick was in 1962 when he first visited Australia, along with Vince McKelvey, another giant of phosphate geology. Together they assessed the phosphate potential of Australia. One of the areas to which they drew particular attention, the Georgina Basin, subsequently proved to host a series of world-class phosphate deposits. At that time I was working on phosphate deposits in central Australia and kept connected with Dick. That connection was further strengthened and became a lifelong friendship when I decided to earn my Ph.D. at the University of Colorado on the Phosphoria Formation. Dick was at that stage at the USGS in Denver and was extraordinarily generous to me, a mere graduate student, offering advice, acting as a thesis adviser, arranging for analyses to be done and even organizing for me to have access to survey pack ponies for some of the higher areas of the Rockies. Our close contacts were maintained over the next decade as we both continued to work on phosphate. I never ceased to be impressed by the incisive and perceptive nature of Dick's research. One example in particular was that in the mid 1960s, when paleomagnetism was regarded as a doubtful science and continental drift a highly questionable hypothesis, Dick threw caution to the wind, and used very uncertain data to demonstrate that phosphorites formed at low palaeolatitudes. He then used this as a basis for prospecting.

Fifteen years later, after an enormous amount of painstaking science, I (together with Mike McIlhinny) was able to show that Dick was absolutely correct in his hypothesis. Our friendship was further strengthened when in the late 1970s John Shergold and I organized Project 156 (Phosphorites) of the International Geological Correlation Programme. Dick was a strong proponent of this program and was instrumental in obtaining support for it both within the United States and internationally. One of the highlights during this period was the western U.S. field trip to the Phosphoria Formation organized and led by Dick, characterized, as always, by a winning combination of camaraderie and outstanding science.

Arthur Notholt was born in 1929 in Scotland. After a B.Sc. (Hons) at the University of Aberdeen he worked for a short period in the coal industry before joining the Colonial Geological Survey (now part of the British Geological Survey) in 1956.

Over the next 30 years he acquired encyclopedic knowledge of fertilizer, raw materials, and especially phosphate rock, in many parts of the world. Although I had met Arthur a few years previously, it was his involvement in IGC Project 156 that brought me and others into close contact with Arthur. His amiable manner and his knowledge of phosphate made Arthur a much valued member of the Project and a close personal friend with many.

Dick and Arthur developed a particularly warm friendship. They were very different people. Dick was a long, lanky Oklahoman who was always somewhere between 40 and 50 years of age even when he was 70! Arthur was short and stocky with a Scottish lilt and a quiet manner. But their differences were no impediment to their mutual respect and the readiness with which they worked together, culminating in the Cambridge University Press volume. Our science is better for their contributions and our lives have been enriched by their friendship. There can be no more appropriate people to dedicate this volume to than Dick Sheldon and Arthur Notholt.

Peter J. Cook
British Geological Survey
Nottingham, U. K.

MARINE AUTHIGENESIS, A HOLISTIC APPROACH

CRAIG R. GLENN*

Department of Geology & Geophysics, University of Hawaii, Honolulu, Hawaii 96822-2285, USA

LILIANE PRÉVÔT-LUCAS AND JACQUES LUCAS

Centre de Géochimie de la Surface (CNRS), EOST-Université Louis Pasteur, 1 rue Blessig, F-67084 Strasbourg Cedex, France

HISTORICAL PERSPECTIVE

This book is a collection of papers focusing on various aspects of authigenic and diagenetic marine minerals and related elemental cycling. In essence, it belongs to a line of specialized volumes marking major steps of continued international cooperation. Closely following the establishment of UNESCO International Geological Correlation Programme (IGCP) Project 156 (Phosphorites) in 1978, *SEPM Special Publication 29* (Marine Phosphorites, edited by Y. K. Bentor) was published in 1980 as a collection of stimulating papers stemming from a symposium on marine phosphorites held at the Xth International Congress on Sedimentology in Jerusalem in July 1978. Nearly simultaneously, a thematic set of special papers on "Phosphatic and Glauconitic Sediments" was published by the *Journal of the Geological Society* (Volume 137, 1980) following a meeting of the Geological Society on the same subject earlier that year. Led by P. J. Cook and J. H. Shergold (1978-1984) and later by W. C. Burnett and S. R. Riggs (1984-1988), IGCP Project 156 conducted 29 international and regional field workshops and symposia.

Over the next two decades, a flourishing of international research on phosphorites and their related facies ensued, with many workers attempting to solve the many "Unsolved Problems" outlined by Bentor (1980) in his seminal introduction to *SEPM Special Publication 29*. A great deal of this research was submitted for Special Publication 52 of the *Journal of the Geological Society of London*, published in 1990, and edited by A. Notholt and I. Jarvis, who organized the final International Symposium of UNESCO Project 156 in Oxford in 1988. In 1991, a successor UNESCO IGCP Project 325 titled "Correlation of Paleogeography with Phosphorites and Associated Authigenic Minerals" was initiated by J. Lucas and L. Prévôt. The title of this research group clearly illustrates the increasing desire of its researchers at that time to expand the scope of the project to the study of the origin and occurrence of other authigenic minerals. In addition to a myriad of independent publications, a number of other important books were published in association with the IGCP Projects during this period. These included a three-volume set of 151 papers published by Cambridge University Press titled "Phosphate Deposits of the World," including *Vol. 1 — Proterozoic and Cambrian Phosphorites* (Cook and Shergold, 1986), *Vol. 2 — Phosphate Rock Resources* (Notholt et al., 1989), and *Vol. 3 — Neogene to Modern Phosphorites* (Burnett and Riggs, 1990). Also published during this time frame were special thematic publications of *Sciences Géologiques* (Lucas and Prevôt, 1979, 1985; Lucas et al., 1989). Later, *Siliceous, Phosphatic and Glauconitic Sediments of the Tertiary and Mesozoic* (Iijima et al., 1994) was published following a symposium of the International Geological Congress in Japan in 1992. In 1993, members of IGCP Project 325 met in Interlaken for a meeting and workshop to discuss concepts and controversies that continued to surround the origin of marine phosphorites and this resulted in a multi-authored set of papers published in *Eclogae geologicae Helvetiae* (Föllmi, 1994).

The present volume contains many papers which were either presented at the final symposium and workshop of UNESCO Project IGCP 325 (led by J. Lucas and L. Prevôt-Lucas) titled "Deposystems of Phosphorites and Related Authigenic Minerals: Processes, Pathways and Products" in Strasbourg (October, 1996), or at the first symposium of its successor, the SEPM Research Group on Marine Authigenesis (convened by C. R. Glenn & W. C. Burnett) titled "Formation of Authigenic Marine Minerals," in Kona, Hawaii (May 1997). As with related predecessor volumes, this special publication contains a number of papers still searching for answers relating to the origin of marine phosphorites, including several papers relating to the biogeochemical cycling of phosphorus, yet it also contains a number of papers that discuss many other aspects of marine authigenic mineralization. The commingling of the various studies of authigenic minerals is partly the result of the increasing awareness that there are many overlaps, even direct associations, between different authigenic minerals, both in time and space. In compiling this volume we have taken the "holistic" attitude towards marine authigenesis, which considers the integrated whole to be more than the simple sum of its parts.

MARINE AUTHIGENESIS: A HOLISTIC APPROACH

Many marine authigenic minerals are more or less directly tied to marine biological activity, both benthic and pelagic. This implies a strong dependence on the chemical characteristics of the environment and on nutrients, which are conducive not only to high biological productivity, but also to variations of predominant species and hence to variation of their chemical composition. In addition, all authigenic minerals may play significant roles in the long-term biogeochemical cycling of elements in the oceans, and many may have been overlooked in terms of their relative importance with regard to providing a sink for reverse weathering reactions. Among nutrients, phosphorus is one of the most important so-called limiting nutrients, which explains and justifies the special interest in the P-cycle during the last decade. *Compton et al.* explore the modern P cycle and its possible variations in the geologic past. These authors gain in

**email: glenn@soest.hawaii.edu*

Marine Authigenesis: From Global to Microbial, SEPM Special Publication No. 66
Copyright 2000 SEPM (Society for Sedimentary Geology), ISBN 1-56576-064-6, p. ix-xi.

refinement on earlier established general principles which interweave climate, weathering of continents, tectonics, rate of oceanic spreading and accompanying mid-oceanic ridge hydrothermal fluxes, etc. *Subramanian* suggests in this volume that the importance of modern Asian riverine P inputs may have been overlooked in some past P-cycle models. Uncertainties in past global P cycle budgets are then further emphasized as *Colman and Holland* provide special attention to the fate of the marine P of which the return flux from its sink in seafloor muds to overlying seawater is higher than the riverine input. These authors emphasize the strong coupling between the redox state of marine sediments and the return flux of P to seawater (increasing with more reducing conditions). As P is a limiting nutrient that plays a dominant control on organic C production, and as organic C burial is a major control on atmospheric O_2 production, *Colman and Holland* also show how the marine P cycle plays an important role in stabilizing atmospheric O_2. *Guidry et al.* further discuss how global tectonics may also play an important role in the removal of phosphate from seawater by carrying a non-negligible part of P into tectonically subducted sediments. In addition, *Rasmussen* suggests that aluminophosphate minerals may be an important underestimated sink for oceanic phosphate.

Sedimentary phosphorite formation, a major global P sink, has occurred episodically over geologic time, and *Shields et al.* discuss what can be discerned regarding the origin of two of the largest grouping of these occurrences using combined sedimentary isotopic modeling (Sr, Nd, S, and C), contrasting results obtained for the Precambrian-Cambrian episode with those for the Late Cretaceous to Recent. A causal connection is put forward between metazoan evolution and widespread phosphogenesis. *Van Houten* re-examines the episodicity of phosphorite formation through time, and contrasts this with new data on the temporal variability of ooidal ironstones. He finds that many ooidal ironstones and phosphorites have many similarities, developing in similar shallow water environments experiencing slow sedimentation rates, being influenced by oceanic upwelling, and exhibiting marked evidence reworking, transport, and condensation. For about 75% of the Phanerozoic, however, these two authigenic facies are dissimilar in their temporal distribution; ironstones appear to correlate with greenhouse phases of global climate, increased continents, and subaerial weathering. Greenhouse conditions thus appear to play a perhaps subordinate control on the timing of major phosphogenesis.

The amount of available nutrients determines the rate of productivity, which utilizes various elements depending on the dominant species, delivering to sediments organic matter, carbonates, and silica in various amounts. *Fonseca* looks at the early diagenetic evolution of this material and how it depends on the ambient redox conditions or on the redox conditions induced by the sediment itself (type, sedimentation rate, etc.). *Lucas and Prévôt-Lucas* find the causes of switching between carbonate-dominated sedimentation and phosphorite-dominated sedimentation are established before deposition. They discuss the role of relative N/P nutrient ratios in determining carbonate-producing phytoplankton versus organic matter-producing naked phytoplankton. They believe when the local morphology of the shelf is favorable, the major cause of the formation of giant phosphate deposits is the type of productivity rather than an episodically enhanced global P-input. *Soudry* explains how micritic limestones built by calcification of bacterial mats and phosphorites may occur closely together in some phosphorite suites, while *Benalioulhaj et al.* conclude that, in the relationship between phosphorites and black shales, the disassociation of the two facies may occur during early diagenesis through differential preservation of organic matter on the shelf; the availability of inorganic clays is one factor favorable to the preservation of organic matter.

Comparative studies of formation and accumulation of phosphorites in various settings provide better knowledge of the required environmental conditions and prove in this volume that it is not reasonable to try to elaborate one single model. Several scenarios are proposed, including phosphatized microbial mats in the shallow water cases studied by *Schwennicke* et al., or diatomaceous, weakly-consolidated organic-rich muds developing phosphatic grains and nodules through a process of progressive phosphatization that involves compaction, crystallization of phosphatic matter along with the dissolution and expulsion of nonphosphatic components, as explained by *Baturin*. *Baturin* convincingly interprets most of the common varieties of carbonate fluorapatite mineral morphologies, including colloform masses, globules, globular aggregates with pronounced radial crystallization, rods, spindles, and dumb-bell shapes, as shown by SEM, to be diagenetic crystallites and not phosphatized (or phosphatizing) bacteria, as previously suggested by other workers. The different scenarios are even further complicated when, as pointed out by *Fountain and McClellan*, originally authigenic pristine minerals change through repeated diagenetic events, from cryptocrystalline carbonate fluorapatite rich in structural CO_3, to hexagonal apatite crystals depleted in CO_3, or to ferruginous CO_3-depleted nodules. Ultimately, more or less intense reworking is able to totally change the character of the primary sediment in which grains, nodules and other types of solid phosphate particles and associated authigenic minerals undergo sorting, displacement (transport), and sedimentological stacking. Phosphate and other authigenic mineral deposits are also subjected to allostratigraphic and sequence stratigraphic controls. As *Grimm* points out, much of the attention paid to phosphorites derives from their hybrid character, including their genesis as biogeochemical precipitates, and the puzzling array of processes that govern their stratigraphic distribution and various facies associations. Utilizing sequence stratigraphy, *Hendrix and Byers* re-examine the processes and products leading to the generation of a heterogeneous collection of sedimentary facies of the "world-class" Permian Phosphoria Formation, which contains about six times the amount of elemental P present in today's oceans. *Riggs et al.* use refined age assessments in interpreting the sea level history and paleoceanographic implications of cyclical phosphogenic events associated with yet another phosphorite giant in the Miocene sections of southeast North America. *Taylor and Macquaker* stress the importance of the variability of accommodation availability as they tackle how early authigenic/diagenetic assemblages (including calcites, siderite, dolomite, bertherine, chamosite, glauconite, and phosphates) vary spatially and temporally across various mudstone-dominated successions.

Authigenic minerals are usually found in periods and places starved of background sedimentation, such as hardgrounds and seamounts, where they accumulate in crusts and various nodules. Millimeter by millimeter isotope studies by *Burnett et al.* of "protocrusts" from the Peru margin upwelling zone indicate an extremely young age for gelatinous phosphatic to phosphoritic crusts, for which a model of authigenic growth from the diffusion of the interstitial P of muds at the sediment-water interface is proposed. Seamounts are favorable settings for phosphorite crusts (*Benninger and Hein*) and/or ferromanganese crusts (*Bertram and Cowen*), which are of similar diagenetic origin, and some of these are associated with iron-manganese and other authigenic mineral associations, including palagonite, smectite, phillipsite, and barite. An experimental investigation *in situ*, on Cross Seamount, suggests that the deposition of organic matter within microenvironments may contribute to the formation of the narrow bands of Mn- and Fe-oxides observed within crusts and nodules. As for other low-productivity phosphorite deposits, the close association between carbonate fluorapatite and Fe-Mn oxyhydroxides suggests that seamount phosphatization is facilitated by Fe-oxyhydroxide redox reactions whereby P adsorbed to Fe-oxyhydroxides is released to bottom waters or pore waters upon encountering reducing conditions. Phosphatization is characterized by mineralization fronts. Its frequent association with Fe-Mn oxy-hydroxides indicates oxic to sub-oxic conditions of genesis.

The importance of microbial activity in marine authigenesis is emphasized by *Konhauser* for a very different reason. He relates the long-debated occurrence and origin of the enigmatic Precambrian Banded Iron Formations (BIFs), one of the most abundant sedimentary deposits of the Precambrian, to bacterial biomineralization associated with hydrothermal processes. The comparison with phosphate deposits is intriguing. The results of REE studies are provided for various authigenic associations: REE in Upper Carboniferous phosphate of Mid-Continent North America by *Cruse et al.*, REE and U in modern phosphatic sediment off Goa, India, by *Nath et al.*, and REE in P-Fe-Mn crusts by *De Carlo et al.* All these results, and even those on Pt in Pacific Fe-Mn crusts by *VonderHaar et al.*, show an obvious diagenetic relationship. *Martín-Algarra and Sánchez-Navas* report on the origin of other examples of crusts in Mesozoic sediments, which they do not consider diagenetic, but biosedimentary microbial accretions (pelagic stromatolites). These pelagic stromatolitic beds, composed of phosphate and iron-manganese nodules, also contain other authigenic minerals such as barite and glauconite, which require similar conditions for formation. These minerals are minor constituents in the previous case but these become the major minerals in other facies, including the bedded barite facies as reviewed by *Jewell*, and the green clay minerals, verdine and glauconite reviewed by *Thamban and Purnachandra Rao*. Both record essential genetic characters revealing their close relationship with phosphatic facies. Hence, the authigenic green clay minerals explored by *Amorosi and Centineo* in the Lowermost Cenomanian of Cape Blanc Nez (Northern France) and by *Kronen & Glenn* in mixed carbonate-siliciclastic forereef sediments (Great Barrier Reef, Australia) play significant sedimentologic, stratigraphic, and sequence-stratigraphic roles.

Dolomite, which plays an important part in ocean-atmosphere interaction, is another authigenic mineral, the genetic conditions of which may seem far different from those of the other authigenic minerals considered above. *Arvidson et al.* revisit the "Dolomite Problem" combining a modified geochemical model of atmospheric CO_2 fluctuations over the past 100 million years with new experimental precipitation rate data for dolomite and calcite. They conclude that the decrease in the dolomite to calcite ratio observed over time is largely a result of changes in the saturation state of the oceans with respect to these minerals and global surface temperature. *Wright* addresses another aspect of the "dolomite problem," the apparent inability to synthesize dolomite in physio-laboratory experiments, and suggests that the lack of convincing hydrologic models requires new exploration of the role of benthic microbial communities in the formation of dolomite, not just in association with the now well-documented "organodolomites," but also linking the role of organic diagenesis played by the interaction between cyanobacteria and sulfate-reducing bacteria in association with the formation of the large platform dolomites of the Proterozoic and Paleozoic. However, the model of microbial dolomite proposed brings this mineral within the fold of the "phosphate family" and supports the idea of a close genetic relationship suggested by the frequent occurrence of dolomite in many phosphate deposits of the world.

University of Hawaii School of Ocean and Earth Science and Technology Contribution Number 5015.

REFERENCES

BENTOR, Y. K., 1980, Phosphorites-the unsolved problems, *in* Bentor, Y. K., ed., Marine Phosphorites-Geochemistry, Occurrence, Genesis: Tulsa, Society of Economic Paleontologist and Mineralogists Special Publication 29, p. 3-18.

BURNETT, W. C., and RIGGS, S. R., eds., 1990, Neogene to Modern Phosphorites: Phosphate Deposits of the World, Volume 3: Cambridge, Cambridge University Press, 464 p.

COOK, P. J., and SHERGOLD, J. H., eds., 1986, Phosphate Deposits of the World, Volume 1—Proterozoic and Cambrian Phosphorites: Cambridge, Cambridge University Press, 386 p.

FÖLLMI, K., ed., 1994, Concepts and Controversies in Phosphogenesis: Eclogae geologicae Helvetiae, v. 87, p. 639-788.

IIJIMA, A., ABED, A. M., and GARRISON, R. E., eds., 1994, Siliceous, Phosphatic and Glauconitic Sediments of the Tertiary and Mesozoic: Zeist, VSP BV, 248 p.

LUCAS, J., and PRÉVÔT, L., eds., 1979, Phosphorites: Strasbourg, Sciences Géologiques Bulletin, v. 32, p. 3-105.

LUCAS, J., and PRÉVÔT, L., eds., 1985, Phosphorites: Strasbourg, Sciences Géologiques Mémoires, 77, 141 p.

LUCAS, J , COOK, P. J., and PRÉVÔT, L., eds., 1989, Apatite and Phosphorites: Strasbourg, Sciences Géologiques Bulletin, 42, p. 139-266.

NOTHOLT, A. J. G, and JARVIS, I., eds., 1990, Phosphorite Research and Development: Oxford, Geological Society of London Special Publication 52, 326 p.

NOTHOLT, A. J. G, SHELDON, R. P., and DAVIDSON, D. F., eds., 1989, Phosphate Deposits of the World Volume 2: Phosphate Rock Resources: Cambridge, Cambridge University Press, 566 p.

SCIENTIFIC REVIEWERS

With deep gratitude, we extend our thanks and appreciation to the following individuals for their excellent work and scientific expertise in reviewing the papers in this volume.

Alessandro Amorosi, Universita' di Bologna
Edouard Bard, Université d'Aix-Marseille
Gleb N. Baturin, Russian Academy of Sciences, Moscow
Amos Bein, Geological Survey of Israel, Jerusalem
David J. Bottjer, University of Southern California, Los Angeles
Jean-Gabriel Bréhéret, Université de Tours
William C. Burnett, Florida State University, Tallahassee
Henry F. Chaftez, University of Houston
Norbert Clauer, Centre de Géochimie de la Surface, CNRS, Strasbourg
Max L. Coleman, University of Reading
Albert S. Colman, Yale University, New Haven
John S. Compton, University of Cape Town
Peter J. Cook, British Geological Survey, Nottingham
David S. Cronan, Imperial College, London
Walter E. Dean, U.S. Geological Survey, Denver
Gabriel M. Filippelli, Indiana University/Purdue University, Indianapolis
Karl B. Föllmi, Universite de Neuchatel
Robert E. Garrison, University of California, Santa Cruz
Ronald J. Gibbs, University of Delaware
Geoff P. Glasby, Universität-Bremen
Craig R. Glenn, University of Hawaii, Honolulu
Karen Graber, Ocean Drilling Program, College Station
Kurt A. Grimm, University of British Columbia, Vancouver
W. Burleigh Harris, University of North Carolina, Wilmington
Marc S. Hendrix, University of Montana, Missoula
James R. Hein, U.S. Geological Survey, Menlo Park
Peter Homewood, ELF-Exploration Production, Pau, France
Ellery D. Ingall, University of Texas, Port Aransas
Ian Jarvis, Kingston University, Kingston upon Thames
Kevin H. Johanesson, University of Nevada, Las Vegas
David Karl, University of Hawaii, Honolulu
David L. Kidder, Ohio University, Athens
Andrea Koschinsky, Freie Universität, Berlin
John D. Kronen, Jr., University of Hawaii, Manoa
H. Roy Krouse, University of Calgary, Canada
Lee R. Kump, Pennsylvania State University, University Park
Jacques Lucas, Université Louis Pasteur, Strasbourg
Timothy W. Lyons, University of Missouri, Columbia
Joe H. S. Macquaker, University of Manchester
Frank T. Manheim, U.S. Geological Survey, Woods Hole
John M. McArthur, University College, London
Judith A. McKenzie, Swiss Federal Institute of Technology, ETH-Zentrum, Zürich
David McConchie, Southern Cross University, New South Wales, Australia
Panagiotis Michalopoulos, State University of New York, Stony Brook
David J. Mallinson, University of South Florida, St. Petersburg
Richard W. Murray, Boston University
Yaacov Nathan, Geological Survey of Israel, Jerusalem
Kenneth Nealson, University of Wisconsin, Milwaukee
Geoffrey W. O'Brien, Australian Geological Survey Organization, Canberra
Gilles S. Odin, Université Pierre et Marie Curie, Paris
Judith T. Parrish, University of Arizona, Tucson
David Z. Piper, U.S. Geological Survey, Menlo Park
Liliane Prévôt-Lucas, Centre de Géochimie de la Surface, CNRS, Strasbourg
Jean-Pierre Prian, Bureau de Recherches Géologiques et Minières, Orléans
Jeffrey E. Richey, University of Washington, Seattle
Stanley R. Riggs, East Carolina University, Greenville
Kathleen C. Ruttenberg, Woods Hole Oceanographic Institution, Woods Hole
John H. Shergold, La Freunie, Benayes, Masseret
David Soudry, Geological Survey of Israel, Jerusalem
Peter J. Sugarman, New Jersey Geological Survey, Trenton
Bjorn Sundby, INRS-Océanologie, Université du Québec
Maurice E. Tucker, University of Durham
Jane S. Tribble, University of Hawaii, Honolulu
Ron T. Watkins, Curtin University of Technology, Perth

OCEAN/ATMOSPHERE HISTORY AND CARBONATE PRECIPITATION RATES: A SOLUTION TO THE "DOLOMITE PROBLEM"?

ROLF S. ARVIDSON

Department of Oceanography, Texas A&M University, College Station, TX 77843-3146

AND

FRED T. MACKENZIE* AND MICHAEL GUIDRY

Department of Oceanography, University of Hawaii, School of Ocean and Earth Science and Technology Honolulu, HI 96822

ABSTRACT: In this paper, we consider how long-term tectonic conditions and their effect on the surface environment of the earth interacted with the global carbonate cycle during the hothouse (greenhouse)–icehouse transition of the past 100 million years. Using the recalculated output of the Berner, Lasaga, and Garrels (BLAG) geochemical model as a template (Berner et al., 1983; Lasaga et al., 1985), we computed changes in seawater carbonate chemistry for the past 100 m.y. Experimental dolomite and calcite precipitation rate data as a function of environmental conditions were used to calculate the ratio of rates of dolomite and calcite precipitation rates during this period of time. We conclude from these model calculations that the observed decrease in the ratio of dolomite to calcite in sedimentary carbonates deposited since the Late Cretaceous transgression was a result of changes in the ocean saturation state with respect to carbonate minerals and global surface temperature. Thus, the solution to the classical "dolomite problem" may lie in relatively small but coupled changes in the composition and temperature of the atmosphere and seawater.

INTRODUCTION

The past 100 m.y. of earth history hold considerable interest for earth scientists, because this interval of time represents the long-term transition between the hothouse (greenhouse) state of the Cretaceous and the icehouse state of the present (Mackenzie and Morse, 1992). It is thought that tectonic processes play a strong role in development of these climate states and the transition from one state to another (Mackenzie and Pigott, 1981; Berner et al., 1983; Fischer, 1984).

Hothouse periods are characterized by high mid-ocean ridge spreading rates, resulting in higher eustatic sealevels, elevated volcanic fluxes of CO_2 and exchange fluxes of components involved in basalt–seawater interaction, and higher average sea surface temperatures and a reduction in latitudinal thermal gradients. These conditions also favor flooding of continental interiors, expansion of shallow water environments, and increased rates of continental weathering and delivery of dissolved components to the ocean. In contrast, icehouse periods are characterized by lower rates of seafloor spreading and CO_2 outgassing, regression of epicontinental seas, continental exposure, and cooler climates with higher poleward temperature gradients. Using data recalculated from the work of Wilkinson and coworkers (Given and Wilkinson, 1987; Wilkinson and Algeo, 1989), Mackenzie and Morse (1992) concluded that these climate states may also be reflected in the geologic record by the mass-age distribution of sedimentary carbonate mineral phases. Hothouse periods such as the mid-Cretaceous and early Paleozoic are characterized by a greater dolomite to calcite ratio than icehouse periods, notably those of the late Paleozoic and Tertiary–Recent.

In this study we examined the Cretaceous to present hothouse–to–icehouse transition in detail. To do so, we apply combined models of long-term biogeochemical cycling and laboratory-derived reaction rate data for carbonate mineral phases. The recalculated output of the Berner, Lasaga, and Garrels (BLAG) geochemical model (Berner et al., 1983; Lasaga et al., 1985) was used to constrain seawater temperature and composition (pCO_2, Ca^{2+}, Mg^{2+}, and total dissolved inorganic carbon, C_T) over the past 100 m.y. Increases in earth surface temperature result directly from increases in atmospheric CO_2 in the BLAG model, and lead to changes in seawater composition through enhanced carbonic acid weathering fluxes on land. Based on laboratory measurements of carbonate precipitation kinetics, Arvidson and Mackenzie (1997) and Arvidson (1998) concluded that the precipitation rate of dolomite increases strongly, relative to calcite, with both increasing temperature and increasing solution saturation index. We used a simple kinetic model, derived from the experimental data and driven by the BLAG model history of temperature and seawater composition to compute precipitation rates for dolomite and calcite over the past 100 m.y.

THE BLAG MODEL AND SEAWATER SATURATION STATE CALCULATIONS

Seawater composition and temperature data were taken directly from the revised BLAG model plots (Lasaga et al., 1985). Calcium, magnesium, and atmospheric CO_2 data at 10 m.y. intervals were converted from total moles to molality (moles/kg-H_2O). The BLAG estimate of dissolved carbon (total moles of HCO_3^-, calculated by charge balance) was converted to total dissolved inorganic carbon concentration. The constraints of pCO_2, total dissolved inorganic carbon, and remaining seawater solutes are sufficient to calculate a complete distribution of species and thermodynamic activities. In the calculations, described below, total calcium, magnesium, and dissolved inorganic carbon are the only bulk components that differ from the standard (modern) seawater composition. The results of our calculations of the composition of seawater using the revised BLAG model are given in Figure 1.

The calculations of saturation index with respect to ideal dolomite and calcite are given by

*email: fredm@soest.hawaii.edu

Marine Authigenesis: From Global to Microbial, SEPM Special Publication No. 66

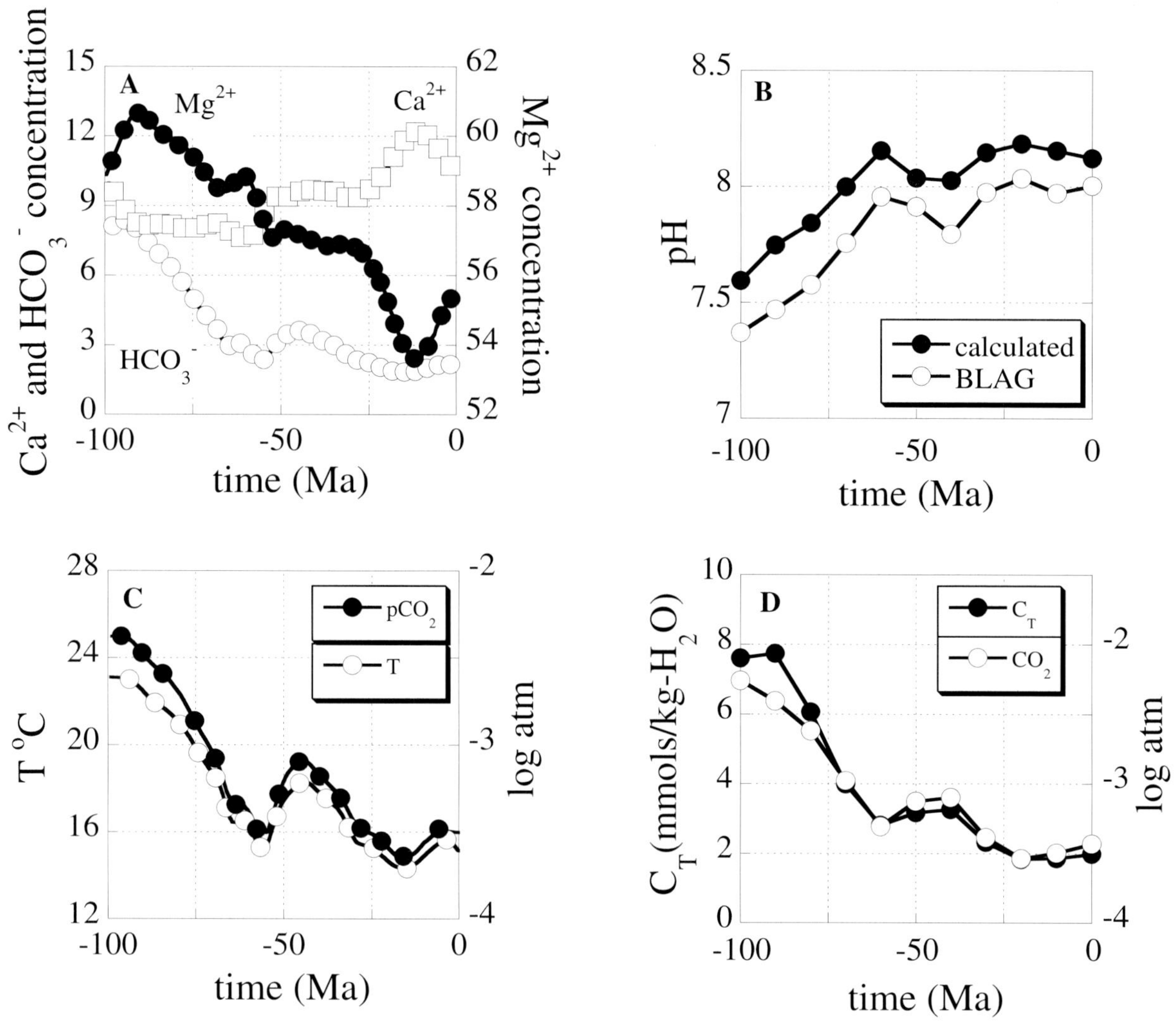

FIG. 1—Model recalculations of seawater composition for the past 100 m.y. based on the BLAG model. (a) Changes in the molal concentration of Ca^{2+}, Mg^{2+}, and HCO_3^-. (b) Comparison of pH taken from the original BLAG model with those recalculated from the BLAG model output (see text). The offset of the two curves arises from differences in the modern ocean pH used in the two models. (c) Changes in pCO_2 and temperature. (d) Comparison of changes in pCO_2 and total dissolved inorganic carbon concentration (C_T).

$$\Omega_{dolomite} = a_{Ca^2} + a_{Mg^2} + a^2_{CO_3^{2-}} / K_{dolomite} \quad (1)$$

and

$$\Omega_{calcite} = a_{Ca^2} + a_{CO_3^{2-}} / K_{calcite} \quad (2)$$

where K is the temperature-dependent equilibrium constant, and a_i is the thermodynamic activity. All thermodynamic calculations were performed using an aqueous model in which aqueous species are represented explicitly and parameterized by dissociation constants and activity coefficients. The latter were estimated using an extended Debye-Hückel formalism (Helgeson, 1967, 1969). This approach was preferred over the more conventional method of "seawater" calculation that uses stoichiometric or apparent equilibrium constants, because the concentration ranges for divalent cations and dissolved inorganic carbon differ substantially from those of normal seawater in our calculations.

Figure 1b shows our calculated pH values which are offset consistently from the estimates of BLAG by ~0.1–0.2 pH units, reflecting our choice of a modern ocean surface seawater pH of ~8.12 versus the whole ocean pH of 8 as used in the BLAG model. This difference is of small significance to our rate calculations because the change in carbonate ion activity affects both calcite and dolomite precipitation rates. The reduction in pH of seawater from the present back to the Cretaceous reflects the simple titration of the system by CO_2. The most important feature of the BLAG model in terms of carbonate mineral precipitation rates is the increase in atmospheric CO_2 with increasing age (Fig. 1c). This increase in CO_2 serves to raise atmospheric temperatures and thus enhance mineral weathering rates on land. The weathering of terrestrial silicate and carbonate rocks through the familiar reactions,

$$2CO_2 + H_2O + CaSiO_3 = Ca^{2+} + 2HCO_3^- + SiO_2 \quad (3)$$

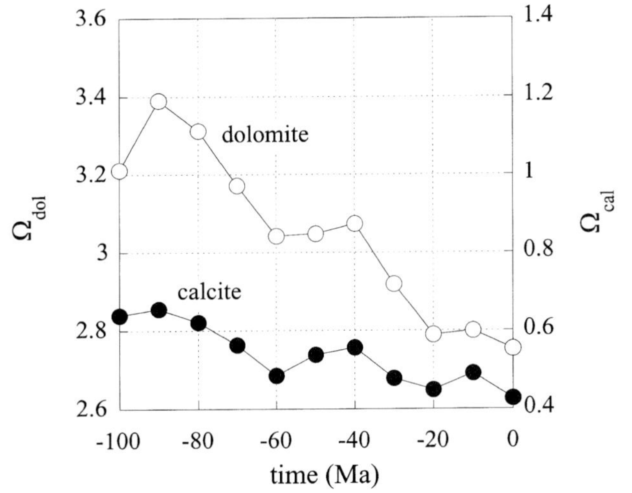

FIG. 2.—Model calculated seawater saturation index (Ω) with respect to ideal calcite and dolomite for the past 100 m.y. Notice the more modest rate of decrease in the saturation state for calcite versus dolomite with decreasing age.

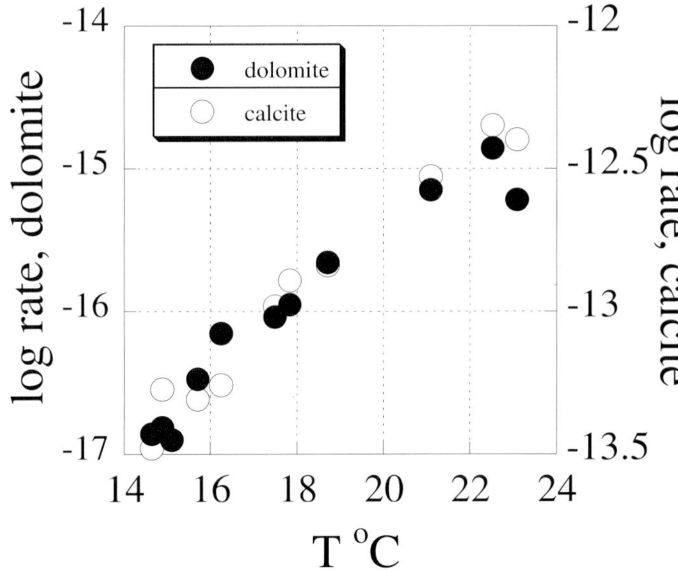

FIG. 3.—Computed dolomite and calcite precipitation rates as a function of temperature. The curves show the log moles/cm^2-sec. Notice the difference in the dolomite scale, on the left, and the calcite scale, on the right. The change in the rate for dolomite is an order of magnitude greater than that for calcite over this temperature range.

$$CO_2 + H_2O + CaCO_3 = Ca^{2+} + 2HCO_3^- \quad (4)$$

results in a transfer of dissolved carbon, mainly as bicarbonate, to the ocean. The impact of this transfer on ocean carbon chemistry can be clearly seen in Figure 1d, which compares BLAG pCO_2 to our computed total dissolved carbon concentration. As atmospheric CO_2 declined over the past 100 m.y., the global average earth surface temperature fell resulting in decreasing amounts of dissolved inorganic carbon entering the ocean via continental water runoff. This in turn has led to a decline in the total dissolved inorganic carbon of ocean water over this period of time.

Figure 2 shows the calculated seawater saturation index with respect to ideal calcite and dolomite, illustrating the general decrease in supersaturation for both phases from Cretaceous to present. Note also that the decrease for calcite is relatively modest compared to that of dolomite (0.23 versus 0.64 log units).

PRECIPITATION RATE LAW FOR CALCITE AND DOLOMITE

The precipitation kinetics of calcite have been studied both in dilute solutions (for a review of the literature, see Morse, 1983; Inskeep and Bloom, 1985) and in seawater (Mucci and Morse, 1983; Burton and Walter, 1987). The calcite rate law used here is taken from Mucci and Morse (1983) and has the simple form

$$r = k(\Omega - 1)^n \quad (5)$$

where n is the order of the overall precipitation reaction having rate r, and the rate constant k has an Arrhenius form ($k = A\exp(-e / RT)$, where e is the activation energy).

Despite much voice given to the notion that reaction kinetics are the key to understanding dolomitization reactions, there are very few data that can be represented in the above form, in which there is an explicit dependence both on saturation index and temperature. Arvidson (1998) fitted experimental data to yield a reaction order, n = 2.26, and temperature-dependent rate constant, log k = –22.3 at 25°C, for the overall precipitation reaction to form dolomite

$$Ca^{2+} + Mg^{2+} + 2CO_3^{2-} = CaMg(CO_3)_2 \quad (6)$$

The key findings from this work are the relatively high order and activation energy for the reaction in terms of equation 5, and therefore a high dependence of the rate on both saturation index and temperature. The strength of the reaction's temperature dependence is critical in the implications for the mass-age distribution of calcite and dolomite in the geologic record.

Figure 3 shows the computed precipitation rates for dolomite and calcite as a function of estimated temperature range for the past 100 m.y. (Savin et al., 1975; Berner et al., 1983; Lasaga et al., 1985). Comparison of axis scales indicates that the dolomite precipitation rate increases by about two orders of magnitude over the temperature range for the 100 m.y. interval, whereas that for calcite increases an order of magnitude less. This large difference, arising from both the higher activation energy of dolomite and its larger change in saturation index, implies a significant temporal variation in the ratio of calcite to dolomite precipitation rates. Thus, the ratio of precipitation rates of these minerals could be proportional to their relative abundance in the sedimentary record. The evolution of the calcite to dolomite rate ratio is plotted in Figure 4. The trend in ratio is very similar in pattern to the trend over the past 100 m.y. in calcite to dolomite ratios observed in sedimentary carbonates. (Given and Wilkinson, 1987; Wilkinson and Algeo, 1989; Mackenzie and Morse, 1992) With decreasing rock age, dolomite progressively becomes less abundant with respect to calcite in carbonate rocks.

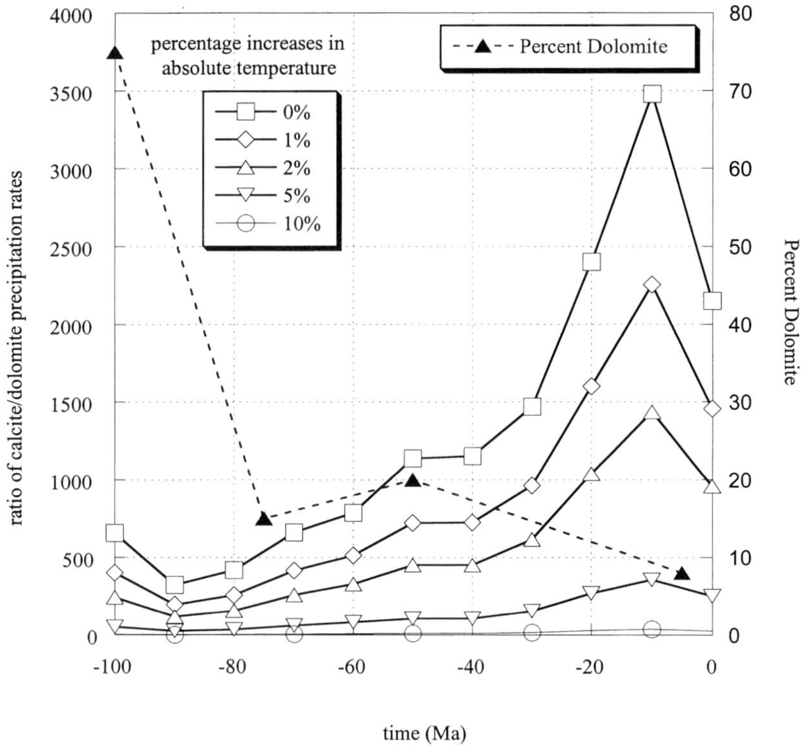

FIG. 4.—Calculated calcite-to-dolomite precipitation rates ratios (solid lines, scale on left) compared with and the relative abundance of dolomite. Large triangles show the epochal averages of dolomite as a percent of total carbonate from Given and Wilkinson (1987) (scale on right). The curve labelled 0% is the ratio calculated using the BLAG model global average temperature. The sensitivity of the ratio to temperature increases was evaluated by calculating the rate ratio at temperatures 1, 2, 5, and 10% higher than the BLAG model global average temperature (solid curves labelled 1%, 2%, 5%, and 10%). The warmer the temperature, the lower the calcite/dolomite precipitation rate ratio and thus the more likely that dolomite will be an important phase in carbonate rocks. Notice the general antithetical relationship between percent dolomite in the rock record and the calculated calcite/dolomite precipitation rate ratio.

DISCUSSION AND CONCLUSION

It is important to note that the temperature record of the BLAG model is globally averaged, and thus does not reflect latitudinal variations in sea surface temperature or the relaxation of icehouse poleward temperature gradients. During Cretaceous sealevel highstands, shallow-water environments expanded because of flooding of continental interiors at low to middle latitudes. The increase in sea surface temperature at these latitudes thus greatly expands the requisite environments of carbonate mineral deposition. Modern dolomite-producing environments are characteristically either warm, restricted regimes in which dolomite growth is promoted by increases in temperature accompanied by evaporative concentration of calcium and magnesium (commonly with an increase in Mg/Ca ratio), or those in which there is a substantial increase in seawater carbonate alkalinity accompanying sulfate reduction. For example, in their case studies of dolomite-producing environments of the Abu Dhabi sabkha of the Persian Gulf, Arvidson and Mackenzie (1997) showed that under conditions of adequate temperature (~40°C) and evaporative concentration, dolomite growth is relatively rapid. It is reasonable to assume that during the Cretaceous the increase in sea surface temperatures in marginal, shallow-water environments of continental interiors was greater than the average global temperature calculated by the BLAG model.

To evaluate the sensitivity of the calcite/dolomite precipitation rate ratio to further temperature increase, we compared the ratio calculated at absolute temperature increases of 1, 2, 5, and 10 percent. As can be seen from Figure 4, under the hothouse maximum, it is possible that the ratio of calcite/dolomite was suppressed well below the value computed using the BLAG model temperature. In other words, significantly warmer regional seas during the Cretaceous (like e.g., Abu Dhabi temperatures) could have enhanced the dolomite precipitation rate relative to that of calcite and increased the probability of a higher dolomite to calcite ratio in deposited sediments. In contrast, the loss of requisite epicontinental environments for carbonate precipitation with decreasing age up to the present makes the BLAG model temperature a more reasonable one with which to estimate the calcite/dolomite precipitation rate ratio. We recognize, of course, that carbonate phases of calcite (including the magnesian calcites) and aragonite are biologically precipitated in the skeletons and tests of organisms and dolomite is not. However, as studies of both modern and ancient carbonates demonstrate, this does not preclude dolomite from being a substantial portion of the sedimentary rock record, either as massive apparently syngenetic dolomite or as dolomite dispersed in other carbonates and in non-carbonates. Indeed, if bacteria do play a strong role in dolomite formation (Folk, 1993), one might anticipate that bacterial activity could increase with increasing temperature, and lead to rates of dolomite precipitation even greater than those calculated from abiogenic laboratory experiments (Arvidson and Mackenzie, 1997; Arvidson, 1998).

We have shown that the ratio of model rates of dolomite and calcite precipitation for the past 100 m.y., as constrained by the BLAG model history of seawater composition and temperature, is in good agreement with the observed ratio of preserved limestone and dolostone masses deposited during the same period of time. If correct, this agreement is important because it suggests that the relative abundance of dolomite in the rock record is a sensitive environmental indicator. Thus, the classical "dolomite problem" may be less of a true paradox than a result of the significant differences in environmental conditions that prevailed in the geologic past.

ACKNOWLEDGMENTS

This work was supported by the National Science Foundation (EAR 9316133) and the donors of The Petroleum Research Fund, administered by the ACS. The authors also wish to thank Albert Colman and Lee Kump for their constructive criticisms and reviews. This is the School of Ocean and Earth Science and Technology (SOEST) contribution number 4837.

REFERENCES

ARVIDSON, R. S., 1998, The kinetics of dolomite precipitation with application to changes in seawater saturation state over the past 100 Ma: Ph.D. Dissertation, University of Hawaii, Manoa, 282 p.

ARVIDSON, R. S., AND MACKENZIE, F. T., 1997, Tentative kinetic model for dolomite precipitation rate and its application to dolomite distribution, Aquatic Geochemistry, v. 2, p. 273-298.

BERNER, R. A., LASAGA, A. C., AND GARRELS, R. M., 1983, The carbonate silicate geochemical cycle and its effect on atmospheric carbon dioxide over the past 100 million years, American Journal of Science, v. 283, p. 641-683.

BURTON, E. A., AND WALTER, L. M., 1987, Relative precipitation rates of aragonite and Mg–calcite from seawater: Temperature or carbonate ion control?, Geology, v. 15, p.111-114.

FISCHER, A. G., 1984, The two Phanerozoic super cycles, *in* Berggren, W. A. and Vancouvering, J. A., eds., Catastrophes in Earth History: Princeton UP, p. 129-148.

FOLK, R., 1993, SEM imaging of bacteria and nannobacteria in carbonate sediments and rocks, Journal of Sedimentary Petrology, v. 63, p. 990-999.

GIVEN, R. K., AND WILKINSON, B. H., 1987, Dolomite abundance and stratigraphic age: Constraints on rates and mechanisms of Phanerozoic dolostone formation, Journal of Sedimentary Petrology, v. 57, p. 1068-1078.

HELGESON, H.C., 1967, Thermodynamics of complex dissociation in aqueous solution at elevated temperatures, Journal of Physical Chemistry, v. 71, p. 3121-3136.

HELGESON, H.C., 1969, Thermodynamics of hydrothermal systems at elevated temperatures and pressures, American Journal of Science, v. 267, p. 729-804.

INSKEEP, W. P., AND BLOOM, P. P., 1985, An evaluation of rate equations for calcite precipitation kinetics at pCO_2 less than 0.01 atm and pH greater than 8, Geochimica et Cosmochimica Acta, v. 49, p. 2165-2180.

LASAGA, A. C., BERNER, R. A., AND GARRELS, R. M., 1985, An improved geochemical model of atmospheric CO_2 fluctuations over the past 100 millions years, *in* Sundquist, E. T., and Broecker, W. S., eds., The Carbon Cycle and Atmospheric CO_2: Natural Variations Archean to Present: Washington, American Geophysical Union, p. 397-412.

MACKENZIE, F. T., AND PIGOTT, J. P., 1981, Tectonic controls of Phanerozoic sedimentary rock cycling, Journal of the Geological Society of London, v. 138, p. 183-196.

MACKENZIE, F. T., AND MORSE, J. W., 1992, Sedimentary carbonates through Phanerozoic time, Geochimica et Cosmochimica Acta, v. 56, p. 3281-3295.

MORSE, J. W., 1983, The kinetics of calcium carbonate dissolution and precipitation: *in* Reviews of Mineralogy: Kinetics of Geochemical Processes, v. 11: Washington, Mineralogical Society of America, p. 227-264.

MUCCI, A., AND MORSE, J. W., 1983, The incorporation of Mg^{2+} and Sr^{2+} in calcite overgrowths: influences of growth rate and solution composition, Geochimica et Cosmochimica Acta, v. 47, p. 217-233.

SAVIN, S. M., DOUGLAS, R. G., AND STEHLI, F. G., 1975, Tertiary marine paleotemperatures, Geological Society of America Bulletin, v. 86, p. 1499-1510.

WILKINSON, B. H., AND ALGEO, T. J., 1989, Sedimentary carbonate record of calcium magnesium cycling, American Journal of Science, v. 289, p. 1158-1194.

BENTHIC MICROBIAL COMMUNITIES AND DOLOMITE FORMATION IN MARINE AND LACUSTRINE ENVIRONMENTS — A NEW DOLOMITE MODEL

DAVID T. WRIGHT*

Department Of Earth Sciences, University Of Oxford, Parks Road, Oxford OX1 3PR

Abstract: Conventional hydrologically-driven models of dolomite formation, though popular, often lack empirical support, and encounter fundamental chemical problems related to kinetic impediments in saline solutions: these include the high hydration energy of the magnesium ion, the extremely low activity of the carbonate ion, and the presence of even very low concentrations of sulfate. Although an organogenic dolomite model exists, it has been mainly limited in application to modern, organic-rich, deep marine sediments. However, growing evidence from modern and ancient sediments points to a greatly enhanced and fundamental role for benthic microbial communities in dolomite formation, linked to anoxic organic diagenesis driven by sulfate reduction during which all kinetic inhibitors to dolomite formation are overcome. Magnesium concentrated in cyanobacterial sheaths may be liberated during degradation into sulfate-free solutions of high ionic strength and carbonate ion activity, to become available for dolomite formation. Another possible source of magnesium in modified saline solutions is from dissociation of complexed ions. These processes, which provide the basis for a new organogenic model for dolomite formation, have operated throughout the geologic record on a range of scales, from relatively minor modern occurrences to the thick, extensive platformal dolostones of the Proterozoic which developed in association with microbially-dominated shallow marine environments.

Analyses of lake waters and sediments from ephemeral Coorong lakes of South Australia reveal seasonally high carbonate alkalinities and magnesium concentrations in association with intense bacterial sulfate reduction and cyanobacterial degradation, indicating a genetic link between microbial mediation of ambient waters and dolomite formation. The ecosystem of the Coorong distal ephemeral lakes during late stages of evaporation provides a small scale analogue for biogeochemical processes operating in microbially-dominated shallow marine environments typical of the Proterozoic, where sulfate reduction would have been a major, shelf-wide phenomenon operating just beneath the sediment surface, continually driving biochemical modification of interstitial waters and sustaining high carbonate alkalinity in an underlying sulfate-free medium.

Thin sections of late Archaean platformal *Conophyton* stromatolites from South Africa reveal progressive disintegration of constituent cyanobacterial sheaths accompanied by the appearance and increasing crystal size of replacive, authigenic dolomite in a calcite matrix, indicating that magnesium was derived from *in situ* degradation of cyanobacterial sheaths. Total loss of sheath material resulted in a coarse dolospar fabric preserving no evidence of its microbial origin. Petrographic analysis of silicified microbial shelf sediments, including cyanobacterial mat, ooids, peloids and intraclasts, from the Cambrian Eilean Dubh Formation of northwestern Scotland reveals a common diagenetic trend in which sequential anoxic biofabric degradation resulted in the appearance and progressive increase in abundance of dolomite, culminating in a dolospar fabric. Structureless dolostones may thus evolve through a sequence of ephemeral fabrics by authigenic precipitation or replacement linked to microbial degradation, but evidence for their origin depends on the preservation of successive stages in their development from organic-rich sediments to massive dolomites. The predominance of stromatolites and other microbialitic sediments in thick dolostone platformal successions throughout much of the Precambrian and early Paleozoic, indicates that organic diagenesis has played a major role in dolomite formation from the microbial to the global scale.

INTRODUCTION

Dolomite is abundant in the rock record, and in Precambrian and Early Paleozoic rocks, is the dominant carbonate, but it becomes much rarer in Tertiary to Quaternary rocks (e.g. Given and Wilkinson, 1987). The mineral occurs in a range of scales, from micromillimetric laminae in limestones, or more rarely in siliciclastic rocks, to thick monomineralic sequences of carbonate platforms. However, after more than two centuries of intensive research and debate, the genesis of dolomite in the natural environment is still surrounded by controversy. Numerous physico-chemical models have been proposed to explain dolomite formation, in both modern and ancient environments. Most models involve dolomitization of limestone by huge volumes of brines or fluids operating in specific hydrologic systems (e.g. Adams and Rhodes, 1960; Badiozamani, 1973; Folk and Land, 1975; Machel and Mountjoy, 1987), but in spite of their popularity, none of these models provides a universal answer for dolomite formation, (for reviews, see Hardie, 1987; Tucker and Wright, 1990), nor do they adequately explain the uneven distribution of dolomite through time.

Modern dolomite is generally microcrystalline, poorly-ordered, non-stoichiometric, and therefore susceptible to diagenetic change, suggesting that geochemical signatures of older dolomites may not reflect the depositional environment. This may explain why studies attempting the integration of geochemistry and sedimentology of dolostones have not produced diagnostic criteria. Surprisingly, dolomite has not been synthesized in physico-chemical laboratory experiments at normal temperatures and pressures, even though thermodynamic considerations dictate that in the marine environment, not only should dolomite precipitate spontaneously, but any solid calcium carbonate should also be immediately dolomitized (Lippmann, 1973). The main obstacles to dolomite formation in normal seawater at sedimentary temperatures and pressures are kinetic factors affecting the behavior of the component ions of dolomite in solution. These factors are related to the tendency of ions in aqueous solutions to form hydration shells. The magnesium cation is attached to the oxygen atoms of six water molecules, which act as electric dipoles whose removal requires energy. In saline solutions, ion pairs form due to short-range interactions of adjacent ions, attracted by coulombic forces. This complexing reduces the ions' activities below their molalities, making precipitation of carbonate minerals unlikely; for example, more than 90% of total CO_3^{2-} is complexed with hydrated metal cations, chiefly magnesium (Garrels and Christ, 1965). Kinetic inhibitors to dolomite formation in surface seawater are:

**email: davew@earth.ox.ac.uk*

Marine Authigenesis: From Global to Microbial, SEPM Special Publication No. 66

1. The high hydration energy of Mg^{2+} (1926 kJ/mole) (Lippmann, 1973; Dasent, 1982; Tucker and Wright, 1990; Slaughter and Hill, 1991);
2. the extremely low activity of CO_3^{2-} anions (e.g. Garrels & Thompson 1962; Leeder 1982);
3. the presence of even very low concentrations of sulfate (e.g. Baker & Kastner 1981; Morrow and Ricketts, 1988), related to the formation of strongly bonded neutral $MgSO_4^0$ ion pairs (Slaughter and Hill, 1991).

The lack of a convincing hydrological model suggests that the search for a solution should be redirected. An appropriate approach is to consider under what conditions the kinetic constraints on dolomite formation might be overcome in aqueous saline environments, and whether those conditions could plausibly have resulted in the thick platformal dolostones abundant in the Precambrian. Most modern dolomite is found in intertidal to supratidal environments (including alkaline lakes), whereas most Precambrian dolomite was deposited in shallow subtidal marine settings. However, a common factor linking both modern and ancient dolomites is their frequently observed association with benthic microbial communities, typically stromatolites and cyanobacterial mats. There is increasing evidence that microbial degradation of organic matter, particularly the activities of sulfate-reducing bacteria (SRB) in anoxic layers beneath cyanobacterial growth surfaces, can play a critical role in the biochemical modification of ambient waters and can supply the conditions necessary for dolomite formation (Lippmann, 1973; Compton, 1988a,b; Last and De Decker, 1990; Slaughter and Hill, 1991; Wright, 1993, 1997; Vasconcelos and McKenzie, 1997, Wright and Robbins, unpub. data.). Moreover, Vasconcelos et al., (1995) have shown experimentally that the rapid formation of dolomite can be achieved in association with cultures of SRB at normal temperatures and pressure.

This paper briefly discusses problems confronting current models for dolomite formation, and presents and examines new evidence from modern lacustrine and ancient platformal marine dolomites that strongly indicates an organogenic origin of dolomite that involves modification of ambient waters by the anoxic microbial degradation of organic matter. Using this data, it is argued that benthic microbial communities have played a major role in dolomite formation throughout the geologic record.

INORGANIC MODELS FOR MODERN DOLOMITE—KINETIC PROBLEMS

When interpreting dolomite formation in the context of conventional models, consideration of fundamental chemical constraints has often been avoided or neglected. For example, it is generally held that massive dolomitization in carbonate platformal sequences requires the long-term addition of magnesium by the circulation of saline dolomitizing fluids derived from seawater. This model is commonly referred to as the *magnesium pump* because seawater is seen as the only sufficiently large reservoir of magnesium. More recently, massive dolomitization by the circulation or tidal pumping of normal seawater through carbonate platforms has also been suggested (e.g. Carballo et al., 1987; Land, 1991). However, it seems most unlikely that unaltered seawater is the dolomitizing solution because of the high enthalpy of hydration of the magnesium ion and the presence of sulfate—both effective inhibitors to dolomite formation. Lippmann (1973) argues that for dolomitization of calcite to even begin in normal seawater, all Ca^{2+} must first be removed by precipitation, and a significant increase in carbonate alkalinity would be required for it to proceed—changes in chemical composition that are impossible to maintain in normal seawater. These models also fail to explain the strong control of facies and environment on dolomite distribution that is commonly seen in dolomitic platform sequences.

A similar case can be made against the mixing zone or *Dorag* model of Badiozamani (1973) in which mixing of marine and meteoric waters produces a solution both undersaturated in calcite and supersaturated in dolomite. However, the subsequent high Mg:Ca ratios in the mixing zone are unlikely to lead to dolomitization because there is no mechanism to breach the high hydration barrier of the magnesium ion. This model also predicts that dolomitization would be concomitant with the dissolution of calcite, but as Hardie (1987) reports, this does not occur in, nor does dolomite form in, modern mixing zones.

Proposed models that invoke an evaporative mechanism, including seepage reflux (Adams and Rhodes, 1960) and evaporative pumping (Hsu and Siegenthaler, 1969) fail to address the kinetic problem. Lippmann (1973) showed that evaporative concentration would not favor dolomitization despite consequent high Mg:Ca ratios, because of the accompanying decrease in the activity of the CO_3^{2-} free ion. The activity of the CO_3^{2-} ion, which is already low, would be further lowered by complexing with cations to form neutral ion pairs, such as $MgCO_3^0$, and Mg^{2+} would also react with the sulfate ion to form $MgSO_4^0$. The nature of the mechanisms for increasing magnesium and carbonate ion activities thus remain problematic.

Hardie (1987) used the presence of sulfate in certain lake waters to argue against the effectiveness of sulfate as an inhibitor to dolomite formation. However, the experiments of Usdowski (1967), Baker and Kastner (1981) and Morrow and Ricketts (1988) provide compelling evidence for sulfate inhibition. Sulfate is of course present in seawater but in modern deep-sea organic-rich sediments, where continual diffusion of seawater SO_4^{2-} feeds and controls SRB populations, dolomite forms in sulfate-free anoxic interstitial waters immediately beneath the zone of bacterial sulfate reduction (Compton, 1988a). In deep-sea samples with no evidence of sulfate reduction, no dolomite was reported (Gieskes et al., 1982). This provides evidence that the removal of sulfate from solution by SRB may lead to dolomitization if other kinetic inhibitors are also overcome.

ORGANOGENIC DOLOMITE

Existing models for marine organogenic dolomite formation in organic-rich sediments in deep-water depositional settings are driven by sulfate reduction (e.g. Irwin, 1980; Baker and Burns,

1985). Both Burns et al. (1988) and Compton (1988b) attribute the nodular dolomite in the Miocene Drakes Bay and Monterey Formations, California to biochemical changes driven by SRB in interstitial waters. The organogenic dolomite is generally distinguished by having a significant proportion of its carbon derived directly from organic carbon (Murata et al., 1969; Claypool and Kaplan, 1974), which is characterized by a widely variable carbon isotope signature that reflects changes in the carbon isotope ratio ($\delta^{13}C$) of dissolved CO_2 caused by sulfate reduction and methanogenesis. Compton (1988a) noted that traces of Holocene dolomite in Deep Sea Drilling Project (DSDP) cores from the Gulf of California (Gieskes et al., 1982) appeared to be restricted to organic-rich hemipelagic sediments, in which very low sulfate concentrations were concomitant with high carbonate alkalinity. He considered that dolomite could precipitate in such conditions without a carbonate precursor, but precluded the application of the organogenic model to thick platformal carbonates on the grounds that bacterial degradation of organic matter could not generate sufficient alkalinity for the formation of significant amounts of dolomite. However, Compton (1988a) did not consider the potential for major changes in water chemistry on ancient shallow-marine shelves with widespread benthic microbial communities.

The potential for radical modification of ambient waters by SRB is likely to be much greater in microbially dominated ecosystems. Such systems were typical of shallow carbonate shelves in Precambrian times. No modern carbonate platform can provide a suitable analogue, but a study of processes operating in the dolomitic distal ephemeral lakes of the Coorong region of South Australia (see below), in which SRB flourish in the late stages of evaporation, provides data applicable to microbially dominated restricted shallow-marine shelves.

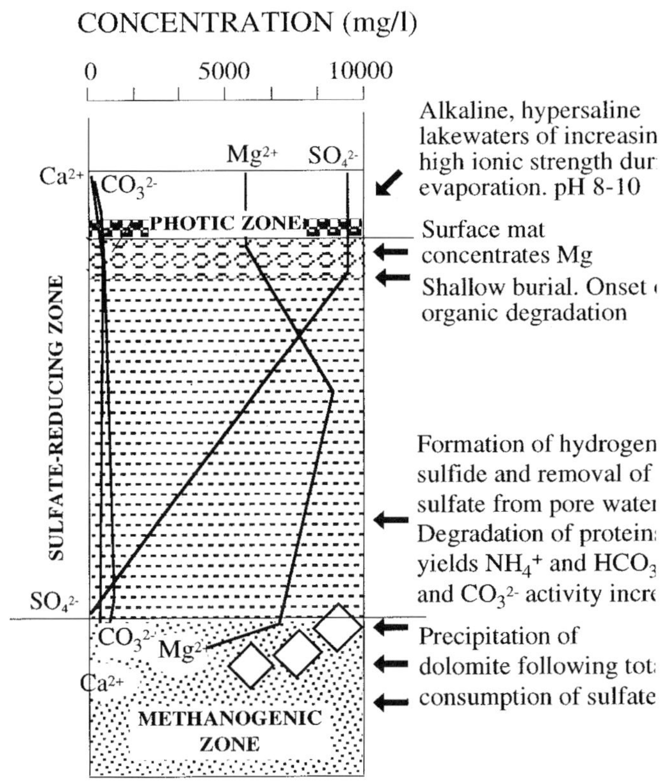

Fig. 1. Schematic representation of generalized trends with depth of calcium, magnesium, carbonate, and sulfate ions. in organic-rich sediments below a modern benthic microbial community (surface mat). Biochemical activity by microbes induces changes in pore water composition. Sulfate concentrations decline sharply within the sulfate-reducing zone, accompanied by a significant increase in free carbonate ion and ammonia, while Mg^{2+} concentrations decrease to zero at the base of the zone, below which dolomite may be found. This schematic is based on data from the Coorong distal lakes (Wright, D. T., unpub. data), from Deep Sea Drilling Project interstitial water profiles in organic-rich sediments (Gieskes et al., 1982), and from Jørgensen (1982).

BACTERIAL SULFATE REDUCTION — CHEMICAL CONSEQUENCES

SRB are among the oldest organisms known, and were widespread in a range of Precambrian environments. Today, they are ubiquitous. They are able to colonize any suitable anoxic environment, and have wide tolerances of temperature, salinity, pH, and ultraviolet radiation (Trudinger, 1992). They are found as constituents of depth-stratified benthic microbial mats, mainly in buried sediments containing organic matter, and in the bottom waters of restricted marine basins and lakes. Mat stratification, which is related to internal community structure, is organized along physical and chemical gradients resulting from the metabolic activities of constituent microbes (Des Marais, 1990). SRB are overlain by autotrophic bacteria which, in turn, are capped by a layer of photosynthetic cyanobacteria; relative depths fluctuate in response to oxygen and hydrogen sulfide concentrations (e.g. Revsbech et al., 1983). As dead material becomes buried by new surface mats, anoxic conditions quickly prevail in the sulfate-reducing zone, which passes down to a zone of methanogenesis (Glenn and Arthur, 1988). SRB are particularly effective in sediments of shallow enclosed seas, where up to 95% of their productivity takes place between 0.1 and 3 cm (Jørgensen, 1982; Canfield and Des Marais, 1991; Wright and Robbins, unpub. data). The consumption of organic matter in anoxic sediments is related to sulfate concentration in ambient waters (Stams, 1985) and to the type of organic matter: prokaryotes and marine algae are richer in amino acids than higher plants (Huc, 1980). In freshwater sediments, organic matter is converted mainly to methane and CO_2. Sulfate is the terminal electron acceptor in the anaerobic oxidation of organic compounds by SRB (Slaughter and Hill, 1991):

$$2CH_2O + MgSO_4^0 \longrightarrow 2HCO_3^- + H_2S + Mg^{2+}.$$

SRB remove sulfate, an important kinetic inhibitor of dolomite formation, in the uppermost layers of anoxic, organic-rich sediments and can anaerobically degrade amino acids completely to CO_2 (Hensgens, 1996). Ammonia is liberated in the process (Stams, 1985). Ammonia concentrations and carbonate alkalinity associated with progressive microbial degradation increase with depth (Fig. 1). Ammonia is highly soluble and is the only natural base which can increase the activity of the carbonate ion on a large scale (Berner, 1980) by raising the pH of ambient waters ($NH_3 + H_2O \longrightarrow NH^{4+} + OH^-$). Bicarbonate buffers the solution, leading to increased carbonate ion concentrations ($OH^- + HCO_3^- \longrightarrow H_2O + CO_3^{2-}$) (Slaughter and Hill, 1991). Continued organic degradation will tend to maintain or increase OH^- and CO_3^{2-} levels until inhibitors to carbonate precipitation are overcome.

The dissociation exponent of the acid (pKa) of the HCO_3^- CO_3^{2-} acid-base system can be used to calculate the concentrations of carbonate species in solutions at a given pH. In normal seawater, pH is less than pKa and the bicarbonate ion (HCO_3^-) is the dominant carbonate species (Fig. 2A). However, in more alkaline solutions, in which pH is greater than pKa, the carbonate ion (CO_3^{2-}) predominates. Moreover, the more saline the solution, the higher the concentration of CO_3^{2-} relative to HCO_3^- at a given pH (Pytkowicz, 1983). This relationship between pH and pKa was found in distal ephemeral lakes of the Coorong area. For example, in hypersaline Pellet Lake (Fig. 3) a pH of 8.46 was measured and a pKa of 7.5 was calculated from titrations of lake water samples on-site (Fig. 2B). Because the pH is greater than the pKa, carbonate ion should be the dominant carbonate species and carbonate ion, as a percent of total carbonate, will be greater than the that shown for pH=8.46 in Fig. 2A.

Stable carbon isotope data can provide information about the carbon source in carbonate rocks. Dissolved carbon derived from organic degradation is typically enriched in ^{12}C because of fractionation by photosynthetic plants, and carbonates precipitated from waters enriched in dissolved organic carbon typically show negative $\delta^{13}C$ values. However, dolomites interpreted to be of microbial origin in ancient platformal environments (e.g. Wright, 1997) carry only a very slightly negative to neutral stable carbon isotope signature. This can be explained by the fact that HCO_3^- is abundant in seawater, and carbon derived from this source is isotopically heavier than carbon derived from organic diagenesis. In response to increased alkalinity driven by sulfate reduction, HCO_3^- dissociates as described above, thereby increasing the proportion of carbonate containing isotopically heavier carbon with respect to the isotopically lighter carbon derived from organic diagenesis. Canfield and Des Marais (1993) report that even in microbial mats growing in evaporative salt ponds, dissolved inorganic carbon (DIC) from diffusion from overlying water is greater than dissolved organic carbon (DOC) produced from sulfate reduction. As a result, organogenic dolomite need not have a strongly negative signature. Furthermore, increased CO_3^{2-} alkalinity from sulfate reduction may be more rapid than dolomite formation, becoming fixed as hydrous magnesium carbonates in pore spaces. This may be later dissolved by porous flow containing a normal marine isotopic composition for carbonate species, so that ensuing dolomitization need not confer a depleted isotopic signal (Lippmann 1973). Another possibility is that the isotopic signature of the dolomite may be inherited from precursor carbonate. It is also worth noting that cyanobacteria in hypersaline brine habitats can record $\delta^{13}C$ values as high as –3‰ (Schidlowski et al., 1984); when mixed with inorganic carbonate, the organic contribution in any precipitated carbonate mineral may be masked.

The activities of SRB can cause profound changes in ambient water chamistry. Because H_2S is a very weak acid, it will have little effect on increased pH, and may be removed from the sediment by bubbling or fixation by iron. Both the removal of the sulfate ion and increased carbonate alkalinity must move the chemical equilibrium towards dolomite formation by diminishing kinetic encumbrances. Evidence is presented below that in the dolomitic Coorong ephemeral lakes of South Australia, SRB totally consume initially high but finite sulfate concentrations prior to dolomite formation during the latest stages of evaporation; where sulfate remains in solution, dolomite does not form.

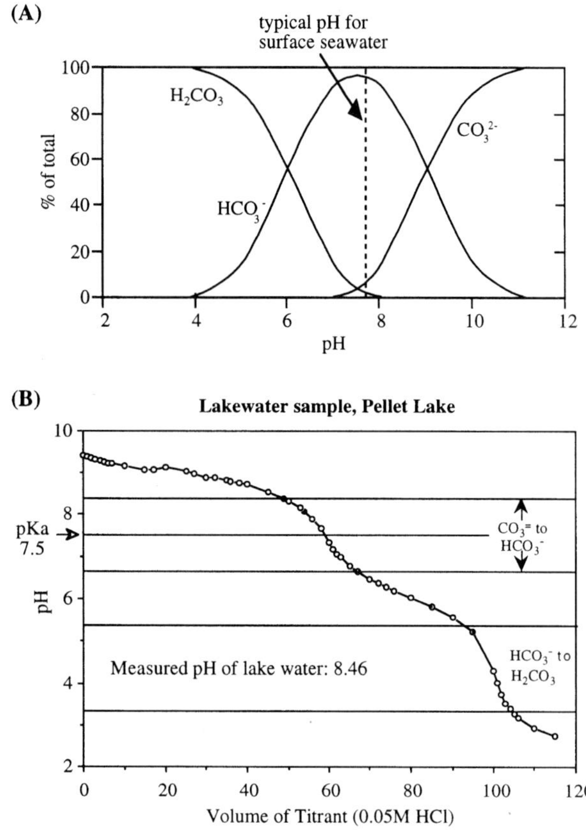

FIG. 2. (A) Graph of carbonate species concentration versus pH for the carbonic acid-carbonate system in seawater (after Pytkowicz, 1983). From this graph, it can be clearly seen that in order to increase CO_3^{2-} ion concentration, the pH of the system must be increased. However, in alkaline solutions of higher ionic strength, the dissociation exponent (pKa) of HCO_3^-, a measure of H+ concentrations, is lower (Slaughter and Hill, 1991). Thus, in solutions of higher ionic strength, the curves in (A) would be shifted to the left and at a given pH, the percent CO_3^{2-} would be higher and the percent HCO_3^- would be lower. (B) A pH-titration plot of a brine sample from the dolomitic distal ephemeral lake, Pellet Lake, from the Coorong, South Australia. Location is shown in Fig. 3. From the titration curve, the pKa of HCO_3-is estimated to be 7.5, which is the midpoint of the steep section of the titration curve that reflects the carbonate-bicarbonate conversion. The pKa (7.5) is an order of magnitude lower than the measured pH of 8.46 so that the carbonate concentration exceeds that of bicarbonate by an amount greater than that calculated from pH alone (using Figure 2a), and will be a factor favoring dolomite precipitation in sulfate-free brines.

COORONG DOLOMITE

Previous Interpretations

Despite many studies (e.g. Alderman, 1959; Von der Borch,

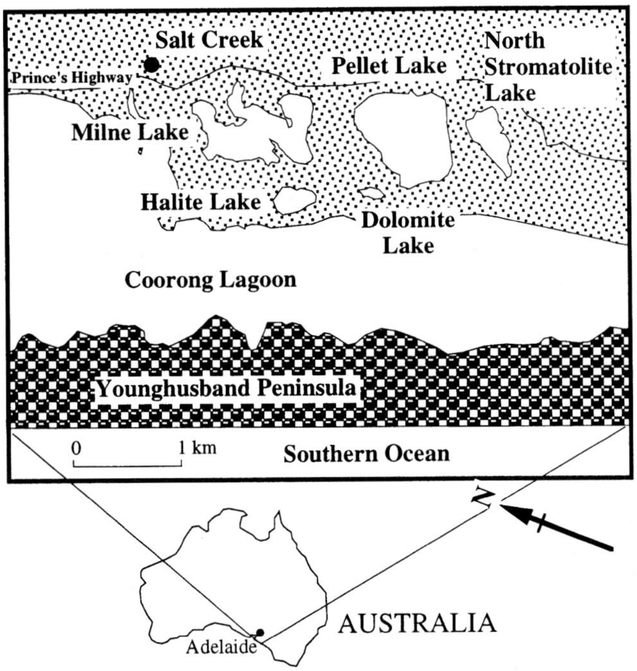

FIG. 3. Locality map showing distal ephemeral lakes near Salt Creek, Coorong area, South Australia. South Stromatolite Lake, also studied, lies in a similar setting 10 km southeast of Salt Creek. Halite Lake is connected to the Coorong Lagoon by a subterranean channel, and is the only lake in the study area that has typical evaporative sedimentary mineralogy—aragonite, gypsum, and halite.

1965, 1976; Von der Borch and Lock, 1979; Botz and Von der Borch, 1984; Warren, 1988) in which the dolomites have generally been interpreted as evaporative in origin, a satisfactory mechanism for dolomite formation in the Coorong lakes remains elusive. The modern sediments of adjacent distal ephemeral lakes in the Coorong may show very different mineralogies. Experiments involving evaporation of the ephemeral lake waters under laboratory conditions have consistently failed to produce dolomite (Von der Borch, pers. commun., 1995, 1996), probably due to the operation of kinetic impediments referred to above (Lippmann 1973). The dominant mineral precipitated in such experiments was aragonite, which was not found in any of the lake sediments in which dolomite was the major mineral (Wright and Robbins, unpub. data).

Methods

In the Coorong, sediment cores were obtained using 4-cm-diameter plastic piping and/or sample tubes. Lake water samples were filtered on site using 0.45-mm filters and preserved in gluteraldehyde in sealed containers; duplicates were analyzed both at CSIRO (Adelaide) and WSP laboratories, Nottingham, England. Lake and pore water pH readings were taken in situ in the field, thus avoiding sample contamination or alteration during conventional laboratory analysis, and in the laboratory for comparison, using an Orion A290 pH meter and Ross Sureflow electrode. Titrations of filtered and unfiltered lakewater samples were performed at the field base at Salt Creek (Fig. 3). Magnesium and calcium concentrations were obtained by atomic absorption; sulfate concentration was measured by ion chromatography. Carbonate alkalinity was determined by titration at CSIRO (Adelaide). The delay during transportation of the samples was likely to have resulted in underestimations of carbonate alkalinity. In addition, semi-quantitative Merck analytical test strips were used for rapid field measurements of sulfate and ammonia in lake and interstitial waters. Mineralogy was determined by XRD. Carefully disaggregated powders were mounted and run on a Philipps HT Generator Model PW 13 20/00 at the Department of Earth Sciences, University of Oxford, using a Diffractor PW 1050.

Microbial Degradation and Modern Dolomite Formation in The Coorong Lakes

In the distal ephemeral lakes (Fig. 3), dolomite sediment is invariably associated with benthic microbial communities, and prior to desiccation, comprises a gelatinous, organic-rich mud that smells strongly of H_2S. Together, H_2S and the presence of authigenic pyrite framboids (e.g. Rosen et al., 1988) indicate active sulfate reduction. SEM analysis confirmed that the dolomite typically consists of spherical aggregates, generally <1μm in size, or 'pellets' which themselves comprise submicron aggregates of grains. TEM and geochemical studies (Rosen et al., 1988) indicate that the dolomite precipitated directly and rapidly from solution, probably from interstitial waters just before total desiccation. The spherical grains are morphologically similar to spherical dolomite grains produced in microbial experiments by Vasconcelos et al., (1995). These grains were

TABLE 1.—CHEMICAL ANALYSES OF COORONG LAKE WATER SAMPLES, DECEMBER 1995

Lake	Mg^{2+} (mg/l)	Ca^{2+} (mg/l)	Mg:Ca Ratio	CO_3^{2-} (mg/l)	SO_4^{2-} (mg/l)	pH	Mineralogy
Pellet	5,280	165	32	21.68	9,650	8.461	dolomite
Milne	5,406	25	216	37.7	9,680	8.749	dolomite, magnesite
S. Stromatolite	28,720	112	256	12.4	73,400	7.363	hydromagnesite, aragonite
Halite	29,620	165	179	10.6	28,300	7.811	aragonite, gypsum, halite
Seawater	**1,313**	**400**	**3.2**	**16.2**	**2,688**	**7.9**	

produced by the mineralization of nannobacteria in a bacterial sulfate-reducing medium and, by analogy, suggests that a similar origin may be considered for the Coorong dolomite. Interestingly, Vasconcelos and McKenzie (1997) proposed a similar anoxic microbial model for dolomite precipitation in Lagoa Vermelha, Brazil.

All lake waters in the Coorong study area (Fig. 3) in late Spring are electrolytic solutions of high conductivity. Analyses of lake waters show major differences between the chemistries of dolomitic and non-dolomitic lakes during late stages of evaporation (Table 1), though it should be recognized that chemical parameters for all lake waters are in a constant state of flux as the system responds to episodic freshening through rainfall and run-off dilution during the evaporative cycle. Lake waters from dolomitic lakes are characterized by higher pH (always >8), elevated concentrations of carbonate and magnesium, and compared to the other lakes, have relatively low levels of sulfate. In the non-dolomitic South Stromatolite Lake, water analyses are characterized by extremely high sulfate and magnesium concentrations coupled with relatively low carbonate concentrations and, on most days, relatively low pH levels (although pH ranged up to 9.7 on some days). Despite high sulfate concentrations, gypsum was not found in any of the studied distal lakes (confirmed by XRD), even though its' precipitation would be expected to accompany straightforward evaporation. The presence of halite crusts directly overlying dolomitic and non-dolomitic muds in desiccated lakes (Fig. 4A, B) negates the possibility that gypsum was somehow dissolved and flushed from the system. Halite Lake is located in the interdunal flat near Salt Creek (Fig. 3), but differs from other lakes in the study area by having a submarine connection to the Coorong Lagoon, and in there being no record of sulfate reduction. Its sedimentary fill is a typical evaporative suite of aragonite, gypsum, and halite.

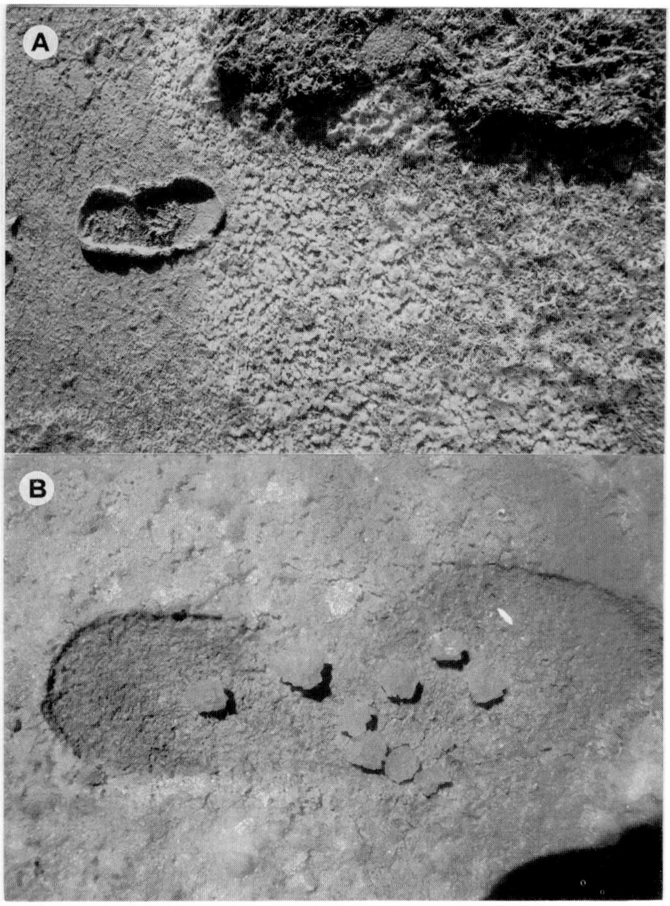

FIG. 4. Desiccated ephemeral lakes with dolomitic mud (A) and hydromagnesite-aragonite mud (B) directly overlain by thin halite crust. The absence of gypsum, despite total evaporation of sulfate-rich lake water, is attributable to complete sulfate reduction. Footprint is about 11 in. (28 cm) long.

Towards a Microbial Dolomite Model

The organogenic model as currently understood (e.g. Compton, 1988a) acknowledges the role of SRB in the removal of sulfate and increased alkalinity on a limited scale, but does not address the kinetic constraints posed by hydration of the magnesium ion. A commonly held view, based on thermodynamic considerations, is that high Mg/Ca ratios promote dolomitization (e.g. Braithwaite 1991), though empirical reports have raised some doubts on this (e.g. Illing et al., 1965; Baker and Kastner, 1981). In the Coorong distal ephemeral lakes, magnesium concentrations in the dolomitic lakes are more than four times greater than in normal seawater. However, the most remarkable feature is the enormous concentration of magnesium in non-dolomitic lakes, reaching almost 30,000 mg/l in South Stromatolite Lake, together with extremely high Mg:Ca ratios (Table 1). Clearly, a high Mg:Ca ratio is insufficient by itself to cause dolomite formation and a process able to dehydrate magnesium ions (or inhibit their hydration) appears to be necessary. A potential source of available magnesium would be magnesium released from complexed ion pairs during sulfate reduction in evaporative solutions of high ionic strength (Slaughter and Hill, 1991):

$$CH_2O + MgSO_4^0 \longrightarrow 2HCO_3^- + H_2S + Mg^{2+}.$$

An alternative source of magnesium, which has not yet been considered in models of dolomitization, is cyanobacteria, their sheaths in particular, which are known to concentrate magnesiumby up to four times its abundance in seawater (Gebelein and Hoffman, 1973). Stromatolites are present in the lakes, and their constituent cyanobacteria will therefore remove magnesium from lake waters throughout their growth. Desiccation of stromatolites during retreat of lake waters near the end of the evaporative cycle occurs in both the dolomitic and non-dolomitic lakes.

The unusual chemical compositions of the Coorong lake waters can be attributed to bacterial sulfate reduction in the uppermost sediments, which through organic degradation promotes high pH values accompanied by high carbonate ion concentrations and activities (Fig. 5). In residual pools of desiccated dolomitic lakes, sulfate was not detected by analytical test strips which are sensitive to the presence of sulfate in solutions containing < 200 mg/l, suggesting that sulfate-reducing bacteria removed all the water sulfate as the lakes dried out. This inter-

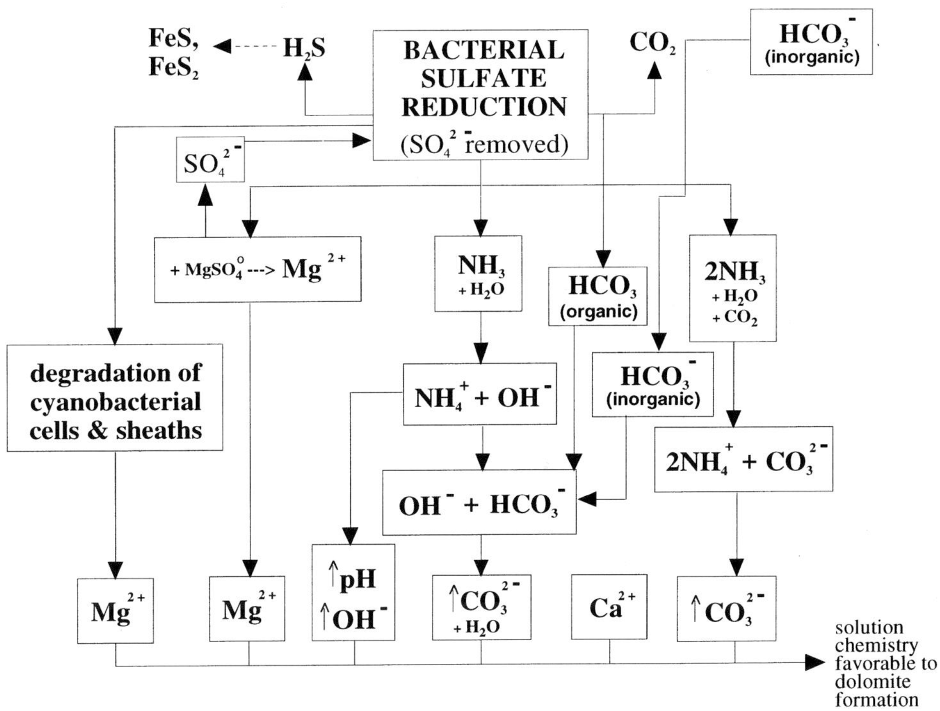

FIG. 5. Flow chart showing modification of ambient waters by bacterial sulfate reduction leading to chemical conditions necessary for dolomite formation.

pretation is supported by the fact that no gypsum ($CaSO_4$) was precipitated. During degradation, the magnesium locked up in cyanobacterial cells and sheaths may be rapidly liberated into the modified lake waters. Because Mg-hydration is not necessarily instantaneous in strong electrolytes, newly released magnesium cations might be available for dolomite formation before forming ligands with water dipoles. No pumping mechanism is required. In non-dolomitic South Stromatolite Lake, sulfate ions persist in lake and interstitial waters throughout desiccation, probably because the sulfate concentration is so high at the beginning of the evaporative cycle (Table 1). The persistence of sulfate may thus explain why aragonite and hydromagnesite are precipitated rather than dolomite (Bathurst, 1975; Berner, 1980; Lyons et al., 1984; Walter, 1986) and the absence of gypsum suggests that the sulfate may be present largely as neutral ion pairs, i.e. $MgSO_4^0$. The data from Halite Lake show calcium concentrations at about the same level as in the dolomitic Pellet Lake (Table 1), suggesting that gypsum would precipitate in the other, isolated, lakes, if mineralogy was solely the result of evaporation.

Bacterial sulfate reduction thus has the potential to provide the chemical conditions necessary for dolomite formation wherever benthic microbial communities, including cyanobacteria, dominate a saline aqueous environment. In alkaline pore waters of high ionic strength, all the kinetic inhibitors to dolomite formation in the immediate environment can be overcome. Raised CO_3^{2-} activities result in high pH, and sulfate-free solutions enable the CO_3^{2-} ion to displace water dipoles from Mg^{2+} ions at dolomite crystal surfaces (Lippmann, 1973). Magnesium may also be supplied in unhydrated form from the degradation of cyanobacterial sheaths (Murata et al., 1972) and from the dissociation of complexed ions following modification of ambient waters by bacterial sulfate reduction.

These microbiogeochemical processes are constrained temporally only in the sense that Metazoans have suppressed microbes in most environments during most of the Phanerozoic. Whenever benthic microbial ecosystems dominated the environment, as they did throughout the Proterozoic and at times during Phanerozoic, the potential for widespread bacterial sulfate reduction and anoxic microbial dolomite formation would have existed. Petrographic evidence that widespread dolomitization was associated with organic degradation in ancient rocks, and that cyanobacterial sheaths were a source of magnesium, is reported below.

PETROGRAPHIC EVIDENCE FOR ANCIENT MICROBIAL DOLOMITE

Relatively deep-water subtidal microbialites from the latest Archean Gamohaan Formation (2521±3 Ma; Sumner and Grotzinger, 1994) of South Africa, include conical stromatolites 10-12 cm high, comprising a calcitic column crossed by irregular, continuous dolomitic laminae ranging in thickness from <1–100 mm. Petrographic examination of thin sections of the upper part of these stromatolites shows that a sparry matrix is traversed by elongate, thin, filamentous structures from 15 µm to >25 µm in diameter and many hundreds of microns in length, commonly intermeshed in strands or bundles traversing the field of view (Fig. 6A). The filaments have been identified as fossil microbial sheaths (*Siphonophycus transvaalensis* Beukes, Klein and Schopf), and they are seen in various states of preservation in both cherts and carbonates (Klein et al., 1987). In the best preserved examples, the outline of the sheath is clearly defined,

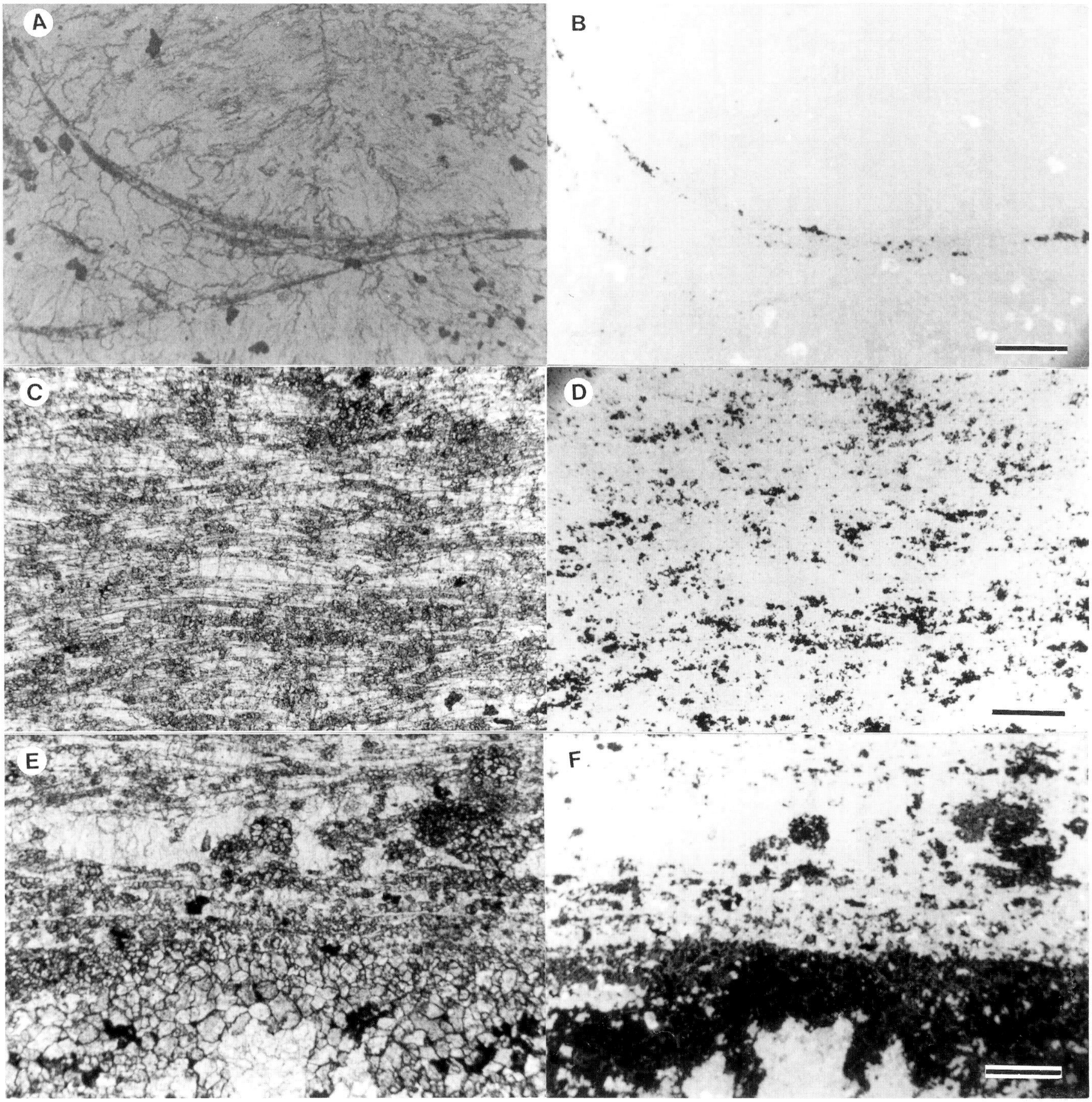

Fig. 6. Thin-section photomicrographs of stromatolites of the Gamohaan Formation (Archaean) in plane-polarized light and the same fields of view under cathodoluminescence. **(A)**. In the upper part of the stromatolite, the sparry matrix is traversed by permineralized filamentous cyanobacterial sheaths. The outline of the sheaths can be clearly seen. Very fine, high-relief, anhedral dolomite crystals are attached to the outer surface of the filaments. Scale bar = 360μm. **(B)** Same field of view as (A). Under cathodoluminescence, dolomite crystals along the filaments appear dark purple (dark gray–black in the photomicrograph) in stark contrast to the pale yellow (light gray in the photomicrograph) of the calcitic matrix. Dolomite is restricted to the surfaces of the cyanobacterial sheaths. No dolomite is found floating in the matrix. Scale bar = 360μm. **(C)** Photomicrograph of sample at a lower level within the stromatolite Filamentous sheaths are more numerous and are closely associated with tiny dolomite crystals. Dolomite crystals are more abundant and slightly larger where sheaths are less well defined. Laterally, discernible sheaths grade into discontinuous laminae and patches of dolomicrite. Plane-polarized light; scale bar = 360μm. **(D)**. Cathodoluminescence of same field of view as in (C). Compare distribution and abundance of the dolomite (purple luminescence, appears dark gray–black in the photomicrograph) with respect to the locations and orientations of cyanobacterial filaments in D. Scale bar = 360μm . (E), (F) Photomicrograph of sample from the lower part of the stromatolite in plane-polarized light. **(E)** and under cathodoluminescence. **(F)**. Scale bar = 200μm. Relatively well-preserved sheaths near the top of the photomicrograph are associated with sparse dolomite. In the center of the photomicrograph, sheaths are recognizable but less well defined, and the dolomite crystals are larger and more abundant. In the lower part of the photomicrograph, sheaths are not discernible, but their former presence is suggested by abundant coarse, turbid dolomite spar forming a hypidiotopic fabric. This sequence represented in Figures 6A–6F is interpreted to reflect progressive dolomitization in association with sheath degradation and illustrates how a structureless dolomitic fabric may develop, bearing no clue to its microbial origin.

and careful inspection may reveal the presence of tiny micritic crystals apparently attached to the outer surface. Under cathodoluminescence, these crystals have a dark purple color (Fig. 6B), which strongly contrasts with the pale yellow of the calcitic matrix, facilitating identification and possibly indicating differences in mineralogy. These crystals are not found dispersed in the matrix where the sheaths are absent. In the interior of the stromatolitic column, the sheaths are less well defined, becoming increasingly encrusted by darkly-luminescing crystals ranging in size from micrite to about 25 μm (Fig. 6C, D). Towards the base of the stromatolite (Fig. 6E, F), sheaths may grade laterally into micritic to microsparry laminae. Lower down, sheaths are barely discernible, but their former presence is indicated by replacive, turbid, brownish, inclusion-rich crystals (Hudson, 1975) up to 100 μm in size, producing a hypidiotopic, coarse crystalline texture. This culminates in a totally replacive, sparry fabric in which no remaining sheath material is visible. It thus appears that the replacive mineral conforms to the original distribution of sheaths of filamentous cyanobacteria involved in stromatolite construction, and that crystal size and abundance increases as preservational detail of the sheaths diminishes.

*Mineralogy.*_
Microprobe and XRD analysis of the Gamohaan dolomite reveals that the darklyluminescing mineral is near-stoichiometric, but rather poorly ordered, dolomite. The Mg:Ca ratio in crystals attached to the sheaths was 45:55 (0.81), whereas in coarser crystals, the ratio is higher 48:52 (0.92). An X-ray diffraction trace (Fig. 7) shows the major reflection peak for dolomite at 30.99° 2θ, corresponding to *d*(104). Subordinate dolomite peaks (110) and (113) appear on the trace in the region of 37.4° and 41.2° 2θ. However, the principal order reflection (015) at 35.3° 2θ is absent, indicating that the dolomite is rather poorly ordered and, in that respect, it resembles modern dolomite.

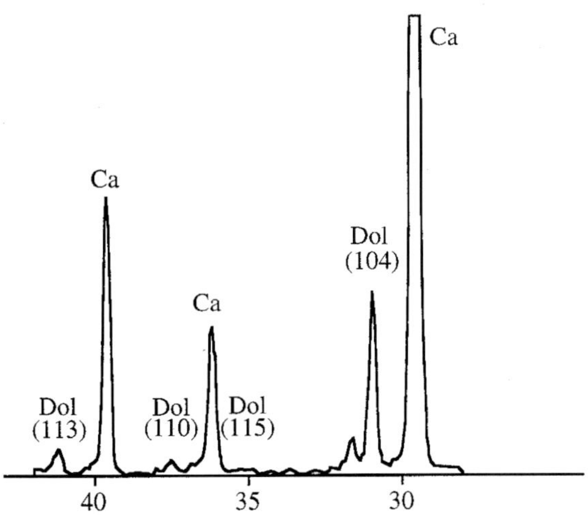

FIG. 7. X-ray diffractogram trace (28° to 42° 2θ) of a bulk powdered sample from a stromatolite from the Gamohaan Formation. The major (104) dolomite reflection peak at 30.9° is accompanied by subordinate (110) and (113) peaks. But the absence of the (015) peak at about 34.8° 2θ indicates that the dolomite is poorly ordered.

*Organic degradational mechanism.*_
The Gamohaan stromatolites display textural and mineralogical features consistent with *in situ* dolomitization of calcium carbonate in association with microbial degradation of cyanobacterial matter. The variable structural detail of the sheaths indicates their differential degradation prior to stabilization of the fabric. The style of encrustation of the sheaths by dolomicrite appears to be analogous to the early diagenetic calcification of partially degraded cells in modern cyanobacterial mats and stromatolites in which mineralization forms molds of felted groups of sheaths (e.g. Horodyski and Von der Haar, 1975; Monty, 1976; Gerdes and Krumbein, 1987). The close physical association between early-formed dolomite and less well preserved cyanobacterial sheaths and the absence of dolomite in the calcitic mineral matrix, strongly indicates that conditions favoring dolomite formation existed within these organic-rich laminae. Support for this hypothesis is provided by Gebelein and Hoffman (1973), who suggested from the known ability of cyanobacteria to concentrate magnesium in their sheaths, and from calculations based on laboratory experiments, that the slow release of magnesium from the long term degradation of the sheaths could result in 1-mm-thick authigenic dolomite laminae in 2-mm-thick cyanobacterial layers .

A genetic link between anoxic sheath degradation and dolomite formation is therefore postulated, whereby sulfate reduction led to the release of unhydrated magnesium into a partially mineralized, chemically modified, fluid microenvironment, resulting in the dolomitization of the calcitic matrix where it was in contact with relict organic-rich laminae. In the early stages of dolomitization, dolomicrite formed on the sheath surface, whereas advanced degradation was accompanied by an increase in the abundance and size of the dolomite, culminating in a coarse, dolospar fabric. The sequence of degradation and dolomite formation as seen in Fig. 6 suggests that massive, structureless dolomite bearing no clue to its organogenic origin may develop because dolomitization is ultimately fabric destructive. This microbially mediated diagenetic dolomitization of calcium carbonate reveals the operation of a process which has the potential to dolomitize significant volumes of limestone, and would be constrained only by the abundance of cyanobacterial matter and the timing of stabilization of the rock fabric.

Widespread Paleozoic Microbial Dolomite

Dolomite is commonly associated with other microbial carbonate facies. Stages in the taphonomic history of microbial mats and ooids have been preserved in fossiliferous cherts of the Cambrian Eilean Dubh Formation of northwestern Scotland (Wright, 1993, 1994, 1997). Better-preserved mats are not associated with carbonate, but the progressive loss of pigmentation and structural detail, indicating sequential degradation, is accompanied by the appearance of and increase in abundance of dolomite. Further evidence of a genetic link between cyanobacterial degradation and dolomite formation is provided by cherts in which oolitic grainstone fabrics are preserved. A diagenetic

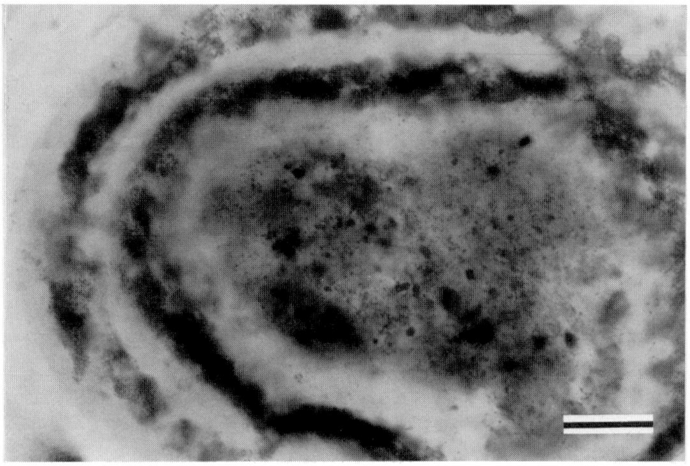

FIG. 8. Photomicrograph in plane-polarized light showing ooid with abundant organic matter (i.e. organic degradation is not well advanced). Dolomite is absent. Dark organic-rich laminae that alternate with pale laminae around a peloidal nucleus reflect periodic colonization of the ooid by epilithic cyanobacteria during accretion. Scale bar = 40 μm.

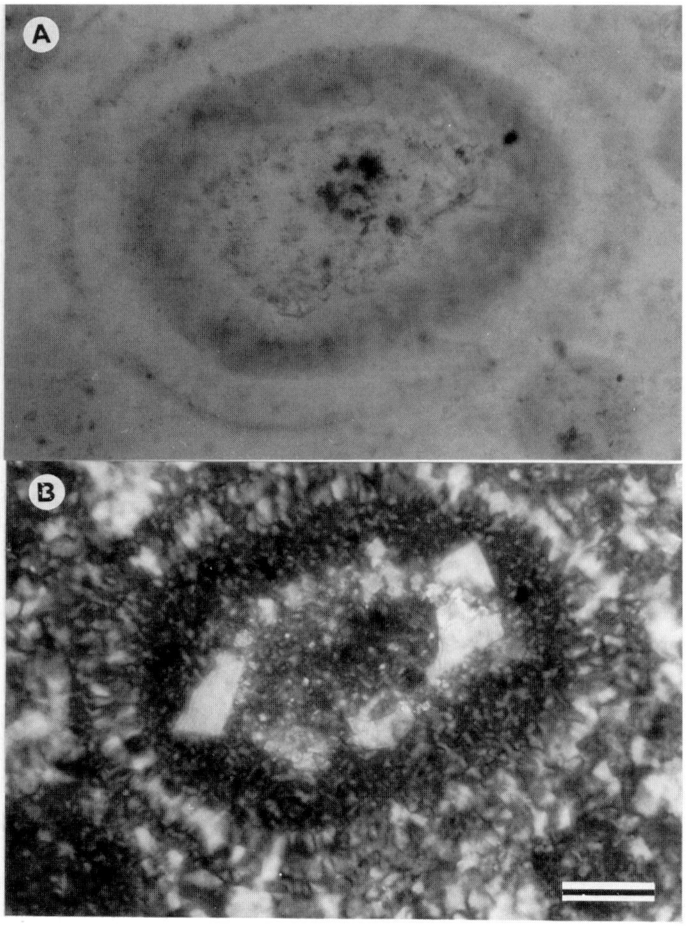

FIG. 9. Photomicrographs of a silicified ooid with intragranular dolomite rhombohedra (large subhedral crystals) at the core-cortex boundary in (A) plane-polarized light and (B) cross-polarized light; scale bar = 40 μm. The pale brown color of the microcrystalline silica in the inner cortex (A) indicates the former presence of organic matter. Organic diagenesis accompanied by cortical leaching produced conditions necessary for dolomite precipitation and crystal growth within the ooid shell.

sequence can be inferred from the various types of organic-rich ooids. Ooids in which organic degradation is not well advanced are not associated with dolomite (Fig. 8). Progressive degradation of organic material, together with dissolution of the ooid shell, is accompanied by an increasing abundance of intragranular dolomite. The unvarying location of dolomite rhombohedra or dolomicrite within ooids at the core-cortex boundary, or in the peloidal nucleus, indicates that the dolomite precipitated and grew there (Fig. 9A, B). Dolomite is not associated with later pore-filling cements. In many cases, clusters of dolomite are enclosed within ooids whose former existence is revealed by a microcrystalline silica replacement visible only under cross-polarized light (Fig. 9A, B). The ooids appear to have provided a microenvironment particularly favorable to dolomite formation, so that each ooid became a "dolomite production unit" in which all the ingredients and conditions necessary for dolomite formation were concentrated. Complete dissolution of the crystal-bearing ooids resulted in the release of a package of dolomite crystals into the sedimentary environment. Upon consolidation, the dolomite sediment became preserved in the geologic record as a dolostone with a structureless, crystalline fabric. In the absence of silicification, only the final product of these processes, a dolomitic rock fabric bearing no clue to its organogenic origin, has survived in the rock record, as seen in the host rock of the cherts (Fig. 10).

MICROBIAL DOLOMITE—A NEW DIMENSION

In the Coorong, lake waters are freshened during the winter and the lakes are colonized by aquatic higher plants such as the *Ruppia maritima*, and Metazoa. These are killed off during summer by increasing salinity, and probably also by H_2S toxicity as microbes, including cyanobacteria and SRB, are alone able to survive. In the absence of burrowing infauna, SRB migrate upwards to near the sediment surface and flourish in the hypersaline conditions (Wright and Robbins, unpub. data). Microbial domination of the lake environment is thus seasonal.

FIG. 10. Photomicrograph of 'structureless' dolostone from the host bed of oolitic chert. The 'structureless' dolostone is interpreted as an accumulation of intragranular (intra-ooid) dolomite grains that were released upon dissolution of host ooids. Scale bar = 100 μm.

Nevertheless, the benthic microbial communities, in which sulfate reduction drives profound changes in ambient water chemistry favorable to dolomite formation, may be considered analogous to cyanobacterially-dominated microbial ecosystems of the past in terms of generic composition and metabolic process. The Eilean Dubh Formation provides an ancient analogue for the abundant, thick, shallow marine Proterozoic dolostone sequences in which stromatolites and other microbially dominated facies reached their maximum development. Microbiogeochemical modification of ambient waters in various organic-rich facies can account for sub-millimeter dolomitic laminations in stromatolites, as well as significant thicknesses of shelf facies including ooidal-peloidal shoals. Silicified 'windows' show that many apparently structureless dolostones may have resulted from authigenic dolomite precipitation or dolomite replacement of precursor carbonate during organic diagenesis in sediments. A model for dolomite formation genetically linked with anoxic microbial degradation is therefore postulated. The model is applicable on a range of scales and environments throughout the biological record, wherever modification of ambient waters was driven by SRB in association with benthic cyanobacteria. Where vast populations of benthic cyanobacteria-dominated ancient, shallow epicratonic seas, ideal conditions for the large-scale biochemical modification of depositional waters may have existed, thus promoting widespread formation of dolomite, especially in inner shelf areas. In the seaward parts of carbonate shelves, where greater exchange with open marine seawater would progressively restore sulfate levels and depress pH and carbonate ion activities, calcium carbonate would dominate and dolomite formation would then be restricted to localized sites of anoxic microbial degradation, such as the organic-rich laminae in the Gamohaan Formation stromatolites. This facies relationship is illustrated by microbialites of the Upper Cambrian Petit Jardin Formation of western Newfoundland: in the inner shelf, the microbialites are dolomitic, whereas seaward, in the mid-shelf area, they become increasingly calcitic (Chow, 1986; Wright, 1993).

The microbial model has the potential to resolve intrinsic contradictions in interpretations of documented dolomitic occurrences in Proterozoic and Phanerozoic rocks (e.g. Tucker, 1982), and to illuminate the environmental effects of microbial modification on ecosystems in the past (Tucker, 1986, 1992). The scarcity of shallow-marine environments suitable for mass colonization by benthic cyanobacteria during most of the Phanerozoic, combined with competitive exclusion by Metazoans, would have reduced the opportunities for widespread dolomite formation, and may thus explain the uneven distribution of dolomite through time. Although the microbial model has not been applied to most Recent dolomite occurrences (with the notable exception of Vasconcelos and McKenzie, 1997), this association would appear to merit closer investigation.

A Review Of Some Modern Dolomite Occurrences

Since the 1950's, dolomite has been reported in a variety of Holocene intertidal and supratidal settings, deep-sea marine locations, and inland alkaline lakes; representative examples are considered below. Most of the occurrences are associated with organic-rich environments, but there is frequently a lack of pore water sulfate and carbonate concentration data .

Abu Dhabi —

One of the clearest modern associations of dolomite and algal mats was reported by Illing et al. (1965), who found dolomite in the subsurface sabkha of the Persian Gulf. Evaporation has led to the precipitation of aragonite and gypsum (Curtis et al., 1963). The prograding sabkha has progressively buried the algal mats of the intertidal environment. The depth at which dolomite occurs increases from near-surface adjacent to the coastline to 0.7 m inland, where it is above the water table, but still at the same level as the decomposing algal laminated sediments beneath the prograding sabkha. At this level, most of the dolomite occurs as rhombs within pellets that range in size from 1 to 5 μm. The pellets are progressively replaced by dolomite. Where the replacement is complete and the original form obliterated, only a microcrystalline texture remains. The abundance of aragonite decreases as the abundance of dolomite increases, leading to the assumptions that the dolomite is a replacement of aragonite (e.g. McKenzie et al., 1980). Petrographic evidence, however, does not unequivocally support that interpretation. In places, euhedral aragonite encrusts dolomite rhombohedra, indicating that the aragonite post-dated the dolomite and elsewhere, dolomite encrusts aragonite, with no evidence for replacement or absorption (Hardie, 1987).

Surface water temperatures range from 40°C to 50°C, reaching 60°C in the porewaters of the sabkha subsurface under dark, porous algal mat, with intense evaporation (Bathurst, 1975 p. 206). Although Illing et al. (1965) considered magnesium-rich subsurface brines of the uppermost tidal flat to be one of the most important factors leading to dolomitization of the original aragonite, sulfate concentrations are high and dolomite is not found. Bathurst (1975) explained the dolomitization as the result of heavy brines percolating downwards through the pore system, with the depth of dolomitization being related to depth of penetration. Apart from kinetic objections to evaporative dolomitization discussed above, if dolomitization was the result of downward percolating brines, dolomite should crosscut the various facies and should be most extensive near the surface. However, dolomite is absent in the sabkha facies, and in the sabkha subsurface it is concentrated in the sub-horizontal layer of buried intertidal algal laminites. Dolomite is also found in the underlying subtidal facies where sulfate is presently pervasive (McKenzie, pers. commun., 1997); however, this dolomite may be related to downward percolation of dolomitizing fluids generated by sulfate-reduction in the overlying algal mat.

McKenzie et al., (1980) invoke a 3-stage hydrological cycle. Brine chemistry evolution, however, can be simply explained by evaporative concentration and precipitation of gypsum, and need not involve dolomite formation at all (Hardie, 1987). In subsurface brines, SO_4^{2-} concentrations decrease landwards, from >15 g/kg to about 1 g/kg with the precipitation of gypsum

(Hardie, 1987). Sulfate concentrations may be further diminished in the subsurface algal laminites by bacterial sulfate reduction and degradation of cyanobacteria, which may produce ideal conditions for dolomite formation. The role of benthic microbial communities should therefore be considered and tested in future research (McKenzie, personal communication, 1997).

Andros, Bahamas.—

In Andros, dolomite is found in the subsurface of palm hammocks. Palm hammocks are supratidal mounds that consist of laminated pelleted muds formed by intermittent flooding and growth of intertidal algal mats. On their lower flanks (30–60 cm above mean high water) are widespread crusts, 2–3 cm thick, containing 20–80% dolomite. Chemical information from these deposits is unavailable, but Bathurst (1975) reported chemical data from dolomite-rich sediments deposited in a similar environment (Shinn, 1968) in the Florida Keys. In the Florida Keys deposit, pore-waters, squeezed from dolomite-rich sediments preferentially colonized by cyanobacteria, were about six times more saline than normal seawater, and had a Mg:Ca ratio greater than 40. Shinn et al. (1965) described the deposit as having a fenestrate fabric, with abundant organic matter and small crystals (1-2 μm) of poorly ordered dolomite with an average composition of Ca_{55}:Mg_{45}. Interestingly, Shinn et al. (1965) comment that dolomite in mollusc shells may not be replacive, but the product of numerous algal borings. They were perplexed by the high pore-water Mg:Ca ratios, stating simply that an ". . .outside source of Mg must exist." A cyanobacterial source would seem plausible.

Bonaire, Netherlands Antilles.—

Deffeyes et al. (1965) described near-surface crusts and unconsolidated supratidal sediments from Bonaire containing about 80–95% well-ordered dolomite. The dolomite grain size was about 2 μm and stoichiometry ranged from Mg_{46} to Mg_{44}. Carbon isotope (^{14}C) dating gave an age of 1,500–2,200 years old. The pore water Mg:Ca ratio was 6.2–11.4 and sulfate levels were reportedly very low. Although Deffeyes et al. (1965) proposed seepage-reflux to explain the dolomite which they interpreted to be a replacement, the proximity of microbial populations in which organic matter can be rapidly converted to organic carbon by high rates of sulfate reduction (Lyons et al., 1994) suggests possible microbial involvement in precipitation of authigenic dolomite at Bonaire.

Lake Hayward, Australia.—

In Western Australia, dolomite is reported from a small permanent hypersaline lake in a setting similar to that of the Coorong of South Australia (Rosen and Coshell, 1992). The euhedral dolomite forms as both a diagenetic cement and as a replacement mineral. It occurs below 60 or 70 cm within a "laminated lacustrine unit" in which algal mats are common and is typically associated with bacterial bodies. The authors (Rosen and Coshell, 1992) acknowledge that the source of magnesium is problematical, but cyanobacterial sheaths would appear to be a possible source. The association of dolomite, bacterial populations, and buried algal mat strongly suggest that a role for microbes in the formation of the Lake Hayward dolomite should be investigated.

These diverse modern examples of dolomite formation all share an association with benthic microbial communities. It can thus be envisaged that the same processes involved in dolomite formation in the past are operating today under similar conditions: the major difference is the scale on which it occurs—a phenomenon which can be related to the comparative abundances and environmental distribution of benthic cyanobacterial communities through time.

CONCLUSION

Bacterial sulfate reduction in modern microbial sediments can cause significant changes to the chemical composition of ambient saline solutions, such that kinetic inhibitors to dolomite formation can be overcome. Dolomite may form as an authigenic phase beneath the zone of sulfate reduction, or as a direct precipitate in alkaline lakes in which sulfate has been entirely consumed. These modern microbiogeochemical processes are analogous to those of ancient microbially dominated environments, but operate on a much more limited scale. In this sense, the present is the key to the past.

It is important to visualize the nature and scale of microbial ecosystems in ancient times. In the absence of higher forms of life during the Precambrian and at times during the Phanerozoic, benthic microbial communities colonized every available environmental niche and were essential constituents of a variety of allochems including ooids, peloids, and intraclasts. Under such conditions, the impact of sulfate-reducing bacteria, cyanobacteria, and other microbes on the chemical and depositional environment was likely to be profound and sustained. Continuous anoxic degradation of organic matter just below the sediment-water interface generated continuously high carbonate alkalinity in underlying sulfate-free pore waters, and maintained massive concentrations of magnesium ions derived from cyanobacterial sheath degradation and possibly from the dissociation of Mg^+ from $MgSO_4^0$. A microbial model for widespread dolomite formation across broad areas of marine shelves in which removal of the kinetic inhibitors to dolomite precipitation is genetically linked to organic degradation, is postulated.

Formation of microbial dolomite involves progressive mineralization during the taphonomic evolution of organic material, and typically culminates in a massive dolomitic fabric preserving no clues to its origin. However, preservational windows allow glimpses at earlier stages in the development and destruction of ephemeral bio-mineralic fabrics and provide evidence for a microbial origin of widespread dolomite deposits formed in a variety of environments throughout the biological record. A review of some modern dolomite occurrences previously ascribed to inorganic, hydrological models suggests that the benthic microbial communities in the depositional environment may be a factor in dolomite formation. Recognition of the potential impact of organic diagenesis on dolomite formation from a microbial to a global scale, combined with an appreciation of the shortcomings of conventional inorganic models, should stimulate renewed research and debate from which a new paradigm may emerge.

ACKNOWLEDGMENTS

I would like to record my sincere thanks to Chris Von Der Borch of Flinders University, Adelaide, for logistical and scientific support during field data acquisition in the Coorong. Thanks also go to Lisa Robbins (University of South Florida), to Robbie Lennard (Flinders University), to Adrian Beach and Megan Easterbrook of CSIRO, Adelaide, and to Adrian Colquoun of WSP, Nottingham, England for laboratory support. I held many fruitful discussions with Matthew Lloyd of Dyson Perrins Laboratory, University of Oxford, who has immeasurably improved my understanding of water chemistry. I wish to record my sincere thanks to Judy McKenzie and Max Coleman for their advice and critical reviews of the manuscript.

I am also grateful to Nic Beukes for providing stromatolite specimens from the Gamohaan Formation, and to Martin Brasier at Oxford University for his encouragement and involvement. The study visit to the Coorong was funded by the Royal Society, whose financial support I gratefully acknowledge.

REFERENCES

ADAMS, J. E. AND RHODES, M. L., 1960, Dolomitization by seepage refluxion: American Association of Petroleum Geologists Bulletin, v. 44, p. 1912–1921.

ALDERMAN, A. R., 1959, Aspects of carbonate sedimentation: Journal of the Geological Society of Australia, v. 6, p. 1–10.

BADIOZAMANI, K., 1973, The Dorag dolomitization model—application to the Middle Ordovician of Wisconsin: Journal of Sedimentary Petrology, v. 43, p. 965–984.

BAKER, P. A. AND BURNS, S. J., 1985, Occurrence and formation of dolomite: American Association of Petroleum Geologists Bulletin, v. 69, p. 1917–1930.

BAKER, P. A. AND KASTNER, M., 1981, Constraint on the formation of sedimentary dolomite: Science, v. 213, p. 214–216.

BERNER, R. A., 1980, Early Diagenesis: a Theoretical Approach: Princeton, Princeton University Press, 241p.

BOTZ, R. W. AND VON DER BORCH, C. C., 1984, Stable isotope study of carbonate sediments from the Coorong area, South Australia: Sedimentology, v. 31, p. 837–849.

BRAITHWAITE, C. J. R., 1991, Dolomites; a review of origins, geometry and textures: Transactions of the Royal Society of Edinburgh: Earth Sciences, v. 82, p. 99–112

BURNS, S. J., BAKER, P. A. AND SHOWERS, W. J., 1988, *in* Shukla, V. and Baker P. A., eds., Sedimentology and Geochemistry of Dolostones: Society of Economic Paleontologists and Mineralogists Special Publication 43, p. 41–52.

CANFIELD, D. E. AND DES MARAIS, D. J., 1993, Biogeochemical cycles of carbon, sulfur, and free oxygen in a microbial mat: Geochimica et Cosmochimica Acta, v. 57, p. 3971–3984.

CARBALLO, J. D., LAND, L. S. AND MISER, D. E., 1987, Holocene dolomitization of supratidal sediments by active tidal pumping, Sugarloaf Key, Florida: Journal of Sedimentary Petrology, v. 57, p. 153–165.

CHOW, N., 1986, Sedimentology and diagenesis of Middle and Upper Cambrian platform carbonates and siliciclastics, Port au Port Peninsula, western Newfoundland: Unpublished Ph.D. Dissertation, Memorial University of Newfoundland, St. John's, 458 p.

CLAYPOOL, G. E. AND KAPLAN, I. R., 1974, The origin and distribution of methane in marine sediments, in Kaplan I. R., ed., Natural Gases in Marine Sediments: New York, Plenum Press, p. 99–140.

COMPTON, J. S., 1988a, Degree of supersaturation and precipitation of organogenic dolomite: Geology, v. 16, p. 318–321.

COMPTON, J. S., 1988b, Sediment composition and precipitation of dolomite and pyrite in the Neogene Monterey and Sisquoc Formations, Santa Maria Basin area, California, in Shukla, V. and Baker P. A., eds., Sedimentology and Geochemistry of Dolostones: Society of Economic Paleontologists and Mineralogists Special Publication 43, p. 53–64.

CURTIS, R., EVANS, G., KINSMAN, D. J. J. AND SHEARMAN, D. J., 1963, Association of dolomite and anhydrite in the Recent sediments of the Persian Gulf: Nature, v. 197, p. 679–680.

DASENT, W. E., 1982, Inorganic Energetics (2nd ed.): New York, Cambridge University Press, 185p.

DEFFEYES, K. S., LUCIA, F. J. AND WEYL, P. K., 1965, Dolomitization of Recent and Plio-Pleistocene sediments by marine evaporite waters on Bonaire, Netherlands Antilles *in* Pray, L. C. and Murray, R. C., eds., Dolomitization and Limestone Diagenesis: Society of Economic Paleontologists and Mineralogists Special Publication 13, p. 71–88.

DES MARAIS, D. J., 1990, Microbial mats and the early evolution of life: Trends in Ecological Evolution, v. 5, p. 140–145.

FOLK, R. L. AND LAND, L. S., 1975, Mg/Ca ratio and salinity: two controls over crystallization of dolomite: American Association of Petroleum Geologists Bulletin, v. 59, p. 60–68.

GARRELS, R. M. AND THOMPSON, M. E., 1962, A chemical model for seawater at 25°C and one atmosphere total pressure: American Journal of Science, v. 260, p. 57–66.

GARRELS, R. M. AND CHRIST, C. L., 1965, Solutions, Minerals and Equilibria: Student Reprint, Tokyo. Harper & Row and John Weatherhill Incorporated, 450 p.

GEBELEIN, C. D. AND HOFFMAN, P., 1973, Algal origin of dolomite laminations in stromatolitic limestone: Journal of Sedimentary Petrology, v. 43, p. 603–613.

GERDES, G. AND KRUMBEIN, W. E., 1987, Biolaminated deposits: Lecture Notes in Earth Science 9, Berlin, Springer-Verlag, 183 p.

GIESKES, J. M., ELDERFIELD, H., LAWRENCE, J. R., JOHNSON, J., MEYERS, B. AND CAMPBELL, A., 1982, Geochemistry of interstitial waters and sediments, Leg 64, Gulf of California, *in* J. R. Curray, D. G. Moore, et al., Initial reports of the Deep Sea Drilling Project, v. 64: Washington D.C., U.S. Government Printing Office, p. 675–694.

GIVEN, R. K. AND WILKINSON, B. H., 1987, Dolomite abundance and stratigraphic age: constraints on rates and mechanisms of Phanerozoic dolostone formation: Journal of Sedimentary Petrology v. 57, 1068–1078.

GLENN, C. R., AND ARTHUR, M. A., 1988, Petrology and major element geochemistry of Peru margin phosphorites and associated minerals: authigenesis in modern organic-rich sediments: Marine Geology, v. 80, p. 231–267.

HARDIE, L. A., 1987, Dolomitization: a critical view of some current views: Journal of Sedimentary Petrology, v. 57, p. 166–183.

HENSGENS, C. M. H., 1996, Alcohol metabolism and nature of electron-dense granules in Desulfovibrio. Unpublished Ph.D. Dissertation, University of Groningen, 112 p.

HORODYSKI, R. J. AND VON DER HAAR, S. P., 1975, Recent calcareous stromatolites from Laguna Mormona (Baja California), Mexico: Journal of Sedimentary Petrology, v. 45, p. 680–696.

HSU, K. J. AND SIEGENTHALER, C., 1969, Preliminary experiments and hydrodynamic movement induced by evaporation and their bearing on the dolomite problem: Sedimentology, v. 12, p. 11–25.

HUC, A. Y., 1980, Origin and formation of organic matter in recent sediments and its relationship to kerogen, in Durand, B., ed., Kerogen: Editions Technip v. 27, p. 445–474.

HUDSON, J. D., 1975, Carbon isotopes and limestone cement: Geology, v. 3, p. 19–22.

ILLING, L V., WELLS, A. J. AND TAYLOR, J. C. M., 1965, Penecontemporary dolomite in the Persian Gulf, *in* Pray, L. C. and Murray, R. C., eds., Dolomitization and Limestone Diagenesis: Society of Economic Paleontologists and Mineralogists Special Publication 13, p. 89–111.

IRWIN, H., 1980, Early diagenetic carbonate precipitation and pore fluid migration in the Kimmeridge Clay of Dorset, England: Sedimentology, v. 27, p. 577–591.

JØRGENSEN, B. B., 1982, Mineralization of organic matter in the sea bed—the role of sulfate reduction: Nature, v. 296, p. 643–645.

KLEIN, C., BEUKES, N. J. AND SCHOPF, J. W., 1987, Filamentous microfossils in the early Proterozoic Transvaal Supergroup: their morphology, significance and palaeoenvironmental setting: Precambrian Research, v. 36, p. 81–94.

LAND, L. S., 1991, Dolomitization models—seawater mixing zones (abs.): Dolomieu Conference on Carbonate Platforms and Dolomitization, p.144.

LAST, W. M. AND DE DECKER, P., 1990, Modern and Holocene carbonate sedimentology of two saline volcanic maar lakes, southern Australia: Sedimentology, v. 37, p. 967–981.

LEEDER, M. R., 1982, Sedimentology: process and product: London, George Allen and Unwin, 344 p.

LIPPMANN, F., 1973, Sedimentary Carbonate Minerals: Berlin, Springer-Verlag, 228 p.

LYONS, W. B., LONG, D. T., HINES, M. E., GAUDETTE, H. E., AND ARMSTRONG, P. B., 1984, Calcification of cyanobacterial mats in Solar Lake, Sinai: Geology, v. 12, p. 623–626.

LYONS, W. B., HINES, M. E., LAST, W. M. AND LENT, R. M., 1994, Sulfate reduction rates in microbial mat sediments of differing chemistries: implications for organic carbon preservation in saline lakes, *in* Renaut, R. W. and Last, W. M., eds., Sedimentology and Geochemistry of Modern and Ancient Saline Lakes: Society of Economic Paleontologists and Mineralogists Special Publication 50, p. 13–20.

MACHEL, H. G. AND MOUNTJOY, E. W., 1987, General constraints on extensive pervasive dolomitization and their application to the Devonian carbonates of western Canada: Bulletin of Canadian Petroleum Geology, v. 35, p. 143–158.

MCKENZIE, J. A., HSU, K. J. AND SCHNEIDER, J. F., 1980, Movement of subsurface waters under the sabkha, Abu Dhabi, UAE, and its relation to evaporative dolomite genesis, *in* Zenger, D. H., Dunham J. B. and Ethington, R. E., eds., Concepts and Models of Dolomitization: Society of Economic Paleontologists and Mineralogists Special Publication 28, p. 11–30.

MONTY, C. L. V., 1976, The origin and development of cryptalgal fabrics, *in* Walter M. R., ed., Stromatolites: Developments in Sedimentology 20, Amsterdam, Elsevier, p. 193–249.

MORROW, D. W. AND RICKETTS, B. D., 1988, Experimental investigation of sulfate inhibition of dolomite and its mineral analogues, *in* Shukla, V. and Baker, P. A., eds., Sedimentology and Geochemistry of Dolostones: Society of Economic Paleontologists and Mineralogists Special Publication 43, p. 25–38.

MURATA, K. J., FRIEDMAN, I. AND MADSEN, B. M., 1969, Isotopic composition of diagenetic carbonates in Miocene marine formations of California and Oregon: U.S. Geological Survey Professional Paper 614-B, 24 p.

MURATA, K. J., FRIEDMAN, I. AND CREMER, M., 1972, Geochemistry of diagenetic dolomites in Miocene marine formations in California and Oregon. U.S. Geological Survey Professional Paper 724-C, 12 p.

PYTKOWICZ, R. M., 1983, Equilibria, Nonequilibria, and Natural Waters: New York, Wiley, 345 p.

REVSBECH, N. P., J¨RGENSEN, B. B., BLACKBURN, T. H. AND COHEN, Y., 1983, Microelectrode studies of the photosynthesis and O2, H2S and pH profiles in a microbial mat: Limnology and Oceanography, v. 28, p. 1062–1074.

ROSEN, M. R. AND COSHELL, L., 1992, A new location of Holocene dolomite formation, Lake Hayward, Western Australia: Sedimentology, v. 39, p. 161–166.

ROSEN, M. R., MISER, D. E. AND WARREN, J. K., 1988, Sedimentology, mineralogy and isotopic analysis of Pellet Lake, Coorong region, South Australia: Sedimentology, v. 35, p. 105–122.

SCHIDLOWSKI, M., MATZIGKEIT, U. AND KRUMBEIN, W. E., 1984, Superheavy organic carbon from hypersaline microbial mats: assimilatory pathway and geochemical implications: Naturwissenschaften v. 71, p. 303–8.

SHINN, E. A, 1968, Selective dolomitization of Recent sedimentary structures: Journal of Sedimentary Petrology v. 38, p. 612–616.

SHINN, E. A., GINSBERG, R. N. AND LLOYD, R. M., 1965, Recent supratidal dolomite from Andros Island, Bahamas: *in* Pray L. C. and Murray R. C. eds., Dolomitization and limestone diagenesis: Society of Economic Paleontologists and Mineralogists Special Publication 13, p. 112–123.

SLAUGHTER, M. AND HILL, R. J., 1991, The influence of organic matter in organogenic dolomitization: Journal of Sedimentary Petrology, v. 61, p. 296–303.

STAMS, A. J. F., 1985, Ecophysical aspects of the electron donor metabolism of sulfate-reducing bacteria: Unpublished Ph.D. Dissertation, University of Groningen, Van Denderen BV Groningen, 111 p.

SUMNER, D. Y. AND GROTZINGER, J. P., 1994, Archaean microbialites and carbonate precipitation, (abs.): *in* Awramik, S. M., ed., Death Valley International Stromatolite Symposium, Laughlin, Nevada, USA, p.75–76.

TRUDINGER, P. A., 1992, Bacterial sulfate reduction: current status and possible origin *in* Schidlowski, M. et al., eds., Early Organic Evolution: Implications for Mineral and Energy Resources: Berlin, Heidelberg, Springer-Verlag, p. 367–377.

TUCKER, M. E., 1982, Precambrian dolomites: petrographic and isotopic evidence that they differ from Phanerozoic dolomites: Geology, v. 10, p. 7–12.

TUCKER, M. E., 1986, Carbon isotope excursions in Precambrian/Cambrian boundary beds, Morocco: Nature, v. 319, p. 48–50.

TUCKER, M. E., 1992, The Precambrian-Cambrian boundary: Seawater chemistry, ocean circulation and nutrient supply in metazoan evolution, extinction and biomineralization: Journal of the Geological Society of London, v. 149, p. 655–668.

TUCKER, M. E. AND WRIGHT, V. P., eds., 1990, Carbonate Sedimentology, London, Blackwell Scientific, 482 p.

USDOWSKI, H.E., 1967, The formation of dolomite in sediments, *in* Muller, G., and Friedman, G.M., eds., Recent Developments in Carbonate Sedimentology in Central Europe: Heidelberg, Springer-Verlag, p. 21–32.

VASCONCELOS, C. AND MCKENZIE, J. A., 1997, Microbial mediation of modern dolomite precipitation and diagenesis under anoxic conditions (Lagoa Vermelha, Rio de Janeiro, Brazil): Journal of Sedimentary Research, v. 67, p. 378–390.

VASCONCELOS, C., MCKENZIE, J. A., BERNASCONI, S., GRUJIC, D. AND TIEN, A., 1995, Microbial mediation as a possible mechanism for natural dolomite formation at low temperatures: Nature, v. 377, p. 220–222.

VON DER BORCH, C. C., 1965, The distribution and preliminary geochemistry of modern carbonate sediments of the Coorong area, South Australia: Geochimica et Cosmochimica Acta, v. 29, p. 781–799.

VON DER BORCH, C. C., 1976, Stratigraphy and formation of Holocene dolomitic carbonate deposits of the Coorong area, South Australia: Journal of Sedimentary Petrology, v. 46, p. 952–966.

VON DER BORCH, C. C., AND LOCK, D., 1979, Geological significance of Coorong dolomites: Sedimentology, v. 26, p. 813–824.

WALTER, L. M., 1986, Relative efficiency of carbonate dissolution and precipitation during diagenesis: a progress report on the role of solution chemistry, *in* Gautier D.L., ed., Roles of Organic Matter in Sediment Diagenesis: Society of Economic Paleontologists and Mineralogists Special Publication 38, p. 1–11.

WARREN, J. K., 1988, Sedimentology of Coorong dolomite in the Salt Creek region, South Australia: Carbonates and Evaporites, v. 3, p. 175–199.

WRIGHT, D. T., 1993, Studies of the Cambrian Eilean Dubh Formation of northwestern Scotland: Unpublished Ph.D. Dissertation, University of Oxford, 165 p.

WRIGHT, D. T., 1994, The role of benthic microbial communities in widespread dolomite formation (abs.), in Awramik, S. M., ed., Death Valley International Stromatolite Symposium, Laughlin, Nevada, USA, p. 95.

WRIGHT, D. T., 1997, An organogenic origin for widespread dolomite in the Cambrian Eilean Dubh Formation, northwestern Scotland: Journal of Sedimentary Research, v. 67, p. 54–64.

VARIATIONS IN THE GLOBAL PHOSPHORUS CYCLE

J. COMPTON*

Dept. of Geological Sciences, University of Cape Town, Rondebosch 7700, South Africa

D. MALLINSON

Dept. of Marine Science, University of South Florida, St. Petersburg, Florida 33701 USA

C. R. GLENN

Dept. of Geology and Geophysics, SOEST, University of Hawaii, Honolulu, Hawaii 96822 USA

G. FILIPPELLI

Dept. of Geology, Indiana University–Purdue University at Indianapolis, Indianapolis, Indiana 46202-5132 USA

K. FÖLLMI

Geological Institute, University of Neuchatel, Ch-2007 Neuchatel, Switzerland

G. SHIELDS

Centre de Geochimie de la Surface, CNRS, 1 rue Blessig, F-67084, Strasbourg Cedex, France

AND

Y. ZANIN

United Institute of Geology, Geophysics and Mineralogy, Siberian Branch of the Russian Academy of Sciences, 630090 Novosibirsk, Russia

ABSTRACT: Phosphorus is a critical element in the biosphere, limiting biological productivity and thus modulating the global carbon cycle and climate. Fluxes of the global phosphorus cycle remain poorly constrained. The prehuman reactive phosphorus flux to the ocean is estimated to range from 0.7—4.8 x 10^{12}g/yr. Uncertainty in the reactive phosphorus flux hinges primarily on the uncertain fate of phosphate adsorbed to iron oxyhydroxide particles which are estimated to constitute 50% or more of the chemically weathered-phosphorus river flux.

Most reactive phosphorus is initially removed from seawater by burial of organic matter and by scavenging onto iron–manganese oxide particles derived from mid-ocean ridge (MOR) hydrothermal activity. Calculation of the oceanic phosphorus burial flux is complicated by early diagenetic redistribution of both oceanic and terrestrial phosphorus. Increased phosphorus input during periods of warm, humid climate is offset to some degree by increased burial rate as productivity shifts to expanded shallow-water estuary and shelf areas where phosphorus is rapidly decoupled from organic matter to form phosphorite. Phosphorus scavenging is greater if high sea levels are associated with increased MOR hydrothermal activity such as during the Late Cretaceous. Less phosphorus is derived from weathering during cool, dry climatic periods but a more direct transportation of phosphorus to the deep ocean, and a shift of productive upwelling regions to deeper water areas allows more phosphorus to be recycled in the water column. Lowered sea level results in less effective trapping of phosphorus in constricted estuary and shelf areas and in an increase in the phosphorus flux to the deep ocean from sediment resuspension. A decrease in MOR spreading rates and the resulting decrease in phosphorus scavenging by iron–manganese oxide particles would result in more phosphorus for the biosphere. Orogeny and glaciation may accelerate chemical weathering of phosphorus from the continents when the increased particle flux is exposed to warm and humid climate. Large, reworked phosphorite deposits may proxy for short-term organic carbon burial and correspond to periods of increased reactive phosphorus input that cannot be accommodated by long-term organic matter and iron-oxide particulate burial.

INTRODUCTION

The weathering of phosphorus and its delivery and fate in the ocean are important aspects of the phosphorus cycle to understand because phosphorus is a limiting nutrient to biological productivity on geologic time scales. Much of the phosphorus delivered to the ocean has the potential to participate in biological production, but whether it does or bypasses the ocean reservoir is poorly understood. Reactive phosphorus, defined here as phosphorus that enters the ocean reservoir and is available for biological uptake, is of particular interest because it links the phosphorus and carbon cycles. Variations in the reactive phosphorus entering the oceans related to changes in climate and tectonism would impact the uptake of CO_2 by primary productivity in the oceans, and the transfer of carbon as organic matter to the deep ocean to be oxidized or buried in sediments. The phosphorus cycle is simplified somewhat by the fact that there are no appreciable gaseous atmospheric phosphorus compounds, but the path of phosphorus from its release by chemical weathering to its transport and burial at sea is complex because of the interaction of phosphorus with the biosphere and iron–manganese oxide particles. A key question of interest to geochemists is how has the amount of reactive phosphorus entering the oceans varied in the past and can these variations be inferred from the geological record of reactive phosphorus sinks?

There have been a number of previous global phosphorus budgets proposed (Froelich et al., 1982; Richey, 1983; Froelich, 1984; Ruttenberg, 1993; Filippelli and Delaney, 1994; Baturin et al., 1995) as well as reviews of our current understanding of the phosphorus cycle (Bentor, 1980; Arthur and Jenkyns, 1981; Glenn et al., 1994b; Föllmi, 1996). This paper presents a brief overview of our quantitative understanding of the reservoirs and fluxes of the global phosphorus cycle with an emphasis on recent results and remaining key uncertainties. The rapid exchange of phosphorus among different reservoirs combined with large anthropogenic influences make it difficult to quantify the amount of phosphorus entering the oceans. Measuring the output of reactive phosphorus from the oceans is hampered by the large area of the seafloor and the redistribution of phosphorus during early burial diagenesis. Processes operating on glacial/interglacial and longer (>1 m.y.) geologic time scales are discussed that might result in variations in the global phosphorus cycle with a particu-

*email address: compton@geology.uct.ac.za

Marine Authigenesis: From Global to Microbial, SEPM Special Publication No. 66

lar focus on the relationship of phosphorite deposits to changes in the amount of phosphorus entering the ocean.

THE GLOBAL PHOSPHORUS CYCLE

Weathering Sources of Phosphorus to the Ocean

The objective is to quantify the principal sources and sinks of marine phosphorus, and to differentiate between prehuman and present-day (polluted) phosphorus fluxes, as well as reactive and nonreactive phosphorus fluxes (Tables 1A, B). A simplified schematic of the global phosphorus cycle indicates that the river flux dominates phosphorus input to the ocean (Fig. 1). The other significant source is eolian particulate phosphorus. River input can be divided into dissolved inorganic phosphorus (DIP), dissolved organic phosphorus (DOP), particulate inorganic phosphorus (PIP), and particulate organic phosphorus (POP).

Dissolved inorganic phosphorus.—

The dissolved inorganic phosphorus (DIP) flux of rivers (orthophosphate, $H_2PO_4^-$, HPO_4^{2-}, PO_4^{3-}) is estimated to be 0.26 to 0.46 x 10^{12}g/yr using a value of 7–10 µg/l as the average phosphorus concentration in unpolluted rivers (Meybeck, 1982; 1993; Savenko and Zakharova, 1995) and a river water flux of between 3.7 to 4.6 x 10^{16}kg/yr (Baumgartner and Reichel, 1975; Holland, 1978). The DIP value of 7-10 µg/l is derived from analyses of the least polluted rivers. The DIP in most rivers is influenced by human activities with DIP concentrations in agricultural and industrial areas as high as 100 to 700 µg/l (Meybeck, 1993; Savenko and Zakharova, 1995). Pollution is estimated to increase the present-day DIP river flux to between 0.8 and 1.4 x 10^{12}g/yr (Meybeck, 1982).

Dissolved organic phosphorus.—

The present-day dissolved organic phosphorus (DOP) river flux can be calculated as 0.55 to 0.68 x 10^{12}g/yr from an average DOP concentration of 15µg/l in rivers (total dissolved phosphorus (TDP) of 25µg/l minus dissolved inorganic phosphorus of 10µg/l) (Meybeck, 1982). Alternatively, the DOP flux can be calculated as 0.2 x 10^{12}g/yr from a dissolved organic carbon flux of 0.22 x 10^{15}g/yr and an average DOC C/P weight ratio of 1000 (Meybeck, 1982). The lower value (0.2 x 10^{12}g/yr) is preferred because high TDP values can be obtained from unfiltered or polluted water samples (Meybeck, 1982).

Particulate organic phosphorus.—

Particulate organic phosphorus (POP) was originally calculated by Meybeck (1982) to be 8 x 10^{12}g/yr using a particulate organic matter (POM) C/P wt ratio of 22, but Ramirez and Rose (1992) determined a much higher POM C/P wt ratio of 193. Using a POM C/P ratio of 193 and 0.17 x 10^{15}g/yr for the terrestrial POM flux (Meybeck, 1993) gives a POP flux of 0.9 x 10^{12}g /yr (Ramirez and Rose, 1992). The POP is associated with the POM flux from soils (94 x 10^{12}g/yr) and shales (76 x 10^{12}g/yr) (Meybeck, 1993). Assuming a POM C/P wt ratio of 193, gives a POP flux of 0.5 x 10^{12}g/yr from soils and 0.4 x 10^{12}g/yr from shales.

Particulate inorganic phosphorus.—

The particulate inorganic phosphorus (PIP) flux includes detrital phosphorus and adsorbed phosphorus. Detrital phosphorus refers to phosphorus that occurs in detrital grains, primarily igneous, metamorphic and sedimentary apatites as well as a trace element in other minerals. Detrital phosphorus is generally covalently bonded in mineral structures and not available for uptake by the biosphere or by adsorption until released by chemical weathering. Adsorbed phosphorus occurs primarily on iron-manganese oxide and oxyhydroxide particle surfaces (Palmer, 1985). Adsorbed phosphorus is potentially available to the biosphere through reversible exchange and is referred to here as iron-bound PIP. The detrital phosphorus and iron-bound PIP fluxes, Table 1A, were estimated from the total and chemically weathered river fluxes of phosphorus. The chemically

TABLE 1A—SUMMARY OF PREHUMAN PHOSPHORUS FLUX TO THE OCEAN (x 10^{12}g/yr)

Riverine Phosphorus	
DIP	0.3–0.5
DOP	0.2 (maximum)
POP (0.5 soil derived; 0.4 shale derived)	0.9 (maximum)
PIP, Fe-bound (P adsorbed to iron-manganese oxide/oxyhydroxides)	1.5–3.0
PIP, detrital	6.9–12.2
Total river phosphorus	9.8–16.8
Eolian phosphorus	1.0 (20% reactive)
Total River + Eolian phosphorus flux	10.8–17.8
Total prehuman potentially reactive phosphorus flux (DIP+DOP+POP+iron-bound PIP+reactive Eolian)	3.1–4.8

TABLE 1B.—SUMMARY OF PRESENT-DAY PHOSPHORUS FLUX TO THE OCEAN (x10^{12}g/yr)

DIP	0.8–1.4 x 10^{12} g/yr
DOP	0.2 (average)
POP (0.5 soil derived; 0.4 shale derived)	0.9 (average)
PIP, Fe-bound (P adsorbed to iron-manganese oxide/oxyhydroxides)	1.3–7.4
PIP, detrital	14.5–20.5
Total river phosphorus	17.7–30.4
Eolian phosphorus	1.05 (20% reactive)
Total River + Eolian phosphorus flux	18.7–31.4
Total present-day potentially reactive phosphorus flux (DIP+DOP+POP+iron-bound PIP+20% Eolian)	3.4–10.1

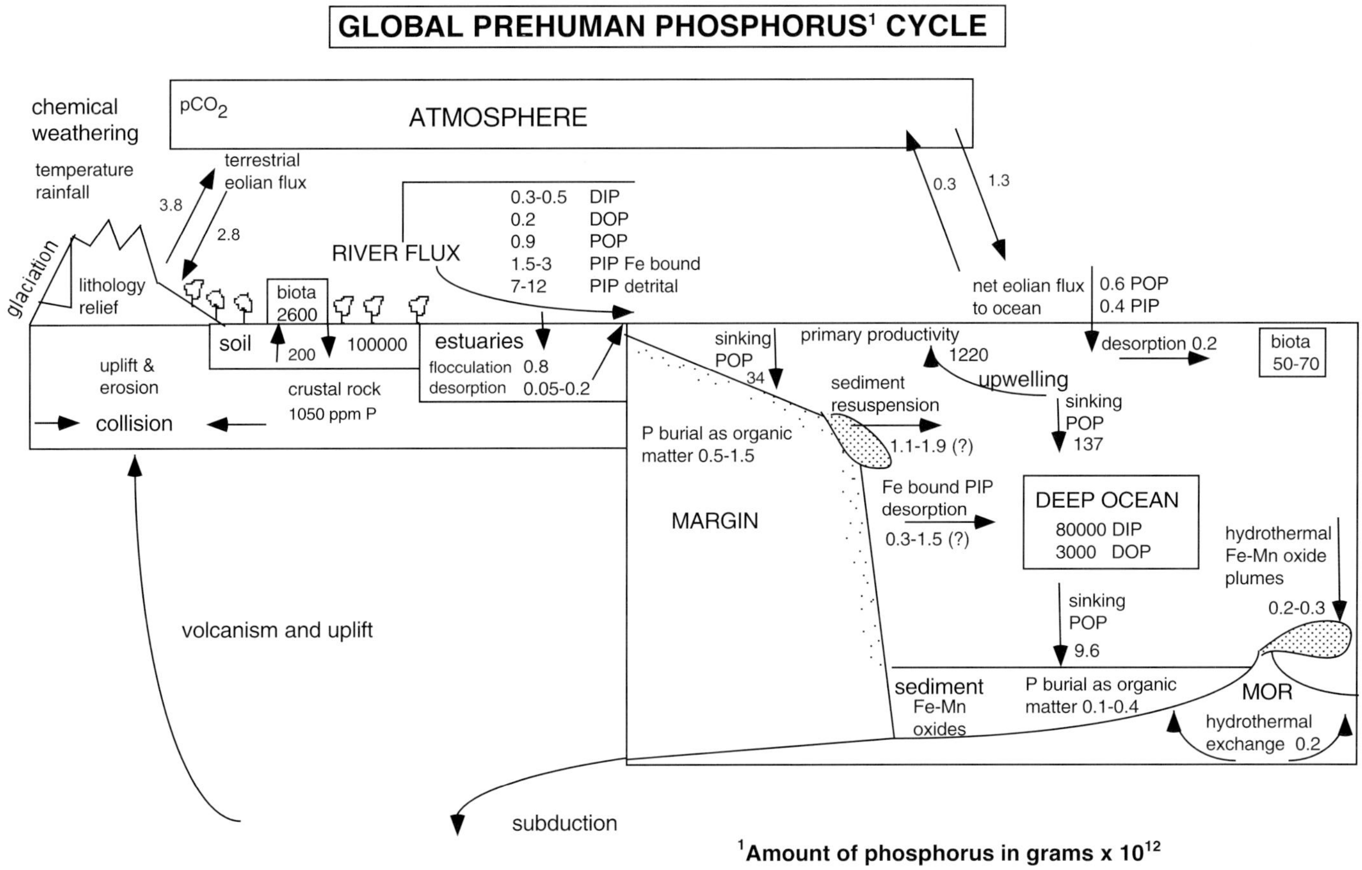

FIG. 1.— The global, prehuman phosphorus cycle with an outline of the major fluxes in units of 10^{12}g/yr (arrows indicate direction of flux) and amount of phosphorus in reservoirs in units of grams x 10^{12}. Sources of data and flux calculations are discussed in the text. Also included are factors important to the total potentially reactive phosphorus flux to the ocean, such as uplift and erosion. Net primary productivity and sinking particulate values assume a Redfield C/P wt ratio of 41 and that approximately 80% of primary production is over the open ocean and approximately 14% of primary production sinks below the photic zone (Martin et al., 1987). The net eolian flux from the continents to the ocean is 1×10^{12}g/yr, of which 20% is estimated to desorb in surface waters (Graham and Duce, 1979). The flux derived from desorption of iron-bound PIP (0.05–0.2 x 10^{12}g) is considered to be a subset of the sediment resuspension flux (1.1–1.9 x 10^{12}g); both fluxes are poorly constrained. DIP = dissolved inorganic phosphorus; DOP = dissolved organic phosphorus; PIP = particulate inorganic phosphorus; POP = dissolved organic phosphorus.

weathered phosphorus flux includes DIP, DOP, POP (soil derived), and iron-bound PIP. The total prehuman phosphorus river flux was calculated to be 9.8–16.8 x 10^{12}g/yr (Table 1A) by assuming that the total present-day sediment load of 1.4–2.4 x 10^{16}g/yr (Milliman and Zyvitski, 1992) is twice that of the prehuman sediment load (Berner and Berner, 1987), that 21–26% of total prehuman weathering is chemical (see arguments below), that crustal rocks have an average phosphorus content of 1050 ppm (Mason, 1967), and that phosphorus is not preferentially leached from crustal rocks. The chemically weathered prehuman phosphorus flux is 2.5–4.2 x 10^{12}g/yr. The PIP flux assumed to occur primarily as exchangeable iron-bound phosphorus is 1.5–3 x 10^{12}g/yr and was calculated as the difference between the total chemically weathered flux and the sum of DIP, DOP and POP fluxes (Table 1A). A similar iron-bound phosphorus input by fluvial particulates of 0.4–4 x 10^{12}g/yr was estimated by Froelich (1988). The river phosphorus flux is dominated by the detrital phosphorus flux (6.9–12.2 x 10^{12}g/yr) which was calculated as the difference between the total river phosphorus flux and the chemically weathered plus shale-derived POP fluxes.

Eolian flux.—

The eolian phosphorus flux to the ocean was estimated to be 1 x 10^{12}g/yr, mostly particulate phosphorus, by Graham and Duce (1979). The eolian flux of terrestrial POM carbon was estimated to be 0.1 x 10^{15} g/yr (Romankevich, 1984; Zafiriou et al., 1985). Assuming that most of the POM is soil derived with a C/P wt ratio of 154 (Ramirez and Rose, 1992) gives an eolian POP flux of 0.6 x 10^{12}g/yr. Therefore, the eolian flux is roughly equally split between PIP and POP. Approximately 20% of the total eolian phosphorus flux is readily soluble in seawater and is an important source of phosphorus for non-upwelling ocean areas (Graham and Duce, 1982). Human activities have increased the eolian dust flux of phosphorus to the ocean by only 0.05 x 10^{12}g/yr (Graham and Duce, 1979).

Prehuman phosphorus input.—

The total (riverine plus eolian) prehuman phosphorus flux to the oceans ranges from 10.8–17.8 x 10^{12}g/yr (Table 1A). The total potentially reactive phosphorus flux, defined as the sum of the chemically weathered flux, shale-derived POP flux, and 20% of the eolian flux, is estimated to be 3.1–4.8 x 10^{12}g/yr.

The ranges in detrital and iron-bound PIP fluxes reflect the range estimated for the prehuman total sediment load of rivers (0.7–1.2 x 10^{16}g/yr) and also depends on the extent of chemical weathering (17–26%), whereas the range in DIP reflects the range in total river water flux (3.7–4.6 x 10^{16}kg/yr). Present-day POP and DOP fluxes of 0.9 and 0.2 x 10^{12}g/yr, respectively, are assumed to be maximum values because the anthropogenic influence is difficult to quantify (Meybeck, 1993). It appears that no more than 20% of the total river flux of POM and DOM is preserved in sediments (Hedges, 1992), which would correspond to around 0.9 x 10^{12}g /yr reactive organic phosphorus, but where and on what time scale this terrestrial organic flux is remineralized remain unclear. The largest pool of potentially reactive phosphorus is iron-bound PIP. Based on the range in potentially reactive phosphorus input and a deep ocean phosphorus reservoir of approximately 8 x 10^{16}g, a residence time of phosphorus in the ocean of 16–27 kyr can be predicted, but a phosphorus residence time on the order of 100 kyr would be predicted if much of the particulate phosphorus flux (POP and iron-bound PIP) is nonreactive during its transit through the ocean. A critical question that remains is: What fraction of the terrestrial iron-bound PIP is released to the oceans?

Present-day phosphorus input.—

For comparison, the total present-day phosphorus river flux (Table 1B) was calculated to be 17.7–30.4 x 10^{12}g/yr from a total sediment load of 1.4–2.4 x 10^{16}g/yr (Milliman and Zyvitski, 1992) and assuming that 17% of present-day total weathering is chemical (Garrels and Mackenzie, 1971). The chemically-weathered phosphorus flux is 3–5.2 x 10^{12}g/yr. If only 10% of present-day total weathering is chemical (Berner and Berner, 1996), then the chemically-weathered phosphorus flux would be 1.8–3.0 x 10^{12}g/yr.

It is difficult to calculate the present-day iron-bound PIP flux, because of the higher DIP flux associated with pollution, and because much of the present-day sediment load of rivers is from accelerated soil erosion (Berner and Berner, 1987). The amount and mineralogy of the iron oxides transported by rivers and their relative stability in estuarine and marine environments remain poorly understood. For example, ferrihydrite is common in young soils as a precursor to hematite and its high surface area should make it a labile source of adsorbed phosphorus. A rough estimate of the iron-bound PIP river flux is calculated to be 1.3–7.4 x 10^{12}g/yr (Table 1B) assuming a 2 wt% average iron oxide content of the river sediment load, an iron oxide surface area of 60–200 m^2/g, and a maximum phosphate adsorption by iron oxide of 2.5 μmol/m^2 (Schwertmann and Taylor, 1989). The calculated iron-bound PIP flux is 8–26% of the total PIP flux, which is similar to the estimated 15–30% iron-bound PIP flux of the St. Lawrence River (Lucotte and d'Anglejan, 1983). The present-day minimum potentially reactive phosphorus flux (3.4 x 10^{12}g/yr) is similar to that of the prehuman flux, but the maximum present-day potentially reactive flux is roughly twice that of the prehuman value (Table 1B).

Chemical weathering.—

The extent of chemical weathering is critical because it determines the flux of potentially reactive phosphorus to the oceans. The percent of total weathering (chemical and mechanical) that is chemical is calculated by dividing the mass of the dissolved load by the mass of the dissolved load plus sediment load carried by rivers. The mass of the dissolved load was calculated from average global river concentrations corrected for human influences and for bicarbonate ion derived from atmospheric and soil CO_2 (Table 2).

The present-day total dissolved load from chemical weathering is 0.32–0.41 x 10^{16}g/yr or 15–19% of the total weathering flux. This range includes the global average of 17% estimated by Garrels and Mackenzie (1971), but is higher than the estimate of 10% by Berner and Berner (1996). Chemical weathering would make up 21–26% of the total weathering flux using a prehuman dissolved load (Table 2) and assuming a doubling of the prehuman sediment load (Berner and Berner, 1987).

The relative amount of reactive phosphorus in rivers derived from chemical weathering of apatite as opposed to other phosphorus-bearing minerals is unknown, but detrital igneous and metamorphic apatites are probably no more susceptible to dissolution than many other minerals containing minor to trace amounts of phosphorus (Zanin, 1989). The average total particulate phosphorus (TPP=PIP+POP) content of present-day rivers is estimated to be 1150 μg/g (Martin and Meybeck, 1979; Meybeck, 1982). Subtraction of soil-derived POP and iron-bound PIP (Table 1B) from TPP indicates that the phosphorus content of the detrital weathering flux (POP from shales and detrital PIP) is comparable to the average phosphorus content of crustal rocks and suggests that phosphorus is not preferentially removed (leached) from fresh rock during chemical weathering.

TABLE 2.— AVERAGE RIVER WATER CONCENTRATIONS (mg/kg)

	Present day Livingstone (1963)	Prehuman (total) Meybeck (1979)	Prehuman Rock-derived
Na	6.3	5.1	5.1
Mg	4.1	3.3	3.3
Ca	15	13.4	13.4
K	2.3	1.3	1.3
SiO_2	13.1	10.4	10.4
Cl	7.8	5.7	5.7
SO_4	11.2	8.2	8.2
HCO_3	58.4	52	21.5*
Fe	0.67		
F	1		
Total	120	99.4	68.9

*The dissolved mass is corrected for CO_2 and H_2O uptake from the atmosphere by assuming one-third of the dissolved inorganic carbon, mostly present as bicarbonate ion, is from the weathering of carbonate rocks and one-third is from weathering of silicate rocks by carbonic acid (Meybeck, 1993).

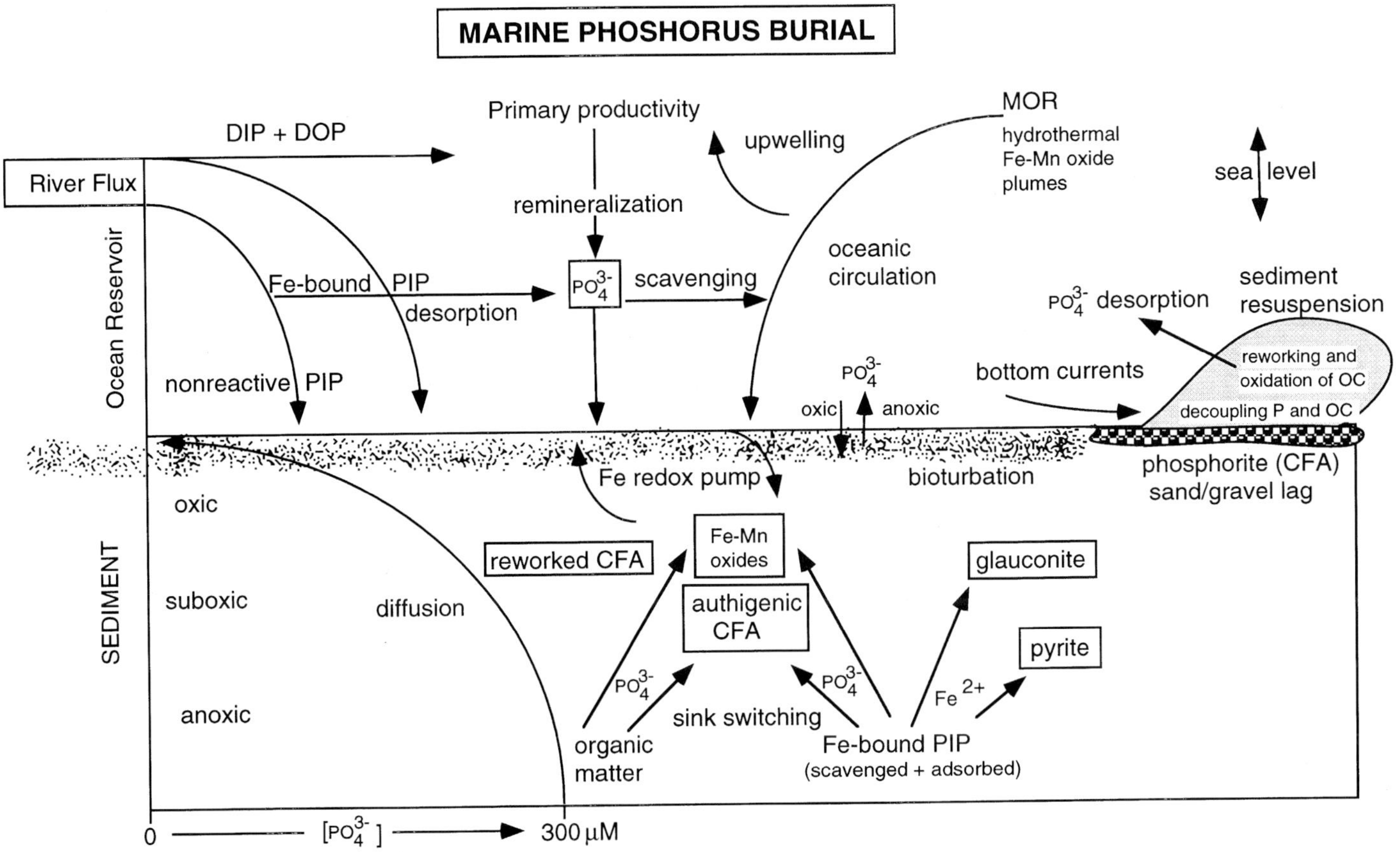

FIG. 2.— Overall schematic of the processes involved in the burial of phosphorus in marine sediments. After burial, most of the organic matter and iron-manganese oxyhydroxide primary phosphorus is released and taken up by authigenic CFA and iron-manganese oxide surfaces in the sediment during diagenesis. This redistribution (sink switching) makes it difficult to determine the net reactive phosphorus burial flux in marine sediments, particularly on continental margins having abundant iron-bound PIP, and possibly reworked CFA.

Reflux of Phosphorus from Sediments to Seawater

Phosphorus buried in sediments can later be remobilized during bacterial degradation of organic matter and reduction of iron oxides, with phosphate ion concentrations commonly increasing up to several hundred micromolar within meters of the sediment/seawater interface (Fig. 2). Most of this remobilized phosphate is removed from the pore water as authigenic carbonate fluorapatite (CFA) or is resorbed by iron oxide particles and does not return to the ocean reservoir (Krom and Berner, 1981; Ruttenberg and Berner, 1993; Ingall and Jahnke, 1994; Lucotte et al., 1994; Filippelli and Delaney, 1996). However, when sediment is resuspended in estuaries and on the shelf, significant amounts of DIP may be released into the water column by desorption (Fox et al., 1986; Sundby et al., 1992). Resuspension of sediment is predicted to result in desorption of PIP because the phosphate ion concentrations in pore waters are much higher than in bottom waters. Berner and Rao (1994) estimate that 5 µg of phosphorus per kg of resuspended sediment is released on the Amazon shelf, which translates to a potentially large global resuspension phosphorus flux of $1.1–1.9 \times 10^{12}$g/yr. Clearly more data are needed to quantify phosphorus flux by sediment resuspension. Significantly more phosphorus may diffuse from near surface sediments back into the ocean reservoir where low oxygen concentration in the bottom waters render iron oxide trapping of phosphorus less efficient (Ingall et al., 1993; Ingall and Jahnke, 1994; Van Cappellan and Ingall, 1994). Although present-day anoxic bottom waters are minor, more widespread ocean anoxia may have occurred during certain periods of Earth history.

Role of Estuaries

The highly variable biological and physical dynamics of estuaries make it difficult to estimate the net phosphorus river flux to the ocean (Wollast, 1983; Froelich, 1984). The conservative mixing behavior of DOC in estuaries would suggest that most DOC and associated DOP is delivered to the sea and perhaps 5–10% is trapped in estuaries by flocculation of primarily humic compounds (Sholkovitz, 1976; Sholkovitz et al., 1978; Burton, 1983). However, most DOC in the ocean appears to be derived from marine rather than riverine sources and suggests that most riverine DOC is photohydrolized to smaller compounds and recycled by bacteria within estuaries and coastal systems (see review by Hedges, 1992). Apparently most riverborne particulate organic matter (POM) and associated POP is removed in estuarine sediments before reaching coastal or shelf areas (Wollast, 1983). However, carbon isotope, C/N ratio, and biomarker data indicate that a significant fraction of the organic matter buried on margins adjacent to large rivers (Prahl et al., 1994) as well as in deep-sea sediments (Druffel et al., 1986) can be terrestrial.

The river water DIP flux appears to be enhanced by desorption of iron-bound PIP in estuaries. Chase and Sayles (1980) observed as much as 3% phosphate desorption associated with an increase in pH during mixing. Froelich (1984) found that the DIP river flux may increase by 50% during passage through estuaries. Extrapolation of these results suggests a global desorption flux to estuaries of 0.05–0.23 x 10^{12}g/yr. A phosphorus mass balance of the Scheldt estuary indicates that much of the phosphate entering the estuary is deposited rather than being directly transported to sea (Wollast, 1983). Considering the above estimates, estuaries may trap approximately one-quarter of the total potentially reactive phosphorus river flux as DOP and POP, as well as a significant but unknown amount of the iron-bound PIP flux. Efficient phosphorus trapping by estuaries is supported by Holocene reactive phosphorus accumulation rates of 1,000–11,000 μmol/cm^2/kyr, which are an order of magnitude greater than the typical phosphorus accumulation rates on continental margins (Filippelli, 1997a; unpublished data).

Oceanic Sinks of Phosphorus

The combined primary sedimentary sinks of reactive phosphorus in the ocean are estimated to range from 1–2.5 x 10^{12}g/yr (Table 3A). Uptake of phosphorus by organisms is the dominant primary sink of reactive phosphorus in the ocean (Table 3A). The burial rate of phosphorus associated with organic matter is estimated to be 0.6–1.9 x 10^{12}g/yr assuming an organic carbon burial flux of 10–30 x 10^{12} mol/yr (Mackenzie et al., 1993) having an average sedimentary C/P molar ratio of 500 (Ramirez and Rose, 1992). In addition to long-term organic carbon burial, short-term (<10^5 yr) variations in organic carbon burial includes organic matter regenerated on or just below the seafloor (Jørgensen, 1983) where released phosphorus can be captured as CFA or adsorbed onto oxyhydroxides, thereby decoupling phosphorus from the nutrient flux back to the ocean. Short-term organic carbon burial is expected to increase the significance of the organic matter primary sink of phosphorus, particularly in shallow-water shelf regions. The range in phosphorus burial estimated by Ruttenberg (1993) and Berner and Berner (1996) (Table 3B) overlaps with estimates of the total potentially reactive phosphorus input (Table 1A), but is near the upper limit of the estimated primary phosphorus burial sinks (Table 3A). The high phosphorus burial estimates of Filippelli and Delaney (1996) (Table 3B) may include a significant amount of nonreactive phosphorus or detrital/reworked CFA especially on continental margins.

TABLE 3A.—OCEANIC SINKS OF REACTIVE PHOSPHORUS (x 10^{12}g/yr)

Organic matter	0.6–1.9
iron oxides (MOR)	0.4–0.5
CFA (direct precipitation)	0–0.1 (?)
TOTAL	1.0–2.5

Nonbiological phosphorus sinks in the ocean are important because they influence the deep water phosphate ion concentration and amount of upwelled phosphorus available for primary productivity. Wheat et al. (1996) estimate that the total amount of phosphorus removed by hydrothermal processes associated with mid-ocean ridges (MOR) is between 0.37–0.5 x 10^{12}g/yr, most of which (0.17–0.31 x 10^{12}g/yr) is DIP removed by iron oxides entrained in nonbuoyant plumes. An additional 0.2 x 10^{12}g/yr is removed during the convective flow of seawater through ridge flanks. Although 30–100% of the phosphate is removed from seawater circulating through basalt, the total amount of phosphorus removed by this process is minor (0.004 x 10^{12}g/yr). The amount of phosphorus uptake by particulate iron oxides in hydrothermal plumes depends on the intensity of hydrothermal activity and the phosphate concentration of the ambient seawater (Feely et al., 1990; 1991). The net amount of phosphorus scavenged from the oceanic water column by iron-manganese oxide-coated eolian or detrital particles is probably less than that removed by fresh MOR iron precipitates, because some phosphorus is adsorbed before entering the ocean. In addition, eolian iron-manganese dust partially desorbs phosphorus in surface waters and may enhance primary production (Graham and Duce, 1982).

CFA precipitated directly on the seafloor is generally not believed to be a significant phosphorus sink. Phosphatic hardgrounds apparently form slowly during extended periods of nondeposition (persistent bottom currents), and because hardgrounds commonly form at the upper surface of carbonate

TABLE 3B.—COMPARISON OF DIFFERENT PHOSPHORUS BURIAL BUDGETS (x 10^{12}g/yr)

Sink	Ruttenberg (1993)*	Berner and Berner (1996)
Org Matter (O-P)	1.3	1.1
iron-oxides (iron-P)	1.2 (0.1)	0.7
CFA	2.8 (0.7)	1.1
loosely-sorbed P	0.4	—
TOTAL	5.7 (2.4)	2.9

*Calculated from sequential extraction data of marine sediments and the flux of riverine suspended material to the margins and deep sea (minimum values in parentheses). CFA includes fish debris and carbonate-bound phosphorus.

Sink Area	Filippelli and Delaney (1996)
Low Productivity	0.2
Moderate to high Productivity	2.4
Continental Margins	3.9
TOTAL	6.5 x 10^{12}g/yr

Calculated as the mean phosphorus accumulation rate over last one million year multiplied by area.

substrate rocks, they are not volumetrically significant (e.g., Föllmi and Garrison, 1991). Another possible mechanism of phosphorus uptake in sediments is by iron redox pumping (Heggie et al., 1990; O'Brien et al., 1990). The iron redox phosphorus pump works by adsorption of phosphate ion from bottom and interstitial waters onto iron oxides, release of this phosphate for CFA formation upon reduction of the iron oxides at depth in the sediment, and then oxidation of the iron as it diffuses across the redox boundary. The amount of reactive phosphorus taken up directly from bottom waters by the iron redox pump mechanism is probably relatively small because most of the iron reduction and oxidation cycle occurs at depth in the sediment where it is overlain by oxygenated bottom waters.

Diagenetic Redistribution of Buried Phosphorus

A major cause of uncertainty in estimates of the reactive phosphorus burial flux in marine sediments is the difficulty in distinguishing between phosphorus removed from the ocean reservoir (a true sink) as opposed to phosphorus that bypasses the ocean reservoir but is released into pore waters and taken up by CFA or iron-oxides during diagenesis (Fig. 2). Sequential extraction techniques have provided accumulation rate data on the various reservoirs of sedimentary phosphorus and have demonstrated the complex dynamics of phosphorus during burial diagenesis (Ruttenberg, 1993; Filippelli and Delaney, 1996). Diagenetic redistribution of phosphorus (sink switching) involves much of the phosphorus that arrives at the seafloor and can complicate the determination of the primary reactive phosphorus output. CFA is not a true sink of reactive phosphorus if it is reworked CFA or phosphorus derived from iron-bound PIP that bypassed the ocean reservoir. The detrital phosphorus flux will include some fraction of reworked sedimentary CFA either locally derived or carried by rivers. The detrital CFA content of river sediment loads is unknown, but is expected to be significant particularly during marine regressions when rivers erode older deposits containing authigenic CFA.

CFA is estimated to make up 30–50% of the total reactive phosphorus output (Ruttenberg, 1993; Table 3B). Fine-grained, disseminated, and hence not easily detected, CFA containing phosphorus derived from both iron-bound PIP and organic phosphorus may occur in detrital-rich continental margin sediments (Ruttenberg and Berner, 1993) and in deep sea sediments (Filippelli and Delaney, 1996). Berner and Berner (1996) using the data of Ruttenberg and Berner (1993) estimate the phosphorus sink associated with finely dispersed authigenic CFA to be 1×10^{12}g/yr, which is comparable to the organic matter phosphorus burial sink. But what percentage of the phosphorus in authigenic CFA is derived from organic matter vs. iron-bound PIP?

The phosphorus in most large phosphorite deposits is derived from the microbial release of organic phosphorus, as indicated by the common association of phosphorite deposits and reworked and condensed sediments below strong upwelling systems (Bentor, 1980). CFA containing phosphorus derived from iron-bound PIP is limited by the amount of iron-rich terrigenous sediment. The lateral or overlapping association of phosphorite deposits with iron-rich authigenic minerals (glauconite, pyrite, and ferruginous crusts) as observed on the Peru-Chile continental margin (Glenn et al., 1994a,b) and southwest African continental margin (Birch et al., 1983; Birch, 1990) suggests that some CFA phosphorus is derived from a large pool of iron-bound PIP. For example, Late Cretaceous Egyptian phosphorite deposits may have formed largely from remobilized iron-bound PIP derived from local rivers (Glenn and Arthur, 1990; Glenn et al., 1994a). The near-absence of glauconite in Neogene phosphorites on the Florida Platform (Riggs, 1984), in contrast to the large glauconite content of Neogene sediments on the New Jersey margin, suggests an increasingly abundant source of iron-bound phosphate associated with terrigenous sediment input to the continental margin. High pore water phosphate concentrations and the presence of authigenic CFA in places (Mountain et al., 1994) suggest that significant amounts of iron-bound PIP are released to form CFA in rapidly deposited, terrigeneous sediment on the New Jersey margin.

The large amount of phosphorus (Froelich, 1984) associated with iron-manganese oxide crusts on biogenic carbonate is probably derived mostly from organic and iron-bound phosphorus in the sediment rather than phosphorus taken up directly from the water column (Palmer, 1985). Precipitation of REE phosphate minerals in marine sandstones has been proposed as a major sink of oceanic phosphorus, on the order of 2.3×10^{12}g/yr (Rasmussen, 1996; also see Rasmussen, this volume), but REE phosphate minerals probably form during sediment diagenesis with phosphorus derived from both reactive and nonreactive phosphorus sources.

DISCUSSION

Most of the phosphorus delivered to the ocean bypasses the marine biosphere. The prehuman reactive phosphorus flux to the oceans could be as small as 0.7×10^{12}g/yr and as much as 4.8×10^{12}g/yr (7–25% of the total phosphorus river flux). The bulk of the phosphorus that bypasses the ocean reservoir is detrital phosphorus ($6.9–12.2 \times 10^{12}$g/yr). In addition, roughly half of the prehuman chemically weathered phosphorus flux of $2.5–4.2 \times 10^{12}$g/yr is apparently adsorbed to iron oxides or is in particulate organic matter and bypasses the ocean reservoir. Assuming the entire dissolved phosphorus flux (DIP+DOP) and 20% of the eolian flux is reactive, the estimated reactive phosphorus burial flux would require that approximately 12–40% of the total potentially reactive particulate phosphorus flux (POP and iron-bound PIP) be released to the oceans as reactive phosphorus. These values are consistent with previous estimates of the soluble-exchangeable fraction of the total river particulate phosphorus of 3–6% (Chase and Sayles, 1980; Hedley et al., 1982; Van Eck, 1982; Tiessen et al., 1984).

What processes might change the reactive phosphorus flux to the ocean? A key assumption is that the reactive phosphorus river flux varies directly with chemical weathering rates. However, the rapid and reversible reactions of DIP, such as adsorption-desorption from iron oxyhydroxide surfaces, uptake by primary producers in river water, and burial in sediments, may buffer changes in the DIP concentration (Wollast, 1983; Froelich, 1988). Although buffering of DIP concentrations is not supported by the high DIP concentration in present-day polluted rivers, these extreme levels of phosphorus entering rivers today may have exceeded the natural buffering capacity of rivers. Natural increases in DIP would be much less than for polluted rivers and may have been moderated by increased uptake as organic phosphorus (DOP and POP) or as iron-bound PIP. The chemically-weathered phosphorus river flux should increase in response to increased chemical weathering rates, but the relative proportion of DIP, DOP, POP, and PIP may vary and impact the amount of reactive phosphorus entering the ocean. Here we consider possible variations in the phosphorus cycle on two time scales: glacial/interglacial cycles and time periods greater than 1 m.y.

Glacial/Interglacial Variations

By increasing the amount of rock and mineral surface area exposed to chemical weathering, mechanical weathering during glacial periods increases the effectiveness of chemical weathering during the transition to warmer and wetter climates of interglacial periods (Anderson et al., 1997). The effect of increased glaciation over the past 2.5 m.y. is predicted to have increased global silicate weathering by 40% and increased the global river cation flux by as much as 8% (Blum, 1997). Increased chemical weathering rates should increase the reactive phosphorus river flux during interglacials, but the concurrent marine transgression is predicted to increase the phosphorus trapping efficiency of estuaries and shelf areas resulting in more rapid phosphorus burial nearshore than in the deep sea (Ruttenberg, 1993). An increase in shelf area from the rise in sea level during deglaciation will result in an increase in primary productivity over shelf areas (Raymont, 1983). The release of regenerated organic phosphorus on or just below the seafloor enhances precipitation of CFA on the shelf because 25% of phosphorus remineralization occurs in shelf sediments compared to only around 1% in pelagic sediments (Jørgensen, 1983). The amount of shelf area underlying highly productive surface waters is therefore an important control on phosphorite formation and on the rapid removal of reactive phosphorus from the ocean (Arthur and Jenkyns, 1981). Pelagic productivity and the biological CO_2 pump are expected to decrease, and atmospheric pCO_2 is expected to increase during interglacials from less phosphorus input to the deep ocean and less intense upwelling (Broecker, 1982).

The decreased rate of chemical weathering in the transition to cooler and drier global climate of glacial periods is predicted to result in less river input of reactive phosphorus. The lower river phosphorus flux may be offset by a greater phosphorus reflux from resuspension of margin sediments, more direct transport of river sediment to the shelf/slope break, and erosion of older interglacial phosphorus-rich estuarine and shelf deposits. During glacial periods, estuaries and shelf areas are less extensive, and therefore, less of the reactive phosphorus river flux is trapped nearshore and more of the river load is deposited in deeper water. As sea level drops, the shelf area decreases and primary productivity shifts to areas overlying deeper water. Primary production over deeper water results in more phosphorus regeneration in the water column where reactive phosphorus is returned by upwelling to sustain high surface water productivity and lower levels of atmospheric CO_2.

The residence time of phosphorus would increase as a consequence of reduced rates of phosphorus input and output (Ruttenberg, 1993). As phosphorus input is reduced towards the minimum values in Table 1A, the residence time approaches 100 kyr making it less likely that variations would be seen on glacial/interglacial timescales. Eolian input also increases during glacial periods as shown in ice core dust records (Petit et al., 1990). The increased dust input further promotes marine productivity by providing phosphorus (Graham and Duce, 1982) and trace elements to surface waters (Berger and Wefer, 1991; Martin, 1991). Trace elements such as iron may in fact be more important than phosphorus in limiting primary production on glacial/interglacial timescales. Greater iron and aluminum inputs during glacial intervals coincide with greater phosphorus concentrations and accumulation rates in two Southern Ocean cores as well as with an increased dust flux in Antarctic ice cores (Latimer and Filippelli, 1997), supporting the possible links between dust, productivity, and nutrient budgets.

Changes in the terrestrial biosphere may also affect the phosphorus flux to the oceans. Interglacial terrestrial carbon storage is estimated to have increased by 750–1050 Gt (10^{15}g) over that of glacial periods (Crowley, 1995). The terrestrial biomass is dominated by soil organic matter with an average C/P wt ratio of 154 (Ramirez and Rose, 1992). The change in the average terrestrial phosphorus storage would be 6×10^{15}g. Uptake or release of this organic phosphorus over a period of 10–20 kyr would result in a significant phosphorus flux of $0.3–0.6 \times 10^{12}$g/yr. The delivery of phosphorus by rivers would be expected to change in response to the terrestrial biomass with a greater flux of POP and DOP relative to iron-bound PIP during interglacials. Uptake of phosphorus by the terrestrial biosphere and trapping of riverine POP in estuaries will act to diminish any increase in the chemically weathered phosphorus flux to the ocean during interglacials.

Variations on Geologic (>1 m.y.) Time Scales

The impact of climate on phosphorus input and oceanic circulation are largely similar to those discussed above for glacial/interglacial cycles but m.y. time scales also include tectonic processes such as MOR spreading rates and associated long-term sea-level fluctuations, variations in uplift and erosion associated with orogenic events, and long-term climate transitions from greenhouse to ice house worlds. The reactive phosphorus input is predicted to be proportional to chemical weathering rates which in turn are related to climate (temperature and rainfall) and mean elevation (tectonism) (Garrels and Mackenzie, 1971; Holland, 1978). Tectonic uplift produces more sediment by mechanical erosion, but does not necessarily increase chemical weathering rates (Garrels and Mackenzie, 1971). A time lag is likely between particle production on continents by glaciation or orogeny and chemical weathering and delivery to oceans for biological uptake (Föllmi, 1995; Van Cappellen and Ingall, 1996). In the case of the uplift and erosion of the Himalayan-Tibetan Plateau (HTP), for example, monsoon rains appear to be an important factor in increasing chemical weathering rates and phosphorus input (Filippelli, 1997b). Phosphorite formation appears to correlate with periods characterized by increased rates of chemical weathering associated with warm, humid climate (Kazarinov, 1966; Zanin, 1989). However, recent studies indicate that there is no simple relationship between climate and chemical weathering rates. Lithology and degree of physical degradation appear more important than climatic factors (temperature and precipitation) in determining the rate of chemical weathering by CO_2 uptake (Edmond and Huh, 1997; Allègre et al., 1997).

Geochemical proxies are improving our understanding of possible variations in chemical weathering rates, but the results remain equivocal. The increase in the marine strontium isotope ratio in the last 40 m.y. is interpreted to represent some combination of increased chemical weathering rates and weathering of increasingly radiogenic source rocks (Raymo et al., 1988; Hodell et al., 1990; Richter et al., 1992). Alternatively, the seawater strontium isotope evolution may reflect slowing of MOR spreading rates rather than uplift and erosion of the HTP (Allègre et al., 1997). An increase in Os isotopes in the last 25 m.y. suggests an increase in the weathering of old black shales (Ravizza, 1993), but a decrease in the Ge/Si ratio in the last 30–40 m.y. suggests an overall decrease in the rate of chemical weathering (Shemesh et al., 1989).

The largest primary sink of reactive phosphorus is organic matter. The marine carbonate $\delta^{13}C$ record is a potentially useful proxy for variations in organic carbon burial, but interpretation of the $\delta^{13}C$ record is complicated by the dependence of $\delta^{13}C$ values on the carbonate ion concentration (Spero et al., 1997). Variations in the carbonate ion concentration, related to changes in atmospheric CO_2 content and ocean chemistry, are sufficient to explain much of the observed variation in the late Pleistocene $\delta^{13}C$ record (Spero et al., 1997). In cases where the $\delta^{13}C$ record can be used to estimate unambiguously changes in long-term organic carbon burial rates, it is still necessary to distinguish between terrestrial and marine organic carbon, as coal deposits do not provide a sink for marine phosphorus. For example, the large positive $\delta^{13}C$ excursion in the late Paleozoic (Carboniferous to early Permian) probably reflects an increase in the formation of coal deposits related to the evolution of vascular land plants (Berner and Canfield, 1989). The large, middle to late Permian, phosphorite deposits of the Phosphoria Formation appear to be related to a major period of glaciation (Sheldon, 1989) and post date the Carboniferous to early Permian positive $\delta^{13}C$ excursion.

Large, reworked phosphorite deposits perhaps proxy for short-term marine organic carbon burial (Compton et al., 1990; 1993). This proxy works by virtue of the rapid decoupling of phosphorus from organic matter during early diagenesis. In addition to phosphorus buried as organic matter, a significant amount of organic phosphorus is remineralized by microbial activity and taken up by precipitation of authigenic CFA while the organic carbon is returned to the oceans by oxidation during early diagenesis (Fig. 2). The decoupling of phosphorus and organic carbon burial fluxes can occur on time scales smaller than the oceanic residence time of carbon (10^5 yr), so that changes in organic carbon burial are not reflected in the $\delta^{13}C$ record. The burial flux of rapidly decoupled phosphorus is predicted to increase during periods of marine transgression and intensified coastal upwelling as productivity shifts to shallow water (shelf) areas where mineralization of phosphorus from organic matter at or below the seafloor is maximized (Jørgensen, 1983). Large phosphorite deposits commonly form on shelf and upper slope areas during marine transgressions and are then reworked by marine regressions. Examples include phosphorite currently forming on the southwest African shelf (Birch et al., 1983; Birch, 1990) and from the Peru-Chile upper continental slope (Glenn et al., 1994a) where upwelling and productivity are particularly strong. Strontium isotope dates of phosphorite grains show that the grains are typically older (by up to several million years) than the sediment in which they occur and indicate extensive reworking of phosphorite deposits (Compton et al., 1993).

Increased MOR spreading rates are expected to increase hydrothermal MOR activity and release of iron-manganese oxyhydroxide particles in plumes to scavenge phosphate from seawater (Wheat et al., 1996). In general, MOR spreading rates have decreased from the middle Cretaceous to the Oligocene and then increased somewhat since the late Oligocene (Kominz, 1984). As phosphate is scavenged by iron oxide particles, the phosphate concentration of the oceans should decrease, thereby providing a feedback to limit the amount of phosphorus scavenged.

The Cretaceous was a period of warm humid climate, high MOR spreading rates, and high sea level. The warm wet climates may have promoted chemical weathering and river input of phosphorus, with 10% greater rainfall estimated for the Late Cretaceous by climate models (Bush and Philander, 1997). A higher phosphorus input may have been offset by increased trapping of phosphorus in estuaries, precipitation of CFA on an expanded shelf area, and scavenging of phosphorus in the water column by iron-manganese oxyhydroxide particles from high MOR spreading rates. The decrease in MOR spreading rates (Kominz, 1984) and the associated decrease in the amount of reactive phosphorus removed by iron-manganese plumes between 100 and 40 m.y. ago may have promoted formation of large Late Cretaceous and early Tertiary phosphorite deposits and increased the organic phosphorus sink by increasing the amount of phosphorus available for biological uptake. Large reworked Late Cretaceous phosphorite deposits (Notholt et al., 1989) together with positive excursions in the marine $\delta^{13}C$ record (Schlanger et al., 1987) may also reflect increased short- and long-term marine organic carbon burial. Although Late Cretaceous oceanic circulation rates were perhaps similar to those of today (Bush and Philander, 1997), lower oxygen-renewal rates may have resulted in expansion of oxygen-minimum zones (Schlanger et al., 1987; Arthur et al., 1987) thereby increasing the total reflux of phosphorus out of sediments into the ocean (Van Cappellen and Ingall, 1994).

In the transition from Eocene to present, climate has become cooler and drier, sea level has dropped, and MOR spreading rates have decreased. The river flux of phosphorus may have increased as a result of uplift and erosion of the HTP (Raymo et al., 1988; Filippelli, 1997b) and increased glaciation (Blum, 1997). A greater phosphorus reflux from sediment resuspension, as well as an increase in reworked CFA, since the late Pliocene are predicted because of high-frequency, high-amplitude sea-level fluctuations concomitant with ice-volume changes. The amount of oceanic phosphorus available for biological uptake and formation of phosphorite deposits would also be enhanced by low MOR spreading rates which reduced the amount of phosphorus scavenged by oxyhydroxides. A positive correlation between sea level and phosphorus accumulation rate from 160 to 32 m.y. ago in the deep sea sediment record is interpreted to result from increased chemical weathering during periods of high sea level. A negative correlation of sea level and phosphorus accumulation rate since 32 m.y. may reflect accelerated chemical weathering from increased glacial erosion (Föllmi, 1995).

Phosphorite Giants

Are large phosphorite deposits ("giants") the result of increased input of reactive phosphorus to the ocean or do they simply represent a focused output of phosphorus from the ocean produced by changes in ocean circulation (large, sustained upwelling sites) and concentrated by later reworking?

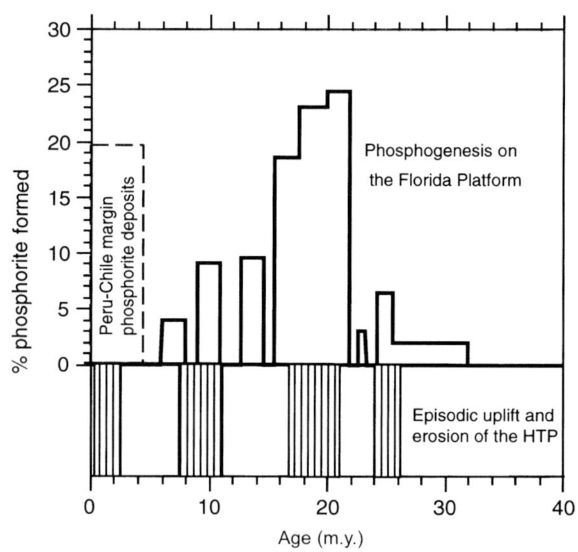

FIG. 3.—Comparison of the timing of proposed periods of increased Himalayan-Tibetan Plateau (HTP) uplift and erosion (Sorkhabi and Stump, 1993) and the timing and relative magnitude of phosphogenesis on the Florida Platform (Compton et al., 1993; Mallinson and Compton, 1997). Percent phosphorite formed is the percent of the total phosphorite on the Florida Platform that formed within the specified age intervals. The vertically hatched areas indicate the proposed time intervals of increased HTP uplift and erosion. Large phosphorite deposits on the Peru-Chile margin (Glenn et al., 1994a) may correspond to increased uplift and erosion of the HTP since 2 m.y. ago.

Phosphorite giants may form when the riverine reactive phosphorus flux exeeds the capacity of the long-term organic matter and particulate iron-oxide fluxes to remove phosphorus from the ocean. Compton and Mallinson (1996) estimate an excess reactive phosphorus river flux to the oceans of between 9–23 x 10^{16} moles from the proposed increase in the strontium river flux that is required to explain the strontium isotope evolution of seawater over the last 30 m.y. (Richter el al., 1992). The reactive phosphorus input is assumed to increase in proportion to the strontium increase. The early to middle Miocene rapid increase in the reactive phosphorus river flux predicted by the strontium model corresponds to an episode of major uplift and erosion of the HTP and phosphogenesis on the Florida Platform (Fig. 3). Uplift and erosion of the HTP has been an ongoing process since 50 m.y. ago, but several studies suggest that uplift and erosion were episodic and particularly rapid in the Miocene from 21–16 m.y. ago, from 11–7 m.y. ago, and possibly from 26–24 m.y. ago (Sorkhabi and Stump, 1993). Early to middle Miocene phosphorite on the Florida Platform can account for roughly 20–60% of the excess reactive phosphorus input between 21–16 m.y. ago. Correlation of the relative magnitude and timing between Neogene uplift and erosion of the HTP and phosphogenesis in the southeast USA suggests that phosphorite deposits of the southeast USA and elsewhere may represent, in part, the burial sink of the proposed increased weathering flux of phosphorus from the HTP to the oceans (Compton and Mallinson, 1996). A greater phosphorus sink from organic matter burial or scavenging is not supported by the Neogene marine $\delta^{13}C$ record (Shackleton and Hall, 1995) or MOR spreading rates (Kominz, 1984).

Uplift and erosion of the HTP continued after the Miocene with a major episode of uplift and erosion from about 2 m.y. ago to the present. Where are the sinks of the phosphorus derived from HTP erosion during the Pliocene and Pleistocene? Large phosphorite deposits have formed in the last 6 m.y. in the Peru-Chile and Namibian margin areas. The Peru/Chile margin is the site of a Modern phosphorite giant containing an estimated 140–746 x 10^{12}g phosphorus (Glenn et al., 1994a). Phosphatic sediment recovered from the Peru margin suggests that this area has been a major phosphogenic province since at least the late Miocene (Garrison and Kastner, 1990). The present-day burial flux of phosphorus as phosphorites on the Peru margin is comparable to that of the Miocene southeast USA burial flux of 0.06–0.4 x 10^{12}g/yr (Mallinson and Compton, 1997). Combined with phosphorite formation on the southwest African shelf as well as an apparent increase in the pelagic phosphorus burial rate from 10 m.y. ago to the present (Föllmi, 1995), appears sufficient to remove the present-day excess phosphorus flux from erosion of the HTP. Alternatively, the phosphorus weathering flux from the HTP may have decreased since the late Pliocene because of an increase in mechanical weathering and decrease in chemical weathering with glaciation (Filippelli, 1997b).

SUMMARY

The pathway of phosphorus from chemical weathering on

the continents to final burial in marine sediments is complex and not well understood. The maximum prehuman potentially reactive phosphorus flux to the ocean is estimated to be 3.1–4.8 x 10^{12}g/yr; the minimum flux could be as low as 0.7 x 10^{12}g/yr depending on the amount of phosphorus released from iron-bound PIP, the fate of dissolved and particulate organic phosphorus, and the influence of estuaries as phosphorus traps. Variations in the rate of burial of primary phosphorus bound to MOR-derived iron-oxide particles can be inferred from MOR spreading rates. Variations in burial rate of phosphorus associated with organic matter are more difficult to evaluate because of the dependence of the $\delta^{13}C$ record on ocean chemistry and on the amount of terrestrial as well as marine organic carbon burial. Much of the organic matter phosphorus and iron-bound PIP is redistributed during early diagenesis to CFA and iron oxides. The use of authigenic CFA as an indicator of the burial flux of reactive phosphorus is complicated by the fact that phosphorus in CFA is also derived from detrital iron-bound PIP (sink switching) and the presence of river-borne (detrital) CFA, particularly in shelf areas adjacent to rivers. Estimated phosphorus output (1–2.5 x 10^{12}g/yr) is less than the phosphorus input flux and suggests that 12–40% of the iron-bound PIP (adsorbed phosphorus) flux is released to the ocean. The uncertainty in the input and output fluxes is large, but could improve with more data on the amount and nature of iron-bound PIP, and the amount of iron-bound phosphorus released by desorption when freshwater mixes with seawater and during sediment resuspension.

The phosphorus cycle has most likely varied in the past in response to changes in chemical weathering rates (climate and orogeny), sea-level fluctuations, MOR spreading rates, glaciation, and oceanic circulation. Major periods of orogeny and glaciation appear to enhance phosphorite formation by increasing the chemical load of rivers. Phosphorite giants are more likely to form during marine transgressions that coincide with increased chemical weathering rates (warm and humid climate combined with increased physical weathering) and less MOR hydrothermal activity. As sea level rises, upwelling and primary productivity shift to expanded shallow-water estuary and shelf areas where a greater portion of the organic phosphorus is released and captured by CFA or iron oxides on or just below the seafloor rather than being recycled in the water column. Shelf deposits are then susceptible to reworking by sea-level fluctuations. A significant but unknown amount of CFA may form by rapid (<10^5 yrs) remineralization of organic matter. Large, phosphorite deposits may provide a proxy for short-term organic carbon burial and a sink of phosphorus when phosphorus input exceeds the long-term organic carbon and MOR iron-oxide burial sinks.

ACKNOWLEDGMENTS

This paper was initiated by discussions of the IGCP 325 workshop on the phosphorus cycle held in October 1996 in Strasbourg. We thank Liliane Prévôt-Lucas and Jacques Lucas for bringing this group together and for their comments on the manuscript. Gleb Baturin and an anonymous reviewer provided useful comments that improved the manuscript. JSC acknowledges support from the Foundation for Research Development and the University Research Committee at UCT. CRG acknowledges support from NSF grant OCE9201305. University of Hawaii, School of Ocean and Earth Science and Technology Contribution No. 4874.

REFERENCES

Allègre, C. J., Gaillardet, J., Levasseur, S. Meynadier, L., and Louvat, P., 1997, The evolution of seawater strontium isotopes in the last hundred million years: Reinterpretation and consequences for erosion and climate models: Seventh Annual V.M. Goldschmidt Conference (Abs): Houston, Lunar and Planetary Institute No. 921, p. 4.

Anderson, S. P., Drever, J. I.,and Humphrey, N. F., 1997, Chemical weathering in glacial environments: Geology, v. 25, p. 399-402.

Arthur, M. A. and Jenkyns, H .C., 1981, Phosphorites and paleoceanography: Oceanologica Acta, Proceedings 26th International Geological Congress, Paris, p. 83-96.

Arthur, M. A., Schlanger, S. O., and Jenkyns, H. C., 1987, The Cenomanian-Turonian oceanic anoxic event, II. Palaeoceanographic controls on organic matter production and preservation, *in* Brooks, J. and Fleet, A. J., eds., Marine Petroleum Source Rocks: London, Geological Society Special Publication, No. 26, p. 401-420.

Baturin, G. N., Lucas, J., and Prévôt-Lucas, L., 1995, Phosphorus behaviour in marine sedimentation: continuous P-behaviour versus discontinuous phosphogenesis: Comptes Rendus Académie des Sciences, Paris, v. 321, p. 263-278.

Baumgartner, A. and Reichel, E., 1975, The world water balance: Amsterdam, Elsevier, 179 p.

Bentor, Y. K., 1980, Phosphorites–the unsolved problems, *in* Bentor, Y.K., ed., Marine phosphorites–geochemistry, occurrence, genesis: Society of Economic Paleontologists and Mineralogists Special Publication, v. 29, p. 3-18.

Berger, W. H. and Wefer, G., 1991, Productivity of the glacial ocean: Discussion of the iron hypothesis: Limnology and Oceanography, v. 36, p. 1899-1918.

Berner, E. K. and Berner, R. A., 1987, The Global Water Cycle, Geochemistry and Environment: Englewood Cliffs, New Jersey, Prentice Hall, 397 p.

Berner, E. K. and Berner, R. A., 1996, Global Environment: Englewood Cliffs, New Jersey, Prentice Hall, 376 p.

Berner, R. A. and Canfield, D., 1989, A new model for atmospheric oxygen over Phanerozoic time: American Journal of Science, v. 289, p. 333-361.

Berner, R. A. and Rao, J.-L., 1994, Phosphorus in sediments of the Amazon River and estuary: Implications for the global flux of phosphorus to the sea: Geochimica et Cosmochimica Acta, v. 58, p. 2333-2340.

Birch, G. F., 1990, Phosphorite deposits on the South African continental margin and coastal terrace, *in* Burnett, W. C. and Riggs, S. R., eds., Phosphate deposits of the world: Vol. 3, Genesis of Neogene to recent phosphorites: Cambridge, Cambridge University Press, p. 153-166

Birch, G. F., Thompson, J., McArthur, J. M., and Burnett, W. C., 1983, Pleistocene phosphorites off the west coast of South Africa: Nature, v. 302, p. 601-603.

Blum, J. D., 1997, The effect of late Cenozoic glaciation on silicate weathering rates and the marine strontium isotope record: Seventh Annual V.M. Goldschmidt Conference (Abs): Houston, Lunar and Planetary Institute No. 921, p. 30.

Broecker, W. S., 1982, Ocean chemistry during glacial time: Geochimica et Cosmochimica Acta, v. 46, p. 1689-1705.

Burton, J. D., 1983, A comment on the behaviour of dissolved organic carbon during estuarine mixing, *in* Bolin, B. and Cook, R.B., eds., The Major Biogeochemical Cycles and Their Interactions: New York, John Wiley and Sons, p. 408-410.

Bush, A. B. G, and Philander, S. G. H., 1997, The late Cretaceous: Simulation with a coupled atmosphere-ocean general circulation model: Paleoceanography, v. 12, p. 495-516.

Chase, E. M. and Sayles, F. L., 1980, Phosphorus in suspended sediments of the Amazon River: Estuarine and Coastal Marine Science, v. 11, p. 383-391.

Compton, J. S. and Mallinson, D. J., 1996, Geochemical consequences of increased late Cenozoic weathering rates and the global CO_2 balance since 100 Ma: Paleoceanography, v. 11, p. 431-446.

COMPTON, J. S., SNYDER, S. W. AND HODELL, D. A., 1990, Phosphogenesis and weathering of shelf sediments from the southeastern United States: Implications for Miocene $\delta^{13}C$ excursions and global cooling: Geology, v. 18, p. 1227-1230.

COMPTON, J. S., HODELL, D. A., GARRIDO, J. R., AND MALLINSON, D. J., 1993, Origin and age of phosphorite from the south-central Florida Platform: Relation of phosphogenesis to sea-level fluctuations and $\delta^{13}C$ excursions: Geochimica et Cosmochimica Acta, v. 57, p. 131-146.

CROWLEY, T. J., 1995, Ice age terrestrial carbon changes revisited: Global Biogeochemical Cycles, v. 9, p. 377-389.

DRUFFEL, E. R. M., HONJO, S., GRIFFIN, S., AND WONG, C. S., 1986, Radiocarbon in particulate matter from the Eastern Sub-Arctic Pacific Ocean: evidence of a source of terrestrial carbon to the deep sea: Radiocarbon, v. 28, p. 397-407.

EDMOND, J. M. AND HUH, Y., 1997, Fixation rates of carbon dioxide by chemical weathering in the continental arc of the western Americas, 15°S to 70°N: Seventh Annual V.M. Goldschmidt Conference (Abs): Houston, Lunar and Planetary Institute No. 921, p. 63.

FEELY, R. A., MASSOTH, G. J., BAKER, E. T., COWEN, J. P., LAMB, M. F. AND KROGSLUND, K. A., 1990, The effect of hydrothermal processes on midwater phosphorus distributions in the northeast Pacific: Earth and Planetary Science Letters, v. 96, p. 305-318.

FEELY, R. A., TREFRY, J. H., MASSOTH G. J., AND METZ S., 1991, A comparison of the scavenging of phosphorus and arsenic from seawater by hydrothermal iron oxyhydroxides in the Atlantic and Pacific oceans: Deep-Sea Research, v. 38, p. 617-623.

FILIPPELLI, G. M., 1997a, Controls on phosphorus concentration and accumulation in oceanic sediments: Marine Geology, v. 139, p. 231-239.

FILIPPELLI, G. M., 1997b. Intensification of the Asian Monsoon and a chemical weathering event in the late Miocene/early Pliocene: Implications for late Neogene climate change: Geology, v. 25, p. 27-30.

FILIPPELLI, G. M. AND DELANEY, M. L., 1994, The oceanic phosphorus cycle and continental weathering during the Neogene: Paleoceanography, v. 9, p. 643-652.

FILIPPELLI, G. M. AND DELANEY, M. L., 1996, Phosphorus geochemistry of equatorial Pacific sediments: Geochimica et Cosmochimica Acta, v. 60, p. 1479-1495.

FÖLLMI, K. B., 1995, 160 m.y. record of marine sedimentary phosphorus burial: Coupling of climate and continental weathering under greenhouse and icehouse conditions: Geology, v. 23, p. 859-862.

FÖLLMI, K.B., 1996, The phosphorus cycle, phosphogenesis and marine phosphate-rich deposits. Earth-Science Reviews, v. 40, p. 55-124.

FÖLLMI, K. B. AND GARRISON, R. E., 1991, Phosphatic sediments, ordinary or extraordinary deposits? The example of the Miocene Monterey Formation (California), *in* Müller, D. W., McKenzie, J. A. and Weissert, H., eds., Controversies in Modern Geology: London, Academic Press, p. 55-84.

FOX, L. E., SAGER, S. L., AND WORFSY, S., 1986, The chemical control of soluble phosphorus in the Amazon estuary: Geochimica et Cosmochimica Acta, v. 50, p. 783-794.

FROELICH, P.N ., 1984, Interactions of the marine phosphorus and carbon cycles, *in* Moore, B. and Dastoor, M. N., eds., The Interaction of Global Biogeochemical Cycles: Pasadena, California Institute of Technology, NASA-JPL Publication 84-21, p. 141-176.

FROELICH, P. N., 1988, Kinetic control of dissolved phosphate in natural rivers and estuaries: a primer on the phosphate buffer mechanism: Limnology and Oceanography, v. 33, p. 649-668.

FROELICH, P. N., BENDER, M. L., LUEDTKE, N. A., HEATH, G. R., AND DE VRIES, T., 1982, The marine phosphorus cycle: American Journal of Science, v. 282, p. 474-511.

GARRELS, R. M. AND MACKENZIE, F. T., 1971, Evolution of Sedimentary Rocks: New York, Norton, 397 p.

GARRISON, R. E. AND KASTNER, M., 1990, Phosphatic sediments and rocks recovered from the Peru margin during Leg 112 of the ocean drilling program, *in* Suess, E., von Huene, R., et al., eds., Proceedings Ocean Drilling Program (College Station, Texas), Scientific Results, v. 112, p. 111-134.

GLENN, C. R., AND ARTHUR, M. A., 1990, Anatomy and origin of a Late Cretaceous phosphorite-greensand giant: Sedimentology, v. 37, p. 123-154.

GLENN, C. R., ARTHUR, M. A., RESIG, J. M., BURNETT, W. C., DEAN, W. E. AND JAHNKE, R. A., 1994a, Are modern and ancient phosphorites really so different?, *in* Iijima, A., Abed, A. M., and Garrison, R. E., eds., Siliceous, phosphatic and glauconitic sediments of the Tertiary and Mesozoic: Proceedings 29th International Geological Congress, Part C: Utrecht, Springer-Verlag, p. 159-188.

GLENN, C. R., FÖLLMI, K. B., RIGGS, S. R., BATURIN, G. N., GRIMM, K. A., TRAPPE, J., ABED, A. M., GALLI-OLIVIER, C., GARRISON, R. E., ILYIN, A., JEHL, C., ROHRLICHV., SADAQAH, R., SCHIDLOWSKI, M., SHELDON, R. E., AND SIEGMUND, H., 1994b, Phosphorus and phosphorites: sedimentology and environments of formation, *in* Föllmi, K. B., ed., Concepts and controversies in phosphogenesis: Eclogae Geologicae Helvetiae, v. 87, p. 747-788.

GRAHAM, W. F. AND DUCE, R. A., 1979, Atmospheric pathways of the phosphorus cycle: Geochimica et Cosmochimica Acta, v. 43, p. 1195-1208.

GRAHAM, W. F. AND DUCE, R. A., 1982, The atmospheric transport of phosphorus to the western North Atlantic: Atmospheric Environment, v. 16 p. 1089-1097.

HEDGES, J. I., 1992, Global biogeochemical cycles: progress and problems: Marine Geology, v. 39, p. 67-93.

HEDLEY, M. J., STEWART, J. W. B., AND CHAUHAN, B. S., 1982, Changes in inorganic and organic soil phosphorus fractions induced by cultivation practices and by laboratory incubations: Soil Science Society of America Journal, v. 46, p. 970-976.

HEGGIE, D. T., SKYRING, G. W., O'BRIEN, G. W., REIMERS, C., HERCZEG, A., MORIARTY, D. J. W., BURNETT, W. C., AND MILNES, A. R., 1990, Organic carbon cycling and modern phosphorite formation on the east Australian continental margin: an overview, *in* Notholt, A. J. G. and Jarvis, I., eds., Phosphorite research and development: London, Geological Society Special Publication, v. 52, p. 87-117.

HODELL, D. A., MEAD, G. A., AND MUELLER, P. A., 1990, Variation in the strontium isotopic composition of seawater (8 Ma to present): Implications for chemical weathering rates and dissolved fluxes to the oceans: Chemical Geology, v. 80, p. 291-307.

HOLLAND, H. D., 1978, The Chemistry of the Atmosphere and Oceans: New York, Wiley Interscience, 351 p.

INGALL, E. D. AND JAHNKE, R., 1994, Evidence for enhanced phosphorus regeneration from marine sediments overlain by oxygen depleted waters: Geochimica et Cosmochimica Acta, v. 58, p. 2571-2575.

INGALL, E. D., BUSTIN, R. M., AND VAN CAPPELLEN, P., 1993, Influence of water column anoxia on the burial and preservation of carbon and phosphorus in marine shales: Geochimica et Cosmochimica Acta, v. 57, p. 303-316.

JØRGENSEN, B.B., 1983, Processes at the sediment-water interface, *in* Bolin, B. and Cook, R.B., eds., The Major Biogeochemical Cycles and Their Interactions: New York, John Wiley and Sons, p. 477-515.

KAZARINOV, V. P., 1966, The problem of exploration of the large deposits of rich phosphorites in Siberia, *in* Bgatov, V.I. and Kazarinov, V.P., eds., Conditions of the Localization of the Phosphate-Bearing Sediments in Siberia: Leningrad (Russia), p. 3-14 (in Russian).

KOMINZ, M. A.,1984, Oceanic ridge volumes and sea-level change: An error analysis: *in* Schlee, J. S., ed., Interregional Unconformities and Hydrocarbon Accumulation: Memoir American Association of Petroleum Geologists, v. 36, p. 109-127.

Krom, M. D. and BERNER, R. A., 1981, The diagenesis of phosphorus in a nearshore marine sediment: Geochimica et Cosmochimica Acta, v. 45, p. 207-216.

LATIMER, J. C., AND FILIPPELLI, G. M., 1997, Phosphorus and iron sedimentation in the Southern Ocean on Glacial/Interglacial time scales: EOS (abs.), American Geophysical Union, v. 78 (17), p. 191.

LIVINGSTONE, D. A., 1963, Chemical composition of rivers and lakes, *in* Fleischer, M., ed., Data of Geochemistry: U.S. Geological Survey Professional Paper 440-G.

LUCOTTE, M. AND D'ANGLEJAN, B., 1983, Forms of phosphorus and phosphorus-iron relationships in the suspended matter of the St. Lawrence estuary: Canadian Journal of Earth Science, v. 20, p. 1880-1890.

LUCOTTE, M., MUCCI, A. HILLAIRE-MARCEL C. AND TRAN S., 1994, Early diagenetic processes in deep Labrador Sea sediments: reactive and nonreactive iron and phosphorus: Canadian Journal of Earth Science, v. 31, p. 14-27

MACKENZIE, F. T., VER, L. M., SABINE, C., LANE, M., AND LERMAN, A., 1993, C, N, P, S global biogeochemical cycles and modeling of global change, *in*

Wollast, R., Mackenzie, F. T., and Chou, L., eds., Interactions of C, N, phosphorus and S Biochemical Cycles and Global Change: NATO ASI Series I, v. 4, Berlin, Springer-Verlag, p. 1-62.

Mallinson, D. J. and Compton, J. S., 1997, Linking phosphogenic episodes on the southeast U.S. margin to marine $\delta^{13}C$ and $\delta^{18}O$ records: Geology, v. 25, p. 103-106.

Martin, J. H., 1991, Iron, Liebig's Law, and the greenhouse: Oceanography, v. 4, p. 52-55.

Martin, J. H., Knauer, G. A., Karl, D. M., and Broenkow, W. W., 1987, VERTEX: carbon cycling in the northeast Pacific. Deep-Sea Research, v. 34, p. 267-285.

Martin, J. M. and Meybeck, M., 1979, Elemental mass-balance of material carried by world major rivers: Marine Chemistry, v. 7, p. 173-206.

Mason, B., 1967, Principles of Geochemistry. New York, John Wiley and Sons, 329 p.

Meybeck, M., 1979, Concentrations des eaux fluviales en éléments majeurs et apports en solution aux océans: Revue de Géologie Dynamique et de Géographie Physique, v. 21, p. 215-246.

Meybeck, M., 1982, Carbon, nitrogen, and phosphorus transport by world rivers: American Journal of Science, v. 282, p. 401-450.

Meybeck, M., 1993, C, N, phosphorus and S in rivers: from sources to global inputs, *in* Wollast, R., Mackenzie, F. T., and Chou, L., eds., Interactions of C, N, phosphorus and S Biochemical Cycles and Global Change: NATO ASI Series I, v. 4, Berlin, Springer-Verlag, p.163-193.

Milliman, J. D. and Zyvitski, J. P. M., 1992, Geomorphic/tectonic control of sediment discharge to the ocean: the importance of small mountainous rivers: Journal of Geology, v. 100, p. 525-544.

Mountain, G. S., Miller, K. G., Blum, P., et al., 1994, Proceedings of Ocean Drilling Program, Initial Reports., v. 150, College Station (Texas), 885 p.

Notholt, A. J. G., Sheldon, R. P. and Davidson, D. F., 1989, Phosphate Deposits of the World, Volume 2, Phosphate Rock Resources: Cambridge, Cambridge University Press, 566 p.

O'Brien, G. W., Milnes, A. R., Veeh, H. H., Heggie, D. T., Riggs, S. R., Cullen, D. J., Marshall, J. F., and Cook, P. J., 1990, Sedimentation dynamics and redox iron-cycling: controlling factors for the apatite-glauconite association on the east Australian continental margin, *in* Notholt, A. J. G. and Jarvis, I., eds., Phosphorite research and development: London, Geological Society Special Publication, v. 52, p. 61-86.

Palmer, M. R., 1985, Rare earth elements in foraminifera tests. Earth and Planetary Science Letters, v. 73, p. 285-298.

Petit, J. R., Mounier, L., Jouzel, J., Korotkevitch, V. S., Kotlyako, V. I., and Lorius, C., 1990, Paleoclimatological and chronological implications of the Vostok core dust record: Nature, v. 343, p. 159-161.

Prahl, F. G., Ertel, J. R., Goni, M. A., Sparrow, M. A., and Eversmeyer, B., 1994, Terrestrial organic carbon contributions to sediments on the Washington margin: Geochimica et Cosmochimica Acta, v. 58, p. 3035-3048.

Ramirez, A. J. and Rose, A. W., 1992, Analytical geochemistry of organic phosphorus and its correlation with organic carbon in marine and fluvial sediments and soils: American Journal of Science, v. 292, p. 421-454.

Rasmussen, B., 1996, Early-diagenetic REE-phosphate minerals (florencite, gorceixite, crandallite, and xenotime) in marine sandstones: A major sink for oceanic phosphorus: American Journal of Science, v. 296, p. 601-632.

Ravizza, G., 1993, Variations of the $^{187}Os/^{186}Os$ record of the past 28 million years as inferred from metalliferous carbonates: Earth and Planetary Science Letters, v. 130, p. 155-167.

Raymo, M. E., Ruddiman, W. F., and Froelich, P. N., 1988, Influence of late Cenozoic mountain building on ocean geochemical cycles: Geology, v. 16, p. 649-653.

Raymont, J. E. G., 1983, Plankton and productivity in the oceans, Volume 2—Zooplankton (2nd Ed.): Oxford, Pergammon Press, 824 p.

Richey, J. E., 1983, Interactions of C, N, phosphorus and S in river systems: a biogeochemical model, *in* Bolin, B. and Cook, R. B., eds., The Major Biogeochemical Cycles and Their Interactions: New York, John Wiley and Sons, p. 365-383.

Richter, F. M., Rowley, D. B., and Depaolo, D. J., 1992, Sr isotope evolution of seawater: The role of tectonics: Earth and Planetary Science Letters, v. 109, p. 11-23.

Riggs, S. R., 1984, Paleoceanographic model of Neogene Phosphorite deposition, U.S. Atlantic continental margin: Science, v. 223, p. 123-131.

Romankevich, E. A., 1984, Geochemistry of Organic Matter in the Ocean: New York, Springer-Verlag, 334 p.

Ruttenberg, K. C., 1993, Reassessment of the oceanic residence time of phosphorus: Chemical Geology, v. 107, p. 405-409.

Ruttenberg, K. C. and Berner, R. A., 1993, Authigenic apatite formation and burial in sediments from non-upwelling, continental margin environments: Geochimica et Cosmochimica Acta, v. 57, p. 991-1007.

Savenko, V. S. and Zakharova, E. A., 1995, Phosphorus in riverine runoff. Doklady Rossiiskoi Akademii Nauk, v. 345, p. 682-685 (in Russian).

Schlanger, S. O., Arthur, M. A., Jenkyns, H. C., and Scholle P. A., 1987, The Cenomanian-Turonian oceanic anoxic event, I. Stratigraphy and distribution of organic carbon rich beds and the marine $\delta^{13}C$ excursion, *in* Brooks, J. and Fleet, A. J., eds., Marine Petroleum Source Rocks: London, Geological Society Special Publication No. 26, p. 371-400.

Schwertmann, U. and Taylor, R. M., 1989, Iron oxides, *in* Dixon, J.B. and Weed, S.B., eds., Minerals in Soil Environments: Madison (Wisconson), Soil Science Society of America, p. 379-438.

Shackleton, N. J., and Hall, M. A., 1995, Stable isotope records in bulk sediment: College Station (Texas), Proceedings Ocean Drilling Program, Scientific Results, v. 138, p. 797-806.

Sheldon, R. P., 1989, Phosphorite deposits of the Phosphoria Formation western United States, *in* Notholt, A. J. G., Sheldon, R. P. and Davidson, D. F., eds., Phosphate Deposits of the World, Volume 2, Phosphate Rock Resources: Cambridge, Cambridge University Press, p. 53-61.

Shemesh, A., Mortlock, R. A., and Froelich, P. N., 1989, Late Cenozoic Ge/Si record of marine biogenic opal: Implications for variations of riverine fluxes to the ocean: Paleoceanography, v. 4, p. 221-234.

Sholkovitz, E. R., 1976, Flocculation of dissolved organic and inorganic matter during the mixing of river water and seawater: Geochimica et Cosmochimica Acta, v. 40, p. 831-845.

Sholkovitz, E. R., Boyle, E. A., and Price, N. B., 1978, The removal of dissolved humic acids and iron during estuarine mixing: Earth and Planetary Science Letters, v. 40, p. 130-136.

Sorkhabi, R. B., and Stump, E., 1993, Rise of the Himalaya: A geochronologic approach: Geological Society of America Today, v. 3, p. 85-92.

Spero, H. J., Bijma, J., Lea, D. W., and Bemis, B. E., 1997, Effect of seawater carbonate concentration on foraminiferal carbon and oxygen isotopes: Nature, v. 390, p. 497-500.

Sundby, B., Gobeil, C., Silverberg, N., and Mucci, A., 1992, The phosphorus cycle in coastal marine sediments: Limnology and Oceanography, v. 37, p. 1129-1145.

Tiessen, H., Stewart, J. W. B., and Cole, C. V., 1984, Pathways of phosphorus transformations in soils of differing pedogenesis: Soil Science Society of America Journal, v. 48, p. 853-858.

Van Cappellen, P. and Ingall, E. D., 1994, Benthic phosphorus regeneration, net primary production, and ocean anoxia: A model of the coupled marine biogeochemical cycles of carbon and phosphorus: Paleoceanography, v. 9, p. 677-692.

Van Cappellen P. and Ingall, E. D., 1996, Redox stabilization of the atmosphere and oceans by phosphorus-limited marine productivity: Science, v. 271, p. 493-496.

Van Eck, T. M., 1982, Forms of phosphorus in particulate matter from the Hollands, Diep/Haringvilet, the Netherlands: Hydrobiologia, v.92, p. 665-681.

Wheat, C. G., Feely, R. A., and Mottl, M. J., 1996, Phosphate removal by oceanic hydrothermal processes: An update of the phosphorus budget in the oceans: Geochimica et Cosmochimica Acta, v. 60, p. 3593-3608.

Wollast, R., 1983, Interactions in estuaries and coastal waters, *in* Bolin, B. and Cook, R.B., eds., The Major Biogeochemical Cycles and Their Interactions: New York, John Wiley and Sons, p. 385-407.

Zafiriou, O. C., Gagosian, R. B., Peltzer, E. T., Alford, J. B. and Loder, T., 1985, Air-to-sea fluxes of lipids at Enewetak Atoll: Geophysical Research, v. 90(D1), p.2409-2423.

Zanin, Y. N., 1989, Phosphate-bearing weathering crusts and their related deposits, *in* Balasubramanian, K. S., et al., eds., Weathering; its products and deposits. Vol. II. Products-deposits-geotechnics: Athens, Theophrastus Publications, p. 321-367.

ROLE OF TECTONICS IN PHOSPHORUS DISTRIBUTION AND CYCLING

MICHAEL W. GUIDRY*, FRED T. MACKENZIE

Department of Oceanography, University of Hawaii, School of Ocean and Earth Science and Technology, Honolulu, HI 96822

AND

ROLF S. ARVIDSON

Department of Oceanography, Texas A&M University, College Station, TX 77843-3146

ABSTRACT: Until very recently, the effect of tectonism on the exogenic phosphorus cycle has been neglected. Currently, the subduction of phosphorus associated with organic matter buried in ocean sediment to the mantle represents a significant flux relative to other fluxes within the exogenic phosphorus cycle. Phosphorus is generally considered to have been the limiting nutrient for oceanic primary production over much of geologic time. Therefore, changes in the total exogenic phosphorus mass control, to a certain extent, the amount of phosphorus available for oceanic production. Over geologic time, the exogenic phosphorus mass is likely governed by plate tectonics; specifically, by the balance between input of newly formed crystalline rock and the output via subduction of oceanic crust and the overlying sediment to the mantle.

INTRODUCTION

In recent years, a great deal of attention has been focused on the global biogeochemical cycle of carbon (e.g., Sundquist and Broecker, 1985). Interest has arisen, in part, because of the connection between the carbon cycle and climate and the issue of an enhanced greenhouse effect and global warming induced by human activities. Part of the current understanding of how the carbon cycle can affect climate is derived from studying the geologic record. On a time scale of tens of millions to hundreds of millions of years, tectonics plays an important role in carbon cycling (e.g., Berner et al., 1983; Holser et al., 1988; Berner, 1990), although there is still controversy concerning the exact nature of tectonic controls (Raymo et al., 1988; Raymo, 1991; Kerrick and Caldeira, 1994; Berner and Caldeira, 1997; Kump and Arthur, 1997). Interestingly, the distributions of carbonate mineral compositions in the geologic record also have been linked to tectonic forcing (Mackenzie and Pigott, 1981; Berner et al., 1983; Fischer, 1984; Mackenzie and Morse, 1992; Arvidson and Mackenzie, 1997). In addition, tectonic forcing affects cycles of elements other than carbon.

Phosphorus availability is thought to limit net primary production in the global ocean on a geologic time scale (e.g. Redfield, 1958; Holland, 1978; Broecker and Peng, 1982; Smith and Mackenzie, 1987; Kump, 1988). However, this conclusion is debatable with arguments being made for nitrogen limitation of production (e.g. Ryther and Dunstan, 1971; Falkowski, 1997). Over geologic time, continental weathering controls the rate of delivery of terrestrial phosphorus to the ocean. Ultimately, the phosphorus is deposited and sequestered in a variety of marine sedimentary pools such as iron-bound phosphorus, organic phosphorus, apatite (e.g., Froelich et al., 1982; Ruttenberg and Berner, 1993; Föllmi, 1996), and aluminophosphates (Rasmussen, 1996; Rasmussen, this volume) and its return to land is then largely dependent on tectonic activity. Lacking a prevalent gas phase, only tectonically linked processes such as the uplift and subduction of marine sediments and production of phosphorus-enriched crystalline rock (i.e. fluorapatite) can return previously sedimented phosphorus to the exogenic cycle for weathering. The exogenic phosphorus cycle has been extensively discussed and modeled (Lerman et al., 1975; Froelich et al., 1982; Mackenzie et al., 1993; Föllmi, 1996; Compton et al., this volume). Nevertheless, until recently, little attention has been paid to the role of tectonics in the phosphorus biogeochemical cycle (Van Cappellen and Ingall, 1996; Kump and Mackenzie, 1996; Colman et al., 1997).

In this paper, we consider the global cycle of organic phosphorus as tied to that of organic carbon and show that subduction may limit the availability of phosphorus in the exogenic cycle and thereby influence long-term, phosphorus-limited productivity in the ocean.

Tectonics and Phosphorus Cycling

The Phanerozoic can be divided into three major tectonic cycles: the Caledonian (Cambrian to early Devonian), Hercynian (early Devonian to early Triassic), and Alpine (early Triassic to present) (Sloss, 1963; Wise, 1974; Ronov, 1980). Tectonic processes can directly affect the biogeochemical cycling and partitioning of elements among the major exogenic reservoirs—crustal rock, ocean, atmosphere, and biosphere (e.g., Garrels and Perry, 1974; Cook and McElhinny, 1979; Holland, 1978; Berner et al., 1983; Lasaga et al., 1985; Raymo et al., 1988; Wilkinson and Walker, 1989; Mackenzie and Morse, 1992). Over geologic time, several geochemical processes associated with tectonic processes are capable of large-scale repartitioning of elements between the principal exogenic reservoirs. The tectonic processes include volcanism, orogeny, seafloor spreading, and subduction. One component of the tectonic cycle, the subduction of oceanic lithosphere, is significant to exogenic element cycling because of the overlying sedimentary component of the lithosphere. The elemental composition of the sedimentary cover is in part a function of the rates of biotic and abiotic processes occurring in the exogenic cycle, such as biological production, sedimentation, diagenesis, and chemical and physical weathering on land. Thus, changes in subduction rates can influence the recycling rate not only of the crystalline lithosphere, but also of elements within the sedimentary cover. In turn, changes in subduction rates can impact the biogeochemical cycles of elements within the exogenic cycle.

Relationships between tectonic processes and the long-term

*email: mguidry@soest.hawaii.edu

Marine Authigenesis: From Global to Microbial, SEPM Special Publication No. 66

geochemical cycle of carbon have been well documented (e.g. Berner, et al., 1983; Lasaga et al., 1985; Arvidson et al., this volume). Atmospheric CO_2 taken up in the weathering of calcium silicates on land is deposited as calcium carbonate in the ocean. The calcium carbonate is subducted or buried to great depth where it is converted to calcium silicate. The net chemical reaction, the Urey reaction (Urey, 1952), involves the release of CO_2 and its return via volcanism and other processes to the Earth's surface:

$$CaCO_3 + SiO_2 = CO_2 + CaSiO_3 \quad (1)$$

In addition, CO_2 and nitrogen gases are released by the volatilization of organic matter buried in the subducted sediments, returning CO_2 and nitrogen to their respective exogenic cycles.

We now discuss the linkages between carbon and phosphorus in marine sediments. There are several mechanisms and processes through which elements are incorporated in sediment riding on oceanic plates. One important set of processes is the production, sedimentation, burial, and diagenesis of organic matter. Sinking detritus from biological production in the euphotic zone carries both carbon and phosphorus to the sediment. Carbon in carbonate shells and tests and organic carbon associated with phosphorus in organic matter are linked initially via the sedimentary flux of organic matter to the sediment-water interface. As the organic matter falls through the water column and is degraded by microbial respiration, the carbon-to-phosphorus ratio (C:P ratio) changes slightly (Copin-Montegut and Copin-Montegut, 1983). This change is the initial phase in the decoupling of the organic carbon and organic phosphorus biogeochemical cycles in the ocean (Van Cappellen and Ingall, 1994). After deposition and burial, microbial and inorganic processes in the oxic, suboxic, and anoxic zones of sediment further modify the C:P ratio (Berner, 1980), enhancing the decoupling of the biogeochemical cycles of the two elements. Organic carbon and phosphorus that remain in the sediments will eventually be subducted as the oceanic plate is recycled into the mantle, unless the sediment is uplifted or becomes part of the accretionary prism adjacent to the subduction zone. Carbon and phosphorus removed from the exogenic cycle via subduction must ultimately be returned to the Earth's surface. Otherwise, the subduction fluxes would lead to an elemental drain from the exogenic cycle. Herein lies a problem with the phosphorus cycle.

During subduction, organic matter in marine sediments undergoes metamorphic reactions at elevated temperatures and pressures. The volatilization of the organic matter releases carbon and nitrogen as gaseous CO_2 and N_2. Interestingly though, on the basis of isotopic and mass-balance constraints, carbon fluxes between the mantle and the exogenic carbon cycle seem to be in quasi- steady-state, but there may be a net flux of nitrogen into the mantle (Javoy, 1997). Because there is no quantitatively important phosphorus gas phase under metamorphic conditions (nor even at surface environmental conditions), it is very likely that the phosphorus released by the volatilization of subducted organic matter is incorporated in apatite in crystalline rocks (Skinner, 1973). Unless phosphine gas (Devai et al., 1988; Gassman, 1994) has been overlooked as an important component of volcanic gases, the subducted organic phosphorus probably cannot be returned to the Earth's surface at the same rate or on the same time scale as subducted organic carbon and nitrogen. In addition, the phosphorus that is recycled in crystalline rocks to the Earth's surface must be chemically weathered to supply the dissolved phosphorus to the ocean.

The amount of time required for the return of the subducted phosphorus to the exogenic system influences its status as a limiting nutrient on geologic time scales. There has been considerable debate concerning whether nitrogen or phosphorus is the limiting nutrient for primary organic production in the global ocean over geologic time. On a time scale of days to decades, local nutrient limitation is complex and temporally variable, and it is likely that several co-limiting factors determine the level of net primary production. Indeed, recent *in situ* experiments have shown that even the availability of trace metals, such as iron, can limit production on a local to regional scale (Martin and Fitzwater, 1989; Coale et al., 1996). In surface waters of some oceanic regions, nitrate has been shown to be limiting (Ryther and Dunstan, 1971), and in some coastal waters, phosphorus (Smith, 1984).

Based on an interpretation of the evolutionary history of enzymatic systems involved in nitrification and denitrification, Falkowski (1997) concluded that nitrogen ultimately limits long-term ocean productivity. This conclusion is in direct contrast to the geochemical argument that over geologic time, phosphorus availability determines the rate of net primary production in the ocean (Redfield, 1958; Holland, 1978; Broecker and Peng, 1982; Smith, 1984; Smith and Mackenzie, 1987; Van Cappellen and Ingall, 1996). The geochemical argument is based on the observation that without a sizable atmospheric reservoir of phosphorus, phosphorus availability is limited by the rate of weathering, whereas in periods of nitrate deficiency, nitrogen-fixing organisms can draw from an enormous atmospheric source to produce fixed nitrogen. Karl et al. (1997) recently demonstrated the validity of this argument on a regional scale in studies of the biogeochemical cycle of nitrogen in the subtropical, oligotrophic North Pacific Ocean. If biotic responses to changing conditions are rapid enough to mitigate the effects of nitrogen deficiencies on short time scales, then arguably, nitrogen is likely not a limiting nutrient on geologic time scales.

If phosphorus is the limiting nutrient in the ocean over geologic time scales, then it is the return of phosphorus to the Earth's surface via crystalline rock production, sediment uplift, and the subsequent release of phosphorus by chemical weathering that control ocean productivity over geologic time by controlling the amount of phosphorus returning to the oceans. By controlling ocean productivity, phosphorus also regulates, to a certain extent, the amount of organic carbon and carbonate shells and tests that are eventually buried and thereby influences the amount of oxygen in the ocean-atmosphere system (Holland, 1994; Van Cappellen and Ingall, 1996).

Other geologic processes that are capable of redistributing the life-essential elements of carbon, phosphorus, and nitrogen are also linked to tectonic activity. First-order changes in sea level, which are related to ridge accretion rates, seafloor spreading rates, and subduction, can be correlated to a certain extent with the areal extent of shallow-water deposition and survival of carbonate rocks (Mackenzie and Morse, 1992) and phosphorites (e.g. Cook and McElhinny, 1979; Glenn et al., 1994; Mallinson and Compton, 1997). Wilkinson and Walker (1989), using continental freeboard as a proxy to model Phanerozoic carbonate partitioning among oceanic environments, demonstrated a genetic link between eustacy and sedimentary carbonate cycling, as did the work of Mackenzie and Morse (1992). In light of these studies, it seems clear that investigations of the geochemical cycling of life-essential elements during the Phanerozoic should incorporate the impacts of tectonic processes.

To investigate the role of tectonic processes in the geochemical cycling of phosphorus, we employed the following approaches: (1) using published sedimentary data, we computed carbonate and phosphorus existing mass for each Phanerozoic period; (2) the mass distributions were used to calculate survival rates and decay constants for sedimentary carbonate rocks and phosphorus in the Phanerozoic sedimentary lithosphere; and (3) through the construction and use of simple box models of the linked geochemical cycles of carbon and phosphorus, we investigated the role of subduction in regulating the availability of phosphorus in the exogenic cycle and hence its role in determining net primary production in the ocean and in regulating atmospheric oxygen.

DATA AND METHODS OF CALCULATION

Carbonate and Phosphorus Mass Distributions

The Phanerozoic mass distributions and survival rates for carbonate in continental and oceanic sediments were calculated using data from Wilkinson and Walker (1989) compiled by Mackenzie and Morse (1992) (Table 1). To maintain consistency with previous work (Wilkinson and Walker, 1989; Wilkinson and Algeo, 1989; Mackenzie and Morse, 1992), we give the carbonate mass data in grams calcium.

The Phanerozoic sedimentary phosphorus mass was divided into the reservoirs of continental and ocean sediments and phosphorite for the respective Phanerozoic Periods (Table 2). Continental phosphorus mass distributions (Table 2) were calculated using data from Ronov (1980). Currently there are no compilations of the phosphorus content of sedimentary rocks on a Period by Period basis for the Phanerozoic, although Föllmi (1995) compiled a record for marine sedimentary phosphorus for the last 160 m.y. and Filippelli and Delaney (1994) determined the phosphorus content of Neogene sediments. Consequently, for each period, the phosphorus mass in continental sedimentary rocks was estimated by multiplying the total continental sedimentary mass by the average phosphorus concentration in sedimentary rocks, 0.126 wt% P_2O_5 (Ronov 1980). Because the phosphorus content of most sedimentary rocks falls within a narrow range, between 0.1 and 0.2% (Poldervaart, 1955), the use of a single value (0.126 wt%) for the global average phosphorus concentration in sedimentary rocks should not affect our interpretations significantly. The Period-averaged mass distributions of phosphorus in ore-grade phosphorite deposits were calculated using data from Cook and McElhinny (1979). The total Phanerozoic phosphorus mass in phosphorite is 1.3×10^{17} grams.

The mass distributions of phosphorus in oceanic sedimentary deposits (Table 2) were calculated using Gregor's (1985) data on the mass-age distribution of Phanerozoic sedimentary rock. To calculate the extant oceanic sediment masses of the Tertiary, Cretaceous, and Jurassic Periods, the sediments were classified as either pelagic or shallow-water deposits. The grouping has a two-fold purpose. First, it enables us to take into account the differences in sedimentary phosphorus concentrations in pelagic versus shallow-water continental margin deposits. Second, in calculations of the total oceanic phosphorus mass, changes from Period to Period could be accounted for in each group's percentage of the total sedimentary mass. Therefore, as we go back in geologic time through the Jurassic, the relative increase in shallow-water deposits versus the concomitant decrease in pelagic deposits is evident.

TABLE 1.—PHANEROZOIC CARBONATE MASS DISTRIBUTION AND SURVIVAL*

Period	Period Duration (10^6 years)	Continental Mass (10^{20} g-Ca)	Continental Survival (10^{20} g-Ca/m.y.)	Ocean Mass (10^{20} g-Ca)	Ocean Survival (10^{20} g-Ca/m.y.)	Total Carbonate Survival (10^{20} g-Ca/m.y.)
Tertiary	56	238.4	4.3	205.1	3.7	7.9
Cretaceous	69	375.6	5.5	101.8	1.5	7.0
Jurassic	70	333.4	4.8	28	4	5.2
Triassic	45	110.5	2.5	–	–	2.5
Permian	40	184.0	4.6	–	–	4.6
Carboniferous	65	289.5	4.5	–	–	4.5
Devonian	50	252.0	5.0	–	–	5.0
Silurian	30	66.0	2.2	–	–	2.2
Ordovician	78	202.8	2.6	–	–	2.6
Cambrian	57	196.7	3.5	–	–	3.5

*data from Wilkinson and Walker (1989) and Mackenzie and Morse (1992)

TABLE 2.—PHANEROZOIC PHOSPHORUS MASS DISTRIBUTION AND SURVIVAL*

Period	Period Duration (10^6 years)	Continental Mass (10^{18}g-P)	Continental Survival (10^{18}g-P/m.y.)	Ocean Mass (10^{18}g-P)	Ocean Survival (10^{18}g-P/m.y.)	P_2O_5 Mass (10^{14}g-P)	P_2O_5 Survival (10^{14}g-P/m.y.)	Total Phosphorus Survival (10^{18}g-P/m.y.)
Tert-Pliocene	66	77.0	1.17	182	2.75	549	8.3	3.9
Cretaceous	66	113.0	1.71	89	1.35	69	1.1	3.1
Jurassic	53	83.6	1.58	33	0.62	69	1.3	2.2
Triassic	50	57.7	1.15	–	–	0.09	0.002	1.2
Permian	45	54.7	1.22	–	–	549	12.2	1.2
Carboniferous	65	71.3	1.10	–	–	0.14	0.002	1.1
Devonian	55	91.6	1.67	–	–	0.005	0.0001	1.7
Silurian	35	35.8	1.02	–	–	0.005	0.002	1.0
Ordovician	55	50.9	.925	–	–	8.7	0.2	0.92
Cambrian	80	82.4	1.03	–	–	44	0.6	1.0

*data from Cook and McElhinny (1979), Ronov (1980), Gregor (1985), Ruttenberg and Berner (1993), and Filippelli and Delaney (1996)

For each Period, the percentage of the pelagic and shallow-water sediments surviving of the total surviving sedimentary mass was determined. The amount of phosphorus contained in the sediment group was obtained by multiplying the percentage for the group by the average phosphorus content of the group. The values used for the average phosphorus content (~0.050 wt% for shallow-water deposits and ~0.031 wt% for pelagic deposits) were computed from phosphorus concentrations in recent near-shore and pelagic sediments in Ruttenberg and Berner (1993), and Filippelli and Delaney (1996). The total phosphorus contained in oceanic sedimentary deposits for each period (Table 2) is the sum of these two products. We did not include data for Triassic and older oceanic deposits in Table 2 because most of the sediment originally deposited has been removed, at least temporarily, from the exogenic cycle by subduction (i.e. differential cycling: Garrels and Mackenzie, 1971; Veizer, 1988).

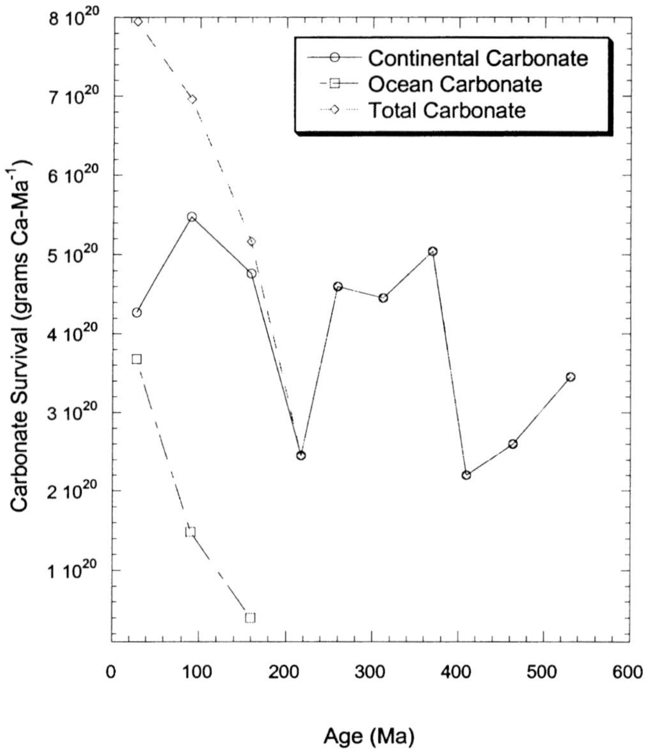

FIG. 1.—Carbonate survival rates (grams calcium/m.y.) for extant Cambrian through Tertiary continental carbonate and Jurassic through Tertiary extant oceanic carbonate. Survival rates for each period are plotted midway between the geochronometric age estimates (in Ma) of the period boundaries. The Total Carbonate curve is the sum of the Continental and Oceanic Carbonate curves.

Survival Rate and Decay Constant Calculations

Phanerozoic survival rates of phosphorus and carbonate reservoirs (Tables 1 and 2) were calculated according to the equation

$$S = m/t \tag{2}$$

where S is the survival rate (in grams/yr), m is the surviving mass in grams of either phosphorus or carbonate, and t (in years) is the duration of the Period (Garrels and Mackenzie, 1971; Garrels et al., 1976; Gregor, 1985). Note that m is the *surviving mass* (i.e. existing mass), not the total mass originally deposited. Carbonate and phosphorus survival rates computed for each reservoir and each period (Tables 1 and 2) are plotted in Figures 1 and 2. Survival rates for each period are plotted midway between the chronometric age estimates of the period boundaries. Differences in the chronometric scales used by various authors resulted in small differences in the ages plotted in Figure 1 and in Figure 2. Any interpretation of the data should be made with this caveat in mind.

In addition to computing the survival rate of oceanic carbonate, oceanic phosphorus, and total phosphorus, we calculated the decay constant, M_1, by fitting exponential curves to the survival rate graphs.

$$Y = M_0 e^{M_1 X} \tag{3}$$

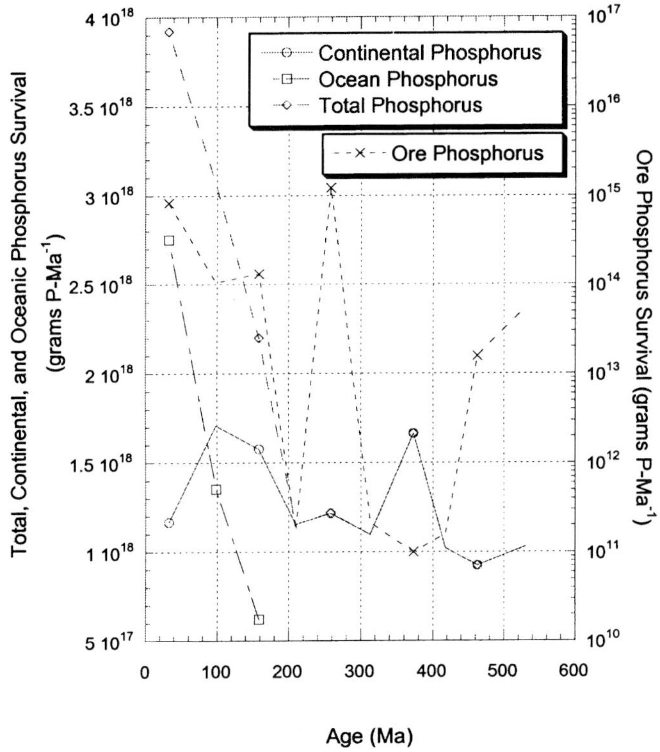

FIG. 2.—Phosphorus survival rates (grams phosphorus per m.y.) for extant Cambrian through Tertiary continental phosphorus (i.e. non-phosphorite phosphorus on continents) and ore-grade phosphorite deposits and Jurassic-Tertiary oceanic phosphorus. The scale for the Ore-Phosphorous curve is on the right. The Total Phosphorus Curve is the sum of the Continental, Ocean, and Ore Phosphorus survival rates (see Table 2). Survival rates for each period are plotted midway between geochronometric age estimates (in Ma) of the period boundaries. Differences in the geochronometric scales used by different authors result in small differences between the midpoints in Figures 1 and 2.

where Y is the survival rate ($g/10^6\ yr^{-1}$) at time X (10^6 yr), M_1 is the decay constant ($10^6\ yr^{-1}$), and M_0 is the rate constant ($g/10^6$ yr) (Wilkinson and Walker, 1989; Garrels and Mackenzie, 1971; Veizer and Jansen, 1979; Veizer and Jansen, 1985). Two implicit assumptions are made using this approach for estimating mass flux relationships. First, the rate of change or cycling is proportional to the mass or volume present (a first-order, constant-mass model of Garrels and Mackenzie, 1971). For this assumption to hold, no preferential preservation or destruction of mass or volume has occurred. Processes of destruction such as erosion are proportional to the mass of material present. Secondly, the rate of change is constant for each period mass unit.

MODELING LONG-TERM CYCLING OF PHOSPHORUS ASSOCIATED WITH ORGANIC CARBON

The objective of this model is to investigate how tectonic processes, especially subduction, influence the cycling of phosphorus associated with carbon. As mentioned previously, sedimentation and burial of organic matter are one mechanism by which phosphorus ultimately becomes sequestered in ocean sediments. To begin with, a first-order, steady-state carbon model was constructed (Fig. 3). Carbon reservoirs and fluxes within this model are those important to long-term carbon cycling. We then constructed a simple first-order, steady-state model of phosphorus associated with organic carbon based on the carbon model (Fig. 4). All fluxes depicted in Figures 3 and 4 are net first order fluxes.

The constraints and provisions of the carbon model are given in Tables 3 and 4 and below. The anthropogenic riverine flows of inorganic and organic carbon are from Meybeck (1982). Estimates of carbon derived from weathering of $CaCO_3$ and $CaSiO_3$ are from the weathering model of Mackenzie (1992). Deposition rates of carbonate and organic matter in the ocean were derived from Milliman (1974), Wollast (1994), Smith and Mackenzie (1987), Mackenzie et al. (1993), Smith and Hollibaugh (1993), and Berner (1982). The metamorphic and volcanic flux of CO_2 is from Des Marais (1985). The CO_2-weathering fluxes to land for weathering of $CaCO_3$ and $CaSiO_3$ are constrained by the stoichiometry of the weathering reactions involving these phases. Subduction fluxes are in accord with estimates in the literature (Berner et al. 1983) for the amount of sediment being subducted, the rate of subduction, and the average carbonate and organic matter content of sediment. The net organic production flux on land is constrained by the fact that the pre-anthropogenic oceans must be net heterotrophic because of (1) organic matter delivery to the oceans by rivers and its respiration in the ocean, (2) a marine burial flux of organic matter on the order of 10^{13} moles carbon per year (Smith and Mackenzie, 1987; Smith and Hollibaugh, 1993), and (3) the estimates of primary production in terrestrial ecosystems (e.g., Woodwell and Whittaker, 1968).

The carbon model has eight reservoirs which are referenced by number in Figure 3: terrestrial organic matter (2), terrestrial silicate minerals (4), terrestrial carbonate minerals (3), atmospheric carbon (1), ocean dissolved inorganic carbon (6), ocean organic carbon (5), sedimentary organic matter (8), and sedimentary carbonate minerals (7). The estimated pre-anthropogenic reservoir sizes and residence times of the steady-state carbon cycle described in this section are given in Table 3. Values for steady-state carbon fluxes and their associated rate constants are given in Table 4. The *Land* reservoir in Figure 3 includes three carbon reservoirs: terrestrial organic matter, terrestrial carbonate minerals, and terrestrial silicate minerals. The *Atmosphere* reservoir is simply the carbon in atmospheric carbon dioxide. The *Ocean* reservoir of carbon in Figure 3 includes dissolved inorganic carbon (DIC), primarily bicarbonate, and carbon in organic matter. The *Sediment* carbon reservoir in Figure 3 includes carbon in organic matter and inorganic carbon in carbonate minerals that have been deposited and buried (Sedimentary CH_2O and Sedimentary $CaCO_3$, respectively, in Table 3).

In the phosphorus model (Fig. 4), the *Land* phosphorus reservoir consists of terrestrial organic and inorganic phosphorus, biota, and humus (Mackenzie et al., 1993). The phosphorus mass in the *Ocean* reservoir of the model (Fig. 4) is the total reactive dissolved inorganic phosphorus (Mackenzie et al., 1993). The *Sedimentary Organic Phosphorus* reservoir (Fig. 4, Table 3) is the phosphorus associated with buried organic matter in sedi-

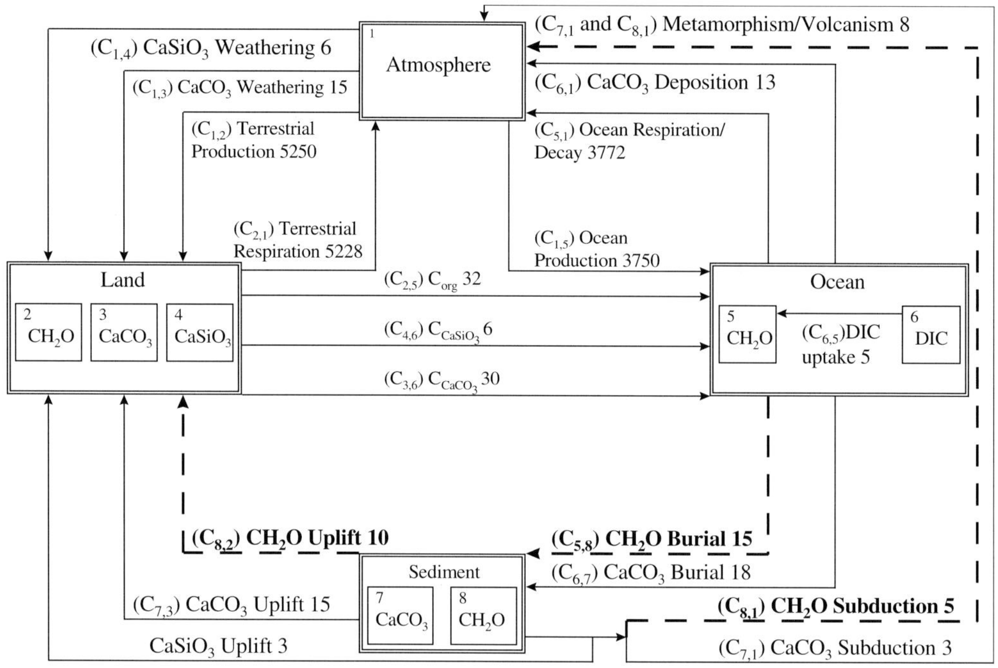

FIG. 3.—Modern steady-state carbon biogeochemical cycle. Flux rates between carbon reservoirs are in units of 10^{12} moles per year. The bold dashed lines indicate the flux rates from which the organic phosphorus model flux rates in Figure 4 were derived. $C_{i,j}$ denotes carbon flux from reservoir "i" to reservoir "j" where i and j represent reservoir numbers 1-8 described in Table 3. Flux rates are listed in Table 4. DIC = dissolved inorganic carbon.

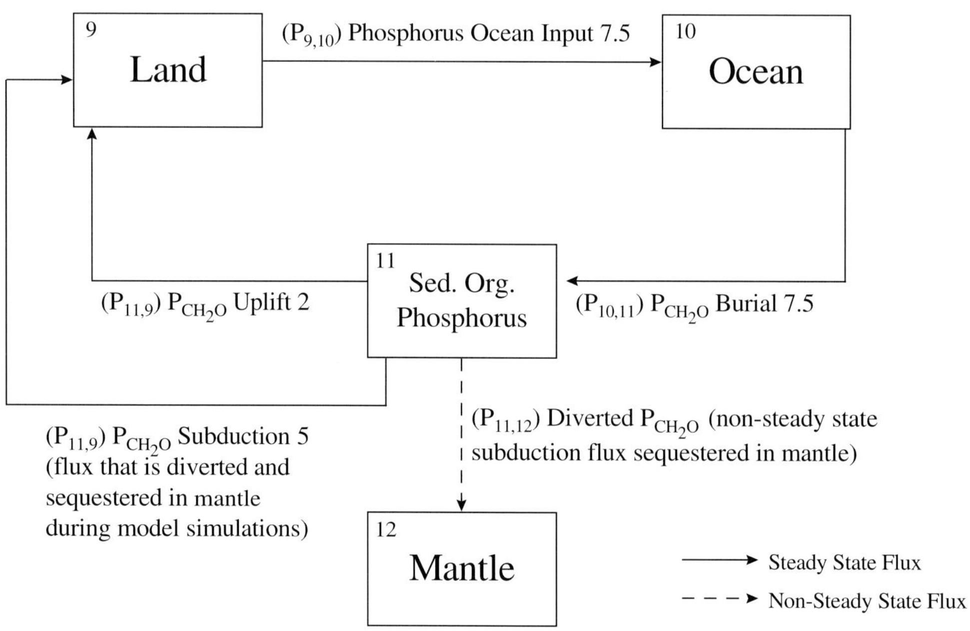

FIG. 4.—Organic phosphorus cycle showing modern steady-state model phosphorus reservoirs—Land, Ocean, and Sedimentary Organic Phosphorus-and the Mantle reservoir. In non- steady-state simulations, the Sedimentary Organic Phosphorus subduction flux to Land was diverted to the Mantle where it was sequestered. Flux rates between reservoirs are in units of 10^{10} moles per year. $P_{i,j}$ denotes phosphorus flux from reservoir "i" to reservoir "j" where i and j represent reservoir numbers 9-12 described in Table 3.

TABLE 3.—CARBON AND PHOSPHORUS RESERVOIR MASSES AND RESIDENCE TIMES AT STEADY STATE*

Reservoir**	Mass (moles)	Residence Time (years)
Terrestrial CH_20 (2)	2.08×10^{17}	39
Terrestrial $CaCO_3$ (3)	$6{,}580 \times 10^{12}$	219
Terrestrial $CaSiO_3$ (4)	$1{,}312 \times 10^{12}$	219
Atmospheric Carbon (1)	$49{,}583 \times 10^{12}$	5.6
Ocean Inorganic Carbon (6)	3.2×10^{18}	89,000
Ocean CH_2O (5)	$83{,}750 \times 10^{12}$	22
Sedimentary CH_2O (8)	13×10^{20}	8.7×10^{7}
Sedimentary $CaCO_3$ (7)	52×10^{20}	2.9×10^{8}
Terrestrial Phosphorus (9)	1.6479×10^{15}	22,000
Ocean Inorganic Phosphorus (10)	3.2395×10^{15}	43,000
Sedimentary Organic Phosphorus (11)	6.5×10^{18}	8.7×10^{7}
Mantle Phosphorus (12)	?	?

* see text for data reference

** the number in parenthesis is the reservoir reference number for Figures 3 and 4 and Table 4.

TABLE 4.—STEADY STATE FLUXES AND RATE CONSTANTS FOR CARBON AND PHOSPHORUS MODELS

Fluxes ($F_{i,j}$) $P_{i,j}$ – Phosphorus $C_{i,j}$ – Carbon	Magnitude (moles/yr)	Rate Constants (k_{ij}) (yr)
$C_{1,2}$	5250×10^{12}	0.106
$C_{1,3}$	15×10^{12}	3.03×10^{-04}
$C_{1,4}$	6×10^{12}	1.21×10^{-04}
$C_{1,5}$	3750×10^{12}	7.56×10^{-02}
$C_{2,1}$	5228×10^{12}	2.51×10^{-02}
$C_{2,5}$	32×10^{12}	1.54×10^{-04}
$C_{3,6}$	30×10^{12}	4.57×10^{-03}
$C_{4,6}$	6×10^{12}	4.57×10^{-03}
$C_{5,1}$	3772×10^{12}	4.50×10^{-02}
$C_{5,8}$	15×10^{12}	1.79×10^{-04}
$C_{6,1}$	13×10^{12}	4.06×10^{-06}
$C_{6,5}$	5×10^{12}	1.56×10^{-06}
$C_{6,7}$	18×10^{12}	5.63×10^{-06}
$C_{7,1}$	3×10^{12}	5.77×10^{-10}
$C_{7,3}$	15×10^{12}	2.88×10^{-09}
$C_{8,1}$	5×10^{12}	3.85×10^{-09}
$C_{8,2}$	10×10^{12}	7.69×10^{-09}
$P_{9,10}$	7.5×10^{10}	4.55×10^{-05}
$P_{10,11}$	7.5×10^{10}	2.31×10^{-05}
$P_{11,9(uplift)}$	2.5×10^{10}	3.84×10^{-09}
$P_{11,9(subduction)}$	5×10^{10}	7.69×10^{-09}

i - denotes supplying reservoir; j - denotes receiving reservoir

mentary deposits which are available to undergo either uplift or subduction. The sedimentary organic phosphorus mass was calculated by multiplying the organic carbon reservoir mass by an atomic ratio of organic carbon to phosphorus in organic matter $(C:P)_{org}$. A value of 200:1 was chosen for reasons given below.

The organic phosphorus fluxes were calculated by multiplying the associated steady-state carbon fluxes by a $(C:P)_{org}$ ratio (Figs. 3, 4). (With the exception of the land-to-ocean flux, all phosphorus fluxes in Fig. 4 are organic phosphorus). The steady state fluxes were computed in the following manner. To determine the burial $(C:P)_{org}$, the average shelf sedimentation rate (cm/kyr) was computed as the ratio of the areal-average shelf sedimentation rate, 12.5 g/cm^2 kyr (Gross, 1972), to the average sediment density of 2.75 g/cm^3. Based on a previously calculated relationship between sedimentation rate and burial $(C:P)_{org}$, the calculated shelf sedimentation rate of 0.0045 cm/yr yields a burial $(C:P)_{org}$ which is approximately 400:1 (Ingall and Van Cappellen, 1990). Because shallow shelf sediments are those most susceptible to uplift, we used the same ratio, 400:1, for the $(C:P)_{org}$ uplift ratio to compute the uplift flux of phosphorus associated with the organic carbon uplift flux. The resultant organic phosphorus uplift flux is 2.5×10^{10} moles/yr.

For pelagic sediments, the ratio of the areal pelagic sedimentation rate, 0.5 g/cm^2-kyr (Worsley and Davies, 1979), to the average sediment density, 2.75 g/cm^3, results in a sedimentation rate of 0.00018 cm/yr. Again, from Ingall and Van Cappellen (1990), the corresponding burial $(C:P)_{org}$ is approximately 100:1. Pelagic sediments are those more likely to be subducted. They have a short residence time compared to sediments deposited elsewhere in the sedimentary lithosphere (Veizer, 1988). We calculated the organic phosphorus subduction flux by using the 100:1 ratio associated with the organic carbon subduction flux. The resultant organic phosphorus subduction flux is 5×10^{10} mol/yr.

To satisfy the steady-state condition, the phosphorus burial rate, i.e. additions to the sedimentary organic phosphorus reservoir, must equal the sum of the two output fluxes, uplift and subduction, which is 7.5×10^{10} mol/yr. The flux of organic phosphorus out of the *Ocean* reservoir, the burial flux, must be balanced by the phosphorus flux from land. We point out that the model does not take into account phosphate removal by hydrothermal processes (Wheat et al., 1996). Therefore, to satisfy steady-state conditions, the flux of phosphorus from land to the ocean was set equal to the organic phosphorus burial flux. The phosphorus flux from land cannot be quantitatively separated into the fluxes associated with various organic and inorganic phases of phosphorus. However, we know that to maintain net heterotrophy within the global oceans, some portion of the weathered terrestrial organic matter entering the oceans must be oxidized (Smith and Mackenzie, 1987). However, even though organic degradation is extensive, not all of the terrestrial matter is oxidized. Therefore, the organic phosphorus burial flux is not entirely due to autochthonous production. In essence, the organic phosphorus buried with ocean sediments can also include "non-reactive" phosphorus associated with refractory terrestrial organic matter. Consequently, we cannot refer to the phosphorus entering the oceans from land as strictly "reactive" phosphorus. The flux of phosphorus from land undoubtedly also includes iron-bound phosphorus, dissolved organic and dissolved inorganic phosphorus, and particulate organic phosphorus. In the water column, particulate organic phosphorus is "potentially reactive" in that it can become mobilized within the water column to stimulate organic production. Ultimately, more production results in burial of more organic matter and burial of more organic phosphorus.

The ratio that couples the organic carbon burial flux to the organic phosphorus burial flux, i.e. the global organic $(C:P)_{org}$ burial ratio, was back calculated by taking the ratio of the organic carbon burial flux to the organic phosphorus burial flux. We arrived at a global organic burial ratio, $(C:P)_{org}$, of 200:1. Other models (e.g. Van Cappellen and Ingall, 1994) have used 250:1 to derive the phosphorus burial rate from the organic carbon burial rate in marine sediments (Froelich et al., 1982). To calculate the mass of sedimentary organic phosphorus, we multiplied the mass of sedimentary carbon by the previously determined global burial $(C:P)_{org}$.

The methodology for determining the uplift, subduction, and global burial $(C:P)_{org}$ ratios and the related organic phosphorus fluxes assumes that the $(C:P)_{org}$ ratios are related to burial rates as demonstrated by Ingall and Van Cappellen (1990). This assumption is subject to debate in light of other studies by Mach et al. (1987) and Ramirez and Rose (1992). Those studies suggest that sedimentary organic carbon and phosphorus content are linearly correlated to one another and are not a function of sedimentation rate. If that were the case, the burial $(C:P)_{org}$ would be independent of sedimentation rate, calling into question the use of the aforementioned relationship between sedimentation rate and burial $(C:P)_{org}$ to calculate phosphorus fluxes. Using a molar burial $(C:P)_{org}$ of 490:1 and organic carbon burial rate of 1.0×10^{13} moles/yr, Mach et al. (1987) calculated the organic phosphorus flux from the oceans to be 2.0×10^{10} moles/yr, which is roughly 25% of our calculated phosphorus burial flux of 7.5×10^{10} (Fig. 4). However, the modeling results and our conclusions are qualitatively the same regardless of the $(C:P)_{org}$ ratio used to determine phosphorus fluxes from organic carbon uplift, subduction, and burial fluxes.

TECTONIC REALMS, PROCESSES, AND PHOSPHORUS CYCLING

Veizer (1988) demonstrated that elements are cycled at different rates in different populations of tectonic realms (e.g. continental platforms, cratons, active and passive margin basins, oceanic crust, etc.). For phosphorus cycling, our model recognizes three tectonic processes with different rates, and therefore time-scales, at which buried organic phosphorus is returned to land. The longest return cycle occurs via subduction of oceanic lithosphere and overlying sediment. Organic phosphorus released from the subducted oceanic sediment during metamorphism is entrained in magma, and eventually returned to or near the Earth's surface in the form of crystalline rock. The medium time-scale return cycle also occurs via subduction but also involves processes of uplift. The sedimentary mass is subducted, metamorphosed, and subsequently uplifted and subjected to weathering. The short-term return cycle results from tectonic uplift at convergent plate margins and the resulting exposure to weathering of previously submerged sedimentary deposits. The inclusion of the subduction and uplift mechanisms in our model recognizes the fact that different tectonic regimes and depositional environments impact element cycles at different time scales (Veizer, 1988).

Upon completion of the steady-state phosphorus model and determination of rate constants, the following perturbation scenarios were investigated to determine the influence of subduction on the cycling of phosphorus. To do this, the subduction flux, which at steady-state returns phosphorus to land, was diverted to a mantle reservoir. Simulations of 200,000 years were then run for four scenarios: (1) current subduction, uplift, and weathering rates, (2) 2 times the subduction rate, 1.5 times the uplift rate, 1.5 times the weathering rate, (3) 2 times the subduction rate, 1.5 times the uplift rate, and 2.0 times the weathering rates, and (4) 0.5 times the subduction, uplift, and weathering rates. A simple and consistent, but not necessarily correct, mechanistic rational for covarying the subduction, uplift, and weathering rates in the various model simulations is as follows. With a scenario of increased global tectonism, i.e. faster rates of seafloor spreading, subduction, uplift, and orogeny, would result in an increase in the rates of physical and chemical weathering even as sea-level rises due to increased ridge volume to inundate low-lying coastal areas. Also, as global tectonism increases, in general, so do the rates of metamorphism and volcanism and the rate of CO_2 release that accompanies those processes and an increase in atmospheric CO_2. This increase in atmospheric CO_2 concentration would lead to both an increased availability of carbonic acid and a rise in temperature and therefore, greater chemical weathering rates. Therefore, the combination of increased uplift and orogeny and increased atmospheric CO_2 leads to an increase in global weathering rates.

In contrast, with a scenario of reduced global tectonism, rates of seafloor spreading, subduction, uplift, and orogeny would decrease. Decreased rates of uplift and orogeny would result in a reduction of weathering rates even if sea level fell due to a reduction in ridge volume. In addition, a decrease in global tectonism would be accompanied by a concomitant reduction in the atmospheric CO_2 contribution from metamorphic and volcanic processes. As atmospheric CO_2 input declined, so would its effects via both temperature and carbonic acid on chemical weathering rates. Therefore, a decrease in global tectonism should be accompanied by a decrease in weathering rates. The mechanistic rationales for the covariance in subduction, uplift, and weathering rates have been hotly debated (Raymo et al. 1988; Berner and Caldeira, 1997; France-Lanord and Derry; 1997; Kump and Arthur, 1997). However, the objective of our model is not to validate the proposed mechanisms for coupling the processes of subduction, uplift, and weathering, but simply to illustrate the importance of the phosphorus subduction flux to the exogenic phosphorus cycle.

RESULTS

Phanerozoic Carbonate Survival Rate Calculations

The total surviving carbonate mass is primarily the extant continental carbonate mass until the last major Phanerozoic tectonic cycle, the Alpine, (Fig. 1). In the Jurassic through Tertiary, oceanic sediments become an increasingly important contributor to late Phanerozoic total carbonate survival mass (Fig. 1). The relative abundance of dolomite with respect to calcite decreases during this period of time (Chilingar, 1956; Sperber et al., 1984; Given and Wilkinson, 1987; Morse and Mackenzie, 1990). In Jurassic-, Cretaceous-, and Tertiary-age rocks, the oceanic mass increases, respectively, from roughly 8%, to 21%, to 47% of the total surviving extant carbonate mass. The exponential curve fit of the ocean carbonate survival data yields a decay constant of -0.017 m.y., an intercept of 6.2 x 10^{20} grams Ca/m.y., and a half-life of 40 m.y. The surviving Triassic- through Cambrian-age carbonate mass consists entirely of *continental carbonate* (i.e. carbonate rocks on continental plates). No appreciable ocean carbonate (i.e. carbonate on oceanic plates) from the Triassic through the Cambrian Periods remains (>180 m.y. in Fig. 1). There is a small amount of Triassic ocean sediment remaining but this does not significantly influence our results. We therefore consider the total sedimentary carbonate extant mass from the Triassic to Cambrian to consist entirely of surviving continental sedimentary carbonate.

Phanerozoic Phosphorus Survival Rate Calculations

As with total carbonate, the total preserved phosphorus mass is primarily the surviving continental phosphorus mass in pre-Jurassic rocks (Fig. 2, Table 2). Except for the Devonian, the total phosphorus survival curve displays an exponential decay trend with increasing age (Fig. 2). The exponential curve fit to the total phosphorus data yields a decay constant of -0.0027 m.y., an intercept of 3.3 x 10^{18} grams P/m.y., and a half-life of 257 m.y. As with the oceanic carbonates, the oceanic phosphorus survival rate increases dramatically with respect to continental phosphorus survival rate from the Jurassic through Tertiary (Fig. 2). The oceanic phosphorus mass in the Jurassic, Cretaceous, and Tertiary deposits represent 28%, 44%, and 71%, respectively, of the total surviving sedimentary phosphorus mass. The extant oceanic phosphorus survival rate data was also fit with an exponential curve. The calculated decay constant is -0.012 m.y., with an intercept of 4.1 x 10^{18} grams P/m.y., and a half-life of 58 m.y. Phanerozoic periods with the greatest phosphorite survival rates are, in decreasing order, Permian, Tertiary, Jurassic, Cretaceous, Cambrian, and Ordovician (Fig. 2). The Triassic, Carboniferous, Devonian, and Silurian Periods have survival rates that are at least one order of magnitude less than those of the aforementioned periods. Note the difference in scale between the survival rates of phosphorus as phosphorite and the other phosphorus reservoirs (Fig. 2).

Exogenic Phosphorus Cycling: Simulation Results

For all model simulations, the changes in phosphorus reservoir mass and fluxes over time are depicted in Figures 5 and 6; respectively. In all simulations, the subduction flux, which returns phosphorus to the *Land* reservoir at steady state, was diverted to the mantle. Therefore, as one would anticipate, without a return flux of phosphorus from the mantle, the combined mass of the *Land, Ocean,* and *Sedimentary Organic* phosphorus reservoirs decreases over time, in spite of changes made to the weathering, subduction, and uplift rates (Fig. 5). Concomitantly, the mantle mass of phosphorus increased with time at a rate controlled entirely by the subduction rate. The rate at which the combined exogenic mass decreases and the manner in which mass is repartitioned between reservoirs depend on the weathering, subduction, and uplift rates. To facilitate discussion, we compare the modeled flux rates and reservoir masses to simulations of current weathering, subduction, and uplift rates.

In the simple first-order phosphorus model presented here, increased tectonism (seafloor spreading, subduction, and uplift) results in increased weathering rates. In response to enhanced tectonism, the ocean reservoir phosphorus mass increases to values greater than the initial value for the first 30,000 years of the simulation. Also, final ocean phosphorus masses are greater for the increased-tectonism scenarios than for the current-condition simulation (Fig. 5B). The increase in the weathering rate is primarily responsible for both of these outcomes. The half-life of the ocean mass under current conditions is 120,000 years and for the increased-tectonism scenarios, 170,000 years. Due to the increased weathering rates, even with increased uplift rates providing more phosphorus to land, the land mass is depleted faster than at current conditions (Fig. 5C). The final land mass of phosphorus for three of the four scenarios (current conditions, decreased tectonism, and increased tectonism with 1.5 x current weathering rate) converges on the same quasi-steady-state value, about 5.5 x 10^{14} moles. However, the land mass in the enhanced-tectonism scenario reaches the quasi-steady-state value much more rapidly. The sedimentary organic phosphorus reservoir is depleted much faster in the increased tectonism scenario due to the elevated subduction flux (Fig. 5D). The mantle reservoir mass grew faster with increased subduction rates compared with the current subduction rate.

With a reduction in tectonism (seafloor spreading, subduction, and uplift), weathering rates decrease. Due to the decrease in the weathering rate in the reduced tectonism scenario, the ocean mass decreased to values below those for the current-condition and enhanced-tectonism scenarios. For the reduced-tectonism scenario, the calculated half-life for the ocean phosphorus mass is 80,000 years. In comparison to current and enhanced tectonic conditions, the land mass phosphorus declined less rapidly than in the current-condition and enhanced-tectonism scenarios, although it reached the same quasi-steady-state value in 200,000 years (Fig. 5C). The sedimentary organic phosphorus mass increased slightly and was greater than the initial mass for most of the simulation. At approximately 60,000 years, the mass reached its peak of 6.501 x 10^{18} moles phosphorus and slowly declined to about 6.499 x 10^{18}, slightly less than the initial mass after 200,000 years. The total increase in mass in the first 60,000 years represents an increase of approximately 0.016%. Nevertheless, it constitutes an increase of 1.0 x 10^{15} moles of phosphorus sequestered in the sediment. Growth of the mantle reservoir mass was slower at the reduced subduction rate than for current and increased subduction rates (Fig. 5A).

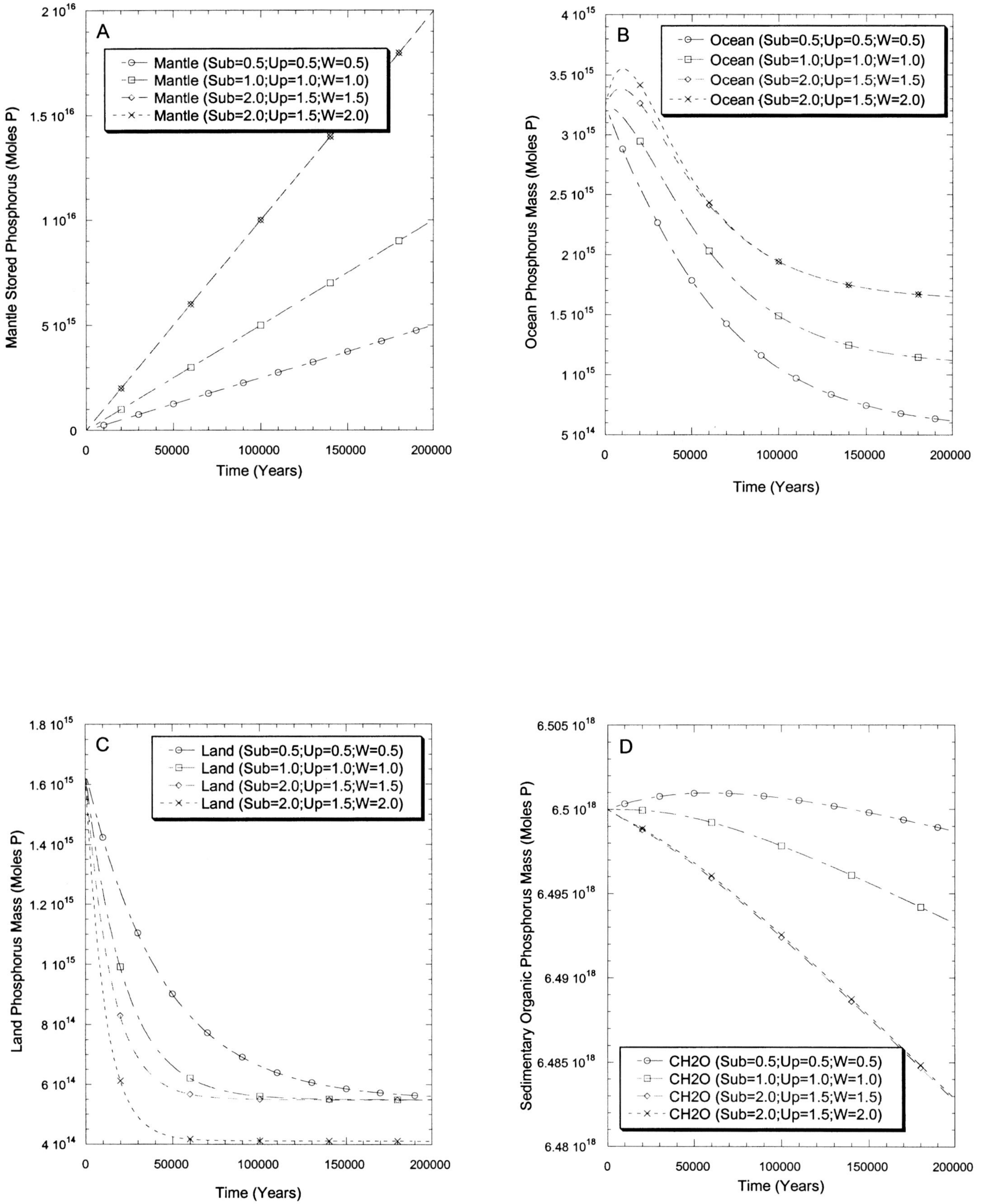

FIG. 5.—The effect of variations in the subduction rate (Sub), uplift rate (Up), and weathering rate (W) on (A) the phosphorus mass sequestered in the mantle, (B) the mass of the ocean phosphorus reservoir, (C) the mass of the land phosphorus reservoir and (D) the mass of the sedimentary organic phosphorus reservoir. Rates are expressed as factors of the modern steady-state flux rates listed in Table 4.

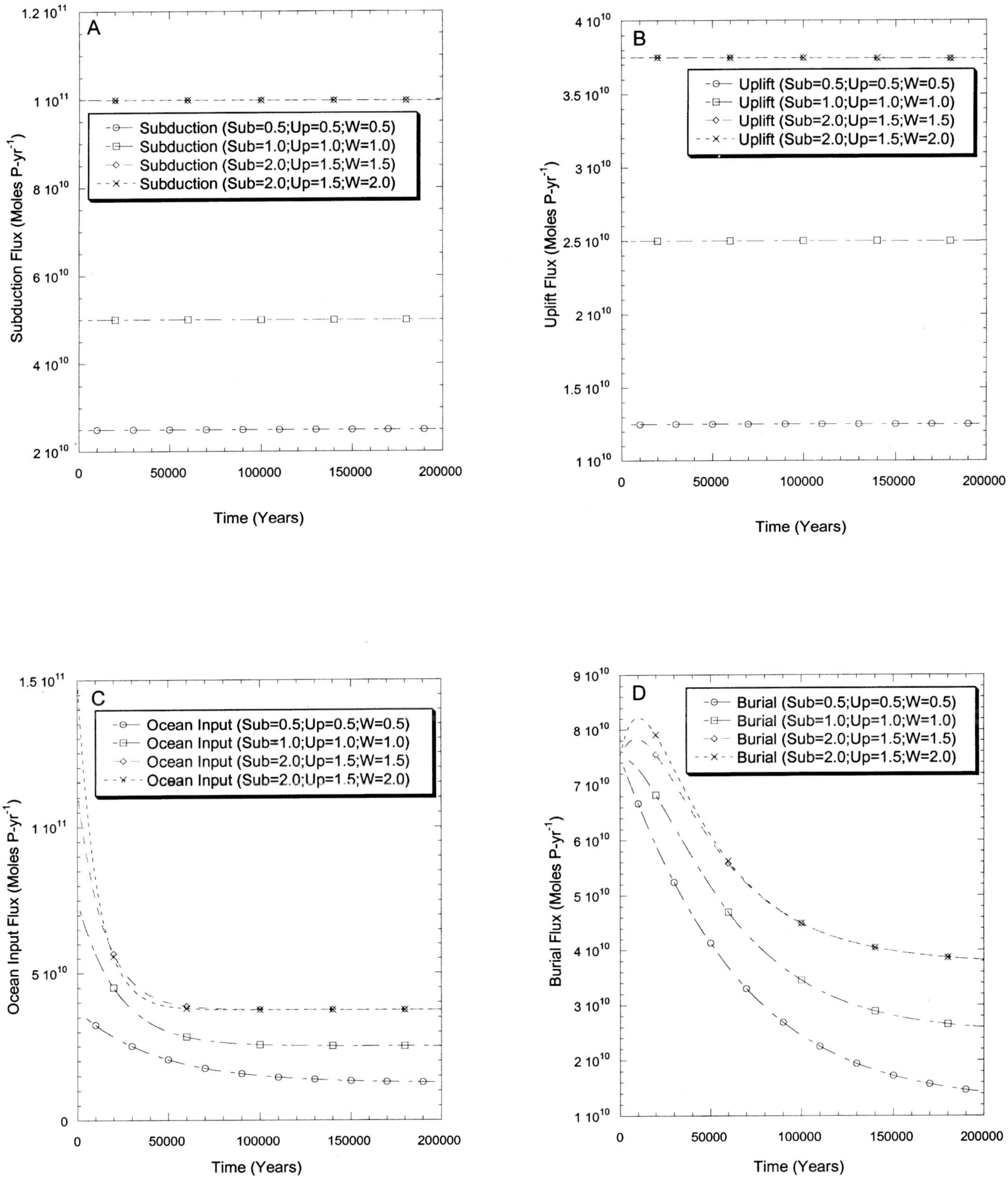

FIG. 6.—The effect of variations in the subduction rate (Sub), uplift rate (Up), and weathering rate (W) on (A) the phosphorus subduction flux, (B) the phosphorus uplift flux, (C) the phosphorus ocean input flux and (D) phosphorus burial flux. Rates are expressed as factors of the modern steady-state flux rates listed in Table 4.

DISCUSSION

Carbonate Cycling and Survival

Phanerozoic cycling of the global carbonate mass has previously been the subject of extensive discussion whose scope far exceeds this work (e.g., Wilkinson and Walker, 1989; Mackenzie and Morse, 1992). However, several points are worth reiterating with regard to Phanerozoic carbonate cycling and its relationship to tectonic processes. Over most of the Phanerozoic, the total preserved sedimentary carbonate mass follows an erratic secular trend (Fig. 1). Only after the beginning of the Triassic, does the trend behave smoothly by increasing exponentially. Fitting the total surviving carbonate mass with an exponential curve (Wilkinson and Walker, 1989) yields a decay constant of –0.002 m.y. or –0.003 m.y. assuming constant mass flux or linear growth models, respectively. The corresponding half-lives are 380 and 240 m.y.

The Cambrian through Triassic total extant sedimentary carbonate mass is composed entirely of continental carbonates (Fig. 1; Table 1). The continental carbonate survival graph in Figure 1 displays three distinct trends-Cambrian through Silurian, Devonian through Triassic, and Triassic to the Tertiary-that correspond to the Caledonian, Hercynian, and Alpine tectonic cycles (Mackenzie and Morse, 1992). The decrease in surviving continental carbonate from the Cretaceous to Tertiary (Fig. 1) has been interpreted as a decline in shallow-water carbonate deposition with a concomitant increase in pelagic carbonate deposition and survival (Wilkinson and Walker, 1989; Mackenzie and Morse, 1992).

The survival rate of Jurassic through Tertiary oceanic carbonates increases (Fig. 1). The decay constant calculated for the ocean carbonates is –0.017 m.y., which is 6 to 9 times the total surviving carbonate decay constants calculated using linear growth and constant-mass flux models by Wilkinson and Walker (1989) . The corresponding half-life of the ocean carbonate reservoir is 41 m.y. The increased survival rate of Tertiary ocean carbonates was apparently at the expense of continental carbonates whose Tertiary survival rate is much lower than the Cretaceous survival rate (Fig. 1). Both biological and eustatic processes were likely responsible (Wilkinson and Walker, 1989; Boss and Wilkinson, 1991; Mackenzie and Morse, 1992). The rise of planktonic carbonate-secreting organisms during the Jurassic may have shifted the locus of carbonate deposition from shallow water to pelagic environments. The decline in sea level since the Mesozoic would also have reduced the areal ratio between shallow-water and pelagic environments.

Phosphorus Cycling and Survival

Until now, investigation of Phanerozoic phosphorus survival has primarily revolved around phosphorus associated with phosphorites (Cook and McElhinny, 1979, Filippelli and Delaney, 1992). Although phosphorites are a concentrated economic source of phosphorus, their contribution to the total Phanerozoic extant phosphorus mass is small (Fig. 2). The total phosphorus in phosphorite is several orders of magnitude less than the total in continental and oceanic phosphorus (Table 2).

Unlike the irregular Cambrian-Triassic continental carbonate and total carbonate survival curves, the total phosphorus mass curve is secular and remarkably smooth. In fact, except for the Devonian survival rate, the survival rate data are fit well by an exponential decay curve decreasing with increasing age. The total phosphorus decay constant calculated for the Phanerozoic, –0.0027 m.y., is within the bounds of the two decay constants, –0.002 and –0.003 m.y., for the global carbonate survival in Wilkinson and Walker (1989). From the Cambrian through the Tertiary, total phosphorus survival is primarily due to the extant continental mass with a minor contribution from the phosphorite mass. Overall, the survival rate of the continental phosphorus mass increases in Cambrian- through Cretaceous-age rocks (Table 2; Fig. 2). Although the rate of increase in the Jurassic and Cretaceous is less with the concomitant increase in oceanic phosphorus surviving mass, the apparent shift of phosphorus mass from the continents to oceans may have been a result of the same two factors that caused the shift in the locus of carbonate deposition during the Alpine tectonic cycle. The calculated half-life for the ocean sedimentary phosphorus mass is 57 m.y., about 40% greater than the calculated half-life of ocean carbonates, which is 41 m.y.

We suggest that the relatively smooth trend of total phosphorus versus the total carbonate survival results, at least in part, from the difference in elemental cycling mechanisms and therefore rates. With a prevalent gas phase, carbon can be transferred in large quantities between reservoirs via the atmosphere over relatively short periods of time. Phosphorus is not afforded such a rapid means of mass transfer. Measured fluxes and the atmospheric reservoir of phosphine gas, which is quickly oxidized, are negligible (Devai et al., 1988). Without a prevalent gas phase, phosphorus must be transferred between reservoirs by mechanisms with slower rates involving either liquid or solid material. Therefore, the long-term cycling of phosphorus is dependent upon the subduction and uplift of sedimentary deposits and the production and weathering of phosphorus-bearing crystalline rocks-processes which occur over geologically significant periods of time.

Tectonism and Phosphorus Cycling

To our knowledge, the importance of phosphorus return to the exogenic phosphorus cycle after having been subducted and perhaps incorporated into the mantle has not been previously quantitatively addressed. The magnitude of the phosphorus subduction flux is dependent on, among other factors, subduction rates. Changes in subduction rates should be accompanied by changes in seafloor spreading rates. Historical spreading rates as calculated by Berner et al. (1983) from data of Pitman (1978) and Southam and Hay (1977) vary by a factor of 0.9–2 from the current spreading rate (Gaffin, 1987). With changing subduction rates, elemental exchange rates between the exogenic system and the mantle change. In terms of carbon, faster subduction rates presumably result in greater CO_2 degassing rates (Javoy et al., 1982).

In the model, as subduction rates increase, the amount of phosphorus transported to and sequestered in the mantle increases. At initial steady-state conditions, the subduction flux of phosphorus is 5.0 x 10^{10} moles/yr, which is two-thirds the magnitude of the phosphorus flux from land to the ocean (7.5 x 10^{10} moles/yr). Doubling the subduction rate resulted in a phosphorus flux to the mantle that is ~33% greater than the initial steady-state land-to-ocean flux and 150% greater than the final quasi- steady-state land-to-ocean flux (Fig. 6A, 6C).

Subduction rates can also decrease and thereby slow the rate of transport of phosphorus to the mantle. Decreasing the subduction rate by one-half, which is equivalent to decreasing the phosphorus subduction flux to 2.5 x 10^{10} moles/yr, resulted in a rate of phosphorus transportation to the mantle that is 33% of the initial steady-state land-to-ocean phosphorus flux. With decreasing subduction rates, we would anticipate that the organic sedimentary phosphorus mass would increase, as it does in the model (Fig. 5D). Regardless of whether or not the subduction rate changes, the magnitude of the phosphorus flux to the mantle is still significant when compared to the magnitude of the flux from land to the ocean.

Changes in seafloor spreading rates presumably are the result of changes in global tectonism (e.g. Gaffin, 1987). Uplift rates may also change if subduction rates change. However, it should be kept in mind that changes in the rate of subduction need not correspond in time with changes in rates of uplift. The uplift process serves to return both continental margin sediments and subducted, metamorphosed sediments to the Earth's surface. At the surface, the uplifted material is exposed to weathering processes and some of the phosphorus it contains is eventually returned to the ocean as reactive phosphorus. Thus, changes in uplift rate change the amount of phosphorus-bearing sedimentary and crystalline rocks available for weathering. Colman et al. (1997) demonstrated that model results can dramatically differ from the simulations given in Van Cappellen and Ingall (1996) if the weathering flux to the oceans is coupled to uplift rates. In our model, uplift rates and subduction rates are both increased. With greater uplift rates more sedimentary and crystalline rock is exposed to weathering, therefore, increased uplift rates lead to an increase in weathering rates. Comparing magnitudes of model fluxes, the steady-state uplift flux rate (2.5 x 10^{10} moles/yr) is 33% that of the ocean input rate (land-to-ocean flux, 7.5 x 10^{10} moles/yr) (Fig. 4). With the increase in uplift and weathering rates, and no replenishment of the phosphorus subduction flux, the land phosphorus mass decreased at faster rates than in the current uplift and weathering scenario (Fig. 5C).

Tectonism and Phosphorus Availability, Ocean Production, and Atmospheric Oxygen

Phosphorus availability in the exogenic system through geologic time has important implications regarding issues of oceanic net primary production and atmospheric oxygen concentrations. What are the implications of restricted phosphorus availability due to sequestration of subducted phosphorus in the mantle? Let us consider the three scenarios and their impacts. The first scenario began with the rates of subduction, uplift, and weathering at current steady-state values but with the subduction flux being sequestered in the mantle rather than being returned to land. Without the return of the subducted phosphorus to land, the land phosphorus mass decreased with time (Fig. 5C). As the phosphorus flux from land to the oceans diminished, so did the oceanic dissolved phosphorus mass. A decrease in the ocean phosphorus mass could result in a decrease in oceanic net primary productivity. With subsequently less organic matter being produced and, presumably, less being buried, the oceanic contribution of oxygen to the atmospheric would decrease. With the decrease in the ocean-to-atmosphere oxygen flux and the continued weathering of reduced materials, atmospheric oxygen levels could decrease.

For the second scenario, we considered increased subduction, uplift, and weathering rates. The increased rate of uplift to land of sedimentary and crystalline rocks resulted in an increase in the weathering flux of phosphorus to the ocean (Fig. 6C). Again, because the subduction flux was shunted to the mantle and not returned to the land reservoir, the land phosphorus mass decreased (Fig. 5C). For the first 30,000 years, the oceanic phosphorus mass increased (Fig. 5B). With the oceanic phosphorus increase, ocean productivity increased and, as a consequence, the organic matter burial rate also increased (Fig. 6D) because of the first-order relationship between the two processes. The increase in the organic matter burial rate should result in an increase in the flux of oxygen from the ocean to the atmosphere. However, at peak organic matter burial rates in the model simulation, only an additional 7 x 10^9 moles O_2/yr could be released into the atmosphere. At that rate, the total amount of additional oxygen that could be generated in 30,000 years is on the order of 10^{14} moles O_2, which is a trivial amount compared to the magnitude of atmospheric oxygen fluctuations during the Phanerozoic (Berner and Canfield, 1989).

However, with increased tectonism comes the potential increase in the amount of reduced material (for example pyrite and organic matter) exposed to weathering and oxidation. The reduced material provides a sink for the increased atmospheric oxygen input during the first 30,000 years. After that short initial interval though, the decrease in the oxygen input and the continued exposure of reduced material could result in a drawdown of atmospheric oxygen.

The third scenario is one with reduced subduction, uplift, and weathering rates. This scenario has interesting implications for atmospheric oxygen concentration. With reduced uplift and the subsequent reduced weathering rates, the organic phosphorus burial flux had dropped from the initial 7.5 x 10^{10} moles/yr to 1.2 x 10^{10} moles/yr, a difference of 6.3 x 10^{10} moles/yr by the end of the simulation. This difference corresponds to a decrease in the organic carbon burial rate of approximately 1.3 x 10^{13} moles/yr. If the organic carbon burial rate was reduced by this amount, in only 300 k.y. 3.8 x 10^{18} moles O_2, or 10% of the current atmospheric O_2 mass, would not be injected to the atmosphere. This reduced input of oxygen to the atmosphere could result in a significant draw down, via oxidation of organic matter and pyrite, of the atmospheric oxygen reservoir over a relatively short geologic interval. On the other hand, the consumption of oxygen by weathering might decrease because of the decreased uplift rate and reduced exposure of organic matter and pyrite to weathering in this scenario and thus counteract this draw down.

In all of our simulations, the oceanic dissolved phosphorus mass decreased over the long term, primarily because the phosphorus subducted was not returned to the exogenic system. This in turn reduced the amount of phosphorus available for primary production and subsequently reduced the rate of organic matter burial. Therefore, the amount of oxygen produced by the degradation of buried organic matter decreased. Other more complicated models (Van Cappellen and Ingall, 1996; Van Cappellen and Ingall, 1994; Colman et al., 1997) have shown that the redox-dependent flux of phosphorus from sediments back into the ocean can also be important in regulating ocean phosphorus availability and, as a consequence, may play a role in the organic matter burial rate and in balancing geologic production and consumption of atmospheric oxygen.

These redox mechanisms (Holland, 1994; Colman and Holland, 1994; Van Cappellen and Ingall, 1994) allow for the "recycling" of phosphorus from sediments back to the overlying water column in the following manner. As tectonism increases, so does the weathering flux of phosphorus from land to ocean (Colman et al., 1997). As the availability of phosphorus in the ocean increases, so does primary production. Increased production results in increased amounts of organic matter settling through the water column and being deposited. Bacterial oxidation of the organic matter, in both the water column and sediments, results in a reduction in the pore water and deep-water oxygen content. As the ocean bottom waters become anoxic, oxidized iron in the near-surface sediment is reduced and the phosphorus sorbed to oxidized iron is released into pore water and the overlying water column. After being returned to the water column, phosphorus is again available for primary production via upwelling. In such a situation, one mole of phosphorus can be responsible for the burial and production, respectively, of more than 106 moles of organic carbon and 138 moles of oxygen before permanent burial and subduction. In this way, oceanic phosphorus redox mechanisms may regulate atmospheric oxygen concentration.

Simulations with increased tectonism and weathering rates resulted in an increase in the oceanic dissolved phosphorus mass for 30,000 years (Fig. 5B) which conceivably would increase primary production and burial of organic matter. However, because our model lacks a redox feedback mechanism, the increased production and burial of organic matter does not trigger the return of phosphorus to the water column as in the model of Colman et al. (1997). Without the recycling of phosphorus from the sediment to the ocean, the dissolved ocean phosphorus mass is reduced by organic matter burial and eventual subduction. For all simulations over the long term, the reduction in dissolved ocean phosphorus and concomitant reduction in primary production and organic matter burial would result in the reduced production of oxygen. Phosphorus recycling might delay such a terminal atmospheric oxygen depletion by reducing the amount of phosphorus lost to subduction. However, ultimately the subducted phosphorus must eventually be returned to the exogenic cycle. If not returned, the impact of subduction on the exogenic phosphorus and oxygen cycles would be dramatic.

The Open Exogenic Phosphorus Cycle: Steady-State Over Geologic Time?

Over geologically short intervals of time, it appears, based on our model, that the removal of phosphorus via subduction is capable of rapidly depleting the exogenic phosphorus mass if the phosphorus is permanently sequestered in the mantle. Because of its biological importance, the depletion of exogenic phosphorus would have critical implications for long-term global productivity. Because of the subduction flux to the mantle, the exogenic phosphorus cycle must be considered to be an open cycle. However, an interesting question is: Has the cycle been in a quasi-steady-state at least in Phanerozoic time?

In spite of the phosphorus drain by subduction, the enormous mass of phosphorus dispersed in crystalline rock could supply enough phosphorus via increased weathering to offset the subduction flux. Presumably, an increase in tectonism accompanied by increases in atmospheric carbon dioxide and temperature would result in increased chemical weathering rates. An increase in weathering of crystalline rock containing phosphorus would release more phosphorus, which could offset the loss of exogenic phosphorus mass sequestered in the mantle via subduction. Eventually though, the crystalline mass would most likely have to be replenished.

The question regarding the return of subducted phosphorus to the exogenic cycle, and the open nature of the cycle bears directly on the current debate regarding mantle mixing. Mounting evidence now suggests that subducted slabs can reach the mantle-core boundary perhaps allowing for whole-mantle mixing and circulation (Grand et al., 1997; van der Hilst et al., 1997). If so, does subducted sedimentary organic and inorganic phosphorus remain in the upper mantle to be entrained by rising magma plumes or is it incorporated in metamorphic rocks and uplifted and thus returned relatively quickly to the exogenic system as crystalline rock? Or does it instead get carried deeper into the mantle where it becomes part of a longer-term return mechanism?

Model Caveats

There are other viable mechanisms to maintain phosphorus in the exogenic system. Our simple model does not include redox-dependent recycling of phosphorus. Phosphorus recycling, which depends on ocean and pore water oxidation states, has been suggested as a means by which permanent burial of reactive phosphorus is prevented (e.g. Berner and Canfield, 1989; Colman and Holland, 1994; Van Cappellen and Ingall, 1994; Van Cappellen and Ingall, 1996; Colman and Holland, this volume). Future modeling efforts of long-term phosphorus cycling should include subduction, uplift, exogenic redox-dependent processes, production of crystalline rock, weathering, and the effects of tectonism on their rates.

In addition, subduction of inorganic phosphorus (i.e. REE-P, iron-bound P, apatite) which is likely an important part of the phosphorus mass delivered to the mantle was not considered in our model. Also, we did not model oceanic hydrothermal processes and water-basalt interactions that result in the uptake of phosphorus from the ocean (Wheat et al., 1996). The (C:P) ratios used by the model to derive phosphorus burial, uplift, and subduction fluxes are debatable. The model results are contingent on these ratios. However, regardless of the $(C:P)_{org}$ ratios used to couple the uplift, burial, and subduction flux of organic carbon and phosphorus, there will be a drain of phosphorus from the exogenic system via subduction and sequestration in the mantle. Is the phosphorus drain offset by a return of phosphorus via weathering of uplifted sediments and crystalline rock? Future modeling efforts should address this question.

CONCLUSIONS

On geologic time scales, variations in rates of tectonism exert a critical control on exogenic elemental cycling. Shifts in the locus of deposition from the continents to deep ocean environments change the cycling pathways that are followed by elements. The depletion of exogenic phosphorus by subduction of organic matter and the associated organic phosphorus is a significant flux in the global biogeochemical cycle of phosphorus. Unlike the bio-essential elements of carbon and nitrogen, which have prominent gas phases, phosphorus is likely to be returned to the Earth's surface in sediments, sedimentary and crystalline rock, and uplifted organic matter. The time-scale of this return certainly differs from that of carbon (as volcanic CO_2) and nitrogen (as nitrogen gases) and is most likely longer. On the geologic time-scale of millions of years, the loss of subducted organic phosphorus is most likely mitigated by changes in chemical weathering rates and/or exogenic redox-dependent processes. However, on an even longer scale of tens of millions to one hundred million years, replenishment of exogenic phosphorus requires the return of subducted phosphorus incorporated in crystalline rock.

ACKNOWLEDGMENTS

This work was supported by the National Science Foundation (EAR 9316133) and the donors of The Petroleum Research Fund, administered by the ACS. We are grateful for the insightful comments on earlier versions of the manuscript by Steve Smith and Daniel Hoover . The authors also wish to thank Albert Colman and Lee Kump for their constructive criticisms and reviews. This is SOEST Contribution Number 4807.

REFERENCES

ARVIDSON, R. S., AND MACKENZIE, F. T., 1997, Tentative kinetic model for dolomite precipitation rate and its application to dolomite distribution, Aquatic Geochemistry, 2, p. 273-298.

BERNER, R. A., 1980, Early Diagenesis, A Theoretical Approach, Princeton, Princeton University Press, 241 p.

BERNER, R. A., 1982, Burial of organic carbon and pyrite sulfur in the modern ocean: Its geochemical and environmental significance, American Journal of Science, 282, p. 451-473.

BERNER, R. A., 1990, Atmospheric carbon dioxide levels over Phanerozoic time, Science, 249, p. 1382-1386.

BERNER, R. A., LASAGA, A. C., AND GARRELS, R. M., 1983, The carbonate silicate geochemical cycle and its effect on atmospheric carbon dioxide over the past 100 million years, American Journal of Science, 283, p. 641-683.

BERNER, R. A., AND CANFIELD, D. E., 1989, A new model of atmospheric oxygen over Phanerozoic time, American Journal of Science, 289, p. 333-361.

BERNER, R. A., AND CALDEIRA, K., 1997, The need for mass balance and feedback in the geochemical carbon cycle, Geology, 25, p. 955-956.

BOSS, S. K., AND WILKINSON, B. H., 1991, Planktonic/eustatic control on cratonic/oceanic carbonate accumulation, Journal of Geology, 99, p. 497-513.

BROECKER, W. S., AND PENG, T. H.,1982, Tracers in the Sea, New York, Eldigio Press, 690 p.

CHILINGAR, G. V., 1956, Relationship between Ca/Mg ratio and geological age,: American Association of Petroleum Geologists Bulletin, 40, p. 2256-2266.

COALE, K. H., JOHNSON, K. S., FITZWATER, S. E., GORDON, R. M., TANNER, S., CHAVEZ, R. P., SAKAMOTO, L. F. C., ROGERS, P., MILLERO, F., STEINBERG, P., NIGHTINGALE, P., COOPER, D., COCHLAN, W.P., LANDRY, M.R., CONSTANTINOU, J., ROLLWAGON, G., AND KUDELA, A. T. R., 1996, A massive phytoplankton bloom induced by an ecosystem-scale iron fertilization experiment in the Equatorial Pacific Ocean:, Nature, 383, p. 495-501.

COLMAN, A. S. AND HOLLAND, H. D., 1994, The global diffusive flux of phosphorus from marine sediments into the oceans (abs): Eos (abs), v. 75, p. 96.

COLMAN, A. S., MACKENZIE, F. T., AND HOLLAND, H. D., 1997, Redox stabilization of the atmosphere and oceans and marine productivity:, Science, 275, p. 406-408.

COOK, P. J., AND MCELHINNY, M. W., 1979, A reevaluation of the spatial and temporal distribution of sedimentary phosphate deposits in the light of plate tectonics,: Economic Geology, 74, p. 315-330.

COPIN-MONTEGUT, C., AND COPIN-MONTEGUT, G., 1983, Stoichiometry of carbon, nitrogen, and phosphorus in marine particulate matter, Deep Sea Research, v. 30, p. 31-46.

DES MARAIS, D. J., 1985, Carbon exchange between the mantle and the crust, and its effect upon the atmosphere: Today compared to Archean time, *in* Sundquist, E. T., and Broecker, W. S., eds., The Carbon Cycle and Atmospheric CO_2: Natural Variations Archean to Present: Washington, American Geophysical Union, p. 602-612.

DÉVAI, I., FELFÖLDY, L., WITTNER, H., AND PLÓSZ, S., 1988, Detection of phosphine: new aspects of the phosphorus cycle in the hydrosphere, Nature, 333, p. 343-345.

FALKOWSKI, P. G., 1997, Evolution of the nitrogen cycle and its influence on the biological sequestration of CO_2 in the ocean, Nature, 387, p. 272-275.

FILIPPELLI, G. M., AND DELANEY, M. L., 1992, Similar phosphorus fluxes in ancient phosphorus deposits and a modern phosphogenic environment, Geology, 20, p.709-712.

FILIPPELLI, G. M., AND DELANEY, M. L., 1994, The oceanic phosphorus cycle and continental weathering during the Neogene, Paleoceanography, 9, p. 642-652.

FILIPPELLI, G. M., AND DELANEY, M. L., 1996, Phosphorus geochemistry of equatorial Pacific sediments, Geochimica et Cosmochimica Acta, 60, p. 1479-1495.

FISCHER, A. G., 1984, The two Phanerozoic super cycles, *in* Berggren, W. A. and Vancouvering, J. A., eds., Catastrophes in Earth History: Princeton, Princeton University Press UP, p. 129-148.

FÖLLMI, K. B., 1995, 160 m.y. record of marine sedimentary phosphorus burial: Coupling of climate and continental weathering under greenhouse and icehouse conditions, Geology, 23, p. 859-862.

FÖLLMI, K. B., 1996, The phosphorus cycle, phosphogenesis and marine phosphate-rich deposits. Earth Science Reviews, 40, p. 55-124.

FRANCE-LANORD, C., AND DERRY, L. A.,1997, Organic carbon burial forcing of the carbon cycle from Himalayan erosion, Nature, 390, p. 65-67.

FROELICH, P. N., BENDER, M. M., LUEDTKE, N. A., HEATH, G. R., AND DE VRIES, T., 1982, The marine phosphorus cycle, American Journal of Science, 282, p. 474-511.

GARRELS, R. M., AND MACKENZIE, F. T., 1971, Evolution of Sedimentary Rocks, New York, W.W. Norton, 397 p.

GARRELS, R. M., AND PERRY, E. A. Jr., 1974, Cycling of carbon, sulfur, and oxygen through geologic time, *in* Goldberg, G. E., ed., The Sea: New York, Wiley-Interscience, p. 303-336.

GARRELS, R. M., LERMAN, A., AND MACKENZIE, F. T., 1976, Controls of atmospheric O_2 and CO_2: Past, present and future, American Journal of Science, 64, p. 306-315.

GASSMAN, G., 1994, Phosphine in the fluvial and marine hydrosphere, Marine Chemistry, 45, p. 197-205.

GAFFIN, S., 1987, Ridge volume dependence on seafloor generation rate and inversion using long term sealevel change, American Journal of Science, 287, p. 596-611.

GIVEN, R. K., AND WILKINSON, B. H., 1987, Dolomite abundance and stratigraphic age: Constraints on rates and mechanisms of Phanerozoic dolostone formation, Journal of Sedimentary Petrology, 57, p. 457-469.

GLENN, C. R., FÖLLMI, K. B., RIGGS, S. R., BATURIN, G. N., GRIMM, K. A., TRAPPE, J., ABED, A. M., GALLI-OLIVIER, C., GARRISON, R. E., ILYIN, A. V., JEHL, C., ROHRLICH, V., SADAQAH, R. M. Y., SCHIDLOWSKI, M., SHELDON, R. E., AND SEIGMUND, H., 1994, Phosphorus and phosphorites: Sedimentology and environments of formation, Eclogae Geologicae Helvetiae, 87, p. 747-788.

GRAND, S. P, VAN DER HILST, R. D., AND WIDIYANTORO, S., 1997, Global seismic tomography: A snapshot of convection in the Earth, GSA Today, 7, p. 1-7.

GREGOR, C. B., 1985, The mass-age distribution of Phanerozoic sediments, *in* Snelling, N.., ed., Geochronology and the Geologic Record, Geological Society of London Memoir, No.10, p. 284-289.

GROSS, M. G., 1972, Oceanography (1st ed.), New Jersey, Prentice Hall. 581 p.

HOLLAND, H. D., 1978, The Chemistry of the Atmosphere and the Oceans (1st ed.): New York, Wiley-Interscience. 351 p.

HOLLAND, H. D., 1994, The phosphate-oxygen connection (abs): Eos (abs), v. 75, p. 96.

HOLSER, W.., SCHIDLOWSKI, M., MACKENZIE, F. T., AND MAYNARD, J. B., 1988, Biogeochemical cycles of carbon and sulfur, *in* Gregor, C. B., Garrels, R. M. Mackenzie, F. T., Maynard, J. B., eds., Chemical Cycles in the Evolution of the Earth: New York:, Wiley-Interscience, p. 105-161.

INGALL, E. D., AND VAN CAPPELLEN, P., 1990, Relation between sedimentation rate and burial of organic phosphorus and organic carbon in marine sediments, Geochimica et Cosmochimica Acta, 54, p. 373-386.

JAVOY, M., 1997, The major volatile elements of the Earth: Their origin, behavior, and fate, Geophysical Research Letters, 24, p. 177-180.

JAVOY, M., PINEAU, F., AND ALLEGRE, C. J., 1982, Carbon geodynamic cycle, Nature, 300, p.171-173.

KARL, D., LETELIER, R., TUPAS, L., DORE, J., CHRISTIAN, J., AND HEBEL, D., 1997, The role of nitrogen fixation in biogeochemical cycling in the subtropical North Pacific Ocean, Nature, 388, p. 533-538.

KERRICK, D. M., AND CALDEIRA, K., 1994, Metamorphic CO_2 degassing and early Cenozoic paleoclimate, Geological Society of America Today, 4, p. 62-65.

KUMP, L.., 1988, Terrestrial feedback in atmospheric oxygen regulation by fire and phosphorus, Nature, 335, p. 152-154.

KUMP, L. R., AND MACKENZIE, F. T., 1996, Regulation of atmospheric oxygen: Feedback in the microbial feedbag, Science, 271, p. 459-460.

KUMP, L. R., AND ARTHUR, M., 1997, Global Chemical Erosion during the Cenozoic: Weatherability Balances the Budgets, in W. Ruddiman, ed., Tectonic Uplift and Climate Change: New York, Plenum Press, p. 399-426.

Lasaga, A. C., Berner, R. A., and Garrels, R. M., 1985, An improved geochemical model of atmospheric CO_2 fluctuations over the past 100 millions years, *in* Sundquist, E. T., and Broecker, W. S., eds., The Carbon Cycle and Atmospheric CO_2: Natural Variations Archean to Present: Washington, American Geophysical Union, p. 397-412.

LERMAN, A., MACKENZIE, F. T. AND GARRELS, R. M., 1975, Modeling of geochemical cycles: phosphorus as an example,: Geological Society of America Memoirs, v. 142, p. 205-218.

MACH, D. L., RAMIREZ, A., AND HOLLAND, H. D., 1987, Organic phosphorus and carbon in marine sediments, American Journal of Science, 287, p. 429-441.

MACKENZIE, F. T., 1992, Chemical mass balance between rivers and oceans, in Encyclopedia of Earth System Science, v. 1: New York, Academic Press, Inc., p. 431-445.

MACKENZIE, F. T., AND PIGOTT, J. P., 1981, Tectonic controls of Phanerozoic sedimentary rock cycling, Journal of the Geological Society of London, 138, p.183-196.

MACKENZIE, F. T., AND MORSE, J. W., 1992, Sedimentary carbonates through Phanerozoic time, Geochimica et Cosmochimica Acta, 56, p.3281-3295.

MACKENZIE, F. T., VER, L. M., SABINE, C., LANE, M., AND LERMAN, A., 1993, C,N,P,S global biogeochemical cycles and modeling of global change, *in* Wollast, R., Mackenzie, F. ., and Chou, L., eds., Interactions of C,N,P and S Biogeochemical Cycles and Global Change: Berlin, Springer-Verlag, p. 1-61.

MALLINSON, D. J., AND COMPTON, J. S., 1997, Linking phosphogenic episodes on the southeast U.S. margin to marine del-13C and del-18O records, Geology, 25, p. 103-106.

MARTIN, J. H., AND FITZWATER, S. E., 1989, Iron deficiency limits phytoplankton growth in the north-east Pacific subarctic, Nature, 331, p. 341-343.

MEYBECK, M., 1982, Carbon, nitrogen and phosphorus transport by world rivers, American Journal of Science, 282, p. 401-450.

MILLIMAN, J. D., 1974, Recent Sedimentary Carbonates 1, Marine Carbonates: New York, Springer-Verlag, 375 p.

MORSE, J. W. AND MACKENZIE, F. T., 1990, Geochemistry of Sedimentary Carbonates. New York, Elsevier. 707 p.

PITMAN, W. C., 1978, Relationship between eustacy and stratigraphic sequences of passive margins, Geological Society of America Bulletin, 89, p.1289-1403.

Poldervaart, A., 1955, Chemistry of the Earth's crust, *in* Poldervaart, A., eds., Crust of the Earth: Special Paper 62, Geological Society of America, Baltimore, Waverly Press, Baltimore, p. 119-144.

RAMIREZ, A. J., AND ROSE, A. W., 1992, Analytical geochemistry of organic phosphorus and its correlation with organic carbon in marine and fluvial sediments and soils, American Journal of Science, 292, p. 421-454.

RASMUSSEN, F., 1996, Early-diagenetic REE-phosphate minerals (florencite, gorceixite, crandallite, and xenotime) in marine sandstones: a major sink for oceanic phosphorus, American Journal of Science, 296, p. 601-632.

RAYMO, M. E., 1991, Geochemical evidence supporting T.C. Chamberlin's theory of glaciation, Geology, 19, p.344-347.

RAYMO, M. E., RUDDIMAN, W. F., AND FROELICH, P. N., 1988, Influence of late Cenozoic mountain building on ocean geochemical cycles, Geology, 16, p. 649-653.

REDFIELD, A.., 1958, The biological control of chemical factors in the environment,American Scientist, 46, p. 205-221.

RONOV, A. B., 1980, The Earth's Sedimentary Shell: Quantitative patterns of its structure, compositions, and evolution: The 20th Vernadskiy Lecture v. I., *in* Yaroshevskiy, A. A., ed., The Earth's Sedimentary Shell: Moscow, Nauka, p.1-80; also, American Geological Institute Reprint Series, v. 5, p. 1-73.

RUTTENBERG, K. C., AND BERNER, R. A., 1993, Authigenic apatite formation and burial in sediments from non-upwelling, continental margin environments, Geochimica et Cosmochimica Acta, 57, p. 991-1007.

RYTHER, J. G., AND DUNSTAN, W. M., 1971, Nitrogen, phosphorus and eutrophication in coastal marine environment, Science, 171, p. 1008-1013.

SKINNER, H. C., 1973, Phase relations in the $CaO-P_2O_5-H_2O$ system from 300 to 600°C at 2 kb H_2O pressure, American Journal of Science, 273, p. 545-560.

SLOSS, L. L., 1963, Sequences in the cratonic interior of North America. Bulletin of the Geological Society of America, 74, p. 93-114.

SMITH, S. V., 1984, Phosphorus versus nitrogen limitation in the marine environment, Limnological Oceanography, 29, p. 1149-1160.

SMITH, S. V., AND MACKENZIE, F. T., 1987, The ocean as a net heterotrophic system: Implications from the carbon biogeochemical cycle, Global Biogeochemical Cycles, v. 1, p. 187-198.

SMITH, S. V. AND HOLLIBAUGH, J. T. 1993, Coastal metabolism and the oceanic organic carbon balance:, Reviews of Geophysics, v. 31, p. 75-89.

SOUTHAM, J. R., AND HAY, W. W., 1977, Time scales and dynamic models of deep-seasedimentation, Journal of Geophysical Research, 82, p. 3825-3842.

SPERBER, C. M., WILKINSON, B. H., AND PEACOR, D. R., 1984, Rock composition, dolomite stoichiometry, and rock/water reactions in dolomitic carbonate rocks, Journal of Geology, 92, p. 609-622.

SUNDQUIST, E. T., and BROECKER, W. S.,(eds.) 1985, The Carbon Cycle and Atmospheric CO_2: Natural Variations Archean to Present, Geophysical Monograph 32, 627 p.

UREY, H., 1952, The Planet, Their Origin and Development: New Haven, Yale University Press, 245 p.

VAN CAPPELLEN, P., AND INGALL, E. D., 1994, Benthic phosphorus regeneration, net primary production, and ocean anoxia: A model of the coupled marine biogeochemical cycles of carbon and phosphorus, Paleoceanography, 9, p. 677-692.

VAN CAPPELLEN, P., AND INGALL, E. D., 1996, Redox stabilization of the atmosphere and oceans by phosphorus limited marine productivity, Science, 271, p.493-496.

VAN DER HILST, R., WIDIYANTORO, S., AND ENGDAHL, E. R.1997, Evidence for deep mantle circulation from global tomography, Nature, 386, p.578-560.

VEIZER, J., 1988, The evolving exogenic cycle, *in* Gregor, C.B., Garrels, R.M.,

Mackenzie, F. T., Maynard, J. B., eds., Chemical Cycles in the Evolution of the Earth: New York: Wiley-Interscience, p. 175-218.

VEIZER, J., AND JANSEN, S.L., 1979, Basement and sedimentary recycling and continental Evolution:, Journal of Geology, 87, p. 341-370.

VEIZER, J., AND JANSEN, S. L., 1985, Basement and sedimentary recycling 2: Time dimension to global tectonics, Journal of Geology, 93, p. 625-643.

WHEAT, C. G., FEELY, R .A. AND MOTTL, M. J.,1996, Phosphate removal by oceanic hydrothermal processes: an update of the phosphorus budget in the oceans. Geochimica et Cosmochimica Acta, 60, p.3593-3608.

WILKINSON, B. H., AND ALGEO, T. J., 1989, Sedimentary carbonate record of calcium magnesium cycling, American Journal of Science, 289, p. 1158-1194.

WILKINSON, B. H., AND WALKER, J. C. G., 1989, Phanerozoic cycling of sedimentary Carbonate:, American Journal of Science, 289, p. 525-548.

WISE, D. U., 1974, Continental margins, freeboard, and the volumes of continents and oceans through time, *in* Burke, K., and Drake, C. L., eds., The Geology of Continental Margins: New York, Springer-Verlag, p. 45-58.

WOLLAST, R., 1994, The relative importance of biomineralization and dissolution of $CaCO_3$ in the global carbon cycle, *in* Doumenge, F., ed., Past and Present Biomineralization Processes. Consideration about the Carbon Cycle: Monaco, Musee Oceanographique, p. 13-35.

WOODWELL, G. M., AND WHITTAKER, R. H., 1968, Primary Production in terrestrial ecosystems, American Zoologist, 8, p. 19-30.

WORSLEY, T. R., AND DAVIES, T. A., 1979, Sea level fluctuations and deep-sea sedimentation rates, Science, 203, p. 455-456.

THE GLOBAL DIAGENETIC FLUX OF PHOSPHORUS FROM MARINE SEDIMENTS TO THE OCEANS: REDOX SENSITIVITY AND THE CONTROL OF ATMOSPHERIC OXYGEN LEVELS.

ALBERT S. COLMAN*

Department of Geology and Geophysics, Yale University, P.O. Box 208109, New Haven, CT 06520-8109

AND

HEINRICH D. HOLLAND

Department of Earth and Planetary Sciences, Harvard University, Cambridge, MA 02138

ABSTRACT: The marine geochemistry of phosphorus links the burial rate of organic carbon in marine sediments to the oxygen content of the atmosphere and may serve as a major component of the system that controls atmospheric oxygen. The return flux of phosphate from marine sediments to seawater is an important part of the marine phosphorus cycle. This paper examines the relationship between the return flux of phosphate and the oxidation state of marine sediments, a necessary preliminary step in defining the efficacy of the oxygen control mechanism.

The diffusive return flux of phosphate from marine sediments to overlying bottom waters was calculated for 193 published pore water phosphate profiles that met a number of stringent criteria for sediment core and pore water recovery and processing. The phosphate return fluxes, scaled to carbon regeneration fluxes, are significantly greater from highly reduced sediments than from highly oxidized sediments. In highly reduced sediments the return phosphate flux from carbon regeneration is frequently augmented by a large phosphate flux released during the reductive dissolution of ferric (oxy)hydroxides. In highly oxidized sediments the return phosphate flux can be somewhat less than the flux to be expected from carbon regeneration. The missing phosphate is probably adsorbed on ferric (oxy)hydroxides in these sediments. The strong coupling between the oxidation state of marine sediments and the return phosphate flux to seawater suggests that the marine phosphate cycle is indeed an important part of the system that stabilizes atmospheric O_2.

The total preagricultural return flux of P from marine sediments was ca. 12 x 10^{11} mol P/yr. This rate is more than an order of magnitude larger than the riverine flux of total dissolved phosphorus to the oceans, ca. 0.3 x 10^{11} mol P/yr. We estimate that the total continental preagricultural flux of reactive P that ultimately appears dissolved in ocean water was ca. 3.5 x 10^{11} mol P/yr. The large flux of continental P to seawater, via direct input of riverine dissolved inorganic and organic P and via the diagenetic return flux from reactive continental particulate P deposited in marine sediments, indicates that the marine residence time of phosphate with respect to terrigenous inputs is ca. 10,000 years. This figure depends quite heavily on the fraction of terrigenous, particulate-phase phosphate that is released to seawater during diagenesis. Variations in this fraction can significantly affect the marine residence time of phosphate and the relative proportion of detrital versus authigenic phosphate phases in marine sediments.

INTRODUCTION

Previous attempts have been made to quantify and model the marine phosphorus cycle, either as an isolated biogeochemical cycle (e.g., Lerman et al., 1975; Froelich et al., 1982; Föllmi, 1996), or coupled with other major biogeochemical cycles (e.g., Mackenzie et al., 1993). In the present study we assess the suggestion that the marine phosphorus cycle and the partial pressure of atmospheric oxygen are coupled parts of a negative feedback mechanism that may have regulated atmospheric oxygen levels throughout the Phanerozoic (Betts and Holland, 1991; Holland, 1994a; Colman and Holland, 1994; Colman, 1994; Van Cappellen and Ingall, 1996; Colman et al., 1997). The proposed coupling of the oxygen and phosphorus cycles occurs during the early diagenesis of marine sediments.

It has long been recognized that a negative feedback control mechanism must be constraining the amount of oxygen in the atmosphere (e.g., Holland, 1973; Walker, 1974). The atmosphere currently contains 20.9% O_2. O_2 fluxes to and from the atmosphere are relatively well defined. The atmospheric oxygen reservoir of 1.2 x 10^{21} g O_2 divided by the total O_2 source or sink fluxes of ca. 3.2 x 10^{14} g O_2/yr yields a residence time of ca. 4 million years for atmospheric O_2 (Holland, 1978). Atmospheric P_{O_2} probably increased from ≤ 0.002 atm to ≥ 0.03 atm between 2.2 Ga and 2.0 Ga and has remained ≥ 0.03 atm since then (Holland, 1994b; Rye and Holland, 1998). Oxygen levels appear to have been fairly stable over the last 350 Ma, not exceeding ca. 0.35 atm nor falling below 0.15 atm (Watson et al., 1978; Cope and Chaloner, 1980, 1985; Robinson, 1989; Berner and Canfield, 1989). Rhoads and Morse (1971) have shown that P_{O_2} must have been > 0.02 atm throughout the Phanerozoic to support the oxygen demand of benthic macrofauna. The relative constancy of P_{O_2} over a span of time two orders of magnitude greater than the residence time of atmospheric O_2 indicates that P_{O_2} has almost certainly been stabilized by a control mechanism that has linked the O_2 content of the atmosphere either to the rate of O_2 production, or to the rate of O_2 consumption, or to both.

Significant sinks for atmospheric O_2 on geologic time scales are the oxidation of organic carbon, ferrous iron, and sulfide sulfur in rocks exposed to weathering, and the oxidation of reduced volcanic gases (e.g., Holland, 1973; Walker, 1974; Holland, 1978). Oxygen is released to the atmosphere on geologic time scales by the long-term burial of reduced carbon, ferrous iron, and sulfide sulfur, largely in marine sediments. The negative feedback required to stabilize atmospheric O_2 is probably generated by the sensitivity of the burial fluxes of reduced species to the amount of oxygen in the atmosphere (Holland, 1978).

Organic carbon burial in marine sediments is quantitatively the most important source of atmospheric oxygen (Holland, 1978, 1984). The worldwide burial rate of organic carbon depends on the supply of organic carbon to marine sediments and the efficiency with which this carbon is buried. The system may be partially stabilized by the coupling of the weathering fluxes to the sedimentation fluxes (e.g., Garrels et al., 1976; Berner and Canfield, 1989). Previous work has explored the

**email: albert.colman@yale.edu*

Marine Authigenesis: From Global to Microbial, SEPM Special Publication No. 66

effect of the oxygen concentration in bottom waters on the burial efficiency of organic carbon (e.g., Demaison and Moore, 1980; Emerson, 1985; Henrichs and Reeburgh, 1987; Canfield, 1989a; Pedersen and Calvert, 1990; Betts and Holland, 1991; Blackburn, 1991; Calvert et al., 1992; Van Cappellen and Canfield, 1993; Hulthe et al., 1998). Recent work by Hartnett et al. (1998, see also Reimers, 1998) indicates that burial efficiency may be correlated with the length of time that organic carbon spends in the oxidized zone of sediments. Although there probably is a link between the carbon burial efficiency and the oxygen content of bottom waters, it is unclear whether or not the connection is strong enough to stabilize atmospheric oxygen levels within the limits inferred from the geologic record during the Phanerozoic. Here we suggest that it is the organic carbon flux from the euphotic zone to marine sediments that is sufficiently sensitive to atmospheric oxygen levels to act as a significant control mechanism (Kump, 1988; Betts and Holland, 1991). We therefore examine the possibility that nutrient availability in the euphotic zone, which should limit export production, is inversely related to atmospheric oxygen levels.

Phosphorus, combined nitrogen, and trace metals have all been proposed as limiting nutrients in various parts of the world ocean today. It is commonly thought that over much of geologic time phosphorus has been the limiting nutrient (e.g., Broecker and Peng, 1982; Holland, 1984; although, see discussions of nitrogen limitation in Codispoti, 1989; Falkowski, 1997). The existence of a negative feedback has been proposed between atmospheric oxygen levels and the diagenetic return flux of phosphorus from marine sediments to the oceans (Holland, 1994a; Colman and Holland, 1994; Colman, 1994; Van Cappellen and Ingall, 1996; Colman et al., 1997). In this paper, we explore the relationship between redox conditions near the sediment-water interface and the return flux of phosphate from marine sediments to seawater.

Phosphorus enters the oceans almost entirely via rivers (Froelich et al., 1982; Meybeck, 1982; Duce et al., 1991) as dissolved inorganic phosphorus (DIP), dissolved organic phosphorus (DOP), particulate inorganic phosphorus (PIP), and particulate organic phosphorus (POP). Traditional models of the marine phosphorus cycle have considered that only DIP and DOP (taken together, total dissolved phosphorus, or TDP) are biogeochemically active. POP, and especially PIP, were thought to be refractory (e.g., Lerman et al., 1975; Froelich et al., 1982). More recent data have shown that particulate phosphorus may contribute significantly to the riverine reactive phosphorus flux (Fox et al., 1986; Froelich, 1988; Berner and Rao, 1994; Ruttenberg and Canfield, 1994). Most of the currently accepted values for the riverine TDP flux are based on Meybeck's (1982) estimate of 3.2×10^{10} moles/yr. Current estimates of the riverine total reactive phosphorus flux range from 4×10^{10} moles/yr (Froelich et al., 1982) to 41×10^{10} moles/yr (derived from Mackenzie et al., 1993). The uncertainty in the total reactive phosphate flux to the oceans is large because the fraction of the particulate phosphate that is labile is poorly constrained.

Sediments are the ultimate sink for marine phosphorus. In freshly sedimented materials, reactive phosphorus is found mainly as a constituent of organic matter and ferric (oxy)hydroxides (e.g., Sherwood et al., 1987). In certain localized environments, biogenic apatite in fish debris is a significant fraction of the total reactive phosphorus (Suess, 1981). During the early diagenesis of marine sediments, both the total phosphorus content and its distribution among various solid phases tend to change. A substantial but highly variable fraction of the particulate reactive phosphorus may be returned to solution and transported out of sediments. Although considerable variability exists from site to site, approximately one-third of the reactive particulate phosphorus that is buried is usually present as a constituent of ferric (oxy)hydroxides, another third as organic phosphorus, and the remaining third as authigenic francolite, a minor dispersed phase in a wide variety of marine environments (Ruttenberg, 1990; Ruttenberg and Berner, 1993; Jensen and Thamdrup, 1993; Berner and Rao, 1994; Vink, 1994; Föllmi, 1996; Slomp et al., 1996a; Vink et al., 1997). The burial of authigenic REE-phosphates may be a previously overlooked globally significant sink for marine phosphate (Rasmussen, 1996). Scavenging of phosphate on hydrothermal plume particles and altered basalts in mid-ocean ridge hydrothermal systems is also a significant global sink (Berner, 1973; Froelich et al., 1977; Feely et al., 1996; Wheat et al., 1996).

The chemical changes that redistribute phosphorus among various solid phases and release phosphorus to pore waters are principally due to the degradation of organic matter during early diagenesis. The connection may be direct, as with the release of metabolites such as phosphate, or indirect, as with the precipitation of francolite, and in rare cases, vivianite and struvite, in response to sufficiently elevated pore water phosphate concentrations and other pore water and sediment conditions.

The degradation of organic matter during the early diagenesis of marine sediments can involve a number of terminal electron acceptors. The most energetically favorable oxidant available is consumed preferentially during the biologically mediated degradation of organic matter. In marine sediments, the sequence in which electron acceptors are normally consumed is oxygen, nitrate, manganese (IV), ferric iron, sulfate, and carbon dioxide.

Oxidants are supplied to sediments and pore waters from the overlying bottom water and by fresh inputs of iron- and manganese-bearing sediments. Oxidants are consumed sequentially with burial leading to depth-based redox zonation in sediments. Pore water chemistry is a sensitive indicator of the successive depletion of oxidants during the burial of organic matter. Overlap between different diagenetic regimes can occur at the same depth horizon in sediments due to heterogeneities in the distribution of organic matter (Jorgensen, 1977; Jahnke, 1985), and the ventilation of burrows (Aller, 1977, 1980a,b, 1988). Kinetic inhibition, microbial ecology, and competing reactions may also create minor deviations from the thermodynamically predicted sequence (Canfield, 1989b; Canfield, 1993; Winters and Buckley, 1986). However, a single terminal electron acceptor normally dominates at any given depth, and the chemistry of pore waters in marine sediments is consistent with the thermodynamically predicted sequence.

Regardless of the electron acceptor, phosphate in decaying organic matter is released to pore waters. The reduction of solid phase ferric iron to ferrous iron in solution liberates additional phosphate because labile ferric (oxy)hydroxide usually contains significant quantities of adsorbed and structurally bound phosphate. Once in solution, phosphate and iron are subject to transport. In most marine settings, transport is dominated by diffusion and benthic infaunal activity, though wave and tidal pumping and episodic resuspension of sediments may be important in shallow shelf environments (Berner, 1980). If phosphate and ferrous iron are transported upwards into oxic environments, the ferrous iron precipitates as ferric (oxy)hydroxide, scavenging phosphate in the process. There may well be multiple crossings of the oxic to suboxic boundary. Ferric (oxy)hydroxide can be buried or mixed below the redox transition, redissolve, diffuse upward, and again be precipitated (e.g., Froelich et al., 1979; Aller, 1980a; Sundby et al., 1992; Canfield et al., 1993; Thamdrup et al., 1994). Such cycling of phosphate across the ferric-ferrous redox boundary in sediments may play an important role in elevating pore water phosphate concentrations so that conditions become favorable for francolite precipitation (e.g., Van Cappellen and Berner, 1988; Heggie et al., 1990; Ruttenberg and Berner, 1993; Glenn et al., 1994; Krajewski et al., 1994; Slomp et al., 1996a). In the zone of sulfate reduction, sulfide may combine with ferrous iron or may react with solid phase ferric oxides and hydroxides to form ferrous monosulfides, which may mature to pyrite (e.g., Canfield, 1989b; Canfield et al., 1992). The sequestration of iron in ferrous sulfides decouples the chemistry and transport of phosphate from those of iron, as phosphate does not coprecipitate with nor adsorb on ferrous sulfides (Krom and Berner, 1980a).

Clearly, a number of factors influence the quantity of phosphate that is released by marine sediments to pore waters and lost to overlying bottom waters. The rate at which reactive particulate phosphorus is deposited in sediments limits the potential return flux. The supply of labile organic matter largely controls the types, rates, and spatial distribution of chemical reactions during diagenesis. The crystallinity of ferric oxides and hydroxides influences their reactivity (Canfield, 1989b) and their phosphorus content (e.g., Lijklema, 1980; Colombo et al., 1994). Temperature influences degradation and diffusion rates. Infaunal community structure influences the redistribution of solid phases in sediments as well as the transport of metabolites out of sediments and the addition of oxidants to sediments. Finally, the oxygen content of bottom water influences the depth at which ferric iron is used as an electron acceptor and the depth at which dissolved ferrous iron precipitates from solution.

In this paper we attempt to define the parameters that control the return flux of phosphate from marine sediments to seawater, to estimate the magnitude of this flux today, and to determine the effect of redox conditions in sediments and bottom water on this flux. To the extent that we are successful, we can begin to define the strength of the negative feedback mechanism that links the level of atmospheric oxygen to the marine phosphorus cycle.

DATA AND ANALYSIS

The data used in this paper for estimating diffusive phosphate fluxes from marine sediments were taken from the literature and appear in Appendix A, where they are listed by study, core ID, and coordinates of the coring station. (Appendices A and B are archived in Microsoft Excel 97 and in tab-delimited ascii formats and are available in digital form at the World Data Center-A for Marine Geology and Geophysics, NOAA/NGDC, 325 Broadway, Boulder, CO 80303, USA; phone 303-497-6339; fax 303-497-6513; URL http://www.ngdc.noaa.gov/mgg/sepm/jsr/; E-mail: wdcamgg@ngdc.noaa.gov.) In a few cases, latitude and longitude coordinates were estimated from map representations of the study site. Water depth, date of coring, and bottom water temperature are all reported. Sedimentation rates are generally based on radionuclide tracer profiles. For a few slowly accumulating sediments, the depth to the Holocene/Pleistocene transition was used to estimate sedimentation rates. For one study (Elderfield et al., 1981), the sedimentation rate was based on diagenetic modeling. Uncertainties in most of the sedimentation rate estimates are a factor of two to three ($\pm 2\sigma$ uncertainty). Porosity was determined by sediment mass loss on drying. Organic carbon content was usually determined by high-temperature combustion of acid-treated sediment samples using CHN elemental analyzers. Data for the behavior of redox-sensitive pore water species were also noted. A discussion of the redox data appears in the calculations section.

Calculations of the phosphate flux require a knowledge of the pore water phosphate gradient at the sediment-bottom water interface (SWI), the surficial sediment porosity, and the temperature (see calculations section). Pore water phosphate concentrations were usually measured with nutrient autoanalyzers. Analytical uncertainties in the phosphate measurements were $\leq \pm 5\%$ ($\pm 2\sigma$ level). Phosphate gradients were calculated assuming a linear change in phosphate concentration between bottom water and the topmost measured pore water concentration,

$$\left(\frac{\partial [PO_4]}{\partial z}\right)_{z=0} = \frac{[PO_4]_{top} - [PO_4]_{BW}}{\Delta z}, \qquad (1)$$

where $[PO_4]$ represents the dissolved orthophosphate concentration. The subscripts *top* and *BW* denote topmost sediment layer and bottom water, respectively; z is the depth coordinate, $z = 0$ signifies the SWI, and Δz represents the distance between the SWI and the mean depth of the topmost sediment layer. Pore water phosphate profiles in surficial marine sediments are commonly nonlinear due to significant pore water phosphate sources and sinks near the SWI. To minimize errors introduced by the assumption of linearity, we have included only profiles for which $\Delta z \leq 1.1$ cm. In most cases, $\Delta z \leq 0.5$ cm (i.e., the surficial layer of sediment from which pore waters were derived was 1 cm thick). Nevertheless, we cannot rule out the possibility that reactions in the top few millimeters of sediment and pore water may cause deviations between actual fluxes and our flux estimates. This topic is addressed further in the discussion of results section. We only used data from cores for which sediments at the SWI showed negligible signs of disturbance due to coring.

Within this set of cores, only those that were processed using methods to avoid oxidation artifacts were included (Bray et al., 1973). Most of the sediments with usable profiles were recovered by means of a diver-held or ship-deployed box core. A few cores were obtained with piston corers, and a few pore water profiles are based on dialyzer or "peeper" equilibrations or on harpoon sampler collections. Pore water was extracted from cores with whole-core squeezers (e.g., Jahnke, 1988; Bender et al., 1987), or by extrusion and processing of individual sediment layers using centrifugation or piston squeezers. Uncertainties in measuring pore water species, in particular particle-reactive species such as phosphate, that arise during coring and pore water separation are discussed in Emerson et al. (1980), Jahnke et al. (1982), Bender et al. (1987), Jahnke (1988), and Jahnke and Knight (1997). The estimated uncertainy in the phosphate gradients reflects the analytical uncertainty and, where applicable, the uncertainty in evaluating data published only in graphical form. No attempt was made to evaluate the accuracy of the reported surficial sediment layer thickness. An additional 5-20% uncertainty in the phosphate gradient would be contributed for each millimeter of uncertainty in the thickness of the surficial sediment layer. Expansion during decompression of pelagic cores may cause pore waters to move relative to the sediment in the core (e.g., Jahnke et al., 1982). We did not attempt to correct for this possible artifact, because it is not clear at present how to do so.

In four deep sea cores, the phosphate gradients within the first centimeter below the SWI were indistinguishable from zero. This is not surprising, since the particulate phosphate flux to these sediments is quite small. Complete remineralization of the flux of labile particulate organic phosphorus would produce a phosphate gradient of ca. 0.01 to 1 μM/cm in typical deep-sea sediments. Due to analytical constraints, the minimum resolvable phosphate gradient is on the order of 0.1 to 0.4 μM/cm when $[PO_4]$ in bottom water is ca. 2 μM. Fortunately, the phosphate gradient in the top decimeter in these cores is essentially linear. This enabled us to estimate the surficial gradients in these cores (Jahnke et al., 1989, Sta. 10; Emerson et al., 1980, cores 9BC26 and 9BC27; Froelich et al., 1979, core 12GC2). The estimates are enclosed in parentheses in Appendix A.

CALCULATIONS

Phosphate Fluxes

The diffusive flux of phosphate across the SWI was calculated in two ways, both based on a standard modified form of Fick's First Law of diffusion (e.g., Berner, 1980). In the first approach we assumed that simple aqueous diffusion is the sole transport mechanism. The second approach incorporated the effects of bioturbation (particle mixing) treated as a diffusive process. Advective burial of pore water exerts a negligible influence on the net flux of phosphate across the sediment-water interface for all sediments encountered in this study and is therefore neglected in our calculations. The flux equations used for transport via aqueous diffusion only and aqueous diffusion plus biological mixing, respectively, were:

$$J_\circ = -\phi_\circ D_s \left(\frac{\partial [PO_4]}{\partial z} \right)_{z=0}, \quad (2)$$

and

$$J_\circ = -\phi_\circ \left(D_s + D_B \right) \left(\frac{\partial [PO_4]}{\partial z} \right)_{z=0}, \quad (3)$$

where J_o = phosphate flux across the sediment-water interface (in mol $PO_4/m^2/yr$);
ϕ_o = porosity of surficial sediment;
D_s = molecular diffusion coefficient in interstitial water (in cm^2/sec)
D_B = biological mixing diffusion coefficient (in cm^2/sec)
z = depth coordinate (with SWI defined as z = 0; z, in cm, increasing with depth)
$(\partial[PO_4]/\partial z)_{z=0}$ = phosphate concentration gradient at the sediment-water interface (in μM/cm).

A unit conversion factor, not shown above, is used to convert the above products into units of mol $PO_4/m^2/yr$.

Molecular Diffusion

D_s in equation (2) is generally related to the aqueous molecular diffusion coefficient, D, via the expression:

$$D_s = \frac{D}{\theta^2} \quad (4)$$

where θ is the tortuosity, which is related to the sediment and pore water electrical resistivity, and to the porosity, ϕ , via the expressions:

$$\theta^2 = \phi F \quad (5)$$

$$F = R_s / R_w = \phi^{-n}. \quad (6)$$

F is the formation factor, defined as the ratio of the bulk sediment resistivity to the interstitial water resistivity (R_s/R_w). The exponent n in equation (6) varies with sediment type, and is higher for unconsolidated, highly porous sediments than for consolidated sands (Berner, 1980). Manheim and Waterman (1974) found that n is ca. 2 for low porosity ($\phi < 0.7$), compacted marine sediments. Data from Andrews and Bennett (1981) and from Ullman and Aller (1982) show that n is ca. 2.5-3 for high porosity ($\phi > 0.7$) muds. The surficial sediments we have reviewed, from coastal settings to deep-sea settings, exhibit high porosities (see Appendix A). We have adopted a value of 2.75 for n in the calculation of bulk sediment diffusion coefficients. Combining equations (4), (5), and (6), and setting n equal to 2.75, we obtain:

$$D_s = D\phi^{1.75} \quad (7)$$

We have used the formulation of Li and Gregory (1974; consistent with the data of Krom and Berner, 1980b) to compute the temperature dependence of the aqueous molecular diffusion coefficient of HPO_4^{2-} :

$$D = 7.34 \times 10^{-6} \frac{cm^2}{\text{sec}} - 0.160 \times 10^{-6} \frac{cm^2}{\text{sec } ^\circ\text{C}} (25^\circ\text{C} - T) \quad (8)$$

Equation (8) is based on measurements in dilute aqueous solutions. Diffusion coefficients calculated in this way may underestimate the effective diffusion coefficient of phosphate, which is enhanced in seawater by ion pairing, primarily with Mg^{2+}, Na^{+}, and Ca^{2+} (Lasaga, 1979; Katz and Ben-Yaakov, 1980; Atlas et al., 1976). Ion pairing of phosphate reduces the net effective charge of the diffusing phosphate species. This, in turn, reduces the hydration number of phosphate and may increase the effective molecular diffusion coefficient of phosphate by up to 50% (Klump and Martens, 1981). Klump and Martens's (1981) estimated enhancement may be too high, since it does not incorporate the effect of the paired cation's mass, which reduces the diffusion rate of the complex. We have neglected the effect of ion pairing. Our flux calculations therefore yield minimum values for the diffusive flux of phosphate.

Bioturbation (Biomixing)

Benthic fauna can influence solute exchange between pore water and bottom water through the physical reworking of sediments or the irrigation of burrows. Field studies (e.g., Callender and Hammond, 1982; Clavero et al., 1991) suggest that when benthic faunal activity is intense in coastal settings, nutrient fluxes can be enhanced by as much as a factor of ten. A variety of techniques have been developed to incorporate bioturbation and bioirrigation into models of early diagenesis (e.g., Schink and Guinasso, 1978; Aller 1980b, 1982, 1988; Berner, 1980; Boudreau, 1984). The simplest and most widely employed, though not necessarily the most accurate, technique treats sediment biomixing and pore water bioirrigation as diffusion processes. However, only a few of the sites in Appendix A have been studied with respect to the faunal enhancement of benthic fluxes.

Boudreau (1994) and Tromp et al. (1995) derived general expressions relating biomixing rates, analogous to diffusion coefficients, to sedimentation rates by examining radiotracer-based estimates of bioturbation rates from a variety of sediment settings:

$$D_b = 4.98 \times 10^{-7} \omega^{0.69} \text{ (Boudreau, 1994)} \quad (9)$$

and

$$D_b = 13.5 \times 10^{-7} \omega^{0.85} \text{ (Tromp et al., 1995)}, \quad (10)$$

where D_b is the biomixing coefficient in units of cm^2/sec and where the sedimentation rate, ω, is in cm/yr. Over the full range of sedimentation rates observed in our study, the two expressions agree within a factor of five. The data sets used by Boudreau (1994) and Tromp et al. (1995) in deriving equations (9) and (10) show a spread of roughly one to two orders of magnitude in bioturbation rates for any given sedimentation rate. However, their data also show that biomixing rarely contributes significantly to pore water transport in areas where the sedimentation rate is < 0.5 cm/yr. Biomixing rates only tend to equal or exceed molecular diffusion rates of solute transport in environments where sedimentation rates are above roughly 1 cm/yr. We have calculated phosphate fluxes with and without biomixing, and we use the mean value from equations (9) and (10) to represent D_b in our flux calculations. The frequent presence of sulfide in the bottom water of Saanich Inlet (Murray et al., 1978) prevents the establishment of a benthic faunal community. Likewise, conditions in the Cape Lookout Bite sediments prevent appreciable faunal activity for most of the year (Klump and Martens, 1981). In both cases we have not calculated an enhanced biomixing flux.

Bioirrigation

Faunal irrigation of burrows often plays a much larger role than biomixing in solute transport between pore water and bottom water. In cases of intense irrigation, total solute fluxes across the SWI may exceed those predicted by simple molecular diffusion by as much as an order of magnitude (e.g., Callender and Hammond, 1982). Bioirrigation has been modeled using radial diffusion (e.g., Aller, 1977, 1980b), non-local exchange (e.g., Boudreau, 1984), and one-dimensional, enhanced-diffusion models (e.g., Berner, 1980). Bioirrigation rates are frequently estimated by comparing fluxes measured using benthic flux chambers, $J_{F.C.}$, with fluxes calculated using diffusion models of SWI solute gradients, J_{diff}. The ratio $J_{F.C.}$: J_{diff} is used to calculate a diffusion enhancement factor due to bioirrigation. Such calculations assume that differences between $J_{F.C.}$ and J_{diff} are solely due to bioirrigation. However, a non-negligible fraction of the flux measured in benthic flux chambers may be due to direct reactions between particles at the SWI and bottom water. Even in the absence of bioirrigation, flux chamber measurements may therefore exceed fluxes calculated from the diffusion modeling of pore water gradients.

Redox-reactive pore water species, such as Fe^{2+}, Mn^{2+}, and species like phosphate that coprecipitate with and/or are adsorbed strongly on freshly precipitated Fe- and Mn- oxides and hydroxides, may not be as efficiently transported as non-redox-reactive species to overlying bottom water via bioirrigation (R. Aller, pers. commun., 1997). Instead, they may become constituents of solid phases on entering oxic microenvironments that surround well-irrigated burrows. This could reduce fluxes below those predicted by vertical diffusion alone. We know of no direct measurements of the radial distribution of particulate phosphate phases surrounding burrow structures, although distinctive color changes accompany the transition of iron from reducing to oxic sediments around burrows. We have chosen not to incorporate bioirrigation in our calculations of phosphate fluxes.

POC Burial Flux, Surface-arriving Flux, and Burial Efficiency

We have calculated the burial rate of particulate organic carbon in units of mol $C/m^2/yr$ using the relationship:

$$\text{POC Burial Flux} = \omega \times (1-\phi) \times \rho_{sol} \times \frac{\%\text{Org.C}_{z=\infty}}{100} \times \frac{1 mol C}{12 g C} \times \frac{10^4 cm^2}{m^2}, \quad (11)$$

where ω = sedimentation rate (in cm/yr), ϕ = porosity, and ρ_{sol} = density of sediment solids (taken to be 2.6 g sed./cm^3 sed). $\%\text{Org.C}_{z=\infty}$ is the weight percentage of organic carbon in dried sediment at a depth where the organic carbon content of the sediment appears to have become constant. This organic carbon consists of compounds that are recalcitrant to decomposition on the time scale of early diagenesis in the particular environment studied. The calculated POC burial fluxes are listed in Appendix A.

The determination of the POC flux arriving in freshly depositing sediments (POC SWI flux) would appear to be straightforward, with the substitution of the %Org.C in surficial sediments for the $\%\text{Org.C}_{z=\infty}$ in equation (11). However, even sediment in the top one or two centimeters of a sediment column is depleted in organic carbon relative to the material arriving at the SWI. There are two principal reasons for this. First, organic matter in freshly arriving sediment tends to be very labile. Second, fresh surficial sediment tends to be mixed rapidly via bioturbation into underlying sediments, and thus sediment at any particular depth is a mixture of sediments of different ages. It is therefore difficult to isolate a surficial sediment layer thin enough to define the composition of sediments as they arrive at the SWI. Accurate benthic sediment-trap data with sufficient temporal resolution to capture seasonal and interannual variability are needed to determine the POC SWI Flux. Unfortunately, such data are lacking for the sites that we studied.

The POC SWI flux can also be determined using the mass-balance relationship: POC SWI flux = (Depth-integrated respiration rate) + (POC burial flux). Respiration rates may be measured using benthic flux chambers or calculated by modeling the concentration of pore water oxidants and/or metabolites (e.g., Canfield, 1989a). However, benthic flux chamber measurements and pore water profiles provide only a "snapshot" of sediment respiration. Within a single year, respiration rates may vary by a factor of four to ten in continental shelf and slope environments (e.g., Martens and Klump, 1984; McNichol et al., 1988) and by a factor of two to four in pelagic sediments (e.g., Smith and Baldwin, 1984; Smith et al., 1992). Variations in respiration rate are due to seasonal blooms and concomitant variations in sedimentation rates and the composition of sedimenting particles. Rapid degradation kinetics (e.g., Cole et al., 1987; Soetaert et al., 1996) are also required for the depositional variations to be manifest in variations of integrated sediment column metabolic rates (Martin and Bender, 1988). In continental shelf environments where the temperature changes seasonally, changes in reaction rates and diffusion kinetics, both of which scale with temperature, also contribute to intra-annual variations in sediment respiration rates (e.g., Aller, 1977). This may lead to erroneous estimates of POC depositional fluxes if single respiration measurements are used to characterize carbon mass balance at individual sites. Time-series measurements of benthic respiration rates could yield accurate descriptions of POC SWI fluxes, but these measurements have not been made for most of the sites that were studied.

We have therefore used the organic carbon burial efficiency trends reported in the literature (e.g., Sarnthein et al., 1987; Henrichs and Reeburgh, 1987; Canfield, 1989a; Betts and Holland, 1991) to estimate POC SWI fluxes at our sites. The burial efficiency (B.E.) is defined as:

$$\text{B.E.} = \frac{\text{POC Burial Flux}}{\text{POC SWI Flux}} \times 100\%. \quad (12)$$

We have used the correlation between burial efficiency and sedimentation rate reported in Betts and Holland (1991), combined with data for the sedimentation rates at the sites used in this study, to determine B.E. and hence the POC SWI flux. Using

$$\text{B.E.} = 2.19 \times 10^{\left(\frac{1.39 Log(\omega)}{Log(7.9+\omega)}\right)}, \quad (13)$$

it follows that,

$$\text{POC SWI Flux} = \frac{\text{POC Burial Flux}}{2.19 \times 10^{\left(\frac{1.39 Log(\omega)}{Log(7.9+\omega)}\right)}} \times 100. \quad (14)$$

One sigma uncertainties in the calculated B.E. in Betts and Holland (1991) are generally a factor of 2 to 2.5.

We also calculated a carbon respiration flux using the difference between POC SWI and POC burial fluxes. The use of a generalized B.E. parameterization for determining surface-arriving and sediment-respiration carbon fluxes at individual sites in our study provides only very rough estimates of carbon fluxes. Nevertheless, these estimates prove useful for comparison with calculated phosphate fluxes, as will be discussed in a later section.

Redox Index

A principal goal of this study was to ascertain whether or not sediment redox conditions, and by extension, bottom water dissolved oxygen content, influence the efficiency with which particulate phosphorus entering sediments is regenerated and returned to the overlying water column. Although both experimental and field evidence indicates that there is a redox-sensitive connection between phosphate and ferric (oxy)hydroxides in sediments (e.g, Klump and Martens, 1981; Sundby et al., 1986; Ingall and Jahnke, 1994, 1997; McManus et al., 1997), it has been difficult to define the quantitative relationship between redox conditions and phosphate fluxes from marine sediments.

To this end, we searched for a widely measured parameter that would serve as an indicator of the redox conditions in pore water-bottom water systems. Bottom water dissolved oxygen concentration and/or oxygen penetration depth might serve as suitable parameters. Unfortunately, such data were unavailable for most of our data set. In their place, we have developed a classification system and a quantification system for redox conditions within surficial marine sediments. An advantage of using sediment redox conditions directly is that, though bottom water oxygen concentrations certainly affect sediment redox conditions, the sediment redox conditions are the direct cause for redox related homogeneous and heterogeneous reactions in pore waters and on particulate surfaces.

We define three sediment redox states in terms of the oxidants used to degrade organic matter. "Oxic" conditions correspond to the regions in which O_2 and NO_3^- are the principal electron acceptors; "suboxic" conditions correspond to regions where Mn and Fe reduction occurs; and "anoxic" conditions correspond to the region in which sulfate reduction and methanogenesis are important. This classification is similar to that suggested by Berner (1981). Both schemes have roughly equivalent definitions of "oxic"; our "suboxic" category corresponds to Berner's "post-oxic", and our "anoxic" category comprises his "sulfidic" and "methanic" categories.

In oxic sediments, phosphate is released during the degradation of organic matter. Ferric (oxy)hydroxides, where present, may scavenge phosphate from pore waters. In suboxic sediments, phosphate is also released during the reductive dissolution of ferric (oxy)hydroxides. In anoxic sediments, remaining ferric (oxy)hydroxides may continue to release phosphate. Sulfide reacts with ferric (oxy)hydroxides in this region and scavenges ferrous iron from pore waters, thereby decoupling iron dynamics from those of phosphate. The highest ratio of phosphate regeneration rate to carbon respiration rate occurs in anoxic and suboxic sediments, and the ratio is lowest in oxic sediments.

Careful analysis of compositional changes in pore water and solid phase profiles allows reasonably accurate determinations of the depths at which transitions from oxic to suboxic and from suboxic to anoxic conditions occur. The transition from oxic to suboxic is taken to be the depth at which $[O_2]$ and $[NO_3^-]$ reach detection limits. The transition to suboxic conditions is also signaled by the appearance of dissolved Mn^{2+}, followed closely by the appearance of Fe^{2+}, in pore waters. NH_4^+ production also becomes significant. The transition from suboxic to anoxic conditions is marked by loss of pore water sulfate. The presence of pore water HS^- and of solid phase AVS and pyrite are also signs of reducing conditions. Depletion of pore water Fe^{2+} with depth, which may result in a subsurface maximum in $[Fe^{2+}]$, is usually due to precipitation of ferrous sulfides, and therefore can be used as an indicator of anoxic conditions. However, subsurface Fe^{2+} maxima can also be due to siderite and vivianite precipitation, non-steady state diagenesis, and preferential bioirrigation of sediment at depth, and care must be taken to distinguish between these possible causes.

Eh is notoriously difficult to measure accurately (e.g., Berner, 1981; Lindberg and Runnells, 1984; Grenthe et al., 1992; Stumm and Morgan, 1996). However, thermodynamically predicted Eh values for Mn, Fe, and S redox transitions correlate roughly with observed pore water chemistry where both Eh and redox-sensitive pore water species were measured. The transition from oxic to suboxic conditions occurs between ca. 0.15 and 0.30 V, and the transition from suboxic to anoxic conditions occurs between ca. 0 and -0.15 V. In a few cases, color transitions in sediments were used to complement other indicators of redox transitions. A brown to green or tan to green color transition with depth is associated with Fe reduction in the oxic to suboxic transition (Hartmann et al., 1976; Lyle, 1983; Hansen et al., 1994; König et al., 1997; Trolard et al., 1997).

To quantify sediment redox conditions, we partitioned the top 10 cm of sediment at each site for which redox indicators were available into 10 1-cm-thick layers. Each layer was assigned a numerical value based on the redox condition prevailing in that layer: oxic = 0, suboxic = 1, and anoxic = 2. We then integrated the redox conditions for the 10 cm of surficial sediment. Sediments in the top layer play a much larger role in determining fluxes across the sediment water interface than do sediments in the bottom layer, due to both the relative proximity of near-surface layers to the SWI and the greater availability of reactive organic matter and labile ferric (oxy)hydroxides in near-surface sediments. After testing a variety of ways of incorporating a depth-weighting scheme into the calculation of the redox parameter, we settled on a linear system. We then normalized the depth-integrated redox parameter to produce a redox index, or *RI*. A sediment core that is fully oxic throughout its top 10 cm has an *RI* value of 0. An *RI* value of 1 characterizes a sediment which is fully anoxic throughout its top 10 cm. The formulation for the *RI* is

$$RI = \frac{\sum_{i=1}^{10} Rx_i \times (11-i)}{\sum_{i=1}^{10} 2 \times (11-i)} , \qquad (15)$$

in which i represents the sediment layer from $(i - 1)$ to i cm below the SWI; Rx_i is the numerical value (0, 1, or 2) of the redox conditions in layer i. The denominator normalizes the *RI* to a range of 0-1.

Annual Mean Phosphate Fluxes and Redox Conditions

At all of the continental shelf sites where cores were collected during the course of several seasons, a pronounced seasonal variation in the phosphate flux and sediment redox conditions was observed (Aller, 1977; Klump and Martens, 1981; McNichol et al., 1988; Jensen et al., 1995). Many factors combine to cause these seasonal variations. Seasonal blooms are followed by increased rates of organic matter sedimentation, bioturbation, and organic matter decomposition. The kinetics of organic matter degradation (e.g., Westrich, 1983) and diffusion (e.g., Li and Gregory, 1974) both scale with temperature. Since there is a distinct and understandable bias in favor of sampling sediments and pore waters during the summer and, to a lesser extent, during the spring and fall, a seasonal correction factor must be applied when using single phosphate pore water profiles to calculate annual phosphate fluxes.

Though seasonal variations in the flux of organic matter to the sediment probably affect the phosphate release rates and

TABLE 1.—SEASONAL VARIATION IN THE DIFFUSIVE PHOSPHATE FLUXES AND REDOX TRANSITIONS AT TIME SERIES SITES

Study-Site	Winter		Spring		Summer		Fall	
	PO_4 Flux (mol $PO_4/m^2/yr$)	Redox Transitions (cm)	PO_4 Flux (mol $PO_4/m^2/yr$)	Redox Transitions (cm)	PO_4 Flux (mol $PO_4/m^2/yr$)	Redox Transitions (cm)	PO_4 Flux (mol $PO_4/m^2/yr$)	Redox Transitions (cm)
Aller (1977)–FOAM	0.014 (±0.003, n=2)	1,4 (n=2)	NA	NA	0.16 (±0.14, n=2)	0,2 (n=2)	0.023 (n=1)	2,4 (n=1)
Aller (1977)–NWC	0.0044 (±0.0036, n=2)	2,3 (n=2)	NA	NA	0.24 (±0.18, n=2)	0,1 (n=2)	0.018 (±0.0080, n=2	1,3 (n=2)
Aller (1977)–Deep	NA	NA	0.0069 (n=1)	2,6 (n=1)	0.025 (n=1)	1,2 (n=1)	0.014 (n=1)	2,4 (n=1)
Jensen et al. (1995)–Aarhus Bay	0.014 (±0.010, n=5)	0,3 (n=5)	0.017 (±0.010, n=2)	0,3 (n=2)	0.037 (±0.013, n=5)	0,2 (n=5)	0.020 (±0.014, n=3)	0,2 (n=3)
Klump and Martens (1981)–CLB	0.028 (±0.018, n=4)	1,4 (n=4)	0.59 (±0.38, n=4)	0,0 (n=4)	0.61 (±0.090, n=4)	0,0 (n=4)	0.17 (±0.11, n=3)	0,1 (n=3)
McNichol et al. (1988)–Buzzards Bay	0.024 (n=1)	0,(8) (n=1)	0.063 (n=1)	0,(4) (n=1)	0.11 (n=1)	0,(4) (n=1)	0.073 (±0.043, n=3)	0,(4) (n=2)

Mean seasonal fluxes ±s.d. are reported. The redox transitions column indicates the mean depth at which the oxic-suboxic transition occurs, followed by the mean depth at which the suboxic-anoxic transition occurs. The redox-based diagenetic regimes and the transitions between those regimes are defined and discussed in the text.

sediment redox conditions more than variations in temperature, seasonal organic matter sedimentation fluxes are rarely reported. However, temperature variations are documented for the time series studies listed above. The temporal patterns in the phosphate fluxes calculated from the above studies suggest that approximate seasonal corrections can be made by binning the data into four seasonal groupings, each of equal duration and based on the seasonality of water temperature. We let the winter season correspond to the period of coldest bottom water and the summer season to the period of warmest water. We do not have any Southern Hemisphere continental shelf data; thus, the winter, spring, summer, and fall seasons correspond, respectively, to January through March, April through June, July through September, and October through December. Table 1 summarizes the seasonal means of the phosphate flux and redox boundaries at the sites for which seasonal data are available. Table 2A shows the fraction of the total annual flux contributed during each season.

We have used the data in Table 2A to estimate the annual phosphate flux at continental shelf sites. We calculate the total flux of phosphate at a site during any given season (in mol P/m^2) to be the mean calculated flux from all the cores collected during that season (in units of mol $P/m^2/yr$) multiplied by the duration of the season (0.25 yr). We then sum the total phosphate fluxes from each season and divide by a correction factor, β, where β is the mean fraction of the total annual flux from the seasons during which fluxes were determined (from the last line in Table 2A). For example, if at a given site, two phosphate fluxes were determined, one in the summer and one in the fall, the mean annual flux calculated using the above method would be 0.25 yr x $(J_{summer} + J_{fall}) / (\beta_{summer} + \beta_{fall})$ where J_{summer} and J_{fall} represent the summer and fall fluxes, and β_{summer} and β_{fall} represent the mean fraction of total annual fluxes owing to the summer and fall seasons (0.50 and 0.17, respectively, from the final row of Table 2A). The full data set in Appendix A is thus converted to mean annual data, which appear in Appendix B.

Table 2B uses the redox transition data presented in Table 1 to document seasonal deviations from annual mean sediment redox conditions. The depths to redox conditions in spring and fall are close to annual mean conditions. Summer redox transitions are generally closer to the SWI than mean annual transitions, and winter transitions are generally deeper in the sediment.

We use the trends evident in Table 2B to estimate annual mean redox conditions for each continental shelf site in Appendix A by means of a three-step calculation. First, seasonal

TABLE 2.—(A) SEASONAL DISTRIBUTION OF THE TOTAL ANNUAL DIFFUSIVE PHOSPHATE FLUX

Study-Site	Annual Flux (mol $PO_4/m^2/yr$)	Fraction of annual flux occurring in each season Winter	Spring	Summer	Fall
Aller (1977)– FOAM[1]	0.068	5%	28%	58%	9%
Aller (1977)– NWC[1]	0.091	1%	28%	66%	5%
Aller (1977)– Deep[1]	0.013	9%	14%	50%	27%
Jensen et al. (1995)– Aarhus Bay	0.022	16%	19%	42%	23%
Klump and Martens (1981)– CLB	0.35	2%	42%	44%	12%
McNichol et al. (1988)– Buzzards Bay	0.068	9%	23%	41%	27%
Unweighted Means:		7%	26%	50%	17%

[1]indicates that for Aller (1977), the "FOAM" and "NWC" Spring fluxes and the "Deep" Winter fluxes (2A) were calculated using the patterns established by the other studies for the % of the total annual flux represented by the Spring and Winter fluxes respectively. Redox transitions (2B) were treated similarly.

(B) SEASONAL VARIATION IN POSITION OF REDOX TRANSITIONS

Study-Site	Mean Redox Transitions (cm)	Seasonal deviations of redox transitions from annual means Winter	Spring	Summer	Fall
Aller (1977)– FOAM[1]	1,4	0,0	NR	+1,+2	-1,0
Aller (1977)– NWC[1]	1,3	-1,0	NR	+1,+2	0,0
Aller (1977)– Deep[1]	2,5	NR	0,-1	+1,+3	0,+1
Jensen et al. (1995)– Aarhus Bay	0,3	0,0	0,0	0,+1	0,+1
Klump and Martens (1981)– CLB	0,1	-1,-3	0,+1	0,+1	0,0
McNichol et al. (1988)– Buzzards Bay	0,4	0,-4	0,0	0,0	0,0
Representative Seasonal Deviations:		-1,-1	0,0	+1,+1	0,0

Positive deviations indicate a shallowing of the redox transition; negative deviations indicate a deepening of the redox transition.

mean redox transition depths are calculated for each season during which cores were collected. Then, for each season represented, an estimate of the annual mean redox transitions is made by subtracting the appropriate seasonal deviation, taken from the final line of Table 2B, from the seasonal mean redox transitions. That is, each seasonal mean provides an estimate of the annual mean transition depths. These estimates, based on corrections to the seasonal means, are then averaged to yield a best estimate for the annual mean redox transition depths. The annual means appear in Appendix B.

Interannual and spatial variability further complicate the calculation of representative annual fluxes and redox conditions. We believe that the effect of interannual and spatial variability on the fluxes and redox conditions will tend to average out in the aggregated data and that our estimates of global fluxes and trends are reasonably unbiased.

DISCUSSION OF RESULTS

General Trends

The calculated diffusive phosphate flux across the SWI for individual sediment pore water profiles ranges from essentially zero to 0.6 mol $PO_4/m^2/yr$, and the estimated mean annual fluxes range from ca. 10^{-5} to 0.5 mol $PO_4/m^2/yr$. The data and calculations in Appendices A and B show that the phosphate fluxes are correlated with several sediment and pore water characteristics, which are themselves correlated with one another. The strong correlation between sedimentation rate and POC depositional flux (Fig. 1A) is largely due to the fact that sedimentation rates in this study vary by over four orders of magnitude, whereas organic carbon concentrations largely cluster within a range of about an order of magnitude. Sediment redox conditions respond to the rate of organic matter degradation, which is in turn related to the POC depositional flux. Consequently, higher POC depositional fluxes would be expected to correlate with more reducing conditions in the sediments, as is shown in Figure 1B. The correlation between sedimentation rate and POC depositional flux (Fig. 1A), coupled with the relationship between the POC depositional flux and sediment redox conditions (Fig. 1B), suggests that sedimentation rates and sediment redox conditions should also be correlated. This is indeed observed in Figure 1C.

Figures 2A, B, and C show the mean annual diffusive phosphate fluxes plotted against the sedimentation rate, the particulate organic carbon flux to the sediment, and the *RI*. The diffusive flux of phosphate across the SWI is larger at higher sedimentation rates and with increasing depositional rates of particulate organic carbon (Figs. 2A, 2B). As discussed previously, the sedimentation rate and the POC depositional flux are rough indicators of the rate at which reactive particulate phosphorus is supplied to sediments. The large scatter in Figures 2A and B is due both to the large uncertainty in the position of each data point and to real differences in the composition of inorganic and organic sediment at the several sites. The POC burial rate, which is much more tightly constrained by our data than the POC depositional flux, correlates with the phosphate return flux

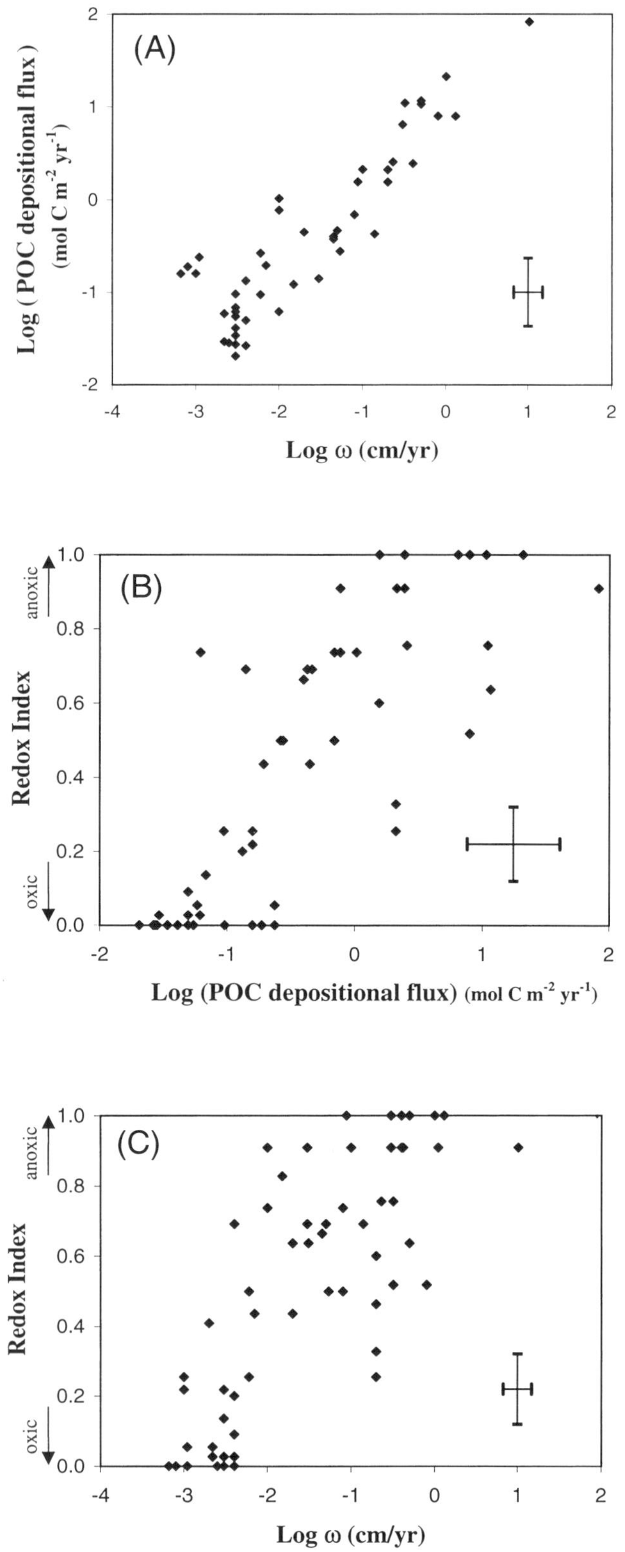

FIG. 1.—(A) Log(POC depositional flux) versus Log(ω); ω represents linear sedimentation rate. (B) Redox Index versus Log(POC depositional flux). (C) Redox Index versus Log(ω). Representative 1-σ error bars, calculated with standard error propagation techniques, are indicated in each plot. A Redox Index value of 0 indicates that a sediment core is oxic throughout its top 10 cm; a Redox Index value of 1 indicates anoxic conditions throughout the top 10 cm. See text for redox classification system and derivation of the RI.

in much the same way as the POC depositional flux. The return phosphate flux is also related to the calculated *RI* (Fig. 2C). Phosphate fluxes from more reducing sediments are generally greater than those from more oxidizing sediments. The data in Figure 2C appear to be more strongly correlated than those in Figures 2A and B. A linear least-squares regression to the data in Figure 2C yields an R^2 value of 0.83, which suggests that redox conditions do, in fact, correlate with phosphate fluxes and that the *RI* captures valuable information linking the redox conditions in sediments to phosphate fluxes.

Carbon Model Predictions of Phosphate Return Fluxes

The fact that the phosphate flux, sedimentation rate, POC depositional flux, and sediment redox conditions are all correlated (Figs. 1 and 2) makes it difficult to unravel the extent to which any one factor controls the rate of phosphate release from marine sediments. We could better understand how these several factors, acting in concert, determine phosphate fluxes by using diagenetic models that follow the degradation of organic matter via the full suite of terminal electron acceptors discussed in the introduction (e.g., models in Rabouille and Gaillard, 1991; Tromp et al., 1995; Wang and Van Cappellen, 1996; Dhakar and Burdige, 1996; Soetaert et al., 1996; Middleburg et al., 1997). However, these models require a good deal more site-specific information on pore water and sediment composition than is available for the cores and sites we studied.

We have therefore modified Berner's (1980) general diagenetic model to predict the phosphate flux from the cores in this study using the sedimentation rate, POC depositional flux, sediment porosity, water depth, and temperature data listed in Appendix B for each site. We then compare this predicted phosphate flux with the calculated diffusive flux and assess the degree to which differences between the two are related to the sediment *RI*.

We begin with Eqn. 6-66 in Berner (1980):

$$[PO_4] = \left(\frac{F \; P_o \, \omega^2}{D_s k + (1+K_p)\omega^2} \right) \times \left(1 - \exp\left\{ \frac{-k}{\omega} z \right\} \right) + [PO_4]_{z=0} \; , \qquad (16)$$

where, $[PO_4]$ = pore water phosphate concentration as a function of depth (in μM),

$F = \rho_{sol}\,(1-\phi)/\phi$, where ρ_{sol} and ϕ retain their previous definitions,

P_o = concentration of metabolizable particulate organic P at SWI (in mol P/g sed solids),

ω and D_s retain their previous definitions,

K_p = equilibrium constant for linear adsorption of phosphate,

k = rate constant for organic matter decomposition (in yr^{-1}),

z = depth below SWI (in cm), and

$[PO_4]_{z=0}$ = bottom water phosphate concentration (in μM).

Equation (16) was derived by Berner (1980) for calculating phosphate concentration profiles in anoxic sediments below the zone of bioturbation. The effects of compaction and the precip-

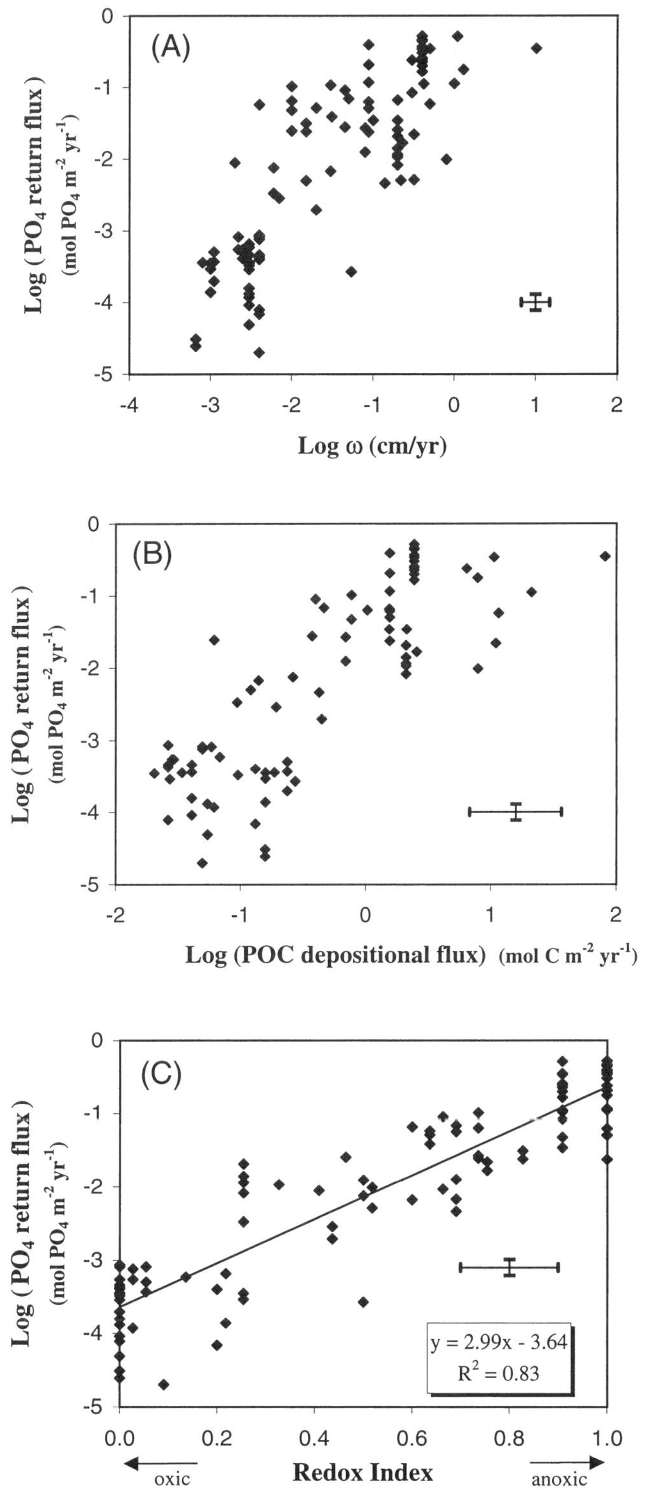

FIG. 2.—(A) Log(Phosphate return flux) versus Log(ω); ω represents linear sedimentation rate. (B) Log(Phosphate return flux) versus Log(POC depositional flux). (C) Log(Phosphate return flux) versus Redox Index. In Figures 2-4, we have used those phosphate fluxes calculated assuming only molecular diffusion as a transport mechanism because the uncertainties in the bioturbation correction term are very large. Furthermore, transport of pore waters due to bioturbation is generally negligible when sedimentation rates are < 0.5 cm yr^{-1}. Essentially identical trends are therefore observed for both the bioturbation-enhanced phosphate fluxes and simple diffusive fluxes. Representative 1-σ error bars are shown in each plot.

itation of authigenic phosphate minerals were assumed to be negligible. The equation was originally parameterized with a K_p appropriate for anoxic sediments, in which ferric (oxy)hydroxides are of minor importance in adsorbing phosphate.

Equation (16) can be modified for predicting the phosphate release flux from surficial marine sediments. We retained the assumption of no compaction. This simplification should not have a significant impact on our calculations. Authigenesis of phosphate minerals was neglected initially. We have assumed that oxic conditions prevail in the surficial sediments, and we reparameterized K_p accordingly.

Differentiating equation (16) with respect to z yields:

$$\frac{\partial [PO_4]}{\partial z} = \frac{kF\ P_o\,\omega\,\exp\left(\frac{-k}{\omega}z\right)}{D_s k + (1+K_p)\omega^2} = \frac{k\left(\frac{1-\phi}{\phi}\right)\rho_{sol}\,P_o\,\omega}{D_s k + (1+K_p)\omega^2} \times \exp\left(\frac{-k}{\omega}z\right). \tag{17}$$

We note that:

$$\mathrm{POC}_{Deposition}\ \mathrm{Flux} = \mathrm{POP}_{Deposition}\ \mathrm{Flux} \times \left(\frac{C}{P}\right)_o = (1-\phi)\rho_{sol}\,P_o\,\omega\left(\frac{C}{P}\right)_o, \tag{18}$$

where $POC_{Deposition}$ and $POP_{Deposition}$ fluxes refer to the depositional fluxes of POC and POP; $(C/P)_o$ is the molar C:P ratio of organic matter in the depositional flux. Combining equations ((17) and ((18) gives:

$$\frac{\partial [PO_4]}{\partial z} = \frac{\left(\frac{k}{\phi}\right)\left(\frac{P}{C}\right)_o \mathrm{POC}_{Deposition}\ \mathrm{Flux}}{D_s k + (1+K_p)\omega^2} \times \exp\left(\frac{-k}{\omega}z\right). \tag{19}$$

Calculating the phosphate flux across the SWI by substituting equation (19) for the phosphate gradient, evaluated at $z = 0$, into equation (16) yields:

$$J_o = \phi D_s\left(\frac{\partial [PO_4]}{\partial z}\right)_o = \frac{k\left(\frac{P}{C}\right)_o}{k + \frac{(1+K_p)}{D_s}\omega^2}\left(\mathrm{POC}_{Deposition}\ \mathrm{Flux}\right). \tag{20}$$

We have dropped the "-", recognizing that the sign is used merely to indicate the direction of the phosphate flux. To simplify the notation, we introduce the parameter A, such that:

$$A = \frac{k\left(\frac{P}{C}\right)_o}{k + \frac{(1+K_p)}{D_s}\omega^2}, \tag{21}$$

and therefore,

$$J_o = A \times \left(\mathrm{POC}_{Deposition}\ \mathrm{Flux}\right). \tag{22}$$

We have used the formulation for k determined by Soetaert et al. (1996). They derived a relationship between k and water depth based on Westrich and Berner's (1984) data for fresh plankton decay and a model for the settling and decay of organic particles through the water column. The Soetaert et al. (1996) model considers organic matter to be composed of two components, one that degrades rapidly and a second that degrades more slowly. The model follows sinking, decaying particles, calculating the composition or reactivity of organic matter as a

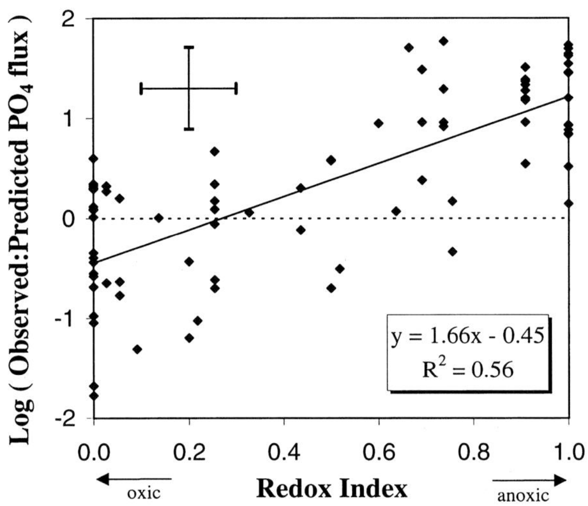

FIG. 3.—Log(Ratio of Observed:Predicted phosphate flux) versus Redox Index. The "Observed" phosphate flux is taken to be the calculated phosphate return flux based on the measured concentration gradient at the SWI; the "Predicted" flux is the phosphate return flux predicted by the POC regeneration model discussed in the text. The horizontal dashed line serves as a reference to denote perfect agreement between observed gradient-based diffusive flux calculations and the POC model predictions of phosphate return flux. Representative 1-σ error bars are shown.

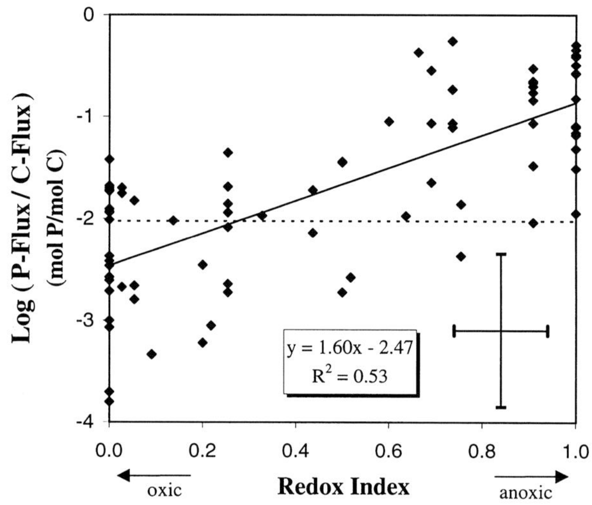

FIG. 4.—Log(P-Flux:C-Flux) versus Redox Index. "P-Flux" is the calculated phosphate return flux based on the measured concentration gradient at the SWI; "C-Flux" is the depth-integrated organic carbon respiration rate predicted by the POC regeneration model discussed in the text. The horizontal dashed line serves as a reference to denote the Redfield P:C ratio of 1:106.

function of depth. Fitting a line to their plot of the bulk organic matter decay constant as a function of water column depth yields:

$$k = 5.8\,yr^{-1} - 4.0\times10^{-4}\,m^{-1}\,yr^{-1}\times H \tag{23}$$

where H is water depth in meters.

We set $K_p = 140$, which falls in the middle of the range (30-240) commonly reported for equilibrium linear adsorption constants for phosphate in oxic sediments (Slomp et al., 1996b; Slomp, 1997, Chapter 5), and we adopted the Redfield ratio (106:1) for the value of $(C/P)_o$.

In Figure 3, we plot the log of the ratio of the gradient-based:POC-model-predicted phosphate fluxes versus the *RI*. Though the data in Figure 3 exhibit much scatter, there is a pronounced trend toward significant underestimation of fluxes in reducing sediments and a slight overestimation of fluxes in oxic sediments. The model's underestimation of phosphate fluxes from reducing sediments indicates that the model includes an inappropriate sink for pore water phosphate or fails to include an important source of pore water phosphate or both. The model's use of a linear adsorption constant appropriate for oxic conditions results in the calculation of storage of phosphate as an adsorbed phase. However, this storage term is small relative to the total phosphate released except at sedimentation rates >1 cm/yr. Recalculation of the model fluxes without adsorption (i.e., $K_p = 0$) results in only slightly better agreement between model and gradient-based phosphate fluxes at these high sedimentation rate sites. Phosphate release from the dissolution of ferric (oxy)hydroxides is not considered in the model. If such release is significant at the sites in our data set, it could explain the large excess phosphate fluxes frequently observed from sediments with *RI* >0.5.

The model also does not take into account phosphate loss from pore waters due to precipitation of authigenic phosphate minerals. In a variety of marine sediments, francolite precipitation has been inferred from solid-phase P sequential extraction analyses and from pore water phosphate and fluoride profiles (e.g., Jahnke et al., 1983; Ruttenberg and Berner, 1993.) Since such precipitation diminishes the phosphate flux out of sediments, the failure to include authigenesis in the model cannot account for the observed large, excess phosphate fluxes. However, intersite variability in the precipitation of francolite, and also in the redox dynamics of the iron-phosphate relationship, are probably responsible for some of the scatter in the flux ratios in reducing settings.

Many of the largest ratios of gradient-based:POC-model-predicted phosphate flux appear to violate phosphorus mass balance in the sediment. We suspect that the largest ratios, which approach 70, are artifacts of uncertainties in the gradient-based flux calculation and in the model input parameters. However, the mean value, ca. 12, should be robust. We also examined the ratio of the POC regeneration flux:phosphate return flux, in molC:molP, and plotted this ratio against the *RI* in Figure 4. The regenerated POC flux was calculated using the burial efficiency relationship discussed earlier. The values of the regenerated POC flux have an uncertainty of roughly a factor of three. Despite the uncertainty for individual data points, the overall trend in the data is strong and indicates that under more oxic conditions, the phosphate flux out of sediments tends to be less than or equal to the flux of phosphate released from organic matter. On the other hand, under more reducing conditions, more phosphate is lost from sediments than would be expected from organic matter degradation alone. Under fully anoxic conditions, the total PO_4 flux is roughly an order of magnitude greater than the quantity of phosphate released by metabolized organic matter.

Mass Balance and Phosphate Regeneration

We can place some constraints on the maximum quantity of phosphate that could be liberated from ferric (oxy)hydroxides to pore waters and lost to overlying waters. The total quantity of phosphate released from ferric (oxy)hydroxides can be scaled to the organic carbon regeneration rate using the reaction stoichiometry:

$$(CH_2O)_{106}(H_3PO_4) + [740-848n]\,CO_2 + 424\,Fe(OOH)_{1-n}(PO_4)_n + 104\,H_2O \rightarrow 424\,Fe^{2+} + [846-848n]HCO_3^- + [1+424n]HPO_4^{2-} \quad (24)$$

where ferric (oxy)hydroxides are represented generically by FeOOH and substitution of phosphate for the "(oxy)hydroxide" group is indicated by the subscript n; where n is equivalent to the molar ratio of Fe:P in the ferric (oxy)hydroxide. Organic matter is assigned Redfield stoichiometry, and the overall reaction stoichiometry derives from the balance between iron reduction, carbon oxidation, and acid consumption (represented by the CO_2 to bicarbonate reaction.) Using the above formulation, we can calculate the molar ratio of regenerated organic carbon to regenerated total phosphate, $(C_{org}{:}PO_4)_{regen}$, according to:

$$(C_{org}:PO_4)_{regen} = 106{:}\left(1+ Y_{Fe} \times 106 \times (Fe{:}C)_{regen} \times n\right) = 106{:}\left(1+ Y_{Fe} \times 424 \times n\right) \quad (25)$$

where $(Fe{:}C)_{regen}$ is the stoichiometric relationship between iron reduction and carbon oxidation and is taken to be four. Organic matter degradation can lead to the reduction of ferric iron to ferrous iron along a more or less direct route, in which iron is the terminal electron acceptor, and along a slightly indirect route, in which sulfide produced by dissimilatory sulfate reduction proceeds to react with and reduce iron. We let Y_{Fe} be the fraction of organic matter remineralization that leads to the reduction of ferric iron, recognizing that the sulfate reduction to sulfide and subsequent reaction with iron may account for a substantial fraction of the total iron reduction (Canfield, 1989b).

Measurements of PO_4 substitution in ferric (oxy)hydroxides in hydrothermal plume particles show that Fe:P ratios typically range from 4 to 10 and correlate with the ambient deep ocean water phosphate concentrations (e.g., Feely et al., 1991). Ingall and Jahnke (1997) have reviewed measurements of Fe:P in reactive ferric iron phases in marine sediments and have settled on an average value of 10.4. These ratios are controlled by the crystallinity of the ferric phase, by the phosphate concentration in the ambient water during precipitation of the ferric (oxy)hydroxide phase, and by pH, among other factors (e.g., Lijklema, 1980). The most reactive ferric (oxy)hydroxide phases in sediments are apt to be those with the highest phosphate concentrations.

Using equation (25), and setting $n = 0.23$ and $Y_{Fe} = 1.0$, we obtain a maximum value for the release of phosphate. Under these conditions $(C_{org}{:}PO_4)_{regen}$ is then 1.1, and roughly 0.9 moles of PO_4 are released per mole of carbon consumed. A more representative parameterization for continental margin sites might be $n = 0.10$ and $Y_{Fe} = 0.5$, yielding a value of 4.9 for $(C_{org}{:}PO_4)_{regen}$.

The availability of reactive iron will also limit the extent of phosphate liberation through iron reduction (e.g., Ingall and Jahnke, 1997). Data from Mississippi Delta and Long Island Sound sediments (Canfield, 1988; Tables VI-1, A-4, A-5, and A-6) indicate that labile organic carbon and reactive ferric iron may be present in freshly accumulating sediment in a 2:1 to 1:4 weight ratio, corresponding to a molar C:Fe ratio of 9:1 to 1.2:1. Using these ratios, we constrain Y_{Fe} to be between 0.03 and 0.20, which is consistent with observations in continental margin settings where organic matter degradation due to iron reduction has been studied (e.g., Slomp et al., 1997). Setting $n = 0.10$, we calculate a range of $(C_{org}{:}PO_4)$ between 12 and 35. This range brackets many of the ratios observed in reducing sediments as plotted in Figure 4. Although this calculation does not prove that the difference between the gradient-based phosphate flux and the flux predicted by the POC-model is due to the dissolution of ferric (oxy)hydroxides, it shows that ferric (oxy)hydroxides can be a sufficiently large source of labile phosphate to account for the difference in many cases.

Sediment Particulate Phosphate Phases

Many sequential and non-sequential extraction techniques have been employed to determine the particulate phases with which phosphorus is associated in marine sediments (e.g., Aspila et al., 1976; Lucotte and d'Anglejan, 1985; Sherwood et al., 1987; Ruttenberg, 1990, 1992; Jensen and Thamdrup, 1993; Vink, 1994; and Vink et al., 1997). Table 3 compares down-core variations of ferric iron-bound P, P in organic matter, and total particulate P for a wide variety of marine sediments. Although the sequential extraction techniques used in the studies reported in Table 3 were not identical, the data have been grouped into equivalent categories.

It would be useful to compare the composition of freshly sedimented material with that of buried material, but we are not aware of sediment trap data at any of the sites in Table 3. We have therefore summarized solid-phase P data from surficial sediments and from a representative downcore sediment layer in each core. The depth resolution of phosphorus measurements varies from study to study. Several cores were sectioned at 0.25 cm to 0.5 cm intervals near the SWI. Even in such thin surficial layers the sediment composition may differ significantly from that of freshly deposited sediment. Dissimilatory reduction of ferric phases liberates Fe-P and Org-P, resulting in a depletion of P in both reservoirs. The mixing of fresh sediment with underlying sediment that is even more depleted in P dilutes the surficial P concentrations. On the other hand, diffusion and coprecipitation of Fe and PO_4 from underlying suboxic to anoxic sediment into a thin, oxic surficial layer would enrich surficial sediment in Fe-P.

Table 3 shows that a substantial fraction of the total P in surficial sediments from a wide variety of marine environments is present as ferric (oxy)hydroxide bound-P. At the Long Island Sound site (Ruttenberg, 1990, 1992; Ruttenberg and Berner, 1993), the use of a 10-cm-thick surficial sample may have masked the presence of Fe-P in the most near-surface sediment.

TABLE 3.—THE CONCENTRATION AND DISTRIBUTION OF PARTICULATE PHOSPHORUS IN MARINE SEDIMENTS

Study	Core ID	Location	Fe-P (in μmol P/g sed. by dry wt.)	Org-P	Other-P	Tot-P	Depth Interval (cm)	Redox Transitions (cm)
Jensen et al. (1995)[1]	10/8/90	Aarhus Bay, Denmark	30.	15.	15.	60.	0-0.25	0,2
			2.	9.	15.	26.	6-8	
	5/27/91	Aarhus Bay, Denmark	12.	12.	14.	38.	0-0.25	0,2
			1.3	7.	14.	22.	6-8	
Jensen and Thamdrup (1993)[1]	11/89	Aarhus Bay, Denmark	24.	12.	12.	48.	0-0.5	(0,3)
			1.3	7.	13.	21.	6-7	
	2/91	Aarhus Bay, Denmark	14.	8.5	18.	40.	0-0.25	1,3
			2.	4.5	16.	22.	6-8	
	5/90	Northern Kattegat	10.	5.0	8.	23.	0-0.25	<5,(<10)
			1.8	4.2	8.	14.	ca. 32-36	
	5/90	Skagerrak	14.	7.5	10.	31.	0-1.0	ca. 5, NR
			3.4	7.3	12.	23.	28-44	
Ruttenberg (1990, 1992); Ruttenberg and Berner (1993)[2]	FOAM-VC D-7/1/86	Long Island Sound, USA	b.d.	2.6	11.0	13.6	0-10	(0,2)
			b.d.	1.8	12.7	14.5	50-60	
	CT15-GC4	Mississippi Delta	5.9	5.4	8.2	19.5	0-2	<5,<5
			4.5	4.7	8.5	17.7	12-14	
	14BC59	Equatorial Atlantic	4.7	3.8	10.0	18.5	0-1	>10,>10
			6.3	3.8	14.4	24.5	5-6	
Slomp et al. (1996a)[3]	OMEX-I	Goban Spur, North Atlantic	3.0	2.7	5.5	11.2	0-0.25	2,5
			1.7	2.2	7.6	11.5	11-15	
	OMEX II	Goban Spur, North Atlantic	4.6	4.7	4.8	14.1	0-0.25	>10,>10
			2.2	3.1	6.6	11.9	11-15	

Sequential extraction data for phosphorus solid phase associations are listed for a variety of sites. At each site, sediment composition in a surficial sediment layer and in a subsurface layer are reported. Measurements of ferric iron-bound P ("Fe-P"), P in organic matter ("Org-P"), and total particulate P ("Tot-P") are reported from the indicated studies. "Other-P" represents the difference between Tot-P and the sum of Fe-P and Org-P, and consists primarily of phosphorus in detrital and authigenic apatite and francolite and in association with $CaCO_3$. The depth intervals chosen for each study provide representative concentrations from surficial and subsurface sediments. In all but the Mississippi Delta and Long Island Sound cores, sediment samples were ≤ 1 cm thick near the surface and somewhat thicker at depth. Parentheses are used to indicate redox transition depths that are uncertain.

[1]Redox transitions for the Aarhus Bay cores in Jensen and Thamdrup (1993) are based on data in Jensen et al. (1995). To be consistent with the Org-P data from other studies, data were binned from two separate organic P pools, one more labile and one less labile, that were identified in the sequential extraction technique employed by Jensen and Thamdrup (1993) and Jensen et al. (1995). In the 5/90 Skagerrak core, the more labile organic P phase decreases by 1.6 μmol P/g sed while the less labile phase increases slightly over the depth interval studies. The increase in the less labile phase, probably due to a change in the composition of the depositional flux, partially offsets the loss in the more labile phase. If we use ΔLabile Org-P in place of ΔOrg-P, we calculate a ΔFe-P:ΔOrg-P (see text) of 6.6, which we use in Figure 5. None of the other cores in these two studies exhibits similar behavior.

[2]Redox transitions for the Long Island Sound core are inferred from Aller (1977). Half of the 0.9 μmol P/g sed increase in Other-P appears to be due to precipitation of francolite and half is due to an increase in the detrital apatite fraction. The Mississippi Delta core shows evidence for francolite precipitation, though most appears to be below 20 cm. The Equatorial Atlantic core exhibits small increases in all inorganic P phases with depth over the top 6 cm. We believe this represents a change in the composition of the depositional flux, though the increase in Fe-P may be due to uptake of phosphate from pore water.

[3]Francolite precipitation is responsible for most of the increase in Other-P at OMEX-I. At OMEX-II, the increase in Other-P is due mainly to an increase in the detritral apatite fraction. Francolite precipitation at OMEX-II appears to be very small or negligible. The O_2 penetration depth at OMEX-II is about 5 cm, and NO_3 penetrates to roughly 14 cm. Pore water Fe, Mn, and NH_4 spikes in the top 3 cm of the OMEX II core imply significant suboxic microenvironments near the surface.

In the remaining sites, for which finer scale sampling was conducted, Fe-P represents between 25% and 50% of the total particulate P. Both Fe-P and Org-P tend to decrease with depth in these sediments. We have introduced a category, Other-P, which is the difference between Tot-P and the sum of Fe-P and Org-P. At many sites, Other-P does not vary significantly with depth, indicating that variations with depth in Fe-P and Org-P are largely diagenetic and not a strong function of changes in sediment source composition through time. In a few cases, Other-P increases slightly with depth; the reasons for this are indicated in Table 3.

The Aarhus Bay studies (Jensen and Thamdrup, 1993; Jensen et al., 1995, listed in Table 3 and Appendices A and B) provide one of the most comprehensive data sets for phosphorus diagenesis. If we take the mean change of particulate phase P with depth as presented in Table 3, coupled with the reported sedimentation rate in terms of a mass flux, we can calculate the rate at which phosphate is lost from the sediment. The mean changes in Fe-P and Org-P concentrations are 18 and 5 μmol/g sed, respectively, and the mean change in Tot-P is 24 μmol/g sed. The sedimentation rate is reported to be 900 g sed m^{-2} yr^{-1}. This corresponds to a total loss rate of PO_4 from the sediment of 0.020-0.022 mol P m^{-2} yr^{-1}. In Appendix B, we calculate the mean annual gradient-supported diffusive flux of phosphate from the Aarhus Bay sediments to be 0.020 mol P m^{-2} yr^{-1}. The agreement is remarkable, given the uncertainties in the pore water and solid phase phosphorus measurements and in the parameters that enter into the diffusive flux calculation. Furthermore, the gradient-supported flux, based on 1-2 years of pore water measurements, and the particulate-phase loss flux, calculated over a depth interval that represents roughly 25 years of deposition, represent averages taken over different periods of time. That the two approaches to calculating the phosphate efflux from the Aarhus Bay site yield such similar fluxes supports the validity of both methods for calculating phosphate loss fluxes, although it is somewhat fortuitous that the two methods yield almost identical flux estimates.

The ratio of the downcore loss of Fe-P to the downcore loss of Org-P, Δ(Fe-P):Δ(Org-P), in the cores of Table 3 is plotted in Figure 5. In all but one of the sites for which fine-scale measurements were made, the loss of Fe-P is greater than the loss of Org-P. The exception is the equatorial Atlantic site. Data for the depth of redox transitions in the sediments were not available for all cores, although at least rough estimates of the transition depths can be made (see Table 3). The calculated *RI* values are in the range of ca. 0.2 to 0.8 for all cores except the equatorial Atlantic core. This is the region of Figure 3 in which the average $PFlux_{observed}$:$PFlux_{POCmodel}$ ranges from ca. 1 to ca. 8. If we attribute the phosphate flux in excess of the POC model-predicted phosphate flux solely to phosphate released from the particulate Fe-P reservoir, then the $P\text{-}Flux_{observed}$:$P\text{-}Flux_{POCmodel}$ range of 1 to 8 corresponds to a Δ(Fe-P):Δ(Org-P) ratio of 0 to 7. Comparison with the data in Figure 5 shows that there is good agreement between the predicted and observed values of Δ(Fe–P):Δ(Org-P). The equatorial Atlantic core, which is fully oxic throughout its top 10 cm (*RI* = 0.00), shows a gain of Fe-P with depth. This could be due to uptake of P by a ferric iron phase during diagenesis or to changes in the composition of the depositional flux at that site.

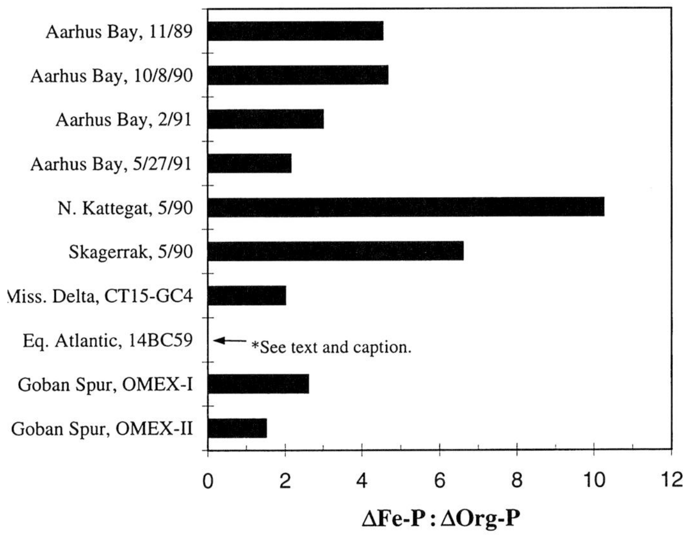

FIG 5. Ratio of downcore loss of Fe-P (ΔFe-P) to downcore loss of Org-P (ΔOrg-P) in surficial marine sediments. All cores show greater loss of Fe-P than Org-P except core 14BC59. Core 14BC59 exhibits slight elevations in Fe-P with depth and shows virtually constant Org-P with depth, as discussed in Table 3 and text. The loss of both Fe-P and Org-P is not compensated for by gains in other particulate P phases (see Table 3.)

Comparison with Flux Chamber Studies

Benthic flux chamber data in Ingall and Jahnke (1997) and McManus et al. (1997) also document large enhancements of the phosphate fluxes from some reducing sediments, but not to the same degree as the data in Figures 3 and 4. As discussed previously, flux chambers measure the combined effect of fluxes across the SWI and production and loss reactions at the sediment surface, whereas pore water gradient-based calculations estimate only the fluxes across the SWI. The degradation of labile organic matter at the sediment surface may increase the phosphate flux measured by a flux chamber substantially without appreciably influencing the pore water phosphate gradient at the SWI. Phosphate release during organic matter degradation at the sediment surface is probably not sensitive to the redox conditions in the underlying sediments if the ambient bottom water is oxygenated. Therefore, we might not expect fluxes measured by flux chambers to show the same level of sensitivity to sediment redox conditions as pore water gradient-based flux calculations. In sediments that are oxic only to a few millimeters in depth, coprecipitation of Fe and phosphate and adsorption of phosphate on ferric (oxy)hydroxides in the top few millimeters might result in artificially high calculated phosphate fluxes if based on apparent SWI diffusion gradients (e.g., McManus et al., 1997).

We believe that the data in Ingall and Jahnke (1997) and McManus et al. (1997) show the pronounced effect of nonsteady state release of phosphate from sediments with shallow oxygen penetration depths (e.g., Gobeil et al., 1997). Both studies (Ingall and Jahnke, 1997; McManus et al., 1997) focus on the relationship between dissolved phosphate fluxes to and from sediments on the western U.S. continental margin in relation to organic carbon metabolism and bottom water oxygen content. Ingall and Jahnke (1997) find that phosphate fluxes from sediments in sites where BW $[O_2]$ <20 μM are consistently higher than those predicted on the basis of the rate of organic matter consumption and Redfield stoichiometry for the organic matter. Phosphate fluxes at sites where BW $[O_2]$ >20 μM are consistent with or fall below the predicted values. Ingall and Jahnke (1997) show that the excess phosphate release from many of the low BW $[O_2]$ sites in their study cannot be explained by steady-state release of iron-bound phosphorus. They suggest that transient uptake and release of phosphate by bacteria in the sediment may be responsible for the relationship between phosphate fluxes and bottom water oxygen concentrations.

McManus et al. (1997) observe patterns in phosphate fluxes that are generally similar to those reported in Ingall and Jahnke (1997). However, the results of McManus et al. (1997) deviate significantly from Ingall and Jahnke's (1997) at low BW $[O_2]$ sites, redefined as BW $[O_2]$ <50 μM, that exhibit carbon regeneration rates below ca. 400 μmol m^{-2} yr^{-1}. At these low BW $[O_2]$, low carbon regeneration sites, McManus et al. (1997) observe phosphate fluxes that are sometimes lower than predicted on the basis of carbon regeneration rates. These fluxes may even be directed into the sediments. McManus et al. (1997) use flux chamber and pore water phosphate, Fe^{2+}, and O_2 measurements and conclude that precipitation of fresh ferric (oxy)hydroxides in the shallow (<1 cm) sediment subsurface is largely responsible for scavenging phosphate from pore water in these low BW $[O_2]$, low carbon regeneration sites, thus reducing or even reversing phosphate exchange fluxes with overlying waters. McManus et al. note that the storage and release of phosphate by bacteria, the process favored by Ingall and Jahnke (1997), may also be important.

Both Ingall and Jahnke (1997) and McManus et al. (1997) briefly consider the potential impact of non-steady state diagenesis on the phosphate fluxes that they report. We believe that this topic merits further attention, and that it may well explain many of their observations. Phosphate fluxes are linked to redox conditions in sediments, as we have shown, and as a result, to bottom water oxygen concentrations, as shown by Ingall and Jahnke (1997) and McManus et al. (1997). McManus et al. (1997) discuss the impact that a seasonally variable depositional flux of organic matter would have on the oxygen penetration depth in sediments. During lulls in organic matter deposition the oxygen penetration depth increases. A deepening of the oxygen penetration depth, corresponding to a lowering of the *RI* in our study, creates a transient period of enhanced ferric (oxy)hydroxide precipitation due to the oxidation of Fe^{2+} in pore water and of particulate ferrous iron phases in the zone between the minimum and maximum depth of oxygen penetration. This process may significantly diminish or even reverse the direction of the phosphate flux across the SWI. In Sundby et al.'s (1986) study, the apparent widening of a thin (1 mm) oxic zone during the onset of winter is responsible for a large discrepancy between the flux chamber and the gradient-supported phosphate fluxes. If significant scavenging in thin, surficial oxic zones is widespread in sediments that are dominantly suboxic or anoxic within the top centimeter, then the fluxes that we calculated may well be fluxes to a solid phase in the near-surface environment.

Following blooms, the oxygen penetration depth decreases, and phosphate bound in ferric (oxy)hydroxides in the zone between the minimum and maximum depth of oxygen penetration may be released. The formation of FeS and pyrite during the onset of anoxic conditions decouples the flux of Fe from that of P and allows P to be released from sediments at rates exceeding those estimated on the basis of steady-state calculations. Bacterial storage of phosphate during oxic conditions and release during reducing conditions would produce essentially the same effect (Ingall and Jahnke, 1997).

We believe that mean phosphate fluxes averaged over an entire year or multiple years are correlated with mean oxygen penetration depths, or analogously, with the mean *RI*. Short term flux measurements may vary considerably about this mean. The variability of the phosphate flux will be a function of changes in the oxygen penetration depth and the depth at which sulfate reduction becomes important. The decoupling of carbon from phosphorus fluxes and of iron from phosphorus fluxes may be important over depth ranges much greater than those observed by Ingall and Jahnke (1997) and McManus et al. (1997). Low BW $[O_2]$ and moderate to large fluxes of labile organic matter at the western U.S. continental margin sites that they studied result in small oxygen penetration depths. Comparable sites with higher BW $[O_2]$ and consequently slightly greater oxygen penetration depths may also be susceptible to large fluctuations in the redox transition depths due to seasonal variations in organic matter deposition. In shallow water settings, temperature dependent variations in the kinetics of microbial degradation of organic matter may also contribute to seasonal changes in redox boundaries. Tables 1 and 2 indicate that seasonal variability in fluxes and the position of redox boundaries is significant. Where non-steady state diagenesis occurs, flux estimates based on a single benthic flux chamber measurement or the measurement of a single pore water gradient may not be representative of average conditions on annual or longer time scales. The seasonal corrections that we apply when calculating mean annual fluxes for continental shelf sites are an improvement over simply equating the observed fluxes with mean annual fluxes. Some of the variability in our data probably results from variability in the timing of coring and pore water sampling with respect to blooms. However, we believe the large number of flux estimates and the wide range of sediment environments represented in our study provide good estimates of mean trends.

The Global Return Flux of Phosphate

The global return flux of phosphate from the oceans is calculated in Table 4 using the relationship between sedimentation rate and phosphate flux in Figure 2A. In Table 4, ocean sediments are divided into two categories: continental margin sediments, i.e., those of the continental shelf and slope, and pelagic sediments, i.e., those of the continental rise and deep-ocean. The

TABLE 4.—THE RETURN FLUX OF PHOSPHATE IN RELATION TO OCEANIC SEDIMENTATION

Ocean Floor Province	Depositional Area* (x 10^6 km^2)	Total Sed. Flux to Province (x 10^{15} g/yr)	Average Sed. Rate (cm/yr)	Return PO_4 Flux from sediments (mol/m^2/yr)	Total Return PO_4 Flux from Province (10^{11} moles/yr)
Case A: Modern sediment fluxes					
Shelf and Slope	8	18.0	0.57	0.15	12.1
Rise and Deep Sea	306	6.9	0.0043	0.0017	5.2
				Oceanwide Total:	17.3
Case B: Preagricultural sediment fluxes					
Shelf and Slope	8	9.0	0.28	0.10	8.4
Rise and Deep Sea	306	5.9	0.0037	0.0013	4.1
				Oceanwide Total:	12.5

Depositional Province	Source	Sediment Mass Fluxes (in 10^{15} g/yr) Case A	Case B
Continental Shelf and Slope	Riverine	18.0	9.0
	Total	18.0	9.0
Continental Rise and Deep Sea	Riverine	2.0	1.0
	Glacial	1.5	1.5
	Aeolian	1.6	1.6
	Biogenic	1.8	1.8
	Total	6.9	5.9

*Depositional Area on Shelf and Slope represents 30% of total continental shelf area (after Kennett, 1982; see text for explanation of calculations.)

hypsometric definitions and surface areas for the continental margin and pelagic provinces are those of Menard and Smith (1966). The total sediment delivery rate of 20 Gt/yr from rivers to the oceans is that estimated by Milliman and Syvitski (1992), which is consistent with the data of Lisitzin (1996, p. 37). We assumed that 90% of the riverine sediment flux is deposited on the continental margins and that the remaining 10% passes to the pelagic realm (Lisitzin, 1996, p. 37). This figure is not well constrained. It has almost certainly fluctuated with changes in sea level, particularly during glacial periods and with the relative extent of active and passive margins (e.g., Milliman, 1991). Sedimentation dynamics on the continental margins are poorly defined. Kennett (1982, pp. 308-309) reports that 70% of exposed continental shelf sediments are relict and that sands and muds are rarely deposited more than 6 km and 30 km, respectively, from shorelines.

We have taken continental margin sediments to be dominantly riverine. The depositional region of the continental margin is taken to be 30% of the 27 x 106 km^2 of continental shelf. The 18 Gt/yr of riverine sediment deposited on the continental margins in our calculation is assumed to be evenly distributed over the 8 x 106 km^2 of depositional area. The mass sedimentation rate is converted to a linear sedimentation rate (from g/cm^2/yr to cm/yr), assuming that sediment solids have a density of 2.6 g/cm^3 and that the porosity of surficial margin sediments is 0.85. Average return phosphate fluxes for sediments in the depositional environment (in units of mol P m^{-2} yr^{-1}) are then calculated as a function of sedimentation rate using the least squares fit of an exponential to the data in Figure 2A. The average return phosphate flux for the depositional margin is then multiplied by the total area of the depositional margin to derive a total phosphate return flux from the continental margins.

The nondepositional shelf and slope regions do not contribute to the phosphate return flux in our calculation, and carbonate sedimentation is not considered. In reality, phosphate fluxes from nondepositional environments are observed, primarily due to inputs of fresh organic matter without appreciable accumulation of other sediment (e.g., Van Raaphorst et al., 1990; data in Appendix A). Neglecting nondepositional environments leads to an underestimate of the total return flux of phosphate.

A more detailed analysis of shelf sediment deposition would show that a substantial fraction accumulates at rates that are higher than the calculated mean rate, e.g., through focused deposition in deltas and estuaries. Some of the inaccuracies introduced by our simplifying assumptions will tend to offset each other. Sensitivity tests of the assumption that fluxes from nondepositional regions can be neglected and that sedimentation in depositional regions is homogeneous produced maximum deviations of ±40% from our best estimates of the return phosphate flux.

The input of riverine terrigenous sediments to the pelagic depositional environment is ca. 2 Gt/yr (Lisitzin, 1996). Glacially derived, aeolian, and biogenic contributions to pelagic sedimentation are taken to be 1.5, 1.6, and 1.8 Gt/yr respectively (Lisitzin, 1996, p. 37). The total pelagic sediment flux, 6.9 Gt/yr, is assumed to be evenly distributed across the continental rises and the deep-sea. The return phosphate flux was calculated as above. The assumption of homogeneous sedimentation rates across the entire continental rise and deep-sea environment introduces an uncertainty of <30% in our calculation. The density of sediment solids was again taken to be 2.6 g/cm^3; the porosity of surficial sediments was taken to be 0.80.

Human land use practices have substantially elevated the riverine flux of terrigenous sediments to the oceans (e.g., Berner and Berner, 1996; Milliman and Syvitski, 1992). Berner and Berner (1996) estimate that the preagricultural riverine flux of terrigenous sediments was about 10 Gt/yr. We have repeated our calculations for the global return flux of phosphate from marine sediments for preagricultural conditions, using the values 9 Gt/yr and 1 Gt/yr, respectively, for the fluxes of river-borne terrigenous sediment to the continental margins and pelagic provinces. The results appear in Table 4. Modern fluxes are considered in Case A; preagricultural fluxes are considered in Case B.

TABLE 5.—PHOSPHATE FLUXES FROM CONTINENTS AND FROM SEDIMENTS

Phosphate Flux		PO_4 flux to the oceans (in 10^{11} mol P/yr)
Riverine (Meybeck 1982)	DIP	0.13
	TDP	0.32
	POP	2.6
	TPP	6.5
	Subtotal (TDP+TPP):	6.8
Aeolian (Duce et al. 1991)	Soluble	0.10
	Insoluble	0.21
	Subtotal:	0.31
Diffusive flux from sediments (from Table 4)	Modern	17.3
	Preagricultural	12.5

DIP = dissolved inorganic phosphorus
TDP = total dissolved phosphorus
POP = particulate organic phosphorus
TPP = total particulate phosphorus

The total return phosphate flux from continental margin sediments is roughly double the total return flux from pelagic sediments. Table 5 compares the global phosphate fluxes calculated in Table 4 with other significant fluxes of phosphate to the oceans. The global return flux of phosphate from marine sediments, roughly 12-17 x 10^{11} mol P/yr, is larger than the combined flux of dissolved and particulate phosphorus to the oceans by riverine and aeolian transport, roughly 7 x 10^{11} mol P/yr . It is important to recognize that a substantial fraction of reactive particulate phosphorus entering the oceans, either bound in organic matter or in ferric (oxy)hydroxides, is probably released to the oceans as orthophosphate only during diagenesis (e.g., Berner and Rao, 1994). The return flux of phosphate from marine sediments to the oceans is therefore a mixture of phosphate from riverine particulates, which is phosphate new to seawater, and phosphate from marine particulates, which is recycled to the oceans by diagenetic release.

At first glance the estimated preindustrial return phosphate flux from sediments (12 x 10^{11} mol/yr) seems rather too large compared to the total flux of riverine and aeolian dissolved and soluble phosphate of ca. 0.4 x 10^{11} mol/yr (see Table 5). However, a second glance shows that the estimated return flux is quite reasonable. Total carbon fixation in marine photosynthesis amounts to ca. 40,000 x 10^{11} mol/yr. This implies an annual rate of phosphate fixation of ca. 400 x 10^{11} mol/yr. Most of the organic matter is metabolized: only ca. 0.3% of the quantity photosynthesized is buried more or less permanently with marine sediments (e.g., Holland, 1978). A somewhat smaller percentage of the 400 x 10^{11} mol of phosphate fixed annually is also buried with marine sediments. Most of the nearly 400 x 10^{11} mol of phosphate released annually from organic matter is regenerated in the water column. Our estimated preagricultural return flux of phosphate from marine sediments represents ≤3% of the estimated annual phosphate release from organic matter and is therefore not unreasonably large, especially if one considers that a substantial fraction of the estimated return flux of phosphate is probably derived from the reductive dissolution of ferric (oxy)hydroxides.

The Oceanic Residence Time of Phosphorus

The calculation of the residence time of phosphorus in the oceans is sensitive to the flux of phosphate that is released from riverine particulates. The data of Fox et al. (1986), Froelich (1988), Berner and Rao (1994), Ruttenberg and Canfield (1994), Vink et al. (1997), and Ruttenberg and Goñi (1997) suggest that a substantial fraction of riverine particulate phosphate is bound in labile phases (e.g., organic matter and ferric (oxy)hydroxides). We define RPP_{riv} as the total flux of reactive particulate phosphate that enters the oceans from the continents, mainly from rivers but with a small aeolian component, and that deposits in marine sediments. The residence time, τ, of phosphate in the oceans with respect to continental inputs is then:

$$\tau = \frac{M_{PO_4}}{(1-BE_{PO_4})RPP_{riv} + TDP_{riv}} , \qquad (26)$$

where M_{PO_4} is the total amount of phosphate in the oceans, and BE_{PO_4} is the burial efficiency in marine sediments of RPP_{riv}.

The sequential extraction data of Berner and Rao (1994) show that 75-85% of the total particulate phosphorus in Amazon River suspended and bottom sediments is in a potentially labile phase (e.g., Fe-P, Org-P, or P associated with biogenic and authigenic apatite or carbonates). The sequential extraction data of Vink (1994) and Vink et al. (1997) show that at least 45-55% of the total particulate phosphorus in Lagunitas Creek (California, USA) suspended and bottom sediments is potentially labile. Ruttenberg and Goñi (1997) identify roughly 25% of the particulate phosphorus in Mackenzie River suspended sediments as organic. Although Ruttenberg and Goñi (1997) do not isolate labile from refractory inorganic particulate phosphate phases, their data on organic versus inorganic phosphorus are consistent with the data for the Amazon and Lagunitas Creek sediments. Taken together, these data suggest that ca. two-thirds, or ca. 4.5 x 10^{11} mol P/yr, of riverine particulate P is potentially reactive.

The total preagricultural flux of P from continents to the oceans is then (1 - BE_{PO_4}) x RPP_{riv} + TDP_{riv}, or ca. 3.5 x 10^{11} mol P/yr for our best estimate of 30% for BE_{PO_4}. This is slightly less than the 4.1 x 10^{11} mol P/yr derived from the model of Mackenzie et al. (1993) for a modern flux and higher than most previous estimates. The residence time for phosphorus in the oceans is calculated to be 3 x 10^{15} mol P/3.5 x 10^{11} mol P yr^{-1} or ca. 10,000 years. Though our estimate for the residence time is slightly less than most other estimates (e.g., see reviews in Ruttenberg, 1993; Filippelli and Delaney, 1996), it is consistent with the recent recognition that authigenic francolite, authigenic

REE-phosphate minerals, and Fe-particle scavenging of phosphate in MOR hydrothermal plumes are potentially large and previously underestimated sinks for reactive phosphorus from the oceans (Ruttenberg and Berner, 1993; Filippelli and Delaney, 1996; Rasmussen, 1996; Feely et al., 1996; Wheat et al., 1996). Such a short residence time for marine reactive phosphorus is also consistent with evidence for variability in phosphate concentrations in the ocean on glacial-interglacial time scales (Boyle, 1986).

CONCLUSIONS

The return flux of phosphate from sediments to the oceans is large, ca. 12 to 17 x 10^{11} mol P/yr, relative to the riverine flux of dissolved inorganic and organic phosphate, ca. 0.3 x 10^{11} mol P/yr. Most of the continental phosphate released in dissolved form to the oceans appears to be riverine particulate phosphate that is solubilized subsequent to deposition and diagenesis in continental margin sediments. The large size of the return flux is due in part to the release of phosphate from riverine reactive particulate phosphate phases, in part to the degradation of marine organic matter, and in part to the release of phosphate accompanying the reductive dissolution of ferric (oxy)hydroxides. The total preagricultural inputs of dissolved phosphate to the oceans from continents including the flux of phosphate that is diagenetically mobilized from reactive particulate phosphorus is ca. 3.5 x 10^{11} mol P/yr, and the residence time of phosphate in the oceans is ca. 10,000 years.

The return flux of phosphate from sediments is redox sensitive. In highly oxidized sediments the return phosphate flux to the ocean is roughly equal to or slightly smaller than the flux generated by the oxidation of organic matter. Where return fluxes of phosphate are less than the quantity of phosphate regenerated during the oxidation of organic matter in surficial sediments, phosphate is probably trapped on ferric (oxy)hydroxides. Francolite precipitation becomes an important sink for pore water phosphate at greater depth in oxic sediments. In highly reduced sediments the return flux of phosphate to the oceans is larger than the flux generated by the destruction of organic matter, largely because this flux is augmented by a flux of phosphate from the dissolution of ferric (oxy)hydroxides.

The modern diffusive phosphate flux from marine sediments originates mainly in regions of active sediment deposition on continental shelves and slopes. In the past, sedimentation patterns may well have been different, and the bulk of phosphate release may have been further offshore. Glacial-interglacial changes in sea level could cause significant shifts in sedimentation. Since the residence time of marine phosphate is ca. 10,000 years, such shifts could cause changes in the size of the marine phosphate reservoir (e.g., Broecker and Peng, 1982; Boyle, 1986; Ruttenberg, 1993). The impact of sedimentation and diagenesis on the transfer of phosphate from the continents to the oceans suggests that proxies used for assessing phosphate concentrations in the ocean may reflect not only changes in weathering fluxes of phosphate to the oceans (e.g., Filippelli and Delaney, 1994, 1996; Föllmi, 1995; Filippelli, 1997) but also changes in the global rate at which sediment phosphate is released diagenetically to the oceans.

The magnitude of the phosphate return flux and its sensitivity to redox conditions suggest that it has the potential to act as a significant factor in controlling the O_2 content of the atmosphere and seawater.

ACKNOWLEDGMENTS

This study was supported by NASA grant NAG5-4174 and by an NSF Graduate Research Fellowship to A.S.C. We greatly appreciate the insightful comments and advice of F. T. Mackenzie offered at many stages in this research. This work has also benefited from many stimulating discussions with R. C. Aller, R. A. Berner, K. C. Ruttenberg, L. M. Ver, and S. Vink. We thank Bjørn Sundby and an anonymous reviewer for their constructive suggestions.

REFERENCES

Aller, R. C., 1977, The influence of macrobenthos on chemical diagenesis of marine sediments: Unpublished Ph.D. Dissertation, Yale University, New Haven, Connecticut, 600 p.

Aller, R. C., 1980a, Diagenetic processes near the sediment-water interface of Long Island Sound. II. Fe and Mn: Advances in Geophysics, v. 22, p. 351-415.

Aller, R. C., 1980b, Quantifying solute distributions in the bioturbated zone of marine sediments by defining an average microenvironment: Geochimica et Cosmochimica Acta, v. 44, p. 1955-1965.

Aller, R. C., 1982, The effects of macrobenthos on chemical properties of marine sediment and overlying water, *in* McCall, P. L., and Tevesz, M. J. S., eds., Animal-Sediment Relations: New York, Plenum, p. 53-102.

Aller, R. C., 1988, Benthic fauna and biogeochemical processes in marine sediments: The role of burrow structures, *in* Blackburn, T. H., and Sørensen, J., eds., Nitrogen Cycling in Coastal Marine environments: New York, Wiley, p. 301-338.

Andrews, D., and Bennett, A., 1981, Measurements of diffusivity near the sediment-water interface with a fine-scale resistivity probe: Geochimica et Cosmochimica Acta, v. 45, p. 2169-2175.

Aspila, K. I., Agemian, H., and Chau, A. S. Y., 1976, A semi-automated method for the determination of inorganic, organic, and total phosphate in sediment: Analyst, v. 101, p. 187-197.

Atlas, E., Culberson, C., and Pytkowicz, R. M., 1976, Phosphate association with Na^+, Ca^{2+}, and Mg^{2+} in seawater: Marine Chemistry, v. 4, p. 243-254.

Bender, M., Martin, W., Hess, J., Sayles, F., Ball, L., and Lambert, C., 1987, A whole-core squeezer for interfacial pore-water sampling: Limnology and Oceanography, v. 32, p. 1214-1225.

Berner, E. K., and Berner, R. A., 1996, Global environment: Water, air, and geochemical cycles: Upper Saddle River, New Jersey, Prentice-Hall, 376 p.

Berner, R. A., 1973, Phosphate removal from sea water by adsorption on volcanogenic ferric oxides: Earth and Planetary Science Letters, v. 18, p. 77-86.

Berner, R. A., 1980, Early diagenesis: A theoretical approach: Princeton, New Jersey, Princeton University Press, 241 p.

Berner, R. A., 1981, A new geochemical classification of sedimentary environments: Journal of Sedimentary Petrology, v. 51, p. 359-365.

Berner, R. A., and Canfield, D. E., 1989, A new model for atmospheric oxygen over Phanerozoic time: American Journal of Science, v. 289, p. 333-361.

Berner, R.A., and Rao, J.-L., 1994, Phosphorus in sediments of the Amazon River and estuary: Implications for the global flux of phosphorus to the sea: Geochimica et Cosmochimica Acta, v. 58, p. 2333-2339.

Betts, J. N., and Holland, H. D., 1991, The oxygen content of ocean bottom waters, the burial efficiency of organic carbon, and the regulation of atmospheric oxygen: Palaeogeography, Palaeoclimatology, Palaeoecology (Global and Planetary Change Section), v. 97, p. 5-18.

Blackburn, T. H., 1991, Accumulation and regeneration: Processes at the benthic boundary layer, *in* Mantoura, R. F. C., Martin, J.-M., and Wollast, R., eds., Ocean Margin Processes and Global Change: New York, Wiley, p. 181-195.

Boudreau, B. P., 1984, On the equivalence of nonlocal and radial-diffusion models for porewater irrigation: Journal of Marine Research, v. 42, p. 731-735.

Boudreau, B. P., 1994, Is burial velocity a master parameter for bioturbation?: Geochimica et Cosmochimica Acta, v. 58, p. 1243-1249.

Boyle, E. A., 1986, Paired carbon isotope and cadmium data from benthic foraminifera: Implications for changes in oceanic phosphorus, oceanic circulation, and atmospheric carbon dioxide: Geochimica et Cosmochimica Acta, v. 50, p. 265-276.

Bray, J. T., Bricker, O. P., and Troup, B. N., 1973, Phosphate in interstitial waters of anoxic sediments: Oxidation effects during sampling procedure: Science, v. 180, p. 1362-1364.

Broecker, W. S., and Peng, T.-H., 1982, Tracers in the Sea: New York, Eldigio, 690 p.

Callender, E., and Hammond, D. E., 1982, Nutrient exchange across the sediment water interface in the Potomac River Estuary: Estuarine, Coastal, and Shelf Science, v. 15, p. 395-413.

Calvert, S. E., Bustin, R. M., and Pedersen, T. F., 1992, Lack of evidence for enhanced preservation of sedimentary organic matter in the oxygen minimun of the Gulf of California: Geology, v. 20, p. 757-760.

Canfield, D. E., 1988, Sulfate reduction and the diagenesis of iron in anoxic marine sediments: Unpublished Ph.D. Dissertation, Yale University, New Haven, Connecticut, 248 p.

Canfield, D. E., 1989a, Sulfate reduction and oxic respiration in marine sediments: Implications for organic carbon preservation in euxinic environments: Deep-Sea Research, v. 36, p. 121-138.

Canfield, D. E., 1989b, Reactive iron in marine sediments: Geochimica et Cosmochimica Acta, v. 53, p. 619-632.

Canfield, D. E., 1993, Organic matter oxidation in marine sediments, *in* Wollast, R., Mackenzie, F. T., and Chou, L., eds., Interactions of C, N, P and S Biogeochemical Cycles and Global Change; NATO ASI Series, Vol. 14: New York, Springer-Verlag, p. 333-363.

Canfield, D. E., Raiswell, R., and Bottrell, S., 1992, The reactivity of sedimentary iron minerals toward sulfide: American Journal of Science, v. 292, p. 659-683.

Canfield, D. E., Thamdrup, B., and Hansen, J. W., 1993, The anaerobic degradation of organic matter in Danish coastal sediments: Iron reduction, manganese reduction, and sulfate reduction: Geochimica et Cosmochimica Acta, v. 57, p. 3867-3883.

Clavero, V., Niell, F. X., and Fernandez, J. A., 1991, Effects of Nereis-diversicolor O F Muller abundance on the dissolved phosphate exchange between sediment and overlying water in Palmones River Estuary, southern Spain: Estuarine, Coastal and Shelf Science, v. 33, p. 193-202.

Codispoti, L. A., 1989, Phosphorus vs. nitrogen limitation of new and export production, *in* Berger, W. H., Smetacek, V. S., and Wefer, G., eds., Productivity of the Ocean: Present and Past: New York, John Wiley and Sons, p. 377-394.

Cole, J. J., Honjo, S., and Erez, J., 1987, Benthic decomposition of organic matter at a deep-water site in the Panama Basin: Nature, v. 327, p. 703-704.

Colman, A. S., 1994, Benthic phosphorus regeneration: The global diffusive flux of phosphorus from marine sediments into the oceans and the control of atmospheric oxygen levels throughout the Phanerozoic: Unpublished Senior Thesis, Harvard University, Cambridge, Massachusetts, 39 p.

Colman, A. S., and Holland, H. D., 1994, The global diffusive flux of phosphorus from marine sediments into the oceans (abs): Eos, Transactions of the American Geophsyical Union, Ocean Sciences Meeting Supplement, v. 75, p. 96.

Colman, A. S., Mackenzie, F. T., and Holland, H. D.; Reply by Van Cappellen, P., and Ingall, E. D., 1997, Redox stabilization of the atmosphere and oceans and marine productivity: Science, v. 275, p. 406-408.

Colombo, C., Barrón, V., and Torrent, J., 1994, Phosphate adsorption and desorption in relation to morphology and crystal properties of synthetic hematites: Geochimica et Cosmochimica Acta, v. 58, p. 1261-1269.

Cope, M. J., and Chaloner, W. G., 1980, Fossil charcoal as evidence of past atmospheric composition: Nature, v. 283, p. 647-649.

Cope, M. J., and Chaloner, W. G., 1985, Wildfire, an interaction of biological and physical processes, *in* Tiffney, B. H., ed., Geological Factors and the Evolution of Plants: New Haven, Connecticut, Yale University Press, p. 257-277.

Demaison, G. J., and Moore, G. T., 1980, Anoxic environments and oil source bed genesis: American Association of Petroleum Geologists Bulletin, v. 64, p. 1179-1209.

Dhakar, S. P., and Burdige, D. J., 1996, A coupled, non-linear, steady state model for early diagenetic processes in pelagic sediments: American Journal of Science, v. 296, p. 296-330.

Duce, R. A., et al., 1991, The atmospheric input of trace species to the world ocean: Global Biogeochemical Cycles, v. 5, p. 193-259.

Elderfield, H., McCaffrey, R. J., Luedtke, N., Bender, M., and Truesdale, V. W., 1981, Chemical diagenesis in Narragansett Bay sediments: American Journal of Science, v. 281, p. 1021-1055.

Emerson, S., 1985, Organic carbon preservation in marine sediments, *in* Sundquist, E. T., and Broecker, W. S., eds., The Carbon Cycle and Atmospheric CO_2 : Natural Variations, Archean to Present; Geophysical Monograph 32: Washington, D.C., American Geophysical Union, p. 78-87.

Emerson, S., Jahnke, R., Bender, M., Froelich, P., Klinkhammer, G., Bowser, C., and Setlock, G., 1980, Early diagenesis in sediments from the eastern Equatorial Pacific, I. Pore water nutrient and carbonate results: Earth and Planetary Science Letters, v. 49, p. 57-80.

Falkowski, P. G., 1997, Evolution of the nitrogen cycle and its influence on the biological sequestration of CO_2 in the ocean: Science, v. 387, p. 272-275.

Feely, R. A., Baker, E. T., Marumo, K., Urabe, T., Ishibashi, J., Gendron, J., Lebon, G. T., and Okamura, K., 1996, Hydrothermal plume particles and dissolved phosphate over the superfast-spreading southern East Pacific Rise: Geochimica et Cosmochimica Acta, v. 60, p. 2297-2323.

Feely, R. A., Trefry, J. H., Massoth, G. J., and Metz, S., 1991, A comparison of the scavenging of phosphorus and arsenic from seawater by hydrothermal iron oxyhydroxides in the Atlantic and Pacific oceans: Deep-Sea Research, v. 38, p. 617-623.

Filippelli, G. M., 1997, Controls on phosphorus concentration and accumulation in oceanic sediments: Marine Geology, v. 139, p. 231-240.

Filippelli, G. M., and Delaney, M. L., 1994, The oceanic phosphorus cycle and continental weathering during the Neogene: Paleoceanography, v. 9, p. 643-652.

Filippelli, G. M., and Delaney, M. L., 1996, Phosphorus geochemistry of equatorial Pacific sediments: Geochimica et Cosmochimica Acta, v. 60, p. 1479-1495.

Föllmi, K. B., 1995, 160 m.y. record of marine sedimentary phosphorus burial: Coupling of climate and continental weathering under greenhouse and icehouse conditions: Geology, v. 23, p. 859-862.

Föllmi, K. B., 1996, The phosphorus cycle, phosphogenesis and marine phosphate-rich deposits: Earth-Science Reviews, v. 40, p. 55-124.

Fox, L. E., Sager, S. L., and Wofsy, S. C., 1986, The chemical control of soluble phosphorus in the Amazon estuary: Geochimica et Cosmochimica Acta, v. 50, p. 783-794.

Froelich, P. N., 1988, Kinetic control of dissolved phosphate in natural rivers and estuaries: A primer on the phosphate buffer mechanism: Limnology and Oceanography, v. 33, p. 649-668.

Froelich, P. N., Bender, M. L., and Heath, G. R., 1977, Phosphorus accumulation rates in metalliferous sediments on the East Pacific Rise: Earth and Planetary Science Letters, v. 34, p. 351-359.

Froelich, P. N., Bender, M. L., Luedtke, N. A., Heath, G. R., and DeVries, T., 1982, The marine phosphorus cycle: American Journal of Science, v. 282, p. 474-511.

Froelich, P. N., Klinkhammer, G. P., Bender, M. L., Luedtke, N. A., Heath, G. R., Cullen, D., Dauphin, P., Hammond, D., Hartman, B., and Maynard, V., 1979, Early oxidation of organic matter in pelagic sediments of the eastern equatorial Atlantic: Suboxic diagenesis: Geochimica et Cosmochimica Acta, v. 43, p. 1075-1090.

Garrels, R. M., Lerman, A., and Mackenzie, F. T., 1976, Controls of atmospheric O_2 and CO_2: Past, present and future: American Scientist, v. 64, p. 306-315.

Glenn, C. R., Föllmi, K. B., Riggs, S. R., Baturin, G. N., Grimm, K. A., Trappe, J., Abed, A. M., Galli-Oliver, C. Garrison, R. E., Ilyan, A.,

JAHL, C., ROHRLICH, V., SADAQAH, R. M. SCHIDLOWSKI, M., SHELDON, R. E., AND SEIGMUND, H., 1994, Phosphorus and phosphorites: Sedimentology and environments of formation: Eclogae geologicae Helvetiae, v. 87, p. 747-788.
GOBEIL, C., MACDONALD, R. W., AND SUNDBY, B., 1997, Diagenetic separation of cadmium and manganese in suboxic continental margin sediments: Geochimica et Cosmochimica Acta, v. 61, p. 4647-4654.
GRENTHE, I., STUMM, W., LAAKSUHARJU, M., NILSSON, A.-C., AND WIKBERG, P., 1992, Redox potentials and redox reactions in deep groundwater systems: Chemical Geology, v. 98, p. 131-150.
HANSEN, H. C. B., BORGGAARD, O. K., AND SØRENSEN, J., 1994, Evaluation of the free energy of formation of Fe(II)-Fe(III) hydroxide-sulphate (green rust) and its reduction of nitrite: Geochimica et Cosmochimica Acta, v. 58, p. 2599-2608.
HARTMANN, M., MÜLLER, P. J., SUESS, E., AND VAN DER WEIJDEN, C. H., 1976, Chemistry of Late Quaternary sediments and their interstitial waters from the NW African continental margin: "Meteor" Froschungsergebnisse, Reihe C, v. 24, p. 1-67.
HARTNETT, H. E., KEIL, R. G., HEDGES, J. I., AND DEVOL, A. H., 1998, Influence of oxygen exposure time on organic carbon preservation in continental margin sediments: Nature, v. 391, p. 572-574.
HEGGIE, D. T., SKYRING, G. W., O'BRIEN, G. W., REIMERS, C., HERCZEG, A., MORIARTY, D. J. W., BURNETT, W. C., MILNES, A. R., 1990, Organic carbon cycling and modern phosphorite formation on the East Australian continental margin: An overview, *in* Notholt, A. J. G., and Jarvis, I., eds., Phosphorite Research and Development: London, Geological Society Special Publication 52, p. 87-117.
HENRICHS, S. M., AND REEBURGH, W. S., 1987, Anaerobic mineralization of marine sediment organic matter: Rates and the role of anaerobic processes in the oceanic carbon economy: Geomicrobiology Journal, v. 5, p. 191-237.
HOLLAND, H. D., 1973, Ocean water, nutrients, and atmospheric oxygen, *in* Ingerson, E., ed., Proceedings of Symposium on Hydrogeochemistry and Biogeochemistry, Vol. I: Washington, D.C., Clarke, p. 68-81.
HOLLAND, H. D., 1978, The Chemistry of the Atmosphere and Oceans: New York, Wiley, 351 p.
HOLLAND, H. D., 1984, The Chemical Evolution of the Atmosphere and Oceans: Princeton, New Jersey, Princeton University Press, 582 p.
HOLLAND, H. D., 1994a, The phosphate-oxygen connection (abs): Eos, Transactions of the American Geophsyical Union, Ocean Sciences Meeting Supplement , v. 75, p. 96.
HOLLAND, H. D., 1994b, Early Proterozoic atmospheric change, *in* Bengston, S., ed., Early Life on Earth; Nobel Symposium 84: New York, Columbia University Press, p. 237-244.
HULTHE, G., HULTH, S., AND HALL, P. O. J., 1998, Effect of oxygen on degradation rate of refractory and labile organic matter in continental margin sediments: Geochimica et Cosmochimica Acta, v. 62, p. 1319-1328.
INGALL, E., AND JAHNKE, R., 1994, Evidence for enhanced phosphorus regeneration from marine sediments overlain by oxygen depleted waters: Geochimica et Cosmochimica Acta, v. 58, p. 2571-2575.
INGALL, E., AND JAHNKE, R., 1997, Influence of water-column anoxia on the elemental fractionation of carbon and phosphorus during sediment diagenesis: Marine Geology, v. 139, p. 219-229.
JAHNKE, R., 1985, A model of microenvironments in deep-sea sediments: Formation and effects on porewater profiles: Limnology and Oceanography, v. 30, p. 956-965.
JAHNKE, R. A., 1988, A simple, reliable, and inexpensive pore-water sampler: Limnology and Oceanography, v. 33, p. 483-487.
JAHNKE, R. A., EMERSON, S. R., REIMERS, C. E., SCHUFFERT, J., RUTTENBERG, K., AND ARCHER, D., 1989, Benthic recycling of biogenic debris in the eastern tropical Atlantic Ocean: Geochimica et Cosmochimica Acta, v. 53, p. 2947-2960.
JAHNKE, R. A., EMERSON, S. R., ROE, K. K., AND BURNETT, W. C., 1983, The present day formation of apatite in Mexican continental margin sediments: Geochimica et Cosmochimica Acta, v. 47, p. 259-266.
JAHNKE, R., HEGGIE, D., EMERSON, S., AND GRUNDMANIS, V., 1982, Pore waters of the central Pacific Ocean: Nutrient results: Earth and Planetary Science Letters, v. 61, p. 233-256.
JAHNKE, R. A., AND KNIGHT, L. H., 1997, A gravity-driven, hydraulically-damped multiple piston corer for sampling fine-grained sediments: Deep-Sea Research I, v. 44, p. 713-718.
JENSEN, H. S., MORTENSEN, P. B., ANDERSEN, F. Ø., RASMUSSEN, E., AND JENSEN, A., 1995, Phosphorus cycling in a coastal marine sediment, Aarhus Bay, Denmark: Limnology and Oceanography, v. 40, p. 908-917.
JENSEN, H. S., AND THAMDRUP, B., 1993, Iron-bound phosphorus in marine sediments as measured by bicarbonate-dithionite extraction: Hydrobiologia, v. 253, p. 47-59.
JØRGENSEN, B. B., 1977, Bacterial sulfate reduction within reduced microniches of oxidized marine sediments: Marine Biology, v. 41, p. 7-17.
KATZ, A., AND BEN-YAAKOV, S., 1980, Diffusion of seawater ions. Part II. The role of activity coefficients and ion paring: Marine Chemistry, v. 8, p. 263-280.
KENNETT, J. P., 1982, Marine Geology: Englewood Cliffs, New Jersey, Prentice-Hall, 813 p.
KLUMP, J. V. AND MARTENS, C. S., 1981, Biogeochemical cycling in an organic rich coastal marine basin — II. Nutrient sediment-water exchange processes: Geochimica et Cosmochimica Acta, v. 45, p. 101-121.
KÖNIG, I., DRODT, M., SUESS, E., AND TRAUTWEIN, A. X., 1997, Iron reduction through the tan-green color transition in deep-sea sediments: Geochimica et Cosmochimica Acta, v. 61, p. 1679-1683.
KROM, M. D., AND BERNER, R. A., 1980a, Adsorption of phosphate in anoxic marine sediments: Limnology and Oceanography, v. 25, p. 797-806.
KROM, M. D., AND BERNER, R. A., 1980b, The diffusion coefficients of sulfate, ammonium, and phosphate ions in anoxic marine sediments: Limnology and Oceanography, v. 25, p. 327-337.
KRAJEWSKI, K. P., VAN CAPPELLEN, P., TRICHET, J., KUHN, O., LUCAS, J., MARTIN-ALGARRA, A., PRÉVÔT, L., TEWARI, V. C., GASPER, L., KNIGHT, R. I., AND LAMBOY, M., 1994, Biological processes and apatite formation in sedimentary environments: Eclogae geologicae Helvetiae, v. 87, p. 701-745.
KUMP, L. R., 1988, Terrestrial feedback in atmospheric oxygen regulation by fire and phosphorus: Nature, v. 335, p. 152-154.
LASAGA, A. C., 1979, The treatment of multi-component diffusion and ion pairs in diagenetic fluxes: American Journal of Science, v. 279, p. 324-346.
LERMAN, A., MACKENZIE, F. T., AND GARRELS, R. M., 1975, Modeling of geochemical cycles: Phosphorus as an example: Geological Society of America Memoirs, v. 142, p. 205-218.
LI, Y.-H., AND GREGORY, S., 1974, Diffusion of ions in sea water and in deep-sea sediments: Geochimica et Cosmochimica Acta, v. 38, p. 702-714.
LIJKLEMA, L., 1980, Interaction of orthophosphate with iron (III) and aluminum hydroxides: Environmental Science and Technology, v. 14, p. 537-541.
LINDBERG, R. D., AND RUNNELLS, D. D., 1984, Ground water redox reactions: An analysis of equilibrium state applied to Eh measurements and geochemical modeling: Science, v. 225, p. 925-927.
LISITZIN, A. P., 1996, Oceanic Sedimentation: Lithology and Geochemistry: Washington, D.C., American Geophysical Union, 400 p.
LUCOTTE, M., AND D'ANGLEJAN, B., 1985, A comparison of several methods for the determination of iron hydroxides and associated orthophosphates in estuarine particulate matter: Chemical Geology, v. 48, p. 257-264.
LYLE, M., 1983, The brown-green color transition in marine sediments: A marker of the Fe(III)-Fe(II) redox boundary: Limnology and Oceanography, v. 28, p. 1026-1033.
MACKENZIE, F. T., VER, L. M., SABINE, C., LANE, M., AND LERMAN, A., 1993, C, N, P, S global biogeochemical cycles and modeling of global change, *in* Wollast, R., Mackenzie, F. T., and Chou, L., eds., Interactions of C, N, P and S Biogeochemical Cycles and Global Change; NATO ASI Series, Vol. 14: New York, Springer-Verlag, p. 1-61.
MANHEIM, F. T., AND WATERMAN, L. S., 1974, Diffusimitry (diffusion constant estimation) on sediment cores by resistivity probe: Initial Reports of the Deep Sea Drilling Project, v. 22, p. 663-670.
MARTENS, C. S., AND KLUMP, J. V., 1984, Biogeochemical cycling in an organic-rich coastal marine basin 4. An organic carbon budget for sediments dominated by sulfate reduction and methanogenesis: Geochimica et Cosmochimica Acta, v. 48, p. 1987-2204.
MARTIN, W. R., AND BENDER, M. L., 1988, The variability of benthic fluxes and sedimentary remineralization rates in response to seasonally variable organic carbon rain rates in the deep sea: A modeling study: American Journal of Science, v. 288, p. 561-574.
MCMANUS, J., BERELSON, W. M., COALE, K. H., JOHNSON, K. S., AND KILGORE,

T. E., 1997, Phosphorus regeneration in continental margin sediments: Geochimica et Cosmochimica Acta, v. 61, p. 2891-2907.

McNichol, A. P., Lee, C., and Druffel, E. R. M., 1988, Carbon cycling in coastal sediments: 1. A quantitative estimate of the remineralization of organic carbon in the sediments of Buzzards Bay, MA: Geochimica et Cosmochimica Acta, v. 52, p. 1531-1543.

Menard, H. W., and Smith, S. M., 1966, Hypsometry of ocean basin provinces: Journal of Geophysical Research, v. 71, p. 4305-4325.

Meybeck, M., 1982, Carbon, nitrogen, and phosphorus transport by world rivers: American Journal of Science, v. 282, p. 401-450.

Middleburg, J. J., Soetaert, K., and Herman, P. M. J., 1997, Empirical relationships for use in global diagenetic models: Deep-Sea Research I, v. 44, p. 327-344.

Milliman, J. D., 1991, Flux and fate of fluvial sediment and water in coastal seas, *in* Mantoura, R. F. C., Martin, J.-M., and Wollast, R., eds., Ocean Margin Processes in Global Change: New York, Wiley, p. 69-89.

Milliman, J. D., and Syvitski, J. P. M., 1992, Geomorphic/tectonic control of sediment discharge to the ocean: The importance of small mountainous rivers: Journal of Geology, v. 100, p. 525-544.

Murray, J. W., Grundmanis, V., and Smethie, W. M., Jr., 1978, Interstitial water chemistry in the sediments of Saanich Inlet: Geochimica et Cosmochimica Acta, v. 42, p. 1011-1026.

Pedersen, T. F., and Calvert, S. E., 1990, Anoxia vs. productivity: What controls the formation of organic-carbon-rich sediments and sedimentary rocks?: American Association of Petroleum Geologists Bulletin, v. 74, p. 454-466.

Rabouille, C., and Gaillard, J.-F., 1991, Towards the EDGE: Early diagenetic global explanation. A model depicting the early diagnosis of organic matter, O_2, NO_3, Mn, and PO_4: Geochimica et Cosmochimica Acta, v. 55, p. 2511-2525.

Rasmussen, F., 1996, Early-diagenetic REE-phosphate minerals (florencite, gorceixite, crandallite, and xenotime) in marine sandstones: A major sink for oceanic phosphorus: American Journal of Science, v. 296, p. 601-632.

Reimers, C. E., 1998, Carbon cycle: feedbacks from the sea floor: Nature, v. 391, p. 536-537.

Rhoads, D. C., and Morse, J. W., 1971, Evolutionary and ecologic significance of oxygen-deficient marine basins: Lethaia, v. 4, p. 413-428.

Robinson, J. M., 1989, Phanerozoic O_2 variation, fire, and terrestrial ecology: Palaeogeography, Palaeoclimatology, and Palaeoecology (Global and Planetary Change Section), v. 75, p. 223-240.

Ruttenberg, K. C., 1990, Diagenesis and burial of phosphorus in marine sediments: Implications for the marine phosphorus budget: Unpublished Ph.D. Dissertation, Yale University, New Haven, Connecticut, 375 p.

Ruttenberg, K. C., 1992, Development of a sequential extraction method for different forms of phosphorus in marine sediments: Limnology and Oceanography, v. 37, p. 1460-1482.

Ruttenberg, K. C., 1993, Reassessment of the oceanic residence time of phosphorus: Chemical Geology, v. 107, p. 405-409.

Ruttenberg, K. C., and Berner, R. A., 1993, Authigenic apatite formation and burial in sediments from non-upwelling, continental margin environments: Geochimica et Cosmochimica Acta, v. 57, p. 991-1007.

Ruttenberg, K. C., and Canfield, D. E., 1994, Chemical distribution of phosphorus in suspended particulate matter from 12 North American rivers: Evidence for bioavailability of particulate P (abs): Eos, Transactions of the American Geophsyical Union, Ocean Sciences Meeting Supplement, v. 75, p. 110.

Ruttenberg, K. C., and Goñi, M. A., 1997, Phosphorus distribution, C:N:P ratios, and $\delta^{13}C_{oc}$ in arctic, temperate, and tropical coastal sediments: tools for characterizing bulk sedimentary organic matter: Marine Geology, v. 139, p. 123-145.

Sarnthein, M., Winn, K., and Zahn, R., 1987, Paleoproductivity of the ocean upwelling and the effect on atmospheric CO_2 and climate change during deglaciation, *in* Berger, W. H., and Labeyrie, L. D., eds., Abrupt Climate Change: Evidence and Implications: Dordrecht, Reidel, p. 311-337.

Schink, D. R., and Guinasso, N. L., Jr., 1978, Redistribution of dissolved and adsorbed materials in abyssal marine sediments undergoing biological stirring: American Journal of Science, v. 278, p. 687-702.

Sherwood, B. A., Sager, S. L., and Holland, H. D., 1987, Phosphorus in foraminiferal sediments from North Atlantic Ridge cores and in pure limestones: Geochimica et Cosmochimica Acta, v. 51, p. 1861-1866.

Slomp, C. P., 1997, Early diagenesis of phosphorus in continental margin sediments: Doctoral Dissertation, Netherlands Institute for Sea Research, Texel, The Netherlands, 178 p.

Slomp, C. P., Epping, E. H. G., Helder, W., and Van Raaphorst, W., 1996a, A key role for iron-bound phosphorus in authigenic apatite formation in North Atlantic continental platform sediments: Journal of Marine Research, v. 54, p. 1179-1205.

Slomp, C. P., Malschaert, J. F. P., Lohse, L., and Van Raaphorst, W., 1997, Iron and manganese cycling in different sedimentary environments on the North Sea continental margin: Continental Shelf Research, v. 17, p. 1083-1117.

Slomp, C. P., Van der Gaast, S. J., and Van Raaphorst, W., 1996b, Phosphorus binding by poorly crystalline iron oxides in North Sea sediments: Marine Chemistry, v. 52, p. 55-73.

Smith, K. L., Jr., and Baldwin, R. J., 1984, Seasonal fluctuations in deep-sea sediment community oxygen consumption: central and eastern North Pacific: Nature, v. 307, p. 624-626.

Smith, K. L., Jr., Baldwin, R. J., and Williams, P. M., 1992, Reconciling particulate organic carbon flux and sediment community oxygen consumption in the deep North Pacific: Nature, v. 359, p. 313-316.

Soetaert, K., Herman, P. M. J., and Middleburg, J. J., 1996, A model of early diagenetic processes from the shelf to abyssal depths: Geochimica et Cosmochimica Acta, v. 60, p. 1019-1040.

Stumm, W., and Morgan, J. J., 1996, Aquatic Chemistry: Chemical Equilibria and Rates in Natural Waters (3rd edition): New York, Wiley, 1022 p.

Suess, E., 1981, Phosphate regeneration from sediments of the Peru continental margin by dissolution of fish debris: Geochimica et Cosmochimica Acta, v. 45, p. 577-588.

Sundby, B., Anderson, L. G., Hall, P. O. J., Iverfeldt, Å., Rutgers van der Loeff, M. M., Westerlund, S. F. G., 1986, The effect of oxygen on release and uptake of cobalt, manganese, iron and phosphate at the sediment-water interface: Geochimica et Cosmochimica Acta, v. 50, p. 1281-1288.

Sundby, B., Gobeil, C., Silverberg, N., and Mucci, A., 1992, The phosphorus cycle in coastal marine sediments: Limnology and Oceanography, v. 37, p. 1129-1145.

Thamdrup, B., Fossing, H., and Jørgensen, B. B., 1994, Manganese, iron, and sulfur cycling in a coastal marine sediment, Aarhus Bay, Denmark: Geochimica et Cosmochimica Acta, v. 58, p. 5115-5129.

Trolard, F., Génin, J.-M. R., Abdelmoula, M., Bourrié, G., Humbert, B., and Herbillon, A., 1997, Identification of a green rust mineral in reductomorphic soil by Mössbauer and Raman spectroscopies: Geochimica et Cosmochimica Acta, v. 61, p. 1107-1111.

Tromp, T. K., Van Cappellen, P., and Key, R. M., 1995, A global model for the early diagenesis of organic carbon and organic phosphorus in marine sediments: Geochimica et Cosmochimica Acta, v. 59, p. 1259-1284.

Ullman, J. W., and Aller, R. C., 1982, Diffusion coefficients in nearshore marine sediments: Limnology and Oceanography, v. 27, p. 552-556.

Van Cappellen, P., and Berner, R. A., 1988, A mathematical model for the early diagenesis of phosphorus and fluorine in marine sediments: apatite precipitation: American Journal of Science, v. 288, p. 289-333.

Van Cappellen, P. and Canfield, D. E.; Reply by Calvert, S. E., Bustin, R. M., and Pedersen, T. F., 1993, Lack of evidence for enhanced preservation of sedimentary organic matter in the oxygen minimun of the Gulf of California: comment and reply: Geology, v. 21, p. 570-572.

Van Cappellen, P., and Ingall, E. D., 1996, Redox stabilization of the atmosphere and oceans by phosphorus limited marine productivity: Science, v. 271, p. 493-496.

Van Raaphorst, W., Kloosterhuis, H. T., Cramer, A., and Bakker, K. J. M., 1990, Nutrient early diagenesis in the sandy sediments of the Dogger Bank area, North Sea: Pore water results: Netherlands Journal of Sea Research, v. 26, p. 25-52.

Vink, S., 1994, Sedimentary phosphorus cycling in Tomales Bay, California: Unpublished Ph.D. Dissertation, University of Hawaii, Honolulu, 169 p.

Vink, S., Chambers, R. M., and Smith, S. V., 1997, Distribution of phosphrous in sediments from Tomales Bay, California: Marine Geology, v. 139, p. 157-179.

Walker, J. C. G., 1974, Stability of atmospheric oxygen: American Journal of

Science, v. 74, p. 193-214.
WANG, Y., AND VAN CAPPELLEN, P., 1996, A multicomponent reactive transport model of early diagenesis: Application to redox cycling in coastal marine sediments: Geochimica et Cosmochimica Acta, v. 60, p. 2993-3014.
WATSON, A., LOVELOCK, J. E., AND MARGULIS, L., 1978, Methanogenesis, fires and regulation of atmospheric oxygen: Biosystems, v. 10, p. 293-298.
WESTRICH, J. T., 1983, The consequences and controls of bacterial sulfate reduction in marine sediments: Unpublished Ph.D. Dissertation, Yale University, New Haven, Connecticut, 530 p.
WESTRICH, J. T., AND BERNER, R. A., 1984, The role of sedimentary organic matter in bacterial sulfate reduction: the G model tested: Limnology and Oceanography, v. 29, p. 236-249.
WHEAT, C. G., FEELY, R. A., AND MOTTL, M. J., 1996, Phosphate removal by oceanic hydrothermal processes: An update of the phosphorus budget in the

TRANSFER OF PHOSPHORUS FROM THE INDIAN SUB-CONTINENT TO THE ADJACENT OCEANS

V. SUBRAMANIAN*

School of Environmental Sciences, Jawaharlal Nehru University, New Delhi, 110 067, India

ABSTRACT: Rivers of South Asia carry dissolved inorganic phosphorus (DIP) at levels far greater than many of the world's rivers such as the Amazon River and European rivers on whose phosphorus data average rates of DIP exported to the oceans are largely based. Similarly, all the major river systems in South Asia have sedimentary phosphorus values that are considerably higher than the world average. Unlike the Amazon or other large river system for which extensive data are available, organic phosphorus contributes very little to the total sedimentary phosphorus pool in South Asian rivers. Trapping of enormous amounts of sediment in the vast alluvial belts in the northern part of South Asia indicate that very little of the sedimentary phosphorus transported by these rivers is likely to be exported to the adjacent marine environment.

INTRODUCTION

Berner and Ji-Long Rao (1994) speculate that levels of dissolved phosphorus in the oceans are largely a function of supply from the continents. Estimates of the global average riverine contribution of phosphorus to the oceans have been, for the most part, based largely on data from the Amazon River and European rivers (Berner and Ji-Long Rao, 1994; Froelich, 1988; Meybeck, 1982). Though the Asian rivers, particularly those draining the densely populated south Asian region, account for the bulk of the global sediment flux to oceans (Milliman et al, 1995; Subramanian, 1996), very little attention has been paid to South Asian rivers in global-scale estimates of fluxes and biogeochemical processes, particularly with respect to phosphorus. This paper attempts to fill the gap so that the understanding of the Phosphorus Cycle is not biased by geographically selective data.

DATA SOURCE

In addition to published phosphorus data for the Indian Subcontinent, new data, primarily for the Bengal Basin, Brahmaputra River and flood plains are presented and discussed. Standard methodology for determining dissolved phosphorus (DIP) and total sediment phosphorus (TDP) were followed (APHA, 1995). Details of the phosphorus-fractionation scheme employed are discussed in Vaithiyanathan et al. (1993). Samples were collected for many of the systems at different seasons in the last two decades and analyzed immediately. Spectrophotometer and Carbon Analyzer were used for phosphorus and carbon, respectively; nitrogen was analyzed by the Kjeldhal method. Accumulation rates were determined by lead-201/radium-226 radiometric dating methods described in Ramesh et al. (1988).

DISSOLVED INORGANIC PHOSPHORUS (DIP)

More than 90% of the discharge from South Asian rivers goes to the Bay of Bengal which connects to the Indian Ocean in the equatorial region (Fig. 1). Table 1 shows the sediment discharge data along with values for dissolved and particle-bound phosphorus for South Asian rivers. Data for the both the Indian and Bangladesh segments of the Brahmaputra and Bengal Basin rivers are shown in Table 1. For comparison, the world average values are also indicated in the table. Meybeck's (1982) world average riverine DIP estimate of 10 µg/l is based on a compilation of data that also includes some extreme values for rain water DIP concentrations (1.3 µg/l over the Rio Negro, South America, to 51 µg/l over Hansen, Sweden). His DIP compilation also includes additional pollution contributions of 16–1,400 µg/l phosphate. Berner and Ji-Long Rao (1994) used value of 12 µg/l for the world average riverine DIP in their calculations. The DIP concentrations in South Asian rivers (Table 1) are generally higher than the world average estimates, perhaps because of pollution from the watershed which has been a region of high population density and extensive agricultural activity and other land use for several thousand years. The ranges shown in Table 1 represent both spatial and temporal variations in DIP. The discharge-weighted average for each location and for each river was used to arrive at the Indian subcontinent average DIP of 40 µg/l given at the bottom of Table 1. Hence, previously published world average DIP values should be revised upwards in the light of our data.

Most of the organic matter is labile in the river sediments of the Indian subcontinent (Gupta et al, 1997; Subramanian and Ittekot, 1991) and is likely to release additional phosphorus into the river water. Therefore, the total dissolved phosphorus (TDP), which is DIP plus dissolved organic phosphorus (DOP), being exported out of South Asia to the Bay of Bengal is likely to be substantially greater than 40 µg/l and a discharge-weighted world average TDP that incorporates the South Asian data is likely to be substantially greater than Meybeck's estimated TDP value of 25 µg/l (Table 1).

For the Ganges estuary (from Calcutta towards the Bay of Bengal, Fig. 1) the DIP data shows highly non-conservative behavior (Subramanian, 1993) whereas the Cauvery estuary, in the peninsular tropical region, DIP data shows conservative behavior right up to the river's junction with the Bay of Bengal (Ramanathan et al., 1993). The DIP values in Table 1 are discharge-weighted averages only down to the fresh water end of the river/location. All along the 5,000-km coastal belt of South Asia, the conservative or non-conservative behavior of phosphorus in estuarine regions will determine the local net export of phosphorus into the adjacent marine environment.

SEDIMENTARY PHOSPHORUS

In comparison to the bed load, suspended-load phosphorus concentrations are consistently more variable even in data from

*email: subrama42@hotmail.com

Marine Authigenesis: From Global to Microbial, SEPM Special Publication No. 66

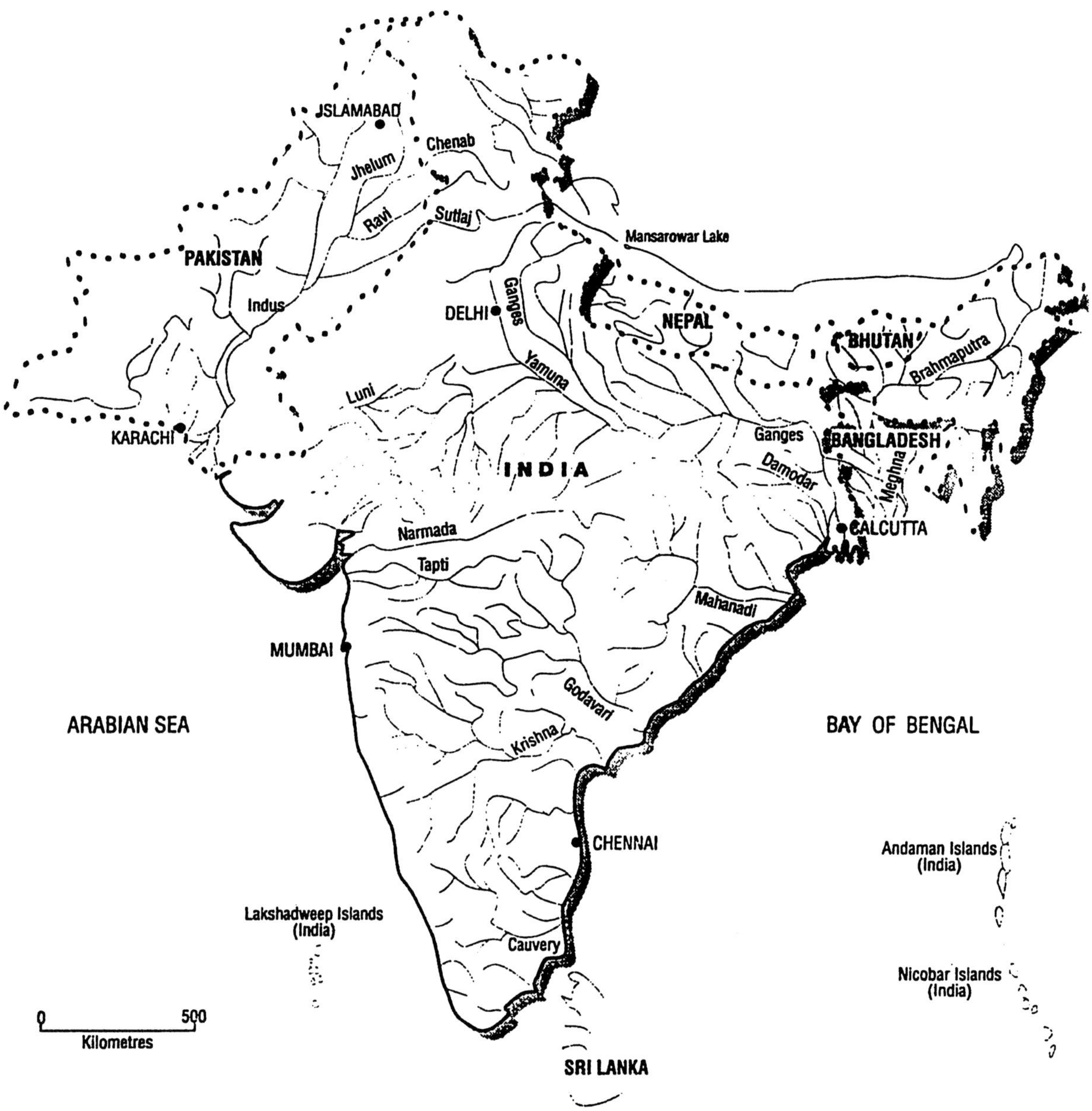

FIG. 1.—Location map showing major rivers of South Asia and urban areas such as Delhi, Calcutta, and Chennai. Dotted lines represent political boundaries between countries.

single rivers (Table 1). For example, sediments of the Yamuna River (passing through Delhi) (Fig. 1) show a wide variation in the sediment-bound phosphorus in both bed load and the suspended load sediments whereas the sediments of the Cauvery River in Peninsular India (Fig. 1) show relatively limited variability in both bed load and suspended load phosphorus reflecting the dynamics of the respective river regimes. Figure 2 shows the spatial variability in the downstream direction of the sediment-bound phosphorus (bed load and suspended load) for two river types represented by the Brahmaputra River, which has a high discharge rate and heavy sediment load of about a billion tons per year (Subramanian, 1996) and the Cauvery River, which has a low discharge rate and low sediment load (Ramanathan et al., 1995). Downstream along the Brahmaputra's 1,000-km stretch in India, there is an increase in the suspended load phosphorus and an increase in suspended load relative to the bed load phosphorus (Fig. 2A). Such downstream enrichment of particulate phosphorus has been reported for many river systems in the world (see for example, de Groot and Golterman, 1990). The downstream profiles for the Cauvery River (Figs. 2B, 2C) are somewhat more erratic, perhaps due to the extremely fine grain size of the particulates and their composition, which is dominantly montmorillonite clay (Subramanian, 1980). The grain size influence on the sedimentary phosphorus is thus interesting in that although the Cauvery has a larger proportion of clay minerals such as montmorillonite, there is no apparent enrichment in phosphorus in the Cauvery relative to the Brahmaputra whose sediment is coarser grained. In fact, unlike the Brahmaputra, the Cauvery's sedimentary phosphorus profile suggests repeated exchange between suspended load and bed load sediments, per-

Table 1.—SUMMARY OF PHOSPHORUS DATA FOR THE RIVERS OF SOUTH ASIA

Rivers	Discharge km^3/yr	$P\text{-}P_4^{-3}$ µg/l	Bed Load Phosphorus in µg/g	Suspended Load Phosphorus in µg/g	Reference
Ganges	393.000	10–1135	109–310	225–2021	Present Study
Gandak	52.000	10–33	310	820–1369	Abbas & Subramanian, 1984
Kosi	22.400	29–33	293	400	Subramanian & Singh, 1998
Sone	21.000	21–39	–	2180–8150	Subramanian & Singh, 1998
Yamuna	96.100	36–1628	22–655	284–7600	Jha et al., 1990
Betwa	0.018	1954	196	2532	Jha et al., 1990
Chambal	0.031	54	262	2314	Jha et al., 1990
Ken	0.011	41	44	1964	Jha et al., 1990
Brahmaputra (Assam)	603.000	16–39	350–1085	926–1195	Present Study
Indus	238.000	140	–	–	Arian, 1989
Tapti	10.000	26–39	–	–	Subramanian, 1979
Gomti	7.390	49–261	870–430	650	Gupta & Subramanian, 1994
Mahanadi	54.000	1–59	483–1163	1837–3502	Chakrapani & Subramanian, 1993
Narmada	41.000	100	–	–	Subramanian, 1979
Krishna	30.000	7–33	587–1273	774	Ramesh et al., 1988
Godavari	119.000	3–202	830	–	Bikshamiah et al., 1991
Cauvery	11.000	21–124	900–1500	800–1400	Ramanathan et al., 1993
Western Ghat Rivers	57.400	3–65	197–340	–	Subramanian, 1997
Brahmaputra (Bangladesh)	744.000	8–29	610–1180	540–6410	Present Study
Surma (Bangladesh)	–	41–121	–	–	Present Study
Meghna (Bangladesh	136.000	1–90	530–940	3780–6410	Present Study
Kushiara (Bangladesh)	–	10–47	–	–	Present Study
Jamuna (Bangladesh)	–	10–42	–	–	Present Study
Ganga (Padma)	450.000	15–98	600–1850	1080–1950	Present Study

Indian Subcontinent Average:	$P\text{-}PO_4^{-3}$	40 µg/l (DIP)	
	Sediment Phosphorus	3456 µg/g (Average based on Ganga-Brahmaputra)	
World Average:	$P\text{-}PO_4^{-3}$	10 µg/g (DIP)	Meybeck, 1982
	Total Dissolved Phosphate	25 µg/l (TDP)	Meybeck, 1982
	Sediment Phosphorus	1150 µg/g	Meybeck, 1982

haps due to re-suspension along its 800-km course from the Western Ghats to its junction with the Bay of Bengal.

In the estuarine region, several factors regulate the fate of sediment-bound phosphorus: fresh water flocculation, relative proportion and type of clay minerals, salinity gradient, extent of re-suspension due to tides, circulation, or dredging as is being done at the mouth of the Ganges Estuary to keep the port at Calcutta open. Redistribution of particulate phosphorus between bed load and suspended load has been reported to be a response to mixing processes in the Ganges estuary (Subramanian, 1993), the Cauvery estuary (Ramanathan et al., 1993), and the Indian west coast estuaries (Balchand and Nair, 1994).

The interaction between the major nutrients, carbon, nitrogen, and phosphorus, in river sediments is well known . However, the graphs in Figure 3 show no strong correlation between carbon and nitrogen and carbon and phosphorus in the two South Asian rivers represented, which suggests that organic processes play a smaller role in phosphorus dynamics in South Asian Rivers than in other rivers.

PHOSPHORUS FRACTIONATION

Sediment-bound phosphorus is fractionated into several compartments and numerous authors have adapted various extraction methods to assess the chemical sites (Balchand and Nair, 1994; Berner and Ji-Long Rao, 1994; Bostrom et al., 1988; de Groot and Golterman, 1991; Lebo, 1991; Vaithyanathan et al., 1993). Figure 4 illustrates the fractionation pattern and the downstream changes in pattern of sediment-bound phosphorus in three South Asian rivers that are characterized by different discharge rates and sediment grain sizes, and different kinds and degrees of modification in the drainage basin. The Brahmaputra River has a high discharge rate, coarse sediment load (Subramanian and Ramanathan, 1996), with low levels of urbanization, and high levels of agricultural use. The Yamuna (a tributary of the Ganges) has a high discharge rate, coarse sediment load, and high levels of urban, industrial, and agricultural use. The Cauvery River has a low discharge rate, fine sediment load, and high levels of urban, industrial, and agricultural activity. The only readily observable differences in the fractionation patterns are the great abundance

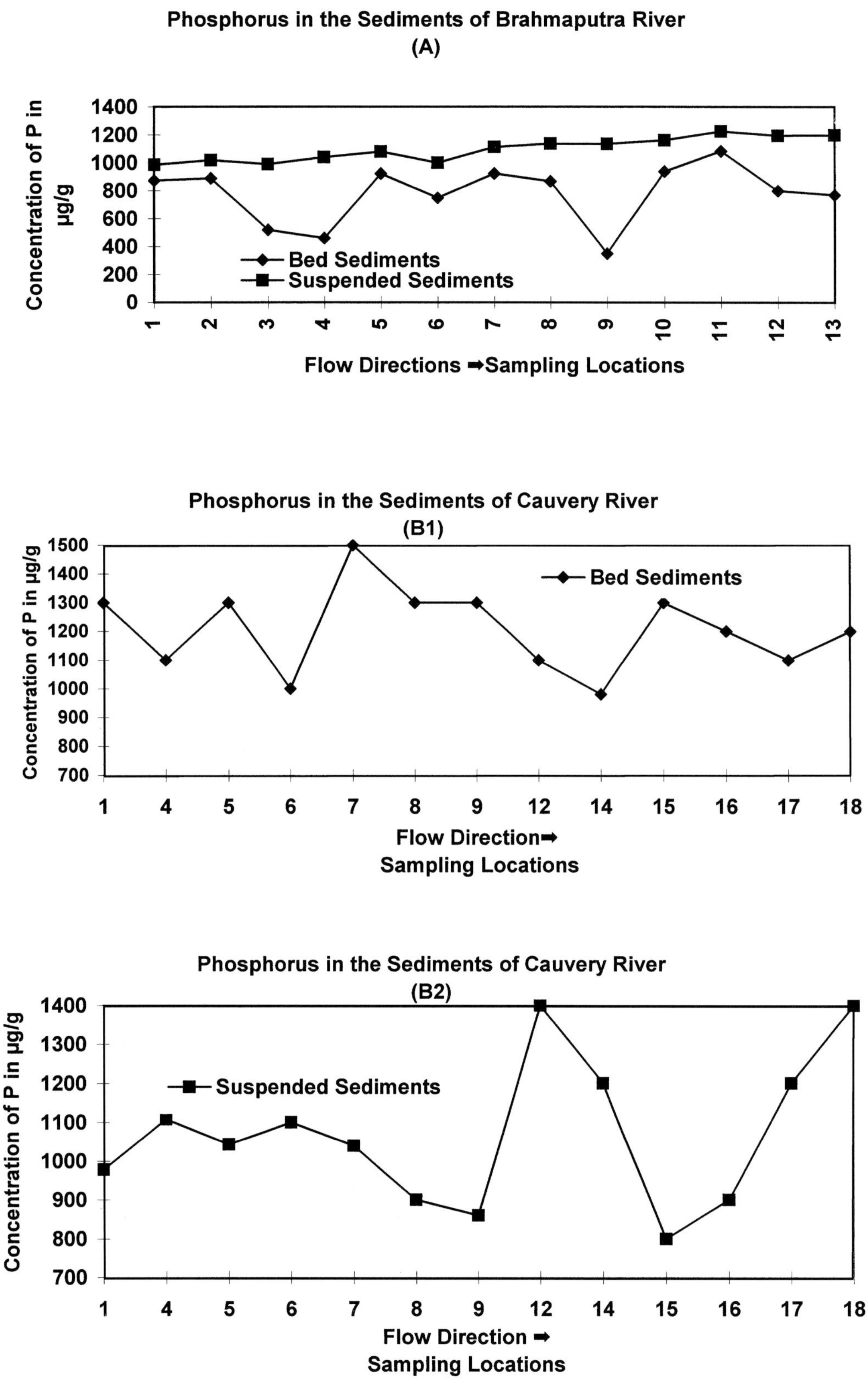

FIG. 2.—Spatial variability of inorganic phosphorus in a river characterized by high sediment loads, the Brahmaputra River, and a river characterized by low sediment loads, the Cauvery River. Phosphorus concentration is in micrograms per gram of sediment. Sample location numbers (horizontal axis) increase in the direction of flow. (A) Inorganic phosphorus concentrations in bedload and suspended-load sediment of the Brahmaputra River. (B1) Inorganic phosphorus concentrations in bedload sediments of the Cauvery River. (B2) Inorganic phosphorus concentrations in suspended-load sediments of the Cauvery River.

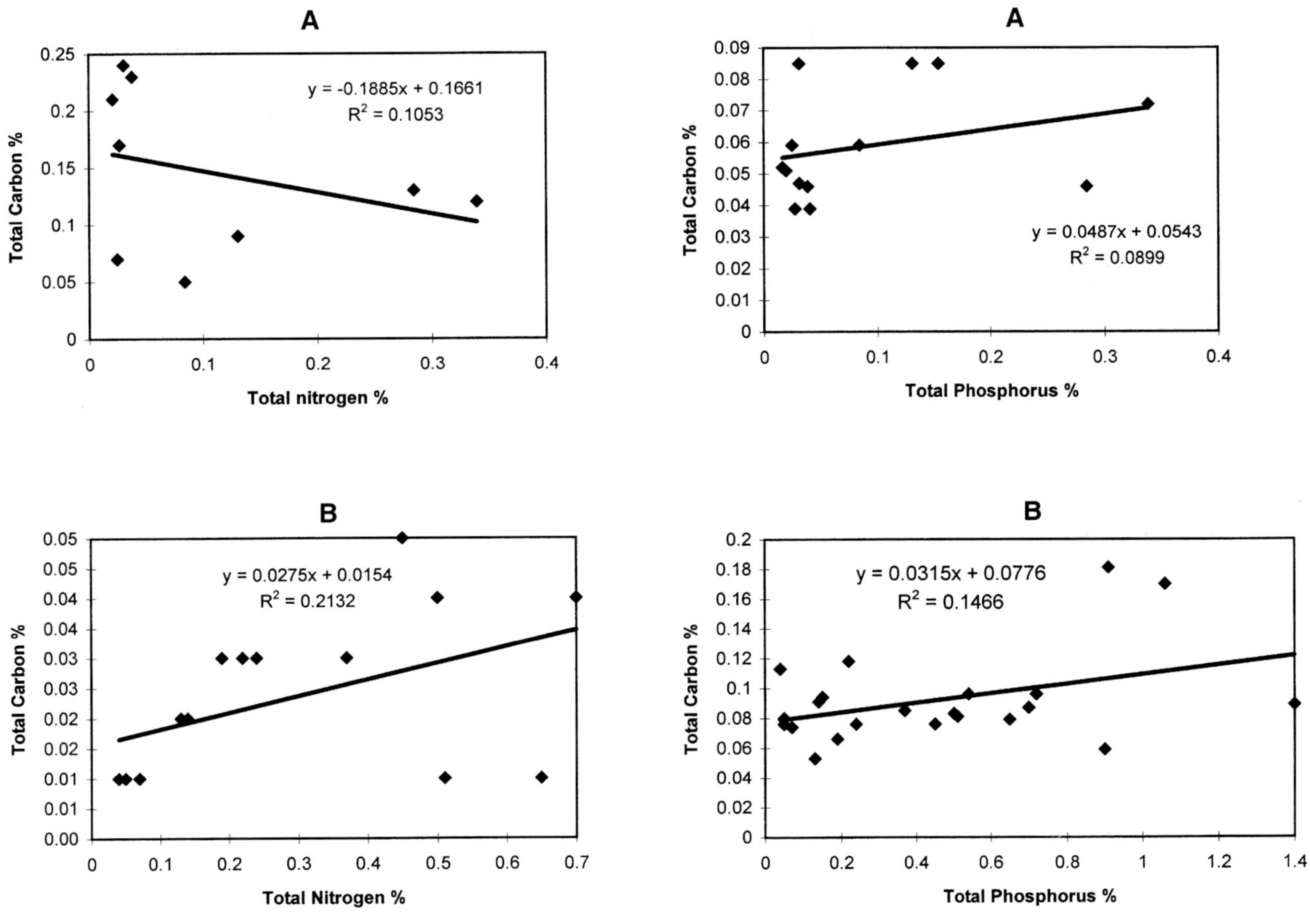

FIG. 3.—Relationship between three major nutrients, carbon, nitrogen, and phosphorus, for two South Asian rivers — the Bengal Basin rivers with high sediment load (A [also Brahmaputra River]) and the Cauvery (B) representing low sediment load for tropical river system.

abundance of organic phosphorus in the Brahmaputra patterns; the dominance of biogenic hydroxyapatite and carbonate-bound phosphorus in the Yamuna patterns, and the dominance of apatitic phosphorus in the Cauvery River patterns. For each of the three rivers, spatial variability in phosphorus fractionation is illustrated by patterns of samples from five or six different locations. Unlike the other two rivers, the Brahmaputra's sediment has a variable percentage of organic phosphorus due to agricultural activity in the watershed. But, in the delta region of the combined Ganga-Brahmaputra-Meghna system, very little organic phosphorus is present. The fractionation patterns from all three systems in the Figure 4 show variable levels of calcium-bound or iron-bound phosphorus and in the Cauvery sediments the detrital apatite-bound phosphorus (i.e. calcium-bound phosphorus) dominates. Berner and Ji-Long Rao (1994) reported the dominance of organic-bound phosphorus in the Amazon River sediments; Vaithiyanathan et al., (1993) showed the dominance of calcium-bound phosphorus for the Ganges sediments. Lebo (1991) suggested that iron and aluminum-bound phosphorus buffers excess sedimentary phosphorus in the Delaware estuary and Balchand and Nair (1994) outlined the importance of exchangeable phosphorus and calcium-bound phosphorus along with the nature of organics in the west coast estuaries. Thus, phosphorus in particulate organic matter may be less important in the South Asian rivers as it is in the Amazon sediments due to the high biodegradability of organic matter in this tropical region, as evidenced by the nature of the organic matter in the Godavari River Sediments (Gupta et. al., 1997).

PHOSPHORUS ACCUMULATION IN FLOOD PLAINS

Post-depositional diagenetic processes re-deploy the sediment-bound phosphorus. Figure 5 illustrates this point using data from two different types of rivers in South Asia: the Krishna River with low sedimentation rates (Ramesh et al. 1988) and the Brahmaputra (Subramanian and Ramanathan, 1996) with variable rates of phosphorus accumulation (Datta, 1996). Such a vertical redeployment is the combined effect of a variety of chemical and biological processes in the closed environment. Note that the graphs have different vertical scales and different horizontal scales. In both cases, the vertical trends are compatible but not the absolute values due to different scales used in the figure. Vertical remobilization of sediment-bound phosphorus is indicated for both the river systems though they are not consistent—increases towards the surface in the upper reaches of Krishna, decreases towards the surface at other places in the same river systems with low sedimentation rates but highly

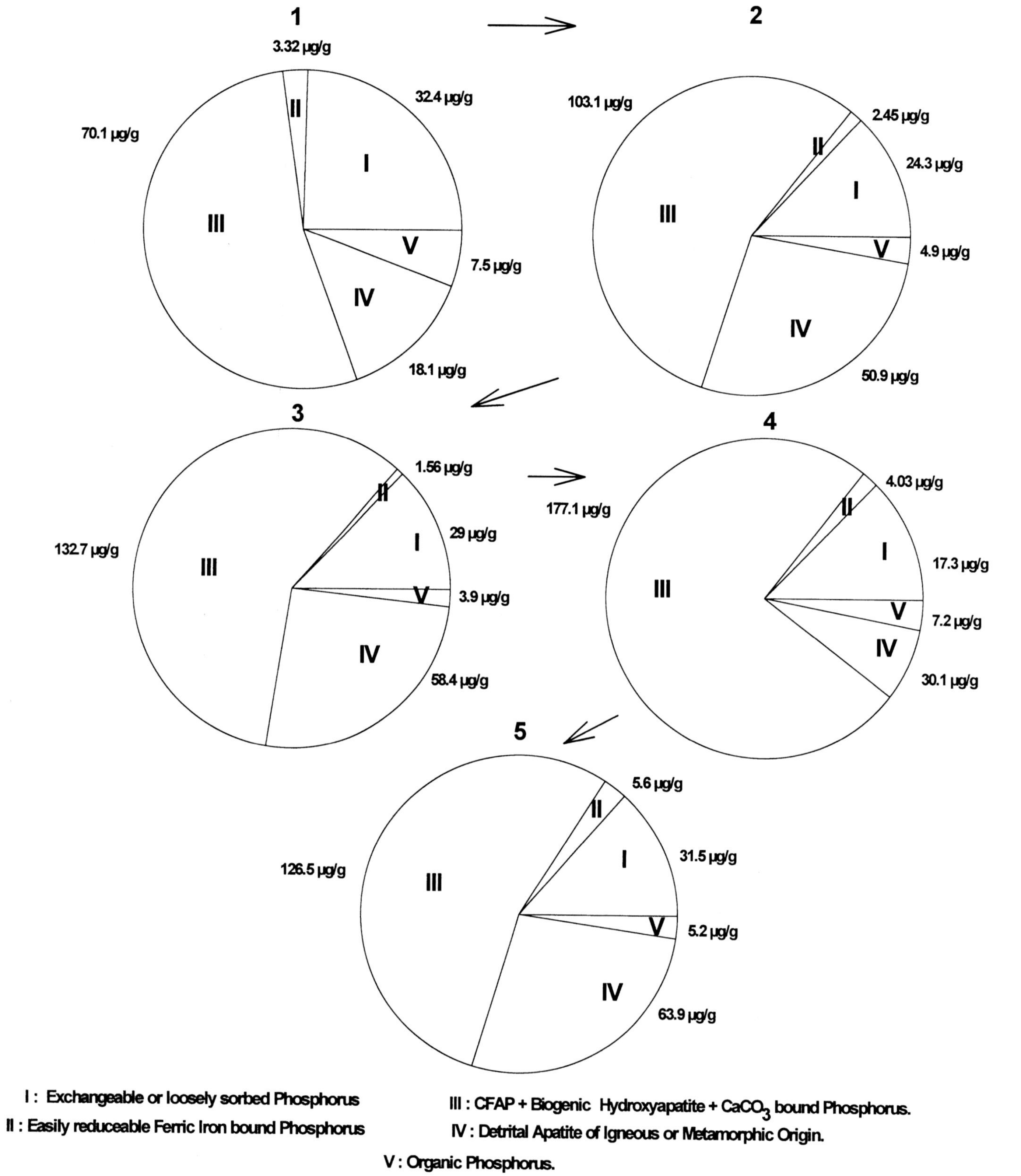

FIG. 4.—Particulate phosphorus-fractionation patterns of samples from Brahmaputra, Yamuna, and Cauvery River sediments. Arrows indicate flow direction between sample sites. Phosphorus-fraction concentration is in micrograms per gram of dry sediment.

(A) Brahmaputra River sediment phosphorus-fractionation patterns. I = iron-bound (CDB-extracted) phosphorus; II = aluminum- and iron–bound, non-occluded (NaOH-extracted) phosphorus; III = calcium-bound (HCl-extracted) phosphorus; IV = organic phosphorus .

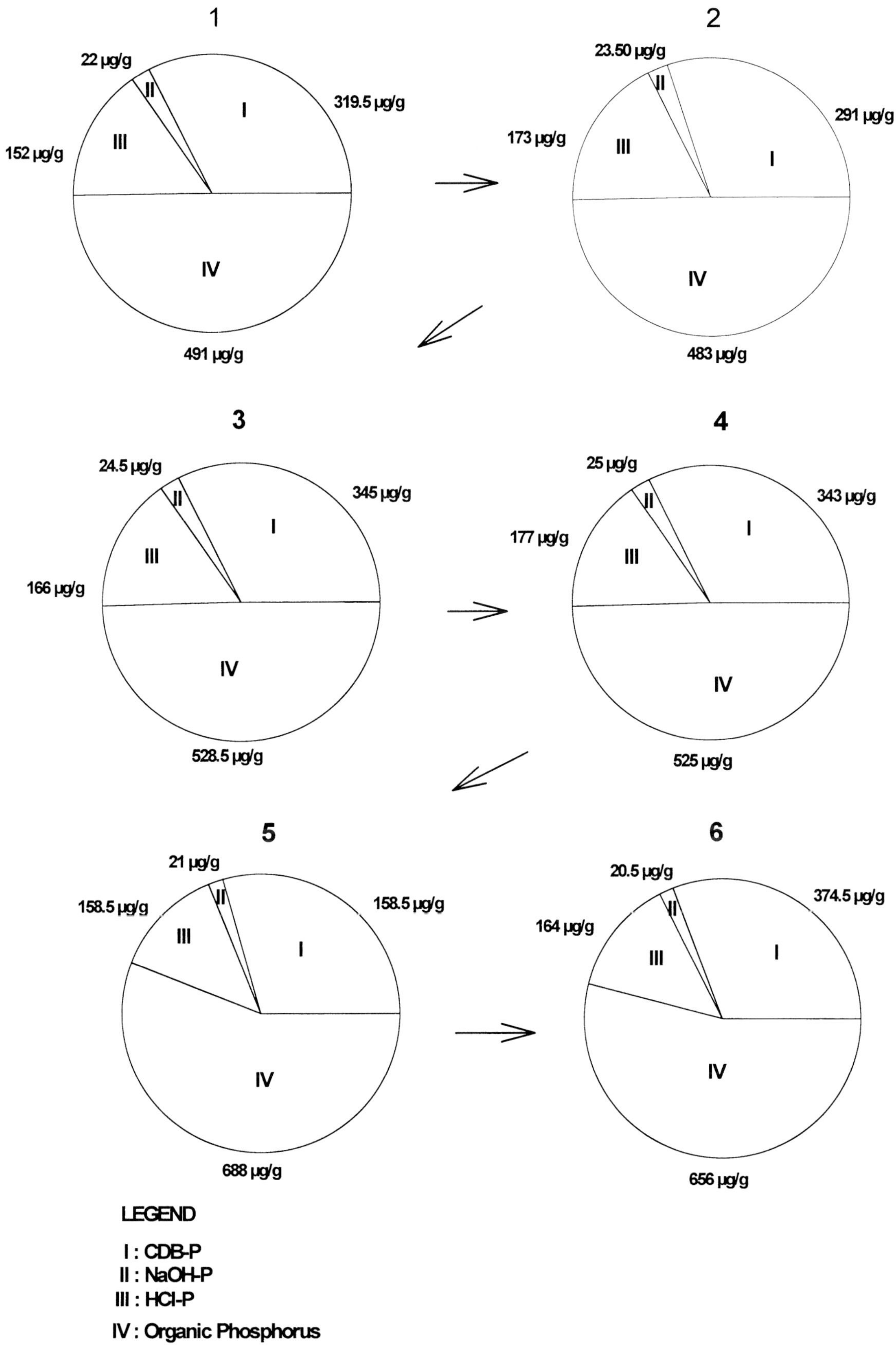

FIG. 4.—Continued. (B) Yamuna River sediment phosphorus-fractionation patterns. I = Exchangeable-phosphorus or loosely sorbed phosphorus; II = Easily reducible ferric iron-bound phosphorus; III = CFA phosphorus + biogenic hydroxyapatite + $CaCO_3$-bound phosphorus; IV = detrital apatite of igneous or metamorphic origin; V = organic phosphorus.

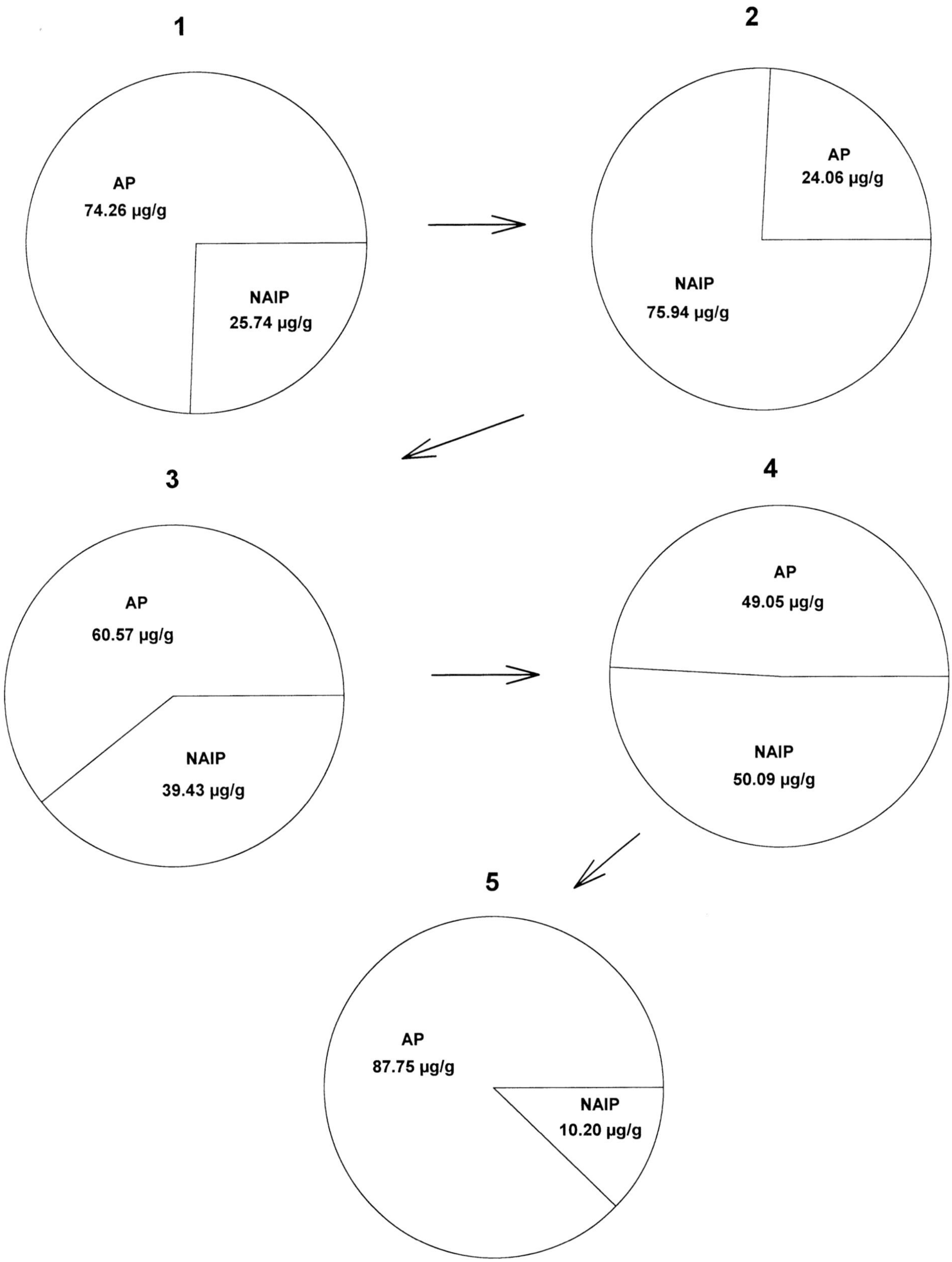

FIG. 4.—Continued. (C) Cauvery River sediment phosphorus-fractionation patterns. AP = apatitic phosphorus; NAIP = non-apatitic inorganic phosphorus.

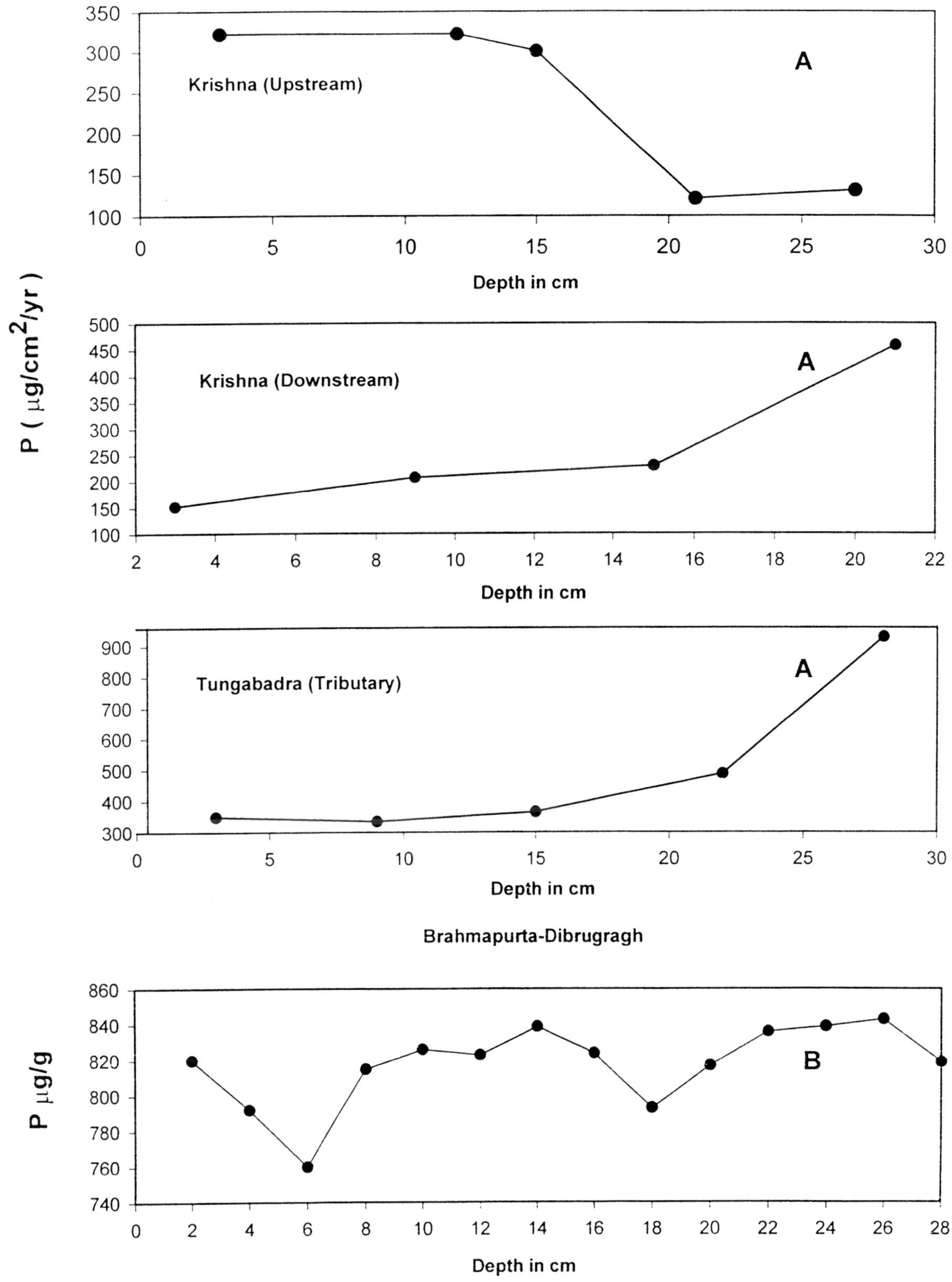

FIG. 5.—Phosphorus accumulation rate versus sample depth from (A) rivers with low sedimentation rates, the Krishna River and one of its tributaries, the Tungabadra River; and (B) from a river with a high sedimentation rate, the Brahmaputra River. Phosphorus accumulation rates were determined by lead-210/radium-226 radiometric dating methods described in Ramesh et al. (1988).

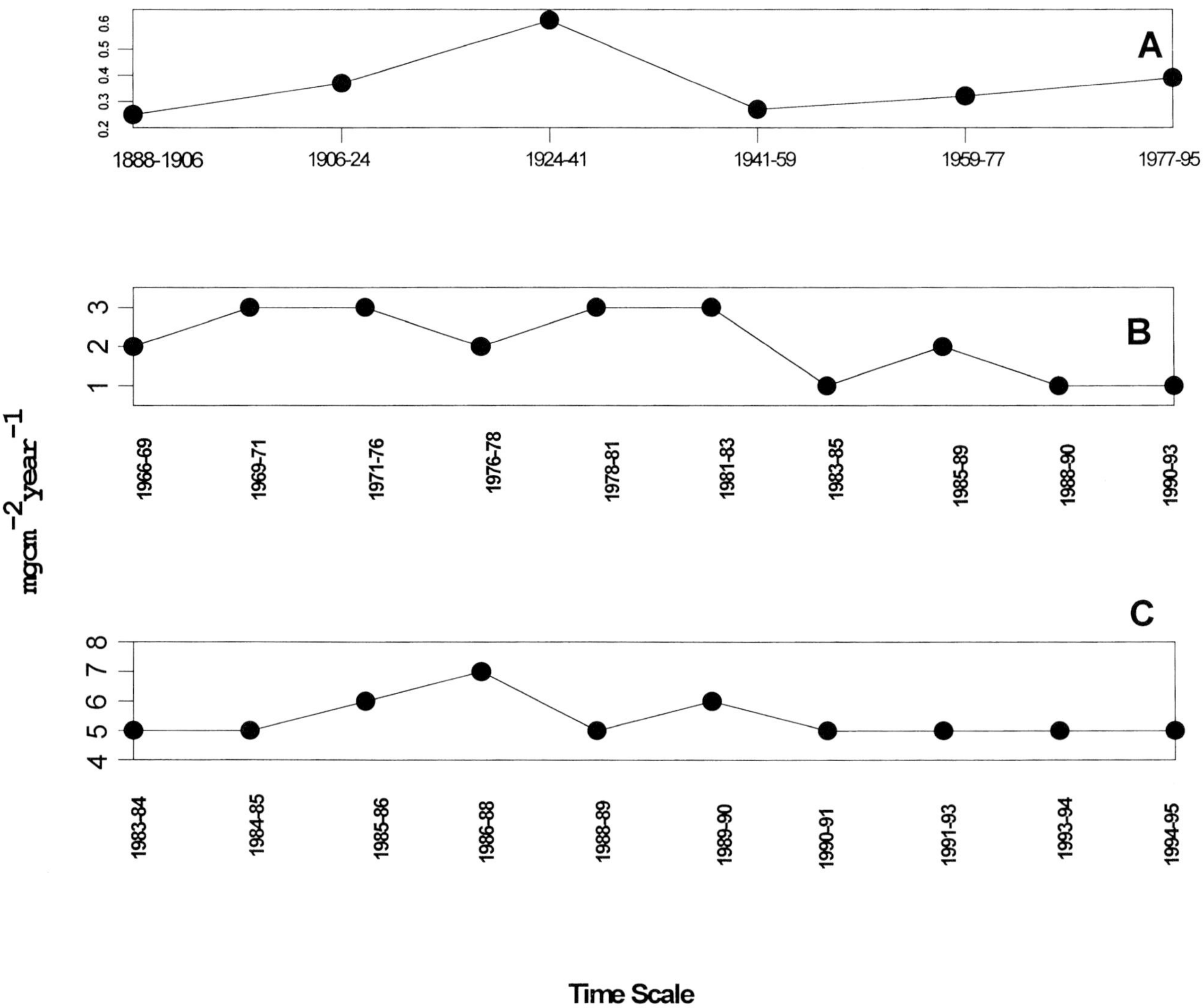

FIG. 6.—Phosphorus accumulation rates in the flood plains of the Yamuna River are illustrated by graphs of the average yearly rate of phosphorus accumulation. (A) Upper reaches of the Yamuna River, in the Himalayas; 1888-1995; mean phosphorus deposition rate is 0.375 mg/cm^2-yr. (B) Middle reaches of the Yamuna River; mean phosphorus deposition rate is 2.1 mg/cm^2-yr. (C) Lower reaches of the Yamuna River; mean phosphorus deposition rate is 5.4 mg/cm^2-yr. Phosphorus accumulation rates were determined by lead-210/radium-226 radiometric dating methods described in Ramesh et al. (1988).

erratic in sediments of the Brahmaputra which has high deposition rates, perhaps representing highly active diagenetic processes.

One of the key aspects of continental riverine environments in South Asia is the presence of extensive flood plains in the entire region. Large rivers such as the Brahmaputra can erode or deposit millions of tons of sediment in a single flood (Coleman, 1969). Thus, the associated phosphorus is either washed away or redeposited. The graphs of phosphorus accumulation rate versus time in Figure 6 illustrate this point. The accumulation rate data in Figure 6 is from three distinct zones of the Yamuna River (a tributary of the Ganges): the upper reaches, near the source in the Himalayas (Saharanpur); the middle reaches, in the urban belt near Delhi; and the lower reaches at the Yamuna's junction with the Ganges (Allahabad). In the upper reaches, the phosphorus accumulation rates are very low; the yearly average over a seventy-year period was 0.375 mg/cm^2-yr (Fig. 6A). The middle reaches had higher phosphorus accumulation rates averaging 2.1 mg/cm^2-yr over a thirty-year period (Fig. 6B). The lower reaches had even higher accumulation rates, averaging 5.4 mg/cm^2-yr over a period of ten years. Thus, downstream along the course of a river, sediment-bound phosphorus accumulates at progressively higher rates and is transported out of the upper mountain regions at a greater rate due to the higher downstream gradient. The amount of sediment-bound phosphorus trapped in the Ganges River alluvium, which occupies an area 2,000 x 1,000 km, whose mean thickness is 1,000 m and maximum thickness is 4 km, must be enormous. Very little of it would have been exported out to the marine environment in the immediate geological past.

One must however remember that after deposition and burial, phosphorus can be remobilized by a variety of diagenetic processes, such as desorption from clay minerals or from iron-aluminum (Fe-Al) hydroxides and break-down of organic matter (Berner and Ji-Long Rao, 1994; Froelich, 1988; Meybeck, 1982). The variations with burial depth in phosphorus accumulation rates observed in river sediments (Fig. 5) and in flood plain, sediments (Fig. 6) may reflect a combination of these diagenetic processes.

CONCLUSION

The study clearly shows that in the Indian Subcontinent, with high population density and increasing use of inorganic fertilizers, the role of organic-bound phosphorus, that has been observed for major rivers such as the Amazon, may be less important in phosphorus dynamics. The database in Table 1 is not yet sufficient for calculating a new global phosphorus budget. However, the South Asian river average values for DIP and sedimentary phosphorus (Table 1) are 2–3 times higher than the global average river values estimated by Meybeck (1982) and Berner and Ji-Long Rao (1994). Because a large proportion of the sediment entering the world's oceans is delivered by Asian rivers (Milliman et al., 1995) by South Asian rivers in particular (Subramanian, 1996), and because those rivers have average phosphorus concentrations significantly greater than the global averages of Meybeck (1982) and Berner and Ji-Long Rao (1994), the total riverine contribution of phosphorus to the oceans is probably significantly higher than the estimates currently in use.

The phosphorus data from South Asian rivers, although limited, indicate an urgent need to re-evaluate global fluxes and processes involved in the global phosphorus cycle. It might be possible to better understand the global biogeochemical cycle of phosphorus by including data for rivers of Southeast Asia and China so that our understanding is based on a database that more realistically represents the world's rivers.

ACKNOWLEDGMENTS

This research was supported by funds from Department of Science and Technology, Department of Ocean Development and Ministry of Environment and Forests, Government of India.

REFERENCES

ABBAS, N. AND SUBRAMANIAN, V. (1984): Erosion and sediment transport in the Ganges River Basin (India). Journal of Hydrology, v. 69, p. 173-182.

APHA, 1995, Standard Methods for the Examination of Water and Wastewater: APHA, Washington, D.C., Section 4, p. 106-144.

ARIAN, A. (1989): Nutrients in Indus river, *in* River Geochemistry, SCOPE, v. 3, Hamburg, University of Hamburg Press, p. 212-219.

BALCHAND, A. N. AND NAIR, S. M. (1994): Fractionation of phosphorus in the sediments of a tropical estuary. Environmental Geology, v. 23, p. 284-294.

BERNER, R. A AND JI-LONG RAO, (1994): Phosphorus in the sediments of Amazon River and estuary: implications for the global flux of phosphorus of sea. Geochimica et Cosmochimica Acta, v. 58, No.10, p. 2333-2339.

BIKSHAMIAH, G., SUBRAMANIAN, V., RAMANATHAN, A. L. AND VAN GRIEKEN, R. (1991): Heavy metal distribution in Godavari River basin. Environmental Geology and Water Sciences, v. 17(2), p. 117-126.

BIKSHAMIAH, G. SUBRAMANIAN, V. (1988): Nature of solute load in the Godavari River basin, India. Journal of Hydrology, v. 103, p. 375-392.

BOSTROM, B., PERSSON, G. AND BROBERG, B. (1988): Bioavailability of different phosphorus forms in fresh water systems. Hydrobiologia, v. 170, p. 133-155.

CHAKRAPANI, G. J. AND SUBRAMANIAN, V. (1993): Rates of erosion and sedimentation in the Mahanadi River basin, India. Journal of Hydrology, v. 140, p. 39-48.

COLEMAN, J. M. (1969): Brahmaputra River channel processes and sedimentation. Sedimentary Geology, v. 3, p. 129-239.

DATTA, D. K. (1996): Particle bound nutrients and heavy metals in the Bengal basin. Unpublished Ph.D. Thesis, Jawaharlal Nehru University, New Delhi, 224 p.

DATTA, D. K. AND SUBRAMANIAN, V. (1997): Nature of solute load in the rivers of Bengal basin, Bangladesh. Journal of Hydrology, v. 198, p. 196-208.

DE GROOT, C. J. AND GOLTERMAN, H. L. (1990): Sequential fractionation of sediment phosphate, Hydrobiologia, v. 192, p. 143-148.

FROELICH, P. N. (1988): Kinetic control of dissolved phosphate in natural rivers and estuaries: a primer on the phosphate buffer mechanism. Limnology and Oceanography, v. 33, p. 649-668.

GARRELS, R. M. AND MACKENZIE, F. T. (1971): Evolution of Sedimentary Rocks. New York, Norton Publishing, 298 p.

GUPTA, L. P. AND SUBRAMANIAN, N, V. (1994): Environmental geochemistry of the River Gomti: a tributary of the Ganga River. Environmental Geology, v. 24, p. 235-243.

GUPTA, L. P., SUBRAMANIAN, V., AND ITTEKOT, V. (1997): Nature of organic matter in the sediments of the Godavari River basin. Biogeochemistry, v. 38(2), p. 103-128.

JHA, P. K., SUBRAMANIAN, V., SITASAWAD, R. AND VAN GRIEKEN, R. (1990): Heavy metals in the sediments of Yamuna River (a tributary of Ganges), India. Science of the Total Environment, v. 95, p. 7-27.

LEBO, M. (1991): Particle bound phosphorus along an urbanised coastal plain estuary, Marine Chemistry, v. 34, p. 225-246.

MARTIN, J. M. AND MEYBECK, M. (1979): Elemental mass balance of material carried by major world rivers. Marine Chemistry, v. 7, p. 173-206.

MEYBECK, M. (1982): Carbon, nitrogen and phosphorus transport by world rivers. American Journal of Science, v. 282, p. 401-450.

MILLIMAN, J. D., RUTKOWSKII, C. AND MEYBECK, M. (1995): River discharge to the sea—A global river index (GLORI). LOICZ Reports and Studies: LOICZ Core Project Office, Texel, The Netherlands, 126 p.

RAMANATHAN, A. L., VAITHIYANATHAN, P., SUBRAMANIAN, V. AND DAS, B. K. (1993): Geochemistry of the Cauvery estuary, east coast of India. Estuaries, v. 16, p. 459-474.

RAMANATHAN, A. L., VAITHIYANATHAN, P., SUBRAMANIAN, V. AND DAS, B. K. (1995): Nature and transport of solute load in the Cauvery River basin, India: Water Research, v. 28, p. 1585-1593.

RAMESH, R. AND SUBRAMANIAN, V. (1988): Nature of dissolved load of the Krishna River basin, India: Journal of Hydrology, v. 103, p. 139-155.

RAMESH, R., SUBRAMANIAN, V. AND PILLAI, K. C. (1988): Sediment and elemental accumulation rates in the Krishna River basin, India. The Science of Total Environment, v. 72, p. 221-226.

SOMAYAJULU, B. L. K., MARTIN, J. M., EISMA, D., THOMAS, A. J., BOROLE, D. V. AND RAO, K. S. (1993): Geochemical studies of the Godavari estuary, India: Marine Chemistry, v.43, p. 83-93.

SUBRAMANIAN, V. AND SINGH, S. (1998): Nature of chemical and sediment load from conference of large rivers—example from Ganges (abs.). *in* International Association of Hydrological Sciences Symposium on Sediment Load, Exeter, UK (July 1998), p. 45.

SUBRAMANIAN, V. (1997): Biogeochemistry of Western Ghat rivers of India (abs.) in European Geophysical Society IGBP-BAHC Session, Vienna, p. 292.

SUBRAMANIAN, V. (1996): The sediment load of Indian rivers: International Association of Hydrological Sciences, v. 236, p. 183-189.

SUBRAMANIAN, V. (1993): Phosphorus, silicon and some trace contaminants in Ganges estuary: Estuaries, v. 16, p. 453-458.

SUBRAMANIAN, V. (1980): Mineralogical input of suspended matter by Indian rivers into the adjacent areas of Indian Ocean: Marine Geology, v. 36, p. 29-34.

SUBRAMANIAN, V. (1979): Chemical and suspended sediment characteristics of rivers of India: Journal of Hydrology, v. 44, p. 37-55.

SUBRAMANIAN, V. AND RAMANATHAN, A. L. (1996): Nature of sediment load in the Ganga-Brahmaputra River system in India, *in* Milliman, J. D. and Haque, B. K., eds., Sea Level Rise and Coastal Subsidence: The Netherlands, Kluwer Academic Press, p. 225-245.

SUBRAMANIAN, V. AND ITTEKOT, V. (1991): Carbon transport in Himalayan rivers, *in* Eisma, D. and Kempe, S., eds., Biogeochemistry of World Rivers, SCOPE Hydrology: Oxford, U.K., Wiley and Sons, v. 42, p. 159-170.

Vaithiyanathan, P., Jha, P. K. and Subramanian, V. (1993): Phosphorus distribution in the sediments of Hooghly (Ganges) estuary, India: Estuarine, Coastal and Shelf Science, v. 37, p. 603-614.

Vaithiyanathan, P., Subramanian, V. and Ramanathan, A L. (1989): Transport and distribution of phosphorus by Indian rivers: Geological Society of India Memoirs, v. 13, p. 127-137.

THE IMPACT OF EARLY-DIAGENETIC ALUMINOPHOSPHATE PRECIPITATION ON THE OCEANIC PHOSPHORUS BUDGET

BIRGER RASMUSSEN*

Centre for Strategic Mineral Deposits, Department of Geology and Geophysics, the University of Western Australia, Nedlands, 6907, Perth, Australia

ABSTRACT: Early-diagenetic aluminophosphate minerals (mainly florencite) are ubiquitous in ancient (Archean to Cretaceous) marine-deposited sandstones. The crystals are <10 μm in diameter and are mostly associated with thin coatings or pockets of detrital clay particles lining quartz grain surfaces. Aluminophosphate crystals are also found in altered detrital aluminosilicate grains (e.g. feldspar).

The aluminophosphate minerals precipitated in sands deposited in shallow marine environments. Diagenetic textures and the presence of structural sulfate in the aluminophosphate minerals suggest that the authigenic crystals formed shortly after deposition, probably in the zones of sulfate reduction and methanogenesis. The aluminophosphate minerals formed as the concentrations of phosphate in the pore water increased and phosphate combined with various cations (e.g. calcium, barium, rare earth elements) to precipitate at sites of aluminum availability (e.g. detrital clay particles and aluminosilicate grains).

Although the aluminophosphate minerals are volumetrically minor constituents (<0.05 wt%), the authigenic crystals, which are present in >90% of samples studied, are spatially and temporally widespread in ancient Australian sandstones. Burial flux estimates show that phosphorus is removed at a rate of ~5.6 x 10^{-7} g/cm^2-yr. This value suggests that aluminophosphate precipitation was probably an important sink for oceanic phosphorus in the past, and possibly significant in present-day coastal environments. The discovery of the magnitude of the authigenic aluminophosphate sink has implications for models of the marine phosphorus cycle.

INTRODUCTION

Phosphorus is an essential nutrient for sustaining life in the oceans (Redfield, 1958; Holland, 1978). Many processes remove phosphorus from the marine cycle. Phosphorite deposition in coastal areas influenced by upwelling currents is generally considered to be a particularly important process volumetrically (Schuffert et al. 1994). Accordingly, most studies of authigenic phosphate minerals have focused on phosphorite deposits, where phosphorus concentrations are locally high (Burnett, 1977; Baturin, 1982; Froelich et al. 1983; Berner, 1990; Glenn et al. 1994; Föllmi, 1996). However, recent work on shallow-marine and shelf sediments show that significant amounts of phosphorus are also deposited in non-upwelling areas (Ruttenberg and Berner, 1993; Lucotte et al. 1994).

In areas influenced by upwelling currents and on shallow-marine shelves, the authigenic phosphate mineral most commonly reported is francolite or carbonate fluorapatite, $Ca_5(PO_4,CO_3,OH)_3F$, (Froelich et al. 1982; Jahnke et al. 1983; Ruttenberg and Berner, 1993). Other authigenic phosphate minerals have also been reported, including viviannite ($Fe_3(PO_4)_2{\bullet}8H_2O$), goyazite ($SrAl_3(PO_4)_2(OH)_5H_2O$), crandallite ($CaAl_3(PO_4)_2(OH)_5H_2O$), xenotime ($YPO_4$), and rhabdophane ((REE, Th)$PO_4{\bullet}nH_2O$) (Berner, 1980; Spötl, 1990; Milodowski and Zalasiewicz, 1991), however, they tend to be present in minute concentrations and spatially restricted. Although solubility data suggests that authigenic REE-phosphates should be abundant (Jonasson et al. 1985; Byrne and Kim, 1993), the literature suggests that such minerals are extremely rare in the sedimentary record.

Recently, authigenic REE-rich aluminophosphate minerals were found in shallow-marine sandstones of all ages from Australian sedimentary basins (Rasmussen, 1996). The minerals appear to be ubiquitous, but volumetrically minor, components of the sedimentary rocks. The aluminophosphates are present as early-diagenetic crystals. Florencite, (REE)$Al_3(PO_4)_2(OH)_6$, is the most abundant mineral type, although minor amounts of crandallite, $CaAl_3(PO_4)_2(OH)_5H_2O$; gorceixite, $BaAl_3(PO_4)_2(OH)_5H_2O$; and even smaller amounts of xenotime, YPO_4; and authigenic apatite, $Ca_5(PO_4)$(F,Cl,OH) are also present. It was proposed that these minerals were an important global sink for oceanic phosphorus in the marine cycle (Rasmussen, 1996). This study documents other occurrences of early-diagenetic aluminophosphate minerals (Table 1) and re-calculates the global flux of phosphorus out of the marine environment as a result of the precipitation of these minerals.

METHODS

Most of the samples studied were from subsurface cores; material from the Grampians Group, Warton Sandstone and Warrawoona Group was from outcrop. Polished thin-sections were examined with a polarizing microscope and a JEOL JSM-6400 Scanning Electron Microscope (SEM) equipped with a LINK Si(Li) PCXA EDS system for energy-dispersive X-ray microanalysis with 15-kV accelerating voltage and 0.3- to 3-nA-beam current. Florencite and xenotime were analyzed on an ARL SEMQ microprobe with 20-kV accelerating voltage and a 5-nA-beam current. Core chips (15–25 mm wide) were examined with a Philips 505 SEM equipped with an integrated EDAX PV9900 system. X-ray diffraction (XRD) analyses to determine major mineral compositions were carried out on a Philips PW1700 automatic diffractometer. The concentrations of authigenic phosphate minerals in the samples were calculated by the methods described by Rasmussen (1996).

*email: brasmuss@geol.uwa.edu.au

Marine Authigenesis: From Global to Microbial, SEPM Special Publication No. 66
 ISBN 1-56576-064-6, p. 89-101.

Table 1.—SEDIMENTARY UNITS WITH EARLY-DIAGENETIC PHOSPHATE MINERALS

Formation	Age	Location	Depositional Environment	Authigenic Phosphate Minerals			
				Florencite	Crandallite & Gorceixite	Xenotime	Apatite
Barrow Group	Early Cretaceous	Carnarvon Basin, W. A.	Deltaic to marine	present	n.o.[1]	present	present
Birdrong Sandstone	Early Cretaceous	Carnarvon Basin, W. A.	Shoreface to nearshore	n.o.	present	n.o.	n.o.
Flag Sandstone	Early Cretaceous	Carnarvon Basin, W. A.	Marine	present	n.o.	n.o.	present
Angel Formation	Early Jurassic	Carnarvon Basin, W. A.	Marine	present	n.o.	n.o.	present
Mungaroo Formation	Mid-Late Triassic	Carnarvon Basin, W. A.	Fluvial to shallow marine	present	n.o.	n.o.	present
Permo-Triassic Sandstones	L Permian-E Triassic	Perth Basin, W. A.	Fluvial to shallow marine	present	n.o.	present	present
Kennedy Group	Permian	Carnarvon Basin, W. A.	Shallow marine	present	present	present	n.o.
Keyling Formation	Early Permian	Bonaparte Basin, W. A.	Fluvial to marine	present	n.o.	present	n.o.
Grampians Group	Late Silurian	Grampians Basin, Vic.	Coastal	present	present	present	n.o.
Tumblagooda Group	L Cam-E Ordovician	Perth Basin, W. A.	Fluvial to shallow marine	present	present	n.o.	present
Goldwyer Formation	Mid-Ordovician	Canning Basin, W. A.	Marine	present	n.o.	n.o.	present
Emanuel Formation	Early Ordovician	Canning Basin, W. A.	Marine	present	n.o.	present	n.o.
Stirling Range Formation	Neoproterozoic	Albany-Fraser Belt	Shallow marine	present	n.o.	present	n.o.
Warton Sandstone	Paleoproterozoic	Kimberley Basin, W. A.	Shallow marine	present	present	present	present
Warrawoona Group	Archean	Pilbara Craton, W. A.	Shallow marine	present	n.o.	present	n.o.

[1] n.o. denotes not observed in this study

RESULTS

Authigenic Phosphate Minerals

Florencite ($(REE)Al_3(PO_4)_2(OH)_6$).—

Florencite is the most abundant of the authigenic phosphate minerals, and is commonly present in trace amounts throughout most of the formations examined (Table 2). The florencite occurs as clusters of minute (<10 μm), euhedral crystals, commonly lining detrital quartz surfaces along with a fine detrital clay (Figs. 1A, 1B). The clay forms a mesh of minute (1–20 μm) randomly oriented anhedral particles composed of Si-Al-K and Si-Al-K-Fe-Mg. Florencite crystals are also present within remnants of altered aluminosilicate grains (e.g. feldspar, muscovite; Figs. 1C, 1D) and amongst pore-filling, recrystallized clay (Fig. 1E).

Most florencite crystals are visible with the optical microscope under high magnification. The crystals are clear to pale yellow to pale green with moderate relief and low birefringence. Florencite crystals are rhombohedral and commonly have a "pseudocubic" habit; some have compositional zoning (Fig. 1F) with up to four distinct zones. The zoning appears to be due to variations in cation (i.e. Ca^{2+}, Ba^{2+}, and REE) and anion (i.e. PO_4^{3-} and SO_4^{2-}) composition, however, geochemical trends are impossible to distinguish due to the minute size of the florencite crystals.

In most samples, florencite is surrounded by diagenetic quartz cement. Locally, clusters of florencite crystals surround early-diagenetic pyrite and siderite, although in places, florencite is embedded within the centers of siderite rhombs (Figs. 2A, 2B). The embedded florencite crystals are smaller than others nearby (Fig. 2B).

Crandallite ($CaAl_3(PO_4)_2(OH)_5H_2O$) and gorceixite ($BaAl_3(PO_4)_2(OH)_5H_2O$).—

Authigenic crandallite and gorceixite were found in several samples as trace constituents (Tables 1 and 2); mostly as minute (<5 μm) rhombohedral crystals, isolated and in dense clusters. The authigenic aluminophosphates were mostly concentrated within pockets of detrital clay that lines detrital quartz surfaces, and within highly altered and leached aluminosilicate grains (e.g. feldspar, mica). Locally, the barium- and calcium-rich aluminophosphates are present within cleavage planes and fractures of partly dissolved feldspar grains and within the framework of highly altered feldspar grains (Fig. 2C). Most of the crystals are anhedral and less than 1 μm in width. Some of the larger crandallite and gorceixite crystals (up to 5 μm in width) display compositional zoning with substantial substitution between Ca^{2+}, Ba^{2+}, Sr^{2+}, and REE, as well as PO_4^{3-} and SO_4^{2-}.

Xenotime (YPO_4).—

Authigenic xenotime is present in most formations (Table 1), however, its abundance is significantly lower than that of the aluminophosphates and apatite (Table 2). Authigenic xenotime is mostly present as diagenetic overgrowths (in apparent optical continuity) on detrital zircon grains. The overgrowths range in size from minute (<0.5 μm) dentate serrations that line the rounded zircon surface, to large pyramidal prisms up to 12 μm long and 20 μm wide (Fig. 2D). Rare clusters of xenotime euhedra (0.1–2 μm) are present around detrital quartz grains and within clay matrix.

Xenotime overgrowths are composed of yttrium and phosphate with minor amounts of thorium, uranium, and REE. Some overgrowths are zoned due to variations in the proportion of uranium and thorium relative to REE. In some samples, xenotime overgrowths are partly engulfed by early-diagenetic siderite, titanium dioxides, cryptocrystalline silica, or quartz cement, suggesting that xenotime crystals probably formed shortly after burial.

Apatite ($Ca_5(PO_4,CO_3,OH)_3F$).—

Authigenic apatite is generally less abundant than the REE-aluminophosphates, although in one core, apatite is more abundant than florencite (Dangara No. 24, Table 2). Authigenic apatite is present as isolated crystals and clusters of minute (<10 μm) crystals in voids in partly dissolved grains of feldspar and mica. In some samples, a thin lining (<1 μm thick) of apatite covers detrital quartz surfaces, and is surrounded by quartz and early-diagenetic siderite cement. Locally, two linings of apatite crystals are separated by quartz and siderite cement.

DISCUSSION

Formation of Phosphate Minerals

The sediment pore-water from which the phosphate minerals precipitated must have been locally enriched in REE, Ca^{2+}, Ba^{2+} and PO_4^{3-}. Calcium and barium ions are relatively abundant ocean water, but concentrations of REE and PO_4^{3-} are very low (Turekian, 1969). For example, the concentration of REE in the ocean ranges from 1 to 6.5 parts per trillion (Elderfield and Greaves, 1982). Several processes are apparently responsible for the low concentrations: adsorption of REE onto ferromanganese nodules (Elderfield et al. 1981a), hydrous iron-oxide particles (Piper, 1974b; Elderfield et al. 1981b; German et al. 1990), and detrital clay aggregates (Piper, 1974a; Dypvik and Brunfelt, 1979); formation of inorganic complexes (e.g. carbonates and phosphates)(Wood, 1990; Byrne et al. 1991), organic adsorption, halmyrolytic reactions between volcanic material and seawater (Fleet, 1984), incorporation into biogenic material (Wright et al. 1987), and incorporation into minerals (e.g. glauconite, phosphorites) by precipitation or fixation (Goldschmidt, 1954; Altschuler 1980).

The concentration of phosphorus in seawater is also very low (about 0.6 parts per million; Krauskopf, 1977), with most phosphorus entering the oceans being removed in the form of organic phosphorus incorporated into phytoplankton in the euphotic zone of the water column. Seawater phosphorus is also removed by adsorption onto hydrous iron-oxide particles and clay minerals, by incorporation into skeletal material and biogenic calcium carbonate, and by mineral precipitation (Berner, 1973; Froelich et al. 1982; Berner et al. 1993).

However, the concentration of REE and phosphorus in pore

TABLE 2.—CONCENTRATION OF EARLY-DIAGENETIC PHOSPHATE MINERALS IN MARINE SANDSTONES

Formation	Well Name/ Outcrop	Concentration in Weight %			
		Florencite	Crandallite & Gorceixite	Xenotime	Apatite
Barrow Group	Flinders Shoal No. 1	0.00118	0	0	0.00004
	Roller No. 1	0.01426	0	0	0
	Saladin No. 7	0.00016	0	0	0
	Skate No. 1	0.00042	0	0	0
Birdrong Sandstone	Rough Range No. 1[1]	0	0.04594	0	0
Flag Sandstone	Harriet No. 1	0.00028	0	0	0.00014
	Tanami No. 1	0	0	0	0
Angel Formation	Cossack No. 1	0.0039	0	0	0.00006
	Wanaea No. 3	0.00083	0	0	0
Mungaroo Formation	North Gorgon No. 1	0.00968	0	0	0.00018
	Goodwyn No. 6	0.00537	0	0	0.00154
Permo-Triassic Sandstones	Dongara No. 4[2]	0.032	0	0	0.00006
	Dongara No. 24[2]	0.025	0	0.00016	0.067
	Mondarra No. 1[1]	0.01344	0	0	0
	Mondarra No. 3[2]	0.0318	0	0.00027	0
	Mount Adam No. 1[2]	0.017	0	0.0006	0.0003
	Strawberry Hill No. 1[2]	0.0055	0	0	0
	West White Point No. 1[2]	0.0011	0	0	0.0005
	Yardarino No. 1[2]	0.0178	0	0	0.00088
Kennedy Group	Dillson No. 1[2]	0.144	0	0	0
	Fennel No. 1	0.00154	0.02134	0	0
Grampians Group	Outcrop[1]	0.0007	0.1175	0.0002	0
Tumblagooda Group	Wandagee No. 1[2]	0.033	0.01085	0	0.00043
Goldwyer Formation	Acacia No. 1	0.001	0	0	0.00034
Emanuel Formation	PHD 1	0.0058	0	0	0
Warton Sandstone	Outcrop	0.03573	0	0.00152	0.00007
	Outcrop	0.00687	0.00393	0	0.00006
Warrawoona Group	Outcrop	0.0005	0	0.0007	0
Average Concentration		0.014	0.007	0.00012	0.0025

[1]Weight percent calculation is based on new sample data and data reported in Rasmussen (1996).
[2]From Rasmussen (1996)

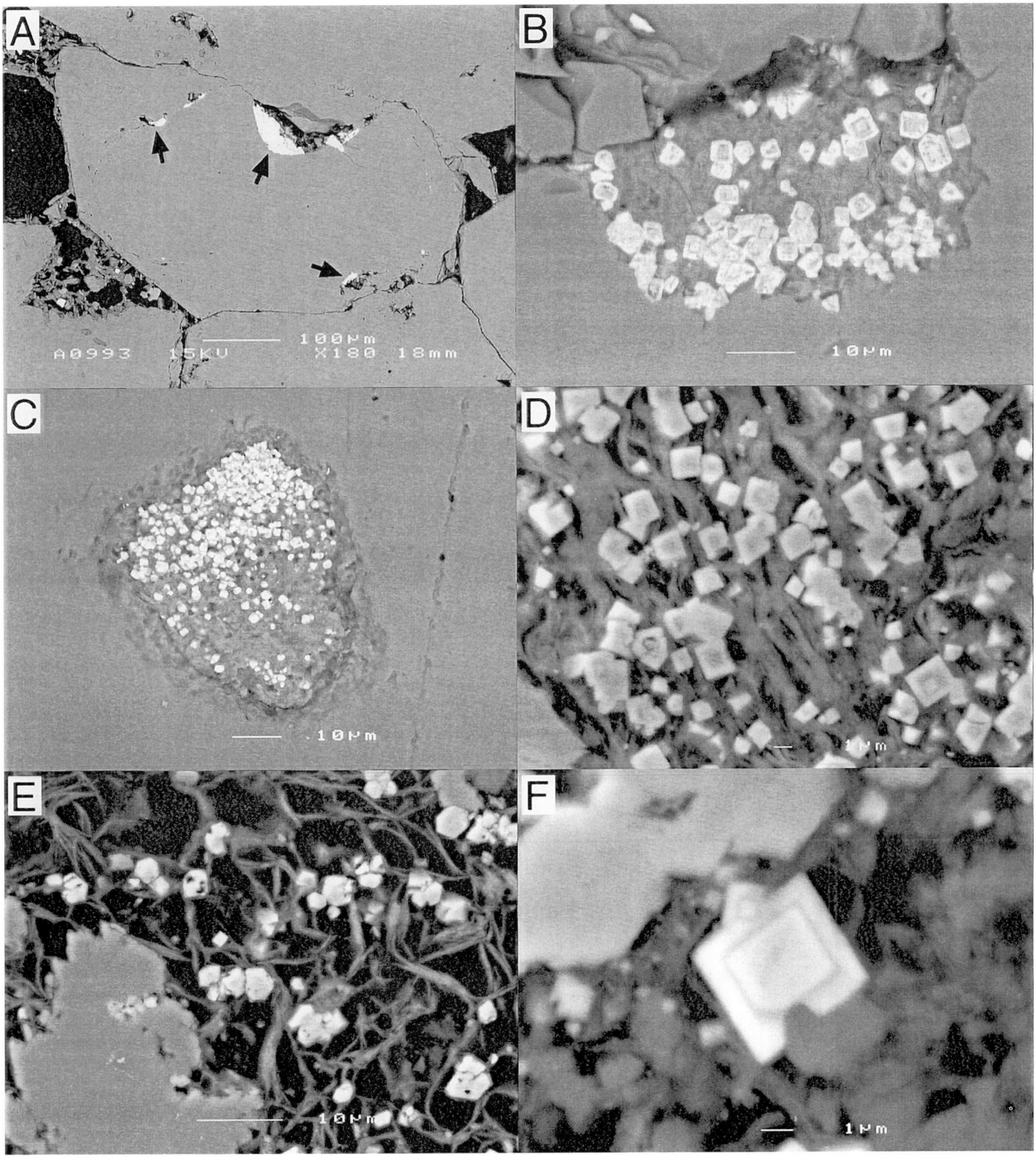

FIG. 1.— (A) Clusters of florencite crystals (white, see arrows) and aggregates of detrital clay lining detrital quartz surfaces. The florencite and clay are partly engulfed by quartz cement. Back-scattered electron (BSE) image, Permo-Triassic sandstone. West White Point No. 1, 2182.4 m. (B) Small dense cluster of florencite crystals with compositional zoning within a pocket of detrital clay. The detrital clay has prevented the full development of syntaxial quartz overgrowth. BSE image, Permo-Triassic sandstone. West White Point No. 1, 2182.4 m. (C) A dense cluster of florencite crystals within an altered aluminosilicate grain that is engulfed by quartz cement. BSE image, Permo-Triassic sandstones. Mondarra No. 1, 2708.5 m. (D) Close-up of euhedral florencite crystals within an altered aluminosilicate grain. Note the compositional zoning of most florencite crystals. BSE image, Kennedy Group. Dillson No. 1, 2395.7 m. (E) Florencite crystals (white) within pore-filling authigenic clay (Si-Al-K with minor Fe and Mg). BSE image, Emanuel Formation. PHD No. 1, 578.2 m. (F) Close-up of two intergrown florencite crystals with compositional zoning. The florencite is surrounded by detrital clay (dark gray) and siderite (gray, top left). BSE image, Kennedy Group. Dillson No. 1, 2395.7 m.

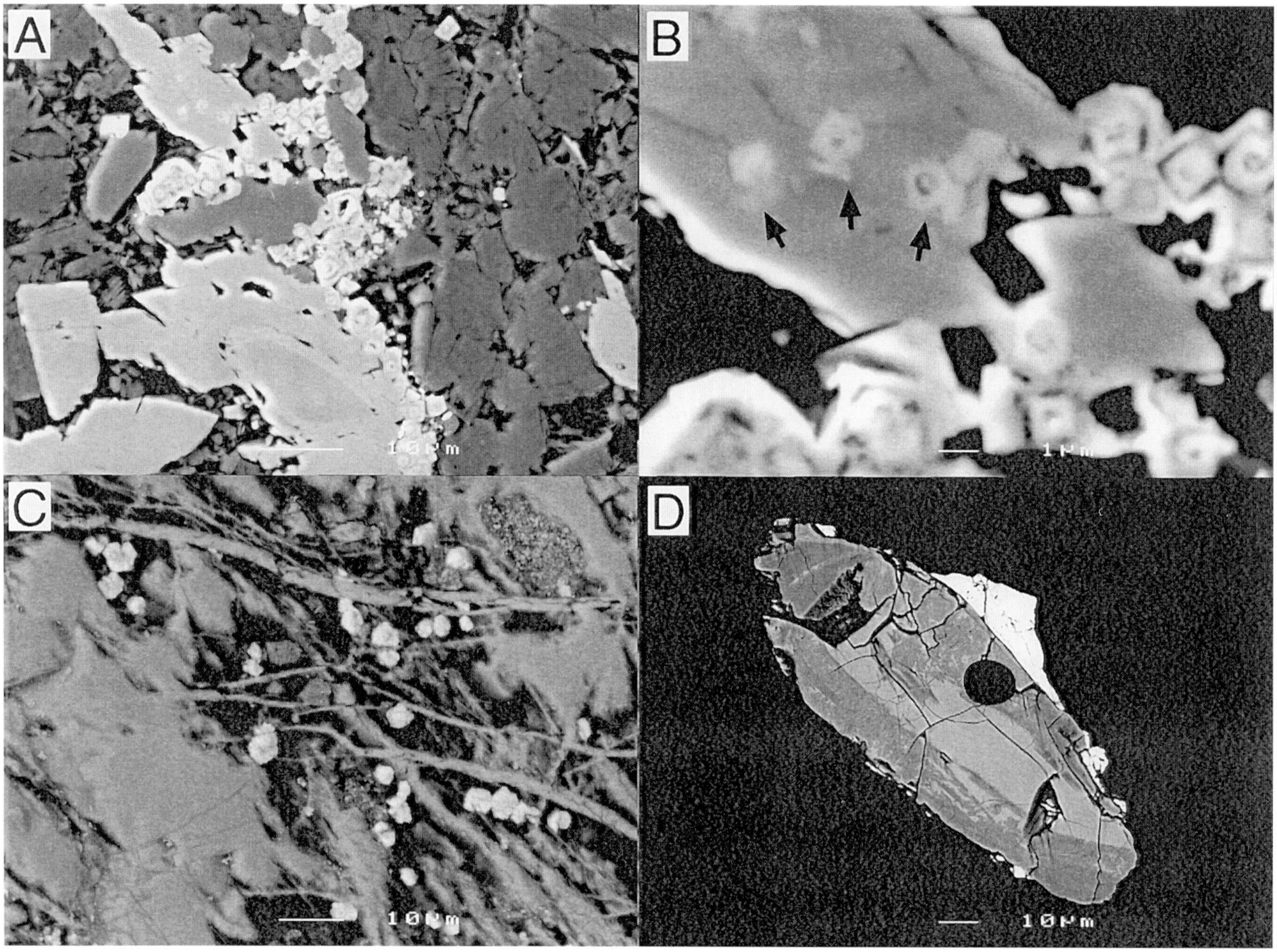

FIG. 2.— (A) A cluster of minute zoned florencite crystals (light gray) adjacent to siderite crystals (gray). The dark gray "groundmass" comprises authigenic kaolinite and quartz. BSE image, Kennedy Group. Dillson No. 1, 2395.7 m. (B) Close-up of BSE image in Figure 2A showing florencite crystals (light gray, see arrows) within early-diagenetic siderite cement (gray). BSE image, Kennedy Group. Dillson No. 1, 2395.7 m. (C) Numerous minute crandallite and gorceixite crystals (white) within an altered feldspar grain. BSE image, Grampians Group. (D) A detrital zircon grain (dark gray) with a "pyramidal" overgrowth of xenotime (white). BSE image, Warton Sandstone.

waters of coastal sediments may be greatly increased during early diagenesis. For instance, REE concentrations immediately below the sediment-water interface may be 10–50 times that of seawater (Elderfield and Sholkovitz, 1987; Sholkovitz et al. 1989). The high concentrations appear to be due to anoxic reduction of iron- and manganese-oxides, in either colloidal suspensions or surface coatings, and the subsequent release of adsorbed REE. The REE concentration may also be increased by the release of REE adsorbed onto clay mineral surfaces (Balashov and Girin, 1969; Roaldset and Rosenqvist, 1971) by the partial dissolution and surface reactions of REE-bearing minerals (e.g. monazite, apatite), and by early-diagenetic dissolution of fish debris. Similarly, the concentration of dissolved phosphate in pore water may also be locally enriched compared to seawater (Sholkovitz, 1973; Martens et al. 1978) as phosphorus is released by microbial decomposition of organic compounds (Berner, 1980; Burnett, 1977; Baturin and Bezrukov, 1979). Phosphorus adsorbed onto the surfaces of clay and iron oxyhydroxide particles may also be re-released shortly after burial during mineral transformation and anoxic reduction, respectively (Froelich et al. 1977; Krom and Berner, 1981; Borgaard, 1983; Watanabe and Tsunogai, 1984).

The sites of aluminophosphate precipitation are aggregates of detrital clay, as well as on feldspar and mica grains, which probably were sources of aluminum (Fig. 3). The aluminum was released during aluminosilicate transformation and alteration within the marine pore water. Because the solubility of aluminum is extremely low in low-temperature aqueous fluids (Garrels and Christ, 1965), the site of aluminum release (e.g. the clay mineral substrate) was probably also the nucleation site, where the aluminum combined with various ions (i.e. REE, Ca^{2+}, Ba^{2+}, PO_4^{3-} and SO_4^{2-}) in the pore water to precipitate as aluminophosphate crystals.

The main source of REE in the aluminophosphates was probably detrital clay, which commonly lines detrital quartz surfaces and locally fills grooves and indentations in quartz grains.

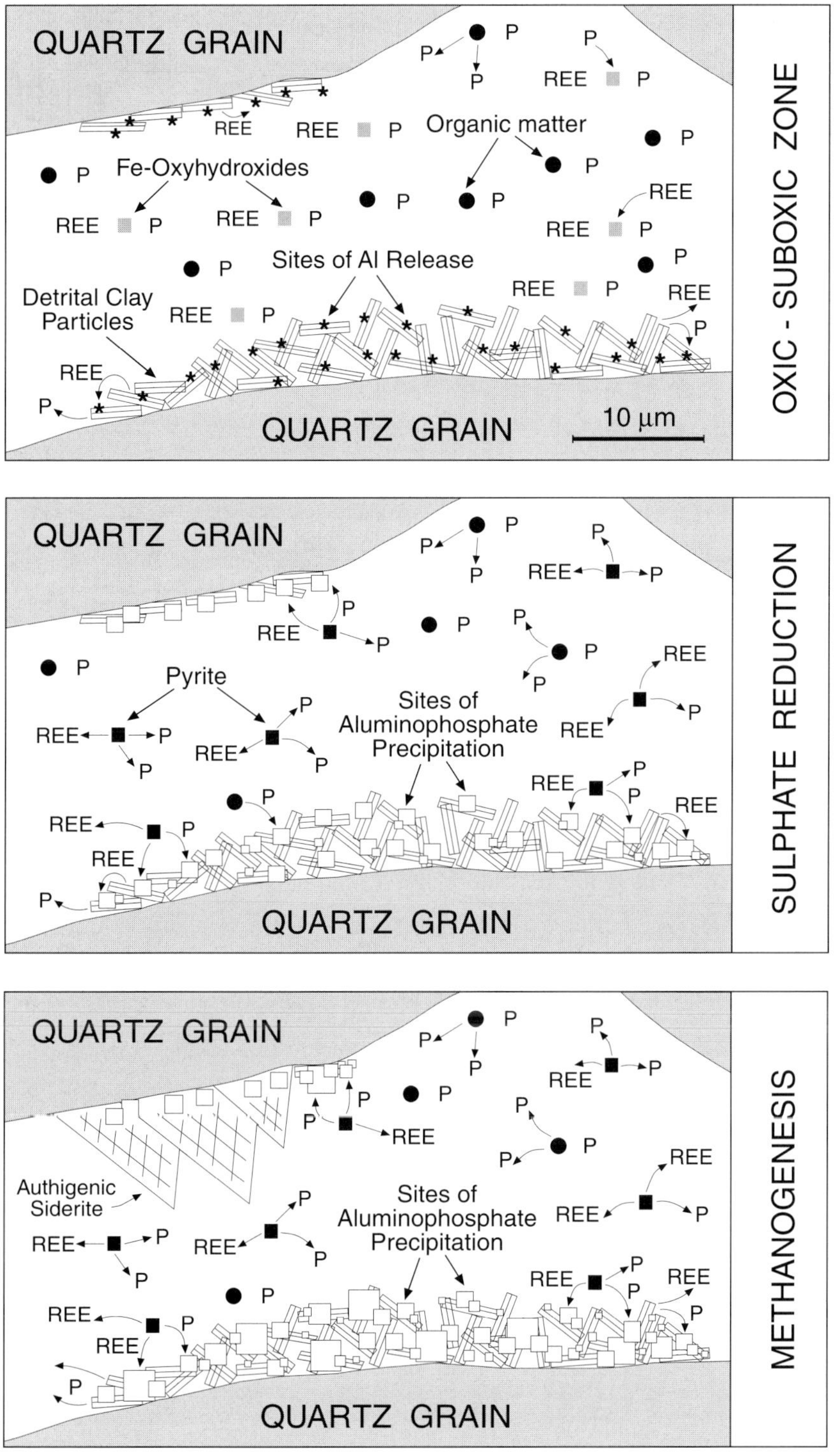

In the oxic zone, organic matter is decomposed by organisms using oxygen dissolved in the pore water, releasing carbon dioxide, water, nitrate and phosphate. The released pore-water phosphate, as well as phosphate migrating from deeper layers, may in part be adsorbed onto the surfaces of Fe-oxyhydroxides. The REE released in underlying geochemical zones and from sea water may also be captured and adsorbed onto the surfaces of Fe-oxyhydroxides. Detrital clay particles undergo transformation and alteration in the marine pore water, releasing adsorbed REE and P, and producing sites of Al release.

In the zone of sulphate reduction, organic matter is decomposed by bacteria using pore water sulphate as an electron acceptor, producing hydrogen disulphide and phosphate, as well as ammonia, carbon dioxide and water. Iron and manganese oxyhydroxides are reduced (releasing REE and P), combining with hydrogen disulphide to form pyrite. REE and P adsorbed onto detrital clay surfaces are released during clay transformation in the marine pore water, and locally nucleate and precipitate on substrates releasing Al (e.g. detrital clay particles, feldspar).

Decomposition of organic matter continues within the zone of methanogenesis, with the release of phosphate, as well as methane and carbon dioxide. If some Fe-oxyhydroxides remain after the depletion of sulphate in the zone bacterial sulphate reduction, then siderite or Fe-rich carbonates may form. In the process, adsorbed P and REE bound to the Fe-oxyhydroxides may be released. Aluminophosphate precipitation and crystal growth continues with the ongoing transformation of detrital clay (releasing adsorbed P and REE) and dissolution and alteration of detrital aluminosilicate grains (releasing Al).

FIG. 3— Schematic diagram showing the main geochemical processes operating below the sediment-seawater interface, resulting in the release of phosphate, rare earth elements, and aluminum, and the precipitation of aluminophosphate minerals.

Iron oxyhydroxides were probably also an important source. REE are commonly adsorbed to and consequently released from surfaces of iron oxyhydroxides during suboxic and/or anoxic reduction. The common occurrence of early-diagenetic pyrite and siderite in most samples attests to the relatively high concentration of iron (probably in the form of oxides/hydroxides) prior to its incorporation into pyrite and siderite.

As mentioned above, the most likely source of phosphorus in the sandstones was organic matter which releases phosphorus during bacterial decomposition. Once organic matter reaches the sediment-seawater interface, it is decomposed by aerobic organisms using dissolved oxygen and releasing various nutrients, including phosphate (see equation below).

$$(CH_2O)_{106}(NH_3)_{16}(H_3PO_4) + 138O_2 \rightarrow 106CO_2 + 122H_2O + 16HNO_3 + H_3PO_4$$

With burial below the oxic zone, phosphate is released from organic matter as it is decomposed by organisms using progressively less efficient oxidants (O_2, NO_3^-, MnO_2, Fe_2O_3, SO_4^{2-}, CO_2) (Froelich et al. 1979). In shallow-marine environments, where sedimentation rates and primary productivity are high, organic matter decomposition by organisms using SO_4^{2-} is dominant (Jørgensen,1982; Canfield et al. 1993). Iron oxyhydroxide particles are another important source of phosphorus (Krom and Berner, 1981). Adsorbed phosphorus is released from iron-oxyhydroxide particles during anoxic reduction. Iron oxyhydroxides in shelf sediments may be present in quantities similar to those of organic phosphorus (Ruttenberg, 1993). As mentioned above, the widespread occurrence of early diagenetic pyrite in the sandstones studied suggests that iron oxyhydroxides were prevalent shortly after burial and before sulfidization.

Timing of Phosphate Mineral Precipitation

Phosphate minerals precipitate in three geochemical zones below the sediment-seawater interface: the sub-oxic zone; the zone of sulfate reduction (associated with the formation of pyrite); and the zone of microbial methanogenesis (commonly associated with siderite precipitation) (Benmore et al. 1983; Coleman, 1985; Morad and Al-Aasm, 1994). In the sandstones studied here, aluminophosphate crystals co-exist with early-diagenetic siderite, and locally, are engulfed by carbonate cement (Figs. 2A, 2B). This indicates that phosphate precipitation predated and was contemporaneous with siderite formation. The presence of significant amounts of SO_4^{2-} (several wt%) in the structure of some of the aluminophosphate crystals suggests that sulfate was still available in the sediment pore-water during phosphate precipitation. Thus, the phosphates probably started forming within the zone of sulfate reduction and continued forming through the zone of microbial methanogenesis.

Further indirect evidence for the timing of aluminophosphate precipitation is their crystal size (<10 μm; mostly <3 μm) which is similar to the size of apatite particles (one to several microns) found in modern phosphorites (Glenn and Arthur, 1988). Crystal growth experiments carried out in modified seawater solutions suggest that fluorapatite crystals that nucleate and precipitate near the sediment-water interface should be restricted in size to 0.1–10 μm during burial (Van Cappellen and Berner, 1991).

Phosphorus Burial Flux

The oceanic flux of an element is commonly calculated assuming that the ocean is in a chemical steady state, with the rate of addition of the element (input) balanced by its rate of removal (output). The major phosphorus input is river water (Froelich et al. 1982; Meybeck, 1982); atmospheric input accounts for less than 10% (Graham and Duce, 1979). For phosphorus, the natural fluvial input is difficult to quantify due to the effect of human activity on river nutrient levels (Meybeck, 1982). Furthermore, it is unknown what proportion of phosphorus eroded from continents is reactive in river water and how phosphorus behaves when the river water mixes with ocean water (Froelich et al. 1982). It is therefore simpler to calculate the phosphorus output, i.e. phosphorus removed from seawater through the deposition and burial of phosphorus-bearing components, than it is to calculate the phosphorus input. The burial flux of phosphorus has been estimated in several studies (Froelich et al. 1982; Ruttenberg and Berner, 1993; Wheat et al. 1996; Filippelli and Delaney, 1996; *see Table 3*). In the study by Froelich et al. (1982), the phosphorus burial flux was calculated for sediments from continental margins influenced by upwelling currents and for deep-sea sediments. The reactive phosphorus burial flux was estimated to be about 3.8×10^{10} mol/yr and the main phosphorus sinks identified were organic matter and biogenic carbonate (Froelich et al. 1982). The later work by Ruttenberg and Berner (1993) showed that along continental margins not affected by upwelling currents the phosphorus output flux is about 18.5×10^{10} moles/yr which is 5 times greater than the output flux in upwelling areas. The higher output flux was mostly due to additional removal of non-volcanogenic iron-bound phosphorus in deltaic environments and burial of carbonate fluorapatite, biogenic apatite, and biogenic carbonate in shelf environments (Ruttenberg, 1993). More recently, Wheat et al. (1996) showed that hydrothermal processes might also be important in removing phosphorus from seawater. Although it is not known which mineral is the main sink (e.g. apatite, carbonate, iron oxides), a large proportion of seawater phosphorus (an amount equivalent to 5–50% of the riverine flux) is thought to be lost during seawater circulation through the oceanic crust.

A study by Filippelli and Delaney (1996) using sequential extraction techniques showed that deep-sea sediment is also an important phosphorus sink. In particular, their study suggests that the geochemical processes operating in continental margin sediments also operate in deep-sea sediments and that carbonate fluorapatite (CFA) is a major phosphorus sink. The authors calculated a global phosphorus burial flux of 21×10^{10} moles/yr by extrapolating the results from their study area to the whole ocean.

In the work by Ruttenberg and Berner (1993) and Filippelli and Delaney (1996), authigenic CFA was identified as a particularly important sink for reactive phosphorus. Ruttenberg

(1993) estimated that about 50% of phosphorus in the ocean is lost through authigenic mineral precipitation. Laboratory experiments by Jonasson et al. (1985) have shown that hydrous REE-phosphates (e.g. rhabdophane and xenotime) should also form in seawater. However, their presence in marine sediments has not been reported in the literature. Recently though, the experimental observations were, in fact, confirmed with the identification of early-diagenetic REE-phosphates (florencite, crandallite, gorceixite, and xenotime) in the majority of marine sandstones studied by Rasmussen (1996).

The apparent absence of REE-phosphates in previous studies may be related to the analytical techniques used to separate the various phosphorus-bearing phases. Many workers use sequential-extraction techniques with various solvents at low temperatures (Lucotte and d'Anglejan, 1985; Ruttenberg, 1992; Filippelli and Delaney, 1996). However, aluminophosphates and xenotime are extremely insoluble at low temperatures, even in strong acids (Norrish, 1968; Sarma and Krishna Murti, 1969). Dissolution of these minerals requires concentrated hydrofluoric acid at temperatures >400°C (Triplehorn and Bohor, 1983). Therefore, sequential-extraction techniques probably did not permit identification of the phosphorus tied up in aluminophosphate minerals as "authigenic phosphorus". It is likely that the REE-phosphates were included with the less soluble fractions—"detrital phosphorus" or "organic phosphorus" or, most likely, that they passed through the extraction cycle undetected. If so, estimates of the proportion of reactive phosphorus lost from seawater through authigenic mineral precipitation may well have been too low.

In the marine sandstones studied here, diagenetic aluminophosphate minerals and xenotime are generally more abundant than authigenic apatite. However, the concentration of apatite in the <0.1-μm-size-fraction, which is difficult to detect using microscopic techniques, may be greater. It is also difficult to determine what proportion of phosphorus was incorporated into the sands as organic phosphorus, calcareous phosphorus, iron-bound phosphorus, and loosely sorbed phosphorus, all of which may have been remobilized during burial as they were subjected to increasing pressure and temperature conditions and changing pore-water chemistry. Hence, direct comparison of the sequential extraction results of Ruttenberg and Berner (1993) and the results of this study is not possible. However, in subsequent calculations, it will be assumed that the sequential extraction technique correctly identified organic phosphorus, calcareous phosphorus, iron-bound phosphorus, loosely sorbed phosphorus, and authigenic apatite phosphorus, and failed to detect REE-phosphate phosphorus. Although some REE-phosphate phosphorus may have been counted twice, that amount was probably offset by the aluminophosphate grains and detrital matrix lost during polished thin-section preparation and the aluminophosphate grains smaller than 0.1 μm that are difficult to detect with microscopic techniques. Therefore, the flux of REE-phosphate was, if anything, underestimated by this method.

The amount of reactive phosphorus removed through authigenic phosphate precipitation (in moles per year) was calculated for each aluminophosphate mineral (ALP) by the following equation:

$$\text{Burial Flux} = \text{SR x (wt\%}_{(ALP)}/100 \text{ x SD)/MW}_{(ALP)}) \text{ x } M_{(ALP)} \text{ x } MS_{(ALP)} \text{ x } A_{(Ocean)}$$

Where SR = sedimentation rate; SD = sand density; $MW_{(ALP)}$ =

TABLE 3.—BURIAL FLUX OF OCEANIC PHOSPHORUS—COMPARISON WITH PREVIOUS STUDIES

	Phosphorus Burial Rate (10^{10} moles/yr)									
Phosphorus Sinks	**Froelich et al. (1982)**	**Ruttenberg (1993)**		**Rasmussen (1996)**		**Wheat et al. (1996)**	**Filippelli & Delaney (1996)**	**This study**	
Organic-P	1.5	4.1		4.1^2		4.1^2		4.1^2	
$CaCO_3$-P	1.5	–		–		–		–	
Fe-Oxyhydroxide-P	0.4	4.0	(0.4)	4.0^2	$(0.4)^2$	–		4.0^2	$(0.4)^2$
Loosely Sorbed-P	–	1.3		1.3^2		1.3^2		1.3^2	
Fish Debris	<0.07	–		–		–		–	
Authigenic Minerals									
– Apatite	<0.4	9.1^1	(2.2)	$9.1^{1,2}$	$(2.2)^{1,2}$	$2.2^{1,2}$		$9.1^{1,2}$	$(2.2)^{1,2}$
– REE Phosphates	–	–		7.56^3		–		6.56^3	
Hydrothermal	–	–		–		1.43		–	
Total Burial Flux	**3.8**	**18.5**	**(8.0)**	**26.06**	**(15.56)**	**9.03**	**21**	**25.06**	**(14.56)**

[1]In Ruttenberg (1993) "Apatite" includes Calcareous-P, biogenic apatite, and authigenic apatite. The values in parenthesis are minimum estimates based on a more restrictive definition of reactive phosphorus. See Ruttenberg (1993) for more detail.
[2]From Ruttenberg (1993).
[3]This value does not include authigenic apatite.

molecular weight; $M_{(ALP)}$ = moles phosphorus/mole ALP; $MS_{(ALP)}$ = the proportion of marine sandstones containing diagenetic ALP; and $A_{(Ocean)}$ = area. Sedimentation rates (SR) in coastal environments, the dominant sites of sand deposition, are generally quite high (>0.3 cm/yr; Gibbs, 1981; Ingall and Van Cappellen, 1990; Alexander et al. 1991); however, a more conservative value (0.1 cm/yr) was used in these calculations. The average concentration of each ALP mineral is shown in Table 2 (e.g. the average concentration for florencite is 0.014 wt%). A sediment grain density of 2.65 g/cm^3 and an initial sand porosity of 40% (Atkins and McBride, 1992) were also assumed, giving a sand density (SD) of about 1.59 g/cm^3. The number of moles of phosphorus in one mole of authigenic phosphate mineral is represented by $M_{(ALP)}$. The estimated proportion of marine sandstones with diagenetic phosphates ($MS_{(ALP)}$), about 13%, was calculated from the following:

1. the abundance of sandstone in the Phanerozoic sedimentary rock record which, by various estimates, range from 14% to 40% (Kuenen, 1941; Krynine, 1948; Ronov and Migdisov, 1971; Holland, 1984); with an average between 20.1% and 34% (Poldervaart, 1955; Horn and Adams, 1966; Ronov, 1982); 20.1% was used here;
2. the proportion of marine sedimentary rocks deposited on continental plates during the Phanerozoic, approximately 70% (Ronov, 1982); and
3. the proportion of marine-deposited sandstones that contain measurable concentrations of authigenic phosphate minerals (about 92% from this study).

To give global flux values in moles phosphorus per year, the estimated burial rate in moles/cm^2-yr was multiplied by the area of the ocean floor, $A_{(Ocean)}$,which is 3.61 x 10^{18} cm^2 (Chester, 1990).

Results from this study (Table 4) show that the phosphorus burial flux (excluding the apatite burial flux) is about 6.56 x 10^{10} moles/yr. The rate calculated in Rasmussen (1996) was 7.56 x 10^{10} moles/yr. The lower phosphorus burial flux rate in this study is due to a lower average concentration of aluminophosphates in the sandstones examined. For example, the average florencite concentration, which was 0.014 in this study, was only 0.038 in the previous study. However, the lower concentration was partly compensated for by the more widespread occurrence of the minerals in this study (in 92% of samples) compared to the previous study in which they were present in only 62% of samples. Nevertheless, the magnitude of the lower rate (6.56 x 10^{10} moles/yr) is similar to that of the authigenic phosphorus (i.e. CFA) sink (between 2.2 and 9.1 x 10^{10} moles/yr) reported by Ruttenberg (1993) and confirms that a significant quantity of phosphorus is removed from the ocean through REE-phosphate precipitation (Table 3).

TABLE 4.—ESTIMATED BURIAL RATES OF PHOSPHORUS IN PHOSPHATE MINERALS

Phosphate Minerals	Phosphorus Burial Flux	
	10^{-7} g/cm^2-yr)	10^{10} moles/yr)
Xenotime	0.04	0.045
Florencite	3.65	4.25
Crandallite & Gorceixite	1.95	2.27
Apatite	1.0	1.15
Total Output	**6.64**	**7.715**

Implications

In order to maintain mass balance of the oceanic phosphorus cycle model, the increase in burial flux attributable to REE-phosphate precipitation must be compensated for by a decrease in output to another sink or by greater input. Recent work on the behavior of phosphorus in river and estuarine waters (Fox et al. 1986; Froelich, 1988) shows that the quantity of reactive phosphorus carried by suspended material (i.e. adsorbed onto hydrous iron and aluminum oxides and associated with organic material) is as much as three times greater than that carried in solution (Rao and Berner, 1993; Berner and Rao, 1994). Extrapolating results from a study of the Amazon River, Berner and Rao (1994) estimated a global solubilized sediment phosphorus flux of 3.5–6 x 10^{10} moles/yr, which translates to a total pre-human phosphorus flux to the oceans of 5–8.5 x 10^{10} moles/yr (after including the natural dissolved phosphorus flux of about 1.5–2.5 x 10^{10} moles/yr; Froelich et al. 1982).

Dissolved elements may also be delivered to the ocean by submarine groundwater discharge, which is estimated locally to be about 40% of the river-water flux (Moore, 1996). If this magnitude of groundwater flux proves to be representative of most coastlines, then the groundwater input may significantly increase the terrestrial flux of dissolved material to the oceans. Assuming a river phosphorus flux of 5–8.5 x 10^{10} moles/yr, then the input of phosphorus through groundwater discharge may be as high as 2–3.4 x 10^{10} moles/yr. However, this potential source of phosphorus remains to be quantitatively tested.

Although there is uncertainty in the estimates of the amount of phosphorus entering the oceans, the total from the various sources is significantly less than the modified phosphorus output (14.5–25 x 10^{10} moles/yr) which suggests that the amount of reactive phosphorus entering the oceans may be significantly higher or that the rate of removal is too high.

Perhaps more important, is the impact of a significant new phosphorus sink (i.e. REE-phosphate precipitation) on our understanding of the carbon cycle, and thereby, of the Earth's climate. Phosphorus is an essential nutrient in sustaining life and is apparently important in controlling primary productivity in the oceans (Holland, 1978). Its concentration in the ocean is thought, by some (e.g. Redfield, 1958; Broecker, 1982), to limit the growth of phytoplankton, which remove CO_2 from surface waters and the atmosphere through photosynthesis. More recently, Falkowski (1997) argued that over much of geological time, nitrogen was more important than phosphorus in limiting primary productivity. However, it seems more likely that nitrogen and phosphorus both play important, and probably interlinked, roles in phytoplankton productivity, with resultant feedback with atmospheric CO_2.

Because atmospheric CO_2 plays an important role in regulating Earth's surface temperature, and has a major impact on climate variation (Berner and Berner, 1987), changes in the volume of phosphorus removed from the oceans and incorporated in relatively inert authigenic minerals, have important feedback relationships with the carbon cycle. Results from this study show that early-diagenetic REE-phosphate minerals, which are ubiquitous in marine sandstones, imply a shorter phosphorus residence time in the ocean. The oceanic residence time can therefore have an impact on the abundance and distribution of marine life and the capacity of the ocean to absorb CO_2, thereby influencing global climates.

CONCLUSIONS

The examination of sandstone units not previously studied has shown that aluminophosphate minerals (i.e. florencite, crandallite, and gorceixite) and xenotime crystals are even more widespread than reported in Rasmussen (1996), but are present in smaller amounts (average 0.014 wt% for florencite). The aluminophosphates are minute euhedral crystals within detrital clay aggregates and within altered aluminosilicate grains, whereas xenotime occurs as syntaxial overgrowths on detrital zircon grains. In the past, the minute crystal size (<10 μm) and extremely low concentrations of the phosphate minerals have probably hindered identification by microscopic techniques, and their insolubility at low temperatures may have prevented their separation and recognition by analytical methods using acid digestion. It is therefore suggested that the REE-phosphates are probably globally widespread in marine-deposited sandstones.

The phosphate minerals in the sandstones studied precipitated shortly after burial, within the zones of sulfate reduction and microbial methanogenesis. The aluminophosphate crystals nucleated on detrital clay and other aluminosilicate particles undergoing transformation in the marine pore water. The detrital clay and aluminosilicate particles were sources of aluminum; phosphate was derived from the decay of organic matter and the release of adsorbed phosphorus from iron oxyhydroxide particles.

The phosphate minerals are present in most (>90%) of the sandstones examined. Estimates of the burial flux of reactive phosphorus by aluminophosphate and xenotime precipitation (6.56×10^{10} moles/yr) is comparable to the combined burial flux of authigenic carbonate fluorapatite, calcareous-phosphorus, and biogenic apatite which is between 2.2×10^{10} and 9.1×10^{10} moles/yr). The magnitude of the REE-phosphate sink has implications for the residence time of phosphorus in the oceans, and through nutrient modulation of the photosynthetic biomass, may also have an impact on oceanic and atmospheric CO_2 levels and hence on the global climate.

ACKNOWLEDGMENTS

The Geological Survey of Western Australia, West Australian Petroleum Pty. Ltd., and Western Mining Corporation (Petroleum Division) Ltd. are acknowledged for providing core material. Samples from the Grampians Group, Warton Sandstone, and Warrawoona Group were provided by Annette George, Rob Ramsay, and Roger Buick, respectively. Roger Buick and Bob Gilkes are thanked for valuable discussions. Karl-Heinz Wyrwohl is thanked for critically reviewing the manuscript. At the Centre for Microscopy and Microanalysis (at UWA), I acknowledge Greg Pooley, Brendon Griffin, and John Hillyer for technical support, and Andrew Johnson and Brendon Griffin (of the same Centre) for access to microscopy and microanalysis equipment. The manuscript benefited from the thoughtful reviews of Gabriel Filippelli, Liliane Prévôt-Lucas, Jacques Lucas, and an anonymous reviewer. This work was carried out while in receipt of an ARC Fellowship.

REFERENCES

Alexander, C. R., Demaster, D. J., and Nittrouer, C. A., 1991, Sediment accumulation in a modern epicontinental-shelf setting: The Yellow Sea: Marine Geology, v. 98, p. 51-72.

Atkins, J. E., and McBride, E. F., 1992, Porosity and packing of Holocene river, dune, and beach sands: American Association of Petroleum Geologists Bulletin, v. 76, p. 339-355.

Altschuler, Z. S., 1980, The geochemistry of trace elements in marine phosphorites, 1. Characteristic abundances and enrichment: Society of Economic Paleontologists and Mineralogists, Special Publication No. 29, p. 19-30.

Balashov, Y. A., and Girin, Y. P., 1969, On the reserve of mobile rare earth elements in sedimentary rocks: Geochemistry International, v. 7, p. 649-659.

Baturin, G. N., 1982, Phosphorites on the Sea Floor: Developments in Sedimentology 33, Amsterdam, Elsevier, 343 p.

Baturin, G. N., and Bezrukov, P. L., 1979, Phosphorites on the seafloor and their origin: Marine Geology, v. 31, p. 317-332.

Benmore, R. A., Coleman, M. L., and McArthur, J. M., 1983, Origin of sedimentary francolite from its sulphur and carbon isotope composition: Nature, v. 302, p. 516-518.

Berner, E. K., and Berner, R. A., 1987, The Global Water Cycle: Geochemistry and Environment: Englewood Cliffs, NJ, Prentice-Hall, 397 p.

Berner, R. A., 1973, Phosphate removal from seawater by adsorption on volcanogenic ferric oxides: Earth and Planetary Science Letters, v. 18, p. 77-86.

Berner, R. A., 1980, Early Diagenesis: A Theoretical Approach: Princeton, NJ, Princeton University Press, 241 p.

Berner, R. A., 1990, Diagenesis of phosphorus in sediments from non-upwelling areas, in Burnett, W. C., and Riggs, S. R., eds., Phosphate Deposits of the World. Volume 3, Neogene to Modern Phosphorites: International Geological Correlation Program, Project 156: Phosphorites. Cambridge, Cambridge University Press, p. 27-32.

Berner, R. A., and Rao, J-L., 1994, Phosphorus in sediments of the Amazon River and estuary: Implications for the global flux of phosphorus to the sea: Geochimica et Cosmochimica Acta, v. 58, p. 2333-2339.

Berner, R. A., Ruttenberg, K. C., Ingall, E. D., and Rao, J.-L., 1993, The nature of phosphorus burial in modern sediments, *in* Wollast, R., Mackenzie, F. T., and Chou, L., eds., Interactions of C, N, P and S Biogeochemical Cycles and Global Change: Berlin, Springer-Verlag, p. 365-378.

Borgaard, O. K., 1983, Effect of surface area and mineralogy of iron oxides on their surface charge and anion-adsorption properties: Clays and Clay Minerals, v. 31, p. 230-232.

Broecker, W. S., 1982, Glacial to interglacial changes in ocean chemistry: Progress in Oceanography, v. 11, p. 151-197.

Burnett, W. C., 1977, Geochemistry and origin of phosphorite deposits from off Peru and Chile: Geological Society of America Bulletin, v. 88, p. 813-823.

Byrne, R. H., Lee, J. H., and Bibgler, L. S., 1991, Rare earth element complexation by PO_4^{3-} ions in aqueous solutions: Geochimica et Cosmochimica Acta, v. 55, p. 2729-2735.

BYRNE, R. H., AND KIM, K-H., 1993, Rare earth precipitation and coprecipitation behaviour: The limiting role of PO_4^{3-} on dissolved rare earth concentrations in seawater: Geochimica et Cosmochimica Acta, v. 57, p. 519-526.

CANFIELD, D. E., JØRGENSEN, B. B., FOSSING, H., GLUD, R., GUNDERSEN, J., RAMSING, N. B., THAMDRUP, B., HANSEN, J. W., NIELSEN, L. P., AND HALL, P. O. J., 1993, Pathways of organic carbon oxidation in three continental margin sediments: Marine Geology, v. 113, p. 27-40.

CHESTER, R., 1990, Marine Geochemistry: London, Unwin Hayman Ltd., 698 p.

COLEMAN, M. L., 1985, Geochemistry of diagenetic non-silicate minerals: kinetic considerations: Philosophical Transactions of the Royal Society of London, v. 315A, p. 39-56.

DYPVIK, H., AND BRUNFELT, A. O., 1979, Distribution of rare earth elements in some North Atlantic Kimmeridgian black shales: Nature, v. 278, p. 339-341.

ELDERFIELD, H., AND GREAVES, M. J., 1982, The rare earth elements in sea water: Nature, v. 296, p. 214-219.

ELDERFIELD H., HAWKESWORTH, C. J., GREAVES, M. J., AND CALVERT, S. E., 1981a, Rare earth element geochemistry of oceanic ferromanganese nodules and associated sediments: Geochimica et Cosmochimica Acta, v. 45, p. 513-528.

ELDERFIELD, H., MCCAFFREY, R. J., LUEDTKE, N., BENDER, M., AND TRUEDALE, V. W., 1981b, Chemical diagenesis in Narraganasett Bay sediments: American Journal of Science, v. 281, p. 1021-1055.

ELDERFIELD, H., AND SHOLKOVITZ, E. R., 1987, Rare earth elements in the pore waters of reducing nearshore sediments: Earth and Planetary Science Letters, v. 82, p. 280-288.

FALKOWSKI, P. G., 1997, Evolution of the nitrogen cycle and its influence on the biological sequestration of CO_2 in the ocean: Nature, v. 387, p. 272-275.

FILIPPELLI, G. M., AND DELANEY, M. L., 1996, Phosphorus geochemistry of equatorial Pacific sediments: Geochimica et Cosmochimica Acta, v. 60, p. 1479-1495.

FLEET, A. J., 1984, Aqueous and sedimentary geochemistry of the rare earth elements, *in* Henderson, P., ed., Rare Earth Element Geochemistry: Amsterdam, Elsevier Science Publications, p. 343-373.

FÖLLMI, K. B., 1996, The phosphorus cycle, phosphogenesis and marine phosphate-rich deposits: Earth Science Reviews, v. 40, p. 55-124.

FOX, L. E., SAGER, S. L., AND WOFSY, S. C., 1986, The chemical control of soluble phosphorus in the Amazon estuary: Geochimica et Cosmochimica Acta, v. 50, p. 783-794.

FROELICH, P. N., 1988, Kinetic control of dissolved phosphate in natural rivers and estuaries: A primer on the phosphate buffer mechanism: Limnology and Oceanography, v. 33, p. 649-688.

FROELICH, P. N., BENDER, M. L., AND HEATH, G. R., 1977, Phosphorus accumulation rates in metalliferous sediments on the East Pacific Rise: Earth and Planetary Science Letters, v. 34, p. 351-359.

FROELICH, P. N., KIM, K. H., JAHNKE, R., BURNETT, W. C., SOUTAR, A., AND DEAKIN, M., 1983, Pore water fluoride in Peru continental margin sediments: Uptake from seawater: Geochimica et Cosmochimica Acta, v. 47, p. 1605-1612.

FROELICH, P. N., BENDER, M. L., LUEDTKE, N. A., HEATH, G. R., AND DEVRIES, T., 1982, The marine phosphorus cycle: American Journal of Science, v. 282, p. 474-511.

FROELICH, P. N., KLINKHAMMER, G. P., BENDER, M. L., LUEDTKE, N. A., HEATH, G. R., CULLEN, D., DAUPHIN, P., HAMMOND, D., AND HARTMAN, B., 1979, Early oxidation of organic matter in pelagic sediments of the eastern equatorial Atlantic: Suboxic diagenesis: Geochimica et Cosmochimica Acta, v. 43, p. 1075-1090.

GARRELS, R. M., AND CHRIST, C. L., 1965, Solutions, Minerals and Equilibria: New York, Harper and Row, 450 p.

GERMAN, C. R., KLINKHAMMER, G. P., EDMOND, J. M., MITRA, A., AND ELDERFIELD, H., 1990, Hydrothermal scavenging of rare-earth elements in the ocean: Nature, v. 345, p. 516-517.

GIBBS, R. J., 1981, Sites of river-derived sedimentation in the ocean: Geology, v. 9, p. 77-80.

GLENN, C. R., AND ARTHUR, M. A., 1988, Petrology and major element geochemistry of Peru margin phosphorites and associated diagenetic minerals: Authigenesis in modern organic-rich sediments: Marine Geology, v. 80, p. 231-267.

GLENN, C. R., FÖLLMI, K. B., RIGGS, S. R., BATURIN, G. N., GRIMM, K. A., TRAPPE, J., ABED, A. M., GALLI-OLIVIER, C., GARRISON, R. E., ILYIN, A. V., JEHL, C., ROHRLICH, V., SADAQAH, R. M., SCHIDLOWSKI, M., SHELDON, R. E., AND SIEGMUND, H., 1994, Phosphorus and phosphorites: Sedimentology and environments of formation: Eclogae Geologicae Helvetiae, v. 87, p. 747-788.

GOLDSCHMIDT, V. M., 1954, Geochemistry: Oxford, Oxford University Press, 733 p.

GRAHAM, W. F., AND DUCE, R. A., 1979, Atmospheric pathways of the phosphorus cycle: Geochimica et Cosmochimica Acta, v. 43, p. 1195-1208.

HOLLAND, H. D., 1978, The Chemistry of the Atmosphere and Oceans: New York, Wiley-Interscience, 351 p.

HOLLAND, H. D., 1984, The Chemical Evolution of the Atmosphere and Oceans: Princeton, Princeton University Press, 582 p.

HORN, M. K., AND ADAMS, J. A. S., 1966, Computer-derived geochemical balances and element abundances: Geochimica et Cosmochimica Acta, v. 30, p. 279-290.

INGALL, E. D., AND VAN CAPPELLEN, P., 1990, Relation between sedimentation rate and burial of organic phosphorus and organic carbon in marine sediments: Geochimica et Cosmochimica Acta, v. 54, p. 373-386.

JAHNKE, R. A., EMERSON, S. R., ROE, K. K., AND BURNETT, W. C., 1983, The present day formation of apatite in Mexican Continental margin sediments: Geochimica et Cosmochimica Acta, v. 47, p. 259-266.

JONASSON, R. G., BANCROFT, G. M., AND NESBITT, H. W., 1985, Solubilities of some hydrous REE phosphates with implications for diagenesis and seawater concentrations: Geochimica et Cosmochimica Acta, v. 49, p. 2133-2139.

JØRGENSEN, B. B., 1982, Mineralization of organic matter in the sea bed - the role of sulphate reduction: Nature, v. 296, p. 643-645.

KRAUSKOPF, K. B., 1977, Introduction to Geochemistry: New York, McGraw-Hill, 617 p.

KROM, M. D., AND BERNER, R. A., 1981, The diagenesis of phosphorus in a nearshore marine sediment: Geochimica et Cosmochimica Acta, v. 45, p. 207-216.

KRYNINE, P. D., 1948, Petrology and genesis of the Third Bradford Sand: Bulletin of the Pennsylvania State College, v. 29, p. 1-134.

KUENEN, P. H., 1941, Geochemical calculations concerning the total mass of sediments in the earth: American Journal of Science, v. 239, p. 161-190.

LUCOTTE, M., AND D'ANGLEJAN, B., 1985, A comparison of several methods for the determination of iron hydroxide and associated orthophosphates in estuarine particulate matter: Chemical Geology, v. 48, p. 257-264.

LUCOTTE, M., MUCCI, A., HILLAIRE-MARCEL, C., AND TRAN, S., 1994, Early diagenetic processes in deep Labrador Sea sediments: reactive and nonreactive iron and phosphorus: Canadian Journal of Earth Science, v. 31, p. 14-27.

MARTENS, C. S., BERNER, R. A., AND ROSENFELD, J. K., 1978, Interstitial water chemistry of anoxic Long Island Sound sediments II: Nutrient regeneration and phosphate removal: Limnology and Oceanography, v. 23, p. 605-617.

MEYBECK, M., 1982, Carbon, nitrogen and phosphorus transport by world rivers: American Journal of Science, v. 282, p. 401-450.

MILODOWSKI, A. E., AND ZALASIEWICZ J. A., 1991, Redistribution of rare earth elements during diagenesis of turbidite/hemipelagite mudrock sequences of Llandovery age from Central Wales, *in* Morton, A. C., Todd, S. P., and Haughton, P. D. W., eds., Developments in Sedimentary Provenance Studies: Geological Society Special Publication, 57, p. 101-124.

MOORE, W. S., 1996, Large groundwater inputs to coastal waters revealed by ^{226}Ra enrichments. Nature, v. 382, p. 612-614.

MORAD, S., AND AL-AASM, I. S., 1994, Conditions of formation and diagenetic evolution of Upper Proterozoic phosphate nodules from southern Sweden: Evidence from petrology, mineral chemistry and isotopes: Sedimentary Geology, v. 88, p. 267-282.

NORRISH, K., 1968, Some phosphate minerals of soils: Transactions of the Ninth Congress of Soil Scientists, v. 2, p. 713-723.

PIPER, D. Z., 1974a, Rare earth elements in the sedimentary cycle: A summary: Chemical Geology, v. 14, p. 285-304.

PIPER, D. Z., 1974b, Rare earth elements in ferromanganese nodules and other marine phases: Geochimica et Cosmochimica Acta, v. 38, p. 1007-1022.

POLDERVAART, A., 1955, Chemistry of the earth's crust, in Poldervaart, A., ed., Crust of the Earth—A Symposium: Geological Society of America Special Paper No. 62, p. 119-144.

RAO, J-L., AND BERNER, R. A., 1993, Phosphorus dynamics in the Amazon river and estuary: Chemical Geology, v. 107, p. 397-400.

RASMUSSEN, B., 1996, Early-diagenetic REE-phosphate minerals (florencite, gorceixite, crandallite and xenotime) in marine sandstones: a major sink for oceanic phosphorus: American Journal of Science, v. 296, p. 601-632.

REDFIELD, A. C., 1958, The biological control of chemical factors in the environment: American Scientist, v. 46, p. 205-222.

ROALDSET, E., AND ROSENQVIST, I. T., 1971, Unusual lanthanide distribution: Natural Physical Sciences, v. 231, p. 153-154.

RONOV, A. B., AND MIGDISOV, A. A., 1971, Geochemical history of the crystalline basement and the sedimentary cover of the Russian and North American platforms: Sedimentology, v. 16, p. 137-185.

RONOV, A. B., 1982, The Earth's sedimentary shell (quantitative patterns of its structure, compositions, and evolution): International Geology Review, v. 24, p. 1213-1388.

RUTTENBERG, K. C., 1992, Development of a sequential extraction method for different forms of phosphorus in marine sediments: Limnology and Oceanography, v. 37, p. 1460-1482.

RUTTENBERG, K. C., 1993, Reassessment of the oceanic residence time of phosphorus: Chemical Geology, v. 107, p. 405-409.

RUTTENBERG, K. C., AND BERNER, R. A., 1993, Authigenic apatite formation and burial in sediments from non-upwelling, continental margin environments: Geochimica et Cosmochimica Acta, v. 57, p. 991-1007.

SARMA, V. A. K., AND KRISHNA MURTI, G. S. R., 1969, Plumbogummite minerals in Indian soils: Geoderma, v. 3, p. 321-327.

SCHUFFERT, J. D., JAHNKE, R. A., KASTNER, M., LEATHER, J., STURZ, A., AND WING, M. R., 1994, Rates of formation of modern phosphorite off western Mexico: Geochimica et Cosmochimica Acta, v. 58, p. 5001-5010.

SHOLKOVITZ, E. R., 1973, Interstitial water chemistry of the Santa Barbara Basin sediments: Geochimica et Cosmochimica Acta, v. 37, p. 2043-2073.

SHOLKOVITZ, E. R., PIEPGRAS, D. J., AND JACOBSEN, B., 1989, The pore water chemistry of rare earth elements in Buzzards Bay sediments: Geochimica et Cosmochimica Acta, v. 53, p. 2847-2856.

SPÖTL, C., 1990, Authigenic aluminium phosphate-sulphates in sandstones of the Mitterberg Formation, Northern Calcareous Alps, Austria: Sedimentology, v. 37, p. 837-845.

TRIPLEHORN, D. M., AND BOHOR, B. F., 1983, Goyazite in kaolinitic altered tuff beds of Cretaceous age near Denver, Colorado: Clays and Clay Minerals, v. 31, p. 299-304.

TUREKIAN, K. K., 1969, The oceans, streams and atmosphere, in Wedepohl, K. H., ed., Handbook of Geochemistry, Volume 1: Berlin, Springer-Verlag, 442 p.

VAN CAPPELLEN, P., AND BERNER, R. A., 1991, Fluorapatite crystal growth from modified seawater solutions: Geochimica et Cosmochimica Acta, v. 55, p. 1219-1234.

WATANABE, Y., AND TSUNOGAI, S., 1984, Adsorption-desorption control on phosphate in anoxic sediments of a coastal sea, Funka Bay, Japan: Marine Chemistry, v. 15, p. 71-83.

WHEAT, C. G., FEELY, R. A., MOTTL, M. J., 1996, Phosphate removal by oceanic hydrothermal processes: An update of the phosphorus budget in the oceans: Geochimica et Cosmochimica Acta, v. 60, p. 3593-3608.

WOOD, S. A., 1990, The aqueous geochemistry of the rare-earth elements and yttrium 1. Review of available low-temperature data for inorganic complexes and the inorganic REE speciation of natural waters: Chemical Geology, v. 82, p. 159-186.

WRIGHT, J., SCHRADER, H., AND HOLSER, W. T., 1987, Paleoredox variations in ancient oceans recorded by rare earth elements in fossil apatite: Geochimica et Cosmochimica Acta, v. 51, p. 631-644.

ISOTOPIC RECORDS ACROSS TWO PHOSPHORITE GIANT EPISODES COMPARED: THE PRECAMBRIAN-CAMBRIAN AND THE LATE CRETACEOUS-RECENT

GRAHAM SHIELDS* AND PETER STILLE

Centre de Géochimie de la Surface, UMR 7517, 1 rue Blessig, 67084, Strasbourg, France

AND

MARTIN D. BRASIER

Department of Earth Sciences, University of Oxford, Oxford, OX1 3PR, UK

ABSTRACT: Sedimentary phosphorite formation has occurred episodically over geologic time. Substantial phosphorite deposits were formed during P-giant episodes of which the most widespread and long-lasting were the Precambrian-Cambrian and the Cretaceous-recent. The aim of this review is to reevaluate these two episodes in the light of sedimentary Sr, Nd, S, and C isotopic records. Some, although not all of these phases of phosphorite formation were accompanied by increases in seawater $^{87}Sr/^{86}Sr$ unrelated to changes in seafloor spreading, which supports a possible link between orogeny and global weathering rates, P-input, and P-giant formation. Assessing the relationship between $\delta^{13}C$ and phases of widespread phosphogenesis is more complicated due to the counterbalancing effects on seawater $\delta^{13}C$ of productivity/organic matter deposition and subsequent phosphogenesis/early diagenetic carbon oxidation. It is recommended that proposed links between phosphogenesis and global changes of $\delta^{13}C$ or P-input continue to be reexamined on a case-to-case basis. The application of secular trends in Nd isotopic ratios to unravelling the often vital role of paleocurrents in phosphorite formation is also discussed. In addition, recent isotopic research is outlined that has led to significant improvements in stratigraphic resolution around the Precambrian-Cambrian boundary. A temporal and causal connection is put forward between metazoan evolution, that is the introduction of bioturbation, fecal pellets, biomineralization, and filter feeders, which would have helped to concentrate mineral phosphate in sediments, and widespread phosphogenesis at this time.

INTRODUCTION

Cook and McElhinney (1979) compiled data for 216 sedimentary phosphorite deposits worldwide, which revealed six broad periods of phosphorite formation during the Phanerozoic era: Precambrian-Cambrian, Ordovician, Permian, Jurassic, Late Cretaceous-Eocene and Miocene-recent (Fig. 1C, 1D). More recent resource data compiled for "major" phosphorite deposits only (Notholt et al., 1989) show a slightly different picture (see Fig. 1A, 1B). Greatest phosphorite resources appear to derive from just two extended episodes that are highlighted in this article: latest Precambrian-Middle Cambrian (<590-510 Ma), and late Cretaceous-Holocene (75-0 Ma). However, major resources are also known from the Middle Cretaceous (Föllmi et al., 1993), the Jurassic of eastern Europe (0.43 x 10^9 metric tons P_2O_5), the mid-late Permian of the western United States (1.8 x 10^9 metric tons), the Carboniferous of the United States, and the Ordovician of the Baltic basin and the United States (0.75 x 10^9 metric tons).

Several researchers have considered the origin of phosphorite episodicity (see summaries in Cook and McElhinney, 1979; and Glenn et al., 1994) and it is still a matter of some controversy (Compton et al., this volume). One proposed link is continental glaciation. Global cooling and increased vertical oceanic overturn may encourage upwelling of phosphate-rich, deep ocean waters similar to those of today (Sheldon, 1980; Sheldon, 1989). High productivity in areas of upwelling helps to transfer P from the zone of photosynthesis to the shelf floor, where early diagenetic processes associated with organic matter decay store this P as francolite. Major resources of the Ordovician period have been used to support a relationship between glaciation and phosphorite formation (e.g., Cook and McElhinney, 1979). However, the largest of these resources, the Baltic basin deposits, are of Early Ordovician age (Krasilnikova and Ilyin, 1989) and cannot be related to the short, latest Ordovician glaciation (Brenchley et al., 1994; Fig. 1A). Moreover, there is little evidence for continental glaciation during the Late Cretaceous and Early Cenozoic, one of the most significant periods of phosphogenesis. Although glacially driven upwelling is a primary factor in phosphogenesis today (Glenn et al., 1994, Fig. 4; and Fig 5 herein), vigorous circulation does not have to be restricted to an icehouse world (Lyle, 1997). Therefore, despite the fact that continental glaciation was absent during several P-giant episodes, upwelling may still have played an important role in phosphogenesis.

Another possible factor could be a warm climate and related ocean stratifiction, which might lead to greatly enhanced storage of nutrients, such as P, in deep, possibly anoxic waters. Upwelling of these P-charged waters could lead to phosphate storage on the more oxic parts of the shelves (Fischer and Arthur, 1977; Donnelly et al., 1990). Although ocean anoxic events cannot explain phosphorite formation since the Miocene, deep, anoxic waters have been proposed for the Cretaceous (Schlanger et al., 1987), Permian (Knoll et al., 1996), Ordovician (Berry and Wilde, 1978), and Precambrian-Cambrian times (Brasier, 1992a; Derry et al., 1992). Oceanic anoxic events (OAEs) should correspond to times of high carbon burial and markedly positive $\delta^{13}C$ values in seawater (Fischer and Arthur, 1977), allowing any connection between P-giants and OAEs to be tested.

It remains unclear whether P-giant episodes relate to times of special oceanographic concentration of phosphorus due to paleoceanographic factors such as upwelling or whether these episodes are related to changes in the global P budget. The most important source of phosphorus in the oceans is continental input through weathering (e.g., Föllmi, 1995). Therefore, if P-giant episodes relate to changes in the global P budget, such episodes may correspond to periods of enhanced continental

*email: shields@illite.u-strasbg.fr

Marine Authigenesis: From Global to Microbial, SEPM Special Publication No. 66

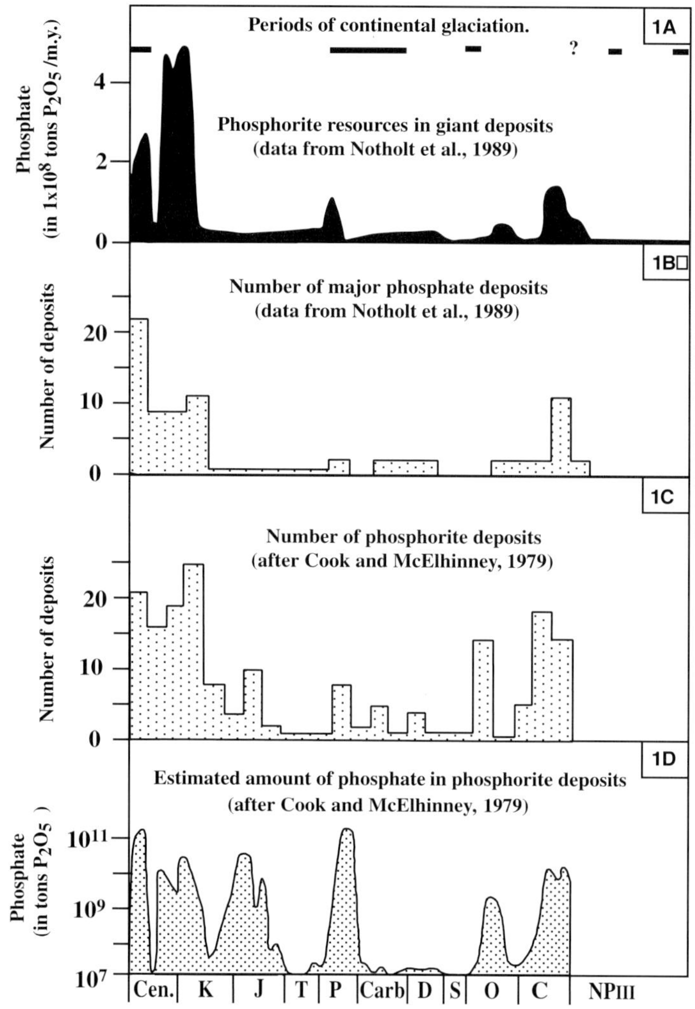

FIG. 1.— (A) Estimated abundance of phosphorite (metric tons P_2O_5) in major deposits based on data in Notholt et al. (1989). N.B. resource data is biased toward the recent due to the depths of older phosphorite reserves, e.g., the Permian Phosphoria Formation (Sheldon, 1989). Black bars represent temporal extent of continental glaciation. (B) Estimated number of major phosphate deposits (evaluation based on resource potential) discussed in Notholt et al. (1989) with a histogram time interval of 25 Ma. Early Cambrian peak is between 550 Ma and 535 Ma. (C) Estimated number of phosphorite deposits through the Phanerozoic with a histogram time interval of 25 Ma (after Cook and McElhinney, 1979). (D) Estimated abundance of phosphorite in deposits expressed in metric tons P_2O_5 on a logarithmic scale (after Cook and McElhinney, 1979). The middle Cretaceous low is possibly misleading (Föllmi et al., 1993).

weathering (e.g., Compton et al., 1993). Uplift and erosion may be especially important for periods of sustained phosphorite formation, such as the Precambrian-Cambrian and Cretaceous-recent episodes discussed here. Another factor might be oceanic hydrothermal circulation, which tends to remove P from the ocean (Wheat et al., 1996). Rates of ocean spreading could have significantly influenced the concentration of P in seawater, although hydrothermal circulation is not likely to have been a major sink (Compton et al., this volume). Local weathering and P input might also be important. Some authors have argued that even major phosphorite deposits, such as those of the late Cretaceous Tethyan realm, could be related more to local river input of phosphorus and less to the global P cycle (Mikhailov et al., 1972; Glenn and Arthur, 1990).

In this article, we concentrate on how the sedimentary isotopic record (Sr, Nd, Os, C, and S isotopes), especially changes in seawater $^{87}Sr/^{86}Sr$ and $\delta^{13}C$, can shed light on the question of P-giant episodicity by looking specifically at the Precambrian-Cambrian and Cretaceous-recent episodes. By looking at what are essentially global parameters (except Nd), the assumption is made that at least some major episodes of phosphorite formation are related to changes in the global P budget. Seawater $^{87}Sr/^{86}Sr$ reacts primarily to three factors: (1) increases in continental weathering rates (which tend to increase $^{87}Sr/^{86}Sr$), (2) changes in the material undergoing weathering (weathering of ancient, basement rocks tends to increase $^{87}Sr/^{86}Sr$, while weathering of juvenile, magmatic rocks will decrease $^{87}Sr/^{86}Sr$, and (3) hydrothermal circulation of Sr at spreading centers also decreases $^{87}Sr/^{86}Sr$. Therefore, if we can relate changes in $^{87}Sr/^{86}Sr$ to continental weathering rates we will be able to determine whether a particular phase of phosphorite formation resulted from an increase in global P input through weathering, which is the major and only significant source of P in the oceans (Föllmi, 1995; Compton et al., this volume). Conversely, in this way we may also ascertain the possible role of changes in seafloor spreading on phosphorite formation. No connection between P-giants and $^{87}Sr/^{86}Sr$ would be expected if phosphorite formation was not linked to the global P budget.

Positive $\delta^{13}C$ excursions have been linked to episodes of widespread phosphogenesis (e.g., Föllmi et al. 1993; Compton et al. 1993), which suggests that an accelerated carbon cycle and increased rates of carbon burial allowed the concentration of P in sediments to rise. Both sets of authors relate changes in carbon burial to global weathering rates and the resultant increase in marine productivity. However, most of the organic carbon associated with giant phosphorite deposits is commonly oxidized later due to reworking, which is essential for the concentration of phosphate in sediments (e.g., Compton and Mallinson, 1996). This carbon oxidation would tend to lower the efficiency of carbon burial and seawater $^{13}C/^{12}C$ enrichment. The temporal relationship between phosphorite formation and seawater $\delta^{13}C$ can be expected, therefore, to be complicated and will be examined later. Present knowledge of secular trends in Nd, Os, and S isotopic ratios through time and their causes renders these isotopic systems of lesser importance in the context of this review. However, all three systems show promise in phosphorite giant research and are discussed alongside Sr and C.

PHOSPHORITE DEPOSITS

Precambrian-Cambrian Boundary Phosphorites

According to data contained in Notholt et al. (1989) around 4.8 x 10^9 metric tons of phosphate (P_2O_5) can be found in phosphorite resources of Precambrian-Cambrian age in China, Vietnam, Mongolia, Kazakhstan, Russia, and Australia. Other deposits of lesser economic importance can be found throughout Asia, Europe, Africa, and Brazil. Cook and Shergold (1986) list 350 discrete deposits of Precambrian-Cambrian age, which

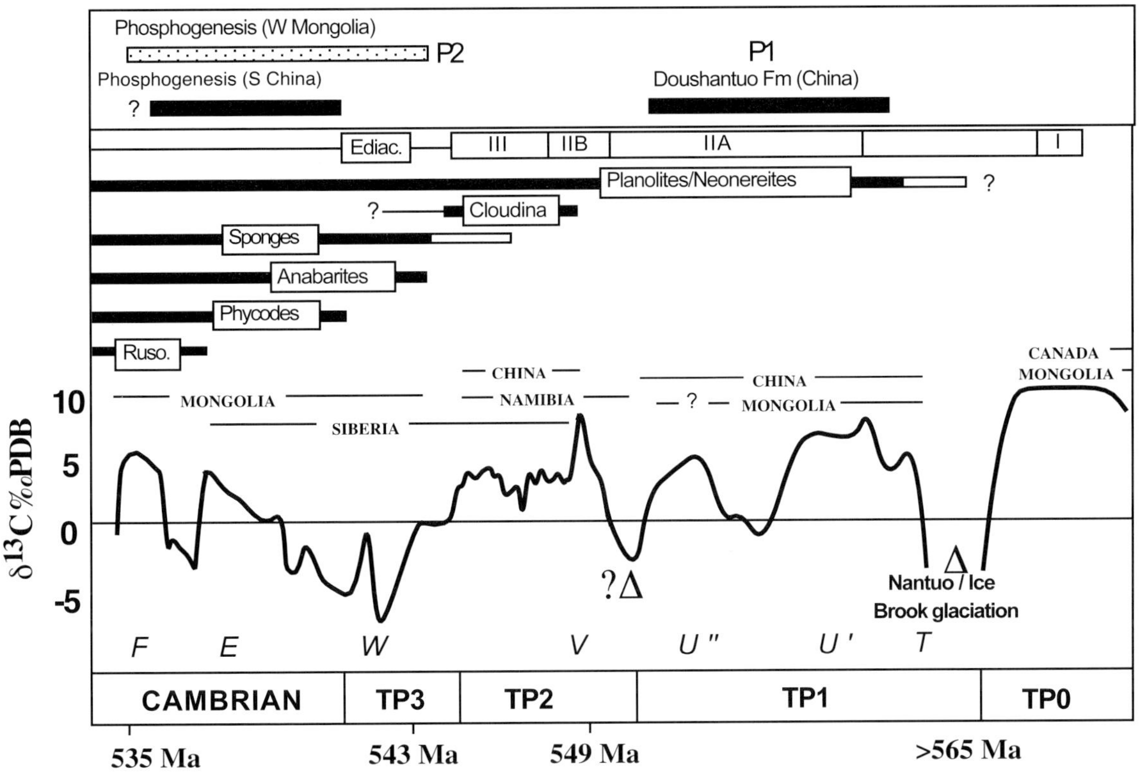

FIG. 2.—Revised calibration of periods of Precambrian-Cambrian phosphogenesis, biostratigraphy, and isotope stratigraphy (Shields, 1999). Filled horizontal bars represent the stratigraphic ranges of fossils: Ruso. = *Rusophycus* or *Cruziana* arthropod traces, Planolites/Neonereites data is from Brasier and McIlroy, 1998. Ediac. = Ediacara-type soft body fossils. Unfilled bars reveal possible extensions to these ranges. Ediacara-type faunal assemblages can be divided into zones: I, small simple discs; IIA, larger, radially symmetrical discs; IIB, complex discs and fronds; III, complex discs, fronds, dickinsoniids, abundant trace fossils, Cloudina, (Kaufman et al., 1997). The C isotope curve has been constructed from the individual records of regions shown above the curve in capital letters (the actual curve is drawn from regions written in bold) by cross-referencing with the Sr isotope record. *F, E, W, V, U'', U',* and *T* are characteristic C-isotope features as in for example, Brasier et al. (1997). Age constraints 535, 543, and 549 Ma are as in Bowring et al. (1993), and Grotzinger et al. (1995). The boundary TP1-TP2 is not chronologically firm and may be older than it appears here. Terminal Proterozoic stage-like divisions TP0 to TP3 can be defined using both the biostratigraphic and the isotopic records. Phosphogenesis phase P1 is confined to time interval TP1, while the more widespread phase P2 begins somewhere in interval TP3 when shelly fossils, e.g., *Anabarites, Cambrotubulus,* and diverse Phanerozoic-type trace fossils first appear in the rock record and continues into the early Cambrian.

together make up 20% of global economic phosphate production. The significance of this P-giant episode is made more remarkable when it is remembered that probably less than half of the phosphorite-bearing rocks of this age have been preserved (Cook and Shergold, 1984; Sengör et al., 1993) and that resource estimates are biased toward the recent due to younger resources being more likely to be buried at shallower depths. There is little evidence for substantial phosphorite formation during the rest of the Precambrian with the possible exception of the early Proterozoic (Melezhik and Fallick, 1996). Economically viable deposits of this age can be found on the Baltic Platform (Scandinavia), and in Korea, Brazil, and Australia.

Shergold and Brasier (1986) reviewed the biochronology of Proterozoic and Cambrian phosphorites. They recognized an episode of phosphogenesis in the Riphean-Vendian interval, <700 Ma ago that was "associated with or immediately after glacial events." Phosphorite deposits of this age, which are not significant resources, can be found in Australia (Adelaide Geosyncline, Amadeus basin, and Pine Creek geosyncline). Other deposits previously thought to be considerably older than the Precambrian-Cambrian boundary, e.g., west African and Mongolian phosphorites, are now believed to be Precambrian-Cambrian in age (Bertrand-Sarfati et al., 1995; Brasier et al., 1996). However, some Brazilian phosphorites could still be of Riphean age (Misi and Kyle, 1994).

The world's oldest, economically significant phosphorite deposits come from the Doushantuo Formation, China (Fig. 2), which lies below Ediacara- and Cambrian fauna-bearing strata and above glaciogenic rocks that are likely of Marinoan glaciation age (e.g., Kaufman and Knoll, 1995). Sr, and C isotope-, and biostratigraphy constrain the age of this formation to between 600 and 550 Ma (Fig. 2; Lambert et al., 1987; Shields, in press). The Doushantuo Formation (and equivalents across South China) is commonly up to 250 m thick with estimated resources of 2 x 10^9 metric tons P_2O_5, although this possibly includes hypothetical resources (Notholt et al., 1989). Ediacarian age (terminal Proterozoic) phosphorite deposits from China are considered to be even more significant economically than lowermost Cambrian deposits of China (Li Yueyan, 1986), We consider that these important deposits represent the beginning of the Precambrian-Cambrian P-giant episode.

The overlying Dengying Formation (up to 700 m thick) only contains phosphorite at its top, in the first member to contain

Cambrian age fossils. This phosphorite is of the same age as the richly phosphatic Meishucun formation and equivalents throughout south China. Lowermost Cambrian phosphorites are found elsewhere in Asia: in Kazakhstan, Mongolia, India, Pakistan, and Vietnam. In China, four Cambrian phosphogenic events can be recognized: in the Meishucun, Canglangpu, and Maochuang stages, and in the middle Cambrian (Shergold and Brasier, 1986). Each major transgression was accompanied by phosphorite deposition (Ye, 1984) whereby the lowermost Cambrian, Meishucun phase appears to have been the most significant from the point of view of resources across Paleotethyan Asia. A lower, middle Cambrian (<520 Ma) phase of phosphorite formation is also evident in Australia (in the Georgina Basin). No major phosphorite deposits are known for the late Cambrian. Shergold and Brasier (1986) propose that there were at least seven sedimentary and phosphatic cycles from the Precambrian-Cambrian boundary to the lowermost Ordovician on the Avalon and Baltic Platforms, although none of these resulted in major phosphorite deposition. One major phosphate deposit is known from the lower Ordovician of the Baltic states, for example, Estonia (ca. 0.65 x 10^9 tons P_2O_5 resource; Krasilnikova and Ilyin, 1989).

In figure 3 and in the text, we divide the Precambrian-Cambrian P-giant episode into three phases. P1 represents the massive phosphorite deposits of the Doushantuo Formation of south China. P2 represents the more widespread lower Cambrian phase of phosphorite deposition. P2 always occurs in the first transgressive unit to contain Cambrian-type shelly fauna and/or diverse trace fossils, i.e., after ca. 545 Ma (Fig. 2). P3 represents the Middle Cambrian deposits of Australia and north China.

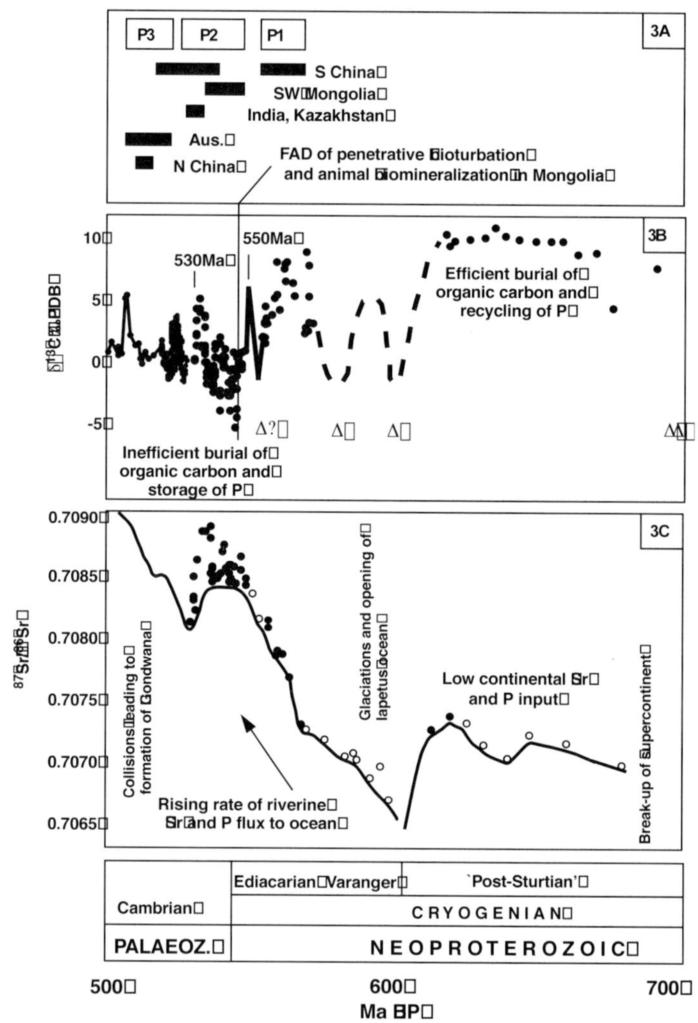

FIG. 3—Phosphorite formation and isotope trends through the latter part of the Neoproterozoic. (A) dark bars represent periods of phosphorite formation, in China and Mongolia as in FIG. 1. P1, P2, and P3 refer to phases of phosphogenesis identified in the text. Biological tie lines and ages are after Brasier et al. (1997), and Grotzinger et al. (1995). (B) C isotope trends are from data in Brasier (1992b) and Brasier et al. (1994; 1996), i.e., mostly from Mongolia and Siberia (filled circles); full line around 550 Ma represents data from Lambert et al., 1987, and Wang et al. (1996) for the Dengying Formation of south China, which we correlate with the post-Marinoan peak of Kaufman and Knoll, 1995. Dashed lines represent general trends based on data for the Varanger glaciations (e.g., Kaufman et al., 1997). Open triangles represent continental glaciation. (C) Sr data (filled circles) are from Shields (1996) (in Brasier et al., 1996) and (empty circles) from Kaufman et al. (1993). The full-line trend has been drawn with the help of data in Nicholas (1996), Derry et al. (1994), and Montanez et al. (1996).

Phosphogenesis Since the Cretaceous

Comprehensive accounts of phosphogenesis since 100 Ma can be found in Burnett and Riggs, 1990; Riggs and Sheldon, 1990; and Notholt et al., 1989. Only a short summary is given here. Most major phosphorite resources were formed during the last 75 Ma, that is, from the Late Cretaceous until today (Fig. 1). This P-giant episode can be divided into two phases: Late Cretaceous-Eocene (75-35 Ma) and Miocene–Recent (25-0 Ma).

Late Cretaceous–Eocene (75-35 Ma)—

In the late Cretaceous (Campanian), phosphorite was deposited in the Middle Eastern Tethys (Egypt, Jordan, Saudi Arabia, Israel, and Syria), Morocco, and northern South America. Especially massive deposits of phosphorite were formed on Africa's northwestern and northern coasts (off Morocco, Tunisia, and Algeria) during the Eocene. Phosphorite formation continued until the Oligocene period in Iraq, Saudi Arabia, and Morocco (Notholt et al., 1989). Resources deriving from this Late Cretaceous-Eocene phase are estimated to make up over 2 x 10^{10} metric tons of P_2O_5 resource (Fig. 1B; Notholt et al., 1989).

Miocene–Recent (25-0 Ma)—

There are no major phosphorite deposits from the Oligocene period, which represents an interval of at least 10 Ma (Fig. 1). At the end of the Oligocene, as continental glaciation began in Antarctica, concentrated phosphogenesis picked up once again and phosphorite formed in the Mediterranean, off Baja California, and in Brazil. Sr isotope evidence suggests that phosphogenesis also began off the eastern United States at this time (Compton et al., 1993). Most modern day continental margin phosphogenesis occurs on the shelves off Peru (east Pacific) and Namibia (west Atlantic), where estimates indicate that we are still in a time of anomalously concentrated phosphogenesis that is related to upwelling (Glenn et al., 1994). Both of these areas

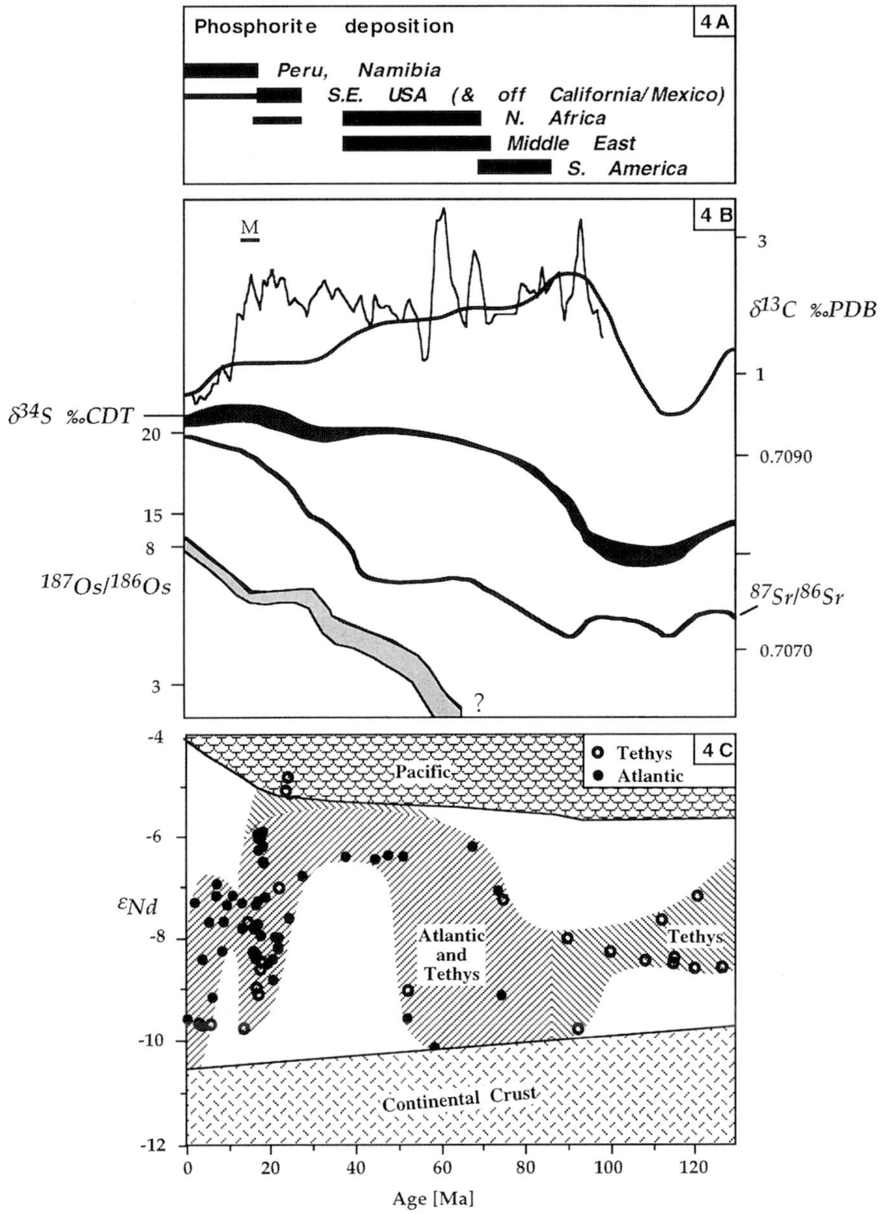

FIG. 4—Phosphorite formation and isotopic trends in seawater for the last 120 Ma. (A) Bars represent the temporal extent of phosphorite deposition in those areas. (B) Seawater $\delta^{13}C$ (from Lasaga, 1989: thick line, and Shackleton and Hall, 1984, 1995; McArthur et al., 1993, for bulk carbonate rock: thin line), M represents the temporal extent of the circa 1‰ positive $\delta^{13}C$ excursion seen in benthic foraminifera: the Monterey event. $\delta^{34}S$ is shown as an uncertainty band (from Burdett et al., 1989; Nielsen, 1989), $^{87}Sr/^{86}Sr$ as a curve (from Hess et al., 1986; McArthur et al., 1994; McArthur, 1994), and $^{187}Os/^{186}Os$ is also shown as an uncertainty band after Peuker-Ehrenbrink et al. (1995). The pre-Cenozoic $^{187}Os/^{186}Os$ trend is not yet clear from the available data. (C) The evolution of Nd isotopic ratio, $^{143}Nd/^{144}Nd$, in the closing Tethys and the newly opening Atlantic oceans (after Stille, 1992; Stille et al., 1996). A lowering of $^{143}Nd/^{144}Nd$ can be explained by increasing influence from the low end member: continental weathering, while rises may represent improved mixing with Pacific seawater, which has a higher Nd isotope ratio. Changes in the $^{143}Nd/^{144}Nd$ ratio of the Tethys and Atlantic reflect changes in paleocurrents, which may help to explain changes in the locii and timing of phosphate deposition in Fig. 5.

have been major provinces of phosphogenesis since the late Miocene (Garrison and Kastner, 1990; Watkins et al., 1995). The switch in location of phosphorite formation from the Middle East to the western seaboards and eastern United States was a response to glacially related upwelling and a change in circulation patterns from predominantly west-east to north-south (Riggs and Sheldon, 1990; see Fig. 5B). Although not all present day phosphogenesis is related to upwelling (Ruttenberg and Berner, 1993), all Miocene-recent major deposits formed in areas of upwelling (Riggs and Sheldon, 1990). The Miocene-recent phase accounts for additional further resources of 6.4 x 10^9 metric tons P_2O_5 (Notholt et al., 1989). The greatest phosphorite deposits of this phase in the southeastern United States were formed during a short, ca. 7 Ma period in the early-mid Miocene, 23-16 Ma (Compton et al., 1993; Compton et al., this volume).

SEDIMENTARY ISOTOPIC RECORDS

The Precambrian-Cambrian Sedimentary Isotopic Record

The transition from the Proterozoic to the Phanerozoic witnessed significant changes in marine isotopic chemistry as well as biological evolution (Donnelly et al., 1990). Recent chronological (Grotzinger et al., 1995) and isotopic studies (Brasier et al., 1996; Kaufman et al., 1997) have helped to improve our knowledge about the relative importance and timing of isotopic events and metazoan evolution (Fig. 2; Shields, in press).

The Sr isotopic record reveals rising seawater $^{87}Sr/^{86}Sr$ from 0.7057 ca. 800 Ma (Veizer et al., 1983; Asmerom et al., 1991) to 0.7092 by middle Cambrian time, ca. 510 Ma (Montanez et al., 1996). The major part of this rise was achieved between ca. 600 Ma and 545 Ma (<0.7066-0.7085; Fig. 3), which represents a steeper and more sustained increase in $^{87}Sr/^{86}Sr$ than has been experienced during the Cenozoic (cf., Fig. 4). Between 540 and 530 Ma, seawater $^{87}Sr/^{86}Sr$ actually decreased from 0.7085 to ca. 0.7080 (Brasier et al., 1996) before renewing its rise in two stages: during the Atdabanian (0.7080-0.7085) and the Middle Cambrian (0.7085-0.7092). Montanez et al. (1996) consider this latter sharp rise to have represented an unroofing event analogous to the uplift and erosion of southern Tibet, which may have caused a similar rise in seawater $^{87}Sr/^{86}Sr$ between 20 Ma and 15 Ma (Richter et al., 1992). Seawater Sr isotope ratio reacts primarily to changes in input of strontium from the continent and submarine hydrothermal exchange (e.g., Chauduri and Clauer, 1986). The sharp rise in seawater $^{87}Sr/^{86}Sr$ through the terminal Proterozoic could thus be interpreted as a relative increase in continental weathering; this is further supported by escalating sedimentation rates and paleomagnetic and tectonic considerations that constrain the final amalgamation of Gondwana to between 555 Ma and 530 Ma (e.g., Bond et al., 1988). Nd isotopes in phosphates tell a similar story (Yang et al., 1995) with ever more continental influence on the Panthalassa and "Chinese" oceans through this time. An increase in the Sr isotopic ratio of material undergoing weathering rather than the rate of overall weathering would also lead to a rise in $^{87}Sr/^{86}Sr$ (Palmer and Elderfield, 1985). However, by comparison with today's dominance of $^{87}Sr/^{86}Sr$ by Himalayan erosion of uplifted basement rocks, these two causes are likely to be coupled during periods of major orogeny and erosion such as at the Precambrian-Cambrian boundary. Another interpretation for the steep rise in seawater $^{87}Sr/^{86}Sr$ could include decreasing rates of ocean spreading (e.g., Chauduri and Clauer, 1986; Allègre et al., 1997). However, this is not supported by the rapidity of plate movements around this time (Torsvik et al., 1996), which would require vigorous spreading. Indeed, these authors consider the period between 600 Ma and 550 Ma as one of maximum latitudunal plate velocity and maximum angular rotation (Torsvik et al., 1996: Fig. 14), a time of intense rifting and drifting associated with the opening of the Iapetus ocean. Therefore, there is strong evidence that the steep rise in $^{87}Sr/^{86}Sr$ from 600 Ma to 550 Ma, and its continuation through to ca. 510 Ma, was caused by greatly increased mountain building activity and weathering input into the oceans.

The Precambrian seawater sulfur isotope record is only poorly constrained (Strauss, 1997). First order changes in seawater $\delta^{34}S$ have paralleled the $^{87}Sr/^{86}Sr$ record as in the Phanerozoic, with low values in the Neoproterozoic, ca. <20‰ (Strauss, 1993) reaching uniquely high values of >33‰ CDT by the early Cambrian, (Claypool et al., 1980; Shields et al., in press). As seawater $\delta^{34}S$ is controlled by the large fractionation involved in the reduction of sulfate to sulfide, such high values in the early Cambrian imply relatively high rates of sulfate reduction and pyrite burial at that time (Donnelly et al., 1990). However, recent work has shown that the Late Precambrian rise in $\delta^{34}S$ may also have been due, in part, to greater isotopic fractionation caused by the evolution of sulfur oxidizing bacteria after 800 Ma (Canfield and Teske, 1996), the effects of which became more significant after the introduction of bioturbation during the terminal Proterozoic (Shields et al., in press).

Consistently high $\delta^{13}C$ values up to +11‰ PDB are characteristic of Sturtian to Marinoan time, that is, ca. 730 Ma to 600 Ma (Fig. 3; Kaufman et al., 1997). Thereafter, average $\delta^{13}C$ values decrease strongly with the beginning of Marinoan time and reach negative values at the Marinoan glaciation. During this period, the Sr isotopic record also reaches lowest isotopic composition values, ca. 0.7066 (Fig. 3). From Marinoan time until the Cambrian the $\delta^{13}C$ record shows strong fluctuations between -5‰ and +8‰, reaching minimal values in the earliest Cambrian. The trend to lower $\delta^{13}C$ values from ca. 700 Ma to 540 Ma implies a sustained decrease in the efficiency of organic carbon (C_{org}) burial, despite Sr isotopic evidence for an increase in nutrient input from weathering.

The Cenozoic Sedimentary Isotopic Record

One of the major isotopic features of the Cenozoic, as with the Precambrian-Cambrian transition, is a rise in seawater Sr isotopic ratio. $^{87}Sr/^{86}Sr$ reached a low point in the late Jurassic (about 150 Ma) of ca. 0.7067 (Burke et al., 1982), rose markedly in the late Cretaceous, ca. 90 Ma (McArthur et al., 1994), and again more steeply from 0.7076 to 0.7091 around 40 Ma in the Eocene, (Fig. 4), This latter steep rise has been related to the uplift and erosion of the Himalayan-Tibetan plateau, or HTP, (Richter et al., 1992), which resulted from the collision of India with Asia at least 50 m.y. ago. The increase in seawater Sr isotope ratio is mirrored by a rise in osmium (Os) isotope ratio (Ravizza, 1993), which provides another possible proxy for chemical weathering rates. However, the major rise in seawater $^{187}Os/^{186}Os$ began more recently (Peuker-Ehrenbrink et al., 1995) than that of $^{87}Sr/^{86}Sr$, which has led some authors to suggest that the effects of the Himalayan uplift and erosion on the radiogenic isotope record can first be seen only 12 m.y. ago (Allègre et al., 1997). Correspondence between the rise in $^{187}Os/^{186}Os$ at this time and a significant fall in $\delta^{13}C$ has been linked to the erosion of organic-rich rocks in the Himalayas as organic matter is commonly enriched in osmium (Ravizza and Esser, 1993). The steep rises in $^{87}Sr/^{86}Sr$ at 90 Ma and 40 Ma are accompanied by sustained falls in the rate of ocean crust production (e.g., Larson, 1991), which implies that these rises were caused at least in part by decreases in hydrothermal Sr exchange near the mid-ocean ridges. Therefore, contrary to the Precambrian-Cambrian rise, it appears unlikely that the pre-Miocene rise in $^{87}Sr/^{86}Sr$, especially that of the Cretaceous, can be related to orogenic uplift and erosion. However, a marked steepening of the $^{87}Sr/^{86}Sr$ curve between 25 Ma and 14 Ma (Miller et al., 1991) is more likely to have been caused by Himalayan erosion (e.g., Raymo, 1994), as there was no significant decrease in ocean crust production at that time (Larson, 1991).

Carbonate C isotope values have tended to decrease over the Cenozoic from around 2‰ to 2.5‰ PDB between 50 Ma and 14 Ma, to around 0‰ today (Fig. 4B). Lasaga's (1989) smoothed curve of marine carbonate $\delta^{13}C$ through the Phanerozoic shows this trend to be part of a longer trend to lower values, which began in the Late Cretaceous (Fig. 4B) and has mirrored sea-level change. One marked increase in seawater $\delta^{13}C$ took place in the Paleocene. Another significant rise in $\delta^{13}C$ occurred during the Early-Middle Miocene (28-14 Ma), which also corresponds to the Cenozoic high in seawater $\delta^{34}S$ (Burdett et al., 1989). This latter rise in $\delta^{13}C$ is best seen in the benthic foraminifera record (e.g., Compton and Mallinson, 1996) and has been attributed to an increase in organic carbon burial that led to global cooling (the Monterey Hypothesis: Vincent and Berger, 1985). Temporal correlation between this $\delta^{13}C$ rise and the progress of glaciation in Antarctica tends to support this hypothesis. The major decrease in $\delta^{13}C$ from ca. 2‰ to 0‰ began after ca. 14 Ma. The interpretation of this fall remains controversial as recent studies indicate that the C_{org} reservoir actually grew over this period (Derry and France-Lanord, 1996). Such a decrease in C_{org} burial efficiency since 14 Ma could be the result of diagenetic phosphogenesis and associated carbon oxidation (Compton and Mallinson, 1996).

The Neodymium isotopic record—

Neodymium (Nd) has a short ocean residence time of around 400 years, which is far shorter than the ocean mixing time of about 1500 years. This means that the oceans are not isotopically homogeneous with respect to Nd as they are to both Sr and Os, which provides an opportunity to separate the effect of tectonics (weathering input and ocean spreading) on radiogenic isotopic ratios from changes in ocean circulation. Stille et al. (1996) traced the evolution of this isotopic parameter in several ocean basins over the past 180 m.y. (see Fig. 4C). The Sr and Nd isotopic records for the Atlantic and Tethys oceans were decoupled from each other during 80-45 m.y. B.P. and again around 20 m.y. B.P. (Stille, 1992). For these periods, Sr isotope ratios imply increased continental influence, while Nd isotope ratios in these ocean basins suggest increased influence from a more primitive source, such as the Pacific, with a more radiogenic isotopic signature. The Nd isotopic evolution of the oceans could be viewed as the result of mixing between two end members: continental crust (weathering) and the influence of the Pacific Ocean. The Pacific Ocean shows an unchanging range of Nd isotope ratios from 200 Ma until today (Stille et al., 1996) that is higher (more mantle-influenced) than that of any other ocean. The reason for this lies in the abundance of volcanic activity and the youthfulness of the rocks surrounding the Pacific Ocean. It seems likely from paleoceanographic and tectonic considerations that the Nd isotopic ratios in the Atlantic and Tethys oceans were partly controlled, when Sr and Nd were coupled, by changes in the balance between continental weathering and submarine hydrothermal exchange, especially between 170 Ma and 90 Ma (Stille et al., 1996) and from 18 Ma to 0 Ma, and partly by exchange with Pacific seawater, when Sr and Nd were decoupled (for example, 80-30 Ma and 20-18Ma). In this case, the rise in seawater Nd isotopic ratio from 90 Ma to 45 Ma suggests that water from the Pacific flowed through the Tethys into the juvenile Atlantic and around the world as a circumequatorial current during the late Cretaceous (Stille et al., 1996; Fig. 5A). Upwelling occurred around the Middle East and led to substantial phosphogenesis there and in Columbia and Venezuela in sediments of late Cretaceous age. During the Eocene (55-35 Ma), the Indian-Tethys seaway became very narrow between the Arabian plate to the south and the Eurasian plate to the north. As the Nd isotopic composition of Tethys seawater remained elevated, it can be assumed that there was still considerable seawater exchange between the two oceans, leading to P deposition where the circumequatorial currents were forced to rise above submerged, continental straits in the Persian Gulf, Morocco, and Mexico (Fig. 5A).

Tectonic changes altered ocean circulation by the close of the Eocene. Coupling of both the Sr and Nd records suggests increasing influence from the continents from 30 Ma until today with a major hiccup at 20-18 Ma, which coincides with the opening of the Drake passage (Stille, 1992; Stille et al., 1996). This opening may also have helped to bring about cold current-related deposition of phosphorite off the coast of Namibia from the late Early Miocene onwards (Fig. 5B). Future isotopic studies on Namibian phosphate will attempt to confirm this temporal connection. It appears plausible that changes in current patterns brought on by the opening of the Drake Passage helped concentrate phosphogenesis off the coasts of the eastern United States, Brazil, and west Africa at the same time. The opening of the Drake Passage may also have helped to instigate the onset of global continental glaciation by establishing the cool circum-Antarctic currents, which thermally insulated Antarctica. Nd isotope studies and in particular their coupling with other isotopic parameters (Sr, Os, Ce) of different ocean residence times (e.g., Stille, 1992; Ravizza and Esser, 1993; Allègre et al., 1997) may help us to understand the part that changing paleocurrents have played in phosphogenesis and environmental change.

DISCUSSION

The Role of Metazoan Evolution in Precambrian-Cambrian Phosphogenesis

Several authors have linked widespread phosphogenesis around the Precambrian-Cambrian boundary with the evolution of mineralized skeletons ca. 550 m.y. ago (e.g., Donnelly et al., 1990). The predominance of phosphatic shelly fauna in rocks of this age has led some to propose that the first shelly animals were phosphatic (Lowenstam and Margulis, 1980; Cook and

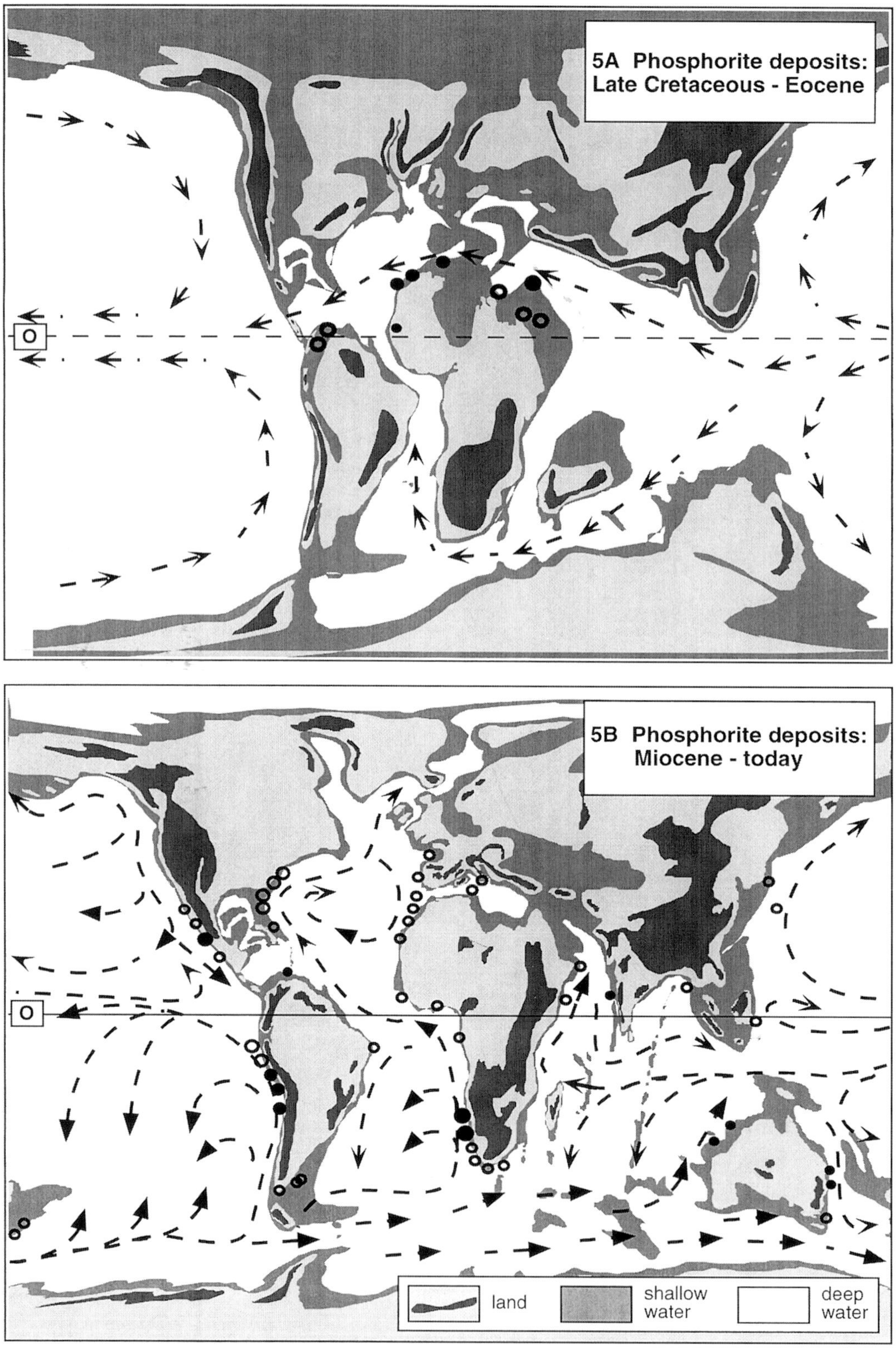

FIG. 5—Evolution of continental positions (after Scotese et al., 1988) and global paleocurrents from the late Cretaceous until today. (5) The Cenomanian, around 94 m.y. ago. Circumequatorial currents flowed east-west through the Middle East and Panama Straits (from Stille et al., 1996), leading to phosphate deposition in these areas between 85 Ma and 35 Ma. Open circles = late Cretaceous phosphogenesis, filled circles = phosphogenesis restricted to the Paleogene (mostly Eocene). (B) The Chattian, around 27 m.y. ago with today's system of currents superimposed: large arrow heads represent cold currents, while small arrows represent warm currents (after Willett et al., 1983). Collision of India with Asia, and Australia with Indonesia has allowed the Indian ocean to develop its own circulation cell. Tectonics helped to close the Indian-Tethys seaway during the Oligocene, which altered ocean circulation from largely east-west to north-south in the Pacific and the newly opening Atlantic. The Drake Passage opened in the early Miocene, allowing Pacific seawater to enter the south Atlantic between Antarctica and South America, leading to upwelling and widespread phosphate deposition in the Atlantic as circum-Antarctic currents and Antarctic glaciation was beginning to take effect. Filled circles = areas of present day phosphogenesis active since the Miocene (from data in Notholt et al., 1989). Open circles = areas of Late Oligocene-Holocene phosphogenesis that have ceased to be active. Large circles represent major deposits (Notholt et al., 1989; Riggs and Sheldon, 1990).

Shergold, 1984; Kempe et al., 1989). Other authors have emphasized the bias in our fossil record due to the taphonomic advantages of secondarily phosphatized, thin calcitic shells (Brasier, 1990). In recent years it has become evident that the very first shelly fauna were not phosphatic but instead siliceous (Brasier et al., 1997) and calcitic (Grotzinger et al., 1995). The development of phosphatic shells in the early Cambrian was simply one of many innovations that arose during the Cambrian explosion and so may tell us little about the mechanisms behind the events of this time (Runnegar, 1987).

The earliest firm evidence for metazoan life is provided by simple trace fossils (Brasier and McIlroy, 1998) and some forms of Ediacara-type, soft-bodied fauna. Paleontological constraints point to the Marinoan glaciation, ca. 600 Ma, as the maximum age for the earliest macrofauna (e.g., Knoll, 1996). Some authors have argued that the introduction of pelagic metazoans would have changed the chemical environment of the late Precambrian oceans. Rapid sinking of fecal pellets would have helped to clear the waters of organic debris, encouraging photosynthesis and allowing oxygen to collect in surface waters (Logan et al., 1995). Filter feeders, e.g., sponges, are among the first animals to appear as body fossils (Fig. 2) and would have contributed to cleaning surface waters and transferring P to the sediment/water interface. We can envisage a change around this time from microbially mediated biodegradation of organic matter within the water column, which would have led to low P concentrations in early Neoproterozoic sediments (Strauss et al., 1997), to a situation more similar to that of today, whereby redox reactions in the top few centimeters of the shelf sediments serve to oxidize organic carbon and release phosphorus during early diagenesis. The earliest signs of sediment digestion known (fecal pellet traces) are younger than 600 Ma (*Neonereites uniserialis,* in Brasier and McIlroy, 1998) and would also have helped to concentrate P in the sediment. In addition, the later introduction of penetrative bioturbation would have helped to mechanically concentrate grains by maintaining phosphatic phases at critical depths in the sediment (Glenn, 1990).

From this short consideration of early metazoan evolution, it is clear that several biological innovations made giant phosphorite deposits a possibility after 600 Ma and even more likely after 545 Ma. Concentration of P in the biological cycle, by way of fecal pellets, filter feeding, bioturbation, and biomineralization led to greater concentration of P in sediments. Increasing oxygenation of the expanding shelf through this time would have encouraged preferential storage of this P and oxidation of C_{org} in sediments during diagenesis (Grill and Richards, 1962). The influence of bioturbation, and related increases in early diagenetic oxygenation of organic matter, on biogeochemical cycling may also be reflected in the continuous decreases in carbon burial efficiency, that is, the trend to low $\delta^{13}C$ values toward the Precambrian-Cambrian boundary (Shields et al., in press).

Phosphorite Formation and Sr and C Isotopic Records

It is important that caution be used when interpreting one phenomenon, phosphorite formation, on the basis of only indirectedly related isotopic trends. Each P-giant episode was accompanied by its own pecularities of climate, evolutionary change, continent configuration, paleocurrent patterns, spreading rates, and orogeny, which must have had their own influences on both phosphogenesis and the isotopic record. Therefore, it is important to interpret each P-giant episode within the framework of what is already known about that particular time in geologic history.

The Precambrian-Cambrian and Neogene were both times of intense mountain building activity and significant increases in seawater $^{87}Sr/^{86}Sr$. Precambrian-Cambrian P-giant phases P1 and P3 occurred during steep rises in $^{87}Sr/^{86}Sr$ (Fig. 3). Based on evidence for rapid plate movements at this time (Torsvik et al., 1996), it was argued earlier that these $^{87}Sr/^{86}Sr$ rises must have been caused by increasing continental weathering during a time of tectonic upheaval rather than by any decrease in mantle influence. The P-giant phase of the Early-Middle Miocene (23-16 Ma) was similarly accompanied by a marked steepening of a $^{87}Sr/^{86}Sr$ rise (28-14 Ma) that was likewise unrelated to significant changes in ocean crust production. Compton and Mallinson (1996) calculate that the proposed increase in the Sr river flux required to explain the Sr isotopic evolution of seawater of Richter et al. (1992) would result in an excess P river flux to the oceans of between 9 and 23 x 10^{16} moles since 30 Ma. However, this must be considered a maximum estimate as some of the $^{87}Sr/^{86}Sr$ rise may be accounted for by the weathering of rocks with more radiogenic isotopic signatures (Palmer and Elderfield, 1985) and slowing spreading rates from ca. 40 Ma to 15 Ma (Kominz, 1984; Larson et al., 1991). Therefore, it can be argued that widespread phosphorite deposition at these times was linked to a global increase in P input from the continent. On the other hand, the Miocene P-giant phase corresponded with a sustained period of minimum ocean crust production (Larson, 1991), which would have helped to keep P concentrations in the oceans high. Phase P2, the most significant and widespread phase of phosphorite formation of the Precambrian-Cambrian episode, was not accompanied by a rise in $^{87}Sr/^{86}Sr$ but instead by a plateau and slight fall. Likewise, the Cretaceous and Paleogene phosphorite deposits of the Tethyan realm cannot be unequivocally linked to changes in uplift and global chemical weathering rates as changes in $^{87}Sr/^{86}Sr$ at this time appear to be associated more with rates of seafloor spreading. This would support the notion that the Tethyan deposits were more closely related to increases in the rates of local weathering and P input from deeply weathered hinterlands to the south (Fig. 5A; Glenn and Arthur. 1990). These authors argue that the great areal extent of the Tethyan P deposits tends to rule out upwelling as the ultimate source of P as productivity is quickly diminished away from the zone of upwelling and some of the deposits were formed up to hundreds of kilometers from the open Tethyan margin (Glenn et al., 1994). However, successive decreases in hydrothermal scavenging of P at this time, which are implied by falls in ocean crust production (Larson et al., 1991), cannot be ruled out as a mechanism that helped to change the local or global P budget during the Cretaceous-Eocene.

The carbon isotopic record would be expected to react to changes in reactive P input from weathering as greater nutrient availability should increase productivity and burial of organic carbon as is considered to have happened in the Middle Miocene (the Monterey Hypothesis: Vincent and Berger, 1985). Following fom this, positive $\delta^{13}C$ excursions would be expected during periods of weathering-driven phosphorite formation. Such a relationship has been proposed for Cretaceous and Tertiary time (e.g., Föllmi et al., 1993), and in particular for the giant phosphorite deposits of the Early-Middle Miocene (Compton et al., 1993). P-giant phases accompanied by increases in the rate of weathering (=P1, P3 and Miocene, from $^{87}Sr/^{86}Sr$) should also be accompanied by increases, therefore, in $\delta^{13}C$ (Delaney and Filipelli, 1994). P1 and P3 phases are indeed associated with positive $\delta^{13}C$ excursions (Figs. 3 and 4), although the lack of resolution offered by present stratigraphic constraints does not allow us to determine the precise temporal relationship between phosphogenesis and $\delta^{13}C$. Where this is possible, for example in the Miocene, it appears that much phosphorite was formed during a period of low rather than high $\delta^{13}C$. Precise correlation between phosphogenesis off southeastern United States and the seawater $\delta^{13}C$ record using Sr isotope stratigraphy has shown that most phosphogenesis occurred during a relative low in $\delta^{13}C$ between 23 Ma and 16 Ma, largely before the major $\delta^{13}C$ peak around 16 Ma (Compton et al., 1993: Fig. 10). Therefore, phosphorite formation was decoupled from global organic carbon burial. This decoupling can be explained by considering the effect that early diagenetic francolite formation has on the carbon cycle. Organic carbon oxidation during diagenesis releases P, which can end up concentrated chemically and mechanically by current action and reworking as phosphate, whereby the more oxidizing the conditions of deposition, the more P will tend to be trapped in the upper centimeters of the sediment relative to organic carbon (e.g., McManus et al., 1997). Greater weathering rates, such as those argued for the Middle Miocene, seem likely to lead to greater productivity, C_{org} burial and P transport to the sediment. However, the formation of phosphate in the sediment and its reworking will itself decrease carbon burial efficiency and lower seawater $\delta^{13}C$. Therefore, the relationship between massive phosphorite formation and $\delta^{13}C$ is more complicated than might be assumed from considering only the effect of nutrient input/higher productivity.

The major phase of Precambrian-Cambrian phosphogenesis (P2) could reflect such a C-P decoupling relationship too. Extremely high $\delta^{13}C$ values in the pre-Marinoan oceans (up to +11‰) imply enhanced organic carbon storage and burial, which may have been the result of deposition beneath anoxic deep waters (e.g., Derry et al., 1992). Under increasingly anoxic conditions P tends to be preferentially released with respect to carbon (Van Capellen and Ingall, 1996; McManus et al., 1997). Phosphorus regeneration in the anoxic or suboxic parts of the water column would have helped to drive productivity in surface waters (Ingall and Jahnke, 1997), allowing high rates of carbon burial (high $\delta^{13}C$) during a time of low weathering rates (low $^{87}Sr/^{86}Sr$) and possibly leading to global cooling (Marinoan glaciation). In this scenario, sporadic switches to low or negative $\delta^{13}C$ values between 750 Ma and 550 Ma represent the return of carbonate species-charged deep waters to the surface as a result of glacially induced oceanic overturn (Fig. 2; Derry et al., 1992). Indeed, negative $\delta^{13}C$ excursions in the Neoproterozoic do coincide with times of glaciation (Kaufman et al., 1997). The trend to lower $\delta^{13}C$ values through the terminal Proterozoic suggests a progressive decrease in the efficiency of organic carbon burial. It is possible that the biological revolutions and the oxygenation of the surface environment envisaged for this time (e.g., Logan et al., 1995) plus the effects of bioturbation were factors in this lowering of carbon burial efficiency as discussed earlier. Just as anoxic bottom waters during the pre-Varanger Neoproterozoic would have led to preferential retention of carbon in sediments and P regeneration (Strauss et al., 1997), oxic/suboxic diagenesis during the subsequent transgression would have favored carbon oxidation and P storage (Grill and Richards, 1962; McManus et al., 1997). At this time $\delta^{13}C$ dropped to uniquely low values with the appearance of the first shelly fauna and penetrative bioturbation in the sedimentary record. Therefore, carbon oxidation efficiency and phosphogenesis reached their respective maximums at the same time and, in a similar fashion to the Middle Miocene, $\delta^{13}C$ recovered gradually to high values of up to +5‰ by ca. 535 Ma (Brasier et al., 1996). However, as with the Miocene phase, it is unclear whether much phosphorite formed during the time of maximum $\delta^{13}C$. It appears that there is no simple relationship between phosphogenesis and $\delta^{13}C$ at the Precambrian-Cambrian boundary as may be the case in the Tethyan Cretaceous (Glenn et al., 1994). Finally, as evidence for Precambrian-Cambrian boundary glaciation is growing (Bertrand-Sarfati et al., 1995), the role of glacial upwelling and reworking caused by sea-level fluctuations at this time should not be discounted in any discussion of the Precambrian-Cambrian boundary phase of phosphogenesis (P2).

The Miocene and the lowermost Cambrian (P2) phosphogenic phases began during times of relatively low $\delta^{13}C$, and arguably ended before a peak in $\delta^{13}C$. Therefore, a connection between maximum global carbon burial and phosphorite formation cannot be supported using the present isotopic record. It could be that the $\delta^{13}C$ record is, at least in part, reacting to early diagenetic phosphogenesis and associated increases in carbon oxidation efficiency. Such a mechanism for controlling seawater $\delta^{13}C$ has already been suggested to cause the post-Miocene fall in $\delta^{13}C$ (Compton and Mallinson, 1996). It may also be the case that these P-giant phases are not responses to increases in the global P budget but rather to palaeoceanographic and more local factors.

SUMMARY

Substantial deposits of sedimentary phosphorite were formed between <590 Ma and 510 Ma and from 75 Ma to 0 Ma. The lack of pre-600 Ma phosphorite can be explained by evolutionary changes, which allowed P to become concentrated in marine sediments during diagenesis for the first time; the initial phase of Precambrian-Cambrian phosphogenesis (P1: <590-555 Ma) corresponds to the introduction of the first macrofauna, fecal pellets, and surface bioturbation, while the major phase (P2: 545-525 Ma) corresponds to the appearance of Phanerozoic-type, diverse, penetrative bioturbation, and a marked increase in metazoan diversity known as the Cambrian explosion. The Precambrian-Cambrian and Cretaceous-recent episodes of phosphorite formation were times of particular domination of seawater $^{87}Sr/^{86}Sr$ by continental weathering. Some individual phases of massive phosphorite formation (P1 = Ediacarian, P3 = Middle Cambrian, and Early-Mid Miocene) were accompanied by significant changes in seawater $^{87}Sr/^{86}Sr$, which were related to pulses of tectonic uplift and weathering. However, there is no Sr isotopic evidence that Middle Cretaceous and Late Cretaceous-Eocene phosphogenesis was related to changes in the global P cycle or budget and may have been the result of local weathering input of P and paleoceanographic factors. Seawater Nd isotope ratio can be used to trace changes in paleoceanography and paleocurrent patterns. In this way, the location and timing of phosphorite formation phases may be more fully understood. Contrary to expectations, major phases of phosphorite formation do not always correlate with periods of maximum $\delta^{13}C$ values (i.e., Middle Miocene and P2), but instead with periods of low $\delta^{13}C$. This decoupling of phosphorite formation and global organic carbon burial may reflect the increase in C oxidation efficiency, which is a result of the early diagenetic transfer of organic P to francolite, or may simply reflect the relative unimportance of global element fluxes. Therefore, no simple relationship between phosphorite formation and $\delta^{13}C$ can be generalized. More high resolution stratigraphic studies are required to assess other periods of phosphogenesis with sufficient precision. For these studies it is recommended that Sr and C isotopic studies be carried out directly on phosphatic successions, which would avoid some of the common problems of stratigraphic correlation.

ACKNOWLEDGMENTS

The authors wish to thank sincerely John Compton, John McArthur and Craig Glenn for their constructive criticism and input throughout the review process.

REFERENCES

ALLEGRE, C. J., GAILLARDET, J., LEVASSEUR, S., MEYNADIER, L., AND LOUVAT, P., 1997, The evolution of seawater strontium isotopes in the last hundred million years: Reinterpretation and consequences for erosion and climate models (abs): Abstract 7th Annual V. M. Goldschmidt Conference, p. 4.

ASMEROM, Y., JACOBSEN, S. B., KNOLL, A. H., BUTTERFIELD, N. J., AND SWETT, K., 1991, Strontium isotopic variations of Neoproterozoic seawater: Implications for crustal evolution: Geochimica et Cosmochimica Acta, v. 55, p. 2883-2894.

BERRY, W. B. N., AND WILDE, P., 1978, Progressive ventilation of the oceans—an explanation for the distribution of Lower Paleozoic black shales: American Journal of Science, v. 278, p. 257-275.

BERTRAND-SARFATI, J., MOUSSINE-POUCHKINE, A., AMARD, B., AND AHMED, A. A. K, 1995, First Ediacara fauna in western Africa and evidence for early Cambrian glaciation: Geology, v. 23, p. 133-136

BOND, G. C., NICKESON, P. A., AND KOMINZ, K. A., 1988, Breakup of a supercontinent between 625 and 555 Ma: New evidence and implictions for continental histories: Earth and Planetary Science Letters, v. 70, p. 325-345.

BOWRING, S. A., GROTZINGER, J. P., ISAACHSEN, C. E., KNOLL, A. H., PELECHATY, S. M., AND KOLOSOV, P., 1993, Calibrating rates of ealy Cambrian evolution: Science, v. 261, p. 1293-1298.

BRASIER, M. D., 1990, Phosphogenic events and skeletal preservation across the Precambrian-Cambrian boundary interval, *in* Notholt, A. F. G., and Jarvis, I., eds., Phosphorite Research and Development: London Geological Society Special Publication 52, p. 289-303.

BRASIER, M. D., 1992a, Nutrient flux and the evolutionary explosion across the Precambrian-Cambrian boundary interval: Historical Biology, v. 5, p. 85-93.

BRASIER, M. D., 1992b, Towards a carbon isotope stratigraphy of the Cambrian System: Potential of the Great Basin succession, *in* Hailwood, E. A., and Kidd, R. B., eds., High Resolution Stratigraphy: Oxford, Geological Society Special Publication 70, p. 341-350.

BRASIER, M. D., GREEN, O., AND SHIELDS, G. A., 1997, Ediacarian sponge spicule clusters from southwestern Mongolia and the origins of the Cambrian fauna: Geology v. 25, p. 303-306.

BRASIER, M. D., AND MCILROY, D., 1998, *Neonereites uniserialis* from c. 600 Ma year old rocks in western Scotland and the emergence of animals: Journal Geological Society London, v. 144, p. 5-13.

BRASIER, M. D., ROZANOV, A. YU., ZHURAVLEV, A. YU, CORFIELD, R., AND DERRY, L. A. 1994, A carbon isotope reference scale for the lower Cambrian series in Siberia: Report of IGCP Project 303: Geological Magazine, v. 131, p. 767-783.

BRASIER, M. D., SHIELDS, G. A., KULESHOV, V. N., AND ZHEGALLO, E. A., 1996. Integrated chemo- and biostratigraphic calibration of early animal evolution: Neoproterozoic-Early Cambrian of southwest Mongolia: Geological Magazine, v. 133, p. 445-85.

BRENCHLEY, P. J., MARSHALL, J. D., CARDEN, G. A. F., ROBERTSON, D. B. R., LONG, D. G. F., MEIDLA, T., HINTS, L. AND ANDERSON, T. F., 1994, Bathymetric and isotopic evidence for a short-lived Late Ordovician glaciation in a greenhouse period: Geology, v. 22, p. 295-298.

BURDETT, J. W., ARTHUR, M. A., AND RICHARDSON, M., 1989, A Neogene seawater sulphur isotope curve from calcareous pelagic microfossils: Earth and Planetary Science Letters, v. 94, p. 189-198.

BURKE, W. H., DENISON, R. E., HETHERINGTON, E. A., KOEPNIK, R. B., NELSON, H. F., AND OTTO, J. B., 1982, Variation of seawater $^{87}Sr/^{86}Sr$ throughout Phanerozoic time: Geology, v. 10, p. 516-519.

BURNETT, W. C., AND RIGGS, S. R., eds., 1990, Neogene to Modern Phosphorites, Phosphate Deposits of the World, Volume 3: Cambridge, Cambridge University Press, 464 p.

CANFIELD, D. E., AND TESKE, A., 1996, Late Proterozoic rise in atmospheric oxygen concentration inferred from phylogenetic and sulphur-isotope studies: Nature, v. 382, p. 127-132.

CHAUDURI, S. AND CLAUER, N., 1986, Fluctuations of isotopic composition of strontium in seawater during the Phanerozoic eon: Chemical Geology, v. 59, p. 293-303.

CLAYPOOL, G. E., HOLSER, W. T., KAPLAN, I. R., SAKAI, H., AND ZAK, I., 1980, The age curves of sulfur and oxygen isotopes in marine sulfate and their mutual interpretation: Chemical Geology, v. 28, p. 199-260.

COMPTON, J. S., HODELL, D. A., GARRIDO, J. R., AND MALLINSON, D. J., 1993, Origin and age of phosphorite from the south-central Florida Platform: Relation of phosphogenesis to sea-level fluctuations and $\delta^{13}C$ excursions: Geochimica et Cosmochimica Acta, v. 57, p. 131-146.

COMPTON, J. S., AND MALLINSON, D. J., 1996, Geochemical consequences of increased late Cenozoic weathering rates and the global CO_2 balance since 100 Ma: Paleoceanography, v. 11, p. 431-446.

COOK, P. J., AND MCELHINNEY, M. W., 1979, A reevaluation of the spatial and temporal deposition of sedimentary deposits in the light of plate tectonics: Economic Geology, v. 74, p. 315-330.

COOK, P. J., AND SHERGOLD, J. H., 1984, Phosphorus, phosphorites and skeletal evolution at the Precambrian-Cambrian boundary: Nature, v. 308, p. 231-236.

COOK, P. J., AND SHERGOLD, J. H., 1986, Proterozoic and Cambrian phosphorites—nature and origin, *in* Cook, P. J., and Shergold, J. H., eds., Proterozoic and Cambrian phosphorites: Phosphate Deposits of the World, Volume 1: Cambridge, Cambridge University Press, p. 369-386.

DELANEY, M. L., AND FILIPELLI, G. M., 1994, An apparent contradiction in the role of phosphorus in Cenozoic mass balances for the world ocean: Paleoceanography, v. 9, p. 513-527.

DERRY, L. A., and FRANCE-LANORD, C., 1996, Neogene growth of the sedimentary organic carbon reservoir: Paleoceanography, v. 11, p. 267-275.

DERRY, L. A., KAUFMAN, A. J., AND JACOBSEN, S. B., 1992, Sedimentary cycling and environmental change in the Late Proterozoic: Evidence from stable and radiogenic isotopes: Geochimica et Cosmochimica Acta, v. 56, p. 1317-1329.

Derry, L. A., Brasier, M. D., Corfield, R. M., Rozanov, A. Yu., and Zhuravlev, A. Yu., 1994, Sr and C isotopes in Lower Cambrian carbonates from the Siberian carbonates from the Siberian craton: A paleoenvironmental record during the "Cambrian explosion": Earth and Planetary Science Letters, v. 128, p. 671-681.

Donnelly, T. H., Shergold, J. H., Southgate, P. N. and Barnes, C. J., 1990, Events leading to global phosphogenesis around the Proterozoic/Cambrian boundary, *in* Notholt, A. F. G., and Jarvis, I., eds., Phosphorite Research and Development: London Geological Society Special Publication 52, p. 273-287.

Fischer, A. G., and Arthur, M. A., 1977, Secular variations in the pelagic realm, *in* Cook, H. E., and Enos, P., eds., Deep-water Carbonate Environments: Tulsa, Society for Sedimentary Geology (SEPM) Special Publication 25, p. 19-50.

Föllmi, K. B., 1995, 160 m.y. record of marine sedimentary burial: Cooling of climate and continental weathering under greenhouse and icehouse conditions: Geology, v. 23, p. 859-862.

Föllmi, K. B., Weissert, H. and Lini, A., 1993, Nonlinearities in phosphogenesis and phosphorus-carbon coupling and their implications for global change, *in* Wollast, R., MacKenzie, F. T., and Chou, L., eds., Interactions of C, N, P and S Biogeochemical Cycles and Global Change, NATO Asi Series 14: Berlin, Springer-Verlag, p. 447-474.

Garrison, R. E., and Kastner, M., 1990, Phosphatic sediments and rocks recovered from the Peru Margin during Leg 112, *in* Suess, E. and von Huene, R., et al., eds., Proceedings of the Ocean Drilling Program Scientific Results 112: College Station, Texas, Ocean Drilling Program, p. 111-134.

Glenn, C. R., 1990, Pore water, petrologic and stable carbon isotopic data bearing on the origin of Modern Peru margin phosphorites and associated authigenic phases, *in* Burnett, W. C., and Riggs, S. R., eds., Neogene to Modern Phosphorites: Phosphate Deposits of the World, Volume 3, p. 46-62.

Glenn, C. R., and Arthur, M. A., 1990, Anatomy and origin of a Cretaceous phosphorite-greensand giant, Egypt: Sedimentology, v. 37, p. 123-154.

Glenn, C. R., Föllmi, K. B., Riggs, S. R., Baturin, G. N., Grimm, K. A., and Trappe, J., 1994, Phosphorus and phosphorites: Sedimentology and Environments of Formation: Eclogae Geologicae Helvetica, v. 87, p. 747-788.

Grill, E. V., and Richards, F. A., 1962, Nutrient regeneration from phytoplankton decomposing in seawater: Journal of Marine Research, v. 22, p. 51-59.

Grotzinger, J., Bowring, S. A., Saylor, B. Z., and Kaufman, A. J., 1995, Biostratigraphic and geochronologic constraints on early animal evolution: Science, v. 270, p. 598-604.

Hess, J. Bender, M. L., and Schilling, J-G., 1986, Evolution of the ratio of strontium-87 to strontium-86 from Cretaceous to Present: Science, v. 231, p. 979-984.

Ingall, E., and Jahnke, R., 1997, Influence of water-column anoxia on the elemental fractionation of carbon and phosphorus during sediment diagenesis: Marine Geology, v. 139, p. 219-229.

Kaufman, A. J., Jacobsen, S. B., and Knoll, A. H., 1993, The Vendian record of Sr- and C-isotopic variations in seawater: Implications for tectonics and paleoclimate: Earth and Planetary Science Letters, v. 120, p. 409-430.

Kaufman, A. J., and Knoll, A. H., 1995, Neoproterozoic variation in the C-isotopic composition of seawater: Stratigraphic and biogeochemical implications: Precambrian Research, v. 73, p. 27-49.

Kaufman, A. J., Knoll, A. H., and Narbonne, G. M., 1997, Isotopes, ice ages, and terminal Proterozoic earth history: Proceedings of the National Academy of Science, v. 94, p. 6600-6605.

Kempe, S., Kazmierczak, J., and Degens, E. T., 1989, The soda ocean concept and its bearing on biotic evolution, *in* Crick, R. E., ed., Origin and Evolution and Modern Aspects of Biomineralization of Plants and Animals: New York, Plenum Press, p. 29-43.

Knoll, A. H., 1996. Daughter of time: Paleobiology, v. 22, p. 1-7.

Knoll, A. H., Bambach, R. K., Canfield, D. E., and Grotzinger, J. P., 1996, Comparative Earth history and Late Permian mass extinction: Science, v. 273, p. 452-457.

Kominz, M. A., 1984, Oceanic ridge volumes and sea-level change: An error analysis, *in* Schlee, J. S., ed., Interregional Unconformities and Hydrocarbon Accumulation: Tulsa, American Association of Petroleum Geologists Memoir 36, p. 109-127

Krasilnikova, N. A. and Ilyin, A. V., 1989, The Ordovician Baltic phosphorite Basin, USSR, *in* Notholt, A. F. G., Sheldon, R. E., and Davidson, D. F., eds., Phosphate Deposits of the World: Phosphate Resources of the World, Volume 2: Cambridge, Cambridge University Press, p. 494-496.

Kump, L.R., 1990, Neogene geochemical cycles; implications concerning phosphogenesis, *in* Burnett, W. C., Riggs, S. R., eds., Phosphate Deposits of the World, Volume 3; Neogene to Modern Phosphorites: Cambridge, Cambridge University Press, p. 273-282.

Lambert, I. B., Walter, M. R., Zang, W., Lu, S., and Ma, G., 1987, Paleoenvironment and carbon isotope stratigraphy of Upper Proterozoic carbonates of the Yangtze Platform: Nature, v. 325, p. 140-142.

Larson, R. L., 1991, Geological consequences of superplumes: Geology, v. 19, p. 963-966.

Lasaga, A., 1989, A new approach to isotopic modeling of the variation of atmospheric oxygen through Phanerozoic time: American Journal of Science, v. 289, p. 411-435.

Li Yueyan, 1986, Proterozoic and Cambrian phosphorites—regional review, *in* Cook, P. J., and Shergold, J. H., eds., Proterozoic and Cambrian Phosphorites, Phosphate Deposits of the World, Volume 1: Cambridge, Cambridge University Press, p. 42-612.

Logan, G. A., Hayes, J. M., Hieshima, G. B., and Summons, R. E., 1995, Terminal Proterozoic reorganization of biogeochemical cycles: Nature, v. 376, p. 53-56.

Lowenstam, H. A., and Margulis, L., 1980, Evolutionary prerequisites for early Phanerozoic calcareous skeletons: Biosystems, v. 12, p. 27-41.

Lyle, M., 1997, Could early Cenozoic thermohaline circulation have warmed the poles?: Paleoceanography, v. 12, p. 161-167.

McArthur, J. M., 1994, Recent trends in strontium isotope stratigraphy: Terra Nova, v. 6, p. 331-358.

McArthur, J. M., Thirlwall, M. F., Gale, A., Kennedy, W. J., Burnett, J. A., Mattey, D., and Lord, A. R., 1993, Strontium isotope stratigraphy for the late Cretaceous: A new curve, based on the English Chalk, *in* Hailwood, E. A., and Kidd, R. B., eds., High Resolution Stratigraphy: Cambridge, The Geological Society Special Publication 70, p. 195-209.

McArthur, J. M., Kennedy, W. J., Chen, M., Thirlwall M. F., and Gale, A., 1994, Strontium isotope stratigraphy for Late Cretaceous time: Direct numerical calibration of the Sr isotope curve based on the US Western Interior seaway: Palaeoceanography, Palaeoclimatology, Palaeoecology, v. 108, p. 95-119.

McManus, J., Berelson, W. M., Coale, K. H., Johnson, K. S., and Kilgore, T. E., 1997, Phosphorus regeneration in continental margin sediments: Geochimica et Cosmochimica Acta, v. 61, p. 2891-2907.

Melezhik, V. A., and Fallick, A. E., 1996, A widespread positive $\delta^{13}C$ anomaly at around 2.33-2.06 Ga on the Fennoscandian Shield: A paradox?: Terra Nova, v. 8, p. 141.157.

Mikhailov, J. A., Baturin, G. N., Kochenov, A. V., Mitrov, Y. V,. and Razvalyagev, A. V., 1972, Environment of phosphogenesis in the Nile syneclise and on the south-west African shelf: Lithological and Mineral Resources, v. 5, p. 3-13 (in Russian).

Miller, K. G., Feigenson, M. D., Wright, J. D., and Clement, B. M., 1991, Miocene isotope reference section, deep sea drilling project site 608: An evaluation of isotope and biostratigraphic resolution: Paleoceanography, v. 6, p. 33-52.

Misi, A., and Kyle, J. R., 1994, Upper Proterozoic carbonate stratigraphy, diagenesis and stromatolitic phosphorite formation, Irece Basin, Bahia, Brazil: Journal of Sedimentary Research, v. A64, p. 290-310.

Montanez, I. P., Banner, J. L., Osleger, D. A., Borg, L. E., and Bosserman P. J., 1996, Integrated Sr isotope variations and sea-level history of Middle to Upper Cambrian platform carbonates: Implications for the evolution of Cambrian seawater $^{87}Sr/^{86}Sr$: Geology, v. 24, p. 917-920.

Nicholas, C, 1996, The Sr isotopic evolution of the oceans during the 'Cambrian explosion': Journal of the Geological Society of London, v. 153, p. 243-254.

Nielsen, H., 1989, Local and global aspects of the Sulphur isotope age curve of oceanic sulphate, *in* Brimblecombe, P., and Lein, A. Yu., eds., Evolution of the Global Biogeochemical Sulphur Cycle, Scope 39: New York, John Wiley & Sons Ltd., p. 57-64.

Notholt, A. J. G., Sheldon, R. P., and Davidson, D. F., eds., 1989, Phosphate

Resources of the World, Phosphate Deposits of the World, Volume 2: Cambridge, Cambridge University Press, 566 p.

PALMER, M. R., AND ELDERFIELD, H., 1985, Sr isotope composition of seawater over the past 75 Myr: Nature, v. 314, p. 526-528.

PEUKER-EHRENBRINK, B., RAVIZZA, G., AND HOFMANN, A.W., 1995, The marine $^{187}Os/^{186}Os$ record of the past 80 million years: Earth and Planetary Science Letters, v. 130, p. 155-167.

RAVIZZA, G., 1993, Variations of the $^{187}Os/^{186}Os$ ratio of seawater over the past 28 million years as inferred from metalliferous carbonates: Earth and Planetary Science Letters, v. 113, p. 569-576.

RAVIZZA, G., AND ESSER, B. K., 1993, A possible link between the seawater osmium isotope record and weathering of ancient sedimentary organic matter: Chemical Geology, v. 107, p. 255-258.

RAYMO, M. E., 1994, The Himalayas, organic carbon burial, and climate in the Miocene: Paleoceanography, v. 9, p. 399-404.

RICHTER, F. M., ROWLEY, D. B., AND DEPAOLO, D. J., 1992, Sr isotope evolution of seawater: The role of tectonics: Earth and Planetary Science Letters, v. 109, p. 11-23.

RIGGS, S. R., AND SHELDON, R. P., 1990, Paleoceanographic and paleoclimatic controls of the temporal and geographic distribution of Upper Cenozoic continental margin phosphorites: *in* Burnett, W. C., and Riggs, S. R., eds., Neogene to Modern Phosphorites, Phosphate Deposits of the World, Volume 3: Cambridge, Cambridge University Press, p. 381-395.

RUNNEGAR, B., 1987, The evolution of mineral skeletons, *in* Crick, R. E., ed., Origin, Evolution and Modern Aspects of Biomineralization in Plants and Animals: New York, Plenum Press, p. 75-94.

RUTTENBERG, K., AND BERNER, R. A., 1993, Authigenic apatite formation and burial in sediments from non-upwelling continental margin environments: Geochimica et Cosmochimica Acta, v. 57, p. 991-1007.

SCHLANGER, S. O., ARTHUR, M. A., JENKYNS, H. C., AND SCHOLLE, P. A., 1987, The Cenomanian-Turonian oceanic anoxic event 1: Stratigraphy and distribution of organic carbon-rich beds and the oceanic $\delta^{13}C$ excursion: Geological Society of London Special Publication 26, p. 371-399.

SCOTESE, C. R., GAHAGAN, L. M., AND LARSON, R. L., 1988, Plate tectonic reconstructions of the Cretaceous and Cenozoic ocean basins, *in* Scotese, C. R., and Sager, W. W., eds., Mesozoic and Cenozoic Plate Reconstructions: Tectonophysics, v. 155, p. 27-48.

SENGÖR, A. M. C., NATAL'IN, B. A., AND BURTMAN, V. S., 1993, Evolution of the Altaid tectonic collage and Paleozoic crustal growth in Eurasia: Nature, v. 364, p. 299-307.

SHACKLETON, N. J., AND HALL, M. A., 1984, Carbon isotope data from Leg 74 sediments, *in* Moore, T. C., Jr., and Rabinowitz, R. D., et al., eds., Initial Reports of the Deep Sea Drilling Project, v. 74: Washington, U.S. Government Printing Office, p. 613-620.

SHACKLETON, N. J., AND HALL, M. A., 1995, Stable isotope records in bulk sediments (Leg 138), *in* Pisias, N. G., Mayer, L. A., Janacek, T. P., Palmer-Julson, A., and van Andel, T. H., eds., Proceedings of the Ocean Drilling Program, Scientific Results, v. 138: College Station, Texas, Ocean Drilling Program, p. 797-805.

SHELDON, R. P., 1980, Episodicity of phosphate deposition and deep ocean circulation—a hypothesis, *in* Bentor, Y. K., ed., Marine Phosphorites: Geochemistry, Occurrence, Genesis: Tulsa, Society for Sedimentary Geology (SEPM) Special Publication 29, p. 239-247.

SHELDON, R. P., 1989, Phosphorite deposits of the Phosphoria Formation, western United States, *in* Notholt, A. F. G., Sheldon, R. P., and Davidson, D. F., eds., Phosphate Resources of the World: Phosphate Deposits of the World, Volume 2: Cambridge, Cambridge University Press, p.53-61.

SHERGOLD, J. H., AND BRASIER, M. D., 1986, Proterozoic and Cambrian phosphorites—specialist studies: Biochronology of proterozoic and Cambrian phosphorites, *in* Cook, P. J., and Shergold, J. H., eds., Proterozoic and Cambrian Phosphorites,Phosphate Deposits of the World, Volume 1: Cambridge, Cambridge University Press, p. 295-326.

SHIELDS, G. A., 1996, Event stratigraphy around the Precambrian-Cambrian boundary with special emphasis on S. China and W. Mongolia: Unpublished Ph.D. Dissertation, ETH Zürich, 133 p.

SHIELDS, G. A., STRAUSS, H., HOWE, S. S. AND SIEGMUND, H., 1999, Sulphur isotope compositions of sedimentary phosphorites from the basal Cambrian of China—Implications for Neoproterozoic-Cambrian biogeochemical cycling: Journal of the Geological Society of London, v. 156, p. 943-955.

STILLE, P., 1992, Nd-Sr isotope evidence for dramatic changes of palaeocurrents in the Atlantic Ocean during the past 80 m.y.: Geology, v. 20, p. 387-390.

STILLE, P., STEIMANN, M., AND RIGGS, S. 1996, Nd isotope evidence for the evolution of the paleocurrents in the Atlantic and Tethys oceans during the past 180 Ma: Earth and Planetary Science Letters, v. 144, p. 9-19.

STRAUSS, H., 1993, The sulfur isotopic record of Precambrian sulfates: new data and a critical evaluation of the existing record: Precambrian research, v. 63, p. 225-246.

STRAUSS, H., VIDAL, G., MOCZYDLOWSKA, M., AND PACZESNA, J., 1997, Carbon isotope geochemistry and palaeontology of Neoproterozoic to early Cambrian siliciclastic successions in the East European Platform, Poland: Geological Magazine, v. 134, p. 1-16.

TORSVIK, T. H., SMETHURST, M. A., MEERT, J. G., VAN DER VOO, R., MCKERROW, W. S., BRASIER, M. D., STURT, B. A., AND WALDERHAUG, H. J., 1996, Continental break-up and collisiosn in the Neoproterozoic and Paleozoic—a tale of Baltica and Laurentia: Earth Science Reviews, v. 40, p. 229-258.

VAN CAPELLEN, P., AND INGALL, E. D., 1996, Redox stabilization of the atmosphere and oceans by phosphorus-limited marine productivity: Science, v. 271, p. 493-496.

VEIZER, J., COMPSTON, W., CLAUER, N. AND SCHIDLOWSKI, M., 1983, $^{87}Sr/^{86}Sr$ in late Proterozoic carbonates: Evidence for a 'mantle' event ca. 900 Ma ago: Geochimica et Cosmochimica Acta, v. 47, p. 295-302.

VINCENT, E., AND BERGER, W. H., 1985, Carbon dioxide and polar cooling in the Miocene: The Monterey hypothesis, *in* Sunquist, E. T., and Broeker, W. S., eds., The Carbon Cycle and Atmospheric CO_2: Natural Variations from Archean to Present: Geophysical Monograph 32: Washington, D.C., American Geophysical Union, p. 455-468.

WANG ZONGZHE, YANG JIEDONG, AND SUN WEIGUO, 1996, Carbon isotope record of Sinian seawater in Yangtze Platform (in Chinese): Geological Journal of Universities, v. 2, p. 112-120.

WATKINS, R. T., NATHAN, Y., AND BREMNER, J. M., 1995, Rare earth elements in phosphorite and associated sediment from the Namibian and South African continental shelves: Marine Geology, v. 129, p. 111-128.

WEISSERT, H., AND LINI, A., 1991, Ice age interludes during the time of Cretaceous greenhouse climate? *in* Müller, D., McKenzie, J. A., and Weissert, H., eds., Controversies in Modern Geology: London, Academic Press, p. 173-191.

WHEAT, C. G., FEELY, R. A., AND MOTTL, M. J., 1996, Phosphate removal by oceanic hydrothermal processes: An update of the phosphorus budget in the oceans: Geochimica et Cosmochimica Acta, v. 60, p. 3593-3608.

WILLETT, B. M., GAYLARD, D. AND ATKINSON, G., eds., 1983, Philips' Modern Atlas, 80th Edition: London, George Philip & Son Ltd., p. 132.

YANG JIEDONG, TAO XIANCONG, AND XUE YAOSONG, 1995, Nd isotopic variations of Chinese seawater during Neoproterozoic through Cambrian: Chemical Geology, v. 135, p. 127-137

YE LIANJUN, 1984, Regularities of the occurrence of China phosphorite deposits: Developments in Geoscience, Contribution to 27th International Geological Congress, Moscow: Beijing, Science Press, p. 191-199.

PHOSPHORITE AND LIMESTONE, TWO INDEPENDENT END-MEMBER PRODUCTS OF THE RANGE OF BIO-PRODUCTIVITY IN SHALLOW MARINE ENVIRONMENTS

JACQUES LUCAS AND LILIANE PRÉVÔT-LUCAS*

Centre de Géochimie de la Surface (CNRS), EOST-Université Louis Pasteur,
1 rue Blessig, F-67084 Strasbourg Cedex, France

ABSTRACT: Calcium carbonate is known to be one of the sedimentary materials occuring with phosphorite in many of the phosphorite deposits. This relationship has been described for various geological periods and at various scales: mineable phosphorite alternating with barren carbonate; calcitic or more often dolomitic phosphorite alternating with phosphatic dolomite or limestone. In the present study, we attempt to explain why there is a link between phosphorite and carbonate. This relationship does not presuppose a genetic link between minerals, although this link does exist where apatite can be recognized as the result of apatitization of pre-existing calcite. Calcite is a true biomineral that is directly produced by living marine organisms as their skeletons. It is deposited as a solid material before various diagenetic recrystallizations add their imprints. Apatite is an authigenic mineral, precipitated during early diagenesis from an organic matter (OM)-rich ooze. The OM must be furnished predominantly by naked and or siliceous organisms and its degradation processes are controlled by dysoxic conditions. The formation of limestones depends on a predominantly carbonate-producing productivity that is common because of the abundance of carbon. Phosphorite formation requires high quantities of OM related to specific noncarbonate-producing productivity that is less common because the availability of phosphorus is minor compared to that of carbon. The "phosphorite factory," which produced the phosphorite deposits, and the "carbonate factory" are independent but complementary rather than competing processes; they do not use the same primary material, nor do they occur at the same phase of the sedimentary cycle. The occurrence of alternations of beds of phosphorite and beds of calcium carbonate is thought to be the record in the sediment of changes in the dominant type of bioproductivity in the water column. Shallow-water shelves without significant detrital influx are the environments where such frequent changes in the type of bioproductivity are known to occur easily in response to slight physical or chemical changes.

INTRODUCTION

Carbon and phosphorus are two essential nutrients in the biosphere. Both are constituents of organic matter (OM), and their ratios vary depending on the species of the organisms (e.g., Baturin, 1982). Both are also skeletal components. Carbonate accumulates in external skeletons and in the shells of marine organisms. Phosphate accumulates in the bones of vertebrates, but is also found in the tests of brachiopods and in fish scales.

This involvement of carbon and phosphorus in biological activity as basic constituents is likely to reflect their similar behavior during the mineralization process. The cycle of carbon and the cycle of phosphorus were often studied (Van Cappellen and Ingall, 1994; Compton et al., this volume) either alone, or in parallel and compared, in spite of the important difference of concentration of both elements in sediments. The high C/P ratio at the Earth's surface leads to very different accumulation rates for calcite (or aragonite) and apatite. Because of the abundance of carbon, carbonates, which have increased proportionally in sediments through geologic time, are omnipresent. They may occur pure, or may be mixed in any proportion with siliciclastic material to form a limestone-marlstone-claystone series. Because the concentration of phosphorus is much lower than that of carbon, apatite, though often not detected in sedimentary deposits because it is not a major constituent and is generally disseminated, is certainly a much more widely distributed mineral than might be supposed from routine analyses, and is found in many *normal* sediments (Ruttenberg and Berner, 1993; Föllmi, 1996). High concentrations of apatite as phosphorite, however, which are the equivalent of high concentrations of calcite as limestone, are rare.

This points to a difference of significance between limestone and phosphorite and the cycles of carbon and phosphorus. Limestone is a normal, common manifestation of the carbon cycle. The equivalent manifestation of the phosphorus cycle is not phosphorite but the mineral apatite, which is disseminated in various marine sediments. Phosphorites are exceptional concentrations of apatite in more-or-less thick beds that form under specific conditions that have been widely discussed (Glenn et al., 1994). Many models of genesis have been proposed, including those by Riggs (1986 and references therein), Cook and Shergold (1986), Lucas and Prévôt-Lucas (1997). Phosphorites are known to occur with many other nonterrigenous sedimentary materials (Riggs, 1986), including carbonates, where continental influx was generally rather low.

The concentration of phosphorus in phosphorite deposits corresponds to a localized increase of several orders of magnitude in the rate of P-burial compared to common sediments. We do not believe that such an increase requires exceptional injections of phosphorus into the global phosphorus cycle (Compton et al., this volume). Instead, it seems to be tied to specific local conditions. In other words, we believe that the episodicity of phosphorite deposits through geological time (Cook and McElhinny, 1979) is not brought about by episodic changes in the phosphorus cycle, but by an episodicity of the conjunction of

*email: jlucas@illite.u-strasbg.fr

Marine Authigenesis: From Global to Microbial, SEPM Special Publication No. 66

a set of locally favorable environmental conditions.

In the present study, we attempt to explain the common though not requisite relationship, which we do not believe to be fortuitous, between phosphorites and limestones. Both resulting from biological activity, their association is not surprising as is not, for the same reason, their common association with chert. The high marine carbon-to-phosphorus ratio accounts for the common occurrences of limestone without apatite and, conversely, for the common occurrence of limestone (or dolomite) in phosphorite deposits. The intimate association of carbonate and phosphorite demands that a mechanism be evoked whereby the repeated transition from conditions favoring one or the other does not require repeated drastic changes in the environment.

Carbonates do appear to be related to phosphorites at various scales in various settings, but in some settings carbonates are either missing or less abundant, so that the carbonate component has not held the attention of many scientists. Carbonates are definitely a major component of the rock suites in most of the phosphorite deposits of the Late Cretaceous along the Tethys margin (Lucas and Prévôt-Lucas, 1995). Belonging to this Tethyan realm, phosphorites shift nonrandomly between Turonian and Lutetian from east to west. In the Moroccan deposits Ganntour and Oulad Abdoun (Boujo, 1976; Belfkira, 1980; Prévôt, 1990; Benalioulhaj et al., this volume), the economically high-grade apatite formations have been shown to contain, on average, as much carbonate as apatite (dolomite being the most frequent and most abundant), even in the mined beds of the Montian, which provide the highest % P_2O_5 (Prévôt, 1990). This deposit has been preserved in a relatively "primary" state, being only slightly reworked, not affected by tectonism, and only very locally weathered or transformed by late diagenesis (Prévôt, 1990, 1991). At this site, the carbonates are either barren (nonphosphatic) beds sharply separated from overlying and underlying phosphorite beds, or within the phosphatic beds.

The example in the Permian Phosphoria Formation given by Sheldon (1989) also shows that "although phosphorites commonly contain a few percent carbonate minerals, and dark-colored carbonate rocks commonly contain a few percent apatite, extensive mixing of these rocks is rare. On the other hand,. . .mudstone and apatite are usually interlaminated." Prian (1980) described the separation of carbonate rocks and phosphorites in Cambrian formations of the Montagne Noire in France.

Most phosphatic formations are described as stratified sedimentary deposits, and we believe that in many of the economic phosphate deposits this interbedding is genetic due to the conditions surrounding the formation, deposition, and preservation of the minerals.

THE MINERALS

Calcium Carbonates

Calcite and aragonite are the main components of the external skeletons of marine organisms. Apart from oolitic or stromatolitic limestones, which are considered chemical and microbiochemical in origin, most limestones were formed by the accumulation of particles derived directly from these skeletons. This is clear for reefal deposits, chalk, and coquinas. It is less evident for fine-grained limestone and marlstone, where sedimentary and diagenetic processes have led to micritization, dissolution, and recrystallization. Nevertheless, there appears little doubt that most limestones are primarily biological in origin.

Thus, most calcite and aragonite in sediments resulted from (1) biomineralization of CO_2 by marine organisms to form their external skeletons, followed by (2) shedding of the calcium carbonate by the organisms, either while living or after death. Most carbonate deposited in particulate form is eventually reworked like all other particles and constitutes the primary calcite or aragonite of the sediment, while early diagenetic processes are responsible for micritization or transformation into secondary calcite.

Apatite

Apatite is thought to form during early diagenesis when organic matter slowly decomposes due to bacterial mediation. Support for this hypothesis comes from both examination of the sedimentary record and experimental work. The association, on various scales, of apatite and OM was frequently recorded. Several of the trace elements associated with apatite are proxies for OM (Prévôt, 1990). There is close resemblance of the OM in phosphorites to the OM in associated black shales (Meunier-Christmann et al., 1989; Benalioulhaj, 1991). Experimental work has shown that microbial action on OM in freshwater and seawater can lead to apatite formation either by direct precipitation from solution or by replacement of a pre-existing carbonate (Lucas and Prévôt, 1985; Lucas et al., 1990; Prévôt et al., 1989). Apatite formation is favored when fluoride is available in a magnesium-depleted environment, which is consistent with a closed marine environment such as in the interstitial pores of freshly deposited marine sediments (Lucas and Prévôt, 1984, 1985; Lucas and Prévôt-Lucas, 1997). The natural and the experimentally formed phosphate mineral is carbonate-fluorapatite, which precludes a primary skeletal origin because vertebrate bones are made of a different apatite, that is, low-fluoride hydroxyapatite, which is diagenetically changed during fossilization (Lucas and Prévôt, 1991).

Apatite is the result of (1) releasing of phosphorus from the OM of living organisms and (2) precipitation in conjunction with the slow, microbially induced degradation of the OM in the uppermost centimeters of the sediment during the very first stages of diagenesis.

Comparison

Comparison of the stories of these two mineral families identifies a major difference. Carbonates are essentially biominerals, in the sense used by Trichet and Defarge (1995), that formed in the seawater column, whereas apatite is the product of a chemical precipitation that occurs below the seawater-sediment interface, within the interstitial fluids that have had their chemical compositions modified by microbial activity. Neither the processes by which these minerals are introduced to the sediment column, nor the environments of deposition, are the same for the two mineral families.

BIOLOGICAL UPTAKE, A PRIMARY CONTRIBUTION TO SEDIMENTATION

Carbonate and phosphate ions are trapped by living organisms. All organisms need phosphorus in their cells, where it is a component of nucleic acids. All organisms also need carbon to form their OM, but many of them utilize a great deal of carbon to make their shells. This is the "carbonate factory " (James, 1979; Homewood, 1996). The relative amounts of carbonate and OM produced must depend on the proportion of organisms having and not having carbonate shells or skeletons (Prévôt, 1990). Moreover, although all organisms need nutrients to develop, not all adopt the same strategy for reproduction and ecological succession (Birkeland, 1987). This point was discussed in detail by Homewood (1996), who suggests a link between the carbonate productivity (forams, coccoliths, etc.) and low nutrient supply. Abundant nutrients may favor the development of phytoplankton, silica-walled organisms (radiolarians, diatoms) or naked organisms (acritarchs, dinoflagellates, and various algae).

Specific nutrients have different effects on the development of particular organisms. The limiting nutrients that are most efficient in the production of OM are phosphorus and nitrogen. The mean ratio of nitrogen to phosphorus in marine organisms, known as the Redfield ratio, is 16. However, these two elements are not necessarily present in this stoichiometric ratio at any given place at the same time. Depending on whether phosphorus or nitrogen is the limiting element, the effect on the assemblage of organisms is different, and the chemical composition of the OM is thus different (Goldman et al, 1979). Karl et al. (1997) have shown that some microorganisms are able to use atmospheric nitrogen, strongly increasing bioproductivity without requiring any additional phosphorus supply. The organisms living in an ecosystem where the ratio of carbon to nitrogen is 20 could sequester appreciable carbon. Falkowski (1997) followed the same idea while considering that the variation of fixed nitrogen has a direct effect on the fixation of CO_2 in the oceans. The massive phytoplankton bloom provoked by the seeding of oceanic water with dissolved iron (experimentally performed *in situ* by Coale et al., 1996) may well have been linked to increased nitrogen fixation, "because nitrogenase requires iron as a co-factor" (Karl et al., 1997). Such enhancement in nitrogen availability can be induced by a supply of iron or other trace elements (Frost, 1996; Tortell et al., 1996) independent of phosphorus supply. Arguing from this, we suggest that an increase of nitrogen in the marine environment results in an increase in the productivity of CO_2-fixing organisms and thus an increase in carbonate production.

On the other hand, areas with an adequate phosphorus supply, such as upwelling zones, are commonly areas of high productivity, where noncalcareous plankton are more prevalent. In this situation, the availabilily of phosphorus is not a limiting factor. Deep water upwelling into the photic zone creates an environment with a nitrogen-phosphorus ratio of less than 16. This situation is not advantageous for the sequestering of CO_2, and the resulting phytoplanktonic population consists either of dinoflagellates or silica-walled organisms (Hutchins and Bruland, 1998).

Indeed, there are other environmental conditions that affect species changes and their relative abundance. These include water transparency, water temperature, and other factors that depend not only on climate, but also on the morphology of basins at the ocean-continent interface, the continental clastic supply, sealevel variations, and other factors that modify the conditions of stratigraphic accommodation (Westbroek et al., 1994; Busson et al., 1997).

Thus, depending on the chemical composition of the environment, a first sorting of the mineral elements is induced by the microorganisms themselves. This sorting is enhanced by the differential behavior of the organic and mineral components of organisms during sedimentation, as documented by many authors dealing with the accumulation of OM and/or carbonates. Tribovillard et al. (1996) discussed the fate of different types of phytoplankton during their sinking and settling. They observed that although all carbonate-shelled organisms have a fairly high ratio of OM to calcite, much calcareous sediment, such as chalk, contains little or no OM. We contend that this is due to differences in the composition of the OM depending on the species assemblages of the plankton. Polysaccharides and polypeptides, which are dominant in carbonate- and silica-walled phytoplankton, are less resistant to decomposition during sedimentation than the lipids, which are more abundant in naked phytoplankton (Goth et al., 1988; Largeau et al., 1990; Derenne et al., 1991; Harvey et al., 1995). However, diatoms, which produce a high quantity of fatty acids during certain stages of their growth, are intermediate in their resistance to decomposition, so that silica-walled plankton also contribute to the production of resistant OM.

Carbonate reservation is also strongly influenced by the $OM/CaCO_3$ ratio in the sediment. As soon as sediment is deposited, especially in shallow water, microbial activity provokes the "respiration of the OM" (Moulin et al., 1985; Walter and Burton, 1987; Emerson and Archer, 1990; Wollast, 1994) leading to dissolution of the carbonate. This is more or less rapid, depending on the redox state of early diagenesis in the environment. The higher the OM content of the sediment, the more easily its calcite is dissolved.

The silica-walled components of the marine phytoplankton play an important and specific role in sedimentation. When carbonate input is low, their OM seems to be resistant enough to contribute, together with the OM of naked plankton, to the accumulation of organic matter in sediments. However, the relative fragility of the OM of silica-walled organisms results in its complete destruction in sediments dominated by carbonate-walled plankton, such as chalk which, for this reason, has very little OM. Opal tests do not interact with carbonates or with OM. Carbonates, which are stable in surface waters, but soluble below the carbonate compensation depth (CCD), behave differently than silica, which dissolves continuously from the surface water to the bottom. Furthermore, within the sediment during early diagenesis, silica dissolution is more active if the sediment is porous (De Wever et al., 1996). Early dissolution of biogenic silica is consistent with the scarcity of siliceous fossils and the abundance of chert and various kinds of silica nodules in both chalk and phosphorites and with the common silicification observed in phosphorites.

We concur with Homewood (1996), in broad view of ecological succession, that the end member products of marine biological activity are mineral biomass (carbonate and silica) and OM. At this stage of the scenario, apatite still has not appeared, because phosphorus is still associated with sedimented organic constituents.

ACCUMULATION AND EARLY DIAGENESIS OF ORGANIC MATTER: THE PHOSPHORITE FACTORY

Mineralization of organic phosphorus requires a chemical environment that is different from that of seawater, but which exists in the pore fluid of organic-rich sediments during early diagenesis. Although disseminated apatite mineralization occurs throughout most oceanic sediments (Krajewski et al., 1994), concentrations of apatite in discrete beds or as giant deposits require the deposition of huge quantities of OM, which may be linked to a primary biomass composed mainly of naked and silica-walled plankton, as discussed above.

The debate over the past decide over the resistance of OM to decay after its deposition on the bottom of the ocean may be stated as "anoxia versus productivity" (Pedersen and Calvert, 1990). Preservation of OM is related to an anoxic environment, but direct evidence suggests that preliminary anoxia of the milieu is not absolutely necessary (Calvert et al., 1992; Ingall et al., 1993; Tribovillard et al., 1994). Anoxia can be the result of a high rate of OM accumulation. In calm water, in which large amounts of OM sink to the sea bottom, renewal of the bottom water oxygen may be insufficient to ensure full oxidation, and so the environment becomes anoxic or at least dysoxic. Thus, an oxygen minimum zone may cause the preservation of OM, but it is not a requisite condition. Up to now, the debate was mainly about deposition of OM in the deep ocean, at depths of several hundreds to several thousands of meters. It seldom dealt with the fate of OM on shallow water shelves, which are known to be the environment in which most of the commercial phosphate deposits formed.

It is obvious that shallow waters are much more sensitive to morphological and climatic variations than the deep water of oceanic basins. In shallow-water systems, local changes in the environment resulting from changes in the bottom morphology or sealevel can induce changes in productivity. Environmental changes are suggested to be linked in a feedback loop to accommodation space and base level (Homewood, 1996). Climatic variations that modify water temperature also have to be taken into account, as well as the influx of continental clastics. The environment is commonly oxic, and the OM is not preserved in the sediment unless it is protected from oxidation and microbial breakdown. A good way for the OM to escape destruction seems to be by its mixing during or soon after deposition with clay material on which it is easily adsorbed. In clay-rich mud, OM is mechanically protected because bacteria, with their hydrolytic enzymes, are too large to have access to the tiny pore spaces of such muds (Benalioulhaj, 1991; Mayer, 1994; Caplan and Bustin, 1996; Lucas and Prévôt-Lucas, 1997). In other words, OM preservation requires a basin with sufficient subsidence for the accumulation of fine siliciclastic material, such as in the black shales deposit of Timahdite (Morocco) (Benalioulhaj et al., this volume). OM with clay may lead to black shales. Black shales are commonly associated with phosphorites. Both are indeed related to the production of OM, yet phosphorites form only during synsedimentary diagenetic destruction of OM.

In the shallowest parts of the shelf, when siliciclastic supply is scarce, marine biogenic material is the main sediment type. These are areas where seawater is oxygenated, and OM is easily destroyed. However, the oxygen available is too low to consume all that OM, and the dysoxic environment promoting apatite formation during the slow decay of OM is induced if the sedimentation rate of OM is high enough (Canfield, 1994).

Noticeable amounts of apatite require the transformation of a large quantity of OM of average composition. A present-day high-productivity zone can provide 2800 $g/m^2/y$ (Parsons et al., 1977). Calculations suggest that in one year 2,800 g/m^2 would be required to form an apatite bed with a maximum thickness of 0.05 mm. (Lucas and Prévôt-Lucas, 1997). This approximation supposes that nothing was lost during the transformation, which is unlikely but does serve to give an idea of the quantities involved. It would take at least 100 years to form a 0.5-cm-thick bed of pure apatite. Considering porosity and the average P_2O_5 content of mineable deposits, 0.5 cm of pure apatite would correspond to 1 cm of phosphorite containing 50% apatite. With these estimates in mind, several questions may be addressed.

1. Does the amount of phosphorus needed to produce a phosphorite deposit require exceptionally high rates of OM deposition? If we take the example of the Oulad Abdoun phosphorite deposit in Morocco, in the area of Khouribga (Benalioulhaj, 1991; Moutaouakil and Giresse, 1993; Benalioulhaj et al., this volume), the time span covered by apatite formation ranged from early Maastrichtian to early Lutetian, a minimum of 25×10^6 years. The cummulative thickness of phosphorites deposited during that period, plus an additional 10% to account for the apatite that has dispersed, is 12 m. Keeping the hypothesis of a 2800 g/m^2yr^1 deposition rate, 12 m would have accumulated in only 1.2×10^5 years, i.e., only 1/200 of the available time. Therefore, even if the OM accumulation rate was much lower than 2800 $gm/cm^2/yr$, the amount of phosphorus supplied by OM was higher by far than the amount needed. This is the same order of magnitude as the 2×10^5 years estimated by Gulbrandsen (1969) for the Phosphoria Formation and is in agreement with the results obtained by the comparison of four ancient phosphorite deposits and one modern phosphogenic environment (Filippelli and Delaney, 1992). Therefore, the formation of a phosphate deposit does not require an exceptionally high rate of phosphorus.
2. Is it possible to accumulate all the OM required in one location? Setting the specific gravity of OM at 1.0 and apatite at 3.0, a 1-cm-thick bed of phosphorite would require about

30 cm of OM. This estimate is in agreement with the calculations by Föllmi (1996). Taking into account the hard parts of the siliceous and calcareous microorganisms, and the interstitial water in the mud, the primary source bed would be approximately 1 m thick. The preservation of a 1-m-thick bed of organic-rich sediment derived from a peculiar type of biomass, i.e. low in carbonate, deposited during a 100-year period of continuously high productivity is extremely unlikely, especially in shallow-water conditions. One meter of sediment per 100 years is a high accumulation rate on the geological time scale, whereas preservation of OM in a reasonably oxygenated environment for several years is a long time on the scale of life. This points to the rarely addressed problem of scale difference between biological time and geological time.

To escape this difficulty and approach both scales at the same time, it is necessary to reexamine the conditions of sedimentation in shallow platform areas that have remained tectonically stable for periods of time on the order of tens of million of years. The sedimentary regime can not have been continuous and stable during such long time spans, especially with respect to phosphate deposits that are clearly the result of discontinuous sedimentation. Phosphate series are known to display repeated lithologic changes and to be condensed series that require more intense diagenetic compaction of the source bed than other sedimentary rocks, due not only to the removal of water, but also to the almost total breakdown of the primary OM.

There might have been discontinuities in the sedimentation of the OM that serves as source bed of phosphorites. We explained that this OM has to be abundant and quickly protected from degradation. Little is known about this topic, and there is disagreement on how OM settles to the bottom. Among several possibilities of rapid sedimentation, the recycling of OM in the form of mechanically resistant fecal pellets produced in large quantities seems to be efficient. Some authors believe this to be a very important mechanism (possibly the sole route) for the formation of apatite grains (Lamboy et al., 1994). This form of OM is likely to be important, but it cannot be the only form of OM, especially in an environment with high productivity and rapid turnover where low-diversity assemblages would rapidly become predominant. Another also frequently invoked mechanism is siliceous or naked phytoplankton blooms that result from physical or chemical changes in the environment. Diatom blooms, which have been studied by many authors (e.g., Parsons et al., 1977; Barron, 1987; Hein and Parrish,1987; Jordan et al., 1991; Grimm et al., 1997), play an important role in the living biosphere that supplies the phosphorite source bed with OM. Kemp and Baldauf (1993) note the occurrence of Neogene oceanic diatom mats correlated for distances of more than 2000 km in the Pacific Ocean at a depth of 2000 to 3000 m. They explain that these diatoms, which lived in surface water, formed thick mats that must have sunk suddenly to the ocean bottom, where they were preserved due to the impenetrability of their meshwork. Repetition of that scenario over a period of 10 million years is suggested to have produced a thick, laminated bed with intercalated calcareous nannofossil oozes (Kemp and Baldauf, 1993). Similar mechanisms may well produce analogous deposits in shallower water where algal mats can settle, through a mechanism of self-sedimentation (Grimm et al., 1997) due to rapid aggregation of a transparent gel formed by the algae.

Diatoms are not found in abundance in the geological record because their opal tests are easily dissoluted, while cysts of dinoflagellates are abundant and well preserved in phosphate horizons and are good indicators of environmental conditions and are useful for stratigraphic correlation (Eshet et al., 1994). Variation in species diversity within one sedimentary sequence often ends with nearly monogeneric assemblages (Prévôt et al., 1979; Rauscher and Doubinger, 1982; Soncini, 1990). The sudden appearances and disappearances of dinoflagellate species in the sedimentary record are thought to reflect repeated blooms. Alternatively, it also may be explained as the result of a continuous supply of OM provided by a continuous development of organisms in a semi-closed environment with restricted exchange with the ocean. Nutrients are enriched in such environments through continuous oceanic supply and their restoration during the decomposition of organisms (Fig. 1). Dinoflagellates become the overwhelmingly dominant and often monospecific phytoplankton. This leads to pollution similar to that caused by present-day red tides, in which organisms consume more oxygen than available, makes the environment euxinic. Dinoflagellates also secrete toxins (saxitoxine) that lead to mass mortality (Brongersma-Sanders, 1957) and to the almost instantaneous sedimentation of a large amount of OM. This discontinuity process is a kind of threshold mechanism, the result of a feedback mechanism that occurs during continuous and regular biological development. Such mechanisms, among others (Grimm et al., 1996), allow the accumulation rate of OM to be high enough at times for the rate of breakdown of OM to be slowed down. Periods of temporary preservation of settled but not sealed OM are one possible cause for "pulsations" in the microenvironments within the fresh sediment, which achieve redox oscillations from oxic to anoxic (Aller, 1994) and opening-closing of the microenvironments (Lucas and Prévôt-Lucas, 1997). These conditions seem required for the precipitation, under the control of microbial activity, of a high amount of apatite. We propose "phosphorite factory," the chain of conditions and mechanisms producing phosphorite, i.e. a rock containing a high content of apatite, as an homologue to "carbonate factory."

The scarce Ca-carb in organic-rich muds is unstable and is commonly transformed into dolomite, since magnesium is not used in the phosphorite factory. The phosphorite factory works discontinuously, because the supply of material is discontinuous. Schematically, when external and internal conditions of the basin remain stable for a sufficient length of time, the repetition of discontinuous supply may produce an accumulation source bed sufficient to form a bed of phosphorite. It has been stressed that apatite precipitation can be rapid. The limiting mechanism for sufficient apatite formation is deposition of OM, which is discontinuous because of the high sensitivity to slight environmental changes. Environmental change can even result in complete cancellation of phosphorus delivery to the factory.

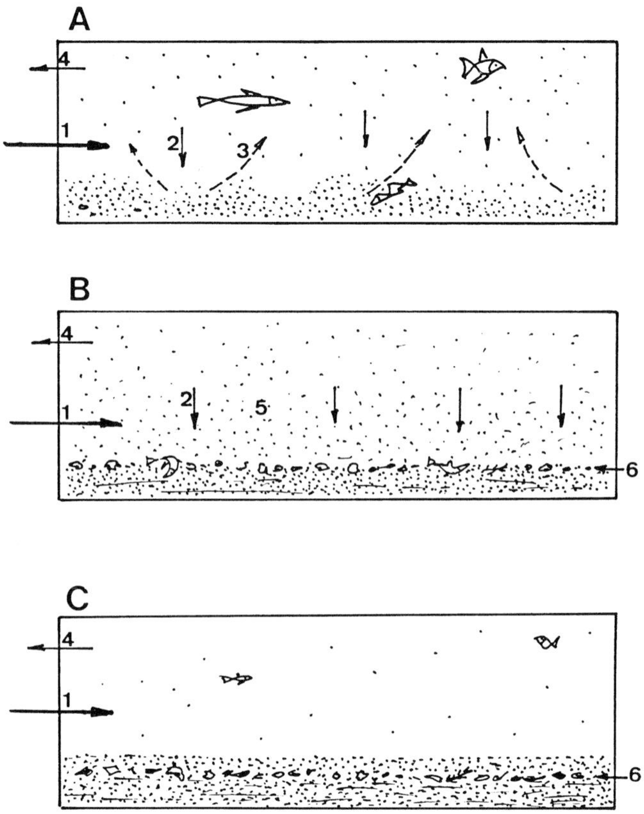

FIG. 1.—A tentative explanation for intermittent rapid OM sedimentation in spite of a regular, continuous supply of nutrients in a restricted shallow-water environment without significant clastic supply. This scenario is for shallow-water platform basins in hot climates, where the amounts of water and nutrients entering the basin (1) exceed the amounts lost by evaporation (4).

A. A constant nutrient supply to the seawater (1) favors bioproductivity. If the nitrogen-to-phosphorus ratio is low, the noncarbonate organisms are favored. Postmortem, the OM sinks (2), and is partly and quickly destroyed by microbial activity. Some of the phosphorus and nitrogen return to the seawater (3)(Colman and Holland, this volume). The phosphorus from the OM adds to that which is supplied from outside the basin and progressively increases the total phosphorus content of the seawater, thus enhancing productivity.

B. The increasing organic productivity (5) consumes more and more oxygen, and produces more and more polluting metabolites. The environment, which becomes dysoxic to anoxic, is characterized by low biodiversity assemblages, which are commonly monospecific blooms. Blooms may end in mass mortality, resulting in sudden deposition of a thick bed of OM on the sea floor. Such mass mortality may be recorded in the sediment by a bone bed (6) (Lucas and Prévôt, 1991).

C. Mass mortality can be considered an ecological process that cleans the water. The oxygenation of the water comes back to normal, allowing biodiversity and productivity to return to "normal" and the OM sedimentation rate to temporarily return to low levels, thus beginning a new cycle.

Sedimentary discontinuities also affect the whole deposit due to changes in productivity. Environmental changes are frequent on continental platforms caused by global changes such as variation in sealevel, astronomical cycles (Kemp and Baldauf, 1993; Giraud et al., 1995), and cyclicity in upwelling currents. Environmental change may induce changes in productivity and the temporary dominance of either carbonate-walled, silica-walled, or naked microorganisms, and thus lead to the deposition of either limestone, chert, or phosphorite, respectively. Repetition of these changes results in the cyclic sequences displayed by many phosphorite deposits over various scales, from the centimeter scale (Schuffert et al., 1998) to the meter scale and larger (Boujo, 1976; Prévôt, 1990). Variations in base level (sealevel variations or tectonics) may result in increasing amounts of coarser clastics which may interrupt phosphorite deposition either temporarily (Benalioulhaj et al., this volume) or permanently (Lucas and Prévôt-Lucas, 1995).

DISCUSSION AND CONCLUSION

Calcite and apatite are two minerals that form due to marine biological activity. It is obvious that the higher the level of productivity, the higher the chances for these minerals to precipitate and settle. However, high rates of undifferentiated productivity are not enough to explain why either calcite or apatite form. Carbonates are directly precipitated by organisms, pelagic (phytoplankton, foraminifera, etc.) and benthic organisms, in which the proportion of OM is relatively low. During and shortly after carbonate sedimentation, the OM is partially or totally destroyed. Greater productivity of such organisms does not change the ratio of carbonate to OM, only the quantity of carbonate. Organic-rich sediments are directly linked to noncalcareous (siliceous or naked) pelagic organisms. Greater productivity of these organisms does not change the ratio of carbonate to OM, but is necessary in order for enough OM to accumulate, even temporarily. OM decay brings about dissolution of the minor carbonates.

Shallow water of shelves or shoals without significant amounts of clay are more favorable than the deep ocean for the formation of apatite, which precipitates after the slow breakdown of OM in a dysoxic environment. In these environments, apatite can accumulate in rock-forming quantities, occurring as phosphorite, whereas it is only dispersed in most oceanic sediments. This uneven distribution of apatite is misleading, because the spectacular apatite accumulations in phosphorite deposits represent only a small proportion of the total mineralized phosphorus in sedimentary deposits of a given age. Thus, the striking apparent episodicity of phosphorite deposition does not reflect rythms fluctuations in the amount of phosphorus available at the earth surface, i.e. a quantitative variation in the phosphorus cycle, rather the episodic conjunct occurrence of favorable paleogeographic and paleoceanographic conditions.

Because their modes of genesis are different, calcite and apatite are totally independant. Most limestones do not contain any visible apatite. Conversely, a number of phosphorite deposits—for example, paleozoic deposits—contain no or scarce calcite. This affords proof, if need be, that phosphorite formation does not require carbonates. Since Mezosoic times, carbonates have become more and more abundant, as a result of enhanced production of carbonate-producing phytoplankton. Hence, the proximity and apparent relationship between two minerals that originate from the living biosphere cannot be considered a wonder. The shift from populations dominated by carbonate-producing phytoplankton to populations dominated by OM-producing siliceous or naked phytoplankton does not require any drastic change of the environment. All depend on the availability of nutrients and seem very sensitive to the nitrogen-phosphorus ratio. Higher nitrogen concentrations favor carbonate-producing plankton, whereas a disequilibrium higher phosphorus favors the predominance of OM-producing naked plankton. Any change achieving such disequilibrium of the nitrogen-phosphorus ratio, even the most unexpected, may play a role. For example, an iron supply, even very low, may increase nitrogen-fixation capacity. Similarly, any slight increase in the supply of phosphorus, either through upwelling currents, or from the continents through river water (Tyrrell and Law, 1997) as suggested by Ilyin (1994, 1997) on the Russian craton, or both the ocean and the continent as proposed for the Egypt deposit (Glenn and Arthur, 1990), may play a role.

Conditions of OM sedimentation and its transformation into apatite are, in any event, transient. The span of time over which favorable living phytoplankton species dominate and for its OM to be suitably preserved is narrow, because the areas favorable for phosphogenesis are those most prone to environmental changes. Global changes (e.g. sealevel or climate) or more regional changes (e.g. tectonic deformation, continental sediment supply, underwater topography) can affect the chemical and/or physical environment (nitrogen-to-phosphorus ratio, temperature, water depth, lighting, and agitation) the living biosphere and redox condition of the sediment. A few successive slight changes, generally hard to detect, may be responsible for the abrupt fluctuations in planktonic assemblages that result in sedimentation gaps or sharply alternating beds of almost pure carbonate and phosphorite. However, just as apatite is dispersed in most marine sediments, usually minor numbers of carbonate tests are deposited with the organic-rich sediment. These tests can be found partially or totally replaced by apatite (Prévôt and Lucas, 1986); however, the apatite replacement of carbonate is a minor mechanism of phosphogenesis.

The hydrodynamic variations related to these environmental changes provoke almost synsedimentary winnowing, which obliterates the characteristics of the pristine sediment by mixing or, on the contrary, by sorting the constituents. Regional changes superimposed on global changes confer upon each phosphate deposit specific characteristics which tend to mask the general conditions of genesis common to all phosphate deposits.

The phosphorite factory, which produced the notable phosphorite deposits, came into play when the necessary environmental and early diagenetic conditions coincided in time and place. The phosphorite factory and the carbonate factory are seen to be independent, but complementary rather than competing, as they do not use the same primary material, nor do they occur during the same phase of the sedimentary cycle.

ACKNOWLEDGMENTS

We are indebted to Guerry McClellan for the helpful revision of our American English and spelling prior to the reviewing process. Very constructive reviews by Peter W. Homewood and Marc S. Hendrix as well as many editorial suggestions and comments improved the manuscript and are greatly appreciated.

REFERENCES

Aller, R. C., 1994, Bioturbation and remineralization of sedimentary organic matter : effects of redox oscillation: Chemical Geology, v. 114, p. 331-345.

Barron, J. A., 1987, Diatomite; environmental and geologic factors affecting its distribution, *in* Hein, J. R., ed., Siliceous sedimentary rock-hosted ores and petroleum: Evolution of ore fields series: New York, Van Nostrand Reinhold Co., p. 164-178.

Baturin, G. N., 1982, Phosphorites on the Ocean floor: Developments in Sedimentology, v. 33: Amsterdam, Elsevier, 343 p.

Belfkira, O., 1980, Evolutions sédimentologiques et géochimiques de la série phosphatée du Maestrichtien des Ouled Abdoun (Maroc): Unpublished Ph.D. Dissertation: University of Grenoble, France, 164p.

Benalioulhaj, N., 1991, Les formations à phosphates et à schistes bitumineux du Bassin des Oulad Abdoun et du Bassin de Timahdit (Maroc): pétrographie, minéralogie, géochimie et environnement de dépôt: Unpublished Ph.D. Dissertation: University of Strasbourg, France, 238 p.

Birkeland, C., 1987, Nutrient availability as a major determinant of differences among coastal hard-substratum communities in different regions of the tropics, *in* Birkeland, C., ed., Comparison between Atlantic and Pacific tropical marine coastal ecosystems: community, structure, ecological processes, and productivity: UNESCO/COMAR workshop, University of the South Pacific, Suva Fiji, 1986, p.45.

Boujo, A., 1976, Contribution à l'étude géologique du gisement de phosphate Crétacé-Eocène des Ganntour (Maroc Occidental): Sciences Géologiques, Mémoires, Strasbourg, France, v. 43, 227p.

Brongersma-Sanders, M., 1957, Mass mortality in the sea: Geological Society of America, Memoirs, v. 67, p. 941-1010.

Busson, G., Noël, D., and Cornée, A., 1997, Les coccolithophoridées, contributeurs essentiels de la fraction calcaire des sédiments marins du Jurassique moyen-supérieur: contraste entre les populations des calcaires et des marnes: Bulletin de la Société Géologique de France: Paris, v. 168, p. 601-610.

Calvert, S. E., Bustin, R. M., and Pedersen, T. F., 1992, Lack of evidence for enhanced preservation of sedimentary organic matter in the oxygen minimum zone of the Gulf of California: Geology, v. 20, p. 757-760.

Canfield, D. E., 1994, Factors influencing organic carbon preservation in marine sediments: Chemical Geology, v. 114, p. 315-329.

Caplan, M. L., and Bustin, R. M., 1996, Factors governing organic matter accumulation and preservation in a marine petroleum source rock from the Upper Devonian to Lower Carboniferous Exshaw Formation, Alberta: Bulletin of Canadian Petroleum Geology, v. 44, p. 474-494.

Coale, K. H., Johnson, K. S., Fitzwater, S. E., Gordon, R. M., Tanner, S., Chavez, F. P., Ferioli, L., Sakamoto, C., Rogers, P., Millero, F., Steinberg, P., Nightingale, P., Cooper, D., Cochlan, W. P., Landry, M. R., Constantinou, J., Rollwagen, G., Trasvina, A. and Kudela, R., 1996, A massive phytoplankton bloom induced by an ecosystem-scale iron fertilization experiment in the equatorial Pacific Ocean: Nature, v. 383, p. 495-501.

Cook, P. J., and McElhinny, M. W., 1979, A reevaluation of the spatial and temporal distribution of sedimentary phosphate deposits in the light of plate tectonics: Economic Geology, v. 74, p. 315-330.

Cook, P. J., and Shergold, J. H., 1986, Proterozoic and Cambrian Phosphorites —nature and origin, *in* Cook, P. J., and Shergold, J. H., eds, Phosphate Deposits of the World, v. 1, Proterozoic and Cambrian Phosphorites: Cambridge University Press, p. 369-386.

Derenne, S., Largeau, C., Casadevall, E., and Rousseau, B., 1991, Chemical evidence of kerogen formation in source rocks and oil shales via selective preservation of thin resistant outer wall of microalgae: origin of ultralaminae: Geochimica et Cosmochimica Acta, v. 55, p. 1041-1050.

De Wever, P., Baudin, F., Azema, J., and Fourcade, E., 1996, Radiolarians and Tethyan radiolarites from primary production to their paleogeography, *in* Nairn, A. E. M., Ricou, L. E., Vrielynck, B., and Dercourt, J., eds, The Ocean Basins and Margins: New York, Plenum, p. 267-318.

Emerson, S., and Archer, D. E., 1990, Calcium carbonate preservation in the ocean: Philosophical Transactions of the Royal Society London, Series A v. 331, p. 29-40.

Eshet, Y., Almogi-Labin, A. and Bein, A., 1994, Dinoflagellate cysts, paleoproductivity and upwelling systems : a Late Cretaceous example from Israël: Marine Micropaleontology, v. 23, p. 231-240.

FALKOWSKI, P. G., 1997, Evolution of the nitrogen cycle and its influence on the biological sequestration of CO_2 in the ocean: Nature, v. 376, p. 272-275.

FILIPPELLI, G. M, AND DELANEY, M. L., 1992, Similar phosphorus fluxes in ancient phosphorite deposits and a modern phosphogenic environment: Geology, v. 20, p. 709-712.

FÖLLMI, K.B., 1996, The phosphorus cycle, phosphogenesis and marine phosphate-rich deposits: Earth-Science Reviews, v. 40, p. 55-124.

FROST, B. W., 1996, Phytoplankton bloom on iron rations: Nature, v. 383, p. 475-476.

GIRAUD, F., BEAUFORT, L., AND COTILLON P., 1995, Contrôle astronomique de la sédimentation carbonatée dans le Crétacé inférieur du Bassin vocontien (SE France): Bulletin de la Société Géologique de France, v. 166, p. 409-421.

GLENN, C. R., AND ARTHUR, M. A., 1990, Anatomy and origin of a Late Cretaceous phosphorus-greensand giant: Sedimentology, v. 37, p. 123-154.

GLENN, C. R., FÖLLMI, K. B., RIGGS, S. R., BATURIN, G. N., GRIMM, K. A., TRAPPE, J., ABED, A. M., GALLI-OLIVIER, C., GARRISON, R. E., ILYIN, A. V., JEHL, C., ROHRLICH, V., SADAQAH, R. M. Y., SCHIDLOWSKI, M., SHELDON, R. E., AND SIEGMUND, H., 1994, Phosphorus and phosphorites: sedimentology and environments of formation: Eclogae Geologicae Helvetiae, v. 87, p. 747-788.

GOLDMAN, J. C., MCCARTHY, J. J., AND PEAVEY, D. G., 1979, Growth rate influence on the chemical composition of phytoplankton in oceanic waters: Nature, v. 279, p. 210-215.

GOTH, K, DE LEEUW, J. W., PÜTTMANN, W., AND TEGELAAR, E. W., 1988, Origin of the Messel oil shale kerogen: Nature, v. 336, p. 759-761.

GRIMM, K . A., LANGE, C. B., AND GILL, A. S., 1996, Biological forcing of hemipelagic sedimentary laminae: evidence from ODP site 893, Santa Barbara Basin, California: Journal of Sedimentary Research, v. 66, p. 613-624.

GRIMM, K. A., LANGE, C. B., AND GILL, A. S., 1997, Self-sedimentation of phytoplankton blooms in the geologic record: Sedimentary Geology, v. 110, p. 151-161.

GULBRANDSEN, R. A., 1969, Physical and chemical factors in the formation of marine apatite: Economic Geology, v. 64, p. 375-382.

HARVEY, H. R., TUTTLE, J. H., AND BELL, J. T., 1995, Kinetics of phytoplankton decay during simulated sedimentation: changes in biochemical composition and microbial activity under oxic and anoxic conditions: Geochimica et Cosmochimica Acta, v. 59, p. 3367-3377.

HEIN, J. R., AND PARRISH, J. T., 1987, Distribution of siliceous deposits in space and time, *in* Hein, J. R., ed., Siliceous sedimentary rock-hosted ores and petroleum: Evolution of ore fields series: New York, Van Nostrand Reinhold Co., p. 12-57.

HOMEWOOD, P. W., 1996, The carbonate feedback system: interaction between stratigraphic accommodation, ecological succession and the carbonate factory: Bulletin de la Société Géologique de France, v. 167, p. 701-715.

HUTCHINS, D. A., AND BRULAND, K. W., 1998, Iron-limited diatom growth and Si/N uptake ratios in a coastal upwelling regime: Nature, v. 393, p. 561-564.

ILYIN, A. V., 1994, Cenomanian phosphorites in the former Soviet Union: Sedimentary Geology, v. 94, p. 109-127.

ILYIN, A. V., 1997, Mid-Cretaceous phosphate platform of the Russian craton: Sedimentary Geology, v. 113, p. 125-135.

INGALL, E. D., BUSTIN, R. M., AND VAN CAPPELLEN, P., 1993, Influence of water column anoxia on the burial and preservation of carbon and phosphorus in marine shales: Geochimica et Cosmochimica Acta, v. 57, p. 303-316.

JAMES, N. P., 1979, Introduction to carbonate facies models, *in* Walker, R. G., ed., Facies Models: Geoscience Canada reprint series, v.1, Waterloo Canada, p. 105-107.

JORDAN, R. W., PRIDDLE, J., PUDSEY, C. J, BARKER, P. F., AND WHITEHOUSE, M. J., 1991, Unusual diatom layers in Upper Pleistocene sediments from the northern Weddell Sea: Deep-Sea Research, v. 38, p. 829—843.

KARL, D., LETELIER, R., TUPAS, L., DORE, J., CHRISTIAN, J., AND HEBEL, D., 1997, The role of nitrogen fixation in biogeochemical cycling in the subtropical North Pacific Ocean: Nature, v. 388, p. 533-538.

KEMP, A. E. S., AND BALDAUF, J. G., 1993, Vast Neogene laminated diatom mat deposits from the eastern equatorial Pacific Ocean: Nature, v. 362, p. 141-144.

KRAJEWSKI, K. P., VAN CAPPELLEN, P., TRICHET, J., KÜHN, O., LUCAS, J., MARTIN-ALGARRA A., PRÉVÔT, L., TEWARI, V. C., GASPAR, L., KNIGHT, R., AND LAMBOY, M., 1994, Biological processes and apatite formation in sedimentary environments: Eclogae Geologicae Helvetiae, v. 87, p. 701-745.

LAMBOY, M., RAO, V. P., AHMED, E., AND AZZOUZI, N., 1994, Nannostructure and significance of fish coprolites in phosphorites: Marine Geology, v. 120, p. 373-384.

LARGEAU, C., DERENNE, S., CASADEVALL, E, BERKALOFF, M., COROLLEUR, B., RAYNAUD, J.F., AND CONNAN, J., 1990, Occurrence and origin of "ultralaminar" structures in "amorphous" kerogens of various source rocks and oil shale: Organic Geochemistry, v. 16, p. 889-895.

LUCAS, J., AND PRÉVÔT, L., 1984, Les synthèses de l'apatite, données nouvelles pour un modèle de genèse des phosphorites sédimentaires, *in* Proceedings of the 27th International Geological Congress, v. 15: Non-metallic mineral ores, VNU Science Press, p.173-186.

LUCAS, J., AND PRÉVÔT, L., 1985, The synthesis of apatite by bacterial activity: mechanism: Sciences Géologiques, Mémoires, Strasbourg, France, v.77, p.83-92.

LUCAS, J., AND PRÉVÔT, L., 1991, Phosphates and fossil preservation, *in* Stehli, F. G., and Jones, D., eds, Taphonomy, releasing the data locked in the fossil record, v. 9, Topics in Geobiology: New York, Plenum, p. 389-409.

LUCAS, J., AND PRÉVÔT-LUCAS, L., 1995, Tethyan phosphates and bioproductites, *in* Nairn, A. E. M., Ricou, L. E., Vrielynck , B., and Dercourt, J. , eds, The Ocean basins and margins, v. 8, The Tethys Ocean: Plenum, p. 367-391.

LUCAS, J., AND PRÉVÔT-LUCAS, L., 1997, On the genesis of sedimentary apatite and phosphate-rich sediments, *in* Paquet, H. and Clauer, N., eds, Soils and Sediments: Springer, p. 249-268.

LUCAS, J., EL FALEH, E. M., AND PRÉVÔT, L., 1990, Experimental study of the substitution of Ca by Sr and Ba in synthetic apatites, *in* Notholt, A. J. G., and Jarvis, I., eds, Phosphorite Research and Development: Geological Society Special Publication, London, v. 52, p. 33-47.

MAYER, L. M., 1994, Surface area control of organic carbon accumulation in continental shelf sediments: Geochimica et Cosmochimia Acta, v. 58, p. 1271-1284.

MEUNIER-CHRISTMANN, C., LUCAS, J., AND ALBRECHT, P., 1989, Organic geochemistry of Moroccan phosphorites and bituminous shales. A contribution to the problem of phosphogenesis: Sciences Géologiques, Bulletin, Strasbourg, France, v. 42, p. 205-222.

MOULIN, E., JORDENS, A., AND WOLLAST, R., 1985, Influence of the aerobic bacterial respiration on the early dissolution of carbonates in coastal sediments: Proceedings of the Progress in Belgium Oceanographic Research, p. 196-208.

MOUTAOUAKIL, D, AND GIRESSE, P., 1993, Pétrologie et environnements sédimentaires des phosphates mésocénozoïques du bassin des Oulad Abdoun (Maroc): Bulletin de la Société Géologique de France, v. 164, p. 473-491.

PARSONS, T. R., TAKAHASHI, M., AND HARGRAVE, B., 1977, Biological Oceanographic Processes (2nd ed.): New York, Pergamon, 332 p.

PEDERSEN, T. F., AND CALVERT, S. E., 1990, Anoxia vs Productivity: What Controls the Formation of Organic-Carbon-Rich Sediments and Sedimentary Rocks?: The American Association of Petroleum Geologists Bulletin, v. 74, p. 454-466.

PRÉVÔT, L., 1990, Geochemistry, Petrography, Genesis of Cretaceous-Eocene Phosphorites. The Ganntour Deposit (Morocco): a type example: Mémoires de la Société Géologique de France, v. 158, 232 p.

PRÉVÔT, L., 1991, Geology of clays in the sedimentary phosphate deposit of Ganntour Basin, Morocco: Proceedings of the 7th Euroclay Conference "Dresden, 1991," Greifswald, p. 859-862.

PRÉVÔT, L., AND LUCAS, J., 1986, Microstructure of apatite replacing carbonate in synthesized and natural samples: Journal of Sedimentary Petrology, v. 56, p. 153-159.

PRÉVÔT, L., EL FALEH, E. M., AND LUCAS, J., 1989, Details on synthetic apatites formed through bacterial mediation. Mineralogy and chemistry of the products: Sciences Géologiques, Bulletin, Strasbourg, France, v. 42, p. 237-254.

PRÉVÔT, L., LUCAS, J., AND DOUBINGER, J., 1979, Une correspondance entre le contenu palynologique et la composition minéralogique et chimique d'une série phosphatée sédimentaire (Ganntour, Maroc): Sciences Géologiques, Bulletin, Strasbourg, France, v. 32, p. 69-90.

PRIAN, J. P., 1980, Les phosphorites cambriennes du versant septentrional de la Montagne Noire, au Sud du Bassin Permien de Camarès (Aveyron): Unpublished Ph.D. Dissertation, Université Pierre et Marie Curie, Paris VI, 407 p.

RAUSCHER, R., AND DOUBINGER, J., 1982, Les dinokystes du Maestrichtien phosphaté au Maroc: Sciences Géologiques, Bulletin, Strasbourg, France, v. 35, p. 97-116.

RIGGS, S., 1986, Proterozoic and Cambrian Phosphorites— Specialist studies: Phosphogenesis and its relationship to exploration for Proterozoic and Cambrian Phosphorites, *in* Cook, P. J., and Shergold J. H., eds, Phosphate Deposits of the World, v. 1 Proterozoic and Cambrian Phosphorites: Cambridge University Press, p. 352-368.

RUTTENBERG, K. C., AND BERNER, R. A., 1993, Authigenic apatite formation and burial in sediments from non-upwelling continental margin environments: Geochimica et Cosmochimica Acta, v. 57, p. 991-1007.

SCHUFFERT, J. D., KASTNER, M., AND JAHNKE, R. A., 1998, Carbon and Phosphorus burial associated with modern phosphorite formation: Marine Geology, v. 146, p. 21-31.

SHELDON, R. P., 1989, Phosphorite deposits of the Phosphoria Formation, western United States, *in* Notholt, A. J. G., Sheldon, R. P., and Davidson D. F., eds, Phosphate Deposits of the World, v. 2, Phosphate Rock Resources: Cambridge University Press, p. 53-61.

SONCINI, M. J. , 1990, Palynologie des phosphates des Ouled Abdoun, Maroc: biostratigraphie et environnements de la phosphatogenèse dans le cadre de la crise Crétacé-Tertiaire: Unpublished Ph. D. Dissertation, University of Strasbourg, France, 243 p.

TORTELL, P D., MALDONADO, M. T., AND PRICE, N. M., 1996, The role of heterotrophic bacteria in iron-limited ocean ecosystems: Nature, v. 383, p. 330-332.

TRIBOVILLARD, N. P., CAULET, J. P., VERGNAUD-GRAZZINI, C., MOUREAU, N. AND TREMBLAY, P., 1996, Lack of organic matter accumulation on the upwelling-influenced Somalia margin in a glacial-interglacial transition: Marine Geology, v. 133, p. 157-182.

TRIBOVILLARD, N. P., RIVIÈRE, M., OUAHDI, R., LALLIER-VERGES, E., AND CAULET, J. P., 1994, L'absence d'un enrichissement marqué en matière organique dans des sédiments déposés en contexte d'upwelling: le courant de Somalie: Bulletin de la Société Géologique de France, v. 165, p. 65-75.

TRICHET, J., AND DÉFARGE, C., 1995, Non-biologically supported organomineralization, *in* Allemand, D., and Cuif, J. P., eds, Biomineralization 1993: 7th International Symposium on Biomineralization special v. 14/2, p. 203-236.

TYRELL, T., AND LAW, C. S., 1997, Low nitrate: phosphate ratios in the global ocean: Nature, v. 387, p. 793-796.

VAN CAPPELLEN, P., AND INGALL, E. D., 1994, Benthic phosphorus regeneration, net primary production, and ocean anoxia: a model of the coupled marine biogeochemical cycles of carbon and phosphorus: Paleoceanography, v. 9, p. 677-692.

WALTER, L. M., AND BURTON, E. A., 1987, Active dissolution in modern shallow marine carbonate sediments: Global implications (abs.): Geological Society of America, Annual Meeting and Exposition, Phoenix, Arizona, p. 880.

WESTBROEK, P., BUDDEMEIER, B., COLEMAN, M., KOK, D. J., FAUTIN, D., AND STAL, L., 1994, Strategies for the study of climate forcing by calcification, *in* Doumenge, F., Allemand, D., and Toulemont A., eds, Past and Present Biomineralization Processes. Considerations about the carbonate cycle: IUCN-COE Workshop, Monaco, Bulletin de l'Institut Océanographique, special v. 13, p. 37-60.

WOLLAST, R., 1994, The relative importance of biomineralization and dissolution of $CaCO_3$ in the global carbon cycle, *in* Doumenge, F., Allemand, D., and Toulemont A., eds., Past and Present Biomineralization Processes. Considerations About the Carbonate Cycle: IUCN-COE Workshop, Monaco, Bulletin de l'Institut Océanographique, special v. 13, p. 13-35.

OOIDAL IRONSTONES AND PHOSPHORITES—A COMPARISON FROM A STRATIGRAPHER'S VIEW.

FRANKLYN B. VAN HOUTEN

Department of Geosciences, Princeton University, Princeton, NJ 08544

ABSTRACT: Chemical components of most ooidal ironstones and phosphorites were initially derived from deeply weathered uplands and deposited in shallow waters of epeiric seas and continental margins that received very little clastic sediments. They were then commonly transformed during diagenesis largely by upwelling seawater and redeposited in winnowed layers.

Ironstones and phosphorites differ, however. Ironstones were first laid down as ferric oxide and kaolinite ooids and peloids, whereas phosphorites were formed first as phosphatic carbonate granules, crusts, and hardground. Ironstones were then transformed to berthierine (or rarely nontronite) during diagenesis, whereas phosphorites were transformed to carbonate fluorapatite (CFA).

In terms of their distribution in the geologic record ironstones and phosphorites are also dissimilar. A few ooidal ironstones were deposited in Early Proterozoic time. Most accumulated during greenhouse phases of the Phanerozoic (Ordovician-Devonian and Jurassic-early Cenozoic). With the exception of the late Cretaceous–early Cenozoic deposits, major phosphorites are notable for their absence in rocks of those ages. In fact, phosphorites were most abundant in the Late Proterozoic and early Cambrian times.

INTRODUCTION

Many geologists have suggested that the chemical environments in which phosphorites and ooidal ironstones form have numerous factors in common (e.g. Pettijohn, 1975,p.432; Kimberley, 1994,p.126). This is true, but it is not the entire story. For this reason these two sedimentary products have been compared in some detail by different researchers, including Cook,1984; Föllmi,1996; Glenn et al., 1994b; McKelvey, 1979; and Petranek and Van Houten, 1997. They have found that ironstones differ from phosphorites in significant ways. Although the origin of the ironstones and phosphorite ooids has not been established, that subject is not considered here.

GEOLOGIC RECORD

The Precambrian ooidal ironstones differ significantly from the phosphorites in stratigraphic history. Banded iron formations are found in Archean and in Early and Late Proterozoic sequences (Trendall and Morris,1983); the Early Proterozoic was clearly the period of greatest iron sedimentation in Earth history (James, 1983,). Ooidal ironstones were present only during the Early Proterozoic time (2500–1600 Ma) in about 7 districts.

Phosphorites, on the other hand, are present more widespread in Precambrian deposits (Cook and Shergold, 1986), but very few are associated with ooidal ironstones. Phosphorites are present in a few Archean formations (3700–2500 Ma), in 8 Early Proterozoic formations (2500–1600 Ma), in 9 Middle Proterozoic formations (1600–1000 Ma), none of which contain banded iron formations or ooidal ironstones, and in 19 Late Proterozoic rocks (1000–570 Ma). However, only 2 of these 36 Proterozoic phosphorites are intimately associated with iron formations.One is the Early Proterozoic phosphorite in the Lake Superior area of the north-central United States and one is the Late Proterozoic Bambui deposit of south-central Brazil. In fact, the Upper Proterozoic phosphorites are most common in stratigraphic intervals in which banded iron formations are the least common. Moreover, the greatest global phosphogenic episode in geological history occurred in Late Proterozoic and Cambrian periods (Notholt and Sheldon, 1986).

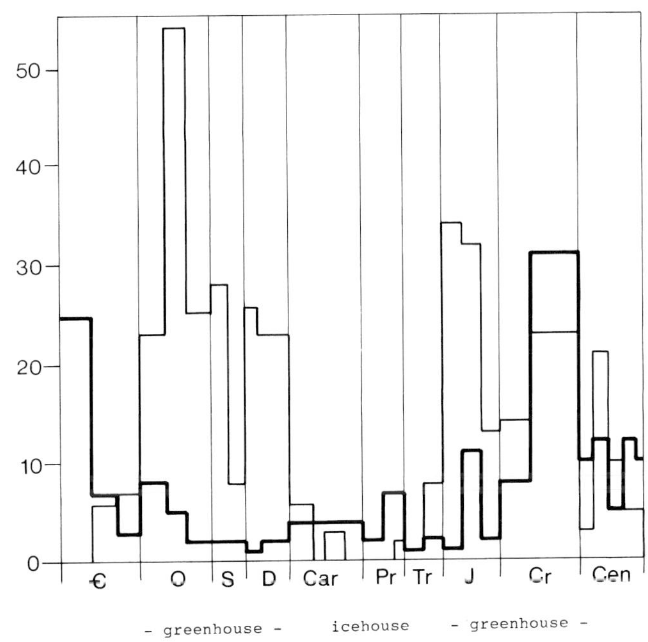

FIG. 1.— Stratigraphic distribution of Phanerozoic phosphorite (thick lines) and ooidal ironstone (thin lines) formations. Greenhouse and icehouse phases indicated at base. Phosphorite data is from Cook, 1992; Cook and McElhinney, 1979; and Lee, 1980. Ironstone data is from Petranek and Van Houten, 1997.

Studies of more than 400 Phanerozoic ooidal ironstones and about 200 Phanerozoic phosphorites (Cook and McElhinney, 1979; Cook, 1992; Lee, 1980) show that they have similar patterns of stratigraphic distribution throughout the Phanerozoic. But when compared in specific stratigraphic terms (Fig.1) only about 90 ironstones (about 25%) are intimately associated with phosphorites (Table 1), so that the phosphorites are not present in more than 300 ironstone sequences (Table 2).

Of the forty Paleozoic sequences that contain both ironstones and phosphorites, 62% also contain carbonates. Of the 17 Jurassic ironstone and phosphorites sequences, 10 (56%) are

Marine Authigenesis: From Global to Microbial, SEPM Special Publication No. 66

TABLE 1.—PHANEROZOIC OOIDAL IRONSTONE DEPOSITS ASSOCIATED WITH PHOSPHORITES

Cambrian Period
- Middle: Canada (Nova Scotia—*Gillis Brook Fm.*), USA (Wyoming, Montana)

Ordovician Period
- Early: central Australia—*Pacoota Fm.*
- Early–Middle: USA (central Oklahoma—*Oil Creek Fm.*)
- Early, Middle, Late: Sweden
- Middle: Siberia, Estonia, Poland, Canada (south Saskatchewan and Manitoba), USA (Iowa, North and South Dakota, Georgia—*Lenoir Fm.)*
- Late: USA (Wisconsin—*Neda Fm.*), south Bolivia—*Jarcas Fm.*

Silurian Period
- Middle: USA (central and western New York—*Clinton Fm.*)

Devonian Period
- Early and Middle: France and Germany
- Middle and Late: south China, USA (central and north Pennsylvania, Maryland, north Virginia—*Needmore Fm.*)
- Late: Nepal, Belgium

Carboniferous Period
- Early: south Ireland
- Late: northeast England—*Robin Hood Fm.*, USA (northwest Texas)

Permian Period
- Late: northwest Australia—*Desert Basin Fm.*., northeast India

Jurassic Period
- Early: west India
- Early–Middle: Belgium
- Early and Middle: southwest Russia—*Malka Fm.*, Bulgaria—*Troyanska Fm.*
- Early, Middle, Late: England— *several fms.*, France and Germany—*several fms.*
- Middle: north Pakistan—*Samana Suk Fm.*, central Spain
- Middle–Late: Canada (Alberta—*Fernie Fm.*)

Cretaceous Period
- Early: Israel—*Ramin Fm.*, central Saudi Arabia
- Early and Middle: France and Germany
- Middle: Siberia
- Late: south-central Russia—*Koper Fm.*, central Kazakhstan—*Ayat Fm.*, north-central Saudi Arabia, north Egypt (western desert), north Colombia and west Venezuela—*Colon Fm.*

Cenozoic Era
- Paleocene: north Pakistan—*Langrial Fm.*
- Eocene: west-central Saudi Arabia—*Wadi Fatima Fm.*, Mali, Niger—*Malbaza Fm.*, Nigeria—*Sokoto Fm.*
- Miocene: northeast Colombia—*Guayobo Fm.*, west Venezuela—*Lagunillas Fm.*

sequences. Of the 16 Cretaceous ironstone and phosphorite sequences, 8 (50%) are associated with carbonates. Of the 7 Cenozoic ironstone and phosphorite sequences, only 1 (14%) is associated with carbonate. Clearly, increasingly more carbonates with associated ironstones and phosphorites are found as the geologic record is traced backward from Cenozoic to Paleozoic time.

Ironstones were rare in Middle and Late Cambrian time, very important in the Ordovician Period, important in the Devonian, Jurassic, and Cretaceous periods, and common in the early Cenozoic Era (Fig.1). In contrast, phosphorites were very important in the Late Proterozoic and Early Cambrian times, common in Early Ordovician Period, less common in the Carboniferous and Permian periods, important in Jurassic, and very important in Cretaceous periods, and important in the Paleocene, Eocene, Miocene, and Pliocene epochs (Fig.1). Accordingly, the relative abundance of ironstones and phosphorites through Phanerozoic time differs considerably.

Cambrian phosphorites without ironstones are found in Australia, central and southeastern Asia, northwestern India, western and northwestern Europe, Newfoundland, and eastern South America. The extensive Ordovician ironstones without phosphorites, on the other hand, are found in north Africa, southern China, northern, central and western Europe, eastern maritime Canada, and southeastern United States (Table 2).

CLIMATE

Most ironstones and phosphorites were laid down in low latitudes, but a few appear to have formed at higher latitudes. Many of the ironstones were deposited during warm periods, greenhouse phases, of earth history: Ordovician-Devonian and Jurassic-early Cenozoic times (Fig. 1). Features associated with some of the ironstones suggest that they were deposited in a rather dry climate during times that were characterized by episodes of large-scale transgressions. During the latest Paleozoic glacial epochs, ironstones were deposited mainly in the Late Carboniferous interglacial period and during the latest Permian post-glacial time (Frakes, 1979). The ironstone in southern Ukraine was deposited in a climate that was relatively mild during Middle Pliocene of the late Cenozoic, a period characterized overall by increasingly colder climates.

Like ironstones, phosphorites were also laid down in warm climates during times characterized by large-scale transgressions. Many were deposited in humid climates (Zanin, 1993,p.9); others in somewhat dryer climates (Cook, 1984). Unlike the ironstones, phosphorites are not clustered in the greenhouse phases (Fig. 1). Apparently, warm climate was not the principal control on phosphorites deposition.

MAJOR FACIES

The chemical components of ooidal ironstones and phosphorites were derived from deep continental weathering of rocks and transported in solution from areas of rather low relief that yielded very little detrital clastic material, (McArthur and

TABLE 2.—PHANEROZOIC OOIDAL IRONSTONES NOT ASSOCIATED WITH PHOSPHORITES

Cambrian Period
- Middle: USA (northwest Arizona)
- Late: USA (Iowa, southeast Arizona, southwest New Mexico, Texas)

Ordovician Period
- Ordovician: south China, Germany, Malaysia
- Early: Norway, Wales, France, Spain, Portugal, Libya, Tunisia, Algeria, Morocco, Canada (Newfoundland, Nova Scotia)
- Middle: Russia, Czech, Wales, France, Spain, Portugal, Libya, Tunisia, Morocco, Algeria
- Late: Turkey, Czech, Wales, England, France, Spain, Libya, Algeria, Morocco, Canada (Newfoundland), USA (Georgia), Venezuela

Silurian Period
- Silurian: Italy
- Early: Turkey, Algeria, Morocco, Canada (New Brunswick), USA (Tennessee, Kentucky, Alabama), Brazil
- Early and Late: USA (Pennsylvania, Ohio, Maryland, Virginia, West Virginia)
- Middle: England, Algeria, Canada (Nova Scotia), Mauritania, Guinea
- Late: Wales, Spain, northwest Argentina, southeast Argentina

Devonian Period
- Devonian: Saudi Arabia
- Early: Mauritania, Guinea, Canada (Nova Scotia), Brazil, northwest Argentina
- Early–Middle: Macedonia, Belgium, Spain
- Middle: Turkey, Poland, Mali, USA (Arizona)
- Middle–Late: USA (New York)
- Late: Czech, USA (Kansas), Iran
- Early, Middle, Late: Libya

Carboniferous Period
- Early: Czech, Wales, Libya, China

Triassic Period
- Middle and Late: Saudi Arabia
- Late: Nepal, Slovakia, USA (Alaska), Canadian Archipelago

Jurassic Period
- Early: Australia (Queensland), Serbia, Poland, Norway, Sweden, Denmark, Scotland, Spain, Iraq, northwest Canada, Canadian Archipelago
- Early–Middle: Luxembourg
- Middle: Nepal, Hungary, Poland, Scotland, Switzerland, Saudi Arabia, Madagascar, Tunisia, Algeria, northwest Canada, Canadian Archipelago
- Late: Switzerland, Canada (Alberta), Canadian Archipelago

Cretaceous Period
- Early: Poland, England, Lebanon, Syria, Libya, Canada (Alberta), USA (New Jersey), Colombia, Venezuela
- Middle: Serbia, England
- Late: Serbia, Macedonia, south Egypt, Oman, Sudan, Nigeria, Canada (Alberta), USA (New Jersey, Utah, New Mexico)

Cenozoic Era
- Paleocene: Venezuela
- Eocene: Tunisia, Algeria, USA (Texas, Louisiana) Colombia, Venezuela, India
- Oligocene: Kazakhstan, Denmark
- Miocene: Malaysia
- Pliocene: north Philippines, south Ukraine

yielded very little detrital clastic material, (McArthur and Walsh, 1984; 1985; Glenn and Arthur, 1990; Bhattacharyya and Crerar, 1993). They were precipitated in rather shallow marine environments at the water-sediment boundary in areas with very little detrital clastic sedimentation. Although some researchers believed that the entire process of ironstones formation took place in shallow seas (see Petranek and Van Houten, 1997), upwelling currents of basin water with large amounts of iron (Maynard, 1986,) and transgression of anoxic conditions into shallow-water regions are thought to be necessary for the formation and diagenesis of many, if not most, ooidal ironstones (Van Houten and Arthur, 1989; Petranek, pers. commun.). These ironstones are commonly in condensed sections or at disconformities and dark gray or black shales are adjacent, in the offshore direction, or below and immediately above the ironstone.

Marine deltaic, nearshore, and offshore (but not the deep-sea) environments have been identified for 217 ooidal ironstone deposits (Table 3). Twelve Paleozoic, 12 Mesozoic, and 7 Cenozoic ironstone deposits were identified as deltaic deposits (Table 3). Considering the relative duration of the Paleozoic, Mesozoic, and Cenozoic eras (about 330, 174, and 66 m.y., respectively), and the numbers of deltaic phosphorites reported for each era, the deltaic setting was apparently a more common site of phosphorite deposition in Cenozoic time than in Mesozoic and Paleozoic times. Based on the reported incidence of offshore marine phosphorites, the offshore environment was a more common site of phosphorite deposition in the Mesozoic Era (21 deposits, 174 m.y.) than in the Paleozoic (24 deposits, 330 m.y.) and Cenozoic Eras (no reports of Cenozoic offshore phosphorites). The apparent incidence of the nearshore phosphorite setting is nearly the same for the Paleozoic (86 deposits, 330 m.y.), Mesozoic (39 deposits, 174 m.y.), and Cenozoic (16 deposits, 66 m.y.) eras.

Transgression or regression of ironstone facies have been identified and reported for a few of the tabulated deltaic, nearshore, and offshore deposits (Table 3). Many more deposits were identified as transgressive than as regressive, which is not surprising in light of the upwelling-transgression model of ironstone formation.

According to Bhattacharyya, (1989), the lean ooidal iron-

TABLE 3.—REPORTED NUMBERS OF PHANEROZOIC OOIDAL IRONSTONES BY DEPOSITIONAL SETTING AND AGE.[1]

Depositional Environment		Paleozoic		Mesozoic		Cenozoic		Total
Alluvial		0		6		7		13
Lacustrine		0		3		3		6
Total Nonmarine		0		9		10		19
Deltaic	Transgressive	12	4	12	5	7	2	31
	Regressive		3		2		1	
Nearshore	Transgressive	86	4	39	9	16	2	141
	Regressive		1		0		3	
Offshore	Transgressive	24	11	21	4	0	0	45
	Regressive		2		1		0	

[1]Source from Petranek and Van Houten, 1997

stones are typically 5–25 cm thick and generally have less than 30% ooids and peloids scattered in iron-rich mudstone. Commonly, these were subtidal mudstones of primary sites. The concentrated ironstones are as much as 3–4 meters thick and contain 60–70% ooids. Many of the concentrated deposits contain ooids that were transported and redeposited in winnowed beds.

Most of the phosphorites were deposited in shallow subtidal environments (Glenn et al., 1994b) on continental margins or epeiric platforms. Only a few were deposited on seamounts and guyots. They occur in lime- and clay-mudstone and sandstone, and they contain crossbedding and burrows. In the landward direction, phosphorite-bearing sequences may contain ironstones instead of phosphorites. Further landward, the sequence then gives way to clastics derived from the nearby land.

Most phosphorites accumulated during rising sea level which was accompanied by transgression of upwelling currents, expansion of shelf areas, and reduced detrital influx; as many modern examples (Glenn et al., 1994a, 1994b). Phosphatic particles were transported from their place of formation to the place of deposition, where they accumulated along disconformities and in condensed strata. Multiple thin phosphorite deposits stack to form beds as much as 150 meters thick.

Many ironstones and phosphorites are found in organic mudstones at the tops of small upward-coarsening transgressive sequences (Southgate and Shergold, 1991). Commonly they grade offshore into black shale that is somewhat enriched in phosphate.

Primary and Diagenetic Products

Ooidal ironstones contain more than 15% iron (by definition) in the form of ooids (<2 mm in diameter) and rare pisoids (>2 mm in diameter). Some ironstones also contain small floccules of clay and peloids.

The main minerals in the Early Proterozoic ooidal ironstones, as in the banded iron formations, are generally hematite and silica. They also contain minor amounts of magnetite, pyrite, and siderite, and locally contain minerals such as greenalite, stilpnomelane, minnesotaite, and chamosite. They have less than a few percent of aluminum and less than 1% phosphate.

Phanerozoic ooidal ironstones commonly contain kaolinite, goethite, hematite, berthierine (rarely nontronite) or chamosite, siderite, and apatite with calcite. Some contain pyrite and magnetite along with authigenic chert. Many contain some apatite in sheaths and matrix of mud. They commonly have more than 5% aluminum and more than 1% phosphate. The phosphatic ironstones are generally associated with phosphorites and are commonly also associated with carbonates (Kearsley, 1989).

The initial deposits of ironstones consisted largely of clay minerals such as goethite and kaolinite. During diagenesis under reducing conditions, the clay minerals were transformed, at least in part, into berthierine. In later diagenesis, the berthierine was transformed into chamosite and apatite and, less commonly, to magnetite. In some deposits, goethite and hematite can also form (Mücke, 1994). Conversely, some ironstones, such as the Cenozoic ironstone deposits of Sudan, have only goethite or hematite ooids and kaolinite. Diagenesis of these deposits may have never gone as far as the diagenetic transformation into berthierine. Although ooidal ironstones and glauconies are similar in many ways, the two are rarely associated in stratigraphic sections (Maynard, 1986), probably, in part, because the precursor clay mineral of ironstones is kaolinite whereas the precursor clay mineral of glauconies is illite-smectite. Glauconies and phosphorites on the other hand are commonly both present, especially in far-offshore realms.

Phosphorites contain rounded granules, commonly about 1–2 mm in diameter; a few are ooids. These granular deposits consist of 5–15% phosphatic granules and phosphatic mudstone, and 10–30% phosphate.

Most phosphorites were formed in moderately reducing marine environments (Zanin, 1993). Composed largely of carbonate fluorapatite (CFA) granules, they also typically contain some phosphatic ooids and nodules along with micritic carbonate, quartz grains, clay minerals, and microfossils such as diatoms and foraminifers. In some phosphorite-bearing rocks, berthierine or chamosite is present (Zanin, 1993). Initially the phosphorus was in organic matter or in particles (Glenn et al., 1994a) associated with micritic or biosiliceous muds. During early diagenesis, the phosphorus became incorporated into CFA. Early-formed phosphatic particles were later concentrated by bottom currents that scoured the seafloor and winnowed away the finer-grained components. Many phosphorites are associated with siderite, dolomite, and pyrite, as well as with peloidal glaucony. As shown by their Phanerozoic record (Fig. 4 in Van Houten and Purucker, 1984), the geologic range of glauconies is much more like that of phosphorites than of ooidal ironstones.

Modern Analogues

Ooidal ironstones have no obvious modern analogue. Because of extensive chemical weathering, most of the sediment delivered to the marine coast of northeast Venezuela clay mud that includes ferric oxide and kaolinite (Kimberley, 1989; 1994). In the nearshore environment, the ferric oxide and kaolinite are transformed into ooids and peloids of berthierine, which is the ferric iron-rich odinite endmember (Bailey, 1988) in greenish-gray mud. Similar ferric iron-rich odinite peloids, or verdine, are found in the muddy deposits of the Orinoco Delta and in other nearshore settings (Odin, 1990; also see Kronen and Glenn, this volume).

Michalopoulas and Aller (1995) have recently found that diagenetic clay minerals, like nontronite and berthierine, can form off the Amazon Delta by "reversed weathering" of the fine oxidized mud along the coast. Presumably, in anoxic sediments the ferrous iron-rich berthierine will persist, whereas in surface muds the ferric iron-rich odinite variety prevails. However, these modern ferric iron-rich ooids and peloids are not found with any of the modern phosphorites. Clearly, examples of the modern ooids and peloids, such as the ones cited briefly here, need to be tested thoroughly.

Phosphorites have numerous modern analogues. Some are introduced by oceanic upwelling of deep, cold anoxic basin water onto aerated shelves where the phosphatic deposit is transformed into phosphorites during diagenesis. Among the modern examples, the phosphorites off Santo Domingo, Baja California (Birch, 1980) are similar to many *ancient ones* deposited in arid climates. They are silty and sandy with phosphatic granules and hardground with a phosphate content of 30% P_20_5. They occur on the middle and outer shelf where the phosphate was derived from upwelling oceanic currents.

Similar phosphorites that are found off the coast of Peru are also related to upwelling currents. Similar processes formed the Miocene phosphorites in California and in the Sechura Desert of Peru and the late Cenozoic phosphorites off the coast of the southeastern United States, and in the Cretaceous of Brazil and Egypt (Glenn et al., 1994b).

Some phosphorites, on the other hand, like the ones on the eastern shelf of Australia (Glenn et al., 1994b, p. 765–766), are not related to upwelling oceanic currents, but they numerous similarities to the offshore Peru phosphorites. In fact, many modern phosphorites are related to the increase in suitable depositional sites, i.e. remote from land and marked by cross-shelf currents that are associated with rising sea level and transgression. Many researchers think that deposition of phosphorites on the continents through time has drifted in low latitudes, favoring upwelling and transgression (Glenn, et al., 1994a). Moreover, these conditions may have resulted from upwelling from basins within a broad shelf rather than only oceanic current upwelling. But there is still a question of the role of upwelling for the phosphorites deposited in epeiric seas, far-removed from upwelling oceanic currents.

CONCLUSIONS

Many ooidal ironstones and phosphorites have several factors in common. They are sedimentary rocks whose chemical components initially derived from deeply weathered uplands and fed by rivers into shallow seas with very little detrital clastic sediments. Their tectonic settings were relatively stable, and generally in low latitudes. They developed in rather similar shallow marine environments with very limited supplies of sediments. They were commonly transported and redeposited in winnowed layers to form condensed beds at disconformities. Like many other sedimentary rocks the ironstones and phosphorites are fine-grained deposits that accumulated in shallow seas.

Ooidal ironstones and phosphorites differ in composition, however. Oceanic or basinal current upwelling probably led to the deposition of most of the ooidal ironstones and phosphorites, but ironstones were first laid down as ferric oxide and kaolinite ooids and peloids in clay mud. Many phosphorites, on the other hand, formed first as phosphatic carbonate in limy mud. Ironstones were then transformed into berthierine, perhaps by a "reverse weathering" pathway during diagenesis, whereas phosphorites were transformed to CFA. In associated deposits, ironstones were commonly mixed with mudstone and sandstone in clastic sequences, whereas the phosphorites were commonly associated with dolomite and chert, as well as glauconies and, in some cases, with carbonates. Accordingly, these generally similar products differ in their mode of initial formation as well as in their diagenesis.

In their stratigraphic distribution, ironstones are also dissimilar to phosphorites. In Archean time, banded iron formations and a few phosphorites were derived largely from anoxic oceanic reservoirs of hydrothermal activity (Bhattacharyya and Crerar, 1993,). In early Proterozoic time, weathering on land began slowly. Some large, shallow platforms were built, and banded iron formations as well as some ooidal ironstones and phosphorites related to upwelling currents were formed. As the areal extent of the shallow platforms increased during Middle and Late Proterozoic time, the reservoir of oceanic iron declined toward its present modest level, and the banded iron formations virtually disappeared. But phosphorite formation grew to a climax in late Proterozoic and early Cambrian times.

Phanerozoic ooidal ironstones and phosphorites also differ. With increased continental area and subaerial weathering, the ooidal ironstones flourished during the greenhouse phases (Fig.1), whereas the phosphorites show no such preference. It is clear that the greenhouse environment that encouraged the development of ironstones was not the principal control of phosphorite accumulation.

Domination by either clastics or carbonates dictated, in large part, whether ironstones or phosphorites resulted, even in similar environments. Perhaps when mixed siliciclastic-carbonate environments prevailed the ironstones and phosphorites formed at the same time, as they did during about 25% of the Phanerozoic Eon.

ACKNOWLEDGEMENTS

D. P. Bhattacharyya and C. R. Glenn helped me sort out comparisons. R. E. Garrison and an anonymous reviewer made great improvements in the subject and style of the paper. This review is a contribution to IGCP Project 277 and Project 325.

REFERENCES

BAILEY, S. W., 1988, Odinite, a new dioctahedral-trioctahedral Fe^{3+}-rich 1:1 clay mineral: Clay Minerals, v. 23, p. 237–247.

BHATTACHARYYA, D. P., 1989, Concentrated and lean oolites: examples from the Nubia Formation at Aswan, Egypt, and significance of the oolite types in ironstone genesis, *in* Young, T. P. and Taylor, W. E. G., eds., Phanerozoic Ironstones: Geological Society Special Publication No. 46, p. 93–103.

BHATTACHARYYA, D. P., AND CRERAR, D. A., 1993, Genetic model for the Phanerozoic oolitic ironstones and some speculations about its implications for the deposition of banded iron formation: Proceedings National Academy Sciences, India, v.63(A),I, p. 48–72.

BIRCH, G. F., 1980, A model of penecontemporaneous phosphatization by diagenetic and authigenic mechanisms from the western margin of southern Africa: Society of Economic Petrology and Mineralogy Special Publication 29, p.79–100.

COOK, P. J., AND SHERGOLD, J. H., 1986, Proterozoic and Cambrian phosphorites—an introduction, *in* Cook, P. J., and Shergold, J. H., eds., Phosphate Deposits of the World, v. 1, Proterozoic and Cambrian Phosphorites: Cambridge, Cambridge University Press, p. 1–8.

COOK, P. J., 1984, Spacial and temporal controls on the formation of phosphate deposits—a review, *in* Nriagu, J. 0. and Moore, P. B., eds., Phosphate Minerals: New York, Springer-Verlag, p. 242–274.

COOK, P. J., 1992, Phosphogenesis around the Proterozoic-Phanerozoic transition: Journal Geological Society, v. 149, p. 615–620.

COOK, P. J., AND MCELHINNEY, M. W., 1979, A reevaluation of the spacial and temporal distribution of sedimentary phosphate deposits in the light of plate tectonics: Economic Geology, v. 74, p. 315–330.

FÖLLMI, K., 1996, The phosphorus cycle, phosphogenesis and marine phosphaterich deposits: Earth Science Reviews, v. 40, p. 55–124

FRAKES, L. A., 1979, Climate through Geologic Time: Amsterdam, Elsevier, 310 p.

GLENN, C. R. AND ARTHUR, M. A.,1990, Anatomy and origin of a Cretaceous phosphorite-greensand giant, Egypt: Sedimentology, v. 37, p. 123–154.

GLENN, C. R., ARTHUR, M. A., RESIG, J. M., BURNETT, W. C., DEAN, W. E., AND JAHNKE, R. A., 1994a, Are modern and ancient phosphorites really so different? *in* Iijima, A., Abed, A. M., and Garrison, R. E.,eds., Siliceous, Phosphatic, and Glauconitic Sediments of the Tertiary and Mesozoic: Utrecht, VSP Publishers, p. 159–188.

GLENN, C. R., FÖLLMI, K. B., RIGGS, S. R., BATURIN, G. V., GRIMM, K. A., TRAPPE, J., ABED, A. M., GALLI OLIVIER, C., GARRISON, R. E., ILYIN, A. V., JEHL, C., ROHRLICH, V., SADAQAH, R. M. Y., SCHIDLOWSKI, M., SHELDON, R. E., AND SIEGMUND, H., 1994b, Phosphorus and phosphorites: sedimentology and environments of formation: Eclogae Geologicae Helvetica, v. 87, p. 747–788.

JAMES, H. L., 1983, Distribution of banded iron-formation in space and time, *in* Trendall, A. F., and Morris, R. G., eds., Iron-formation Facts and Problems: Amsterdam, Elsevier, p. 471–490.

KEARSLEY, A. T., 1989, Iron-rich ooids, their mineralogy and microfabric: clues to their origin and evolution, *in* Young, T. P., and Taylor, W E. G., eds. Phanerozoic Ironstones: Geological Society Special Publication No. 46, p. 141–164.

KIMBERLEY, M. M., 1989, Exhalative origin of iron formations: Ore Geology Reviews, v. 5, p. 13–145.

KIMBERLEY, M. M., 1994, Debate about ironstones: has solute supply been surficial weathering, hydrothermal convection, or exhalation of deep fluids?: Terra Nova, v. 6, p. 116–132.

LEE, A. I. N., 1980, Fertilizer mineral occurrence in the Asia-Pacific region: Honolulu, EastWest Centre, p. 124–139.

MAYNARD, J. B., 1986, Geochemistry of oolitic ironstones, an electron microprobe study: Economic Geology, v. 81, p. 1473–1483.

MCARTHUR, J. M., AND WALSH, J. N., 1984/1985, Rare-earth geochemistry of phosphates; Chemical Geology, v. 49, p. 191–220.

MCKELVEY, V. E., 1979, Phosphate in sediments, *in* Fairbridge, R. W., and Bourgeois, J., eds., The Encyclopedia of Sedimentology: Stroudsburg, Pennsylvania, Dowdon, Hutchinson and Ross, p. 574–578.

MICHALOPOULOS, P., AND ALLER, R. C., 1995, Rapid clay formation in Amazon Delta sediments: reversed weathering and oceanic elemental cycles: Science, v. 270, p. 614–617.

MÜCKE, A., 1994, Postdiagenetic ferrugination of Phanerozoic oolitic ironstones: a contribution to their genesis, *in* Wolf, K. H. and Chilingarian, G. V., eds., Development in Sedimentology, Diagenesis IV: Amsterdam, Elsevier, p. 396–423.

NOTHOLT, A. J., AND SHELDON, R. P., 1986, Proterozoic and Cambrian phosphorites—regional review; world resources, *in* Cook, P. J. and Shergold, J. H., eds., Phosphate Deposits of the World, Volume 1, Proterozoic and Cambrian Phosphorites: Cambridge, Cambridge University Press, p. 9–19.

ODIN, G. S., 1990, Clay mineral formation at the continental-ocean boundary: Clay Minerals, v. 25, p. 477–483.

PETRANEK, J., 1998, Transgression, anoxic processes, upwelling, and ooidal ironstones a genetic relationship?: manuscript in review.

PETRANEK, J., AND VAN HOUTEN, F. B., 1997, Phanerozoic ooidal ironstones: Czech Geological Survey, Special Papers, v. 7, 71 p.

PETTIJOHN, F .J., 1975, Sedimentary Rocks, 3rd edition: New York, Harper and Row, 628 p.

SOUTHGATE, P N., AND SHERGOLD, J. H., 1991, Application of sequence stratigraphic concepts to Middle Cambrian phosphogenesis, Georgina Basin, Australia: BMD Journal of Australian Geology and Geophysics, v. 12(2), p. 119–144.

TRENDALL, A. F. AND MORRIS, R. G., eds., 1983, Iron-formation Facts and Problems: Amsterdam, Elsevier, 558 p.

VAN HOUTEN, F. B., AND ARTHUR, M. A., 1989, Temporal patterns among Phanerozoic oolitic ironstones and oceanic anoxia, *in* Young, T. P., and Taylor, W. E. G., eds., Phanerozoic Ironstones: Geological Society Special Publication No. 46, p. 33–49.

VAN HOUTEN, F. B., AND PURUCKER, M. E., 1984, Glauconitic peloids and chamositic ooids: favorable factors, constraints and problems: Earth Science Reviews, v. 20, p. 211–243.

ZANIN,Y. N., 1993, Geochemical conditions of formation of mineral parageneses of phosphate-bearing rocks: Russian Geology and Geophysics, v. 34, p. 1–14.

HYDROTHERMAL BACTERIAL BIOMINERALIZATION: POTENTIAL MODERN-DAY ANALOGUES FOR BANDED IRON-FORMATIONS

KURT O. KONHAUSER*

School of Earth Sciences, University of Leeds, Leeds, UK LS2 9JT

ABSTRACT: Precambrian banded iron-formations (BIFs) are chemical sediments of hydrothermal origin and consist of Fe-rich minerals with alternating layers of chert. Because microorganisms potentially played a role in their precipitation, the study of bacterial-mineral interactions at modern hydrothermal environments may provide small-scale analogues to those conditions under which they accumulated. Interestingly, microbial populations currently growing at hot springs and deep-sea vents are commonly encrusted in iron and silicate minerals. Iron biomineralization occurs either passively through interaction between the reactive sites of the cell and dissolved cationic iron from the hydrothermal fluid, or actively through chemolithotrophic iron-oxidation by bacteria such as *Gallionella* genera. Amorphous silica precipitates on individual bacteria through hydrogen bonding between hydroxy groups in the extracellular polymers and hydroxyl groups in dissolved silica, with some colonies becoming completely cemented together within a siliceous matrix up to several micrometers thick. Iron-silicates form due to reactions between dissolved silica and cell-bound iron. In these predominantly nonspecific processes, bacterial cells simply catalyze reactions that are rendered possible by the supersaturated conditions created by the sudden physical and chemical changes induced through venting. Diagenetic reactions, some of which are also catalyzed by microorganisms growing in the sediment, can further alter the mineralogy of these primary precipitates, leading to the formation of secondary magnetite and siderite. In this way, all of the main mineralogical components of BIFs can be associated with microbial activity.

INTRODUCTION

Banded iron-formations (BIFs) are some of the most abundant sedimentary deposits of the Precambrian, with major accumulations having formed during the late Archean (2.7–2.5 Ga) and early Proterozoic (2.5–1.8 Ga) (Isley, 1995). These thin-bedded or laminated rocks contain an anomalously high content of iron, and commonly, layers of chert (James, 1954). Most BIFs have similar bulk compositions and mineral assemblages, although Archean BIFs are not as thick or laterally extensive as their Proterozoic counterparts (Gole and Klein, 1981). Iron and silica contents are typically in the range of 20–35 wt. % and 40–50 wt. %, respectively (James, 1983), while the mineralogy consists of quartz, magnetite, hematite, various carbonates, ferrous iron-silicates, and pyrite; some carbonaceous material is also present (McConchie, 1987; Klein and Beukes, 1992). The banding in BIFs generally comprises alternating layers of Fe-rich minerals with layers of chert (and carbonate) that can be observed at three different scales: macro, meso, and micro (Trendall, 1965; Trendall and Blockley, 1970). Macrobands are coarse alterations of contrasting Fe-oxide facies (magnetite, hematite) and shale facies (referred to as S-macrobands, with chert, siderite, and iron-silicates) decimeters to tens of meters in thickness; mesobands are centimeter thick units; and microbands usually range from submillimeter to 5 mm (Morris, 1993). The lateral continuity of mesobands is a characteristic feature of some iron-formations, and in the Dales Gorge Member, Hamersley Group BIF, Western Australia, for example, they can be traced for more than 130 km through the basin (Ewers and Morris, 1981).

It is generally accepted that BIFs are chemical precipitates (Morris, 1993), with the iron primarily derived from a hydrothermal source (Simonson, 1985; Klein and Beukes, 1992; Isley, 1995; Simonson and Hassler, 1996). This model is supported by (i) estimates of higher surface heat flow and rapid hydrothermal cycling rates for the late Archean-early Proterozoic (Turcotte, 1980; Jacobsen and Pimentel-Klose, 1988), (ii) rare earth element (REE) distribution patterns in BIFs (Jacobsen and Pimentel-Klose, 1988; Derry and Jacobsen, 1990; Bau and Möller, 1993), (iii) sulfur and neodymium isotopic compositions of Archean and early Proterozoic iron-formations (Cameron, 1983; Jacobsen and Pimentel-Klose, 1988), and (iv) Sr^{87}/Sr^{86} ratios and ^{18}O depletion in Archean and early Proterozoic carbonates (Veizer et al., 1982; Beukes et al., 1990). Interestingly, the transition from hydrothermal-dominated ocean chemistry of the Archean to continental-dominated modern seawater was established by approximately 0.7–0.8 Ga, coinciding with the last major BIF deposits of the Rapitan Group, N.W.T., Canada (Klein and Beukes, 1992).

Although the origin of iron in BIFs has been clarified, mechanisms for their deposition are still under debate (see McConchie, 1987 and Morris, 1993, for detailed discussions). Interactions between microorganisms and dissolved ferrous iron and silica in the Precambrian ocean have been suggested as one plausible means for mineral precipitation (Cloud, 1965, 1973; LaBerge, 1973; Baur et al., 1985; Birnbaum and Wireman, 1985; Holm, 1987, 1989; Nealson and Myers, 1990; Widdel et al., 1993; Ehrenreich and Widdel, 1994; Konhauser and Ferris, 1996). Cloud (1965, 1973) originally proposed that primitive oxygen-releasing photosynthetic bacteria, which may have lacked suitably advanced oxygen-mediating enzymes, attached any free O_2 that was generated to convenient oxygen acceptors in the immediate vicinity of the bacterial population. The abundant ferrous iron in solution at that time would have served as such an acceptor, thereby maintaining the reducing environment necessary for their survival. This process of transferring oxygen

*email: kurt@earth.leeds.ac.uk

Marine Authigenesis: From Global to Microbial, SEPM Special Publication No. 66

to nonbiological oxygen acceptors may have later evolved into a source of energy for chemolithotrophic microaerophilic bacteria such as *Gallionella ferruginea*. Holm (1987, 1989) speculated that slow biological oxidation of hydrothermal solutions by these microorganisms, in an ocean with limited free photosynthetic oxygen, was primarily responsible for the precipitation of ferrous to ferric iron over wide ocean basins. Alternatively, the discovery of purple bacteria that couple ferrous iron oxidation to anoxygenic photosynthesis has been suggested as a means for early BIF deposition in the Archean, before free oxygen became available as an oxidant (Widdel et al., 1993; Ehrenreich and Widdel, 1994).

Simultaneously, saturation of ocean waters with respect to silica due to the absence of silica-secreting microorganisms (e.g., diatoms, radiolarians) at that time provided ideal conditions for the precipitation of amorphous silica (Siever, 1992). The sudden cooling and pressure drop of hydrothermal effluent after venting on the ocean floor (Williams and Crerar, 1985), evaporation on continental shelves (Morris, 1993), tidal flats and restricted lagoons (Knoll, 1985), and coprecipitation with iron (Ewers, 1983) could each have led to silica supersaturation and precipitation. During diagenesis, these primary silica precipitates subsequently may have undergone a dissolution-reprecipitation pathway with a characteristic sequence of mineralogical transformations progressing from amorphous silica to low-temperature cristobalite to quartz, the phase of lowest entropy and slowest nucleation and growth rates (Williams and Crerar, 1985). A biological role for amorphous silica precipitation has also been considered, since bacteria are known to create favorable geochemical conditions for silicification through their metabolic activities (Birnbaum and Wireman, 1985) and they provide highly reactive surfaces for silica nucleation and mineral growth (Urrutia and Beveridge, 1993). The presence of Precambrian microfossils in carbonaceous cherts (Awramik et al., 1983; Walsh and Lowe, 1985; Schopf and Packer, 1987; Schopf, 1993) and BIFs (Barghoorn and Tyler, 1965; Cloud, 1965; LaBerge, 1973), as well as the presence of silicified stromatolites (Lowe, 1980; Walter et al., 1980) would certainly seem to suggest that microorganisms were, in some manner, inextricably linked to silica precipitation.

Because Precambrian banded iron-formations are chemical sediments of hydrothermal origin, and microorganisms potentially played a role in their precipitation, I propose here that the study of bacterial-mineral interactions at modern hydrothermal environments may provide small-scale analogues to those conditions under which they accumulated. This paper will show (i) that bacteria directly form iron and silicate minerals in modern hydrothermal environments and (ii) that these biomineralization processes correlate well with conditions that may have existed in the Precambrian, resulting in banded iron-formation.

HYDROTHERMAL BIOMINERALIZATION

Bacteria are found today throughout the hydrosphere, commonly inhabiting environments that cannot sustain other life forms. At hot springs, photosynthetic cyanobacteria (e.g., *Synechococcus* sp.) and filamentous bacteria (e.g., *Chloroflexus* sp.) frequently form well-defined laminated mats adjacent to hydrothermal vents and in their outflow channels (Walter et al., 1972; Walter, 1976; Doemel and Brock, 1977). In Yellowstone National Park, USA, for example, these mats can reach several millimeters in thickness with both *Synechococcus* and *Chloroflexus* cells growing to depths where sufficient light exists, below which anaerobic decomposition by heterotrophic bacteria takes place (Doemel and Brock, 1977; Walter et al., 1992). In some high-sulfide springs, cyanobacterial growth is suppressed, and pure *Chloroflexus* populations are developed via anoxygenic photosynthesis (Castenholz, 1973). In groundwater from geothermal areas, such as those in Iceland, high ferrous iron concentrations encourage the growth of iron-oxidizing bacteria such as *Gallionella ferruginea* (Holm, 1987).

In deep-sea hydrothermal environments, bacterial productivity is highest in the vicinity of warm vents, where reducing fluids mix with downwelling cold, oxygenated seawater (Jannasch and Mottl, 1985). Bacterial communities growing at these sites have concentrations that range from 10^6 to 10^9 cells per milliliter (Jannasch and Wirsen, 1981). Reduced sulfur compounds appear to represent the major electron donors for aerobic microbial metabolism, although methane-, hydrogen-, iron-, and manganese-oxidizing bacteria are also locally important (Jannasch and Mottl, 1985). Methanogenic, acetogenic, and sulfur-reducing bacteria commonly occupy anoxic niches where CO_2 and H_2 are present (Baross et al., 1982; Harmsen et al., 1997).

These mixed bacterial communities interact diversely with their surrounding aqueous environment. Passive interactions occur when microorganisms behave as a solid-phase sorbents for dissolved ions. This behavior stems from bacterial cells having high surface area to volume ratios (Beveridge, 1988) and exceptionally reactive surfaces comprising exposed anionic functional groups (Beveridge and Fyfe, 1985). Once bound to the cells, these metals reduce the activation energy barriers to heterogeneous nucleation by providing sites where additional ions can be sorbed and authigenic minerals are formed (Ferris, 1997; Fortin et al., 1997; Konhauser, 1997, 1998). Conversely, some bacteria actively promote mineral formation by producing metabolic end products (e.g., OH^-, CO_2, H^+, NH_3, and H_2S) that react with dissolved metal ions, leading to supersaturation (Thompson and Ferris, 1990; Fortin and Beveridge, 1997), or they produce enzyme-mediated changes in redox state (e.g., oxidation of Fe^{2+} to Fe^{3+}) (Ghiorse and Ehrlich, 1992). In this regard bacteria catalyze reactions that are kinetically slow, or even thermodynamically unfeasible prior to their intervention, and hence induce mineral growth under potentially unfavorable conditions (Ehrlich, 1996). The biogenically formed minerals have crystal habits and chemical compositions similar to those produced by precipitation from inorganic solutions since they are governed by the same equilibrium principles that control mineralization of their inorganic counterparts.

Iron Hydroxides

In terrestrial hot springs, the presence of iron hydroxide, in

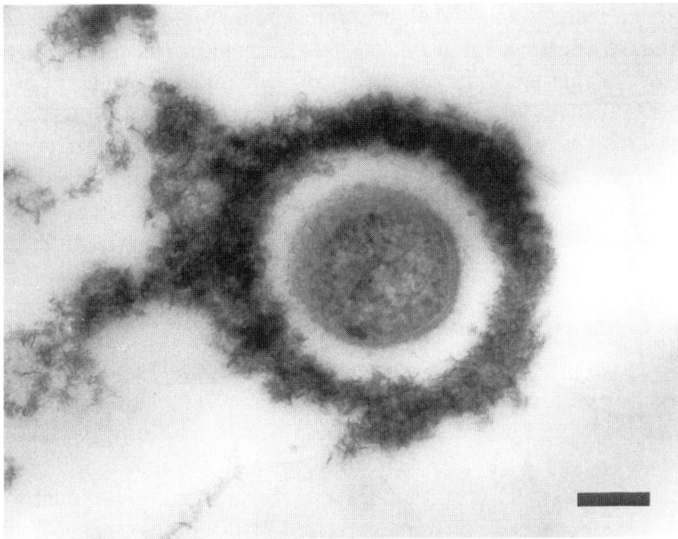

FIG. 1.—Transmission electron micrograph (TEM) of an epilithic bacterial cell from a hot spring effluent channel at Lýsuhóll, Iceland. The bacterial cell is surrounded by a dense, iron-rich capsule consisting of amorphous ferric hydroxide. Scale bar = 140 nm.

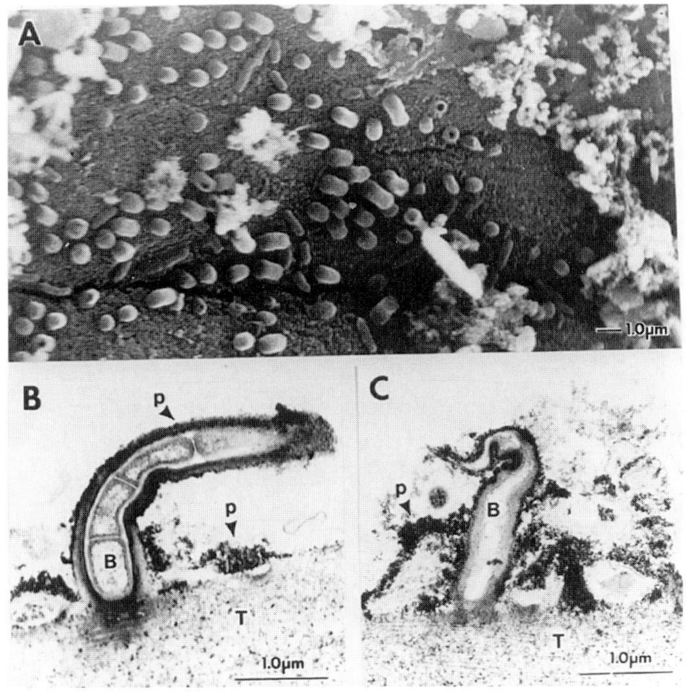

FIG. 2.—Electron micrographs of vestimentiferan tubes from a site of Fe-oxide encrustation on the southern Juan de Fuca Ridge. (A) Scanning electron micrograph (SEM) of the tube surface with light colonization by sheathed bacteria. (B) TEM photo showing Fe particle (p) accumulation on sheath of bacterium (B) attached to the tube surface (T). (C) Extensive Fe particle accumulation in unorganized extracellular slime surrounding the bacterium. Reprinted with permission of the author and CRC Press (Juniper and Tebo, 1995).

association with microbial mats, has been reported from geothermal areas in Iceland (Holm, 1987; Konhauser and Ferris, 1996) and Yellowstone National Park (Ferris et al., 1986). At Húsatóttir, Iceland, samples of Fe(III) precipitate were taken from the bottom of a drilled well (80 m in depth) used to pump low-temperature (12°C) geothermal water (Holm, 1987). Microscopic analysis revealed that the iron precipitate consisted mainly of living cells, stalks, and other remains of *Gallionella ferruginea*. On 1 mg of dry precipitate dispersed in 1 ml of water, the total living cell count of *Gallionella ferruginea* was 2 x 10^8 cells (Holm, 1987). Transmission electron microscopy (TEM) and energy dispersive X-ray spectroscopy (EDS) of microbial mats growing in hot spring outflow channels of Lýsuhóll, Iceland, indicated a highly mineralized bacterial community (Konhauser and Ferris, 1996). Individual bacterial cells (Fig. 1) or microcolonies were frequently observed surrounded by an iron-rich capsule (a hydrated polysaccharide material that is naturally anionic and has great potential for scavenging metals). In Lýsuhóll, precipitation of iron phases was not restricted to cells with encapsulating material; the partial and complete encrustation of some bacterial cells by fine-grained (<100 nm in diameter) spheroids composed of amorphous ferric hydroxide were also observed. Other cells underwent intracellular mineralization so that the cytoplasm was completely replaced with amorphous iron hydroxide once the cell had lysed.

In marine hydrothermal environments, Fe-hydroxides are commonly precipitated in the region immediately surrounding the vent, the peripheral sediments, and the hydrothermal plume (Karl et al., 1988; Juniper and Tebo, 1995). Baross and Deming (1985) provided evidence that at an active black smoker from the East Pacific Rise, iron was deposited in significantly greater amounts on rock surfaces covered with microbial mats than on uncolonized rocks. Electron microscopic examination of vestimentiferan tubes at two chimneys on the southern Juan de Fuca Ridge (Fig. 2) revealed diverse unicellular and filamentous bacteria colonizing the tube surfaces where they accumulated iron hydroxide (as Fe-rich spheres 6–8 μm in diameter) around their sheaths (Tunnicliffe and Fontaine, 1987). These spheres were shown to eventually coalesce to form a continuous crust in both vertical and horizontal dimensions. The sequence of formation suggests that iron was initially adsorbed to reactive organic sites on the cells, but that further iron accumulation may have resulted autocatalytically (Tunnicliffe and Fontaine, 1987). Extensive deposits of biogenic ferric hydroxide have also been described from the surface sediments of the ancient caldera of Santorini, Aegean Sea. Hanert (1973) and Holm (1987) both found Fe-encrusted stalks of *Gallionella ferruginea* occurring in such masses that the activity and importance of this bacterium in iron sedimentation was unquestionable. In deep hydrothermal areas, iron hydroxides were found to form on bacterial filaments growing in mud deposits at the Larson and Red Seamounts, near 21°N, East Pacific Rise (Alt, 1986, 1988) and on the Loihi Seamount, Hawaii (Karl et al., 1988, 1989). Because the mud consisted essentially of bacterial filaments, and the hydrothermal waters in which these deposits formed were slightly acidic (pH 5 to 6) and low in dissolved O_2 (spontaneous oxidation of ferrous to ferric iron would be very slow under these conditions), biologically catalyzed precipitation was clearly favored over abiotic oxidation processes (Juniper and Tebo, 1995).

The microbial precipitation of ferric hydroxide (e.g., ferrihydrite) can occur both passively and actively. In the first instance, the oxidation and hydrolysis of cell-bound ferrous iron or the binding of ferric species [e.g., $Fe(OH)^{2+}$, $Fe(OH)_2^+$] and cationic colloidal species [e.g., $Fe(H_2O)_5(OH)_5(OH)^{2+}$] to negatively charged polymers (MacRae and Edwards, 1972; Ferris et al., 1989) can result in ferric iron deposition. Alternatively, ferrous iron transported into an oxygenated environment spontaneously reacts with dissolved oxygen (at circumneutral pH) to precipitate rapidly as ferric hydroxide on available nucleation sites. Bacteria passively act as such sites, and over a short period of time the microbial mats can become completely encrusted in amorphous iron as abiological surface catalysis accelerates the rate of mineral precipitation (Ghiorse, 1984). It has even been suggested that under circumneutral pH any bacterium that produces acidic, extracellular polymers will nonspecifically adsorb positively charged Fe-hydroxides (Ghiorse, 1984). This is due to the point of zero charge (pH where the mineral has zero charge) of amorphous Fe-hydroxides (e.g., ferrihydrite) being in the range of 5.3 to 7.5 for natural samples (Schwertmann and Fechter, 1982), which thereby allows reactive organic sites to scavenge ferric iron from the surrounding waters. The growth of iron-depositing microbial populations in environments where they are constantly covered in amorphous iron hydroxide suggests that these sites must also provide the bacteria with favorable growth conditions. Quite possibly the continual supply of iron, as well as the trace metals and anions that adsorb or coprecipitate directly onto iron hydroxides (German et al., 1991), may serve as an ideal nutrient source in close proximity to the cells.

The active process by which iron hydroxides form stems from the metabolism of Fe(II)-oxidizing bacteria that derive energy from the biological oxidation of Fe(II) to Fe(III) for the fixation of carbon from carbon dioxide dissolved in water (Ehrlich, 1990). At neutral pH, Fe(II) oxidation by *Gallionella* genera occurs under partially reduced conditions of 0.1–1 mg of O_2 per liter (Ehrlich, 1990). The existence of these microaerophilic bacteria relies on their oxidizing efficiency relative to abiotic oxidation at low-oxygen fugacity. Because very little energy is generated in the oxidation of ferrous to ferric iron, these microaerophilic bacteria must oxidize large quantities of iron to grow (Lees et al., 1969). Consequently, even a small number of bacteria can be responsible for precipitating vast amounts of iron (Brock et al., 1984). Ferrous iron has also been observed to undergo microbial oxidation under anoxygenic conditions. Widdel et al. (1993) and Ehrenreich and Widdel (1994) have described purple, nonsulfur bacteria that combined ferrous iron oxidation with CO_2 fixation for cellular material, with light as the energy source. Ferrous iron oxidation took place on the outer membrane so that the insoluble ferric hydroxide remained outside of the cell (Ehrenreich and Widdel, 1994). Straub et al. (1996) also demonstrated that the biological oxidation of ferrous iron, in the absence of oxygen, was possible by light-independent, chemotrophic activity with nitrate as the electron acceptor.

Once the primary iron hydroxides are precipitated and incorporated into the sediment, several diagenetic reactions can subsequently alter their mineralogies, textures, and structures. The transformation of ferric hydroxide into hematite takes place through internal aggregation and rearrangement (Cornell et al., 1987), and in marine sediments, its formation is favored under high temperatures and low organic matter content (Schwertmann and Fitzpatrick, 1992). However, the observed presence of a mixed ferric hydroxide/hematite assemblage on bacterial cells, in present-day biofilms, suggests that this process may also occur rapidly at surface conditions (Ferris et al., 1989; Sawicki et al., 1995). Dissimilatory Fe(III)-reducing bacteria have been shown to produce magnetite under anaerobic conditions, as a by-product during the oxidation of organic compounds coupled with the reduction of poorly crystalline, Fe(III) oxide (Lovley, 1991). For example, in hydrothermal sediments located 25 km from the East Pacific Rise, authigenic magnetite was recovered at the redoxcline between Fe-oxidizing to Fe-reducing conditions (Karlin et al., 1987). The formation of siderite in marine sediments has also been shown to result from the ability of bacteria (e.g., *Desulfovibrio* sp.) to release ferrous iron, via bacterial Fe(III)-reduction, into waters with excess seawater bicarbonate (Coleman et al., 1993). In deep sediment from Santorini, beneath the iron hydroxide layers, siderite was identified as a by-product of microbial diagenesis (Holm, 1987).

Amorphous Silica

The close association of microorganisms with amorphous silica has been reported from a variety of hot springs, including Yellowstone National Park (Walter et al., 1972; Walter, 1976; Ferris et al., 1986; Hinman and Lindstrom, 1996); Taupo Volcanic Zone, New Zealand (Jones et al., 1997a, b); Kenya Rift (Jones and Renaut, 1996); El Tatio, Chile (Jones and Renaut, 1997); and Geysir, Iceland (Schultze-Lam et al., 1995; Konhauser and Ferris, 1996). The role of microorganisms in silica precipitation at hot springs generally has been considered a passive process (Walter et al., 1972), in that rapid cooling at ambient air temperatures, evaporation and/or steam loss, and a change in pH of the hydrothermal waters following discharge can all induce silicification of growing cells (Fournier, 1985). However, several recent studies using electron microscopy have shown that microorganisms provide favorable nucleation sites that serve as templates for silicification. In sediments from Yellowstone, Ferris et al. (1986) observed individual bacterial cells preserved in successive stages of amorphous silica encrustation; some cells were thoroughly embedded in a mineralized matrix. In New Zealand, bacterial filaments were similarly encrusted by amorphous silica (Jones et al., 1997a, b). That microorganisms catalyze silica precipitation was confirmed through analyses of microbial mats growing around a hot spring vent in Geysir, Iceland, where filamentous phototrophic bacteria resembling *Chloroflexus aurantiacus,* and the unicellular cyanobacterium *Synechococcus* sp., underwent rapid silicification (Schultze-Lam et al., 1995; Konhauser and Ferris, 1996). The mineralization associated with the cells occurred as spheroidal grains (200–500 nm in diameter) both extracellularly on the sheaths of living bacteria and intracellularly within the cytoplasm presumably after the cells had lyzed. In many filaments, the silica crystallites appeared to have coalesced such that individual precipitates were no longer distinguishable (Fig. 3); entire colonies sometimes became cemented together in a

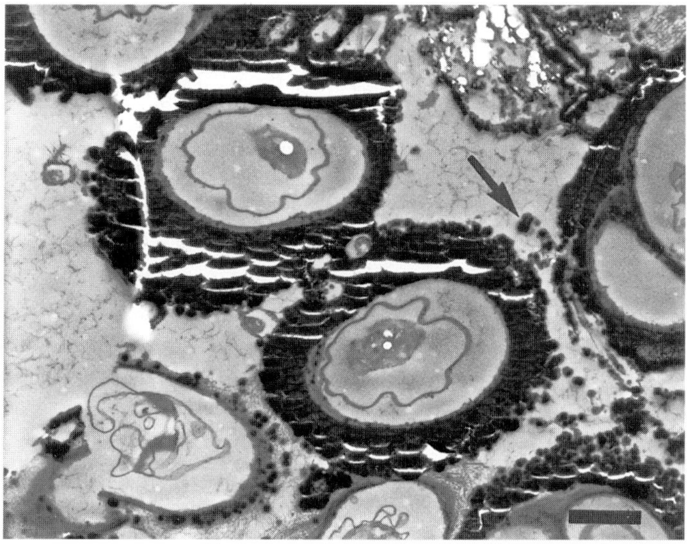

FIG. 3.—TEM image of a colony of filamentous bacteria (from Geysir, Iceland) with epicellular amorphous silica on outer sheath surfaces. Note the presence of a few discernible spheres on the outer edges on the mineralized matrix (arrow). Scale bar = 1.1 μm.

siliceous matrix several micrometers thick (Konhauser and Ferris, 1996). Eventually only the sheath and cell wall of the original organic framework remained recognizable. Thus, it appears that silicification begins when the microorganisms are living (i.e., they serve as nucleation sites) and continues for some time after their death. In the latter stages of silicification, the development of amorphous silica cement between silicified microbial filaments is presumably abiogenic (Jones et al., 1997a). Silicification has also been shown to occur in thermal springs in the Kenya Rift where water temperatures reach above 90°C, temperatures too high to support cyanobacteria and most bacteria. Instead, hyperthermophilic bacteria were shown to survive in this extreme environment, preferentially fixing silica onto secreted microbial mucus (Jones and Renaut, 1996).

In a hydrothermal venting site with talus piles on the wall of a graben at 21°30'S, East Pacific Rise, hot water mixes with seawater in the talus piles producing sulfide and amorphous silica deposits (Juniper and Fouquet, 1988). Upon magnification, the silica crust revealed a filamentous microtexture in which individual filaments were cemented together. Figure 4 shows a section of such a filament heavily coated in concentric layers of deposited silica, with the hollow space likely representing the original bacterial cell (Juniper and Fouquet, 1988).

Based on findings attained from studies on the silicification of wood (Leo and Barghoorn, 1976; Sigleo, 1978), the association of silica with microbial surfaces is probably established by hydrogen bonding of the hydroxyl groups in silicic acid [$Si(OH)_4$] to available hydroxy groups of the bacterial sheath (Ferris et al., 1988). Continued reaction of the bound silica with more silicic acid, from the surrounding waters, results in polymerization and eventual dehydration, which leads to the formation of silica crystallites (Iler, 1980). Results from artificial fossilization studies on microorganisms have confirmed that bacteria can be completely silicified (Oehler and Schopf, 1971; Francis et al., 1978; Westall et al., 1995), with the preferential

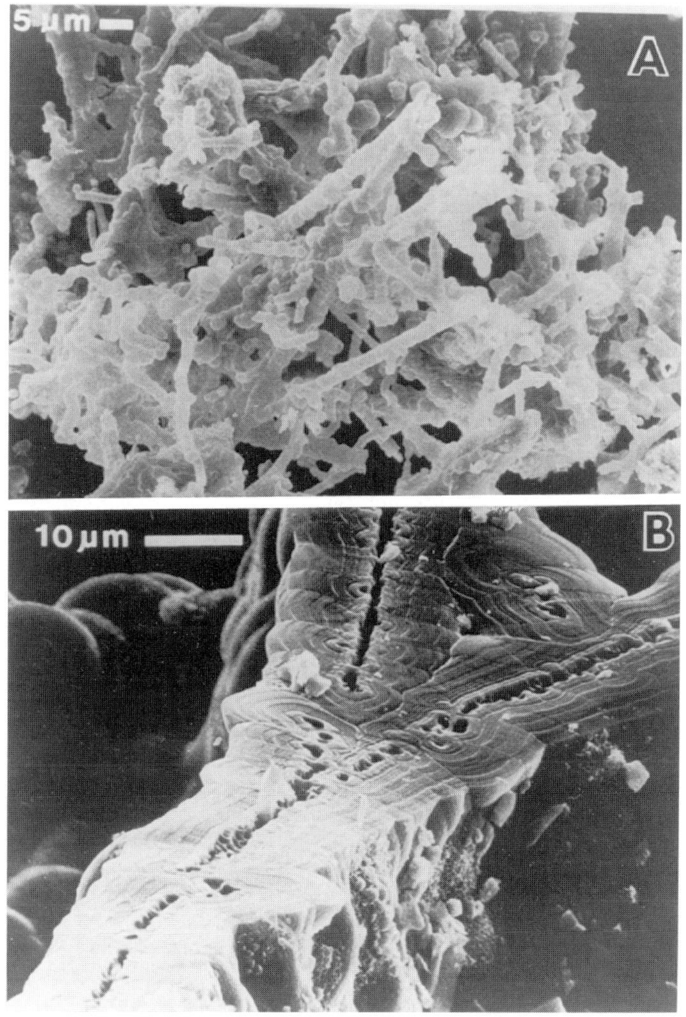

FIG. 4.—(A) Low-magnification SEM photo showing filamentous microtexture of hydrothermal deposits (from 21°30'S, East Pacific Rise). (B) Section of a filament heavily coated in silica. The hollow internal space likely represents the original filament that was covered in concentric layers of deposited silica. Reprinted with permission of the author and *Canadian Mineralogist* (Juniper and Fouquet, 1988).

preservation of cell walls and sheaths (Oehler, 1976). Ferris et al. (1988) have also shown the rapid development of silica crystallites on cellular debris following cell lysis. Evidently, lyzed microorganisms continue to serve as templates for silica deposition, with cellular degradation enhancing the mineralization process by increasing the availability of functional groups capable of forming hydrogen bonds with dissolved silica (Knoll, 1985; Ferris et al., 1988). General features of increasing silicification with time include thickening of silica crusts around the bacteria, increasing crystallization of the silica from loose aggregates into dense spheres, and eventually the formation of a mammillated external crust (Westall et al., 1995). Interestingly, pretreatment of cells with iron appeared to denature their autolytic enzymes, thus maintaining intact organic residues during silicification (Ferris et al., 1988).

Iron Silicates

Bacterial cells, associated with granular and spheroidal crystallites composed exclusively of iron and silica, have been examined from acidic hot spring sediments from Yellowstone National Park, USA (Ferris et al., 1986) and from mats growing in hot spring outflow channels, Krisuvik, Iceland (Konhauser and Ferris, 1996). In the Yellowstone samples, both intact and lyzed cells were mineralized, with grains typically smaller than 100 nm (Ferris et al., 1986); one cell that appeared to have undergone an extended episode of iron-silica growth and compaction had an interlocking mosaic that completely encrusted the microorganism. In Krisuvik, bacterial cells were associated with diverse mineral assemblages (Konhauser and Ferris, 1996). In some cells several spheroidal grains (~200 nm) were shown to be embedded in dense Fe-rich capsular material. In other cells from the same sample site, larger amorphous precipitates (~500 nm), consisting of iron and silica in approximately equal proportions, were evident directly on the cell wall (Fig. 5).

Iron-silica deposits have also been identified from various deep-sea hydrothermal sites. One active site at the Philosopher Vent on Explorer Ridge in the northwest Pacific had iron-silica deposits forming with a filamentous morphology (Juniper and Fouquet, 1988). The resemblance of these filaments to those produced by microorganisms, along with the presence of organic carbon (1.3%) and filamentous bacteria (from one vent sample), suggested that bacteria provided substrata for mineral nucleation. Transmitted light microscopy of these iron-silica deposits further identified intense iron accumulation on the filaments, upon which silica precipitated (Fig. 6). Extensive accumulation of microbial-mineral flocs (dominated by iron and silica oxides) were also observed in association with abundant diffuse venting in new lava flows and around less extensive, new venting on older basalt in the CoAxial Segment of the Juan de Fuca Ridge (Juniper et al., 1995). The presence of filaments (resembling *Leptothrix* sp.) in nontronite deposits (Fe-rich smectite) from seamounts in the East Pacific Rise (Alt, 1988), and the distinct microtube-like morphology in nontronite deposits from white smoker chimneys in the Galapagos Rift and Mariana Trough (Köhler et al., 1994), also suggest that bacteria are directly involved in catalyzing clay mineral formation.

The precipitation of iron-silicate phases closely follows the two-step mineralization process originally described by Beveridge and Murray (1976), in that iron is initially bound to anionic cellular sites, after which dissolved silica is added to the growing mineral via hydrogen bonding between hydroxyl groups. The Fe-rich sites on the cell surface presumably serve as precursors to the more complex surface precipitates, with the iron-silicates using some fraction of the precursor surface as a template for their own growth, in effect circumventing the need for direct nucleation (Steefel and Van Cappellen, 1990). This sequence of reactions appears to correspond with previous theories of hydrothermal nontronite formation that suggested these clays formed either by sorption of silica onto Fe-oxyhydroxides or directly through precipitation from hydrothermal fluids (Köhler et al., 1994).

Experimental work has also confirmed the ability of bacterial cells to form iron-silicate minerals. Silicate anion binding to bacteria was enhanced when the cell walls were preloaded with Fe(III), indicating that the metal participated in a cationic bridging mechanism for silicate binding to the bacterial surface through the formation of ternary complexes (e.g., wall-metal-silicate) (Urrutia and Beveridge, 1993, 1994). Growth of the precipitates then continued until complex silicate structures were formed.

DISCUSSION

There are three similarities between present-day hydrothermal systems and Precambrian oceans that support this paper's contention that modern bacterial biomineralization may provide small-scale analogues to the conditions under which BIFs were

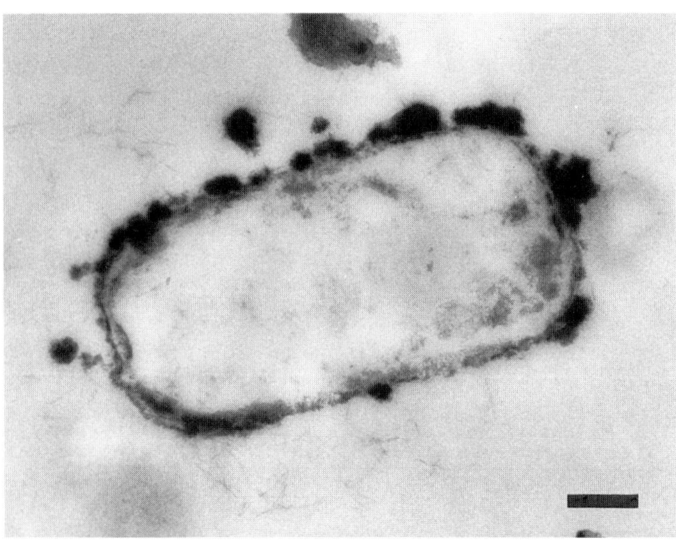

FIG. 5.—TEM image of a bacterial cell from Krisuvik, Iceland, with amorphous iron-silicate grains on the outer cell wall. Scale bar = 200 nm.

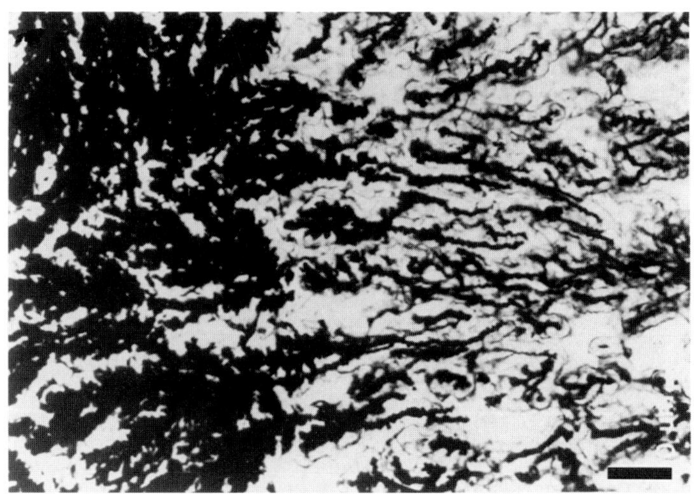

FIG. 6.—Transmitted light microscopy image of the filament-mineral association in iron-silica deposits in a chimney fragment from 12°50'N, East Pacific Rise. The transition from iron-rich (dark) to silica-rich (light) over a distance of 2 mm is shown. Scale bar = 2 μm. Reprinted with permission of the author and *Canadian Mineralogist* (Juniper and Fouquet, 1988).

deposited. First, modern deep-sea vents and hot springs are characterized by the release of warm to hot reducing solutions with high elemental concentrations (Corliss et al., 1979; Henley, 1984). For example, effluent collected from vents along the Mid-Atlantic Ridge and the East Pacific Rise contained dissolved silica and iron (averaging between the two sites) approximately 1,100 ppm and 100 ppm, respectively (Campbell et al., 1988). These values are approximately 1,000 times and 50,000 times greater than concentrations for silica and iron, respectively, in background marine waters (Drever, 1988). Similarly, hot spring effluents (from a variety of sites globally) have dissolved silica concentrations typically exceeding 200 ppm, with some geothermal wells having up to 1,100 ppm (Henley, 1984), while iron concentrations range from 0.8 ppm (Lýsuhóll, Iceland) (Konhauser and Ferris, 1996) to 2.4 ppm (Mammoth Springs, Yellowstone) (Fournier, 1989). The combination of high dissolved iron and silica are unique to hydrothermal systems, and are analogous to conditions in much of the Precambrian, when ocean chemistry was controlled by the interaction of warm seawater with the basaltic oceanic crust (Veizer et al., 1982). In the Archean and early Proterozoic, hydrothermal effluent was transported from the deep sea into the oxic surface waters by unusual geostrophic or wind-driven upwelling (Lowe, 1994) or directly by hydrothermal plumes generated at mid-ocean ridge crests with substantially shallower depths than present (Isley, 1995). This led to surface seawater with high metal concentrations, perhaps up to 250 ppm iron (Veizer, 1983), and silica concentrations close to or at saturation (~120 ppm) with respect to amorphous silica (Siever, 1992).

Secondly, a consortium of microorganisms are currently present in hot springs and deep-sea vents, some of which resemble those found as Precambrian microfossils, including cyanobacteria, anoxygenic phototrophic bacteria, and iron-depositing bacteria. There are several morphological lines of evidence that suggest that cyanobacteria existed at 3.5 Ga (Awramik, 1992), including the presence, in rocks from the Warrawoona Group (~3.5 Ga), of trichome-like filaments (Awramik et al., 1983); microfossils with widths greater than most filamentous bacteria (Schopf, 1993); and coccoid forms enclosed in a wall-like structure (Schopf and Packer, 1987). The existence of stromatolites (presumed to have formed in shallow, near-shore marine settings) further implies the presence of cyanobacteria, since the microorganisms that formed them were phototactic, photoautotrophic, and produced mucus sheaths, all attributes found in present-day cyanobacteria (Awramik, 1992). In addition to cyanobacteria, mats formed by sulfide-dependent, phototrophic bacteria (e.g., *Chloroflexus* sp.), and possibly sulfur-oxidizing bacteria such as *Beggiatoa* sp., may also have been locally abundant in the Precambrian (Doemel and Brock, 1977; Walter et al., 1972, 1992). Lastly, the presence of thread-like structures in chert from the Gunflint Iron Formation, Ontario (aged ~ 2.0 Ga) that resemble filamentous iron-depositing bacteria such as *Crenothrix* sp., *Sphaerotilus* sp., and *Leptothrix* sp., additionally suggests the presence of bacteria that were involved in the oxidation of ferrous to ferric iron (Barghoorn and Tyler, 1965; Cloud, 1965).

The resemblance of a wide consortia of early Precambrian microorganisms to those existing today is not surprising. During that period in Earth's history, when major BIFs were accumulating, microorganisms were already flourishing, with nearly all prokaryotic phyla having by that time evolved and diversified (Awramik, 1992). This view is supported by the presence of fossilized microbial remains, stromatolites, chemical fossils, and biogenic minerals reported from early Archean rocks (Lowe, 1980; Awramik et al., 1983; Hayes et al., 1983; Ohmoto et al., 1993), some at least 3.8 Ga (Mojzsis et al., 1996). In fact, the construction of stromatolites by 3.5 Ga implies that these microbes were morphologically well-developed by that time (Awramik, 1992; McNamara and Awramik, 1992).

The third similarity reflects the common occurrence of iron and silicate minerals around modern hydrothermal vents and their presence as primary mineralized constituents of BIFs. Iron hydroxides and amorphous silica are the predominant mineral phases that precipitate around low-temperature, sulfide-deficient, deep-sea vents, where they can form centimeter-thick layers (by diffuse venting through underlying basalts or sulfides), oxide muds, and chimney structures (Juniper and Tebo, 1995). At hot springs, iron and silicate biominerals are also observed, sometimes displaying crude microbanding (K.O. Konhauser, in prep., 2000) reminiscent of the very finely laminated iron-chert beds in, for example, the Hamersley BIF (Morris, 1993).

As discussed above, bacteria contribute to the precipitation of authigenic minerals in modern hydrothermal systems. Herein lies the premise of this paper. If microorganisms are present today in hydrothermal environments, and if they precipitate iron and silicate minerals, then it seems reasonable to argue that Precambrian bacterial populations, also growing in hydrothermal waters, could have formed similar biominerals, i.e., the primary minerals in BIFs. I speculate that previous models for BIF deposition (discussed below) have underestimated the role of microorganisms in the deposition of BIFs. In this new hypothesis, bacteria, growing in the Precambrian oceans, may have served as nucleation sites that catalyzed primary iron and silica mineralization.

In the early Precambrian, biological activity was probably limited to the flat margins of the microcontinents, where microorganisms grew as benthic mats or in the open ocean above areas of dynamic upwelling of deep, nutrient-rich waters or hydrothermal plumes (Lowe, 1994). The planktonic microorganisms presumably flourished when ferrous iron and nutrients were available, allowing them to form iron hydroxides, but they declined in numbers (no mineralization period) when these components became depleted (Cloud, 1973; Baur et al., 1985). It is unclear what proportion of the ferric iron was initially produced in the Archean via photochemical processes (Cairns-Smith, 1978; Braterman et al., 1983; François, 1986), anoxygenic photosynthesis (Widdel et al., 1993; Ehrenreich and Widdel, 1994), photodissociation of water vapor in the atmosphere, generating free oxygen to react with ferrous iron (Walker et al., 1983), or oxygenic photosynthesis (Cloud, 1965, and many others). In a large part it will depend on when cyanobacteria came into existence. If they existed as early as 3.5 Ga (Awramik, 1992), then these microorganisms would certainly have had a major impact on the initial production of oxygen

(Cloud, 1965, 1973). Because the initial O_2 partial pressures may have been extremely low (Walker et al., 1983), bacteria would not have been able to gain energy from oxidation of Fe(II) to Fe(III) with O_2 as an electron acceptor, using instead energy sources such as sunlight to oxidize Fe (Ehrlich, 1990). As O_2 partial pressures increased beyond some threshold concentration, oxidation by microaerophilic bacteria such as *Gallionella* sp. would have become possible and perhaps even favored (under low pO_2) compared to inorganic oxidation of Fe(II) (Holm, 1987).

The end result of any of these modes of biomineralization would be the precipitation of ferric hydroxide in the water column, its sedimentation, and the systematic accumulation of iron-rich sediment on the ocean floor (Walker et al., 1983). Along with ferric iron, the remains of the microorganism would be deposited. This is suggested by the common presence of spheroidal structures in BIFs, which have been interpreted as relict microfossils (Cloud and Licari, 1968; LaBerge, 1973). Once deposited, the ferric hydroxide eventually would have consolidated as (i) hematite, where oxidizing conditions were maintained in the sediment or organic material was limited; (ii) as magnetite, where partial Fe-reduction occurred in the sediments by microbial organic matter mineralization; or (iii) as siderite/ferrous silicate when substantial Fe-reduction occurred (Morris, 1993). Magnetite and siderite are both major components in iron-oxide facies and S-macrobands, respectively (Ewers and Morris, 1981), and their presence suggests that a significant fraction of BIFs were diagenetically modified under suboxic to anoxic conditions (Walker et al., 1983). That the latter diagenetic reactions were microbially catalyzed is suggested by isotopically light $\delta^{13}C$ values in BIF carbonate minerals (Perry et al., 1973; Baur et al., 1985), which implies that isotopically light organic carbon in the bottom sediments was oxidized to bicarbonate with the concomitant reduction of some electron acceptor, presumably ferric iron (Walker, 1984). This oxidation may also explain the absence of organic matter in BIFs. The by-product of Fe-reduction could be either magnetite, commonly evident as overgrowths on hematite (Morris, 1980), or siderite, which is present in the Gunflint Iron Formation, Canada, for example, as uniform, spherical grains (approximately 30–40 μm in diameter) with organic-rich cores (Goodwin, 1956). This correlates well with the suggestion that during the hundreds of millions of years when ferric iron was being deposited and organic matter accumulated in the sediment, that a simple respiratory chain to ferric iron developed and eventually served as a precursor to more complex respiratory chains associated with extant organisms (Nealson and Myers, 1990).

The fundamental differences between mineralization in present-day hydrothermal environments and the Precambrian oceans are (i) the size of the deposits and (ii) the relationship between the microorganisms and the chemocline (where anoxic, ferrous-rich waters meet oxygenated surface waters). In the first instance, the oxygenated conditions on the ocean floor today preclude the formation of large-scale deposits in modern hydrothermal environments. Mineral precipitation occurs peripheral to the vent, where anoxic hydrothermal fluids are mixed with oxygenated deep-sea waters (Juniper and Tebo, 1995). In the presence of oxygen and at pH~8 (the half-life of ferrous iron being only be a few minutes, Millero et al., 1987), ferrous iron is rapidly oxidized to ferric iron and precipitated as ferric hydroxide (Schwertmann and Fitzpatrick, 1992). Even if the hydrothermal effluent is hot (giving low oxygen solubility) and acidic, and the surrounding ambient ocean bottom waters are low in dissolved O_2, ferric hydroxides still accumulate in relative proximity to the vents, with a gradual depletion outwards (Juniper and Tebo, 1995). Thus, high concentrations of ferrous iron are not found in modern marine waters. Similarly, in hot spring effluents there is an almost immediate precipitation of ferric hydroxide near the geyser. In contrast, the anoxic bottom waters of the Precambrian oceans allowed for iron to dissipate greater distances away from the vent, and subsequently precipitate out of solution in the deep oceans during the Archean or in huge basinal areas (e.g., 100,000 km^2 for the Hamersley Group, Trendall and Blockley, 1970) during the Proterozoic (Beukes and Klein, 1992; Simonson and Hassler, 1996). Precipitation occurred over hundreds of millions of years as the oceans became increasingly oxygenated, and it was only when the rate of photosynthetic oxygen production exceeded the rate of supply of ferrous iron that it became substantially depleted in marine waters and free oxygen accumulated (Walker et al., 1983). Based on the oxidation state of paleosols, the nature of uranium ores, the presence of red-beds, trace metal content in black shales, and the evolution of eukaryotes, this may not have occurred until after 2.2 Ga (Holland, 1994).

The second key difference between present-day hydrothermal environments and those under which BIFs were deposited is that today, extant microorganisms, growing as mats in hot spring effluent or on vent chimneys and in surrounding marine sediments, are found at the chemocline. In the Precambrian, however, biomineralization was removed from the site of hydrothermal emissions, allowing the "biogenic" deposits to form thick accumulations of great lateral extent. The disparity is due to radically different conditions on the early Earth. The Archean (at the time 3.5–3.2 Ga) was characterized by high rates of mantle outgassing, limited chemical weathering, elevated atmospheric CO_2 levels and surface temperatures of perhaps 30–50°C that prevented the accumulation of polar ice, cold-water sinking, and deep ocean stirring (Lowe, 1994). The Archean oceans were thus strongly stratified with a shallow, wind-mixed surface layer containing dissolved CO_2 and small amounts of O_2, which were derived from oxygenic photosynthesis in areas of high planktonic productivity (Cloud, 1965, 1973; Schidlowski, 1976), and from photochemical processes that dissociate water vapor (Walker et al., 1983). The deeper, reducing waters were enriched in ferrous iron (Klein and Beukes, 1989; Beukes et al., 1990; Beukes and Klein, 1992).

The distribution of Archean iron-formations suggests that these relatively small deposits (compared to Proterozoic BIFs) were ubiquitous in deep water, not overwhelmed by coarse clastic debris (Lowe, 1994), and presumably represented local accumulations close to marine vents (Simonson and Hassler, 1996). The relative absence of shallow-water BIFs indicates that the upper, wind-mixed layer of the ocean contained little dissolved

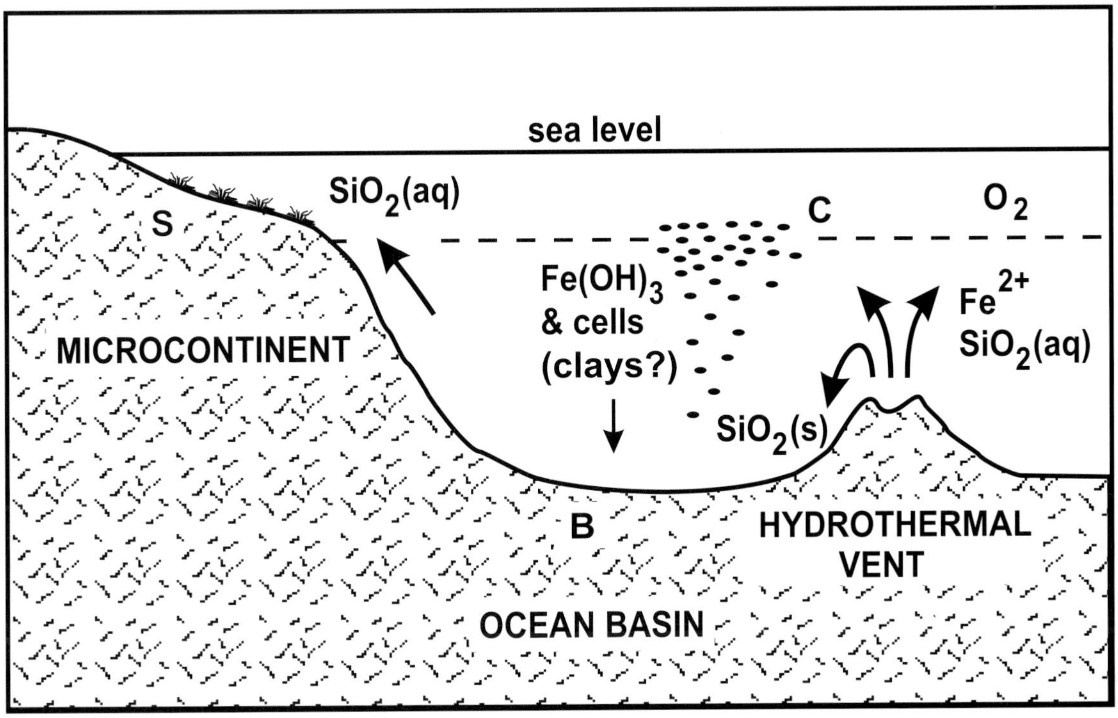

FIG. 7.—Schematic diagram of the Archean environment, modified from Lowe (1994). The Archean oceans were strongly stratified with a shallow surface layer containing small amounts of oxygen, while the deeper reducing waters were enriched in ferrous iron. Benthic mats grew on the shallow sediments of the microcontinental shelf (S), while bacterial plankton grew in the open ocean above areas of dynamic upwelling or hydrothermal plumes. At the chemocline (C), ferric hydroxide precipitation was catalyzed by plankton through binding of cationic iron onto their cells; the source of the ferric iron was a combination of photochemical processes, anoxygenic or oxygenic photosynthesis. Thus, all iron was removed from the deep waters before it spread onto the shelf. Upon death, these iron-rich cells accumulated in the bottom sediments (B), where diagenetic reactions associated with burial led to the transformation of ferric hydroxide into hematite, magnetite, or siderite. Iron deposition may have been episodic due to fluctuating nutrient supplies for the overlying plankton. Precipitation of silica occurred on the ocean floor due to pressure/temperature changes after venting. At the chemocline, some silica may also have reacted with cell-bound iron to form clay-like phases. The remaining dissolved silica was transported onto the shelf, where silica supersaturation (via evaporation) allowed microorganisms to induce silicification, leading to the formation of silicified stromatolites.

iron, quite possibly due to iron being removed from upwelling deep waters before it spread onto adjacent continental shelves (Lowe, 1994). This is reflected in Archean shallow water deposits that are depleted in iron (Beukes and Klein, 1992). Thus, it seems likely that in the Archean major BIFs would have formed above oceanic hydrothermal vents or in areas of upwelling, where ferrous iron and dissolved silica were brought into contact with oxygenated water at the chemocline (Beukes and Klein, 1992), yet below the level of the shallow continental shelf (Lowe, 1994). Plankton growing in the surface waters of the open ocean could have provided reactive sites to catalyze iron mineralization (Fig. 7). Quite possibly some of the cell-bound iron may also have reacted with dissolved silica, forming iron-silicate phases reminiscent of those formed today (Siever and Woodford, 1973; Siever, 1992). Such reactions would have kept dissolved silica concentrations in the open ocean below saturation with respect to amorphous silica (Knoll, 1985), although amorphous silica could have precipitated on the ocean floor due to the supersaturated conditions created by the sudden physical and chemical changes induced through venting. As iron was removed from the upwelled waters, the remaining Si-rich waters may have been transported onto the flat margins of the microcontinents where supersaturated conditions (via evaporation) would have allowed benthic microorganisms to induce silicification. That microorganisms may have facilitated this form of mineralization is evident from the associations of silicified microfossils with ancient microbial mats and organic-rich mud, as represented by silicified stromatolites and carbonaceous cherts (Knoll, 1985). The lack of extant bacteria that actively control silica precipitation (i.e., in the manner of diatoms), and the natural and experimental studies that show silicate minerals commonly associated with microbial cells (when subject to high dissolved silica) suggest that it is possible to invoke a passive biological interpretation for Precambrian cherts, and hence BIFs.

The end of the Archean witnessed significant environmental changes, beginning with the formation of several large blocks of continental crust about 3.3–3.1 Ga, increased physical and chemical weathering, removal of CO_2 from the atmosphere, and the subsequent deposition of large carbonate platforms by 2.6 Ga (Lowe, 1994). The shallow sea hosted enormous bacterial mats, which both consumed more CO_2 through photosynthesis and released O_2 to the atmosphere. The depletion of atmospheric CO_2 by weathering, biological activity, and sedimentation on these new continental blocks was accompanied by the Huronian glaciation 2.4–2.1 Ga ago, which led to the formation of polar shelf ice, cold water sinking, and inevitably the mixing of ocean waters (Lowe, 1994). As the stratified ocean began to break down, ferrous iron and silica were actively transported onto the continental slopes and shallow shelves, leading to thick, laterally extensive BIF deposits (Simonson and Hassler, 1996), in waters ranging from 10–700 m (Isley, 1995). The lateral grading of BIFs into stromatolites or oolites, along with sedimentary

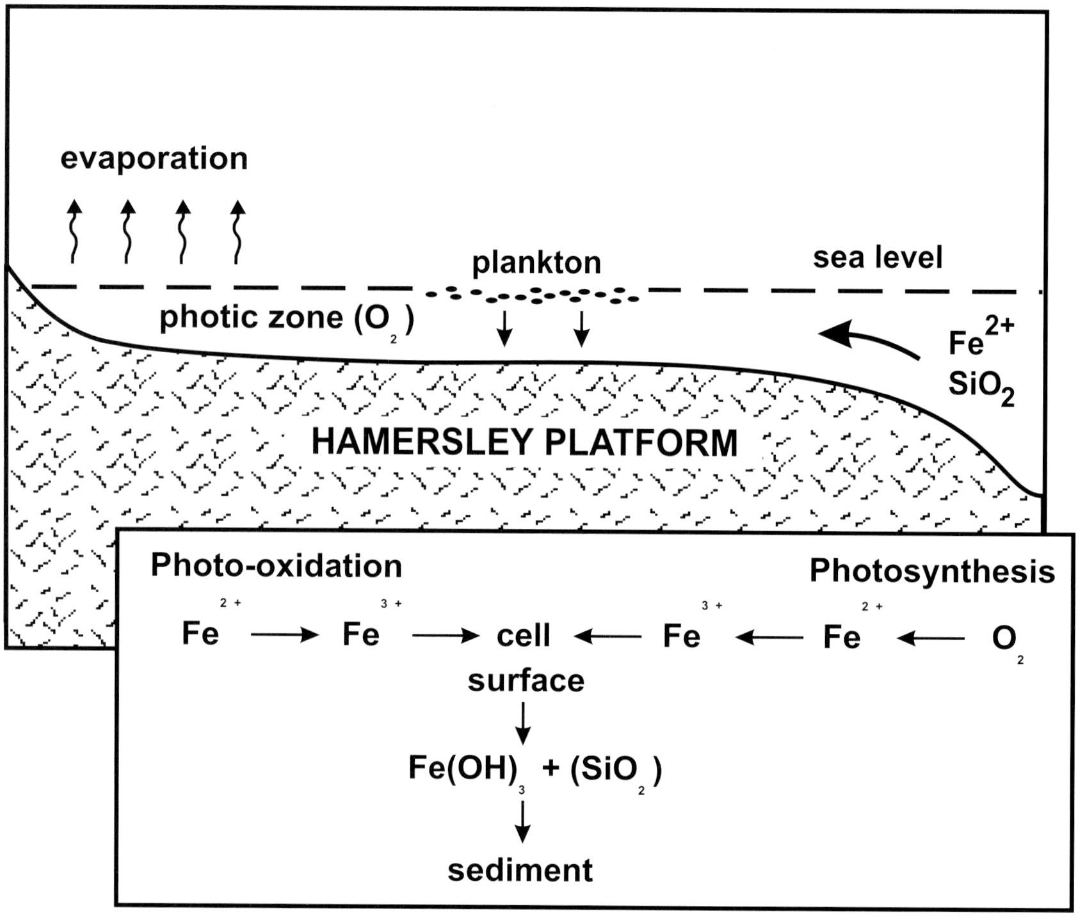

FIG. 8.—Schematic diagram of BIF deposition in the Proterozoic, modified from a genetic model of the Hamersley Group (~2.5 Ga) from Morris (1993). When the stratified ocean began to break down, ferrous iron and silica were transported onto the continental shelf. Upwelling also supplied nutrients to the depository, triggering a parallel growth of microorganisms. These bacterial communities subsequently served as passive nucleation sites for ferric iron deposition (as iron-rich mesobands) by catalyzing reactions rendered possible by supersaturated conditions developed at the chemocline (see inset). Within these iron oxide mesobands, the alternating Fe-rich and Si-rich microbands signified seasonal and climatic changes, with iron precipitated when upwelling currents recharged the depository with Fe(II) and nutrients, and amorphous silica precipitated due to evaporation on the shelf when iron was depleted.

structures consistent with wave agitation and shallow water deposition (Gole and Klein, 1981; Simonson, 1985), also confirms that paleodepositional depths had decreased. This transition may have been imposed by a transgressive sequence that inhibited siliciclastic input and shifted primary organic matter and carbonate production shoreward (Klein and Beukes, 1989; Beukes and Klein, 1992). The chemocline thus moved from the open oceans onto the continental shelves, thereby facilitating BIF deposition in these relatively shallow waters (Schissel and Aro, 1992; Simonson and Hassler, 1996).

According to a genetic model for the banded-iron formation of the Hamersley Group (~2.5 Ga), Morris and Horwitz (1983) and Morris (1993) proposed that during periods of vigorous and sustained mid-ocean ridge activity, which allowed high levels of Fe(II) to reach the continental shelf, the combined effects of photo-oxidation (which possibly included anoxygenic photosynthesis, Widdel et al., 1993; Ehrenreich and Widdel, 1994) and oxygenic photosynthesis in the Precambrian oceans caused ferric iron deposition and the formation of iron-rich mesobands (Fig. 8). Upwelling also supplied nutrients to the depository, triggering a parallel growth of microorganisms. If, as modern hydrothermal systems suggest, bacteria have an important role as nucleation sites for the precipitation of minerals (Juniper and Tebo, 1995; Konhauser and Ferris, 1996), then it is possible that in the Precambrian, these bacterial communities also catalyzed the mineralization process. Within iron oxide mesobands, the alternating Fe-rich and Si-rich microbands further signified seasonal and climatic changes. Iron precipitated when upwelling currents recharged the depository with Fe(II) and nutrients, while amorphous silica precipitated due to evaporation on the shelf when iron was depleted. Si-levels likely reflected background surface ocean concentrations from a composite of sources including continental weathering and hydrothermal input. Superimposed on this background deposition were sustained periods of volcanicity which resulted in the Si-rich macrobands gradually returning to the normal Fe-oxide macrobands as volcanic input waned. Once the chemical stratification in the oceans broke down completely, they became completely mixed and depleted in iron, as is indicated by the general absence of major iron-formations after 1.8 Ga (Beukes and Klein, 1992).

ACKNOWLEDGMENTS

Funding for parts of this work was provided by a Royal Society Research grant (574005.G501). I thank Simon Bottrell, Joe Cann, Sarah Gleeson, Rob Raiswell, Sue Bowler, and two anonymous reviewers for their critical comments during the preparation of this manuscript.

REFERENCES

ALT, J. C., 1986, A hydrothermal/bacterial iron deposit from a seamount near 21°N, East Pacific Rise (abs.): Geological Society of America Abstracts, v. 18, p. 526.

ALT, J. C., 1988, Hydrothermal oxide and nontronite deposits on seamounts in the eastern Pacific: Marine Geology, v. 81, p. 227-239.

AWRAMIK, S. M., SCHOPF, J. W., AND WALTER, M. R., 1983, Filamentous fossil bacteria from the Archean of Western Australia: Precambrian Research, v. 20, p. 357-374.

AWRAMIK, S. M., 1992, The oldest records of photosynthesis: Photosynthesis Research, v. 33, p. 75-89.

BARGHOORN, E. S., AND TYLER, S. A., 1965, Microorganisms from the Gunflint chert: Science, v. 147, p. 563-577.

BAROSS, J. A., AND DEMING, J. W., 1985, The role of bacteria in the ecology of black-smoker environments, in Jones, M. L., ed., Hydrothermal vents of the Eastern Pacific: An overview: Bulletin of the Biological Society of Washington, v. 6, p. 355-371.

BAROSS, J. A., LILLEY, M. D., AND GORDON, L. I., 1982, Is the CH_4, H_2 and CO venting from submarine hydrothermal systems produced by thermophilic bacteria?: Nature, v. 298, p. 366-368.

BAU, M., AND MÖLLER, P., 1993, Rare earth element systematics of the chemically precipitated component in Early Precambrian iron formations and the evolution of the terrestrial atmosphere-hydrosphere-lithosphere system: Geochimica et Cosmochimica Acta, v. 57, p. 2239-2249.

BAUR, M. E., HAYES, J. M., STUDLEY, S. A., AND WALTER, M. R., 1985, Millimeter-scale variations of stable isotope abundances in carbonates from banded iron-formations in the Hamersley Group of Western Australia: Economic Geology, v. 80, p. 270-282.

BEUKES, N. J., AND KLEIN, C., 1992, Models for iron-formation deposition, in Schopf, J. W., and Klein, C., eds., The Proterozoic Biosphere: Cambridge, Cambridge University Press, p. 147-151.

BEUKES, N. J., KLEIN, C., KAUFMAN, A. J., AND HAYES, J. M., 1990, Carbonate petrography, kerogen distribution, and carbon and oxygen isotope variations in an early Proterozoic transition from limestone to iron-formation deposition, Transvaal Supergroup, South Africa: Economic Geology, v. 85, p. 663-690.

BEVERIDGE, T. J., 1988, The bacterial surface: General considerations towards design and function: Canadian Journal of Microbiology, v. 34, p. 363-372.

BEVERIDGE, T. J., AND MURRAY, R. G. E., 1976, Uptake and retention of metals by cell walls of Bacillus subtilis: Journal of Bacteriology, v. 127, p. 1502-1518.

BEVERIDGE, T. J., AND FYFE, W. S., 1985, Metal fixation by bacterial cell walls: Canadian Journal of Earth Sciences, v. 22, p. 1893-1898.

BIRNBAUM, S. J., AND WIREMAN, J. W., 1985, Sulfate-reducing bacteria and silica solubility: A possible mechanism for evaporite diagenesis and silica precipitation in banded iron formations: Canadian Journal of Earth Sciences, v. 22, p. 1904-1909.

BRATERMAN, P. S., CAIRNS-SMITH, A. G., AND SLOPER, R. W., 1983, Photo-oxidation of hydrated Fe^{2+}-Significance for banded iron formations: Nature, v. 303, p. 163-164.

BROCK, T. D., SMITH, D. W., AND MADIGAN, M. T., 1984, Biology of Microorganisms (4th Edition): Englewood Cliffs, Prentice-Hall, 847 p.

CAIRNS-SMITH, A. G., 1978, Precambrian solution photochemistry, inverse segregation, and banded iron formations: Nature, v. 276, p. 807-808.

CAMERON, E. M., 1983, Genesis of Proterozoic iron-formation: Sulphur isotope evidence: Geochimica et Cosmochimica Acta, v. 47, p. 1069-1074.

CAMPBELL, A. C., PALMER, M. R., KLINKHAMMER, G. P., BOWERS, T. S., EDMOND, J. M., LAWRENCE, J. R., CASEY, J. F., THOMPSON, G., HUMPHRIS, S., RONA, P., AND KARSON, J. A., 1988, Chemistry of hot springs on the Mid-Atlantic Ridge: Nature, v. 335, p. 514-519.

CASTENHOLZ, R. W., 1973, The possible photosynthetic use of sulfide by the filamentous phototrophic bacteria of hot springs: Limnology and Oceanography, v. 18, p. 863-876.

CLOUD, P. E., Jr., 1965, Significance of the Gunflint (Precambrian) microflora: Science, v. 148, p. 27-35.

CLOUD, P., 1973, Paleoecological significance of the banded iron-formation: Economic Geology, v. 68, p. 1135-1143.

CLOUD, P., AND LICARI, G. R., 1968, Microbiotas of the banded iron formations: National Academy of Sciences Proceedings (United States), v. 61, p. 779-786.

COLEMAN, M. L., HEDRICK, D. B., LOVLEY, D. R., WHITE, D. C., AND PYE, K., 1993, Reduction of Fe(III) in sediments by sulphate-reducing bacteria: Nature, v. 361, p. 436-438.

CORLISS, J. B., DYMOND, J., GORDON, L. I., EDMOND, J. M., VON HERZEN, R. P., BALLARD, R. D., GREEN, K., WILLIAMS, D., BAINBRIDGE, A., CRANE, K., AND VAN ANDEL, T. H., 1979, Submarine thermal springs on the Galápagos Rift: Science, v. 203, p. 1073-1083.

CORNELL, R. M., GIOVANOLI, R., AND SCHINDLER, P. W., 1987, Effect of silicate species on the transformation of ferrihydrite into goethite and hematite in alkaline media: Clays and Clay Minerals, v. 35, p. 21-28.

DERRY, L. A., AND JACOBSEN, S. B., 1990, The chemical evolution of Precambrian seawater: Evidence from REEs in banded iron formations: Geochimica et Cosmochimica Acta, v. 54, p. 2965-2977.

DOEMEL, W. N., AND BROCK, T. D., 1977, Structure, growth, and decomposition of laminated algal-bacterial mats in alkaline hot springs: Applied and Environmental Microbiology, v. 34, p. 433-452.

DREVER, J. I., 1988, The Geochemistry of Natural Waters (2nd Edition): Englewood Cliffs, Prentice Hall, 437 p.

EHRLICH, H. L., 1990, Geomicrobiology (2nd Edition): New York, Marcel Dekker, 646 p.

EHRLICH, H. L., 1996, How microbes influence mineral growth and dissolution: Chemical Geology, v. 132, p. 5-9.

EHRENREICH, A., AND WIDDEL, F., 1994, Anaerobic oxidation of ferrous iron by purple bacteria, a new type of phototrophic metabolism: Applied and Environmental Microbiology, v. 60, p. 4517-4526.

EWERS, W. E., 1983, Chemical factors in the deposition and diagenesis of banded iron-formation, in Trendal, A. F., and Morris, R. C., eds., Banded Iron-Formation: Facts and Problems: Amsterdam, Elsevier, p. 491-512.

EWERS, W. E., AND MORRIS, R. C., 1981, Studies of the Dales Gorge Member of the Brockman Iron Formation, Western Australia: Economic Geology, v. 76, p. 1929-1953.

FERRIS, F. G., 1997, Formation of authigenic minerals by bacteria, in McIntosh, J. M., and Groat, L. A., eds., Biological-Mineralogical Interactions: Ottawa, Mineralogical Association of Canada Short Course Series 25, p. 187-208.

FERRIS, F. G., BEVERIDGE, T. J., AND FYFE, W. S., 1986, Iron-silica crystallite nucleation by bacteria in a geothermal sediment: Nature, v. 320, p. 609-611.

FERRIS, F. G., FYFE, W. S., AND BEVERIDGE, T. J., 1988, Metallic ion binding by Bacillus subtilis: Implications for the fossilization of microorganisms: Geology, v. 16, p. 149-152.

FERRIS, F. G., TAZAKI, K., AND FYFE, W. S., 1989, Iron oxides in acid mine drainage environments and their association with bacteria: Chemical Geology, v. 74, p. 321-330.

FORTIN, D., AND BEVERIDGE, T. J., 1997, Microbial sulfate reduction within sulfidic mine tailings: Formation of diagenetic Fe sulfides: Geomicrobiology Journal, v. 14, p. 1-21.

FORTIN, D., FERRIS, F. G., AND BEVERIDGE, T. J., 1997, Surface-mediated mineral development by bacteria, in Banfield, J. F., and Nealson, K. H., eds., Geomicrobiology: Interactions Between Microbes and Minerals: Washington, D.C., Mineralogical Society of America, p. 161-180.

FOURNIER, R. O., 1985, The behavior of silica in hydrothermal solutions, in Berger, B. R., and Bethke, P. M., eds., Geology and Geochemistry of Epithermal Systems: Reviews in Economic Geology, v. 2, p. 45-61.

FOURNIER, R. O., 1989, Geochemistry and dynamics of the Yellowstone National Park hydrothermal system: Annual Review of Earth and Planetary Sciences, v. 17, p. 13-53.

FRANCIS, S., MARGULIS, L., AND BARGHOORN, E. S., 1978, On the experimental silicification of microorganisms. II. On the time of appearance of eukaryotic organisms in the fossil record: Precambrian Research, v. 6, p. 65-100.

François, L. M., 1986, Extensive deposition of banded iron formations was possible without photosynthesis: Nature, v. 320, p. 352-354.

German, C. R., Campbell, A. C., and Edmond, J. M., 1991, Hydrothermal scavenging at the Mid-Atlantic Ridge: Modification of trace element dissolved fluxes: Earth and Planetary Science Letters, v. 107, p. 101-114.

Ghiorse, W. C., 1984, Biology of iron- and manganese-depositing bacteria: Annual Review of Microbiology, v. 38, p. 515-550.

Ghiorse, W. C., and Ehrlich, H. L., 1992, Microbial biomineralization of iron and manganese, in Skinner, H. C. W., and Fitzpatrick, R. W., eds., Biomineralization. Processes of Iron and Manganese: Cremlingen, Catena Supplement 21, p. 75-99.

Gole, M. J., and Klein, C., 1981, Banded iron-formations through much of Precambrian time: Journal of Geology, v. 89, p. 169-183.

Goodwin, A. M., 1956, Facies relations in the Gunflint iron Formation: Economic Geology, v. 51, p. 565-595.

Hanert, H. V., 1973, Rezente marine Eisenerze auf Santorin, Griechenland. II. Bakterogenese von Eisenhydroxidsedimenten: Geologische Rundschau, v. 62, p. 803-813.

Harmsen, H. J. M., Prieur, D., and Jeanthon, C., 1997, Distribution of microorganisms in deep-sea hydrothermal vent chimneys investigated by whole-cell hybridization and enrichment culture of thermophilic subpopulations: Applied and Environmental Microbiology, v. 63, p. 2876-2883.

Hayes, J. M., Kaplan, I. R., and Wedeking, K.W., 1983, Precambrian organic geochemistry, preservation of the record, *in* Robertson, J. M., ed., Fluid-Mineral Equilibria in Hydrothermal Systems. Volume 1: El Paso, Society of Economic Geologists, p. 9-29.

Henley, R. W., 1984, Chemical structure of geothermal systems, in Robertson, J. M., ed., Fluid-Mineral Equilibria in Hydrothermal Systems. Volume 1: El Paso, Society of Economic Geologists, p. 9-29.

Hinman, N. W., and Lindstrom, R. F., 1996, Seasonal changes in silica deposition in hot spring systems: Chemical Geology, v. 132, p. 237-246.

Holland, H. D., 1994, Early Proterozoic atmospheric change, in Bengtson, S., ed., Early Life on Earth: New York, Columbia University Press, p. 237-244.

Holm, N. G., 1987, Biogenic influences on the geochemistry of certain ferruginous sediments of hydrothermal origin: Chemical Geology, v. 63, p. 45-57.

Holm, N. G., 1989, The $^{13}C/^{12}C$ ratios of siderite and organic matter of a modern metalliferous hydrothermal sediment and their implications for banded iron formations: Chemical Geology, v. 77, p. 41-45.

Iler, R. K., 1980, Isolation and characterization of particle nuclei during the polymerization of silicic acid to colloidal silica: Journal of Colloid and Interface Science, v. 75, p. 138-148.

Isley, A. E., 1995, Hydrothermal plumes and the delivery of iron to banded iron formation: Journal of Geology, v. 103, p. 169-185.

Jacobsen, S. B., and Pimentel-Klose, M. R., 1988, A Nd isotopic study of the Hamersley and Michipicoten banded iron formations: The source of REE and Fe in Archean oceans: Earth and Planetary Science Letters, v. 87, p. 29-44.

James, H. L., 1954, Sedimentary facies of iron-formations: Economic Geology, v. 49, p. 235-293.

James, H. L., 1983, Distribution of banded iron-formation in space and time, in Trendall, A. F., and Morris, R. C., eds., Iron-Formation: Facts and Problems: Amsterdam, Elsevier, p. 471-490.

Jannasch, H. W., and Wirsen, C. O., 1981, Morphological survey of microbial mats near deep-sea thermal vents: Applied and Environmental Microbiology, v. 41, p. 528-538.

Jannasch, H. W., and Mottl, M. J., 1985, Geomicrobiology of deep-sea hydrothermal vents: Science, v. 229, p. 717-725.

Jones, B., and Renaut, R. W., 1996, Influence of thermophilic bacteria on calcite and silica precipitation in hot springs with water temperatures above 90°C: Evidence from Kenya and New Zealand: Canadian Journal of Earth Sciences, v. 33, p. 72-83.

Jones, B., and Renaut, R. W., 1997, Formation of silica oncoids around geysers and hot springs at El Tatio, northern Chile: Sedimentology, v. 44, p. 287-304.

Jones, B., Renaut, R. W., and Rosen, M. R., 1997a, Biogenicity of silica precipitation around geysers and hot-spring vents, North Island, New Zealand: Journal of Sedimentary Research, v. 67, p. 88-104.

Jones, B., Renaut, R. W., and Rosen, M. R., 1997b, Vertical zonation of biota in microstromatolites associated with hot springs, North Island, New Zealand: Palaios, v. 12, p. 220-236.

Juniper, S. K., and Fouquet, Y., 1988, Filamentous iron-silica deposits from modern and ancient hydrothermal sites: Canadian Mineralogist, v. 26, p. 859-869.

Juniper, S. K., and Tebo, B. M., 1995, Microbe-metal interactions and mineral deposition at hydrothermal vents, in Karl, D. M., ed., The Microbiology of Deep-Sea Hydrothermal Vents: Boca Raton, CRC Press, p. 219-253.

Juniper, S. K., Martineu, P., Sarrazin, J., and Gélinas, Y., 1995, Microbial mineral floc associated with nascent hydrothermal activity on CoAxial Segment, Juan de Fuca Ridge: Geophysical Research Letters, v. 22, p. 179-182.

Karl, D. M., Brittain, A. M., and Tilbrook, B. D., 1989, Hydrothermal and microbial processes at Loihi Seamount, a mid-plate hot-spot volcano: Deep-Sea Research, v. 36, p. 1655-1673.

Karl, D. M., McMurtry, G. M., Malahoff, A., and Garcia, M. O., 1988, Loihi Seamount, Hawaii: A mid-plate volcano with a distinctive hydrothermal system: Nature, v. 335, p. 532-535.

Karlin, R., Lyle, M., and Heath, G. R., 1987, Authigenic magnetite formation in suboxic marine sediments: Nature, v. 326, p. 490-493.

Klein, C., and Beukes, N. J., 1989, Geochemistry and sedimentology of a facies transition from limestone to iron-formation deposition in the early Proterozoic Transvaal Supergroup, South Africa: Economic Geology, v. 84, p. 1733-1774.

Klein, C., and Beukes, N. J., 1992, Time distribution, stratigraphy, and sedimentologic setting, and geochemistry of Precambrian iron-formations, in Schopf, J. W., and Klein, C., eds., The Proterozoic Biosphere: Cambridge, Cambridge University Press, p. 139-146.

Knoll, A. H., 1985, Exceptional preservation of photosynthetic organisms in silicified carbonates and silicified peats: Royal Society of London Philosophical Transactions, v. B311, p. 111-122.

Köhler, B., Singer, A., and Stoffers, P., 1994, Biogenic nontronite from marine white smoker chimneys: Clays and Clay Minerals, v. 42, p. 689-701.

Konhauser, K. O., 1997, Bacterial iron biomineralisation in nature: FEMS Microbiology Reviews, v. 20, p. 315-326.

Konhauser, K. O., 1998, Diversity of bacterial iron mineralization: Earth-Science Reviews, v. 43, p. 91-121.

Konhauser, K. O., and Ferris, F. G., 1996, Diversity of iron and silica precipitation by microbial mats in hydrothermal waters, Iceland: Implications for Precambrian iron formations: Geology, v. 24, p. 323-326.

Laberge, G. L., 1973, Possible biological origin of Precambrian iron-formations: Economic Geology, v. 68, p. 1098-1109.

Lees, H., Kwok, S. C., and Suzuki, I., 1969, The thermodynamics of iron oxidation by ferrobacilli: Canadian Journal of Microbiology, v. 15, p. 43-46.

Leo, R. F., and Barghoorn, E. S., 1976, Silicification of wood: Harvard University Botanical Museum Leaflets, v. 25, p. 1-46.

Lovley, D. R., 1991, Magnetite formation during microbial dissimilatory iron reduction, in Frankel, R. B., and Blakemore, R. P., eds., Iron Biominerals: New York, Plenum Press, p. 151-166.

Lowe, D. R., 1980, Stromatolites 3,400-Myr old from the Archean of Western Australia: Nature, v. 284, p. 441-443.

Lowe, D. R., 1994, Early environments: Constraints and opportunities for early evolution, in Bengtson, S., ed., Early Life on Earth: New York, Columbia University Press, p. 24-35.

MacRae, I. C., and Edwards, J. F., 1972, Adsorption of colloidal iron by bacteria: Applied Microbiology, v. 24, p. 819-823.

McNamara, K. J., and Awramik, S. M., 1992, Stromatolites: A key to understanding the early evolution of life: Science in Progress, v. 76, p. 345-364.

McConchie, D., 1987, The geology and geochemistry of the Joffre and Whaleback shale members of the Brockman Iron Formation, Western Australia, in Appel, P. W. U., and LaBerge, G. L., eds., Precambrian Iron-Formations: Athens, Theophrastus, p. 541-597.

Millero, F. J., Sotolongo, S., and Izaguirre, M., 1987, The oxidation kinetics of Fe(II) in seawater: Geochimica et Cosmochimica Acta, v. 51, p. 793-801.

Mojzsis, S. J., Arrhenius, G., McKeegan, K. D., Harrison, T. M., Nutman, A. P., and Friend, C. R. L., 1996, Evidence for life on Earth before 3,800 million years ago: Nature, v. 384, p. 55-59.

Morris, R. C., 1980, A textural and mineralogical study of the relationship of iron ore to banded iron-formation in the Hamersley Iron Province of Western Australia: Economic Geology, v. 75, p. 184-209.

MORRIS, R. C., 1993, Genetic modelling for banded iron-formation of the Hamersley Group, Pilbara Craton, Western Australia: Precambrian Research, v. 60, p. 243-286.

MORRIS, R. C., AND HORWITZ, R. C., 1983, The origin of the iron-formation-rich Hamersley Group of Western Australia-Deposition on a platform: Precambrian Research, v. 21, p. 273-297.

NEALSON, K. H., AND MYERS, C. R., 1990, Iron reduction by bacteria: A potential role in the genesis of banded iron formations: American Journal of Science, v. 290A, p. 35-45.

OEHLER, J. H., 1976, Experimental studies in Precambrian paleontology: Structural and chemical changes in blue-green algae during simulated fossilization in synthetic chert: Geological Society of America Bulletin, v. 87, p. 117-129.

OEHLER, J. H., AND SCHOPF, J. W., 1971, Artificial microfossils: Experimental studies of permineralization of blue-green algae in silica: Science, v. 174, p. 1229-1231.

OHMOTO, H., KAKEGAWA, T., AND LOWE, D. R., 1993, 3.4-billion-year-old biogenic pyrites from Barberton, South Africa: Sulfur isotope evidence: Science, v. 262, p. 555-557.

PERRY, E. C., JR., TAN, F. C., AND MOREY, G. B., 1973, Geology and stable isotope geochemistry of the Biwabik iron formation, northern Minnesota: Economic Geology, v. 68, p. 1110-1125.

SAWICKI, J. A., BROWN, D. A., AND BEVERIDGE, T. J., 1995, Microbial precipitation of siderite and protoferrihydrite in a biofilm: Canadian Mineralogist, v. 33, p. 1-6.

SCHIDLOWSKI, M., 1976, Archean atmosphere and evolution of the terrestrial oxygen budget, in Windley, B. F., ed., The Early History of the Earth: London, John Wiley, p. 525-535.

SCHISSEL, D., AND ARO, P., 1992, The major early Proterozoic iron and manganese deposits and their tectonic setting: Economic Geology, v. 87, p. 1367-1374.

SCHOPF, J. W., AND PACKER, B. M., 1987, Early Archean (3.3-billion to 3.5-billion-year-old) microfossils from Warrawoona Group, Australia: Science, v. 237, p. 70-72.

SCHOPF, J. W., 1993, Microfossils of the early Archean Apex chert: New evidence of the antiquity of life: Science, v. 260, p. 640-646.

SCHULTZE-LAM, S., FERRIS, F. G., KONHAUSER, K. O., AND WIESE, R. G., 1995, In situ silicification of an Icelandic hot spring microbial mat: Implications for microfossil formation: Canadian Journal of Earth Sciences, v. 32, p. 2021-2026.

SCHWERTMANN, U., AND FITZPATRICK, R. W., 1992, Iron minerals in surface environments, in Skinner, H. C. W., and Fitzpatrick, R. W., eds., Biomineralization. Processes of Iron and Manganese: Cremlingen, Catena Verlag, p. 7-30.

SIEVER, R., 1992, The silica cycle in the Precambrian: Geochimica et Cosmochimica Acta, v. 56, p. 3265-3272.

SIEVER, R., AND WOODFORD, N., 1973, Sorption of silica by clay minerals: Geochimica et Cosmochimica Acta, v. 37, p. 1851-1880.

SIGLEO, A. C., 1978, Organic geochemistry of silicified wood, Petrified Forest National Park, Arizona: Geochimica et Cosmochimica Acta, v. 42, p. 1397-1405.

SIMONSON, B. M., 1985, Sedimentological constraints on the origins of Precambrian iron-formations: Geological Society of America Bulletin, v. 96, p. 244-252.

SIMONSON, B. M., AND HASSLER, S. W., 1996, Was the deposition of large Precambrian iron formations linked to major marine transgressions?: Journal of Geology, v. 104, p. 665-676.

STEEFEL, C. I., AND VAN CAPELLEN, P., 1990, A new kinetic approach to modeling water-rock interaction: The role of nucleation, precursors, and Ostwald ripening: Geochimica et Cosmochimica Acta, v. 54, p. 2657-2677.

STRAUB, K. L., BENZ, M., SCHINK, B., AND WIDDEL, F., 1996, Anaerobic, nitrate-dependent microbial oxidation of ferrous iron: Applied and Environmental Microbiology, v. 62, p. 1458-1460.

THOMPSON, J. B., AND FERRIS, F. G., 1990, Cyanobacterial precipitation of gypsum, calcite, and magnesite from natural alkaline lake water: Geology, v. 18, p. 995-998.

TRENDALL, A. F., 1965, Progress report on the Brockman Iron Formation in the Wittenoom-Yampire area: Western Australia Geological Survey Annual Report, 1964, p. 55-65.

TRENDALL, A. F., AND BLOCKLEY, J. G., 1970, The iron formations of the Precambrian Hamersley Group, Western Australia; with special reference to the associated crocidolite: Western Australia Geological Survey Bulletin, v. 119, 336 p.

TUNNICLIFFE, V., AND FONTAINE, A. R., 1987, Faunal composition and organic surface encrustations at hydrothermal vents on the southern Juan de Fuca Ridge: Journal of Geophysical Research, v. 92, p. 11303-11314.

TURCOTTE, D. L., 1980, On the thermal evolution of the earth: Earth and Planetary Science Letters, v. 48, p. 53-58.

URRUTIA, M. M., AND BEVERIDGE, T. J., 1993, Mechanism of silicate binding to the bacterial cell wall in Bacillus subtilis: Journal of Bacteriology, v. 175, p. 1936-1945.

URRUTIA, M. M., AND BEVERIDGE, T. J., 1994, Formation of fine-grained metal and silicate precipitates on a bacterial surface (Bacillus subtilis): Chemical Geology, v. 116, p. 261-280.

VEIZER, J., 1983, Geologic evolution of the Archean-early Proterozoic earth, in Schopf, J. W., ed., Earth's Earliest Biosphere. Its Origin and Evolution: Princeton, Princeton University Press, p. 240-259.

VEIZER, J., COMPSTON, W., HOEFS, J., AND NIELSEN, H., 1982, Mantle buffering of the early oceans. Naturwissenschaften, v. 69, p. 173-180.

WALKER, J. C. G., 1984, Suboxic diagenesis in banded iron formations: Nature, v. 309, p. 340-342.

WALKER, J. C. G., KLEIN, C., SCHIDLOWSKI, M., SCHOPF, J. W., STEVENSON, D. J., AND WALTER, M. R., 1983, Environmental evolution of the Archean-early Proterozoic earth, in Schopf, J.W., ed., Earth's Earliest Biosphere. Its Origin and Evolution: Princeton, Princeton University Press, p. 260-290.

WALSH, M. M., AND LOWE, D. R., 1985, Filamentous microfossils from the 3,500-Myr-old Onverwacht Group, Barberton Mountain Land, South Africa: Nature, v. 314, p. 530-532.

WALTER, M. R., 1976, Hot-spring sediments in Yellowstone National Park, in Walter, M. R., ed., Stromatolites: Amsterdam, Elsevier, p. 489-498.

WALTER, M. R., BAULD, J., AND BROCK, T. D., 1972, Siliceous algal and bacterial stromatolites in hot spring and geyser effluents of Yellowstone National Park: Science, v. 178, p. 402-405.

WALTER, M. R., BUICK, R., AND DUNLOP, J. S. R., 1980, Stromatolites 3,400-3,500 Myr old from the North Pole area, Western Australia: Nature, v. 284, p. 443-445.

WALTER, M. R., BAULD, J., DES MARAIS, D. J., AND SCHOPF, J. W., 1992, A general comparison of microbial mats and microbial stromatolites: Bridging the gap between the modern and the fossil, in Schopf, J. W., and Klein, C., eds., The Proterozoic Biosphere: Cambridge, Cambridge University Press, p. 335-338.

WESTALL, F., BONI, L., AND GUERZONI, E., 1995, The experimental silicification of microorganisms: Palaeontology, v. 38, p. 495-528.

WIDDEL, F., SCHNELL, S., HEISING, S., EHRENREICH, A., ASSMUS, B., AND SCHINK, B., 1993, Ferrous iron oxidation by anoxygenic phototrophic bacteria: Nature, v. 362, p. 834-836.

WILLIAMS, L. A., AND CRERAR, D. A., 1985, Silica diagenesis, 11. General mechanisms: Journal of Sedimentary Petrology, v. 55, p. 312-321.

BEDDED BARITE IN THE GEOLOGIC RECORD

PAUL W. JEWELL*

Department of Geology and Geophysics, University of Utah, Salt Lake City, UT 84112

Abstract: Barite occurs throughout the geologic record as massive beds, laminations, rosettes, and nodules. The most important scientific and economic occurrences of barite are stratabound and stratiform massive beds from the Early Archean and the Early to Middle Paleozoic. Paleozoic bedded barites are by far the most volumetrically significant deposits in the geologic record. Additional occurrences have been documented in some Middle Proterozoic, Late Proterozoic, and Mesozoic rocks and in several localities on the modern ocean floor. Bedded barite is believed to have formed as emanations from seafloor sediments, as diagenetic replacements of preexisting minerals, or as direct precipitants due to biological fixation of barium in the water column. Direct field evidence to differentiate between these theories is often lacking or contradictory. Geochemical studies, particularly those that have employed $\delta^{34}S$ and $^{87}Sr/^{86}Sr$ analyses of the barite, have proven very useful in understanding bedded barite genesis.

The low solubility of barite relative to other natural salts has helped barite survive as a pseudomorph of stratiform evaporite minerals in some Archean sedimentary sequences. Other examples of Archean barite appear to have a shallow water detrital or authigenic origin. Very low $\delta^{34}S$ values of Archean barite are interpreted as indicating a low-sulfate ocean. Large deposits of Paleozoic bedded barite are typically found in fine-grained, organic-rich siliciclastic sequences and are associated with massive and disseminated sulfides, cherts, phosphorites, and less frequently limestones and volcanic rocks. $\delta^{34}S$ analyses indicate that almost all bedded barite had a seawater sulfate source. The genetic link between Paleozoic bedded barites and sedimentary submarine exhalative Pb-Zn sulfide deposits has been established by field and geochemical study of deposits in western Canada and western Europe. $^{87}Sr/^{86}Sr$ analyses suggest that these bedded barites have a continental barium source. Economically important bedded barites in China, Arkansas, and Nevada have no significant sulfides or other hydrothermal manifestations. The clear association of dissolved barium and barite with biological cycles in the modern ocean, associations with phosphorites and cherts, and $^{87}Sr/^{86}Sr$ analyses that are comparable to contemporaneous seawater suggest that the Chinese, Nevada, and Arkansas barite deposits formed as biological precipitates on the seafloor. Bedded barite formed by this mechanism holds promise as an indicator of high paleoproductivity and open ocean sulfate reduction during selected periods of the Paleozoic. The lack of world class examples of bedded barite in Mesozoic and Cenozoic black shale sequences indicates a lack of open ocean sulfate reduction during these periods of geologic time.

INTRODUCTION

Barite ($BaSO_4$) is found throughout the geologic record and is widely documented in modern oceanic environments. Several factors make barite an economically and scientifically interesting mineral. The high specific gravity (4.48 g/cm^3) and widespread occurrence of almost pure barite make it economically useful, particularly in the manufacture of drilling muds for the oil and gas industry. Barite is a common gangue mineral in a variety of hydrothermal ore deposits including volcanic- and sediment-hosted deep-sea hot springs associated with seafloor volcanism. These associations are used as a modern analog for bedded barite in sedimentary exhalative ("sedex") deposits, which are common in many fine-grained siliciclastic sedimentary sequences in the geologic record.

Barium in modern seawater has long been recognized as following biological cycles in a manner similar to elements such as phosphorous and silicon (Chow and Goldberg, 1960; Turekian and Johnson, 1966). Barite and elevated barium concentrations in modern deep water marine sediments have been noted by oceanographers for decades (Goldberg and Arrhenius, 1958). Careful research has shown that barite forms discrete particles in the water column of the open ocean as a result of biological processes (Dehairs et al., 1980; Bishop, 1988). This has led to the development of barite as an important paleoproductivity tracer (e.g., Schmitz, 1987; Dymond et al., 1992).

The solubility of barite is dependent on the redox state of water, which means that barite dissolves in reducing open water and diagenetic environments. This feature has been used to argue that transitions between anoxic, sulfate-reducing, and oxygenated waters are necessary for the formation of bedded barite and associated minerals in a variety of marine settings, including a sulfate-poor Archean ocean (Perry et al., 1971), deep sea hydrothermal emanations (Turner, 1992), and coastal upwelling zones (Jewell and Stallard, 1991; Jewell, 1994).

Barite has the useful characteristic of containing two elements (barium and sulfur) that can be interpreted within the context of well-established isotopic techniques. Stable isotopes of sulfur have a long history of being used to interpret both the source and genesis of a wide variety of sulfide and sulfate minerals. Within this context, barite has played a role in understanding the global history of the sulfur cycle (e.g., Schidlowski et al., 1983). Also, the chemistry of barium is sufficiently similar to that of strontium that $^{87}Sr/^{86}Sr$ techniques have been extensively used to understand the parentage (for example, continental versus mantle versus seawater source) of barite. For instance, detailed geochemical analysis of the biologically produced barite fraction of deep-sea modern to Miocene sediments shows that barite tracks the $^{87}Sr/^{86}Sr$ composition of contemporaneous seawater (Paytan et al., 1993; Martin et al., 1995). On the other hand, barites that precipitate from fluids discharging on the seafloor tend to reflect the $^{87}Sr/^{86}Sr$ values of minerals that reacted with the formation fluids (e.g., Aquilina et al., 1997). These studies provide confidence that the large number of $^{87}Sr/^{86}Sr$ analyses of older bedded barites accurately reflect the parentage of the fluids from which they formed.

All of these features have combined to produce a wealth of studies on barite geology and geochemistry. In spite of this, many aspects of bedded barite genesis remain poorly under-

*email: pwjewell@mines.utah.edu

Marine Authigenesis: From Global to Microbial, SEPM Special Publication No. 66

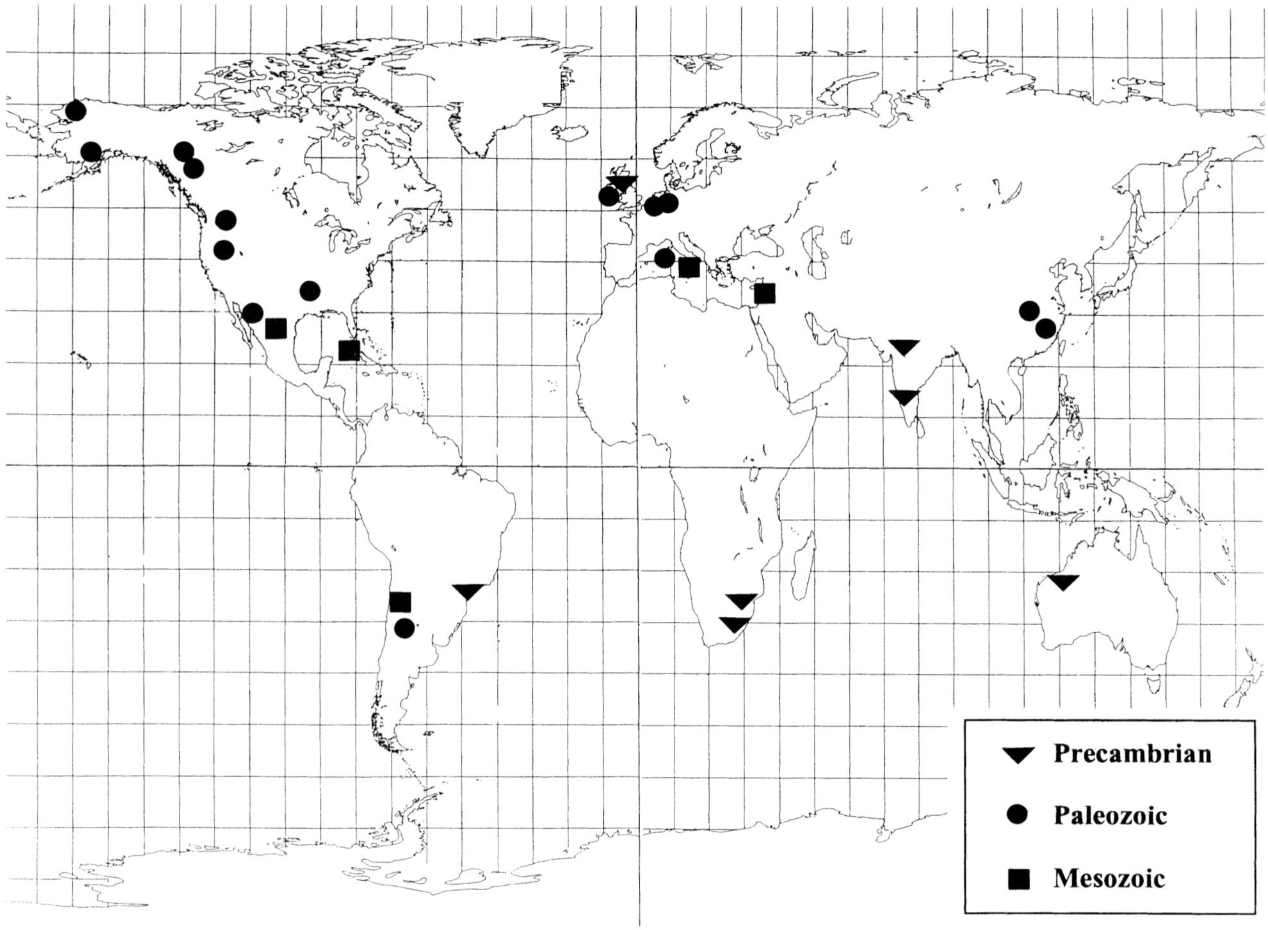

FIG. 1.—Geographic distribution of major bedded barite deposits on the continents.

stood. Although summary articles of some aspects of barite formation have been undertaken (e.g., Brodtkorb, 1989; Maynard and Okita, 1991), no systematic attempt has been made to synthesize theories of barite genesis throughout the geologic record. This paper will focus on massive bedded or stratiform barite in sedimentary rocks in an effort to understand the genesis of this important worldwide economic and scientific resource. Although the discussion is not meant to include all bedded barite occurrences, specific examples are discussed from all of the most important geologic eras and localities where bedded barite has been found. Particular attention is paid to Paleozoic bedded barites, which constitute the largest deposits in the world. The emphasis in this paper is on deposits that are well described in mainstream scientific literature. Lesser emphasis is given to bedded barite occurrences that have only been documented in abstracts or less accessible publications. It should also be noted that nodular barite occurs with many of the massive, bedded barites that are discussed. In some cases, these two contrasting barite morphologies have been used to make specific inferences about environments of deposition (e.g., Graber and Chafetz, 1990) and the isotope chemistry of the two morphologies is often distinctly different (e.g., Rye et al., 1978). A comprehensive discussion of nodular barite would have significantly expanded the scope of this review and for the sake of brevity, the emphasis is placed on bedded barite.

PRECAMBRIAN BEDDED BARITE

Isolated bedded barite occurrences are described from a small number of Precambrian shields throughout the world (Fig. 1, Table 1). Since many Precambrian barite deposits have been extensively metamorphosed, it is often difficult to ascertain their precise genesis. Nevertheless, these barites have sparked considerable scientific interest because of their possible role in interpreting Archean evaporites and the evolution of the redox state of the Archean atmosphere and ocean. The best documented Archean barite occurrences are in Australia, South Africa, and India, all of which have ages >3200 m.y. The paucity of documented bedded barite in rocks from the Late Archean through the Late Proterozoic is puzzling and has been noted by at least one author (Thorpe, 1979).

Archean Barite

Archean bedded barite from the Pilbara block of western Australia (Lambert et al., 1978; Barely et al., 1979; Dunlop and Buick, 1981) and the Barberton Mountain Land of South Africa (Perry et al., 1971; Lowe and Knauth, 1977) represent the oldest bedded barites in the geologic record (Table 1). Measurement of interfacial angles of individual crystals of the Australian barite demonstrate that they are pseudomorphs of

TABLE 1.— SUMMARY OF AGE LITHOLOGIC AND DEPOSITIONAL CHARACTERISTICS OF IMPORTANT PRECAMBRIAN AND PALEOZOIC BEDDED BARITE DEPOSITS

Location and geologic formation	Age	Lithologic associations	Depositional setting
Western Australia			
Warrawoona Group[1]	~3450 m.y.	Cherts, fine-grained metasiliciclastic rocks	Shallow water to subaerial evaporitic shelf
South Africa			
Onverwacht Group[2,3]	~3450 m.y.	Cherts, quartzitic micaceous amphibolites	Shallow water evaporitic shelf
Fig Tree Group[3,4]	~3200 m.y.	Cherts, dolomitic conglomerates	Submarine ridge
India			
Sagur Group[5,6]	>3300 m.y	Quartzites, talc schists, phyllites, metavolcanics	Shallow water evaporitic shelf
Aravalli Supergroup[6]	~2000 m.y.	Quartzites, metabasalts	Alluvial, shallow marine shelf
Scotland			
Argyll Group[7]	610-640 m.y.	Quartzites, phyllites, schists	Epicratonic rift basin
Southern China			
Niutitang, Lujiaping, Longmaxi Formations[8]	Early Cambrian	Black shales, cherts, phosphorites	Continental shelf (south), deep water siliciclastic basin (north)
Western Canada			
Mount Mye Unit[9]	Cambrian	Noncalcareous phyllites, schists	Epicratonic rift basin
Vangorda Unit[9]	Cambro-Ordovician	Calcereous phyllites, schists	Epicratonic rift basin
Road River Formation[10]	Ordovician-Silurian	Black shales, cherts	Epicratonic rift basin
Lower Earn Group[10]	Devonian	Black shales, cherts	Epicratonic rift basin
Nevada, U.S.A.			
Vinini Formation[11]	Ordovician	Shales, cherts, siltstones	Deep water continental margin
Slaven Chert[11]	Late Devonian	Cherts, shales	Deep water continental margin
Germany			
Meggan Beds[12]	Middle Devonian	Black shales, siltstones	Epicratonic starved basin
Wissenbach Shales[13]	Middle Devonian	Black shales, siltstones	Shallow water shelf
Belgium			
Unnamed[14]	Late Devonian	Shales, argillaceous limestones	Shallow water evaporitic shelf
Arkansas, U.S.A.			
Stanley Shale[15]	Late Mississippian	Shales, siltstones, limestones	Deep water slope
Ireland			
Unnamed[16]	Mississippian	Limestones, dolomites	Shallow water shelf

[1] Barley et al. (1979); Dunlop and Buick (1981)
[2] Lowe and Knauth (1977)
[3] Reimer (1980)
[4] Heinrichs and Reimer (1977)
[5] Radhakrishna and Sreenivasaiya (1974)
[6] Deb et al. (1991)
[7] Willan and Coleman (1983); Hall et al. (1991)
[8] Wang and Li (1991)
[9] Jennings and Jilson (1986)
[10] Abbott et al. (1986)
[11] Stewart (1980)
[12] Krebs (1981)
[13] Hannak (1981)
[14] Dejonghe and Boulvain (1993)
[15] Morris (1974)
[16] Taylor and Andrew (1978)

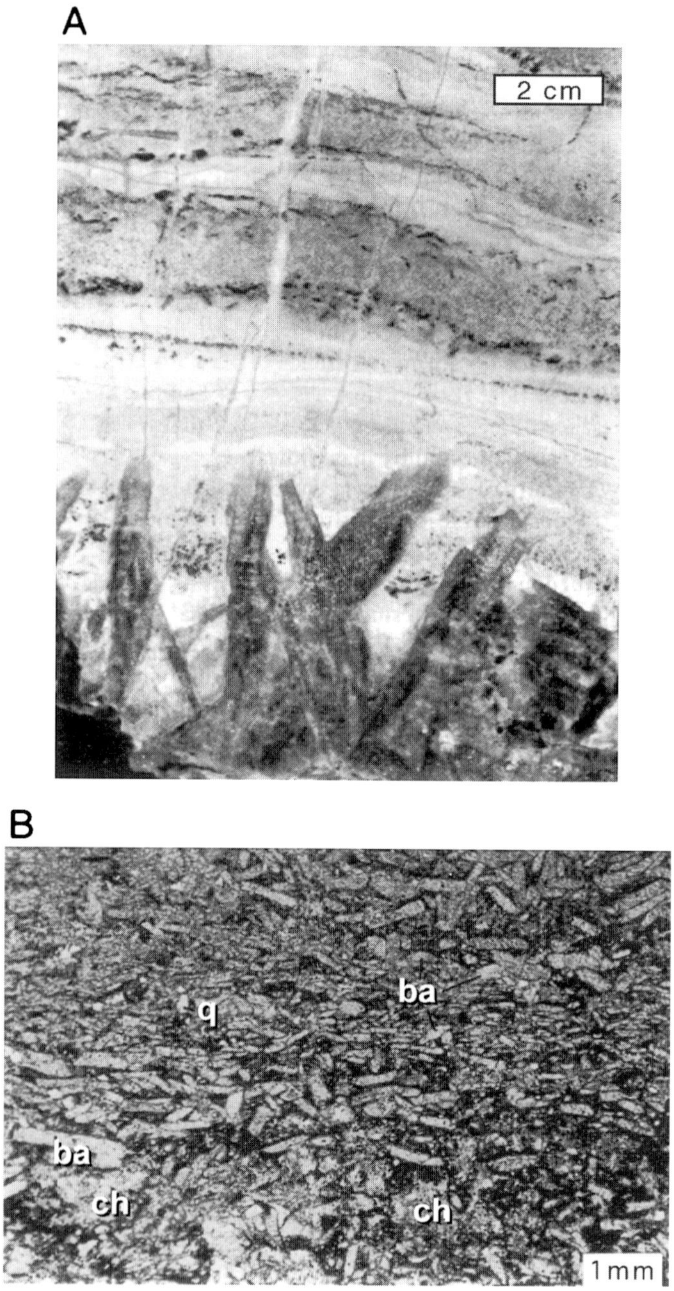

FIG. 2.—Field examples of Precambrian bedded barite. (A) Stratiform barite-chert horizon from North Pole deposit of Western Australia. Universal stage measurement of barite crystals show that they have interfacial angles of gypsum (from Lambert et al., 1978). (B) Detrital barite of the Archean Fig Tree Group, Swaziland, South Africa (from Heinrichs and Reimer, 1977).

gypsum (Lambert et al., 1978) (Fig. 2A) and that a portion of this sedimentary sequence was originally composed of evaporites. The high solubility of all common evaporite minerals relative to barite has allowed these barites to survive subsequent tectonism, metamorphism, and surface exposure. The Australian Archean barites appear to be the oldest remnants of evaporites in the geologic record (Lambert et al., 1978).

There are differences of opinion regarding the origin of barite in the ~3450 m.y. Onverwacht Group of South Africa. Petrographic and mineralogical analysis suggests an evaporite origin similar to that of the Australian barites (Lowe and Knauth, 1977; Lambert et al., 1978). Reimer (1980) advocates a volcanic source of the barium on the basis of close stratigraphic proximity to large volumes of volcanic rocks and minor amounts of associated sulfides. Detailed work on younger (~3200 m.y.) bedded barite in the overlying Fig Tree Group shows that these deposits are largely detrital with some diagenetic recrystallization rather than being pseudomorphs of evaporites (Heinrichs and Reimer, 1977; Reimer, 1980) (Fig. 2B). An obvious source of the detrital barite would, of course, be the older Onverwacht barite.

Archean bedded barite from central India is highly metamorphosed and occurs with a variety of Ba-bearing metamorphic minerals. As with the South African and Australian barites, no significant sulfide mineralization is noted. Whereas the Indian bedded barites are clearly synsedimentary in origin, and their exact depositional environment is difficult to interpret (Radhakrishna and Sreenivasaiya, 1974; Deb et al., 1991).

The Australian, South African, and Indian Archean bedded barite have all been studied isotopically. Assuming that the primary Rb-Sr signature of these barites has been retained, the Western Australian barite has the most primitive Sr signature of all the bedded barites (Table 2). Comparison of the $^{87}Sr/^{86}Sr$ composition of these barites with lunar and meteorite $^{87}Sr/^{86}Sr$ analyses have allowed constraints to be placed on the origin of the Earth and the Moon (McCulloch, 1994). South African and Indian Archean bedded barites are on average slightly more radiogenic than the Australian barite (Table 2), although both are consistent with mantle buffering of Archean seawater composition (Perry et al., 1971; Deb et al., 1991; Strauss, 1993).

Sulfur isotopic analysis of Archean bedded barite has been a critical piece of evidence for understanding the redox state and sulfate concentrations of Archean seawater. $\delta^{34}S$ compositions of the Australian, South African, and Indian bedded barites are all low relative to the minor amounts of coexisting sulfides (Table 2), which suggests that the biologic fractionation of sulfide sulfur observed throughout the post-Archean geologic record was not operative in the Archean ocean. Rather, the low sulfide-sulfate fractionation is attributed to inorganic oxidation of volcanic H_2S or SO_2 in an ocean with very small sulfate concentrations (Perry et al., 1971; Cameron, 1982; Schidlowski et al., 1983; Strauss, 1993).

Proterozoic Barite

Examples of bedded barite from the Proterozoic are rare (Table 1). The lone example from the Early to Middle Proterozoic is the ~2000 m.y. bedded barite of the Aravalli Supergroup of central India. These bedded barites are associated with supracrustal quartzites, metavolcanics, and some carbonates but no known massive sulfides (Deb et al., 1991). The Late Proterozoic Dalradian Supergroup in Scotland has an exceptional number of bedded barite occurrences with a clear association to exhalative massive sulfide deposits as well as some barite deposits with only minor associated sulfides (Willan and Coleman, 1983; Hall et al., 1991). These deposits are found in

TABLE 2.— SUMMARY OF SELECTED $^{87}Sr/^{86}Sr$ and $\delta^{34}S$ DATA FOR BEDDED BARITE. VALUES REPORTED ARE MEAN + ONE STANDARD DEVIATION OF ALL ANALYSES REPORTED IN THE LITERATURE REFERENCE.

Location and geologic formation	$^{87}Sr/^{86}Sr$ (barite)	$^{87}Sr/^{86}Sr$ (seawater)[1]	$\delta^{34}S$ (‰)
Western Australia,			
Warrawoona Group (Early Archean)	.70053± 3 (n=4) [2]		3.6± 0.5 (n=18) [3]
	.70052 [4]		5.4±0.9 (n=2) [4]
South Africa			
Onverwacht Group (Early Archean)			3.0–6.9 (n=11) [5]
Fig Tree Group (Early Archean)	.7009-.7017 (n=8) [6]		3.8±0.4 (n=7) [3]
	.7014 [3]		3.4±0.3 (n=11) [6]
			4.1–8.4 [7]
			2.0–4.3 [8]
India			
Sagur Group (Early Archean)	.7018±7 (n=7) [9]		5.4±0.7 (n=12) [9]
			4.3+0.2 (n=6) [10]
Aravalli Supergroup (Middle Proterozoic)	.7234±3 (n=6) [9]		18.8±0.5 (n=16) [9]
Scotland			
Argyll Group (Late Proterozoic)	.7139±6 (n=6) [11]	.7071	36.5±2.0 (n=16) [11]
			32.3±2.7 (n=20)[12]
Southern China			
Qinling Region (Early Cambrian)	.7084±3 (n=7) [13]	.7095	22.9±0.8 (n=17) [14]
Jiangnan Region (Early Cambrian)	.7085+2 (n=5) [13]		34.9±2.2 (n=3) [14]
Western Canada			
Faro deposit (Late Cambrian)			25.2±1.2 (n=6) [15]
Grum deposit (Late Cambrian)			31.8±2.2 (n=10) [15]
DY deposit (Late Cambrian)			31.7±7.2 (n=13) [15]
Vulcan deposit (Late Silurian - Early Devonian)			43.5 (n=13) [16]
Jason deposit (Middle Devonian)	.7134±5 (n=10) [17]	.7077	24.4±1.2 (n=17) [18]
Lower Earn Group (Late Devonian)			27.0±3.4 (n=3) [19]
Nevada, U. S. A.			
East Northumberland Canyon (Late Devonian)	.7083 [20]	.7083	24.4±2.2 (n=10) [20]
Mountain Springs Deposit (Late Devonian)	.7097±8 (n=7) [21, 22]		
Argenta (Late Devonian)	.7107±8 (n=2) [21, 22]		29.1±13.6 (n=6) [24]
Miller (Late Devonian)	.7091 [23]		32.6±`8.7 (n=4) [24]
Queen Lode (Late Devonian?)	.70810 [21]		
Western Europe			
Meggen deposit (Middle Devonian)	.71072 [21]	.7077	23.1±2.0 (n=72) [25]
Ramselberg deposit (Middle Devonian)	.71100 [21]		22.7±3.1 (n=58) [26]
Chaudfontaine deposit (Late Devonian)	.7112±2 (n=11) [27]		28.0±1.5 (n=9) [27]
Arkansas, U. S. A.			
Fancy Hill deposit (Late Mississippian)	.7084±2 (n=2) [21]	.7082	
Millchem deposit (Late Mississippian)	.7085±2 (n=2) [21]		
Chamberlain deposit (Late Mississippian)	.7084±7 (n=7) [28]		
Italy (Mid-Late Triassic)	.70707±4 (n=4) [29]	0.7077	
Cuba (Jurassic)	.7120±6 (n=5) [21]		
Mexico (Early Cretaceous)	.7084±6 (n=8) [30]	0.7072	24.3±12.0 [30]
Paita (Holocene, Peru middle shelf)	.7104±3 (n=4) [31]		
Chiclayo Canyon (Holocene, Peru continental slope)	.7109±4 (n=6) [31]		

[1] Burke et al. (1982)
[2] McCulloch (1994); primary barite only
[3] Lambert et al. (1978)
[4] Strauss (1993)
[5] Reimer (1980)
[6] Perry et al. (1971)
[7] Reimer (1980); detrital barite
[8] Reimer (1980); diagenetic barite
[9] Deb et al. (1991)
[10] Hoering (1988)
[11] Hall et al. (1991)
[12] Willan and Coleman (1983)
[13] Wang and Chu (1994); excludes one value (0.713661) from Jiangnan area and another (0.716710) from Qinling area
[14] Wang and Li (1991); excludes analyses of concretionary barite
[15] Shanks et al. (1987); all samples described as baritic massive sulfides
[16] Mako and Shanks (1984); no reported standard deviations of data
[17] Turner (1986)
[18] Gardner and Hutcheon (1985)
[19] Cecile et al. (1983)
[20] Rye et al. (1978)
[21] Maynard et al. (1995)
[22] Barbieri and Masi (1983)
[23] Jewell, unpublished
[24] Mitchell (1980)
[25] Buschendorf et al. (1963)
[26] Anger et al. (1966)
[27] Dejonghe et al. (1989)
[28] Hulen (1978)
[29] Barbieri et al. (1984)
[30] Kesler and Jones (1981)
[31] Aquilina et al. (1997)

siliciclastic sedimentary sequences deposited during a period of extensional tectonics along a continental margin. In this sense, the Proterozoic Scottish deposits are similar to the well-studied Paleozoic barite of western Canada and western Europe. It is noteworthy that most large, well-known, Proterozoic sediment-hosted massive sulfide deposits (for example, Sullivan in British Columbia and the McArthur River area in northeastern Australia) are devoid of bedded barite (Gustafson and Williams, 1981).

$^{87}Sr/^{86}Sr$ analyses of the central Indian and Scottish Proterozoic bedded barites show clear continental affinities (Table 2) and probably reflect the circulation of hydrothermal fluids through continental crust. All Proterozoic bedded barites have $\delta^{34}S$ compositions that are heavier than Archean barite, which suggests that extensive biogenic fractionation and high sulfate oceans were common in the Proterozoic (Willan and Coleman, 1983; Hall et al., 1991; Deb et al., 1991) (Table 2).

Additional occurrences of Proterozoic bedded barite from southern Brazil (Cassedanne, 1989) and Zimbabwe (Thorpe, 1979) are poorly documented with respect to age, geologic setting, and genesis. The latter is particularly intriguing because of its similarities to the South African and Australian Archean bedded barites and its occurrence in rocks that could be anywhere from Late Archean to Middle Proterozoic in age.

PALEOZOIC BEDDED BARITE

Paleozoic bedded barites constitute the most widespread and volumetrically significant amounts of barite in the geologic record. Major deposits are pre-Pennsylvanian in age and found throughout the world (Fig. 1). The preponderance of Paleozoic bedded barites occurs in black shale sequences with associated cherts and phosphorites as well as lesser amounts of volcanic rocks, limestones, and sandstones (Table 1). Significant amounts of barite occur with Paleozoic sedimentary rock-hosted massive sulfide deposits, although many large, economically important bedded barites are not associated with significant amounts of sulfide. Like Precambrian barite, the sulfur isotopic composition of Paleozoic barite has been effectively used to track open and closed ocean chemistry for selected periods of time (e.g., Goodfellow and Jonasson, 1984) and strontium isotopes have proven effective in documenting the source of barium in a variety of settings (e.g., Maynard et al., 1995). Paleozoic bedded barites have $\delta^{34}S$ signatures that are close to but not precisely the same as the Paleozoic evaporite curve (Fig. 3).

Southern China

Bedded barite deposits in southern China are among the largest barite reserves of the world. In spite of this, the Chinese deposits are among the most poorly documented of all barite provinces with respect to regional geologic setting and detailed studies of specific deposits.

Although Late Sinian (Late Proterozoic) rocks host concretionary barite and witherite, all of the major Chinese bedded barite deposits are found in Early Cambrian black shales and cherts of the Qinling and Jiangnan regions in the south-central portion of the country (Wang and Li, 1991). Barite deposits of the more southern Jiangnan area are believed to have formed on the interior of the Yangtze craton, while the Qinling deposits are found in the suture zone that formed between the Yangtze and Sino-Korean craton to the north during the early Paleozoic (Maynard and Okita, 1991). Exact age determinations and stratigraphic correlations of barite-bearing strata are lacking, although the massive barite beds appear to occur in stratigraphically higher intervals in deep water facies to the northeast (Wang and Li, 1991).

The Chinese barite deposits occur in massive beds, laminations, rosettes, and nodules and are apparently unique among major barite provinces of the world in being associated with major amounts of witherite ($BaCO_3$) and barytocalcite ($CaBa(CO_3)_2$). High concentrations of vanadium (up to 1% V_2O_5) have been documented in associated sedimentary rocks. Like many barite provinces, phosphatic concretions and organic carbon-rich shales (up to 16% total organic carbon) are commonly associated with the barite (Wang and Li, 1991). $^{87}Sr/^{86}Sr$ values from most of the Chinese deposits are very close to those of contemporaneous Cambrian seawater (Wang and Chu, 1994) (Table 2). However, single samples from the Sanjiang deposit of the Jiangnin area and Liulin deposit of the Qinling area are anomalously radiogenic. $\delta^{34}S$ values of the barite grade from relatively heavy (+35‰) in the platform settings to somewhat lighter values (+23‰) in the geosynclinal or continental margin deposits (Table 2). These observations are explained within the context of a euxinic interior basin that depleted seawater sulfate concentrations leading to relatively heavy sulfate that was then incorporated into the barite (Wang and Li, 1991).

Western Canada

Western Canada (the Yukon, Northwest Territories (NWT), and northern British Columbia) are host to a large number of economically important sedimentary rock-hosted massive sulfide (sedex) and barite deposits. These deposits are found in thick sequences of fine-grained siliciclastic rocks and chert deposited in an epicratonic rift basin that was active from the Cambrian through the Late Devonian (Table 1) and bounded to the east and west by carbonate platforms (Abbott et al., 1986).

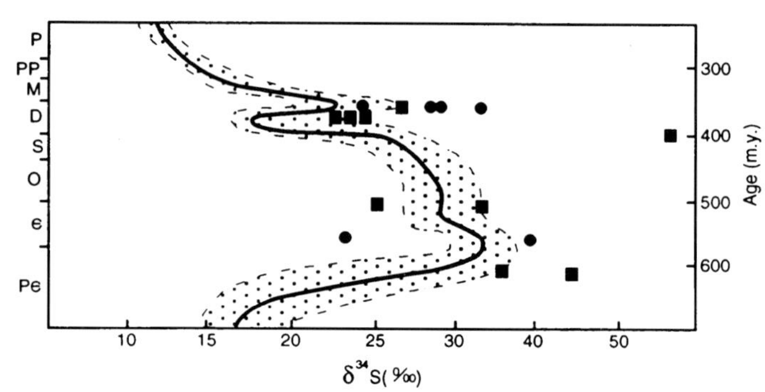

FIG. 3.—Mean values of $\delta^{34}S$ (‰) of Late Proterozoic through Paleozoic bedded barite from Table 2 plotted against the $\delta^{34}S$ curve of Claypool et al. (1980). Circles refer to barite-only deposits; squares refer to Pb-Zn-Ba deposits.

Along the Yukon-NWT boundary, this rift is known as the Selwyn Basin and to the southeast in British Columbia it is known as the Kechika Trough. The wide range of bedded barite and sulfide deposit ages in western Canada has allowed determination of secular changes of both sulfide and sulfate $\delta^{34}S$ values during the Middle Paleozoic. This work suggests that the Selwyn basin was anoxic, stratified, and hydrographically isolated from the open ocean during much of its existence (Cecile et al., 1983; Goodfellow and Jonasson, 1984).

Within the Selwyn Basin, three major districts of bedded barite and massive sulfide mineralization have been recognized (Carne and Cathro, 1982). The Anvil Camp on the western side of the basin hosts a number of important massive sulfide deposits at the contact of Cambrian noncalcareous metapelites of the Mount Nye Unit and calcareous metapelites of the Vangorda Unit (Jennings and Jilson, 1986). Barite mineralization appears to be minor, although a number of barite $\delta^{34}S$ analyses have been reported (Shanks et al., 1987) (Table 2). The Howards Pass Camp deposits along the Yukon-NWT boundary are hosted by Upper Ordovician-Lower Silurian carbonaceous shales and siltstones of the Road River Group and also have relatively small amounts of bedded barite (Carne and Cathro, 1982; Goodfellow et al., 1983). Deposits of massive sulfide and bedded barite are reported in the Upper Silurian to Lower Devonian portion of the Road River Formation south of the Howards Pass Camp (Mako and Shanks, 1984) (Fig. 4A). The MacMillan Pass Camp located immediately north of the Howards Pass Camp hosts several economically important sedex Pb-Zn-Ba deposits (Gardner and Hutcheon, 1985; Turner, 1986), as well as several bedded barite deposits with no associated sulfides (Lydon et al., 1985b). The number of "barite-only" deposits from the MacMillan Pass is greater than those of either the Anvil or Howards Pass areas. The massive sulfide and bedded barite deposits are found in Upper Devonian carbonaceous shales of the Lower Earn Group.

A

B

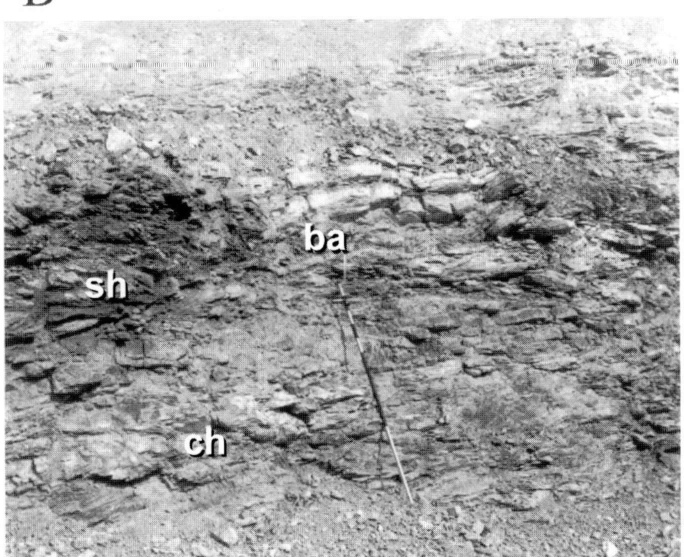

FIG. 4.—Field examples of Paleozoic bedded barite. (A) Laminated barite-chert from the base of the upper Earn Group (Late Devonian to Mississippian) near the Vulcan deposit, Northwest Territories, Canada (from Mako and Shanks, 1984). Note pencil for scale. (B) Sequence of deformed barite, chert, and organic-rich shales of the Slaven Chert (Late Devonian) in the Argenta mine of north central Nevada (from Jewell and Stallard, 1991). Jacob staff is 1.5-m long.

Bedded barite and shale hosted Pb-Zn-Ba deposits in the Kechika Trough of northeastern British Columbia are more spatially and temporally dispersed than those in the Selwyn basin. Deposits are found in host rocks that range in age from Ordovician through Devonian. Bedded barite deposits with no associated massive sulfides include the Ordovician Sika and the Devonian Kwadacha deposits (MacIntyre, 1982).

Research in the Selwyn Basin and Kechika Trough has emphasized an exhalative origin for the bedded barites (e.g., Large, 1983; Lydon et al., 1985a). In addition to association with massive sulfide mineralization, evidence of an exhalative origin of the bedded barite includes hydrothermal breccias, contemporaneous volcanism, and $^{87}Sr/^{86}Sr$ values, which indicate that the barite was derived from continental crust (Table 2). It is noteworthy that some of the bedded barites (particularly from the Devonian section) with no associated massive sulfide mineralization resemble other barite-only provinces of the world (China, Nevada, Arkansas) in terms of morphology and association with cherts and organic-rich shales (e.g., Mako and Shanks, 1984; Lydon et al., 1985a). It should be further noted that $^{87}Sr/^{86}Sr$ data have been reported for only one deposit (the Devonian Jason Pb-Zn deposit) (Table 2).

Alaska

Alaska has a number of bedded barite occurrences (Clark and Poole, 1989; Schmidt, 1997). Interestingly, shale-hosted stratiform deposits in Alaska appear to be either massive sulfide deposits with no barite or deposits composed only of barite (Schmidt, 1997). An exception is the Red Dog deposit in the western Brooks Range, a world class (77 million metric ton) Pb-Zn-Ba deposit with significant amounts of bedded barite hosted by Lower Mississippian to Permian black shales and chert. Genetic relationships of the barite to sulfide mineralization at Red Dog are similar to sedex deposits of western Canada and Germany (Moore et al., 1986).

Deposits composed only of bedded barite are hosted by Devonian through Triassic black shales, cherts, and minor amounts of limestones and fine-grained siliciclastic rocks. Most of these deposits are found in the western Brooks Range or in southwestern Alaska. Exact genetic relationships between the barite and various rocks types are poorly known due to lack of exposure or detailed geologic investigations. Reported $\delta^{34}S$ values of the barite range from 21‰ to 28‰ (Schmidt, 1997).

Nevada

Central Nevada contains some of the largest bedded barite deposits in the world. Barite is found in more than 100 individual mines and prospects that have reserves of approximately 90 million metric tons. An overview of the Nevada barite province can be found in Mitchell (1980) and Papke (1984).

Most of the central Nevada bedded barite is found in Ordovician through Late Devonian cherts and shales of the Roberts Mountain allochthon. These deep-water facies rocks were thrust over shallow-water facies rocks during the Late Devonian through Early Mississippian Antler Orogeny. Tectonic overprinting during the Mesozoic and Cenozoic has greatly complicated the host rocks of these barite deposits. The most common bedded barite bearing strata are the Ordovician Vinini Formation and the Late Devonian Slaven Chert (Papke, 1984). The general lithologic similarities and extreme structural complications of these two units make them difficult to distinguish in the field. Both are dominantly chert and shale, although the Vinini has higher amounts of coarse-grained siliciclastics (Stewart, 1980). Barite deposits in both formations are associated with cherts and organic-rich shales with lesser amounts of limestone and igneous rocks (Dubé, 1988; Graber and Chafetz, 1990; Jewell and Stallard, 1991) (Fig. 4B). Interestingly, all deposits that have been studied in detail and produced reliable paleontological data are Middle or Late Devonian in age. Sulfur isotopic data (Mitchell, 1980; Rye et al., 1978) show that massive bedded barite has a seawater sulfate source, while concretionary and rosette barite have relatively heavy sulfur ratios that are characteristic of a closed sulfur system. Analyses (Rye et al., 1978; Maynard et al., 1995) show that $^{87}Sr/^{86}Sr$ ratios of massive barites likewise are close to or slightly heavier than those of contemporaneous seawater (Table 2).

The relatively well-studied nature of the Nevada bedded barite has not yielded consensus about its origin. Structural complexity, poor stratigraphic control, and paucity of dateable fossils contribute to this situation. Ketner (1963) advocated diagenetic replacement of limestone on the basis of rare brachiopods found within some of the barite beds. Most subsequent workers have argued for a synsedimentary origin, with deposition by exhalative processes being a popular mechanism (e.g., Poole, 1988; Dubé, 1988). Shawe et al. (1969) and Jewell and Stallard (1991) argue for deposition by biological processes, citing a lack of significant associated volcanism and massive sulfides, analogs of barite deposition in the modern ocean, and major and minor geochemical systematics. Jewell (1994) details a model for the formation of a sulfate-reducing coastal upwelling system as a depositional mechanism for the Nevada bedded barite.

Arkansas

The Arkansas bedded barite province is hosted by the Late Mississippian Stanley Shale and has been an important producer of barite for much of this century. The deposits are found in an east-west-trending syncline in the central part of the state; the principle zones of bedded barite occur at Chamberlin Creek in the east and Fancy Hill to the west. The Stanley Shale consists of organic-rich black shales interbedded with turbiditic siltstones and sandstones that were deposited on the southern edge of a continental margin that received sediment from the Appalachian Mountains to the north (Morris, 1974; Maynard and Okita, 1991). Devonian novaculite (metamorphosed chert), shales, and volcanic tuffs underlie the barite deposits. The barite occurs in a variety of morphologies, including rhythmic bands of shale, barite, and carbonate (Zimmerman and Amstutz, 1989).

The Arkansas bedded barites differ from the Nevada barites in having dominantly shale and siltstone host rocks, larger concentrations of carbonate, and trace amounts of anhydrite. A detailed trace element study of the Fancy Hill bedded barite deposit has also shown that the minor sulfide present was diagenetic with little input from hydrothermal sources (Howard and Hanor, 1987). $^{87}Sr/^{86}Sr$ analyses of the Arkansas barite are very close to those of contemporaneous seawater (Maynard et al., 1995) (Table 2).

Other Occurrences in North and South American

Lesser known Paleozoic bedded barite deposits have been noted in Sonora, Mexico (Poole, 1988; Clark and Poole, 1989), and northeastern Washington state (Mills et al., 1971). Neither area has massive sulfide deposits that can be directly related to the bedded barite. A small number of Paleozoic nodular and lensoid deposits have been described in west-central Argentina in a gray and black shale sequence of Ordovician age (Brodtkorb et al., 1982).

Western Europe

Well-documented bedded barite is associated with sedimentary rock-hosted Pb-Zn deposits in Germany (Meggen and Rammelsberg), Belgium (Chaudfontaine), and Ireland (Silvermines and a number of other deposits). Lesser known barite deposits occur throughout Europe. The German deposits have many of the same characteristics as the sedex deposits in western Canada, that is, a central core of massive stratabound Pb-Zn mineralization surrounded by an apron of bedded barite (Krebs, 1981; Hannak, 1981). Reported tonnages of barite at Meggen are much higher than at Rammelsberg (10 million tons versus 0.2 million tons) (Fuchs, 1989). Deposition of the

Meggen and Rammelsberg ore bodies also appears to have occurred in sediment-starved rift basins, although these basins were not nearly as large as the Selwyn and Kechika Basins in Canada. $\delta^{34}S$ analysis of these barite deposits indicate a seawater source for the sulfur and $^{87}Sr/^{86}Sr$ analyses suggest a continental source for the barium (Table 2).

The Chaudfontaine deposit is a relatively recent discovery in the Late Devonian shales and carbonates of western Belgium (Dejonghe, 1979). The deposit has a high percentage of bedded barite relative to Pb-Zn massive sulfides and the barite has a wide variety of crystal morphologies (Dejonghe, 1990) that appear to have grown in a semi-restricted evaporite basin (Dejonghe and Boulvain, 1993). The source of the sulfur was seawater, while other metals including barium have a continental source (Dejonghe et al., 1989) (Table 2).

Pb-Zn-Ba deposits are also associated with Carboniferous carbonate rocks in a variety of localities in Ireland (Morrisey et al., 1971; Coomer and Robinson, 1976; Taylor and Andrew, 1978). Stratiform bedded barite deposits have been described in Sardinia from Middle Cambrian dolomites (Barbieri et al., 1984b; Padalino et al., 1989) and in association with stratiform Pb-Zn mineralization in Devonian black shales, siltstones, and limestones in the Pyrenees of France (Lhegu et al., 1988). These deposits have relatively low tonnage and have not attracted extensive scientific scrutiny.

MESOZOIC TO MODERN BEDDED BARITE

Noteworthy bedded barite occurrences are relatively rare in Mesozoic and younger rocks. None have the large size and association with black shale sequences typical of Paleozoic bedded barites (Table 1). As discussed below, this may be attributed to the possibility that sulfate reducing ocean conditions were present during the Paleozoic that were not present during the Mesozoic.

A number of massive barite occurrences have been documented in deep sea drilling sites and on the seafloor of the modern ocean (Dean and Schrieber, 1977). Some of these deposits have been used as analogs for a variety of older bedded barite deposits (Torres et al., 1996).

Mesozoic Bedded Barite

Barite and a small amount of Pb-Zn mineralization is found in Middle to Late Triassic rocks of southern Italy (Barbieri et al., 1984a). $^{87}Sr/^{86}Sr$ analyses of this deposit suggest that the barite precipitated from seawater (Table 2). Early Jurassic dolomites host small bedded barite deposits in France whose paragenesis appears related to evaporite basins (Fuchs, 1989). A number of similar Early Jurassic barite deposits are found in evaporite sequences in central Argentina (Brodtkorb et al., 1989). Pb-Zn-Ba deposits from Jurassic sedimentary rocks in Cuba have been described by Maynard et al. (1995). $^{87}Sr/^{86}Sr$ data from these deposits suggest a continental source (Table 2). An early Cretaceous bedded barite deposit from the Atacama region of Chile occurs as mantos in a sequence of limestones and calcarenites that were extensively altered during subaerial diagenesis (Diaz, 1990). The geology and isotopic geochemistry of similar manto-type barite and celestite ($SrSO_4$) deposits are described in Middle Cretaceous limestones of northeastern Mexico (Kesler and Jones, 1981). Interestingly, this barite has $^{87}Sr/^{86}Sr$ that is distinctly more radiogenic than the celestite. $\delta^{34}S$ of the celestite is also lighter than that of the barite. Bedded barite from a Cretaceous chalk-chert sequence in the Judean Desert of Israel has an association with biologically produced sediments (cherts, phosphorites) that is similar to that of Paleozoic deposits (Bogoch and Shirav, 1978).

Bedded Barite on the Seafloor and from Deep-Sea Drilling Sites

Bedded barite associated with seafloor spreading on the modern ocean floor has been documented in a number of localities (Church, 1979; Lonsdale, 1979; Koski et. al., 1985, 1988; Peter and Scott, 1988). Other localities such as the California borderlands (Cortecci and Longinelli, 1972) and the Peru convergent margin (Aquilina et al., 1997) are not associated with seafloor spreading. The latter setting appears to be an example of fluid circulation through an accretionary prism.

Deep-sea drilling over the past several decades has led to the discovery of a number of bedded barite occurrences (Dean and Schrieber, 1977). Unfortunately, it is generally impossible to determine the morphology and areal extent of barite beds intersected by these cores. Nevertheless, analysis of the barite and associated pore fluids in deep-sea drilling sites holds the promise of providing considerable insight into how bedded barite may have formed in similar environments throughout the geologic record. A particularly interesting analysis of this sort is given by Torres et al. (1996) who model dissolution and reprecipitation of barite along diagenetic fronts where sulfate concentration of the pore fluids shows large gradients.

THEORIES OF BEDDED BARITE GENESIS

The wide variety of bedded barite and associated lithologies has led to several theories of bedded barite genesis. For the purposes of this review, these are placed into three categories: diagenetic replacement, biologic precipitation, and exhalative. All three are supported to greater or lesser degrees by field relationships, geochemical data, and comparisons to modern analogs. Of the three categories, exhalative is perhaps the most diverse. In the present context, sedimentary exhalative (sedex) is meant to include all fluids expelled from the underlying sediment column. No distinction is made between fluids heated by magma, by circulation along a deep seated fault, or simply by geothermal gradients in an overpressured basin.

Diagenetic Replacement

Replacement of evaporite sequences by barite in some Archean bedded barite occurrences is well-documented on the basis of crystal morphologies of the barite and surrounding sedimentary associations (Lambert et al., 1978; Reimer, 1980).

Diagenetic replacement of limestone has also been suggested for Paleozoic bedded barite deposits of Nevada (Ketner, 1963), although the evidence is far less convincing because of the stratabound nature of the barite as well as the fact that limestone and barite are seldom found in close spatial proximity (Papke, 1984; Poole, 1988). Most subsequent researchers have adopted a synsedimentary mechanism for the Nevada as well as virtually all other Proterozoic through Paleozoic bedded barite deposits. In spite of the lack of evidence for the formation of bedded barite deposits by wholesale diagenetic replacement, there is considerable evidence that diagenesis is important on local scales (Papke, 1984). In fact, diagenetic reorganization of bedded barite would be a logical consequence of any early sediment diagenesis that included sulfate reduction. Sulfate reduction has been documented in organic-rich marine sediments associated with diagenetic barite deposits in the modern ocean (Torres et al., 1996). Diagenesis is also considered the primary genetic mechanism for bedded barites in the carbonate-hosted Mesozoic mantos deposits of Mexico (Kesler et al., 1981).

Exhalative

A sedimentary exhalative origin of bedded barite has been applied to virtually all deposits that have temporal and spatial associations with sediment-hosted massive sulfide deposits as well as to many bedded barites without these associations. This model evolved following the discovery from the 1960s through the 1980s of hydrothermal vents at seafloor spreading centers and continental margins. These discoveries led to a consensus that many ancient volcanic-hosted massive sulfide deposits were syngenetic rather than hydrothermal replacement deposits (Franklin et al., 1981). This exhalative model subsequently became a popular although by no means universally accepted model for sedimentary rock-hosted massive sulfide and bedded barite deposits throughout the geologic record (Gustafson and Williams, 1981). The most widely recognized modern analog for sedex deposits is the Guyamas Basin in the Gulf of California, where a large number of submarine hot springs and massive sulfide deposits have been documented in sedimentary basins near seafloor spreading zones (Koski et al., 1985; Peter and Scott, 1988). However, in many ancient sedex deposits a clear link with volcanism is lacking (Carne and Cathro, 1982; Large, 1983; Lydon et al., 1985a). For these deposits, the source of hydrothermal fluids is believed to be circulation along the normal faults of continental rift basins (Fig. 5). In this sense, no modern analog has been documented for sedex deposits, such as those of western Canada and Europe.

A variation of the sedex model that does have documented modern analogs involves excess sediment pore pressure rather than hydrothermal heating driving the exhalative processes. These seafloor discharges are characterized by relatively modest temperatures and tend to form in basins with relatively high sedimentation rates (for example, the Gulf of Mexico) or in those that are undergoing subduction (for example, the Peru margin). Fluids that precipitate barite along the Peru margin are reduced, contain high concentrations of dissolved barium, and have elevated $^{87}Sr/^{86}Sr$ values, which indicate derivation from a continental source (Aquilina et al., 1997). This form of the sedex deposition has been advocated for barite-only deposits in Arkansas (Howard and Hanor, 1987), although $^{87}Sr/^{86}Sr$ signatures would seem to indicate that the barium was derived from contemporaneous seawater rather than continental material (Maynard et al., 1995; Table 2).

Biological Precipitation

A biological origin for marine barite has been advocated since oceanographic surveys documented the association between marine biogoechemistry, elevated concentrations of barium in modern ocean sediments, and the occurrence of discrete barite particles in biogenic matter of the water column (Goldberg and Arhennius, 1958; Chow and Goldberg, 1960; Church, 1979; Dehairs et al., 1980). Barite formation has been documented in the microenvironments of organic particles falling through the water column (Bishop, 1988), which thereby accounts for the correlation between surface productivity and barite accumulation in the underlying sediments (Dymond et al., 1992). Seawater is the source of the barium as well as the $^{87}Sr/^{86}Sr$ signature of these particles (Martin et al., 1995). Barite accumulations of several weight percent are observed below zones of high biological productivity and low siliciclastic sedimentation rates (e.g., Dymond, 1981).

Biologically produced barite particles that fall into sulfate reducing waters of restricted basins of the modern ocean dissolve and enrich those waters in dissolved barium (Falkner et al., 1993). If sulfate-reduction were present in the open waters of ancient oceans, then the barite would dissolve and be reprecipitated as bedded barite in the ocean's oxic-anoxic transition zone (Fig. 6). Formation of massive bedded barite would require high surface productivity, significant water mass flux (that is, strong currents), and open ocean sulfate reduction to exist in the same locality (Jewell, 1994). In modern coastal upwelling zones such as the coast of Peru, these first two criteria are met while the third (sulfate reduction) has only been observed under rare circumstances (Dugdale et al., 1977). As discussed in more detail below,

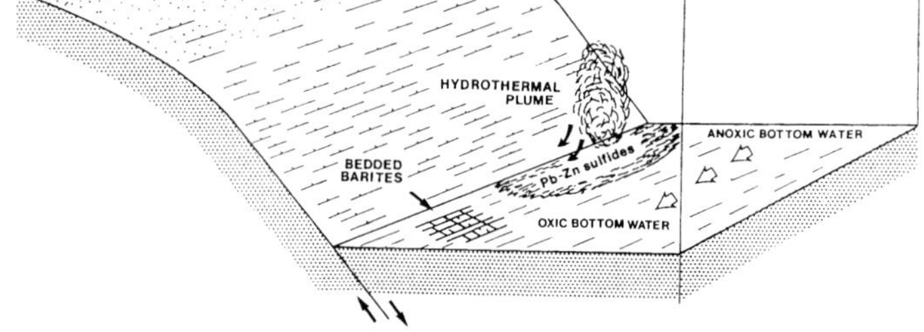

FIG. 5.—Diagram of sedimentary exhalative model of bedded barite deposition. Massive sulfide mineralization occurs near seafloor hydrothermal vents that are often associated with a major fault. Restricted circulation in the submarine basin promotes anoxic water, allowing dissolved barium to migrate to the site of bedded barite deposition at the O_2-H_2S transition zone.

open ocean sulfate reduction was probably more common in Paleozoic oceans (Berry and Wilder, 1978) and, in fact, could have been a natural consequence of the relatively large size of Middle to Early Paleozoic oceans (Jewell, 1995). Under these circumstances, bedded barite sedimentation rates could have been comparable to those of bedded chert (Jewell, 1994).

Associations with biologically produced sediments such as cherts and phosphorites and parallels with biogenic barite in the modern ocean were noted for the Northumberland Canyon bedded barite deposits of central Nevada (Shawe et al., 1969). A biological origin was later advocated for all of the Nevada deposits (Jewell and Stallard, 1991). The biological precipitation mechanism is appealing for barite-only provinces such as Nevada, China, and Arkansas because it explains the lack of massive sulfide deposits in the host rocks as well as why these three barite-only provinces have a dominantly seawater $^{87}Sr/^{86}Sr$ signature (Table 2).

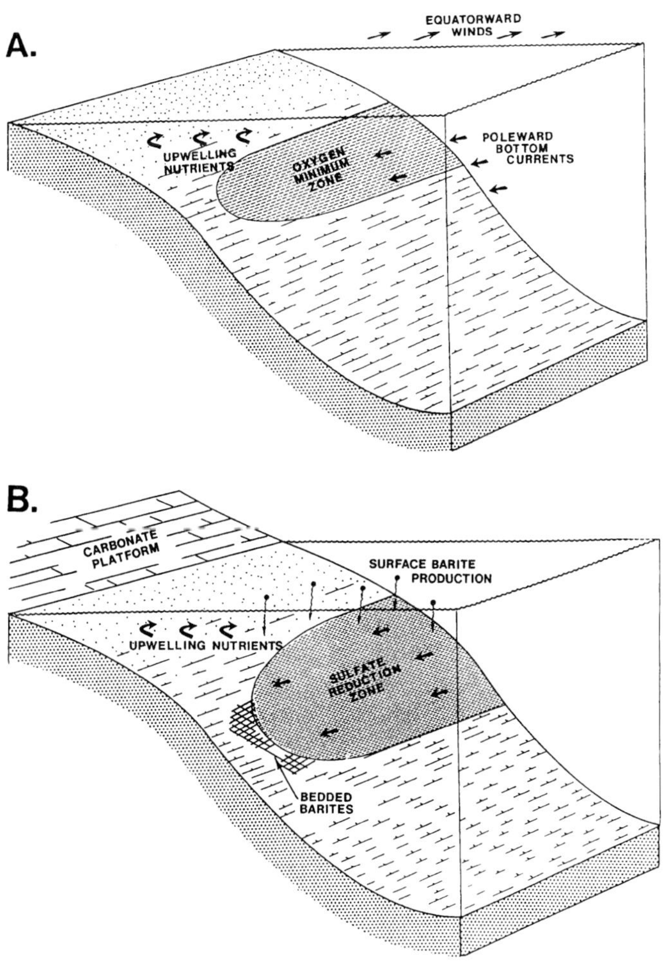

FIG. 6.—Diagrammatic representation of modern and ancient upwelling zones and their relationship to bedded barite formation. (A) Modern upwelling in which equatorward winds induce upwelling of nutrient-rich water. (B) Hypothesized Late Devonian upwelling zone. Disareobic or anoxic deep ocean causes water to be sulfidic below the photic zone. Biogenically produced barite particles settle into the H_2S-rich water. Barite is deposited at the O_2-H_2S boundary due to the interaction of barium-rich water with sulfate-rich water (from Jewell and Stallard, 1991).

DISCUSSION

Despite being a minor element in most crustal rocks, beds of essentially pure barite with thicknesses of tens of meters and strike lengths of several kilometers are found in selected sedimentary sequences throughout the world. Pure bedded barite represents an upgrade of ~1000 times over concentrations of elemental barium in average continental and marine sediments. What peculiar circumstances mobilized and transported large quantities of dissolved barium over long distances before it could react with sulfate to deposit beds of pure barite up to tens of meters thick at various times in the geologic record in a manner not observed in the modern ocean?

Sulfate reduction of seawater and concurrent dissolution of barium-bearing mineral phases is the most logical way of transporting elevated concentrations of barium to a suitable site of deposition. In seafloor hydrothermal and diagenetic systems, seawater is reduced and barium is leached during circulation of seawater through crustal rocks. H_2S-bearing fluids with elevated concentrations of dissolved barium discharge into oxygenated water to form barite-rich deposits near seafloor vents in the modern ocean (e.g., Koski et al., 1985; Peter and Scott, 1988; Aquilina et al., 1997). If similar discharges occurred in sulfate-reducing waters of ancient oceans, then the dissolved barium could migrate long distances from the exhalative source and produce the barite "apron," which is a common feature of sedex deposits (e.g., Large, 1983; Lydon et al., 1985a) (Fig. 5). In a similar fashion, if barite-bearing particles produced by biogenic processes in the upper portions of the water column entered sulfate-reducing water, they would dissolve and be reprecipitated at the edges of the sulfate-reducing zone (Fig. 6). Three critical questions thus emerge. (1) What conditions would lead to areally extensive sulfate reduction in ancient oceans given modern oceanographic conditions in which sulfate reduction is rare? (2) Why are massive bedded barite deposits most common in Early and Middle Paleozoic rocks? (3) How can a distinction be made between the sedimentary exhalative and biogenic models of barite formation?

In modern marine settings, sulfate reduction occurs when oxidation of organic matter exhausts the supply of oxygen and nitrate in the water column. This can occur as a result of physical limitations on the exchange of oxygenated water with the open ocean, high fluxes of organic matter, or some combination of these two factors. The most common sulfate reducing environments in the modern ocean involve settings where the exchange of water with the open ocean is restricted (for example, silled basins such as the Black Sea, Baltic Sea, and fjords). The sulfate-reducing waters of these basins have relatively high dissolved barium concentrations (Falkner et al., 1993). The Cariaco Trench is a fault-bounded graben off the coast of Venezuela that has restricted circulation and anoxic bottom waters (Richards, 1975). This setting may be a good modern analog for the ancient Selwyn, Meggen, and Rammelsberg basins, where active extensional tectonics might have produced seafloor hydrothermal circulation along normal faults as well as restricted the exchange of water with the open ocean.

Extensional tectonic basins and the subsequent restriction of water column circulation on a localized scale is a less appealing

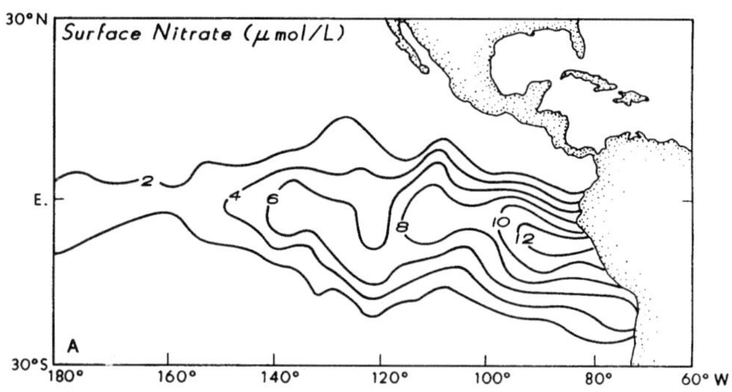

FIG. 7.—Surface nitrate concentrations (μmols/L) in the equatorial Pacific ocean (from Toggwieler et al., 1991)

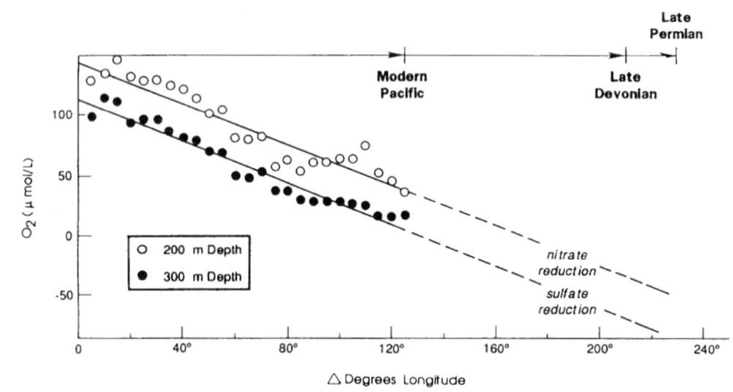

FIG. 8.—Oxygen concentrations as a function of ocean basin width (in changes in degrees of longitude in eastern direction for Pacific Ocean (data from Levitus, 1982). Dashed lines represent linear extrapolations of modern data to ocean widths which are believed to have existed in the Late Paleozoic. Negative oxygen concentrations refer to nitrate and sulfate reduction (from Jewell, 1995).

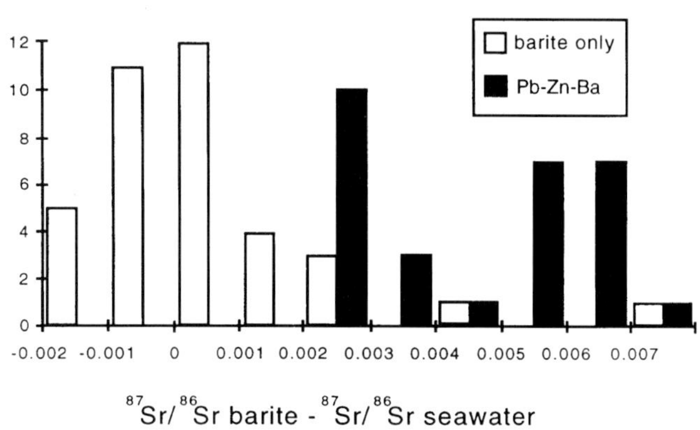

FIG. 9.—Histogram of measured $^{87}Sr/^{86}Sr$ of bedded barite relative to contemporaneous seawater $^{87}Sr/^{86}Sr$ values (modified from Maynard et al., 1995). Contemporaneous seawater values are from Burke et al. (1982). Barite-only data are from China, Nevada, and Arkansas and Pb-Zn-Ba values are from Scotland, Germany, Belgium, and western Canada (Table 2).

mechanism for the genesis of barite-only deposits such as those of Nevada and Arkansas. Although evidence of extensional tectonics has been found in some of these bedded barite deposits (e.g., Dubé, 1988), rift basins capable of restricting seawater circulation have not been systematically documented in the same fashion as the important sedex provinces of western Canada and Europe. Various arguments for widespread sulfate reduction in the open ocean during the Early and Middle Paleozoic have been put forward, however. Widespread evidence of anoxia on continental shelves suggests that the global ocean was oxygen deficient during the early Paleozoic (Berry and Wilde, 1978). Jewell (1995) outlined a mechanism whereby sulfate reduction was a commonplace feature of the equatorial regions of the very large ocean that existed concurrently with the large, closely spaced continents of the Middle Paleozoic through the Middle Mesozoic. Elevated concentrations of surface nutrients along the equatorial axis of the modern Pacific Ocean have long been noted (Fig. 7). This is a result of equatorial upwelling, which causes high surface productivity, in conjunction with the very strong, eastward-flowing equatorial undercurrent. Sinking organic matter is remineralized in the undercurrent and then recirculated to the surface in the equatorial upwelling system. The result is a "nutrient-trap" in which nutrients become enriched in an eastward direction (Fig. 7). Equatorial water at intermediate depths also shows pronounced oxygen depletion in an eastward direction, for example, near the South American coast, where it is close to anoxic (Fig. 8). Sulfate-reducing water has been observed during rare occasions in this area (e.g., Dugdale et al., 1977). The much larger size of the ocean during the Middle and Late Paleozoic would have enhanced this nutrient-trapping effect, thereby leading to widespread sulfate-reducing conditions in the equatorial ocean during this period (Fig. 8). These equatorial waters, in turn, would feed coastal upwelling at low latitudes along the west-facing portions of continents, further enhancing the possibility of sulfate reduction and the formation of bedded barite deposits (Fig. 6).

The biogenic model for bedded barite formation requires a seawater source of barium and strontium within the barite. This is supported by much of the data from China, Nevada, and Arkansas, which tend to have $^{87}Sr/^{86}Sr$ values close to that of contemporaneous seawater (Fig. 9, Table 2). It is also significant that paleogeographic reconstructions of these three localities plot very close to the paleoequator (Fig. 10), where the nutrient-trapping (and barium-enrichment) effect would have been greatest (Fig. 7).

In summary, understanding the genesis of bedded barite throughout the geologic record continues to be a useful and interesting avenue of research. Clear association of bedded barites with sedimentary exhalative deposits holds out the promise of finding new base metal deposits in the geographic vicinity. Likewise, a biologic origin of bedded barite allows the possibility of reconstructing productivity and redox conditions in ancient oceans. Several additional research possibilities are also possible to realize these potentials, including more detailed comparison of areas that contain both Pb-Zn-Ba and barite-only deposits (for example, Alaska and the Selwyn Basin) to understand specific differences between the two. In this regard, $^{87}Sr/^{86}Sr$ studies would be particularly useful in understanding sources of the barium (Fig. 9). More detailed descriptions of the morphology and form of barite occurrences on the modern ocean for which the genesis is clearly understood would also be useful for making meaningful comparisons with ancient deposits.

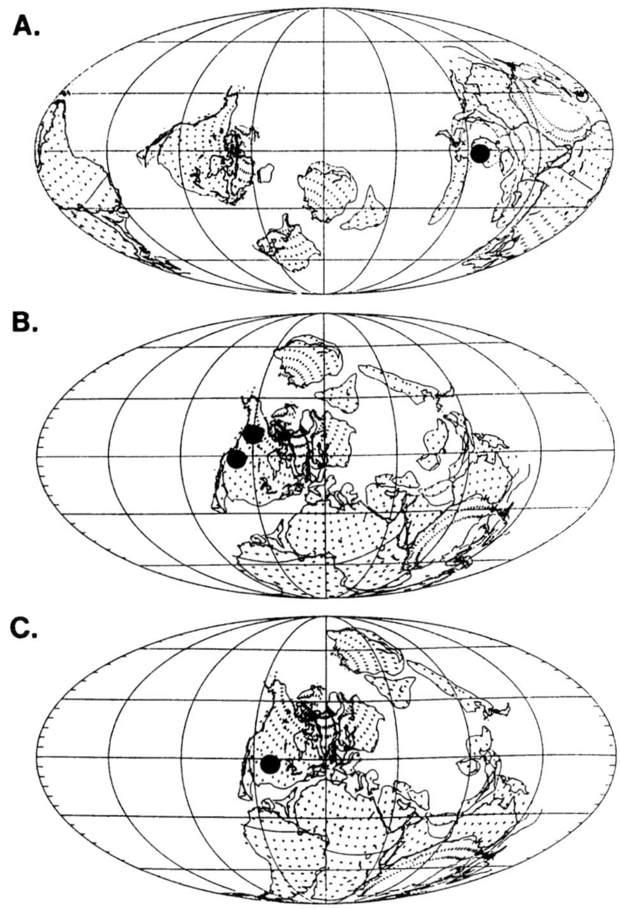

FIG. 10.—Paleogeographic distribution of important bedded barite provinces. (A) Central China (Early Cambrian). (B) Nevada and Selwyn Basin (Late Devonian). (C) Arkansas (Late Mississippian) (paleogeographic configurations from Scotese and McKerrow, 1990).

ACKNOWLEDGMENTS

The paper was greatly improved by the reviews of Karen Graber and Roy Krouse. Robert Stallard was helpful in formulating my ideas about bedded barite formation. Discussion of bedded barite with Karen Graber over the years has been much appreciated. Acknowledgment is made to the Donors of the Petroleum Research Fund, which is administered by the American Chemical Society, for partial support of this research.

REFERENCES

ABBOTT, J. G., GORDEY, S. P., AND TEMPELMAN-KLUIT, D. J., 1986, Setting of stratiform, sediment hosted lead-zinc deposits in Yukon and northeastern British Columbia, *in* Morin, J., ed., Mineral Deposits of the Northern Cordillera: Canadian Institute of Mining and Metallurgical Engineering Special Volume 37, p. 1-18.

ANGER, G., NIELSEN, H., PUCHLET, H., AND RICKE, W., 1966, Sulfur isotopes in the Rammelsberg ore deposit (Germany): Economic Geology, v. 61, p. 511-536.

AQUILINA, L., DIA, A. N., BOULEGUE, J., BOURGOIS, J., AND FOUILLAC, A. M., 1997, Massive barite deposits in the convergent margin of Peru: Implications for fluid circulation within subduction zones: Geochimica et Cosmochimica Acta, v. 61, p. 1233-1245.

BARBIERI, M., AND MASI, U., 1983, Strontium geochemistry of barite from two lower Paleozoic deposits from Tennessee and Nevada (U.S.A): Minerologica Petrographica Acta, v. 27, p. 251-254.

BARBIERI, M., DE VIVO, B., PERRONE, V., AND TURCO, E., 1984a, Strontium geochemistry of the San Donato barite mineralization (Calabria, Italy): Chemical Geology, v. 45, p. 279-288.

BARBIERI, M., MASI, U., AND TOLOMEO, L., 1984b, Strontium geochemical evidence for the origin of the barite deposits from Sardinia (Italy): Economic Geology, v. 79, p. 1360-1365.

BARLEY, M. E., DUNLOP, J. S. R., GLOVER, J. E., AND GROVES, D. I., 1979, Sedimentary evidence for an Archean shallow water volcanic sedimentary facies, eastern Pilbara Block, Western Australia: Earth and Planetary Science Letters, v. 43, p. 74-84.

BERRY, W. B. N., AND WILDE, P., 1978, Progressive ventilation of the oceans — an explanation for the distribution of the lower Paleozoic black shales: American Journal of Science, v. 278, p. 257-275.

BISHOP, J. K. B., 1988, The barite-opal-organic carbon association in oceanic particulate matter: Nature, v. 332, p. 341-343.

BOGOCH, R., AND SHIRAV, M., 1978, Petrogenesis of a Senonian barite deposit, Judean Desert, Israel: Mineralium Deposita, v. 13, p. 383-389.

BRODTKORB, M. K., 1989, Nonmetalliferous Stratabound Ore Fields: New York, Van Nostrand Reinhold, 332 p.

BRODTKORB, M. K., RAMOS, V., BARBIERI, M., AND AMETRANO, S., 1982, The evaporitic celestite-barite deposits of Neuquen, Argentina: Mineralium Deposita, v. 17, p. 423-436.

BRODTKORB, M. K., SCHALAMUK, I. B. A., AND AMETRANO, S., 1989, Barite and celestite stratabound ore fields in Argentina, *in* Brodtkorb, M. K., ed., Nonmetalliferous Stratabound Ore Fields: New York, Van Nostrand Reinhold, p. 41-68.

BURKE, W. H., DENISON, R. E., HETHERINGTON, E. A., KOEPNICK, R. B., NELSON, H. F., AND OTTO, J. B., 1982, Variation of seawater $^{87}Sr/^{86}Sr$ throughout Phanerozoic time: Geology, v. 10, p. 516-519.

BUSCHENDORF, F., NIELSEN, H., PUCHLET, H., AND RICKE, W., 1963, Schwefelisotopen-untersuchungen am pyrit-sphalerit-baryt-lager Meggen/Lenne (Deutschland) und an verschieden Devon-evaporiten: Geochimica et Cosmochimica Acta, v. 27, p. 501-523.

CAMERON, E. M., 1982, Sulfate and sulfate reduction in early Precambrian ocean: Nature, v. 296, p. 145-148.

CARNE, R. C., AND CATHRO, R. J., 1982, Sedimentary exhalative (sedex) zinc-lead-silver deposits of the northern Canadian Cordillera: Canadian Institute of Mining and Metallurgy Bulletin, v. 75, p. 66-78.

CASSEDANNE, J., 1989, Brazilian stratabound barite ore fields, *in* Brodtkorb, M. K., ed., Nonmetalliferous Stratabound Ore Fields: New York, Van Nostrand Reinhold, p. 69-92.

CECILE, M. P., SHAKUR, M. A., AND KROUSE, H. R., 1983, The isotopic composition of western Canadian barites and the possible derivation of oceanic $\delta^{34}S$ and $\delta^{18}O$ curves: Canadian Journal of Earth Sciences, v. 20, p. 1528-1535.

CHOW, T. J., AND GOLDBERG, E. D., 1960, On the marine geochemistry of barium: Geochimica et Cosmochimica Acta, v. 20, p. 192-198.

CHURCH, T. M., 1979, Marine Barite, *in* Burns, R.G. ed., Marine Minerals: Reviews in Mineralogy, v. 6: Washington, D.C., Mineralogical Society of America, p. 170-210.

CLARK, S. H. B., AND POOLE, F. G., 1989, Stratabound ore fields of North America (excluding Arkansas), *in* Brodtkorb, M. K., ed., Nonmetalliferous Stratabound Ore Fields: New York, Van Nostrand Reinhold, p. 93-117.

CLAYPOOL, G. E., HOLSER, W. T., KAPLAN, I. R., SAKAI, H., AND ZAK, I., 1980, The age curves of sulfur and oxygen isotopes in marine sulfate and their mutual interpretations: Chemical Geology, v. 28, p. 199-260.

COOMER, P. G., AND ROBINSON, B. W., 1976, Sulphur and sulphate oxygen isotopes and the origin of the Silvermines deposit, Ireland: Mineralium Deposita, v. 11, p. 155-169.

CORTECCI, G., AND LONGINELLI, A., 1972, Oxygen isotope variations in a barite slab from the sea bottom off southern California: Chemical Geology, v. 9, p. 113-117.

DEAN, W. E., AND SCHRIEBER, B. C., 1977, Authigenic barite, Leg 41, deep sea drilling project, in Lancelot, Y., Seiboll, E., et al., eds., Initial Reports of the Deep Sea Drilling Project, v. 41: Washington, D.C., United States Government Printing Office, p. 915-934.

DEB, M., HOEFS, J., AND BAUMANN, A., 1991, Isotopic composition of two

Precambrian stratiform barite deposits from the Indian shield: Geochimica et Cosmochimica Acta, v. 55, p. 303-308.

DEHAIRS, F., CHESSELET, R., AND JEDWAB, J., 1980, Discrete suspended particles of barite and the barium cycle in the open ocean: Earth and Planetary Science Letters, v. 49, p. 528-550.

DEJONGHE, L., 1979, Discovery of a sedimentary Ba (Fe, Zn, Pb) ore body of Frasnian age at Chaudfontaine, Province of Liege, Belgium: Mineralium Deposita, v. 14, p. 15-20.

DEJONGHE, L., 1990, The sedimentary structures of barite: examples from the Chaudfontaine ore deposit, Belgium: Sedimentology, v. 37, p. 303-323.

DEJONGHE, L., AND BOULVAIN, F., 1993, Paleogeographic and diagenetic context of a baritic mineralization enclosed within Frasnian peri-reefal formations: Case history of the Chaudfontaine mineralization, Belgium: Ore Geology Reviews, v. 7, p. 413-431.

DEJONGHE, L., DOULEGUE, J., DEMAIFFE, D., AND LETOLLE, R., 1989, Isotope geochemistry (S, C, O, Sr, Pb) of the Chaudfontaine mineralization (Belgium): Mineralium Deposita, v. 24, p. 132-140.

DIAZ, L. L, 1990, The Mamina barite mine, Atacama region, Chile, *in* Fontbote, L., Amstutz, G. C., and Cardozo, M., eds., Stratabound Ore Deposits in the Andes: New York, Springer Verlag, p. 523-536.

DUBÉ, T. E., 1988, Tectonic significance of Upper Devonian igneous rocks and bedded barite, Roberts Mountains Allochthon, Nevada, U. S. A., *in* McMillan, N. J., Embry, A. F., and Glass, D. J., Devonian of the World, V. 2: Sedimentation (Proceedings of the 2nd International Symposium of the Devonian System): Calgary, Canadian Society of Petroleum Geology Memoir 14, p. 235-247.

DUGDALE, R. C., GEORING, J. J., BARBER, R. T., SMITH, R. L., AND PACKARD, T. T., 1977, Denitrification and hydrogen sulfide in the Peru upwelling region during 1976: Deep Sea Research, v. 24, p. 601-608.

DUNLOP, J. S. R., AND BUICK, R., 1981, Archean epiclastic sediments derived from mafic volcanics, North Pole, Pilbara block, Western Australia: Adelaide, Geological Society of Australia Special Publication 7, p. 225-233.

DYMOND, J., 1981, Geochemistry of Nazca plate surface sediments: An evaluation of hydrothermal, biogenic, detrital, and hydrogenous sources, *in* Kulin, L., Dymond, J., Dasch, and E., Hussong, D., eds., Nazca Plate: Crustal Formation and Andean Convergence: Boulder, Geological Society of America Memoir 154, p. 133-174.

DYMOND, J., SUESS, E., AND LYLE, M., 1992, Barium in deep sea sediment: A geochemical proxy for paleoproductivity: Paleoceanography, v. 7., p. 163-181.

FALKNER, K. K., KLINKHAMMER, G. P., BOWERS, T. S., TODD, J. F., LEWIS, B. L., LANDING, W. M., AND EDMOND, J. M., 1993, The behavior of barium in anoxic waters: Geochimica et Cosmochimica Acta, v. 57, p. 537-554.

FRANKLIN, J. M., LYDON, J. W., AND SANGSTER, D. F., 1981, Volcanic-hosted massive sulfide deposits, *in* Skinner, B. F., ed., Economic Geology 75th Anniversary Volume: El Paso, Texas, Economic Geology Publishing Company, p. 485-627.

FUCHS, Y. A., 1989, Stratiform barite ore fields in western Europe, *in* Brodtkorb, M. K., ed., Nonmetalliferous Stratabound Ore Fields: New York, Van Nostrand Reinhold, p. 149-164.

GARDNER, H. D., AND HUTCHEON, I., 1985, Geochemistry, mineralogy, and geology of the Jason Pb-Zn deposits, Macmillan Pass, Yukon, Canada: Economic Geology, v. 80, p. 1257-1276.

GOLDBERG, E. D., AND ARRHENIUS, G. O. S., 1958, Chemistry of Pacific pelagic sediments: Geochimica et Cosmochimica Acta, v. 13, p. 153-212.

GOODFELLOW, W. D., AND JONASSON, I. R., 1984, Ocean stagnation and ventilation defined by $\delta^{34}S$ secular trends in pyrite and barite, Selwyn Basin, Yukon: Geology, v. 12, p. 583-586.

GOODFELLOW, W. D., JONASSON, I. R., AND MORGANTI, J. M., 1983, Zonation of chalcophile elements about the Howards Pass (XY) Zn-Pb deposit, Selwyn Basin: Journal of Geochemical Exploration, v. 19, p. 503-542.

GRABER, K. K., AND CHAFETZ, H. S., 1990, Petrography and origin of bedded barite and phosphate in the Devonian Slaven Chert of central Nevada: Journal of Sedimentary Petrology, v. 60, p. 879-911.

GUSTAFSON, L. B., AND WILLIAMS, N., 1981, Sediment-hosted stratiform deposits of copper, lead, and zinc, *in* Skinner, B. F., ed., Economic Geology 75th Anniversary Volume: El Paso, Texas, Economic Geology Publishing Company, p. 139-178.

HALL, A. J., BOYCE, A. J., FALLICK, A. E., AND HAMILTON, P. J., 1991, Isotopic evidence of the depositional environment of Late Paleozoic stratiform barite mineralisation, Aberfeldy, Scotland: Chemical Geology, v. 87, p. 99-114.

HANNAK, W. W., 1981, Genesis of the Rammelsberg ore deposits near Goslar/Upper Horz, Federal Republic of Germany, *in* Wolf, K H., ed., Handbook of Strata-bound and Stratiform Ore Deposits, V. 9: Amsterdam, Elsevier, p. 551-642.

HEINRICHS, T. K., AND REIMER, T. O., 1977, A sedimentary barite deposits from the Archean Fig Tree Group of the Barberton Mountain Land (South Africa): Economic Geology, v. 72, p. 1426-1441.

HOERING, T. C., 1988, The isotopic composition of bedded barites from the Archean of southern India: Carnegie Institute of Washington Annual Report, v. 87-88, p. 122-127.

HOWARD, K. W., AND HANOR, J. S., 1987, Compositional zoning in the Fancy Hill stratiform barite deposit, Ouachita Mountains, Arkansas, and evidence for the lack of associated massive sulfides: Economic Geology, v. 82, p. 1377-1385.

HULEN, P. L, 1978, Determination of the source of barite in the Chamberlain Creek deposit of Arkansas: Sr isotopic evidence: Unpublished M.S. Thesis, University of Kansas, Lawrence, 76 p.

JENNINGS, D. S., AND JILSON, G. A., 1986, Geology and sulfide deposits of the Anvil Range, Yukon, *in* Morin, J., ed., Mineral Deposits of the Northern Cordillera: Canadian Institute of Mining and Metallurgical Engineering Special Volume 37, p. 319-361.

JEWELL, P. W., 1994, Paleoredox conditions and the origin of bedded barites along the Late Devonian North American continental margin: Journal of Geology, v. 102, p. 151-164.

JEWELL, P. W., 1995, Geologic consequences of globe-encircling equatorial currents: Geology, v. 23, p. 117-120.

JEWELL, P. W., AND STALLARD, R. F., 1991, Geochemistry and paleoceanographic setting of central Nevada bedded barites: Journal of Geology, v. 99, p. 151-170.

KESLER, S. E., AND JONES, L. M., 1981, Sulfur and strontium isotopic geochemistry of celestite, barite, and gypsum from the Mesozoic basins of northeastern Mexico: Chemical Geology, v. 31, p. 211-224.

KETNER, K. B., 1963, Bedded Barite Deposits of the Shoshone Range, Nevada: Washington, D.C., U. S. Geological Survey Professional Paper 475, p. B38-B41.

KOSKI, R. A., LONSDALE, P. F., SHANKS, W. C., BERNDT, M .E., AND HOWE, S. S., 1985, Mineralogy and geochemistry of a sediment-hosted hydrothermal sulfide deposit from the southern trough of the Guaymas Basin, Gulf of California: Journal of Geophysical Research, v. 90, p. 6695-6707.

KOSKI, R. A., SHANKS, W. C., BOHRSON, W. A., AND OSCARSON, R. L., 1988, The composition of massive sulfide deposits from the sediment-covered floor of the Escanaba Trough, Gorda Ridge: Implications for depositional processes: Canadian Mineralogist, v. 26, p. 655-673.

KREBS, W., 1981, Geology of the Meggen ore deposit, *in* Wolf, K H., ed., Handbook of Strata-bound and Stratiform ore Deposits, V. 9: Amsterdam, Elsevier, p. 551-642.

LAMBERT, I. B., DONNELLY, T. H., DUNLOP, J. S. R., AND GROVES, D. I., 1978, Stable isotope studies of early Archean sulfate deposits of probable evaporitic and volcanic origin: Nature, v. 276, p. 808-811.

LARGE, D. E., 1983, Sediment-hosted massive sulfide lead-zinc deposits: An empirical model, *in* Sangster, D. F., ed., Short Course in Sediment-hosted Stratiform Lead-Zinc Deposits: Victoria, Mineralogical Association of Canada, p. 1-30.

LEVITUS, S., 1982, Climatological Atlas of the World Ocean: Washington, D.C., National Oceanic and Atmospheric Administration Professional Paper 13, 172 p.

LHEGU, J., JEBRAK, M., AND TOURAY, J. C., 1988, Fluorite and barite deposits of France, *in* 7th International Association on the Genesis of Ore Deposits Symposium Proceedings, p. 297-307.

LONSDALE, P., 1979, A deep-sea hydrothermal site on a strike-slip fault: Nature, v. 281, p. 531-534.

LOWE, D. R., AND KNAUTH, L. P., 1977, Sedimentology of the Onverwacht Group (3.4 billion years), Transvaal, South Africa, and its bearing the characteristics and evolution of the early Earth: Journal of Geology, v. 85, p. 699-723.

LYDON, J. W., GOODFELLOW, W. D., AND JONASSON, I. R., 1985a, A general model for stratiform baritic deposits of the Selwyn Basin, Yukon Territory and District of MacKenzie *in* Current Research, Part A: Ottawa, Geological Survey of Canada, Paper 85-1A, p. 651-660.

LYDON, J. W., JONASSON, I. R., AND HUDSON, K. A., 1985b, The distribution of gold in the Tea barite deposit, Yukon Territory, *in* Current Research, Part A:

Ottawa, Geological Survey of Canada, Paper 85-1A, p. 661-667.

MacIntrye, D. G., 1982 Geologic setting of recently discovered stratiform barite-sulfide deposits in northeast British Columbia: Canadian Institute of Mining and Metallurgy Bulletin, v. 75, p. 99-113.

Mako, D. A., and Shanks, W. C., 1984, Stratiform sulfide and barite-fluorite mineralization of the Vulcan prospect Northwest Territories: Exhalation of basinal brines along a faulted continental margin: Canadian Journal of Earth Sciences, v. 21, p. 78-91.

Martin, E. E., Macdougall, J. D., Herbert, T. D., Paytan, A., and Kastner, M., 1995, Strontium and neodymium isotopic analyses of marine barite separates: Geochimica et Cosmochimica Acta, v. 59, p. 1353-1361.

Maynard, J. B., and Okita, P. M., 1991, Bedded barite deposits in the United States, Canada, Germany, and China: Two major types based on tectonic setting: Economic Geology, v. 86, p. 364-376.

Maynard, J. B., Morton, J.., Valdes-Nordarse, E. L., and Diaz-Carmona, A., 1995, Sr isotopes of bedded barites: guide to distinguishing basins with Pb-Zn mineralization: Economic Geology, v. 90, p. 2058-2064.

McCulloch, M. T., 1994, Primitive $^{87}Sr/^{86}Sr$ from an Archean barite and conjecture on the Earth's age and origin: Earth and Planetary Science Letters, v. 126, p. 1-13.

Mills, J. W., Carlson, C. L., Fewkes, L. W., Handlen, G. P., Jayprakash, G. P., Johns, M. A., Morganti, J. M., Neitzel, T. W., Ream, L. R., Sanford, S. S., and Todd, S. G., 1971, Bedded barite deposits of Stevens County, Washington: Economic Geology, v. 66, p. 1157-1163.

Mitchell, A. W., 1980, Barite deposits of Nevada: American Institute of Mining, Metallurgical, and Petroleum Engineering Transactions, v. 270, p. 1868-1873.

Moore, D. W., Young, L. E., Modene, J. S., and Plahuta, J. T., 1986, Geologic setting and genesis of the Red Dog zinc-lead-silver deposit, western Brooks Range, Alaska: Economic Geology, v. 81, p. 1696-1727.

Morrisey, C. J., Davis, G. R., and Steed, G. M., 1971, Mineralization in the Lower Carboniferous of central Ireland: Transactions of the Institute of Mining and Metallurgy, Section B, v. 80, p. B174-185.

Morris, R. C., 1974, Sedimentary and tectonic history of the Ouachita Mountains, *in* Dickinson, W. R., ed., Tectonics and Sedimentation: Tulsa, Society for Sedimentary Geology (SEPM) Special Publication 22, p. 120-142.

Padalino, G., Tocco, S., and Viola, M., 1989, Stratabound and karstic deposits of barite in Sardinia, *in* Brodtkorb, M. K., ed., Nonmetalliferous Stratabound Ore Fields: New York, Van Nostrand Reinhold, p. 165-200.

Papke, K. G., 1984, Barite in Nevada: Reno, Nevada Bureau of Mines and Geology Bulletin 98, 125 p.

Paytan, A., Kastner, M., Martin, E. E., MacDougall, J. D., and Herbert, T. D., 1993, Barite: a new monitor of the oceanic Sr isotopic ratio: Nature, v. 366, p. 445-449.

Perry, E. C., Jr., Monster, J., and Reimer, T., 1971, Sulfur isotopes in Swaziland System barites and the evolution of the Earth's atmosphere: Science, v. 171, p. 1015-1016.

Peter, J. M., and Scott, S. D., 1988, Mineralogy, composition, and fluid inclusion microthermometry of seafloor hydrothermal deposits in the southern trough of Guaymas Basin, Gulf of California: Canadian Mineralogist, v. 26, p. 567-587.

Poole, F. G., 1988, Stratiform barite in Paleozoic rocks of the western United States, *in* 7th International Association on the Genesis of Ore Deposits Symposium Proceedings, p. 309-319.

Radhakrishna, B. P., and Sreenivasaiya, C., 1974, Bedded barytes from the Precambrian of Karnataka: Journal of the Geological Society of India: v. 15, p. 314-317.

Reimer, T. O., 1980, Archean sedimentary baryte deposits of the Swaziland Supergroup (Barberton Mountain Land, South Africa): Precambrian Research, v. 12, p. 393-410.

Richards, F. A., 1975, The Cariaco Basin (Trench): Oceanography and Marine Biology Annual Reviews, v. 13, p. 11-67.

Rye, R. O., Shawe, R., and Poole, F. G., 1978, Stable isotopic studies of bedded barite at East Northumberland Canyon in Toquima Range, central Nevada: U. S. Geological Survey Journal of Research, v. 6, p. 221-229.

Schidlowski, M., Hayes, J. M., and Kaplan, I. R., 1983, Isotopic inferences of ancient biochemistries: Carbon sulfur, hydrogen, and nitrogen, *in* Schopf, J. W., ed., Earth's Earliest Biosphere: Its Origin and Evolution: Princeton, Princeton University Press, New Jersey, p. 149-186.

Schmidt, J. M., 1997, Shale-hosted Zn-Pb-Ag and barite deposits of Alaska in Mineral Deposits of Alaska, *in* Goldfarb, R. J., and Miller, L. D., eds., Economic Geology Monograph 9: El Paso, Texas, Economic Geology Publishing Company, p. 35-65.

Schmitz, B., 1987, Barium, equatorial high productivity, and the northward wandering of the Indian continent: Paleoceanography, v. 2, p. 63-77.

Scotese, C. R., and McKerrow, W. S., 1990, Revised world maps and introduction, *in* McKerrow, W. S., and Scotese, C. R., eds., Paleozoic Paleogeography and Biogeography: London, Geological Society of London Memoir 12, p. 1-21.

Shanks, W. C., III, Woodruff, L. G., Jilson, G. A., Jennings, D. S., Modene, J. S., and Ryan, B. D., 1987, Sulfur and lead isotope studies of stratiform Zn-Pb-Ag deposits, Anvil Range, Yukon: Basinal brine exhalation and anoxic bottom water mixing: Economic Geology, v. 82, p. 600-634.

Shawe, D. R., Poole, F. G., and Brobst, D. A., 1969, Newly discovered bedded barite in East Northumberland Canyon, Nye County, Nevada: Economic Geology, v. 64, p. 245-254.

Stewart, J. H., 1980, Geology of Nevada: [AUTHOR: PROVIDE CITY OF PUBLISHER HERE], Nevada Bureau of Mines and Geology Special Publication 4, 136 p.

Strauss, H., 1993, The sulfur isotopic record of Precambrian sulfates: new data and a critical evaluation of the existing record: Precambrian Research, v. 63, p. 225-246.

Taylor, S., and Andrew, C. J., 1978, Silvermine orebodies, County Tipperary, Ireland: Transactions of the Institute of Mining and Metallurgy, Section B, v. 87, p. B111-124.

Thorpe, R. I., 1979, A sedimentary barite deposit from the Archean Fig Tree Group of the Barberton Mountain Land (South Africa)—A discussion: Economic Geology, v. 74, p. 700-702.

Toggweiler, J. R., Dixon, K., and Broecker, W. S., 1991, The Peru upwelling system and ventilation of South Pacific thermocline: Journal of Geophysical Research, v. 96, p. 6907-6924.

Torres, M. E., Brumsack, H. J., Bohrmann, G., and Emeis, K. C., 1996, Barite fronts in continental margin sediments: a new look at barium remobilization in the zone of sulfate reduction and formation of heavy barites in diagenetic fronts: Chemical Geology, v. 127, p. 125-139.

Turekian, K. K., and Johnson, D. G., 1966, The barium distribution in sea water: Geochimica et Cosmochimica Acta, v. 30, p. 1153-1174.

Turner, R. J. W., 1986, The genesis of the stratiform lead-zinc deposit, Jason property, Macmillan Pass, Yukon: Ph.D. Thesis, Stanford University, Stanford, Calif., 205 p.

Turner, R. J. W., 1992, Formation of Phanerozoic stratiform sediment-hosted zinc-lead deposits: Evidence for the critical role of ocean anoxia: Chemical Geology, v. 99, p. 165-188.

Wang, Z., and Chu, X., 1994, Strontium isotopic composition of the Early Cambrian barite and witherite deposits: Chinese Science Bulletin, v. 39, p. 52-55.

Wang, Z., and Li, G., 1991, Barite and witherite deposits in Lower Cambrian shales of South China: Stratigraphic distribution and geochemical characterization: Economic Geology, v. 86, p. 354-363.

Willan, R. C. R., and Coleman, M. L., 1983, Sulfur isotope study of the Aberfeldy barite, zinc, lead deposit and minor sulfide mineralization in the Dalradian metamorphic terrain, Scotland: Economic Geology, v. 78, p. 1619-1656.

Zimmerman, R. A., and Amstutz, G. C., 1989, Barite deposits of Arkansas, *in* Brodtkorb, M. K., ed., Nonmetalliferous Stratabound Ore Fields: New York, Van Nostrand Reinhold, p. 119-148.

U-SERIES, ^{14}C, AND STABLE ISOTOPE STUDIES OF RECENT PHOSPHATIC "PROTOCRUSTS" FROM THE PERU MARGIN

W. C. BURNETT*[1], C. R. GLENN[2], C. C. YEH[1], M. SCHULTZ[1], J. CHANTON[1], AND M. KASHGARIAN[3]

[1]*Department of Oceanography, Florida State University, Tallahassee, FL 32306-3048*

[2]*Department of Geology & Geophysics, University of Hawaii, Honolulu, Hawaii 96822-2285*

[3]*Center for Accelerator Mass Spectrometry, Lawrence Livermore National Laboratories, 7000 East Avenue, Livermore, CA 94551-0808*

ABSTRACT: Phosphorites on the Peru shelf occur mainly as dark and dense conglomerate ("D"-phosphates) and occasionally as soft, friable ("F"-phosphates) protocrusts, which appear as laterally extensive, centimeter-thick plates of bored, phosphatized sediment. Several undisturbed and well-documented samples of these protocrusts were collected by box coring and submersible operations during an expedition of the R/V *Seaward Johnson* in 1992. Three of these crusts were analyzed in layer-by-layer fashion for U-series isotopes, while two were analyzed for AMS-^{14}C, and stable carbon isotopes. Unlike the D-phosphates, which may show a progressive decrease in their δ^{13}C values (from 0 to –5‰) with a co-linear increase in structural CO_2 contents (from 2–6 wt. %), the protocrusts show a more restricted range with most values from 0 to –2‰ δ^{13}C and 4.2–4.7 wt.% CO_2, indicative of their precipitation directly on or slightly below the seafloor.

U-series disequilibrium results (^{230}Th/^{234}U, ^{231}Pa/^{235}U) indicate that the protocrusts are very young. They are so young, in fact, that corrections for "common" ^{230}Th and ^{231}Pa (that is, those amounts not associated with decay of the incipient uranium) are very important, and different interpretations (of growth direction, for example) are possible depending upon which assumptions are applied. AMS ^{14}C results are very consistent with the U-series ages in the topmost, highest-phosphate portion of these crusts. A few millimeters into the less phosphatic portions of the crusts, uncorrected ^{230}Th and ^{231}Pa ages diverge from the ^{14}C ages, probably because of contamination of these daughter nuclides from associated sediment grains. An isochron approach confirms that there is a minimal age difference between the upper and lower layers of the protocrusts. All protocrusts display a distinct trend of decreasing phosphate content with depth into the crust. Based on these observations, we propose a growth model that suggests that phosphatic protocrusts grow upward at rates of 2–9 mm/ky in response to downward diffusion of phosphate from an interstitial phosphate pore water spike just below the sediment-seawater interface. Later exhumation of protocrusts resulted in erosion into the more common and more complicated phosphatic hardgrounds, conglomerates, and nodules found scattered throughout the sediments of the Peru shelf upwelling zone.

INTRODUCTION

Uranium-series disequilibrium dating of sea-floor phosphorites in the 1970s by Baturin et al. (1972), Veeh et al. (1973), and Burnett and Veeh (1977) showed that sediments in the Peru upwelling zone host modern phosphorites, making that area one of only a few worldwide where contemporary phosphatization has been confirmed by dating techniques. Later studies, emphasizing physical-chemical research of the principal mineral phase, authigenic carbonate fluorapatite (CFA), as well as studies of the associated sediments and pore solutions, established an early diagenetic origin for these modern phosphorites (Froelich et al., 1988; Glenn et al., 1988). CFA has now been recognized as forming either through replacement of biogenic calcium carbonate (e.g., Manheim et al., 1975; Jarvis, 1992) or via direct precipitation within pore solutions near the sediment-seawater interface (Burnett, 1977; Froelich et al., 1988; Glenn and Arthur, 1988). Time scales for generating individual phosphatic pellets, one of the main phosphatic morphologies recognized in the geologic record, were estimated at less than a few years. The accumulation of CFA pellet beds, however, may extend over much longer time scales and be limited to localized areas of low to zero sediment accumulation rates (Burnett et al., 1988).

Prior uranium-series studies of phosphatic crusts and nodules, found at or near the sediment-seawater interface on the Peru shelf, suggested that centimeter-scale crusts form over a period of a few thousand years (Burnett et al., 1982; Kim and Burnett, 1986). Burnett et al. (1982), in the first attempt to determine the growth rates of marine phosphatic nodules by uranium disequilibrium dating, reported that phosphatic nodules from the Peru seafloor grew on the order of a few millimeters per thousand years, and the direction of growth was interpreted as *downward* (towards the sediment) based on an assumed orientation for these dredge samples. Surprisingly, these growth rates are up to four orders of magnitude slower than measured accumulation rates of near-by sediment. The Peru shelf phosphate nodules, therefore, display the same type of apparent growth rate paradox as displayed by deep-sea manganese nodules; that is, one would expect that up to several meters of sediment might completely cover a nodule in the amount of time (10^3–10^4 yrs) it takes for growth. Yet, nodules are often observed to occur at the sediment surface. The resolution of this apparent inconsistency is most likely related to either: (1) biological mixing that causes upward movement of these concretions relative to the associated sediment; or (2) an effect of nodule growth in localized areas of very slow to no sediment accumulation. A later, far more detailed study of a single phosphatic crust, sampled by a box core in its apparent growth position, showed a growth rate comparable to those measured previously but with *upward* growth (Kim and Burnett, 1986). This observation was not considered inconsistent, however, as that crust (GS-1) was buried beneath about 14 cm of sediment and thus was apparently growing upward toward the sediment-water interface and into a high diffusional source of dissolved interstitial phosphate. Detailed pore water profiles of Peru shelf sediments reported by Froelich et al. (1988) showed a distinct maximum in the dissolved PO_4^{3-} concentration just below the sediment-seawater interface. Based on

email: wburnett@mailer.fsu.edu

Marine Authigenesis: From Global to Microbial, SEPM Special Publication No. 66

the measured growth rates for nodules (Burnett et al., 1982) and calculated rates for pellets (Burnett et al., 1988), it was found that authigenic uptake rates of phosphorus into these various forms of CFA generally agreed with the estimated diagenetic uptake rates based on pore-water models (Froelich et al., 1983; Froelich et al., 1988).

During the Ocean Drilling Program (ODP) Leg 112 on the Peru margin in 1986, phosphorites were sampled from six sites in forearc basins on the shelf and upper slope. These phosphorites were classified by Garrison and Kastner (1990) according to their petrography, morphology, and other characteristics into three main types: (1) "F-phosphorites," which represent **friable**, light-colored micro-nodules and **peloids** of CFA; (2) "P-phosphorites" for phosphatic peloids; and (3) "D-phosphorites" for **dense** and **dark**-colored nodules, gravels, and hardgrounds. These various types of phosphorites are thought to record a time and energy spectrum, with F-phosphorites at the lower end and D-phosphorites at the upper. The overall occurrence and petrography of F-phosphorites indicated to Garrison and Kastner (1990) that they may form at the earliest stage of phosphatization. The F-types may be particularly significant, therefore, in terms of describing the environmental conditions that initially lead to precipitation of CFA. Garrison and Kastner (1990) described the F-phosphorites sampled during the ODP expedition as friable, sand-size peloids, micronodules, and laminae of cryptocrystalline CFA, which is often present in laminated diatomaceous muds. Burnett (1980) had indicated previously that this kind of phosphorite, which he had termed "collophane mudstone," is most abundant in Quaternary muds located at the intersection of the oxygen-minimum zone with the outer shelf and upper slope off Peru and Chile.

Since F-phosphorites are thought to represent the initial phosphatization of diatomaceous muds, application of precise dating tools to oriented F-phosphatic crusts should provide growth direction and rate information, which is important for understanding the interrelationships between the F-phosphorite formation and the diagenetic process occurring within the sediment. This study was initiated in an attempt to connect the sedimentary diagenetic processes with observed variations in the fine-scale chemical composition of phosphatic crusts. To this end, three oriented F-type phosphatic "protocrusts" were subsampled and analyzed on a millimeter-scale, layer-by-layer basis for uranium-series isotopes and compositional parameters. In addition, two of these crusts were also analyzed for radiocarbon to constrain further possible age interpretations based on the U-series disequilibrium results.

SAMPLING AND EXPERIMENTAL METHODS

Samples

F-phosphatic crusts BC-33, BC-57, and BC-123 were collected with a Soutar-type box core during an expedition (1092) of the R/V *Seaward Johnson* in October 1992 (Table 1). Submersible operations using the Johnson Sea Link enabled the close observation and documentation, for the first time, of the environmental setting where modern phosphorites have been found. We call these particular samples "protocrusts" because of their extremely friable nature and observed lateral continuity on the sea floor. Our observations suggested that these crusts are very immature, that is, at a very early stage in their evolution. Crust BC-33 was about 14 mm thick, light green, significantly softer on the bottom surface and relatively more indurated (cemented) on the top. This crust is flat on the top but bulbous on the bottom, and is riddled with numerous vertically oriented worm holes (Fig. 1). A gradual increase in the hardness of this

TABLE 1.—SOME CHARACTERISTICS, LOCATIONS AND WATER DEPTHS FOR THE PHOSPHATIC PROTOCRUSTS SAMPLED DURING THE 1992 R/V *SEAWARD JOHNSON* EXPEDITION TO THE PERU MARGIN AND ANALYZED IN THIS STUDY.

Name	Sample Type	XRD Mineralogy[1]	Latitude (°S)	Longitude (°W)	Depth (m)
JSL3362-Bucket9	bulk protocrust	cfa, qz	11°59.9′	77°50.0′	466
SJ1092-BC33	phosphate crust	cfa, qz, cal	13°30.6′	76°51.1′	305
SJ1092-BC57	phosphate crust	qz, dolo, cfa, cal	13°30.7′	76°41.7′	172
SJ1092-BC123	phosphate crust	qz, cfa, cal	11°59.4′	77°49.2′	376
RC23-06-GS-1[2]	phosphate crust	cfa, qz, cal	7°36.3′	80°38.0′	480
SJ1092-BC123 0–1 cm	benthic forams	—	11°59.4′	77°49.2′	376
JSL3356 SPWC-1 0–1 cm	benthic forams	—	13°30.0′	76°52.2′	385

[1]Abbreviations: cal = calcite; cfa = carbonate fluorapatite; dolo = dolomite; and qz = quartz
[2]From Kim and Burnett (1986)

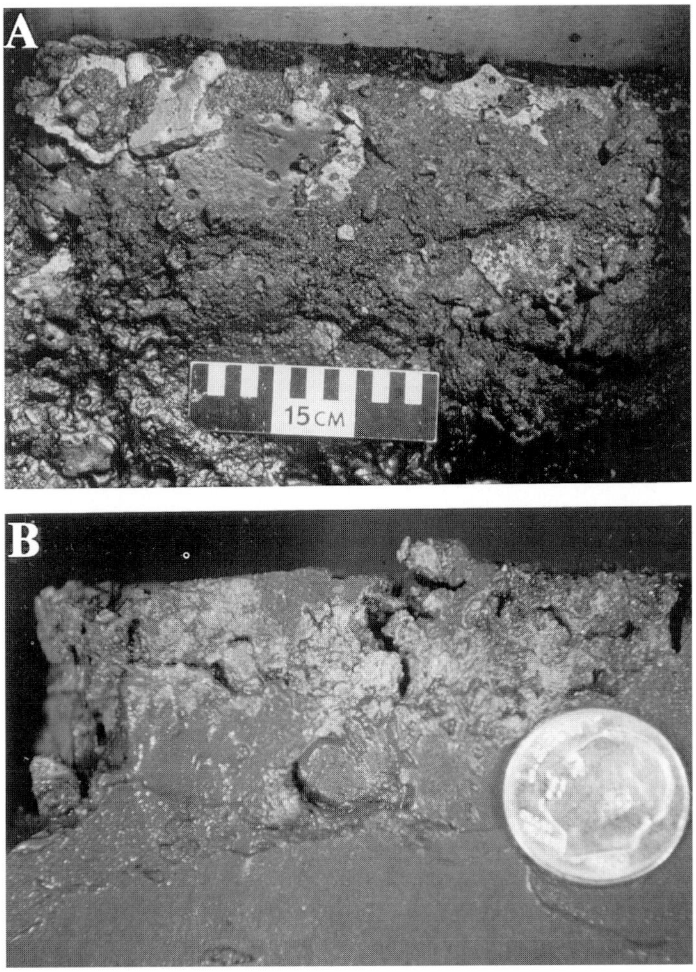

FIG. 1.—(A) Photograph looking down on light olive-brown colored friable BC-33 phosphatic protocrust, resting on gelatinous muds, as seen immediately after recovery by box coring. The crust has been bored and is rounded along its edges, presumably by bottom current erosion. (B) Side view of light olive-brown, friable and water-saturated phosphatic protocrust BC-63, extending about 2 cm into underlying green clays. Although this sample was not included in the present study, it is representative of the vertical nature of protocrust morphologies. Diameter of coin is 18 mm.

crust from bottom to top was observed during subsampling. Crust BC-57 was 25 mm in thickness and consisted of a hard rind that was brownish-black on the surface with underlying grayish-colored dolomite that was significantly softer than the surface rind. The hardness of crust BC-57 was also observed to gradually increase from the bottom towards the top of the crust. BC-123 was the thickest of the crusts reported here at about 30 mm and was also the most friable throughout, although somewhat more indurated on the top surface.

One other F-type crust investigated during this study, JSL3362 Bucket 9, was sampled directly by the submersible from an area of laterally extensive crust material. The crust was collected from a large plate of what appeared to be a continuous phosphatic pavement, 10–15 ft wide by 20 ft long. The crust is green-gray in color and very friable throughout. Since the extremely irregular shape of this sample made layer-by-layer sampling nearly impossible, only bulk analyses were attempted for this sample. Crusts BC-33, BC-57, and BC-123 were sub-sampled for isotopic analysis by scraping 1–3 mm contiguous layers from the crusts using a small knife or razor. Each interval of the crust was then powdered using a mortar and pestle, dried in a dessicator until the sample reached a constant weight, and stored in plastic bottles until analysis.

Two samples of hand-separated benthic foraminifers were also analyzed for radiocarbon for purposes of making a ^{14}C reservoir correction. One sample (JSL3356SPWC-1) was taken from surface sediment of a core collected by the submersible at a site transitional between soft and hard bottom sediments and close to the site where BC-33 was collected (Table 1). The final sample consisted of the monospecific benthic foraminifers *Cancris carmenensis Natland.* The foraminifers showed excellent preservation and no evidence of significant abrasion. The second sample was from surface sediment (0–1 cm) from the same box core that collected protocrust BC-123. The separated sample consisted of mixed benthic species including: *Cancris carmenensis Natland, Cancris inflatus (d'Orbigny), Bolovina plicata d'Orbigny, Bolivina seminuda* var. *foraminata RE & KC Stewart,* and *Bolivina interjuncta Cushman.* All of these foraminifers also showed excellent preservation.

Uranium-Series Disequilibrium Measurements

Activities of ^{238}U, ^{234}U, ^{232}Th, ^{230}Th, and ^{231}Pa for these samples were determined by isotope dilution alpha spectrometry. All samples were analyzed for a 2M HCl soluble fraction and selected samples were run after total dissolution. The chemical separation and purification of U, Th, and Pa isotopes was accomplished by utilizing an approach based on a combination of extraction chromatographic resin (TRU•Resin®, manufactured by Eichrom Industries, Inc., Darien, IL) and standard AG-50 cation exchange resin (Burnett and Yeh, 1995). Briefly, after the sample was loaded onto the TRU•Resin® column in 2M HNO_3, common ions were rinsed out with 2M HNO_3, the column converted to a chloride medium, lanthanide elements removed with 4M HCl, and Th eluted with 1M HCl. U and Pa were then eluted together in 20 mL of a 0.1M HCl-0.1M HF solution. A small amount (~5 mL) of concentrated $HClO_4$ was added and the solution was heated to fumes of $HClO_4$ to drive off all F^- before loading onto the cation exchange column. The solution was then picked up in 0.1M HCl and loaded onto the AG-50 column for the final and complete separation of Pa, U, and residual Th. All sources were prepared by electrodeposition onto stainless steel disks using a rotating cathode electrodeposition unit.

Activities of ^{238}U (^{234}Th), ^{226}Ra, ^{210}Pb, ^{137}Cs, and ^{40}K were determined in the same samples by nondestructive gamma spectrometry utilizing a well-type intrinsic germanium (IG) detector. Powdered samples were packed in plastic counting vials, covered with an approximately 3 mm thick epoxy layer to prevent gaseous ^{222}Rn from escaping, and aged for more than 3 weeks to ensure establishment of secular equilibrium between ^{226}Ra and ^{222}Rn before counting. Salient features of these results are given in this paper, and all measurements have been reported previously in Yeh (1994).

Radiocarbon Studies

Accelerator mass spectrometry (AMS) ^{14}C determinations for carbonate fluorapatite were performed on bulk samples of BC-33, BC-57, BC-123, and GS-1, a protocrust described earlier by Kim and Burnett (1986). In addition, layer-by-layer AMS ^{14}C analyses were performed on several layers from two of these crusts (BC-33 and BC-123), determined to be the youngest of the available samples based on the bulk ^{14}C analyses. All radiocarbon samples were pretreated with 0.5M triammonium citrate, known as "Silverman's solution" (Silverman et al., 1952; Kolodny and Kaplan, 1970), to remove associated calcite. All samples were reacted with this solution for approximately 50 hours at 30°C on a shaker table, rinsed several times with deionized water, and dried. Kolodny (1969) has shown that this treatment completely solubilizes calcite while leaving the CFA intact. While there is no known way to separate dolomite chemically from CFA, dolomite was not detected in BC-33 nor BC-123 by x-ray diffraction procedures (detection limit ≈5%). After pretreatment, samples were reacted with degassed H_3PO_4 to evolve the structural CO_2 from the carbonate fluorapatite structure. Using a helium gas stream, the CO_2 was degassed from the reaction vials, collected and purified cryogenically on a vacuum line, transferred to break-seal ampoules, and sent to Lawrence Livermore National Laboratories (LLNL) where the CO_2 was converted to graphite and run by AMS. Blanks (calcite spar) were run with each batch of samples and were shown to be within the range usually observed in the AMS laboratory at LLNL.

Subsamples of the purified CO_2 were collected from the vacuum line for analysis of stable carbon isotopes. These samples were run on the dual inlet of a Finnigan MAT Delta S isotope ratio mass spectrometer at Florida State University. The $\delta^{13}C$ results are reported relative to VPDβ, and are calibrated relative to NBS 19. The reported uncertainties represent instrumental analytical error.

All radiocarbon results are reported following the conventions in Stuiver and Polach (1977), that is, results are reported as fraction modern and the conventional radiocarbon ages are calculated using the Libby half-life of 5568 y. We also show calibrated radiocarbon ages that use the CALIB 3.0 program of Stuiver and Reimer (1993) to correct for reservoir and other effects. The reservoir corrections were made by assigning the benthic foraminifera results to zero age using 540 as the delta-R value (940–400 ^{14}C years). Sample preparation backgrounds have been subtracted for each sample based on the measurements of ^{14}C-free calcite and scaling the backgrounds relative to the individual sample sizes.

Chemical and Mineralogical Studies

The weight percent (±10%) of the structurally bound carbonate in the CFA was determined in these samples by a semi-quantitative X-ray diffraction (XRD) peak-pair method utilizing an empirically derived relationship between the CO_2 content of CFA and the a-cell dimension of apatite (Gulbrandsen, 1970). Differences in 2θ spacing between peaks at 51.6° (410) and 53.1° (004) were measured from X-ray diffractograms scanned at slow speed under Ni-filtered Cu K_α radiation. The principal mineralogical constituents of these samples were also identified using x-ray diffraction techniques. Phosphorus concentrations (±2%) were determined following procedures detailed in Walsh et al. (1991). Weighed homogenized samples were placed in pre-weighed crucibles, ashed in a muffle furnace for at least 4 hours, acidified with 25 mL mixed reagent containing 2 mL of 6M HCl and deionized water, and neutralized with 2 mL of 0.5M $NaHCO_3$. They were then analyzed on a spectrophotometer for PO_4 concentration within the range of 0.1-10 μM PO_4.

RESULTS AND DISCUSSION

Compositional and Depth Trends

One of the most striking features of these phosphatic protocrusts is the strong depth dependencies of several compositional features (Fig. 2). Crusts BC-33 and BC-123 both show very systematic linear decreases in phosphate content from the top to the bottom of the crust. Crust BC-57 also has the highest concentration of phosphate at the top surface although the trend is not nearly as systematic. This is a reflection of the dolomitic nature of the internal portion of this crust, that is, unlike the other crusts, BC-57 is a phosphatized dolomite crust with a rind of phosphatic material on its top and bottom. Crust GS-1 (not shown) also showed a very systematic trend of steadily decreasing phosphate towards the bottom of the crust (Kim and Burnett, 1986).

Plots of uranium and thorium concentrations versus crust thickness display nearly opposite patterns with uranium generally decreasing from top to bottom, as does the phosphate, while thorium increases. These trends most likely reflect the differing locations of U and Th in separate mineral phases within these F-phosphatic crusts. Uranium concentrations closely follow the distribution of P_2O_5 with correlation coefficients as high as 0.948 (BC-33) while the distribution of thorium in each crust demonstrates negative relationships to P_2O_5 (Fig. 3). It is interesting, and perhaps significant, that the two crusts that have the highest uranium and phosphate contents (BC-33 and GS-1) both have significant negative intercepts on the U-axis of their respective U versus P_2O_5 plots. These trends are quite different than the thorium versus phosphate relationships, for example, which extrapolate to Th intercepts (0% P_2O_5) of about 4–6 ppm, reasonable concentrations for the detrital fraction of these crusts. In general, the Th concentrations tend to follow the concentration of insoluble residue in the 2M HCl digests.

These elemental trends reinforce the concept that uranium is primarily associated with the CFA fraction while thorium primarily resides in non-CFA phases such as detrital mineral grains. Fission-track mapping had shown previously that U is intimately associated with CFA on a microscopic scale (Burnett and Veeh, 1977; Kress and Veeh, 1980; Veeh, 1985). However, more recent studies by Lucas and Abbas (1989) have shown that while uranium is linked to the phosphate, the concentrations are dependent upon pigmentation (organic matter) and grain structure. Strongly pigmented grains and the cortex of coated grains consistently showed higher concentrations of uranium in the

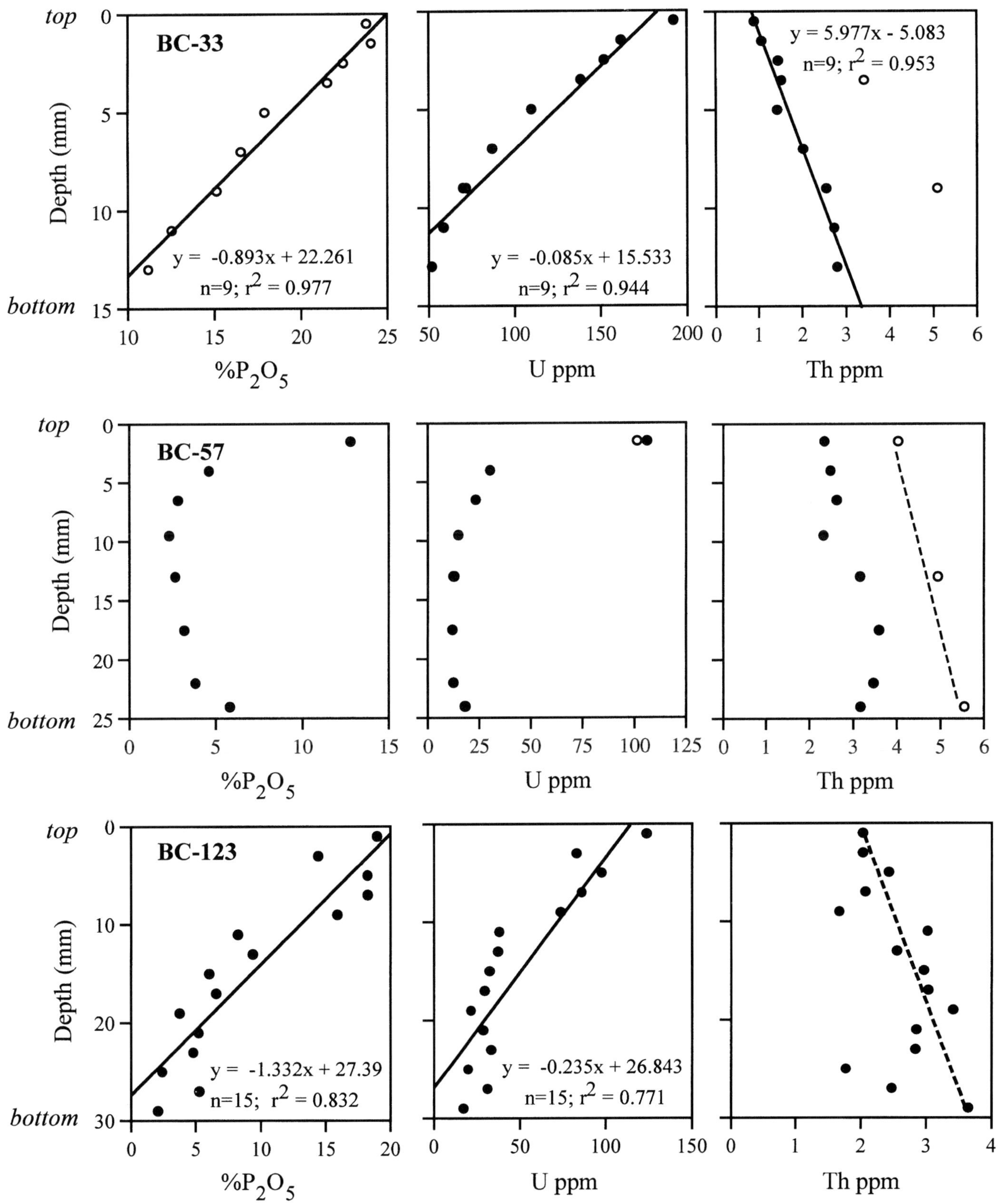

FIG. 2.—Content of P_2O_5 (%), U (ppm), and Th (ppm) versus crust depth for three phosphatic protocrusts, BC-33 (top), BC-57 (middle), and BC-123 bottom). Open symbols in Th plots for BC-33 and BC-57 represent total dissolution while all other results based on the 2M HCl soluble fractions.

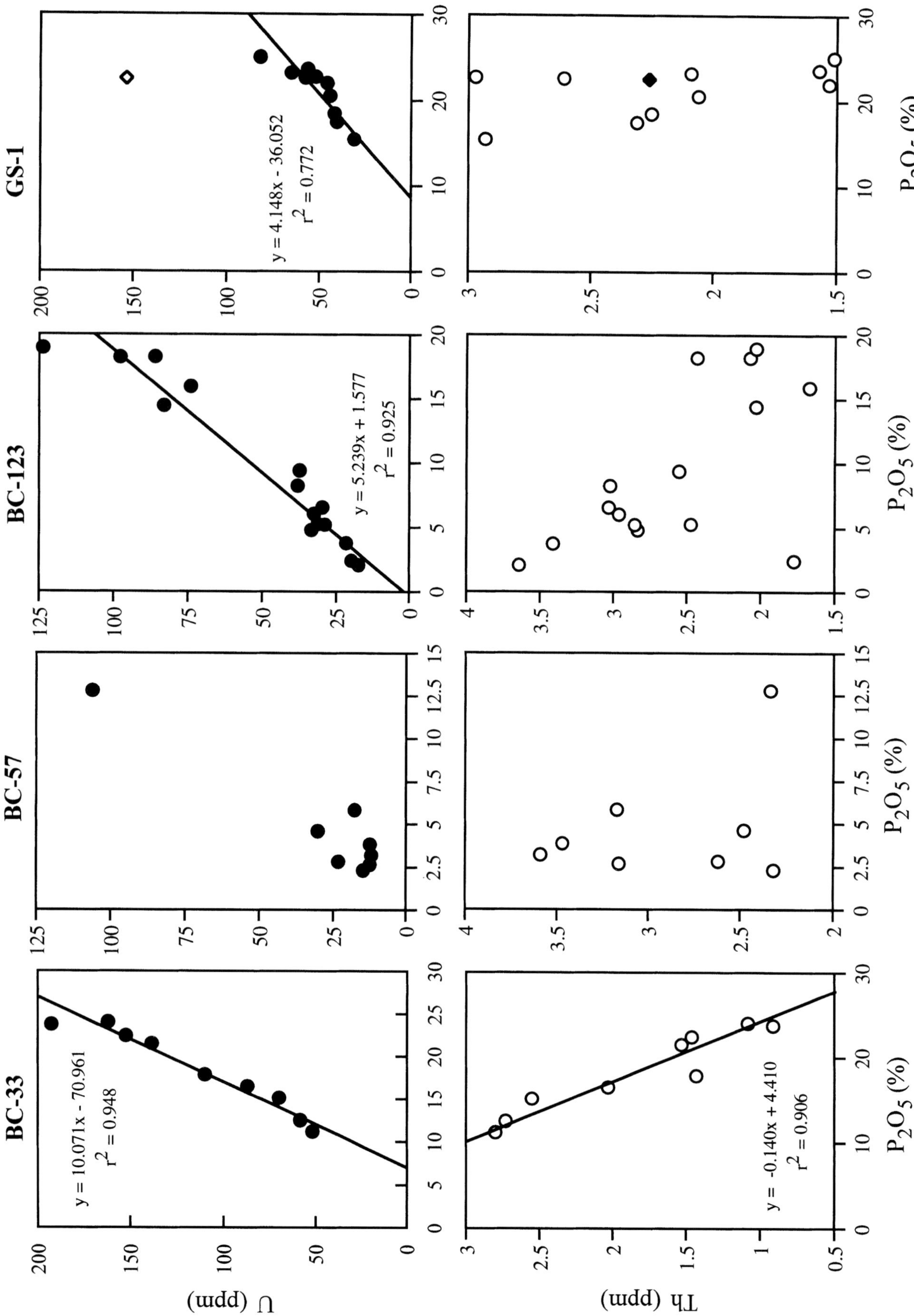

Fig. 3.—Plots of U (ppm) and Th (ppm) versus P_2O_5 (%) for four phosphatic protocrusts from the Peru shelf. Data for GS-1 taken from Kim and Burnett (1986).

grains they examined. Lucas and Abbas suggest that their observations imply that some uranium is not part of the CFA mineral lattice but is adsorbed at the surface of the crystallites. We examined the U/P_2O_5 ratios from over 120 Peru shelf phosphorites divided into three groups: nodules (D-phosphates); protocrusts (F-phosphates); and pellets (P-phosphates). Frequency plots (Fig. 4) of these three groups show them to have very similar distributions, the only exception being an occasional nodule with a very high ratio. All protocrusts display the highest U/P_2O_5 ratios on the uppermost surface with systematic decreases in the lower layers. Thus, the decrease of uranium and increase in thorium from crust top to bottom is apparently controlled primarily by decreasing CFA (reflecting in the decreasing P_2O_5 concentrations) and increasing detrital material percentages towards the bottom of the sample. Since most Th concentrations shown are based on the 2M HCl soluble fraction, there is either some non-phosphatic soluble phase (perhaps iron oxides) that is supplying the bulk of the measured Th or the 2M HCl tends to leach a constant fraction of the total Th from the insoluble residue.

Although it is not entirely clear what may have caused the negative U-intercepts on the uranium versus phosphate plots, it could be that there is a certain amount of phosphate that initially precipitates in a non-CFA form. The $U-P_2O_5$ plots imply that no uranium is structurally incorporated into the apatite lattice (or its precursor phase) until the phosphate content has reached about 7–9% P_2O_5. It has been proposed by several investigators that phosphate may precipitate directly from the dissolved phase as a non-crystalline precursor phase before recrystallizing to CFA (Baturin, 1971; Jahnke, 1984; Froelich et al., 1988; Glenn et al., 1994). It is also possible that phosphate could precipitate initially as a crystalline, yet unstable phase such as struvite ($MgNH_4PO_4 \cdot 6H_2O$), which cannot accommodate uranium (L. Prévôt-Lucas, pers. comm., 1997). Although such phases have never been identified from an actual sample, it may be that the uranium-phosphate relationships observed in BC-33 and GS-1 provide indirect evidence for an initial, high-phosphate, low-uranium phase that may have since recrystallized to CFA or is still present in these samples.

The activities of ^{238}U determined by α-spectrometry, on either the 2M HCl soluble fraction or samples that were totally digested, agree with the determinations made by γ-spectrometry (which measures total U) within statistical error in almost every case (Yeh, 1994). The internally consistent results for ^{238}U activities by these different approaches and the strong relationship between U and P_2O_5 provides further support that most of the U is associated with an acid-soluble phosphatic phase. Since CFA was the only phosphatic phase identified by XRD procedures, it appears that CFA must be where the bulk of the U resides. The presence of a non-CFA, low uranium, crystalline or non-crystalline phosphatic phase (perhaps as a precursor to CFA) must also remain a possibility.

Most $\delta^{13}C$ values of the structural carbonate within CFA from Peru protocrusts are in a relatively narrow range from 0 to –2‰. If D phosphates (indurated nodules) are included, $\delta^{13}C$ may reach values as low as –5‰ and dolomitic phosphates are characterized by a very broad range from +7 to –11‰ (Glenn et al., 1988). The same authors also showed that the amount of structural carbonate in CFA generally increases as $\delta^{13}C$ decreases. The structural-CO_2 in CFA is generally thought to reflect the substitution of PO_4^{3-} in the lattice by CO_3^{2-} with a coupled substitution of F^- to make up the charge difference (McClellan, 1980; Nathan, 1984; McClellan and Van Kauwenbergh, 1990). Glenn et al. (1988) speculated that the

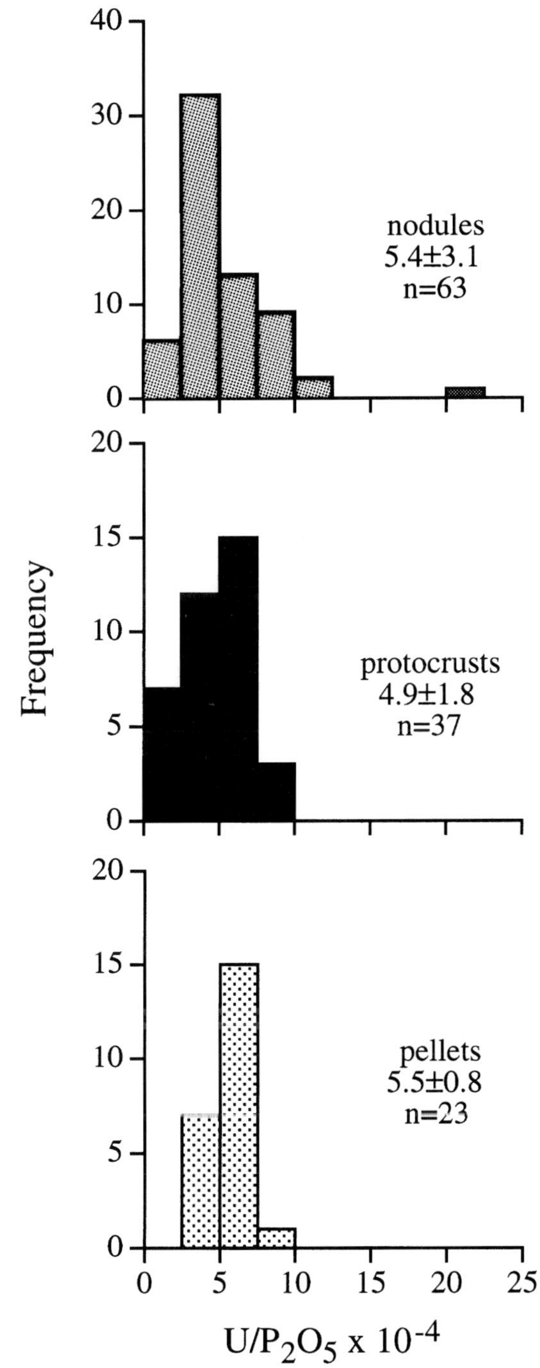

FIG. 4.—Frequency plots showing the distribution of U/P_2O_5 ratios in nodules (D-phosphates), protocrusts (F-phosphates), and pellets (P-phosphates). Ratios calculated from data in Kim and Burnett (1985), Burnett et al. (1988), and this study.

observed relationship between the amount of carbonate and its isotopic composition was a consequence of increasing carbonate and accompanying charge-balancing substitutions in the CFA lattice in response to increasing pore-water carbonate ion concentrations with depth below the sediment-water interface.

A rough inverse relationship between CFA structural-CO_2 and the phosphate content in crust GS-1 was observed earlier by Kim and Burnett (1986). Their results showed phosphate increasing towards the top crust surface (as we now show for crusts BC-33 and BC-123), and CFA-CO_3^{2-} increasing towards the bottom of the crust. This observed association may be a primary feature reflecting the varying CO_3^{2-} and PO_4^{3-} concentrations of the sediment pore solutions from which the CFA precipitated, or it could be due to some type of secondary effect, perhaps similar to the loss of CO_2 by burial diagenesis or weathering reported from other areas (Zanin and Krivoputskaya, 1976; Matthew and Nathan, 1977; Lucas et al., 1980; McArthur, 1980 and 1985; McClellan, 1980; Panczer et al., 1989). If it is a primary feature, which is likely in this modern environment of phosphate precipitation, it is particularly important in terms of mineral genesis. For example, Glenn et al. (1988) suggested that in addition to other constraints such as phosphate and fluoride ion availability, the lower limit of CFA precipitation in suboxic to anoxic sediments may be controlled by lattice poisoning due to excessive dissolved carbonate concentrations.

Although we did not observe trends between phosphate and structural carbonate in the crusts examined here, perhaps because of the inaccuracy of the semi-quantitative XRD technique employed for the determination of CFA-CO_3^{2-}, our stable isotope results showed a clear trend of lighter carbon (from the CFA-CO_3^{2-}) with depth in crust BC-123 (Fig. 5). A very similar pattern was shown earlier for crust GS-1 (Glenn et al., 1988). These trends and isotopic compositions are consistent with an in situ model of CFA precipitation that occurred while the crust was in its present position, that is, the CFA precipitated within the upper few centimeters of the sediment-seawater interface, cementing the other sediment components into these observed "protocrusts."

Ages, Growth Rates, and Directions

Uranium-series disequilibrium studies.—

The basic principle of the uranium-series disequilibrium dating method as applied to authigenic minerals is to measure the activity ratio of certain parent–daughter pairs in systems that (1) contain measurable and preferably high uranium; (2) have a low content of detrital thorium; and (3) behave as a closed geochemical system. One may apply the ingrowth of the daughter nuclides ^{230}Th and ^{231}Pa, which approach secular equilibrium with their respective parents, ^{234}U and ^{235}U, over a time span of a few hundred thousand years. If an authigenic marine mineral acquires U at its time of formation, with no accompanying ^{230}Th or ^{231}Pa, then the ^{230}Th/^{234}U activity ratio will be a function of time as given by Broecker (1963):

$$\frac{^{230}\mathrm{Th}}{^{234}\mathrm{U}} = \frac{1-e^{-\lambda_{230}t}}{^{234}\mathrm{U}/^{238}\mathrm{U}} + \left[1-\frac{1}{^{234}\mathrm{U}/^{238}\mathrm{U}}\right]\frac{\lambda_{230}}{\lambda_{230}-\lambda_{234}}\left[1-e^{-(\lambda_{230}-\lambda_{234})t}\right]$$

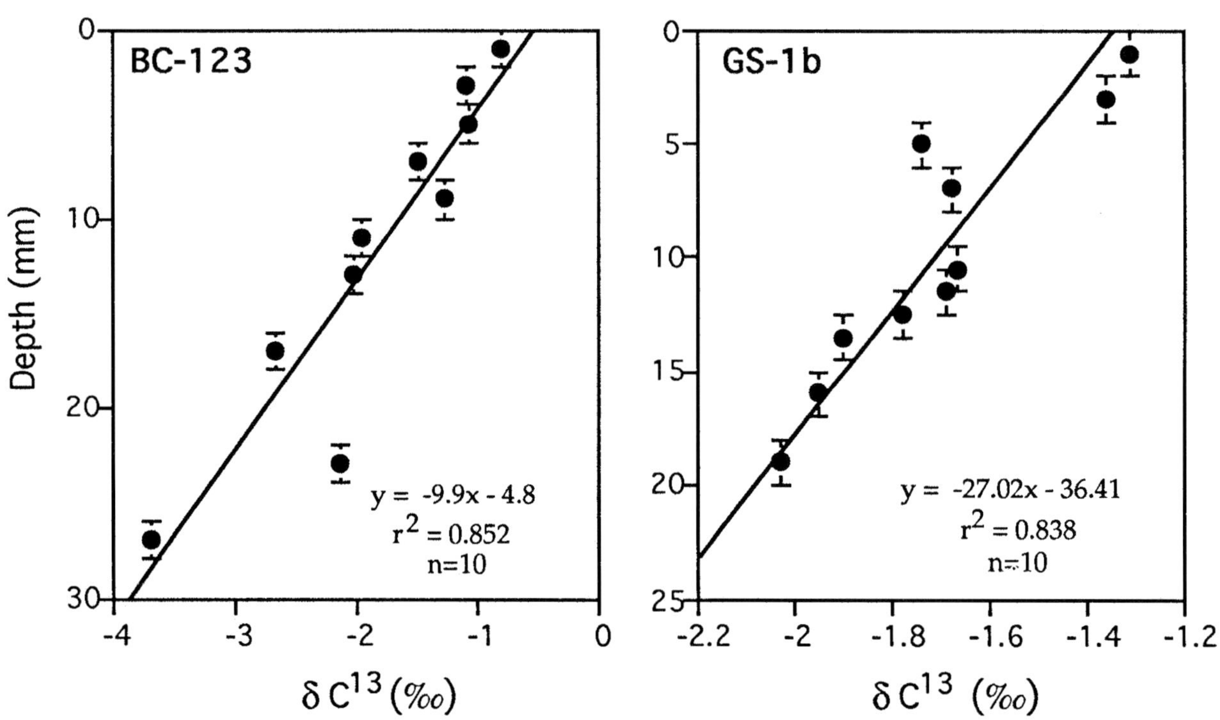

FIG. 5.—Plots of the stable carbon isotopic composition (expressed as δ^{13}C) versus depth in two phosphatic protocrusts, BC-123 (left) and GS-1 (right). Data for GS-1 taken from Glenn et al. (1988).

Because of recoil and other processes, there is an "excess" amount of ^{234}U dissolved in seawater over its long-lived parent ^{238}U. Because U behaves conservatively in the ocean and has a long residence time, its isotopic composition appears to be uniform both in space and in time (Chen et al., 1986). After incorporation into a marine mineral, the decay of the excess ^{234}U towards secular equilibrium with ^{238}U may be described by the equation:

$$\left(\frac{^{234}U}{^{238}U}\right)_t = 1-\left[\left(\frac{^{234}U}{^{238}U}\right)_0 - 1\right]e^{-\lambda_{234}t}$$

where $(^{234}U/^{238}U)_t$ and $(^{234}U/^{238}U)_0$ are the present and initial activity ratios, respectively. Unfortunately, the method is not very precise because there is only a 14% maximum change in the isotope ratio over time. Since the alpha spectrometric measurements have typical counting errors of at least 1–2%, the resulting age range is quite broad. We thus only display $^{234}U/^{238}U$ results for confirmation purposes and rely solely on ^{230}Th and ^{231}Pa ingrowth ages for U-series dating.

Since there is no other U isotope besides the parent in the ^{235}U decay chain, the equation that describes the ingrowth of ^{231}Pa is straight-forward:

$$\frac{^{231}Pa}{^{235}U} = 1-e^{-\lambda_{231}t}$$

The symbols λ_{230}, λ_{231}, and λ_{234} in the above equations represent the decay constants for ^{230}Th, ^{231}Pa, and ^{234}U, respectively.

Uranium-series disequilibrium methods have proven particularly useful for determination of the absolute ages of marine phosphorites because sedimentary apatite contains relatively high uranium (~100 ppm) and generally low thorium (<5 ppm). Previous studies have shown that the activity ratios of $^{234}U/^{238}U$-$^{230}Th/^{234}U$ and $^{230}Th/^{234}U$-$^{231}Pa/^{235}U$ for sea-floor phosphorites display a good degree of concordance, at least as good as that shown for unrecrystallized corals (Burnett and Veeh, 1992).

Ocean-floor phosphate grains, crusts, and nodules always contain at least small amounts of impurities including detrital phases, glauconite, biogenic opal, pyrite, carbonates, and/or organic matter in addition to CFA (e.g., Glenn and Arthur, 1988). Since much of the natural thorium (^{232}Th) contained in phosphorites is thought to be associated with the non-CFA fraction, a correction for detrital ^{230}Th (termed "common" or "nonradiogenic" ^{230}Th) contamination is necessary in order to obtain accurate uranium-series ages. This correction may be particularly important for very young samples, such as those reported here, since the ratio of contaminating ^{230}Th to that derived from decay of U associated with CFA, could be very high. One approach to correct for common thorium is to use the long-lived ($>10^{10}$ y) ^{232}Th concentration in each sample as a "stable" tracer of contamination. The nonradiogenic ^{230}Th may then be assessed if an initial activity ratio of $^{230}Th/^{232}Th$ can somehow be determined or assumed (Kaufman and Broecker, 1965). The following equation would then apply:

$$^{230}Th_c = {}^{230}Th_m - \left(\frac{^{230}Th}{^{232}Th}\right)_0 {}^{232}Th \times e^{-\lambda_{230}t}$$

where $^{230}Th_c$, $^{230}Th_m$, and $(^{230}Th/^{232}Th)_o$ represent the activities of corrected-^{230}Th, uncorrected (measured) ^{230}Th, and the initial $^{230}Th/^{232}Th$ activity ratios, respectively. Previously suggested initial values of $^{230}Th/^{232}Th$ for Peru margin phosphorites have ranged from 1 to 4 based on measurements of very young phosphorites and associated sediments, nodules, and recent sediments from this area (Burnett and Veeh, 1977; Jahnke et al., 1983; Kim and Burnett, 1985). An important point is that there is no *a priori* reason to suspect that there is one unique initial ratio for the entire Peru shelf. Indeed, depending upon environmental circumstances, it would be possible to have a fluctuating $(^{230}Th/^{232}Th)_0$ at any one site. Thus, each case must be evaluated independently. Unfortunately, there is no long-lived or stable isotope of Pa to normalize $^{231}Pa/^{235}U$ results.

The U-series results for the Peru protocrusts are given in Table 2. All individual slices of the crusts examined contain an excess amount of ^{234}U over ^{238}U, with most $^{234}U/^{238}U$ activity ratios falling within the range of 1.144±0.002, the value for modern seawater (Ku et al., 1977; Chen et al., 1986). Several other isotopic characteristics are shared by these samples. For example, the activity of ^{230}Th generally decreases from the top of the crust to the bottom, even though the content of ^{232}Th systematically increases over this same interval. This apparently divergent behavior of two thorium isotopes is a reflection of their different sources, that is, ^{230}Th is mainly from ingrowth from U in the CFA (more concentrated at the top) while ^{232}Th is mainly associated with detrital mineral grains (more concentrated at the bottom). The $^{230}Th/^{232}Th$ activity ratios thus decrease systematically from top to bottom as a result of these trends.

Since ^{230}Th measurements uncorrected for common ^{230}Th will produce $^{230}Th/^{234}U$ activity ratios that are too high, ages derived from these activity ratios must be considered maximum. Although the $^{231}Pa/^{235}U$ ages are internally consistent with the uncorrected ^{230}Th ages, the ^{231}Pa ages may also be affected by initial Pa from detrital phases and should be considered maximum ages as well. The seemingly consistent ages between these two geochronometers could simply be a consequence of the release of contaminating $^{230}Th/^{231}Pa$ in a fixed ratio from non-CFA phase(s) into the fraction being analyzed. In fact, most $^{230}Th/^{231}Pa$ activity ratios in these samples are in a relatively narrow range from 10.8 to 12.6, close to the theoretical production ratio (11.3) from the U dissolved in seawater.

Since there is no way to be certain what the exact initial value was of the $^{230}Th/^{232}Th$ activity ratio, we ran a series of corrections with increasing values of $(^{230}Th/^{232}Th)_0$ until these corrections resulted in a zero age intercept on the respective depth versus age plot for each protocrust (Fig. 6). This approach of forcing a zero age on one surface assumes that the crust was actively growing when sampled. Thus, the corrected age trends shown should be considered minimum ages—greater assumed values of $(^{230}Th/^{232}Th)_0$ would produce negative ages. We thus are left with a situation in which we have constrained the possi-

TABLE 2.—CONCENTRATIONS AND ACTIVITY RATIOS OF URANIUM-SERIES RADIONUCLIDES TOGETHER WITH DATING RESULTS FOR PHOSPHATIC CRUSTS SJ1092 BC-33, BC-57, BC-123, AND JSL3362-BUCKET 9.
All analyses are based on the 2M HCl soluble fraction unless otherwise noted. Errors shown are ±1σ based on counting statistics. See text for discussion of distinction between corrected and uncorrected ^{230}Th ages.

Interval (mm)	P_2O_5 (%)	U (ppm)	Th (ppm)	$^{234}U/^{238}U$	$^{230}Th/^{234}U$	$^{230}Th/^{232}Th$	$^{231}Pa/^{235}U$	Pa-Age (kyr)	Th-Age (kyr) uncorrected	Th-Age (kyr) corrected[a]
JSL3362-Bucket9										
Bulk	—	58.2	3.07	1.14±0.01	0.030±0.002	2.0±0.1	—	—	3.4±0.2	1.7±0.2[c]
Bulk[b]	—	55.4	6.02	1.14±0.01	0.038±0.002	1.2±0.1	—	—	4.2±0.2	0.7±0.2[c]
SJ1092 BC-33										
top										
0-1	25.19	192.6	0.91	1.15±0.01	0.026±0.001	19.5±1.1	0.069±0.005	3.4±0.2	2.9±0.1	2.3±0.1
1-2	24.05	161.9	1.08	1.13±0.01	0.027±0.001	14.0±0.6	0.057±0.005	2.8±0.3	3.0±0.1	2.1±0.1
2-3	22.43	152.3	1.46	1.16±0.01	0.029±0.001	10.6±0.6	0.063±0.004	3.1±0.2	3.2±0.2	2.0±0.2
3-4	21.51	138.4	1.53	1.15±0.01	0.030±0.002	9.6±0.7	0.070±0.006	3.4±0.3	3.3±0.2	1.9±0.2
3-4[b]	—	138.7	3.42	1.16±0.01	0.032±0.001	4.6±0.4	0.100±0.008	5.0±0.4	3.6±0.1	—
4-6	17.89	110.0	1.43	1.14±0.02	0.033±0.002	8.8±0.5	0.062±0.004	3.0±0.2	3.7±0.2	2.0±0.2
6-8	16.51	87.1	2.03	1.15±0.01	0.038±0.001	5.7±0.2	0.088±0.007	4.3±0.4	4.2±0.2	1.2±0.2
8-10	15.12	69.9	2.55	1.16±0.02	0.047±0.002	4.5±0.2	0.096±0.009	4.8±0.5	5.2±0.2	0.5±0.3
8-10[b]	—	71.7	5.09	1.14±0.01	0.054±0.002	2.6±0.2	—	—	6.0±0.3	—
10-12	12.51	58.6	2.73	1.17±0.02	0.057±0.003	4.4±0.2	0.124±0.011	6.3±0.6	6.4±0.3	0.4±0.3
12-14	11.16	51.8	2.80	1.15±0.02	0.068±0.003	4.4±0.2	0.113±0.011	5.6±0.6	7.7±0.4	0.6±0.4
bottom										
SJ1092 BC-57										
Bulk	—	52.0	2.58	1.11±0.01	0.080±0.004	5.5±0.2	—	—	9.1±0.4	7.5±0.4[c]
Bulk[b]	—	52.6	6.01	1.16±0.01	0.088±0.004	2.7±0.1	0.172±0.019	8.9±1.1	10.0±0.5	6.4±0.5[c]
top										
0-3	12.79	106.1	2.34	1.15±0.01	0.073±0.003	11.4±0.5	0.131±0.008	6.6±0.5	8.2±0.4	7.2±0.4
0-3[b]	—	101.3	4.03	1.17±0.01	0.075±0.003	6.7±0.2	0.157±0.014	8.1±0.8	8.5±0.3	6.7±0.3
3-5	4.60	30.2	2.48	1.16±0.01	0.089±0.005	3.8±0.2	—	—	10.1±0.6	6.5±0.6
5-8	2.82	23.2	2.62	1.10±0.01	0.104±0.006	3.1±0.2	—	—	11.9±0.7	6.7±0.7
8-11	2.30	14.6	2.32	1.10±0.01	0.109±0.006	2.3±0.1	—	—	12.5±0.7	5.0±0.7
11-15	2.66	12.2	3.16	1.13±0.01	0.135±0.008	1.8±0.1	—	—	15.7±1.0	3.8±1.0
11-15[b]	—	12.7	4.94	1.10±0.01	0.191±0.009	1.6±0.1	—	—	23.0±1.2	9.6±1.2[c]
15-20	3.19	11.7	3.59	1.14±0.01	0.135±0.010	1.5±0.1	—	—	15.7±1.3	2.4±1.3
20-23	3.83	12.2	3.47	1.19±0.01	0.136±0.009	1.7±0.1	—	—	15.8±1.1	3.1±1.1
23-25	5.82	17.6	3.17	1.08±0.01	0.095±0.006	1.7±0.1	—	—	10.8±0.7	2.1±0.7
23-25[b]	—	18.3	5.55	1.13±0.01	0.148±0.007	1.7±0.1	0.242±0.026	13.1±1.6	17.4±0.8	7.3±0.8[c]
bottom										

TABLE 2.—CONTINUED

Interval (mm)	P_2O_5 (%)	U (ppm)	Th (ppm)	$^{234}U/^{238}U$	$^{230}Th/^{234}U$	$^{230}Th/^{232}Th$	$^{231}Pa/^{235}U$	Pa-Age (kyr)	Th-Age (kyr) uncorrected	Th-Age (kyr) corrected[a]
SJ1092 BC-123										
top										
0-2	18.94	123.7	2.03	1.12±0.03	0.026±0.001	5.3±0.32	0.057±0.006	2.8±0.3	2.8±0.1	1.72±0.08
2-4	14.43	83.1	2.03	1.14±0.03	0.028±0.002	4.0±0.44	0.073±0.012	3.6±0.6	3.1±0.2	1.57±0.18
4-6	18.21	97.8	2.43	1.12±0.05	0.021±0.002	2.8±0.47	0.154±0.042	2.2±0.6	2.4±0.3	0.70±0.25
6-8	18.24	86.1	2.07	1.14±0.04	0.025±0.001	3.6±0.21	0.053±0.010	2.6±0.5	2.7±0.1	1.20±0.11
8-10	15.91	74.0	1.67	1.11±0.04	0.030±0.002	4.5±0.46	0.082±0.018	4.0±0.9	3.3±0.2	1.88±0.18
10-12	8.24	38.1	3.02	1.11±0.04	0.065±0.003	2.8±0.15	0.135±0.023	6.9±0.3	7.4±0.3	2.14±0.31
12-14	9.40	37.4	2.55	1.16±0.04	0.057±0.002	2.9±0.17	0.205±0.056	10.9±3.5	6.4±0.3	2.07±0.27
14-16	6.05	32.5	2.96	1.12±0.05	0.068±0.003	2.6±0.14	0.194±0.056	10.2±3.4	7.7±0.3	1.72±0.35
16-18	6.57	29.7	3.03	1.11±0.07	0.084±0.005	2.8±0.23	0.171±0.040	8.9±2.3	9.5±0.6	2.71±0.62
18-20	3.76	21.6	3.41	1.16±0.03	0.108±0.004	2.4±0.12	0.277±0.046	15.3±3.1	12.4±0.5	2.20±0.48
20-22	5.22	28.9	2.85	1.15±0.07	0.093±0.005	3.3±0.21	0.223±0.038	11.9±2.3	10.7±0.6	4.32±0.59
22-24	4.80	33.4	2.83	1.13±0.07	0.081±0.005	3.3±0.28	0.172±0.034	8.9±2.0	9.2±0.6	3.70±0.57
24-26	2.40	19.8	1.77	1.06±0.07	0.124±0.014	4.4±1.01	0.037±0.009	7.9±2.4	13.4±1.6	7.50±1.60
26-28	5.28	31.3	2.47	1.12±0.04	0.108±0.005	4.7±0.46	0.206±0.031	10.9±1.9	12.5±0.7	7.29±0.66
28-30	2.08	17.2	3.64	1.12±0.03	0.163±0.007	2.6±0.16	0.354±0.048	20.7±3.6	19.2±0.9	4.91±0.88
bottom										

Notes:

[a]Common thorium correction based on initial activity ratios $(^{230}Th/^{232}Th)_0$=4.1, 1.5, and 2.0 for crusts BC-33, BC-57, and BC-123 , respectively. See text for discussion.

[b]Total dissolution.

[c]Based on an initial activity ratio $(^{230}Th/^{232}Th)_0$=1.

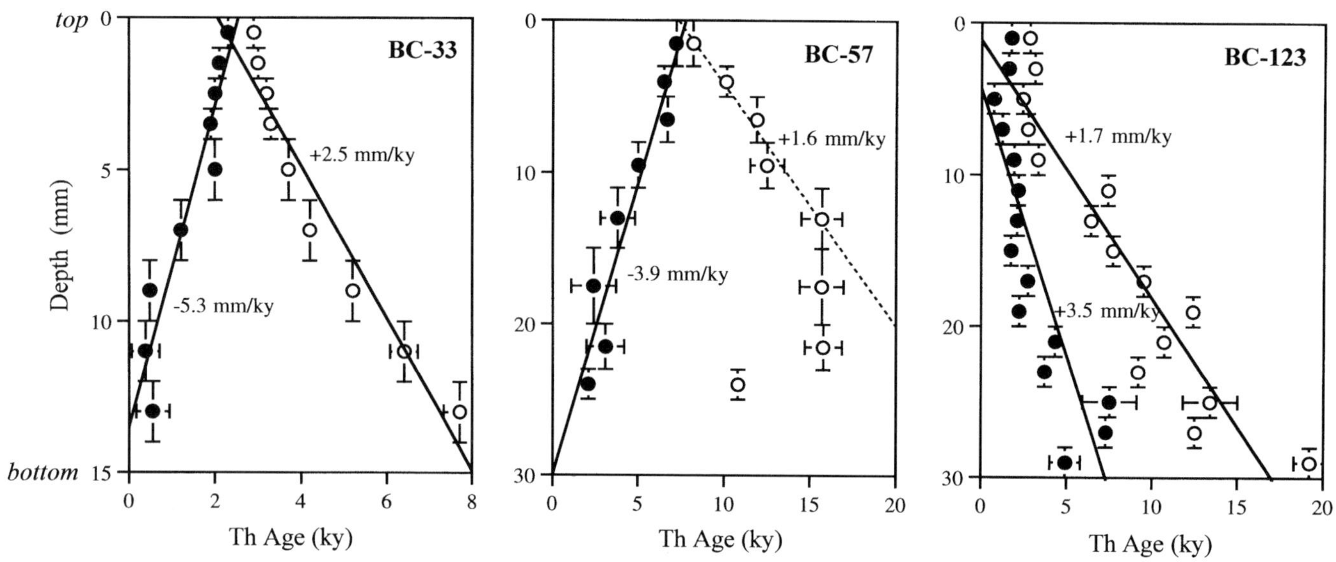

FIG. 6.—Corrected and uncorrected ^{230}Th ages (ky) plotted against crust thickness for three phosphatic protocrusts, BC-33 (left), BC-57 (middle), and BC-123 (right). Growth rates are calculated from the slopes of the regression lines and referred to as positive (+) for upward growth and negative (–) for apparent downward growth.

ble age interpretations to these upper and lower bounds. The true ages must lie somewhere between these two extremes. These plots illustrate how important the correction for common ^{230}Th is for these very young samples. In crusts BC-33 and BC-57, for example, the corrected age trends display a slope that is opposite to the uncorrected, maximum ages. Uncorrected ages increase downward, implying upward growth, while corrected ages decrease with depth, suggesting downward growth. Thus, the interpretation of the direction of crustal growth depends upon which model is applied. Note that the age trends converge at the top surface of each crust, that is, the difference between the corrected and uncorrected ages is least at the surface. This was also the case for crust GS-1 which was examined in a similar manner (Kim and Burnett, 1986). The divergence of ^{230}Th ages with depth in the crust is the result of more impurities (and thus more common ^{230}Th) in that direction. The upper layer of all these crusts is purest with respect to CFA concentration.

Thus, completely different age scenarios may be obtained for protocrusts depending upon what assumptions are used to estimate the extent of the initial contaminating ^{230}Th. For BC-33, an upper-limit $(^{230}Th/^{232}Th)_0$ activity ratio of 4.1 results in a calculated growth rate of 5.3 mm/ky with apparent downward growth. On the other hand, if uncorrected (maximum) ^{230}Th ages are used, the growth rate would be 2.5 mm/ky with upward growth. The age-versus-crust-thickness profile for BC-57 shows that the uncorrected Th-age distribution has a concave shape, that is, the middle layers have ages apparently older than those on either surface. Using an initial $^{230}Th/^{232}Th$ activity ratio (1.5) that forced a zero age intercept, the corrected Th-age-versus-crust-thickness plot displays linear downward growth with a calculated growth rate of 3.9 mm/ky. A value of 2.0 was the highest $(^{230}Th/^{232}Th)_0$ activity ratio that could be applied to the BC-123 data without producing negative ages. Both corrected and uncorrected age trends in BC-123 suggest upward growth at 1.7 mm/ky (uncorrected) to 3.5 mm/ky (corrected).

The apparent $^{231}Pa/^{235}U$ ages from crusts BC-33 and BC-123 are most consistent with the uncorrected ^{230}Th ages (Fig. 7). Note that the trend lines for both the corrected and uncorrected ^{230}Th ages are shown for reference in the same diagram. Thus, both the uncorrected ^{230}Th and ^{231}Pa ages support an upward growth model for these crusts at rates of a few millimeters per thousand years. Only a few ^{231}Pa analyses were performed for crust BC-57 and these also agreed better with the uncorrected ^{230}Th ages from the same subsamples (Table 2).

Since these crusts were recovered in their *in situ* growth positions on the surface of the seafloor, it was hoped that these protocrusts would provide unambiguous information regarding their growth directions. In the past, we have felt that the downward growth model seemed more plausible. A crust such as BC-33, for example, has an extrapolated zero age at the bottom surface when the upper-limit $(^{230}Th/^{232}Th)_0$ correction is applied, indicating that the crust was actively growing when sampled. Downward growth for a crust positioned at the sediment-seawater interface seems reasonable since the typical distribution of dissolved phosphate and fluoride in sediment pore waters from the Peru margin is such that it could supply a downward-growing crust with these components by upward diffusion (Suess, 1981; Froelich et al., 1983; Froelich et al., 1988). On the other hand, downward growth does not appear to be a possible interpretation in the case of BC-123 because larger values of the $(^{230}Th/^{232}Th)_0$ activity ratio will produce negative ages near the top surface rather than the bottom as in the other crusts. Thus, the possibility of upward growth of crusts located at the sediment surface may need to be re-evaluated.

During our routine gamma spectrometric analysis of these

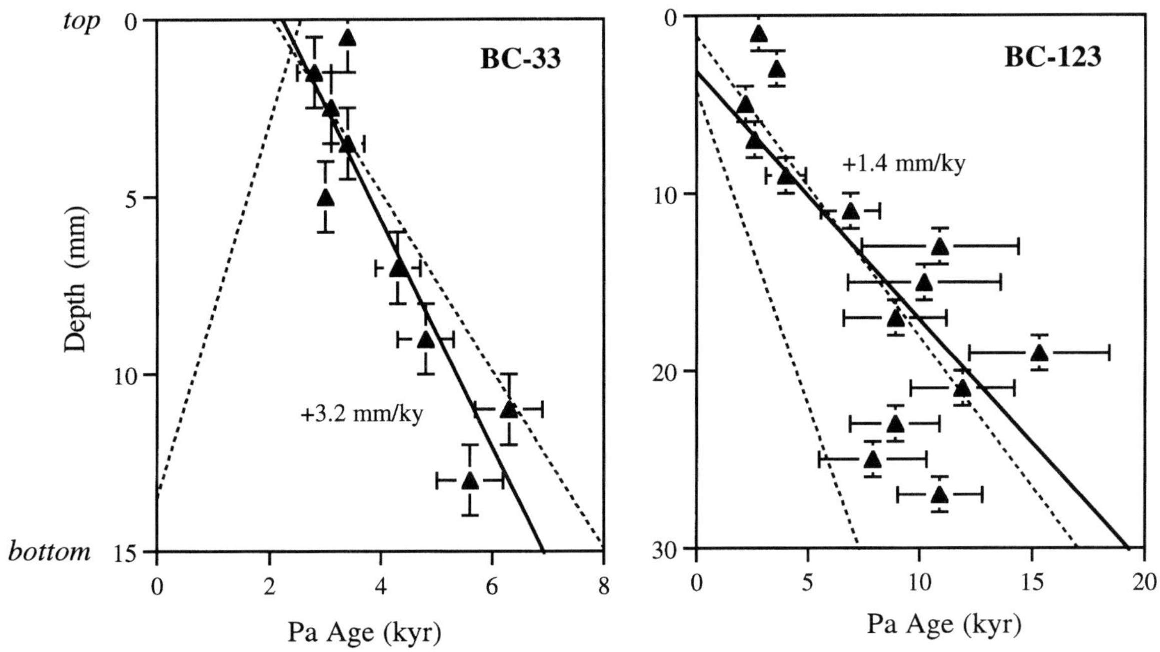

FIG. 7.—Pa-231 ages (ky) plotted against crust thickness for two phosphatic protocrusts, BC-33 (left) and BC-123 (right). Age-depth trends based on corrected and uncorrected ^{230}Th ages are shown for comparison by dashed lines.

samples, one crust (BC-33) was discovered to contain low but measurable amounts of ^{137}Cs (Fig. 8). Since radiocesium exists in the environment only as a fallout-derived nuclide, this could imply that the crust must have formed within the last few decades. This seems extremely unlikely in view of the U-series results (and the ^{14}C results we present below). The ^{137}Cs displays a depth-dependent pattern in the crust with the highest concentrations in the deeper sections where the other impurities are also highest. We also note that the ^{137}Cs follows the distribution of potassium (as shown by ^{40}K) quite closely, perhaps because they are both contained in K-bearing clay minerals such as illite, a common clay in Peru shelf sediments (Rosato and

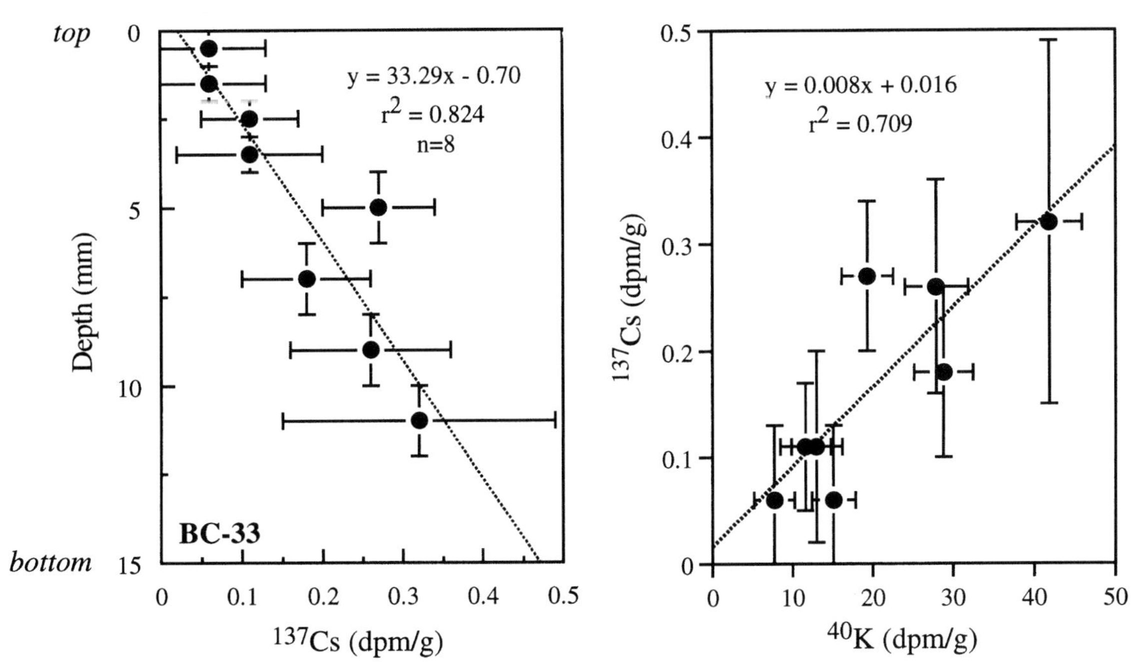

FIG. 8.—Activity (dpm/g) of ^{137}Cs plotted against crust thickness in protocrust BC-33 (left). A plot of the ^{137}Cs activity versus ^{40}K activity (right) shows that the radiocesium is probably incorporated into contaminating clay minerals that may have infiltrated the protocrust during burrowing.

Kulm, 1981). Although these crusts are indeed very young, none of our results are consistent with an age of a few decades. The low activities of ^{137}Cs measured in BC-33 must be a result of infiltration of modern sediment into the interior of the crust, probably via the worm burrows.

All growth rates determined for this group of protocrusts are based on simple linear regressions of the crust thickness versus age relationships. These are thus estimates of long-term average growth rates, not instantaneous rates of growth. While these rates are all in the relatively narrow range of 2–5 mm/ky, we recognize that actual growth rates could fluctuate markedly from these averages during the history of the protocrusts. In addition, the U-series results for two of the three crusts are ambiguous in terms of their direction of growth. This occurs because the precise interpretation of the ^{230}Th ages depends critically upon the correction for detrital ^{230}Th in these very young samples. A completely different explanation arises depending upon the contamination model used. Further refinement of these interpretations requires dating by a completely independent method such as via radiocarbon techniques.

Radiocarbon studies.—

Our results for the AMS ^{14}C analyses of four homogenized samples (GS-1, BC-33, BC-57, and BC-123) of phosphatic protocrusts, two samples of separated benthic foraminifers, and two sets of layer-by-layer analyses of crusts BC-33 and BC-123 are given in Table 3. Since both sets of benthic foraminifers produced results within experimental error of each other, the mean value of the two runs were used for the carbon reservoir correction. When the calibrated ^{14}C ages are plotted against depth, the results show overall good agreement with the U-series ages (Fig. 9). Note that both the corrected and uncorrected ^{230}Th age trends are also shown on these plots for comparison. In crust BC-33, the ^{14}C ages in the topmost layers are virtually the same as the U-series ages and the deeper layers in the crust have carbon ages intermediate between the upper and lower bounds of the Th age trends. The overall agreement is also quite good in crust BC-123, although the carbon ages tend to agree better with the common thorium corrected ages in this case, perhaps because of the larger detrital contribution (less CFA) in this crust compared to BC-33. Significantly, the two ^{14}C age versus crust depth trends are quite similar. Both trends display a systematic increase in age with depth; that is, the carbon ages imply upward growth, in agreement with the uncorrected ^{230}Th and ^{231}Pa age trends. In both crusts, the ^{14}C age interpretations display more rapid growth than the U-series interpretations with calculated growth rates of 9.1 mm/ky and 7.4 mm/ky for BC-33 and BC-123, respectively.

Another way to compare these independent radiometric geochronometers is to construct a type of "concordia" curve that shows the theoretical progression in time of the two age-dependent parameters, that is, the ^{14}C activity (expressed as the "fraction modern") and the $^{230}Th/^{234}U$ activity ratio. Samples that have both measurements can then be superimposed on such a diagram to evaluate the degree of concordancy and to examine any trends that may indicate the presence of possible artifacts. This was done earlier by Burnett et al. (1988) for groups of small (~250 μm) phosphatic pellets that had been separated from organic-rich sediment on the Peru shelf. The ^{14}C ages in that case were much younger than either the corrected or uncorrected U-series ages and showed a distinct trend on the concordia plot that they interpreted as a mixing line between a modern phosphatic fraction and one that was significantly older. Since every analysis of these small pellets consisted of hundreds to

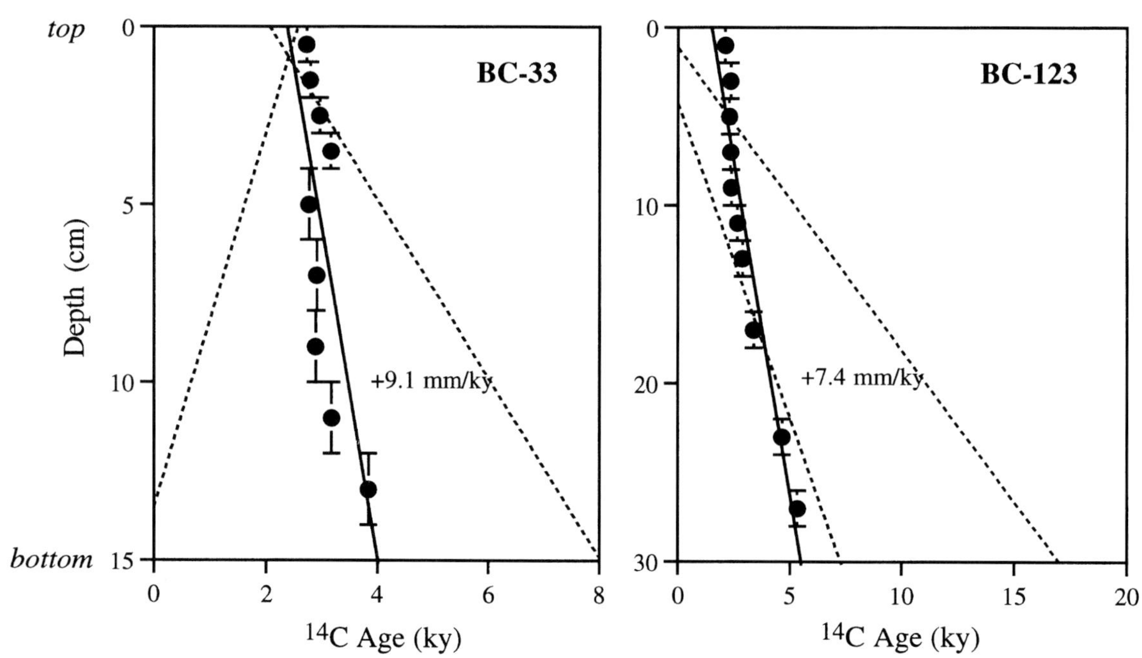

FIG. 9.—AMS ^{14}C ages (ky) plotted against crust thickness for phosphatic protocrusts BC-33 (left) and BC-123 (right). Age-depth trends based on corrected and uncorrected ^{230}Th ages are shown for comparison by the dashed lines.

TABLE 3.—ESTIMATED CARBONATE CONTENT OF THE CARBONATE FLUORAPATITE, STABLE CARBON ISOTOPIC COMPOSITION, AND AMS ^{14}C RESULTS FROM BULK HOMOGENIZED SAMPLES OF 4 PHOSPHATIC PROTOCRUSTS, SEPARATED BENTHIC FORAMINIFERA, AND DISCRETE SECTIONS THROUGH PHOSPHATE PROTOCRUSTS BC-33 AND BC-123.

Sample Type/ Name (CAMS)[1]	CFA-CO_2 (%)	$\delta^{13}C$	Fraction Modern	$D^{14}C$ (‰)	^{14}C Age Conventional yrs.	^{14}C Age Calibrated yrs.
bulk phosphate crusts						
GS-1 bulk (28540)	—	-1.76±0.02	0.2120±0.0020	-787.95±2.05	12460±80	13440±134
BC-33 bulk (28541)	—	0.56±0.02	0.6794±0.0047	-320.61±4.66	3110±60	2270±103
BC-57 bulk (28542)	—	-10.32±0.02	0.0906±0.0020	-909.44±1.99	19290±180	21930±280
BC-123 bulk (28543)	—	-0.03±0.02	0.6419±0.0042	-358.06±4.24	3560±60	2760±68
benthic forams (28538/28539)						
JS1092 BC-123, 0-1 cm	—	-0.20±0.02	0.8921±0.0061	-107.90±6.08	920±60	0[2]
JS-3356 SPWC-1	—	-0.16±0.02	0.8874±0.0061	-112.62±6.05	960±60	0[2]
SJ1092-BC-33 protocrust						
0-1 mm (33818)	4.3	0.01±0.03	0.6494±0.0033	-350.61±3.31	3470±50	2720±45
1-2 mm (33821)	4.4	-0.27±0.02	0.6404±0.0038	-359.64±3.80	3580±50	2780±70
2-3 mm (33822)	4.4	0.71±0.01	0.6296±0.0042	-370.39±4.24	3720±60	2950±115
3-4 mm (33820)	4.6	-1.70±0.02	0.6193±0.0042	-380.68±4.17	3850±60	3150±120
4-6 mm (33817)	4.4	-0.28±0.02	0.6427±0.0043	-357.28±4.27	3550±60	2760±68
4-6 mm[3] (33819)	—	-0.30±0.02	0.6401±0.0032	-359.87±3.16	3580±40	2780±63
6-8 mm (33823)	4.4	-0.38±0.02	0.6327±0.0042	-367.33±4.16	3680±60	2900±113
8-10 mm (33824)	4.5	-0.29±0.08	0.6330±0.0045	-367.03±4.48	3670±60	2880±110
10-12 mm (33816)	4.5	-0.36±0.02	0.6186±0.0034	-381.38±3.36	3860±50	3160±108
12-15 mm (33815)	4.5	-2.19±0.02	0.5763±0.0039	-423.70±3.88	4430±60	3840±118
SJ1092-BC-123 protocrust						
0-2 mm (35113)	4.4	-0.81±0.01	0.6886±0.0045	-311.36±4.50	3000±60	2100±105
2-4 mm (35115)	4.3	-1.07±0.01	0.6664±0.0044	-333.63±4.36	3260±60	2360±103
4-6 mm (35114)	4.3	-1.09±0.01	0.6769±0.0044	-323.13±4.43	3140±60	2290±98
6-8 mm (35116)	4.3	-1.49±0.01	0.6680±0.0038	-332.03±3.80	3240±50	2350±83
8-10 mm (35117)	4.5	-1.28±0.01	0.6655±0.0038	-334.48±3.80	3270±50	2380±105
10-12 mm (35118)	4.6	-1.96±0.01	0.6550±0.0043	-344.99±4.29	3400±60	2670±128
14-16 mm (35119)	3.9	-2.03±0.01	0.6326±0.0042	-367.45±4.15	3680±60	2890±113
16-18 mm (35120)	4.4	-2.67±0.02	0.6014±0.0031	-398.59±3.12	4080±50	3400±73
22-24 mm (35121)	4.5	-2.14±0.01	0.5349±0.0036	-465.11±3.59	5030±60	4650±140
26-28 mm (35122)	4.2	-3.69±0.02	0.5001±0.0027	-499.95±2.71	5570±50	5330±90
blanks (mean±sd, n=7)	—	-3.42±0.11	0.0112±0.0101	-988.80±10.10	38611±6672	N/A

[1]CAMS = Center for Accelerator Mass Spectrometry (LLNL).
[2]Assumed modern for reservoir correction.
[3]Replicate analysis of the 4-6 mm layer of BC-33 including all pretratments, gas transfers, etc.

thousands of discrete grains, this mixing relationship was very possible and was consistent with the observed results. In the crust samples reported here, most samples fall reasonably close to the theoretical concordia line (Fig. 10). Bulk samples of BC-33 and BC-123 both plot to the high side of the concordia (high ^{230}Th, ^{14}C), GS-1 is on the line, and the dolomitic BC-57 is the only sample that plots significantly on the low (low ^{230}Th or low ^{14}C) side. Since these ^{230}Th/^{234}U activity ratios are uncorrected for common thorium, one would expect contaminating thorium to drive the data points to the high side on this type of diagram. The location of BC-57 on the diagram may indicate that CO_2 was evolved from the dolomite in this sample as well as the CFA (the pretreatment cannot distinguish between these two phases) and that the dolomite has a significantly older carbon age than the CFA (where most of the uranium would have been derived). In crusts BC-33 and BC-123, where layer-by-layer analyses were performed, the data plot slightly on the low side of the concordia for the topmost (highest phosphate content) layers and then trend across the line and toward the higher values with depth into these crusts. The most reasonable explanation would be that these trends are an effect of increasing contamination by thorium in the deeper crustal layers, which are higher in detrital mineral content. Thus, the carbon ages support an upward growth model for these protocrusts and suggest growth rates higher than those calculated on the basis of Th and Pa ages, which are subject to contamination from associated sediment in these very young crusts.

An isochron approach.—

Another way of handling the U-series data is to assume that all layers in each examined crust are the same age within the resolution of the ^{230}Th dating method and that apparent differences are due solely to various levels of contamination. This possibility may be examined by construction of a ^{230}Th/^{232}Th-^{234}U/^{232}Th "isochron" diagram similar to that used to date volcanic rocks (Taddeuci et al., 1967; Kigoshi, 1967; Allegre, 1968) and has been used before for dating phosphatic sediments from islands (Roe et al., 1983) and the Peru shelf (Kim and Burnett, 1986). The following equation applies:

$$\frac{^{230}\mathrm{Th}}{^{234}\mathrm{U}} = \frac{^{234}\mathrm{U}}{^{232}\mathrm{Th}}\left[1 - e^{-\lambda t}\right] + \left(\frac{^{230}\mathrm{Th}}{^{232}\mathrm{Th}}\right)_o e^{-\lambda t}$$

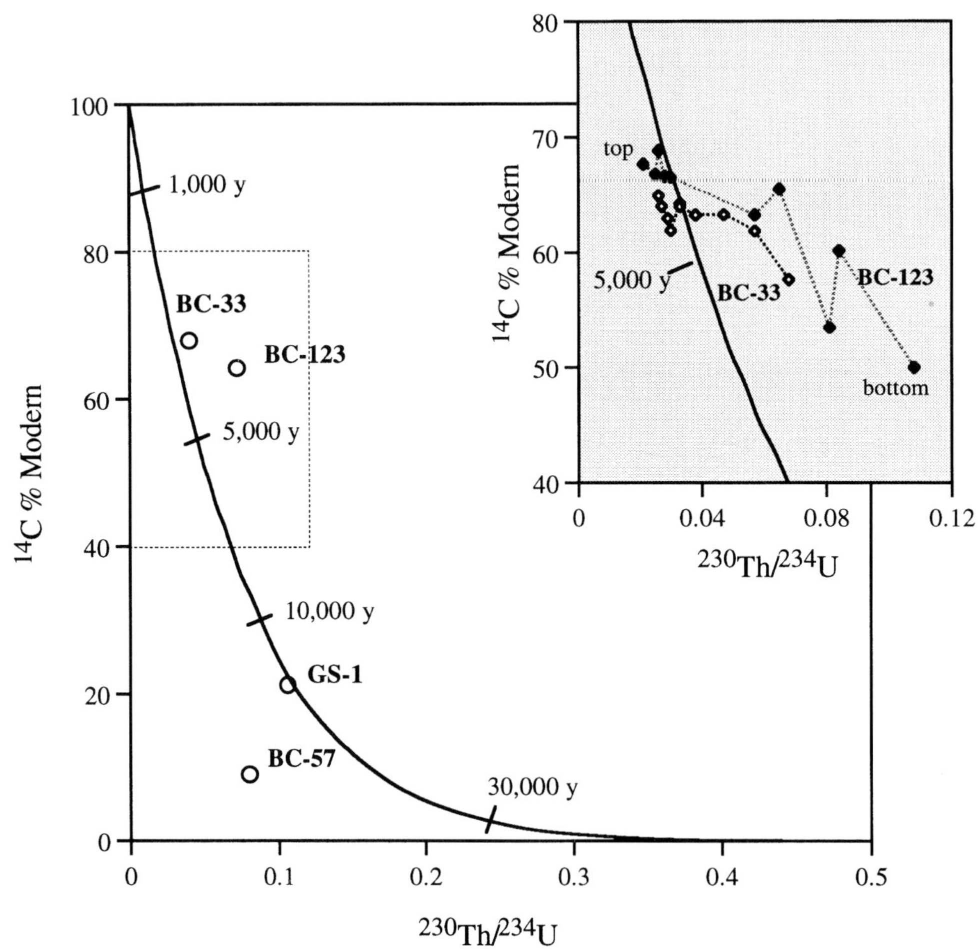

FIG. 10.—Curve showing the theoretical progression of the calibrated ^{14}C activity (expressed as percent modern) and the ^{230}Th/^{234}U activity ratio. Results based on bulk analyses of four phosphatic protocrusts are shown by the circles while the layer-by-layer results from crusts BC-33 and BC-123 are shown in the expanded section. The tick marks show the progression of these parameters in time.

Where (^{230}Th/^{232}Th) and (^{234}U/^{232}Th) are the observed activity ratios; λ is the decay constant of ^{230}Th; and t is time. Straight line segments on this type of plot have a slope of $(1-e^{-\lambda t})$ and are thus an indication of age. With time, the slope increases from 0 (zero age) to 1 (infinity).

When the U-series results from protocrusts are plotted in this manner, the data fit along very well-defined linear trends and suggest that the actual top-to-bottom age gradients must be very slight as was also indicated by the ^{14}C ages (Fig. 11). The slope of the best-fit line for BC-33 suggests an isochron age of 2480±30 y, in excellent agreement with the calibrated ^{14}C age for the bulk sample of 2270±103 y. Crust BC-57, on the other hand, has a much lower isochron age of 7520±200 y compared to the ^{14}C age of 21,930±280 y. This is very likely a reflection of a real age difference between the authigenic CFA, which hosts the U, and the dolomite, which contains the bulk of the analyzed carbon. Protocrust BC-123 has the youngest isochron age at 960±360 y while the bulk ^{14}C age was higher at 2760±68 y. This disagreement is probably not significant in view of the very low slope of the isochron line and correspondingly large uncertainty. Protocrust GS-1 has an isochron age of 5540±220 y while its bulk AMS ^{14}C age is much older at 13,440±134 y. We are not certain what causes this discrepancy, although it may be that a small amount of older dolomite has influenced the carbon ages. X-ray diffraction analysis (detection limit ≈5%) did not detect dolomite in this sample although only a few percent of old

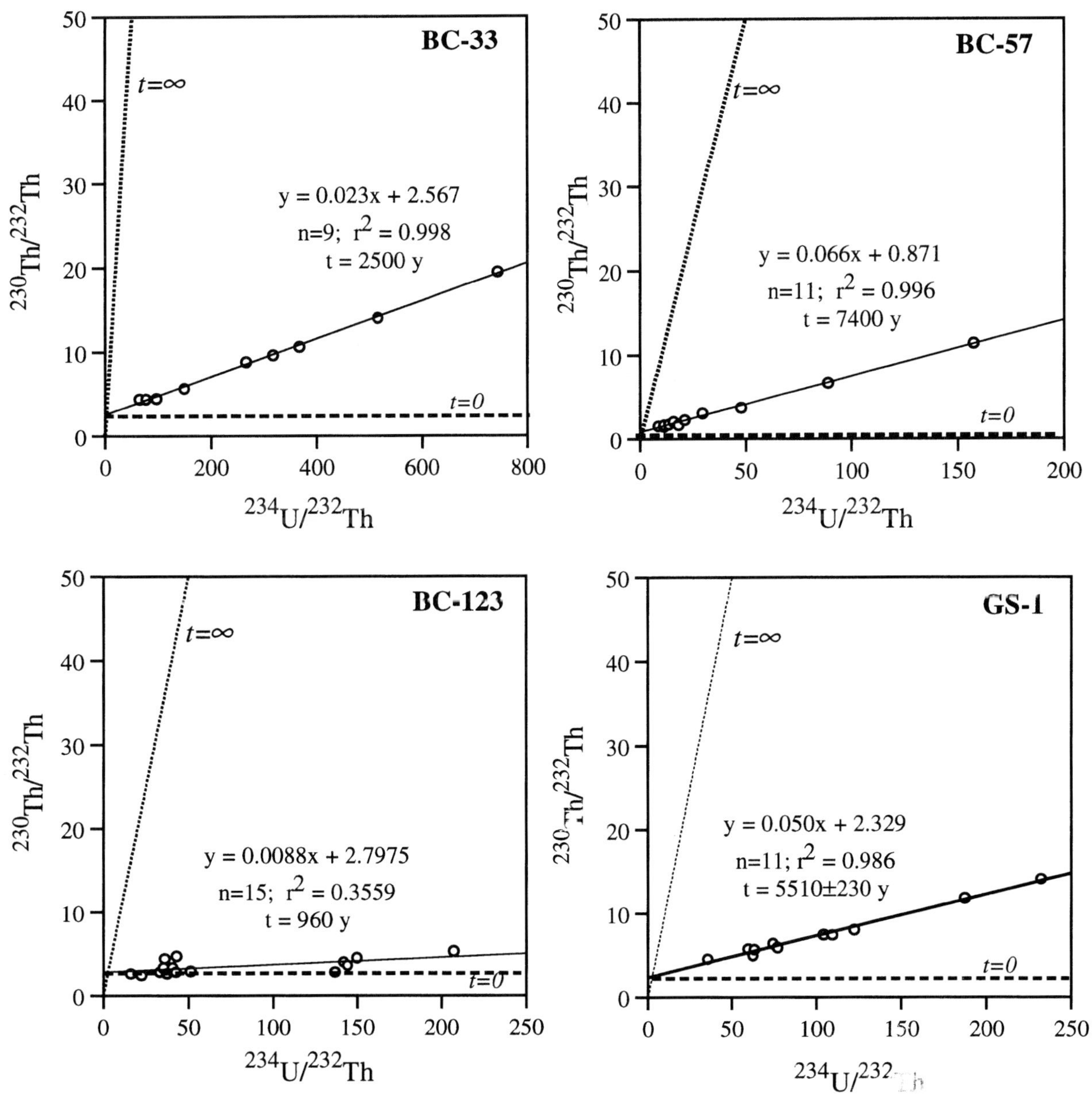

FIG. 11.—Plots of the acivity ratios ^{230}Th/^{232}Th versus ^{234}U/^{232}Th for four phosphatic protocrusts from the Peru shelf: BC-33, BC-57, BC-123, and GS-1. Ages are calculated from the slopes of the best-fit linear regression lines ("isochrons") drawn through the data in each case. Results from GS-1 taken from Kim and Burnett (1986).

dolomite could have produced this effect. We are certain that the apparently high radiocarbon result is not an analytical problem. Another large (~200 g) piece of the same crust was analyzed by conventional ^{14}C dating several years ago in a different laboratory and yielded an age of 12,300±300 y (Burnett and Kim, 1986), in excellent agreement with the conventional (non-calibrated) age of 12,460±80 y by the AMS dating method. We also note that the U-series isochron age (5540±220 y) fits the ^{14}C ages of the enclosing sediment while the bulk ^{14}C age does not. The organic carbon in the sediment just above GS-1 dated at 2965±270 y, while the sediment immediately below this buried crust yielded a conventional ^{14}C age of 9245±530 y (Kim and Burnett, 1986).

A Downward-Flux/Upwards-Growth Model of Protocrust Formation

All four of the phosphatic protocrusts examined in the detailed manner presented in this paper have the following features in common: (1) fairly uniform thicknesses of 2–3 cm; (2) systematic increases in their phosphorus and uranium concentrations from bottom to top; (3) systematic decreases in thorium (^{232}Th) concentrations from bottom to top; (3) ^{230}Th (uncorrected), ^{231}Pa, and ^{14}C ages that decrease from bottom to top with maximum age spreads of only a few thousand years; and (4) $\delta^{13}C$ values for CFA-CO_3 from 0 to –4‰. In addition, some more evolved (D-type) morphotypes have displayed increasing structural carbonate concentrations in the CFA lattice with depth (age) and an apparent decrease in crystallinity was observed from bottom to top of protocrust GS-1 (Kim and Burnett, 1986). Lamboy (1990) reported that overgrowth and/or recrystallization in the lower part of crust GS-1, which he studied by scanning electron microscopy, had transformed ovoidal particles into elongated hexagonal CFA crystals. Two crusts (BC-33 and GS-1) also showed evidence for the presence of a low-uranium phosphatic phase, perhaps indicating the presence of a residual non-CFA precursor phase such as "amorphous calcium phosphate" (ACF; Froelich et al., 1988), an "octacalcium phosphate" (OCP_p; Van Cappellen and Berner, 1991), or some as yet unidentified phase.

Any model for the formation of these protocrusts must be consistent with these observations. Taken together, these findings suggest that these crusts grew upward with CFA precipitating at or just below the sediment-seawater interface at growth rates ranging from about 7 mm/ky to 9 mm/ky. It is possible that CFA precipitation was actually initiated throughout the entire crust more-or-less simultaneously, but the upper layers grew longer as permeability was progressively occluded. Since upward growth indicates that the source of the dissolved phosphate is from above, we are left with the question of whether the overlying bottom water is that source. Although possible, we favor the interpretation that the actual growth of these crusts occurred while they were buried under a few centimeters or so of organic-rich sediment (Fig. 12). Froelich et al. (1988) showed that a very distinctive maximum in the porewater PO_4^{3-} versus depth profiles typically occurs in these sediments just below the sediment-seawater interface. Thus, this interfacial PO_4^{3-} spike creates a very large concentration gradient to drive a diffusional flux both upward (into the overlying water column) as well as downward into the sediment. We hypothesize that this downward flux drives the observed upward protocrust growth. It seems more realistic to us that the downward pore water fluxes would be responsible for the apparent continuous CFA deposi-

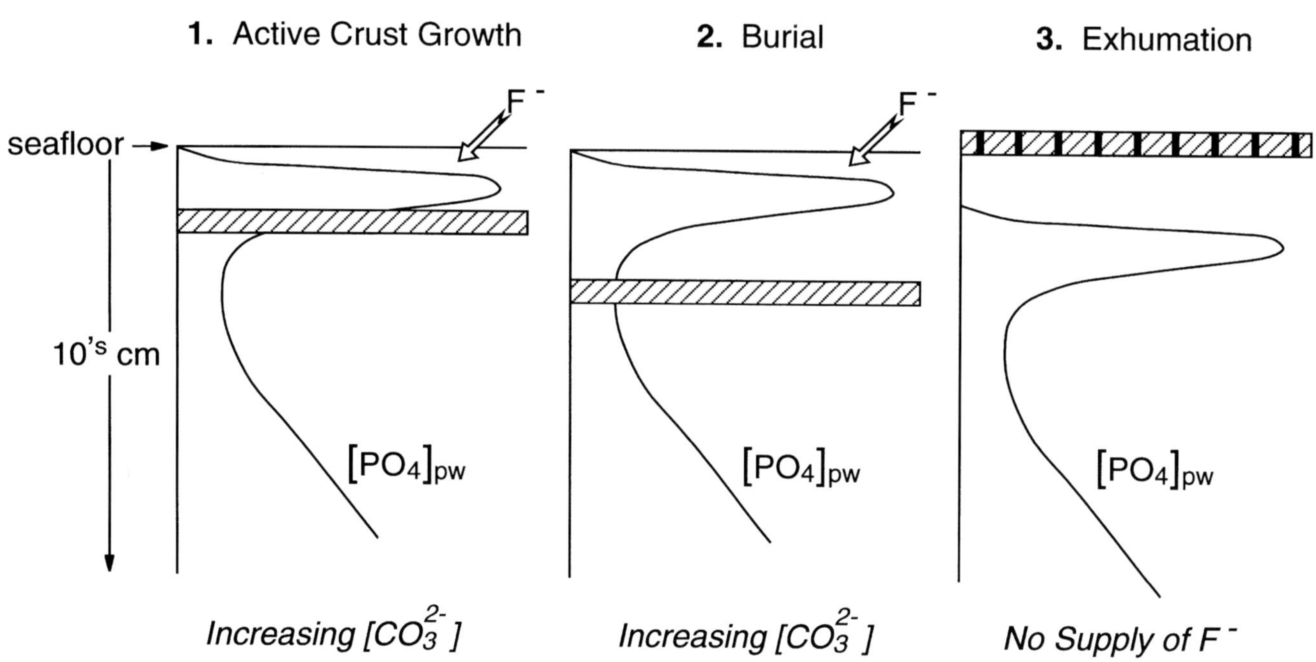

FIG. 12.—Model depicting a possible sequence of events leading to the development of a phosphatic protocrust. Active growth occurs when CFA precipitates out of solution in response to the downward diffusion of PO_4^{3-} and F^-. Increased porosity and burial below the reaction zone stops growth. Later exhumation by bottom currents may expose some of these crusts at the sediment-seawater interface.

tion (over a time scale of a few thousand years) than postulating a process (such as an iron-phosphate pump mechanism; Krom and Berner, 1981; Froelich et al., 1988) to return a fraction of the phosphate lost into the overlying water column as a source. Uptake (burial) fluxes calculated for phosphatic pellets (0.5–4.4 μmol-P/cm^2•yr; Burnett, 1990) and nodules (0.3–6.6 μmol-P/cm^2•yr; Burnett et al., 1982) agree well with estimated downward diagenetic fluxes of phosphate calculated from pore-water models (0.1–4.5 μmol-P/cm^2•yr; Froelich et al., 1988). Estimated upward benthic fluxes of phosphate into the overlying bottom water are much higher at 2–33 μmol-P/cm^2•yr (Froelich et al., 1988).

If these crusts were buried while forming, they would be analogous to the buried phosphatic layers observed in cores from off Baja California, Mexico (Schuffert et al., 1994). These investigators calculated downward diffusive fluxes (0.1–1.3 μmol-P/cm^2•yr) from pore-water profiles that are also similar to our calculated uptake estimates. Since three of the protocrusts studied here (BC-33, BC-57, and BC-123) were sampled in an oriented fashion at the sediment surface (GS-1 was found buried below about 14 cm of sediment), they must have been exposed, presumably by currents sweeping sediments away after something caused a change in current intensity or direction. This would explain why the crusts do not appear to have been actively growing at the time of sampling, that is, the top surface does not have a zero age but appears to be a few thousand years old (see Figs. 6 and 9). The extrapolated ages for the topmost (youngest) surfaces may thus indicate the times when these crusts stopped forming either because of burial beneath the reaction zone or by exhumation.

The "downward-flux/upwards-growth" model of crust formation may also explain why the phosphate concentration is consistently higher on the top surface of these crusts. The actual CFA precipitation may occur at first in a more-or-less continuous fashion throughout the crust thickness, but since the top is closer to the interfacial phosphate spike, it receives the bulk of the dissolved phosphate. As the CFA forms and fills in the sediment porosity, diffusion of phosphate (from the interfacial spike) and fluoride (from overlying seawater) to the lower layers becomes more limited and eventually ceases while precipitation continues on the topmost surface. This concept is somewhat similar to the proposal by Lucas and Prévôt-Lucas (1997) of several stages of crystallization to account for aggregates of cryptocrystalline CFA. This apparently continuous CFA deposition will continue until the flux dissipates (perhaps by burial away from the phosphate spike) and/or the crust is exhumed and exposed at the surface. At that point, the protocrusts are subject to erosion from boring organisms and other processes such as current scouring and may break the crusts into smaller units, turn over, be re-cemented, buried, etc., and occur as the familiar "D"-phosphates, hardgrounds, and nodules.

SUMMARY AND CONCLUSIONS

Our detailed studies of oriented phosphatic protocrusts collected from the Peru shelf show that these 2–3 cm crusts are characterized by strong compositional gradients. Crusts have the highest concentration of phosphate and uranium at the uppermost surface with systematic decreases downward. Thorium and detrital mineral concentrations have opposite trends. Exact interpretations of U-series disequilibrium ages are difficult because of the young age of these samples and the presence of initial daughter nuclides (^{230}Th, ^{231}Pa). AMS ^{14}C ages of CFA-carbon are within the range set by the upper- and lower-bounds of the uncorrected and corrected ^{230}Th ages. Furthermore, the ^{14}C age versus depth trends suggest relatively rapid (7–9 mm/10^3 yrs) upward growth in agreement with U-series isochrons that show minimal age gradients within each crust.

These results, together with additional information about the phosphatizing environment on the Peru shelf, led us to suggest that these protocrusts form at a very shallow depth below the sediment-seawater interface in response to the downward diffusion of dissolved PO_4^{3-} from an interstitial phosphate maximum and downward diffusion of F^- from overlying seawater. CFA precipitation is probably more-or-less continuous throughout the crust but continues longer in the uppermost layers as the crust is buried. Eventually, as the existing sediment grains are cemented by precipitation of CFA, porosity is lowered to the point where diffusion to the lower layers is no longer possible and growth ceases entirely when the top surface is buried beneath the reaction zone. Later exhumation may expose some of these crusts at the sediment surface where they may become burrowed and eroded into the commonly observed phosphatic nodules, conglomerates, and hardgrounds.

ACKNOWLEDGMENTS

The authors wish to acknowledge the assistance of Johanna Resig (UH), who hand-separated the foraminiferal samples used for the ^{14}C reservoir correction. S. Rajan (UH) assisted in the sample preparation for XRD and stable isotope measurements, and Alan Baker and Laura Latham (FSU) performed the sample pretreatments for the AMS ^{14}C analyses. We are indebted to the crew and scientific party of RV *Seward Johnson* and Johnson Sea Link I Cruise SJ10-92 for their roles in sample and data collection, and express particular thanks to co-chief scientists Rick Jahnke and Michael Arthur for organizing and obtaining funding for the expedition. An earlier version of this manuscript benefited from the careful reviews of Norbert Clauer, Liliane Prévôt-Lucas, Jacques Lucas, and an anonymous reviewer. Financial support for this study was provided by grants from NSF (Marine Geology and Geophysics Program) to W.C.B. (OCE9214493) and C.R.G. (OCE9201305). AMS radiocarbon analyses were performed under the auspices of the U.S. Department of Energy by Lawrence Livermore National Laboratory under contract W-7405-ENG-48. University of Hawaii, School of Ocean and Earth Science and Technology Contribution #4790.

REFERENCES

ALLEGRE, C. J., 1968, ^{230}Th dating of volcanic rocks: A comment: Earth and Planetary Science Letters, v. 5, p. 209-210.

BATURIN, G. N., 1971, Stages of phosphorite formation on the ocean floor:

Nature (London), Phys. Sci., v. 232, p. 61-62.

Baturin, G. N., Merkulova, K. I., and Chalov, P. I., 1972, Radiometric evidence for recent formation of phosphatic nodules in marine shelf sediments: Marine Geology, v. 13, p. M37-M41.

Broecker, W. S., 1963, A preliminary evaluation of uranium series inequilibrium as a tool for absolute age measurement on marine carbonates: Journal Geophysical Research, v. 68, p. 2817-2834.

Burnett, W. C., 1977, Geochemistry and origin of phosphorite deposits from off Peru and Chile: Geological Society of America Bulletin, v. 88, p. 813-823.

Burnett, W. C., 1980, Apatite–glauconite associations off Peru and Chile: Paleo-oceanographic implications: Journal of the Geological Society of London, v. 137, p. 764-767.

Burnett, W. C., 1990, Phosphorite growth and sediment dynamics in the modern Peru shelf upwelling system, *in* Burnett, W. C., and Riggs, S. R., eds., Phosphate Deposits of the World, V. III: Neogene to Modern Phosphorites: London, Cambridge University Press, p. 62-72.

Burnett, W. C., Baker, K. B., Chin, P. A., McCabe, W. and Ditchburn, R., 1988, Uranium-series and AMS ^{14}C studies of modern phosphatic pellets from Peru shelf muds: Marine Geology, v. 80, p. 215-230.

Burnett, W. C., Beers, M. J., and Roe, K. K., 1982, Growth rates of phosphate nodules from the continental margin off Peru: Science, v. 215, p. 1616-1618.

Burnett, W. C. and Veeh, H. H., 1977, Uranium-series disequilibrium studies in phosphorite nodules from the west coast of South America: Geochimica et Cosmochimica Acta, v. 41, p. 755-764.

Burnett, W. C. and Veeh, H. H., 1992, Uranium-series studies of marine phosphates and carbonates, *in* Ivanovich, M., and Harmon, R. S., eds., Uranium-Series Disequilibrium: Applications to Earth, Marine, and Environmental Sciences: London, Oxford University Press, p. 487-512.

Burnett, W. C. and C.-C. Yeh, 1995, Separation of protactinium from geochemical materials via extraction chromatography: Radioactivity & Radiochemistry, v. 6, p. 20-28.

Chen, J. H., Edwards, R. L., and Wasserburg, G. L., 1986, ^{238}U, ^{234}U, and ^{232}Th in seawater: Earth and Planetary Science Letters, v. 80, p. 241-251.

Froelich, P. N., Arthur, M. A., Burnett, W. C., Deakin, M., Hensley, V., Jahnke, R., Kaul, L., Kim, K. H., Soutar, A., and Vathakanon, C., 1988, Early diagenesis of organic matter in Peru continental margin sediments: Phosphorite precipitation: Marine Geology, v. 80, p. 309-346.

Froelich, P. N., Kim, K. H., Jahnke, R., Burnett, W. C., Soutar, A. and Deakin, M., 1983, Pore water fluoride in Peru continental margin sediments: Uptake from seawater: Geochimica et Cosmochimica Acta, v. 47, p. 1605-1612.

Garrison, R. E., and Kastner, M., 1990, Phosphatic sediments and rocks recovered from the Peru Margin during ODP Leg 112, *in* Suess, E., von Huene, R., et al., eds., Proceedings of the Ocean Drilling Program, Scientific Results: v. 112, p.111-134.

Glenn, C. R., and Arthur, M. A., 1988, Petrology and major element geochemistry of Peru margin phosphorites and associated diagenetic minerals: Authigenesis in modern organic-rich sediments: Marine Geology, v. 80, p. 231-267.

Glenn, C.R., Arthur, M.A., Yeh, H.W., and Burnett, W.C., 1988, Carbon isotopic composition and lattice-bound carbonate of Peru-Chile margin phosphorites: Marine Geology, v. 80, p. 287-307.

Glenn, C. R., Arthur, M. A., Resig, J. M., Burnett, W. C., Dean, W. E. and Jahnke, R. A., 1994, Are modern and ancient phosphorites really so different? *in* Iijima, A., Abed, A. M., and Garrison, R. E., eds., Siliceous, Phosphatic and Glauconitic Sediments of the Tertiary and Mesozoic: The Netherlands, VSP BV, p. 159-188.

Gulbrandsen, R. A., 1970, Relation of Carbon Dioxide Content of Apatite of the Phosphoria Formation to Regional Facies: Washington, D.C., U.S. Geological Survey Professional Paper, 700(B), p. B9-B13.

Jahnke, R. A., 1984, The synthesis and solubility of carbonate fluorapatite: American Journal of Science, v. 284, p. 58-78.

Jahnke, R. A., Emerson, S. R., Kim, K. H., and Burnett, W. C., 1983, The present day formation of apatite in Mexican continental margin sediments, Geochimica et Cosmochimica Acta, v. 47, p. 259-266.

Jarvis, I., 1992, Sedimentology, geochemistry and origin of phosphatic chalks: The Upper Cretaceous deposits of NW Europe: Sedimentology, v. 39, p. 55-97.

Kaufman, A., and Broecker, W. S., 1965, Comparison of ^{230}Th and ^{14}C ages for carbonate materials from Lakes Lahontan and Bonneville: Journal of Geophysical Research, v. 70, p. 4039-4054.

Kigoshi, K., 1967, Ionium dating of igneous rocks: Science, v. 156, p. 932-934.

Kim, K. H., and Burnett, W. C., 1985, ^{226}Ra in phosphate nodules from the Peru/Chile seafloor: Geochimica et Cosmochimica Acta., v. 49, p. 1073-1081.

Kim, K. H., and Burnett, W. C., 1986, Uranium-series growth history of a Quaternary phosphatic crust from the Peruvian continental margin: Chemical Geology, v. 58, p. 227-244.

Kolodny, Y., 1969, Are marine phosphorites forming today? Nature, v. 224, p. 1017-1019.

Kolodny, Y., and Kaplan, I. R., 1970, Uranium isotopes in sea-floor phosphorites: Geochimica et Cosmochimica Acta, v. 34, p. 3-24.

Kress, A. G., and Veeh, H. H., 1980, Geochemistry and radiometric ages of phosphatic nodules from the continental margin of northern New South Wales, Australia: Marine Geology, v. 36, p. 143-157.

Krom, M. D., and Berner, R. A., 1981, The diagenesis of phosphorus in a nearshore sediment: Geochimica et Cosmochimica Acta, v. 45, p. 207-216.

Ku, T.-L., Knauss, K. G., and Mathieu, G. G., 1977, Uranium in open ocean: Concentration and isotopic composition: Deep-Sea Research, v. 24, p. 1005-1017.

Lamboy, M., 1990, Microstructures of a phosphate crust from the Peruvian continental margin: Phosphatized bacteria and associated phenomena: Oceanologia Acta, v. 13, p. 439-451.

Lucas, J., and Abbas, M., 1989, Uranium in natural phosphorites: The Syrian example: Sciences Geol. Bulletin, v. 42, p. 223-236.

Lucas, J., and Prévôt-Lucas, L., 1997, On the genesis of sedimentary apatite and phosphate-rich sediments, *in* Paquet, H., and Clauer, N., eds., Soils and Sediments: Berlin, Springer-Verlag, p. 249-268.

Lucas, J., Flicoteaux, R., Nathan, Y., Prévôt, L., and Shahar, Y. 1980, Different aspects of phosphorite weathering, *in* Bentor, Y. K., ed., Marine Phosphorites: Geochemistry, Occurrence, Genesis: Tulsa, Society for Sedimentary Geology (SEPM), v. 29, p. 41-51.

Manheim, F. T., Rowe, G. T., and Jipa, D., 1975, Marine phosphorite formation off Peru: Journal of Sedimentary Petrology, v. 45, p. 243-251.

Matthew, A., and Nathan, Y., 1977, The decarbonation of carbonate fluorapatite (francolite): American Mineralogist, v. 62, p. 565-573.

McArthur, J. M., 1980, Post-depositional alteration of the carbonate-fluorapatite phase of Moroccan Phosphates, *in* Bentor, Y. K., ed., Marine Phosphorites: Geochemistry, Occurrence, Genesis, v. 29: Tulsa, Society for Sedimentary Geology (SEPM), p. 53-60.

McArthur, J. M., 1985, Francolite geochemistry—compositional controls during formation, diagenesis, metamorphism and weathering: Geochimica et Cosmochimica Acta, v. 49, p. 23-25.

McClellan, G. H., 1980, Mineralogy of carbonate fluorapatites: Journal of the Geological Society of London, v. 137, p. 675-681.

McClellan, G. H., and Van Kauwenbergh, S. J., 1990, Mineralogy of sedimentary apatites, *in* Notholt, A. J. G., and Jarvis, I., eds., Phosphorite Research and Development: London, Geological Society of London Special Publication 52, p. 23-31.

Nathan, Y., 1984, The mineralogy and geochemistry of phosphorites, *in* Nriagu, J. O., and Moore, P. B., eds., Phosphate Minerals: Berlin, Springer-Verlag, p. 275-291.

Panczer, G., Nathan, Y., and Shiloni, Y., 1989, Trace element characterization of phosphate nodules in Israel: Sciences Geol. Bulletin, v. 42, p. 173-184.

Roe, K. K., Burnett, W. C., and Lee, A. I. N., 1983, Uranium disequilibrium dating of phosphate deposits from the Lau Group, Fiji: Nature, v. 302, p. 603-606.

Rosato, V. J., and Kulm, L. D., 1981, Clay mineralogy of the Peru continental margin and adjacent Nazca Plate: Implications for provenance, sea level changes, and continental accretion: Boulder, Geological Society of America Memoir 154, p. 545-568.

Schuffert, J. D., Jahnke, R. A., Kastner, M., Leather, J., Sturz, A., and Wing, M. R., 1994, Rates of formation of modern phosphorite off western Mexico: Geochimica et Cosmochimica Acta, v. 58, p. 5001-5010.

Silverman, S. R., Fuyat, R. K., and Weiser, J. D., 1952, Quantitative determination of calcite associated with carbonate-bearing apatites: American Mineralogist, v. 37, p. 211-222.

STUIVER, M., AND POLACH, H. A., 1977, Discussion: Reporting of ^{14}C data: Radiocarbon, v. 19, p. 355-363.

STUIVER, M., AND REIMER, P. J., 1993, Extended ^{14}C data base and revised CALIB 3.0 ^{14}C age calibration program: Radiocarbon, v. 35, p. 215-230.

SUESS, E., 1981, Phosphate regeneration from sediments of the Peru continental margin by dissolution of fish debris: Geochimica et Cosmochimica Acta, v. 45, p. 577-588.

TADDEUCI, A., BROECKER, W. S., AND THURBER, D. L., 1967, ^{230}Th dating of volcanic rocks: Earth and Planetary Science Letters, v. 3, p. 338-342.

WALSH, T. W., SMITH, S. V., AND VINK, S., 1991, Review of particulate C:N:P methods used by UH/HIMB analytical services, *in* Hurd, C. D., and Spencer, D. W., eds., Marine Particles: Analysis and Characterization, Geophysical Monograph 63: Washington, D.C., American Geophysical Union, p. 91-96.

VAN CAPPELLEN, P., AND BERNER, R. A., 1991, Fluorapatite crystal growth from modified seawater solutions: Geochimica et Cosmochimica Acta, v. 55, p. 1219-1234.

VEEH, H. H., BURNETT, W. C. AND SOUTAR, A., 1973, Contemporary phosphorites on the continental margin of Peru: Science, v. 181, p. 844-845.

VEEH, H. H., 1985, Uranium-series dating applied to phosphate deposits on coral reef islands, in Proceedings of the Fifth International Coral Reef Congress, Tahiti, v. 3, p. 463-469.

YEH, C.-C., 1994, Uranium-series disequilibrium dating of ODP Leg 112 and other phosphorites from the Peru Margin: Unpublished M.S. Thesis, Florida State University, Tallahassee, 99 p.

ZANIN, YU. N., AND KRIVOPUTSKAYA, L. M., 1976, Transformation of apatite material of phosphorite in catagenesis: Doklady of the Academy of Sciences USSR Earth Science Section, v. 230, p. 200-202.

FORMATION AND EVOLUTION OF PHOSPHORITE GRAINS AND NODULES ON THE NAMIBIAN SHELF, FROM RECENT TO PLEISTOCENE

G. N. BATURIN*

Institute of Oceanology, Russian Academy of Sciences, 36 Prospect Nakhimova, Moscow, 117 851, Russia

ABSTRACT: Phosphatic grains and nodules related to recent diatom oozes on the inner Namibian shelf as well as Pleistocene grains and nodules from the phosphorite deposit on the adjacent outer shelf have been studied by means of scanning electron microscopy combined with several analytical methods.

Recent grains and nodules are being formed simultaneously in the same diatom oozes. Both consist of poorly, moderately, and well-lithified varieties that reflect the process of progressive phosphatization, compaction, and crystallization of phosphatic matter along with dissolution and expulsion of nonphosphatic components. Similar processes are going on during the phosphatization of macro- and microcoprolites. The common varieties of apatite mineral morphologies are colloform masses without distinct texture, globules and globular aggregates with more or less pronounced radial crystallization, and micrometer-sized elongated particles and their aggregates of various shape, which are interpreted as diagenetically formed crystallites. All of these textures are related to each other, with the globular one becoming predominant in most lithified accretions. Reworked Pleistocene grains reveal textural features quite similar to the recent grains. They contain similar organic remains, which proves their common origin in the same environment. The chemical evolution of recent accretions in the course of their lithification consists of gradual increases of phosphatic components and decreases of nonapatite components. The most striking feature of their geochemistry is the extremely low concentration of rare Earth elements in recent grains in contrast to their high concentration in Pleistocene grains. This might mean that phosphate absorbs these elements during prolonged contact with the ambient seawater.

INTRODUCTION

The Namibian shelf is the part of the east Atlantic phosphogenic belt that stretches along the western continental margin from the Pyrenean peninsula in the north to the Agulhas Bank in the south. It was famous for its fish resources as early as in the middle of the 19th century. Later it drew the attention of oceanographers as the site of coastal upwelling and associated phenomena such as high biologic productivity, red tides, and hydrogen sulfide excursions, with consequent mass mortality of marine fauna (Copenhagen, 1953; Brongersma-Sanders, 1957; Hart and Currie, 1960).

The first sediment samples recovered in this "azoic" zone in the Walvis Bay turned out to be soft diatomaceous oozes hosting enormous populations (10^7 cells per cm^3) of filamentous sulfate-reducing bacteria without any signs of other marine life (Butlin, 1949).

From 1960 to 1970 the fishery resources of this area were investigated by several expeditions of the Ministry of Fisheries of the former Soviet Union. Along with oceanographic data, these expeditions recovered a series of sediment samples, including phosphatic and glauconitic sands. Phosphatic components were also found in terrigenous, calcareous, and diatomaceous sediments (Senin, 1968, 1970; Emelyanov and Senin,1969; Emelyanov, 1973 a,b).

In 1968 the location near the Walvis Bay was investigated during the CIRCE expedition of the Scripps Institution of Oceanography on board RV *Argo,* and by Dutch oceanographers working with the Marine Research Laboratory, Walvis Bay, in 1968-1969. Both expeditions provided collections of bottom samples for geochemical investigations (Calvert and Price, 1970, 1983; Brongersma-Sanders et al., 1980).

In 1968 and 1972 this area was revisited by the Soviet research vessels *Akademic Kurchatov* and *Mikhail Lomonosov*, which recovered representative collections of phosphatic accretions, including the recent samples, thus renewing attention in this area (Baturin, 1969, 1971, 1982; Baturin et al., 1970; Veeh et al., 1974).

Simultaneously, several expeditions organized by Cape Town University and the Geological Survey of South Africa continued the detailed investigation of this area, where one of the greatest phosphate deposits on the seafloor was discovered on the outer shelf (Bremner and Willis, 1975; Bremner and Rickard, 1977; Rogers, 1977; Bremner, 1978, 1980; Bremner and Rogers, 1990).

Subsequently, in 1992, the Russian RV *Professor Logachev*, working under contract with the South African Geological Survey, recovered several cores and two large volume samples of the phosphatic sands in this area for future detailed analysis (Bremner, 1992).

MATERIALS AND METHODS

The materials used for this study include recent phosphatic accretions and phosphatized coprolites, enclosing oozes recovered by the author from five stations on the inner Namibian shelf, and two large volume samples of phosphatic sands from the outer Namibian shelf recovered by the *Professor Logachev* expedition and provided by its chief geologist, O. N. Kolosov. The locations of these stations are shown in Table 1.

The larger nodules were handpicked, while the sand fractions from diatomaceous oozes were first concentrated onboard the ship by washing and later separated in the laboratory by wet sieving. Since these fractions contained components other than phosphatic grains (fish debris, quartz, etc.), the material for scanning electron microscope (SEM) and microchemical analysis was handpicked, while the materials for bulk chemical analyses were sieved and handpicked, leaving only minor amounts of impurities.

*email: batur@geo.sio.rssi.ru

Marine Authigenesis: From Global to Microbial, SEPM Special Publication No. 66
 ISBN 1-56576-064-6, p. 185-199.

TABLE 1.—LOCATION OF SAMPLES

Station Number	Latitude S	Longitude E	Depth M	Research Vessel
152	22°40.8'	14°20.6'	76	*Akademik Kurchatov*
209	23°30.0'	14°05'	120	-"-
2046	22°40'	14°15.6'	87	*Mikhail Lomonosov*
2047	22°27'	14°10.5'	78	-"-
2048	22°08'	13°18.4'	85	-"-
L4-12	25°15.9'	14°03'	199	*Professor Logachev*
L4-17	24°54.9'	13°54'	190	-"-
215	17°25'	11°29'	150	*Sozidatel*

The material was studied under the SEMs coupled with microprobe analyzer using conventional techniques at the Geological Institute of the Strasbourg University, the Institute of Paleontology, and the Institute of Mineral Resources in Moscow.

The chemical composition of phosphorites was determined by the following methods: wet chemistry (P, CO_2, Corg, F, Stot) was used at the Institute of Oceanology, Moscow, and at the Geological Institute, Strasbourg; atomic absorption (major and minor elements) was used at the Institute of Oceanology; microchemistry (P, Si, Al) was used at the Institute of Oceanology; and ICP-MS (rare earth elements and yttrium) was used at both the Geological Institute and the Institute of Oceanology. The standard samples used for these analyses include B.C.R. N 32 (Moroccan phosphorite) and SDO-8 (oceanic diatomaceous ooze).

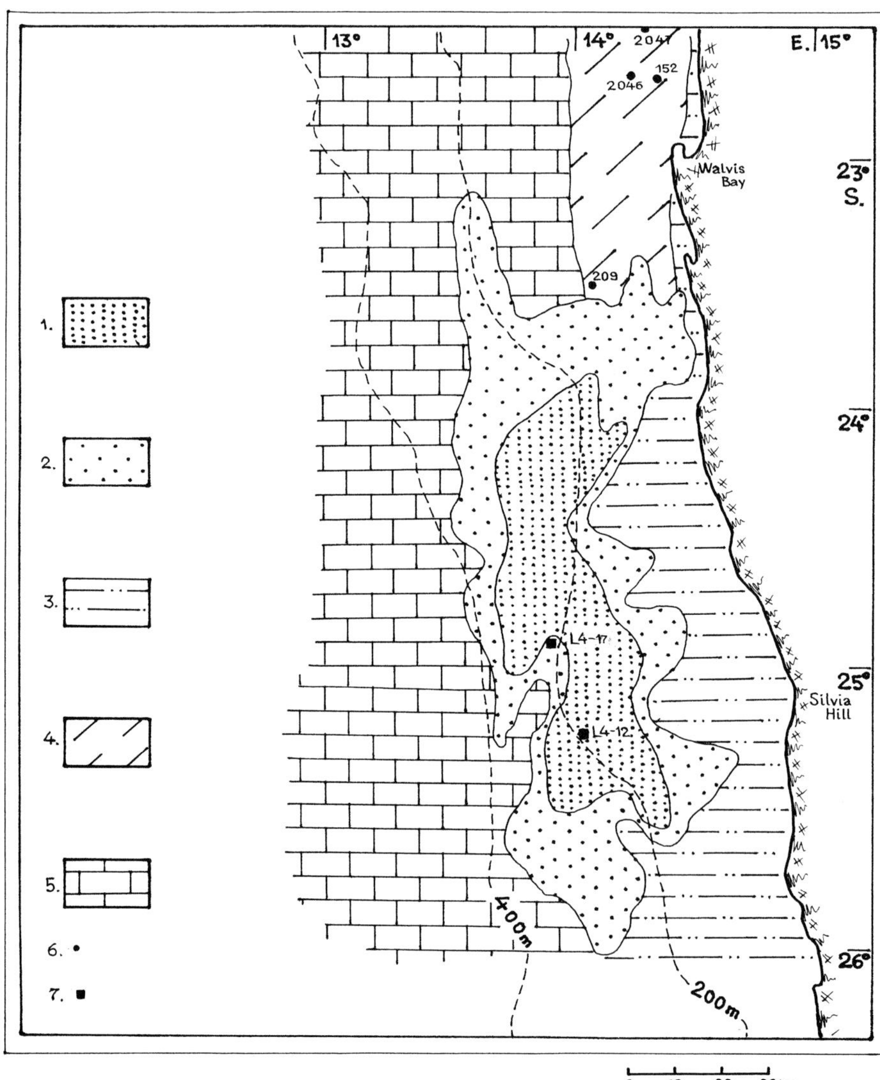

FIG. 1.—The scheme of sediment type distribution on the southern part of the Namibian shelf (based on data from Emelyanov and Senin, 1969; Senin, 1970; Emelyanov, 1973a, b; Rogers, 1977; Bremner, 1978; Bremner and Rogers, 1990; Bremner, 1992; and author's materials). (1) Phosphatic sands; (2) Shelly phosphatic sands; (3) Terrigenous sediments; (4) diatomaceous oozes; (5) calcareous sediments; (6) stations studied by RV Akademik Kurchatov and Mikhail Lomonosov; (7) stations studied by RV Professor Logachev.

TABLE 2.—AVERAGE CHEMICAL COMPOSITION OF SEDIMENT SAMPLES (%) FROM THE NAMIBIAN SHELF

Sediment type	Number of samples	$CaCO_3$	SiO_2	Corg	Fe_2O_3	MnO	TiO_2	P_2O_5
Terrigenous	11	8.43	2.59	2.46	3.30	0.02	0.43	0.41
Calcareous	14	60.49	0.50	3.55	1.39	0.01	0.17	1.69
Glauconitic	1	20.51	1.11	0.89	18.60	0.02	0.45	0.58
Slightly diatomaceous	6	9.24	19.75	7.01	3.03	0.01	0.37	0.50
Diatomaceous	10	4.49	39.9	5.35	1.31	0.01	0.01	0.82
Phosphatic sands	2	39.8	3.7	2.87	2.3	0.01	0.09	17.3

Based on data from Emelyanov and Senin, 1969; Senin, 1970; Emelyanov, 1973a, b.

SEDIMENTS

The sediments of the Namibian shelf are varied and include terrigenous, calcareous, diatomaceous, phosphatic, and glauconitic sediments, as well as sediments of mixed composition.

The scheme of the major sediment type distributions in the southern part of the Namibian shelf is shown in Figure 1.

The belt of diatomaceous oozes stretches along the shore approximately from 18° to 23°30'S between 50-60 m depth to 120-140 m depth. An area of phosphatic and shelly phosphatic sands lies to the south that extends roughly along the 200-m isobath from 23° to 26°S, and is 70-90 km wide.

The sediments outside the area of phosphatic sands are more or less enriched in phosphorus, as indicated by their average chemical composition (Table 2), owing to inclusions of phosphatic material. The latter is represented by phosphatic grains, fish debris, phosphatized coprolites, and occasionally the debris of recent phosphatic brachiopod *Disciniscus tenius*, which contains, on the average, 23.5% P_2O_5 (Bremner, 1980).

Phosphatic grains, which will be described later, have been found in all types of shelf sediments. Fish debris is most abundant in diatomaceous oozes where more than the half of the sand traction consists of fish remnants, including on average, 35% rib and skull debris, 11% vertebrae, 4,5% scales, and less than 1% teeth (Bremner, 1980). However, especially near the Walvis Bay, our observations show some sediment samples that contain up to 10% or even more of fish scales and bones that were possibly left after repeated episodes of mass mortality.

Phosphatized coprolites, which are also most abundant in diatom oozes, are from about 1 mm to several centimeters long and were derived from various organisms, including fish and sea lions. The bigger coprolites contain fish bones and have a rough surface while the small coprolites contain diatom remains and are relatively smooth.

The age of diatomaceous oozes and associated accretions determined by organic remains and isotope methods is recent (Baturin et al., 1972; Veeh et al., 1974), while the age of phosphatic sands from the outer shelf determined by the Sr method is predominantly Pleistocene (McArthur et al., 1990).

TEXTURES OF RECENT GRAINS AND NODULES

Morphology

Both small and large accretions, or grains and nodules associated with diatom oozes, are roughly divided into nonlithified and lithified varieties, with an intermediate variety represented by moderately lithified accretions. The nonlithified varieties are soft to friable, predominantly whitish or yellowish, with rough surfaces owing to inclusions of diatom frustules. The moderately lithified accretions are more compact, gray to light brown, and the well lithified varieties are quite hard and brown to black with relatively smooth surfaces. The morphology of all accretions varies, and includes relatively isometric, flat, oval, and irregular shapes, with elongated shapes evidently belonging to coprolites. The morphologies of some grains are shown in Plate 1.

Textures of Nonlithified Accretions

The inner texture of the softer accretions is strongly influenced by the abundance of organic remains, mainly diatom frustules and their debris; these are of the same recent species that are present in the enclosing diatomaceous oozes. In some samples, coccoliths belonging to *Emiliania Huxleyi* sp. are also present.

Mainly colloform, globular, and microcrystalline varieties represent phosphatic matter itself.

Colloform phosphate appearing as porous irregular clumps is seen on the surface of diatom frustules and in interstices between the biogenic particles of poorly consolidated sediments (Plate 2A, B). Occasionally, this mass assumes the form of globular aggregates; some globules seem to be hollow inside with open apertures (Plate 2C).

Microcrystalline phosphate is represented by micrometer-sized, elongated particles that appear individually or in the form of intergrowths or more complex aggregates on the surface of diatom frustules and in voids (Plate 2D, F). The ends of these particles may assume acicular, rounded, or botryoidal form. In the latter case their ends may grow thicker than the central part of their bodies so that they look like dumbbells. Their intergrowths are highly variable, forming twins, trillings, crosses, druses, rosettes, asterisks, and globular bodies. In some places they are densely packed together in a parallel manner to form aggregates with a botryoidal appearance (Plate 2F).

PLATE 1.—The forms of recent phosphatic grains from diatom oozes. (A, B) Friable gray grains with rough surface owing to embedded diatom frustules. (C, D) Hard black grains, rounded (C) and irregular (D). (E, F) Soft white coprolites.

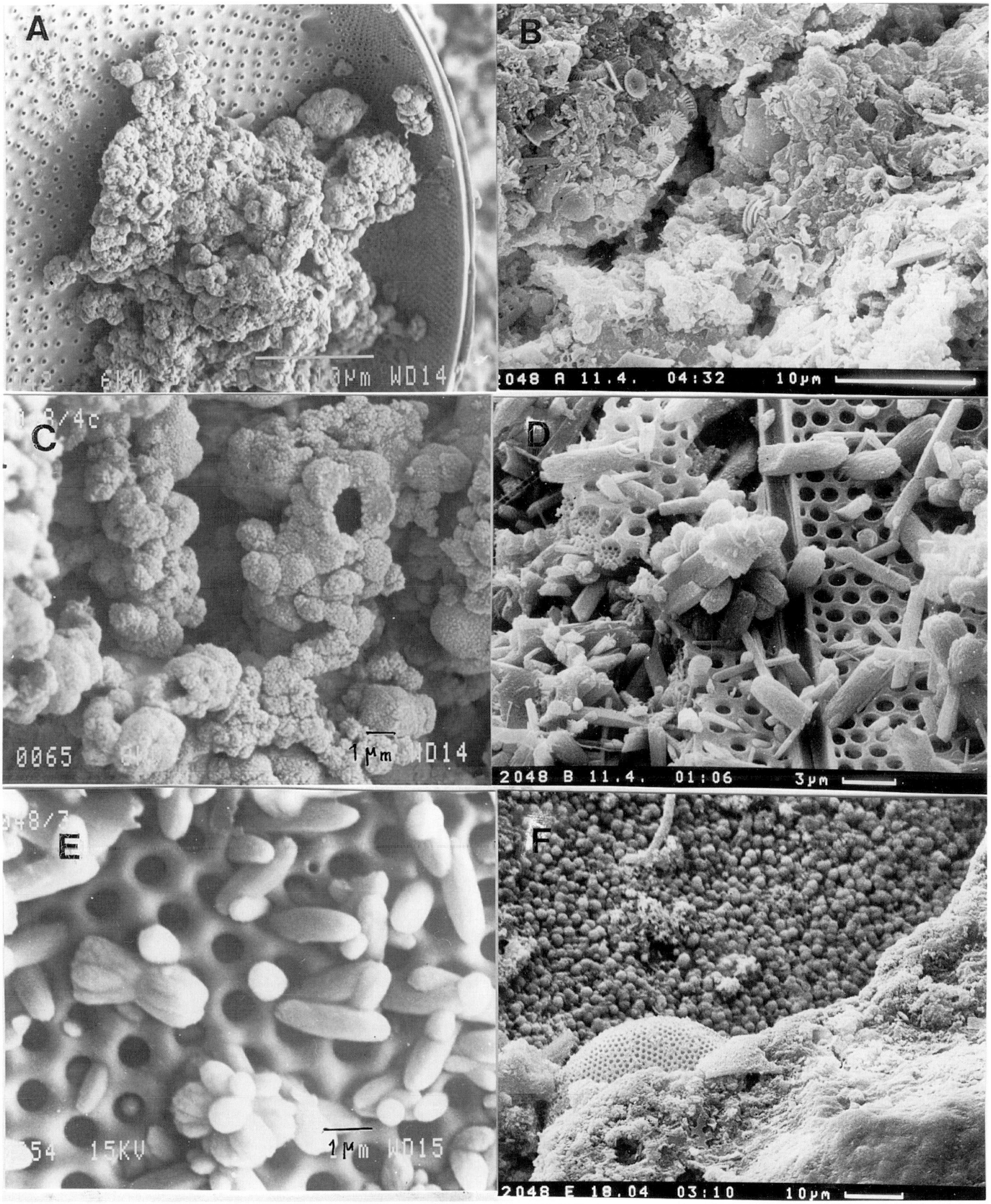

PLATE 2.—Textures of soft and friable phosphatic accretions from diatom oozes. (A) Colloform porous phosphatic aggregate on the surface of recent diatom (*Actinocyclus* sp.) frustule in the friable nodule. (B) Colloform mass of phosphate with inclusions of diatom detritus and tests of coccoliths *Emiliania huxleyi* (Lohm) Hay, Mokler in the soft grain. (C) Irregular globular aggregate with botryoidal surface in the friable nodule; some globules are hollow with visible apertures. (D) Rod-shaped phosphatic particles and their aggregates growing on diatom debris in the soft white grain. The ends of rods reveal botryoidal texture. Some rods are broken, showing their compact structure. (E) Rod-shaped particles and their multiform aggregates on the gridded diatom surface in the soft nodule. The rods coalesce in a parallel manner to form a thicker dumbbell-shaped aggregate seen in the central part of the photograph and a stellar-form aggregate below. (F) Densely packed aggregate of rods in the cavity among the mass of colloform phosphate in the soft white grain.

Textures of Lithified Accretions

Lithified grains and nodules contain much less biogenic material owing to its partial dissolution. Meanwhile, their phosphatic matter is represented by essentially the same morphologies found in nonlithified accretions.

Colloform phosphate appears as irregular clumps that often assume more or less globular shapes, especially where the rock shows significant voids. Some of these globules, whose diameter is about 10 μm, have central holes (Plate 3A). In other places the colloform mass is aggregated into individual isometric or irregular smaller clumps some of which also have central holes (Plate 3B). Generally, in the lithified accretions a much greater part of the phosphate assumes the globular form than in nonlithified accretions. Most globules acquire botryoidal surfaces and show a radial crystallization that may reach their central parts; in other globules, such crystallization is limited to their outer layer while the central part remains structureless (Plate 3C, D).

Rod-shaped particles in lithified accretions are rather common but not predominant, and they occur individually or as aggregates in association with globules, colloform matrix, or on diatom surfaces (Plate 3C, E, F). The size of these particles varies, from about 0.5 μm to 5-6 μm. Some of these are only half developed, with their bases resting on diatom frustules. The surfaces of the bigger particles are roughly striated, as if they consisted of bundles of finer needles (Plate 3E, F).

Textures of Phosphatized Coprolites

The nonlithified coprolites contain abundant organic remains, mainly diatom detritus embedded in porous colloform matrix of knobby appearance (Plate 4A). In other parts, phosphate takes the form of irregular porous clumps among which globules are being formed, some with holes inside (Plate 4B). The well-formed globular aggregates are also present, consisting of globules with smooth surfaces stacked together (Plate 4C).

The moderately and well-lithified coprolites, both big and small, contain fewer organic remains and are characterized by the predominance of more developed globular textures.

Individual globules and globular masses have botryoidal surfaces and reveal partial radial crystallization of their outer parts. The spindle-like particles are not abundant and are found on the surfaces of botryoidal aggregates or in pores (Plate 4D, F). In some samples, the partly dissolved diatom frustules are filled by densely packed aggregate of these particles that stands upright and is overgrown by a later generation of similar particles (Plate 4F).

The coupled chemical and thermogravimetric investigations carried out earlier revealed that the phosphatic matter forming the bulk of both accretions and coprolites consists of carbonate-fluorapatite (CFA) containing on average 32% P_2O_5, 4% F, and 4% CO_2 (Bliskovskiy et al.,1975).

TEXTURES OF PLEISTOCENE GRAINS

Like recent grains and nodules, those of the Pleistocene consist of colloform, globular, and microcrystalline phosphate. The following description is based on observations of typical grains.

Colloform phosphate has a massive appearance and contains disseminated remnants of wholly phosphatized diatoms and rarer coccoliths belonging to the same *Emiliania* sp. as those found in recent accretions (Plate 5A, B).

Globular phosphate seems to be less abundant. Some globules with poorly defined contours having small holes in their centers are seen among the colloform matrix (Plate 5C). Another variety, aggregates of multidimensional globular pellets with well-defined contours, are found in void space inside the colloform matrix. Broken globules reveal complete radial crystallization (Plate 5D).

Microcrystalline phosphate occurs as individual particles on colloform surfaces (Plate 5E), inside phosphatized diatom frustules, or in voids left after their dissolution (Plate 5A) as aggregates in voids in the collophane matrix (Plate 5C), and in particular cases in association with pyrite globules (Plate 5F). The predominant shape of individual crystallites is dumbbell-like, with their central parts slimmer than their ends, which have a botryoidal appearance. These particles are mainly 4-8 μm long and 1-2 μm thick, with some thinner particles appearing more rod-like with less bulbous ends (Plate 5C). Some of these bodies are half-developed, with one of their thick ends emerging from a colloform matrix or pyrite surface, which makes them look like globules; intergrowths of these particles forming rosettes are also common (Plate 5F).

CHEMICAL COMPOSITION

The concentrations of major and minor elements in phosphatic materials described above and recent diatom ooze and calcareous shells from phosphatic sand are shown in Table 3.

The sequence beginning with the recent diatom ooze and ending with the recent well-lithified nodules reveals a gradual increase in P_2O_5, CaO, CO_2, and F. An enrichment in Sr is seen only at the beginning of this sequence, and its content is nearly constant in moderately and well-lithified nodules, while the behavior of Ba is erratic, showing occasional enrichment or depletion that is unrelated to P_2O_5 content.

Among the minor elements, Zn, Pb, V, Cr, Li, and U reveal no distinct trends. All other major and minor elements are more or less depleted in the course of nodule lithification.

The behavior of major elements during the lithification of phosphatized coprolites is the same as in nodules, but some of the minor elements behave differently. The fresh coprolites are essentially enriched in Zn and V, but their content decreases in lithified samples, in contrast to U, which increases. The microelements Mn, Cu, Ni, Co, Pb, Cr, Zn, and Li show no significant changes in concentration during the lithification process.

The sand fractions of diatom ooze, which are enriched in phosphatic grains, reveal an appropriate enrichment in major phosphate-building elements. This is most pronounced in the coarser fractions, which also show substantial enrichment in organic carbon, a trend that is not observed in other phosphatic

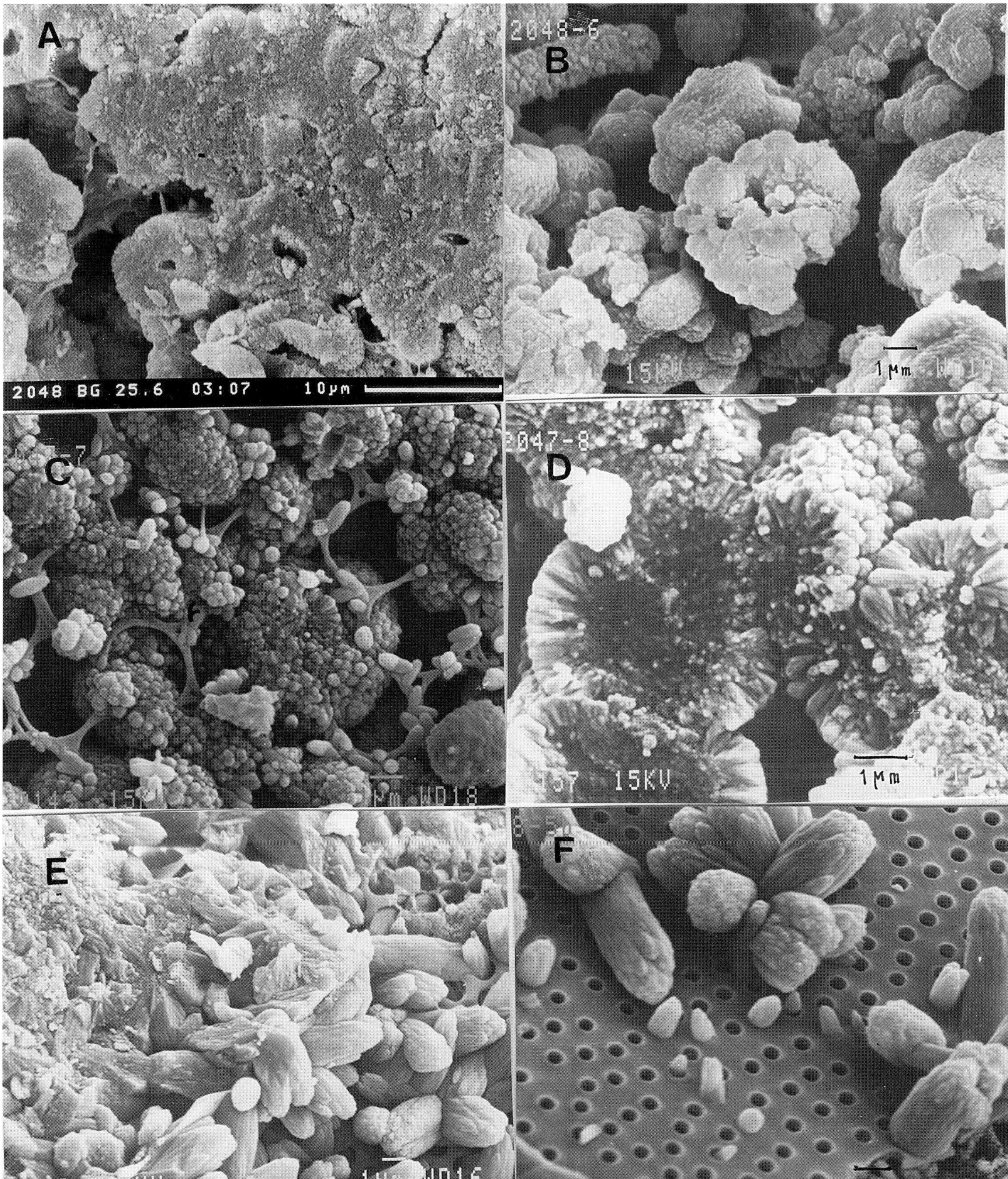

PLATE 3.—Textures of hard accretions from diatom oozes. (A) Formation of globules in the compact mass of colloform phosphate in the hard brown grain. Some globules have hollows in their centers. (B) Irregular globular clumps with microbotryoidal surface in moderately lithified nodule. (C) Globular aggregate with botryoidal surface in association with remains of dissolved diatom frustule and rod-shaped particles in the lithified nodule. (D) Broken globules in the lithified nodule revealing partial or complete radial crystallization. (E) Aggregate of rod-shaped particles in association with colloform phosphate in the moderately lithified nodule. (F) Isolated and aggregated rod-shaped particles growing on the surface of diatom frustule in the moderately lithified nodule. Some rods standing upright are half-developed. The surface of bigger rods is roughly striated.

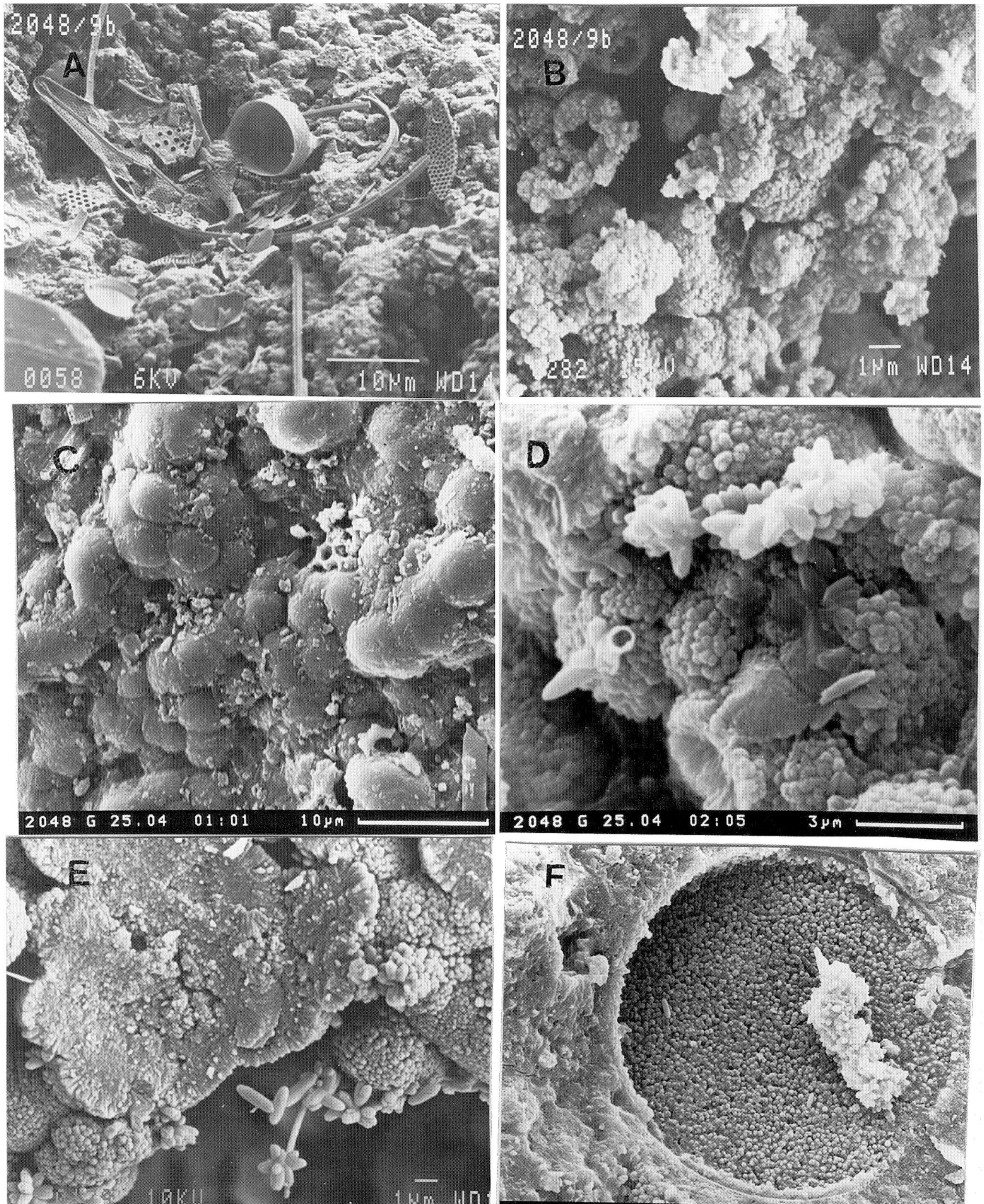

PLATE 4.—Textures of phosphatized coprolites from diatom oozes. (A) Colloform phosphate in association with remains of recent diatoms (*Chaetoceros* and *Gyrosigma* sp.) in the soft coprolite. (B) Globular and irregular aggregates of phosphate in the soft coprolite. Some globules reveal smaller or bigger apertures. (C) The smooth globular surface appearance of the soft microcoprolite. (D) Globular aggregate with botryoidal surface in association with aggregate of rod-shaped particles in the soft microcoprolite. (E) Globular aggregate with botryoidal surface in association with rod-shaped particles in the lithified coprolite. The broken globules reveal radial crystallization of their outer parts. (F) Wholly phosphatized diatom filled with aggregate of densely packed, rod-shaped particles overgrown by another aggregate of similar particles in the microcoprolite. The surrounding mass of phosphate reveals no signs of crystallization.

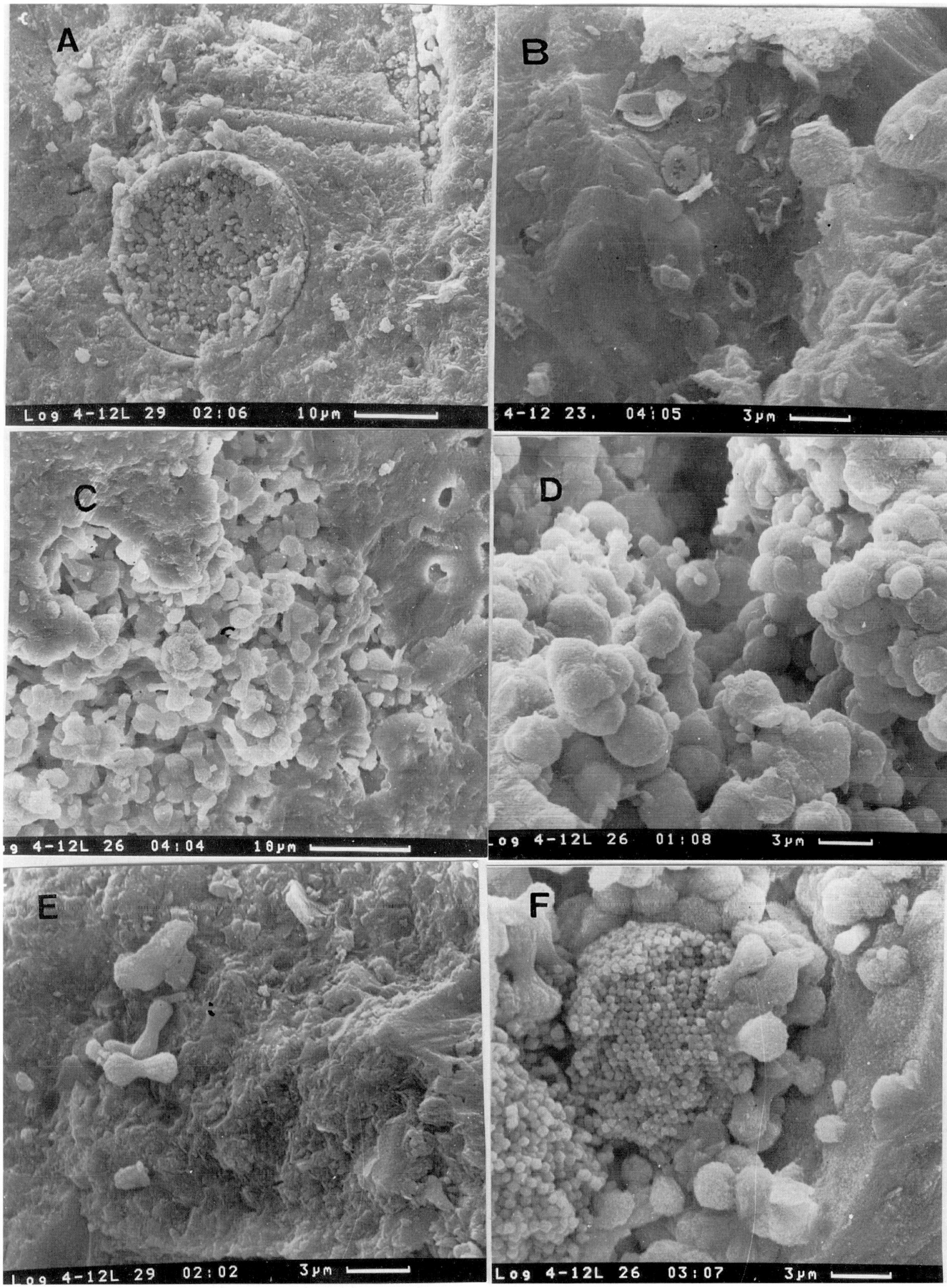

PLATE 5.—Textures of Pleistocene grains from phosphatic sands. (A) Colloform phosphate with inclusions of phosphatized diatom frustules. In the upper right corner the void left after dissolution of a diatom is filled with microcrystals. (B) Colloform phosphate with inclusions of intact and partly dissolved coccoliths (*Emiliania* sp.) (C) Aggregate of rod- and dumbbell-shaped crystallites in association with colloform phosphate. In the upper right corner two poorly defined globules with voids in their centers are seen. (D) Globular aggregate consisting of intergrowth of well-defined globules. The broken globules reveal radial crystallization. (E) Individual dumbbell-like crystallites in the colloform matrix. (F) Assemblage of individual and aggregated dumbbell-like crystallites in association with pyrite globules inside the void in the colloform matrix. Some half-developed crystallites are growing on the pyrite surface or emerging from the matrix. Below the center a rosette of crystallites is seen.

TABLE 3.—CHEMICAL COMPOSITION OF PHOSPHORITES AND ASSOCIATED SEDIMENT

Element	1	2	3	4	5	6	7	8	9
Major elements, %									
P_2O_5	1.68	7.49	24.48	29.56	31.71	28.82	30.90	11.90	25.45
CaO	4.1	12.6	41.2	45.8	48.2	46.1	50.1	19.42	40.55
MgO	0.85	1.62	1.54	1.56	1.562	1.52	1.35	n.d.	n.d.
Na_2O	1.73	3.47	2.45	1.46	1.20	2.09	1.17	2.66	2.22
K_2O	0.52	0.41	0.17	0.09	0.05	0.09	0.06	1.53	0.48
SiO_2	71.1	37.8	9.0	2.6	0.3	0.5	0.2	49.6	8.96
Al_2O_3	2.3	1.4	0.4	0.17	<0.02	0.1	<0.02	5.68	0.98
Fe_2O_3	1.23	0.58	0.54	0.32	0.12	0.10	0.11	2.38	0.57
TiO_2	0.13	0.08	0.03	0.02	0.01	0.03	0.01	0.20	0.62
SrO	0.03	0.10	0.25	0.33	0.30	0.27	0.29	0.14	0.33
BaO	0.03	0.09	0.07	0.23	0.06	0.03	0.03	n.d.	n.d.
CO_2	1.66	1.56	5.11	5.54	6.28	5.12	6.28	2.19	1.06
Corg	4.37	3.97	1.30	1.13	0.92	1.26	0.75	2.05	3.90
Stot	2.15	0.76	1.28	0.93	0.73	0.87	0.71	n.d.	n.d.
F	0.019	0.34	2.76	3.09	3.36	2.67	3.28	n.d.	n.d.
Minor elements, ppm									
Mn	40	53	20	17	19	21	24	217	84
Cu	20	24	12	25	9	20	20	44	37
Zn	36	31	37	40	59	155	90	100	108
Ni	71	37	31	25	28	41	30	35	40
Co	10	11	7	1	1	3	3	1	1
Pb	<10	40	20	30	60	45	30	21	54
V	168	35	75	53	53	130	42	30	90
Cr	41	39	47	45	35	38	31	122	188
Zr	46	10	12	8	7	10	12	n.d.	n.d.
Sc	3.1	1.7	0.3	0.2	0.2	0.1	0.1	n.d.	n.d.
Li	11	9	15	14	14	17	15	30	19
Cd	33	25	3	1	1	4	2	21	27
U	44	51	56	61	53	18	51	n.d.	n.d.

1-11 - Recent samples from five stations taken within the area of diatomaceous oozes; 12-15- Pleistocene samples from station L4-12; 16 - Miocene sample.
1 - diatomaceous ooze containing phosphorite accretions, fish debris, and phosphatized coprolites;
2 - very soft clump of diatomaceous ooze among the semi-liquid mass of enclosing ooze;
3-5 - slightly, moderately, and well-lithified phosphatic nodules;
6, 7 - soft and hard coprolites;
8-11 - silt (0.05-0.1 mm), fine sands (0.1-0.25 mm), medium sand (0.25-0.5 mm), and coarse sand (0.5-2 mm) fractions of diatomaceous oozes;
12-14 - silt, fine sand, and medium sand fractions of phosphorite sands;
15 - gravel-sized nodules from phosphatic sands;
16 - large Miocene nodule from the northern part of the Namibian shelf;
17 - shell debris from phosphatic sand.

materials from this area. As for microelements, the coarser fractions are enriched in Zn, V, and Pb, while Mn and Cu are depleted.

Pleistocene sands (samples 12-15, Table 3) show highest phosphate concentration in coarser fractions along with enrichment in Fe and S, which reflects the high pyrite content. They are also depleted in lithic elements and some microelements, like Mn, V, Zr, and Li. In contrast, Ni, Pb, and Cd are slightly enriched, while the remainder of the microelements shows no consistent trends.

The relatively high concentration of CO_2 in two fractions of Pleistocene sands (7.06% and 8.76%) might be related to admixtures of biogenic calcareous material left after incomplete separation of sands during the sample preparation procedure.

The Pleistocene nodules are higher in phosphatic components as compared to associated sands and lower in all major lithic and some microelements (Cr, Zr, Sc, and Li), while other microelements reveal similar concentrations, namely Mn, Cu, Zn, Ni, Co, Pb, V, and U.

The large Miocene nodule (sample 16, Table 3) from the northern part of the Namibian shelf is less phosphatic as compared to smaller Pleistocene nodules, but more pyritized as evidenced by high Fe and S concentrations, and lower in most of the microelements, including U. Lower or even the lowest concentrations of all microelements with the exception of Mn are found in calcareous shells associated with the phosphatic sands.

TABLE 3.—CONTINUED

Element	10	11	12	13	14	15	16	17
				Major elements, %				
P_2O_5	29.46	30.40	10.91	25.50	25.77	30.74	22.17	0.56
CaO	49.50	51.36	25.7	44.1	45.7	49.1	35.3	55.7
MgO	n.d.	n.d.	1.39	1.22	1.12	0.93	1.11	0.06
Na_2O	1.88	1.90	0.92	1.26	0.98	0.45	1.30	0.14
K_2O	0.18	0.15	0.92	0.33	0.10	0.05	0.72	0.05
SiO_2	3.64	2.18	39.9	6.2	4.7	1.7	16.2	0.2
Al_2O_3	0.22	0.25	2.8	1.2	1.1	0.5	2.0	0.2
Fe_2O_3	0.37	0.57	4.8	3.9	3.1	1.6	5.6	0.1
TiO_2	0.005	<0.005	0.34	0.10	0.08	0.03	0.14	0.02
SrO	0.39	0.39	0.14	0.26	0.27	0.30	0.20	0.19
BaO	n.d.	n.d.	0.01	0.02	0.05	0.04	0.01	0.01
CO_2	1.45	1.23	8.78	6.03	7.06	4.86	3.65	44.00
Corg	3.43	3.50	1.05	1.30	1.43	1.42	0.76	0.92
Stot	n.d.	n.d.	0.95	2.69	2.49	1.63	3.71	0.15
F	n.d.	n.d.	1.17	2.91	2.88	3.45	2.41	n.d.
				Minor elements, ppm				
Mn	36	30	125	55	70	102	72	254
Cu	16	27	16	15	13	18	6	3
Zn	164	175	38	33	33	46	30	3
Ni	33	23	56	102	90	56	70	34
Co	7	6	11	10	12	10	7	<5
Pb	119	65	48	77	67	62	40	n.d.
V	98	100	83	47	38	64	62	5
Cr	168	120	157	184	140	66	98	10
Zr	n.d.	n.d.	159	57	47	73	97	6
Sc	n.d.	n.d.	14	19	13	3	5	0.4
Li	13	14	60	48	48	33	18	18
Cd	22	24	13	18	17	10	<1	12
U	n.d.	n.d.	70	100	100	110	68	n.d.

n.d.=not determined

Special attention must be given to the behavior of REEs which are extraordinarily depleted relative to the enclosing diatom oozes in all Recent nodules, in particular in moderately lithified ones (Table 4). The Pleistocene nodule and sands, especially their finer fraction (0.25-0.1 mm), are two to three orders of magnitude higher in REE as compared to recent ones.

The concentration of REE in the Miocene nodule (sample 10, Table 4) is intermediate between Pleistocene grains (sample 8) and Pleistocene nodules (sample 9), while calcareous shell material is depleted relative to diatom ooze and enriched relative to Recent nodules.

DISCUSSION

The Mode of Phosphate Deposition

Textures of recent phosphatic grains and nodules demonstrate that both are diagenetic products intimately related to the enclosing sediments. The earliest forms of phosphate deposition from dissolved phosphate-enriched interstitial water inside this sediment are micrometer-sized crystallites if the free space allows for crystallization or irregular colloform mass if the space is scarce and concentration of dissolved phosphate is higher. Further dehydration and compaction of this mass promotes globular aggregation accompanied by partial or complete radial crystallization of globular pellets.

Biogenic debris, especially gridded diatom surfaces, seems to be a preferred site for such deposition, and this is followed by subsequent dissolution of this debris, as evidenced by the presence of nearly intact frustules overgrown by phosphatic particles. That phosphate deposition is supported by dissolved species of biogenic origin as demonstrated by diatom-shaped voids that are partly filled by phosphatic material.

Phosphatization of coprolites presents a special problem. The coprolites of living organisms such as sea lions or birds contain only a few percent of P_2O_5 (Hutchinson, 1960). Even fresh coprolites recovered from the sea floor in rapidly accreting sediment areas are nearly entirely phosphatized. This requires an extremely rapid phosphatization process, the nature of which is unclear.

TABLE 4.—CONCENTRATION OF REE AND Y (IN PPM) IN PHOSPHORITES

Element	Samples										
	1	2	3	4	5	6	7	8	9	10	11
La	9.24	4.80	1.72	0.101	0.15	0.89	209.6	133.4	12.4	59.3	2.5
Ce	17.71	8.80	3.26	0.219	0.26	0.85	502.9	310.1	24.7	139.2	4.6
Pr	2.31	1.16	0.42	0.023	0.028	0.21	51.1	31.0	2.5	14.5	0.5
Nd	9.01	4.37	1.65	0.107	0.104	0.41	214.5	130.0	10.1	60.4	2.2
Sm	1.70	0.86	0.29	0.008	0.13	0.11	44.5	27.7	2.0	11.7	0.35
Eu	0.32	0.23					10.3	6.3	0.6	2.7	0.07
Gd	1.49	0.81	0.32	0.026	0.037	0.07	43.0	26.1	2.0	11.2	0.33
Tb	0.25	0.11	0.044	0.0031	0.0037	0.102	6.9	4.2	0.3	1.7	0.04
Dy	1.54	0.75	0.28	0.018	0.032	0.09	42.8	26.4	2.2	11.0	0.28
Ho	0.32	0.16	0.061	0.005	0.008	0.08	8.9	5.5	0.5	2.3	0.06
Er	0.87	0.47	0.18	0.015	0.030	0.020	22.5	14.1	1.3	5.7	0.13
Tm	0.17	0.08	0.036	0.003	0.008	0.09	4.0	2.6	0.2	1.0	0.02
Yb	0.93	0.59	0.23	0.039	0.080	0.11	21.7	13.8	1.3	5.0	0.14
Lu	0.14	0.08	0.033	0.0033	0.0104	0.08	3.3	2.1	0.3	0.7	0.02
Y	10.19	5.48	2.39	0.29	0.51	1.11	327.0	237.3	20.7	80.6	2.1

1-6–Recent samples; 7-9–Pleistocene samples; 10- Miocene sample.
1-2–diatomaceous ooze.
2–very soft clump of slightly phosphatized ooze.
3–soft nodules.
4 somewhat denser nodule.
5–hard brownish nodule.
6–hard black nodule.
7–0.1-0.25 mm fraction of phosphatic sand.
8–0.25-0.5 mm fraction of phosphatic sand.
9–nodule from phosphatic sand.
10–nodule from the northern part of the Namibian shelf.
11–shell debris from phosphatic sand.

The Role of Bacteria

The problem of bacterial participation in phosphorite formation has a long history, beginning with the pioneering work of Cayeux (1936) and continuing with the work of numerous investigators up to present. The major argument in favor of bacterial action in phosphogenesis is the presence of more or less abundant bacteria-like particles in some phosphorites of various ages recovered on the continents or from the seafloor. Their most common morphologies are rounded, oval, rod-, spindle-, and dumb-bell-shaped, which are interpreted as phosphatized bacterial bodies (Mirtov et al., 1987; Lamboy, 1990, 1994). However, the latter three varieties of such bodies seen in our samples are probably not bacteria, but rather prophosphatic crystallites. Moreover, the large variations in their dimensions are not typical for bacteria, nor are their intergrowths, their crystalloform appearance, and the attachment of their half-developed bodies to hard surfaces.

As shown by Van Cappellen and Berner (1991), the fluorapatite crystallites grown in modified seawater solutions have the same dimensions as seen in our samples. Moreover, some other synthetic mineral species grown in experiments likewise have the same forms and dimensions, for example, the iron hydroxide crystals shown on the cover of the book edited by Halbach et al. (1988).

Another argument against the bacterial origin of these particles is the fact that diatom oozes of the Namibian shelf are heavily populated by filamentous sulfate-reducing bacteria without any other vital species (Butlin, 1949; Novozhilova and Baturin, 1973). Filaments are rarely seen in our samples. This is logical because most of the bacterial bodies are too delicate and ephemeral to survive the diagenetic processes, as demonstrated by Westall and Rince (1994).

Phosphatization seems impossible without bacterial participation. However, the role of bacteria in recent diatomaceous oozes is not as an agent in the direct precipitation of phosphate at concentrations of up to about 30% P_2O_5. Rather, the bacteria act to decompose organic matter, reduce sulphate, and mobilize soluble phosphate. Enrichment of dissolved phosphate is the main prerequisite for phosphate deposition and diagenetic accretion. According to Ruttenberg and Berner (1993) mini-scale authigenic apatite formation proceeds along continental margins in various sediment types irrespective of the type of bacterial populations present. In such cases the role of bacteria again appears to be indirect.

The Interrelations of Phosphate Morphologies and Textures

The presence of unreworked grains and nodules in recent

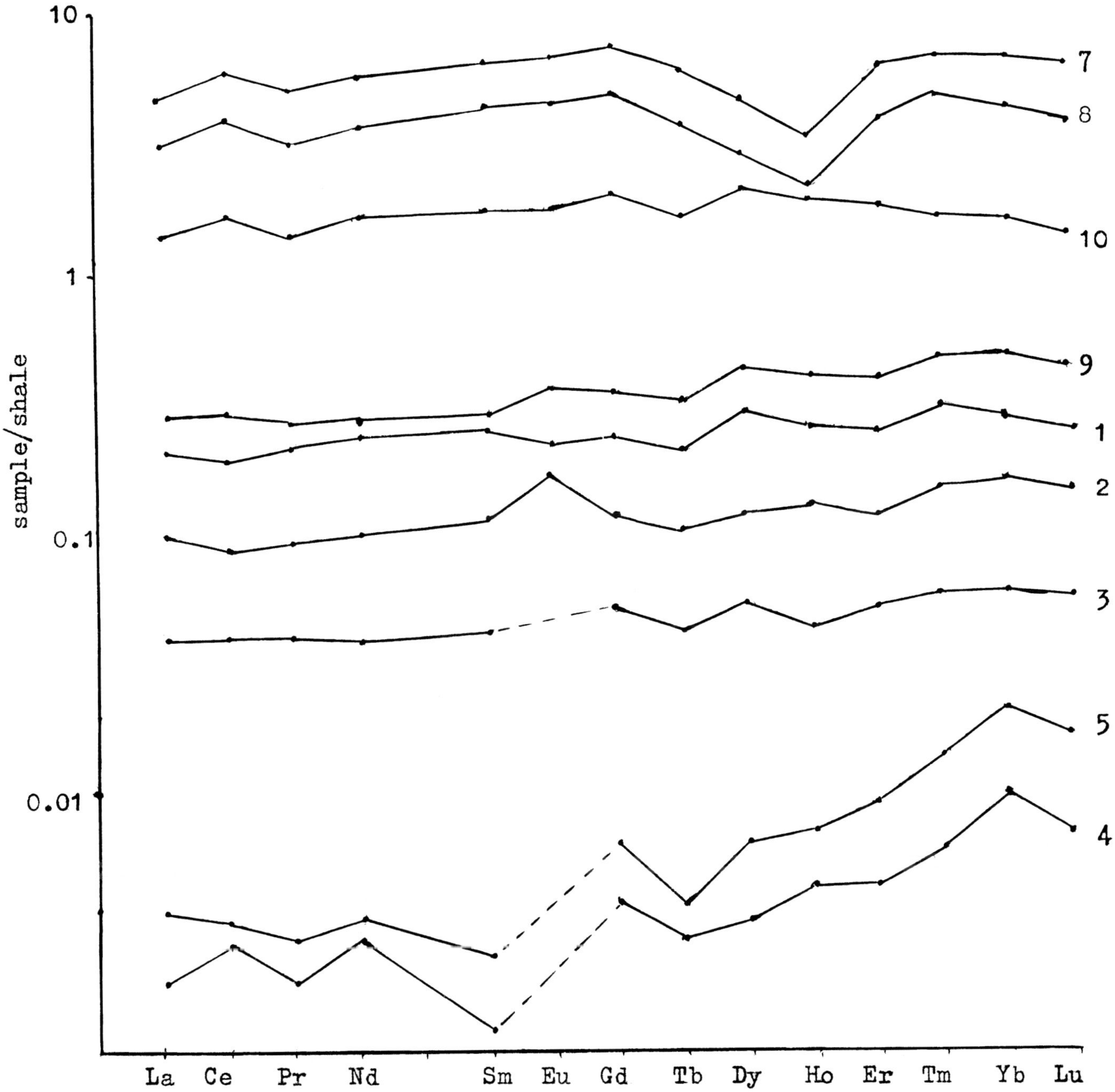

FIG. 2.—Distribution patterns of shale-normalized REE in Namibian shelf phosphorites, diatomaceous ooze, and shell debris. The number of samples is the same as in Table 3.

diatomaceous oozes provides evidence that both are forming simultaneously and their distribution depends on microchemical variabilities. Of course, this does not exclude the possibility of future aggregation of grains into nodules or disintegration of nodules into granular aggregates.

The observed sequences of phosphate accretions beginning from soft, slightly phosphatized diatomaceous ooze and ending with hard well-lithified bodies may mean that they represent the gradual evolution of their fabrics and compositions in the course of diagenesis rather than the one-stage formation of softer or harder bodies. A similar process of gradual lithification accompanies the phosphatization of coprolites. But it is possible that the process of lithification might be interrupted at some early stage, in which case the soft accretion would be buried in sediment, as was observed in Miocene deposits on subsea mountains in the Sea of Japan (Bersenev and Shkolnik, 1984). But if the sediment undergoes reworking, only hard accretions have a chance at surviving, as demonstrated by the Pleistocene phosphatic sands consisting only of well-lithified components.

The inner texture of recent and Pleistocene grains, nodules, and coprolites reveals the three interrelated forms of phosphate: micrometer-sized elongated crystallites, colloform matrix, and globules. Of these, the first and the second seem to be primary and may form simultaneously.

Colloform masses may be transformed into globules that are undergoing more or less developed crystallization. On the other hand, elongated crystallites may form tightly packed rosettes and other aggregations that mimic the appearance of globules growing in the colloform matrix. In any event, the close textural resemblance between recent and Pleistocene grains and nodules implies their origin in common processes.

Geochemical Considerations

The geochemical data confirm the lithologic and SEM observations, which reveal the formation and evolution of recent grains and nodules that develop from more or less phosphatized weakly consolidated ooze into well-lithified accretions. The process of lithification increases the concentration of phosphatic components and reduces nonphosphatic concentrations. This applies to most microelements, with a few exceptions.

Comparison of compositions of recent diatom oozes, nodules, grains, and coprolites with the compositions of older materials (Table 3) yields the following trends for microelement enrichment relative to diatom ooze: Zn 2 to 5 times in recent grains and coprolites; Pb 2 to 10 times in all analyzed materials; Cr 2 to 3 times enriched in recent and Pleistocene grains; U 2 to 3 times; and Li about 4 times enriched in Pleistocene grains and nodules.

The average values of most microelement concentrations in phosphorites given by Altschuler (1980) are close to ours, but his enrichment factors based on average shale composition are quite different, generally higher, since diatom ooze is depleted in most microelements relative to shale, with the exception of Cd, U, Ni, and V.

The behavior of REE (Fig. 2 and Table 4) is much more interesting, since phosphorites are widely known as sites of concentration of these elements in sedimentary sequences. In contrast to previous observations of ancient phosphorites, the recent accretions are much lower in REE than are the enclosing sediments, which in turn are depleted in REE relative to the average shale that was first established by Baturin et al. (1972) and confirmed by later works (McArthur et al., 1984; Watkins et al., 1995).

The Pleistocene sands (samples 7 and 8, Table 4), which are somewhat enriched in REE as compared to the average phosphorite values given by Altschuler (1980), might be considered as a lag deposit formed during reworking of source sediments similar to recent diatom oozes hosting analogous phosphate accretions. Hence, the only possible source of REE in phosphatic sands is the shelf bottom water, whose composition of these elements is close to the shale pattern (Balashov, 1976). The time of exposure of phosphorite in association with water movement may play an important role for their REE enrichment, as indicated by higher content of REE in older phosphorites on the Pacific subsea mountains (Baturin et al.,1972; Tambiyev and Baturin, 1982).

CONCLUSIONS

Diatom oozes accumulating on the Namibian shelf within the east Atlantic phosphorite-bearing belt are producing recent phosphatic accretions of sand to gravel size that are growing by aggregation and cementation of material from the same source sediment. The textural and chemical evolution of these accretions leads to their transformation into well-lithified grains and nodules similar to those found in older deposits. The dominant textures for the phosphate in both recent and Pleistocene accretions are colloform, globular, and microcrystalline. These are all intimately related to each other.

The resemblance of lithified recent accretions to their Pleistocene counterparts from the adjacent shelf area in terms of their appearance, texture, bulk chemical composition, and the character of the organic remains implies that both originated in similar environmental conditions. These conditions were governed by coastal upwelling associated with high biologic productivity, the accumulation of biogenous sediments enriched in phosphorus, and diagenetic processes.

The depletion of recent accretions in REE, in contrast to their high concentration in reworked Pleistocene nodules and especially in finer sands, implies that the process of REE accumulation in phosphorites is postdepositional. That is, it is controlled by the duration of phosphorite exposure, the intensity of bottom water movement, and the surface area of exposed phosphorite.

ACKNOWLEDGMENTS

I thank J. M. Bremner, University of Cape Town, and O. N. Kolosov, chief geologist of RV Professor Logachev, for supplying samples of Pleistocene phosphorites; J. Lucas and L. Prévôt-Lucas, who arranged my stay at Strasbourg University from 1993 to 1994 as a visiting scientist and kindly afforded me analytical facilities and all kinds of assistance; P. Karcher, the Strasbourg University, and E. A. Zhegallo, Institute of Paleontology, Moscow, for their invaluable assistance while working with SEM; J. Krempp, University of Strasbourg, and M. V. Baturina, my spouse, for performing F analyses; A. B. Isaeva, A. V. Dubinin, V. P. Kazakova, A. N. Rudakova, N. N. Zavadskaya, and N. P. Tolmacheva, Institute of Oceanology, Moscow, for analyzing samples by ICP-MS, atomic absorption, wet chemical, microelemental, and mineralogical methods; V. V. Mukhina, G. K. Kazarina, and O. B. Dmitrenko, Institute of Oceanology, for micropaleontological determinations; and O. N. Chaikina, Institute of Oceanology, for technical assistance in preparing the manuscript.

Special thanks are due to two reviewers, R. E. Garrison and F. T. Mannheim, who made numerous valuable suggestions that helped to improve the manuscript.

This work was completed with financial support from the Russian Foundation of Fundamental Research, Grant No. 96-05-65236.

REFERENCES

ALTSCHULER, Z. S., 1980, The geochemistry of trace elements in marine phosphorites, part 1. Characteristic abundances and enrichment, in Bentor, Y. K., ed., Marine Phosphorites—Geochemistry, Occurrence, Genesis: Tulsa, Society for Sedimentary Geology (SEPM) Special Publication 29, p. 19-30.

BALASHOV, Y. A., 1976, Geochemistry of rare earth elements: Moscow, Nauka, 268 p. (in Russian).

BATURIN, G. N., 1971, Stages of phosphorite formation on the oceanic floor: Nature. Physical Science, v. 232, p. 61-62.[

BATURIN, G. N., 1969, Authigenic phosphorite concretions in Recent sediments of the South-Western Africa shelf: Moscow, Doklady Academii Nauk SSSR, v. 183, p. 1359-1367 (in Russian).

BATURIN, G. N., 1982, Phosphorites on the sea floor: origin, composition, and distribution: Amsterdam, Elsevier, 343 p.

BATURIN, G. N., BLISKOVSKIY, V. Z., AND MINEEV, D. A., 1972, Rare earth elements in phosphorites from the sea floor: Moscow, Doklady Academii Nauk SSSR, v. 207, p. 954-957 (in Russian).

BATURIN, G. N., KOCHENOV, A. V., AND PETELIN, V. P.,1970, Phosphorite formation on the South Western Africa shelf: Lithology and Mineral Resources, n. 3, p. 15-26 (in Russian).

BATURIN, G. N., MERKULOVA, K. I., AND CHALOV, P. I., 1972, Radiometric evidence for recent formation of phosphatic nodules in marine shelf sediments: Marine Geology, v. 13. p. 437-441.

BERSENEV, I. I., AND SHKOLNIK, E. L., 1984, Phosphorites from the bottom of the Sea of Japan, in Zimin,S. S., ed., Phosphates of Eastern Asia and Adjacent Seas: Vladivostok, Far-East Scientific Center of USSR, p. 162-179 (in Russian).

BLISKOVSKIY,V. Z., BATURIN, G. N., AND KUZMINA, T. S., 1975, About the phosphatic matter of some phosphorites from the sea floor: Lithology and Mineral Resources, n. 4, p.4-9 (in Russian).

BREMNER, J. M., 1978, Sediments on the continental margin off South West Africa between latitudes 17° and 25°S: Unpublished Ph.D.Dissertation, University of Cape Town, South Africa, 300 p.

BREMNER, J. M., 1980, Concretionary phosphorite from SW Africa: Journal of the Geological Society of London, v. 137, p. 773-786.

BREMNER, J. M., 1992, Brief description of a west-coast phosphorite deposit: Report N. 1992-0 100, Pretoria ,Geological Survey of South Africa, 17 p.

BREMNER, J. M., AND RICKARD, R. S., 1977, On the formation of phosphorite pellets: Proceedings of the Electron Microscopy Society of South Africa, v. 7, p. 83-84.

BREMNER, J. M., AND ROGERS, J., 1990, Phosphorite deposits on the Namibian continental shelf, in Burnett, W. C., and Riggs, S. R., eds., Phosphate Deposits of the World, V. 3, Neogene to Recent Phosphorites: Cambridge, Cambridge University Press, p. 143-152.

BREMNER, J. M., AND WILLIS, J. P., 1975, Glauconite and phosphorites from South West Africa: Procedings of the Electron Microscopy Society of South Africa, v. 5, p. 127-128.

BRONGERSMA-SANDERS, M., 1957, Mass mortality in the sea, in Hedgpeth, J. M., ed., Treatise on Marine Ecology and Paleoecology: Geological Society of America Memoir, v. 67, p. 941-1010.

BRONGERSMA-SANDERS, M., STEPHAN, K. M., KWEE, T. G., AND DE BRUIN, M., 1980, Distribution of minor elements in cores from the South-West Africa shelf with notes on plancton and fish mortality: Marine Geology, v. 37, p. 91-132.

BURNETT, W. C., 1977, Geochemistry and origin of phosphorite deposits from off Peru and Chile: Bulletin of the Geological Society of America, v. 88, p. 813-823.

BUTLIN, K. P., 1949, Some maladorous activity of sulfate reducing bacteria: Proceedings of the Society of Applied Bacteriology, v. 12, p. 39-42.

CAYEUX, L., 1936, Existence de nombreuses bacteries dans les phosphates sedimentaires de tout age: Comptes Rendus de l'Academie de Sciences Paris, v. 23, p. 1198-1200.

CALVERT, S. E., AND PRICE, N. B.,1970, Minor metal contents of recent organic-rich sediments off South West Africa: Nature, v.227, p. 593-595.

CALVERT, S. E., AND PRICE, N. B., 1983, Geochemistry of Namibian shelf sediments, in Suess, E., and Thiede, J., eds.,Coastal Upwelling. Its Sediment Record, Part A: New York, Plenum Press, p. 337-375.

COPENHAGEN, W. J., 1953, The periodic mortality of fish in the Walvis region. A phenomenon within the Benguela current: Investigation Report No. 14, Cape Town, Department of Commerce and Fisheries, 35 p.

EMELYANOV, E. M., 1973a, The distribution and composition of muds on the South Western African shelf: Transactions of the Institute of Oceanology of the USSR, v. 95, p. 211-238 (in Russian).

EMELYANOV, E. M., 1973b, The composition of slightly phosphatic and phosphatic sediments on the South Western african shelf: Ibidem, v. 95, p. 239-261 (in Russian).

EMELYANOV, E. M., AND SENIN, Y. M., 1969, Composition of the sediments of the South West African shelf: Lithology and Mineral Resources, n. 2, p. 10-25 (in Russian).

HALBACH, P, FRIEDRICH, G., AND VON STACKELBERG, U., eds., 1988, The Manganese Nodule Belt of the Pacific Ocean: Stuttgart, F.Enke Verlag, 264 p.

HART, T. J., AND CURRIE, R. J., 1960, The Benguela current: Discovery Report, v. 31, p. 123-298.

HUTCHINSON, G. E., 1950, Survey of contemporary knowledge of biochemistry. 3. The biochemistry of vertebrate excretion: Bulletin of the American Museum of Natural History, v. 96, 554 p.

LAMBOY, M., 1990, Microstructures of a phosphatic crust from the Peruvian continental margin: Phosphatized bacteria and associated phenomena: Oceanologica Acta, v. 13, p. 439-451.

LAMBOY, M., 1994, Nannostructure and genesis of phosphorites from ODP Leg 112, the Peru margin: Marine Geology, v. 118, p. 5-22.

MCARTHUR, J. M., AND WALSH, J. N., 1984, Rare Earth Geochemistry in phosphorites: Chemical Geology, v. 47, p. 191-220.

MCARTHUR, J. M., SAHAMI, A. R., THIRWALL, M., HAMILTON, P. J., AND OSBORN, A. O., 1990, Dating phosphogenesis with strontium isotopes: Geochimica et Cosmochimica Acta, v. 54. p. 1343-1351.

MIRTOV, Y. V., ZANIN, Y. N., KRASILNIKOVA, N. A., GUREVICH, B. G., KRIVOPUTSKAYA, L. M., KRASILNIKOVA, I. G., AND SUKHOV, Y. K., 1987, Ultramicrostructures of phosphorites. Atlas of Photographs: Moscow, Nauka, 224 p. (in Russian).

NOVOZHILOVA, M. I., AND BATURIN, G. N., 1973, Some data about the bacterial microflora in sediments of the Eastern part of the Atlantic Ocean: Proceedings of the Institute of Oceanology, Academy of Sciences of the USSR, v. 95, p. 268-273 (in Russian).

ROGERS, J., 1977, Sediments on the continental margin off the Orange River and the Namib Desert: Unpublished Ph.D. Dissertation, University of Cape Town, South Africa, 212 p.

RUTTENBERG, K. C., AND BERNER, R. A., 1993, Authigenic apatite formation and burial in sediments from non-upwelling, continental margin environments: Geochimica et Cosmochimica Acta, v. 57, p. 991-1007.

SENIN, Y. M., 1968, Peculiarities of sediment formation on the South Western Africa shelf: Lithology and Mineral Resources, no. 4, p. 108-111 (in Russian).

SENIN, Y. M., 1970, Phosphorus in bottom sediments of the South Western Africa shelf: Lithology and Mineral Resouces, n. 1, p. 11-26 (in Russian).

TAMBIEV, S. B., AND BATURIN, G. N., 1982, Rare earth elements in the process of oceanic phosphorite formation: Lithology and Mineral Resources, n. 3, p. 3-15 (in Russian).

VAN CAPPELLEN, P., AND BERNER, R. A., 1991, Fluorapatite crystal growth from modified seawater solutions: Geochimica et Cosmochimica Acta, v. 55, p. 1219-1234.

VEEH, H. H., CALVERT, S. E., AND PRICE, N. B., 1974, Accumulation of uranium in sediments and phosphorites on the southwest African shelf: Marine Chemistry, v. 2, p. 189-202.

WATKINS, P. T., NATHAN, Y., AND BREMNER, J. M., 1995, Rare earth elements in phosphorite and associated sediment from the Namibian and South African continental shelves: Marine Geology, v. 129, p. 111-128.

WESTALL, F., AND RINCE, Y., 1994, Biofilms, microbial mats and microbe-particle interactions: Electron microscope observations from diatomaceous sediments: Sedimentology, v. 41, p. 147-162.

MINERALOGICAL AND GEOCHEMICAL EVIDENCE FOR THE ORIGIN OF PHOSPHORITE NODULES ON THE UPPER WEST FLORIDA SLOPE

KENDALL B. FOUNTAIN* AND GUERRY H. MCCLELLAN

Department of Geology, University of Florida, Gainesville, Florida 32611, USA

ABSTRACT: Phosphorite nodules recovered from the upper west Florida slope provide an analog for phosphogenesis under conditions of marginal upwelling that are significantly different from the regional, continental shelf-upwelling models applied to the majority of the southeastern United States. We believe that ferruginous (AI and AII) and nonferruginous (BI and BII) phosphorite nodules record the episodic precipitation of francolite since the mid-Miocene (12–15 Ma) in response to positioning of the Gulf of Mexico Loop Current. It is likely that during sea level highstands, deflection of the Loop Current landward increases the C_{org} flux, establishing conditions suitable to the concentration of dissolved inorganic phosphorous (DIP) within redox stratified sediments. Francolite containing approximately 6.2% CO_2 (a-values: 9.316-9.324 Å) apparently precipitates in response to the early diagenesis of organic matter and/or Fe-P shuttling under suboxic to anoxic conditions (δ^{13}C:–6.55–0.47‰ PDB, δ^{18}O: –4.92–2.07‰ PDB), forming nonferruginous nodules (< 6% Fe_2O_3) and hardgrounds. Anhedral to subhedral francolite (ovoids, globules, and botryoids) encountered in BI and BII nodules is mineralogically consistent with primary nucleation, and appears to record a microbial component. Subsequent recrystallization of francolite, in response to submarine diagenesis, obliterates such fabrics, producing euhedral, hexagonal crystallites (AI and AII) possessing a decrease in carbonate substitution to 4.9–5.5% CO_2 (a-values: 9.322-9.328 Å).

During sea level lowstands, oxic conditions prevail, favoring Fe-enrichment of nodules (up to 22% Fe_2O_3) at the expense of carbonate minerals and glaucony. This is consistent with the evolution of primary, BII nodules into ferruginous lithotypes (AI and AII). Aragonite dissolution and the dissolution and/or reprecipitation of both calcite and francolite (δ^{13}C: –1.09–0.67‰ PDB, δ^{18}O: –0.61–1.38‰ PDB) within ferruginous nodules occurs during such intervals in response to changes in pore water redox and pH. Both petrographic observations and major element trends illustrate Fe-enrichment processes associated with loss of fine micritic cements and biogenic carbonate as a result of dissolution and/or replacement by FeOOH precipitates, as well as the oxidation of glaucony.

INTRODUCTION

Periods of phosphogenesis indicate significant changes in both paleoenvironment and sedimentation within a local depositional system, often in response to global factors such as sea level fluctuation and plate tectonics (Cook and McElhinny, 1979; Sheldon, 1980). Of the regions in the world where conditions in the geologic past have favored phosphogenesis, the southeastern United States, and in particular, Florida, is one of the most extensive and widely studied (Riggs, 1979a, b, 1984a, b; Van Kauwenbergh et al., 1990; Compton et al., 1993). However, onshore deposits of significant commercial value in Florida have been the focus of research, whereas offshore deposits have received only restricted study. As a result, there is a limited understanding of the depositional and diagenetic processes responsible for the development of offshore phosphorite deposits along the Florida continental margin. Neogene to Quaternary deposits in this region formed in response to diverse geologic and paleoceanographic conditions, including nearshore to shallow-shelf environments, and along the outer shelf and upper slope of now drowned marginal terraces (Gorsline and Milligan, 1963; Mullins and Neumann, 1979; Riggs, 1984b; Doyle and Holmes, 1985). Previous research has shown that most of the deep-water (>200 m) phosphorites located south and east of Florida record a temporal transition from limestone phosphatization and nodule formation to precipitation of seafloor ferromanganese and phosphatic crusts, with subsequent redistribution as lag deposits (Burnett and Gomberg, 1977; Mullins and Neuman, 1979; Manheim et al., 1980). Paleoceanographic changes, particularly increased current strengths, are interpreted to be the cause for such changes in phosphogenesis. A similar transition has been documented along the upper west Florida slope (WFS) in association with Neogene phosphorite nodules (Doyle and Holmes, 1985), and has been correlated to paleoceanographic changes in the Middle Miocene that produced intensification of Loop Current flow at about 12–15 Ma (Mullins et al., 1987).

WFS phosphorites, associated with submarine outcrops of Middle Miocene to Pleistocene sediments in water depths of 450 to 600 m, consist of ferruginous and nonferruginous (< 6% Fe_2O_3) nodules and hardgrounds that apparently developed in response to Loop Current positioning in the eastern Gulf of Mexico (Mullins et al., 1987). This phosphogenic system is unique in that ferruginous and nonferruginous nodules coexist under marginal upwelling conditions significantly different from the regional, continental shelf-upwelling models applied to the majority of the southeastern United States (Riggs, 1979a, b, 1984a, b) in which large C_{org} fluxes are considered essential for phosphorite accumulation (Compton et al., 1993). Although WFS surface sediments are C_{org}-rich (6-8% total organic matter), concentrations are significantly less than other offshore regions of Neogene to recent nodule formation characterized by intense upwelling.

Representative samples of phosphorite collected from the WFS have been analyzed by mineralogical, petrographic, and geochemical methods to develop a model of phosphorite nodule formation and subsequent submarine diagenesis. The results presented here focus on the relationship between sea level and nodule lithotype, including evidence that alternating Loop Current conditions driven by eustatic sea level perturbations since the mid-Miocene (12–15 Ma) have controlled nodule formation and the development of complex diagenetic textures. This, in turn, suggests an evolutionary link between nonferruginous and ferruginous nodule lithotypes.

*email: kendall@ufl.edu

Marine Authigenesis: From Global to Microbial, SEPM Special Publication No. 66

GEOLOGICAL SETTING

The general geology of the upper west Florida slope and the details of the Neogene stratigraphy have been summarized by Mullins et al. (1987, 1988a, b) and references therein. The study area is located south of the Desoto Canyon, along the shelf-slope transition, in a complicated series of breaks with steps in between (Doyle and Holmes, 1985). Between the steps, which represent outcrops or reefs, lie gently sloping areas of unconsolidated carbonate sediments. In this region, the continental margin forms a distally steepened carbonate ramp with gentle slopes (1°–2°) extending to water depths as great as 2000 m (Uchupi, 1967). Below 2000 m, the sea floor drops at a sharp angle (20°–30°) to the abyssal depths of the Gulf of Mexico basin along the Florida Escarpment.

WFS phosphorites are oriented nearly north-south along the shelf-slope transition in water depths of 450 to 600 m (Fig. 1). Incorporated planktonic foraminifera within nodule and hardground fragments suggest an outer shelf (100–200 m) to upper bathyal (>200 m) setting for phosphogenesis (Mullins et al., 1987). In the northern portion of the study area (OTB5, OTB8, and WFS8), ferruginous and nonferruginous phosphorites cap Lower to Middle Miocene shelf-margin carbonates and clays that comprise seaward-prograding clinoforms (Mullins et al., 1988b). Sediments located between local outcrops of these phosphorites are a relatively coarse-grained, winnowed foram-sand facies (Mullins et al., 1988a). A basal unconformity underlies the phosphorites, and is the result of a depositional hiatus of Middle Miocene (Serravallian; 12–15 Ma) age, whereas a landward unconformity bounding the top of the layer in some areas is associated with a Late Miocene (Tortonian) age eustatic sea-level lowstand beginning at 10.2 Ma (Mullins et al., 1987). The Middle Miocene basal unconformity corresponds to intensification of the Gulf of Mexico Loop Current (Mullins et al., 1987), which, in turn, correlates to the onset of deposition dominated by the vertical accumulation of pelagic sediments, resulting in a gently dipping aggradational ramp that has withstood Quaternary climatic fluctuations (Gardulski et al., 1991). It is with these Middle Miocene to Quaternary sediments that nonferruginous phosphorites in the southern portion of the study area (OTB10 and WFS3) are associated (Fig. 1).

Oceanic circulation in the eastern Gulf of Mexico is dominated by clockwise flow of the Loop Current, which enters the Gulf through the Straits of Yucatan as the Caribbean Current and

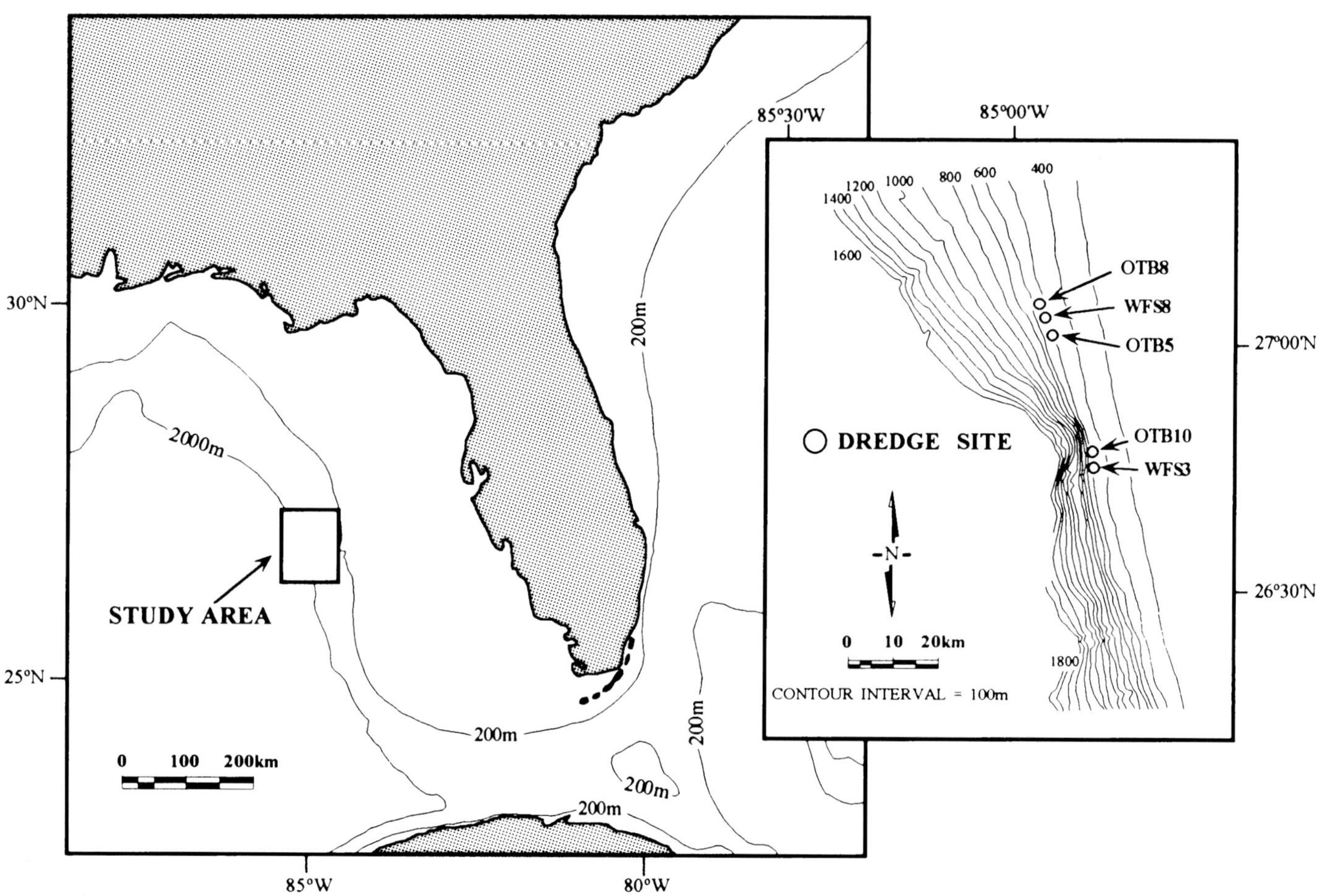

FIG. 1.— Location of the study area on the upper west Florida slope illustrating the bathymetric characteristics of the shelf-slope transition and the location of dredge sites employed in this study.

exits via the Straits of Florida as the Florida Current (Mullins et al., 1988b). The Loop Current presently extends to the sea floor and has produced a swath of winnowed foraminiferal sand at 400–600 m, reportedly inhibiting off-platform transport of shallow-water grains (Mullins et al., 1987; Gardulski et al., 1991). Upwelling occurs along the margin of the Loop Current in association with eddies that commonly break away from the current and migrate either westward or eastward over the outer shelf of west Florida (Elliot, 1982), enhancing primary production (Jones, 1973) and the resulting C_{org} flux to WFS sediments.

METHODS

Three cruises aboard the *RV Suncoaster* were conducted in December 1989, March 1991, and May 1993, approximately 250 km west-southwest of Tampa, Florida (Fig. 1). Twenty-six samples were examined from 19 phosphorite nodule/hardground specimens recovered by dredge hauls (OTB5, OTB8, OTB10, WFS3, and WFS8) made during the December 1989 and May 1993 cruises (Table 1). Samples for thin section preparation and SEM analyses were separated prior to pulverizing the remaining portions to minus 250 mesh with a ceramic mortar and pestle. Carbonate minerals (calcite, dolomite, and aragonite) were removed from portions of the minus 250 mesh powder via a sodium acetate-acetic acid (NaOAc) buffer at pH=5 as outlined by Compton et al. (1993). Sample SR-18 retained minor dolomite after repeated treatment.

The mineralogy of both bulk and NaOAc-treated samples was determined using a Philips Electronics Instruments X-ray diffractometer with Cu Kα radiation operated at 40 KV and 20 mA with a step size and time constant of 0.01° and 2 s. NaOAc-treated samples were analyzed to determine peak locations to a precision of 0.005°2Θ or better in order to calculate the unit cell a- and c- dimensions of francolite using least-squares cell refinement software. Commercial 1 µm α-corundum was added to samples as an internal standard to correct for sample displacement and instrument errors. Following the method of McClellan and Lehr (1969), francolite carbonate content (%CO_2) was calculated using unit cell a-values. The molar concentrations of other major substituting species in the francolite structure were determined using the models developed by McClellan and Gremillion (1980).

Both polarized light microscopy (PLM) and scanning electron microscopy (SEM) were used for petrographic analysis. Thin sections were examined using an Olympus Model BHS System binocular petrographic microscope. Exterior and frac-

TABLE 1.— DESCRIPTION AND LITHOTYPE CLASSIFICATION OF UPPER WEST FLORIDA SLOPE PHOSPHORITE SAMPLES.

Sample	Specimen	Lithotype	Description
SR-1	OTB8-A	AI	Ferruginous, laminated phosphatic crust with cryptalgal structures (microstromatoloids)
SR-2	OTB8-A	AI	Ferruginous, phosphorite nodule/hardground section
SR-3	OTB8-A	AI	Ferruginous, MnO-rich (Mn-dendrites) phosphorite nodule/hardground section
SR-4	OTB8-A	AI	Semi-ferruginous to ferruginous, phosphorite nodule/hardground section
SR-5	OTB10-D	BII	Nonferruginous, friable phosphorite nodule
SR-6	OTB10-B	BII	Nonferruginous, friable phosphorite nodule
SR-7	OTB10-A	BII	Nonferruginous, friable phosphorite nodule
SR-8	OTB5-H	AII	Ferruginous portion of a composite (conglomeratic) phosphorite nodule
SR-9	OTB5-H	AII	Semi- to nonferruginous portion of a composite (conglomeratic) phosphorite nodule
SR-10	OTB5-E	BI	Nonferruginous, well-indurated phosphorite nodule
SR-11	OTB5-D	BI	Nonferruginous, well-indurated phosphorite nodule section
SR-12	OTB5-D	BI	Nonferruginous, well-indurated phosphorite nodule section
SR-13	OTB5-A	AII	Conglomeratic, ferruginous phosphorite nodule
SR-14	OTB5-AA	AI	Ferruginous, phosphorite nodule/hardground section
SR-15	OTB5-AA	AI	Ferruginous, laminated phosphatic crust with cryptalgal structures (microstromatoloids)
SR-16	OTB5-AA	AI	Ferruginous, MnO-rich (Fe-Mn crusts) phosphorite nodule/hardground section
SR-17	WFS8-M	AII	Ferruginous portion of a composite (conglomeratic) phosphorite nodule
SR-18	WFS8-K	BI	Nonferruginous, well-indurated phosphorite nodule
SR-19	WFS8-J	AII	Conglomeratic, ferruginous phosphorite nodule
SR-20	WFS8-H	AII	Conglomeratic, ferruginous phosphorite nodule
SR-21	WFS8-A	AII	Conglomeratic, ferruginous phosphorite nodule
SR-22	WFS8-F	AII	Conglomeratic, ferruginous phosphorite nodule
SR-23	WFS8-C	AII	Conglomeratic, ferruginous phosphorite nodule
SR-24	WFS8-L	BI	Nonferruginous, well-indurated phosphorite nodule
SR-25	WFS3-A	BII	Nonferruginous, friable phosphorite nodule
SR-26	WFS3-B	BII	Nonferruginous, friable phosphorite nodule

tured surfaces examined by SEM were analyzed using a JSM-6400 scanning electron microscope. Major element analyses were carried out on the acid soluble (HCl + HNO_3) fraction of bulk samples by inductively coupled plasma analysis (ICP) at Activation Laboratories LTD, in Ancaster, Ontario.

Carbon and oxygen isotopic values were determined for structurally bound francolite carbonate using NaOAc-treated samples reacted with 100% phosphoric acid *in vacuo* at 90°C for 4 hours to extract purified CO_2 gas. Isotopic ratios were measured on-line by a triple-collector VG Isogas PRISM mass spectrometer. All carbon and oxygen isotopic results are reported in standard delta notation corrected to PDB using a reference gas generated from Carrara Marble and corrected to international standards. Experiments performed on powdered carbonate and phosphorite standards (Carrara Marble and NBS-120c) to estimate analytical precision indicate values of ±0.18‰ for $\delta^{18}O$ and ±0.14‰ for $\delta^{13}C$ (1-sigma; n=30).

RESULTS

Phosphorite Nodule Textures

Two phosphorite nodule types were initially identified by external appearance and macroscopic textures: ferruginous and nonferruginous (Fig. 2). The ferruginous nodules were associated with dredge sites OTB5, OTB8, and WFS8, while nonferruginous nodules were recovered at OTB5, OTB10, WFS3, and WFS8 (Fig. 1). The nodules range in size from approximately 2–20 cm in diameter, and commonly possess a flattened shape. Ferruginous nodules exhibit extensive diagenetic textures and appear darker in color than nonferruginous nodules.

The ferruginous nodules are subdivided into two lithotypes: AI and AII (Fig. 2, Table 1). AI nodules possess complex sedimentary and diagenetic textures including secondary porosity, secondary carbonate- and francolite-filled borings and fractures, Fe-Mn crusts, microstromatoloids, and pelagic limestone-filled borings with well developed geopetal fabrics. AII nodules are

	FERRUGINOUS		NONFERRUGINOUS	
	AI	AII	BI	BII
Macroscopic Description	Tan to gold-brown, well-indurated, with a rough, earthy, oxidized surface and complex sedimentary and diagenetic textures including secondary porosity and microstromatoloids	Dark brown to gray well-indurated, polished, ferruginous lithoclasts in a brown to tan, well-indurated, semi- to nonferruginous matrix	Tan to gray-brown, well-indurated, with a dull to earthy surface texture and extensive bioturbation with partially infilled borings	Light brown to light tan, friable, with a dull to earthy surface texture and extensive bioturbation with little to no infilling with pelagic debris
Lithology	Phosphatic biomicrite to fossiliferous micrite	Phosphatic intramicrudite with ferruginous biomicrite to fossiliferous micrite intraclasts in a semi- to nonferruginous fossiliferous micrite to biopelmicrite matrix	Phosphatic biomicrite	Phosphatic biomicrite
Mineralogy	Francolite Low-Mg Calcite Goethite Glaucony	Francolite Low-Mg Calcite Goethite Glaucony Quartz Dolomite	Francolite Low-Mg Calcite Aragonite Quartz Goethite Glaucony	Francolite Low-Mg Calcite High-Mg Calcite Aragonite Quartz Goethite Glaucony
Francolite Microtextures	Euhedral, prismatic hexagonal crystallites 1-3 μm in diameter and 0.5-2 μm in length	Euhedral, prismatic hexagonal crystallites 1-3 μm in diameter and 0.5-2 μm in length within intraclasts and globose or pustular crystallites in the matrix	Pustulose sheets consisting of closely packed botryoids and subhedral crystallites transitional between a globular and euhedral, hexagonal morphology	Ovoids, globules, and botryoids consistent with primary precipitation of francolite
Francolite a-values	9.323-9.328 Å	9.322-9.328 Å	9.316-9.322 Å	9.316-9.324 Å
Francolite % CO_2	4.9-5.4%	4.9-5.5%	5.5-6.2%	5.3-6.2%
Fe_2O_3	2.48-22.45%	1.23-13.83%	2.17-3.65%	2.15-5.32%
P_2O_5	11.34-28.62%	16.20-23.94%	5.70-17.76%	3.44-11.73%
$\delta^{13}C$	0.67‰ to -1.09‰	-0.13‰ to -0.79‰	0.41‰ to -6.55‰	0.47‰ to -0.70‰
$\delta^{18}O$	1.32‰ to -0.61‰	1.38‰ to -0.56‰	1.98‰ to -4.92‰	2.07‰ to -0.76‰

FIG. 2.— Chart summarizing select characteristics of phosphorite nodule lithotypes.

conglomeratic, with some nodules composed of a single polished lithoclast associated with later accreted nonferruginous material. The nonferruginous nodules also are subdivided into two lithotypes: BI and BII (Fig. 2, Table 1). BI nodules are associated with ferruginous lithotypes in the northern section of the study area (OTB5, OTB8, and WFS8), while BII nodules are restricted to southern sites (OTB10 and WFS3) (Fig. 1). Both lithotypes are lighter in color than ferruginous species and have a dull to earthy surface texture.

Mineralogy

Bulk mineralogy.—

XRD analyses confirmed macroscopic observations, indicating four sample populations with francolite (carbonate fluorapatite) as the principal apatite mineral (Fig. 2, Table 2). Mineralogy of the AI nodules consists of francolite, low-Mg calcite, and goethite as the dominant phases, with trace glaucony evident in all but one of the samples (SR-1). Low-Mg calcite is not present in sample SR-1, and samples SR-2 through SR-4 display extreme differences in goethite content from the same ferruginous nodule. The mineralogy of the AII nodules is similar to that of AI species, with the addition of quartz and dolomite to most of the samples. Samples SR-9 and SR-13 contain no dolomite, with SR-9 containing trace aragonite.

BI nodules consist of francolite, low-Mg calcite, aragonite, and quartz, with traces of goethite and glaucony. An exception to this is sample SR-18, which contains an extremely high dolomite content and no aragonite. The presence of aragonite differentiates BI nodules from either of the ferruginous lithotypes. The mineralogy of BII nodules is similar to that of BI nodules, consisting of francolite, low-Mg calcite, high-Mg calcite, aragonite, and quartz, with trace glaucony and goethite. The presence of high-Mg calcite (> 4 mol % $MgCO_3$) differentiates the two nonferruginous lithotypes.

Francolite crystal chemistry.—

Francolite unit cell a- and c-values, calculated as a proxy for

TABLE 2.— BULK MINERALOGY OF UPPER WEST FLORIDA SLOPE PHOSPHORITE SAMPLES.

Sample	Lithotype	Francolite	Low-Mg Calcite	Goethite	High-Mg Calcite	Aragonite	Quartz	Glauconite	Dolomite
SR-1	AI	***		**					
SR-2	AI	**	**	**				*	
SR-3	AI	**	**	**				*	
SR-4	AI	***	**	*				*	
SR-5	BII	**	**	*	**	*	*	*	
SR-6	BII	*	**		***	*	*	*	
SR-7	BII	*	**	*	***	*	*		
SR-8	AII	**	**	*					*
SR-9	AII	**	**			*	*		
SR-10	BI	**	**	*		*	*	*	
SR-11	BI	**	**	*		*	*		
SR-12	BI	**	**			*	*	*	
SR-13	AII	***	**	*			*		
SR-14	AI	**	**	**				*	
SR-15	AI	***	*	**				*	
SR-16	AI	**	**	**				*	
SR-17	AII	***	*	**					*
SR-18	BI	**	**				*		***
SR-19	AII	***	**	*			*	*	*
SR-20	AII	***	*	*			*	*	*
SR-21	AII	***	*	*			*	*	*
SR-22	AII	***	*	*			*	*	*
SR-23	AII	**	**	*			*	*	*
SR-24	BI	**	***			*	*		
SR-25	BII	**	**	*	***	*	*	*	
SR-26	BII	**	**		**	*	*	*	

Note: Mineral component designations correspond to the following semi-quantitative scheme: *** = major (> 50%), ** = minor (20-50%), and * = trace (< 20%).

CO_3 substitution, calculated values of $\%CO_2$ and $\%CO_3$, and francolite model compositions are given in Table 3. The unit cell a-values of ferruginous nodules (AI and AII) range between 9.322 Å and 9.328 Å ($\%CO_2$: 5.5–4.9), while nonferruginous nodule (BI and BII) unit cell a-values vary between 9.316 Å and 9.324 Å ($\%CO_2$: 6.2–5.3) (Fig. 2, Table 3). Unit cell *c*-values determined for ferruginous and nonferruginous nodules range between 6.893 Å to 6.907 Å and 6.889 Å to 6.906 Å, respectively.

Statistical models based on unit cell a-values predict a limited substitution series for carbonate fluorapatite with end member empirical formulas $Ca_{10}(PO_4)_6F_2$ (fluorapatite) and $Ca_{10-X-Y}Na_XMg_Y(PO_4)_{6-Z}(CO_3)_ZF_{0.4Z}F_2$ (francolite) (McClellan, 1980; McClellan and Gremillion, 1980). Application of these models indicates that francolite from ferruginous nodules possesses CO_3/PO_4 ratios between 0.22 and 0.25, while nonferruginous samples possess higher CO_3/PO_4 ratios between 0.25 and 0.29 (Table 3). The calculated theoretical francolite composition for both ferruginous and nonferruginous samples are:

ferruginous:

$Ca_{9.60}Na_{0.29}Mg_{0.11}(PO_4)_{4.92}(CO_3)_{1.08}F_{2.43}$ –
$Ca_{9.53}Na_{0.34}Mg_{0.13}(PO_4)_{4.79}(CO_3)_{1.21}F_{2.48}$

nonferruginous:

$Ca_{9.54}Na_{0.33}Mg_{0.13}(PO_4)_{4.82}(CO_3)_{1.18}F_{2.47}$ –
$Ca_{9.47}Na_{0.38}Mg_{0.15}(PO_4)_{4.66}(CO_3)_{1.34}F_{2.54}$

Petrography and Scanning Electron Microscopy

Ferruginous nodules.—

Nodule/hardground lithologies are summarized in Fig. 2. Allochems within AI nodules are primarily fragmented and nonfragmented planktonic (globigerinids and *Orbulina* sp.) and benthic foram tests with trace glaucony and detrital quartz in a matrix of francolite and micrite. Secondary porosity due to bioturbation is common in these nodules (Fig. 3A). Burrows are variably infilled with fossiliferous micrite, mixtures of micrite

TABLE 3.— UNIT CELL VALUES, CARBONATE CONTENT, AND MODEL COMPOSITIONS OF FRANCOLITE FROM THE UPPER WEST FLORIDA SLOPE.

Sample	Lithotype	a-value*	c-value*	$\%CO_2$	$\%CO_3$	CO_3	PO_4	CO_3/PO_4	Na	Mg
SR-1	AI	9.326	6.902	5.1	7.0	1.14	4.86	0.23	0.31	0.12
SR-2	AI	9.326	6.906	5.1	7.0	1.14	4.86	0.24	0.31	0.12
SR-3	AI	9.325	6.907	5.2	7.1	1.15	4.85	0.24	0.31	0.12
SR-4	AI	9.327	6.901	5.0	6.8	1.12	4.88	0.23	0.30	0.12
SR-5	BII	9.324	6.896	5.3	7.2	1.18	4.82	0.25	0.33	0.13
SR-6	BII	9.323	6.899	5.4	7.4	1.21	4.80	0.25	0.33	0.13
SR-7	BII	9.321	6.898	5.6	7.6	1.24	4.76	0.26	0.35	0.13
SR-8	AII	9.328	6.898	4.9	6.7	1.09	4.91	0.22	0.29	0.11
SR-9	AII	9.325	6.897	5.2	7.1	1.16	4.84	0.24	0.32	0.12
SR-10	BI	9.319	6.897	5.9	8.1	1.28	4.72	0.27	0.36	0.14
SR-11	BI	9.318	6.897	6.0	8.2	1.31	4.70	0.28	0.37	0.14
SR-12	BI	9.322	6.898	5.5	7.5	1.21	4.79	0.25	0.34	0.13
SR-13	AII	9.325	6.901	5.2	7.1	1.16	4.84	0.24	0.32	0.12
SR-14	AI	9.327	6.893	5.0	6.8	1.12	4.88	0.23	0.31	0.12
SR-15	AI	9.323	6.900	5.4	7.4	1.20	4.80	0.25	0.33	0.13
SR-16	AI	9.328	6.893	4.9	6.7	1.08	4.92	0.22	0.29	0.11
SR-17	AII	9.326	6.897	5.1	7.0	1.14	4.86	0.23	0.31	0.12
SR-18	BI	9.321	6.906	5.6	7.6	1.23	4.77	0.26	0.34	0.13
SR-19	AII	9.327	6.897	5.0	6.8	1.11	4.89	0.23	0.30	0.12
SR-20	AII	9.328	6.898	4.9	6.7	1.10	4.91	0.22	0.30	0.12
SR-21	AII	9.325	6.895	5.2	7.1	1.16	4.84	0.24	0.32	0.12
SR-22	AII	9.328	6.900	4.9	6.7	1.08	4.92	0.22	0.29	0.11
SR-23	AII	9.322	6.893	5.5	7.5	1.21	4.79	0.25	0.34	0.13
SR-24	BI	9.316	6.889	6.2	8.5	1.34	4.66	0.29	0.38	0.15
SR-25	BII	9.318	6.896	6.0	8.2	1.29	4.71	0.27	0.36	0.14
SR-26	BII	9.316	6.891	6.2	8.5	1.34	4.66	0.29	0.38	0.15

Note: Calculation of weight $\%CO_2$ based on the method of McClellan and Lehr (1969) – error of estimate ± 0.15. Weight $\%CO_3$ = 1.3636 × $\%CO_2$. Francolite model compositions based on the method of McClellan and Gremillion (1980). Unit cell a- and c-values are measured in Angstroms (Å), with calculated component concentrations measured in moles.
* Estimated std. dev. = 0.002 Å or less.

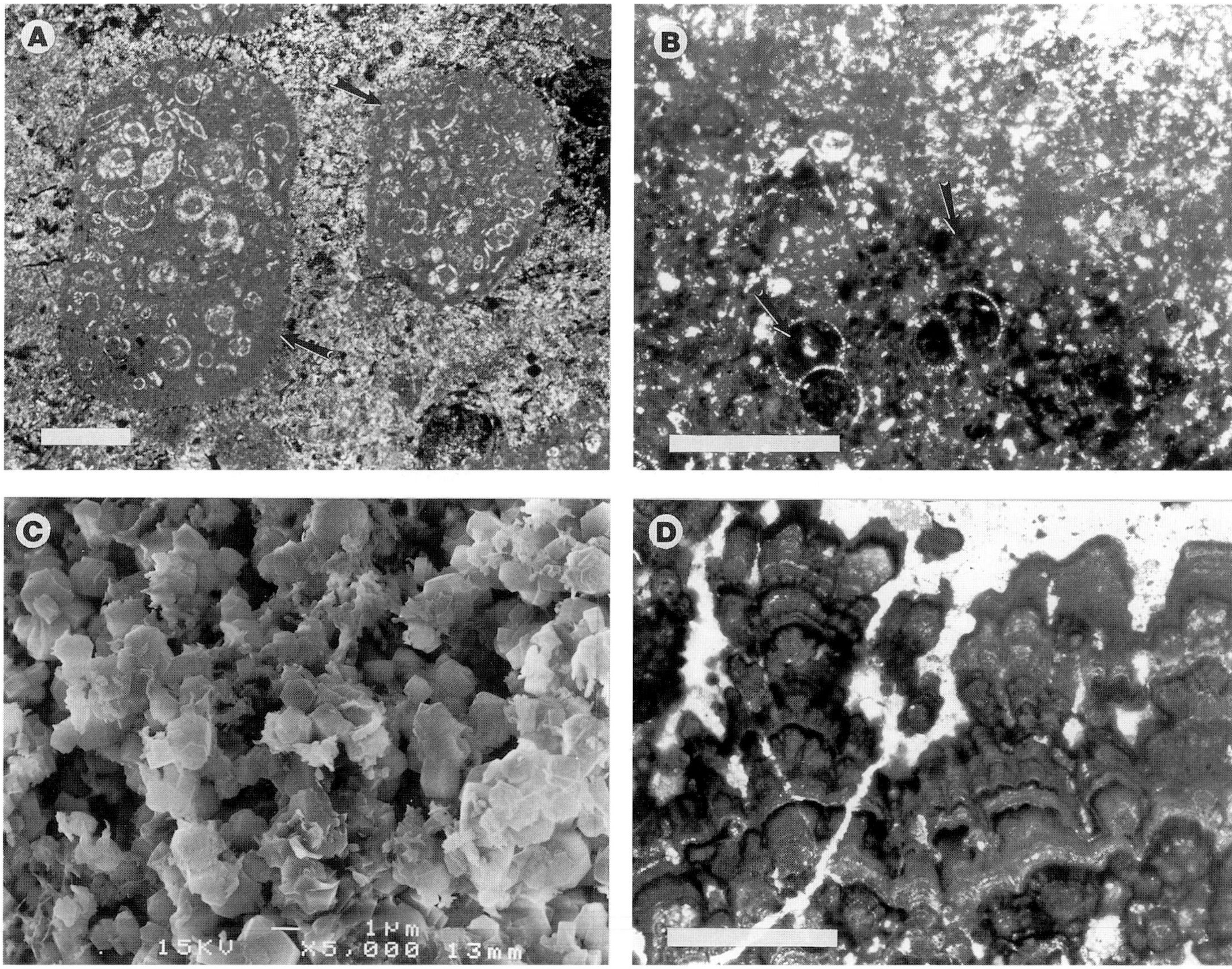

FIG. 3.— Photomicrographs of AI ferruginous nodules (scale bars in PLM photomicrographs = 0.5 mm). (a) Phosphatic biomicrite to fossiliferous micrite nodule (SR-4) with *Thalassinoid* burrows infilled by fossiliferous micrite. (b) Goethite and FeOOH replacing and infilling foram tests and replacing micrite and francolite matrix (SR-2). (c) Euhedral, hexagonal francolite crystallites (SR-2) approximately 1–3 µm in diameter (a-axis direction) and 0.5–2 µm in length (c-axis direction). (d) Example of a cryptalgal microstromatoloid structure (SR-1) possessing a laterally linked hemispheroid (LLH) morphology composed of alternating francolite and goethite laminae.

and francolite, or laminated and/or dendritic Fe-Mn crusts, and commonly exhibit secondary replacement of the surrounding matrix by francolite. Sparry low-Mg calcite is present within voids, infilling and/or replacing planktonic and benthic foram tests, echinoid fragments, gastropod tests, and micritic cement/matrix, and as thin veins (SR-14 through SR-16). Goethite and iron-oxyhydroxides (FeOOH) exhibit heterogeneous concentration, imparting a mottled appearance on both a macroscopic and microscopic scale (Fig. 3B). They are present as an alteration product of incorporated glaucony grains, and replace and/or infill foraminiferal tests, matrix, and cement. Void-infilling and matrix francolite occur as euhedral, prismatic hexagonal crystallites (Fig. 3C), while glaucony occurs in a granular form, where it is highly altered to goethite, and in "rosette" overgrowths of francolite and matrix. Features unique to the AI nodules are phosphatic, cryptalgal microstromatoloids (SR-1 and SR-15) commonly found within surface irregularities and bioturbation features (Fig. 3D). Similar in appearance to features documented in Albian limestones of the Polish Tatra Mountains (Krajewski, 1981), the Transvaal Supergroup, South Africa (Lanier, 1988), and offshore Late Pliocene phosphatic hardgrounds on the Queensland Plateau (Glenn and Kronen, 1993), these features possess a geopetal fabric consisting of finely laminated, botryoidal francolite/goethite believed to precipitate in response to microbial mediation near the sediment-water interface (Flugel, 1982) and/or a pore water Fe-P shuttling mechanism (Glenn and Kronen, 1993).

AII nodules are phosphatic intramicrudites (Fig. 2) composed of ferruginous intraclasts similar in appearance to AI nodules in a semi- to nonferruginous matrix (Fig. 4A). Matrix allochems are primarily fragmented and nonfragmented foram tests, glaucony granules, and detrital quartz, with trace echinoid

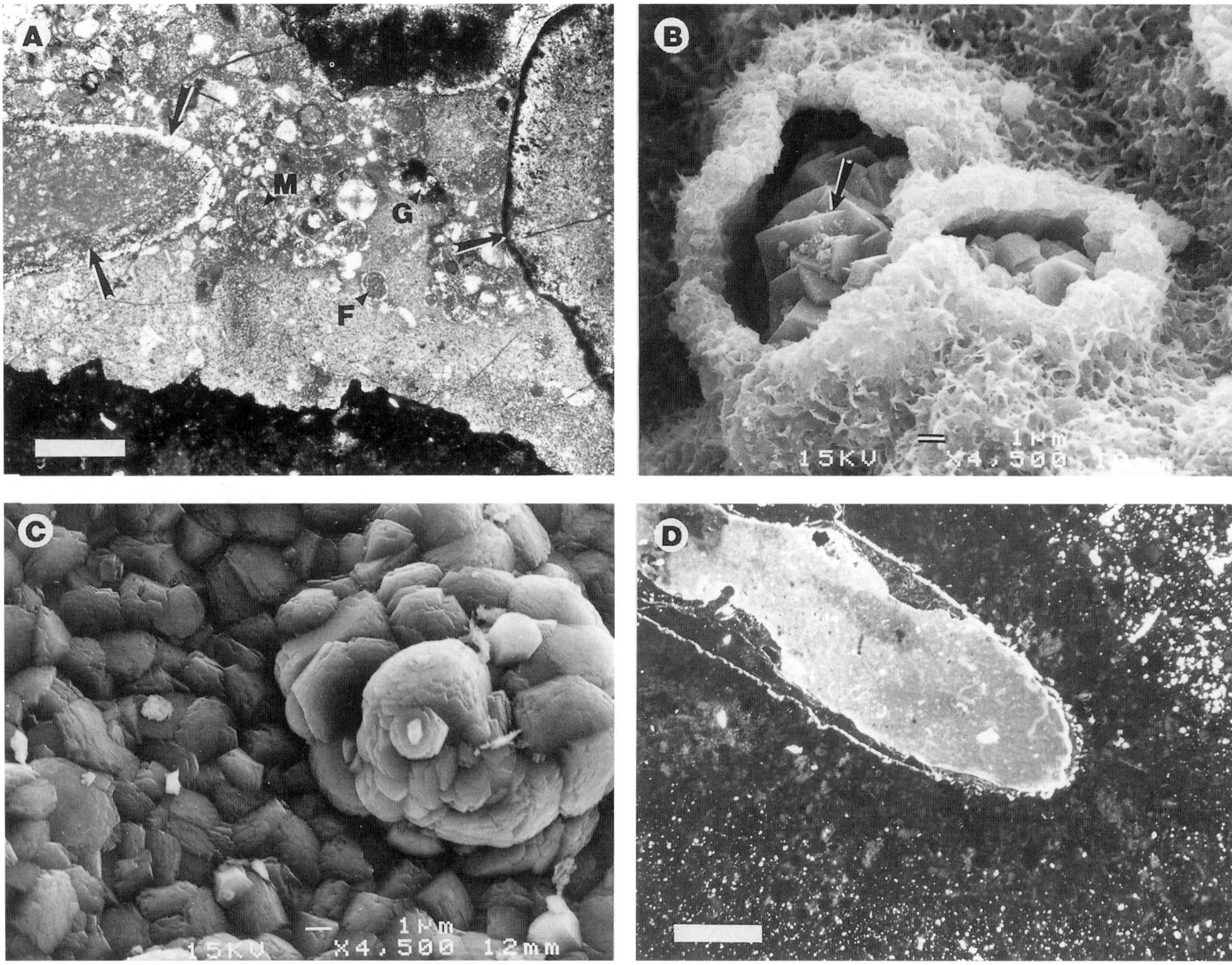

FIG. 4. Photomicrographs of AII ferruginous nodules (scale bars in PLM photomicrographs = 0.5 mm). (a) Phosphatic intramicrudite nodule (SR-21) illustrating intraclast surfaces with secondary veneers (arrows) of glaucony, francolite, and goethite. Note francolite (F), glaucony (G), and micrite (M) infilling foram tests. (b) "Rosette" or flake-like glaucony veneer on intraclast (SR-20), which occurs over a secondary francolite veneer (arrow). (c) Francolite replacing and infilling a foram test in the semi- to nonferruginous matrix of an AII nodule (SR-19). (d) *Thalassinoid* burrow infilled by fossiliferous micrite showing secondary replacement of micritic nodule matrix by francolite (SR-13).

fragments and gastropod tests in a matrix of micrite and francolite. Glaucony occurs as dark green to yellow-green granules and veneers (Fig. 4B), often with goethite alteration, and granules commonly retain the external shape of foram chambers. Francolite occurs in two distinct morphologies, intraclast crystallites, which are similar to those observed in AI nodules, and globose or pustular crystallites within the matrix (Fig. 4C). As with the AI nodules, AII nodules exhibit extensive evidence of bioturbation and secondary porosity in both the intraclasts and the surrounding matrix. Within the matrix, burrows are variably infilled with either fossiliferous micrite or mixtures of micrite and francolite, and commonly exhibit secondary phosphatization of surrounding nodule matrix (Fig. 4D). Dolomite is dispersed throughout the matrix as rhombohedral crystals averaging approximately 10 μm in size.

Nonferruginous nodules.

BI nodules are phosphatic biomicrites (Fig. 2) with allochems of fragmented and nonfragmented planktonic (globigerinids and *Orbulina* sp.) and benthic forams, echinoid fragments, and trace gastropods, with granular and test infilling glaucony and trace detrital quartz in a micrite and francolite matrix (Fig. 5A). Goethite and FeOOH show variable enrichment, replacing micrite, francolite, and glaucony test infillings, and replacing and/or infilling matrix and foram tests. Some nodules show evidence of bioturbation, with partial to no infilling by pelagic carbonate (Fig. 5B). Void-infilling and matrix francolite morphologies are pustulose (Fig. 5C) to subhedral (Fig. 5D), and often retain an intertwining, porous structure.

BII nodules are phosphatic biomicrites (Fig. 2) similar to BI specimens with the addition of bivalve fragment allochems. Goethite and FeOOH enrichment, essentially restricted to the replacement of glaucony, micrite, and francolite test infillings, is more limited than that observed with BI nodules (Fig. 6A). Glaucony infills foram tests (Fig. 6B), with minor accumulations of granules retaining the shape of skeletal chambers. In most instances, the glaucony exhibits little to no alteration.

The BII nodules possess a complex assemblage of francolite microtextures, including ovoids, globules, and botryoids. Ovoids are pervasive within foram tests, and resemble nonfilamentous, bacilli bacteria approximately 3–4 μm in length and approximately 1–2 μm in width (Fig. 6C). Globular francolite, found primarily in the sample matrix, consists of subspherical, isolated globules or agglomerates, and appears to act as a cement, binding clays as well as coccolithophorid plates to allochems. Botryoids, morphologically similar to francolite observed in BI nodules, are the most common, and usually occur as a pore-lining or matrix cement (Fig. 6D).

Geochemistry

Major elements.—
Major element data for bulk nodule samples are reported in Table 4. Due to nodule heterogeneity and the use of both ferruginous and nonferruginous samples in this study, correlation coefficient values ($r > 0.6$) were considered significant.

Total Fe (reported as Fe_2O_3) exhibits overlapping ranges for ferruginous (1.23%–22.45%) and nonferruginous (2.15%–5.32%) samples (Fig. 2). Fe_2O_3 shows a correlation to MnO ($r = 0.61$) and CaO for the total data set, and to MnO ($r = 0.60$), CaO ($r = -0.73$), MgO, and K_2O (excluding SR-16 and SR-3) for ferruginous samples (Fig. 7). P_2O_5 also exhibits overlapping ranges for ferruginous (11.34%–28.62%) and nonferruginous (3.44%–17.76%) samples (Fig. 2). The P_2O_5 contents correlate with Na_2O and MgO (excluding SR-18) for the total data set (Fig. 8), and with Fe_2O_3 (excluding SR-3) for ferruginous samples (Fig. 7). SiO_2, commonly used as an indicator of detrital flux, correlates with K_2O ($r = 0.74$) and Al_2O_3 for the total data set (Fig. 8). Al_2O_3, also a detrital flux indicator, correlates well with TiO_2 ($r = 0.85$) for the total data set, and with K_2O for nonferruginous samples (Fig. 8).

Carbon and oxygen isotopes.—
The $\delta^{13}C$ and $\delta^{18}O$ values for francolite-CO_3 of ferruginous nodules range from –1.09‰ PDB to 0.67‰ PDB and –0.61‰ PDB to 1.38‰ PDB, respectively (Fig. 2, Table 4). The $\delta^{13}C$ and $\delta^{18}O$ of francolite-CO_3 from nonferruginous nodules range from –6.55‰ PDB to 0.47‰ PDB and –4.92‰ PDB to 2.07‰ PDB, respectively (Table 4, Fig. 2). Excluding SR-18, nonferruginous nodules exhibit a strong correlation between $\delta^{13}C$ and $\delta^{18}O$ ($r = 0.95$), with a 1.5‰ spread in $\delta^{13}C$ and a 3‰ spread in $\delta^{18}O$. Examined separately, BII nodules show an even better correlation between $\delta^{13}C$ and $\delta^{18}O$ ($r = 0.97$).

DISCUSSION

Significance of Francolite Microtextures and Crystal Chemistry

Francolite in nonferruginous nodules is of a highly substituted variety, possessing CO_2 contents near the francolite structural limit of 6.3% (McArthur, 1985) (Fig. 2, Table 3). The $\delta^{13}C$ and $\delta^{18}O$ values of structural CO_3 are similar to other areas of offshore phosphorite accumulation (Fig. 9, Table 4), and indicate francolite precipitation under suboxic to oxic redox conditions (Benmore et al., 1983; McArthur et al., 1986). The only anomalous $\delta^{13}C$ and $\delta^{18}O$ values are associated with dolomite-rich SR-18 (Fig. 9), and may reflect deeper burial under anoxic conditions suitable to dolomite precipitation.

Under suboxic conditions in post mid-Miocene WFS sediments (> 90% $CaCO_3$) (Mullins et al., 1988a), pore water alkalinity was likely buffered by metastable carbonate mineral dissolution (high-Mg calcite and aragonite), as indicated petrographically, and by upward diffusion of CO_3 produced in the zone of sulfate reduction (Glenn et al., 1988; Compton et al., 1993). Given that the resulting pore waters would retain a relatively constant CO_3^{2-} concentration (Froelich et al., 1979; Sayles, 1981; Emerson et al., 1982), a narrow range of primary francolite CO_2 concentration would be expected, independent of burial depth. This condition differs significantly from environments lacking the buffering effects of carbonate mineral dissolution (Peru-Chile, Agulhas Bank), where initial francolite CO_2 contents are variable (McArthur, 1985; Glenn, 1990). As a consequence, variations observed for francolite unit cell a-values must be caused by the decarbonation of the francolite structure toward a fluorapatite composition (Chien and Black, 1976; McClellan and Van Kauwenbergh, 1990), a feature well documented in the geologic record (McClellan and Van Kauwenbergh, 1991). This appears to occur on the WFS through the episodic recrystallization of primary francolite.

Francolite microtextures are consistent with recrystallization in response to changes in pore water pH (Fig. 2). Both BI (botryoids) and BII (ovoids, globules, and botryoids) francolite microtextures correspond to small unit cell a-values (9.316 – 9.324 Å) compatible with a primary francolite composition (Table 3). The samples are representative of authigenic precipitation and retain an anhedral to subhedral crystallite morphology, commonly preserving microtextural similarities to both solitary and colonial benthic bacterial communities (Figs. 5C-D, 6C-D). Ovoid crystallites (Fig. 6C), similar in appearance to nonfilamentous, botuliform bacilli bacteria, are analogous to those identified on the East Australian margin as *Pseudomonas* spp. (O'Brien et al., 1981). Such organisms can accumulate elevated concentrations of nucleic acids, phospholipids, and poly-P, possibly acting as a nucleating site for francolite precipitation. According to the model of Nathan et al. (1993), bacteria can act to sequester phosphorous as intracellular poly-P during biological activity under oxic surface sediment conditions, either to hydrolyze poly-P during cycling to suboxic to anoxic conditions or liberate dissolved inorganic phosphorous (DIP) during post-mortem decomposition. This would effectively aid in the concentration

TABLE 4.— MAJOR ELEMENT AND STABLE ISOTOPIC COMPOSITION OF UPPER WEST FLORIDA SLOPE PHOSPHORITE SAMPLES.

Sample	Lithotype	SiO_2	TiO_2	Al_2O_3	Fe_2O_3	MnO	MgO	CaO	Na_2O	K_2O	P_2O_5	LOI	$\delta^{13}C$‰ (PDB)	$\delta^{18}O$‰ (PDB)
SR-1	AI	3.93	0.13	1.99	11.15	0.41	1.33	41.37	1.18	0.34	28.62	9.52	0.67	0.85
SR-2	AI	2.03	0.01	0.33	20.28	0.39	1.34	39.90	0.63	0.16	11.64	22.00	–0.77	0.55
SR-3	AI	0.81	0.01	0.29	5.19	0.91	1.24	51.39	0.56	0.10	11.68	27.78	0.44	0.93
SR-4	AI	2.38	0.01	0.59	2.48	0.07	0.96	53.19	1.22	0.36	25.80	11.64	–0.60	1.18
SR-5	BII	4.74	0.04	1.58	5.32	0.12	2.66	46.01	0.70	0.50	10.58	27.54	0.29	1.16
SR-6	BII	4.84	0.04	1.41	2.90	0.21	2.98	46.47	0.71	0.52	6.24	32.43	0.47	2.07
SR-7	BII	2.92	0.02	0.76	2.91	0.62	2.49	49.42	0.32	0.30	3.44	37.13	–0.05	1.07
SR-8	AII	3.39	0.03	0.97	13.83	0.20	1.54	42.91	0.74	0.38	16.20	17.54	–0.31	–0.56
SR-9	AII	2.14	0.02	0.66	1.23	0.05	1.15	52.30	1.15	0.26	20.72	19.15	–0.40	0.34
SR-10	BI	2.81	0.04	1.01	3.65	0.10	1.51	48.97	0.93	0.30	16.34	22.59	–0.80	–0.85
SR-11	BI	4.49	0.06	1.49	2.87	0.09	1.38	48.45	0.99	0.44	17.76	19.52	0.41	1.10
SR-12	BI	4.03	0.05	1.29	2.46	0.06	1.25	49.66	0.99	0.28	17.60	19.71	0.25	1.98
SR-13	AII	2.70	0.03	0.77	7.78	0.37	1.18	47.44	1.20	0.40	23.94	12.06	–0.79	0.85
SR-14	AI	1.42	0.02	0.47	20.79	1.05	1.47	38.72	0.64	0.14	14.64	18.47	–0.12	0.85
SR-15	AI	3.63	0.08	1.98	14.00	1.22	2.08	38.55	1.01	0.34	24.32	12.43	0.08	1.32
SR-16	AI	3.37	0.05	1.66	22.45	5.54	2.31	31.60	0.65	0.46	11.34	18.84	–1.09	–0.61
SR-17	AII	3.88	0.04	1.20	11.75	0.11	1.52	39.65	1.04	0.26	21.26	13.28	–0.57	–0.36
SR-18	BI	3.91	0.04	0.80	2.61	0.16	7.18	39.45	0.50	0.14	7.89	34.17	–6.55*	–4.92*
SR-19	AII	6.42	0.05	1.28	3.40	0.04	1.47	42.52	1.19	0.62	21.00	15.14	–0.48	1.17
SR-20	AII	5.21	0.05	1.34	6.39	0.09	1.32	40.72	1.11	0.62	20.30	15.65	–0.42	0.56
SR-21	AII	4.32	0.05	1.04	8.30	0.10	1.21	42.07	1.24	0.50	21.21	14.56	–0.23	1.38
SR-22	AII	6.61	0.06	1.69	8.27	0.07	1.37	39.32	1.10	0.45	20.67	13.89	–0.32	0.16
SR-23	AII	4.52	0.05	1.03	5.45	0.17	1.30	44.12	1.01	0.42	17.16	19.95	–0.13	0.22
SR-24	BI	4.51	0.03	0.78	2.17	0.10	2.13	47.34	0.57	0.29	5.70	35.15	–0.70	–0.55
SR-25	BII	4.24	0.04	1.27	2.86	0.08	2.75	46.06	0.64	0.35	6.55	33.86	0.70	–0.76
SR-26	BII	3.60	0.04	1.13	2.15	0.05	2.18	47.67	0.83	0.27	11.73	28.05	0.35	1.58

Note: Major element concentrations are in weight percent of air dry material, measured on bulk samples. Samples for isotopic analysis were treated by NaOAc.
* Isotopic values are an average based on three separate analysis (sample contains trace dolomite).

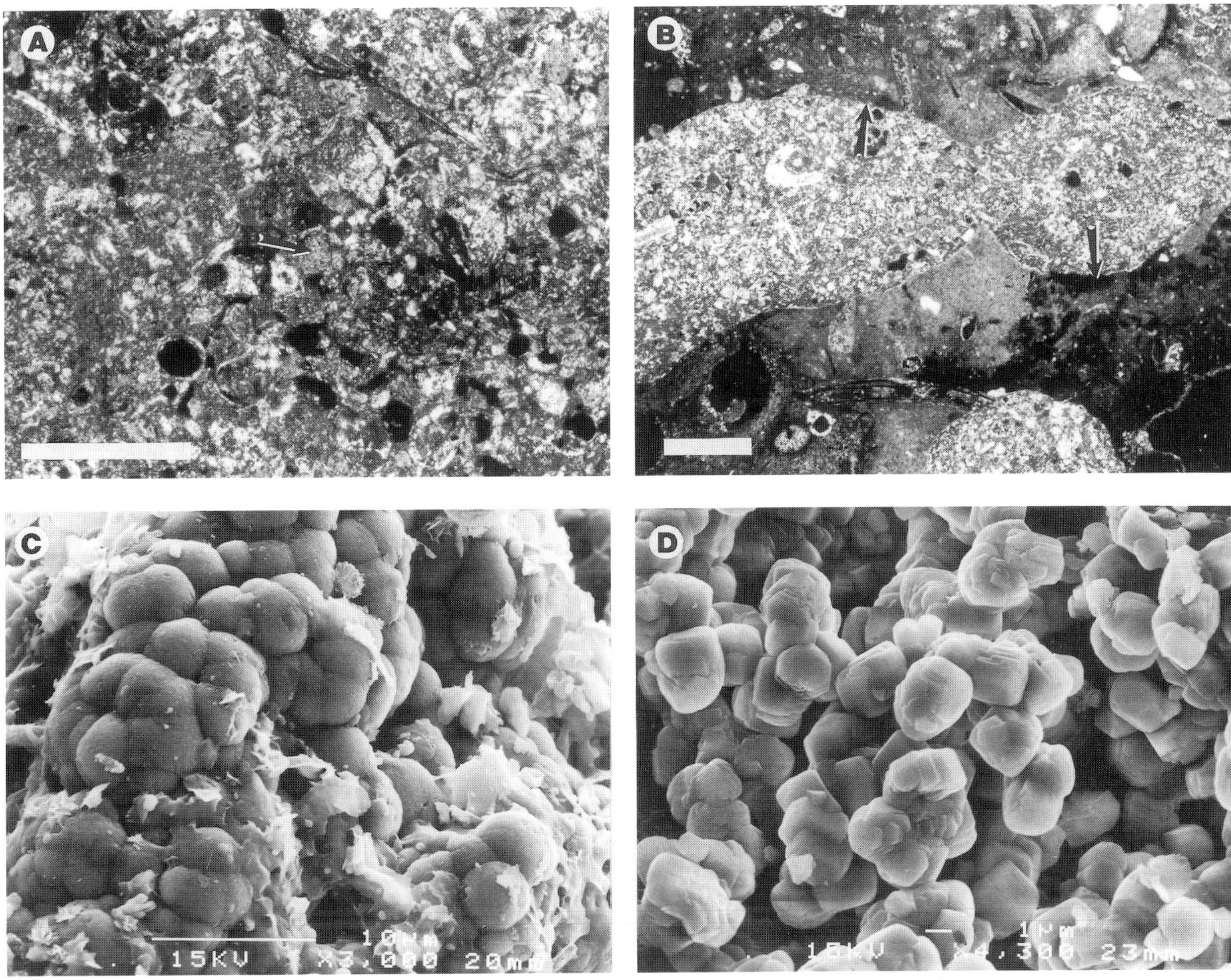

FIG. 5.— Photomicrographs of BI nonferruginous nodules (scale bars in PLM photomicrographs = 0.5 mm). (a) Phosphatic biomicrite nodule (SR-11) possessing minor Fe-enrichment. Note goethite alteration of glaucony foram infilling (arrow). (b) Example of a burrow infilled by pelagic sediments (SR-10) with considerable Fe-enrichment and glauconitization of surrounding burrow walls and nodule matrix (arrows). (c) Pustulose francolite morphology consisting of closely packed botryoids (SR-24). (d) Subhedral francolite crystallites transitional between a globular and euhedral, hexagonal morphology (SR-11).

of DIP for francolite precipitation. Although it is uncertain whether bacteria play an active or passive role in francolite precipitation, the preservation of microbial structures suggests little to no francolite recrystallization in BII nodules, with subsequent obliteration of such fabrics indicative of submarine diagenesis.

Under oxic conditions, interstitial waters can decrease in pH due to H^+ release during aerobic oxidation and cessation of manganese and ferric iron reduction (Froelich et al., 1979). If this occurs on the WFS, it may cause preferential micro-scale local recrystallization of francolite toward a more thermodynamically stable species with decreased CO_3 substitution. With a decrease in the CO_3/PO_4 ratio, the unit cell a-value increases, producing francolite that is less susceptible to early diagenetic recrystallization and possesses a lower free energy of formation (Chien and Black, 1976; McClellan and Van Kauwenbergh, 1991). Variations in porosity and bioturbation can play an important role in mediating these early diagenetic effects by inhibiting or enhancing the migration of solutions. With increased francolite and goethite content, nodule porosity would decrease, reducing the effectiveness of pore waters in altering francolite.

The development of crystal faces on BI francolite morphologies similar to filamentous *Beggiatoa* bacterial mat structures (Fig. 5D), and euhedral, hexagonal francolite crystallites in AI and AII nodules (Figs. 3C and 4C) illustrate subsequent stages of recrystallization, paralleling a decrease in francolite CO2 (Fig. 2, Table 3). These observations agree with those of LeGeros et al. (1967), which showed that decreasing CO_3 concentrations in synthetic apatites increased their crystallinity. The result for the WFS is ferruginous nodule francolite possessing unit cell a-values (9.322–9.328 Å), indicative of reduced substitution and greater thermodynamic stability.

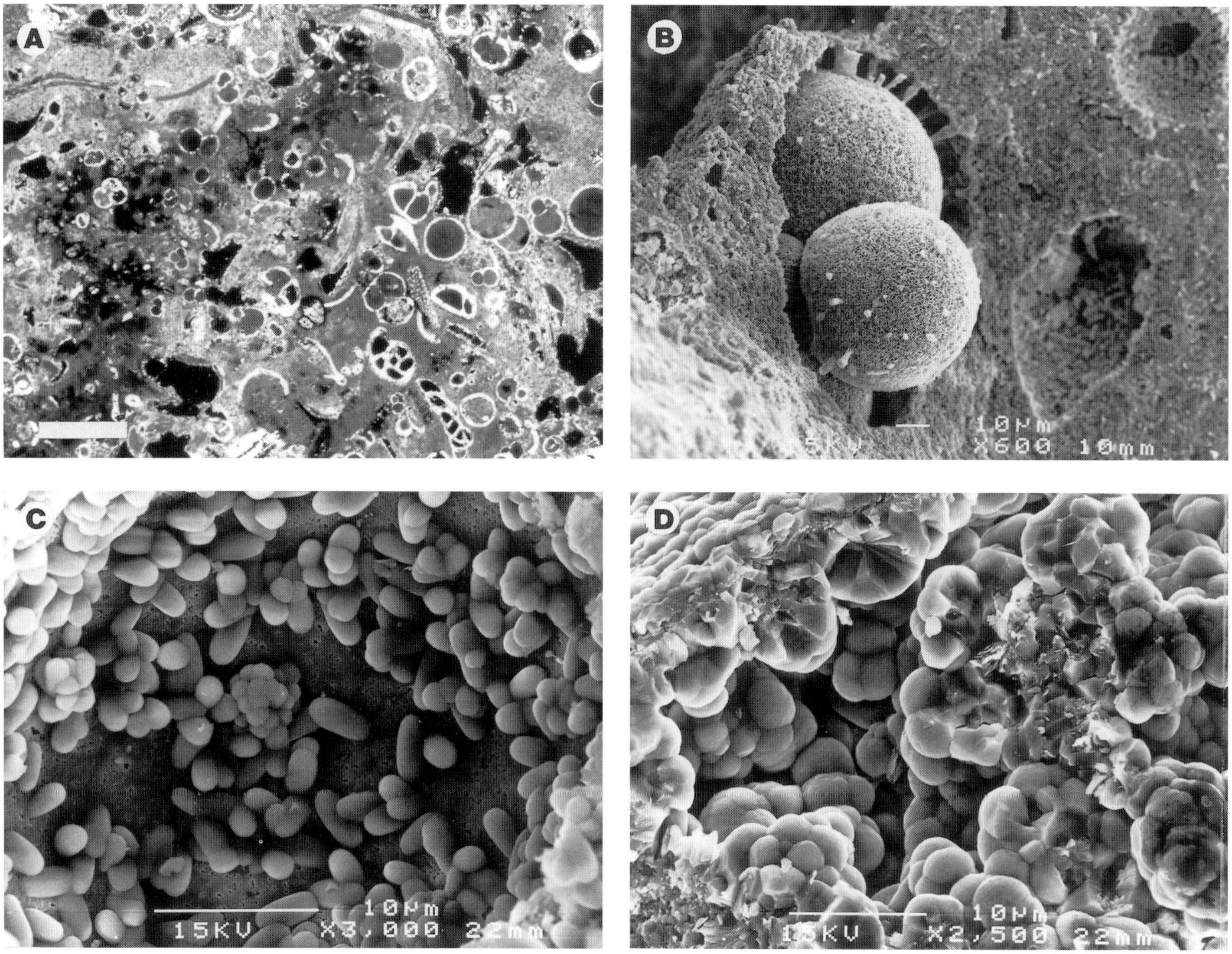

FIG. 6.— Photomicrographs of BII nonferruginous nodules (scale bars in PLM photomicrographs = 0.5 mm). (a) Phosphatic biomicrite nodule (SR-5) exhibiting little alteration of matrix or test infillings by goethite and/or FeOOH. (b) Interpenetrating crystallites of glaucony filling the chambers of a foram test (SR-5). (c) Ovoid francolite morphologically similar to botuliform bacilli bacteria (SR-26). (d) Pore-lining botryoid francolite exhibiting a radial fibrous internal structure and botryoidal to finely granular surface (SR-26).

Mineral Paragenesis

Mineralogical characteristics of WFS nodules are consistent with the evolution of primary, nonferruginous nodules (BII) into ferruginous lithotypes (AI and AII). Through an episodic process of nodule/hardground growth, associated with periods of submarine diagenesis and reworking, nodules evolve from light-colored, friable species toward well-indurated, Fe-enriched species with complex textural characteristics. Pelagic sedimentation appears to be the source of high-Mg calcite, low-Mg calcite, and aragonite within BII nodules. High-Mg calcite is derived from benthic forams, echinoid fragments, and micritic inter- or intragranular cement, and is lost through transformation to low-Mg calcite in BI and ferruginous nodules (Fig. 10). Petrographic observations of sparry low-Mg calcite replacing foram tests and echinoid fragments, along with XRD data, support this conclusion. The trace aragonite found in nonferruginous nodules and the semi- to nonferruginous matrix of some AII nodules is derived from gastropod tests and bivalves and is lost by dissolution/recrystallization to low-Mg calcite as well (Fig. 10). Low-Mg calcite is also derived from planktonic forams and coccoliths.

Glaucony is the most common Fe-silicate associated with phosphate deposits (Glenn and Arthur, 1988), and is an indicator of low sedimentation rate, particularly under current conditions suitable to the winnowing of fine terrigenous particles (Odin, 1985). Since glaucony contains iron in both reduced and oxidized states (Fe^{2+}:Fe^{3+} approximately 1:7), its development most likely occurs during the initial stages of nodule evolution (Fig. 10), taking place within suboxic sediments after partial reduction of ferric iron (Glenn and Arthur, 1988). Within the suboxic zone, soluble reduced iron may be reoxidized and incorporated with small quantities of Fe^{2+} during glauconitization. Exhibiting little to no alteration, glaucony in BII nodules

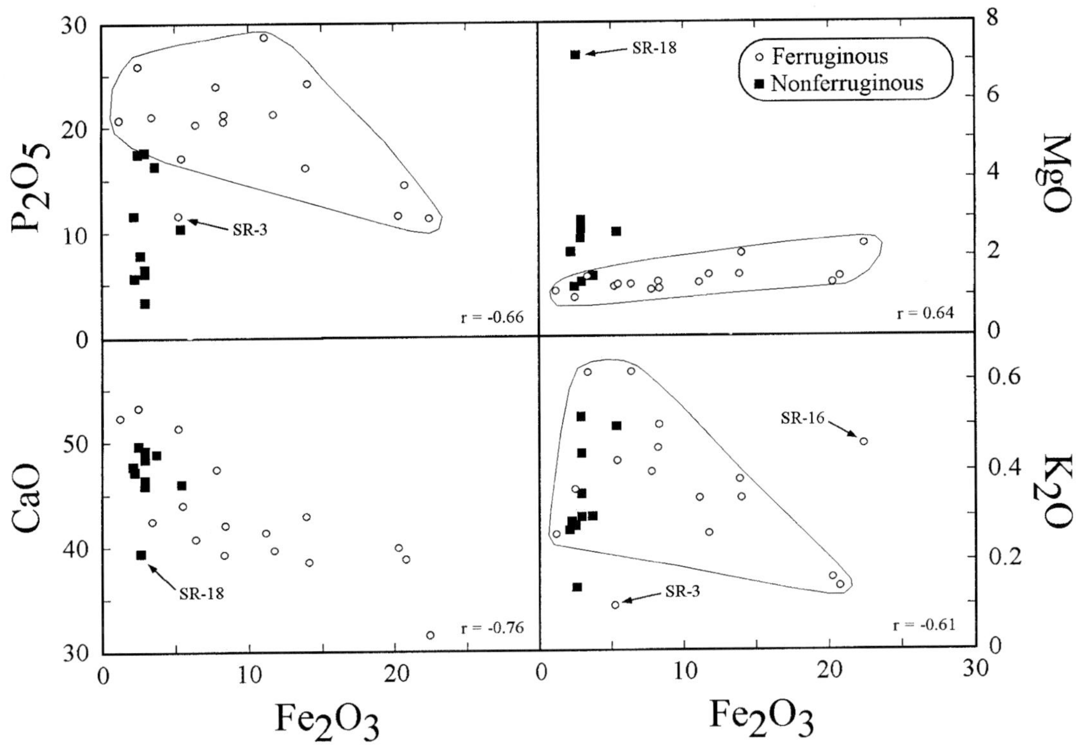

FIG. 7.— Relationship between Fe_2O_3 and the major elements P_2O_5, CaO, MgO, and K_2O for upper west Florida slope phosphorite samples (concentrations are in weight percent of air dry material; r values correspond to delineated data when not total data set).

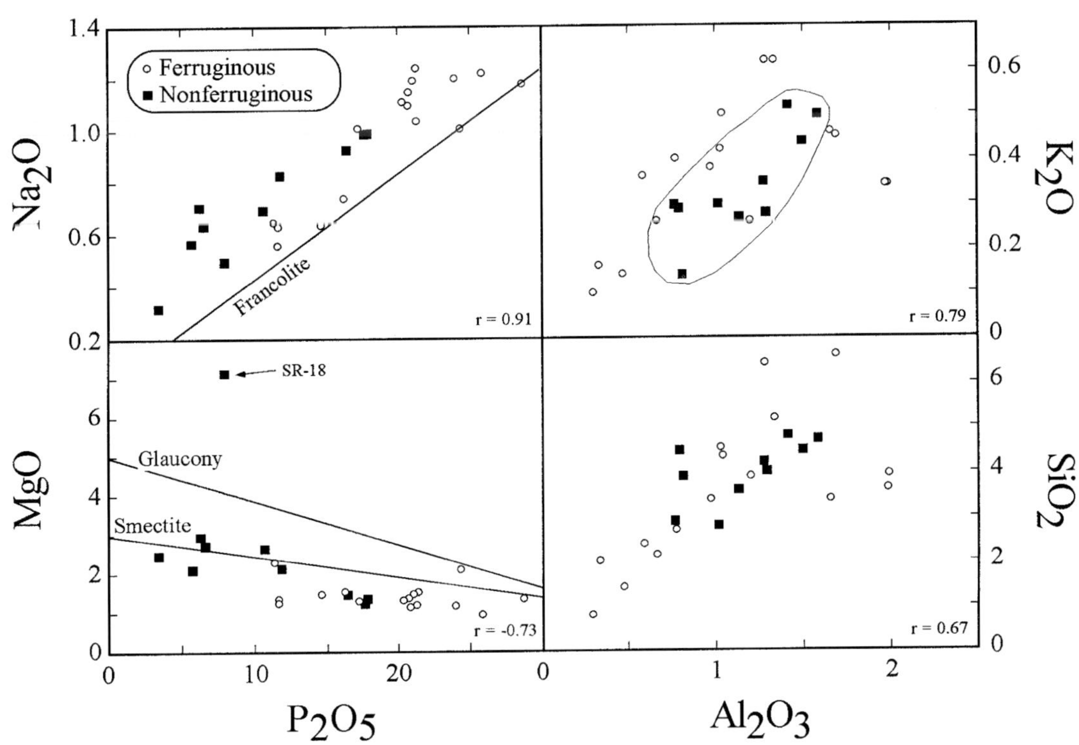

FIG. 8.— Relationship between P_2O_5 and the major elements Na_2O and MgO, and between Al_2O_3 and the major elements K_2O and SiO_2 for upper west Florida slope phosphorite samples (concentrations are in weight percent of air dry material; r values correspond to delineated data when not total data set).

(Fig. 6B) is representative of recent authigenic mineralization, with the progressive alteration to goethite observed for BI (Fig. 5A) and ferruginous nodules consistent with repeated exposure to oxic conditions.

Dolomite, a trace authigenic phase in WFS nodules, precipitates at deeper anoxic levels in association with microbial sulfate reduction and/or the deeper zone of methanogenesis (Glenn and Arthur, 1988; Glenn, 1990) as indicated by the $\delta^{13}C$ (+6.55‰ PDB) and $\delta^{18}O$ (–4.92‰ PDB) values of SR-18 (Fig. 9). Although experiments indicate that the Mg ion can inhibit the precipitation of francolite (Martens and Harris, 1970; Nathan and Lucas, 1976), aragonite dissolution coupled with dolomite precipitation may decrease the effects of Mg inhibition by decreasing the pore water Mg/Ca ratio, particularly in the suboxic zone (Bentor, 1980). The resulting decrease in alkalinity caused by diffusion of carbonate ions decreases calcium carbonate solubility, permitting francolite precipitation from interstitial waters (Baturin and Savenko, 1985). Although precipitation of dolomite may be contemporaneous with francolite precipitation in shallower, suboxic sediments, dolomite accumulation within AII and BI (SR-18) nodules would most likely post-date initial francolite accumulation, occurring after continued burial or after episodes of increased current velocity. With increased current activity, sediment mixing and winnowing could rework dolomite into shallower sediments, permitting incorporation into nodules during subsequent periods of nodule growth as observed in AII specimens.

The lack of pyrite in WFS samples can be explained by several processes: degradation during oxic sediment diagenesis and reworking, consumption of Fe^{2+} during glaucony precipitation in the suboxic zone with recycling to surficial oxic sediments, or the development of nodules above the zone of sulfate reduction. The latter theory is supported by the $\delta^{13}C$ and $\delta^{18}O$ values of structural francolite-CO_3, which indicate francolite precipitation under suboxic conditions.

Significance of Phosphorite Nodule Geochemistry

Complex textures associated with ferruginous phosphorites (e.g., East Australian margin) are thought to originate from the effects of well-oxygenated waters and erosive bottom currents interacting with nonferruginous nodules susceptible to oxic submarine diagenesis (O'Brien et al., 1990). Trends between Fe_2O_3 and the major element species CaO and K_2O illustrate several Fe-enrichment processes associated with such alteration (Fig. 7). The negative Fe_2O_3-CaO correlation is due to the loss of fine micritic cements and biogenic carbonate as a result of dissolution and/or replacement by FeOOH precipitates, as well as the dilution of CaO during direct FeOOH incorporation. Petrographic observations indicate that fine-grained micritic cements and biogenic materials, including foram tests, are the most susceptible to dissolution and/or replacement (Fig. 3B). This is illustrated by increases in goethite and FeOOH observed in BI and BII nodule porosity and cements (Figs. 5A and 6A) and the subsequent replacement of allochems observed in ferruginous nodules (Fig. 3B). The Fe_2O_3-K_2O trend for ferruginous samples (Fig. 7) is attributed to the alteration of glaucony to FeOOH and goethite in response to conditions suitable to the oxidation of structurally bound iron in glaucony. Ferruginous samples exhibit a wide range in Fe_2O_3/K_2O (5.48–148.50), with lowest values near the accepted Fe_2O_3/K_2O of 3 to 5 for average glaucony (Odin and Matter, 1981). Interpretation is complicated by AII samples (SR-19 and SR-20), which exhibit K_2O concentrations as great as 0.62%. These values are related to the samples containing both unaltered glaucony grains incorporated with semi- to nonferruginous matrix and glauconitic veneers on ferruginous intraclasts (Fig. 4A-B). Other samples (SR-3 and SR-16) show poor correlation due to intraspecimen heterogeneity. As a means of Fe-enrichment, petrographic observations of glaucony alteration indicate that it is subordinate to both carbonate mineral dissolution and/or replacement, and the direct assimilation of FeOOH. Nonferruginous samples show variability in K_2O concentrations (0.14%–0.52%) reflecting variations in glaucony sedimentation (Fig. 7).

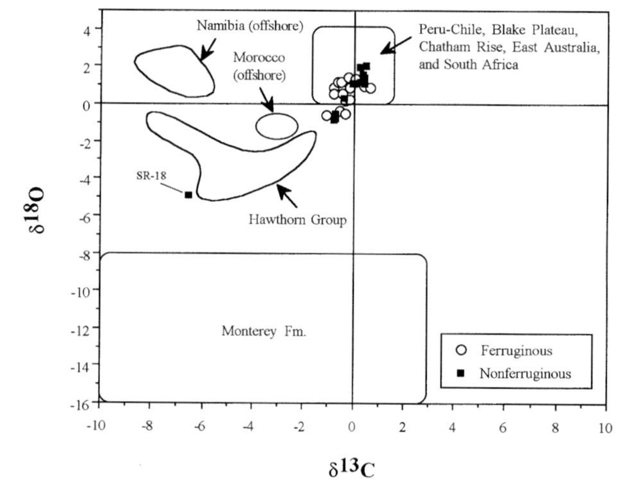

FIG. 9.— Carbon ($\delta^{13}C$) versus oxygen ($\delta^{18}O$) isotopic composition of structurally bound francolite CO_3 from the upper west Florida slope (modified after Compton et. al., 1993). The isotopic composition of francolite from other phosphorite occurrences is plotted for comparison (McArthur et. al., 1986; Glenn et. al., 1988; Kastner et. al., 1990; Compton et. al., 1993).

The Fe_2O_3-MgO distribution for nonferruginous samples can be attributed to high-Mg calcite in BII nodules and dolomite in SR-18 causing elevated MgO concentrations, with the transformation of high-Mg calcite to low-Mg calcite resulting in reduced MgO values for BI nodules (Fig. 7). Increasing MgO concentrations associated with ferruginous samples are more problematic, but probably indicate the assimilation of MgO as a result of adsorption by goethite and/or FeOOH (Parker and Siesser, 1972). This probably occurred after the initial transformation of high-Mg calcite to low-Mg calcite in nonferruginous nodules as indicated by a lack of metastable carbonate phases (except SR-9) in ferruginous nodules (Table 2), and petrographic evidence for their replacement by sparry, low-Mg calcite. The Fe_2O_3-MnO correlation is probably due to the adsorption of Mn on FeOOH during Fe-enrichment, similar to correlations observed for the East Australian margin (O'Brien et al., 1990).

The source of iron necessary for the accumulation of goethite and FeOOH in ferruginous nodules is problematic. Although the oxidation of glaucony, pyrite, and other iron monosulfide phases during oxic conditions would offer a concentrated source of Fe to the sediments, it seems unlikely that this is of major importance. It seems more likely that most of the iron concentrated in the nodules originates from the large continuous supply delivered by rivers to the Gulf of Mexico (Mississippi, Mobile, and Apalachicola Rivers) as dissolved Fe, particulate Fe, and Fe associated with suspended particles.

Fe_2O_3 contents for both nonferruginous (1.23–5.32%) and ferruginous (2.48–22.45%) samples indicate that nonferruginous nodules form with only 1–6% Fe_2O_3, with ferruginous nodules accumulating up to 22% Fe_2O_3 during diagenesis (Fig. 2). The P_2O_5-Fe_2O_3 distribution for nonferruginous samples can be attributed to the preferred accumulation of P_2O_5 (3.44–17.76%) during suboxic conditions suitable to francolite precipitation, with little to no Fe-enrichment (Fig. 7). During periods favorable to Fe-enrichment, francolite precipitation would be limited as a result of reduced porosity and oxic conditions, while colloidal FeOOH could continue to precipitate in micro-pores. This is illustrated by the negative P_2O_5-Fe_2O_3 correlation for ferruginous nodules that results from preferential Fe-enrichment, diluting the francolite signature (Fig. 7). However, during alternating episodes of francolite precipitation, P_2O_5 concentrations would increase, resulting in as much as 28.62% P_2O_5 in ferruginous samples (Fig. 2).

Francolite content is the primary factor influencing the P_2O_5-Na_2O correlation (Fig. 8). Although the WFS data parallel the theoretical end-member francolite trend possessing 1.4% Na_2O at 34.0% P_2O_5 (Van Kauwenbergh et al., 1990), the displacement above the trend suggests the presence of other Na_2O-bearing components, such as glaucony and/or smectite, a minor accessory phase in WFS sediments (Mullins et al., 1987, 1988a).

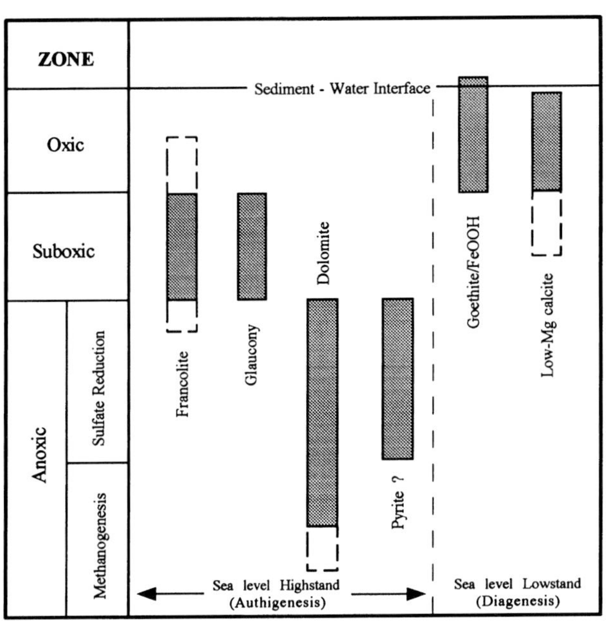

FIG. 10.— Proposed paragenetic sequence and associated geochemical zones for mineral species observed in upper west Florida slope phosphorite nodules.

The negative P_2O_5-MgO correlation likely results from the dilution of a detrital or authigenic MgO signature due to increased francolite content (Fig. 8). A Mg-rich clay species such as smectite and/or glaucony would be the most probable MgO source, possessing maximum MgO concentrations of approximately 3% and 5%, respectively (Neuman and Brown, 1987). Dilution of this signature by francolite would reduce the MgO concentration to 1.4% at 34.0% P_2O_5 for a theoretical francolite end-member composition (McClellan, 1980; Van Kauwenbergh et al., 1990). Values that extend above these trends are probably due to MgO adsorption by goethite (Parker and Siesser, 1972), or the presence of dolomite (SR-18) or high-Mg calcite (Fig. 8). The Al_2O_3-K_2O and Al_2O_3-SiO_2 correlations are related to the presence of authigenic glaucony and detrital clays in WFS sediments (Fig. 8).

Proposed Model for Upper West Florida Slope Phosphogenesis

Organic carbon oxidation, iron and phosphorous recycling between oxic and anoxic sediments, and sediment mixing by bioturbation are considered to be key processes in the development of modern phosphorites (Bentor, 1980; O'Brien et al., 1990; Compton et al., 1993). Although large C_{org} and DIP fluxes are commonly employed to explain phosphogenesis in regions of strong coastal upwelling (Peru-Chile and Namibia), the WFS possesses only weak seasonal upwelling (Elliot, 1982) with total organic matter values between 6–8%, with a range of 3–12% over the whole shelf-slope transition (Mullins et al., 1988a). As a result, DIP concentration within sediments may result not only from the early diagenesis of organic matter, but also as a result of redox driven Fe-P shuttling similar to that proposed for East Australian margin phosphorites (Heggie et al., 1988; O'Brien et al., 1990). With this model, phosphorous is directly extracted from overlying waters, as well as from the shallow oxidation of C_{org} (Schuffert et al., 1994). Thus, DIP enrichment within suboxic interstitial waters is not limited by C_{org} accumulation rates, and is augmented by the oceanic phosphorous reservoir through phosphorous adsorption by particulate FeOOH.

Based on stratigraphic constraints and Loop Current positioning over the WFS (Mullins et al., 1987; 1988b), phosphorite nodule development appears to exhibit a cyclicity corresponding to eustatic sealevel perturbations. During sealevel highstands, the Loop Current would be deflected landward, drawing nutrient-rich upwelling waters onto the shelf-slope break and restricting detrital sediment flux from the east (Fig. 11). Increased productivity of surface waters, in turn, would increase the C_{org} flux to WFS sediments under the influence of reduced current velocities. This would establish conditions suitable to both francolite and glaucony authigenesis through the oxidation of organic matter and/or a Fe-P shuttling mechanism. DIP released from the oxidation of organic matter could be scavenged by Fe^{3+} near the oxic sediment-water interface, with shuttling of both FeOOH and adsorbed P to the suboxic zone accomplished through mechanical winnowing of surficial sediments, or through the activity of infaunal bioturbation (Heggie et al., 1990; Schuffert et al., 1994). Fe^{2+} released under suboxic

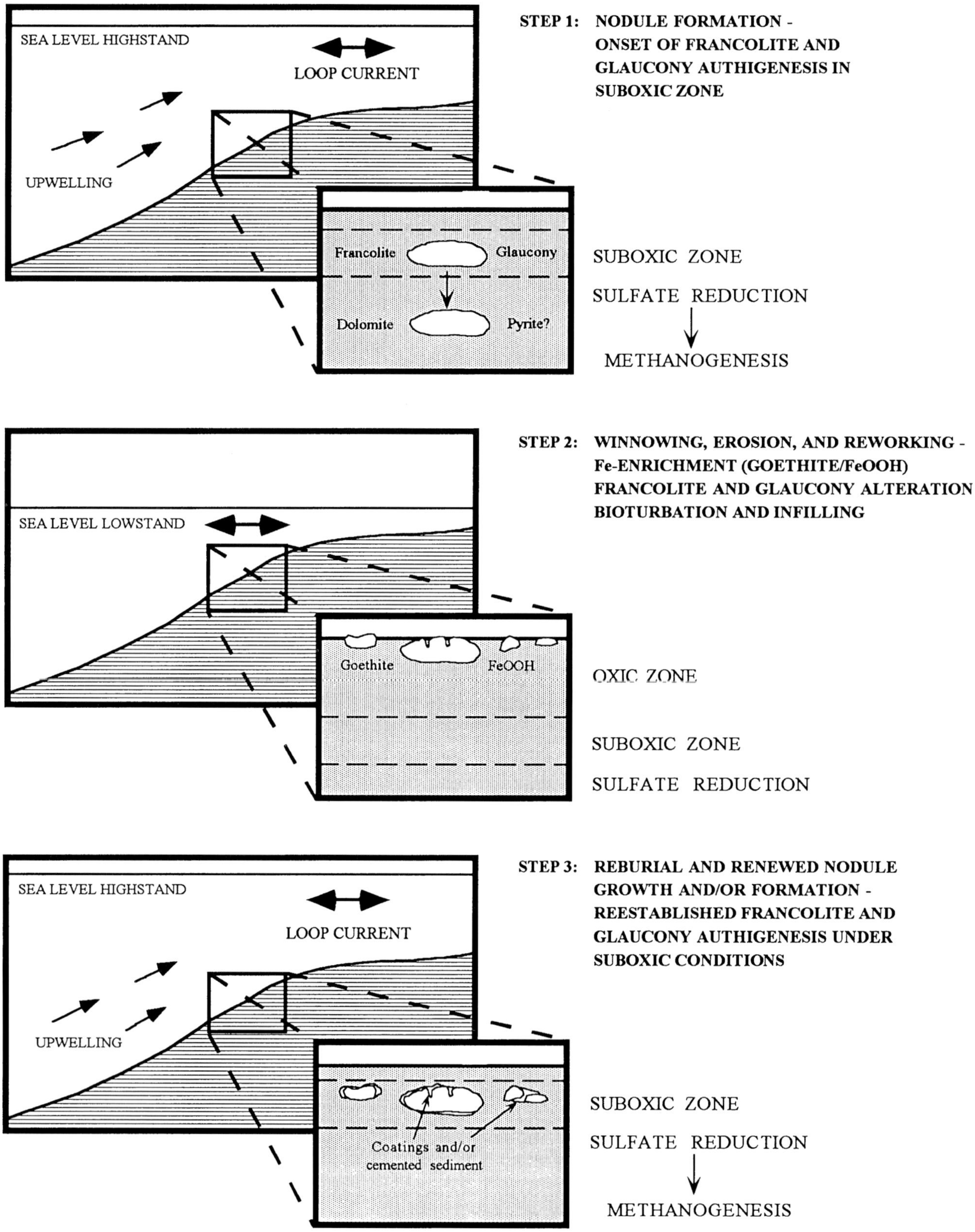

FIG. 11.— Proposed diagenetic model for phosphorite nodule evolution in sediments of the upper west Florida slope. In step 1, francolite and glaucony are precipitated in the suboxic zone in response to an increased C_{org} flux and DIP concentration, forming primary nonferruginous nodules. In step 2, winnowing and erosion expose nodules to oxic conditions, resulting in preferential Fe-enrichment at the expense of glaucony and carbonate minerals, with recrystallization of francolite. Reworking of nodules into lag deposits with subsequent reburial produces AII nodules. In step 3, reduced current velocity and DIP concentration reestablish conditions suitable to renewed francolite and glaucony precipitation as matrix cement, veneers, and replacement textures (Fig. 4A), promoting further nodule growth.

TABLE 5.— COMPARISON OF THE AVERAGE MAJOR ELEMENT COMPOSITION OF UPPER WEST FLORIDA SLOPE PHOSPHORITE NODULES TO OTHER OFFSHORE PHOSPHORITES.

	WFS[a]	WFS[b]	EAS[c]	EAS[d]	Agulhas[e] Bank	Peru-Chile[f]	Namibia[g]	Blake[h] Plateau
SiO_2	3.55	4.01	17.6	24.0	14.79	22.13	1.4	3.7
TiO_2	0.04	0.04	0.26	0.32	0.10	—	0.06	0.15
Al_2O_3	1.08	1.15	3.46	3.53	2.13	5.15	0.37	1.1
Fe_2O_3	10.17	2.99	29.5	5.00	6.17	2.85	0.87	4.1
MnO	0.67	0.16	0.06	0.04	0.01	—	—	0.45
MgO	1.42	2.65	2.59	1.31	1.38	1.07	0.6	1.2
CaO	42.86	46.95	19.6	31.9	36.87	33.93	51.3	50.0
Na_2O	0.98	0.72	0.6	0.9	0.76	0.85	—	0.60
K_2O	0.36	0.34	1.04	0.89	1.55	1.30	0.13	0.30
P_2O_5	19.41	10.38	9.6	11.0	17.53	22.61	32.1	22.2
LOI	16.37	29.02	14.7	19.6	16.40	8.78	—	14.9
CaO/P_2O_5	2.20	4.52	2.04	2.90	2.10	1.50	1.60	2.25
Fe_2O_3/K_2O	28.25	8.79	28.37	5.62	3.98	2.19	6.69	13.67

Note: Major element concentrations are in weight percent of air dry material, measured on bulk samples. Total iron reported as Fe_2O_3.
[a] Upper west Florida slope composite average based on 16 ferruginous phosphorite nodule samples.
[b] Upper west Florida slope composite average based on 10 nonferruginous phosphorite nodule samples.
[c] East Australian shelf composite average based on 12 ferruginous phosphorite nodule samples (Kress and Veeh, 1980; Cook and Marshall, 1981).
[d] East Australian shelf composite average based on 4 nonferruginous phosphorite nodule samples (Kress and Veeh, 1980; O'Brien and Veeh, 1980).
[e] Composite average based on 12 analyes (Parker, 1975).
[f] Composite average based on 15 analyses (Burnett, 1977).
[g] Composite average based on 2 analyses (Price and Calvert, 1978).
[h] Blake Plateau composite phosphorite (Manheim et al., 1980).

conditions can diffuse upward, where it may be reoxidized to Fe^{3+} to form FeOOH compounds, or may take part in glaucony and/or pyrite precipitation (Figs. 10 and 11). DIP released to pore waters may diffuse upward only to be re-adsorbed by FeOOH at the oxic-suboxic boundary or take part in francolite precipitation above the zone of sulfate reduction (Heggie et al., 1990). Employing both processes, DIP tends to be trapped and concentrated under suboxic conditions suitable to nodule formation (Fig. 11).

During sealevel lowstands, deflection of the Loop Current westward would increase current velocity over the study site, imparting oxic conditions to WFS sediments through intensified winnowing and reduction of the C_{org} flux (Fig. 11). This, in turn, would cease conditions suitable for nodule formation, resulting in preferential ferruginization of nodules at the expense of glaucony and carbonate minerals, with coeval recrystallization of francolite. With Fe-P shuttling, iron would be concentrated at the oxic-suboxic boundary (Heggie et al., 1990), preventing iron concentration in the nodules from being limited by FeOOH supply to the sediments. Conditions associated with a sealevel lowstand also would be responsible for reworking of some ferruginous nodules into lag deposits (AII nodules) as a result of induced erosional conditions, and perhaps, infaunal bioturbation and epifaunal boring (Fig. 11). These mechanisms would tend to periodically fracture or rupture primary hardground layers, explaining the flattened shape of most nodules. Those portions of localized hardgrounds and nodules not reworked would subsequently evolve into the AI lithotype through continued Fe-enrichment during periods of very slow sedimentation, nondeposition, and submarine erosion. Goethitic surface veneers observed in AII samples (Fig. 4A) likely precipitated during hiatal episodes, and may have formed contemporaneously with goethitic laminae in AI microstromatoloids (Fig. 3D) near the sediment-water interface. Francolite veneers on intraclasts (Fig. 4A), francolite replacement textures associated with burrow structures (Fig. 4D), and francolite laminae in AI microstromatoloids (Fig. 3D) also probably formed during hiatal periods, but under conditions of high bottom water DIP concentration.

The dissolution of aragonite and the precipitation of sparry calcite (low-Mg calcite) are features commonly interpreted as the result of freshwater diagenesis (Folk, 1974; Longman, 1980). However, combined regional tectonic subsidence and eustatic sealevel change estimates of about 380 m since the Middle Miocene (Mullins and Neumann, 1979) indicate that even during the Late Miocene (Tortonian) age sealevel lowstand beginning 10.2 million years ago (Haq et al., 1987) the outer shelf remained submerged (Mullins et al., 1988b). This makes it difficult to invoke meteoric diagenesis as a means of explaining these features, despite sealevel fluctuations. However, precipitation of sparry calcite and the dissolution of aragonite have been recorded under marine conditions (Mullins et al., 1985; Dix and Mullins, 1988; Freeman-Lynde et al., 1988; Glenn and Kronen, 1993). Thus, increases in the aragonite saturation level coincident with positioning of the aragonite lysocline at or above the shelf-slope transition can be used to explain aragonite dissolution and subsequent sparry calcite precipitation (Glenn and Kronen, 1993) on the WFS. Excess alkalinity generated by aragonite dissolution would promote precipitation of low-Mg calcite, particularly within zones of secondary porosity. Low-Mg calcite mobility also may be explained by periods of oxic diagenesis (Figs. 10 and 11). In the absence of reduced iron, carbonate dissolution may occur (Coleman, 1985), promoted by H^+ release and decreased pH accompanying an increase in sediment depth of the oxic-suboxic boundary (Froelich et al., 1979). Thus, periods of francolite and calcite dissolution/reprecipitation may be contemporaneous during episodes of near-surface submarine exposure and winnowing.

Phosphorite nodules from the WFS are very similar in composition to other offshore deposits of Neogene to recent age (Table 5). The nearest analogs are likely the East Australian margin, where both ferruginous "Type II" and nonferruginous "Type I" nodules form in response to Fe-P shuttling under conditions that lack prominent coastal upwelling and high C_{org} fluxes (O'Brien and Veeh, 1980; O'Brien et al., 1990), and the Peru-Chile margin, a classic example of an eastern boundary current upwelling site, where ferruginous "D-phosphates" and nonferruginous "F-phosphates" (Garrison and Kastner, 1990) have formed in response to organic matter oxidation under conditions of high biological productivity and large C_{org} fluxes (Glenn et al., 1994). A comparison of these sites is summarized by Glenn et al. (1994). The WFS appears to be transitional between these two environments, possessing weak seasonal upwelling and C_{org} fluxes intermediate between the two (East Australia: TOC < 0.5%, Peru-Chile: TOC = 1–20%), further evidence of our belief that both organic matter oxidation and Fe-P shuttling may be necessary to elevate pore water DIP concentrations to levels necessary for francolite precipitation.

CONCLUSIONS

Marginal upwelling conditions on the upper west Florida slope have resulted in the accumulation of ferruginous and nonferruginous phosphorite nodules in water depths of 450 to 600 m. These phosphorites differ significantly from onshore deposits in Florida and record the episodic precipitation of highly substituted francolite with approximately 6.2% CO_2 since the mid-Miocene (12–15 Ma). Periods of increased C_{org} flux associated with deflection of the Gulf of Mexico Loop Current landward are interpreted to have reduced both current velocity over the shelf-slope break and the detrital sediment flux from the east, creating conditions suitable to the concentration of DIP within redox-stratified sediments. The early diagenesis of organic matter and/or Fe-P shuttling result in francolite and glaucony authigenesis within suboxic sediments, involving an as yet uncertain microbial component (*Beggiatoa* bacterial mats and botuliform bacilli bacteria). This leads to the formation of nodules similar to the BII lithotype, which contain 3–12% P_2O_5 and 1–6% Fe_2O_3.

With deflection of the Loop Current westward, increased current velocity would winnow WFS sediments, creating conditions suitable to exposure and reworking of nodules and/or hardgrounds. As indicated by petrographic and geochemical data, the resulting oxic conditions favor preferential enrichment of FeOOH and goethite (up to 22% Fe_2O_3) at the expense of carbonate minerals (aragonite and high-Mg calcite) and glaucony. Furthermore, highly substituted francolite (BI and BII) undergoes recrystallization toward a more thermodynamically stable composition characterized by decreased CO_2 (4.9–5.5 %) and larger unit cell a-values (9.322–9.328 Å). This obliterates anhedral to subhedral crystallite fabrics, resulting in the euhedral, hexagonal crystallites found in ferruginous nodules. Further sealevel oscillations would result in the complex reworking and reburial of preexisting nodules, evolving AI and AII lithotypes from nonferruginous nodules (BI and BII) and hardgrounds.

ACKNOWLEDGMENTS

The authors gratefully acknowledge the service and financial support of the Florida Institute of Oceanography, the Southeastern Section of the Geological Society of America, the Division of Sponsored Research of the Graduate School of the University of Florida, and the Florida Geological Survey. We also acknowledge the crew of the RV *Suncoaster* and the staff of the University of Florida's Major Analytical Instrumentation Center for their assistance in both data collection and analysis. We are indebted to these colleagues. This paper also benefitted from the insights of the Syracuse group, and in particular, George R. Dix, whose initial observations related to WFS phosphogenesis were invaluable, as well as the reviews of Geoffrey W. O'Brien and David L. Kidder and the editorial comments of Craig R. Glenn.

REFERENCES

Baturin, G. N., and Savenko, V. S., 1985, Mechanism of formation of phosphorite nodules: Oceanology, v. 25, p. 747-750.

Benmore, R. A., Coleman, M. A., and McArthur, J. M., 1983, Origin of sedimentary francolite from its sulphur and carbon isotope composition: Nature, v. 302, p. 516-518.

Bentor, Y. K., 1980, Phosphorites—The unsolved problems, *in* Bentor, Y. K., ed., Marine Phosphorites—Geochemistry, Occurrence, Genesis: Tulsa, Society for Sedimentary Geology (SEPM) Special Publication 29, p. 3-18.

Burnett, W. C., 1977, Geochemistry and origin of phosphorite deposits from off Peru and Chile: Geological Society of America Bulletin, v. 88, p. 813-823.

Burnett, W. C., and Gomberg, D. N., 1977, Uranium oxidation and probable subaerial weathering of phosphatic limestone from the Pourtales Terrace: Sedimentology, v. 24, p. 291-302.

Chien, S. H., and Black, C. A., 1976, Free energy of formation of carbonate apatites in some phosphate rocks: Journal of the Soil Science Society of

America: v. 40, p. 234-239.

Coleman, M. L., 1985, Geochemistry of diagenetic nonsilicate minerals: Kinetic considerations: Philosophical Transactions of the Royal Society of London, Series A, v. 315, p. 39-56.

Compton, J. S., Hodell, D. A., Garrido, J. R., and Mallinson, D. J., 1993, Origin and age of phosphorite from the south-central Florida platform: Relation of phosphogenesis to sea-level fluctuations and $\delta^{13}C$ excursions: Geochimica et Cosmochimica Acta, v. 57, p. 131-146.

Cook, P. J., and Marshall, J. F., 1981, Geochemistry of iron and phosphorous-rich nodules from the East Australian continental shelf: Marine Geology, v. 41, p. 205-221.

Cook, P. J. and McElhinny, M. W., 1979, A re-evaluation of the spatial and temporal distribution of sedimentary phosphate deposits in the light of plate tectonics: Economic Geology, v. 74, p. 315-330.

Dix, G. R., and Mullins, H. T., 1988, Rapid burial diagenesis of deep water carbonates: Exuma Sound Bahamas: Geology, v. 16, p. 680-683.

Doyle, L. J., and Holmes, C. W., 1985, Shallow structure, stratigraphy, and carbonate sedimentary processes of west Florida upper continental slope: American Association of Petroleum Geologists Bulletin, v. 69, p. 1133-1144.

Elliot, B. A., 1982, Anticyclonic rings in the Gulf of Mexico: Journal of Physical Oceanography, v. 12, p. 1292-1309.

Emerson, S., Grundmanis, V., and Graham, D., 1982, Carbonate chemistry in marine pore waters: MANOP sites C and S: Earth and Planetary Science Letters, v. 61, p. 220-232.

Flugel, E., 1982, Microfacies Analysis of Limestones: Berlin, Springer-Verlag, 633 p.

Folk, R. L., 1974, The natural history of crystalline calcium carbonate: Effects of magnesium content and salinity: Journal of Sedimentary Petrology, v. 44, p. 40-53.

Freeman-Lynde, R. P., McClain, W. R., and Lohmann, K. C., 1988, Deep-marine origin of equant spar cements in Oligocene-Miocene Perireef Boundstones, Leg 101, Site 635, Northeast Providence Channel, Bahamas, *in* Austin, J. A., Jr. et al., eds., Proceedings of the Ocean Drilling Program, Scientific Results, v. 101, p. 255-261.

Froelich, P. N., Klinkhammer, G. P., Bender, M. L., Lvedtke, N. A., Heath, G. R., Cullen, D., Dauphin, P., Hammond, D., Hartman, B., and Maynard, V., 1979, Early oxidation of organic matter in pelagic sediments of the eastern equatorial Atlantic: Suboxic diagenesis: Geochimica et Cosmochimica Acta, v. 43, p. 1075-1090.

Gardulski, A. F., Gowen, M. H., Milsark, A., Weiterman, S. D., Wise, S. W., Jr., and Mullins, H. T., 1991, Evolution of a deep-water carbonate platform: Upper Cretaceous to Pleistocene sedimentary environments on the west Florida margin: Marine Geology, v. 101, p. 163-179.

Garrison, R. E. and Kastner, M., 1990, Phosphatic sediments and rocks recovered from the Peru margin during ODP Leg 112, *in* Suess, E., et al., eds., Proceedings of the Ocean Drilling Program Scientific Results, v. 112, p. 111-134.

Glenn, C. R., 1990, Pore water, petrologic and stable carbon isotopic data bearing on the origin of Modern Peru margin phosphorites and associated authigenic phases, in Burnett, W. C., and Riggs, S. R., eds., Phosphate Deposits of the World, v. 3: Neogene to Modern phosphorites: Cambridge, Cambridge University Press, p. 338-352.

Glenn, C. R., and Arthur, M. A., 1988, Petrology and major element geochemistry of Peru margin phosphorites and associated diagenetic minerals: authigenesis in modern organic-rich sediments: Marine Geology, v. 80, p. 231-267.

Glenn, C. R., Arthur, M. A., Yeh, H.-W., and Burnett, W. C., 1988, Carbon isotopic composition and lattice-bound carbonate of Peru-Chile margin phosphorites: Marine Geology, v. 80, p. 287-307.

Glenn, C. R., and Kronen, J. D., Jr., 1993, Origin and significance of Late Pliocene phosphatic hardgrounds on the Queensland Plateau, northeastern Australian margin, *in* McKenzie, J. A., et al., eds., Proceedings of the Ocean Drilling Program, Scientific Results, v. 133, p. 525-534.

Glenn, C. R., Föllmi, K. B., Riggs, S. R., Baturin, G. N., Grimm, K. A., Trappe, J., Abed, A. M., Galli-Olivier, C., Garrison, R. E., Ilyin, A. V., Jehl, C., Rohrlich, V., Sadaqah, R. M. Y., Schidlowski, M., Sheldon, R. E., and Siegmund, H., 1994, Phosphorous and phosphorites: Sedimentology and environments of formation: Eclogae geol. Helv., v. 87, p. 747-788.

Gorsline, D. S. and Milligan, D. B., 1963, Phosphatic deposits along the margin of the Pourtales Terrace: Deep-Sea Research, v. 10, p. 259-262.

Haq, B.U., Hardenbol, J., and Vail, P. R., 1987, Chronology of fluctuating sea levels since the Triassic: Science, v. 235, p. 1156-1167.

Heggie, D. T., O'Brien, G. W., Blanks, A., Reimers, C. E., Burnett, W. C., and McArthur, J. M., 1988, E. Australian modern marine phosphorites: Geochemical cycling through pore waters (abs.): American Geophysical Union/American Society of Limnology and Oceanography Ocean Sciences Meeting, New Orleans, January 1988, Eos Transactions, AGU, v. 68, p. 1711.

Heggie, D. T., Skyring, G. W., O'Brien, G. W., Reimers, C., Herczeg, A., Moriarty, D. J. W., Burnett, W., and Milnes, A. R., 1990, Organic carbon cycling and modern phosphorite formation on the East Australian continental margin: An overview, *in* Notholt, A. J. G., and Jarvis, I., eds., Phosphorite Research and Development: Oxford, Geological Society Special Publication 52, p. 87-117.

Jones, J. I., 1973, Physical oceanography of the northeastern Gulf of Mexico and Florida continental shelf area: Tallahassee, Florida State University, Summary of Knowledge of the Eastern Gulf of Mexico, p. IIB-1-IIB-41.

Kastner, M., Garrison, R. E., Kolodny, Y., Reimers, C. E., and Shemesh, A., 1990, Coupled changes of oxygen isotopes in PO_4^{3-} and CO_3^{2-} in apatite, with emphasis on the Monterey Formation, California, *in* Burnett, W. C., and Riggs, S. R., eds., Phosphate Deposits of the World 3: Neogene to Modern Phosphorites: Cambridge, Cambridge University Press, p. 312-324.

Kress, A. G., and Veeh, H. H., 1980, Geochemistry and radiometric ages of phosphatic nodules from the continental margin of northern New South Wales, Australia: Marine Geology, v. 36, p. 143-157.

Krajewski, K., 1981, Phosphate microstromatolites in the High-Tatric Albian limestones in the Polish Tatra Mountains: Bulletin de Academie polonaise des Sciences, Serie des Sciences de la Terre, v. 29, p. 175-183.

Lanier, W., 1988, Structure and morphogenesis of microstromatolites from Transvaal supergroup, South Africa: Journal of Sedimentary Petrology, v. 58, p. 89-99.

Legeros, R. Z., Trautz, O. R., Legeros, J. P., and Klein, E., 1967, Apatite crystallites: Effects of carbonate on morphology: Science, v. 155, p. 1409-1411.

Longman, M. W., 1980, Carbonate diagenetic textures from nearsurface diagenetic environments: American Association of Petroleum Geologists Bulletin, v. 64, p. 461-487.

Manheim, F. T., Pratt, R. M., and Mcfarlin, P. F., 1980, Composition and origin of phosphorite deposits of the Blake Plateau, *in* Bentor, Y. K., ed., Marine Phosphorites—Geochemistry, Occurrence, Genesis: Tulsa, Society for Sedimentary Geology (SEPM) Special Publication 29, p. 117-138.

Martens, C. S., and Harris, R. C., 1970, Inhibition of apatite precipitation in the marine environment by Mg-ions: Geochimica et Cosmochimica Acta, v. 34, p. 621-625.

McArthur, J. M., 1985, Francolite geochemistry—compositional controls during formation, diagenesis, metamorphism and weathering: Geochimica et Cosmochimica Acta, v. 49, p. 23-35.

McArthur, J. M., Benmore, R. A., Coleman, M. L., Soldi, C., Yeh, H.-W., and O'Brien, G. W., 1986, Stable isotope characterization of francolite formation: Earth and Planetary Science Letters, v. 77, p. 20-34.

McClellan, G. H., 1980, Mineralogy of carbonate fluorapatite: Journal of the Geological Society, London, v. 137, p. 675-682.

McClellan, G. H., and Lehr, J. R., 1969, Crystal chemical investigation of natural apatites: American Mineralogist, v. 54, p. 1374-1391.

McClellan, G. H., and Gremillion, L. R., 1980, Evaluation of phosphatic raw material, *in* Khasaureh, et al., eds., The Role of Phosphorous in Agriculture: Madison, Wisconsin, American Society of Agronomy, p. 43-80.

Mcclellan, G. H., and Van Kauwenbergh, S. J., 1990, Mineralogy of sedimentary apatites, *in* Notholt, A. J. G., and Jarvis, I., eds., Phosphorite Research and Development: Oxford, Geological Society Special Publication 52, p. 23-31.

McClellan, G. H., and Van Kauwenbergh, S. J., 1991, Mineralogical and chemical variation of francolites with geological time: Journal of the Geological Society, London, v. 148, p. 809-812.

Mullins, H. T., and Neumann, A. C., 1979, Geology of the Miami Terrace and its paleoceanographic implications: Marine Geology, v. 30, p. 205-236.

Mullins, H. T., Wise, S. W., Jr., Gardulski, A. F., Hinchey, E. J., Masters, P.

M., AND SIEGEL, D. I., 1985, Shallow subsurface diagenesis of Pleistocene periplatform ooze: Northern Bahamas: Sedimentology, v. 32, p. 473-494.

MULLINS, H. T., GARDULSKI, A. F., WISE, S. W., JR., AND APPLEGATE, J., 1987, Middle Miocene oceanographic event in the eastern Gulf of Mexico: Implications for seismic stratigraphic succession and Loop Current/Gulf Stream circulation: Geological Society of America Bulletin, v. 98, p. 702-713.

MULLINS, H. T., GARDULSKI, A. F., HINCHEY, E. J., AND HINE, A. C., 1988a, The modern carbonate ramp slope of central west Florida: Journal of Sedimentary Petrology, v. 58, p. 273-290.

MULLINS, H. T., GARDULSKI, A. F., HINE, A. C., MELILLO, A. J., WISE, S. W., Jr., AND APPLEGATE, J., 1988b, Three-dimensional sedimentary framework of the carbonate ramp slope of central west Florida: a sequential seismic stratigraphic perspective: Geological Society of America Bulletin, v. 100, p. 514-533.

NATHAN, Y. and LUCAS, J., 1976, Experiments on the direct precipitation of apatite in sea water; implication in the genesis of phosphorites: Chemical Geology, v. 18, p. 181-186.

NATHAN, Y., BREMNER, J. M., LOEWENTHAL, R. E., AND MONTEIRO, P., 1993, Role of bacteria in phosphorite genesis: Geomicrobiology Journal, v. 11, p. 69-76.

NEUMAN, A. C. D. AND BROWN, G., 1987, The chemical constitution of clays, *in* Neuman, A.C.D., ed., Chemistry of Clays and Clay Minerals: New York, John Wiley & Sons, p. 1-128.

O'BRIEN, G. W., AND VEEH, H. H., 1980, Holocene phosphorite on the East Australian continental margin: Nature, v. 288, p. 690-692.

O'BRIEN, G. W., HARRIS, J. R., MILNES, A. R., AND VEEH, H. H., 1981, Bacterial origin of East Australian continental margin phosphorites: Nature, v. 294, p. 442-444.

O'BRIEN, G. W., MILNES, A. R., VEEH, H. H., HEGGIE, D. T., RIGGS, S. R., CULLEN, D. J., MARSHALL, J. F., AND COOK, P. J., 1990, Sedimentation dynamics and redox iron-cycling: controlling factors for apatite-glauconite association on the East Australian continental margin, *in* Notholt, A. J. G., and Jarvis, I., eds., Phosphorite Research and Development: Oxford, Geological Society Special Publication 52, p. 61-86.

ODIN, G. S., 1985, Significance of green particles (glaucony, berthierine, chlorite) in arenites, *in* Zuffa, G. G., ed., Provenance of Arenites: Dordrecht, D. Reidel Publishing Company, p. 279-307.

ODIN, G. S., AND MATTER, A., 1981, De glauconiarum origine: Sedimentology, v. 28, p. 611-641.

PARKER, R. J., 1975, The petrology and origin of some phosphorites from the South African continental margin: Journal of Sedimentary Petrology, v. 45, p. 230-242.

PARKER, R. J., AND SIESSER, W. G., 1972, Petrology and origin of some phosphorites from the South African continental margin: Journal of Sedimentary Petrology, v. 42, p. 434-440.

PRICE, N. B., AND CALVERT, S. E., 1978, The geochemistry of phosphorites from the Namibian shelf: Chemical Geology, v. 23, p. 151-170.

RIGGS, S. R., 1979a, Petrology of the Tertiary phosphorite system of Florida: Economic Geology, v. 74, p. 195-220.

RIGGS, S. R., 1979b, Phosphorite sedimentation in Florida—a model phosphogenic system: Economic Geology, v. 74, p. 285-314.

RIGGS, S. R., 1984a, Paleoceanographic model of Neogene phosphorite deposition, U.S. Atlantic continental margin: Science, v. 223, p. 123-131.

RIGGS, S. R., 1984b, Patterns of Miocene phosphate sedimentation on the southeastern United States continental margin: Proceedings of the 27th International Geological Congress, v. 15, p. 201-222.

SAYLES, F. L., 1981, The composition and diagenesis of interstitial solutions—II. Fluxes and diagenesis at the water-sediment interface in the high latitude North and South Atlantic: Geochimica et Cosmochimica Acta, v. 45, p. 1061-1086.

SCHUFFERT, J. D., JAHNKE, R. A., KASTNER, M., LEATHER, J., STURZ, A., AND WING, M. R., 1994, Rates of formation of modern phosphorite off western Mexico: Geochimica et Cosmochimica Acta, v. 58, p. 5001-5010.

SHELDON, R. P., 1980, Episodicity of phosphate deposition and deep ocean circulation—a hypothesis, *in* Bentor, Y. K., ed., Marine Phosphorites—Geochemistry, Occurrence, Genesis: Tulsa, Society for Sedimentary Geology (SEPM) Special Publication 29, p. 239-247.

UCHUPI, E., 1967, Bathymetry of the Gulf of Mexico, in Symposium on the geological history of the Gulf of Mexico, Antillean-Caribbean region: Gulf Coast Association of Geological Societies Transactions, v. 17, p. 161-172.

VAN KAUWENBERGH, S. J., CATHCART, J. B., AND MCCLELLAN, G. H., 1990, Mineralogy and Alteration of the Phosphate Deposits of Florida: Washington, D.C., United States Geological Survey Bulletin 1914, 46 p.

RARE-EARTH ELEMENTS AND URANIUM IN PHOSPHATIC NODULES FROM THE CONTINENTAL MARGINS OF INDIA

B. NAGENDER NATH,* B. RAMALINGESWARA RAO

Geological Oceanography Division, National Institute of Oceanography, Dona Paula, Goa-403 004, India

K. MOHANA RAO

Regional Centre, National Institute of Oceanography, 176 Lawsons Bay Colony, Visakhapatnam-530 017, India

AND

CH. M. RAO

Geological Oceanography Division, National Institute of Oceanography, Dona Paula, Goa-403 004, India

ABSTRACT: A sediment core collected from a bathymetric high off Goa on the western continental margin of India has yielded phosphatic nodules at various subsurface depths (at 110, 150, 305, 355, 435, 500, 505, and 525 cm). The nodules are hosted by sediments of Pleistocene age. They are <1–5 cm in size, have carbonate flourapatite (CFA) as a single authigenic mineral phase, are free of detrital inclusions, and have very high P_2O_5 contents (>30%). In addition, soft and hard phosphatic nodules have also been recovered from the eastern continental margin of India. Eight nodules from various subsurface depths of the sediment core from the western continental margin and two nodules from the eastern continental margin were analyzed for uranium and rare-earth elements (REEs) by inductively coupled plasma-mass spectrometry (ICP-MS).

Total REE contents are very low (8–21 ppm) in western continental margin nodules and only slightly higher in eastern continental margin nodules (maximum is 42 ppm). REE abundances relative to shale (ΣREE in sample/ΣREE in shale) are less than 0.22 and are comparable to phosphatic nodules from the Namibian continental margin, and are slightly lower than the nodules from the Peruvian continental margin. Light REEs are depleted, with La_n/Yb_n [$(La_{sample}/La_{shale})/(Yb_{sample} / Yb_{shale})$] values lower than unity, indicating a minimal contribution from terrestrial sources. The cerium anomaly (Ce/Ce*), which is a measure of Ce fractionation relative to the neighboring REEs, ranges between 0.64 and 1.38 with most close to 1, indicating very little Ce fractionation. Uranium concentrations are very high in nodules from both margins.

Low REE contents and LREE depletion collectively indicate either a seawater or porewater source for these elements. Since Ce is mostly stable in its trivalent state and U precipitates under reducing conditions, the absence of Ce fractionation in association with U enrichment indicates that the nodules may have formed by authigenic precipitation in reducing porewaters. High productivity in association with upwelling might have driven the accumulation of organic matter, which in turn would help enrich phosphate in porewaters, eventually leading to the formation of high-grade phosphorites.

Ce anomaly values in nodules do not correlate with paleoredox conditions in the water column, thus the REEs for these nodules are interpreted to mainly reflect porewater conditions.

INTRODUCTION

Rare-earth elements (REEs) have been used to trace many oceanographic processes because their oceanic residence times (10^2–10^3 years) are shorter than the mixing time of the ocean (1600 years) (Goldberg et al., 1963; Elderfield and Greaves, 1982; Grandjean et al., 1987). Their distributional trends in marine minerals have also been used to establish the origin of different phases (e.g., Piper et al., 1974; Elderfield et al., 1981; Taylor and McLennan, 1985; Hein et al., 1993; Nath et al., 1997a). Further, REE abundances preserved in authigenic deposits and marine sediments were often considered to indicate precisely the primary depositional environment of the precursor sediment (Shimizu and Masuda, 1977; Steinberg et al., 1983; Murray et al., 1990, 1991). Moreover, changes in the Ce anomaly in biogenic apatites have been used to interpret the paleoredox variations (e.g., Wright et al., 1984). REE abundances in phosphorites are highly variable, with very high contents in deep-sea biogenic phosphates (e.g., Elderfield and Paggett, 1986; Toyoda and Tokonami, 1990; Nath et al., 1994) and very low levels in newly formed, pure authigenic phosphatic precipitates (e.g., McArthur and Walsh, 1984; Kidder and Eddy-Dilek, 1994; Watkins et al., 1995). Conversely, uranium is highly enriched in the apatite of sedimentary phosphorites (e.g., Altschuler, 1980). Studies of U distributions in phosphorites assume particular importance given the potential for radiometric dating (Burnett and Veeh, 1977) and the possibility of economic U recovery from phosphorites (Starinsky et al., 1982, and references therein).

Phosphatic coatings on algal limestones of Holocene age from the outer continental shelf of the west coast of India have been reported by Nair (1985) and have been linked to the coastal upwelling prevalent in the area. Subsequent detailed SEM and EDAX studies have addressed some of these phosphatized algal nodules (Rao and Nair, 1988). These authors observed that outer laminations within the nodules contained higher P_2O_5 contents (9.5%) than those measured for the inner layers (5.4%). They further found endolithic filamentous fungi and cellular structures resembling bacteria within the phosphatic grains and, based on this observation, preferred a microbial origin over the upwelling processes for phosphate enrichment. Borole et al. (1987) carried out radiometric studies ($^{230}Th/^{234}U$) on the phosphatic coatings of the nodules studied by Rao and Nair (1988) and found that growth rates were lower (0.1 mm/1000 years) than those observed in the Peru/Chile area (Kim and Burnett, 1985). Lower growth rates were attributed by Borole et al. (1987) to weak upwelling in the region. Recently Rao and Ramalingeswara Rao (1996) reported the occurrence of very high-grade phosphorites (33.9%) from a topographic high along the western continental margin of India and related this to upwelling phenomena. While REE studies have been carried out on phosphatic grains, pellets, and phosphatized limestones at

*email: nagender@csnio.ren.nic.in

Marine Authigenesis: From Global to Microbial, SEPM Special Publication No. 66

TABLE 1.—COMPARISON OF TOTAL REE CONCENTRATIONS IN PHOSPHATIC MATERIAL FROM VARIOUS OFFSHORE REGIONS

Location	Type	ΣREE (ppm)	Reference
	Average Apatite	483	Altschuler et al.(1967)
Off California	Phosphorite	295	Goldberg et al.(1963)
Off Namibia	Concretion	566	Baturin et al.(1972)
Namibian Margin	Authigenic (concretionary) phosphorite	3.5 to 7.3	McArthur and Walsh (1984)
	Purified pelletal phosphorite	821 to 1177	McArthur and Walsh (1984)
Namibian Continental Shelf	Glauconized phosphatic pellets	1146-1282	Watkins et al.(1995)
	Phosphatic fish debris	7.28	Watkins et al.(1995)
	Phosphatic concretions	8.55 to 10.1	Watkins et al.(1995)
	Phosphatic pellets	799 to 980	Watkins et al.(1995)
	Phosphatic nodules	4.41 to 73.7	Watkins et al.(1995)
	Phosphatic crusts	30	Watkins et al.(1995
Blake Plateau	Concretion	342	Baturin et al.(1972)
Blake Plateau	Manganiferous phosphatic foraminiferal limestone	233 to 466	McArthur and Walsh (1984)
Agulhas Bank	Concretion	256	Baturin et al.(1972)
Agulhas Bank	Phosphatic limestones	45 to 113	McArthur and Walsh (1984)
	Slightly Phosphatised limestones	7.7 to 67	McArthur and Walsh (1984)
Peruvian Margin	Phosphatic quatzofeldspathic packstones and wackestones	23 to 41	McArthur and Walsh (1984)
Peru Shelf	Apatite pellets	159-202	Piper et al.(1988)
	Fish debris	29.9	Piper et al.(1988)
	Crusts	37 to 62	Piper et al.(1988)
Moroccan Shelf	Phosphatic intraclastic limestone	209	McArthur and Walsh (1984)
	Phosphatic foraminiferal limestone	23 to 41	McArthur and Walsh (1984)
South African Continental shelf	Phosphatised glauconite	366 to 400	Watkins et al.(1995)
	Conglomeratic rock phosphate	124 to 229	Watkins et al.(1995)
	Conglomeratic phosphatised limestone	116	Watkins et al.(1995)
Pacific Seamount	Phosphorite	845	Baturin et al.(1972)
S.P.Lee Guyot (Central Pacific)	CFA replaced white foraminiferal sand	22	Hein et al.(1993)
South Johnston Island Ridge (Pacific)	Partly CFA replaced and cemented brown volcanoclastic siltstone	261	Hein et al.(1993)
	CFA replaced white clastic limestone	644	Hein et al.(1993)
	CFA replaced bioclastic limestone	147	Hein et al.(1993)
Dana Seamount (Pacific)	CFA replaced brown limestone	446 to 574	Hein et al.(1993)
Western continental margin of India Marginal High off Goa	Phosphorite concretions and pellets	8.1 to 20.9	This study
Eastern continental margin of India off Chennon	Hard Phosphorite concretion	41.6	This study
	Soft concretion	33.5	This study

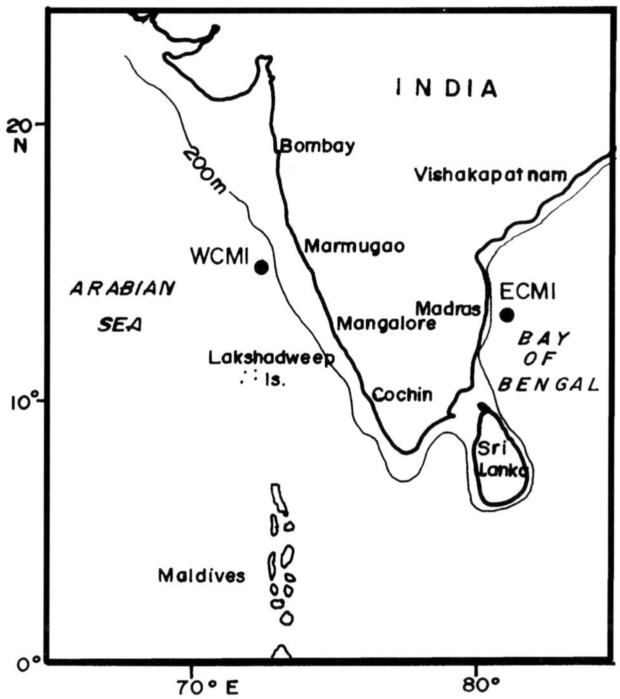

FIG. 1.—Location map. Dots represent the sediment core sites from which the phosphatic nodules have been recovered.

many locations in the Pacific and Atlantic (Table 1 and references therein), such analyses have not been performed on phosphorites of the Indian margin. The present contribution focuses on REE and U concentrations in phosphatic pellets and nodules from the western and eastern continental margins of India (WCMI and ECMI, respectively). More specifically, the present work is an attempt to relate REE and U contents and their distributional patterns to the depositional and diagenetic factors controlling the formation of phosphorites.

MATERIALS, METHODS, AND OCEANOGRAPHIC SETTINGS

A sediment core (#MH-1) measuring 5.85 m was collected from the summit of a marginal topographic high (water depth 320 m) from the WCMI during cruise 87A of *O.R.V. Sagar Kanya* in 1993. The sediments are light olive gray to olive gray in color, silty clay and clay in texture, and are hemipelagic calcareous ooze in character. An additional core was collected during cruise 119 of *O.R.V. Sagar Kanya* in 1996 from a flat terrain between 150 and 200 m bathymetric contours along the ECMI. Core locations are shown in Fig. 1. Both cores have yielded phosphatic nodules. The sediment cores were split and subsampled onboard. Phosphatic nodules observed during subsampling and wet sieving in the laboratory were selected by hand.

These nodules were powdered and digested using open acid digestion method. Samples were moistened with few drops of water. A mixture of 7 ml of HF, 3 ml of HNO_3, and 1 ml of $HClO_4$ was added to the beakers containing samples, and they were swirled until completely wet. The beakers were then covered with lids and kept overnight for digestion after adding 1 ml each of 10 µg/ml Rh and Bi (to act as internal standards). The next day the beakers were heated on a hot plate at 200°C for about 1 hour, after which the lids were removed and the contents were evaporated almost to dryness. The evaporation process was repeated three times after adding a mixture of 3 ml HF and 1 ml $HClO_4$ each time. Then the contents were eventually dissolved using 10 ml of 1:1 HNO_3 The detailed procedure is described in Balaram et. al., (1995). The samples were analyzed for trace element and REE contents using inductively coupled plasma–mass spectrometry (ICP-MS) at the National Geophysical Research Institute, Hyderabad, India.

Isotopes ^{139}La, ^{140}Ce, ^{141}Pr, ^{145}Nd, ^{147}Sm, ^{151}Eu, ^{157}Gd, ^{159}Tb, ^{163}Dy, ^{165}Ho, ^{166}Er, ^{169}Tm, ^{173}Yb, and ^{175}Lu were measured for REE analyses. Potential polyatomic and doubly charged ion interferences are from ^{123}TeO, ^{123}SbO (for La), ^{124}TeO, ^{124}SbO (for Ce), ^{125}TeO (for Pr), ^{128}TeO (for Nd), $^{130}BaOH$ (for Sm), ^{135}BaO (for Eu), ^{141}PrO (for Gd), ^{143}NdO (for Tb), ^{147}SmO (for Dy), ^{149}SmO (for Ho), ^{150}SmO, ^{150}NdO (for Er), ^{153}EuO (for Tm), ^{155}GdO (for Yb), and ^{159}TbO (for Lu). The oxide formation levels were monitored by barium, as barium oxide creates maximum oxide ion interference. The measurement conditions were optimized for a minimum oxide ion ratio (BaO^+/Ba^+ <0.1%). Because of the low abundance of barium levels in these samples, no significant interference effects on Eu were noticed and the oxide ion interference effects of various rare earth elements was found to be insignificant. Further, major elements such as P and Al were found earlier by Balaram (1992) to cause the least interference or no interference with the estimation of trace elements. Therefore, the phosphate matrix of the samples such as those studied here would not create any serious analytical interferences for REEs, U, and Th. The precision of analysis was better than 5% RSD.

Seasonal upwelling prevails along the WCMI (Wyrtki, 1973). The intensity and duration of upwelling vary from south to north (Shetye et al., 1985). A pronounced oxygen minimum zone (OMZ) exists at intermediate depths of the Arabian Sea (Wyrtki, 1973). Hydrochemical data collected in 1984 showed that the OMZ in the WCMI occurs at median water depth between 150 and 1200 m (S. N. D'Souza, pers. commun., 1993) and that the OMZ in the ECMI is less intense. Weak upwelling off Madras, with waters from 30 m depth reaching the surface, was reported in ECMI (Murty and Varadachari, 1968).

RESULTS AND DISCUSSION

Phosphatic nodules in the WCMI core are dark gray, yellowish gray, and light yellow in color, sub-rounded, oval, or irregular in shape. They occur in the size range of <1-5 cm and have rough surface texture with small depressions and cracks (Rao and Ramalingeswara Rao, 1996). Among the two nodules collected from ECMI, the surface texture of one is soft and the other is hard. Soft nodule is more friable and loosely bound, while the hard one is more compact. The ECMI nodules are also dark brownish gray colored and irregular in shape (Vaz, 1995).

The WCMI nodules are composed almost exclusively of carbonate flourapatite (CFA) and do not show detectable concen-

Table 2.—TRACE AND RARE-EARTH ELEMENT CONTENTS (in ppm) IN PHOSPHATIC NODULES OF INDIAN CONTINENTAL MARGINS

	← *Western Continental Margin of India* →								*Eastern Margin*	
Element	CFA1 (110-115)	CFA2 (150-155)	CFA3 (305-310)	CFA4 (355-360)	CFA5 (435-440)	CFA6 (500-505)	CFA7 (505-510)	CFA8 (525-530)	Hard	Soft
Sc	2.39	3.83	3.19	3.3	1.3	3.44	2.17	3.47	1.74	2.98
Y	9.61	5.79	6.83	7.54	2.83	7.52	2.99	5.65	6.92	7.03
Zr	6.02	13.86	8.98	7.54	3.57	7.68	4.18	6.89	29.34	71.09
La	2.41	1.42	2.89	3.65	1.2	3.33	1.51	71.43	10.19	7.96
Ce	2.95	2.74	5.8	7.27	2.89	5.93	3.1	6.43	15.79	12.84
Pr	0.47	0.35	0.59	0.77	0.23	0.6	0.44	0.8	1.66	1.5
Nd	1.51	1.42	3.38	3.96	1.52	2.92	1.3	2.49	6.3	5.58
Sm	2	1.61	1.34	1.74	1.15	1.77	1.51	1.34	1.24	0.92
Eu	0.12	0.1	0.16	0.22	0.04	0.19	0.15	0.26	0.36	0.31
Gd	0.59	0.35	0.87	0.79	0.18	1.22	0.46	1.4	1.55	1.23
Tb	0.11	0.01	0.09	0.12	0.06	0.09	0.04	0.09	0.16	0.13
Dy	0.47	0.4	0.34	0.87	0.32	0.78	0.3	0.65	1.11	0.92
Ho	0.12	0.07	0.15	0.2	0.06	0.15	0.05	0.14	0.3	0.25
Er	0.39	0.27	0.24	0.55	0.15	0.7	0.07	0.31	1.08	0.83
Tm	0.08	0.04	0.06	0.06	0.02	0.04	0.03	0.04	0.17	0.1
Yb	0.6	0.41	0.51	0.64	0.28	0.6	0.35	0.88	1.54	0.83
Lu	0.08	0.06	0.08	0.05	0.03	0.05	0.02	0.06	0.12	0.08
Hf	0.13	0.38	0.29	0.21	0.01	0.32	0.1	0.17	0.45	0.98
Th	0.23	0.13	0.36	0.23	0.08	0.32	0.28	0.47	1.78	2.1
U	23.44	110	88	151	21.44	73	29.5	29.1	119	73
ΣREE	11.99	9.23	16.5	20.89	8.13	18.37	9.33	16.32	41.57	33.48
Ce/Ce*	0.64	0.89	1.02	1	1.26	0.96	0.87	1.38	0.87	0.85
La_n/Yb_n	0.30	0.26	0.42	0.42	0.32	0.41	0.32	0.12	0.49	0.71

Numbers in parentheses for the samples from Western margin indicate the sub-surface depth (in cm) of core at which the phosphorites were recovered.

trations of detrital minerals such as quartz, feldspar, or clay. Furthermore, microscopic studies have not revealed the presence of detrital minerals specifically in or around apatite crystallites (Rao and Ramalingeswara Rao, 1996). However, ECMI nodules are reported to contain calcite and quartz along with carbonate flourapatite (Vaz, 1995). The WCMI samples contain 30.9–33.9% P_2O_5 and 52.8–55.2% CaO. The CaO/P_2O_5 ratios vary between 1.56 and 1.78, with an average of 1.65. These values are slightly higher than the most cation and anion substituted francolite (1.621) end of flourapatite (1.318) – CFA range (Manheim and Gulbrandsen, 1979; McClellan and Van Kauwenbergh, 1990; Hein et al., 1993). CaO/P_2O_5 ratios would be higher (2.26; Birch, 1979) if the phosphorites were formed by replacement (of calcium carbonate, for example) processes. Phosphatic nodules and rocks from the ECMI were found to contain lower CaO/P_2O_5 (0.38; Vaz, 1995), which is probably controlled by lower calcium contents. P_2O_5 contents in the nodules under consideration are much higher than those in the phosphatized algal nodules of the western shelf of India (Rao and Nair, 1988), the Quaternary East Australian phosphatic nodules (mean 11.2%; O'Brien et al., 1990), and nodules of the Agulhas Plateau (10–18%; Parker and Seisser, 1972). But they are comparable to the phosphate nodules from Peru-Chile, Namibia, Morocco, Chatham Rise, the southern California Borderland and the Phosphoria formation (average >20%; O'Brien et al., 1990, and references therein).

Rare-earth Element Abundances

Total ΣREE concentrations in all the phosphatic nodules, from both WCMI and ECMI, are very low, with all samples containing ΣREE <52 ppm. ΣREE in WCMI does not vary widely, ranging between 8 and 21 ppm (with a mean of ~14 ppm). These concentrations are less than those of 33–52 ppm in ECMI nodules (Tables 1 and 2). The REE concentrations are compared with published REE data for other regions in Table 1. ΣREEs measured here are much less than those in average apatite (Altschuler et al., 1967) and many forms of Neogene phosphatic material from offshore locations of California (Goldberg et al., 1963), Namibia (Baturin et al., 1972; McArthur and Walsh, 1984; Watkins et al., 1995), the Blake Plateau (Baturin et al., 1972; McArthur and Walsh, 1984); Peru (Piper et al., 1988), South Africa (Watkins et al., 1995), Morocco (McArthur and Walsh, 1984), and many Pacific seamounts (Hein et al., 1993). Apart from the low concentrations in limestones listed for several locations worldwide in Table 1, which are most likely due to carbonate dilution (e.g., Murray and Leinen, 1993), the WCMI concentrations are comparable only to the authigenic phosphate concretions, crusts, and nodules of Holocene age from the offshore regions of Namibia and Peru (McArthur and Walsh, 1984; Piper et al., 1988; Watkins et al., 1995). Like the samples from Namibia and Peru, nodules of the present study are non-ferruginous and occur associated with organic-rich biogenic sediments (Veeh et al., 1974; Burnett and Veeh, 1977). When compared to shales (PAAS of Taylor and McLennan, 1985; later modified by McLennan, 1989), ΣREE contents in WCMI phosphatic nodules are 0.11 times less than those in average shale and range between 0.18 and 0.22 in ECMI nodules. Such low concentrations are consistent with the suggestion of McArthur and Walsh (1984) that francolite characteristically forms with extremely low ΣREE contents. Most of the biogenic phosphates likely acquire their REE enrichments postdepositionally (Kolodny, 1983; McArthur and Walsh, 1984; Elderfield and Paggett, 1986), sometimes reaching eight-fold enrichment relative to shales. Based on these studies in Pennsylvanian phosphates, Kidder and Eddy-Dilek (1994) have suggested that MREE enrichment is a primary geochemical signature that could be enhanced or diminished, depending upon burial conditions. The present samples are, however, depleted in ΣREEs, although they appear to be of Pleistocene age.

Shale-normalized REE patterns of WCMI and ECMI nodules are shown (Figs. 2 and 3). The shale values used here are from McLennan (1989) and represent post-Archaean Australian shales (PAAS), which were originally proposed by Taylor and McLennan (1985). WCMI phosphorites reveal a gently sloping pattern with slight enrichment of HREEs (Fig. 2). No significant variation is seen for Ce and Eu, which are sensitive to redox changes. Eu however, responds to redox at higher temperatures and remains stable at Earth-surface P-T conditions. The patterns for ECMI phosphorites (Fig. 3) show an enrichment in La compared to other LREEs, strong positive Eu anomaly, and a gradual increase from Tb to Lu indicating HREE enrichment.

The La_n/Yb_n [$(La_{sample}/La_{shale})/(Yb_{sample}/Yb_{shale})$] ratio is used here as a measure of light to heavy REE fractionation. La_n/Yb_n ratios for all the samples are lower than unity, indicating LREE depletion. Ratios vary between 0.12 to 0.42 in the WCMI core (Table 2), with a minimum value for the deepest sample. There is no clear relationship between ΣREE and La_n/Yb_n ratios. In ECMI, the La_n/Yb_n ratio is higher in soft concretions (0.71) than the harder concretions (0.49) (Table 2).

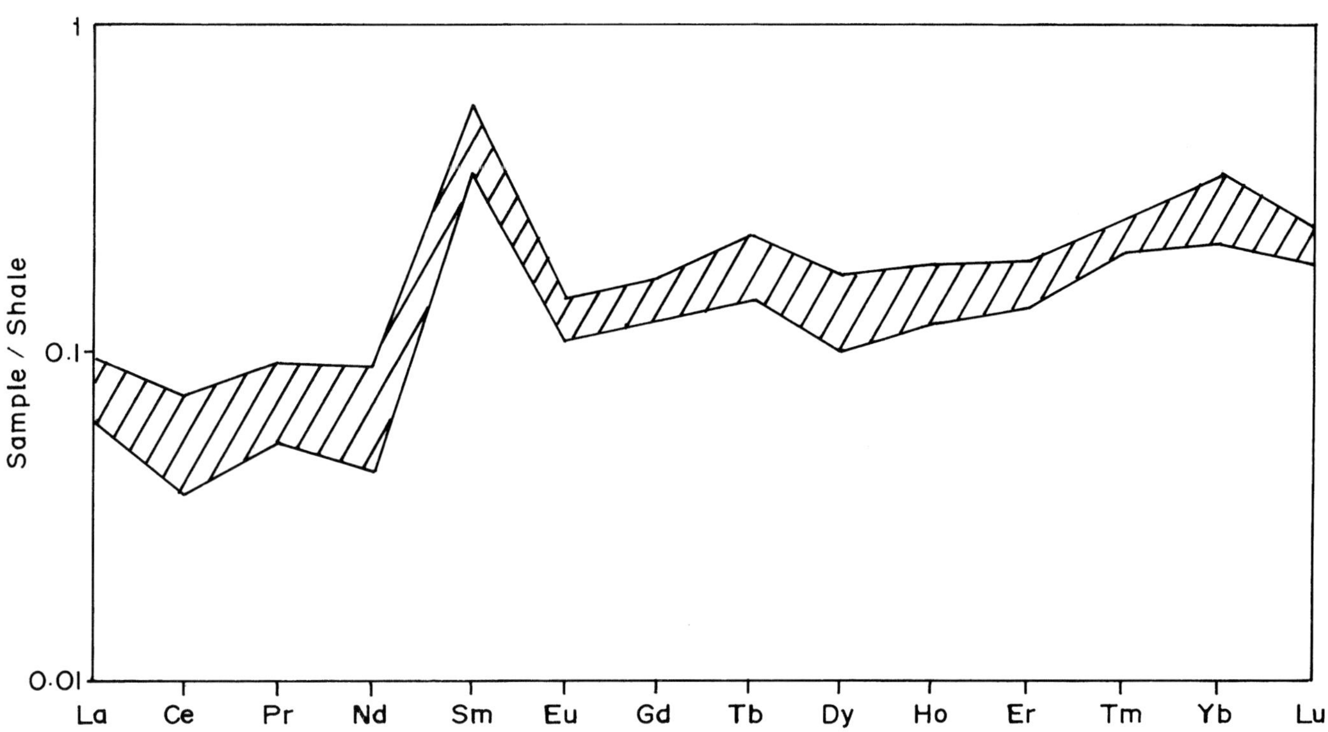

FIG. 2.—The range of shale-normalized REE patterns for nodules (#CFA1 to CFA8) from Western Continental Margin of India (WCMI)

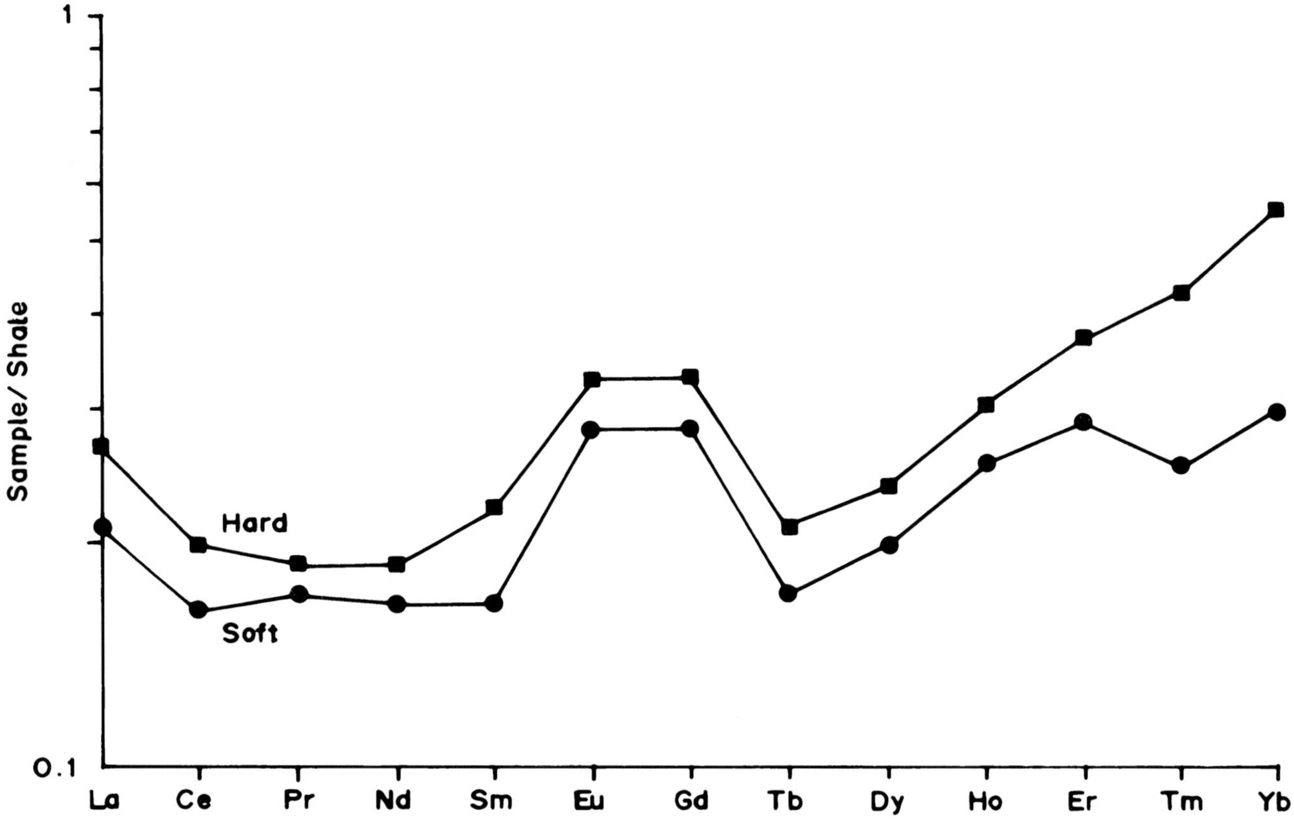

FIG. 3.—Shale-normalized REE patterns for the two nodules from the Eastern Continental Margin of India (ECMI)

The Ce/Ce* ratios, which are a measure of Ce fractionation relative to its trivalent neighbors, can be used here to define distinct groups of samples. Ce/Ce* values are <1 in samples #CFA1, CFA2, CFA7 (see Table 2 for corresponding depths in the core) from WCMI, and both hard and soft nodules of ECMI; approximately 1 in #CFA3, CFA4, and CFA6, and >1 in #CFA5 and CFA8. There is no clear variation of Ce/Ce* either with respect to sediment depth or ΣREE contents.

Although the concretions of the present study range in size from <1 to 5 cm, the ΣREE contents are low and relatively invariant. There seems to be no relationship between concretion size and the ΣREE contents in our samples. This differs from the earlier findings of McArthur and Walsh (1984) and Watkins et al. (1995) that the high surface area/mass ratio of phosphatic concretions promotes REE enrichment.

Sources of REE

Rare-earth elements in phosphorites can be contributed from either authigenic or detrital sources (Piper et al., 1988), or a combination of both. Authigenic component includes material supplied by biogenic, seawater, or porewater sources. Therefore, it is important to ascertain the relative proportion of the phases containing REEs, which may in turn provide constraints on the intensity of terrestrial input (detrital) or the chemical properties of the depositional environment (authigenic). It is possible to discern the relative importance of these sources from REE patterns. Shale-normalized REE patterns of sedimentary phases derived from detrital silicates would be flat and lack any fractionation of LREE to HREE. Shale-normalized patterns of WCMI phosphorites are not flat and slope upward towards HREE. La_n/Yb_n values for both WCMI and ECMI are much lower than 1 to 1.3 ratios, characteristic of material delivered from the upper-continental crust to the oceans (Piper, 1974; Taylor and McLennan, 1985; Sholkovitz, 1990). Further, the shale-normalized patterns of sediments from the central Western continental margin of India derived from Indian peninsular sources are almost flat (except at Eu, where a positive anomaly is seen) and show no Ce fractionation (Nath et al., 1997b). The WCMI phosphorites contain extremely low concentrations of detrital elements (Al_2O_3 is <0.1% and SiO_2 is <0.02%) and lack minerals such as aluminosilicates and quartz that are detrital in nature. Such low Al_2O_3 concentrations would only support ΣREE contents of less than 1 ppm ($Al_2O_{3sample}$ x ΣREE_{shale} / Al_2O_{3shale} = ΣREE in detrital phases of sample) and indicate that only a small portion of REEs in these phosphorites could have been derived from terrestrial sources. Positive Eu anomalies in ECMI phosphates suggest that they have relatively higher detrital (from terrestrial sources) REE contribution. Positive Eu anomalies cannot result from low-temperature water column conditions (Elderfield, 1988) and the anomalies might have been derived from the feldspars and other minerals rich in europium.

Biogenic sources may also contribute REEs to the phos-

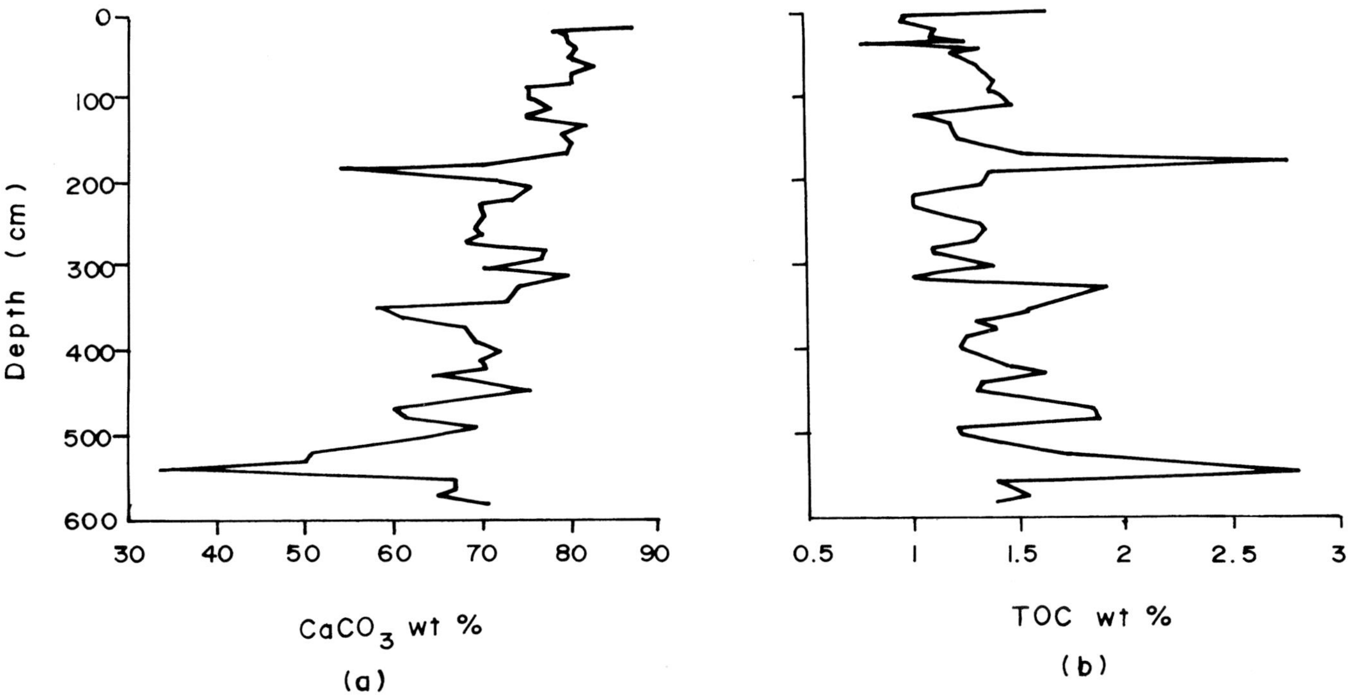

FIG. 4.—Variation of a) calcium carbonate and b) organic carbon in WCMI sediment core with depth

phatic grains. Major biogenic components of sediments are calcium carbonate, opaline silica, and organic matter. $CaCO_3$ dominates over biogenic silica in transporting REEs in the marine environment (Elderfield and Greaves, 1982; McArthur and Walsh, 1984; Piper et al., 1988). $CaCO_3$ is high in the WCMI core (Fig. 4A) and may be responsible for a portion of REEs found in phosphorites. The sediments from the WCMI contain <2% biogenic silica and contain negligible quantities of opal-bearing skeletons such as radiolaria and diatoms (Naidu et al., 1992). Moreover, biogenic silica contains very low REE contents (<1% of shale composition; Piper, 1974) and exerts a secondary and perhaps minor influence on REE distribution in the oceans (Byrne and Kim, 1990). Middle REE enrichment was found in phosphatic nodules from mid-continent (Kansas and northeastern Oklahoma) Pennsylvanian cyclothems (Kidder and Eddy-Dimek, 1994), which have been attributed to a biogenic origin. Kidder and Eddy-Dimek (1994) have suggested that the phosphate acquired this signal by accumulation of fecal material excreted by organisms that consumed the MREE-enriched algae. Their argument is supported by the experimental work of Stanley and Byrne (1990), which showed that the green alga *Ulva lactuca* is capable of selectively incorporating MREE. The absence of MREE enrichment in Indian margin phosphates suggests that the coprolites or other fecal material are not contributing to the formation of the phosphates studied here.

Organic coatings on particles and the organic matter in the sediments may be additional sources. From their experiments on a variety of surfaces, Byrne and Kim (1990) found that the affinity of the REEs for surfaces can be strongly affected by the organic coatings. Murray and Leinen (1993) invoked an increase in scavenging capacity of biogenic and terrigenous particles when coated with organic films. Organic matter may itself be a source of REEs, as it is a source of PO_4^{3-} of the apatite (Piper et al., 1988). On decomposition of organic matter, PO_4^{3-} may be released into porewaters and mobilization takes place in diffused form until suitable sites are available for the precipitation and formation of francolite or apatite. Similarly, REEs, which are found in organic matter, may also be available to the porewater. However, there is no correlation between ΣREE contents in phosphorites and sediment TOC profile. For example, nodules CFA6, 7, and 8 recovered at 500–505, 505–510, and 525–530 cm have variable ΣREE concentrations (18.4, 9.3, and 16.3 ppm, respectively); however, the sediment TOC values are scattered but gradually increase with sediment depth (Fig. 4B).

Thus, it is clear that the ΣREE contributions to phosphate nodules (studied here) from aluminosilicates, carbonate, opal, or organic carbon are less significant. This leaves the possibility to two dissolved sources, that is, seawater and porewater. It is generally believed that REEs in phosphorites are, to a great extent, contributed by seawater (Goldberg et al., 1963; Altschuler et al., 1967; Piper, 1974; Altschuler, 1980). This is supported by the presence of two features in many phosphorites, a marked negative Ce anomaly and HREE enrichment patterns, both of which are characteristic of oxic seawater (e.g., Altschuler, 1980; Kolodny, 1983). With the exception of nodules from the shallowest depth (#CFA1 from 110–115 cm; Ce/Ce* 0.64; Table 1), all other samples from both WCMI and ECMI are either characterized by Ce/Ce* values close to 1 (#CFA2, 3, 4, 6, 7, and ECMI nodules) or greater than 1 (#CFA5 and 8 having Ce/Ce* values of 1.26 and 1.38, respectively). Ce/Ce* values either close to 1 or >1 and the HREE enrichment, as recorded by phosphorites of the present study, are characteristic features of suboxic to anoxic seawater (e.g., German and Elderfield, 1990) or anoxic porewaters (Elderfield and Sholkovitz, 1987).

The direct uptake of REEs from seawater raises serious mass balance problems, as one gram of phosphate would need to adsorb REE from hundreds of tons of seawater to account for the REE levels observed in many phosphorites, including those of the present study (Grandjean et al., 1987). Phosphatic grains having a small surface area, such as those of the present study, may not be able to adsorb/acquire significant amounts of REEs in continental shelf and slope environments with large sedimentation rates. On the other hand, REEs in nearshore porewaters are highly mobile during early diagenesis, and the distribution coefficient (K_d) between REEs in sediment relative to porewater (~10^6) is much lower than in seawater [~10^{12}] (Elderfield and Sholkovitz, 1987; Sholkovitz and Elderfield, 1988; Sholkovitz et al., 1989). Furthermore, Nozaki and Horibe (1983) suggest that there is a substantial benthic flux of ^{227}Ac, which may be taken as an analog for the REEs (Staudigel et al., 1985/86). Therefore, phosphorites would require lower porewater-rock ratios than seawater-rock ratios for the incorporation of REEs. Zero or positive Ce anomalies have been found in phosphatic samples containing sulphur, which is isotopically light (light sulfur is an indicator of bacterial sulphate reduction under anaerobic conditions that can occur both in reducing porewaters or an anoxic water column), compared to coeval seawater sulfate (Claypool et al., 1980). Oxidative precipitation of insoluble Ce (IV) oxide was found to occur throughout the well-oxygenated portion of the water column, while the reductive dissolution of Ce (IV) to Ce (III) was found to occur in the suboxic and anoxic waters of the Black Sea (German et al, 1991). Therefore, Ce is unfractionated or stable in the 3+ state under suboxic conditions (McArthur and Walsh, 1984) in both suboxic porewaters or anoxic water column. But the mass balance considerations shown above indicate that the REEs in these phosphates are acquired from pore fluids. However, we are not able to resolve whether they are acquired during early (Bernat, 1975; Grandjean et al., 1987; Grandjean and Albarede, 1989) or late diagenesis (Toyoda and Tokonami, 1990).

Uranium Incorporation

Uranium concentrations are high in WCMI samples and range between 21 and 151 ppm; hard and soft nodules of ECMI have 119 and 73 ppm respectively. These concentrations are 7–50 times the crustal abundance, yet are close to the range of 30–260 ppm generally reported for phosphorites (e.g., Altschuler, 1980). Although it was argued from earlier studies that U does not enter apatite phases and instead exists as a separate mineral phase (Serebryakova and Razumnaya, 1962; Baturin and Dubinchuk, 1978, 1979), later studies have proved that the bulk of U is included in the apatite lattice (Burnett and Veeh, 1977; Kolodny, 1983). The degree of U incorporation into apatite can be expressed by the U/P_2O_5 ratio. This ratio varies between 6 x 10^{-5} and 4 x 10^{-4} in our samples. Higher values, found in phosphorites of many locations, are close to a ratio of (2–4) x 10^{-4} (Kolodny, 1983).

The WCMI sediments were deposited within the OMZ, and sediments exposed to the OMZ are characterized by high C_{org} contents. In general, high C_{org} can contain tens of ppm uranium, and upon decomposition, C_{org} contributes several mg of U per liter of interstitial solution (Starinsky et al., 1982). A linear relationship between U and C_{org} was found in OMZ sediments of the Arabian Sea (Nath et al., 1997b). The decomposition of C_{org} can contribute both P and U for the newly formed francolite/carbonate flourapatite. The high correlation between organic matter and U has been demonstrated in many high productivity regions of the ocean (Calvert, 1976; Starinsky et al., 1982; Kolodny, 1983) and are typical of reducing sediments (Veeh, 1967; Mo et al., 1973; Glenn et al., 1995; Nath et al., 1997b). However, lack of significant reduction from U(VI) to U(IV) has been noticed in the suboxic and anoxic waters of the Cariaco Trench (Anderson, 1987) and Black Sea (Anderson et al., 1989). A significant increase in porewater and solid phase U enrichment is found at shallow depths in sediments underlying suboxic and anoxic waters of the Black Sea (Barnes and Cochran, 1990) and off California (Klinkhammer and Palmer, 1991). Inorganic precipitation of U in sulfate reducing environments is possible (e.g., Klinkhammer and Palmer, 1991; Lovley et al., 1993). Further, in the presence of $CaCO_3$, transition of hydroxy to carbonate UO_2^{2+} can occur. During this transition, UO_2^{2+} is removed from solution onto carbonate mineral surfaces. Once adsorbed, U may be more susceptible to mineralization by other common chemogenic phases such as phosphate, which have high affinities for carbonate surfaces and form relatively less soluble uranyl compounds (Morse et al., 1984). Low U contents in seamount phosphorites (e.g., Hein et al., 1993) with high bottom water oxygen contents also suggest that anoxic/suboxic conditions are essential for U enrichment. U contents are lower than 10 ppm in phosphorites or phosphotized sediments of several Pacific seamounts (Hein et al., 1993) likely because seamount environments are characterized by a lack of both abundant organic matter to transport U to the bottom and the reducing conditions to fix such U once present (Kolodny, 1983).

In an earlier study U contents were found to be different in friable and consolidated nodules, with an increase in U/P_2O_5 during lithification (Baturin et al., 1974). ECMI nodules show clear indications of variability in U contents, with hard concretion having a higher concentration than the softer one. Harder, more consolidated, nodules may represent more advanced diagenesis (Kolodny, 1983).

U and REE Constraints on the Phosphorite Formation

Based on the lower U contents in the phosphate fraction in phosphatized algal nodules from the western continental shelf off Goa (70–150 m water depth) compared to the associated carbonate material, Rao and Nair (1988) assigned a microbial origin and ruled out the possibility of upwelling as a principal source for their formation.

However, the nodules from WCMI considered here have higher U contents, and therefore, upwelling could be the driving force for the phosphatization. Upwelling leads to fresh supplies of phosphate and nutrients to the photic zone, which in turn enhances the productivity of the surface waters. The Arabian Sea is characterized by summer monsoons. Associated with summer monsoons are southwesterly winds that lead to the development of coastal and open-ocean upwelling systems (Clemens and Prell, 1990; Prell et al., 1992). Upwelling systems dominate the surface area properties such as temperature and nutrient content, mixed layer thickness, and productivity of the euphotic zone (e.g., Wyrtki, 1971). Higher productivity leads to increased accumulation of biogenic material and is partly responsible for the formation of the OMZ, another characteristic feature of the Arabian Sea. Such higher sedimentation rates and lower oxygen levels in the water column lead to more efficient burial of organic carbon in the sediments (e.g., Calvert et al., 1996 and references therein); and the formation and maintenance of anoxic conditions in the sediments. Anoxia in bottom waters and within the sediments facilitates precipitation of the reduced form of U (Nath et al., 1997b, and references therein). When organic carbon in anoxic subsurface sediments is degraded or decomposed, phosphate is released into the porewater and leads to the formation of apatite (e.g., Burnett, 1977; Ruttenberg and Berner, 1993).

Most of the modern marine sedimentary phosphorites are associated with the upwelling of cold, nutrient-rich waters onto continental shelves such as those of Peru-Chile (e.g., Burnett, 1977; Burnett and Veeh, 1977), Namibia (Baturin et al., 1972), and Mexico (Jahnke et al., 1987). Formation of ECMI phosphorites, however, appears not to be related to upwelling. The characteristics of ECMI nodules are similar to those of ancient and modern phosphorite deposits found in areas with little coastal upwelling and low productivity, such as off the coast of South Africa (e.g., Birch et al., 1983) and East Australia (O'Brien and Veeh, 1980). The ECMI has lower productivity compared to the WCMI. Furthermore, as discussed above, WCMI sediments are high in carbonate content. Carbonate increases the stability field of francolite (Nathan and Sass, 1981) and prevents the alteration of francolite to varieties containing less carbonate (Van Kauwenbergh and McClellan, 1990). Dissolution of carbonate may lead to the alteration or disappearance of francolite (Altschuler, 1973; Lucas et al., 1980). Shale-normalized REE patterns of ECMI phosphorites show positive Eu anomaly. A positive Eu anomaly is not a characteristic feature of a low-temperature environment, and therefore could not have been developed during the authigenic formation of phosphorites. This may have been derived from the adjoining continental rocks and may suggest a contribution from terrestrial sources.

If we consider that higher U contents in hard nodules relative to softer varieties are due to diagenetic enrichment, nodules (CFA2, CFA3, CFA4, and CFA6 in the WCMI core) that have higher U contents (>70 ppm) than others studied here may have undergone advanced diagenesis, suggesting an important role for porewater as a source of phosphate. Furthermore, higher ΣREE contents are found in the nodules having higher U contents (except #CFA2), and hard concretion of ECMI has higher ΣREE (Table 2) contents than the softer concretion, indicating a contribution of REEs from diagenetic pore fluids.

Downcore Variation of REEs in Phosphate Nodules and Paleoceanographic Implications

REE variations and, in particular, variation in Ce anomaly changes in biogenic apatites have been used to infer paleoredox and paleoclimatic conditions (e.g., Wright et al., 1984). Ce anomalies are important in low-temperature environments, since Ce has multiple oxidation states and responds to oxidation-reduction reactions (e.g., Piper, 1974; Nath et al., 1994; 1997b, and references therein). This character of Ce response to changing redox conditions has been used by some workers to determine the secular variations of paleoredox conditions of ancient oceans (Wright et al, 1984, 1987; see review by Holser, 1997). Since direct measurements of REE content of ancient seawater is not possible, analyses of REE in authigenic and biogenic materials have been widely adopted by some to study the redox and water masses of ancient oceans (Grandjean et al., 1987, and references therein). In a study on rocks of the Upper Cretaceous Niobrara formation of Colorado, MacLeod and Irving (1996) found a statistically significant correlation between Ce/Ce* and indicators (such as organic carbon and intensity of bioturbation) of changes in the bottom water oxygen contents. Ce/Ce* values above 1, or positive Ce anomalies, have been used to infer anoxic conditions during the precipitation of biogenic phosphates. Ce anomaly values (Ce/Ce*) range from 0.64 to 1.38 in WCMI samples. No clear correlation exists between the Ce/Ce* values and sample depth within the core, which may indicate the relative age of phosphatization; however, there is a general increase in Ce/Ce* values from <1 (a negative anomaly) at the shallowest depths (100–150 cm) to a significantly positive anomaly (1.26) at 435 cm (Fig. 5). The largest positive Ce anomaly is found in the nodule recovered at a depth of 525 cm.

Sediment samples from different intervals of the WCMI core were measured for ^{14}C, but no activity could be detected, which indicates that these sediments are older than ^{14}C detectable age (i.e., >35 ky; Rao and Ramalingeswara Rao, 1996). Furthermore, micropaleontological studies have shown that all the sediment sections contain *G. truncatulinoides*, which indicates that they are of Pleistocene age (D. Naidu, pers. commun., 1997). An attempt has also been made to date the nodules using

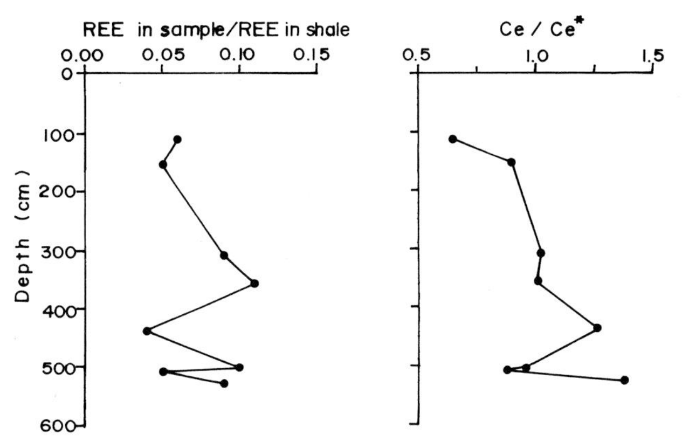

FIG. 5.—Variation of shale-normalized total REE contents and Ce anomalies (represented as Ce/Ce*) with depth in WCMI sediment core.

the U/Th disequilibrium method. The ages are in excess of the dating limit of the method (>250,000 yrs), as the $^{230}Th/^{234}U$ activity ratio is 0.95±0.03. Further, the $^{234}U/^{238}U$ activity ratio (0.96±0.02) is near secular equilibrium, indicating that these nodules are older than 0.8 m.y. (early Pleistocene; Rao and Ramalingeswara Rao, 1996). As the sampling location is on a topographic high, strong currents could prevail, leading to erosion and exposure of older sediment. Thus, the phosphorites under consideration may not be modern.

To ascertain whether the Ce anomaly in biogenic phosphates reflects paleoredox variations or not, it is important to know the bottom water oxygenation conditions of the past. The marginal high from which the sediment core was recovered has a water depth of 320 m and therefore comes under the direct influence of the modern oxygen minimum zone (OMZ). Incidentally, the sediments are grayish green in color, which indicates a low-oxygen environment. The OMZ extends from the median depth of 150–1250 m in this area. An OMZ is a characteristic feature of the Arabian Sea, which encloses one of the thickest of the major oxygen-poor layers of the world ocean, bearing suboxic concentrations (<0.1 ml/l; Wyrtki, 1971; Qasim, 1982; Slater and Kroopnick, 1984). Ce anomaly values observed in the present study at WCMI are negative at 110–115, 150–155, and 505–515 cm, and positive or close to 1 at all other depths. Most of the studies on paleoredox interpretations from biogenic phosphates have interpreted the positive Ce anomalies to reflect suboxic to anoxic conditions (e.g., Wright et al., 1984) as seen in present day anoxic basins (e.g., DeBaar et al., 1988; German and Elderfield, 1990). If we assume that the periods during which the positive anomalies were recorded reflect suboxic to anoxic conditions due to the impingement of OMZ as in the present day, negative anomalies should be due to oxic conditions. For oxic conditions to exist, the position of OMZ must fluctuate. Major fluctuations in the OMZ can be driven by eustatic sea-level changes. The maximum sea-level fall/rise during the Pleistocene in the northern Indian Ocean has been reported to be between 100 m and 125 m (Cochran, 1990). Assuming a shift in OMZ, by 100 m due to changes in sea-level, the area would be still within the core of OMZ and does not explain the negative Ce anomalies. Therefore, Ce fractionation in biogenic phosphates does not seem to be influenced by redox variations in water column, at the WCMI site. The reason could be due to contribution of REEs by porewaters. Therefore, the REE patterns in the sedimentary phases may be affected by changing redox state within sediments. Availability of REEs in porewaters, depends on the supply of organic material, sedimentation rates, dissolution and diffusion kinetics and availability of microreducing environments for the precipitation.

CONCLUSIONS

Among the phosphatic nodules from the Western and Eastern continental margins of India studied here, ΣREE abundances are very low in pure authigenic precipitates of WCMI compared to those of ECMI. Although the nodules are interpreted to belong to Pleistocene age, the REE contents are very low. This is different to the earlier findings that the REE contents in phosphates increase with age.

Shale-normalized REE patterns and other fractionation indices suggest that WCMI phosphorites are pure authigenic precipitates and have negligible contribution from terrestrial sources. In contrast, ECMI phosphates are characterized by positive Eu anomalies on one hand and features indicating low temperature processes on the other suggesting both terrestrial and authigenic sources. High uranium contents in nodules studied here, compared to other seamount phosphorites from Pacific, are consistent with the characteristics of the area i.e., oxygen depleted bottom waters and high organic carbon supply.

Although the REE fractionation indices may not constrain the contrast between seawater and porewater sources, mass balance considerations tend to favour the porewater as a major authigenic source for formation of WCMI phosphates. A simple sequence of formation is envisaged i.e., from the supply of higher organic matter during enhanced upwelling leads to the formation of suboxic to anoxic conditions in the sedimentary column. Subsequently, during the degradation of organic matter, phosphate and U are released into porewater and tend to precipitate as concretions/nodules at favourable sites.

Consequent to the formation of these nodules from porewaters, Ce anomalies found in these concretions are not related to the bottom water oxygenation conditions of the past. Analogous to the enrichment of U during advanced diagenetic stage found elsewhere and here as well, REE contents also seem to increase with consolidation and increased diagenesis. This may further complicate the use of cerium anomaly in biogenic phosphates as a paleoredox proxy.

ACKNOWLEDGMENTS

We thank E. Desa, Director, NIO, for his kind permission to publish this work. The ICP-MS analyses were kindly carried out by Dr. V. Balaram, NGRI, Hyderabad. Both the reviewers Rick Murray and Tim Lyons have thoroughly reviewed the paper and their comments and corrections were very constructive. Many thanks to Craig Glenn for his editorial comments, references on uranium and for bearing with our delay. This is NIO contribution number 2131.

REFERENCES

ALTSCHULER, Z. S., 1973, The weathering of phosphate deposits–geochemical and enrvironmental aspects, *in* Griffith, E, J., ed., Environmental Phosphorus Handbook: New York, John Wiley & Sons, p. 33-96.

ALTSCHULER, Z. S., 1980, The geochemistry of trace elements in marine phosphorites—part I, Characteristics, abundances and enrichment, *in* Bentor., Y. K., ed., Marine Phosphorites: Tulsa, Society for Sedimentary Geology (SEPM) Special Publication 29, p. 9-30.

ALTSCHULER, Z. S., BERMAN, S., AND CUTTITTA, F., 1967, Rare earths in phosphorites—geochemistry and potential recovery: Washington, D.C., U.S. Geological Survey Professional Paper 575 B, p. 1-8.

ANDERSON, R. F., 1987, Redox behaviour of uranium in an oxic marine basin: Uranium, v. 3, p. 145-164.

ANDERSON, R.F., FLEISCHER, M.Q. AND LEHURRAY, A.P., 1989, Concentration, oxidation state and particulate flux of uranium in the Black Sea: Geochimica et Cosmochimica Acta, v. 53, p. 2215-2224.

BALARAM, V., 1992, Effect of concomitant elements on the analyte signals in ICP-MS: A case study: Proceedings of the. National Seminar on Modern Trends in Analytical Techniques, Applications in Chemistry, p. 162-175.

BALARAM, V., RAMESH, S. L., AND ANJAIAH, K. V., 1995, Comparative study of the sample decomposition procedures in the determination of trace and rare earth elements in anorthosites and related rocks by ICP-MS: Fresenius Journal Analytical Chemistry, v. 353, p. 176-182.

BARNES, U.C., AND COCHRAN, J.R., 1990, Uranium removal in oceanic sediments and the oceanic U balance: Earth and Planetary Science Letters, v. 97, p. 94-101.

BATURIN, G. N., AND DUBINCHUK, V.T., 1978, On uranium forms in oceanic phosphorites: Okeanologiya, v. 18, p. 1036-1041.

BATURIN, G. N., AND DUBINCHUK V. T., 1979, Microstructures of oceanic phosphorites, *in* Atlas of Microphotographs: Moscow. Nauka, 198 p.

BATURIN, G. N., MERKULOVA K. I., AND CHALOV P. I., 1972, Radiometric evidence for recent formation of phosphatic nodules in marine shelf sediments: Marine Geology, v. 13, M37-M41.

BATURIN G. N., MERKULOVA K. I., AND CHALOV P. I., 1974, Absolute dating of oceanic phosphorites by disequilibrium uranium: Geokhimiya 5, p. 801-807 (English translation Geochemistry, v. 11, p. 568-574.

BERNAT, M., 1975, Les isotopes de L'uranium et du Thorium et Les terres Rares Dans L'environment: Paris, Marin Cahiers ORSTROM, Série Géologie, VII, p 65-83.

BIRCH, G. F., 1979, Phosphatic rocks on the western margin of South Africa: Journal Sedimentary Petrology, v. 49, p. 93-110.

BIRCH, G. F., THOMSON, J., MCARTHUR, J. M., AND BURNETT, W. C., 1983, Pleistocene phosphorites off the west coast of South Africa: Nature, v. 302, p. 601-603.

BOROLE, D. V., RAJAGOPALAN, G., AND SOMAYAJULU, B. L. K., 1987, Radiometric ages of phosphorites from the West coast of India: Marine Geology, v. 78, p. 161-165.

BURNETT, W. C., 1977, Geochemistry and origin of phosphorite deposits from off Peru and Chile: Geological Society of America, v. 88, p. 813-823.

BURNETT, W. C., AND VEEH, H. H., 1977, Uranium-series disequilibrium studies in phosphorite nodules from the west coast of South America: Geochimica et Cosmochimica Acta, v. 41, p. 755-764.

BYRNE, R. H., AND KIM, K. H., 1990, Rare earth element scavenging in seawater: Geochimica et Cosmochimica Acta, v. 54, p. 2645- 2656.

CALVERT, S. E., 1976, The mineralogy and geochemistry of nearshore sediments. *in* Riley, J. P., and Chester, R. eds., Chemical Oceanography: New York, Academic Press, p. 187-280.

CALVERT, S. C., BURTIN, R. M., AND INGALL, E. D., 1996, Influence of water column anoxia and sediment supply on the burial and preservation of organic carbon in marine shales: Geochimica et Cosmochimica Acta, v. 60, 1577-1593.

CLAYPOOL, G. E., HOLSER, W. T., KAPLAN, I. R., SAKAI, H., AND ZAK, I., 1980, The age curves of sulfur and oxygen isotopes in marine sulfate and their mutual interpretations: Chemical Geology, v. 28, p. 199-260.

CLEMENS, S. C., AND PRELL, W. L., 1990, Late Pleistocene variability of Arabian Sea summer monsoon winds and continental aridity: Eolian records from the lithogenic component of deep-sea sediments: Paleoceanography, v. 5, p. 109-145.

COCHRAN, J. R., 1990, Himalayan uplift, sea level, and the record of Bengal fan sedimentation at the ODP Leg 116 sites, *in* Cochran, J. R., Curray, J. R., Sager, W. W., and Stow, D. A. V., eds.: College Station, Texas, Proceedings of the Ocean Drilling Program Scientific Results 116, p. 397-414.

DE BAAR, H. J. W., GERMAN, C. R., ELDERFIELD, H., AND GAANS, P. V., 1988, Rare earth element distribution in anoxic waters of the Cariaco Trench: Geochimica et Cosmochimica Acta, v. 52, p. 1203-1219.

ELDERFIELD, H., 1988, The oceanic chemistry of the rare-earth elements: London, Philosophical Transactions of the Royal Society of London A325, p. 105-126.

ELDERFIELD, H., AND GREAVES, M. J., 1982, The rare earth elements in seawater: Nature, v. 296, p. 214-219.

ELDERFIELD, H., AND PAGGETT, R., 1986, REE in icchthyoliths: variations with redox conditions and depositional environment: Science of the Total Environment, v. 49, p. 175-197.

ELDERFIELD, H., AND SHOLKOVITZ, E., 1987, Rare earth elements in the porewaters of reducing nearshore sediments: Earth Planetary Science Letters, v. 82, p. 280-288.

ELDERFIELD, H., HAWKESWORTH, C. J., GREAVES, M. J., AND CALVERT, S. E., 1981, Rare earth element geochemistry of oceanic ferro-manganese nodules and associated sediments: Geochimica et Cosmochimica Acta, v. 45, p. 513-528.

GERMAN, C. R,. AND ELDERFIELD, H., 1990, Rare-earth elements in the NW Indian Ocean: Geochemica et Cosmochimica Acta, v. 54, p. 1929-1940.

GERMAN, C. R., HOLLIDAY, B. P., AND ELDERFIELD, H., 1991, Redox cycling of rare earth elements in the suboxic zone of the Black Sea: Geochimica et Cosmochimica Acta, v. 55, p. 3553-3558.

GLENN, C. R., RAJAN, S., MCMURTRY, G. M., AND BENAMAN, J., 1995, Geochemistry, Mineralogy and Stable Isotopic Results from Ala Wai Estuarine Sediments: Records of hypereutrophication and abiotic whitings: Pacific Science, v. 49, p. 367-399.

GOLDBERG, E. D., KOIDE, M., SCHMITT, R. A., AND SMITH, R. H., 1963, Rare-earth distributions in the marine environment: Journal Geophysical Research, v. 68, p. 4209-4217.

GRANDJEAN, P,. AND ALBAREDE, F., 1989, Ion probe measurement of rare earth elements in biogenic phosphates: Geochimica et Cosmochimica Acta, v. 53, p. 3179-3183.

GRANDJEAN, P., CAPETTA, H., MICHARD, A., AND ALBAREDE, F., 1987, The assessment of REE patterns and ^{143}Nd/^{144}Nd ratios in fish remains: Earth Planetary Science Letters, v. 84, p. 181-196.

HEIN, J. R., YEH, H. W., GUNN, S. H., SLITER, W. V., BENNINGER, L. M., AND WANG, C. H., 1993, Two major Cenozoic episodes of phosphogenesis recorded in equitorial Pacific seamount deposits: Paleoceanography, v. 8, p. 293-311.

HOLSER, W. T., 1997, Evaluation of the application of REEs to paleoceanography: Paleogeography, Paleoclimatology and Paleoecology, v. 132, p. 309-323.

JAHNKE, R. A., EMERSON, S. R., ROE, K. K., AND BURNETT, W. C., 1987, The present day formation of apatite in Mexican continental margin sediments: Geochimica et Cosmochimica Acta, v. 47, p. 259-266.

KIDDER, D. L., AND EDDY-DILEK, C. A., 1994, Rare-earth element variation in phosphate nodules from midcontinent Pennsylvanian cyclothems: Journal of Sedimentary Research, v. A64, p. 584-592.

KIM, H. H., AND BURNETT, W. C., 1985, ^{226}Ra in phosphate nodules from the Peru/Chile seafloor: Geochimica et Cosmochimica Acta, v. 49, p. 1073-1081.

KLINKHAMMER, G. P., AND PALMER, M. R., 1991, Uranium in the oceans: where it goes and why: Geochimica et Cosmochimica Acta, v. 55, p. 1794-1806.

KOLODNY, Y., 1983, Phosphorites, *in* Emiliani, C., ed., The Sea, v. 7: New York, John Wiley & Sons, p. 981-1023.

LOVLEY, D. R., RODEN, E. E., PHILLIPS, E. J. P., AND WOODWARD, J. C., 1993, Enzymatic iron and uranium reduction by sulfate-reducing bacteria: Marine Geology, v. 113, p. 41-53.

LUCAS, J., FLICOTEAUX, R., NATHAN, Y., PRÉVÔT, L., AND SHAHOR, Y., 1980, Differential aspects of phosphorite weathering: Tulsa, Society for Sedimentary Geology (SEPM) Special Publication 29, p. 41-57.

MACLEOD, K. G., AND IRVING, A. J., 1996, Correlation of cerium anomalies with indicators of paleoenvironment: Journal of Sedimentary Research, v. 66, p. 948-955.

MANHEIM, F. T., AND GULBRANDSEN, R. A., 1979, Marine phosphorites, *in* Burns, R. G., ed., Marine Minerals: Washington, D.C., Mineralogical Society of America, v. 6, p. 151-173.

MCARTHUR, J. M., AND WALSH, J. N., 1984, Rare earth geochemistry of phosphorites: Chemical Geology, v. 47, p. 191-220.

MCCLELLAN, G. H., AND VAN KAUWENBERGH, S. J., 1990, Mineralogy of sedimentary apatites, *in* Notholt, A. J. G., and Jarvis, I., eds., Phosphorite Research and Development: London, Geological Society Special Publication 52, p. 23-31.

MCLENNAN, S. M., 1989, Rare earth elements in sedimentary rocks: Influence of provenance and sedimentary processes, *in* Lipin, B. P., and McKay, G. A., eds., Geochemistry and mineralogy of rare earth elements: Reviews in Mineralalogy, v. 21, p. 169-200.

MO, T., SUTTLE, A. D., AND SACKETT, W. M., 1973, Uranium concentrations in marine sediments: Geochimica et Cosmochimica Acta, v. 37: p. 35-51.

MORSE, J. W., SHANBHAG, P. M., SATTO, A., AND CHOPPIN, G. R., 1984, Interaction of uranyl ions in carbonate media: Chemical Geology, v. 42, p. 85-99.

MURRAY, R. W., AND LEINEN, M., 1993, Chemical transport to the seafloor of the equatorial Pacific Ocean across a latitudinal transect at 135°W: Tracking sedimentary major, trace, and rare earth element fluxes at the Equator and the Intertropical Convergence Zone: Geochimica et Cosmochimica Acta., v. 57, p. 4141-4163.

MURRAY, R. W., BUCHHOLTZ TEN BRINK, M. R., GERLACH, D. C., RUSS, G. P., III., AND JONES, D. L., 1991, Rare earth, major and trace elements in chert from the Fraciscan Complex and Monterey Group, California: Assessing REE sources to fine grained marine sediments: Geochimica et Cosmochimica Acta, v. 55, p. 1875-1895.

MURRAY, R.W., BUCHHOLTZ TEN BRINK, M. R., JONES, D. L., GERLACH, D. C., AND RUSS G. P. III., 1990, Rare earth elements as indicators of different marine depositional environments in chert and shale: Geology, v. 18, p. 268-271.

MURTY, C. S., AND VARADACHARI, V. V. R., 1968, Upwelling along the east coast of India: Bulletin National Institute of Sciences, India, v. 36, p. 80-86.

NAIDU, P. D., PRAKASH BABU, C., AND RAO, CH. M., 1992, The upwelling record in the sediments of the western continental margin of India: Deep Sea Research, v. 39, p. 715-723.

NAIR, R. R., 1985, Holocene phosphorites of the Western continental margin of India: Mahasagar Bulletin of the National Institute of Oceanography, India, v. 18, p. 273-279.

NATH, B. N., PLÜGER, W. L., AND ROELANDTS, I., 1997a, Geochemical constraints on the hydrothermal origin of ferromanganese encrustations from the Rodriguez Triple Junction, Indian Ocean, *in* Nicholson, K., Hein, J. R., Buhn, B., and Dasgupta, S., eds., Manganese Mineralization, Geochemistry and Mineralogy of Terrestrial and Marine Deposits: London, Geological Society of London Special Publication, v. 119, p. 199-211.

NATH, B. N., BAU, M., RAMALINGESWARA RAO, B., AND RAO, CH. M., 1997b, Trace and rare earth elemental variation in Arabian Sea sediments through a transect across the oxygen minimum zone: Geochimica et Cosmochimica Acta, v. 61, p. 2375-2388.

NATH B. N., ROELANDTS I., SUDHAKAR M., PLUEGER, W. L., AND BALARAM, V., 1994, Cerium anomaly variations in ferromanganese nodules and crusts from the Indian Ocean: Marine Geology, v. 120, p. 385-400.

NATHAN, Y., AND SASS, E., 1981, Stability relations of apatites and calcium carbonates: Chemical Geology, v. 34, p. 103-111.

NOZAKI, Y,. AND HORIBE, Y., 1983, Alpha emitting thorium isotopes in northwest Pacific deepwaters: Earth Planetary Science Letters, v. 65, p. 39-50.

O'BRIEN, G. W., AND VEEH, H. H. 1980, Holocene phosphorite on the East Australian continental margin: Nature, v. 288, p. 690-692.

O'BRIEN, G. W., MILNES, A. R., VEEH, H. H., HEGGIE, D. T., RIGGS, S. R., CULLEN, D. J., MARSHALL, J. F., AND COOK, P. J., 1990, Sedimentation dynamics and redox iron-cycling: controlling factors for the apatite-glauconite association on the East Australian continental margin, *in* Notholt, A. J. G., and Jarvis, I., eds., Phosphorite Research and Development: London, Geological Society Special Publication 52, p. 61-86.

PARKER, R. J., AND SIESSER, W. G., 1972, Petrology and origin of some phosphorites from the South African continental margin: Journal of Sedimentary Petrology, v. 42, p. 434-440.

PIPER, D. Z., 1974, Rare earth elements in the sedimentary cycle, a summary: Chemical Geology, v. 14, p. 285-304.

PIPER, D. Z., BAEDECKER, P. A., CROCK, J. G., BURNETT, W. C., AND LOEBNER, B. J., 1988, REE in the phosphatic-enriched sediment of the Peru shelf: Marine Geology, v. 80, p. 269 -285.

PRELL, W. L., MURRAY, D. W., CLEMENS, S. C., AND ANDERSON, M., 1992, Evolution and variability of the Indian Ocean summer monsoon: Evidence from the Western Arabian Sea Drilling Program, *in* Duncan, R. A., Rea, D. K., Kidd, R. B., Von Rad, U., and Weissel, J. K., eds., Synthesis of Results from scientific drilling in the Indian Ocean, Geophysical Monograph, v. 70: Washington, D.C., American Geophysical Union, p. 447-470.

QASIM, S. Z., 1982, Oceanography of the northern Arabian Sea: Deep Sea Research, v. 29, p. 1041-1068.

RAO, CH. M., AND RAMALINGESWARA RAO, B., 1996, Phosphorite nodules in a sediment core from a bathymetric high off Goa, western continental margin of India: Current Science, v. 70, p. 308-312.

RAO, V. P., AND NAIR, R. R., 1988, Microbial origin of the phosphorites of the western continental shelf of India: Marine Geology, v. 84, p. 105-110.

RUTTENBERG, K. C., AND BERNER, R. A., 1993, Authigenic apatite formation and burial in sediments from non-upwelling, continental margin environments: Geochimica et Cosmochimica Acta, v. 57, p. 991-1007.

SEREBRYAKOVA, M. B., AND RAZUMNAYA, Y. G., 1962, Form of occurrence of uranium in apatite: Academy Nauka, USSR Dokladi, v. 143, p. 153-156. (English translation)

SHETYE, S. R., SHENOI, S. S. C., ANTONY, M. K., AND KUMAR, K., 1985, Monthly-mean wind stress along the coast of the north Indian Ocean: Proceedings of the Indian Academy of Science, v. 94, p. 129-137.

SHIMIZU, H., AND MASUDA, A., 1977, Cerium in chert as an indicator of marine environment of its formation: Nature, v. 266, p. 346-348.

SHOLKOVITZ, E. R., 1990, Rare earth elements in marine sediments and geochemical standards: Chemical Geology, v. 88, p. 333-347.

SHOLKOVITZ, E. R., AND ELDERFIELD, H., 1988, Cycling of dissolved rare earth elements in Chesapeake Bay: Global Biogeochemical Cycles, v. 2, p. 157-176.

SHOLKOVITZ, E. R., PIEPGRAS, D. J., AND JACOBSEN, S. B., 1989, The porewater chemistry of rare earth elements in Buzzards Bay sediments: Geochimica et Cosmochimica Acta, v. 53, p. 2847-2856.

SLATER, R. D., AND KROOPNICK, P., 1984, Controls on dissolved oxygen distribution and organic carbon deposition in the Arabian Sea, *in* Milliman, J. D., and Haq, B. U., eds., Marine Geology and Oceanography of Arabian Sea and coastal Pakistan: New York, Nostrand Reinhold, p. 305-313.

STANLEY, J. K., JR., AND BYRNE, R. H., 1990, The influence of solution chemistry on REE uptake by *Ulva lactuca L.* in seawater: Geochimica et Cosmochimica Acta, v. 54, p. 1587-1595.

STARINSKY, A., KATZ, A., AND KOLODNY, Y., 1982, Incorporation of Uranium into diagenetic phosphorite: Geochimica et Cosmochimica Acta, v. 46, p. 1365-1374.

STAUDIGEL, H., DOYLE, P., AND ZINDLER, A., 1985/86, Sr and Nd isotope systematics in fish teeth: Earth Planetary Science Letters, v. 76, p. 45-56.

STEINBERG, M., BONNOT-COURTOIS, C., AND TLIG, S., 1983, Geochemical contribution to the understanding of bedded chert, *in* Iijima, A., Hein, J.R., and Siever, R., eds., Siliceous Deposits in the Pacific Region. Developments in Sedimentology, v. 36: Amsterdam, Elsevier, p. 193-210.

TAYLOR, S. R., AND MCLENNAN, S. M., 1985, The Continental Crust, Its Composition and Evolution: London, Blackwell, 312 p.

TOYODA, K., AND TOKONAMI, M., 1990, Diffusion of rare earth elements in fish teeth from deep-sea sediments: Nature, v. 345, p. 607-609.

VAN KAUWENBERGH, S. J., AND MCCLELLAN, G. H., 1990, Comparative geology and mineralogy of the southeastern United States and Togo-phosphorites, *in* Notholt, A. J. G., and Jarvis, I., eds., Phosphorite Research and Development: London, Special Publication of the Geological Society, v. 52, p.139-159.

VAZ, G. G., 1995, Phosphatic nodules in the outer continental shelf off Madras, Bay of Bengal: Indian Journal of Marine Science, v. 24, p. 8-12.

VEEH, H. H., 1967, Deposition of uranium from the oceans: Earth and Planetary Science Letters, v. 3, p. 145-150.

VEEH, H. H., CALVERT, S. E., AND PRICE, N. B., 1974, Accumulation of uranium in sediments and phosphorites on the southwest African shelf: Marine Chemistry, v. 2, p.189-202.

WATKINS, R. T., NATHAN, Y., AND BREMNER, J. M., 1995, Rare earth elements in phosphorites and associated sediment from the Namibian and South African continental shelves: Marine Geology, v. 129, p. 111-128.

WRIGHT, J., SEYMOUR, R. S., AND SHAW, H. F., 1984, REE and Nd isotopes in conodont apatite: Variations with geological age and depositional environment: Boulder, Geological Society of America Special Paper, v. 196, p. 325-340.

WRIGHT, J., SHRADER, H. AND HOLSER, W. T., 1987, Paleoredox variation in ancient oceans recorded by REEs in fossil apatite: Geochimica et Cosmochimica Acta, v. 51: 631-644.

WYRTKI, K., 1971, Oceanographic Atlas of the Indian Ocean Expedition: Washington, D.C., National Science Foundation, 531 p.

WYRTKI, K., 1973, Physical Oceanography of the Indian Ocean, *in* Zeitzchel, B., ed., The Biology of the Indian Ocean: New York, Springer, p.18-36.

DISTRIBUTION AND COMPOSITION OF VERDINE AND GLAUCONY FACIES FROM THE SEDIMENTS OF THE WESTERN CONTINENTAL MARGIN OF INDIA

M. THAMBAN AND V. PURNACHANDRA RAO*

National Institute of Oceanography, Dona Paula - 403 004, Goa, India

ABSTRACT: Investigations on green grains from sediments of the western continental margin of India, between Ratnagiri and Cape Comorin, (water depth 37-330 m) indicate the presence of verdine and glaucony facies. Verdine facies occurs over an area of about 100,000 km^2, representing the largest sedimentary basin in the world associated with low fluvial input. Green grains occur as irregular grains, fecal pellets, and infillings/internal molds of microfossils. They are abundant on the shelf off river mouths and their distribution varies with sediment type. Grains vary from dark green to pale green or brownish green. Green grains studied here are a mixture of predominant authigenic green clay and detrital clay minerals and are altered. Both phyllite C and suspected phyllite V- (verdine minerals) associated green grains occur on the continental shelf (between 37 m and 100 m), the former being associated with the transition zone between inner and outer shelf and the latter with relict sands and reefs on the outer shelf. On the continental slope, suspected phyllite V occurs at depths between 100 m and 235 m, followed by phyllite C at 280 m depth and glauconitic smectite of the glaucony facies at 330 m. As these grains are composed of a mixture of predominant authigenic clay, detrital clay, and some altered products, their major element composition differs from those of the green grains (reported elsewhere) that contain pure authigenic clay. The low REE contents and flat shale-normalized REE patterns suggest that the REEs were inherited from the substrate.

We suggest that the size of the verdine deposit is related to the influx of iron rather than the amount of fluvial discharge. The color and morphology of the grains do not reflect the authigenic mineral or its evolution. Green grains on this margin formed at different times when the sea level was at different depths during the Late Quaternary. The distribution of verdine and glaucony facies on the southwestern margin of India is different from those of the distribution along the east coast of India, Senegal, and French Guiana margins, suggesting different paleogeography and subsidence history of the western Indian margin during the Late Quaternary.

INTRODUCTION

Verdine and glaucony are two authigenic green clay facies that occur in the shallow marine environment. Verdine is a newly distinguished facies and it has been reported to occur only in 15 locations in the tropical oceans of the world (Odin, 1985, 1988, 1990; Odin et al., 1989; Rao et al., 1993, 1995). Of these, 10 are from the Atlantic, three from the Indian, and two from the western Pacific Oceans. Except for the verdine from Miocene sediments of the Niger delta, all other occurrences are from recent marine sediments <15 ka B.P. (Odin, 1988). It develops at the continent-ocean boundary near an abundant source of iron at depths between 5 m and 60 m (Odin, 1988). On the other hand, glaucony develops in the open ocean, farther from continental input and at relatively large depth (>60 m). As the verdine facies is replaced by glaucony at greater depths, its distribution may be used for the reconstruction of paleogeography of the margins.

Mallik et al. (1976) and Thrivikramji and Machado (1982) reported the occurrence of green grains (internal molds of microfossils) at a few locations at depths <50 m on the continental shelf of western India. They referred to them as glauconite without analyzing their mineralogy. Subsequently, Rao et al. (1993) conducted detailed studies of the distribution of verdine and glaucony on the Kerala continental shelf and slope. In this paper, we report the distribution, mineralogy, and geochemistry of the green grain facies for a large region between Ratnagiri and Cape Comorin on the western margin of India which represents one of the largest basins containing verdine facies. It almost equals that off the Amazon river mouth.

*email: vprao@csnio.ren.nic.in

GEOLOGY AND PHYSIOGRAPHY OF THE REGION

The study area lies in the tropical belt between 17° and 7°N and experiences a humid tropical climate (Fig. 1). The Western Ghats are the major physiographic features along the coast and are composed of the Deccan Trap basalt in the northern part (up to Goa) and Precambrian gneisses, schists, and charnockites in the southern part. The Ghat rocks are extensively lateritized. The Ghats tend to occur farther inland 50-80 km from the coast, south of Bhatkal to Quilon (Krishnan, 1968). The coastal region here consist of Miocene Warkala and Quilon beds (mainly ferruginized sandstones with clay intercalations) and recent alluvium. There are several medium (~85-250 km long) and minor rivers and numerous streams of seasonal nature. They originate from and drain through the steeper slopes of the Western Ghats, covering an area of about 54,000 km^2, and in total discharge about 95.6 km^3 of water annually into the Arabian Sea (Rao, 1979).

The width of the continental shelf is broader at about 130 km off Ratnagiri and narrows to 60 km off Cochin. It is again wider (120 km) at the southern tip, off Cape Comorin. Accordingly, the width of the inner shelf also varies. It is relatively wide and extends to 60 m water depth in the northern part and narrows down to 30 m water depth in the southern part off Cochin (Rao and Wagle, 1997). The shelf break occurs at about 120 m water depth in the northern part and at about 90 m in the southern part. The sediments on the inner shelf are terrigenous muds between Ratnagiri and Quilon and carbonate sands between Quilon and Cape Comorin (Nair and Pylee, 1968; Hashimi et al., 1982). Outer shelf sediments are relict sands and are carbonate-dominated with reefs (Rao, 1989) between Ratnagiri and Mangalore and terrigenous sand-dominated between Mangalore and Quilon. Biogenic sands again occur on the outer shelf south of Quilon. The surficial sediments from the continental slope are mostly silty clays.

Marine Authigenesis: From Global to Microbial, SEPM Special Publication No. 66

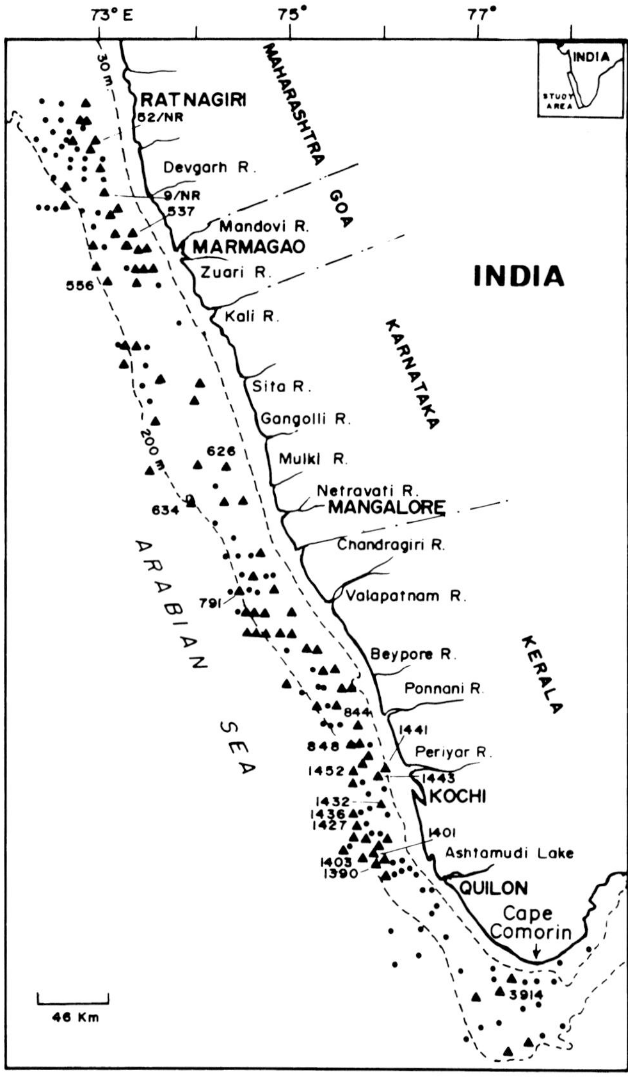

FIG. 1.—Sample location map. Triangles indicate the samples that contain more than 5% green grains in the coarse fraction. Station numbers are given for those samples that were used for detailed mineralogy and geochemistry.

MATERIALS AND METHODS

For the present study, 195 surficial sediment samples collected during different cruises of *R. V. Gaveshani* and *M. V. Nand Rachit* were selected from the western continental shelf and upper slope of India between Ratnagiri in the north and Cape Comorin in the south (Fig. 1). Coarse fraction (>63 μm) of the sediments was separated by wet sieving, split into different sizes, from which the 125–250 μm and 250–500 μm size fractions were used for the present study. Coarse fraction constituents were identified under a binocular microscope and distribution of major sediment constituents and green grains (Table 1) have been quantified by counting about 400 grains in each sample.

Green grains were separated by using a Frantz Isodynamic Separator with a longitudinal slope of 25° and a lateral slope of 17° using 0.6–0.8 mA current. Pure green grains were obtained by hand picking from the magnetically separated fraction under a binocular microscope. The separated green grains were powdered to fine size and scanned from 3° to 15° 2θ and 3° to 40° 2θ (selected samples) at 1° 2θ/min. on an X-ray diffractometer using nickel-filtered Cu Kα radiation. The samples were subjected to ethylene glycol (100°C/1 hour), 1N acetic acid (70°C/2 hours), and heat (250°C/1 hour, 490°C/2 hours and 600°C/4 hours) treatments, and X-ray diffractograms were obtained at each stage.

Geochemistry of the green grains was carried out on only 13 samples, owing to the difficulty in obtaining enough material. Silica was analyzed on a spectrophotometer, following Shapero and Brannock (1962), Na and K by atomic absorption spectrophotometer (AAS), and other major elements by using ICP-AES. REE analyses were carried out on ICP-MS. The accuracy of the analyses as assessed by using International standards (MAG 1 – Marine Mud; SO 3 – Soil-Canadian; and GL O – Glauconite) is better than ± 2% for Si, Al, Fe, Mg, K, and Ca, better than ± 15% for REEs, except for Gd and Ho, where it is >15%. Sodium (Na) showed a higher uncertainty, which may be due to some systematic error.

RESULTS

Nature and Distribution of Green Grains

Of the 195 samples studied, 116 (59%) showed the presence of green grains (Fig. 1). The content of green grains in the coarse fraction varied from 2.4% to 51.2%. The higher percentages (23-51%) of green grains were found diametrically opposite the Periyar, Ponnani, Netravati, Mandovi-Zuari, and other river mouths (Fig. 1; Table 1). Green grains were either absent or present in very low quantities in muddy sediments of the inner shelf. They were less abundant (< 8%) in the terrigenous sand-dominated sediments and abundant (up to 51%) in bioclast-dominated sediments. In general, the green grain facies covered a larger area on the shelf and slope off Kerala than off Karnataka, Goa, and Maharashtra (see Fig. 1). During magnetic separation, higher amounts of heavy minerals were also obtained along with green grains from the shelf sediments off the southern tip of India.

The green grains were present in three morphologies: i) irregular free grains, ii) glossy and rough surface-textured fecal pellets, and iii) infillings and internal molds of predominant microfossils. The color of the grains varied from dark green to pale dull green and even to brown ochre. On the shelf, irregular dark green grains with cracked outer surfaces were predominant and infillings and molds were rare (Table 1). Associated grains include quartz, shell fragments, foraminifers, and ostracodes, which were sometimes stained brown. Highly polished brown (ochre) pellets occur on the shelf (at 49 m depth) off the southern tip of India (G-167/ 3914); even irregular grains showed highly smoothened surfaces. Abundant mollusk fragments and benthic foraminifers followed by quartz (stained) and heavy minerals were associated with them (Table 1).

The green grains from the upper continental slope at depths between 100 m and 235 m were mostly dark green pellets and internal molds of microfossils; irregular free grains were rare. Molds were also dark green and shiny. Green grains were

TABLE 1.—DISTRIBUTION, MORPHOLOGY, AND MINERALOGY OF THE GREEN GRAINS IN SELECTED SAMPLES.

Sl. No	Sample No.	Depth (m)	Distribution of grains (%) Bio.	Terri.	G. grn	Morphology of the green grains	Predominant authigenic green minerals
1	G-71/ 1441	37	nd	nd	nd	Irregular dark green grains; few infillings	Phyllite C
2	G-71/ 1443	38	39.0	37.0	24.0	Irregular dark green grains	Phyllite C
3	G-71/ 1432	41	48.0	25.0	27.0	Irregular dark green grains, cracked outer surface	Phyllite C
4	G-71/ 844	42	34.9	14.4	50.7	Irregular dark green grains, cracked outer surface	Phyllite V*
5	G-18/ 626	44	48.4	28.2	23.4	Irregular dark green grains; few infillings; molds	Phyllite V*
6	G-167/ 3914	49	80.1	14.7	5.2	Brown glossy pellets	Goethite, Phyllite V*
7	G-18/ 537	61	52.5	34.9	12.6	Dark green internal molds and casts of microfauna	Phyllite V
8	NR-2A/ 52	65	76.0	6.2	17.8	Dark green molds and pellets; few brownish grains	Phyllite C
9	NR-2A/ 9	70	56.9	28.6	14.5	Irregular dark green grains; few pellets; stained grains	Phyllite V*
10	G-18/ 556	75	67.2	10.0	22.8	Dark green shiny molds; also irregular grains	Phyllite V
11	G-30/ 848	117	46.9	25.6	27.5	Irregular dark green grains; few dark molds	Phyllite V*
12	G-71/ 1390	117	74.1	3.3	22.6	Dark green infillings and molds of foraminifers	Phyllite V
13	G-30/ 791	150	70.2	—	29.8	Dark green molds and infillings of microfossils	Phyllite V*
14	G-30/ 634	150	73.6	—	26.4	Dark green molds of foraminifers and ostracods	Phyllite V*
15	G-71/ 1436	170	73.0	—	27.0	Dark green glossy pellets; also molds and infillings	Phyllite V
16	G-71/ 1452	190	62.0	—	38.0	Dark shiny molds of foraminifers and pteropods	Phyllite V
17	G-71/ 1427	235	77.0	—	23.0	Dark green glossy pellets and also irregular grains	Phyllite V
18	G-71/ 1401	280	48.8	—	51.2	Dark green pellets; infillings of foraminifers	Phyllite C
19	G-71/ 1402	330	58.0	—	42.0	Infillings of foraminifers; few internal molds	Glauconitic smectite

nd - not determined ; Bio.- Biogenic particles ; Terri.- Terrigenous particles; G.grn- Green grains; * suspected phyllite V.

mostly found as infillings of foraminifers at depths >280 m. The outer tests of many green grains were still intact and appear light to pale green in color; when the outer tests were removed, the green pigment was found to be dark in color.

Mineralogy of the Green Grains

Based on geographic location, depth, and availability of sufficient material, 14 samples were selected for X-ray diffraction studies. Besides, there were five samples in the study area whose mineralogy has been reported (Rao et al., 1993). Both oriented and unoriented slides were analyzed. The peaks were, however, at the noise level in unoriented slides. The peaks are slightly better in oriented slides, which were used throughout the study. Here, we emphasize the results of the samples that showed different mineralogy than those reported by Rao et al. (1993).

Green grains from the continental shelf.—
Irregular dark grains occurring at two stations (G-71/ 1441 and G-71/ 1443) at about 40 m depth were analyzed. Phyllite C was found to be the verdine mineral and was identical to that which Rao et al. (1993) reported from a nearby station (G-71/ 1432) (see Fig. 2A). Comparison of green grain and detrital clay X-ray diffractograms (Fig. 2A&B) suggest that the larger 7Å peak in green grains might be due to kaolinite inherited from the substrate.

X-ray diffractograms of the untreated samples (irregular dark green grains) at 62–65 m water depth off Ratnagiri (NR-2A/52) and Goa (NR-2A/9, 537) exhibit a large 14Å peak and a small dome around 7Å (Fig. 3). Quartz peaks and a peak at 10Å probably due to illite were present in all of the diffractograms. During glycolation, the 14Å peak was shifted to about 16Å. Acetic acid treatment reduced the 7Å peak and marginally enhanced the peak at 14Å. During heat treatment, the peaks were initially lowered (at 250ºC/1 hour) and a large dome developed around 10Å (at 490ºC/2 hours), which remained there even after the sample was heated at 600ºC for 4 hours. The large 14Å peak in untreated X-ray diffractograms, its shifting to higher d-spacings during ethylene glycol solvation, and its behavior on heat treatment confirmed the typical properties of phyllite C.

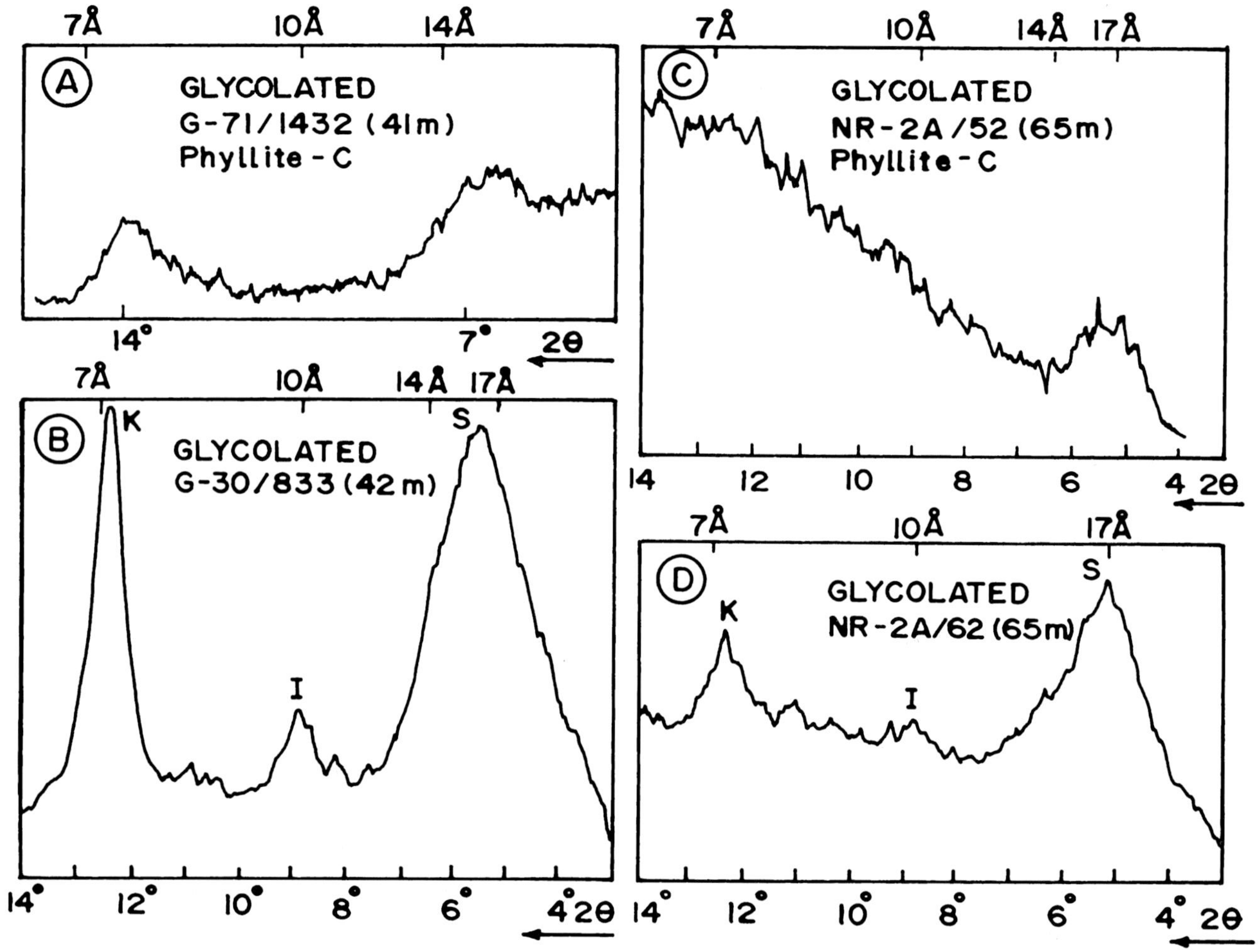

FIG. 2.—Representative x-ray diffractograms of authigenic green grains (a & c) and detrital clays (b & d) (in the vicinity of green grains) on the continental shelf (Fig. 1 for location). Diffractogram (A) was obtained using CoKα radiation. S—smectite; I—illite; K—kaolinite.

When compared with detrital clay X-ray diffractograms from the same region, the 7Å and 14Å reflections of the green grains were broad and poorly crystalline, perhaps due to a mixture of authigenic and detrital clay minerals and immaturity of the authigenic clay (see Fig. 2C&D).

Irregular dark green grains were also present on the outer continental shelf (see Fig. 1) at depths between 42 m and 75 m in the study area. The X-ray diffractograms of the green grains from six stations revealed a large 7Å peak and a small dome around 14Å (Fig. 4). Ethylene glycol treatment had no significant effect on these peaks. However, acetic acid treatment lowered the 7Å peak and enhanced the peak at 14Å. Both peaks were destroyed after heating the samples to 250ºC/1 hour. A peak at about 10Å appears in glycolated and acetic acid-treated samples. The behavior of the green grains differs from sample to sample on heat treatment at higher temperatures. In some samples (see sample 556, Fig. 4A), the 14Å peak was slightly enhanced at 490ºC/2 hours and a portion of it shifted toward lower d-spacings with a broad dome at about 10Å at 600ºC/4 hours. In others (samples 626, 844, Fig. 4B), the 14Å peak was completely collapsed and a broad dome developed at around 10Å (to 490ºC/2 hours), which intensified on further heating (at 600ºC/4 hours).

Smectite and kaolinite were the dominant detrital clay minerals in the surficial sediments of this region (Rao and Rao, 1995). The properties of smectite (expansion of 15Å reflection on glycolation and its collapse to 10Å on heat treatment at relatively low temperatures) were not seen in the X-ray diffractograms. Kaolinite is a nonmagnetic mineral. Since we have analyzed magnetically separated green grains, the 7Å reflection may predominantly be due to authigenic clay. Moreover, the large 7Å and a small 14Å peak, absence of shift in 14Å peak after glycolation, and enhancement of 14Å after acetic acid treatment seen in all of the X-ray diffractograms (Fig. 4) are characteristic of young phyllite V (Odin, 1988), which has been named odinite (Bailey, 1988). Development of the 14Å peak and a portion of it shifting to lower d-spacings (after heat treatment) recorded in some samples (556, Fig. 4A) are also characteristic of phyllite V. The complete collapse of the 14Å peak to lower d-spacings at higher temperatures seen in other samples (626, Fig. 4B) does not confirm phyllite V. In fact, these samples were lying off the river mouths and closer to the inner shelf (see Fig. 1). Verdine can be destroyed quickly upon recrystallization (Odin, 1988). It could be that the authigenic clay in these green

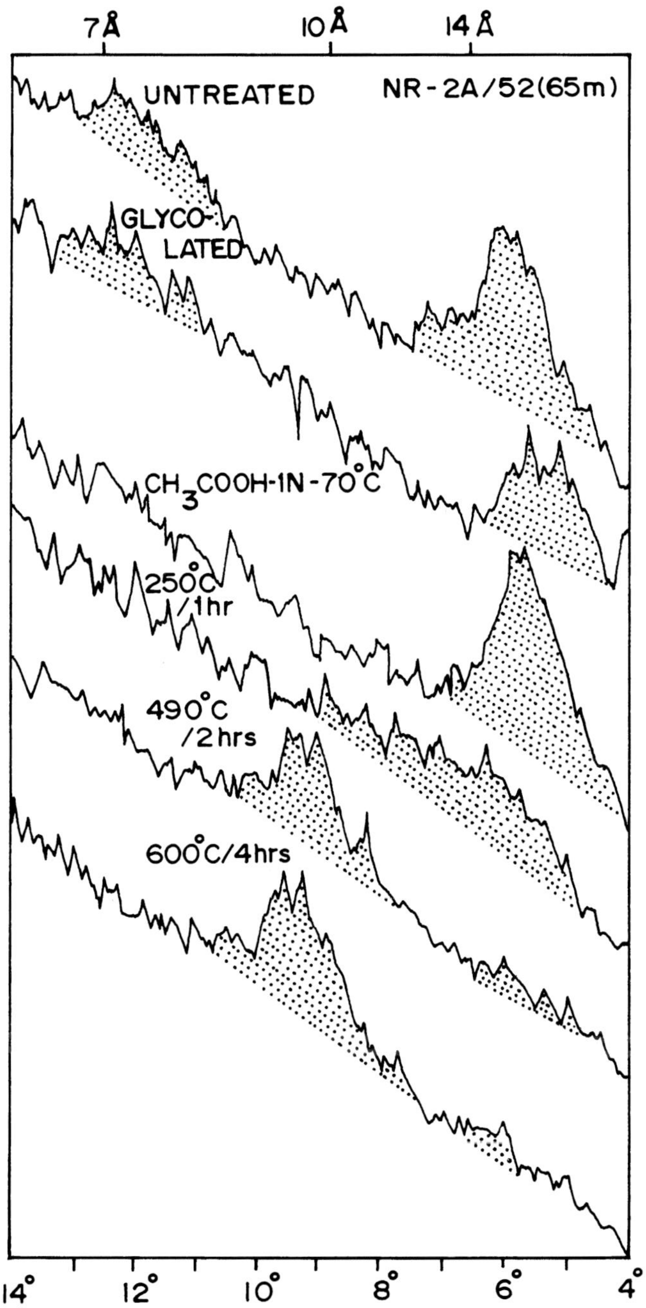

FIG. 3.—X-ray diffraction patterns of irregular dark green grains off Ratnagiri. Shaded portions are probable areas for authigenic clay reflections.

grains was subsequently oxidized and rapidly altered, destroying some original properties of authigenic minerals. We therefore suspect that the authigenic clay is phyllite V. The 10Å peak in glycolated and acetic acid-treated samples and its enlargement after heat treatment may be due to the presence of detrital clays, including illite, inherited from the substrate. Therefore, the green grains represent a mixture of predominant authigenic and detrital clay minerals and their alteration products.

X-ray diffractograms of untreated samples of shiny brown pellets at station G-167/ 3914 (see Fig. 1) did not show significant peaks at 7Å and 14Å, but showed distinct reflections (at 4.18Å, 2.45Å, and 2.69Å) of goethite (Fig. 5). The 7Å and 14Å peaks were slightly enhanced after glycolation without shifting to higher d-spacings. The peak at 14Å was enhanced after acetic acid treatment, but centered around 16.7Å. A broad dome was developed at about 10Å, when heated to 490°C for 2 hours.

The behavior of these grains after glycolation and acetic acid treatment (Fig. 5) is more or less similar to those of the grains from the outer shelf (Fig. 4), which suggests the presence of verdine minerals, but heat treatment did not confirm this. Abundant goethite indicates that the authigenic clay minerals in these grains probably were altered considerably following a low sea level, and thus are difficult to identify at present. Since the characteristics of glycolated and acid-treated X-ray diffractograms resemble those of the grains of phyllite V, we suspect that the authigenic green clay is phyllite V of the verdine facies. The poor 7Å and 14Å reflections in these grains may be due to strong alteration.

Green grains from the continental slope.—

Four samples were analyzed from the continental slope of Karnataka and Kerala, at depths between 117 m and 150 m. Green grains occured as dark green glossy pellets, shiny molds and infillings of foraminifers, and pteropods. X-ray diffractograms showed a well-developed 7.2Å peak and a poorly developed dome at about 14Å (Fig. 6). The dome at about 14Å shifted marginally to higher d-spacings upon glycolation and intensified following acetic acid treatment. The 14Å peak completely collapsed and a large dome was developed at about 10Å up on heat treatment. The large 7Å and small 14Å reflections in untreated X-ray diffractograms and the decrease of 7Å and increase of 14 Å reflections upon acetic acid treatment are characteristic of phyllite V. However, the collapse of the 14Å peak on heating does not confirm phyllite V. The peak at 10Å and the marginal shift of the 14Å peak upon glycolation suggest the presence of detrital illite and expandable mineral like smectite, inherited from the substrate. Although the morphology of the grains differs from that of the outer shelf, the X-ray diffractograms obtained after different treatments are similar, suggesting that these grains contain a mixture of authigenic and detrital clays and are altered. The authigenic green clay is most probably young phyllite V. Rao et al. (1993) also reported phyllite V in the green grains (light green rugose pellets and dark green glossy pellets) from two samples collected at 100 m and 205 m water depth. Although the color and morphology of the green grains vary at these depths (between 100 m and 235 m), the verdine mineral was always young phyllite V (odinite). Pale green pellets and infillings from 280 m depth (G-71/ 1401) were separately analyzed by Rao et al. (1993) and showed that phyllite C was the dominant mineral phase in both morphologies. Pale green infillings from the 330 m depth terrace off Cochin (G-71/ 1403) were analyzed by Rao et al. (1993), who reported glauconitic smectite in these grains.

Distribution of the different mineral phases.—

With wide geographic coverage and high sampling frequency, it is possible to reach the following conclusions: all of the green grains studied here were a mixture of predominant

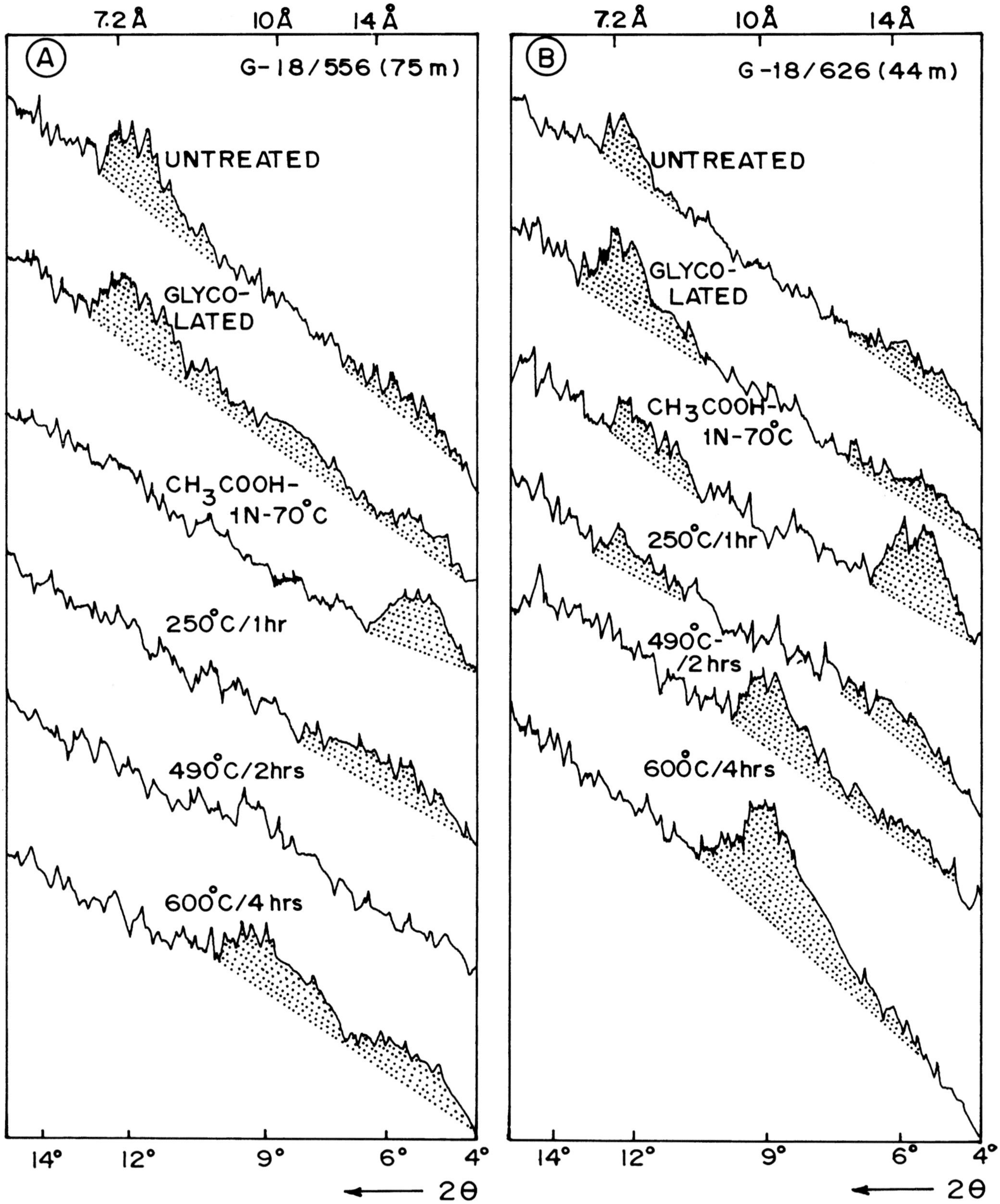

FIG. 4.—X-ray diffraction patterns of irregular dark green grains from the outer shelf off Goa (A), and Karnataka (B).

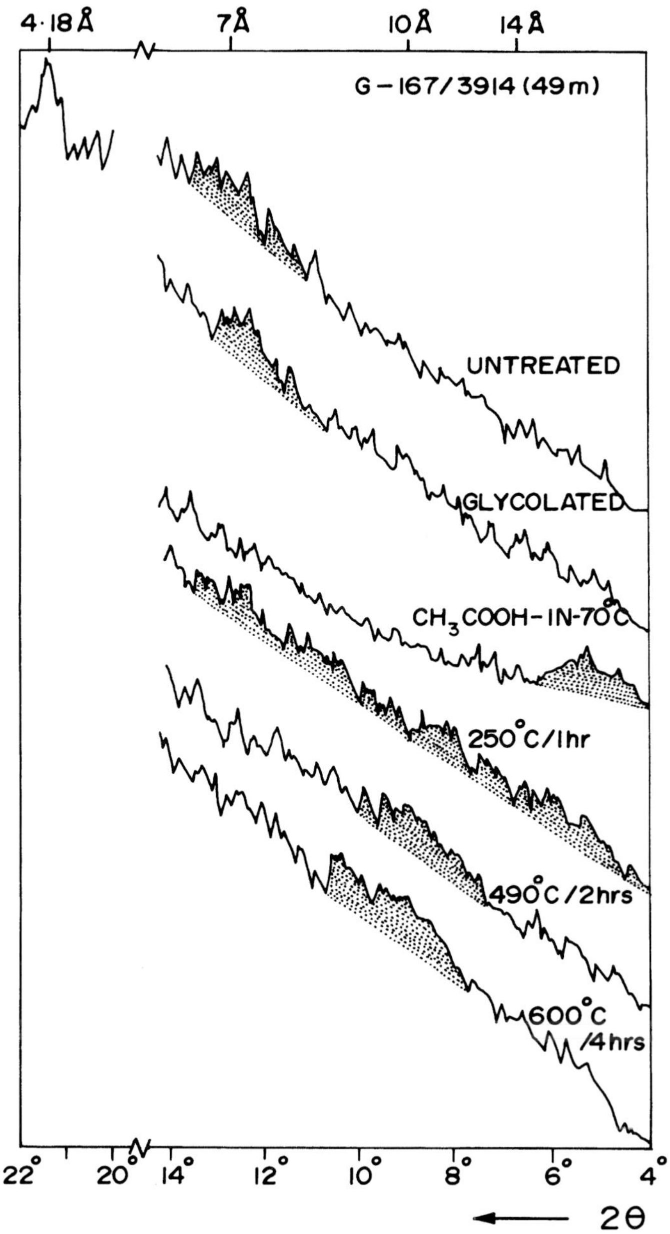

FIG. 5.—X-ray diffraction patterns of brown glossy pellets off Cape Cormorin.

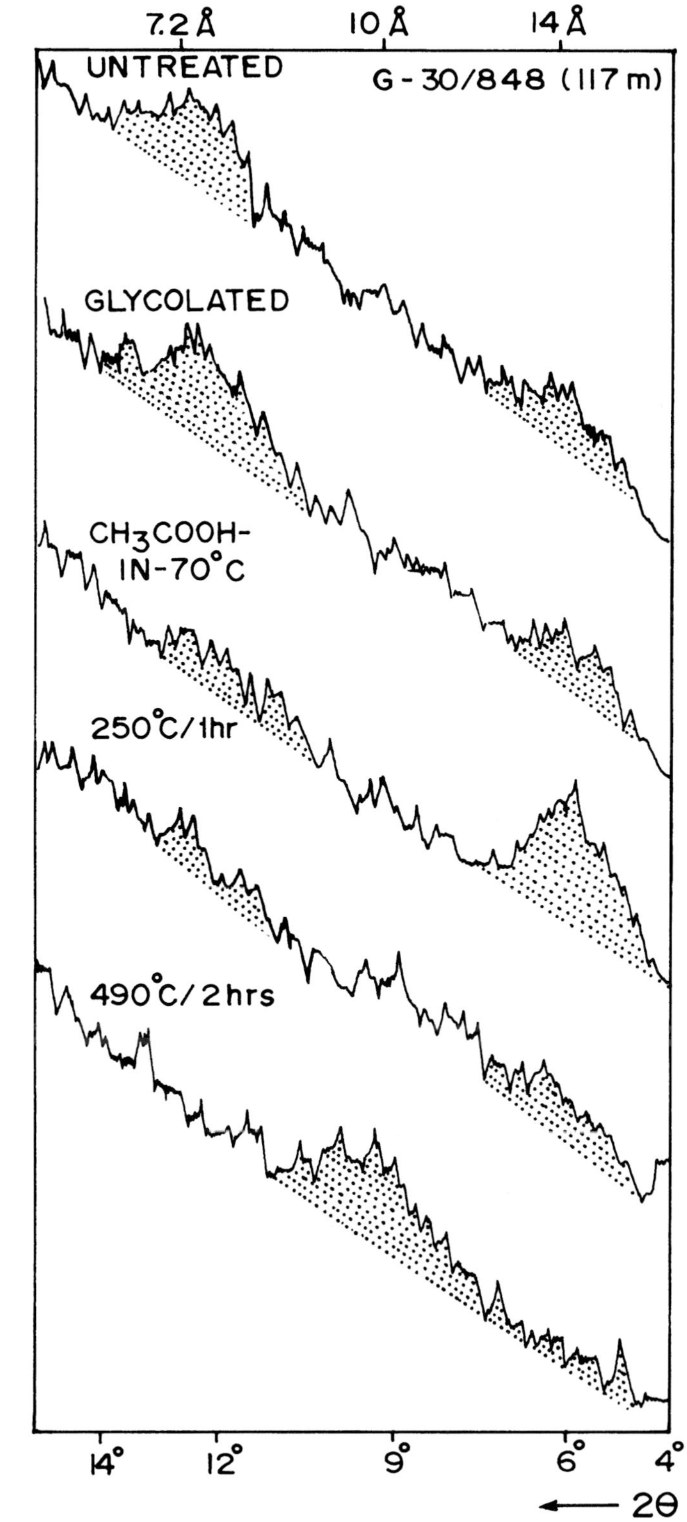

FIG. 6.—X-ray diffraction patterns of dark green grains from the continental slope off Kerala.

authigenic and detrital clay minerals and were intensely altered. Both phyllite C- and suspected phyllite V-associated green grains were present in the shelf sediments; samples showing phyllite C were located off river mouths and at the transition zone between the clastic inner shelf and relict outer shelf sediments; the green grains associated with relict sands and in places with reefs of the outer shelf most probably contain phyllite V. Brownish ochre grains contain abundant goethite and authigenic clay, which was difficult to identify at present. The associated sediment constituents (abundant carbonate fragments, quartz, and heavy minerals) (Table 1) indicate a high-energy oxidizing environment, which was not conducive to the formation of authigenic green clays (Odin, 1988). This implies that the initially formed metastable authigenic clay (maybe phyllite V) probably recrystallized to goethite under high-energy conditions. Although the authigenic clay in the green grains on the continental slope was altered, it is most likely phyllite V down to a depth of 235 m, followed by phyllite C at 280 m water depth. Glauconitic smectite of glaucony facies was

TABLE 2.—DISTRIBUTION OF MAJOR ELEMENTS (%) AND TOTAL REEs (ppm) IN GREEN GRAINS FROM THE STUDY AREA.

Sample No.	Depth (m)	Pred. Authi. Mineral	SiO_2	Al_2O_3	Fe_2O_3	MgO	K_2O	Na_2O	MnO	CaO	ΣREE
G-71/ 1441	37	Phyllite C	32.2	7.6	14.4	11.0	0.6	0.5	0.08	7.3	110.0
G-71/ 1443	38	Phyliite C	37.0	8.8	14.3	10.8	0.9	0.8	0.06	4.1	113.0
G-71/ 1432	41	Phyllite C	39.19	7.7	17.8	11.4	0.6	0.9	0.04	2.8	142.6
G-30/ 844	42	Phyllite V*	39.5	8.9	23.5	12.7	1.1	0.5	0.03	2.2	145.6
G-167/ 3914	49	Goethite, Phyllite V*	27.4	8.9	17.3	10.4	0.5	0.3	0.13	4.1	111.0
NR-2A/ 9	70	Phyllite V*	38.0	9.0	21.8	10.9	0.7	0.3	0.14	3.9	51.2
G-30/ 848	117	Phyllite V*	nd	9.3	22.6	12.1	1.0	0.4	0.16	2.7	111.6
G-30/ 791	150	Phyllite V*	30.0	4.8	14.7	7.6	0.5	0.6	0.04	6.9	47.6
G-71/ 1436	170	Phyllite V	24.9	6.6	15.0	7.9	0.4	0.5	0.04	13.5	75.8
G-71/ 1452	190	Phyllite V	nd	7.9	18.6	8.2	0.6	0.4	0.08	4.8	70.9
G-71/ 1427	235	Phyllite V	27.6	7.1	16.4	7. 1	0.6	0.4	0.02	12.9	54.2
G-71/ 1401	280	Phyllite C	36.1	8.9	21.9	8.0	2.0	0.3	0.01	4.5	45.4
G-71/ 1402	330	Glauconitic smectite	22.1	5.3	23.0	3. 1	2.0	0.4	0.01	15.5	77.3
French Guiana+		Phyllite V	36.9-39.1	10.8-12.0	17.9-19.5	8.3-11.0	1.1-1.4	0.2-0.3	na	0. 3-0. 7	na
New Caledonia+		Phyllite V	31.2, 34.0	5.2, 6.1	22.0, 23.1	10.8, 13.7	0.18, 0.16	0.02	0.02, 0.03	0.7, 1.1	na
Senegal Shelf +		Phyllite V	32.6, 34.8	5.9, 9.3	21.7, 25.1	12.0, 13.2	0.5, 0.17	0.2	0. 02	0. 4	na
GL - $O^@$			52. 13	7. 47	20. 73	4. 39	8.48	0.14	0.005	0.98	na
GL - O^S			50. 90	7. 55	19. 40	4. 45	7.95	0.05	na	0.95	na

+ Odin (1988); nd–not determined; na–data not available; * suspected phyllite V
GL-$O^@$–Glauconite standard values obtained in the present study; GL-O^S–certified values (Odin and Matter, 1981).
Note that these green grains are a mixture of predominant authigenic clay, detrital clay, and some altered products.

present at 330 m water depth.

Geochemistry

Major elements.—

As mentioned above, the green grains studied here are a mixture of predominant authigenic green clay and detrital clay minerals and altered. Table 2 shows how much the composition of the green grains of the western margin of India deviates from that of green grains reported elsewhere containing pure authigenic green clay.

SiO_2, Al_2O_3 , and MgO contents of the verdine-associated green grains vary widely and range from 24.9 to 39.5%, 4.8% to 9.3%, and 7.6% to 12.7%, respectively (Table 2). Higher percentages of these elements occur in green grains from the shallow shelf and their concentrations decreased with increasing distance from the coast (see Table 2). In general, silica content was higher in phyllite C-associated green grains than in those of phyllite V. Al_2O_3 contents of the verdine-associated green grains from the western margin of India were lower (~7–9%) than Guianese verdine (Al_2O_3 ~11–12%), but slightly higher than those of New Caledonia and Senegal deposits (~6%; Odin, 1988). The silica (22.1%) content of the glaucony sample (G-71/1403) at 330 m water depth was extremely low compared to any other glaucony occurrences in the world oceans (SiO_2 ~40–50%; Odin and Matter, 1981; Odin, 1988). Glaucony developed mainly within microfossil tests. The very high Ca (11.1 %, mainly as carbonate) and high moisture content of the sample may be responsible for low SiO_2 values.

Total Fe is reported here as Fe_2O_3. Because the synsedimen-

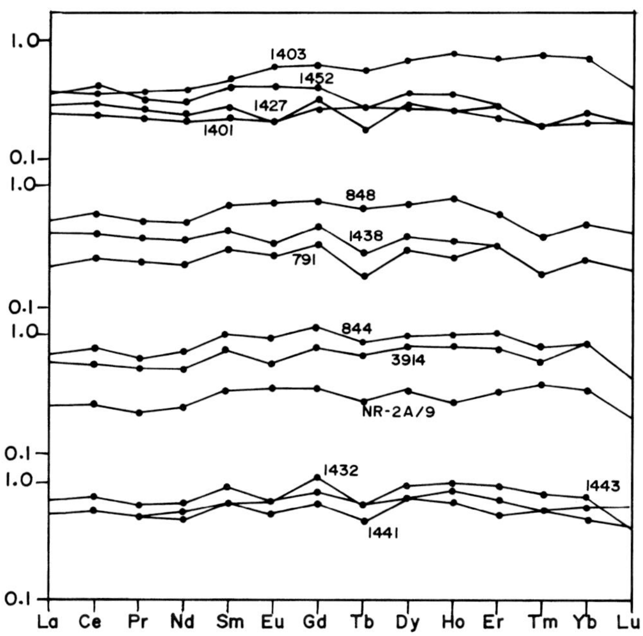

FIG. 7.—Shale-normalized REE patterns of the green grains

tary iron-rich clay minerals are more likely to be ferric than ferrous (Odin, 1988), the reported values may approximate the total iron. Fe_2O_3 varied from 14.3% to 23.6%. The green grains off the river mouths contain more Fe. Iron concentrations do not vary with water depth (Table 2). Calculations based on aluminium-normalized [$(Fe/Al)_{SAMPLE}/(Fe/Al)_{SHALE}$ ~ 8.0; Murray et al., 1992] distribution indicate that the green grains contain "excess Fe" that is not associated with the aluminosilicate matrices.

K_2O concentrations in verdine-associated green grains range from 0.38% to 1.98% and are comparable to those reported for verdine from elsewhere. The low K_2O (~2.0%) content in the glaucony sample suggests that the authigenic clay mineral is in an early stage of formation (Odin, 1988), and mixture of authigenic minerals, which is in agreement with the mineralogy. Na_2O contents range from 0.27% to 0.91% and are uniformly higher than those reported from other regions.

CaO content varies from 2.2% to 15.5%, being highest in grains associated with the tests of microfossils and lowest in free green grains. This indicates that the CaO analyzed here may not be a part of the green mineral, but reflects the type of substrate in which the authigenic clay minerals formed and thus is considered an impurity (Odin, 1988).

The interelement relationships indicate that Si showed strong positive correlations with Al (r = 0.74) and Mg (r = 0.80). Fe and K showed strong correlation with eachother (r = 0.65), but did not show significant correlations with Si, Al, Mg, and Mn. Ca showed negative correlations with Si (r = -0.86), Al (r = -0.73), Mg (r = -0.85), Mn (r = -0.5), and REEs (r = -0.5). It thus appears that elements such as Si, A!, Mg, and Mn were contributed by the initial substrate. Most of the K and Fe may have derived from the water column. Ca is bio-detrital and indicate the type of substrate.

Rare earth elements.—

Total REE content (ΣREE) in verdine grains varied from 45 ppm to 146 ppm. These values were higher in the shallow shelf region and decreased away from the coast. The ΣREE contents were lower than they were in average shale (NASC- Piper,1974) and associated detrital clays but are comparable with that of glaucony (glauconite) (Fleet et al., 1980). Shale-normalized REE plots of the samples were relatively flat (Fig. 7). The LREE to HREE enrichment, as calculated by the La_n/Yb_n ratio (Murray et al., 1992), varied widely between 0.9 and 2.0 for the verdine-associated grains. This ratio was relatively high (La_n/Yb_n ~1-2.0) in the green grains that occur closer to the land, and associated with clayey sediments of the continental slope, but lower (0.88-1.0) in some green grains that are associated with the abundant microfossils of the outer shelf/shelf break. A HREE enrichment (La_n/Yb_n = 0.6) was observed in glauconitic smectite sample from the terrace wherein the green pigment was associated with microfossil tests.

The low ΣREE concentrations and flat shale-normalized REE patterns in verdine-associated green grains indicate that they retain the REE pattern of weathered aluminosilicate rocks. Verdine minerals, unlike other authigenic minerals (Fe-Mn nodules, barite, authigenic montmorillonite, phillipsite, and phosphorites), form in shallow marine environments where high concentrations of dissolved/particulate elements are available and relatively high rates of sedimentation prevail. It is apparent that under these conditions the REEs are not preferentially scavenged during authigenic clay mineral formation and thus do not act as a "sink" for REEs.

The low REE contents compared to the NASC shale and associated detrital muds and moderate correlation of REEs with Mg (r = 0.65) implies that the sources for REEs are the detrital particles associated with the initial substrate and REE concentrations are diluted by the addition of REE-depleted authigenic clay. "Excess Fe" indicates its source other than aluminosilicates, most probably the dissolved iron transported from the tropical weathering of laterites. No correlation of REEs with Fe indicates two different sources. Iron in the nearshore mostly occurs as Fe^{3+}, which is not a stable crystalline form and is readily available for the formation of synsedimentary minerals. Since verdine formation is a relatively fast process, Fe^{3+} must have been scavenged rapidly to form iron-rich clay minerals. Fleet (1984) suggested that a substantial portion of the REEs can be loosely held onto clays and therefore are available to take part in exchange reactions. During authigenic green clay mineral formation and recrystallization of the host sediment (initial substrate) some REEs may have been lost to sea water, a likely factor contributing to lower REE contents in the verdine grains.

DISCUSSION

Relationship of Verdine to Fluvial Influx and Coastal Geology

Iron is a crucial element for verdine formation and is supplied to the sea by streams as detrital iron (particulate and dissolved) and by exhalations primarily from spreading ridges and island arcs. Earlier studies demonstrated that the size of the verdine deposit generally reflects the amount of the river runoff (Odin, 1988). Higher percentages of green grains on the shelf off river mouths (Fig. 1) and their higher Fe_2O_3 contents indicate that iron is largely contributed through rivers. Abundant green grains in the southern part of the study area, especially on the Kerala continental shelf and slope, may be due to the large number (44) of streams cutting across the coastal belt and flow to the sea through numerous lagoons and backwaters. The total annual discharge of the major and minor rivers in the study area is only about 95.6 km^3 (Rao, 1979), but verdine covers an area of about 100,000 km^2. In contrast, the Amazon, the second largest river in the world, discharges about 3,187.5 km^3 of water (Lisitzin, 1972) into the Atlantic and the verdine-bearing basin off the river mouth is only about 100,000 km^2 (Odin, 1988). This implies that the verdine basin off the west coast of India represents an exceptionally large basin when compared to the annual (low) fluvial input. Thus, the suggestion that the size of the verdine deposit is related to the amount of river discharge (Odin, 1988) is not valid. This also means that the amount of iron released from continental waters is more important than the volume of water discharged by the rivers, and hinterland geology plays a crucial role. The intense chemical weathering of the extensive lateritic cappings of Western Ghat rocks and Warkala and Quilon beds with ferruginous sandstones in the coastal region (Krishnan, 1968) may have contributed abundant iron which in turn was utilized for the formation of authigenic green clay minerals.

Nature of Verdine Grains and its Relationship to Sediment Type

The inner shelf sediments are clayey with <10% coarse fraction, and high organic carbon and clay accumulation rates are expected to be higher. Absence of green grains in these sediments may be due to the lack of suitable substrate, which provides necessary microenvironments for authigenic green clay formation. High rates of sedimentation may also have inhibited the growth of authigenic clay.

The presence of phyllite C closer to the land and at the transition between the inner and outer shelf suggests that the phyllite C forms where there is some contribution of fine detrital material. Manjunatha and Shankar (1992) reported minor sediment contribution at about 40–50 m water depth off Karnataka, as evidenced by the clay accumulation rates that vary from 0.56 to 0.89 mm/year. Odin (1988) recorded that phyllite C is an estuarine and nearshore facies and occurs in areas of fine-sediment contribution. The green grains that occur on the outer shelf are associated with relict detrital sands and/or reefs and contain suspected phyllite V. The reefs generally grow at shallow depths in association with warmer and clear waters. The presence of reefs at 75 m water depth and the radiocarbon dates of the corals and algal limestones (8.3 ka B.P.) (Rao and Lamboy, 1996) indicate their submerged and relict nature. The green grains associated with the reefs may have formed more or less at the same time

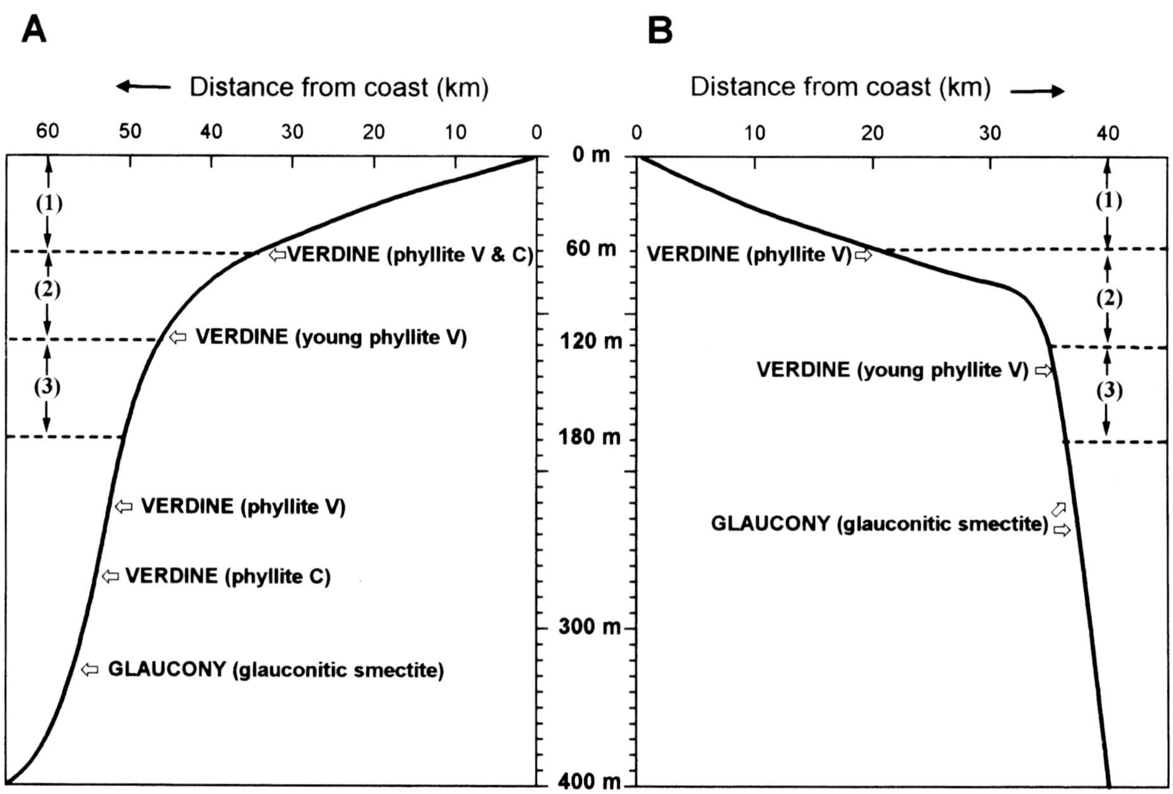

FIG. 8.—Distribution of verdine and glaucony facies from the western continental margin (off south of Cochin) (A), and eastern continental margin (off Kalingapatnam) (B), of India. Zone (1)—Depth range (0–60 m) for the formation of present day verdine; zone (2)—depth range (60–120 m) for verdine formed during the Holocene transgression; zone (3)—depth range (120–180 m) for verdine formed during the last glacial maximum (LGM).

and thus are considered relict and relate to the low sea level stands during the Holocene. Phyllite V-associated green grains occur on the continental slope at depths between 100 m and 235 m in clayey silt sediments. These grains probably formed during the Last Glacial Maximum and/or during the subsequent sea-level transgression. Phyllite C also occurs at 280 m depth in our study area. The different initial substrates of phyllite C for the shelf and slope occurrences rules out lateral transport. It is difficult to explain the formation of phyllite C at this depth (280 m) unless tectonic activity and subsidence of the terrace is involved (see below). The mineralogical properties of green grains on the shelf are more distinct in the northern part of the study area (sample nos. NR-2A/52 &556; Figs. 3 and 4A) than those in the southern part (626 & 3914; Figs. 4B and 5) and the green grains closer to the shelf edge (848; Fig. 6), indicating that the latter are more altered, probably due to high-energy conditions subsequent to their formation.

Bailey (1988) and Odin (1988) interpreted that the color of verdine grains is related to its mineralogy and morphological evolution. Verdine grains on the shelf are dark green and mostly occur as irregular grains. Both phyllite C and phyllite V are associated with them (Table 1). Brown glossy pellets contain poorly crystalline verdine mineral. On the other hand, young phyllite V (odinite) is invariably associated with light green rugose and dark green shiny pellets and dark green infillings on the continental slope. These results thus do not support Bailey (1988) and Odin (1988), but confirm our earlier findings (Rao et al., 1993; Rao et al., 1995) that color and morphology of the grains do not reflect the type of authigenic clay mineral or its evolution. The better crystallized phyllite V-associated samples with microfossil tests and the poorly crystallized verdine with abundant goethite in brown-colored pellets indicate preservation of authigenic clay in the former and its alteration to neoformed minerals in the latter. It is therefore suggested that the processes leading to differences in color may be very complex and not necessarily be related to mineralogy.

Distribution of Verdine and Glaucony Facies on the East and West Coasts of India

Although the samples consisting of green grains are from water depths between 37 m and 330 m, they were recovered at depths <150 m on the continental margin of Maharashtra, Goa, and Karnataka. It has been sufficiently documented from the present study and from Rao et al. (1993) that the mineralogy of the green grains up to 150 m is similar throughout the study area. As our studies on the Kerala margin extend up to 330 m depth and the green grains represent both verdine and glaucony facies minerals, the distribution of these facies on this margin is compared with that of the east coast of India (Fig. 8). According to Odin (1988), phyllite V and phyllite C are two expressions of the verdine facies that occurs at depths <60 m in recent marine sediments. However, verdine occurs at depths between 40 m and 280 m followed by glaucony at 330 m on the Kerala continental shelf and slope. It has been suggested that verdine and glaucony formation on this margin is diachronous (Rao et al., 1993), slope facies formed during the Last Glacial Maxima (LGM) when the sea-level was at 120 m below the present level, and the shelf facies formed during the Holocene transgression. So the expected depth ranges for verdine and glaucony facies on the slope during the LGM are 120-180 m and >180 m, respectively. However, on the continental slope of Kerala, verdine occurs down to a depth of 280 m. The verdine mineral at 100–235 m (phyllite V) and at 280 m (phyllite C) is different and rules out the possibility of lateral transport on the slope.

On the other hand, verdine facies (only phyllite V) occurs at depths between 18 m and 170 m, followed by glaucony down to 247 m on the east coast of India (Rao et al., 1995) (Fig. 8). The change from verdine to glaucony deeper than 170 m on this continental slope is well in agreement with the suggested depth ranges of these facies during the LGM. Moreover, their distribution is similar to that of the Senegal and French Guiana margins (Odin, 1988).

The distribution of verdine and glaucony facies on the Kerala continental shelf and slope therefore appears anomalous as it does not coincide with the suggested depth range for these facies. This suggests that the paleogeography of the southwest coast of India was different during the Late Quaternary, pointing to the subsidence of the terrace. Rao et al. (1996) reported evidence of Late Quaternary neotectonic activity and subsidence along the western continental margin of India.

ACKNOWLEDGMENTS

We thank E. Desa, Director, and R. R. Nair, Former Head, Geological Oceanography Division of the National Institute of Oceanography, for their encouragement. We are grateful to S. Krishnaswami, M. M. Sarin, and S. K. Singh, Physical Research Laboratory, Ahmedabad, and V. Balaram, National Geophysical Research Institute, Hyderabad, for extending analytical facilities. J. N. Pattan helped at various stages during analytical work and in discussions and B. N. Nath helped in discussions. G. Prabhu assisted in XRD analyses of the samples. Thamban thanks CSIR, New Delhi, for granting a Research Fellowship (CSIR-NET). This is NIO contribution No. 2556.

REFERENCES

Bailey, S. W., 1988, Odinite: A new dioctahedral-trioctahedral Fe-rich 1:1 clay mineral: Clay Minerals, v. 23, p. 237-247.

Fleet, A. J., 1984, Aqueous and sedimentary geochemistry of the rare earth elements, *in* Henderson, P., ed., Rare Earth Element Geochemistry (Developments in Geochemistry, 2): Amsterdam, Elsevier, p. 343-369.

Fleet, A. J., Buckley, H. A., and Johnson, L. R., 1980, The rare earth element geochemistry of glaucony and celadonites: Journal of Geological Society of London, v. 137, p. 683-688.

Hashimi, N. H., Nair, R. R., Kidwai, R. M., and Rao, V. P., 1982, Carbonate mineralogy and faunal relationship in tropical shallow water marine sediments: Sedimentary Geology, v. 32, p. 89-98.

Krishnan, M. S., 1968, Geology of India and Burma: Madras, Higginbothoms Limited, 536 p.

Lisitzin, A. P., 1972, Sedimentation in the world oceans: Tulsa, Society for Sedimentary Geology (SEPM) Special Publication 17, 218 p.

Mallik, T. K., Venkatesh, K. V., Sengupta, R., Rao, B. R., and Ramamurty, M., 1976, Heavy mineral distribution patterns in the northern part of the

Arabian Sea, Western continental shelf of India: Indian Journal of Earth Science, v. 3, p. 178-187.

MANJUNATHA, B. R., AND SHANKAR, R., 1992, A note on the factors controlling the sedimentation rate along the western continental shelf of India: Marine Geology, v. 104, p. 219-224.

MURRAY, R. W., BUCHHOLTZ-TEN BRINK, M. R., GERLACH, D. C., PRICE RUSS III, G., AND JONES, D. L., 1992, Interoceanic variation in the rare earth, major and trace element depositional chemistry of chert: Perspectives gained from the DSDP and ODP records: Geochimica et Cosmochimica Acta, v. 56, p. 1897-1913.

NAIR, R. R., AND PYLEE, A., 1968, Size distribution and carbonate content of the sediments of the western shelf of India: Bulletin of National Institute of Sciences India, v. 38, p. 411-420.

ODIN, G. S., 1985, La verdine, faciès granulaire vert, marin et côtier, distinct de la glauconie: distribution actuelle et composition: Comptes Rendus de l' Académie des Sciences de Paris, v. 301, p. 105-108.

ODIN, G. S., 1988, Green Marine Clays (Developments in Sedimentology, 45): Amsterdam, Elsevier, 445 p.

ODIN, G. S., 1990, Clay mineral formation at the continent-ocean boundary: The verdine facies: Clay Minerals, v. 25, p. 477-483.

ODIN, G. S., and Matter, 1981, De glauconiarum origine: Sedimentology, v. 28, p. 611- 641.

ODIN, G. S., AMOURIC, M., BAILEY, S. W., AND MACKINNON, D. R., 1989, Minéralogie du faciès verdine: Comptes Rendus de l' Académie des Sciences de Paris, v. 308, p. 395-400.

PIPER, D. Z., 1974, Rare earth elements in the sedimentary cycle: a summary: Chemical Geology, v. 17, p. 287-297.

RAO, K. L., 1979, India's water wealth: Its assessment, uses and projections (2nd ed.): New Delhi, Orient Longman Limited, 267 p.

RAO, P. S., 1989, Sonograph patterns of the central western continental shelf of India: Journal of Coastal Research, v. 9, 85-90.

RAO, V. P., AND LAMBOY, M. 1996, Genesis of apatite in the phosphatized limestones of the western continental shelf of India: Marine Geology, v. 136, p. 41-53.

RAO, V. P., LAMBOY, M., AND DUPEUBLE, P. A., 1993, Verdine and other associated authigenic (glaucony, phosphate) facies from the surficial sediments of the southwestern continental margin of India: Marine Geology, v. 111, p. 133-158.

RAO, V. P., AND RAO, B. R. 1995. Provenance and distribution of clay minerals in the sediments of the western continental shelf and slope of India: Continental Shelf Research, v. 15, p. 1757-1771.

RAO, V. P., THAMBAN, M., AND LAMBOY, M., 1995, Verdine and glaucony facies from surficial sediments of the eastern continental margin India: Marine Geology, v. 127, p. 105-113.

RAO, V. P., VEERAYYA, M., THAMBAN, M., AND WAGLE, B. G., 1996, Evidences of Late Quaternary neotectonic activity and sealevel changes along the western continental margin of India: Current Science (India), v. 71, p. 213-219.

RAO, V. P., AND WAGLE, B. G. 1997, Geomorphology and surficial geology of the western continental shelf and slope of India: A Review: Current Science (India), v. 73, p. 330-350.

SHAPERO, L., AND BRANNOCK, W. W., 1962, Rapid analysis of silicate, carbonate and phosphate rocks: United States Geological Survey Bulletin, v. 1144, p. A21-A 22.

THRIVIKRAMJI, K. P., AND MACHADO, T., 1982, Autochthonous glauconite in the sediments off Trivandrum coast, Kerala: Indian Journal of Marine Sciences, v. 11, p. 336-337.

DIAGENETIC EVOLUTION OF SEAMOUNT PHOSPHORITE

LAURA M. BENNINGER*
Geology Department, University of California, Davis, California, 95616, USA
AND
JAMES R. HEIN**
U.S. Geological Survey, MS 999, 345 Middlefield Rd., Menlo Park, California, 94025, USA

ABSTRACT: Phosphorite, limestone, basalt, and breccia/conglomerate occur on the flanks of central Pacific Cretaceous seamounts. Phosphatization occurred during the Tertiary, at or immediately below the sediment-water interface, within thin bodies of porous carbonate sediment and within all rock types that occur on the seamounts. The phosphorites are composed of carbonate fluorapatite (CFA), which lines pores and replaces carbonate and silicate protoliths; protolith fabric is locally preserved in fine detail. Phosphatization is typically most pervasive adjacent to pores, fractures, and the outer surfaces of samples.

In order of decreasing abundance, diagenetic minerals associated with seamount phosphorites (SMP) include: CFA, Fe-Mn oxyhydroxides, palagonite, smectite, phillipsite, and barite. This mineral assemblage indicates that CFA formed under conditions consistent with low-temperature, open-circulation seafloor alteration of basalt. The common association between CFA and precipitation of Fe-Mn oxyhydroxides indicates that SMP formed under oxic to suboxic conditions.

Phosphatization is characterized by mineralization fronts; the angular superposition of those fronts (successive mineralizing fluids coming from different directions) locally leads to irregularly shaped domains that differ in pervasiveness of phosphatization, degree of fabric preservation, and abundance of secondary minerals. Phosphatization is, therefore, controlled by a variety of factors including protolith mineralogy and fabric, degree of pore-fluid saturation with respect to CFA, rate of pore-fluid/seawater circulation (rock-water ratio), and number of previous episodes of phosphatization.

INTRODUCTION

Despite the widespread occurrence of phosphorite on seamounts (Baturin, 1982; Cullen and Burnett, 1986; Burnett et al., 1987; Bersenev et al., 1990; Hein et al., 1993), little is known about the distribution, petrology, or genesis of these unique deposits. Study of sedimentary marine phosphorite has traditionally focused on large, economically important deposits such as the Permian Phosphoria Formation, and on contemporary continental margin deposits such as those forming off the Peru-Chile margin. As a result, current phosphogenic models are based on the study of organically rich deposits that formed in highly productive environments of intense and continuous upwelling. In contrast, seamount phosphorites (SMP) occur in moderate- to low-productivity, organically poor settings (Hein et al., 1993; mean organic carbon content (TOC) is 0.12%); consequently, it is unlikely that upwelling-based models fully explain seamount phosphogenesis.

Recent discoveries of phosphatic nodules and hardgrounds from organic-poor settings, such as those along the east Australian margin, have led to the recognition that phosphogenesis within low-productivity environments may be more widespread than previously thought (O'Brien and Heggie, 1988; O'Brien et al., 1990). Studies of those deposits suggest that bacterial mediation and redox cycling of Fe-Mn oxyhydroxides may be crucial mechanisms for concentrating phosphorus during formation of phosphorite in nonupwelling zones (Jarvis, 1992; Schaffer, 1986; Lucas and Prévôt, 1985). Through the study of mid-Pacific SMP, we evaluate the broad applicability of these mechanisms to phosphogenesis within moderate- to low-productivity environments.

SMP predominantly formed during two time periods in the Cenozoic, one spanning the Eocene-Oligocene boundary and the other spanning the Oligocene-Miocene boundary (Hein et al., 1993; J. R. Hein and L. H. Chan, unpublished data, 1997). Other, however less extensive, episodes of phosphogenesis also occurred at various times during the Cenozoic. Compositional and paragenetic analyses preclude a hydrothermal origin for these phosphatized (carbonate and volcanic) rocks and sediments. Additionally they indicate that phosphatization occurred under fluctuating oxic/suboxic conditions (Hein et al., 1993). These results suggest that even though the depositional settings of seamount phosphorites contrast with phosphorites from other low-productivity settings, the geochemical constraints on their formation are similar.

In this paper the term seamount is used to describe submarine highs that include seamounts, guyots, ridges, and edifices that support atolls. Phosphorite is used as a general term and is

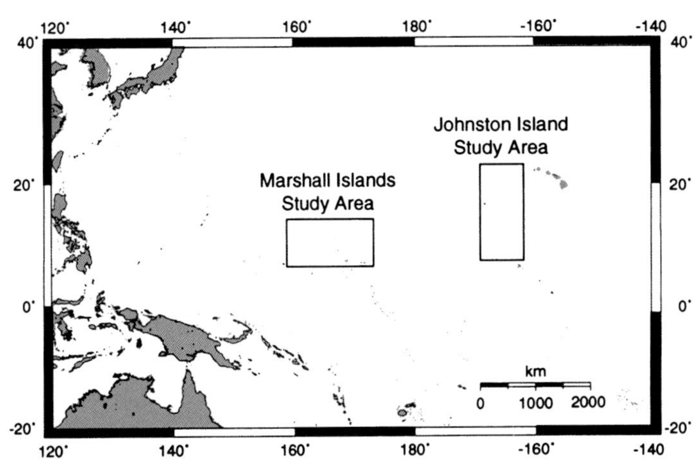

FIG. 1.—Locations of Johnston Island and Marshall Islands regions, central Pacific.

*email: benninger@geology.ucdavis.edu
**email: jhein@octopus.wr.usgs.gov

Marine Authigenesis: From Global to Microbial, SEPM Special Publication No. 66

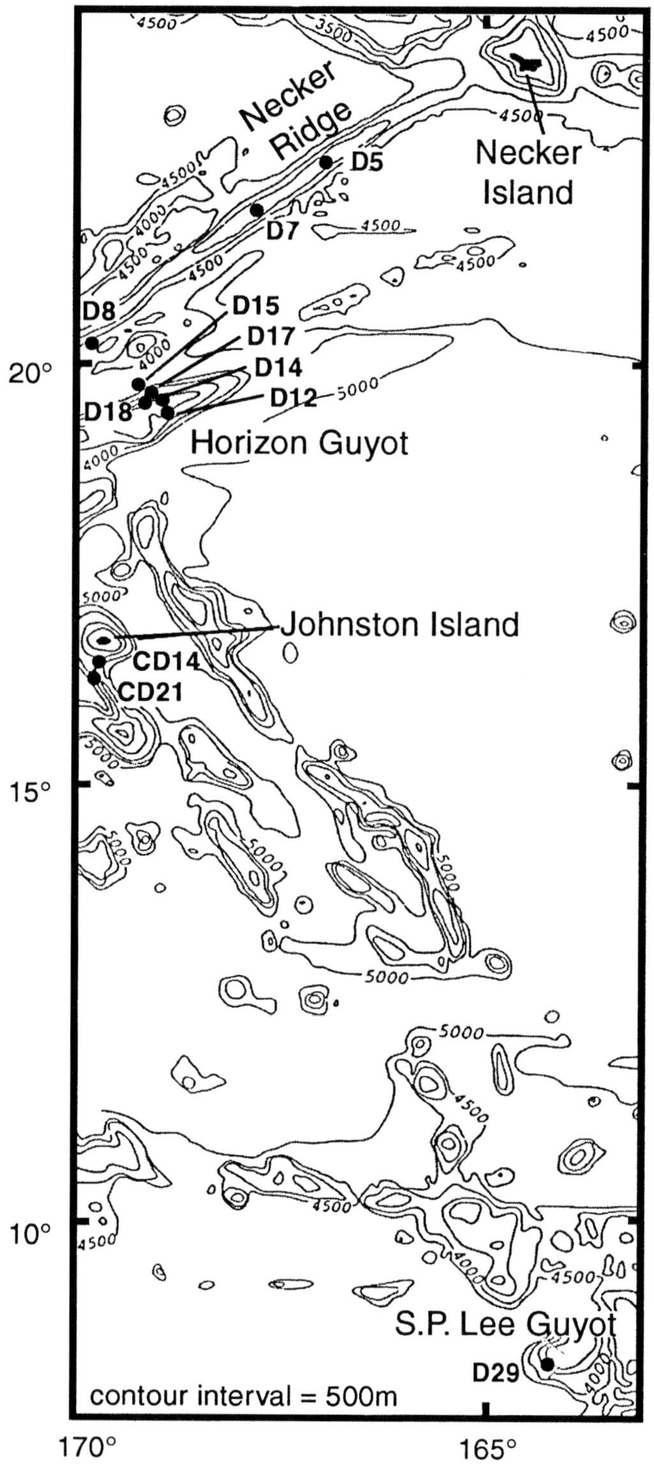

FIG. 2.—Bathymetry of the Johnston Island region showing locations of dredge stations (filled circles).

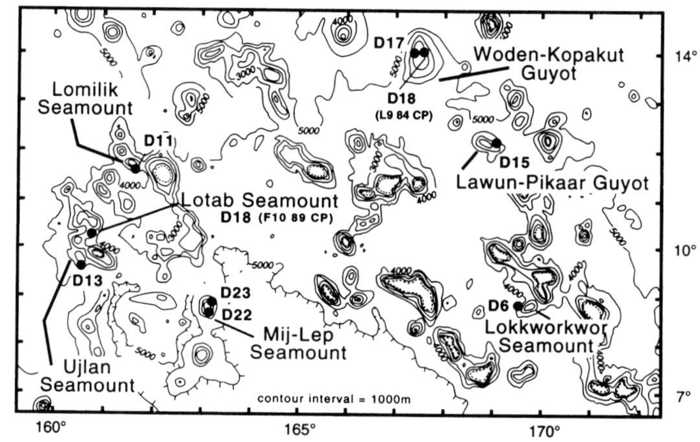

FIG. 3.—Bathymetry of the Marshall Islands region showing locations of dredge stations (filled circles).

not meant to imply an economic value for these deposits. The terms replacement and alteration are used interchangeably to describe dissolution of one mineral and precipitation of another in its place. These terms are not meant to imply replacement or alteration through molecular or ionic substitution.

METHODS AND SAMPLES

This study is based on thin section examination of 28 phosphorite samples dredged from 10 equatorial Pacific seamounts (Fig. 1; Table 1). Seventeen samples are from the Johnston Island region (Fig. 2) and 11 are from the Marshall Islands region (Fig. 3). This suite of samples was selected as representative of the hundreds of Pacific seamount phosphorite samples that were collected by U.S. Geological Survey chain-bag dredge operations over an 8-year period (1983-1991). Samples from the Johnston Island region were collected from water depths between 1750 m and 4800 m and those from the Marshall Islands region were taken from depths between 1000 m and 2900 m (Table 1). Water depths of most samples are accurate to within 100 m because of short dredge intervals. Generally, prior to dredging, single-channel seismic profiles, 10 kHz bathymetry, and 3.5 kHz sub-bottom profiles were taken; during dredging, the latter two systems were in operation. Shipboard navigation used the Global Positioning System (GPS) and Rhorho, a software program that allows direct ranging on Loran C.

Mineralogy of powdered samples was determined using an X-ray diffractometer (XRD) with CuKα radiation and curved-crystal carbon monochromator. Petrographic analyses of doubly polished thin sections (0.3 mm thick) were conducted under transmitted and reflected light. Selected thin sections were subsequently carbon coated to 200Å and then analyzed with a scanning electron microscope (SEM) with back-scattered electron (BSE) capabilities and an X-Ray analyzer (EDS).

DEPOSITIONAL SETTING

SMP occur on seamounts in the Indian, Atlantic, and Pacific Oceans (Baturin, 1982), and are apparently most abundant within equatorial Pacific regions. A compilation of SMP occurrences reported in the literature indicates that deposits range in latitude from 26°N to 33°S, while most occur at latitudes between 8°N and 26°N (Benninger, 1998).

Nearly all equatorial Pacific seamounts host at least minor occurrences of phosphorite. Phosphorites were recovered from about 90% of the approximately 100 equatorial Pacific

TABLE 1.—SAMPLE DESCRIPTIONS AND LOCATIONS

Sample	Edifice	Latitude (on & off bottom)	Longitude (on & off bottom)	Depth (m)	General Thin Section Description
L5 83 HW					
D5-A3-1	Necker Ridge	22° 18.98′ 22° 18.30′	166° 53.93′ 166° 54.12′	2350–2100	Phosphatized, poorly sorted, breccia-conglomerate with crust of Fe-Mn oxyhydroxides.
D5-A3-10	Necker Ridge	22° 18.98′ 22° 18.30′	166° 53.93′ 166° 54.12′	2350–2100	Phosphatized, poorly sorted, breccia-conglomerate cemented by phosphatized foraminiferal carbonate.
D5-A3-11	Necker Ridge	22° 18.98′ 22° 18.30′	166° 53.93′ 166° 54.12′	2350–2100	Phosphatized volcaniclastic breccia encrusted by Fe-Mn oxyhydroxides.
D5-A3-12	Necker Ridge	22° 18.98′ 22° 18.30′	166° 53.93′ 166° 54.12′	2350–2100	Intensely phosphatized volcaniclastic breccia overlain by ~1.5 cm of phosphatized limestone.
D5-A3-15	Necker Ridge	22° 18.98′ 22° 18.30′	166° 53.93′ 166° 54.12′	2350–2100	Moderately altered vesicular basalt encrusted by Fe-Mn oxyhydroxides. Vesicles in zeolitized and phosphatized basalt are partly to completely filled by carbonate fluorapatite (CFA).
D5-A3-18	Necker Ridge	22° 18.98′ 22° 18.30′	166° 53.93′ 166° 54.12′	2350 - 2100	Phosphatized and brecciated glassy basalt encrusted by Fe-Mn oxyhydroxides. Fe-Mn oxyhydroxide crust supports pebbles of phosphatized basalt and phosphatized breccia-conglomerate.
D7-A1	Necker Ridge	21° 47.89′ 21° 48.01′	167° 37.41′ 167° 36.92′	2100	Partly phosphatized and palagonitized brecciated glassy basalt with crust of Fe-Mn oxyhydroxides.
D8-A4	Necker Ridge	20° 12.57′ 20° 11.88′	169° 30.36′ 169° 31.15′	4800–4500	Fe-Mn oxyhydroxide nodule with a nucleus of phosphatized carbonate mudstone.
D12-A10	Horizon Guyot	19° 22.4′ (on bottom)	168° 41.69′ (on bottom)	2000	Phosphatized, poorly-sorted, glassy basalt breccia encrusted by ~0.75 mm of Fe-Mn oxyhydroxides.
D14-A2	Horizon Guyot	19° 30.19′ (on bottom)	168° 46.67′ (on bottom)	1800–1750	Intensely phosphatized volcaniclastic breccia with CFA matrix and crust of Fe-Mn oxyhydroxides.
D14-A6	Horizon Guyot	19° 30.19′ (on bottom)	168° 46.67′ (on bottom)	1800–1750	Palagonitized basalt with 1 cm thick crust of Fe-Mn oxyhydroxides. Thin (0.5 mm) layer of phosphatized foraminiferal carbonate separates Fe-Mn oxyhydroxides from basalt substrate.
D15-B1	Horizon Guyot	19° 41.64′ 19° 40.54′	168° 59.647′ 168° 59.02′	4500–4300	Fe-Mn oxyhydroxide encrusted phosphatized limestone containing phosphatic peloids, altered hyaloclastite fragments, rare phosphorite, fish fragments, and Fe-Mn oxyhydroxide crust fragments.
D17-A1	Horizon Guyot	19° 33.20′ (on bottom)	168° 49.94′ (on bottom)	2400	Non-vesicular basalt clast with thin palagonite rim overlain by layered phosphatized limestone and encrusted by porous Fe-Mn oxyhydroxides.
D18 B3-7	Horizon Guyot	19° 30.89′ 19° 29.69′	168° 51.28′ 168° 51.91′	1799–1780	Phosphatized non-vesicular basalt overlain by 2-3 mm of Fe-Mn oxyhydroxides and topped by partly phosphatized foraminiferal limestone.
D29 1A-A	S.P. Lee Guyot	8° 17.82′ 8° 18.69′	164° 22.11′ 164° 21.04′	1400	Phosphatized planktonic foraminiferal grainstone in which CFA lines and partly fills molds of sand- to silt-sized foraminiferal tests.
L9 84CP					
D6-2	Lokkwor-kwor	8°45.28′ (on bottom)	169° 49.53′ (on bottom)	2900	Intensely phosphatized breccia-conglomerate encrusted by Fe-Mn oxyhydroxides.
D6-3	Lokkwor-kwor	8° 45.28′ (on bottom)	169° 49.53′ (on bottom)	2900	Moderately phosphatized clastic breccia overlain by 1.5 cm of phosphatized foraminiferal wackestone and encrusted by Fe-Mn oxyhydroxides.
D6-15C	Lokkwor-kwor	8° 45.28′ (on bottom)	169° 49.53′ (on bottom)	2900	Partly phosphatized foraminiferal limestone encrusted by 2 mm of Fe-Mn oxyhydroxides.
D15-4	Lawun-Pikaar	12° 10.97′ 12° 12.71′	168° 59.74′ 168° 58.30′	1000–1300	Partly phosphatized foraminiferal wackestone with abundant disseminated Fe-Mn oxyhydroxides and encrusted by porous Fe-Mn oxyhydroxides.
D17-2	Woden-Kopakut	13° 54.28′ 13° 54.24′	167° 37.64′ 167° 38.25′	1600	Phosphatized massive foraminiferal limestone encrusted by Fe-Mn oxyhydroxides.
D18 3B-B	Woden-Kopakut	13° 53.98′ (on bottom)	167° 39.22′ (on bottom)	1600	Moderately altered basalt clast with common fractures infilled by phosphatized foraminiferal-rich carbonate and locally lined by smectite.

TABLE 1.—(CONTINUED)

Sample	Edifice	Latitude (on & off bottom)	Longitude (on & off bottom)	Depth (m)	General Thin Section Description
F10 89 CP					
D11-3E	Lomilik	11° 37.0′ 11° 39.0′	161° 41.1′ 161° 39.8′	1870–1465	Phosphatized volcaniclastic siltstone with interlaminations of volcaniclastic mudstone.
D13-3A (2)	Lotab	09° 37.2′ 09° 37.3′	160° 32.6′ 160° 33.0′	2130–2280	Phosphatized breccia-conglomerate containing phosphatized clasts encrusted by Fe-Mn oxyhydroxides. Within phosphatized carbonate matrix, CFA-rich and CFA-poor regions are separated by sharp, irregular boundaries that locally cross-cut foraminifera tests.
D18-1C	Lotab	10° 15.5’ 10° 15.15′	160° 40.4′ 160° 40.7′	1975–1940	Partly phosphatized foraminiferal wackestone.
D22-1A	Mij-Lep	08° 43.7′ 08° 43.6′	163° 11.4′ 163° 12.9′	1270–1245	Fe-Mn oxyhydroxide crust containing common voids, possibly burrows. Phosphatized carbonate lines and fills some voids.
D22-1C	Mij-Lep	08° 43.7′ 08° 43.6′	163° 11.4′ 163° 12.9′	1270–1245	Phosphatized hyaloclastite with sparse phosphatized pumice clasts and glassy basalt fragments.
D23-6	Mij-Lep	08° 52.0′ 08° 51.7′	163° 10.3′ 163° 10.6′	1405–1305	Phosphatized basalt breccia encrusted by Fe-Mn oxyhydroxides.
F7 86 HW					
CD 14-2B	S. Johnston Island Ridge	16° 26.8′ 16° 26.2′	169° 32.3′ 169° 33.6′	2530–2340	Phosphatized foraminiferal mudstone
CD 21-7-2	S. Johnston Island Ridge	16° 26.6′ 16° 27.2′	169° 31.7′ 169° 31.8′	2550–2370	Non-vesicular basalt encrusted by 3.5 cm of Fe-Mn oxyhydroxides.

seamounts that were dredged over a 10-year period, and the estimated overall mean phosphorus content is 2.3% P (5.2% P_2O_5) for those deposits (Hein et al., 1993). Some equatorial areas, including the Johnston Island and Marshall Islands areas, appear to be especially phosphorite rich, with from 5% to >25% CFA within 60–70% of all rocks dredged (Hein et al., 1993).

Dredge and drill core recoveries indicate that SMP occurs at, or immediately below, the sediment-water interface. Drill core from the Emperor Seamount Chain contains phosphatic grains and fracture fillings in the uppermost 2 m of core (Jackson et al., 1980) and core from Resolution Guyot, Mid Pacific Mountains, contains phosphorite within 15 cm of the seafloor (Sager et al., 1993). Moreover, while phosphorite was recovered many meters below the seafloor at both Allison Guyot in the Mid Pacific Mountains (Sager et al., 1993) and Wodejebato Guyot in the Marshall Islands (Premoli-Silva et al., 1993), those phosphorites co-occur with Fe-Mn oxyhydroxide crusts, implying formation within a few meters of the paleo-sediment-water interface.

SMP occurs preferentially on seamount flanks (Hein et al., 1990, 1993) and the deposits are not concentrated within any particular water depth interval. SMP is also found on seamount summits (Sager et al., 1993; Premoli-Silva et al., 1993) and redeposited SMP is found within turbidite and debris flows (Sager et al., 1993). Preferential occurrence of SMP on seamount flanks (rather than summits) may reflect phosphogenesis during times when seamount summits were emergent and therefore unavailable to phosphate-rich seawater. Alternatively, rapid sedimentation rates on seamount summits may have precluded phosphatization. In addition, SMP formation is likely associated with the oxygen minimum zone (Benninger, 1998), which generally intersects the seamount flanks rather than their summits.

In the Marshall and Johnston Islands areas, seamounts exhibit from 3 km to 5 km of relief and have moderately steep flanks that vary in slope from 9° to 27° (Kayen et al., 1989; Hein et al., 1990). Seamount flanks, which host most SMP deposits, are generally draped by volcaniclastic breccias and conglomeratic limestones. Typically, flanks are free of fine-grained sediment except for local sediment ponds. Seamount summits range from rugged, sediment-starved basaltic ridge tops to terraced platforms covered by up to 200 m of fine-grained carbonate sediment locally interbedded with volcaniclastic talus (Schwab et al., 1988; Hein et al., 1990).

SEAMOUNT PHOSPHORITE LITHOTYPES

SMP are classified here according to protolith type since phosphorites form through alteration of preexisting seamount rocks and sediments. In approximate order of decreasing abundance, CFA occurs as an authigenic pore-lining phase and/or diagenetic replacement within three main rock types: 1) volcaniclastic breccia-conglomerate, 2) limestone, and 3) basalt; these are described in detail below. To a lesser extent, CFA also

FIG. 4.—A) Volcaniclastic breccia-conglomerate containing partly to completely phosphatized lithic fragments within a matrix (light gray) of CFA-replaced, massive, foraminiferal limestone. Angular and subrounded clasts are very poorly sorted; grains range from cobble- (bottom of photo) to silt-sized. A thin (< 0.5 cm) Fe-Mn oxyhydroxide crust (black) covers the upper surface of the rock. *F10-89-CP* D19-3. Scale bar = 2 cm. B) Bedded phosphatic foraminiferal- and nannofossil-rich wackestone. In layers A and D, cryptocrystalline CFA (locally intergrown with dendritic Fe-Mn oxyhydroxides) supports abundant molds of foraminifera and nannofossil tests. Layer A is topped by 1 cm of Fe-Mn oxyhydroxides (black) with phosphatic interlayer (light gray). Within layers B and C, CFA matrix supports pebble- to silt-sized intraclasts of volcaniclastic breccia-conglomerate, partly to unaltered basalt and glassy basalt, Fe-Mn oxyhydroxide crust fragments, and phosphorite. Burrows are present in layer D and along contact between layers D and C. *L9-84-CP* D6-3. Scale bar = 2 cm. C) Massive, completely phosphatized foraminiferal grainstone. CFA cement surrounds sand-sized foraminifera molds and minor silt-sized nannofossil molds. Patina of Fe-Mn oxyhydroxides (black) encrusts outer surface of sample (left). *L5-83-HW* D29-A1-A. Scale bar = 2 cm. D) Massive phosphatized basalt (light gray) with common fractures and minor vugs filled by phosphatized detrital carbonate (white). Basalt groundmass is locally CFA-replaced. Basalt block is topped by 4-cm thick Fe-Mn oxyhydroxide crust (darker gray) that contains common vugs filled with geopetal CFA-replaced foraminiferal limestone (dark cream). *L9-84-CP* D18-3. Scale bar = 2 cm.

lines and/or replaces: 4) detrital carbonate within fractures and vugs, 5) detrital carbonate within Fe-Mn oxyhydroxide crusts or separating crusts from substrate, and 6) preexisting phosphorite. Rare CFA of possible hydrothermal origin is also suggested by rare veins lined by CFA, phillipsite, and barite. Hydrothermal SMP from the central Pacific were previously identified by oxygen isotope analyses (Hein et al., 1993).

Phosphatized Volcaniclastic Breccia-Conglomerate

Most SMP occurs within volcaniclastic breccia-conglomerate, and secondarily within volcaniclastic sandstone, volcaniclastic siltstone, and hyaloclastite. This group of phosphatized clastic rocks (hereafter referred to as volcaniclastic breccia-conglomerate) typically contains partly to completely phosphatized lithic fragments within a matrix of partly to completely CFA-replaced, massive, foraminiferal limestone (Fig. 4A). Angular to rounded clasts are commonly encrusted by Fe-Mn oxyhydroxide layers (patina to 3 cm) and are typically poorly sorted; clasts range up to cobble size. Commonly, large fragments are grain-supported and smaller clasts are matrix-supported.

Lithic fragments within phosphatized breccia-conglomerate largely consist of vesicular glassy basalt (containing minor plagioclase, pyroxene, and/or olivine microphenocrysts), and nonvesicular glassy basalt (containing abundant plagioclase micro-laths and minor pyroxene micro-phenocrysts). To a lesser extent, clasts also include preexisting phosphorite, limestone, glass shards, and Fe-Mn oxyhydroxide crust fragments. Minor broken plagioclase laths, fish debris, and ironstone are locally present. The abundance of preexisting phosphorite is poorly known because of the difficulty in distinguishing between preexisting phosphorite clasts and clasts phosphatized after inclusion within breccia-conglomerate. The abundance of volcanic glass is also poorly known because highly altered glass shards are difficult to distinguish from massive, CFA-replaced limestone matrix.

Phosphatized Limestone

The second most abundant SMP rock type consists of massive to well-bedded phosphatized foraminifera- and nannofossil-rich wackestone (Fig. 4B), packstone, and rare grainstone (Fig. 4C). Cryptocrystalline CFA matrix (commonly intergrown with dendritic Fe-Mn oxyhydroxides) supports intraclasts and foraminifera and nannofossil tests that are typically represented as molds. In decreasing order of abundance, intraclasts are composed of extensively altered glassy basalt and hyaloclastite, fresh to partly altered basalt and glassy basalt, Fe-Mn oxyhydroxide crust fragments, and phosphorite. Rarely, intraclasts also include fish debris, euhedral barite, plagioclase, and pyroxene. Vugs within phosphatized limestones are commonly lined by Fe-Mn oxyhydroxides.

Within bedded phosphatized limestone (Fig. 4B), millimeter- to centimeter-thick layers are distinguished by differences in allochem and intraclast abundance, degree of phosphatization, and differences in crystallinity of the CFA matrix. Soft-sediment deformation features (including flame structures and convolute laminae) are locally present. Rarely, layers are separated by laminated Fe-Mn oxyhydroxides.

Phosphatized Basalt

The third most abundant protolith is composed of either plagioclase-phyric Hawaiite with abundant randomly oriented plagioclase microlites, minor Ti-rich clinopyroxenes, minor Fe-Ti oxides, and minor olivine; or alkalic basalt with sparse olivine, clinopyroxene, and plagioclase phenocrysts. Massive to vesicular phosphatized basalt is typically encrusted by Fe-Mn oxyhydroxides (1 mm-3 cm thick). Within phosphatized basalts CFA lines vugs; replaces basalt groundmass; and replaces vug, void, and fracture-filling detrital carbonate (Fig. 4D).

PARAGENESIS OF SEAMOUNT PHOSPHORITE

Paragenetic diagrams are presented for phosphatized volcanic material (Fig. 5) and phosphatized limestone (Fig. 6). A separate paragenetic diagram is not presented for phosphatized breccia-conglomerates because those rocks reflect phosphatization processes and paragenetic histories that are characteristic of both phosphatized volcanic rocks and limestone. The paragenetic sequences are generalized from observation of many thin sections and do not necessarily represent mineralization sequences observed in all samples. Unidentified and rarely occurring phases are not included in the paragenetic diagrams.

Phosphatization of Volcanic Material

Numerous workers (Alt, 1995; Thompson, 1983; Honnorez, 1981) reported that open-circulation, low-temperature alteration of basalt is typified by early dissolution and/or palagonitization of glass, followed by partial replacement of volcanic material by zeolites, smectites, and Fe oxides. Thin-section and SEM analyses presented below indicate that this same sequence of diagenetic alteration and replacement operates within phosphatized seamount volcanic material (Fig. 5).

As discussed in detail in the following section, phosphatization of volcanic material occurs through lining of vugs and fractures by CFA, smectite, zeolite, and Fe-Mn oxyhydroxides, and through replacement of volcanic material by intergrown CFA,

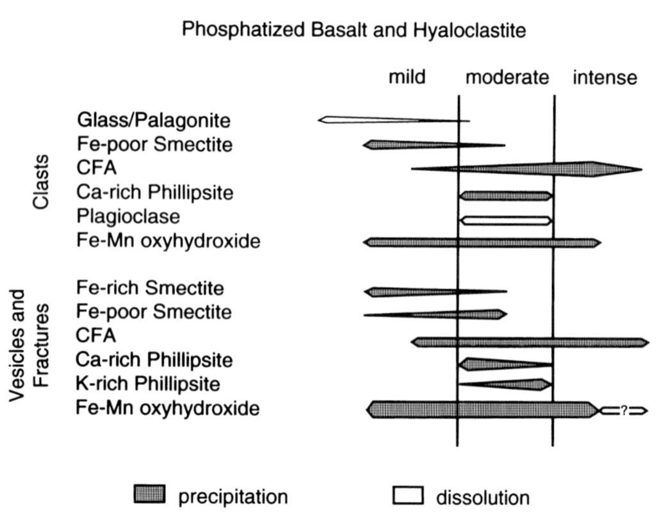

FIG. 5.—Paragenesis diagram for phosphatization of volcanic material. Thickness of bars is schematically proportional to degree of mineralization.

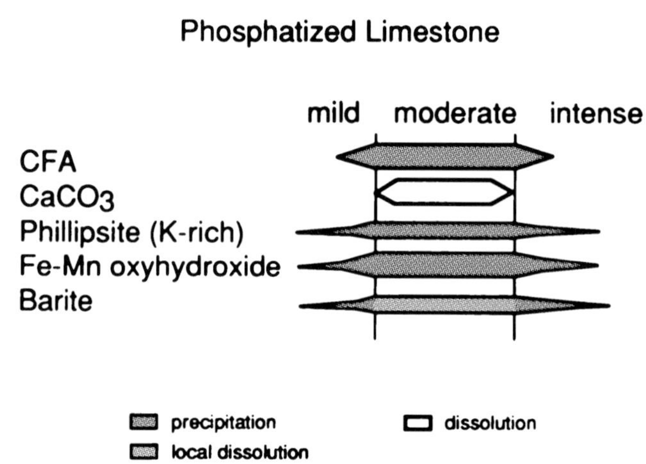

FIG. 6.—Paragenetic diagram for phosphatization of limestone. Thickness of bars is schematically proportional to degree of mineralization.

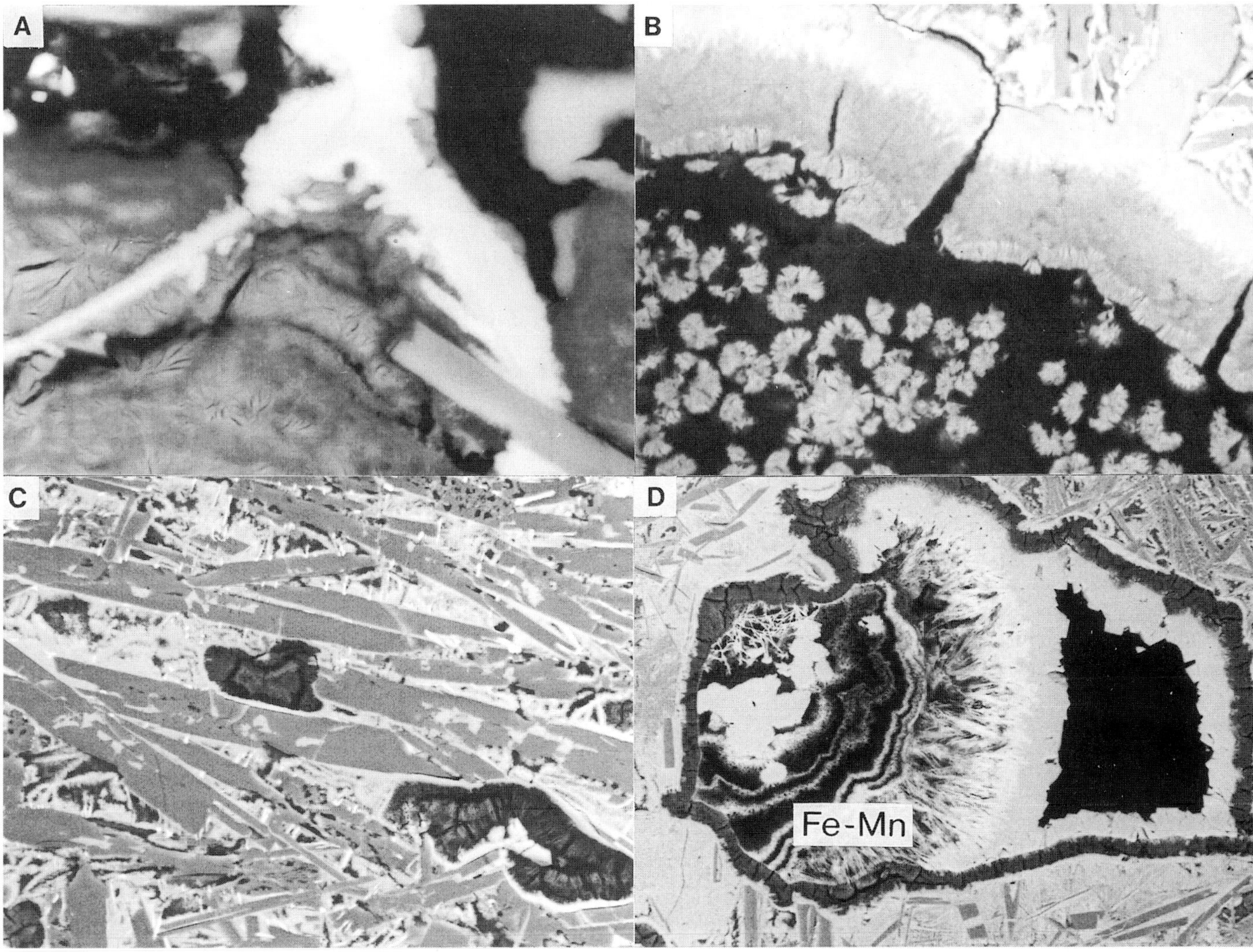

FIG. 7.—A) Back-scattered electron (BSE) image of mildly phosphatized glassy basalt. CFA (white) cross-cuts and embays smectite-replaced volcanic glass (mottled gray). Radial fibrous smectite (by EDS) ranges from Fe-poor (light gray regions) to Fe-rich (darker gray). Plagioclase lath (lower right, medium gray) is unaltered as indicated by its smooth, regular margins. *F7-86-HW* D23-6. Image width = 60 μm. B) BSE image of smectite-lined vesicle within phosphatized basalt. Fe-rich smectite (cream) lines vesicle and is in turn lined by two layers of Fe-poor smectite (dark and medium gray). Clusters of radial-fibrous Fe-poor, Mg-rich smectite (medium gray) partly fill vesicle (black). Within phosphatized basalt (upper right), glassy matrix is replaced by CFA (white) while plagioclase phenocrysts (gray laths) are unaltered. *L5-83-HW* D5-A3-10. Image width = 125 μm. C) BSE image of highly phosphatized basalt. Major CFA (white) and moderate fibrous smectite (dark gray) replace glassy groundmass within basalt, overprinting original basalt fabric. Plagioclase laths (medium gray) are Ca-rich, showing no sign of albitization. Note relict, subparallel Fe-Ti oxides (bright white north/south-oriented streaks). *L5-83-HW* D5-A3-10. Image width = 200 μm. D) BSE image of highly phosphatized basalt. CFA-replaced glass (white) surrounds partly dissolved plagioclase laths (light gray). Reticulate fabric within CFA (lower middle) reflects primary quench texture inherited from the volcanic glass during fabric-preservative CFA replacement. Vesicle is lined by a thin layer of CFA (white), which is in turn lined by phillipsite (dark gray). Botryoidal-fibrous Fe-Mn oxyhydroxides (black and white fibers and lamina filling left half of vesicle, labeled Fe-Mn) surround a nucleus of granular CFA (white). A final layer of blocky CFA (white) lines the remaining pore and passes into open pore space (black). *L5-83-HW* D5-A3-10. Image width = 450 μm

smectite, and zeolite. Phosphatization is commonly more intense adjacent to fractures, large vugs, or sample surfaces and the intensity of mineralization diminishes gradually with distance from pore space. Locally, the pervasiveness of CFA precipitation changes abruptly across linear to undulating boundaries that roughly parallel fractures and/or vug boundaries. These observations suggest that the pervasiveness of phosphatization is partly dependent on rates of fluid exchange and seawater/rock ratios. In addition, diagenetic alteration of volcanic clasts to CFA can be fabric-preserving; thus primary silicate textures locally play a crucial role in determining CFA replacement fabrics.

Mild phosphatization: Formation of palagonite, smectite, and CFA.—

Based on cross-cutting relationships, the earliest diagenetic stage in weakly phosphatized volcanic rocks is local palagonitization of volcanic glass followed by partial alteration of volcanic glass to Fe-poor smectite and CFA (Fig. 7A). Within regions of very mild phosphatization, CFA-filled fractures

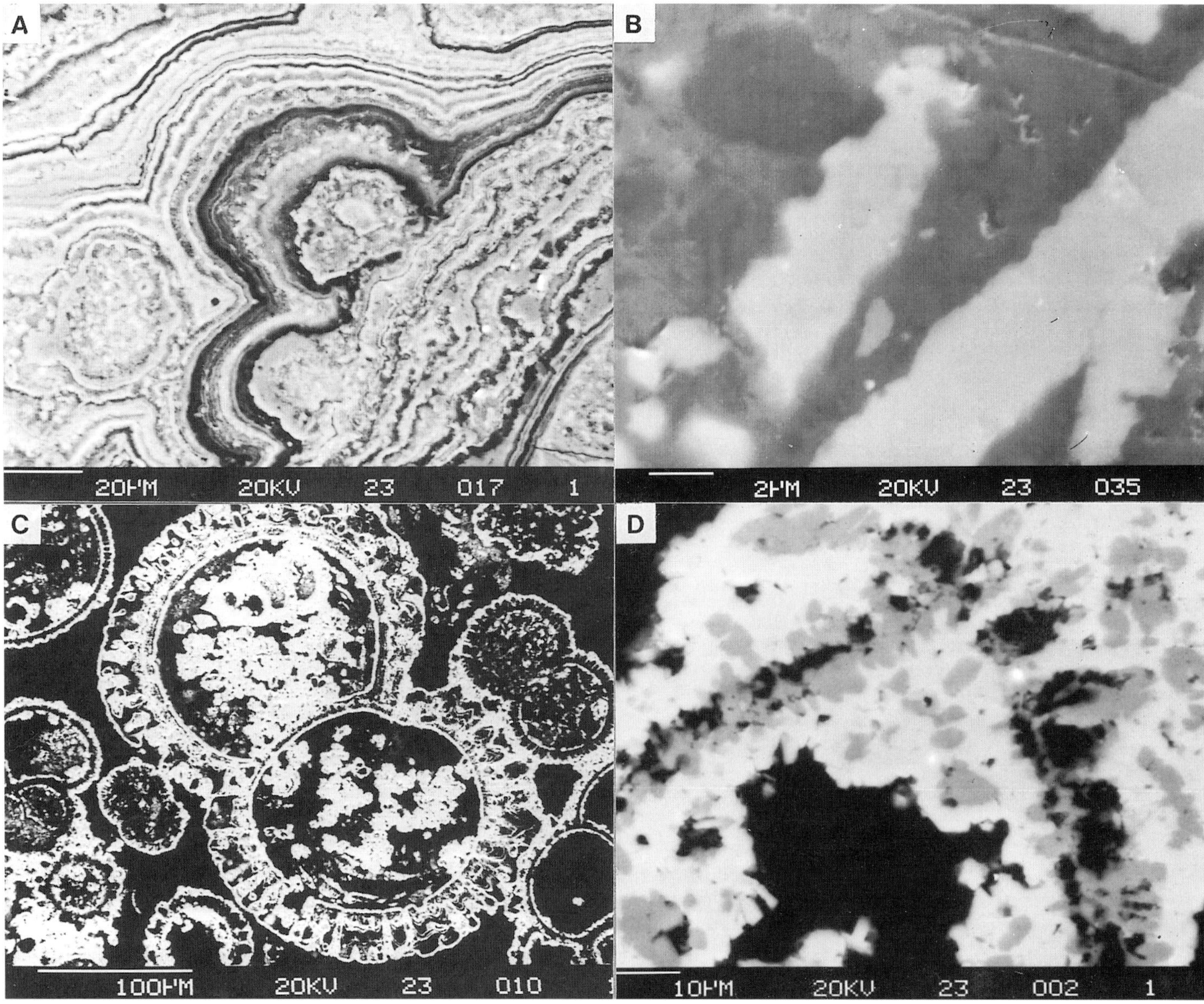

FIG. 8.—A) Back-scattered electron (BSE) image of vug-filling botryoidal CFA within phosphatized basalt. CFA (white and gray) forms botryoidal laminations separated by pore space (black) and surrounds minor Fe-Ti oxides (small bright white blebs). This atypical CFA texture may indicate CFA replacement of Fe-Mn oxyhydroxides or primary precipitation of CFA into pore space. *L5-83-HW* D18-B3-7. Image width = 200 μm. B) BSE image showing euhedral barite crystals (white) embayed by CFA (medium and dark gray). Medium and dark gray regions within CFA are compositionally indistinguishable by EDS analysis. *L5-83-HW* D18-B3-7. Image width = 22 μm. C) BSE image showing CFA molds (white) of foraminifera tests. Margins of molds are smooth and uniform, mirroring fine details of original test morphology and preserving original limestone fabric. *L5-83-HW* D29-1A-A. Image width = 500 μm. D) BSE image showing CFA coating (white) on partly replaced foraminifera test. Irregular and indistinct margins of CFA coating suggest partial CFA-replacement of carbonate (gray) before test partly dissolved to pore space (black). *L5-83-HW* D18-B3-7. Image width = 70 μm.

cross-cut smectite-replaced glass, indicating that smectite precipitation preceded precipitation of CFA.

Mild phosphatization is also typified by lining and partial filling of vugs and vesicles by smectite, CFA, and Fe-Mn oxyhydroxides. Vesicles are commonly lined by isopachous Fe-rich smectite (nontronite by EDS), followed by a layer of Fe-poor, Mg-rich smectite (Fig. 7B). Less commonly, Fe-poor smectite is followed by Fe-rich smectite; rarely multiple layers of isopachous smectite alternate between Fe-poor and Fe-rich. Many smectite-lined vugs and vesicles are subsequently lined or infilled by botryoidal Fe-Mn oxyhydroxides and/or microcrystalline to granular CFA. Sequences of vug-lining mineralization are highly variable both among and within samples.

Moderate to intense phosphatization: Smectite, zeolite, and CFA mineralization.—

In contrast to mildly phosphatized volcanic material where CFA commonly cross-cuts and embays smectite-replaced glass, moderately to intensely phosphatized volcanic glass is replaced by intergrown CFA and smectite. Relative amounts of CFA and smectite precipitation vary at the thin-section scale. In regions of increased phosphatization, glass alters primarily to CFA and only secondarily to smectite (Fig. 7C). Within plagioclase-rich material, glass is also replaced by Ca-rich phillipsite (by EDS) and plagioclase is partly replaced by CFA.

Moderate to intense phosphatization can be either fabric-destructive or fabric-preservative. Within regions of complete replacement by fabric-destructive cryptocrystalline CFA, protolith identity is commonly discernible only through the presence of relict Fe-Ti oxide and/or local concentrations of cryptocrystalline smectite and zeolite. In regions of fabric-preservative phosphatization, CFA pseudomorphs primary protolith textures. Regions of reticulate CFA (Fig. 7D) and CFA growths on plagioclase laths resemble characteristic quench textures observed in MORB glass (Lofgren, 1974). Fine-scale preservation of primary silicate textures is surprising since CFA cannot replace silicate material through solid solution. Although the exact mechanism of replacement is unknown, fabric-preservative replacement of glass by CFA suggests coeval CFA precipitation and glass dissolution. Alternatively, phosphatization may have occurred through dissolution of glass to pore space (leaving an intact boxwork of microphenocrysts), infilling of resultant pores by CFA, later dissolution of microphenocrysts to pore space, and final infilling of phenocryst molds by CFA.

Vugs, vesicles, and fractures within moderately to intensely phosphatized volcanics are typically lined by abundant radial to euhedral phillipsite; cryptocrystalline or micro-granular CFA; and colloform or dendritic Fe-Mn oxyhydroxides (Fig. 7D). Sequences of void-filling phases vary widely, both among samples and within a single sample. In general, however, smectite, phillipsite, or Fe-Mn oxyhydroxides line pores, while Fe-Mn oxyhydroxides or CFA fill them. EDS analyses indicate that phillipsite occurring within vugs is largely K-rich, whereas phillipsite within glassy groundmass is most commonly Ca-rich. Local void-filling botryoidal CFA suggests that CFA may replace Fe-Mn oxyhydroxides (Fig. 8A), although CFA precipitation into open space can also form botryoids (D. Soudry, pers. commun., 1997).

Rarely occurring minerals.—
In addition to the major alteration phases discussed above, some phosphatized volcanic rocks also contain local occurrences of barite and, rarely, sparry calcite. Barite typically occurs within vugs and fractures as tiny euhedral laths disseminated in phosphatized detrital carbonate. Less commonly, veins are filled by a box work of pitted and embayed euhedral barite crystals within a CFA/smectite/phillipsite matrix. Those barite crystals commonly have scalloped edges, which suggests local CFA-replacement of barite (Fig. 8B).

Sparry calcite occurs within two breccia-conglomerates included in this study. Within those samples, clasts are cemented by partly phosphatized carbonate matrix. Dissolution cavities (< 75 mm across) within this matrix are partly to entirely filled by sparry calcite.

Two unidentified phases are locally present within phosphatized volcanic rocks. One breccia-conglomerate sample contains abundant CFA pseudomorphs of an unidentified mineral that is morphologically similar to "dog tooth" calcite. In that sample, Fe-Mn oxyhydroxide encrusted CFA pseudomorphs extend isopachously from clast surfaces into CFA-replaced carbonate matrix. A second unidentified mineral, pseudomorphed by CFA, calcite, and phillipsite, is locally common. The unidentified crystals are twinned, chevron-shaped, and typically occur sandwiched between substrate and an enveloping Fe-Mn oxyhydroxide crust; rarely they fill voids. Chevron twinning and a square cross-section of crystals (when viewed down the apparent c-axis) suggest an original mineralogy of gypsum. Alternatively, while chevron twinning is not typical of zeolites, the mineral occurrence and square cross-section are indicative of phillipsite.

Phosphatization of Limestone

Phosphatized limestone is commonly associated with abundant Fe-Mn oxyhydroxides, phillipsite, and minor barite. Fe-Mn oxyhydroxides and phillipsite are especially common in limestone that is proximal to, or hosts, volcanic material. As discussed in detail below, while diagenetic mineralization sequences vary greatly within and among thin sections, petrographic and SEM analyses indicate that barite, Fe-Mn oxyhydroxides, and phillipsite generally precipitated both before and after phosphatization of the carbonate, and that the latter two also precipitated during phosphatization (Fig. 6). During phosphatization, calcite was dissolved and early-stage barite underwent local dissolution.

Early precipitation of accessory phases is indicated by their common dissemination within a massive CFA matrix, suggesting formation prior to occlusion and infilling of pore space by CFA. SEM images show CFA coatings on euhedral barite and phillipsite. Furthermore, phosphatized limestone commonly contains lithic fragments that are encrusted by Fe-Mn oxyhydroxides and phillipsite. These clasts were clearly encrusted prior to inclusion within the limestone, and therefore prior to phosphatization. The relative timing of precipitation among accessory phases is not determined due to lack of cross-cutting relationships.

Paragenesis during phosphatization.—
Within mildly phosphatized limestone, cryptocrystalline CFA occurs as a pore-lining phase and/or cement. In those rocks, massive CFA surrounds minor allochems and locally abundant foraminifera tests. Microcrystalline relict carbonate commonly imparts a sparkling birefringence to the CFA matrix and tests are partly to fully dissolved. Molds of foraminifera tests preserve original limestone fabric and their smooth margins mirror fine details of original test morphology (Fig. 8C). Fine-scale preservation of original limestone fabric precludes embayment or contemporaneous dissolution of carbonate and indicates that within regions of mild phosphatization, carbonate was not replaced by CFA.

In contrast, within moderately and intensely phosphatized limestone most carbonate dissolved and phosphatization appears to have occurred through both pore-lining and replace-

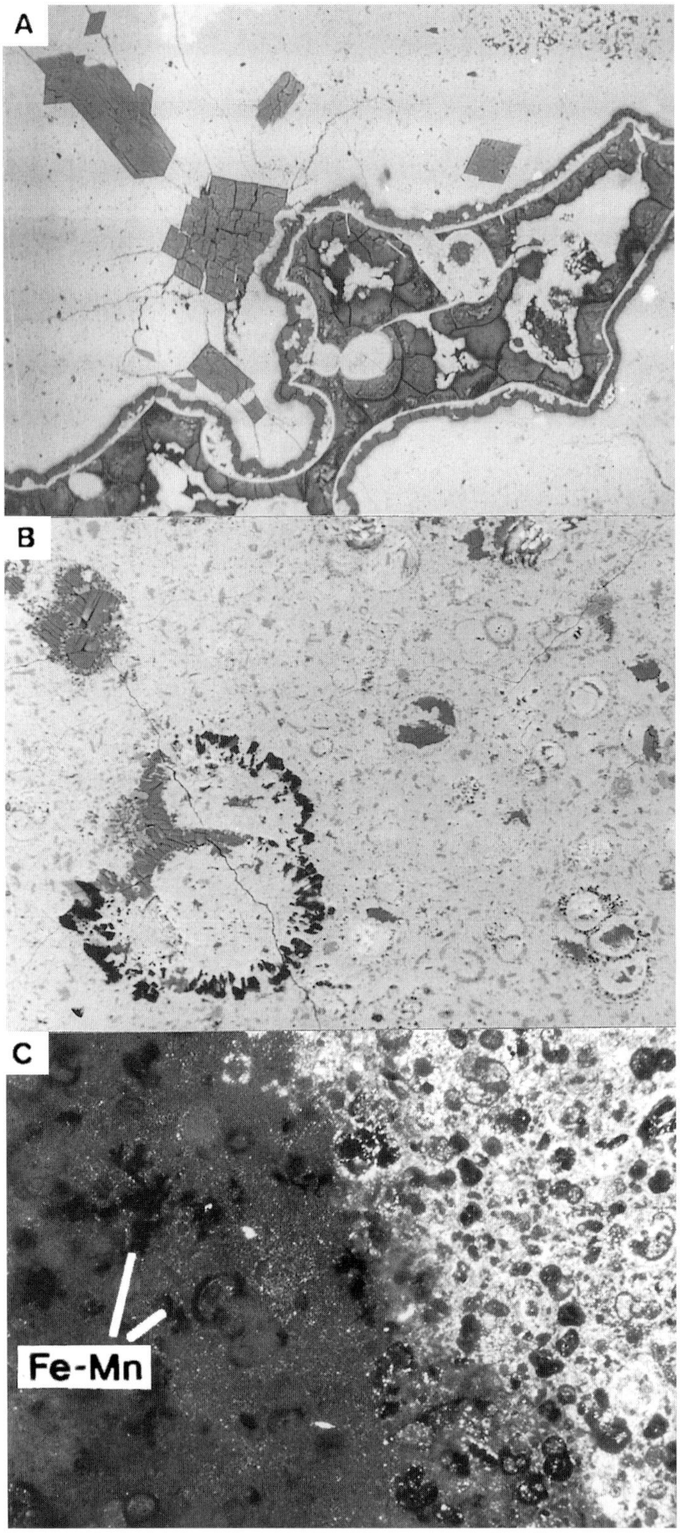

FIG. 9.—A) Back-scattered electron (BSE) image of K-rich phillipsite within phosphatized limestone proximal to an altered hyaloclastite fragment. Euhedral K-rich phillipsite (light gray, rhombohedrons and prisms) is supported by a massive CFA matrix (white). Hyaloclastite fragment (lower left corner extending to upper right) is replaced and lined by smectite (medium and dark gray) and CFA (white). *F7-86-HW* D22-1C. Image width = 500 μm. B) BSE image of CFA-replaced limestone with abundant microfossil tests. Relict carbonate tests (light gray) are surrounded by cryptocrystalline CFA (white), which commonly embays, and therefore replaces carbonate grain margins. Foraminiferal test (upper left) is infilled by phillipsite (dark gray); foraminiferal mold (black, lower left) is partly filled by phillipsite. *L9-84-CP* D6-3. Image width = 400 μm. C) Photomicrograph of phosphatized limestone under cross-polarized light. Dark-colored massive CFA contains common foraminiferal molds (very dark gray) and common disseminated dendritic Fe-Mn oxyhydroxides (labeled Fe-Mn). Unphosphatized carbonate on right is separated from the phosphatized region by an abrupt contact that cross-cuts foraminiferal tests, indicating a mineralization front. Minor relict microcrystalline carbonate (tiny white blebs at middle left) is more abundant within CFA that is closer to the unphosphatized region. *L5-83-HW* D18-B3-7. Image width = ~1.5 cm.

ment processes. Dissolution of carbonate is indicated by foraminifera test molds or tests pseudomorphed by CFA. In addition, SEM images indicate that CFA matrix typically contains little relict carbonate.

Within severely to moderately phosphatized limestone, CFA reflects both pore-lining and replacement textures; one part of a test can be replaced while another is dissolved. Pore-lining is indicated by CFA pseudomorphs of foraminifera tests (commonly distinguishable from CFA matrix by color differences), which typically form through partial CFA-lining of moldic pores. Lamboy (1993) described this same sequence (CFA-lining of tests, carbonate-dissolution, CFA-infilling of molds) for the formation for CFA pseudomorphs in phosphatic carbonates from the Paris Basin and Peru shelf. In addition, CFA replacement of carbonate is indicated by local CFA embayments into allochem margins. Margins of molds within moderately and highly phosphatized limestone are commonly irregular and indistinct, suggesting partial CFA replacement prior to carbonate dissolution (Fig. 8D). Whether phosphatization occurs through authigenic (pore-lining) or diagenetic (alteration) processes, the resultant CFA fabric can preserve or overprint original protolith fabric.

Fe-Mn oxyhydroxides and phillipsite are common within moderately to highly phosphatized carbonates, especially in phosphatized limestone that is proximal to, or supports, volcanic material. Fe-Mn oxyhydroxides form abundant disseminated botryoids and dendrites; common euhedral K-rich phillipsite occurs as rosettes and solitary prisms (Fig. 9A). Less commonly, microcrystalline phillipsite and Fe-Mn oxyhydroxides pseudomorphed foraminifera by lining molds (Fig. 9B). The relative timing of precipitation between Fe-Mn oxyhydroxides and phillipsite is poorly constrained due to the lack of cross-cutting relationships. However, while phillipsite and CFA intergrowths are not observed, coprecipitation is likely because they coprecipitated within nearby phosphatized volcanic rocks.

Phosphatized volcanic-rich carbonate locally contains microcrystalline Fe-poor smectite disseminated within CFA groundmass and/or lining molds from dissolved hyaloclastite and basalt fragments. Within CFA-replaced groundmass, coprecipitated CFA and smectite are intergrown.

Moderately and highly phosphatized limestone is locally barite-rich. Barite typically occurs as small euhedral laths that are disseminated within a CFA (after carbonate) matrix. Common CFA embayments of barite indicate dissolution of barite during phosphatization.

Post-phosphatization paragenesis.—

Post-phosphatization precipitation of barite, phillipsite, and Fe-Mn oxyhydroxides is indicated by SEM and petrographic analyses. Euhedral phillipsite nucleates on CFA and fills molds within CFA that are also commonly infilled by barite and Fe-Mn oxyhydroxides. Sparry calcite is rarely present within molds and large secondary dissolution vugs within CFA. Again, relative timing of precipitation between those phases could not be determined because cross-cutting relationships are lacking.

Mineralization fronts.—

CFA-replaced carbonate locally contains compositionally distinct domains that are separated by sharp boundaries. Domains are distinguishable by differences in crystallinity and by differences in abundances of CFA, Fe-Mn oxyhydroxides, barite, and/or relict carbonate. Boundaries between domains locally cross-cut foraminifera tests, suggesting that the boundaries are mineralizing fluid fronts rather than of depositional origin (Fig. 9C). These mineralization fronts are similar to those reported to occur in seamount silicified carbonates (Hein et al., 1981).

Locally, mineralization fronts broadly parallel voids, fractures, and/or outer surfaces of samples. Typically, regions adjacent to voids and surfaces are relatively enriched in CFA and Fe-Mn oxyhydroxides, indicating that mineralization fronts began at permeable openings and seawater entryways. Domains commonly have complex amoebae-like shapes. Those shapes most likely reflect multiple, overlapping fronts of mineralization and dissolution.

CONCLUSIONS

1. Phosphatization of both silicate and carbonate seamount protoliths occurs through precipitation in voids and through replacement. Regardless of protolith rock type, phosphatization can be either fabric-destructive or fabric-preserving. Within seamount phosphorites, direct molecular replacement of carbonate is not the most important process leading to fabric-preserving replacement by CFA. Rather, fabric-preserving CFA mineralization commonly occurred indirectly when CFA precipitated around microfossils, microfossil carbonate subsequently underwent dissolution, and finally further CFA precipitated in voids. This fabric-preserving phosphatization of carbonate was also described by Soudry and Lewy (1990) and by Lamboy (1993).
2. Fabric-preserving phosphatization of volcanic material is indicated by fine-scale replication of silicate fabrics within CFA. The precise mechanism of replacement is unknown, however solid-state substitution between CFA and silicate minerals clearly cannot occur. Fabric-selective phosphatization may have occurred through coeval CFA precipitation and glass dissolution, or alternatively through a process of cementation-dissolution-cementation analogous to that described for fabric-preserving phosphatization of limestone.
3. Phosphatization is most pervasive near sample surfaces, large pores, and fractures. Increased pervasiveness of phosphatization with increased proximity to fluid access routes suggests that fluid exchange rates and rock/water ratios are important controls on seamount phosphogenesis. It is likely that phosphatizing fluids were seawater-dominated. This interpretation is supported by diagenetic mineral assemblages that are typical of open-circulation, low-temperature seafloor alteration.
4. Breccia-conglomerates clearly show that CFA formed near the sediment-water interface because that is where reworked clasts can be shed from seamount flanks and redeposited within carbonate sediment. Occurrences of reworked phosphorite within a partly phosphatized matrix further indicate that phosphatization occurs over a relatively long time span, or occurs during several episodes separated by large time spans. Phosphatization on seamounts is a long-term process because phosphogenesis must encompass (1) initial phosphatization of limestone and/or volcanic material, (2) mass wasting and redeposition of phosphatized clasts, and (3) further phosphatization of carbonate matrix.
5. The mineral assemblage found in SMP deposits is characteristic of low-temperature seafloor weathering and implies that CFA precipitated hydrogenetically from seawater of ambient temperature. This interpretation is supported by microprobe analyses that indicate that dioctahedral, low-temperature smectite is intergrown with CFA (Benninger, 1998). In addition, oxygen isotopic analyses of CFA confirm a low-temperature origin for most Marshall and Johnston Island phosphorites (Hein et al., 1993). This finding is not surprising since the SMP deposits formed during Tertiary time when the Cretaceous seamounts were no longer hydrothermally active.
6. The close association between CFA and Fe-Mn oxyhydroxide precipitation suggests that similar to other low-productivity phosphorite deposits, SMP formation may have been facilitated by Fe-oxyhydroxide redox reactions. In seawater, Fe-oxyhydroxides have a weakly positive surface charge and can complex with phosphate ion (Roy, 1981). We suggest that phosphorus adsorbs onto surfaces of Fe-oxyhydroxide particles as they fall though the water column. Once within the oxygen minimum zone of suboxic Fe reduction, dissolution of Fe-Mn oxyhydroxides can release P and thereby increase phosphate concentration. Analogous to the Fe-P shuttle described for other low-productivity phosphorites (Jarvis, 1992; Schaffer, 1986; Lucas and Prévôt, 1985), Fe^{2+} released during Fe reduction may migrate to more oxic waters, reform oxyhydroxides, and again be available to shuttle P to (suboxic) phosphatization sites.
7. SMP formed within shallow sediment ponds and within sediment-free breccia and basalt where bacterially mediated anoxic diagenetic zones cannot have developed. Therefore, unlike other continental margin phosphorites that formed in bacterially mediated diagenetic zones within reducing sediments, SMP must have formed below a low-oxygen water column, probably within the oxygen minimum zone that invariably intersects the flanks of equatorial Pacific seamounts.

ACKNOWLEDGMENTS

This research was made possible through financial assistance from the U.S. Geological Survey Graduate Intern Program. Additional funding was provided by Sigma Xi, The Scientific Research Society (Grant-in-Aid of Research award #7143), and by the Geological Society of America and National Science Foundation (research award #4626-91). We are grateful to Leanne Roberts for providing invaluable technical support. We thank David Soudry and an anonymous reviewer for helpful comments.

REFERENCES

ALT, J. C., 1995, Subseafloor processes in mid-ocean ridge hydrothermal systems, *in* Humphries, S., Zierenberg, R. A., Mullineaux, L. S., and Thompson R. E., eds., Seafloor Hydrothermal Systems: Physical, Chemical, Biological and Geological Interactions, Geophysical Monograph 91: Washington D.C., American Geophysical Union, p. 85-113.

BATURIN, G. N., 1982, Phosphorites on the Sea Floor-Origin, Composition and Distribution, *in* Developments in Sedimentology, v. 33: New York, Elsevier Scientific Publishing Company, 343 p.

BENNINGER, L. M., 1998, Distribution, origin, and diagenetic evolution of Pacific seamount phosphorite: M.S. Thesis, University of California, Davis, 82 p.

BERSENEV, I. I., BATURIN, G. N., LELIKOV, E. P., AND GUSEV, V. V., 1990, Neogene phosphorites of the Sea of Japan, *in* Burnett, W. C., and Riggs, S. R., eds., Phosphate Deposits of the World, Volume 3: Neogene to Modern Phosphorites: Cambridge, Cambridge University Press, p. 167-176.

BURNETT, W. C., CULLEN, D. J., AND MCMURTRY, G. M., 1987, Open-ocean phosphorites—in a class by themselves?, *in* Teleki, P. G., Dobson, M. R., Moore, J. R., and von Stackelberg, U., eds., Marine Minerals: Advances in Research and Resource Assessment: Dordrecht, D. Reidel Publishing Company, p. 119-134.

CULLEN, D. J. AND BURNETT, W. C., 1986, Phosphorite associations on seamounts in the tropical southwest Pacific Ocean: Marine Geology, v. 71, p. 215-236.

HEIN, J. R., VALLIER, T. L., AND ALLAN, M. A., 1981, Chert petrology and geochemistry, Mid-Pacific Mountains and Hess Rise, Deep Sea Drilling Project Leg 62, *in* Thede, J., Vallier, T. L., eds., Initial Reports of the Deep Sea Drilling Project, v. 62: Washington, U. S. Government Printing Office, p. 711-748.

HEIN, J. R., ET AL., 1990, Geological, geochemical, geophysical, and oceanographic data and interpretations of seamounts and Co-rich ferromanganese crusts from the Marshall Islands, KORDI-USGS, R.V. *Farnella,* Cruise F10-89-CP, U.S. Geological Survey Open File Report 90-407, 246 p.

HEIN, J. R., YEH, H. W., GUNN, S. H., SLITER, W. V., BENNINGER, L. M., AND WANG, C. H., 1993, Two major Cenozoic episodes of phosphogenesis recorded in equatorial Pacific seamount deposits: Paleoceanography, v. 8, p. 293-311.

HONNOREZ, J., 1981, The aging of the oceanic crust at low temperature, *in* Emiliani, E., ed., The Sea: the Oceanic Lithosphere, v. 7: New York, John Wiley and Sons, p. 525-587.

JACKSON, E. D., ET AL., 1980, Proceedings of the Ocean Drilling Program, Initial Reports, v. 55: Washington, U.S. Government Printing Office, Ocean Drilling Program, 868 p.

JARVIS, I., 1992, Sedimentology, geochemistry and origin of phosphatic chalks: The Upper Cretaceous deposits of NW Europe: Sedimentology, v. 39, p. 55-97.

KAYEN, R. E., SCHWAB, W. C., LEE, H. J., TORRESAN, M. E., HEIN, J. R., QUINTERNO, P. J., AND LEVIN, L. A., 1989, Morphology of sea-floor landslides on Horizon Guyot: application of steady-state geotechnical analysis: Deep-Sea Research, v. 36, p. 1817-1839.

LAMBOY, M., 1993, Phosphatization of calcium carbonate in phosphorites: microstructure and importance: Sedimentology, v. 40, p. 53-62.

LOFGREN, G. E., 1974, An experimental study of plagioclase crystal morphology: isothermal crystallization: American Journal of Science, v. 274, p. 243-273.

LUCAS, J., AND PRÉVÔT, L., 1985, The synthesis of apatite by bacterial activity: mechanism, *in* Lucas, J., and Prévôt, L., eds., Phosphorites: Sciences Géologiques Mémoire, v. 77, p. 83-92.

O'BRIEN, G. W., AND HEGGIE, D., 1988, East Australian continental margin phosphorites: EOS, v. 69, p. 2.

O'BRIEN, G. W., MILNES, A. R., VEEH, H. H., HEGGIE, D. T., RIGGS, S. R., CULLEN, D. J., MARSHALL, J. F., AND COOK, P. J., 1990, Sedimentation dynamics and redox iron-cycling: controlling factors for the apatite-glauconite association of the east Australian continental margin, *in* Notholt, A. J. G., and Jarvis, I., eds., Phosphorite Research and Development: London, Geological Society of London Special Publication 52, p. 61-86.

PREMOLI-SILVA, I., ET AL., 1993, Proceedings of the Ocean Drilling Program, Initial Reports, v. 144: College Station, Texas, Ocean Drilling Program, p. 145-312.

ROY, S., 1981, Manganese Deposits: London, Academic Press, 458 p.

SAGER, W. W., WINTERER, E. L., FIRTH, J. V., ET AL., 1993, Proceedings of the Ocean Drilling Program, Initial Reports, v. 143: College Station, Texas, Ocean Drilling Program, p. 273-288.

SCHAFFER, G., 1986, Phosphatic pumps and shuttles in the Black Sea: Nature, v. 321, p. 515-517.

SCHWAB, W. C., LEE, H. J., KAYEN, R. E., QUINTERNO, P. J., AND TATE, G. B., 1988, Erosion and slope instability on Horizon Guyot, Mid Pacific Mountains: Geo-Marine Letters, v. 8, p. 1-10.

SOUDRY, D., AND LEWY, Z., 1990, Omission–surface incipient phosphate crusts on early diagenetic calcareous concretions and their possible origin, upper Campanian, southern Israel: Sedimentary Geology, v. 66, p. 151-163.

THOMPSON, G., 1983, Basalt-seawater interaction, *in* Rona, P. A., Bostrom, K., and Smith, K. L., eds., Hydrothermal Processes at Seafloor Spreading Centers: New York, Plenum, p. 225-278.

DIAGENESIS OF FERROMANGANESE CRUSTS: CHEMICAL AND BIOLOGICAL ALTERATION OF ARTIFICIAL SUBSTRATES ON CROSS SEAMOUNT

MIRIAM A. BERTRAM* AND JAMES P. COWEN

Department of Oceanography, School of Ocean and Earth Science and Technology, University of Hawaii at Manoa, Honolulu, Hawaii, 96822-2285, USA

ABSTRACT: An experimental approach was used to investigate in-situ metal-oxide accumulation on artificial substrates. Mn-containing (ferromanganese-oxide, rhodochrosite) and Mn-free ($CaCO_3$, basalt) substrates were deployed on Cross Seamount (18°40′N, 158°17′W) at a number of depths (800–2000 m) and time intervals (1–42 months). Chemical and biological alterations to the experimental surfaces were documented by scanning electron microscopy and energy dispersive X-ray fluorescence spectroscopy.

Geochemical alterations of the experimental substrates included the formation of Mn-oxide coatings on some Mn-containing substrates and an apparent loss of Mn from ferromanganese-oxide substrates. Temporal and spatial patterns in Mn-oxide deposition were related to environmental changes and the presence of a particular benthic foraminifera. These patterns, as well as morphological features of the Mn-oxide deposits, suggest that diagenetic alteration of ferromanganese crusts can occur and may be associated with surface water productivity and benthic organisms. These data suggest that suboxic diagenesis, within a microenvironment created by the deposition of organic-rich particulate matter, may contribute to the formation of the narrow bands of Mn and Fe oxides observed within naturally occurring ferromanganese-oxide crusts and nodules.

INTRODUCTION

The chemical composition and growth rate of ferromanganese crusts and nodules is thought to be regulated by the composition of seawater (Hein et al., 1992; McMurtry et al., 1994) and the rate of supply of metals to the oxide surface (Dymond et al., 1984). Nodules occur in the sediments, where particulate metals can be mobilized during the degradation of biogenic matter derived from the surface oceans or during sedimentary diagenesis through diffusive/advective transport of dissolved metals. Crusts form pavements of oxyhydroxide minerals on basaltic and other solid surfaces and are thought to accrete by the slow deposition of metals from seawater (Aplin and Cronan, 1985). Fine-scale analyses of distinct laminae within ferromanganese crusts reveal distinct morphological and chemical boundaries (McMurtry et al., 1994; Wen et al., 1997) as well as gradations in composition from the crust surface to the underlying substrate (Hein et al., 1992; Wen et al., 1997). The chemical composition of these layers has been used to estimate the age and growth rate of an oxide deposit and the paleoceanographic conditions during deposition (e.g., McMurtry et al., 1994; Wen et al., 1997), a valid approach if the original chemical composition is preserved through time.

Ferromanganese nodule composition and rapid accretion rate reflects the supply of metals from three processes: suboxic diagenesis of the underlying sediments, oxic diagenesis of particles at the sediment-water interface, and hydrogenetic deposition from seawater. In contrast, ferromanganese crusts are thought to accrete hydrogenetically, scavenging metals at the oxide-seawater interface (Aplin and Cronan, 1985). Because crusts are thought to accrete by a single process, the composition of ferromanganese crusts (outside of hydrothermal influence) is used to model hydrogenetic oxide accretion (Dymond et al., 1984). However, organic matter can collect in topographic lows that occur on millimeter to meter scales, and create a chemically isolated environment where oxygen levels drop. It is hypothesized here that under these conditions the mobilization of Fe, Mn, and other components of the underlying substrate could be coupled to the oxidation of organic matter. Diagenetic processes in sediments overlying ferromanganese crusts would then contribute to the observed chemistry and accumulation rate of the underlying oxide mineral.

The flux of organic-rich particles to the seafloor is an important food source for benthic communities. Small organisms, primarily bacteria and benthic foraminifera, thrive within organic carbon-rich deposits on the seafloor (e.g., Turley et al., 1988), and the flux of particulate organic carbon may support the bacterial (Burnett and Nealson, 1983) and foraminiferal (Mullineaux, 1987) communities observed on ferromanganese crusts and nodules. Attached foraminifera have long been thought to facilitate oxide accretion (e.g., Graham and Cooper, 1959) because Fe and Mn are found in foraminiferal chamber walls (Graham and Cooper, 1959; Dudley, 1976; Margolis et al., 1979; Gooday and Haynes, 1983; Bertram and Cowen, 1994) and waste products (Riemann, 1983). Bacteria are known to catalyze the oxidation of Mn in many marine environments (Nealson et al., 1988), are documented on the surfaces of dredged crusts and nodules (Burnett and Nealson, 1983), and are also thought to contribute to the accumulation of ferromanganese-oxides. If benthic foraminifera and bacteria provide surfaces that facilitate metal uptake from seawater, their presence could both initiate the deposition and promote the accumulation of Fe and Mn oxides on the seafloor. This influence would be pronounced where microbiota are concentrated, for example, within organic carbon-rich particulate matter.

Despite our long-time awareness of the physical association between oxide deposits and benthic foraminifera and bacteria, their role in the accumulation of Mn and Fe minerals on the seafloor remains largely unknown. The binding of metals onto bacterial surfaces, with subsequent oxidation, accelerates the deposition of Mn-oxides relative to that of abiotic Mn-oxidation rates (Hastings and Emerson, 1986). Thus, the growth of such bacteria on nodule and crust surfaces could accelerate oxidation

*Current address: Pacific Marine Environmental Laboratory, NOAA, 7600 Sand Point Way, N.E., Seattle, WA 98115-0070, USA.
email: bertram@pmel.noaa.gov

Marine Authigenesis: From Global to Microbial, SEPM Special Publication No. 66

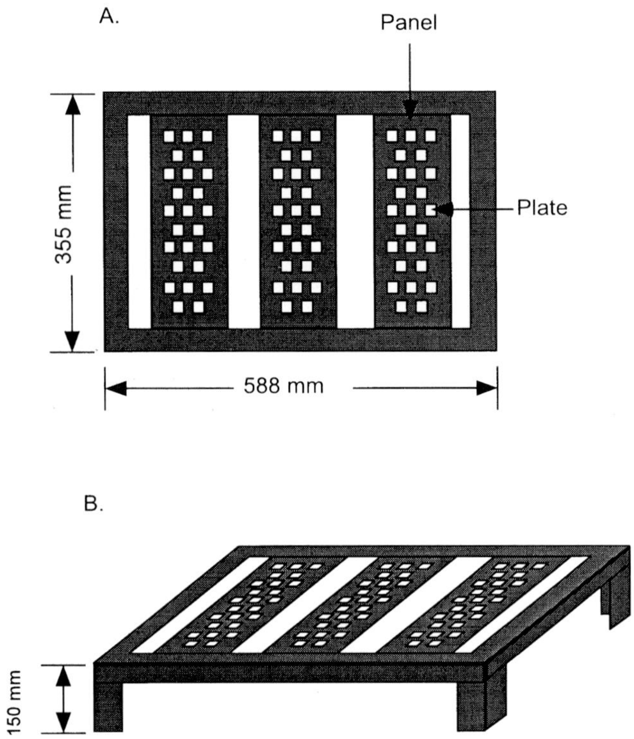

FIG. 1.—Experimental array. A) Top view illustrating panels and plates. B) Side view.

of metals from seawater. The difference in accretion rates for suboxic and hydrogenetic processes (200 and 1 mm/10^6 yr, respectively) is thought to reflect the different sources of metals (Dymond et al., 1984). If metal oxide accretion is limited by the supply rate of metals to the surface, organic surfaces associated with bacteria and foraminifera can provide a mechanism for rapid deposition of oxides when metals become available.

An experimental approach is used in this in-situ study to investigate the accumulation rates and compositions of metal-oxides accreting on Cross Seamount. Experimental plates were deployed at a number of depths (800 m, 975 m, 1285 m, and 1985 m) and time intervals (between 1 month and 42 months). Biological structures and chemical properties of the minerals observed on plate surfaces were documented by scanning electron microscopy and X-ray fluorescence spectroscopy.

Study Area

Cross Seamount (18°40′N, 158°17′W) was formed approximately 83 million years ago just south of the equator and has traveled with the oceanic plate to its current location (Sager and Pringle, 1987). The basaltic flanks of the seamount are covered by layers of ferromanganese-oxides, and the summit at 400 m is covered with a thick sediment cap (Kelly, 1988). A broad oxygen minimum occurs in this region of the Pacific and was seen in dissolved oxygen profiles obtained during this study. Oxygen concentrations gradually increased from a low of approximately 30 μM O_2 at 700 m to 80 μM near 2000 m.

TABLE 1.—NUMBER OF PLATES ANALYZED FOR MN AND FE DETRITAL PARTICLES AND COATINGS.

Experiment			Substrate Composition				
Depth (m)	Time (months)	Year Deployed	FeMn Oxide	$MnCO_3$	$CaCO_3$	Basalt	Total
800	19	1989	3 (3)‡	(3)	6 (0)	3 (0)	15 (6)†
800	30	1989	1 (2)	1 (2)	1 (0)	1 (0)	4 (4)
975	1	1992	3 (2)	0 (0)	1 (0)	0 (0)	4 (2)
975	11	1991	3 (3)	0 (0)	4 (0)	1 (0)	8 (3)
975	19	1989	1 (2)	1 (3)	1 (0)	1 (0)	4 (5)
975	42	1989	1 (3)	3 (3)	2 (0)	1 (0)	7 (6)
1285	24	1991	3 (2)	0 (0)	4 (0)	2 (0)	9 (2)
1985	3	1993	1 (1)	2 (2)	1 (0)	1 (0)	5 (3)
1985	22	1991	2 (2)	0 (0)	3 (0)	1 (0)	6 (2)
		Total	19 (20)	10 (13)	23 (0)	11 (0)	62 (33)

†Systematic surface scans done at 500x for FeMn oxide, basalt and $CaCO_3$ plates, results presented in Bertram and Cowen (1994). Three $MnCO_3$ plates were scanned at 200x, as were the other plates included in this data set.

‡( ) # of plates analyzed by point count.

EXPERIMENTAL METHODS

Experimental arrays containing 50–100 plates were deployed on Cross Seamount for variable durations lasting from 1 month to 42 months at depths of 800 m, 975 m, 1285 m, and 1985 m (Bertram and Cowen, 1994, 1999). Study sites were accessed by the *PISCES V* submersible operated by the Hawaii Undersea Research Laboratory. The depths and sites were chosen to most effectively utilize submersible time. Two sites could be visited during a single dive, and experiments deployed at 800 m and 975 m were initiated during a dive of opportunity. Submersible operations are restricted to waters shallower than 2000 m, which determined the deployment depths of the second set of experiments. Four of the arrays were deployed in 1989, three in 1991, one in 1992, and one in 1993 (Table 1). The arrays (Fig. 1) consisted of a fiberglass rack and three to five Plexiglas panels, each of which contained plates with embedded fine-grained minerals. Individual plates were secured to panels with nylon screws. The fiberglass rack held plates parallel 150 mm above the seafloor.

Experimental Substrates

Ferromanganese-oxide and rhodochrosite were embedded in a polyester resin and attached to 4 cm^2 Plexiglas plates. Ferromanganese-oxide material was obtained from a crust that was previously chemically and mineralogically characterized (4RD 7-5, DeCarlo et al., 1987) and rhodochrosite ($MnCO_3$) was obtained commercially. Both materials were ground and wet-sieved to obtain a < 500 μm size fraction. The ground ferromanganese oxide was rinsed with diluted HCl and then with NaOCl (both at 25°C) to remove $CaCO_3$ and organic matter, respectively. Both the ground ferromanganese-oxide and

TABLE 2.—FERROMANGANESE-OXIDE SUBSTRATES ANALYZED FOR CHEMICAL COMPOSITION.

Plate[†]	Depth meters	Deployment Time (months)	Treatment[‡]	Number of chemical analyses used to determine average composition
Control Type I	0	0	None	10
Control Type I	0	0	DI only	10
Control Type II	1000	6 hours	DI only	10
FeMnA19-1	800	19	DI only	9
FeMnA19-2	800	19	DI only	10
FeMnA19-3	800	19	DI only	9
FeMnA30-1	800	30	DI only	10
FeMnA30-2	800	30	DI only	13
FeMnB1-1	975	1	DI only	10
FeMnB11-1	975	11	DI only	10
FeMnB19-1	975	19	DI only	7
FeMnB19-2	975	19	DI only	5
FeMnB42-1	975	42	Fix + DI	10
FeMnB42-2	975	42	Fix + DI	10
FeMnC24-1	1285	24	Fix + DI	10
FeMnC24-2	1285	24	Fix + DI	10
FeMnD3-1	1985	3	Fix + DI	10
FeMnD22-1	1985	22	Fix + DI	10

[†]Codes used in plate identification: FeMn = ferromanganese-oxide, etc.; A = 800 m, B = 975 m, C = 1285 m, D =1985 m; 19, 30 etc. = deployment time; -1, -2 = replicate plates.
[‡]DI water was adjusted to pH of 7.5–8 using concentrated NaOH. Fixed samples were submerged in a 2% formaldehyde in filtered seawater solution then desalted using solutions of increasing proportions of DI water.

$MnCO_3$ were rinsed in deionized water and dried at 60°C prior to plate construction. $CaCO_3$ and basalt substrates were constructed as described elsewhere (Bertram and Cowen, 1994). Two types of controls were used and prepared as described above. Control type I was never exposed to seawater and control type II was attached to the submersible basket during transit to and from the seafloor.

Upon recovery some plates were rinsed only in pH-adjusted deionized water prior to dessication. To preserve organic structures, other plates were first fixed in a 2% formalin/seawater solution and then desalted using solutions containing increasing proportions of pH-adjusted-distilled water in 0.2 μm filtered seawater (Table 2). After the substrates received their final distilled water rinse and were returned to the laboratory (within 48 hours of recovery), they were dried overnight at 60°C and stored in desiccators prior to being carbon coated for analysis by scanning electron microscopy. No detectable differences in the abundance or state of preservation of the biological structures or in the composition of the minerals were seen between rinsed and fixed plates.

Analytical Procedures

The composition, morphology, and abundance of mineral deposits on the plate surfaces were documented using a scanning electron microscope (SEM) interfaced to an energy dispersive X-ray fluorescence spectrometer (EDS). Chemical discrimination between mineral phases of differing average atomic number was done using a backscattered electron (BSE) detector. Particles and precipitates not significantly obstructed by other particles were analyzed for elemental composition and these data were used to approximate the mole percent of each element in the whole-mounted samples.

Three groups of minerals, distinguished by their surface texture, relief, and size are classified here as detrital particles, the substrate, or coatings. Detrital particles include skeletal material (for example, coccoliths and radiolarian tests), fine silicates (for example, clays), and other material of pelagic origin. Minerals embedded in resin are referred to as the substrate. Minerals that coated these detrital particles, substrate surfaces, or benthic foraminifera are classified as coatings.

Plates were scanned in parallel tracts at 200x magnification to examine 60–80% of plate surfaces from all experiments. Morphological and compositional data for coatings and detrital particles, and compositional data for the substrate, were collected. Minerals largely composed of Mn are classified as Mn-deposits and those with [Fe]≥[Mn] are Fe deposits. Deposits were analyzed three to 15 times, depending on the size and availability of unobstructed regions, to obtain a representative chemical composition. Numerical abundance and mean composition of these minerals were compiled (Fig. 2A, B). These data are useful in describing the composition, morphology, and relative abundances or accumulation rates of mineral deposits.

Spatial Coverage by Detrital Particles and Coatings

A point count procedure was used to quantify the area occupied by commonly observed deposits. The constituent at each of 500 points uniformly distributed across the substrate surface was recorded. The number of times a Mn or Fe deposit was encountered was converted to a percentage of the total points counted, which is a procedure for distinguishing the abundance of features randomly distributed across surfaces (Meese and Tomich, 1992).

RESULTS

The surfaces of plates deployed in 1989 were altered significantly. Alterations included the accumulation of detrital particles, the appearance of extensive coatings of Mn oxide on ferromanganese oxide and $MnCO_3$ substrates, and the loss of Mn from the surrounding ferromanganese oxide substrate. Coatings of Mn-oxide were either not seen on plates deployed in 1991-1993, or were numerically insignificant, particularly relative to that observed on plates deployed in 1989.

Detrital Particle Accumulation

The relative rate of particle deposition and retention was calculated from the abundance of Fe and Mn detrital particles observed during the systematic scans of the plate surfaces. The

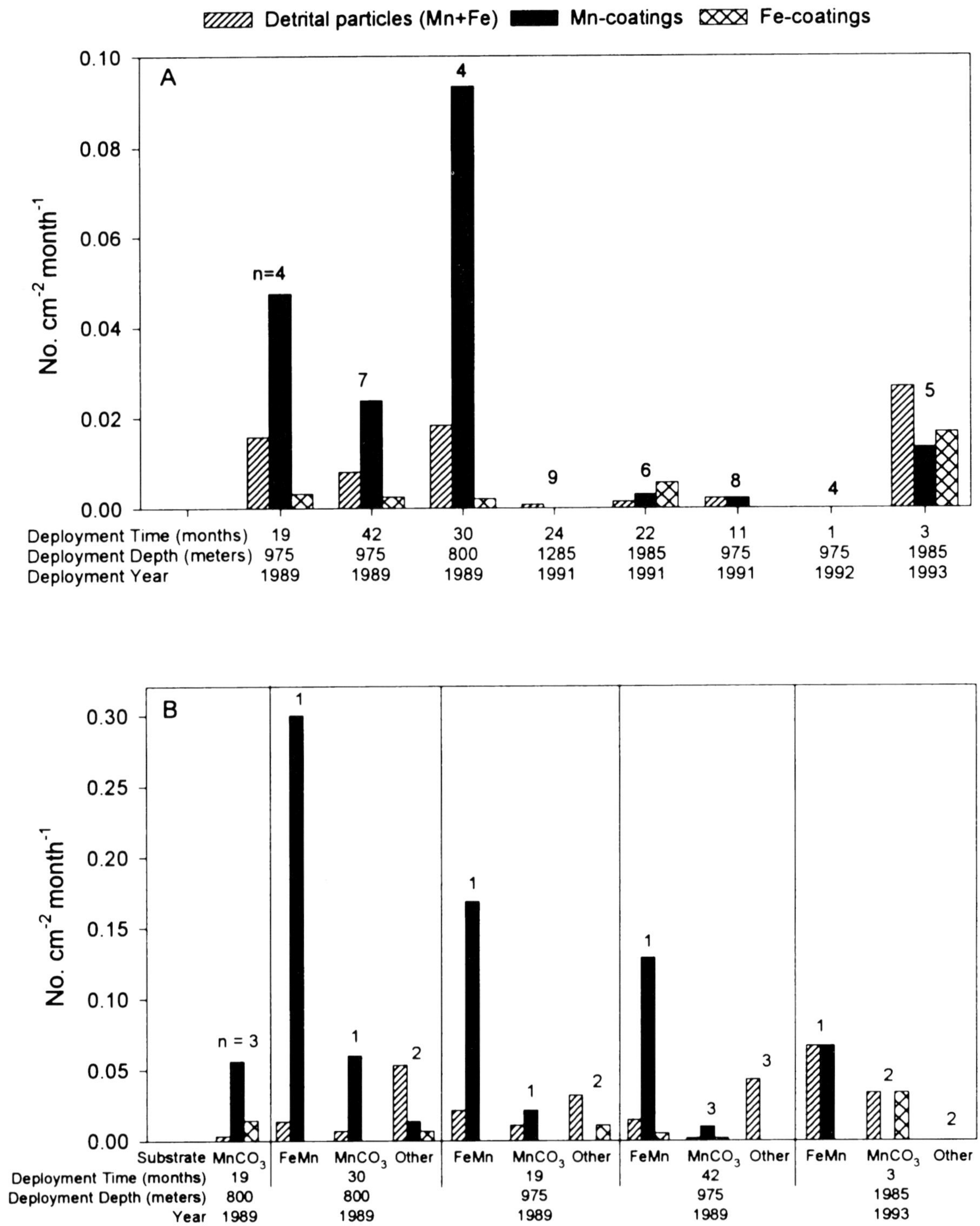

FIG. 2.— A) Average accumulation rate of Fe- and Mn-rich detrital particles and precipitated coatings on plates from all experiments except the 800-m, 19-month deployment. Much of the later data were collected using different SEM survey conditions and are therefore not directly comparable. Both detrital particles and Mn coatings accumulated most rapidly on plates deployed in 1989. B) Similar data divided according to substrate types for experiments with the most rapid accumulation rates of Fe- and Mn-rich detrital particles and coatings. Data from the $MnCO_3$ plates deployed for 19 months at 800 m were collected under the same conditions as data from other experiments and are included here. Mn coatings accumulated most rapidly on the ferromanganese-oxide substrates and are nearly absent on the basalt and $CaCO_3$ substrates (represented by "other"). In both A and B "n" is the number of plates analyzed to obtain the average accumulation rate.

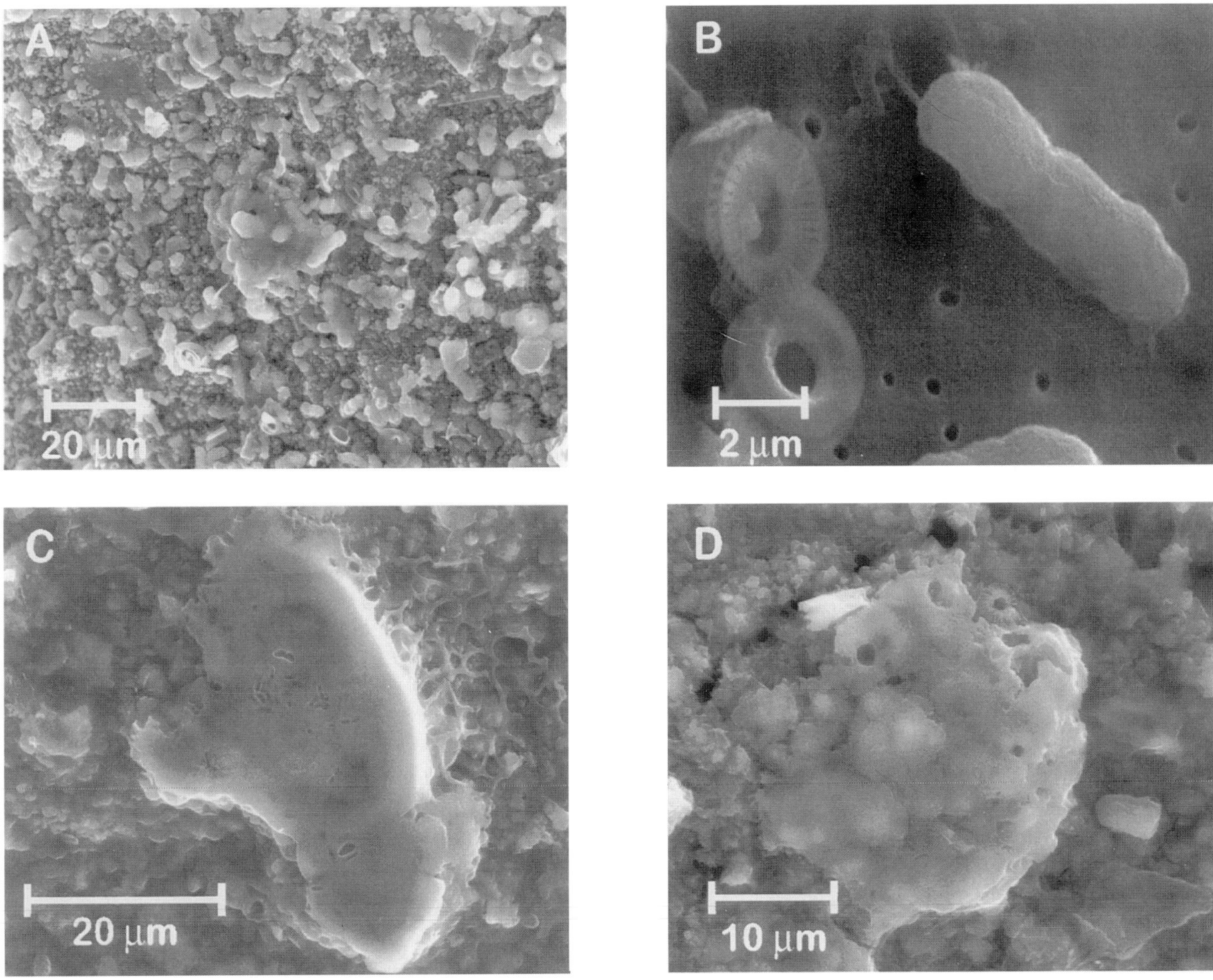

FIG. 3.— Scanning electron micrographs showing: A) Oblong structures encrusted in Mn, and a small coating of Mn oxide at the center of this micrograph are both thought to be of bacterial origin (ferromanganese-oxide substrate, 42 months at 975 m). B) Mn-coated bacterium attached to a siliceous detrital particle (rhodochrosite substrate, 19 months at 800 m). C) Imprints of bacterial size and shape in a Mn coating (ferromanganese-oxide substrate, 19 months at 800 m). D) Other Mn coating (substrate as in "C").

Fe and Mn detrital particle data from a single plate were combined because of their low abundances, and the sum was normalized to the area scanned (≈3.20 cm^2/plate) and deployment time (1–42 months). Detrital Mn and Fe particles accumulated at a faster rate on plates deployed in 1989, and on plates deployed for 3 months in 1993, than on those deployed in 1991 or 1992 (Fig. 2A). These detrital particles did not occupy enough surface area to be quantified using the point count procedure.

Coating Morphology, Abundance, and Composition

Mn-coatings covered the substrate surface (Fig. 3A, C, D), detrital particles (Figs. 3B and 4D), and the outer chamber walls of particular attached agglutinated foraminifera (Fig. 4A,B). Layers up to 7 µm thick covered the underlying substrates, detrital particles, or benthic foraminifera. These Mn-coatings were observed almost exclusively on ferromanganese-oxide and $MnCO_3$ substrates (Fig. 2B), and covered enough of the plate surface (> 4 mm^2) to be detected during the point count procedure. In addition, the coatings contained encrusted capsules (Fig. 3A, B) and bacterial size cylindrical imprints (Fig. 3C) suggestive of bacteria.

Ferromanganese-oxide plates deployed in 1989 were covered by variable abundances of Mn-coatings (up to 12% of the surface) and $MnCO_3$ plates contained scattered Mn-coatings that occupied little surface area. Spatial coverage by Mn-coatings was greatest on plates deployed for the longest period of time (30 months or 42 months at 800 m or 975 m, all deployed in 1989, Table 3). The accumulation rate of Mn-coatings estimated from numbers of deposits observed was greatest on the ferromanganese-oxide plates deployed in 1989 for 19 months at 800 m (Fig. 2A).

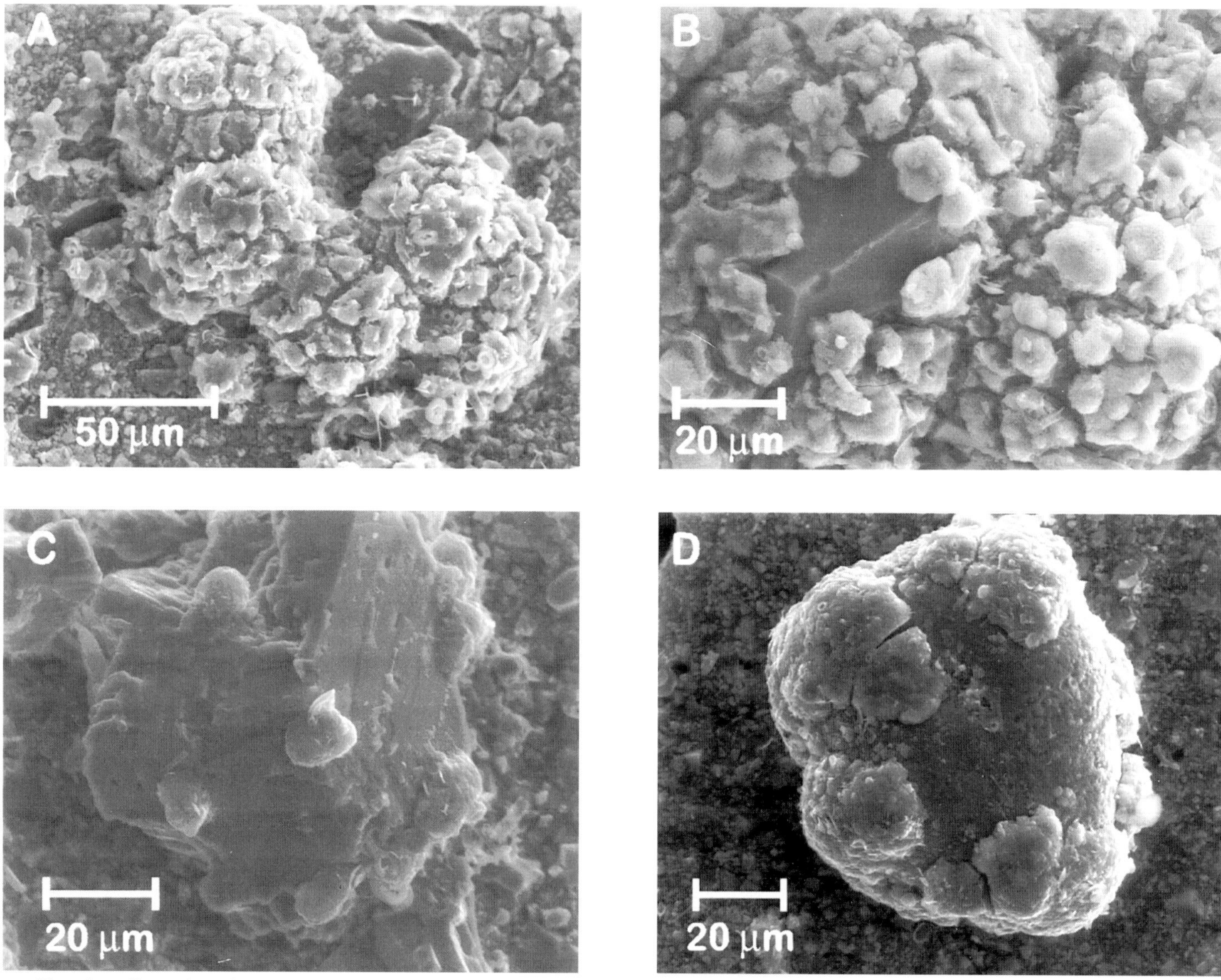

FIG. 4.— Scanning electron micrographs showing A) Agglutinated foraminifera resembling Placopsilina sp. completely encrusted in Mn oxide. The coating only covers the foraminifera and does not extend over the substrate (ferromanganese-oxide substrate, 42 months at 975 m). B) Mn coating surrounding a detrital aluminosilicate particle agglutinated within the walls of a benthic foraminifera (substrate as in "A"). C) Bulby Mn coating on a substrate grain (rhodochrosite substrate, 42 months at 975 m). D) Well-developed Mn oxide coating on a detrital calcareous foraminifera (ferromanganese-oxide substrate, 42 months at 975 m).

Twenty-four Mn-coatings that were observed on nine ferromanganese-oxide and rhodochrosite plates were chemically analyzed by EDS. The coatings on plates recovered after 3 months at 1985 m were too small and infrequently observed to obtain representative compositional data. The average composition of Mn-coatings observed on ferromanganese-oxide substrates deployed in 1989 was Mn (16±6.8%) > Fe (2.0±1.9%) > Mg (1.9±1.5%) > Ca (1.5±1.0%) > Si (1.1±1.2%) > Al (0.34±0.36%) > K (0.27±0.17%) > P (0.24±0.18%) > Co (0.16±0.13%) > Ni (0.073±0.088%; C and O made up the balance. The only notable difference in the composition of coatings observed on rhodochrosite substrates and those on ferromanganese-oxide substrates was the absence of Ni. This Mn-coating is an oxide or oxyhydroxide as inferred from the environment of formation and absence of significant amounts of S or other anion.

Fe-coatings were rarely observed and occurred only on the substrates (not covering detrital particles or foraminifera) (Fig. 2A, B). They were observed on plates from experiments deployed in 1989, 1991, and 1993, at all depths, for both short and long times. These coatings were not extensive enough to be detected during the point count analysis. Three Fe coatings were chemically analyzed: one from a ferromanganese-oxide plate deployed for 22 months at 1985 m, one from a calcite plate deployed for 19 months at 800 m, and one from a calcite plate deployed for 19 months at 975 m. The average composition of these coatings was Fe (46±32%) > Ca (3.4±2.7%) > Mn (3.1±4.6%) > Si (2.8±2.0%) > Mg (0.54±0.90%) > P (0.46±0.42%) > K (0.24±0.24%) > Al (0.19±0.33%) > Ni (0.06±0.11%) > Co (0.047±0.063%).

Ferromanganese-Oxide Substrate Composition

The average composition of the control ferromanganese-oxide substrates (types I and II, combined) was Mn (16±6%) > Fe (9.4±3.0%) > Si (3.2±1.4%) > Ca (1.2 ±0.97%) > P (0.88±0.41%) > Mg (0.50±0.29%) > Al (0.49±0.13%) > Co (0.37±0.14%) > Ni (0.27±0.09%) >K (0.16±0.10 %). The average composition of the 15 deployed ferromanganese-oxide substrates was Mn (11±4%) > Fe (10±2.7%) > Si (2.6±1.2%) > Ca (1.8±0.66%) > Mg (1.0±0.39%) > P (0.65±0.24%) > Al (0.48±0.15%) > Co (0.30±0.16%) > K (0.22±0.07%) > Ni (0.09±0.04%). Deployed ferromanganese-oxide plates had 30% and 67% less Mn and Ni, respectively, than the control, while Fe concentration remained unchanged. The Fe/Mn ratio obtained from analyses of each substrate shows that the three plates with the most altered Fe/Mn ratios were those deployed in 1989 at 800 m or 975 m and recovered after 30 months or 42 months (Fig. 5C, D).

DISCUSSION

Ferromanganese-oxide crusts and nodules generally accrete in layers, and assuming continuous growth, the average net-growth rate of a hydrogenous ferromanganese-oxide crust is 1 μm per 1000 years (Barnes and Dymond, 1967; Segl et al., 1984; Cowen et al., 1993). When metals are made available by oxic and suboxic diagenesis of detrital particles, accretion of metal oxides can occur much faster than when metals are supplied from seawater (hydrogenous deposition) (Dymond et al., 1984) and produce a 1 μm layer of new material in only 5 years. Many of the coatings observed on these substrates were deposited 1000 times faster than hydrogenous accretion, with discrete deposits forming a 2–7-μm thick layer in 1.5–3.5 years. Hydrogenous accretion of Mn and Fe from deep-ocean waters onto the oxide surface is not consistent with these growth rates, although extremely fast oxic or suboxic accretion could be.

Environmental Variability

The accumulation of detrital particles and Mn-oxide coatings on substrates deployed in 1989, but not those deployed in 1991 or later, suggests a shift in environmental conditions. A comprehensive time-series data set is available from an open-ocean site in the central north Pacific Ocean, station ALOHA (22°45'N, 158°W) (Chiswell et al., 1990; Winn et al., 1991, 1993; Tupas et al., 1993, 1994). Biogeochemical measurements, including the measurement of particulate carbon flux, have been collected by the Hawaii Ocean Time-series Program (HOT) since 1988 (Chiswell et al., 1990). Approximately monthly, the HOT program deploys free-drifting sediment traps for 4 days at 150 m, 300 m, and 500 m. Carbon content is determined from CHN analyses of the trapped particles and is converted to particulate carbon flux (Winn et al., 1991). These data show that water column fluxes dropped by 30% between 1990 and 1991 at station ALOHA, a shift that has been correlated with an El Niño-Southern Oscillation (ENSO) event (Karl et al., 1995, 1996). Although station ALOHA and Cross Seamount are separated by over 300 km, biogeochemical responses to ENSO events and other large-scale physical oceanographic changes likely impact both stations similarly. A change in particle flux to the plates is reflected in the higher deposition and retention rate of detrital particles on plates deployed in 1989 than on those deployed in later years (Fig. 2A, B). The greater abundances of benthic foraminifera seen on plates deployed in 1989 relative to those on plates deployed in subsequent years likely reflect the greater availability of food and other particles derived from the surface waters (Bertram and Cowen, 1999). It seems likely that a shift in the magnitude and perhaps composition of the particle flux between 1990 and 1991 contributed to the rapid accumulation of Mn-oxides in 1989.

Diagenetic Alteration of Ferromanganese-Oxide Substrates

The four ferromanganese-oxide substrates deployed in 1989 for 30 months or 42 months have an Fe/Mn ratio that is greater than 1; the 11 other deployed substrates, and the three controls, have Fe/Mn ratios less than 1 (Fig. 5C, D). If we assume that the Fe and Mn content of the controls represents the initial Fe/Mn ratio of the deployed substrates, an increase in the Fe/Mn ratio represents a change in the composition of the substrate. This change in the Fe/Mn ratio may be the result of: 1) a real decrease in the Mn and proportional increase in Fe content (all data are as a percent of the total), 2) a real increase in percent Fe, or 3) a combination of real Fe uptake and Mn loss. The most extensive Mn-oxide coatings were observed on ferromanganese surfaces with the greatest Fe/Mn ratio, and substrates that had Fe/Mn ratios of less than one contained few if any Mn coatings (compare Table 3, Fig. 5C, D). This suggests that Mn from the substrate was made available for Mn-precipitation on the plate surface.

Mobilization of Mn from the particulate (Mn IV) to dissolved form (Mn II) would occur if oxygen concentrations at the substrate surface were depleted. Environments where Mn (IV) is reduced, migrates to a more oxygenated environment, and is reoxidized and deposited, are documented in sediments (e.g., Froehlich et al., 1979) and particular ocean basins (e.g., Cowen and Bruland, 1985; Tebo, 1991), and proposed for Mn-nodules (Dymond et al., 1984). δ-MnO_2 (Mn IV) reduction is associated with low Eh conditions that can occur in a restricted environment during organic carbon (Corg) oxidation:

$$(C_{org}) + 2Mn^{4+} + 2H_2O = 2Mn^{2+} + CO_2 + 4H^+$$

Just outside the oxygen-depleted microzone, and just above the mineral surface, Mn (II) can be oxidized to Mn (IV), and form an oxidized Mn mineral:

$$2Mn^{2+} + O_2 + 2H_2O = 2MnO_2 + 4H^+$$

The suggested migration and redox chemistry of Mn at these substrate surfaces parallel those described by Froehlich et al.

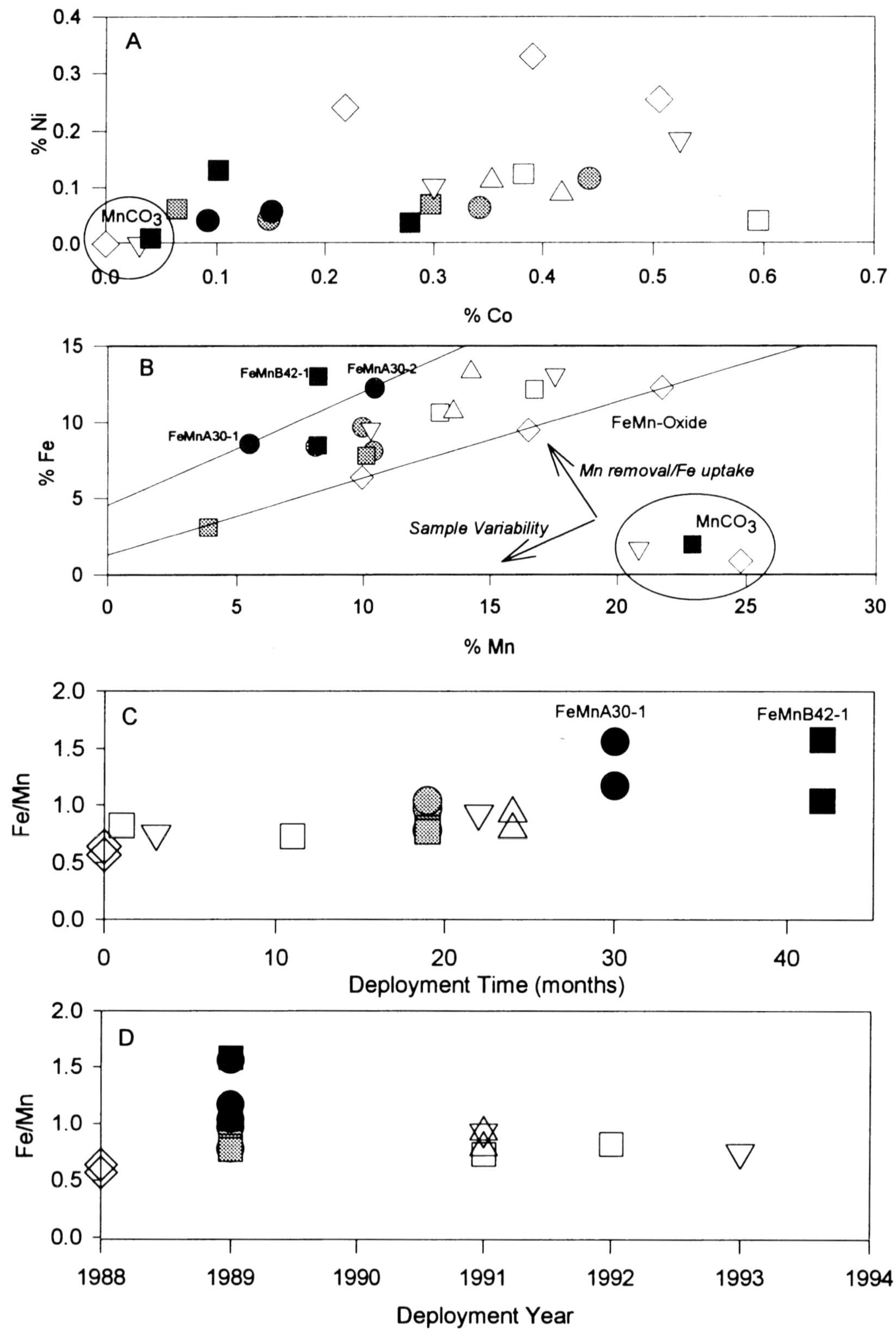

FIG. 5.—Chemical composition of ferromanganese-oxide and rhodochrosite substrates from all experiments. A) %Co versus %Ni. B) %Mn versus %Fe. See table 2 for codes used to identify plates. C) Deployment time versus Fe/Mn ratio in ferromanganese-oxide substrate. D) Deployment year versus Fe/Mn ratio in ferromanganese-oxide substrate. Symbols used: ◇ = controls, ○ = 800 m deployments, □ = 975 m deployments, △ = 1285 m deployments, ▽ = 1985 m deployments. Gray fill=19 month deployment, Black fill = 30-month or 42-month deployment, white = other deployments.

TABLE 3.— PERCENT OF PLATE SURFACE COATED BY MN.

Depth (m)	Experiment Time (months)	Year Deployed	Plate †	% Cover‡
800	19	1989	FeMnA19-1	<0.2
			FeMnA19-2	<0.2
			FeMnA19-3	<0.2
			$MnCO_3$A19-1	0.4
			$MnCO_3$A19-2	<0.2
			$MnCO_3$A19-3	<0.2
800	30	1989	FeMnA30-1	2.3
			FeMnA30-2	<0.2
			$CaCO_3$30-1	<0.2
975	19	1989	FeMnB19-1	0.8
			FeMnB19-2	<0.2
975	42	1989	FeMnB42-1	12
			FeMnB42-2	1
			FeMnB42-3	0.3
			$MnCO_3$B42-1	<0.2
1985	3	1993	FeMnD3-1	<0.2
			$MnCO_3$ D3-1	<0.2

†See Table 2 for codes used in plate identification.
‡<0.2 indicates coatings observed during the systematic scans of the plate surface, but not detected by the point count procedure. Mn coatings were not observed on any of the other plates.

(1979) in suboxic zones within the sediments. Support for this hypothesis is seen in the pattern of Mn-accumulation on different substrates during different intervals. There was an absence of Mn accretion on Mn-free substrates ($CaCO_3$ and basalt substrates) (Table 3, Fig. 2B), and on substrates deployed when particle flux was inferred to be relatively low. Significant Mn accretion was seen on ferromanganese-oxide substrates deployed in 1989. These data are consistent with the interpretation that the accumulation of carbon-rich particles at the ferromanganese-oxide surfaces deployed in 1989 produced conditions conducive to Mn reduction and reoxidation.

A comparison between alterations to the ferromanganese-oxide and rhodochrosite substrates is useful in deducing the conditions that may have persisted on the seafloor during deployment. Unlike δ-MnO_2, rhodochrosite ($MnCO_3$) should dissolve when exposed to oxygenated seawater. In the waters surrounding Cross Seamount, the ion concentration product for $MnCO_3$,

$$[Mn^{2+}] \times [CO_3^{2-}] = (10^{-9}) \times (7 \times 10^{-5}) = 7 \times 10^{-14},$$

is much less than that needed for saturation ($3.6 \pm 0.7 \times 10^{-9}$, Wartell et al., 1990). Rhodochrosite is therefore expected to provide a continuous source of Mn (II) to the overlying microlayer of seawater. The extensive Mn-coatings on ferromanganese-oxide substrates and relatively few Mn-coatings on rhodochrosite substrates could reflect differential rates of dissolution of the two substrates, particularly if the accumulation rate of oxides reflects the availability of appropriate metal ions (Dymond et al., 1984).

Plates deployed in 1989 for 19 months should be compositionally similar to plates deployed in 1989 for 30 months and 42 months if most of the alteration to the substrate occurred as a result of deposition of particulate organic matter on substrate surfaces. Yet, the Fe/Mn ratio of ferromanganese-oxide substrates deployed in 1989 for 19 months is similar to that of substrates deployed in later years. This apparent conflict has no satisfactory explanation, although three possibilities are considered here. First, anoxic conditions may have developed on these plates during organic carbon degradation, mobilizing both Fe and Mn in similar proportions during the first 19 months. The Fe/Mn ratio would remain approximately the same, although mass would be lost. This is a possibility, although it is difficult to confirm given the nature of the data and highly unlikely in the open ocean. Second, given the variability in the chemical composition of ferromanganese-oxide minerals, and the presumed patchy deposition of particulate organic matter, the data may not be representative of the substrate composition if foo few analyses were made. Finally, Fe uptake over time, although unlikely, could explain these data given the low Fe content of the Mn-coatings.

The high accumulation rate of detrital particles and coatings on plates deployed for 3 months in 1993 is more difficult to explain because particle flux was still relatively low (although the interval of deployment was not well-sampled by the HOT program). If Mn coatings are formed in conjunction with the degradation of organic carbon, as proposed above, then the accumulation of a single marine aggregate at the plate surface could produce the observed data. Alternatively, Mn and Fe coatings may be deposited hydrogenetically from seawater.

Co and Ni

Diagenetic processes could also impact the Co and Ni content of Fe and Mn oxides in the original substrate and the Mn coatings. Ni is depleted in all exposed ferromanganese-oxide substrates relative to the original substrates (Fig. 5A), suggesting that Ni was mobilized from the substrate. This Ni could be incorporated into a Mn oxide outside the microenvironment that resulted in its mobilization. Interestingly, Ni is relatively abundant in Mn-coatings observed on plates deployed in 1989 for 19 months, but not in coatings on plates deployed in 1989 for longer times (Fig. 6B). Either the Ni-rich layer deposited during the first 19 months was subsequently 1) coated with a layer poor in Ni, 2) depleted in Ni by exchange between Ni and major seawater cations (e.g., Mg, see Fig. 6B), or 3) was never present in the Mn coatings found on the 30- and 42-month plates. The later is unlikely, if the coatings analyzed from the 30- and 42-month plates began to accumulate during the first 19 months (all were deployed in 1989).

In contrast, neither the coatings on the $MnCO_3$ substrate, or the $MnCO_3$ substrate itself, contain detectable Ni concentrations. This suggests that the Ni in the coatings on the ferromanganese-oxide plates is supplied by the underlying ferromanganese oxide or that different processes control the Ni

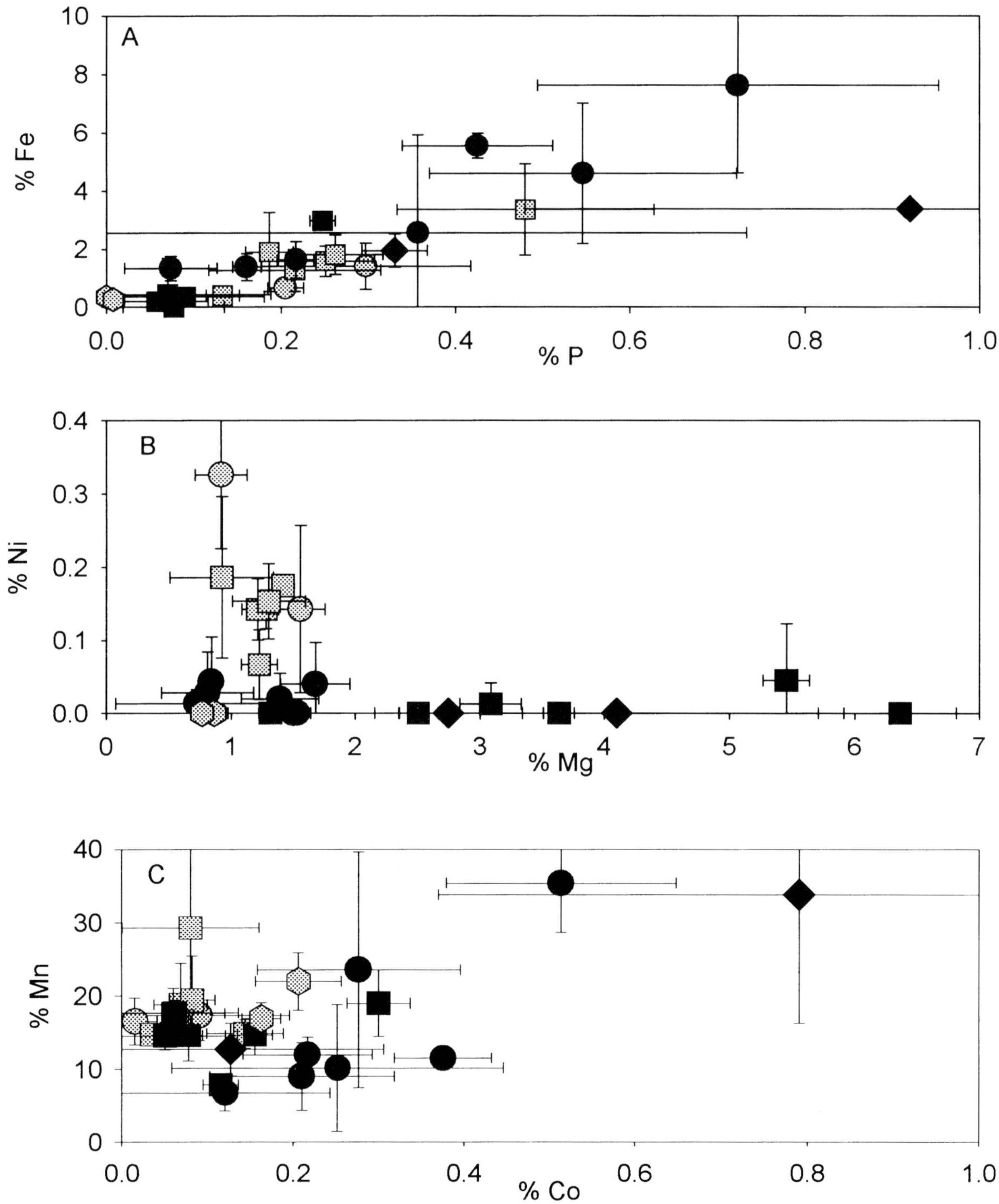

FIG. 6. Chemical composition of Mn coating. Concentration data (mole percent) are from coatings observed on ferromanganese-oxide (FeMn-oxide) and rhodochrosite ($MnCO_3$) plates deployed in 1989; coverage of plates from other deployment dates or of other compositions was not extensive enough to obtain representative and reliable data on coating composition. Mean ± 1 standard deviations are plotted for each element. A) %P versus %Fe. B) %Mg versus %Ni. C) %Co versus %Mn. Symbols used: circle = FeMn-oxide substrate deployed at 800 m, square = FeMn-oxide substrate deployed at 975 m, hexagon = $MnCO_3$ substrate deployed at 800 m, diamond = $MnCO_3$ substrate deployed at 975 m. Gray fill = 19-month deployment, Black fill = 0- or 42-month deployment.

content of the precipitates on these two substrate types.

The rate of Mn oxide formation influences the Mn-mineralogy of the precipitate and thus the types of cations that can be accommodated in the crystal structure for charge balance (e.g., Usui, 1979; Dymond et al., 1984; Usui et al., 1989). The relatively high percentage of Ni in the Mn oxides formed on the ferromanganese plates deployed for 19 months in 1989, and the absence of Ni in coatings on $MnCO_3$ plates, may reflect the rate of Mn-oxide formation. $MnCO_3$ and ferromanganese oxide dissolve under different conditions and at different rates, thus supplying Mn to the region of precipitation at a different rate. This explanation can be extended to explain the absence of Ni in the coatings observed on ferromanganese-oxide plates deployed for 30 months or 42 months. Because a decrease in the Mn-oxide dissolution rate is expected after the most reactive organic matter is consumed, the supply of dissolved Mn is diminished. Alternatively, the absence of Ni in the Mn-oxide coatings observed on ferromanganese-oxide plates deployed for 30 months or 42 months (that were also in place for the first 19 months) could reflect the diminished source of Ni from dissolving detrital particles and exchange between Ni and Mg or another major seawater cation in the later 11–23 months of the experiment.

Co concentrations in the oxidized Mn-mineral appear to be independent of environmental conditions and inferred Mn-mineralogy. Co is present in greatest abundance in the coatings observed on both the ferromanganese-oxide and $MnCO_3$ substrates deployed for 30 months or 42 months. The higher percent Co in Mn-coatings on plates deployed for the longest time suggests time-dependent oxidation of dissolved Co from seawater. This time-dependence may reflect the slow kinetics of Co uptake onto surfaces and change in oxidation state (Co^{2+} to Co^{3+}, Murray, 1975; Li, 1991, and references therein). This observation is consistent with the expectation that a rapidly accreting oxide should contain less Co than a slowly accreting one (Halbach and Puteanus, 1984).

Fe and P

The presence of an Fe P-phase is suggested by the positive association between the two elements in the Mn oxide (Fig. 6A). Phosphate is commonly adsorbed to FeOOH from seawater in surface sediments (e.g., Jarvis et al., 1994). The positive association between the two elements in the coatings (Fig. 6A) likely reflects adsorption of P (as phosphate) onto Fe oxides that are a minor component of these Mn coatings.

Biologically Mediated or Geochemically Controlled Precipitation?

The morphology and rapid deposition rate of the secondary Mn oxides may indicate a biological role in the accretion of these deposits. Morphological support for bacterially mediated formation is seen in the oblong, bacterial-sized structures that are encrusted in Mn (Fig. 4A,B). In addition, these precipitates accreted at a rate 1000 x faster than inorganic Mn oxides precipitated in the laboratory (e.g., Hem, 1981; Sung and Morgan, 1981 as discussed in Bertram and Cowen, 1994), a rate comparable to that of Mn uptake onto the surfaces of bacterial spores (Hastings and Emerson, 1986). The combination of Mn-encrusted bacteria-like particles and greatly accelerated precipitation rates relative to expected inorganic geochemical rates is not proof of biological mediation but highly suggests it.

Interestingly, the chambers of a particular agglutinated foraminifera, one that resembles *Placopsilina sp.*, were typically coated with Mn oxides on the ferromanganese and $MnCO_3$ substrates. These foraminifera utilize the underlying substrate as a portion of their chamber walls and cement detrital particles with an organic cement to form the remainder of the test. New chambers are agglutinated as these foraminifera age, with each chamber larger than the preceding one. The manganese oxides were observed on the outer walls of the chambers, which formed first and occasionally over the entire foraminifera. Because similar foraminifera on Mn-containing plates deployed after 1990 are not coated, it is unlikely that the foraminifera alone are responsible for the presence of the coating. The presence of bacteria that facilitate Mn-deposition may also be required. Whatever the mechanism, foraminiferal chamber walls and other particles are focal points for the deposition of Mn-oxides.

The manganese-oxide coatings that form on agglutinated foraminifera clearly integrate aluminosilicates, coccoliths, barite, and other detrital particles into the ferromanganese crust by surrounding them in a layer of oxide material. One result of the accumulation of ferromanganese oxide around benthic foraminifera is illustrated by the persistence of foraminifera-associated minerals, including agglutinated detrital particles and the precipitated skeletons of calcareous forms within ferromanganese crusts and nodules. Distinct layers in ferromanganese-oxide crusts include abundant molds of tests, while other layers are free of such molds. Wendt (1974) hypothesized that this layering reflected "changes in the microbiological, sedimentary, and/or geochemical environment". The formation of Mn oxides on the benthic foraminifera observed in this study illustrate one of the ways in which foraminiferal tests and detrital material can be rapidly incorporated and preserved in a ferromanganese-oxide deposit.

IMPLICATIONS

The rapid accumulation of manganese oxides documented here provides a straightforward mechanism for the fossilization of the liths produced by coccolithophorids. Coccoliths are produced in abundance in surface ocean waters of the oligotrophic ocean and are among the detrital particles that become incorporated into ferromanganese-oxide deposits. As coccolithophores evolved, the microarchitecture of the lith changed (e.g., Bolli et al., 1985) and thus the imprints that are left behind in the ferromanganese oxides can be used as time markers for dating the age of discrete layers within the crust (Janin, 1987; Cowen et al., 1993). Suboxic mobilization and redeposition of Mn facilitates the rapid incorporation of liths from coccolithophorids as well as other detrital particles within the growing oxide. A detailed imprint of the lith remains within the oxide after dissolution of the calcite.

Ferromanganese crusts are thought to be free of significant postdepositional alteration (Wen et al., 1997) yet this study shows that the distribution of Mn and other trace elements can reflect diagenetic processes. The formation of lith imprints within crusts, as described above, clearly requires that postdepositional alteration of incorporated detrital particles does occur for some elements, and studies show that the ferromanganese oxides are open systems with respect to Sr (VonderHaar et al., 1995). Detailed chemical and biostratigraphic analyses are available for a crust recovered from Schumann Seamount in the Hawaiian Islands (RD50S1B, Cowen et al., 1993; McMurtry et al., 1994; Wen et al., 1997). Rapid accretion rates (> 2 mm/Myr) are calculated using coccolith biostratigraphy, for a layer approximately 6 mm in thickness within this crust (Cowen et al., 1993). This layer is rich in detrital elements (Fe, Al, Si; McMurtry et al., 1994) and overlain by an apatite-rich layer interpreted to have accumulated during a period of elevated productivity (Cowen et al., 1993).

The covariance of Co with Mn, and the minimum concentrations (with respect to the composition of the analyzed crust) of both elements at the base of this rapidly accreting layer (McMurtry et al., 1994) are consistent with diagenetic reworking promoted by periodic suboxic conditions at the crust surface. The hypothesized pathway for the observed chemical signature in the rapidly accreting layer is as follows. An initial Fe-Mn layer deposited purely by hydrogenetic processes should have a hypothetical Fe/Mn ratio of less than 1. Suboxic mobilization of Mn would lower the Mn content, increasing the Fe/Mn ratio to greater than 1. Cobalt associated with the oxidized Mn minerals would also become depleted in the ferromanganese layer. Rapid deposition of a Mn-rich layer over the altered crust would then produce areas with very low Fe/Mn ratios mixed with hydrogenetic deposits of more Fe-rich oxides. Continued periodic deposition of organic-rich particles at the crust surface would continue to promote the leaching/redeposition of Mn until the flux of particles from the surface oceans became too rapid for oxide minerals to continue to accrete, leading to the formation of an apatite layer. A gradual shift to increased productivity in the surface oceans thus promotes reductive dissolution of the ferromanganese oxides in a crust. Further detailed stratigraphic studies are necessary before such a pathway can be fully substantiated.

ACKNOWLEDGMENTS

We thank W. Burnett, D. Cronan, and J. Resing for their careful reviews of this manuscript. We are most grateful to the pilots and crews of the *PISCES V* submersible and *R/V Kila* for their assistance in the deployment and recovery of these experiments.

The work was partially funded by the University of Hawaii Sea Grant College Program (SOEST) project number R/OM-6 under Institutional Grant No. NA36RG0507 from NOAA Office of Sea Grant, Department of Commerce. Additional support was provided by the State of Hawaii's Department of Business and Economic Development and NSF OCE-9217514. Valuable submersible time was granted by NOAA National Undersea Research Program and the Hawaii Undersea Research Laboratory.

This is Sea Grant publication UNIHI-SEAGRANT-JC-98-06, and contribution no. 4827 from the School of Ocean and Earth Science and Technology, University of Hawaii.

REFERENCES

Aplin, A. C. and Cronan, D. S., 1985, Ferromanganese oxide deposits from the Central Pacific Ocean, I. Encrustations from the Line Islands Archipelago: Geochimica et Cosmochimica Acta, v. 49, p. 427-436.

Barnes, S. S. and Dymond, J. R., 1967, Rates of accumulation of ferromanganese nodules: Nature, v. 213, p. 1218-1219.

Bertram, M. A. and Cowen, J. P., 1999, Temporal variations in the deep-water colonization rates of small benthic foraminifera: the results of an experiment on Cross Seamount: Deep-Sea Research, v. 46, p. 1021-1050.

Bertram, M. A. and Cowen, J. P., 1994, Testate rhizopod growth and mineral deposition on experimental substrates from Cross Seamount: Deep-Sea Research, v. 41, p. 575-601.

Bolli, H. M., Saunders, J. B. and Perch-Nielsen, K., 1985, Plankton Stratigraphy: Cambridge Earth Science Series, Cambridge University Press, 1032 p.

Burnett, B. R. and Nealson, K. H., 1983, Energy dispersive X-ray analysis of the surface of a deep-sea ferromanganese nodule: Marine Geology, v. 53, p. 313-329.

Chiswell, S., Firing, E., Karl, D., Lukas, R., Winn, C., 1990, Hawaii Ocean Time-Series Program Data Report 1, 1988-1989: School of Ocean and Earth Science and Technology, Univ. of Hawaii, Honolulu, HI, 175 p.

Cowen, J. P. and Bruland, K.W., 1985, Metal deposits associated with bacteria: implications for Fe and Mn marine biogeochemistry: Deep-Sea Research, v. 32, p. 253-272.

Cowen, J. P., DeCarlo, E. H. and McGee, D. L., 1993, Calcareous nannofossil biostratigraphic dating of a ferromanganese crust from Schumann Seamount: Marine Geology, v. 115, p. 289-306.

DeCarlo E. H., P. A. Pennywell, C. M. and Fraley, 1987, Geochemistry of ferromanganese deposits from the Kiribati and Tuvalu region of the West Central Pacific Ocean: Marine Mining, v. 6, p. 301-321.

Dudley, W. C., 1976, Cementation and iron concentration in foraminifera on manganese nodules: Journal of Foraminiferal Research, v. 6, p. 202-207.

Dymond, J., Lyle, M., Finney, B., Piper, D. Z., Murphy, K., Conard, R., Pisias, N., 1984, Ferromanganese nodules from MANOP Sites H, S, and R—Control of mineralogical and chemical composition by multiple accretionary processes: Geochimica et Cosmochimica Acta, v. 48, p. 931-949.

Froehlich, P. N., Klinkhammer, G. P., Bender, M. L., Luedtke, N. A., Heath, G. R., Hammond, D., Hartman, B., and Maynard, V., 1979, Early oxidation of organic matter in pelagic sediments of the eastern equatorial Atlantic: suboxic diagenesis: Geochimica et Cosmochimica Acta, v. 43, p. 1075-1090.

Gooday, A. J. and Haynes, J. R., 1983, Abyssal foraminifers, including two new genera, encrusting the interior of Bathysiphon rusticus tubes: Deep-Sea Research, v. 30, p. 591-614.

Graham, J. W. and Cooper, S. C., 1959, Biological origin of manganese-rich deposits of the sea floor: Nature, v. 183, p. 1050-1051.

Halbach, P. and Puteanus, D., 1984, The influence of the carbonate dissolution rate on the growth and composition of Co-rich ferromanganese crusts from Central Pacific seamount areas: Earth and Planetary Science Letters, v. 68, p. 73-87.

Hastings, D. and Emerson, S., 1986, Oxidation of manganese by spores of a marine bacillus: Kinetic and thermodynamic considerations: Geochimica et Cosmochimica Acta, v. 50, p. 1819-1824.

Hein, J. R., Bohrson, W. A., Schulz, M. S., Noble, M. and Clague, D. A., 1992, Variations in the fine-scale composition of a central Pacific ferromanganese crust: paleoceanographic implications: Paleoceanography, v. 7, p. 63-77.

HEM, J. D., 1981, Rates of manganese oxidation in aqueous systems: Geochimica et Cosmochimica Acta, v. 45, 1369-1374.

JANIN, M., 1987, Micropaléontologie de concrétions polymétalliques du Pacifique central: zone Clarion-Clipperton, Chaine Centre-Pacifique, Ile de la Ligne et archipel des Touamotou (Eocene-Actuel): Societe Geologique de France, Paris, 135 p.

JARVIS, I., BURNETT, W. C., NATHAN, Y., ALMBAYDIN, F. S. M., ATTIA, A. K. M., CASTRO, L. N., FLICOTEAUX, R., HILMY, M. E., HUSAIN, V., QUTAWNAH, A. A., SERJANI, A. AND ZANIN Y. N., 1994, Phosphorite geochemistry: state-of-the-art and environmental concerns: Eclogae Geologicae Helvetiae, v. 87, p. 643-700.

KARL, D. M., CHRISTIAN, J. R., DORE, J. E., HEBEL, D. V., LETELIER, R. M., TUPAS, L. M. AND WINN, C. D. , 1996, Seasonal and interannual variability in primary production and particle flux at Station ALOHA: Deep-Sea Research, v. 43, p. 539-568.

KARL, D. M., LETELIER, R., HEBEL, D., TUPAS, L., DORE, J., CHRISTIAN, J. AND WINN, C., 1995, Ecosystem changes in the North Pacific subtropical gyre attributed to the 1991-1992 El Niño: Nature, v. 373, p. 230-234.

KELLY, K. M., 1988, The effects of mass wasting on the development of cobalt-rich ferromanganese crusts on Cross Seamount, Hawaii: Unpublished M.S. Thesis, University of Hawaii, Honolulu, 405 p.

LI, Y. H., 1991, Distribution patterns of the elements in the ocean: A synthesis: Geochimica et Cosmochimica Acta, v. 55, p. 3223-3240.

MARGOLIS, S. V., DUGOLINSKY, B. K., AND DUDLEY, W. C. , 1979, Microchemistry and morphology of biogenic detrital and authigenic phases in Pacific manganese nodules: La Genèse des Nodules de Manganèse, v. 289, p. 179-190.

MCMURTRY, G. W., VONDERHAAR, D. L., EISENHAUER, A., MAHONEY, J. J. AND YEH, H. -W., 1994, Cenozoic accumulation history of a Pacific ferromanganese crust: Earth and Planetary Science Letters, v. 125, p. 105-118.

MEESE, R. J. AND TOMICH, P. A., 1992, Dots on the rocks: a comparison of percent cover estimation methods: Journal of Experimental Marine Biology and Ecology, v. 165, p. 59-73.

MULLINEAUX L. S., 1987, Organisms living on manganese nodules and crusts: distribution and abundance at three North Pacific sites: Deep-Sea Research, v. 34, p. 165-184.

MURRAY, J. W., 1975, The interaction of cobalt with hydrous manganese dioxide: Geochimica et Cosmochimica Acta, v. 39, p. 635-647.

NEALSON, K. H., TEBO, B. M., AND ROSSON, R. A., 1988, Occurrence and mechanisms of microbial oxidation of manganese: Advances in Applied Microbiology, v. 33, p. 279-318.

RIEMANN, F., 1983, Biological aspects of deep-sea manganese nodule formation: Oceanologica Acta, v. 6(3), p. 303-311.

SAGER, W. W. AND PRINGLE, M. S., 1987, Paleomagnetic constraints on the origin and evolution of the Musicians and South Hawaiian Seamounts, Central Pacific Ocean, *in* Keating, B. H., Fryer, P., Batiza, R. and Boehlert, G. W., eds., Seamounts, Islands, and Atolls: Geophysical Monograph 43, p. 133-164.

SEGL, M., MANGINI, A., BONANI, G., HOFMANN, J., NESSI, M., SUTER, M., WOLFLI, W., FRIEDRICH, G., PLUGER, W. L., WIECHOWSKI, A., AND BEER, J., 1984, Be-dating of a manganese crust from the central North Pacific and implications for ocean paleocirculation: Nature, v. 309, p. 540-543.

SUNG, W. AND MORGAN J. J., 1981, Oxidative removal of Mn(II) from solution catalyzed by the γ-FeOOH (lepidocrocite) surface: Geochimica et Cosmochimica Acta, v. 45, p. 2377-2383.

TEBO, B. M., 1991, Manganese(II) oxidation in the suboxic zone of the Black Sea: Deep-Sea Research, v. 38, p. S883-S905.

TUPAS, L., SANTIAGO-MANDUJANO, HEBEL, D., LUKAS, R. KARL, D. AND FIRING, E., 1993, Hawaii Ocean Time-Series Program Data Report 4: School of Ocean and Earth Science and Technology, Univ. of Hawaii, Honolulu, HI, 248 p.

TUPAS, L., SANTIAGO-MANDUJANO, HEBEL, D., FIRING, E., BINGHAM, F., LUKAS, R., AND KARL, D., 1994, Hawaii Ocean Time-Series Program Data Report 5: School of Ocean and Earth Science and Technology, Univ. of Hawaii, Honolulu, HI, 156 p.

TURLEY, C. M., LOCHTE, K. AND PATTERSON, D. J., 1988, A barophilic flagellate isolated from 4500 m in the mid-North Atlantic: Deep-Sea Research, v. 35, p. 1079-1092.

USUI, A., 1979, Minerals, metal contents and mechanism of formation of manganese nodules from the Pacific Basin (GH76-1 and GH77-1 areas), *in* Bischoff, J. L. and Piper D. Z., eds., Marine Geology and Oceanography of the Pacific Manganese Nodule Province: Plenum Press, p. 651-679.

USUI, A., MELLIN, T. A., NOHARA, M. AND YUASA, M., 1989, Structural stability of marine 10Å manganates from the Ogasawara (Bonin) arc: implication for low-temperature hydrothermal activity: Marine Geology, v. 86, p. 41-56.

VONDERHAAR, D. L., MAHONEY, J. J. AND MCMURTRY, G. M., 1995, An evaluation of strontium isotopic dating of ferromanganese oxides in a marine hydrogenous ferromanganese crust: Geochimica et Cosmochimica Acta, v. 59, p. 4267-4278.

WARTELL, M., SKIKER, M., AUGER, Y., AND BOUGHRIET, A., 1990, Interaction of manganese (II) with carbonates in seawater: Assessment of the solubility product of $MnCO_3$ and Mn distribution coefficients between the liquid phase and $CaCO_3$ particles: Marine Chemistry, v. 29, p. 99-117.

WEN, X., DECARLO, E. H. AND LI, Y. H., 1997, Interelement relationships in ferromanganese crusts from the central Pacific ocean: Their implications for crust genesis: Marine Geology, v. 136, p. 277-297.

WENDT, J., 1974, Encrusting organisms in deep-sea manganese nodules, *in* Hsu K. J. and H. C. Jenkyns, eds., Pelagic sediments on land and under the sea: Blackwell Scientific, Great Britain, p. 437-447.

WINN, C., CHISWELL, S., FIRING, E., KARL, D., LUKAS, R., 1991, Hawaii Ocean Time-Series Program Data Report 2: School of Ocean and Earth Science and Technology, Univ. of Hawaii, Honolulu, HI, 175 p.

WINN, C., LUKAS, R. KARL, D. AND FIRING, E., 1993, Hawaii Ocean Time-Series Program Data Report 3: School of Ocean and Earth Science and Technology, Univ. of Hawaii, Honolulu, HI, 228 p.

RARE EARTH ELEMENT FRACTIONATION IN HYDROGENETIC FE-MN CRUSTS: THE INFLUENCE OF CARBONATE COMPLEXATION AND PHOSPHATIZATION ON SM/YB RATIOS

E. HEINEN DE CARLO*, X. Y. WEN, AND J. P. COWEN

Department of Oceanography, School of Ocean and Earth Science and Technology, University of Hawaii, Honolulu, HI 96822-2285

ABSTRACT: Hydrogenetic Fe-Mn crust layers accrete sequentially over millions of years and incorporate many trace elements by particle scavenging processes. By regulating the chemical environment at the aqueous/particle interface, conditions in the oceanic water column can thus influence the composition of Fe-Mn crusts. Because crusts are subject to much less diagenesis and post-depositional reworking than most other sedimentary deposits, they represent condensed stratigraphic sequences that reflect oceanic conditions through geologic time. Unfortunately many past efforts to reconstruct growth histories of crusts suffered from limitations inherent in dating techniques. Recently it has been demonstrated that Fe-Mn crust laminae can be dated reliably using nannofossil biostratigraphy; and a combined approach using biostratigraphy and other dating methods now provides better chronological control on crust stratigraphy than previously available.

In this paper we present high-resolution (<3 mm) rare earth element (REE) profiles in hydrogenetic Fe-Mn crusts from two Pacific seamounts. We evaluate the concentrations of these trace elements in light of results of laboratory scavenging experiments and attempt to resolve the influence of solid-phase mineralogy and seawater chemistry on the extent of REE fractionation in growth layers of crusts. Because the chemistry of the REE in seawater is dominated by complexation with dissolved carbonate we hypothesize that temporal variations in REE fractionation in Fe-Mn crusts reflect primarily changes in the dissolved carbonate content of seawater. The use of the Sm/Yb ratio is proposed as a convenient measure of REE fractionation in Fe-Mn crusts and variations thereof are examined with respect to changes in oceanic carbonate content during the Cenozoic. Shale-normalized Sm/Yb ratios in nonphosphatized sections of crusts from the Pacific are believed to represent unaltered signals of conditions in the oceans at the time of accumulation and correlate inversely with the CCD throughout the Cenozoic.

INTRODUCTION

Early research efforts on marine ferromanganese (Fe-Mn) crusts focused primarily on the regional distribution and abundance of potentially economic mineral deposits and the main factors governing their formation. Little emphasis was placed on detailed studies of Fe-Mn crusts, although they are typically characterized by a high degree of compositional heterogeneity as examination of the broad data compilation by Manheim and Lane-Bostwick (1989) quickly reveals.

Over the past decade or so, however, many authors have conducted detailed textural, mineralogical, and elemental analyses, studied partitioning of elements between seawater and Fe-Mn oxides, and applied dating techniques and/or empirical relationships designed to infer the growth history of Fe-Mn crusts (Halbach and Puteanus, 1984; von Stackelberg et al., 1984; Janin, 1988; Pichoki and Hoffert, 1987; Futa et al., 1988; Segl et al., 1989; Alvarez et al., 1990; Ingram et al., 1990; De Carlo, 1991; Neumann and Stueben, 1991; De Carlo and McMurtry, 1992; Hein et al., 1992; Cowen et al., 1993; McMurtry et al., 1994; Chabaux et al., 1995; Koschinsky and Halbach, 1995; Bau et al., 1996; Wen et al., 1997). Unfortunately, many of the earlier detailed studies only examined the compositional variability of Fe-Mn crusts on a relatively coarse scale (>5-10 mm).

It is now generally accepted that Fe-Mn crusts represent a relatively well-preserved, albeit highly condensed, geologic record of biogeochemical conditions in the oceans extending back to the onset of Fe-Mn oxide accumulation onto the substrate material. Published crust growth rates in the range of ~0.1–5 mm/Ma (e.g., Segl et al. 1984, 1989; Cowen et al., 1993) suggest that 5–10-cm thick crusts represent at least several tens of million years of growth history. Some of the more important factors that influence the composition and structure of crusts include bottom current activity; the range of the calcite compensation depth (CCD) and lysocline; biological productivity; the depth and intensity of the oxygen minimum zone; the supply of dissolved metals and particulate matter to seawater; hydrothermal activity; and, perhaps to a much less significant extent, post-depositional and diagenetic reworking (von Stackelberg et al., 1984; Segl et al., 1984, 1989; Cowen and Taylor, 1987; Alvarez et al. ,1990; De Carlo 1991; Hein et al. 1992; De Carlo et al., 1994; Koschinsky and Halbach, 1995; Wen, 1995; Bau et al., 1996; Wen et al., 1997; Bertram and Cowen, this volume).

The slow growth rate of crusts presents ideal conditions for scavenging of trace elements from seawater onto sequentially deposited Fe and Mn oxide layers. Hence, stratigraphic distributions of trace elements potentially provide the means of examining changing geochemical conditions in the oceanic water column over time. The constraints imposed by the very slow growth rates of crusts, however, necessitate that fine-scale analyses be conducted to determine how the compositional variations and morphological boundaries in crusts correlate to changes in environmental conditions during the course of Fe-Mn oxide accumulation. For example, De Carlo (1991) and McMurtry et al. (1994) studied a 9.5-cm-thick Fe-Mn crust recovered from Schumann Seamount, near Hawaii, and independently proposed that strata of this crust recorded a series of geological events and changes in depositional conditions in the water column of the Pacific Ocean. Varying textural characteristics of the crust, for example, were proposed to correlate with changes in the patterns of deep oceanic water flow. Unfortunately an absolute chronology that could validate these interpretations was lacking until successful application of nannofossil biostratigraphy constrained much of the inferred growth history (Cowen et al., 1993).

*email: edecarlo@soest.hawaii.edu

Marine Authigenesis: From Global to Microbial, SEPM Special Publication No. 66

Rare Earth Elements

The rare earth elements (REE) are highly enriched in marine Fe-Mn crusts (e.g., Aplin, 1984; Elderfield, 1988; De Carlo and McMurtry, 1992) and provide an excellent means of examining relationships between oceanic chemistry and the composition of Fe-Mn crusts. The REE comprise a particularly coherent group with a preferred trivalent oxidation state, except for Ce and Eu, which can also exist in the tetra- and di-valent oxidation states, respectively (Henderson, 1984; and references therein). Only Ce, however, cycles between two oxidation states under typical oceanic conditions.

Shielding of 4f electrons by electrons in the 6s and 6p subshells and an increasing effective nuclear charge across the group lead to a steady decrease in ionic radius with increasing atomic number. The preferred trivalent oxidation state for the group combined with sequential filling of the 4f electron subshell also lead to gradual changes in chemical properties across the REE family that are particularly evident in their aqueous chemistry (e.g., Choppin, 1984; Cantrell and Byrne, 1987; Elderfield, 1988; Byrne and Kim, 1990; Wood, 1990; Lee and Byrne, 1993; Johannesson et al., 1995; Byrne and Sholkovitz, 1996).

The REE as a group can provide unique insights into geochemical processes in the marine environment (Elderfield, 1988). These elements are influenced by riverine, atmospheric, and hydrothermal input, and participate in scavenging, complexation, and redox reactions that regulate trace element abundances. Studies of the oceanic chemistry of the REE (Elderfield, 1988; Elderfield et al., 1990; Bertram and Elderfield, 1993; Sholkovitz et al., 1994), modeling (Byrne and Kim, 1990; Lee and Byrne, 1992, 1993), and experimental work (Byrne and Kim, 1990; Koeppenkastrop and De Carlo, 1992, 1993; De Carlo et al., 1997) all indicate that the marine abundance of REE is regulated by scavenging processes. In the oceanic water column these often involve coatings of Mn and Fe oxides and organic matter on particles (e.g., Balistrieri et al., 1981; Elderfield, 1988).

Probably the single most important manifestation of the slight variations in chemical properties across the REE family is the greater propensity for heavy rare earth elements (HREE) to form complexes than their light counterparts (LREE). Because of this and the fact that REE are particle reactive elements, their partitioning between seawater and Fe-Mn crusts is the product of competitive complexation between ligands at the surface of crusts and those dissolved in seawater (Cantrell and Byrne, 1987; Stanley and Byrne, 1990; Koeppenkastrop et al., 1991; Koeppenkastrop and DeCarlo, 1992, 1993; De Carlo et al., 1997). Solution chemistry thus exerts a strong control on REE uptake by Fe and Mn oxides and is typically manifested in fractionation between HREE and LREE (Cantrell and Byrne, 1987; Elderfield, 1988; DeCarlo, 1991; DeCarlo and McMurtry, 1992; Piepgras and Jacobsen, 1992).

Four parameters appear to be particularly important in regulating the concentration of REE in seawater and in particulate matter (for example, crusts): 1) water depth, 2) the source (provenance) of the REE, 3) the type and abundance of solution ligands (that is, solution speciation), and 4) the nature and abundance of reactive particulate phases (Cantrell and Byrne, 1987; Elderfield, 1988; Byrne and Kim, 1990; De Carlo, 1991; Piepgras and Jacobsen, 1992; De Carlo and McMurtry, 1992; Koeppenkastrop and De Carlo, 1992). These parameters are all interrelated inasmuch as they involve the supply of dissolved REE, their solution chemistries, and their relative affinities for surfaces.

Although REE abundances have been used to infer modes of crust genesis (Aplin, 1984; Fleet, 1984; De Carlo, 1991; De Carlo and McMurtry, 1992), a more incisive tool for interpreting paleoceanographic conditions may be a measure of the extent of fractionation between the light (LREE) and heavy (HREE) members of the REE family. Fractionation of the REE during removal from seawater is influenced by both the solution chemistry and by the abundance and composition of the scavenging particles (Cantrell and Byrne, 1987; Byrne and Kim, 1990; Koeppenkastrop and De Carlo, 1992, 1993; De Carlo et al., 1997). Although the flux of particulate matter through the water column will exert an influence on the total amount of REE scavenged from seawater, the fractionation between any two REE (expressed as a ratio of these REE) should remain independent of particulate flux for any given mineral surface as long as the sorptive capacity of the particulate phase is not exceeded and the REE content of seawater remains constant.

Cantrell and Byrne (1987) and Lee and Byrne (1993) demonstrated that CO_3^{2-} complexation is the primary process that controls the speciation of REE in seawater. These authors reported that stability constants of REE dicarbonato complexes (calculated for 25°C and zero ionic strength) vary by more than two orders of magnitude between La and Lu, whereas stability constants for complexes with other ligands in solution typically vary by less than one order of magnitude, excepting for those with phosphate. Hence, during removal of dissolved REE from seawater by particles, the fractionation that results in the typical HREE-enriched shale-normalized pattern of seawater (e.g., de Baar et al., 1985; Piepgras and Jacobsen, 1992; Zhang et al., 1994) is primarily the result of differences in the stability of LREE and HREE carbonate complexes. We have previously demonstrated in laboratory studies that uptake of dissolved REE by particles is a competition between solution and surface complexation, and that different mineral phases (for example, FeOOH versus δ–MnO_2) result in a slightly differing extent of REE fractionation during removal from solution. However, the distribution coefficients (K_d) determined in our work vary at most by one order of magnitude across the REE family (Koeppenkastrop and De Carlo, 1992, 1993; De Carlo et al., 1997). Thus, the concentration of CO_3^{2-} and its effect on REE speciation in seawater remain the dominant controlling factors in REE fractionation during scavenging by solid phases.

In this paper, we present high spatial resolution REE profiles in Fe-Mn crusts recovered from central Pacific seamounts near Hawaii and the Phoenix Islands, Kiribati (Wen, 1995; Wen et al.,1997). We evaluate HREE/LREE fractionation (that is, Sm/Yb) in these crusts by focusing on the relationships between the concentration of CO_3^{2-} in seawater and the uptake of REE by carrier phases previously elucidated through experimental work. We postulate that changes in trace element profiles within crusts, particularly those of the REE, are correlated with changes in paleoceanic conditions. A corollary is that individual crusts from different areas of the Pacific Ocean should be correlatable if they recorded evidence of basin-wide events within their strata.

METHODS

FeMn crusts selected for this study were recovered at depths of 2200-2500 m from seamounts in two areas of the central Pacific Ocean (Fig. 1). Sampling was carried out by chainbag dredging with guidance from 3.5 kHz echo sounding and SeaMARC II sidescan acoustic imaging. Sample 6RD0808 consists of a maximum of 60 mm of FeMn oxide underlain by approximately 20 mm of carbonate fluorapatite admixed with volcanic fragments and was recovered from McKean Seamount near the Phoenix islands, Kiribati, during the Moana Wave cruise 8602 (DeCarlo et al., 1987a). Other samples recovered in the same dredge haul were similar in crustal thickness and substrate, and are also briefly discussed in the context of their correlation to crust 6RD0808. Sample RD50S1B was collected from Schumann Seamount, north of the main Hawaiian Islands, during the Kana Keoki cruise 84082402 (DeCarlo et al., 1987b). The Schumann Seamount crust exhibits a variable thickness with a maximum of 95 mm of FeMn oxides on a large basalt fragment substrate, although we report here only results from a 44-mm-thick section taken from an area where the crust is about 65 mm thick. It is estimated, on the basis of biostratigraphic dating, that approximately two-thirds of the growth history of the

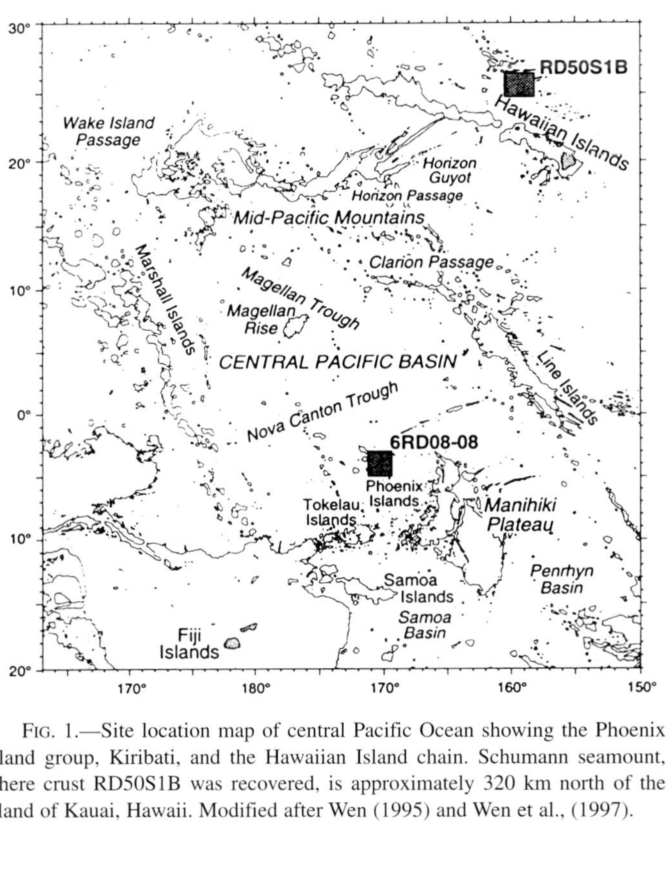

FIG. 1.—Site location map of central Pacific Ocean showing the Phoenix Island group, Kiribati, and the Hawaiian Island chain. Schumann seamount, where crust RD50S1B was recovered, is approximately 320 km north of the island of Kauai, Hawaii. Modified after Wen (1995) and Wen et al., (1997).

crust is represented by this section (Cowen et al., 1993).

Analytical methods have been previously described by Wen (1995) and Wen et al. (1997) and are summarized here. Approximately 10–15 mg of subsamples, collected with a microauto drill at 2-mm to 3-mm intervals along the growth axis of each crust, were weighed on a microbalance and transferred to 3-ml Savillex Teflon vials. To each vial was added 250 μl each of concentrated HNO_3 and HCl, and samples were subsequently digested in a microwave oven as described by DeCarlo (1991). After cooling, 100 μl of HF was added to solubilize any remaining refractory aluminosilicates. Samples were quantitatively transferred to preweighed 8-ml bottles, diluted with 0.5 M H_3BO_3, and the final solution weight was recorded.

Major and minor elements (Fe, Mn, Ca, P, Mg, Al, Si, Ti) in the crusts were determined by inductively coupled plasma-optical emission spectrometry (ICP-OES). REE (Y, La, Ce, Pr, Nd, Sm, Eu, Gd, Dy, Ho, Er, Tm, Yb, Lu) and other trace elements (Cr, Ba, Ir, Pt, Au, Bi, Th, U, Pb) were determined by inductively coupled plasma-mass spectrometry (ICP-MS). Indium was added as an internal standard to dilutions of the original digested samples prior to analysis. All samples were run in duplicate. Precision of trace element analysis was typically 3% except for Ir and Pt, where unoptimized conditions led to low count rates and the precision was limited by counting statistics to no better than 10%. The accuracy of determinations was ascertained by simultaneous analysis of USGS standard nodule A1 (Wen, 1995).

Laboratory experiments were also designed to evaluate systematically the partitioning of REE between the aqueous and synthetic mineral particles under varying pH and ionic strength conditions. Studies with synthetic systems were deemed necessary to establish trends in REE distribution coefficients and other important parameters that must be known to apply quantitatively the REE fractionation model proposed in this work. Methods followed those described by Koeppenkastrop and De Carlo (1992) and are only summarized below.

Batch equilibrations were carried out with 1000 g of liquid phase (seawater or $NaNO_3$ solutions) in Teflon bottles spiked with 125 μg of each of 14 naturally occurring REE (for example, 893 nM La to 715 nM Lu). Spiked solutions were adjusted to within 0.05 pH units of the desired value with dilute NaOH and/or HNO_3, after which was added 10 mg of a pure solid phase (FeOOH or MnO_2) taken from a well-stirred stock suspension. The particles used in our studies display characteristics (for example, specific surface area, particle size range, residual surface charge) typical of fresh, hydrous ferric oxide (Dzombak and Morel, 1990). Thus, the mass of suspended FeOOH and initial dissolved REE concentrations in our experiments result in approximately a two-fold higher concentration of surface sites on the solid phases than dissolved REE in solution. Slight pH shifts after addition of the solid phase to the REE spiked solution usually required readjustment of the pH to the desired value. The pH was measured again after 24 hours, readjusted as necessary, and the system was allowed to equilibrate for an additional 24 hours. No further pH adjustments were carried out. All experiments were conducted at room temperature (~23°C) and stirred continuously. The liquid and the solid phases were sepa-

TABLE 1: RARE EARTH ELEMENT ABUNDANCE (ppm) IN CRUST 6RD08-08 FROM KIRIBATI

Depth (mm)	La	Ce	Pr	Nd	Sm	Eu	Gd	Tb	Dy	Ho	Er	Tm	Yb
4	192	444	29.8	130.1	23.3	6.28	34.2	5.47	31.4	7.6	23.4	3.6	23.6
6	159	398	22.0	96.2	17.6	4.80	25.5	4.04	26.2	6.5	20.1	3.0	20.5
8	116	327	14.8	63.6	11.5	3.13	18.6	2.83	18.9	4.8	15.6	2.5	16.1
10	121	343	14.8	66.6	11.6	3.11	18.7	2.90	20.3	5.2	16.5	2.6	17.5
12	133	419	17.1	75.2	13.0	3.75	21.0	3.20	21.2	5.5	17.4	2.7	18.3
14	118	414	15.3	66.7	12.1	3.22	19.2	2.96	20.1	5.2	17.1	2.6	16.9
16	140	475	19.4	84.8	15.1	4.15	23.9	3.61	23.8	6.0	18.4	2.8	18.3
18	127	581	18.2	75.7	14.2	3.72	22.2	3.28	20.0	4.9	15.9	2.5	16.7
20	126	564	19.4	82.3	15.2	3.91	22.5	3.34	19.8	4.8	15.4	2.4	15.2
22	127	644	20.7	85.1	16.7	4.28	23.3	3.48	20.4	4.8	14.8	2.3	14.9
24	145	685	23.7	98.1	18.6	4.72	26.1	3.86	22.7	5.2	15.5	2.4	15.5
26	143	711	24.9	101	19.6	4.84	25.7	3.92	22.3	4.9	14.7	2.2	14.1
28	156	781	27.7	112	21.7	5.30	30.3	4.32	23.4	5.2	14.7	2.2	14.6
30	144	834	25.5	102	20.1	4.95	26.3	3.89	21.2	4.6	13.3	2.1	12.8
32	155	906	28.8	110	21.1	5.34	28.2	3.94	22.7	4.7	13.6	2.1	12.8
34	145	782	27.2	107	20.7	4.84	27.2	3.91	21.6	4.5	13.2	1.9	12.0
36	171	987	31.4	122	23.3	5.73	31.5	4.40	23.8	4.9	14.3	2.1	13.7
38	146	913	26.5	102	19.4	4.64	26.2	3.75	20.4	4.3	12.9	2.0	12.3
40	90	763	17.1	67.1	13.2	3.29	18.7	2.63	14.4	3.1	8.8	1.4	9.1
42	131	646	22.0	91.6	17.7	4.55	26.0	3.92	23.4	5.5	17.1	2.5	15.5
44	175	910	25.9	105	18.5	4.77	31.2	4.45	27.2	6.7	20.6	3.2	20.7
48	140	920	19.4	80.1	14.3	3.99	25.9	3.50	21.7	5.7	17.3	2.7	17.9
50	206	871	30.5	130	23.8	6.67	40.0	5.66	35.2	8.9	27.6	4.0	25.7
52	177	844	22.0	92.2	16.5	4.55	29.5	4.13	27.2	7.1	23.8	3.6	23.4
54	202	926	26.5	110	20.4	5.56	34.5	4.79	30.6	8.2	27.0	4.1	27.2
58	126	1100	16.0	66.0	11.8	3.02	22.1	2.60	16.2	3.9	12.1	1.9	12.6
60	661	468	112	482	94.1	24.6	129	19.6	122	29.7	86.4	12.0	73.4
62	179	77	30.2	131	25.3	6.74	37.1	5.78	34.8	8.3	24.9	3.5	22.2
64	326	39	59.0	252	47.4	13.2	65.1	11.0	68.7	16.5	49.2	7.2	44.5
66	192	19	17.4	68.0	10.4	3.00	18.7	2.99	23.0	6.5	22.0	3.5	23.5
68	107	12	7.1	27.1	3.9	1.06	7.0	1.17	8.5	2.9	10.8	1.7	12.8
70	86	15	12.3	48.5	8.9	2.67	9.9	1.47	8.6	2.2	7.6	1.3	8.7
72	61	17	10.5	44.1	9.2	2.30	11.8	1.93	12.1	2.9	8.9	1.3	9.0
74	37	10	5.0	20.8	3.7	1.07	5.1	0.82	5.2	1.3	4.1	0.6	4.4
76	38	11	6.6	27.7	5.5	1.56	7.3	1.21	7.6	1.9	5.6	0.8	5.7
78	27	9	5.0	20.5	4.0	1.10	5.4	0.88	5.7	1.4	4.5	0.7	4.6
80	77	33	10.4	46.0	8.6	2.60	13.9	2.20	15.9	4.2	13.9	2.2	14.3

rated by vacuum filtration through 0.22 μm Millipore membranes upon completion of the experiments and subsequently analyzed for their REE content.

RESULTS

REE Analyses in Crusts

Concentrations of strictly trivalent REE, exemplified in Fig. 2 by Sm and Yb, do not appear to display strong trends with depth in the Kiribati crust (Table 1). These elements are in the range of 11.5–23.8 mg/kg and 9.1–27.2 mg/kg, respectively, except at the substrate boundary, where they increase sharply (Sm:94.2 mg/kg; Yb:73.4 mg/kg). In the Schumann Seamount crust (Table 2, Fig. 2B), Sm and Yb concentrations are typically greater (33–108 mg/kg Sm, 19–130 mg/kg Yb) than in the Kiribati crust.

The concentration of Ce (Table 1, Fig. 2A), unlike those of the strictly trivalent REE, increases nearly monotonically from less than 400 mg/kg a few millimeters below the surface of Kiribati crust 6RD08-08 to approximately 1100 mg/kg just above the substrate boundary where a sharp decrease to 470 mg/kg Ce coincides with previously noted increases in Sm and

TABLE 2. RARE EARTH ELEMENT ABUNDANCE IN CRUST RD50S1B FROM SCHUMANN SEAMOUNT

Depth (mm)	La	Ce	Pr	Nd	Sm	Eu	Gd	Tb	Dy	Ho	Er	Tm	Yb
3	215	914	48.9	199	43.5	10.4	52.3	7.8	41.2	8.2	22.1	3.2	20.5
5.5	281	840	69.9	280	60.8	14.6	66.4	10.3	52.6	10.2	27.8	4.0	25.8
8	300	1090	75.4	297	63.0	15.2	69.4	10.6	53.0	10.3	26.8	4.1	25.6
10.5	252	1115	63.1	245	53.0	12.6	59.5	8.9	44.6	8.5	23.8	3.5	22.8
13	254	1185	67.6	267	56.7	13.5	62.3	9.5	47.7	9.4	25.4	38.0	24.0
15.5	228	1030	55.4	219	48.2	11.8	56.9	8.2	41.4	8.3	22.7	3.3	20.1
18	270	1178	66.0	234	57.8	14.0	66.7	9.8	50.4	9.8	28.0	3.9	26.9
20	262	1030	38.1	286	62.5	15.4	73.0	10.8	55.3	11.3	30.5	4.3	26.8
22.5	424	989	111	486	408	27.3	122	18.2	98.8	20.4	54.7	7.4	45.0
25.5	287	963	74.2	310	68.6	16.9	79.7	11.9	60.6	12.5	32.8	4.5	27.8
28	344	828	89.2	388	82.4	20.5	92.4	13.6	71.4	14.2	37.8	5.0	29.7
31	267	1162	56.7	232	46.5	11.3	56.8	8.1	41.9	8.8	23.7	3.4	21.2
33.5	248	1063	48.0	193	38.1	9.0	47.3	6.6	33.7	7.0	20.0	2.9	19.0
36.5	226	1162	43.3	169	32.8	7.8	42.8	5.9	30.7	6.2	17.8	2.6	16.4
39	268	1220	53.3	212	41.3	9.9	52.9	7.5	38.3	7.7	21.3	3.1	20.4
44	288	1277	59.0	234	46.7	11.2	57.8	8.1	41.3	8.4	23.4	3.4	20.4

Yb. Ce abundances in the Schumann Seamount crust are variable between 830 mg/kg and 1270 mg/kg (Table 2, Fig 2B) with little apparent trend except in the lowermost section, where concentrations of Ce also appear to increase steadily.

Shale-normalized patterns of intervals within the two crusts (Figs. 3 and 4) show that : (1) Positive Ce and Gd anomalies are nearly ubiquitous in both crusts, and a slight Ho anomaly also appears in many patterns. (2) Except for Ce, REE patterns are relatively flat with a slight middle REE (MREE) enrichment in crust RD50S1B, whereas fractionation throughout crust 6RD0808 is variable and more extensive near the top and bottom of the crust. (3) The shale-normalized LREE/HREE ratio in crust RD50S1B tends to be near or greater than one, whereas in crust 6RD0808 this ratio is always less than one and fluctuates significantly throughout the various sections of the crust. (4) REE patterns within the massive carbonate fluorapatite (CFA) substrate (6080 mm) of crust 6RD0808 are similar in shape to that of seawater (de Baar et al., 1985; Piepgras and Jacobsen, 1992; Zhang et al., 1994), although highly enriched relative to seawater. The extent of REE enrichment varies throughout the phosphatized substrate with the greatest REE concentrations observed near the Fe-Mn crust-substrate boundary. Major element profiles throughout each crust are shown in Fig. 5.

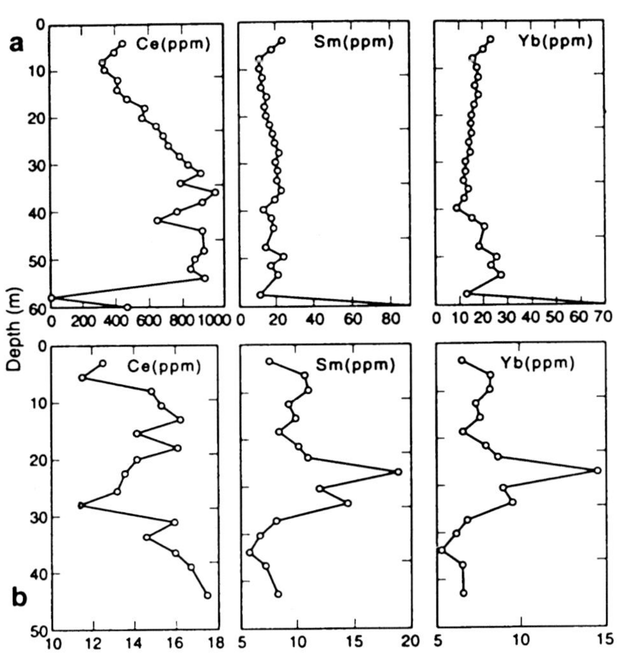

FIG. 2.—Profiles of Ce, Sm, Yb in (a) crust 6RD08-08 and (b) RD50S1B; elemental concentrations are expressed in ppm (mg/kg).

Laboratory Experiments

Uptake curves of individual REE shift toward lower pH as a function of increasing atomic number in both $NaNO_3$ and in seawater below pH 7 (Fig. 6). In seawater (Fig. 6B), the REE adsorption edges for La, Gd, and Yb are better defined and occur over a narrower pH range than in $NaNO_3$ solution (Fig.6A), suggesting that the REE have an overall greater affinity for FeOOH particles suspended in seawater than in $NaNO_3$ solution. Additionally, the order and extent of REE sorption reverse above pH 7 in seawater, a feature not observed in $NaNO_3$ solution lacking significant amounts of carbonate species.

Differences in the extent of scavenging of individual REE by the solid phase are evident at nearly every pH (Fig. 7). However, at pH 6, a strongly convex upward pattern of REE uptake between La and Gd, a slightly less convex upward pattern between Gd and Ho, and a steady increase in uptake between Ho and Lu are particularly apparent. These types of variations have

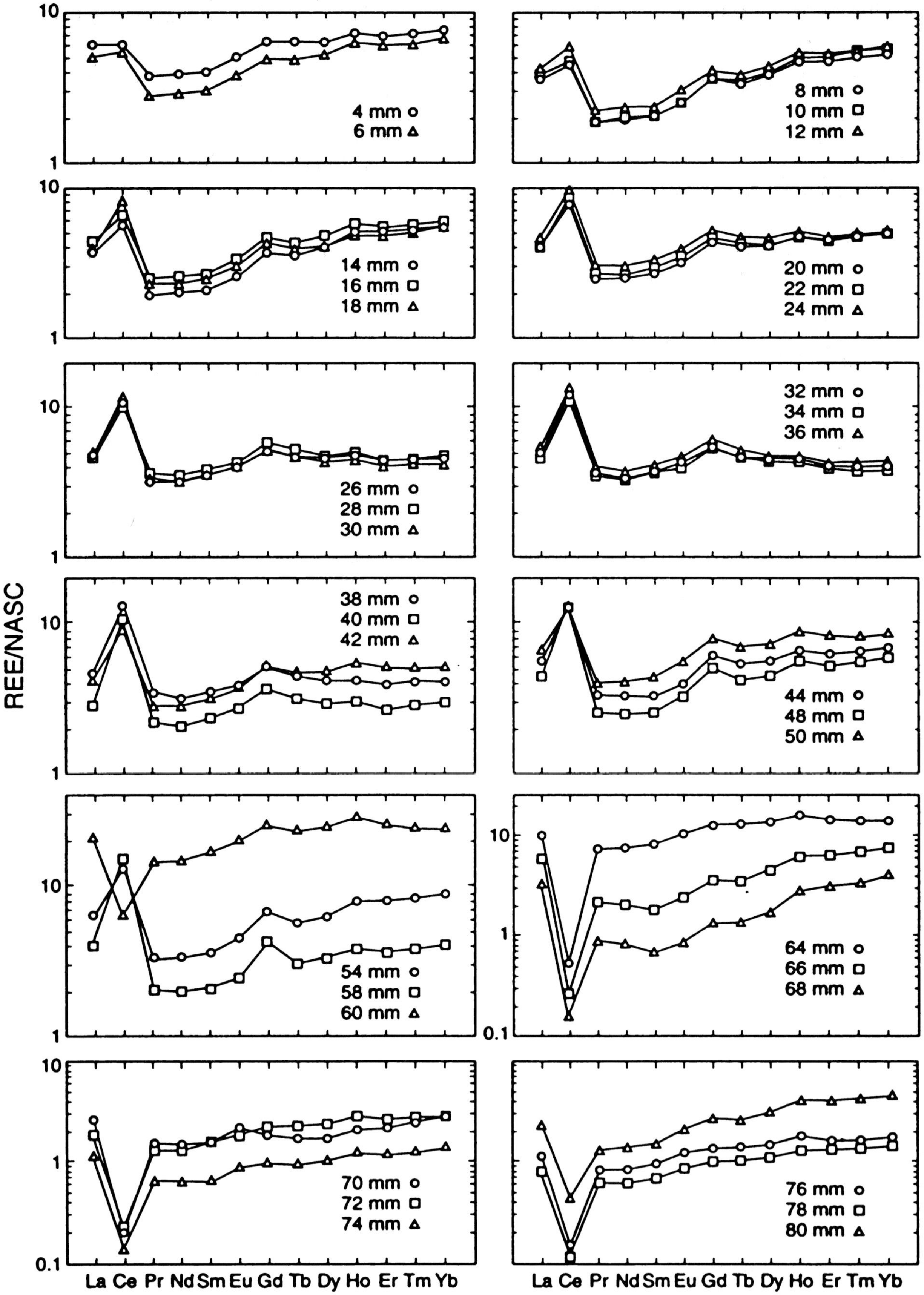

FIG. 3:—Shale-normalized REE patterns for individual Fe-Mn oxide layers (0–58 mm) and the phosporite substrate (60–80 mm) of crust 6RD08-08 from McKean Seamount, Kiribati, showing variable fractionation between LREE and HREE as well as the Ce anomalies relative to strictly trivalent REE. (Modified after Wen, 1995).

been referred to as "tetrad effects" (e.g., Masuda and Ikeuchi, 1979; Akagi et al., 1993; Kagi et al., 1993; McLennan, 1994) and will be referred to as such here. Subtle tetrad effects also occur at pH 4-5 (Fig. 7B) and are also evident but less pronounced above pH 7, where a combination of high efficiency of REE uptake from solution and carbonate complexation (e.g., Cantrell and Byrne, 1987) overshadow tetrad effects.

DISCUSSION

REE abundances as well as interelement relationships, such as those between Ce and Mn (e.g., Fig. 10 in Wen et al., 1997) and between the trivalent REE and Fe differ in the crusts from the two sampling areas. Additionally REE fractionation differs in the two crusts. A first-order interpretation of these observations is that the two crusts accumulated under different oceanic water column conditions. Below we discuss differences observed in these two crusts with a focus on the chemical properties of the REE and their main carrier phases (Fe and Mn oxides and CFA) in crusts.

Fractionation between HREE and LREE in Fe-Mn crusts, as observed in the various layers (Figs. 3 and 4), should reflect differences in the propensity of individual REE to form complexes (both with ligands in seawater and with the principal solid phases in crusts). Complexation reactions with ligands at the surface of Fe and Mn oxides are controlled by fundamental properties of the solid phases and are unlikely to have changed over geologic time. Data in Table 3 substantiate this assumption. Theoretical $K_{d(REE)}$ calculated from adsorption isotherms and $K_{d(REE)}$ observed in kinetic sorption experiments (Koeppenkastrop and De Carlo, 1993) are consistent with, although slightly lower than, those calculated for Fe-Mn crusts based on concentrations of REE in crusts from the exclusive economic zone of the Hawaiian Archipelago and in Pacific Ocean seawater (De Carlo and McMurtry, 1992). The latter $K_{d(REE)}$ exhibit the same range of fractionation between Sm and Yb as those calculated by Bau et al. (1996) using average concentrations of REE in central Pacific crusts and average seawater concentrations reported by Piepgras and Jacobsen (1992). Thus we hypothesize that differences in LREE/HREE fractionation throughout the Fe-Mn crusts, illustrated in Fig. 8 by variations in the Sm/Yb ratio, result primarily from accumulation in waters where either carbonate ion concentrations differ or where different supplies/sources of dissolved REE occur.

Table 3.— COMPARISON OF THEORETICAL AND OBSERVED $K_{d(REE)}$ IN SEAWATER EXPERIMENTS WITH THOSE CALCULATED FOR Fe-Mn CRUSTS

	Sm	Yb	Lu
Theoretical#	7.57	6.57	6.54
Experiment#	7.34	6.69	6.60
Sigma($K_{d\text{-expt}}$.)	0.05	0.05	0.05
Fe-Mn Crusts@	7.7	7.1	6.9
Fe-Mn Crusts&	7.94	7.34	—

Theoretical $K_{d(REE)}$ are calculated from the slope of isotherms (Koeppenkastrop, 1992); experimental values are from previous studies (Koeppenkastrop and De Carlo, 1993)

@ De Carlo and McMurtry (1992)

& Bau et al. (1996)

The uptake of REE from seawater and solutions of differing ionic strength displays consistent trends that reflect parameters key to our hypothesis (Figs. 6 and 7). First, the propensity of HREE to complex to a greater extent than LREE is clearly shown in the order of uptake (Fig. 6) of the REE from either seawater or $NaNO_3$ solution. In the absence of strong solution ligands, the HREE are scavenged preferentially owing to surface complexation with hydroxyl groups on the FeOOH particle surface. This has also been demonstrated through a different approach in prior work by our group and others (e.g., Byrne and Kim, 1990; Koeppenkastrop and De Carlo, 1992). Second, the paramount influence of CO_3^{2-} on REE uptake becomes quite apparent above pH 7 (Fig. 6), where there is a greater propensity of HREE to complex than their LREE counterparts. Complexation with CO_3^{2-}- leads to a lessened uptake of REE by FeOOH, although the overwhelming majority of the REE in solution is still scavenged by the particle surface. Our results support the hypothesis that uptake of REE from solution is primarily a competitive complexation process. These findings are important because they demonstrate that the observed tetrad effects reflect fundamental properties of the REE, although the low pH at which the experiments were conducted is not representative of what typically occurs in the oceanic water column. Thus, the REE behave coherently irrespective of the phases actually involved and their behavior in natural waters can be predicted from fundamental properties, with the primary variable of CO_3^{2-} complexation exerting an overwhelming control in seawater. Other parameters involved in REE scavenging in the oceanic water column (for example, properties of FeOOH and MnO_2) are not as likely to change over geologic time as might the abundance of CO_3^{2-}.

We describe the equilibrium partitioning of REE between seawater and the surface of the metal oxide where:

$$REE^{3+}_{(aq)} + 3M\text{-}OH \Longleftrightarrow REE\text{-}(O\text{-}M)_3 + 3H^+ \quad (1)$$

The equilibrium constant for this reaction is:

$$K_{d(REE)} = [REE\text{-}(O\text{-}M)_3][H^+]3/[REE_T]_{(aq)}[M\text{-}OH]_3 \quad (2)$$

where [M-OH] is the concentration of surface hydroxyl sites on the metal oxide particles, $[REE\text{-}(O\text{-}M)_3]$ is the concentration of REE complexed at the metal oxide surface and

$$[REE_T]_{(aq)} = [REE_f] + [REE\text{-}CO_3^+] + [REE\text{-}(CO_3)_2^-] \quad (3)$$

where $[REE_f]$ is the combined concentrations of free REE^{3+} as well as hydroxy, chloro, and sulfato complexes. The latter three occur at nearly the same concentrations as free REE^{3+} in seawater at pH 7.9 (Cantrell and Byrne, 1987) and should not vary

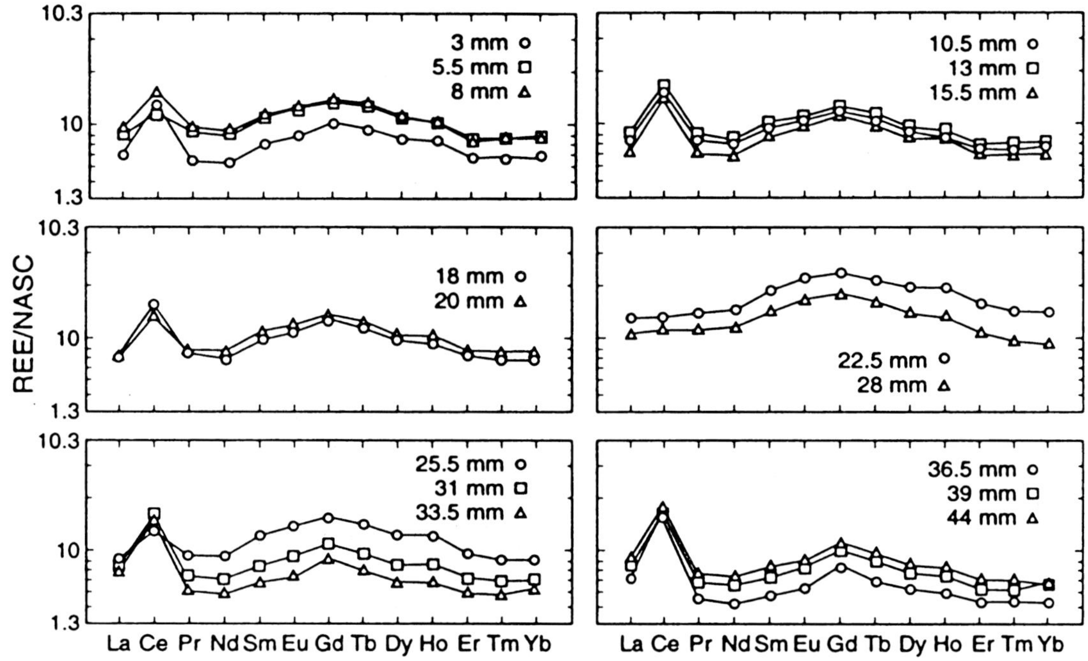

FIG. 4.—Shale-normalized REE patterns for layers of crust RD50S1B from Schumann Seamount, near the Hawaiian Archipelago, showing MREE enrichment, less HREE/LREEE fractionation than observed in Kiribati crusts, and patterns with no Ce anomalies for detrital-rich intervals. (Modified after Wen, 1995)

significantly owing to the near invariance of the concentrations of these ligands in open ocean water. Assuming that the concentration of surface sites on Fe-Mn oxide particles is in large excess compared to concentration of REE in solution, and that the pH of the seawater will not change during sorption reactions, Equation (2) then reduces to:

$$K_{d(REE)} = [REE\text{-}(O\text{-}M)_3]/[REE_T]_{(aq)} \quad (4)$$

Owing to a lack of knowledge of the exact bonding mechanism that takes place at the metal oxide, we will henceforth simplify the above expression and ignore the stoichiometry involving the number of surface hydroxyls needed to complex a given REE, and describe the surface bound REE as REE-O-M. This assumption has no effect on the subsequent theoretical development and is warranted as it has been suggested previously that the REE are scavenged by FeOOH through a complexation mechanism rather than by ion-exchange reactions (Koeppenkastrop and De Carlo, 1993).

Combining Equations (3) and (4) yields:

$$K_{d(REE)} = [REE\text{-}O\text{-}M]/([REE_f] + [REE\text{-}CO_3^{+}] + [REE\text{-}(CO_3)_2^{-}]) \quad (5)$$

The complexation of REE with CO_3^{2-} in seawater is described as follows (e.g., Cantrell and Byrne, 1987):

$$REE^{3+} + CO_3^{2-} <====> REE\text{-}CO_3^{+} \quad (6)$$

$$\beta_1 = [REE\text{-}CO_3^{+}]/[REE^{3+}][CO_3^{2-}] \quad (7)$$

$$REE^{3+} + 2CO_3^{2-} <====> REE\text{-}(CO_3)_2^{-} \quad (8)$$

$$\beta_2 = [REE\text{-}(CO_3)_2^{-}]/[REE^{3+}][CO_3^{2-}]^2 \quad (9)$$

Substitution of Equations (7) and (9) into Equation (5) and rearrangement leads to:

$$K_{d(REE)} = [REE\text{-}O\text{-}M]/(\beta_2[REE_f][CO_3^{2-}]^2 + \beta_1[REE_f] + [CO_3^{2-}] + [REE_f]) \quad (10)$$

Equation (10), a similar development to that of Erel and Stolper (1993), relates the distribution coefficient of a given REE to its concentration at the particle surface, its concentration in solution, and the stability constants for mono- and di-carbonato complexes. Taking the ratio of Equation (10) for any two REE allows us to develop a relationship between their K_d, the dissolved concentration of these REE, and their concentrations at the particle surface. The latter is considered here to be the concentrations of REE in Fe-Mn crusts. Choosing Sm and Yb as representative LREE and HREE, respectively, we obtain:

$$\frac{K_{d(Sm)}\,[Sm^{3+}](\beta_{2(Sm)}[CO_3^{2-}] + \beta_{1(Sm)} + 1/[CO_3^{2-}])}{K_{d(Yb)}\,[Yb^{3+}](\beta_{2(Yb)}[CO_3^{2-}] + \beta_{1(Yb)} + 1/[CO_3^{2-}])} = \frac{[Sm\text{-}O\text{-}M]}{[Yb\text{-}O\text{-}M]} \quad (11)$$

where [Sm-O-M] and [Yb-O-M] are the concentrations of the two REE in the Fe-Mn crusts. The only variables remaining in Equation (11) are the dissolved concentrations of REE and CO_3^{2-} in seawater. If we assume that the concentration of REE in seawater has not changed significantly over time, then the concentration of CO_3^{2-} in seawater remains the only unknown and can be calculated from concentrations of REE observed in the individual layers in Fe-Mn crusts.

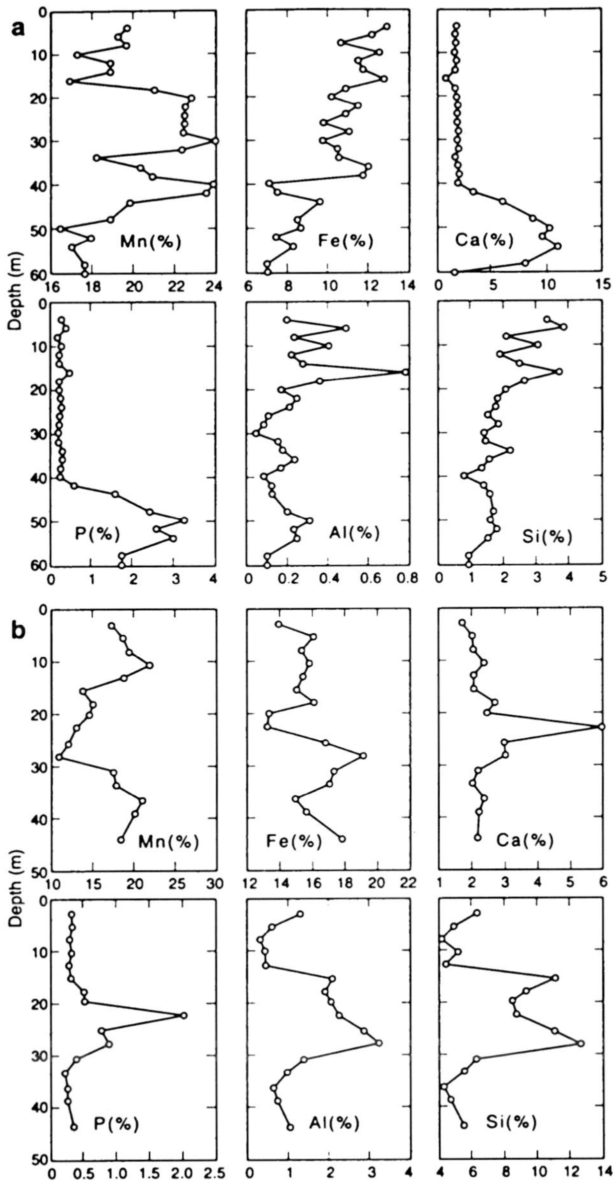

FIG. 5.—Depth profiles of major constituents Mn, Fe, Ca, P, Al, and Si in (a) crust 6RD08-08 from Kiribati and (b) crust RD50S1B from Hawaii. (Modified after Wen et al., 1997)

If one accepts that CO_3^{2-} exerts the primary control on REE speciation in seawater, it becomes apparent that a greater concentration of dissolved CO_3^{2-} should lead to less free REE^{3+} in solution. Previously calculated fractions of free Sm and Yb (Fig. 9 in Lee and Byrne, 1992) are reproduced in Table 4 to demonstrate the changes in the relative amounts of free metal ion of these elements at dissolved CO_3^{2-} concentrations within the range commonly expected for seawater. Thus, HREE (for example, Yb), which form much stronger CO_3^{2-} complexes than LREE (for example, Sm), will be complexed to a greater extent when dissolved CO_3^{2-} concentrations are greater, and will be subjected to a slightly diminished attraction for the carrier phases of Fe-Mn crusts. If, on the other hand, the concentration of CO_3^{2-} in seawater decreases, although LREE will still be more prone to scavenging by the carrier phases of Fe-Mn crusts than HREE (Koeppenkastrop and De Carlo, 1992), they will be complexed by CO_3^{2-} in seawater to a lesser extent, thus be relatively more available for surface complexation with hydroxyl groups on the Fe and Mn oxide particles. Consequently, the concentration of CO_3^{2-} in seawater should alter the proportions of the REE scavenged by solid phases owing to significant differences in the affinity of individual REE for ligands. Therefore, varying Sm/Yb ratios in crusts should result from competitive complexation effects. Low Sm/Yb ratios in crusts should reflect low concentrations of CO_3^{2-} in seawater whereas higher concentrations of dissolved CO_3^{2-} should result in greater Sm/Yb ratios in crusts.

Confident association of changing Sm/Yb ratios in crusts with variations in seawater CO_3^{2-} over time requires the following assumptions: 1) the primary geochemical signature of crusts is not diagenetically modified over geologic time; 2) the provenance of dissolved REE to crusts has remained nearly constant; 3) the ratio of total Sm to Yb in seawater does not vary with water depth where the crusts accumulated, and 4) variations in fluxes of particles can be identified and accounted for.

Under such conditions, dissolved CO_3^{2-} concentration, pH, and the resultant effects on REE speciation will play dominant roles in determining the primary REE signature of Fe-Mn crusts. Fortunately, interelement relationships in the hydrogenetic Fe-Mn crusts described here show little evidence of postdepositional processes, except possibly within the lower (older) section of Kiribati crust 6RD08-08, in which the abundance of CFA (that is, P in Fig. 5A) increases with age (Wen, 1995; Wen et al., 1997). In Hawaii crust RD50S1B the section between 20 mm and 28 mm depth contains abundant detrital matter (for example, Al, Si in Fig. 5B) that also potentially compromises the use of the primary hydrogenetic Sm/Yb signature. The contribution of detrital matter to this interval is clearly evident in Fig. 4, where shale-normalized REE patterns are rather flat except for a slight MREE enrichment. The paleopath of the seamount on which the crust accreted (e.g., McMurtry et al., 1994; Wen, 1995) carried it through the tradewind belt where aeolian input of continental dust contributed significantly to its total and REE composition. The paleopath of the Kiribati crust, however, never passed through that zone and it is estimated that more than half of its growth history occurred within the subtropical belt south of the Equator (Yan and Kroenke, 1993, in Wen, 1995).

Two effects likely result from post-depositional phosphatization of the crust. First, if one assumes that phosphatization occurs purely by precipitation of fine-grained apatite within the ample pore space of Fe-Mn crusts, the process will contribute REE whose ratios are different from and not representative of the primary Fe-Mn oxide. The REE signature of carbonate fluorapatite, the most likely phase to form, will be HREE enriched (for example, layers below 60 mm shown in Fig. 3). Hence, a decrease in the bulk Sm/Yb ratio is to be expected by adding proportionally more Yb than Sm to the already existing REE inventory. Although it is tempting to correct for this REE input through normative calculations, a second effect likely precludes this option. Bau et al. (1996) recently reported that phosphatiza-

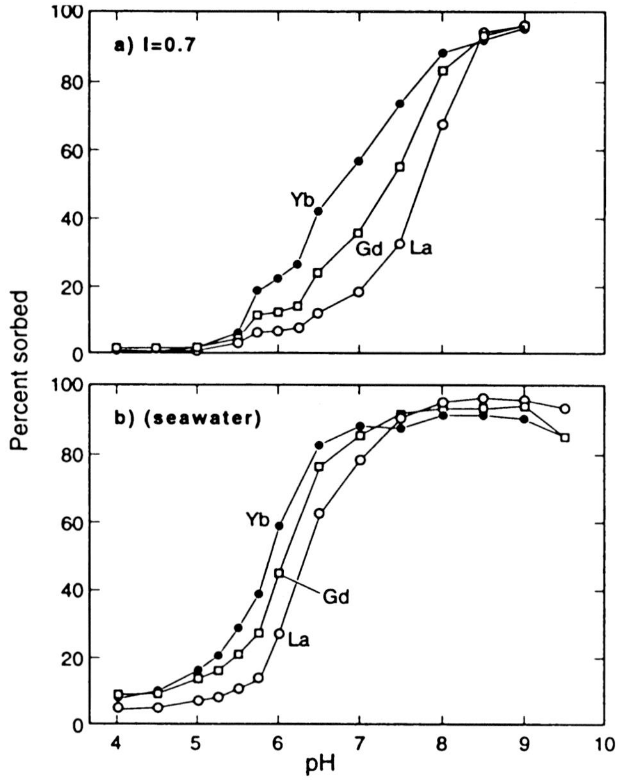

FIG. 6.—Uptake of dissolved REE from solution by synthetic FeOOH as a function of pH in experiments conducted with (a) 0.7M $NaNO_3$ and (b) seawater.

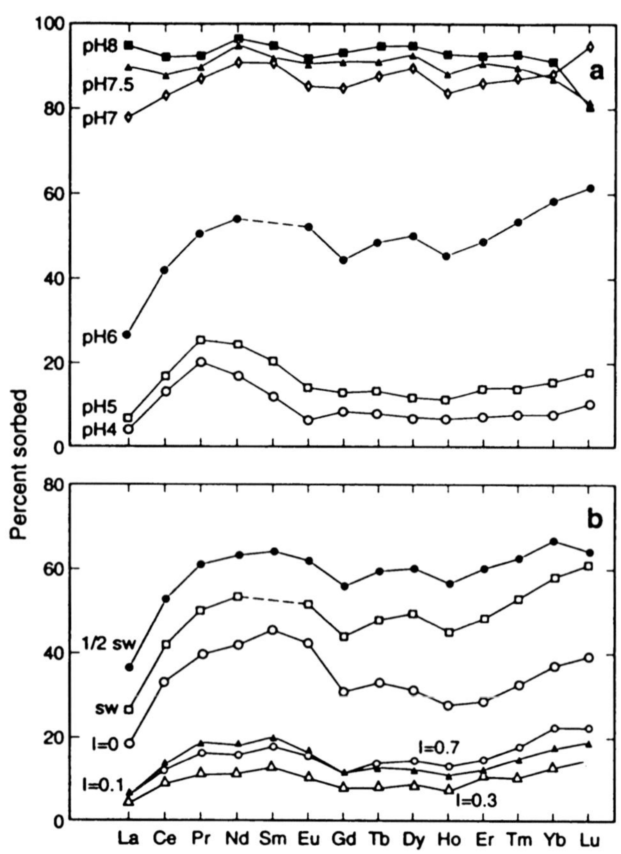

FIG 7.—Uptake of dissolved REE from solution by synthetic FeOOH for individual REE at (a) pH 4-8 in seawater and (b) at pH 6 for solutions of varying ionic strength and composition (sw = seawater, 1/2sw = 50% seawater).

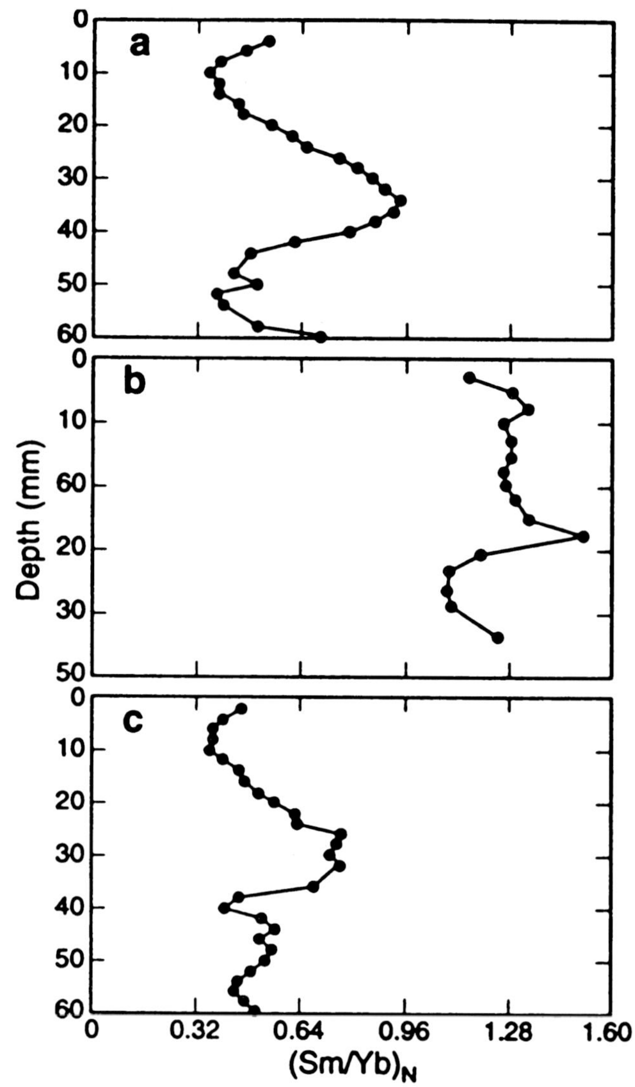

FIG. 8.—Plot of the shale-normalized Sm/Yb ratio as a function of: (a) depth within crust 6RD08-08 from McKean Seamount, Kiribati; (b) depth within crust RD50S1B from Schumann Seamount near Hawaii; (c) depth within for crust 6RD08-03 from McKean Seamount, Kiribati (modified after De Carlo and Pruszkowski, 1995)

tion of Fe-Mn crusts not only leads to the addition of REE but also results in localized redistribution of the REE, presumably owing to the existence of reducing microenvironments during formation of the apatite in the pores of the Fe-Mn oxide. Thus, primary REE signatures can be altered by remobilization and re-equilibration of REE originally scavenged by Fe and Mn oxides in addition to a separate REE contribution from the newly formed apatite. Incorporation of large amounts of (aeolian) detrital matter, such as are inferred to have occurred when Schumann Seamount passed through the tradewind belt, also potentially alters the hydrogenetic REE signature. However, in this case there is no remobilization of the REE associated with Fe and Mn oxides; rather a composite exists that includes detrital REE with a shale-like pattern. The latter can be subtracted by calculating proportions of detrital matter in the interval in question and then subtracting the shale-like contribution to obtain the hydrogenetic REE signature. This approach was used by Wen (1995) to calculate proportions of detrital and hydrogenetic contributions to metals in crusts. Thus, whereas it is theoretically possible to correct for contributions to crusts of REE derived from detrital matter, attempts to correct for the alteration of primary REE signatures during post-depositional phosphatization are unlikely to be successful.

The assumed integrity of the REE source over geologic time can be assessed by examining Nd isotopic compositions, from which the provenance of the REE and variations thereof can be evaluated (e.g., Piepgras and Wasserburg, 1980; Piepgras and Jacobsen, 1992). Nd isotopic data for three subsamples corresponding to the troughs and crest of the Sm/Yb cycle (10, 34, and 54 mm) of crust 6RD08-08 shown in Fig. 8 are consistent with our hypothesis, that the variations in Sm/Yb in our crusts result from changes in dissolved CO_3^{2-} in seawater rather than from large differences in ancient REE sources. The ε_{Nd}, a measure of the deviation of the $^{143}Nd/^{144}Nd$ signature of a sample relative to a chondritic reservoir at a given time, can be used to examine source provenances. Values of $\varepsilon_{Nd}(0)$ for the three intervals (–3.6, 5.2, and –5.0, respectively) are in agreement with previously published data for Fe-Mn nodules and crusts (Piepgras and Wasserburg, 1980; Aplin et al., 1986; Ling et al., 1997) and also are consistent with contemporary seawater $\varepsilon_{Nd}(0)$ values below 2000 m depth reported by Piepgras and Jacobsen (1988). They reported an average Pacific deep water (PDW between 2000–4000 m depth) $\varepsilon_{Nd}(0)$ of –3.4 with a range of –3.0 to –3.8 (2σ of 0.3–0.6). Antarctic bottom water (AABW) contributes to the $\varepsilon_{Nd}(0)$ range of –4.4 to –5.0 in deeper waters of the western Pacific Ocean (Piepgras and Jacobsen, 1988). Surface waters were found to have more variable compositions likely owing to localized geographic influence on REE inputs to surface water.

The small but analytically significant difference in $\varepsilon_{Nd}(0)$ between the intervals at 10 mm and 34 mm suggests that, during the period in which this portion of the crust accumulated, a slight change in the source of REE occurred with increased input of material exhibiting more radiogenic (i.e., less negative) $\varepsilon_{Nd}(0)$ values. Similar differences in $\varepsilon_{Nd}(0)$ between shallow and deep intervals in crusts and nodules have also been observed by others (Piepgras and Wasserburg, 1980; Aplin et al., 1986; Ling et al., 1997). Additional isotopic analyses throughout intervals of particular interest in this crust would help constrain further changes in REE input to mid-depth waters of the central Pacific Ocean over geologic time. For example, the detailed work of Ling et al. (1997), although unconstrained temporally for crust horizons older than about 10–15 Ma, generally shows smaller variations in $\varepsilon_{Nd}(0)$ in the temporally constrained intervals (and even lesser changes in older undated sections of the crust) than were observed in this study. Their findings support our assumption that the source of REE to the oceans has not changed greatly during the approximately 50 m.y. period represented in the section of the crust sequences uncomplicated by phosphatization, although this assumption may not be entirely accurate for the outermost (youngest) few millimeters of the crusts.

The assumption that Sm/Yb ratios in seawater do not vary as a function of depth in the ocean requires more careful examination. Calculations using water column REE data from Piepgras and Jacobsen (1992) and Zhang and Nozaki (1996) show that Sm/Yb ratios vary with depth in Pacific Ocean seawater (Fig. 9). Furthermore, the various sites have different surface water Sm/Yb signatures that are attributable to different local input sources. A large decrease in the Sm/Yb ratio in the first 500–1000 m of the water column is observed at all stations and reflects REE fractionation owing to preferential LREE uptake during scavenging by settling particles. Subsequent reversals in the Sm/Yb ratio at depth are likely due to a similar fractionation process during redissolution of settling particles in the deep ocean.

If Fe-Mn crusts accumulated at a constant water depth throughout their growth history the variations in seawater Sm/Yb ratio with depth could be ignored unless one were comparing crusts recovered from different water depths. However, because the seamounts have subsided in some fashion related to the Sclater curve and have also been periodically uplifted during transit over lithospheric bulges, the effect of seamount depth changes needs to be considered.

Careful examination of the data underlying Fig. 9 reveals that the Sm/Yb ratio varies by only 22% between a depth of 2000 m and 4000 m at the three western North Pacific stations. If we include the two western South Pacific stations the variability reaches 26% over the same depth range. Interstation variability in the Sm/Yb ratio at a given depth is on the order of 5–15% over a latitudinal range comparable to that needed to include the paleopaths of the seamounts on which the crusts grew (Wen, 1995). In contrast, however, the range of variations in the Sm/Yb ratios in crusts shown in Fig. 8 is nearly a factor of three. If a portion of a seamount initially at a water depth of 1000 m or more when Fe-Mn crusts began accumulating subsided 1000–2000 m further during the time corresponding to a half-cycle in the Sm/Yb ratio cycles shown in Fig. 8, the depth contribution to the change in signal would be expected to be on the order of approximately 25%, with other parameters held constant. This constitutes a small portion of the overall Sm/Yb change observed (Fig. 8). Nonetheless, a fully quantitative treatment of the relationship between the Sm/Yb ratio and carbonate concentrations in seawater will require an accounting of variations in the seamount depth through time.

In summary, exclusive of sections of crusts that have been

TABLE 4.— FREE FRACTIONS OF Sm AND Yb IN SEAWATER AT DIFFERENT TOTAL CARBONATE CONCENTRATIONS

Total $[CO_3^{2-}]$ (μM)	Sm	Yb
44.7	0.066	0.019
79.4	0.032	0.0071
151	0.013	0.0025
257	0.0054	0.0009

Data adapted from Figure 9, Lee and Byrne (1992)

post-depositionally phosphatized or that contain abundant detrital matter, Sm/Yb ratio fluctuations should primarily reflect changes in the chemistry of oceanic water, especially its dissolved CO_3^{2-} concentration, during the period in which the crusts grew. Biostratigraphic dating of the samples in the current study suggests this to be a period of at least 55 Ma, or most of the Cenozoic (Cowen et al., 1993). Thus, age-dating of key intervals in crusts RD50S1B and 6RD08-08 temporally constrain variations in the REE concentrations and Sm/Yb ratios discussed here (Cowen et al., 1993, and pers. comm., 1997).

REE and Sm/Yb profiles in the crusts from Kiribati, Schumann Seamount, and those collected from other regions suggest that conditions controlling the incorporation of REE (and other metals) into crusts have varied over geologic time as well as spatially within the ocean (for example, Fig. 8). The abrupt decrease in the Sm/Yb ratio near 40 mm depth in crust 6RD08-08 (Fig. 8a) coincides with an unconformity observed by optical microscopy (Wen, 1995) and is interpreted to reflect an erosional event that occurred near the Eocene-Oligocene boundary. Alternatively, this feature could represent chemical dissolution of the crust rather than mechanical erosion. The net result is the same as the crust record for said period does not exist. A textural feature similar to that present in crust 6RD08-08 was also observed in crust RD50S1B, and is believed to be contemporaneous with that in the Kiribati crust (De Carlo, 1991; McMurtry et al., 1994). When replotted as a function of age (Fig. 10), the large changes in the Sm/Yb ratio near 40 mm depth in the crusts seem less dramatic and are consistent with our first order interpretation that variations in this ratio reflect more gradually changing carbonate concentrations in seawater during the Cenozoic.

According to the following chemical equilibrium applied to seawater, when the CO_2 concentration increases, the concentration of CO_3^{2-} must decrease.

$$CO_2 + CO_3^{2-} + H_2O \quad <====> \quad 2HCO_3 \qquad (12)$$

The relationship between CO_3^{2-} and CO_2 implies that relatively low concentrations of CO_3^{2-} in seawater are a response to relatively high pCO_2 in seawater. Elevated aqueous CO_2 concentrations will then cause shoaling of the lysocline and the CCD.

When changes in the CCD over the past 50 Ma (van Andel et al., 1975) are superimposed on the age reconstructed Sm/Yb curve for crust 6RD08-08 (Fig. 10), the appearance of an inverse relationship between the Sm/Yb and the CCD lends support to our hypothesis that changes in seawater CO_3^{2-} are reflected in REE fractionation in Fe-Mn crusts. The lack of any relationship between the Sm/Yb ratio and major elements (Fe, Mn, Ca, or P; Wen et al., 1997), except perhaps within the lower phosphatized section of the Kiribati crust (see Fig. 5), is consistent with this hypothesis.

Cyclicity in the Sm/Ym ratio has also been observed in other crusts. A similar Sm/Yb pattern exists in crust 6RD08-03 recovered from the same dredge haul on McKean Seamount, Kiribati (Fig 8C, and De Carlo and Pruszkowski, 1995) and in several crusts from the Ellice Islands chain and the Marshall Islands (G. M. Mc Murtry, pers. comm., 1998). Thus, we expect that the crusts recovered from the west central Pacific Ocean should all be correlatable. Two crusts from the northern Line Islands and another from the western Indian Ocean (X. Y. Wen and E. H. De Carlo, pers. comm., 1998) also display Sm/Yb ratio fluctuations throughout the crust laminae. We believe these crusts also record fluctuations in seawater CO_3^{2-} throughout the Tertiary, although their chronology remains undetermined, and the different Sm/Yb ratio patterns in these crusts (data not shown here) imply that they were exposed to different oceanic conditions and/or contain sections that have been subject to post-depositional reworking. Alternatively these crusts experienced periods of nondeposition or more likely erosion or dissolution that destroyed some of the original/primary Sm/Yb record.

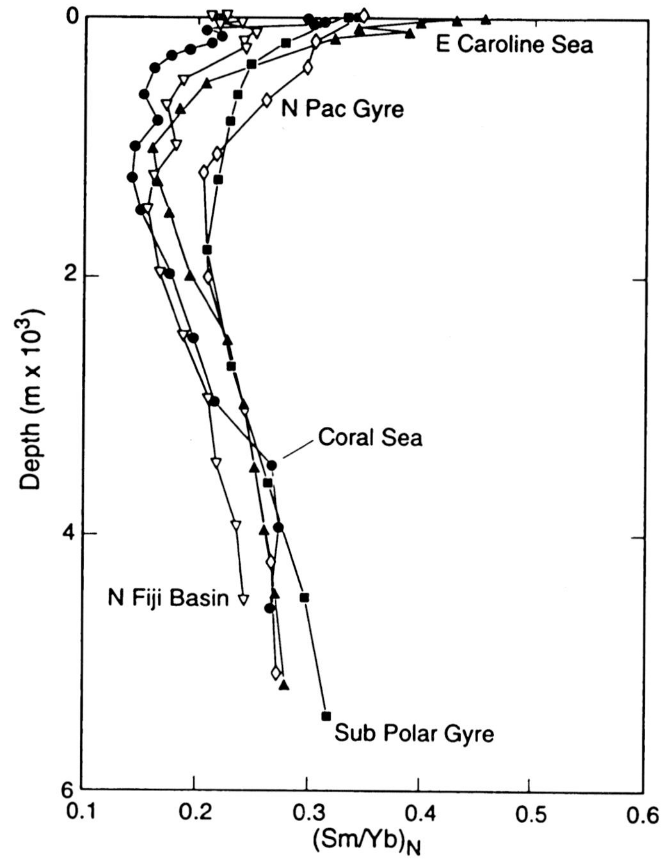

FIG. 9.—Depth profiles of shale-normalized Sm/Yb ratios of seawater from the Pacific Ocean. Data taken from Piepgras and Jacobsen (1992) for stations TPS47-39-1 (western north Pacific Subpolar gyre) and TPS47-271-1 (western N. Pacific Gyre) and from Zhang and Nozaki (1996) for stations SA5 (E. Caroline Basin), SA7 (Coral Sea) and SA12 (N. Fiji Basin). Note the tight fit of the Sm/Yb ration for the various stations in the depth range of 2000-3000 m.

Crust RD50S1B displays a different Sm/Yb pattern (Fig. 8B) than those from the west central Pacific region, implying growth under different conditions. This interpretation is consistent with results presented by Wen (1995) and Wen et al. (1997) on the basis of principal component factor analysis and interelement relationships. For example, Ce anomalies in the Schumann Seamount crust are strongly correlated with Mn, an element with a strong hydrogenetic signature, whereas in the Kiribati crust, Ce anomalies are strongly correlated wtih ele-

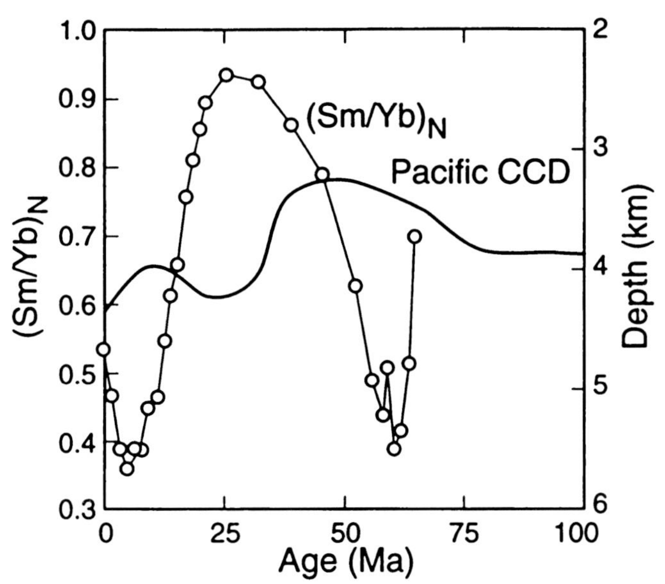

FIG. 10.—Plot of the shale-normalized Sm/Yb ratio in crust 6RD08-08 versus age reconstructed from nanofossil biostratigraphy. Variations in the depth of the CCD during the Cenozoic (from van Andel et al., 1975) are also shown, revealing the inverse relationship between the Sm/Yb ratio and the CCD (and by extension the lysocline). The inverse correlation appears to fail in the lower section of the crust likely owing to post-depositional phosphatization that alters Sm/Yb ratios, which then do not follow the trends predicted on the basis of the theory developed in this paper.

ments that are believed to display a biological influence (Wen, 1995; Wen et al., 1997). The different Ce associations in these two crusts, one consistent with Ce anomalies representing a good paleoredox indicator (that is, correlated with strongly oxidizing conditions in the water column, e.g., De Carlo, 1991) and the other indicating a strong biological influence on this parameter, are reasonable when one considers the history of the seamounts on which the crusts accreted.

Because the paleotracks of the two seamounts traversed different regions, it was expected that differences in the depositional environments to which the crusts were exposed would be reflected in their compositions. For example, crust RD50S1B from the Schumann Seamount (~26°N, ~160°W) spent a significant portion of its early growth history in the equatorial zone of high productivity (McMurtry et al., 1994). Later, passage within the tradewind belt, where continental eolian dust is an important source of sediments, led to a greater influence of detrital matter on crust accumulation. McKean Seamount in Kiribati (~4°S, ~170°W) is currently south of the equator and never passed through the equatorial zone of high productivity, thus the composition of crusts from this seamount should reflect significantly different environmental conditions (for example, a shallower CCD; van Andel et al., 1975) during the corresponding time period. The more "western" paleotrack of the Kiribati Seamounts also likely led to a greater influence of mid-depth water circulation from the western Pacific Ocean.

In the Equatorial Pacific Ocean, the depth and shape of the calcite lysocline appears to have changed significantly in response to climate forcing (Archer, 1991). Farrell and Prell (1989, 1991) also demonstrated that the depth of the calcite lysocline changed significantly in response to glacial/interglacial climate fluctuations. However, how carbonate ion concentrations change in the water column during glacial/interglacial periods remains unclear. Sm/Yb patterns such as those presented here for crusts 6RD0808 and 6RD08-03 (Fig. 8), revealing a systematic variation in REE fractionation as a function of time, may then be quite useful in reconstructing changes in the calcite lysocline during the Tertiary. Additional work with other crusts from other regions should help elucidate how mid-depth oceanic circulation varied throughout different regions of the world's oceans.

CONCLUSIONS

We provide further evidence that Fe-Mn crust laminae record changes in depositional conditions over geologic time and demonstrate a potential for correlation between various areas of the Pacific Ocean. Furthermore, we show that evidence of regional and widespread oceanic conditions coexist in crusts. Fortunately, by judiciously employing REE ratios and abundance patterns, we believe that these contributions can be resolved so as to reconstruct accurately the growth history of crusts. Biostratigraphic dating placed reliable temporal constraints on individual crust strata needed to effectuate paleoreconstructions that cannot be achieved by other methods.

The Cenozoic evolution of the Pacific Ocean resulted from a combination of tectonic events and climatic episodes (Van Andel et al., 1975), with the latter strongly influencing depositional conditions over time. The temporal changes in oceanic circulation, climatic cycles (CO_2 flux and reservoirs), ensuing variations in the geographical extent of the equatorial zone of high productivity, as well as major episodes of phosphogenesis represent some of the more important phenomena that are reflected in the geochemical signature of Fe-Mn crusts.

We conclude from this study and the results presented by Wen et al. (1997) that:

- Changes in crustal characteristics and chemical composition associated with paleoceanographic changes appear correlatable in crusts from different locations within the west central Pacific Ocean.
- Fractionation of REE during uptake from seawater can be modeled from a knowledge of the partitioning behavior of these elements (determined in experimental studies) and from the major element composition of Fe-Mn crusts.
- The fractionation of REE during uptake from seawater by crusts is controlled by solution composition to a greater extent than by solid phase composition.
- REE fractionation, as exemplified here by the Sm/Yb ratio, reflects variations in oceanic chemical conditions (e.g., $[CO_3^{2-}]aq$) during their growth histories.
- The Sm/Yb ratio, in absence of reliable temporal constraints, may also be useful as a relative dating tool if the chronology of other crusts from the given region is available and Sm/Yb fractionation patterns can be correlated between individual crusts.
- The Sm/Yb ratio can only be applied to infer paleoceanic

fluctuations in CO_3^{2-} in crust intervals devoid of post-depositional reworking such as phosphatization.

ACKNOWLEDGMENTS

The work presented in this paper is derived partly from the M.S. thesis of XiYuan Wen. We acknowledge Khalil J. Spencer for providing Nd isotopic data. We thank C. R. Glenn, Y. H. Li, and F. T. Mackenzie for their encouragement and numerous insightful discussions. Helpful reviews of this work and further insights were provided by K. Johannesson and J. R. Hein. This research was sponsored by the NOAA/Sea Grant College Program through Institutional Grant NA89AA-D-SG063, Project R/OM-7 to EHDC. This is SOEST contribution No. 4799 and Sea Grant Contribution JC-98-12.

REFERENCES

Akagi, T., Shabani, M. B., and Masuda, A., 1993, Lanthanide tetrad effect in kimuraite [$CaY_2(CO_3)_4*6H_2O$]: Implication for a new geochemical index: Geochimica et Cosmochimica Acta, v. 57, 2899-2905.

Alvarez, R., De Carlo, E. H., Cowen, J. P., and Andermann, G., 1990, Unusual morphological characteristics of marine ferromanganese crusts: Marine Geology, v. 94, p. 239-249.

Aplin, A.C., 1984, Rare earth element geochemistry of Central Pacific ferromanganese encrustations: Earth and Planetary Science Letters, v. 71, p. 13-22.

Aplin, A. C., Michard, A., and Albarede, F., 1986, $^{143}Nd/^{144}Nd$ in Pacific ferromanganese encrustations and nodules: Earth and Planetary Science Letters, v. 81, p. 7-14.

Archer, D. E. 1991. Equatorial Pacific calcite preservation cycles: Production or dissolution? Paleoceanography, v. 6, p. 561-571.

Balistrieri, L., Brewer, P. G., and Murray, J. W., 1981, Scavenging residence times of trace metals and surface chemistry of sinking particles in the deep ocean: Deep Sea Research, v. 28A, p. 101121.

Bau, M. Koschinsky, A., Dulski, P., and Hein, J. R., 1996, Comparison of the partitioning behaviour of yttrium, rare elements, and titanium between hydrogetic marine ferromanganese crusts and seawater: Geochimica Cosmochimica Acta, v. 60, p. 1709-1725.

Bertram, M. H., and Cowen, J. P., 1994, Testate rhizopod growth and mineral deposition on experimental substrates from Cross Seamount: Deep Sea Research, v. 41, p. 575-601.

Bertram, C. J. and Elderfield, H., 1993, The geochemical balance of the rare earth elements and neodymium isotopes in the oceans: Geochimica et Cosmochimica Acta, v. 57, p. 1957-1986.

Byrne, R. H., and Kim, K.-H., 1990, The influence of surface and solution chemistry on rare earth distribution in seawater: Geochimica et Cosmochimica Acta, v. 54, p. 2645-2656.

Byrne, R. H., and Sholkovitz, E., 1996, Marine Chemistry and Geochemistry of the Lanthanides, *in* Gschneidner, K. A., Jr., and Eyring, L., eds., Handbook on the Physics and Chemistry of Rare Earths: Amsterdam, Elsevier, v. 23, p. 497-593.

Cantrell, J. and Byrne, R. H., 1987, Rare earth element complexation by carbonate and oxalate ions: Geochimica et Cosmochimica Acta, v. 51, p. 597-605.

Chabaux, F., Cohen, A. S., O'Nions, R. K., and Hein, J. R., 1995, ^{238}U-^{234}U-^{230}Th chronometry of Fe-Mn crusts: Growth processes and recovery of thorium isotopic ratios of seawater: Geochimica et Cosmochimica Acta, v. 59, p. 633-638.

Choppin, G. R., 1984, Lanthanide complexation in aqueous solutions: Journal of the Less Common Metals, v. 100, p. 141-151.

Cowen, J. P., and Taylor, G. T., 1987, The impact of micro-organisms on the physical growth and chemical composition of cobalt-rich manganese crusts of Pacific seamounts, in Proceedings of the 8th International Symposium on Environmental Biogeochemistry, Nancy, France.

Cowen, J. P., De Carlo, E. H., and McGee, D., 1993, Calcareous nannofossil biostratigraphic dating of a ferromanganese crust from Schumann Seamount: Marine Geology, v. 115, p. 289-306.

De Baar, H. J. W., Bacon, M. P., Brewer, P. G., and Bruland, K. W., 1985, Rare earth elements in the Pacific and Atlantic Oceans: Geochimica et Cosmochimica Acta, v. 49, p. 1943-1959.

De Carlo, E. H., 1991, Paleoceanographic implications of intrasample rare earth element variability in a marine Fe-Mn crust from the Hawaiian Archipelago: Marine Geology, v. 98, p. 449-467.

De Carlo, E. H., and McMurtry, G. M., 1992, Rare Earth Element geochemistry of seamount ferromanganese deposits from the Hawaii Archipelago: Chemical Geology, v. 95, p. 235-250.

De Carlo, E. H., and Pruszkowski, E., 1995, Laser ablation ICP-MS determination of alkaline and rare earth elements in marine ferro-manganese deposits: Atomic Spectroscopy, v. 16, p. 65-73.

De Carlo, E. H., Pennywell, P., and Fraley, C. M., 1987a, Geochemistry of ferromanganese deposits from the Kiribati and Tuvalu region of the West Central Pacific Ocean: Marine Mining, v. 6, p. 301321.

De Carlo, E. H., McMurtry, G. M., and Kim, K. H., 1987b, Geochemistry of ferromanganese crusts from the Hawaiian Archipelago—I. Northern survey areas: DeepSea Research, v. 34, p. 441467.

De Carlo, E. H., Wen, X. Y, and Cowen, J. P., 1994, Rare earth element fractionation in Fe-Mn crusts: A potential paleoceanographic indicator of seawater carbonate (abs.): Eos, Transactions, American Geophysical Union, v. 75, p. 398.

De Carlo, E. H., Wen, X. Y, and Irving, M., 1997, The influence of redox reactions on the uptake of dissolved Ce and other rare earth elments by suspended Fe and Mn oxide particles: Aquatic Geochemistry, v. 3, p.1-33.

Dzombak, D. A., and Morel, F. M. M., 1990, Surface Complexation Modeling: Hydrous Ferric Oxide: New York, John Wiley and Sons.

Elderfield, H., 1988, The oceanic chemistry of the rare-earth elements: Philosophical Transactions Royal Society of London, v. A325, p. 105-126.

Elderfield, H., and Pagett, R., 1986, Rare earth elements in ichthyoliths: Variations with redox conditions and depositional environment: Science of the Total Environment, v. 49, p. 175-197.

Elderfield, H., Upstill-Goddard, R., and Sholkovitz, E. R., 1990, The rare earth elements in rivers, estuaries, and coastal seas and their significance to the composition of seawater: Geochimica et Cosmochimica Acta, v. 54, p. 971-991.

Erel, Y., and Stolper, E. M., 1993, Modeling of rare earth element partiioning between particles and solution in aquatic environments: Geochimica et Cosmochimica Acta, v. 57, p. 513-518.

Farrel, J. W., and Prell, W. L., 1989, Climatic change and $CaCO_3$ preservation: An 80,000 year bathymetric reconstruction from the central equatorial Pacific Ocean: Paleoceanography, v. 4, p. 447-466.

Farrel, J. W., and Prell, W. L., 1991, Pacific $CaCO_3$ preservation and $\delta^{18}O$ since 4Ma: Paleoceanic and paleoclimatic implications: Paleoceanography, v. 6, p. 485-498.

Fleet, A. J., 1984, Aqueous and sedimentary geochemistry of the rare earth elements, *in* Henderson, P., ed., Rare Earth Element Geochemistry: Amsterdam, Elsevier, p. 343-374.

Futa, K., Peterman, Z. E., and Hein, J. R., 1988, Sr and Nd isotopic variations in ferromanganese crusts from the central Pacific: Implications for age and source provenance: Geochimica Cosmochimica Acta, v. 52, p. 2229-2233.

German, C. R., and Elderfield, H., 1990, Application of the Ce Anomaly as a paleoredox indicator: The ground rules: Paleoceanography, v. 5, p. 823-833.

Halbach, P., and Puteanus, D., 1984, Influence of the carbonate dissolution rate on the growth and composition of Corich ferromanganese crusts from Central Pacific seamount areas: Earth and Planetary Science Letters, v. 68, p. 7387.

Hein, J. R., Bohrson, W. A., Schulz, M. J., Noble, M., and Clague, D. A., 1992, Variations in the fine-scale composition of a central Pacific ferromanganese crust: Paleoceanographic implications: Paleoceanography, v. 7, p. 63-77.

HENDERSON, P., ed., 1984, Rare Earth Element Geochemistry: Amsterdam, Elsevier, 510 p.

INGRAM, B. L., HEIN, J. R., AND FARMER, G. L., 1990, Age determinations and growth rates of Pacific ferromanganese deposits using strontium isotopes: Geochimica et Cosmochimica Acta, v. 54, p. 1709-1721.

JANIN, M.-C., 1988, Biostratigraphie comparée de nodules et sédiments du Pacifique Nord-Equatorial: Bulletin de la Société Géologique de France, v. 8t-II, p. 373-380.

JOHANNESSON, K. H., LYONS, W. B., STETZENBACH, K. J., AND BYRNE, R. H., 1995, The solubility control of rare earth elements in natural terrestrial waters and the significance of PO_4^{3-} and CO_3^{2-} in limiting dissolved rare earth concentrations: A review of recent information: Aquatic Geochemistry, v. 1, p. 157-173.

KAGI, H., DOHMOTO, Y., TAKANO, S., AND MASUDA, A., 1993, Tetrad effect in lanthanide partition between calcium sulfate crystal and its saturated solution: Chemical Geology, v. 107, p. 71-82.

KOSCHINSKY, A., AND HALBACH, P., 1995, Sequential leaching of marine ferromanganese precipitates: Genetic implications: Geochimica et Cosmochimica Acta, v. 59, p. 5113-5132.

KOEPPENKASTROP, D., AND DE CARLO, E. H., 1992, Sorption of rare earth elements from seawater onto pure mineral phases: An experimental approach: Chemical Geology, v. 95, p. 251-263.

KOEPPENKASTROP, D., AND DE CARLO, E. H., 1993, Uptake of rare earth elements from solution by metal oxides: Environmental Science and Technology, v. 27, p. 1796-1802.

KOEPPENKASTROP, D., DE CARLO, E. H., AND ROTH, M., 1991, A method to investigate the interaction of rare earth elements with metal oxides in aqueous solution: Journal of Radioanalytical and Nuclear Chemistry, v. 151, p. 337-346.

LEE, J.-H., and BYRNE, R. H., 1992, Examination of comparative rare earth element complexation behavior using linear free-energy relationships: Geochimica et Cosmochimica Acta., v. 56, p. 1127-1138.

LEE, J.-H., AND BYRNE, R. H., 1993, Complexation of trivalent rare earth elements (Ce, Eu, Gd, Tb, Yb) by carbonate ions: Geochimica et Cosmochimica Acta., v. 57, p. 295-302.

LE SUAVE, R., PICHOKI, C., PAUTOT, G., HOFFERT, M., MOREL, Y., VOISSET, M., MONTI, S., AMOSSE, J., AND KOSAKEVITCH, A., 1989, Geological and mineralogical study of Co-rich ferromanganese crusts from a submerged atoll in the Tuamotu Archipelago (French Polynesia): Marine Geology, v. 87, p. 227-247.

LING, H. F., BURTON, K. W., O'NIONS, R. K., KAMBER, B. S., VON BLANKENBURG, F., GIBB, A. J., AND HEIN, J. R., 1997, Evolution of Nd and Pb isotopes in Central Pacific seawater from ferromanganese crusts: Earth and Planetary Science Letters, v. 146, p. 1-12.

MANGINI, A., SEGL, M., KUDRASS, H., WIEDICKE, M., BONANI, G., HOFMANN, H. J., MORENZONI, E., NESSI, M., SUTTER, M., AND WOLFLI, W., 1986, Diffusion and supply rates of ^{10}Be and ^{230}Th radioisotopes in tow manganese encrustations from the South China Sea: Geochimica et Cosmochimica Acta, v. 50, p. 149-156.

MANHEIM, F.T., AND LANE-BOSTWICK, C. M., 1989, Chemical composition of ferromanganese crusts in the world ocean: A review and comprehensive database. Washington, D.C., U.S. Geological Survey Open File Report, 200 p.

MASUDA, A., AND IKEUCHI, Y., 1979, Lanthanide tetrad observed in marine environment: Geochemical Journal, v. 14, p. 19-22.

MCLENNAN, S. M., 1994, Rare earth element geochemistry and the "tetrad" effect: Geochimica et Cosmochimica Acta, v. 58, p. 2025-2033.

MCMURTRY, G. M.., VONDERHAAR, D. L., EISENHAUER, A., MAHONEY, J. J., AND YEH, H. W, 1994, Cenozoic accumulation history of a Pacific ferromanganese crust: Earth and Planetary Science Letters, v. 125, p. 105-118.

MOFFETT, J., 1990, Microbially mediated cerium oxidation in seawater: Nature, v. 345, p. 421-423.

NEUMANN, T. AND STEUBEN, D., 1991, Detailed geochemical study and growth history of some ferromanganese crusts from the Tuamotu Archipelago: Marine Mining, v. 10, p. 29-48.

PICHOKI, C. AND HOFFERT, M., 1987, Characteristics of Co-rich ferromanganese nodules and crusts sampled in French Polynesia: Marine Geology, v. 77, p. 109-119.

PIEPGRAS, D. J., AND WASSERBURG, G. J., 1980, Neodymium isotopic variations in seawater: Earth and Planetary Science Letters, v. 50, p.128-138.

PIEPGRAS, D. J., AND JACOBSEN, S. B., 1988, The isotopic composition of neodymium in the North Pacific: Geochimica et Cosmochimica Acta, v. 52, p. 1373-1381.

PIEPGRAS, D. J., AND JACOBSEN, S. B., 1992, The behavior of rare earth elements in seawater: Precise determination of variations in the North Pacific water column: Geochimica et Cosmochimica Acta, v. 56, p. 1851-1862.

PUTEANUS, D., AND HALBACH, P., 1988, Correlation of CO concentration and growth rate: A method for age determination of ferromanganese crusts: Chemical Geology , v. 69, p. 73-85.

SHOLKOVITZ, E. R., LANDING, W. M., AND LEWIS, B. L., 1994, Ocean particle chemistry: The fractionation of rare earth elements between suspended particles and seawater: Geochimica et Cosmochimica Acta., v. 58, p. 1567-1579.

SEGL, M., MANGINI, A., BONANI, G., HOFMANN, H. J., NESSI, M., SUTER, M., WOLFLI, W., FRIEDRICH, G., PLUGER, W. L., WIECHOWSKI, A., AND BEER, J., 1984, Be-10 dating of a manganese crust from Central North Pacific and implications for palaeocirculation: Nature, v. 309, p. 540-543.

SEGL, M., MANGINI, A., BEER, J., BONANI, G., SUTER, AND WOLFLI, W., 1989, Growth rate variation of manganese nodules and crusts induced by paleoceanographic events: Paleoceanography, v. 4, p. 511-530.

STANLEY, J. K., AND BYRNE, R. H., 1990, The influence of solution chemistry on REE uptake by *Ulva lactuca L.* in seawater: Geochimica et Cosmochimica Acta., v. 54, p. 1587-1595.

VAN ANDEL, T. H., HEATH, G. R., AND MOORE, T. C., 1975, Cenozoic History and Paleoceanography of the Central Equatorial Pacific Ocean: Boulder, Geological Society of America Memoir 143.

VON STACKELBERG, U., KUNZENDORF, H., MARCHIG, V., AND GWODZ, R., 1984, Growth history of a large ferromanganese crust from the equatorial north Pacific nodule belt: Geologisches Jarbuch, v. A7, p. 213235.

WEN, X. Y., 1995, Geochemical characteristics of Pacific Seamount ferromanganese crusts: Evidence for their genesis. Unpublished M.S. Thesis, University of Hawaii, Honolulu, p. 131.

WEN, X. Y., AND DECARLO, E. H., 1994, A comparative study of the geochemistry and internal structure of seamount ferromanganese crusts (abs.): Eos, Transactions, American Geophysical Union, v. 74, p. 78.

WEN, X. Y., DE CARLO, E. H., AND LI, Y. H., 1997, Interelement relationships in ferromanganese crusts from the central Pacific Ocean: Their implications for crust genesis: Marine Geology, v. 136, p. 277-297.

WOOD, S. A., 1990, The aqueous geochemistry of the rare earth elements and yttrium: 1 Review of available low-temperature data for inorganic complexes and the inorganic REE speciation of natural waters: Chemical Geology, v. 82, p. 159-186.

YAN, C. Y., AND KROENKE, L. W., 1993, A plate tectonic reconstruction of the Southwestern Pacific, 0100 Ma: Proceedings of the Ocean Drilling Program: Scientific Results, v. 130.

ZHANG, J., AMAKAWA, H., AND NOZAKI, Y., 1994, The comparative behaviors of Yttrium and Lanthanides in the seawater of the North Pacific: Geophysical Research Letters, v. 21, p. 2677-2680.

ZHANG, J., AND NOZAKI, Y., 1996, Rare earth elements and yttrium in seawater: ICP-MS determinations in the East Caroline, Coral Sea, and South Fiji basins of the western South Pacific Ocean: Geochimica et Cosmochimica Acta., v. 60, p. 4631-4644.

PLATINUM AND OTHER RELATED ELEMENT ENRICHMENTS IN PACIFIC FERROMANGANESE CRUST DEPOSITS

DENYS L. VONDERHAAR* AND GARY M. MCMURTRY
School of Ocean and Earth Science and Technology, University of Hawaii, Honolulu, Hawaii 96822, USA
DIETER GARBE-SCHÖNBERG
Geologisch-Paläontologisches Institut, Universität Kiel, Olshausenstrasse 40, 2300 Kiel 1, Germany
DORIS STÜBEN
Institut für Petrographie und Geochemie, Universität Karlsruhe, Kaiserstrasser 12, 76128 Karlsruhe, Germany
AND
BRADLEY K. ESSER
Isotope Sciences Division, Lawrence Livermore National Laboratory, Livermore, CA 94551, USA

ABSTRACT: Many of the oldest and most depositionally complete hydrogenetic ferromanganese crusts are recognized to be enriched in Pt, but the source of this enrichment is debated. Despite the presence of numerous Fe-, Ni-, and Pt-rich spherules in the ferromanganese crusts studied, ratios of the platinum group elements (PGE) in the ferromanganese material from one sample indicate that Pt is greatly enriched compared to the other PGE and slightly enriched (about two times) with respect to seawater ratios. Osmium isotopic ratios also indicate a nonchondritic source for the PGE. Much of the crust Pt and Ir correlates with Fe-rich layers. Results of seawater-ferromanganese oxide adsorption experiments using ^{191}Pt and ^{192}Ir suggest that Pt is preferentially scavenged by FeOOH when compared to Ir, suggesting a seawater source for the enrichment. Platinum, Fe, Mn, Ni, and Co all co-vary within the Pt-rich zone in stratigraphic transect 1 of this crust. The more detailed transect 2 results differ somewhat from those of transect 1. These data are compared with Pt and Ir data obtained by laser ablation-inductively coupled plasma-mass spectrometry (LA-ICP-MS) analyses at ~1-mm intervals. Redox-sensitive elements such as Fe, Mn, Co, Cr, the derivative Ce and Eu anomalies, and Sm/Yb ratios from transect 2, when compared with the Pt profiles, suggest that localized redox conditions associated with either O_2 minimum zone processes or with the formation of carbonate fluorapatite (CFA) are responsible for the Pt enrichments. Analyses of two other crust samples from the same seamount and several other ferromanganese crusts from other areas of the Pacific suggest that the PGE enrichment is more closely tied to the formation of the CFA than to O_2 minimum zone processes, carbonate ion variations in the water column, or cosmogenic sources.

INTRODUCTION

It has often been suggested that the variability in elemental composition among layers of ferromanganese crusts reflects differences in the depositional environment (e.g., Halbach and Puteanus, 1984; Hein et al., 1992; Segl et al., 1984; Mangini et al., 1990). Because hydrogenous ferromanganese crusts occur on relatively sediment-poor seamount slopes and are characterized by slow rates of growth, they have the potential to reveal the long-term history of oceanic midwater column processes. The accumulation of ferromanganese crusts has been proposed to be associated with the oxic-suboxic history of the water in which they formed (Frakes and Bolton, 1984). Bioproductivity, bottom current activity, depth of the CCD and lysocline, rate of carbonate dissolution (Halbach and Puteanus, 1984), and position of the seamount in response to plate tectonics and subsidence are all factors that may be reflected in the chemical composition of individual ferromanganese layers. Post-depositional changes in conditions that control redox interactions within the ferromanganese crust-sediment-water system may also have a profound effect on the distribution of the transition elements and platinum group elements (PGE) in these deposits. Despite abundant evidence for these changes, they have largely been ignored.

Commonly, both carbonate fluorapatite (CFA) and ferromanganese crust deposits are ubiquitous and closely associated on the upper slopes of Pacfic seamounts (e.g., Burnett et al., 1987). Changes in major and minor element depth profiles through these deposits may be correlated with changes observed in ocean chemistry, possibly since the time of seamount formation, if initiation of ferromanganese accumulation occurs soon after seamount formation. From the time the seamounts in the Hawaiian Archipelago formed in the late Cretaceous (75–90 Myr), there have been ocean-wide environmental changes, and the seamounts have moved from their area of formation in the equatorial Pacific to their present positions in the north central Pacific Ocean. This movement is associated with both horizontal plate motion and vertical motion due to isostatic subsidence of the underlying lithosphere upon which the seamounts formed. The seamounts have therefore changed their positions relative to the CCD, the O_2 minimum zone, and the equator. Major events should be recorded in the chemistry of the microlayers of ferromanganese crusts, which vary in thickness from less than a millimeter to several millimeters, assuming a quasi-continuous accumulation. There have also been major changes in ocean chemistry during the tens of millions of years over which the seamounts have existed. During this time the ocean has changed from a relatively warm, unstratified ocean to the ocean of today, characterized by strong AABW currents and mid-ocean gyres, with upwelling regimes at continental margins and along the equator (e.g., Kennett, 1982).

Platinum Group Element Geochemistry

Much work has been done in the past decade on Pt and other PGE in sediments and ferromanganese deposits (Esser and Turekian, 1988; Palmer et al., 1988; Halbach et al., 1989; Colodner et al., 1992; Ravizza and McMurtry, 1993), but knowl-

*email:denys@iniki.soest.hawaii.edu

Marine Authigenesis: From Global to Microbial, SEPM Special Publication No. 66

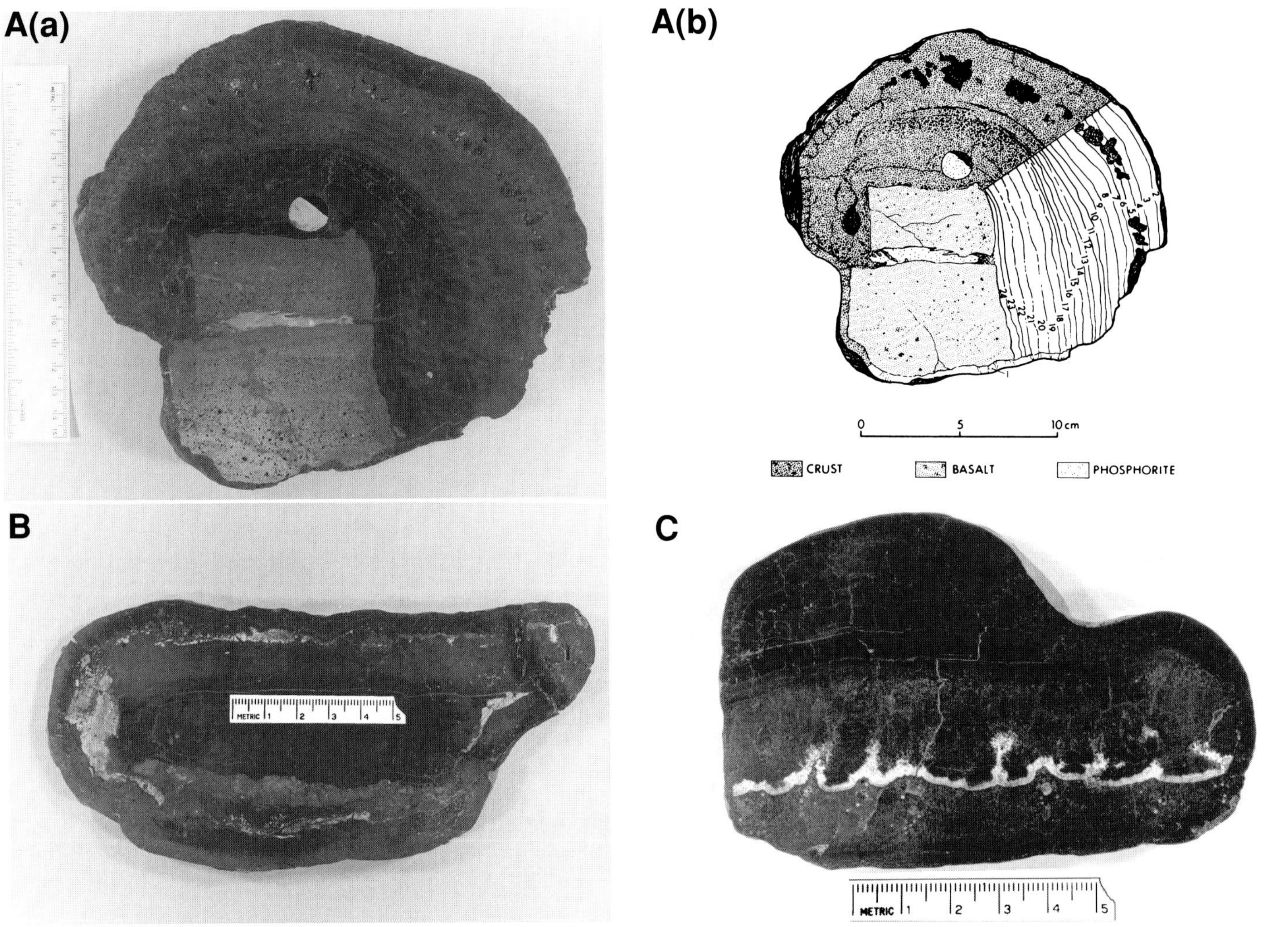

FIG. 1. (A)(a) Photograph of polished vertical section of crust sample KK84-RD50 S1B dredged from Schumann Seamount in the Hawaiian Archipelago. The ferromanganese crust grew around a basalt fragment that contains large cracks filled with carbonate fluorapatite. (b) Schematic diagram showing sampling locations for chemical and isotopic data from transect 1. (B) Photograph of polished section of crust sample KK84-RD50 S2 from Schumann Seamount. This sample differs from RD50 S1B in that it is not attached to a substrate, and has an encircling layer of carbonate fluorapatite and an outer rind of ferromanganese oxide surrounding the sample. (C) Photograph of sample KK84-RD50 S5, also from Schumann Seamount. The sample is most likely oriented upside down in this photograph, and like sample RD50 S2, is a "pseudo-nodule" with an outer rind of ferromanganese oxide.

(D) Photograph of sample KK84-PD8 S2 from an unnamed seamount within the Hawaiian Archipelago. The ferromanganese crust lies on top of an altered hyaloclastite substrate with carbonate fluorapatite within the substrate's matrix. (E) Photograph of crust sample MW88 RD33 S6 from Lo-En Guyot in the Marshall Islands Archipelago, dredged from an area close to the base of the guyot at a water depth of 2400 m. This sample consists of a thick ferromanganese crust with a considerable amount of infiltrated CFA on top of a brecciated basalt substrate. (F) Photograph of crust sample MW 88 RD35 S4 from an area on the flank, close to the summit of Lo-En Guyot. This sample was dredged from a water depth of 1775 m and is similar in appearance to the other Lo-En Guyot sample.

edge remains limited of any of the PGE's behavior in seawater, felsic rocks, and atmospheric aerosol concentrations because of their extremely low concentrations (Hodge et al., 1985; Goldberg et al., 1986; Palmer et al., 1988; Colodner, 1991; Koide et al., 1991). Meteorites are highly enriched in PGE as compared to the Earth's crust. Enrichments in Pt and Ir particularly have been interpreted as reflecting changes in the amount of cosmic material in marine sediments. Because marine hydrogenous ferromanganese deposits preferentially accumulate Pt when compared to sediments, these enrichments have long been attributed to the concentration of cosmogenic material within these deposits (Castaing and Fredriksson, 1958; Finkelman, 1970; Jedwab, 1975; Esser and Turekian, 1988). As early as 1975, spherules and micron-sized metallic flakes found in ferromanganese nodules were considered to be of a cosmic origin (Jedwab, 1975). These Fe-rich spherules and flakes were commonly associated with large amounts of Ni. High levels of Pt were also found in spherules interspersed within a ferromanganese crust sample from Tuamotu in French Polynesia (Kosakevitch, 1987). The spherules found in this crust were considered to be contemporaneous with the Eocene-Oligocene transition and were thought to have been deposited during a hiatus in the growth of the ferromanganese crust. The spherules were trapped in holes, possibly due to erosion of the original crust layer, and were overlain by a younger crust generation (Kosakevitch, 1987). High Pt values in the lower part of the fer-

D

E

F

romanganese crust were attributed to Pt-rich nuclei found in the spherules in this part of the crust. The cosmogenic input was not quantified in this study, however, and it is not clear that these enrichments primarily result from included cosmogenic debris.

Halbach et al. (1989) first tried to quantify the contribution of cosmic debris to Pt concentrations in ferromanganese crusts, but this work was inconclusive. The number of spherules collected in the size class between 100 μm and 350 μm was unable to account for the anomalously high Pt values; but it is likely that much of the material was ignored because the lower limit on the size class is too large, and oxic diagenesis of the spherules renders them insensitive to separation by magnetic techniques. Magnetic techniques would also be ineffective in separating a large proportion of the spherules if a greater percentage of spherules were stony rather than iron. Therefore, estimates that have used the number of magnetically separated spherules to determine cosmogenic concentrations and flux rates are likely to be underestimates (e.g., Esser, 1991). Because even less is known about the behavior of PGE in seawater (Hodge et al., 1985; Goldberg et al., 1986; Jacinto and van den Berg, 1989; Colodner, 1991), PGE source determinations are difficult.

Most of the recent literature suggests seawater is the primary source for the Pt enrichment in ferromanganese deposits because the ratios of Pt to the other PGE are nonchondritic (e.g., Hodge et al., 1986; Goldberg et al., 1986; Palmer et al., 1988; Koide et al., 1991; Colodner et al., 1992). Iridium and Pt contents in ferromanganese nodule deposits in the Pacific Ocean were found to have a strong co-variance with Co, whereas Ni did not co-vary with the PGE. This relationship was attributed to the simultaneous oxidation of Ir, Pt, and Co to higher oxidation states from the dissolved form, where the higher oxidation states are preferentially accommodated in the solid phases in the minerals (Goldberg et al., 1986). It is therefore probable that much of the PGE enrichment is due to enhanced adsorption of the PGE from seawater onto the ferromanganese flocs in these deposits.

Rampino (1981) and, more recently, Colodner et al. (1992) suggested that PGE enrichments in sediments at biostratigraphic boundaries, excepting possibly the K-T boundary, may be in response to changing redox conditions (degradation of organic matter) during early diagenesis, rather than the result of a cosmogenic or volcanic source. Colodner et al. proposed that a redox gradient is created when a turbidite unit, relatively rich in organic carbon (up to 1%), is deposited on a relatively oxic pelagic sediment. These authors found that Pt may be scavenged by both oxyhydroxide and organic-rich sediments, whereas Ir uptake is only observed in organic-rich sediments. The difference between the Pt and Ir ratios in the ferromanganese deposits may also be because of the ~50-fold lower abundance of Ir in seawater compared to Pt. Other studies have proposed Ir enrichment in sediments as a result of reduction and removal of Ir from seawater under anoxic conditions (Schmitz, 1985; Holser et al., 1989), and bacterial scavenging of Ir via Fe-oxides (Playford et al., 1984; Orth et al., 1988). Rhenium/Ir and Pt/Ir ratios in sediments are sensitive to redox changes, however (Colodner et al., 1992), as are most transition elements, and therefore ratios of the PGE may not indicate their origin.

Goals of This Study

Correlation of microlayers within samples of the ferromanganese encrustations from the same seamount should reveal a chemical signature consistent with the conditions encountered along the paleopath and the chemical signatures associated with major oceanic events. The overall aim of this study was to investigate the stratigraphic distribution of the PGE in ferromanganese crusts to establish whether the enrichment of these

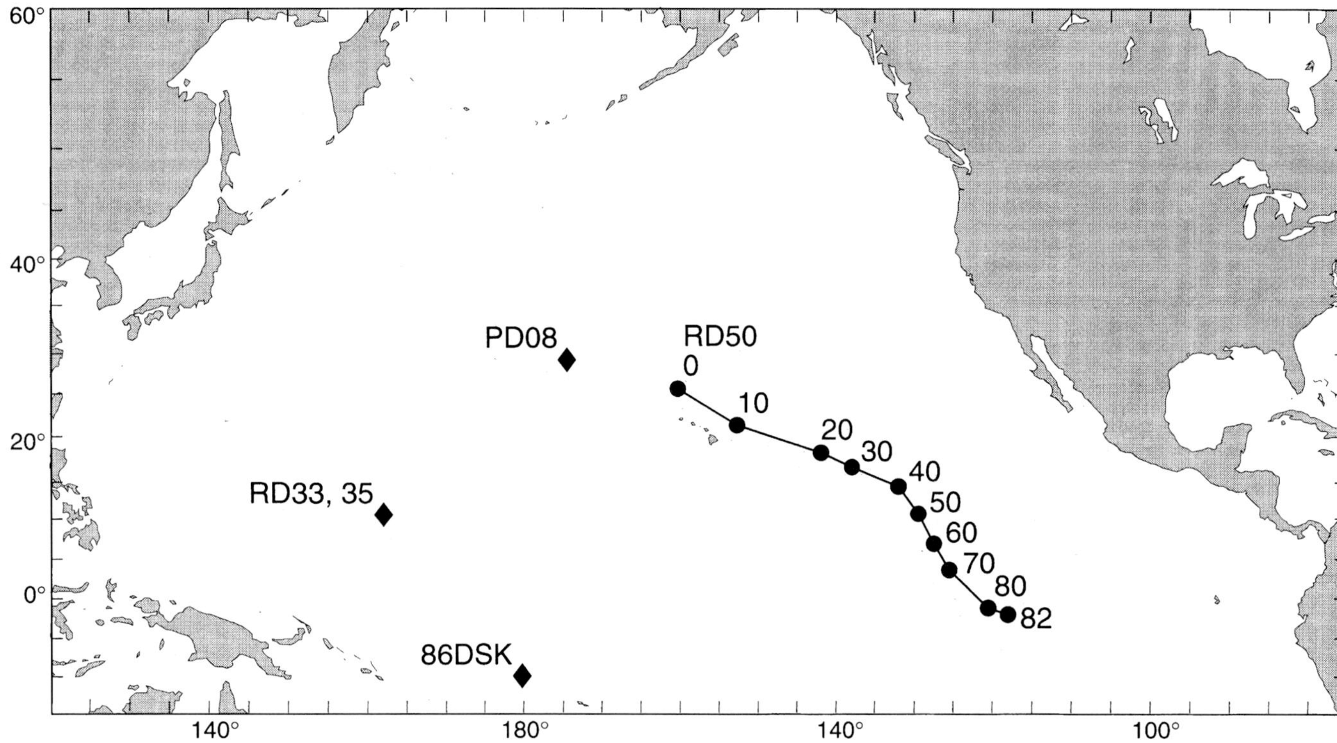

FIG. 2.—Paleopath of Schumann Seamount (solid dots) from a position south of the equator, where it formed at 82 Ma, to its present position ~700 km north of Kauai, Hawaii. Crust samples KK84-RD50 S1B, RD50 S2, and RD50 S5 were dredged from 2250 to 2600-m water depth near the seamount summit. Numbers refer to paleoposition ages in millions of years (Ma). Also shown are the locations of dredge samples PD08 S2, RD33 S6 and RD35 S4, and 86 DSK.

TABLE 1.—LOCATION AND DEPTH OF SAMPLES.

Sample	Thickness	Seamount	Latitude	Longitude	Water Depth
KK84 RD50 S1B	96 mm	Schumann	25°55.50′N	159°57.50′W	2250–2600 m
KK84 RD50 S2	80 mm	Schumann	25°55.50′N	159°57.50′W	2250–2600 m
KK84 RD50 S5	70 mm	Schumann	25°55.50′N	159°57.50′W	2250–2600 m
KK84 PD8 S2	60 mm	Unnamed	29°11.08′N	174°01.27′W	1100–1475 m
SO66 86 DSK	50 mm	Tuvalu area	9°56.24′S	179°56.55′E	1865 m
MW88 RD33 S6	133 mm	Lo-En Guyot	9°56′N	152°52′E	2400 m
MW88 RD35 S4	111 mm	Lo-En Guyot	10°00′N	162°41′E	1775 m

elements is primarily due to (1) input from cosmogenic or from mantle sources, (2) enhanced adsorption of PGE onto Mn and Fe flocs as a result of local changes in redox conditions as they fall below the O_2 minimum zone, (3) post-depositional scavenging by Fe-oxides, or (4) post-depositional enrichment along redox fronts associated with carbonate fluorapatite (CFA) deposition.

METHODS

Sample Selection

Seven samples were chosen for Pt analyses; three of these samples (KK84 RD50 S1B, Fig. 1A; RD50 S2, Fig. 1B; and RD50 S5, Fig. 1C) were dredged from Schumann in the Musician's Group (Fig. 2). Another sample, KK84 PD8 S2, was dredged from an unnamed seamount in the Hawaiian Archipelago (Fig. 1D). Sample SO66 86 DSK was recovered from a seamount in the Tuvalu area of the South Pacific; two other crusts were dredged from Lo-En Guyot in the Marshall Islands Archipelago; MW8805 RD33 S6 (Fig. 1E) and RD35 S4 (Fig. 1F). These samples were chosen because they exhibited the most complete depositional sequence and mineralogy consistent with little or no diagenetic alteration of the major phases, primarily amorphous FeOOH and vernadite. These hydrogenous ferromanganese-oxide crusts ranged in thickness from 70 mm to 96 mm (Fig. 1, Table 1).

Two of the crusts, RD50 S1B and PD8 S2, were analyzed for major and minor elements as well as Pt concentrations by various methods. Details of the analysis are given below. Samples KK84 RD50 S1B, RD50 S2, (in cooperation with A. Eisenhauer), and PD8 S2, (in cooperation with P. Sharma), were analyzed by ^{10}Be techniques. Growth rates were also determined for RD50 S1B and PD8 S2 by a modified "Co chronometer" (Manheim, 1986). In addition, Sr isotopes were determined for the CFA contained within the crust for both RD50 S1B and RD50 S2 (McMurtry et al., 1994 and J. A. Cooper et al., pers. commun., 1988). Finally, microfossil evidence (Cowen et al., 1993) and Os isotopes (VonderHaar, 1997 and B. K. Esser and D. L. VonderHaar, in prep., 1999) were used to date the ferromanganese layers below those that are effectively dated by ^{10}Be methods in the RD50 S1B sample.

We compared the microdistributions of Pt with redox-sensitive rare earth elements (REE) (for example, Ce and Eu). The REE were determined by laser ablation-inductively coupled plasma-mass spectrometry (LA-ICP-MS) at the University of Kiel in all but the RD50 S1B sample. The REE in the RD50 S1B sample were determined by solution ICP-MS at the University of Kiel and the Pt and Ir concentrations by LA-ICP-MS at VG Ltd. in England.

Subsampling of the Ferromanganese Crusts

Crusts RD50 S1B and PD8 S2 were manually separated along visibly distinct layers. Sample PD8 S2 was subdivided into eight layers. RD50 S1B was subdivided twice; initially, the slab closest to the middle of the crust sample was divided into 25 subsamples with an average thickness of 4 mm using a high-speed, hand-held tool with a metal saw (Fig. 1A). Subsequently, a second, adjacent slab was divided into 59 subsamples with an average thickness of 1.5 mm using the same tool but with a corundum grinding stone attachment. When subdividing the samples care was taken to follow visible growth boundaries and to clean the equipment and area between each subsampling interval to ensure minimal sample mixing. A split of powdered sample from the first subsampling of RD50 S1B was used for Pt analyses by Pb-cuppellation fire assay ICP-MS, with other trace and major elements determined by X-ray fluorescence spectrometry (XRF) by XRAL Laboratories, Don Mills, Canada. If more than 0.5 g of sample remained, these layers were also analyzed for Os concentration and isotopic composition by solution ICP-MS and neutron thermal ionization mass spectrometry (NTIMS). Half of the layers sampled in the second transect of RD50 S1B were used for Os isotopic and concentration determinations, with alternate layers reservd for additional Pt-group element (PGE) and rare earth element (REE) concentration analyses. In all cases, the subsamples were ground with a porcelain mortar and pestle to pass a 100-mesh screen and were oven-dried at 110°C for 24 hours. The PD8 S2 subsamples were measured for major and minor elemental concentrations by inductively coupled plasma-atomic emission spectrometry (ICP-AES) and for growth rate determinations by ^{10}Be analyses. The three remaining samples, the two Lo-En Guyot samples from the Marshall Islands archipelago, and the sample from the Tuvalu area of the South Pacific, as well as PD8 S2, RD50 S1B, RD50 S2, and RD50 S5, were cut into thick petrographic-slide-sized pieces and impregnated with epoxy for LA-ICP-MS analyses of Pt and REE.

Platinum and Ir Adsorption Experiment Methods

Seawater collected from ~1000 m water depth off Keahole Point, island of Hawaii, was filtered to <0.2 μm and irradiated with ultraviolet light for 24 hours. Three different particle concentrations (400 mg/L, 800 mg/L, and 1700 mg/L) of sample KK84 PD6 S5 were prepared from powders that were ground and sieved to pass 200 mesh. These powders were put into an opaque 1-L size Nalgene bottle, and filled with 1 L of the prepared seawater. A solution of synthetic, amorphous FeOOH with a particle concentration of 2000 mg/L and a similar solution of synthetic δ-MnO_2 with a particle concentration of 1200 mg/L were also put into the treated seawater in the same type of bottles and kept at typical deep seawater pH (7.5-7.8). These solutions were mixed constantly for a week on an orbital shaker in a dark room with a temperature of ~2°C. Known amounts of ^{191}Pt ($t_{1/2}$ = 2.90 days) and ^{192}Ir ($t_{1/2}$ = 73.8 days) radiotracers were then added to the solution and shaken for the duration of the experiment. At time intervals of 1 hour, 1 day, 1, 2, and 3 weeks, 100 mL of solution was removed and filtered using a filtration rig with two 0.2 μm polycarbonate filters. Activities of the radiotracers in the filter, filter blank, and filtrate were counted using a high-purity Ge gamma detector with a 4096-channel pulse-height analyzer. K_d's were calculated using the relationship of Li et al. (1984):

$$K_d = \frac{A_{Filter} - A_{Filter\ Blank}}{A_{Water}} \times \frac{V \times 10^3\ (mL/g)}{m \times f}$$

where A = activity of a radiotracer (cpm/100 mL water or cpm/filter), m = dry weight of solid particles on the filter (mg), f = factor to convert filter geometry to 100 mL plastic bottle geometry.

RESULTS AND DISCUSSION

Schumann Seamount Study

Most work concentrated on crust sample RD50 S1B (McMurtry et al., 1987, 1989, 1994; VonderHaar and McMurtry, 1988). Major element analyses (Fig. 3) and XRD mineralogy reveal three major components included in this crust: ferroman-

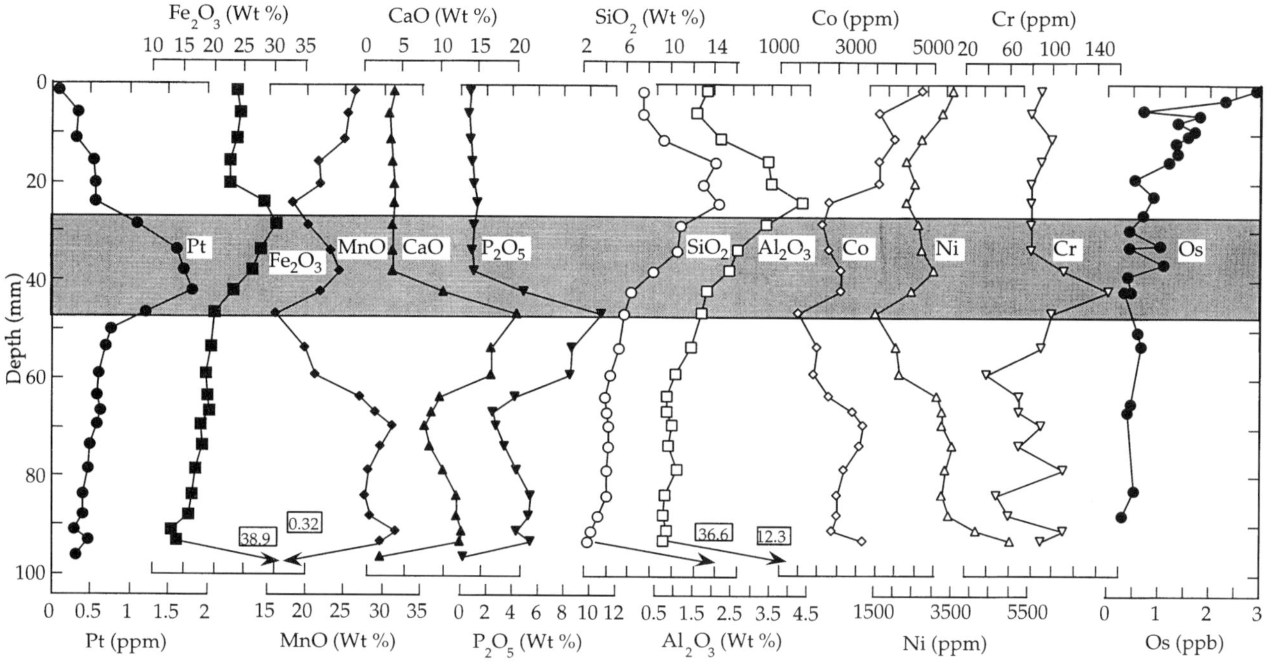

FIG. 3.—Platinum concentrations obtained by Pb-cuppellation fire assay ICP-MS are compared to the XRF analyses of the transition elements on layers from RD50 S1B transect 1. Osmium results determined by ICP-MS and NTIMS on samples from both transects 1 and 2 are also shown. Iron oxide and Pt concentrations are similar throughout the crust, whereas Pt, Mn, Ni, and Co co-vary only in the Pt-rich zone. Chromium is most concentrated at the base of the highest peak and just above the highest P_2O_5 concentrations.

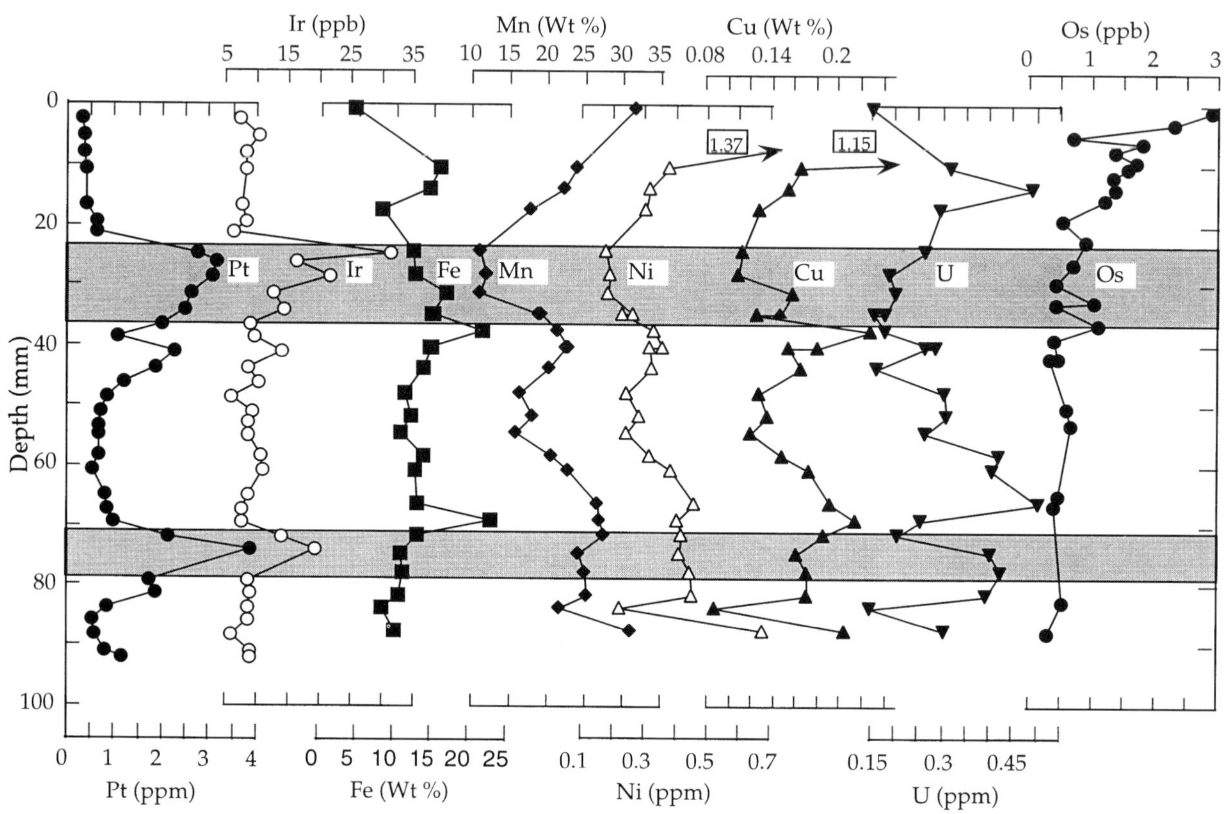

FIG. 4.—Platinum and Ir concentrations determined by LA-ICP-MS and Fe, Cu, Mn, Ni, and U concentrations by solution ICP-MS on the transect 2 samples. These elements show similar correlations to those found by Colodner et al. (1992) in Pt-rich zones within pelagic sediments interrupted by a turbidite unit and interpreted by these authors as a "reductive-bleaching" type of diagenetic process.

ganese oxides, CFA, and aluminosilicate debris. Mineralogically, these components consist primarily of vernadite, amorphous Fe oxides, minor goethite, CFA, quartz, and feldspar. Ferromanganese crusts have been differentiated by Halbach and Puteanus (1984) into two growth generations, younger and older, which are often separated by an episode of phosphatized carbonate or CFA deposition. Although this crust also consists of an older and younger crust generation, there is more than one episode of phosphatic deposition. Both episodes of phosphatic deposition, however, apparently occurred during the 30 Myr hiatus in ferromanganese crust accumulation (McMurtry et al., 1994). This crust exhibits three distinct groupings according to the elemental distributions determined by XRF and Pb-cuppellation fire assay analyses: the "older generation" (>50 mm) impregnated by CFA, the CFA-poor "younger generation" (<20 mm), and the Pt-rich layers (20–50 mm) just above the layer richest in CFA. Iron, generally more abundant in the outermost layers of the crust, has a maximum between 20 mm and 40 mm (Fig. 3) when analyzed by XRF on the transect 1 samples, whereas Mn is more variable but generally increases with depth into the crust, showing dilution effects with both the CFA and the aluminosilicate debris. The Mn and Fe data on the more finely divided transect 2 (Fig. 4) from this sample show trends similar to those of the original transect 1; however, the Fe peak is slightly narrower and there is an additional Fe peak in the lower part of the crust between 65 mm and 70 mm. The Mn data are virtually the same between the two transects with transect 2 again showing a slightly lower displacement of the peaks within the crust and a greater amount of Mn in the outermost layer. The XRF-derived Ca and P concentrations from transect 1 are highly intercorrelated and have a maximum from about 40 mm to almost 65 mm in the interior of the crust, whereas aluminosilicates are concentrated in a layer 10-30 mm into the crust. Calcium, P, Si, and Al concentrations were not determined for the transect 2 data.

Pt Profiling of RD50 S1B

A stratigraphic profile of platinum concentrations obtained by Pb-cuppellation fire assay ICP-MS is compared in Figure 3 with the XRF analyses of the transition element concentrations in layers with an average thickness of 4 mm in transect 1 of sample RD50 S1B. On the same crust sample but on a different subsample, LA-ICP-MS Pt concentrations every 3 mm are compared with the transition element concentrations obtained by solution ICP-MS on layers with an average thickness of 1.5 mm from transect 2 (Fig. 4). The Pt profile obtained on transect 1 exhibits a well-defined maximum of almost 2 ppm from 25 mm to 50 mm into the crust and is similar to the Fe profile (Fig. 3), particularly below 25 mm in this sample. Microscopic and SEM-EDX examination of the Pt and Fe-rich layer revealed Fe-Ni-rich spherules that are known to contain elevated levels of Pt and other PGE (i.e., Kyte et al., 1980; Esser, 1991). The presence of spherules in this layer originally led us to hypothesize that the Pt enrichments were likely from a cosmogenic source. The SEM image in Figure 5 shows a solid spherule with a 60-μm diameter and a portion of a hollow one. The spheres are found along with detrital material between ferromanganese growth columns and tend to be somewhat common in this layer (located at 38-42 mm depth), although they are not always easily discernible from other detritus if they are broken or hollow. Most of the Fe-rich spherules in this layer appear to be completely altered to FeOOH, which renders them inseparable using magnetic techniques.

The PGE ratios do not substantiate a meteoritic component in this crust. Iridium profiles were obtained on this crust at the same time as the Pt profiles, using LA-ICP-MS. Both elements exhibit a broad peak from ~25–55 mm (Fig. 4) and another, much sharper peak from ~70–80 mm depth. Unlike the bulk Pt data in transect 1, the laser ablation data in transect 2 show a second, deeper Pt (and Ir)-rich layer in this crust, probably because the laser is site-specific to a 100-μm spot diameter. The site-specific nature of the analyses also contributes to scatter in a heterogeneous medium, much like the results of electron microprobe profiling. Laser ablation data, although not as quantitative as solution data by ICP-MS, do produce reliable comparisons of relative amounts of Pt and Ir, so were used to calculate the Pt/Ir ratios. The average Pt/Ir ratio for this sample transect (117) is similar to those reported by Goldberg et al. (1986) and Koide et al. (1991), 100 and 103 respectively, for ferromanganese nodules.

However, the Pt and Ir concentrations, measured by LA-ICP-MS for RD50 S1B, are $>10^4$ more enriched than those found in nodules by Goldberg et al. (1986). Although the Os/Ir ratios in pelagic sediments studied by Goldberg et al. were very similar to those found in the ferromanganese nodules with an average of 0.51, (range from 0.13–1.43), the Pt/Ir ratios in the sediments averaged ~7, (range from 1.33–19), indicating a marked enrichment of Pt in the ferromanganese nodules over pelagic sediments (Goldberg et al., 1986). An oxidation-reduction reaction was suggested as the mechanism for enhancement of Pt, Ru, Ir (Goldberg et al., 1988), and Os (Koide et al., 1991) in ferromanganese oxide minerals.

Pt concentrations are highly enriched over Os in the RD50 S1B sample (Figs. 3 and 4) when compared to unaltered chondritic sources, which would have an ~1:1 Pt/Os ratio. The Pt data collected by laser ablation ICP-MS were obtained from total, unseparated samples, whereas the Os measurements collected by solution ICP-MS and NTIMS were made on leachates of subsamples of the bulk hydrogenous material. Correlations between these elements can be considered reliable only if there is a negligible cosmogenic component. Indeed, Os concentration data from the 6N HCl leach—when compared to the 1.1N H_2SO_4 plus 3% H_2O_2 leach—are indistinguishable, indicating a negligible meteoritic component or that the 6N HCl leach is incapable of dissolving this component (B. K. Esser and D. L. VonderHaar, in prep., 1999). Koide et al. (1991) found Os to covary with Pt, Ru, and Ir in ferromanganese nodules; however, factor analysis results on the transect 1 concentrations of this crust sample show hydrogenous Os to positively correlate with total Co and Ti but to negatively correlate with total Pt and Ir (VonderHaar, 1997). The PGE measured in this crust have ratios more similar to those found in pelagic sediments and other fer-

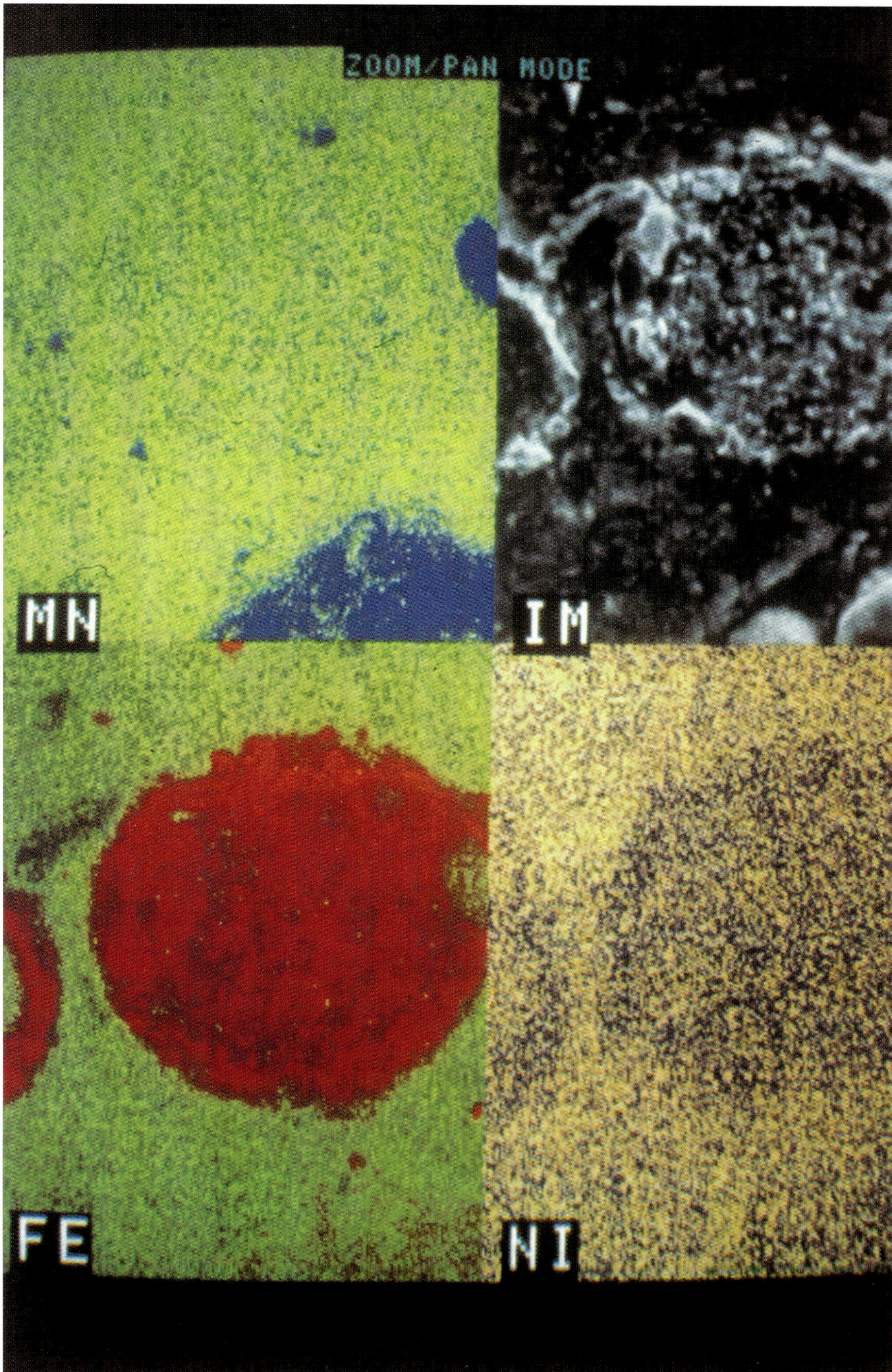

FIG. 5.—SEM images of spherules found in a layer at 38 to 42 mm depth in sample RD50 S1B. A 60 μm-diameter solid spherule with a hollow spherule next to it is shown in the image (IM) in the upper right. The Fe (in red on the lower left) and Ni (in purple on the lower right) x-ray maps show the spherules to be Fe- and Ni-rich, whereas the Mn map (in blue on the upper left), shows them to be Mn-poor, indicating a cosmogenic origin.

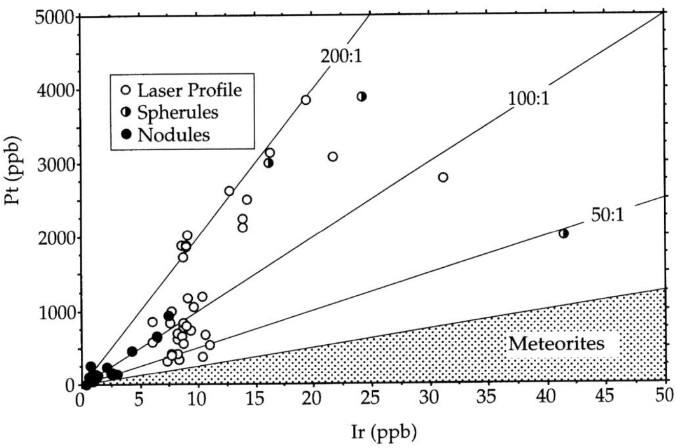

FIG. 6.—Plot of Pt vs. Ir for three individual spherules measured by LA-ICP-MS, along with laser profile data on the ferromanganese oxides for the RD50 S1B crust, and Mn nodule data (Goldberg et al., 1986). Two spherules have Pt/Ir ratios close to 200:1, whereas one spherule shows a ratio of 50:1, closer to the range for unaltered meteorite ratios.

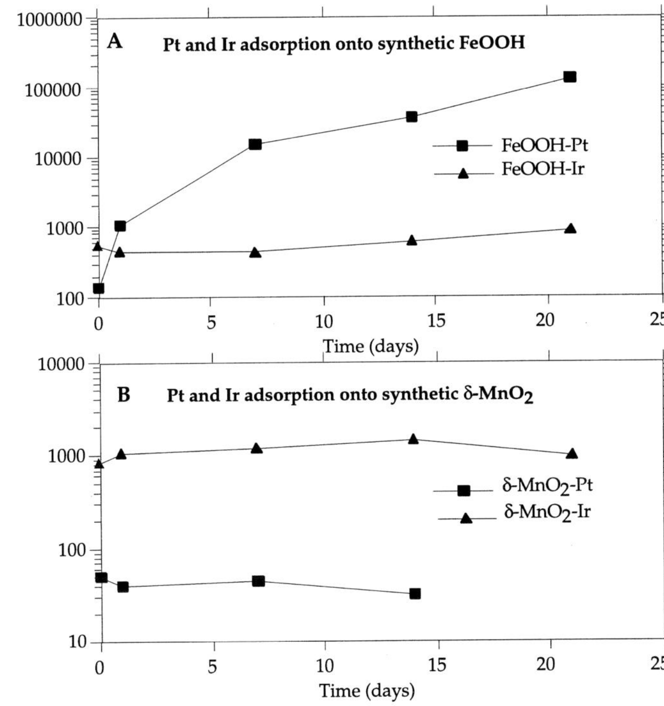

FIG. 7.—(A) The results from a Pt and Ir adsorption radio-tracer experiment onto a synthetic, amorphous FeOOH phase are shown. Adsorption is expressed as Kd, or distribution coefficient. The first data point is at 1 hour, the next point is one day, and the next three points are 1 week, 2 weeks, and 3 weeks, respectively. (B) Platinum and Ir adsorption onto a synthetic δ-MnO_2 phase. Note the Pt value was not available for the third week due to the short half-life of ^{191}Pt (2.90 days); there was not enough activity to measure this value.

romanganese deposits than in pristine meteoritic material, although this does not necessarily preclude a meteoritic origin. Because there is evidence for alteration of the spherules (Fig. 5), effects—if any—of this alteration are unknown on the PGE.

Individual spherules analyzed at the same time as the laser ablation profiling show Pt/Ir ratios that are up to two orders of magnitude higher than those found in C-1 chondrites and twice as high as average seawater values reported by Goldberg et al. (1986) (Fig. 6). Lower Ir values have been reported for a population of spinel-bearing spherules that were compared to irregular fragments of cosmic ablation debris from the same K-T boundary sediment (Robin et al., 1993). The Ir values reported for this crust are within the range of the K-T boundary spherule values, though the average Ir concentration (84 ppb) is higher for the boundary sediment spherules (Robin et al., 1993). Unfortunately, platinum was not measured in the boundary sediment spherules. The trend in Figure 6 suggests that many of the spherules in this crust have either lost Ir, or have preferentially adsorbed Pt onto Fe-rich spherules relative to Ir (and Os), either as they fell through the water column or post-depositionally.

Platinum and Ir Adsorption Experiments

To examine the relative contribution of dissolved Pt and Ir in seawater to crusts, a multiple radiotracer adsorption experiment modeled after Li et al. (1984) was performed using a natural ferromanganese phase, a synthetic FeOOH phase, and a synthetic δ-MnO_2 phase. Deep seawater collected at 1000 m was filtered and UV-treated for these experiments. The results indicate that Pt uptake, reported as K_d or the distribution coefficient, is several orders of magnitude greater onto FeOOH than δ-MnO_2 surfaces (Fig. 7). Iridium adsorption, expressed in the same units, is initially much greater onto the δ-MnO_2 phase than the FeOOH, but after 3 weeks, these values approach one another. The natural ferromanganese phase showed similar adsorption characteristics to the synthetic δ-MnO_2 phase after the initial week. This sample contained nearly twice the amount of MnO_2 as Fe_2O_3 (40.0 wt. % and 23.4 wt. %, respectively), so the results are expected. For FeOOH, the experimental Pt/Ir ratios approach the seawater value of ~100 in about 3 weeks, and Pt/Ir quickly attains a steady value of about 20 for δ-MnO_2 (Fig. 7). Further work needs to be done in this area, including desorption studies and adsorption/desorption experiments for other Pt-group elements such as Os and Re. These results indicate that the Pt/Ir ratios of settling particulate FeMn oxide flocs (i.e., Halbach et al., 1989; Koschinsky and Halbach, 1995) may already have acquired the seawater value through adsorption in the water column before incorporation onto the ferromanganese crust surface. These results also indicate that alteration of chondritic Pt/Ir ratios upon formation of Fe oxides in cosmogenic flakes and spherules in both the water column and in the ferromanganese crust itself may render source determinations by PGE ratios totally meaningless.

Relationship Between Pt and the First-row Transition Elements

Hodge et al. (1985) found a correlation between Co and Pt in ferromanganese nodules, which they attributed to the incorporation of these elements in higher valence states following oxidation of the dissolved divalent phase. This would lead one to expect that Mn would correlate with Pt because Mn and Co are generally positively correlated in the hydrogenous phases.

However, Mn and Pt are associated only within the Pt-rich layer and not throughout the RD50 S1B crust. The Pt-rich layer lies just above a CFA-rich layer in this sample, and in this layer the Pt, Fe, Ni, Co, and Mn all correlate (Fig. 3). Nickel and Co show a strong covariance with Mn throughout the "older generation" (depth range of 50–94 mm) of this sample, as well as within the Pt-rich layer in transect 1 (depth range of 22–50 mm). Hydrated and chloro-complexed metal cations like Co and Ni were shown by Koschinsky and Halbach (1995) to be sorbed onto the negatively charged δ-MnO_2 surfaces.

The Pt and Fe covary throughout the "older generation" and the Pt-rich layer, but the Pt concentrations decrease in the "younger generation," while the corresponding Fe values increase slightly before leveling off. Both the "younger generation" subsamples and the Pt-rich layer are enriched in Ni and Co relative to their Fe_2O_3 content. In the Pt-rich zone, there is an excess of Pt over Fe in the transect 1 data. When Pt/Fe is compared with the Fe profile, the Pt enrichment is two to four times as high in this part of the crust, suggesting that there must be another mechanism, in addition to adsorption, which is responsible for the most Pt-rich part of the ferromanganese crust (Fig. 8). Cobalt enrichment in the outer layers of crusts has been attributed by several authors (Heye and Marachig, 1977; Halbach et al., 1983; Manheim, 1986) to slower ferromanganese growth rates. Cobalt and Os correlate well within the "younger generation," and somewhat in the Pt-rich zone, indicating that Os concentrations may also represent slower growth with variable enrichment within the Pt-rich zone (Fig. 3).

In transect 2 Pt and Ir concentrations obtained by LA-ICP-MS and Fe, Cu, Mn, and Ni obtained by solution ICP-MS, show a juxtaposed relationship between the Pt and transition metals, with the Fe, Mn, Cu, and Ni peaks shifted either slightly above or below the Pt and Ir peaks (Fig. 4). These results are similar to those exhibited in the sediment cores studied by Colodner et al. (1992), and were shown to be the result of diagenetic processes. As organic material decomposes in the sediments, it consumes a well-established series of oxidants (Froelich et al., 1979). Trace metals are subsequently partitioned between the pore waters and solid phases as a result of these redox transitions. Elements such as Co and Ni follow Mn; in general, they are lost from the Mn-reducing zone and fixed under more oxic conditions in the zone of MnO_2 precipitation. Other elements, such as Cr, V, Mo, and U, possess stable soluble forms under oxic conditions and are scavenged or precipitated under more reducing conditions (Colodner, 1991). Chromium was measured in transect 1 and is plotted with the Pt concentrations in Figure 3. Indeed, the relative positions of these elements in the hypothesized redox gradient resemble those exhibited between Cr, V, and Pt in the cores studied by Colodner et al. (1992) that surround the oxic-suboxic transition zone. This and other work by Colodner (1991) clearly demonstrates redistribution of Pt, Ir, and Re along redox boundaries in oxic pelagic sediments interrupted by organic-rich reducing sediments. Uranium was measured in transect 2 but the results are less clear than the Cr and Pt results from transect 1, probably because the Pt and U in transect 2 were measured on different subsamples.

The ferromanganese crust and the turbidite-interrupted sediment sequence are obviously different from one another, particularly regarding the time scales involved, and the amount of organic carbon present during the phosphatization is not known. However, the basic redox associations between Pt and the transition elements still should apply. So though the CFA found within the ferromanganese oxides is probably a suboxic rather than an anoxic deposit (e.g., Glenn et al., 1994), the relative change in oxidation between the organic-rich turbidite and the pelagic sediments and between the CFA and ferromanganese oxides may be similar. Koschinsky and Halbach (1995) also recognized that phosphatization can cause changes in the mineral associations of some elements. They found that the Fe-bearing vernadite phase was partially dissolved by the slightly reducing conditions that occur during the phosphatization episode, mobilizing all elements associated with the Mn phase. The CFA appears to behave analogously to the turbidite in the pelagic sediment unit, as evidenced by the inverse relationship between the Pt and P concentrations shown in Figure 3. CFA deposition is comparatively more reducing and organic-rich (i.e., Glenn et al., 1994) than the surrounding highly oxic ferromanganese material. From Sr isotope-derived ages of the CFA contained within the ferromanganese oxides (McMurtry et al., 1994), and textural clues found in the photomicrographs of this sample (VonderHaar, 1997, Appendix A, pp. 265 -271), at least the layer of CFA at a depth from ~40–65 mm was not deposited concordantly with the ferromanganese oxides in which it is contained, but apparently formed from relatively reducing pore waters (when compared to the ferromanganese oxide surfaces), either during a period of higher productivity and increased sedimentation or when the seamount subsided into the low-O_2 or O_2-minimum zone. This area of the crust contains the greatest amount of CFA material and is just below the most mottled, porous section, which contains the greatest amount of Pt. The mottled tex-

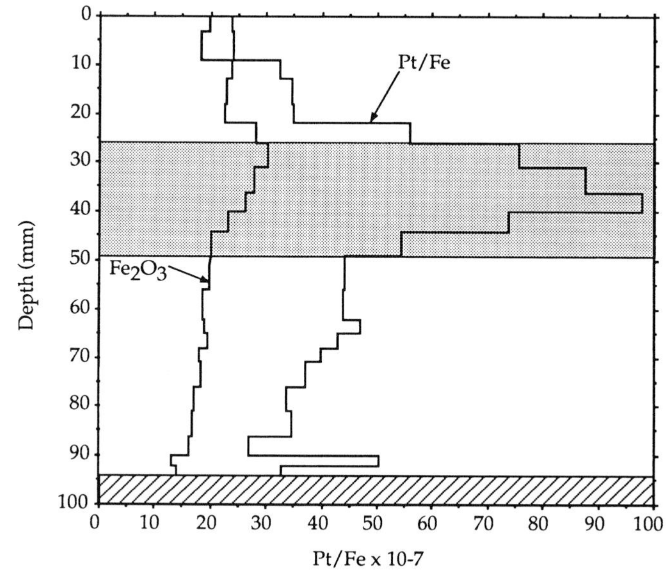

FIG. 8.—Iron concentations and Pt/Fe ratios of the transect 1 subsamples from RD50 S1B show Pt enrichment 2-4 times higher in the stippled area from ~25-50 mm. This suggests another mechanism in addition to seawater adsorption for the areas with the greatest amount of Pt enrichment.

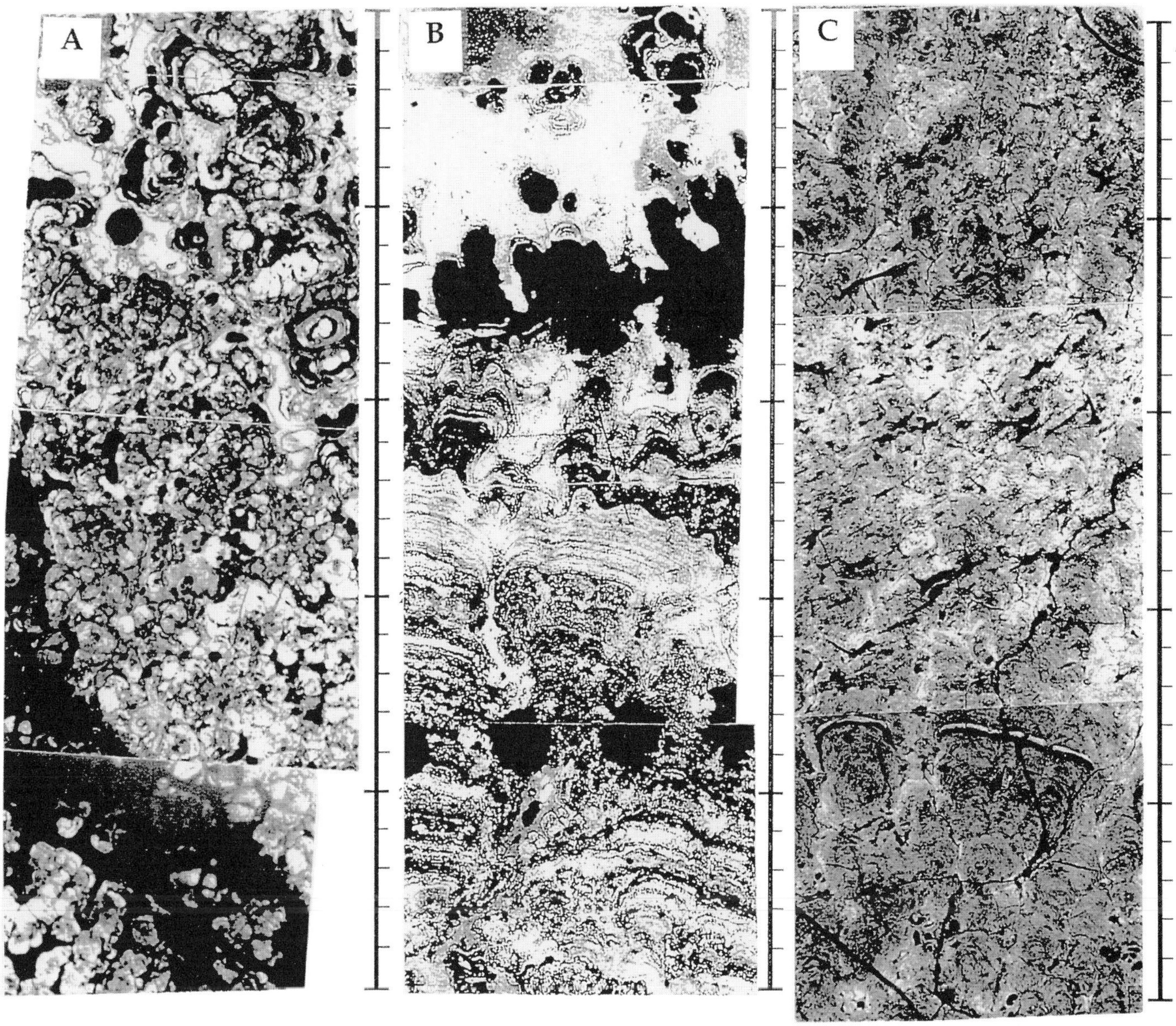

FIG. 9.—(A) Photomicrograph of RD50 S1B from 25-30 mm depth interval showing mottled texture with dissolution features and ferromanganese replacement by CFA. (B) Photomicrograph of RD50 S2 from 15-20 mm depth interval in the todorokite-rich layer of the ferromanganese material just below the contact with the CFA and within the upper Pt peak for this crust. The large holes between 16 and 17 mm and the more well-defined holes between 18 and 19 mm are artifacts of the laser ablation process. (C) Photomicrograph of PD8S2 from 60-65 mm depth interval showing a change in texture indicative of a change in growth rate in the area just above the substrate between 63 and 64 mm.

ture also continues into the CFA-rich zone, but the porosity is lessened because the interstitial spaces are filled with the fine-grained CFA. The uppermost layer of CFA in this crust occurs at a depth of ~18 mm into the crust and extends down to ~27 mm depth (VonderHaar, 1997, Appendix A, pp. 265-267), although the amount of CFA is not substantial enough that it may be clearly recognized in the Ca and P data from the broad subsampling in transect 1. It appears that there was dissolution of some of the ferromanganese oxide material before or during the time of the CFA deposition, and that the CFA was deposited in at least two episodes (Fig. 9A).

Pt and the Rare Earth Elements in RD50 S1-B Crust

The 70–90 mm interval of transect 2 in this crust exhibits a strong positive Ce anomaly relative to the other layers, with the highest enrichment just above the apex of the lower Pt peak (Fig. 10). There is another Ce anomaly peak between 15–20 mm, a few millimeters above the upper Pt peak in transect 2. A large Ce anomaly together with peaks in Mn, Cu, and Co has also been observed in discrete deep layers of ferromanganese crusts recovered from the Mid-Pacific Mountains region west of Hawaii (Puteanus, 1986). These enrichments could imply either depositional or post-depositional redox fluctuations that have affected the dissolved Ce uptake onto ferromanganese oxides. Enhanced uptake may reflect increased Mn found in the O_2 min-

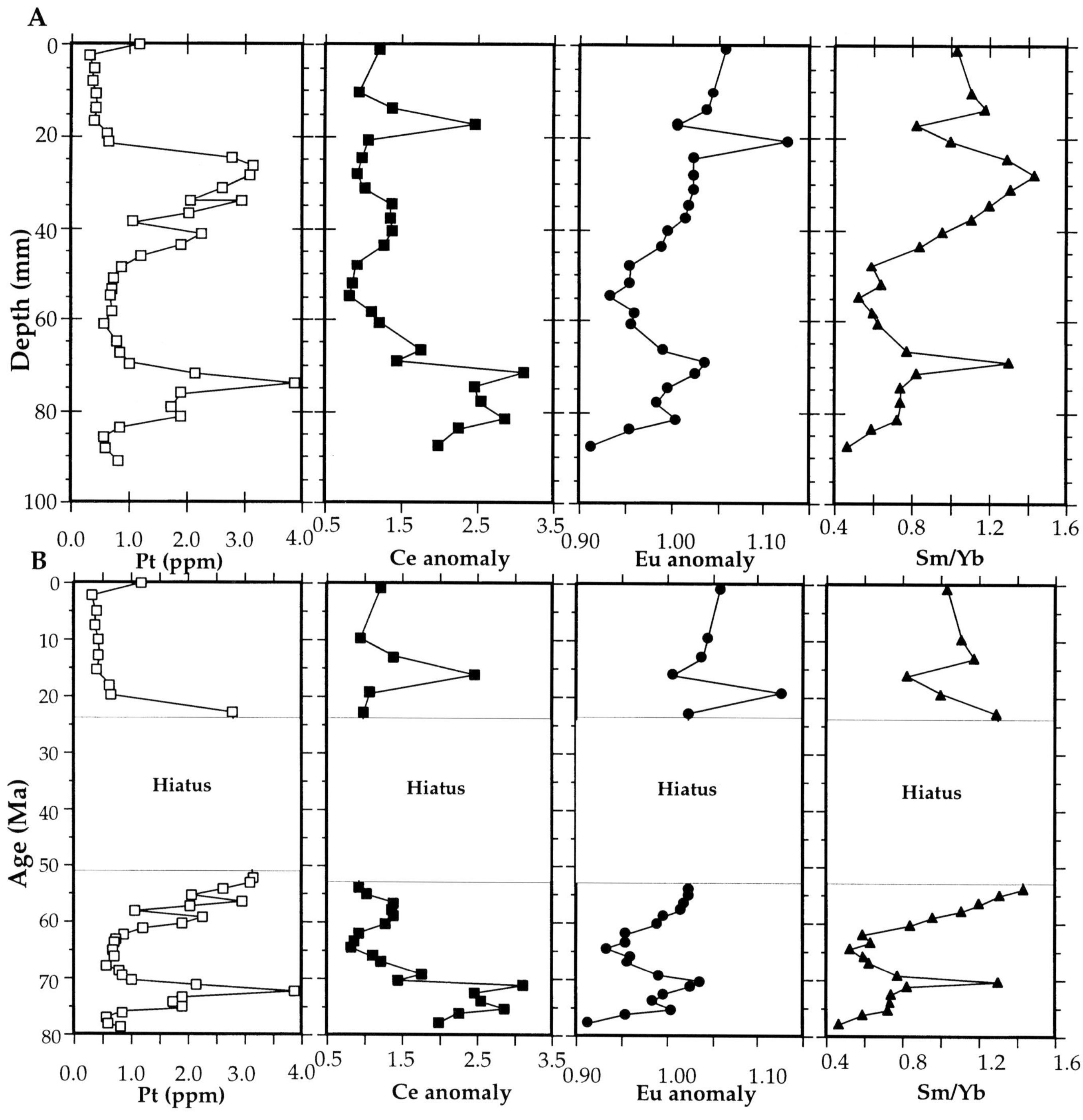

FIG. 10.—(A) LA-ICP-MS Pt concentrations are shown together with ICP-MS Ce anomaly, Eu anomaly, and Sm/Yb on transect 2 subsamples for crust RD50 S1B. (B) Age assignments based on ^{10}Be, Co-chronometry, microfossil evidence, and Os isotopes for the LA-ICP-MS Pt data and the ICP-MS Ce and Eu anomalies, and Sm/Yb ratios.

imum zone with deposition as particles fall into a more O_2-rich environment directly below this zone. Suboxic conditions encountered as the seamount sinks within the O_2 minimum zone with increased organic C preservation and input to the ferromanganese crusts could produce changing redox conditions similar to those described for CFA deposition. Cerium exists in the oceanic environment either in the trivalent, dissolved state, or as highly insoluble CeO_2; the preferential removal of dissolved Ce occurs as a consequence of redox reactions (Henderson, 1984). Redox sensitive REE such as Ce (which can be oxidized) and Eu (which cannot be oxidized) are potentially indicative of paleoredox conditions within the ferromanganese crust (e.g., Elderfield 1988; DeCarlo, 1991). Manganese concentrations and the Ce anomaly are positively correlated within the RD50 S1B crust (Fig. 10), indicating that in this crust, the Ce anomaly appears to be a good oxidative redox indicator.

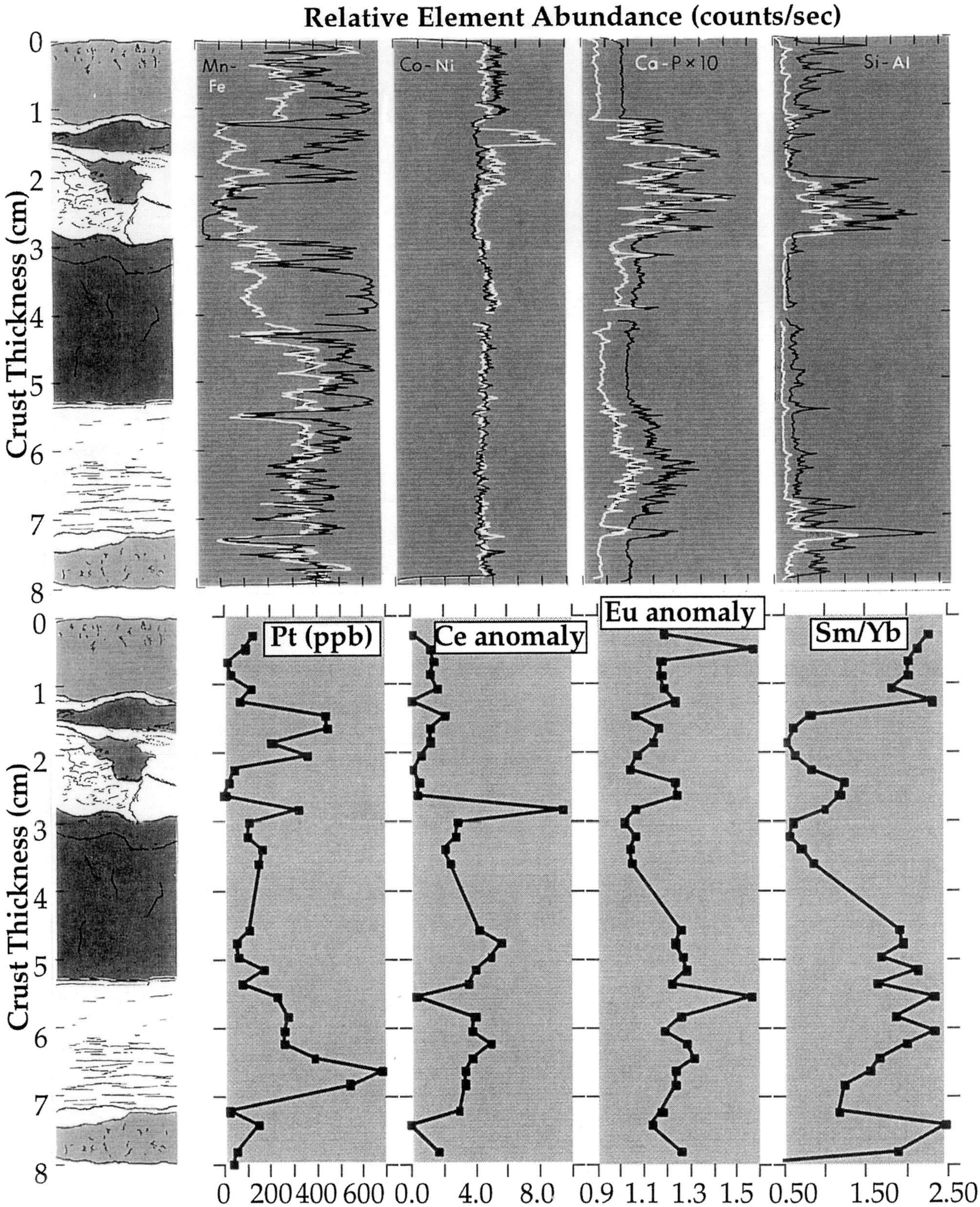

FIG. 11.—Semi-quantitative major element data of crust sample RD50 S2 collected by electron microprobe are compared to quantitative Pt and REE data collected on the same subsample by LA-ICP-MS.

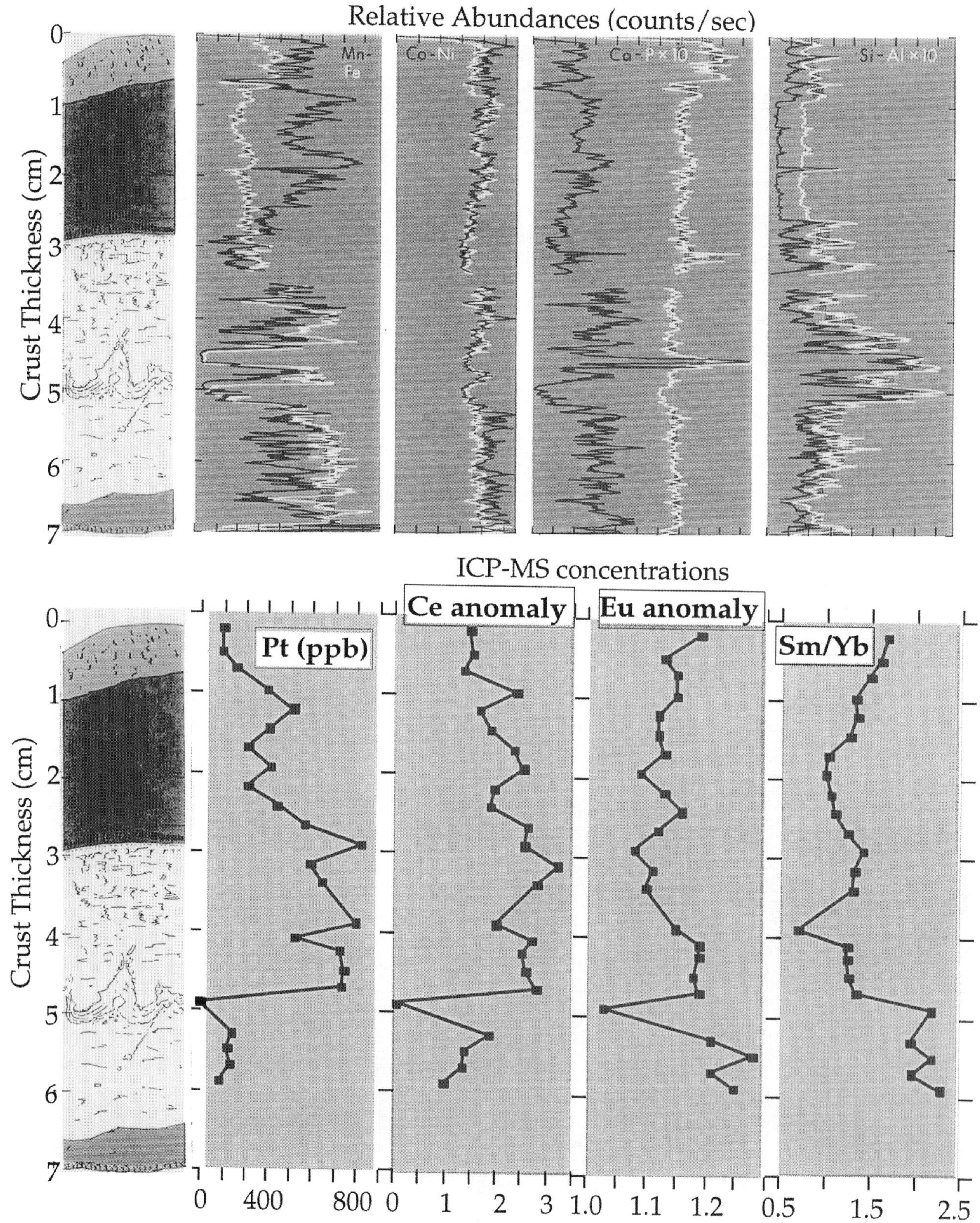

FIG. 12.—Semi-quantitative major element data of crust sample RD50 S5 collected by electron microprobe are compared to quantitative Pt and REE data collected on the same subsample by LA-ICP-MS.

TABLE 2.—TYPICAL DISTRIBUTIONS OF SHALE-NORMALIZED CE AND EU ANOMALIES AND SM/YB RATIOS IN THE MAJOR PHASES FOUND WITHIN THE FERROMANGANESE CRUST SAMPLES, ORGANIC MATTER AND SEAWATER

Phase	Ce anomaly	Eu anomaly	Sm/Yb
Fe-Mn Oxides	(+)	none	(+)
CFA	(-) to none, rarely (+)	none	generally (+), sometimes shale-like, rarely (-)
Alumino-silicates	none	none to slightly (-)	none
Organic material	none	none, (-) Gd	(+)
Seawater-surface	slightly (-) to none	(-)	slightly (-)
Seawater - 1000 m	(-)	slightly (-)	(-)

Moffett (1990) has experimentally shown that the oxidation of trivalent Ce to the more insoluble CeO_2 can result from biologically mediated oxidation in the upper water column. Coatings of both organic matter and FeMn oxides play a major role in the binding of trace elements to particle surfaces (e.g., Hunter, 1983).

The Pt concentration, Ce anomaly, Eu anomaly, and Sm/Yb ratio are all positively correlated in RD50 S1B, but with an offset of the peaks in the upper part of the crust above 40 mm. The Pt and Sm/Yb peaks are very broad from 25 mm to 45 mm, with apexes at ~28 mm, and are the most closely correlated. The Eu and Ce anomalies have a sharp peak consisting of a single data point, with apexes at 20 and 15 mm, respectively, and broader peaks at the lower interval from 70 mm to 85 mm (Fig. 9). Typical distributions of Pt, Ce, and Eu anomalies, and shale-normalized Sm/Yb ratios are given in Table 2 for the major phases in the ferromanganese crusts and in seawater. An interpretation of these correlations is discussed after presentation of further data in other ferromanganese crusts.

Laser Ablation ICP-MS Profiles of Pt and the REE in Other Ferromanganese Crusts

Detailed Pt profiles of seven ferromanganese crust samples were determined by LA-ICP-MS. Laser profiling of two other thick crust samples (RD50 S2 and RD50 S5) from the same dredge haul as sample RD50 S1B revealed that the Pt-rich layers are not unique to one sample (VonderHaar et al., 1990) (Figs. 11 and 12). However, the Pt concentrations obtained by LA-ICP-MS for the RD50 S2 and RD50 S5 crusts are much lower than those for either the bulk Pb-cuppellation, fire assay values, or the LA-ICP-MS values for RD50 S1-B. All of the crusts from Schumann Seamount are similar in that the Pt enrichments exhibit an inverse relationship with the P concentrations (Figs. 3, 11, and 12).

TABLE 3.—SUMMARY OF CRUST ANOMALY CORRELATIONS

Crust Sample	Pt concentration	Ce anomaly	Eu anomaly	Sm/Yb
KK84-RD50 S1B	+	+	+	+
KK84-RD50 S2	+	+	+	-
KK84- RD50 S5	+	+	+	-
KK84-PD8 S2	+	+	+	-
SO66 86 DSK	+	+	+	-
MW88 RD33 S6	+	-	-	-
MW88 RD35 S4	+	-	-	-

In addition to the three crust samples investigated from Schumann Seamount, four thick crusts from three other Pacific seamounts were also measured for Pt and REE by LA-ICP-MS (Figs. 1 and 2). We investigated the detailed REE profile in these crusts to ascertain if (1) any changes in the redox potential could indicate environmental changes such as subsidence of the seamount through the O_2 minimum zone, which may be reflected in variations of the REE distribution (e.g., Liu et al., 1988; Elderfield et al., 1990; DeCarlo, 1991; DeCarlo and McMurtry, 1992); (2) if these variations are correlated with Pt enrichment; or (3) if the changes in the REE distribution are more reflective of basin-wide conditions. Inspection of Figure 13A-D shows that there are indeed multiple layers of Pt enrichment in these crusts, which confirms that the Pt enrichment occurs in discrete layers within ferromanganese deposits throughout the Pacific Ocean. In the companion crust RD50 S5 from Schumann Seamount and in the sample from the Tuvalu area of the South Pacific, 86 DSK (Figs. 12 and 13B), there appears to be a good positive correlation of Pt with the Ce anomaly. The RD50 S5 sample (Fig. 12) also shows a negative correlation with the Sm/Yb ratio, and the RD50 S2 sample (Fig. 11) shows a good positive correlation between the Eu anomaly and the Sm/Yb ratio. The Eu data are not as well calibrated as the other REE because of potential interference by BaO^+ formation from oxygen in the ferromanganese oxides during the laser ablation process (Garbe-Schönberg and McMurtry, 1994). A summary of the correlations of Pt concentrations, Ce and Eu anomalies, and the Sm/Yb ratios in the crusts studied is presented in Table 3.

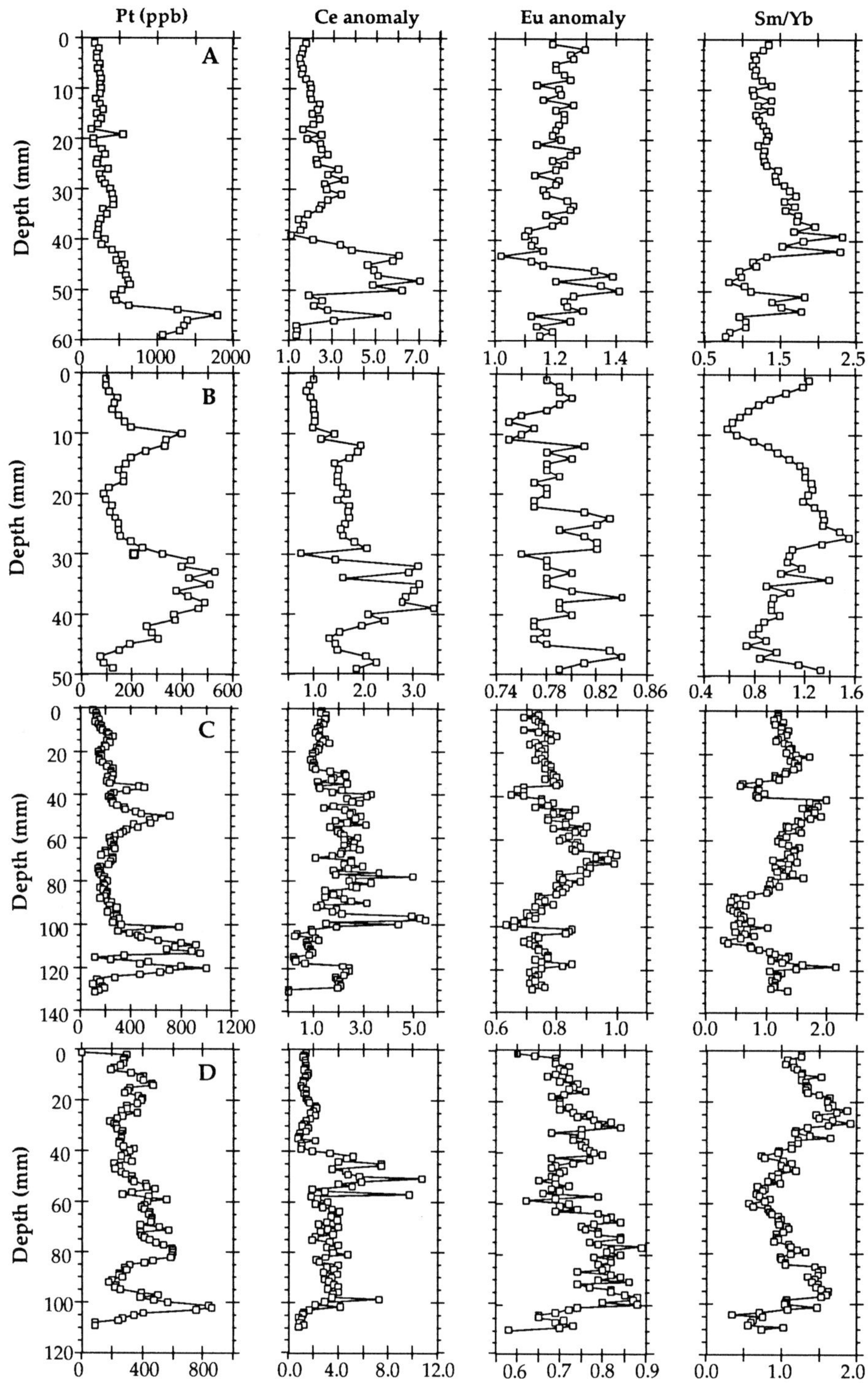

FIG. 13.—(A) Platinum, Ce and Eu anomalies, and Sm/Yb ratios for sample PD8 S2, from an unnamed seamount in the Hawaiian Archipelago, show very small variations in the upper ~40 mm. Platinum concentrations and the Ce anomaly have a definite positive correlation from 50-60 mm but the correlation above this is much less distinct. Greater variations in the Ce anomaly, Eu anomaly, and Sm/Yb ratio exist below 40 mm, with the Ce anomaly and Sm/Yb ratio showing a negative correlation in this area. (B) Platinum, Ce and Eu anomalies, and Sm/Yb ratios for sample 86 DSK from the Tuvalu area in the South Pacific. This crust shows a good positive correlation between the Pt concentration and the Ce anomaly data. The Sm/Yb ratio also shows a systematic variation with depth in the uppermost 20 mm. (C) Platinum, Ce and Eu anomalies, and Sm/Yb ratios for sample RD33 S6 from Lo-En Guyot in the Marshall Islands area. This sample does not show the positive correlation between the Pt concentration and Ce anomaly shown in the two Schumann Seamount crusts and the Tuvalu area crust. This crust seems to show several hiatuses marked by jumps in the REE anomalies, and is also complicated by carbonate fluorapatite filling many vugs and veins within the crust. (D) Platinum, Ce and Eu anomalies, and Sm/Yb ratios for another sample from Lo-En Guyot, RD35 S4, show similarities to RD33 S6 such as no correlation between Pt concentration and the Ce anomaly data. There is a more systematic variation in the Sm/Yb ratio in this crust but it is not as distinct as in the Tuvalu area crust, 86 DSK.

The positive correlation of Pt with the Ce anomaly in these crusts suggests Pt uptake under relatively oxidizing conditions, in agreement with the oxidative scavenging mechanism proposed for Pt enrichment in ferromanganese deposits by Hodge et al. (1985). Because of the close association of the Pt enrichments with CFA material, however, the correlation between the Pt and Ce anomaly is often damped due to the presence of CFA, which generally does not exhibit a Ce anomaly or has a negative Ce anomaly. However, the negative correlation of Pt with the Sm/Yb ratio, a measure of the slope of the heavy rare earth elements (HREE) in shale-normalized patterns, suggests an enrichment of the HREE in some of the Pt-rich zones. Because the light rare earth elements (LREE) have been shown to preferentially sorb onto synthetic Fe and Mn oxides relative to HREE (Koeppenkastrop and De Carlo, 1992) due to the greater stability of aqueous HREE carbonate complexes in seawater, the fractionation of the HREE to LREE may reflect the close association of Pt with CFA. Because of the inherent heterogeneity of the samples, it is difficult to determine the significance of the Pt, Ce, and Eu anomalies and the LREE/HREE enrichment as reflected by the Sm/Yb ratios from these data. The detailed Pt and REE profiles of the two Marshall Islands seamount crusts (Fig. 13C, D) do not show the correlation between Pt enrichment and Ce anomalies seen in the two Schumann Seamount crusts and the Tuvalu crust. Although the Ce anomaly has often been used as an indicator of paleo-redox conditions, caution must be used when applying it to bulk samples consisting of variable amounts of phases that preferentially absorb trivalent REE, such as FeOOH and CFA, because of the sensitivity of the calculation of the anomaly to increases in the trivalent REE. The lack of an association between the Pt concentrations and Ce anomaly in some of the crusts is probably due to the more pervasive impregnation of CFA in these crusts, or the positive correlation of Fe with Pt concentrations, which would dampen the Ce anomaly (Table 3).

Dating Results and Discussion

To help constrain whether the PGE enrichment is primarily due to cosmogenic input, O_2 minimum zone redox processes, or post-depositional redox processes, determination of the age of the particular layers is required. If the Pt-rich horizon occurs at the same time in all samples throughout the Pacific Ocean, then it is likely to be due to cosmogenic input, which may reflect an accumulation of such debris due to a hiatus in deposition or meteoritic input, particularly if it coincides with a known impact time like the Cretaceous-Tertiary boundary. If the Pt-rich horizon occurs within the same stratigraphic unit in samples from the same area of a particular seamount, and the age of the horizon correlates with a time period when the seamount would be subsiding through the O_2 minimum zone, then redox gradients associated with increased organic input within the O_2 minimum zone are likely to be responsible for the Pt enrichment. Finally, if the Pt-rich zones exhibit a "halo" type redox gradient consistently associated with CFA, post-depositional redox processes are likely to control the concentration of Pt and possibly Ir in the deposit.

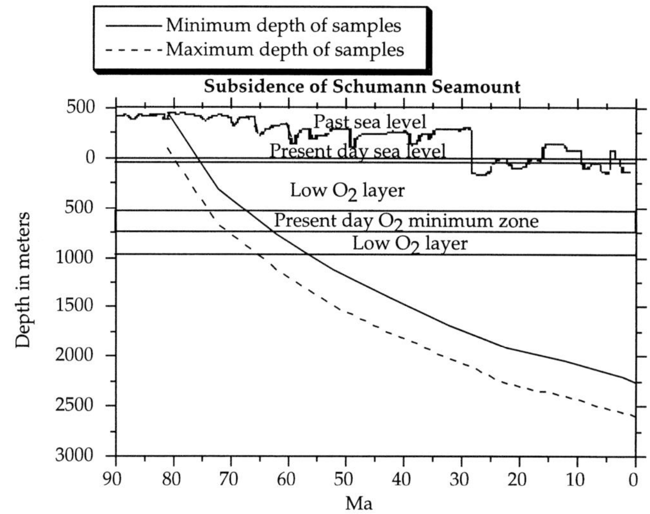

FIG. 14.—Paleo-depth curve showing the possible range of water depths of Schumann Seamount from the time of seamount formation at 82 Ma (Sager and Pringle, 1987). This curve was calculated using the general relationships of Parsons and Sclater (1977) for seamounts greater than 70 million years old. The formula was adjusted for the current depth of the seamount summit. Variations in past sealevel are also indicated (Kennett, 1982).

The RD50 S1B sample was well-dated by excess ^{230}Th profiling (Murphy, 1989), biostratigraphy (Cowen et al., 1993), ^{10}Be profiling (Fig. 15A), a modified "Co chronometer" (McMurtry et al., 1994), and comparison of the ferromanganese stratigraphic $^{187}Os/^{186}Os$ isotopic ratio with the Os seawater curve over time (B. K. Esser and D. L. VonderHaar, in prep., 1999). These methods indicate that this crust formed at an average rate of 1.1 mm/Myr in the uppermost 23 mm, growth was interrupted by a 30 Myr hiatus, and the lower 23–95 mm formed at a slightly accelerated average rate of ~2.5 mm/Myr. These dating results indicate that the ferromanganese oxides began to accrete within a few million years after the formation of the seamount around 82 ± 2 Ma (Sager and Pringle, 1987), during the late Cretaceous.

Age assignments based on these methods yield an age of ~72 Myr at the apex of the lower Pt peak and a gradual but sporadic Pt enrichment beginning at about 62 Myr for the upper Pt peak continuing up and possibly through the hiatus at about 52 Ma (Fig. 10). In this crust, therefore, it does not appear that the Pt enrichment can be attributed to cosmogenic or volcanogenic input at the Cretaceous-Tertiary boundary between 64–65 Ma (Kyte et al., 1980). The Os isotopic and concentration values also do not indicate a cosmogenic or volcanic input at the Cretaceous-Tertiary boundary (VonderHaar, 1997). These results, along with the absence of PGE ratios that could substantiate a cosmogenic source, indicate that the Pt enrichment in this crust is not from a cosmogenic or mantle source. Subsidence of the seamount does indicate, however, that the seamount would have passed below the O_2 minimum zone coincident with the time that the lower Pt enrichment was being deposited at ~70 Ma (Fig. 14). The Ce anomaly is most positive from 76 Ma to 71 Ma and reaches negative values (below 1.0) at

65 Ma, indicating a change in redox conditions between 71 Ma to 65 Ma, if the redox conditions are preserved with the deposition of the ferromanganese oxides. The Eu anomaly is also slightly positive from 70 Ma to 71 Ma, and reaches its lowest value at ~65 Ma. The shale-normalized Sm/Yb ratio peaks just above the Pt and Ce anomaly, where there is also a peak in the Mn and Fe concentrations. The LREE are associated with the Fe and Mn phase, so this peak appears to be following variations in the major element trends.

The RD50 S1B crust has higher concentrations of Pt than the other two crusts collected from the same dredge. This suggests that this crust contains additional PGE enrichment from either (1) a larger amount of older, more Pt-rich material contained within this crust and not in the others; (2) more Pt enrichment because a larger amount of the older crust was affected by the impregnation of the CFA material, accessing more Pt to increase its concentration within the redox gradient; or (3) the analyses obtained on RD 50 S2 and RD 50 S5 were collected on different machines at different times than those obtained on RD50 S1B. The last explanation is unlikely, however, because the PGE in all samples were calibrated against USGS A-1 and P-1 ferromanganese nodule standards (Garbe-Schönberg and McMurtry, 1994).

The outermost centimeter of the companion crust sample, RD50 S2, was also dated by ^{10}Be methods (Fig. 15B); results show a faster rate of accumulation than that of the RD50 S1B crust. The upper 3.5 mm grew at an apparent rate of 3.3 mm/Myr and the lower 6.5 mm grew at a slower apparent rate of 1.5 mm/Myr, closer to the rate observed for RD50 S1B. The CFA layer which surrounds the crust just below this outer rind of ferromanganese oxide material was dated by Sr isotopic methods. The apparent age of the CFA determined by the Sr isotopes is 28 Myr. Both the outer rind of ferromanganese oxide and the CFA are younger than the interior part of the crust, which must be older than 28 Myr. This crust also shows two Pt-rich areas in the laser ablation data, like the RD50 S1B crust, but the RD50 S2 crust differs in that it was detached from the substrate, and both the CFA and outer rind of ferromanganese oxide surround the whole sample. The aluminosilicate material between 20–30 mm in Figure 11 may actually be remnants of the original substrate material; it has a coarser texture than the aluminosilicate debris found in the RD50 S1B sample and looks much like hyaloclastite with a ferromanganese and CFA matrix, so the oldest part of the crust is the highly reflective, Mn-rich layer just below the upper CFA and aluminosilicate-rich material in Figure 11. The texture of the ferromanganese oxide in this section is the same as the ferromanganese oxide just above the substrate in RD50 S1B in that it consists of Mn-rich, wavy undulating layers. The ferromanganese material containing the Pt spike from 14–22 mm in sample RD50 S2 was analyzed by XRD and found to contain the diagenetic ferromanganese phase, todorokite. This area also shows increased concentrations of Ni. This part of the crust was probably affected by a low-temperature diagenetic process; the Pt enrichment in this area is therefore likely to be due to post-depositional processes (Fig. 9B). The Mn-rich area above this, from 30 mm to 42 mm, corresponds to the Mn-rich area in RD50 S1B found between 60–88

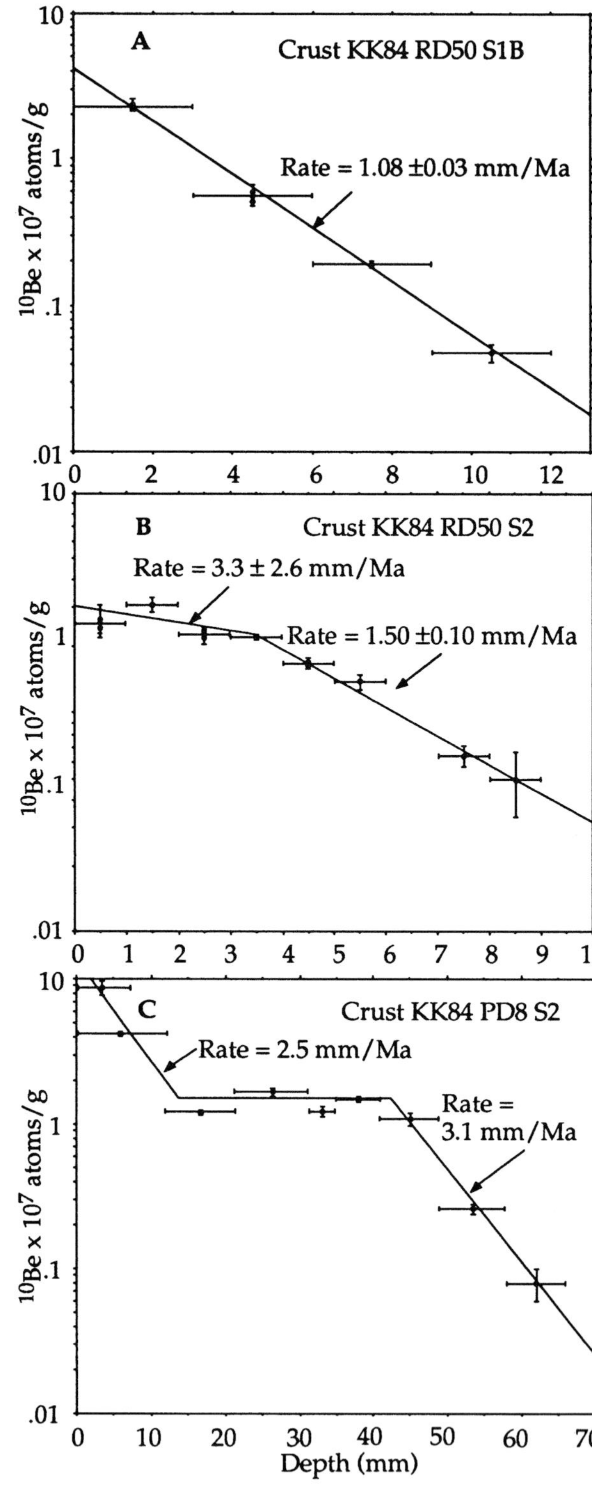

FIG. 15.—(A) ^{10}Be results for crust sample RD50 S1B yielding an average rate of accumulation of 1.08 mm/million years for the outermost 12 mm of this sample. (B) ^{10}Be results for the upper 10 mm of crust sample RD50 S2. The rate of 3.3 mm/million years determined for the outermost 3.5 mm is much faster than that determined for the companion sample, RD50 S1B. The average rate from 3.5-8.4 mm depth of 1.5 mm/million years is much closer to the RD50 S1B rate. (C) ^{10}Be results for the upper 65 mm of crust sample PD8S2. The average rate of 2.5 mm/million years determined for the outermost 20 mm is followed by a period of nearly instantaneous growth from 20-45 mm depth and a rate of 3.1 mm/million years for the layer from 45-65 mm depth.

mm. There, as in the RD50 S1B crust, the Fe phase becomes enriched relative to Mn. However, the Fe enrichment is not as pronounced in the RD50 S2 sample, and the sample has not been as affected by impregnation of the CFA as has the RD50 S1B sample. The CFA in this sample is for the most part confined to the discrete layer surrounding the sample. The Pt spike from 66 mm to 72 mm in Figure 11 lies within the CFA-rich crust but just below the discrete layer of CFA and within the upper aluminosilicate-rich layer, as in the upper Pt spike in the RD50 S1B crust (Fig. 4). The upper Pt spike is unlikely to be due to redox conditions as the seamount subsided through the low O_2 layer, because this part of the sample is more likely to have been deposited at the same time as the upper Pt peak in the RD50 S1B sample, between 52 Ma and 62 Ma. Again, this Pt peak is in close proximity to the CFA. Platinum enrichment as a result of post-depositional processes associated with the migration of a redox front through the sample after deposition of the CFA is the hypothesis most consistent with the data. The lesser degree to which the Pt is enriched in this sample, despite its proximity to the RD50 S1B sample, is therefore more likely because the CFA impregnated the RD50 S2 sample to a lesser degree.

Another ferromanganese crust, PD8 S2 from an unnamed seamount in the Hawaiian archipelago, has also been dated by ^{10}Be profiling and a modified "Co-chronometer" based on the ^{10}Be growth rates in this crust (Fig. 15C). This crust differs from the Schumann Seamount crusts in that it precipitated relatively rapidly, with ^{10}Be-determined rates of 2.5 mm/Myr for the first 20 mm, followed by nearly instantaneous growth from 20 mm to 45 mm, and a growth rate of 3.1 mm/Myr from 45 mm to 60 mm. The growth rates determined by the modified "Co-chronometer" vary from 3.5 mm to 11.5 mm/Myr and in general are faster than those measured by ^{10}Be, except in the 20–45 mm depth range (VonderHaar, 1997). Changes in the texture of this sample also indicate a change in growth rate in the lowermost 63.6–63.8 mm. However, the ^{10}Be data were collected on broad subsamples, and this change is not indicated by the lowermost data point (Fig. 9C). The Pt peak in this sample is between 52 mm and 60 mm, just above the textural change between CFA-poor and CFA-rich ferromanganese material.

The PD8 S2 sample was collected from a water depth of 1100–1475 m. It is therefore unlikely that the Pt enrichment between 52 and 60 mm, just above the substrate, is due to redox processes associated with the O_2 minimum zone, because the Pt enrichment is in the lowermost part of the sample and would have been above the O_2 minimum zone when the Pt-rich ferromanganese layers were deposited. However, age determinations for the included CFA and for the lowermost ferromanganese layers are necessary before any definitive conclusions can be drawn for this sample. Nevertheless, the relatively shallow depth of this seamount has important implications for the model proposed by Wen (1995) and DeCarlo et al. (this volume), as discussed below.

Paleoclimate Implications

Because productivity, ocean chemistry, atmospheric CO_2, continental weathering, and climate are all interrelated, dramatic changes in ocean circulation and climate such as those that occurred during the Cenozoic (c.f. Barron and Baldauf, 1989) have been shown to be reflected in climate-forcing biogeochemical cycles through the regulation of nutrient fluxes exhibited in long-term variation in P, N, Si, and possibly Fe and Mn accumulation (Mangini et al., 1990). Other climate-sensitive time-series such as changes in sea level, depth, and range of the CCD and lysocline, Sr isotopic records, $\delta^{18}O$ records, and differences between planktonic and benthic $\delta^{13}C$ compositions have all been used as proxies for paleoproductivity (e.g., Berger et al., 1989). Recent work by Wen (1995) and DeCarlo et al. (this volume) has found apparent cyclical variations of the shale-normalized Sm/Yb ratio in a ferromanganese crust from Kiribati that the authors attribute to past climatic conditions in terms of the CO_2 variations in seawater during the Tertiary. There are a number of ways the ocean can change its ability to hold CO_2, but the most likely route is through a change in its productivity (Broeker and Peng, 1986). Cantrell and Byrne (1987) have demonstrated that CO_3^{2-} complexation largely controls the speciation of REE in seawater, the proportion of each REE complexed with carbonate increases with increasing atomic number, and the degree of complexation influences the availability of the REE for scavenging onto surfaces. The change in speciation from the LREE to the HREE results in fractionation, whereby the LREEs are preferentially adsorbed to surfaces and the HREEs are preferentially retained in seawater (Koeppenkastrop and DeCarlo 1992). This fractionation results in the HREE-enriched, shale-normalized pattern of seawater (Piepgras and Jacobsen, 1992). Systematic fractionation between the HREE and LREE not only reflects the competition between solution and surface complexation but also the extent of the REE complexation in seawater by carbonate ions through geologic time (Cantrell and Byrne, 1987). Therefore, secular variations in the CO_2-HCO_3-CO_3^{2-} budget in seawater should also be reflected in the HREE to LREE ratios in seawater over the same timescale.

Wen (1995) and DeCarlo et al. (this volume) attribute high normalized Sm/Yb ratios to lower concentrations of carbonate ion in seawater. Relatively low concentrations of carbonate ion in seawater are a response to relatively high CO_2 levels in seawater, which create conditions that would be reflected in a shallower lysocline and CCD level. The Kiribati crust they investigated shows an inverse relationship between the normalized Sm/Yb ratio and the CCD from 50 Ma to the present. However, because the Kiribati crust in that study and the Schumann Seamount crust in this study formed above the CCD and lysocline, a better proxy of the carbon system may be found in the differences between planktonic and benthic $\delta^{13}C$ compositions. The Schumann Seamount crust also exhibits a cyclical variation in the shale-normalized Sm/Yb ratio, but because there is a 30 Myr hiatus from ~22 Ma to 52 Ma, when the Sm/Yb ratio is plotted with age, there is not a clear inverse relationship with the CCD, sea level, or differences between planktonic and benthic $\delta^{13}C$ compositions (e.g., Kennett, 1982). The 86DSK crust from the Tuvalu area of the South Pacific also shows an apparent cyclical variation of the normalized Sm/Yb ratio, but in the opposite sense from either the Kiribati or Schumann Seamount crust, being similar only to the upper part of the Kiribati crust, from the surface to 20-mm depth (Fig. 13B). Although the work by Wen (1995) does not show an apparent correlation between the major phases and the Sm/Yb ratio, it is still possible that the ratio is merely reflecting a shift from greater amounts of CFA below the Sm/Yb peak to greater amounts of Fe (and possibly Mn) oxides at the peak. Because the CFA is able to impregnate older ferromanganese oxides post-depositionally, and apparently has in the case of the RD50 S1B Schumann Seamount crust, much of the trace element variation seems to reflect this factor rather than the paleoceanographic conditions at the time of the ferromanganese oxide deposition.

CONCLUSIONS

Our conclusions can be summarized as follows:

- The RD50 S1B ferromanganese crust sample consists of three major groupings based upon elemental composition: the older generation (>50 mm) impregnated by CFA, the CFA-poor younger generation (<20 mm) and the Pt-rich layers (20–50 mm) just above the layer richest in CFA.
- Platinum and Fe profiles are similar in both transect 1 and transect 2 of the RD50 S1B crust sample. The Pt- and Fe-rich layers in this ferromanganese crust are associated with a mottled texture and Fe-Ni-rich spherules.
- The PGE show a marked enrichment of Pt over Ir and Os compared to unaltered chondritic sources. However, the spherules incorporated within the ferromanganese layers show signs of oxidation, and the effects of seawater alteration on the PGE are not known.
- Adsorption experiments indicate that Pt uptake, as measured by the Kd, is several orders of magnitude greater onto FeOOH surfaces than δ-MnO_2 surfaces. Iridium adsorption is initially much greater onto the δ-MnO_2 phase than onto the FeOOH phase, but after 3 weeks, these values approach one another. For FeOOH, the experimental Pt/Ir ratios approach the seawater value of ~100 in about 3 weeks, and Pt/Ir quickly attains a steady value of about 20 for δ-MnO_2.
- In the Pt-rich grouping from ~20–50 mm depth into the crust, there is an excess of Pt over Fe in the transect 1 data. When the Pt/Fe ratio is compared with the Fe profile, the Pt enrichment is two to four times greater in this part of the crust, suggesting that there must be another mechanism in addition to adsorption from seawater that is responsible for the excess Pt in the part of the crust that is richest in Pt.
- The consistent stratigraphic affiliation of the Pt enrichments with CFA within all the hydrogenous ferromanganese crusts studied, sometimes corresponding with positive Ce anomalies, negative Eu anomalies, and positive Sm/Yb ratios, suggests that the enrichment process is linked to either enhanced absorption of Pt in the O_2 minimum zone or is most likely a result of post-depositional enrichment associated with the influx of CFA into the ferromanganese crusts.
- From the samples we have studied, it does appear that irregardless of the age of the sample, the conditions necessary for Pt enrichment include the post-depositional impregnation of older ferromanganese material by CFA, and that greater Pt enrichment is found in those samples in which the CFA has impregnated a greater amount of ferromanganese material.
- Because of the close association of the CFA with Pt, it is possible that changes in the shale-normalized Sm/Yb ratio reflect post-depositional changes rather than basin-wide phenomenon associated with the oceanic carbon system. More carefully dated samples, such as the Schumann Seamount RD50 S1B crust, need to be analyzed for major elements, REE, and PGE before any meaningful paleoceanographic results can be drawn.
- From this work it appears that most PGE in crusts originate from seawater, and that post-depositional redox gradients during impregnation of ferromanganese crusts by CFA are a more important factor in the Pt enrichment process than cosmogenic sources or changing paleoceanographic conditions.

Areas worth further investigation are the relative post-depositional mobility of the Pt-group elements during diagenesis, including uptake from seawater onto ferric oxides; further age correlations with subsidence of the seamounts to see if these enrichments commonly occur when the seamount is passing through the O_2 minimum zone; and the apparent inverse relationship of the layers rich in Pt to those containing CFA. More detailed analyses should be done on physically or chemically separated phases so that the contributions of the various phases may be resolved.

ACKNOWLEDGMENTS

Much assistance was received with the analyses. We thank A. Mangini and A. Eisenhauer of the Universität Heidelberg, Germany, and P. Sharma of Prime Lab, Purdue University, for ^{10}Be analyses, and Maximilian Cremer for PD-8 subsampling. We are grateful to G. Ehrhardt of the University of Missouri Research Reactor Facility for his assistance in preparing the ^{191}Pt and ^{192}Ir radiotracers. The manuscript was greatly improved by formal reviews from G. P. Glasby and A. Koschinsky, and we have benefited from comments and encouragement given by F. T. Manheim. This publication is funded in part by a grant from the National Oceanic and Atmospheric Administration, project R/OM-11, which is sponsored by the University of Hawaii Sea Grant College Program, SOEST, under institutional grant No. NA36RG0507 from the NOAA Office of Sea Grant, Department of Commerce. The views expressed herein are those of the authors and do not necessarily reflect the views of NOAA or any of its sub-agencies. This is UNHI-SEAGRANT-JC-98-25, State of Hawaii, DBEDT, Ocean Resources Branch Contribution No. 138 and SOEST contribution 4840.

REFERENCES

Barron, J. A., and Baldauf, J. G., 1989, Tertiary cooling steps and paleoproductivity, *in* Berger, W. H., Smetacek, V. S., and Wefer, G., eds., Productivity of the Ocean: Present and Past: Chichester, Wiley, p. 341-354

Berger, W. H., Smetacek, V. S., and Wefer, G.,1989, Ocean productivity and paleoproductivity-An overview, *in* Berger, W. H., Smetacek, V. S., and Wefer, G., eds., Productivity of the Ocean: Present and Past: Chichester, Wiley. p. 1-34.

Burnett, W. C., Cullen, D. J., and McMurtry, G. M., 1987, Open-ocean phosphorites—In a class by themselves?, *in* Teleki, P. G., ed., Marine Minerals: Reidel, p. 119-134.

Cantrell, K. J., and Bryne, R. H., 1987, Rare earth element complexation by carbonate and oxylate ions: Geochimica Cosmochimica Acta, v. 51, p. 597-605.

Castaing, R., and Fredriksson, K., 1958, Analyses of cosmic spherules with an X-ray microanalyser: Geochimica Cosmochimica Acta, v. 14, p. 114-117.

Colodner, D. C., 1991, The marine geochemistry of rhenium, iridium, and platinum: Unpublished Ph.D. Dissertation, Massachusetts Institute of Technology and the Woods Hole Oceanographic Institution, Woods Hole, Mass., 269 p.

Colodner, D. C., Boyle, E. A., Edmond, J. M., and Thomson, J., 1992, Post-depositional mobility of platinum, iridium, and rhenium in marine sediments: Nature, v. 358, p. 402-404.

Cowen, J. P., DeCarlo, E. H., and McGee, D. L., 1993, Calcareous nannofossil biostratigraphic dating of a ferromanganese crust from Schumann seamount: Marine Geology, v. 115, p. 289-306.

DeCarlo, E. H., 1991, Paleoceanographic implications of rare earth element variability in a marine Fe-Mn crust from the Hawaiian Archipelago: Marine Geology, v. 98, p. 449-467.

DeCarlo, E. H., and McMurtry, G. M., 1992, Rare earth element geochemistry of Fe-Mn deposits from the Hawaiian Archipelago: Chemical Geology, v. 95, p. 235-250.

Elderfield, H., 1988, The oceanic chemistry of the rare-earth elements: Philosphical Transactions of the Royal Society of London, v. A325, p. 104-126.

Elderfield, H., Upstill-Goddard, R., and Shokovitz, E. R., 1990, The rare earth elements in rivers, estuaries, and coastal seas and their significance to the composition of ocean waters: Geochimica Cosmochimica Acta, v. 54, p. 971-991.

Esser, B. K., 1991, Osmium isotope geochemistry of terrigenous and marine sediments: Unpublished Ph.D. Dissertation, Yale University, New Haven, Conn., 308 p.

Esser, B. K., and Turekian, K. K., 1988, Accretion rate of extraterrestrial particles determined from osmium isotope systematics of Pacific pelagic clay and manganese nodules: Geochimica Cosmochimica Acta, v. 52, p. 1383-1388.

Finkelman, R. B., 1970, Magnetic particles extracted from manganese nodules: Suggested origin from stony and iron meteorites: Science, v. 167, p. 982-984.

Frakes, L. A., and Bolton, B. R., 1984, Origin of manganese giants: sea-level change and anoxic-oxic history: Geology, v. 12, p. 83-86.

Froelich, P. N., Klinkhammer, G. P., Bender, M. L., Luedtke, N. A., Heath, G. R., Cullen, D., Dauphin, P., Hammond, D., Hartman, B., and Maynard, V., 1979, Early oxidation of organic matter in pelagic sediments of the eastern equatorial Atlantic: Suboxic diagenesis: Geochimica et Cosmochimica Acta, v.43, p. 1075-1090.

Garbe-Schonberg, C. D., and McMurtry, G. M., 1994, In-situ micro-analysis of platinum and rare earths in ferromanganese crusts by laser ablation-ICP-MS (LAICPMS): Fresenius Journal of Analytical Chemistry, v. 350, p. 264-271.

Glenn, C. R.., Föllmi, K. B., Riggs, S. R., Baturin, G. N., Grimm, K. A., Trappe, J., Abed, A. M., Galli-Olivier, C., Garrison, R. E., Ilyin, A., Jehl, C., Rohrlich, V., Sadaqah, R. M., Schidlowski, M., Sheldon, R. E., and Siegmund, H., 1994, Phosphorus, sedimentology, and paleoceanography: Eclogae Geologicae Helveticae, v. 87, p. 747-788.

Goldberg, E. D., Koide, M., Yang, J. S., and Bertine, K. K., 1988, Comparative marine chemistries of platinum group metals and their periodic table neighbors, *in* Kramer, J. R., and Allen, H. E., eds., Metal Speciation, Theory Analysis and Application: Chelsea, Mich., Lewis Publishers, Inc., p. 201-217.

Goldberg, E. D., Hodge, V., Kay, P., Stallard, M., and Koide, M., 1986, Some comparative marine chemistries of platinum and iridium: Applied Geochemistry, v. 1, p. 227-232.

Halbach, P., and Puteanus, D., 1984, Influence of the carbonate dissolution rate on the growth and composition of Co-rich ferromanganese crusts from Central Pacific seamount areas: Earth and Planetary Science Letters, v. 68, p. 73-82.

Halbach, P., Kriete, C., Prause, B., and Puteanus, D., 1989, Mechanisms to explain the platinum concentration in ferromanganese seamount crusts: Chemical Geology, v. 76, p. 95-106.

Halbach, P., Segl, M., Puteanus, D., and Mangini, A., 1983, Co-fluxes and growth rates in ferromanganese deposits from Central Pacific seamount areas: Nature, v. 304, p. 716-719.

Hein, J. R., Borson, W. A., Schulz, M. S., Noble, M., and Clague, D. A., 1992, Variations in the fine-scale composition of a Central Pacific ferromanganese crust: Paleoceanographic implications: Paleoceanography, v.7, p. 63-77.

Henderson, P., 1984, Rare earth element geochemistry, *in:* Henderson, P., ed., Elsevier Developments in Geochemistry 2: Amsterdam, Elsevier, 510 p.

Heye, D., and Marchig, V., 1977, Relationship between the growth rate of manganese nodules from the Central Pacific and their chemical constitution: Marine Geology, v. 23, p. 19-25.

Hodge, V. F., Stallard, M., Koide, M., and Goldberg, E. D., 1985, Platinum and the platinum anomaly in the marine environment: Earth and Planetary Science Letters, v. 72, p. 158-162.

Hodge, V. F., Stallard, M., Koide, M., and Goldberg, E. D., 1986, Determination of platinum and iridium in marine waters, sediments, and organisms: Analytical Chemistry, v. 58, p. 616-620.

Holser, W. T., et al., 1989, A unique geochemical record at the Permian/Triassic boundary: Nature, v. 337, p. 39-44.

Hunter, K. A., 1983, The adsorptive properties of sinking particles in the deep ocean: Deep-Sea Research, v. 30, p. 669-675.

Jacinto, G. S., and Van Den Berg, C. M. G., 1989, Different behavior of platinum in the Indian and Pacific Oceans: Nature, v. 337, p. 332-334.

Jedwab, J., 1975, Cosmic flakes in manganese nodules: Meteoritics, v. 10, p. 191-207.

Kennett, J. P., 1982, Marine Geology: Englewood Cliffs, N.J., Prentice-Hall, 813 p.

Koeppenkastrop, D., and DeCarlo, E. H., 1992 Sorption of rare-earth elements from seawater onto synthetic mineral particles: an experimental approach: Chemical Geology, v. 95, p. 251-263.

Koide, M., Goldberg, E. D., Niemeyer, S., Gerlach, D., Hodge, V., Bertine, K. K., and Padova, A., 1991, Osmium in marine sediments: Geochimica et Cosmochimica Acta, v. 55, p. 1641-1648.

Kosakevitch, A., 1987, Presence de spherules cosmiques ferro-nickeliferes a platinoides dans un encoutement polymetallique sous-marin de Tuamotu (Polynesie francaise): Paris, C.R. Academy of Science, v. 305, p. 105-108.

Koschinsky, A., and Halbach, P., 1995, Sequential leaching of marine ferromanganese precipitates: Genetic implications: Geochimica et Cosmochimica Acta, v. 59, p. 5113-5132.

Kyte, F. T., Zhou, Z., and Wasson, J. T., 1980, Siderophile enriched sediments from the Cretaceous-Tertiary boundary: Nature, v. 288, p. 651-656.

Li, Y.-H., Burkhardt, L., Buchholtz, M., O'Hara, P., and Santschi, P. H., 1984, Partition of radiotracers between suspended particles and seawater: Geochimica et Cosmochimica Acta, v. 48, p. 2011-2019.

Liu, Y. G., Miah, M. R. U., and Schmitt, R. A., 1988, Cerium; a chemical tracer for paleo-oceanic redox conditions, *in* Zartman, R. E., Papanastassiou, D. A., and Depaolo, D. J., eds., Conference on Isotope Tracers in Geochemistry and Geophysics: Oxford, Pergamon, p. 1361-1371.

Mangini, A., Eisenhauer, A., and Walter, P., 1990, Response of manganese in the ocean to the climatic cycles in the Quaternary: Paleoceanography, v. 5, p. 811-821.

Manheim, F. T., 1986, Marine cobalt resources: Science, v. 232, p. 553-564.

McMurtry, G. M., VonderHaar, D. L., and Mahoney, J. J., 1987, Ferromanganese crust stratigraphy, the Cretaceous-Tertiary and Eocene-Oligocene boundaries, and Pacific paleoceanography from 82 Myr BP (abs.): Chemical Geology, v. 70, p. 120.

McMurtry, G. M., VonderHaar, D. L., Eisenhauer, A., Mahoney, J. J., and

YEH, H.-W., 1994, Cenozoic accumulation history of a Pacific ferromanganese crust: Earth and Planetary Science Letters, v. 125, p. 105-118.

MCMURTRY, G. M., VONDERHAAR, D. L., MAHONEY, J. J., GREGSON, D., AND TYNE, C., 1989, Platinum-group metals in ferromanganese crusts from Pacific Seamounts (abs.): Bulletin of the Geological Society of Finland, v. 61, p. 47.

MURPHY III, W. E., 1989, Geochemistry and U-series geochronology of hydrothermal and hydrogenous metal-rich deposits from the Pacific Ocean: Unpublished M.S. Thesis, University of Hawaii, Honolulu, 141 p.

MOFFETT, J. W., 1990, Microbially mediated cerium oxidation in seawater: Nature, v. 345, p. 421-423.

ORTH, C. J., QUINTANA, L. R., GILMORE, J. S., BARRICK, J. E., HAYWA, J. N., AND SPESSHARDT, S. A., 1988, Pt-group metal anomalies in the Lower Mississippian of southern Oklahoma: Geology, v. 16, p. 627-630.

PALMER, M. R., FALKNER, K. K., TUREKIAN, K. K., AND CALVERT, S. E., 1988, Sources of osmium isotopes in manganese nodules: Geochimica et Cosmochimica Acta, v. 52, p. 1197-1202.

PARSONS, B., AND SCLATER, J. G., 1977, An analysis of the varition of ocean floor bathymmetry and heat flow with age: Journal of Geophysical Research, v. 82, p. 803-827.

PIEPGRAS, D. J., AND JACOBSEN, S. B., 1992, The behavior of rare earth elements in seawater: Precise determination of variations in the North Pacific water column: Geochimica et Cosmochimica Acta, v. 56, p. 1851-1862.

PLAYFORD, P. E., MCLAREN, D. J., ORTH, C. J., GILMORE, J. S., AND GOODFELLOW, W. D., 1984, Iridium anomaly in the Upper Devonian of the Canning Basin, Western Australia: Science, v. 226, p. 437-439.

PUTEANUS, D., 1986, Geologie und geochemie von manganerzkrusten auf submarinen gebirgen des Zentralpazifiks: Unpublished Ph.D. Dissertation, Technische Universitat Clausthal, Clausthal-Zellerfield, 242 p.

RAMPINO, M. R., 1981, A noncatastrophic explanation for the iridium anomaly at the Cretaceous/Tertiary boundary, *in* Geological Implications of Impacts of Large Asteroids and Comets: Boulder, Geological Society of America Special Paper, v. 190, p. 455-461.

RAVIZZA, G., AND MCMURTRY, G. M., 1993, Osmium isotopic variations in metalliferous sediments from the East Pacific Rise and the Bauer Basin: Geochimica et Cosmochimica Acta, v. 57, p. 4301-4310.

ROBIN, E., FROGET, L., JEHANNO, C., AND ROCCHIA, R., 1993, Evidence for a K/T impact event in the Pacific Ocean: Nature, v. 363, p. 615-617.

SAGER, W. W., AND PRINGLE, M. S., 1987, Paleomagnetic constraints on the origin and evolution of the Musicians and South Hawaiian seamounts, Central Pacific Ocean, *in* Keating, B. H., Fryer, P., Batiza, R., and Boehlert, G. W., eds., Seamounts, Islands, and Atolls, Geophysical Monograph 43: Washington, D.C., American Geophysical Union, p. 133-162.

SEGL, M., MANGINI, A., BONANI, G., HOFMAN, H. J., NESSL, M., SUTER, M., WOLFLI, W., FRIEDRICH, G., PLUGER, W. L., WIECHOWSKI, A., AND BEER, J., 1984, ^{10}Be-dating of a manganese crust from the Central North Pacific and implications for ocean palaeocirculation: Nature, v. 309, p. 540-543.

SCHMITZ, B., 1985, Metal precipitation in the Cretaceous-Tertiary boundary clay at Stevns Klint, Denmark: Geochimica et Cosmochimica Acta, v. 49, p. 2361-2370.

VONDERHAAR, D. L., 1997, Geochemical and isotopic interpretations of ferromanganese crusts from the Pacific Ocean: Unpublished Ph.D. Dissertation, University of Hawaii, Honolulu, 349 p.

VONDERHAAR, D. L., AND MCMURTRY, G. M., 1988, Chemical and isotopic stratigraphy of a ferromanganese crust from the Hawaiian Archipelago (abs.): Eos, Transactions of the American Geophysical Union, v. 69, p.1501.

VONDERHAAR, D. L., PUTEANUS, D., AND MCMURTRY, G. M., 1990, Detailed Pt profiles and their stratigraphic interpretation in ferromanganese crusts (abs.), Fifth Circum-Pacific Energy and Mineral Resources Conference: Tulsa, American Association of Petroleum Geologists Bulletin, v. 74, p. 1008.

WEN, X. Y., 1995, Geochemical characteristics of Pacific seamount ferromanganese crusts: Evidence for their genesis: Unpublished M.S. Thesis, University of Hawaii, Honolulu, 131 p .

SPATIAL AND TEMPORAL DISTRIBUTION OF AUTHIGENIC MINERALS IN CONTINENTAL SHELF SEDIMENTS: IMPLICATIONS FOR SEQUENCE STRATIGRAPHIC ANALYSIS

KEVIN G. TAYLOR*

Department of Environmental and Geographical Sciences, Manchester Metropolitan University, Chester Street, Manchester, M1 5GD, United Kingdom

AND

JOE H. S. MACQUAKER

Department of Earth Sciences, University of Manchester, Oxford Road, Manchester, M13 9PL, United Kingdom

ABSTRACT: Early diagenetic minerals are common within mudstone-dominated shallow marine sediments. However, our understanding of how these early diagenetic assemblages vary in a spatial and temporal sense across sedimentary basins is poorly developed. Such an appreciation is important, as early diagenetic mineral assemblages show a clear relationship with stratigraphic architecture and stratal surfaces.

In this study we present integrated stratigraphic and geochemical data for four differently aged mudstone-dominated successions (the Lower Jurassic Scunthorpe and Brant Mudstone Formations of eastern England, the Lower Jurassic Cleveland Ironstone Formation of eastern England, the Triassic Westbury Formation of South West England, and the Upper Cretaceous Mesaverde Group of Book Cliffs, Utah). Both spatial and temporal variability in early diagenetic mineral assemblages can be recognized in these successions, taking the form of ooidal ironstones, phosphates (apatite-rich units and bone-beds), concretionary carbonates, and pyritic mudstones. These variations are interpreted to result from spatial and temporal variations in physical and geochemical conditions in the sediments close to the sediment-water interface. Specifically, the physical and geochemical conditions required for the formation of these cemented units occur at key stratal surfaces (marine flooding surfaces, maximum flooding surfaces, and down-dip equivalents of sequence boundaries).

The physical conditions that occur at sequence boundaries (low accommodation, high energy, and frequent sediment reworking) can lead to the predominance of suboxic diagenesis and the formation of berthierine- and siderite-rich ooidal ironstones. Conditions of low net sediment accumulation and oxic/suboxic diagenesis upon major transgressive surfaces and maximum flooding surfaces leads to the formation of phosphate-rich units. Breaks in sediment accumulation at marine flooding surfaces lead to enhanced early diagenesis and the formation of laterally extensive carbonate cements and concretionary horizons. Sulfide-dominated early diagenesis (that dominated by sulfate reduction) is predominant during "normal" periods of shallow marine sedimentation and appears to mark the bulk of sediments within systems tracts. This study emphasizes the importance of integrating stratigraphic and geochemical data in developing quantitative and predictive models for early diagenesis.

INTRODUCTION

Authigenic minerals (for example, calcite, siderite, dolomite, berthierine, chamosite, glauconite, and phosphates) are commonly present in sediments deposited in shallow marine environments. These minerals occur in both ancient and modern sediments and are particularly common in fine (mud and mudstone) successions (e.g., Coleman, 1985), although they can also be important in coarse (sands and sandstones) successions (e.g., Bjorkum and Walderhaug, 1990a,b; Prosser et al., 1993). There is widespread consensus that these minerals precipitate during early diagenesis via a range of bacterially mediated reactions (e.g., Claypool and Kaplan,; Irwin et al., 1977; Coleman, 1985; Curtis, 1987a). Numerous studies of these reactions in recent (e.g., Froelich et al., 1979; Aller et al., 1986; Ruttenberg and Berner, 1993; Canfield et al., 1993) and ancient sediments (e.g., Gautier, 1982; Curtis et al., 1986; Raiswell, 1987, 1988; Coleman and Raiswell, 1995; Mozley, 1996) have shed light upon the mechanisms likely to be responsible for their precipitation and have led to a better understanding of how the compositions of the pore fluids evolve during these processes. In spite of all these studies, however, our understanding of how these early diagenetic assemblages vary in a spatial and temporal sense across sedimentary basins is much less well-developed. Such an understanding is of key importance in basin analysis and stratigraphic interpretation as many of the key factors that control authigenic cement precipitation are themselves linked to changes in relative sea level and accommodation availability (*sensu.* Jervey, 1988). It is only recently, with the development of sequence stratigraphic methodology (e.g., Van Wagoner et al., 1990), that the tools to interpret diagenetic variability in a temporal and spatial way have become available.

Given this preamble the intention of this contribution is to describe how the sedimentological, stratigraphic and geochemical characteristics of units enriched in early diagenetic cements vary in four different stratigraphic settings (the Lower Jurassic Scunthorpe and Brant Mudstone Formations of eastern England, the Lower Jurassic Cleveland Ironstone Formation of eastern England, the Triassic Westbury Formation of South West England, and the Upper Cretaceous Mesaverde Group of Book Cliffs, Utah). From these data the key relationships between early diagenesis and stratigraphic architecture (especially stratal surfaces) on a basin-scale within these successions will be discussed. This study will also emphasize the importance of integrating stratigraphic and geochemical data in developing quantitative and predictive models for early diagenesis.

EARLY DIAGENETIC REACTIONS CONTROLLING AUTHIGENIC MINERAL FORMATION

Studies on recent sediments have shown that in addition to aerobic respiration reactions, a range of anaerobic bacterially mediated organic matter oxidation reactions (e.g., nitrate reduction, sulfate reduction, methane oxidation, and methanogenesis) also occur during early diagenesis (e.g., Coleman, 1985; Aller et al., 1986; Canfield et al., 1993; Table 1). Early researchers com-

*email: k.g.taylor@mmu.ac.uk

Marine Authigenesis: From Global to Microbial, SEPM Special Publication No. 66

TABLE 1.—EARLY DIAGENETIC ORGANIC MATTER OXIDATION REACTIONS TAKING PLACE WITHIN ORGANIC-RICH SEDIMENTS.

	Diagenetic Reaction	Reactants		Products
Aerobic Reaction	Aerobic Oxidation	$CH_2O + O_2$	$\longrightarrow$	$HCO_3^- + H^+$
Anaerobic Reactions	Manganese Reduction	$CH_2O + 2MnO_2 + H_2O$	$\longrightarrow$	$2Mn^{2+} + HCO_3^- + 3OH^-$
	Iron Reduction	$CH_2O + 2Fe_2O_3 + H_2O$	$\longrightarrow$	$4Fe^{2+} + HCO_3^- + 7OH^-$
	Sulphate Reduction	$CH_2O + SO_4^{2-}$	$\longrightarrow$	$HS^- + 2HCO_3^- + H^+$
	Methanogenesis	$CH_2O + H_2O$	$\longrightarrow$	$CH_4 + HCO_3^- + H^+$

monly assumed that aerobic respiration and sulfate reduction accounted for the bulk of organic matter oxidation (e.g., Bender and Heggie, 1984; Henrichs and Reeburgh, 1987; Canfield, 1989). Recently, however, workers (e.g., Tugel et al., 1986; Lovley and Phillips, 1988; Myers and Nealson, 1988; Lovley and Lonergan, 1990; Sørensen and Jørgensen, 1987; Hines et al., 1991; Canfield et al., 1993) have begun to recognize that bacterial reduction of metal oxides (Fe and Mn), linked to organic matter oxidation, are also very important. Indeed, Aller (1990) and Canfield et al. (1993) have shown that in some sediments dissimilatory metal oxide reduction is the only significant pathway for anaerobic organic matter oxidation. Moreover, the dominance of metal oxide reduction in some settings within particular ancient sediments has now also been recognized (e.g., Okita et al., 1988; Minoura et al., 1991; Taylor and Curtis, 1995).

These reactions involving organic matter oxidation release products into sediment pore-water (for example, bicarbonate, sulfide, Fe^{2+}, and Mn^{2+}; Table 1) which, where supersaturation is reached, will give rise to mineral precipitation during early diagenesis. Different organic matter oxidation reactions commonly give rise to minerals and mineral assemblages with characteristic elemental and isotopic compositions (Coleman, 1985). Indeed, such characteristic compositions have been recognized in ancient sediments and attributed to specific organic matter oxidation pathways (e.g., Irwin et al., 1977; Curtis et al., 1986; Raiswell, 1988; Taylor and Curtis, 1995). Consequently, changes in the relative importance of these reactions within sediment pore waters during early diagenesis will cause characteristic diagenetic mineral assemblages to precipitate.

SPATIAL AND TEMPORAL VARIABILITY IN AUTHIGENIC MINERAL ASSEMBLAGES

Due to the generally poor exposure and limited high-resolution stratigraphic control available for most ancient mudstone-dominated shelf sediment successions, recognition of temporal and spatial variability in early diagenesis, and the scale of any such variability, is generally lacking. In spite of this, such variability in early diagenesis, and resulting authigenic mineral assemblages, should be expected. For instance, studies on modern sediments (e.g., Berner, 1980; Canfield, 1989; Canfield et al., 1993) have shown that, within modern shallow marine sediments, the relative importance of different early diagenetic reactions can vary temporally and spatially over small distances. These studies have shown that changes in both rate of sediment accumulation and substrate composition were controlling such variability. Similar primary controls have also been postulated in ancient sedimentary successions (e.g., Curtis et al., 1986; Raiswell, 1997). A major control on sediment accumulation rate is change in relative sea-level and hence accommodation availability within sedimentary basins (Jervey, 1988). Sequence stratigraphers have commonly modeled how changes in accommodation availability can be used to constrain basin stratigraphy. Most of these studies, however, have concentrated on up-dip sand-dominated successions and there have been relatively few that have concentrated on mud-dominated shallow marine shelf sediments. Recently, however, a number of workers have attempted to incorporate mudstone-dominated successions into these models (e.g. Loutit et al., 1988; Wignall, 1991, 1994; Leithold, 1994; Macquaker, 1994a; Macquaker et al., 1998). From these it is clear that cemented units commonly occur in association with the major stratal surfaces and that the characteristics of these cemented units vary in different stratigraphic settings.

In order to understand the fundamental sedimentological and geochemical controls on early diagenetic mineral assemblages, sediment successions must be studied for which good stratigraphic control is available. With these data it will be possible to place the diagenetic controls within a wider spatial and temporal context. To this end we will discuss four mudstone-dominated successions whose stratigraphic context is well

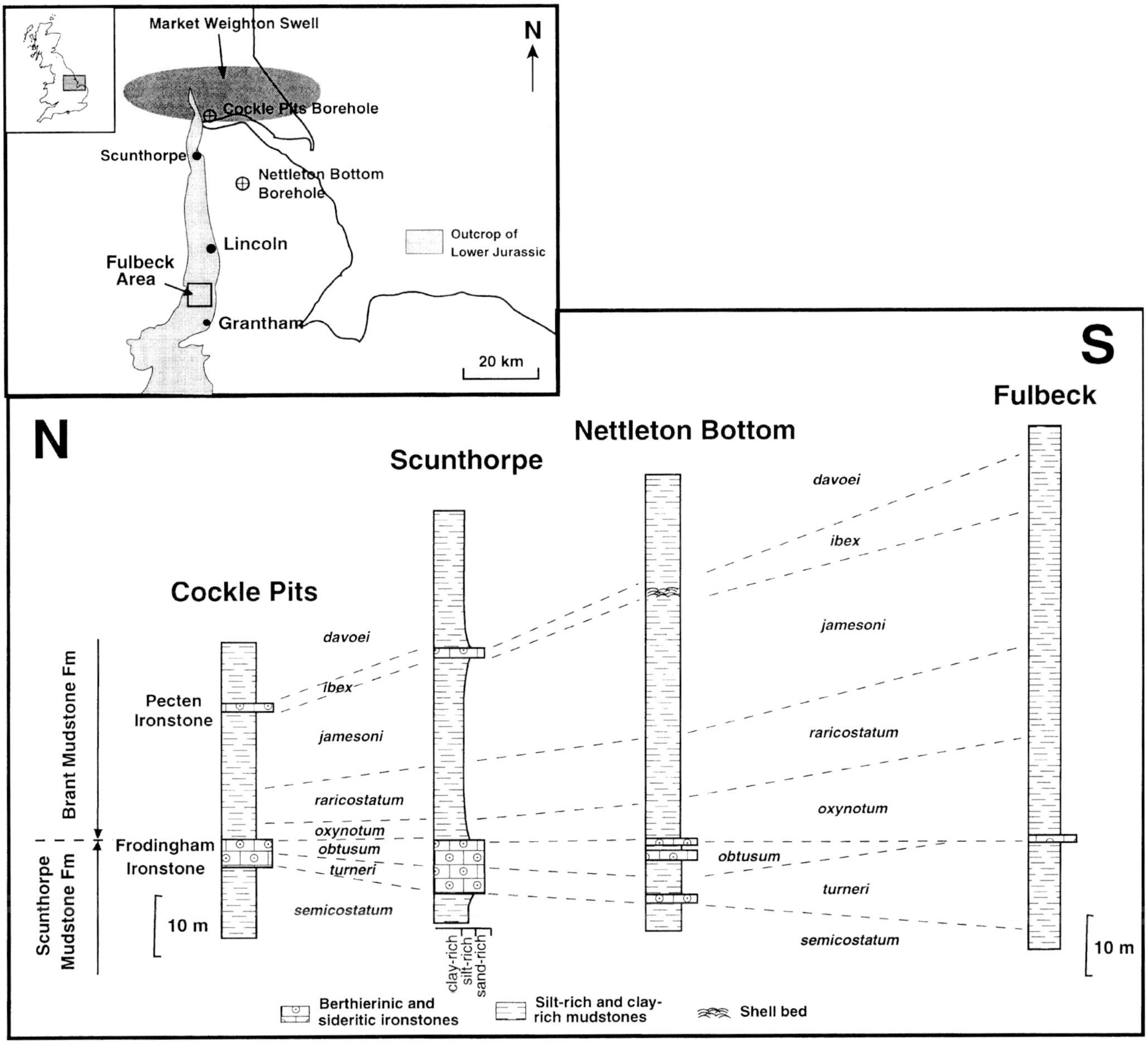

FIG. 1.—Location map and stratigraphic correlations of the Lower Liassic ironstones and mudstones of the Scunthorpe and Brant Mudstone Formations on the East Midlands Platform. Note the stratigraphic thickening in a southward direction. Stratigraphic correlation based on (Brandon et al., 1990). Information on grain-size variation within the mudstones is only included for the Scunthorpe area due to the nature of the core in the other areas.

constrained: i) a mudstone-dominated succession with subordinate ironstones from the Lower Liassic (Lower Jurassic) Scunthorpe and Brant Mudstone Formations of Eastern England; ii) a mudstone-dominated succession with subordinate ironstones and apatite-rich horizons from the Cleveland Ironstone (Lower Jurassic) of Eastern England; iii) a mudstone-dominated succession with subordinate ironstones from the Upper Cretaceous of Colorado and Utah; and iv) a mudstone-dominated succession with subordinate "bone-beds," shell-beds, and concretionary carbonates from the Upper Triassic, Westbury Formation of South West England.

METHODS

Precise details of the methods employed to characterize both the stacking patterns and the composition of the diagenetic mineral assemblages present within these fine-grained successions can be found in Taylor and Curtis (1995) and Macquaker and Taylor (1996). In general the mineralogy of the bulk rock and the clay fraction of samples was determined by X-ray diffraction utilizing a Philips PW1050 X-ray diffractometer operating at 40Kv 20 mA. The textures present, mineralogy, and grain size of the detrital fraction of each sample were determined by first preparing polished thin sections 20–25 μm thick, and then analyzing each utilizing combined optical and electron optical (backscattered electron imagery) petrographic methods. Electron optical analyses were performed in a JEOL 6400 scanning electron microscope equipped with a Link 4-quadrant solid-state backscattered electron detector and Link Energy dispersive spectrometer operating at 20 KV, 9 mm, and 2.0 mA.

Finally, the trace element (Sr, Mn) contents of carbonate minerals were determined on the thin sections using wave dispersive analysis on a Cameca electron microprobe.

RESULTS

Scunthorpe and Brant Mudstone Formations, Eastern England

The Lower Liassic sediments in the northern part of the East Midlands Shelf, Lincolnshire, U.K., (Fig. 1) comprise a succession of marine mudstones, within which two ooidal ironstones, the Frodingham Ironstone and the Pecten Ironstone, are locally developed. These ironstones were deposited upon the Market Weighton Swell structure, which formed a positive topographic feature during the Lower Lias (Fig. 1; Sellwood and Jenkyns, 1975; Bott, 1988). The presence of this intrabasinal high led to the deposition of stratigraphically condensed Lower Liassic ooidal ironstones and mudstones over this area. The ammonite zones from the base of the Frodingham Ironstone to the top of the Pecten Ironstone are represented by 30 m of sediment (Brandon et al., 1990). The age equivalent sediments to the north and south of the swell axis are markedly expanded (the same ammonite zones are represented by 100 m of marine mudstones in the south; Brandon et al., 1990; and 115 m of marine mudstones in the north; Van Buchem et al., 1992). The two ironstones are most markedly stratigraphically condensed compared to age-equivalent marine mudstones to the south (Fig. 1). Taylor and Curtis (1995) have described in detail the stratigraphy and sedimentology of the succession. The presence of an abundant diverse marine fauna, a dominance of infaunal suspension feeders, thick-walled bivalves, trough cross-bedding, and storm scours (Hallam 1963; Sellwood 1972; Young et al. 1991; Taylor and Curtis, 1995) indicate that the ironstones were deposited under well-oxygenated, shallow, high energy conditions. Sediments deposited both between times of ironstone deposition and lateral to ironstones are silt-rich mudstones (terminology of Macquaker & Gawthorpe, 1993). These mudstones contain an abundant macrofauna, again indicating deposition in oxygenated conditions. However, the lack of high energy bedding, coupled with the faunal affinities (Sellwood, 1972), suggests deposition in deeper, lower energy environments than the ironstones.

Both ironstones are dominated by authigenic berthierine and siderite, with glauconite present in some beds within the Pecten Ironstone. Where present, glauconite occurs as irregular peloids commonly acting as nuclei to berthierine ooids (Fig. 2A) and has an average composition of $K_{1.28}$ $(Fe_{2.2}Al_{1.44}Mg_{0.51})$ $(Si_{6.8}Al_{1.2})$ O_{20} $(OH)_4$. Berthierine within the ironstones takes the form of both ooids and pore-lining and pore-filling cements (e.g., Fig. 2A) and has an average composition of $(Fe_{6.81}Mg_{1.27}Al_{2.92})$ $(Si_{6.39}Al_{1.61})$ O_{20} $(OH)_{16}$. Siderite generally takes the form of isolated rhombs within pore-filling berthierine, but where present as the dominant cement individual crystals are less apparent (Fig. 2A). This siderite has a Mn-poor, Mg-rich composition (average composition $Fe_{0.69}Mg_{0.2}Ca_{0.1}Mn_{0.01}CO_3$). Where the three above minerals occur together, the berthierine can be seen to post-date the glauconite (for example, glauconite peloids acting as nuclei to berthierine ooids; Fig. 2A). Siderite can be seen to replace berthierine, and hence, post-date it (Fig. 2A). Pyrite is absent within most parts of the ironstone, but where it is present it can clearly be seen to replace berthierine and siderite and, therefore, precipitated after these minerals.

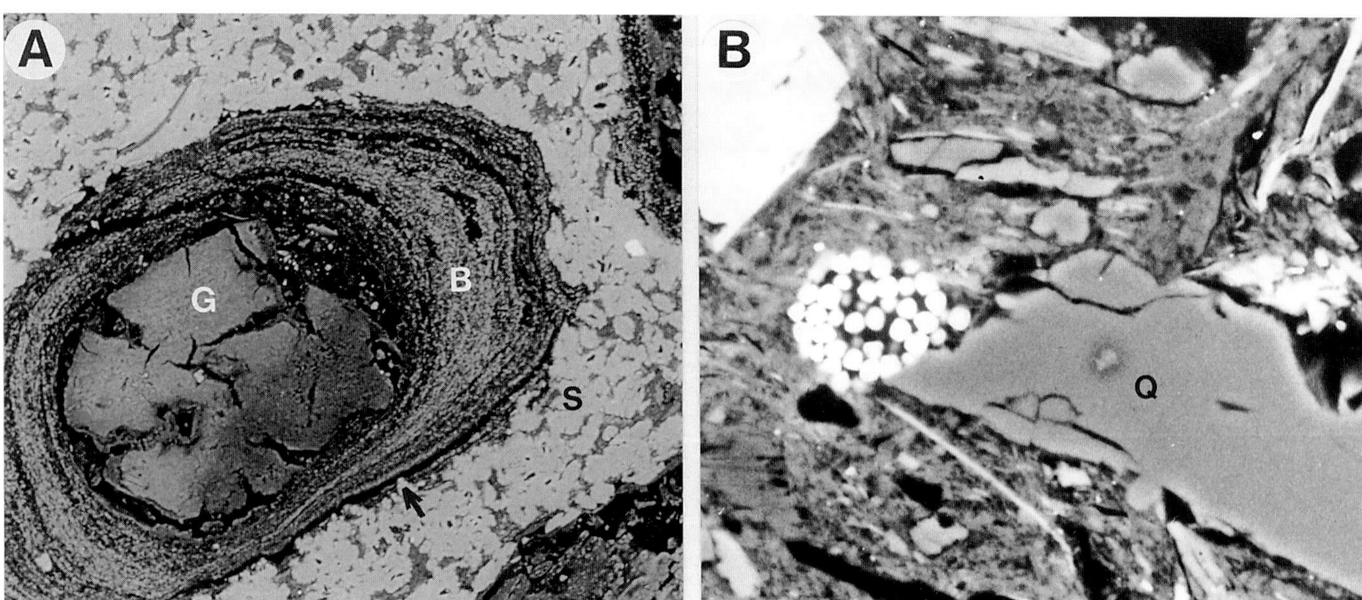

FIG. 2.—(A) Backscattered electron image of ooidal ironstone from the Lower Liassic Brant Mudstone Formation of the East Midlands Platform (Pecten Ironstone, sample from Scunthorpe area). Note the presence of glaucony peloids (G), surrounded by a berthierine cortex (B). Grain-rimming and pore-filling siderite cement (S) post-dates glaucony and siderite, replacing berthierine in places. Horizontal field of view = 500 μm (B). Backscattered electron image of mudstone from the Brant Mudstone Formation of the Scunthorpe area. Note the presence of detrital quartz (Q) and a clay mineral matrix. Note also the presence of framboidal pyrite (high backscattered coefficient). Horizontal field of view = 30 μm.

The mudstones are comprised almost entirely of detrital material, predominantly quartz and clay (chlorite and mica flakes, along with finer-grained clay matrix containing illite, kaolinite and chlorite). Within all the mudstone samples studied, pyrite was the predominant authigenic mineral, generally occurring as framboidal pyrite (Fig. 2B), but euhedral pyrite was also present. Details on the variations in grain size within the mudstones are limited as a result of core quality. However, in the Scunthorpe area, outcrop sections show that the ooidal ironstones cap a coarsening-upward mudstone unit and are overlain by an upward-fining mudstone unit (Fig. 1). These upward-coarsening and upward-fining units are on the scale of 1 m to 5 m.

Lower Jurassic Cleveland Ironstone Formation, UK

The Lower Jurassic Cleveland Ironstone Formation of northeastern England (Fig. 3) is composed predominantly of marine mudstones with subordinate ooidal ironstone beds (up to 1.0 m thick) and apatite-rich units (up to 0.15 m thick) (Fig. 3) (Rawson et al., 1982; Howard, 1985). The mudstones are composed of detrital clay and quartz grains, with clay-rich, silt-rich and sand-rich mudstones all present (Fig. 3). The early diagenetic mineral assemblage in the mudstones is dominated by pyrite, either in the form of framboids or euhedral crystals (Fig. 4). The ooidal ironstones contain abundant reworked intraclasts, abraded shell material and internal erosion surfaces. They are predominantly composed of berthierine-rich ooids (Fig. 4A), grain rimming and pore-filling berthierine cements, multiple generations of rhombohedral siderite cements, and non-ferroan calcite cements. Pyrite is rare within these units. The apatite-rich units are composed of pervasive, early microcrystalline apatite and berthierine cement (Fig. 4E). As in the ironstones, pyrite is only present in small amounts within these units.

Based on grain-size observations within the mudstones of the Cleveland Ironstone Formation, both upward-coarsening and upward-fining successions are present (Figs. 3 and 5A). These features are present at three scales: upward-fining on a 0.001-0.05-m small-scale (Fig. 5A), intermediate-scale upward-coarsening on a decimeter to meter scale (Fig. 3), and both upward-coarsening and upward-fining on a 1-3-m large scale

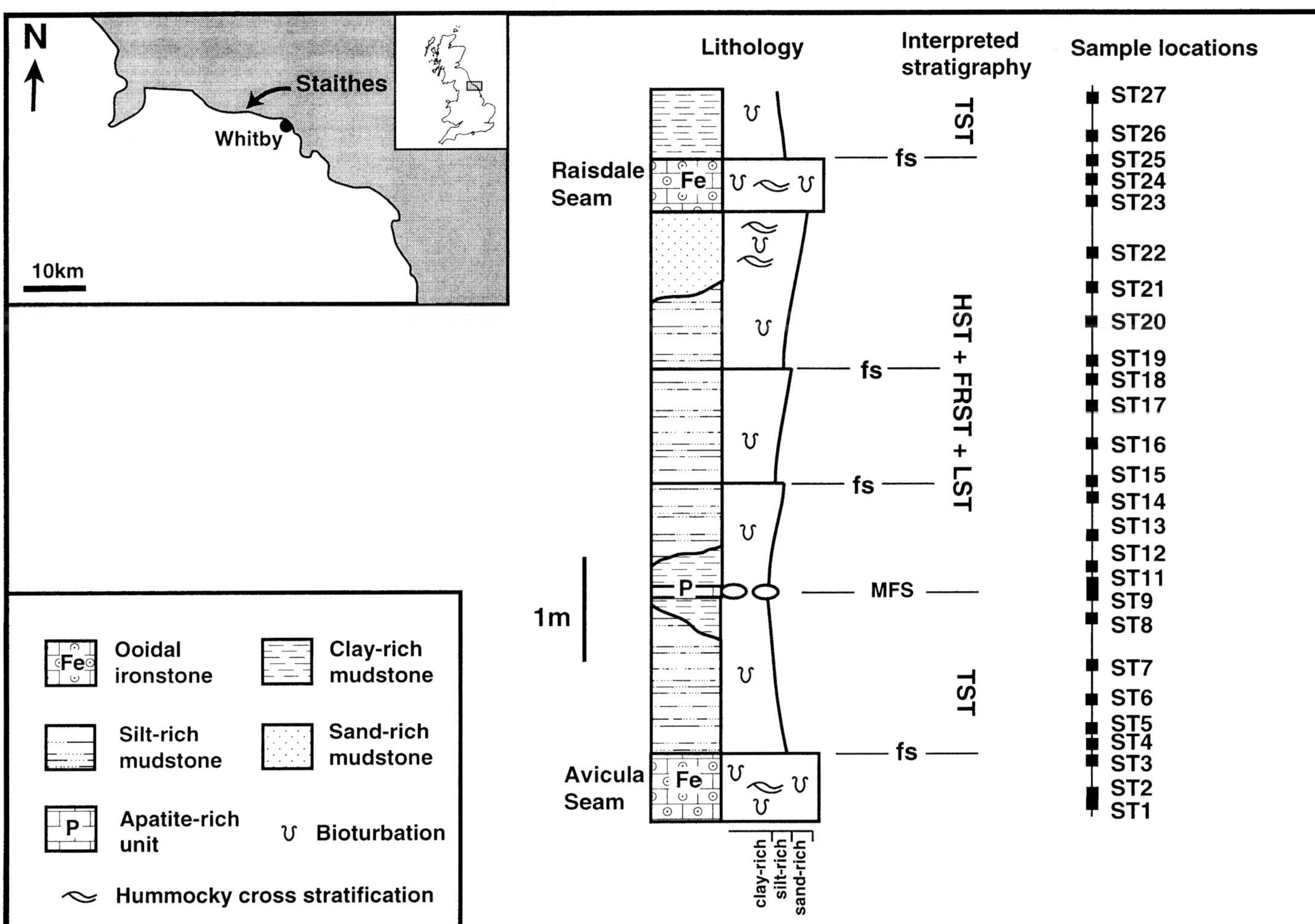

FIG. 3.—Sedimentary log, stacking patterns, sample locations and sequence stratigraphic interpretation of one cycle within the Cleveland Ironstone Formation exposed at the Yorkshire coast. Note the presence of large-scale coarsening-upward and fining-upward packages (on the scale of 1–3-m) and small-scale upward-coarsening packages (on a scale of 1 m). (HST = highstand systems tract; LST = lowstand systems tract; TST = transgressive systems tract; FRST = forced regressive systems tract).

(Fig. 3; Macquaker and Taylor, 1996, 1997). The small-scale upward-fining packages commonly comprise laminated silt-rich/sand-rich mudstones at their bases and bioturbated clay-rich/silt mudstones toward their tops (Fig. 5A).

Between the apatite-rich concretionary horizon and the overlying ironstone (the Raisdale Seam) there are three separate decimeter- to meter-scale upward-coarsening packages. The basal units in these intermediate-scale, upward-coarsening couplets commonly comprise clay-rich and silt-rich mudstones, in contrast to their tops which are coarser and predominantly comprises silt- and sand-rich mudstones and muddy sandstones. These intermediate scale upward-coarsening successions are separated from one another by a thin unit (<0.2 m) over which there is a marked decrease in grain size (Macquaker and Taylor, 1997).

The 1-3-m large-scale, upward-coarsening and fining successions are superimposed over the intermediate and smaller scale variability. These larger scale units can be recognized by comparing the grain-size in the samples at the tops of the intermediate-scale upward-coarsening units (Fig. 3). Large scale, upward-coarsening successions are present between the apatite-rich concretionary horizons and the overlying ooidal ironstone horizon (Fig. 3), while upward-fining successions are present between the ooidal ironstones and the overlying apatite-rich concretionary horizon (Fig. 3). Thus, the ironstones occur where the large-scale patterns change from overall upward-coarsening to upward-fining, while the apatite-rich units occur where the stacking patterns change from large-scale upward-fining to upward-coarsening.

Upper Triassic Westbury Formation, UK

The Upper Triassic Westbury Formation of southwest England and south Wales (Fig. 6) is predominantly composed of marine siliciclastic mudstones with subordinate wakestones, bioclastic packstones, calcareous sandstones, and "bone-beds" (Macquaker, 1994b; Warrington and Ivimey Cook, 1995). The siliciclastic-rich mudstones are variously composed of detrital clay and quartz with minor K-feldspar (Figs. 7B and C). Petrographic analyses reveal that the following mudstone microfacies are present: clay-rich, silt-rich, sand-rich, and concretionary carbonate mudstones (Fig. 7). In the mudstones the early diagenetic mineral assemblage is simple and dominated by →

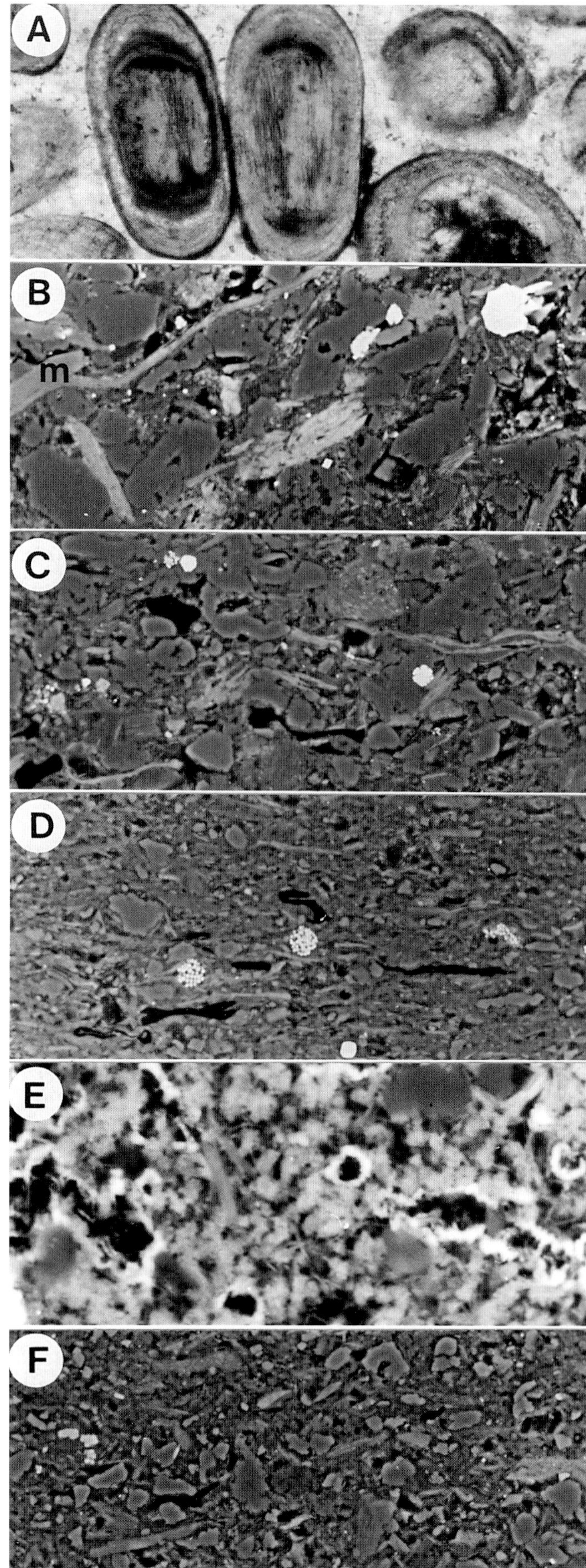

FIG. 4.—Optical photomicrograph and backscattered electron images (BSEI) of lithofacies present within the Cleveland Ironstone Formation exposed at Staithes, Yorkshire, eastern England. Note the presence of systematic grain-size variations through the succession (see Fig. 3 for sample locations). Horizontal field of view: optical photomicrographs = 2 mm; BSEIs = 150 μm. (A) Optical photomicrograph of ooidal ironstone (sample ST23). Sample contains ooids composed of berthierine with a siderite pore-filling matrix cement. (B) BSEI of sand-rich mudstone (sample ST22). Note the presence of detrital quartz, K-feldspar, and muscovite (m). (C) BSEI of silt-rich mudstone (sample ST16). (D) BSEI of clay-rich mudstone (sample ST11). Note the preponderance of fine-grained material. (E) BSEI of apatite-rich unit (sample ST9). Note the presence of both berthierine cement (laths) and apatite cement (intermediate backscatter coefficient) throughout the matrix. (Bright rims in the image are edge effects). (F) BSEI of silt-rich mudstone (sample ST7).

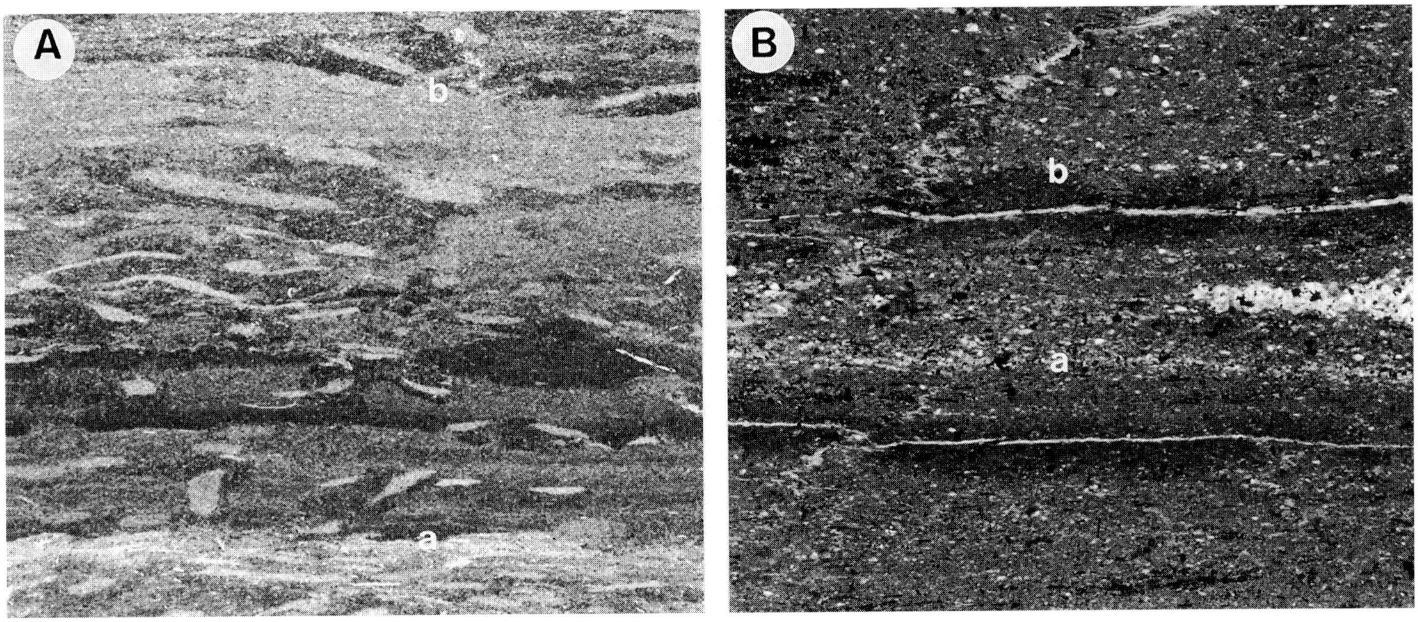

FIG. 5.—Small-scale, upward-fining units within mudstones. (A) Silt-rich mudstone within the Cleveland Ironstone Formation (sample ST14). This image is a negative image and therefore finer layers are represented by paler parts of the image. Note the presence of upward-fining (for example, from a to b) on a 1–3-mm scale, and the Phycosiphon isp burrowing in the upper parts piping material down into lower parts of the unit. Vertical field of view = 9 mm. (B) Silt-rich mudstone from the Westbury Formation (sample Lav 7; positive photographic image). Note the presence of upward fining units (for example, from a to b) on the scale of 2–3 mm. Vertical field of view = 7.5 mm.

pyrite, either in the form of framboids or euhedral crystals (Figs. 7B and E). In the carbonate-rich units the early diagenetic assemblages are much more complicated and a number of zoned non-ferroan calcite, ferroan calcite, pyrite, and glaucony cements are present. These are commonly associated with reworked bone debris (Figs. 7A and F).

Comparison of successive mudstones within the Westbury Formation reveals the presence of both upward-coarsening and upward-fining units (Fig. 6). These features are present at three scales: upward-fining on a 0.001-0.01-m scale (Fig. 5B), upward-coarsening on a decimeter to meter-scale (Fig 6) and both upward-coarsening and upward-fining on a 1-3-m scale (Fig. 6). The base of the Westbury Formation unconformably overlies the top of the Blue Anchor Formation. Intraclasts and reworked bone debris are commonly present (unit 1) on this unconformity (Figs. 6 and 7). Following deposition of this lag deposit, the mudstone units overall fine upwards (unit 2) before being capped by a recrystalized wackestone (unit 3, Fig. 6). This wackestone is then overlain by two intermediate scale upward-coarsening successions, which together form a large-scale upward-coarsening succession (units 4-6). Each of the intermediate-scale upward-coarsening units are separated from one and another by calcareous sandstone units. The carbonate-cemented units in the lower part of this succession have varying petrographic characteristics but share the fact that reworked bone debris is particularly concentrated toward their tops (see Figs 6 and 7). Above unit 6 the stacking patterns change and unit 7 progressively upward-fines until it is capped by a bioclastic wackestone (unit 8). Following deposition of unit 8 the mudstones form two upward-coarsening successions (separated by unit 9) before being overlain by interbedded bioclastic wackestones, ooidal grainstones and mudstones of the Lilstock Formation. Thus, bone-rich units occur where the large-scale stacking patterns change from overall upward-coarsening to upward-fining, while carbonate-cemented units occur at the tops of intermediate-scale coarsening-upwards units.

Within individual mudstones units of the Westbury Formation the small-scale variability commonly takes the form of upward-fining couplets on a millimeter-scale. These couplets commonly comprise sharp-based silt and fine, sand-sized laminae overlain by finer-grained, bioturbated clay-rich laminae (Fig. 5B). In contrast, in the coarser units (sandstones, shell-beds, and conglomerates), the upward-fining couplets are thicker (centimeter-scale) and variously comprise poorly sorted fine sand, reworked shell debris, and intraclasts at their bases and more mud-rich laminae toward their tops (see Macquaker, 1994b). In both the fine and coarser parts of the succession the internal structure of the couplets is commonly modified by bioturbation.

Upper Cretaceous Mesaverde Group Sediments, Book Cliffs, Utah-Colorado.

Ooidal ironstone.—

The Upper Cretaceous Blackhawk Formation and Castlegate Sandstone exposed in the Book Cliffs of Utah and Colorado form a succession of fluvial and shallow marine sediments. These were deposited on the western margin of the Western Interior Seaway as part of a clastic wedge shed from the Siever Thrust Belt. The excellent exposure in the Book Cliffs permits detailed sedimentological and sequence stratigraphic models to be developed (e.g., Swift et al., 1987; Pattison, 1995; Van Wagoner, 1995; O'Byrne and Flint, 1995). The existence of this

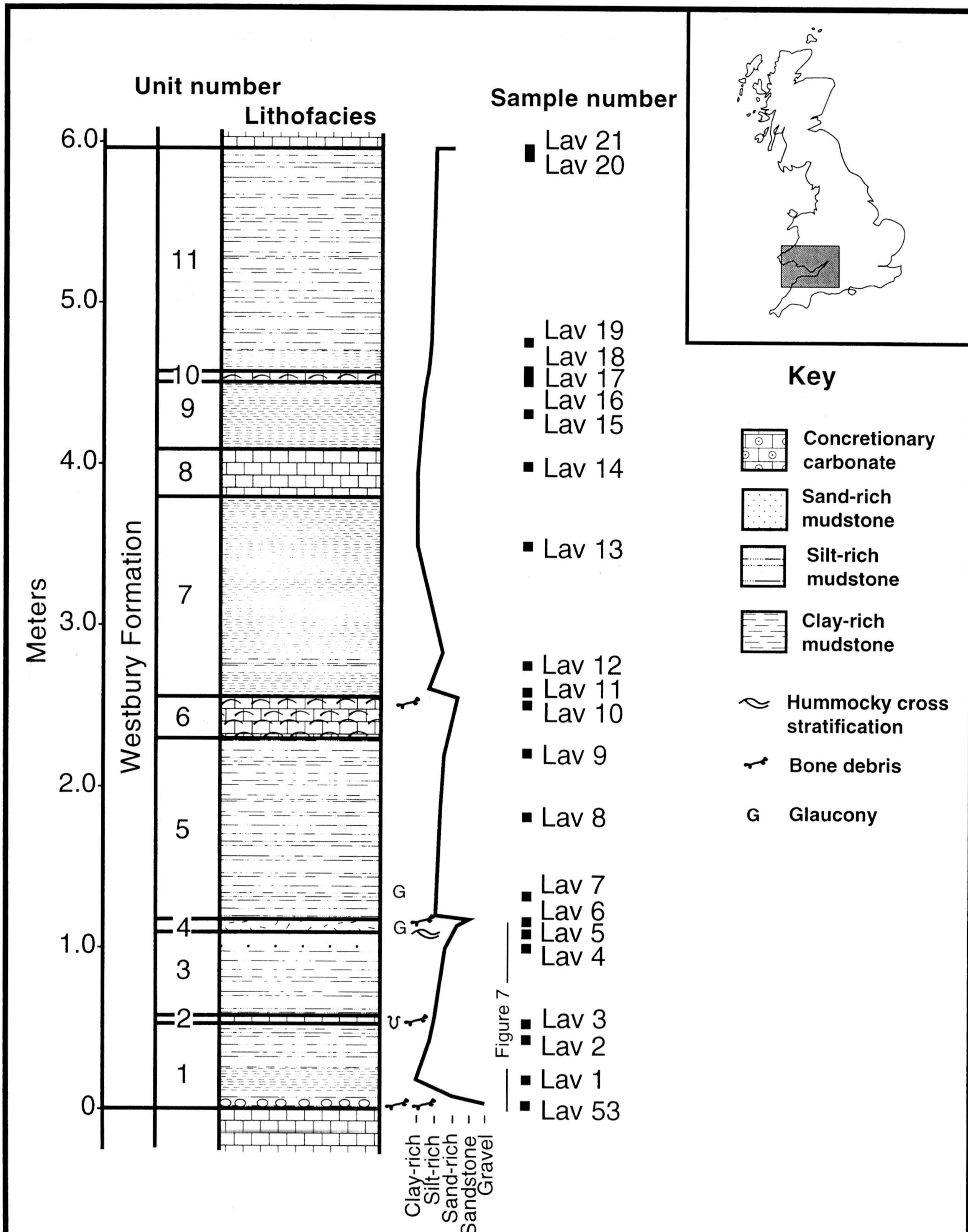

FIG. 6.—Sedimentary log, stacking patterns, and sample locations of the Triassic Westbury Formation exposed at Lavernock, in South West England. Note the presence of large-scale, coarsening-upward and fining-upward packages (on the scale of 1–3 m), and the presence of bone-rich sediments capping the coarsening-upward units.

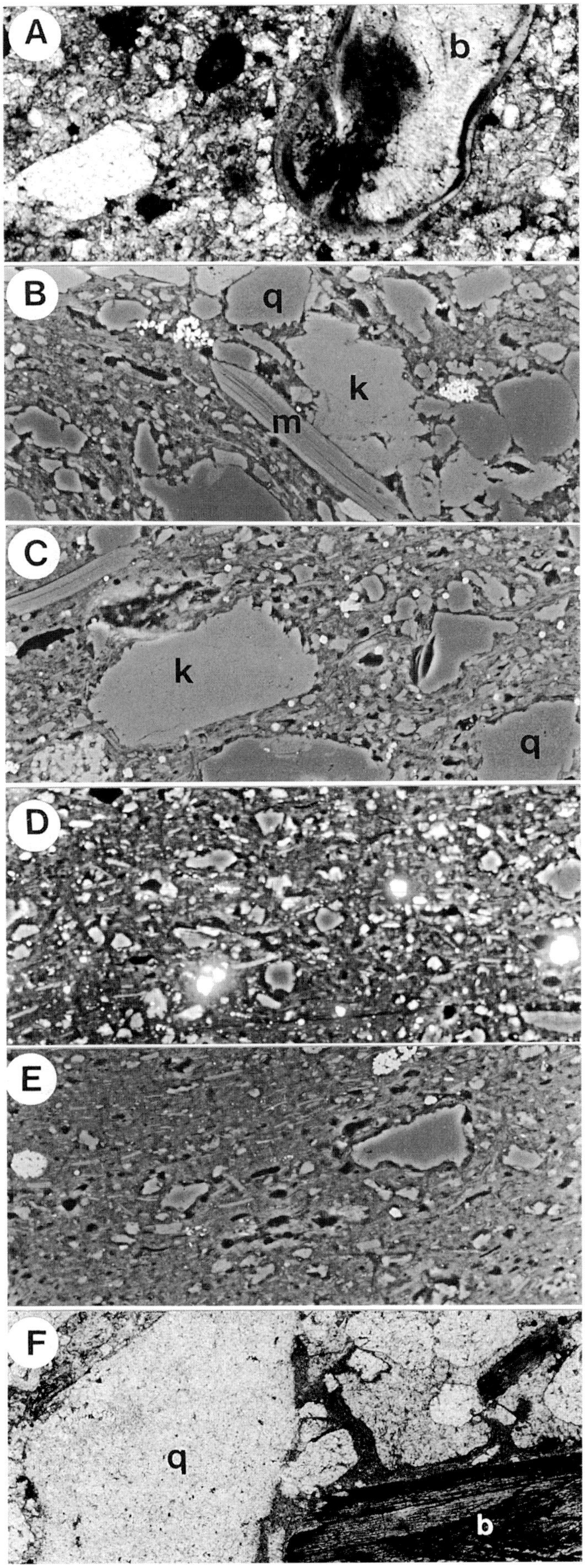

framework has allowed spatial and temporal variations in early diagenetic processes to be placed into a detailed sequence stratigraphic context for both mudstones and sandstones (Taylor et al., 1995, 1997; Yocum, 1996). Three ooidal ironstones have been recognized within the Mancos Shale of the Book Cliffs (Chan, 1992; Van Wagoner, 1995). One of these is associated with the Castlegate Sandstone and has been the subject of detailed sequence stratigraphic and geochemical study (Van Wagoner, 1995; Yocum, 1996.

The ooidal ironstone in question sits upon the Castlegate Sequence Boundary, a major sequence boundary within the succession exposed within the Book Cliffs (see Van Wagoner, 1995). The Castlegate Sequence Boundary can be traced unequivocally from proximal environments, where it forms the base, to incised channel fills to distal environments where it occurs beneath the ooidal ironstone. Stratigraphic correlation, based on field analysis, shows that the up-dip equivalent of the ooidal ironstone is a thick succession (up to 100 m) of fluvial-dominated sediments representing the Castlegate lowstand systems tracts (Fig. 8; Van Wagoner, 1995). The ironstone itself is localized to an area of approximately 5 km in up-dip to down-dip extent (Van Wagoner, 1995). It rests sharply on a thin succession of marginal-marine to non-marine silts and muds, which themselves sharply overlie lower shoreface fine sandstones and siltstones (Fig. 8). The ironstone has a maximum thickness of 2 m and is overlain by fine-grained mudstones of the Mancos Shale. This upper surface can be traced throughout the area and represents a major transgressive surface (the Buck Tongue Flooding Surface), which itself marks the base of the overlying transgressive system tract. The ironstone itself is composed of berthierine and kaolinite ooids, with grain-rimming and pore-filling siderite and calcite cement. Pyrite is rare to absent. Many of the berthierine ooids have been partially replaced by siderite, and most ooids have undergone later replacement by ferroan dolomite.

Carbonate cemented horizons.—

Within the shallow marine sandstones of Book Cliffs, laterally extensive carbonate cements are present beneath major marine flooding surfaces. These cements are laterally extensive for tens of kilometers and show evidence of firmground formation following deposition (for example, unpelleted Skolithos isp. borrows; Taylor et al., 1995). Petrographically, these cement zones comprise zoned ferroan dolomite and calcite cements overgrowing detrital non-ferroan dolomite grains. These later-

FIG. 7.—Optical photomicrographs and backscattered electron images (BSEI) of lithofacies present within the Westbury Formation at Lavernock in southern England. Note the presence of systematic grain size variations through the succession (see Fig. 5 for sample locations). (Horizontal field of view: optical photomicrographs = 2 mm; BSEIs = 150 μm. (A) Optical photomicrograph of calcareous sandstone with abundant bone-debris (b). Note the presence of detrital quartz sand and opaque pyrite (sample Lav 6). (B) BSEI of sand-rich mudstone. This sample contains detrital quartz (q), K-felspar (k) and muscovite (m). (Sample Lav 5). (C) BSEI of sand-rich mudstone (k = detrital K-feldspar, q = detrital quartz. (Sample Lav 4). (D). BSEI of silt-rich mudstone (sample Lav 2). (E) BSEI of clay-rich mudstone (sample Lav 1). (F) Optical photomicrograph of bone-bed. (b = bone material, q = detrital quartz, sample Lav 53).

commonly been recognized in hemipelagic environments (Froelich et al. 1979; Aller, 1990), such conditions have recently been recognized within shallow marine sediments also (Aller et al., 1986; Sørensen and Jørgensen, 1987; Aller, 1993). Although no unequivocal modern environments of ironstone formation are known, authigenic iron rich clays are known from a number of shallow marine environments (e.g., Porrenga, 1965; Odin, 1988). In addition, minor amounts of authigenic Fe-clays have been detected within modern Amazon Shelf sediments dominated by suboxic diagenetic processes (Rude and Aller, 1989; Aller, 1993). Aller et al. (1986) suggested such a setting as a possible modern analogue for ancient ironstones.

Sedimentological and faunal evidence within the ooidal ironstones indicates that net sediment accumulation rates were slower and water depths were shallower than for the pyrite-dominated mudstones within the successions. These sediments would, therefore, most likely have experienced a high degree of physical and biological reworking. The extent of sediment reworking has been proposed as a major control on the degree to which suboxic (metal oxide reduction) reactions take place within marine sediments (Aller, 1990; Raiswell, 1997). Within the ooidal ironstones, suboxic processes would, therefore, be expected to have been more prominent. This is indeed observed; within the ironstones Fe(III) reduction appears to have been the only significant pathway for organic matter oxidation. The predominance of Fe(III) reduction within highly reworked sediments may result from a combination of low organic matter quality and enhanced iron oxide contents (as a result of the prolonged sediment reworking). Under such conditions it has been shown that Fe(III) reduction can inhibit sulfate reduction (Coleman, 1985; Lovley and Phillips, 1987; Chapelle and Lovley, 1992). A similar explanation has been invoked to explain high rates of Fe(III) reduction within Amazon shelf sediments (Aller, 1993). Similar effects have been observed by Canfield et al. (1993), whereby enhanced levels of manganese oxides have led to manganese reduction dominating organic matter oxidation in some recent shallow marine sediments.

As was discussed in the above section, we interpret all the ooidal ironstone units described here to have formed either upon sequence boundaries as correlative equivalents of sequence boundaries or as forced regressive units. The extensive sediment reworking and slow sediment accumulation associated with these stratigraphic settings led to the predominance of suboxic diagenesis within these units, and thereby the generation of the geochemical conditions required for the formation of berthierine and siderite.

Apatite-rich and bone-rich units.—

Phosphate precipitation has been described here from two different stratigraphic horizons: at maximum flooding surfaces and at major transgressive surfaces. Both of these stratal surfaces are characterized by periods of reduced sediment accumulation (Loutit et al., 1988; Jervey, 1988). In addition, both of these surfaces were broadly associated with iron-rich sediments, as evidenced by their current mineralogy (berthierine-siderite-phosphate, glaucony-calcite-phosphate for the major transgressive surfaces on top of the ironstones and beneath the bone beds, and berthierine-phosphate at the maximum flooding surfaces within the Cleveland Ironstone Formation). The high contents of these iron minerals, characteristic of suboxic diagenesis (see discussion above), indicate that iron reduction was a major process within these sediments. The iron-redox cycling of phosphorus has been suggested to be important in the precipitation of phosphate cement within Quaternary phosphorite-bearing sediments of the East Australian Shelf (e.g., Heggie et al., 1990; O'Brien et al., 1990) and in shelf environments (Ruttenberg and Berner, 1993; Macquaker, 1994b). Given the evidence for the importance of iron reduction in the apatite-rich units of the Cleveland Ironstone Formation, early diagenetic liberation of P adsorbed onto iron oxides was potentially a major process, leading to enhanced phosphate levels in early diagenetic pore-waters and the precipitation of apatite. The slow net sediment accumulation rates characteristic of the phosphate horizons would have led to greater sediment residence time in oxic and suboxic zones. As a result, enhanced iron-redox cycling of P should have occurred (due to both bioturbation of sediment and upward chemical diffusion of Fe(II) to the surface oxic zone). Given these comments, we believe that the occurrence of phosphate-rich sediments at maximum flooding surfaces and major transgressive surfaces in the Cleveland Ironstone Formation was most likely a result of slow sedimentation rates coupled with enhanced iron-redox cycling of P at these horizons. Phosphate precipitation within the bone beds may also have been enhanced as a result of decreased pore-water pH due to low detrital iron input and increased sulfide oxidation as a result of marine transgression (Macquaker, 1994b).

Carbonate-rich units.—

The carbonate-rich units described in this paper all occur beneath marine flooding surfaces. A major factor controlling the degree of cement precipitation during early diagenesis is the residence time spent by the sediment package within early diagenetic zones (Curtis, 1987b). This residence time is a function of sediment accumulation (burial) rate. At low rates of sediment burial, residence times within individual zones are high. During such times the increased residence time of the sediment within early diagenetic zones allows greater localized intensity of early diagenetic reactions, for example, through the diffusive replenishment of reactants (sulfate and methane through diffusion, organic matter through bioturbation) and the enhanced production of bicarbonate (Berner, 1980; Curtis, 1987b). Consequently, low sediment accumulation rates allow the potential for enhanced localized cementation during early diagenesis. Raiswell (1987) has argued for, from geochemical evidence, just such a mechanism for formation of carbonate concretions within some Jurassic marine mudstone successions in eastern England.

From field and stratigraphic evidence, coupled with petrographic analysis, we conclude that carbonate-rich units formed during early diagenesis as a result of marine flooding events. During marine flooding events, rate of accommodation increase may outpace sediment supply resulting in a break, or significant reduction, in sedimentation (Jervey, 1988; Fig. 9). During such

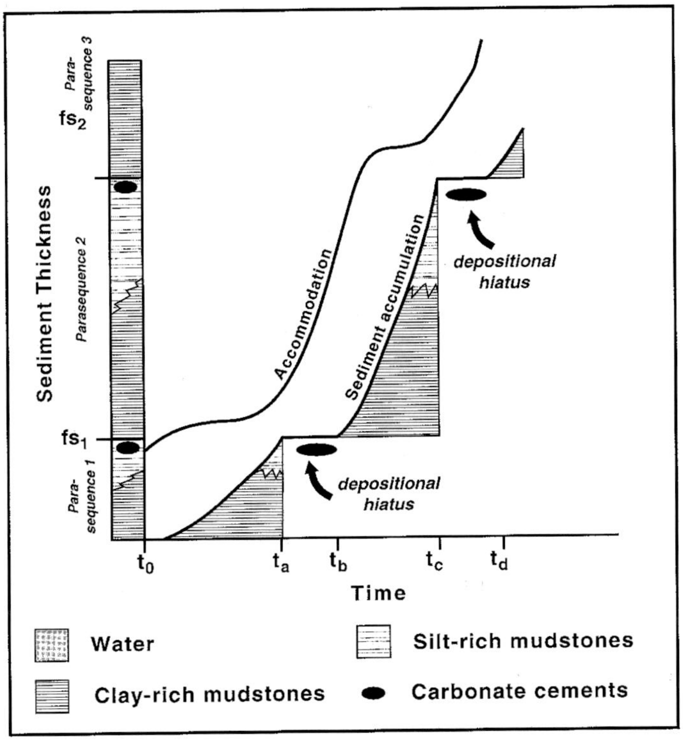

FIG. 9.—The formation of concretionary cements during marine flooding events. Between times t_a and t_b and times t_c and t_d the rate of accommodation development exceeds that of sediment supply, resulting in a depositional hiatus and a marine flooding surface. During this time enhanced sediment residence time within early diagenetic zones leads to the formation of concretionary cements at these horizons. (Modified after Taylor et al., 1995).

times the increased residence time of the sediment within early diagenetic zones led to greater localized intensity of early diagenetic reactions (Taylor et al., 1995). Since marine flooding surfaces are of significant regional lateral extent within sedimentary basins (Van Wagoner et al., 1990), diagenetic cemented horizons associated with these surfaces are potentially important markers for stratigraphic analysis of mudstone-dominated sediments

CONCLUSIONS

The specific conclusions of the data and discussion presented in the paper are:

1. Both spatial and temporal variability in early diagenetic mineral assemblages can be recognized in the four studied mudstone-dominated shallow marine successions (the Lower Jurassic Scunthorpe and Brant Mudstone Formations of eastern England, the Lower Jurassic Cleveland Ironstone Formation of eastern England, the Triassic Westbury Formation of South West England, and the Upper Cretaceous of Book Cliffs, Utah). This variation results from spatial and temporal variations in physical and geochemical conditions, which have a direct control on the early diagenetic reactions in the sediments.
2. The physical conditions that occur at sequence boundaries (low accommodation space, high energy, common sediment reworking) can lead to the predominance of suboxic diagenesis and the formation of berthierine- and siderite-rich ooidal ironstones.
3. Conditions of low net sediment accumulation and enhanced oxic/suboxic diagenesis on major transgressive surfaces and maximum flooding surfaces leads to the formation of apatite-rich units and bone-rich units.
4. Breaks in sediment accumulation at marine flooding surfaces lead to enhanced early diagenesis and the formation of laterally extensive carbonate cements and concretionary horizons.
5. Sulfide-dominated early diagenesis (that dominated by sulfate redution) is predominant during "normal" periods of shallow marine sedimentation and appears to mark the bulk of sediments within systems tracts.

The recognition that spatial and temporal variability in early diagenetic minerals is linked to variations in physical and geochemical environmental conditions, and that these conditions can be interpreted within a sequence stratigraphic framework, has major implications for the interpretation of shallow marine sedimentary successions. Commonly, the style of diagenetic mineralogy is the most prominent feature within these successions, and recognizing their relevance is of key importance for basin and stratigraphic analysis. While the examples discussed here illustrate the formation of specific diagenetic assemblages at specific stratigraphic locations, the exact nature of diagenesis and its spatial variability within individual stratigraphic settings will also depend, to an extent, on factors independent of stratigraphic development (for example, productivity, sediment composition).

The results of this work highlight the importance of integrating geochemical and sequence stratigraphic analysis in both basin analysis and authigenic marine mineral analysis. Characteristic diagenetic mineral assemblages have the potential to act as important sequence stratigraphic marker horizons in what are often otherwise macroscopically homogenous successions.

ACKNOWLEDGMENTS

Aspects of this research have been funded by the Natural Environment Research Council (Grant Nos. GR3/7887 and GR3/8928), British Gas, Union Pacific Resources Co., which is gratefully acknowledged. The authors also thank Charles Curtis, Rob Gawthorpe, Dave Hunt, and John Van Wagoner for stimulating discussions over the research period. Kurt Grimm, Craig Glenn, and an anonymous reviewer are thanked for constructive comments on an earlier version of this paper.

REFERENCES

ALLER, R. C., 1990, Bioturbation and manganese cycling in hemi-pelagic sediments: Philosophical Transactions of the Royal Society of London, v. A 331, p. 51-68.

ALLER, R. C., 1993, Influence of terrestrial weathering on early diagenetic reactions in continental shelf sediments: Chemical Geology, v. 107, p. 437-438.

ALLER, R. C., MACKIN, J. E., AND COX, R. T., 1986, Diagenesis of Fe and S in Amazon inner shelf muds: Apparent dominance of Fe reduction and implications for the genesis of ironstones: Continental Shelf Research, v. 6, p. 263-289.

BENDER, M. L., AND HEGGIE, D. T., 1984, Fate of organic carbon reaching the deep sea floor: a status report: Geochimica et Cosmochimica Acta, v. 48, p. 977-986.

Berner, R. A., 1980, Early Diagenesis: A Theoretical Approach: Princeton, New Jersey, Princeton University Press.

Berner, R. A., 1981, A new geochemical classification of sedimentary environments: Journal of Sedimentary Petrology, v. 51, p. 359-365.

Bjorkum, P. A., and Walderhaug, A., 1990a, Geometrical arrangement of calcite cementation within shallow marine sandstones: Earth-Science Reviews, v. 29, p. 145-161.

Bjorkum, P. A., and Walderhaug, A., 1990b, Lateral extent of calcite cemented zones in shalow marine sandstones, *in* Buller, A. T., et al., eds, North Sea Oil and Gas Reservoirs II: Graham & Trotman, London, p. 331-336.

Bott, M. H. P., 1988, The Market Weighton gravity anomaly — granite or graben?: Proceedings of the Yorkshire Geological Society, v. 47, p 47-53.

Brandon, A., Sumbler, M. G., and Ivimey-Cook, H. C., 1990, A revised lithostratigraphy for the Lower and Middle Lias (Lower Jurassic) east of Nottingham, England: Proceedings of the Yorkshire Geological Society., v. 48, p. 121-141.

Canfield, D. E., 1989, Sulfate reduction and oxic respiration in marine sediments: Implications for organic carbon preservation in euxinic environments: Deep-Sea Research, v. 36, p. 121-138.

Canfield, D. E., Thamdrup, B., and Hansen, J. W., 1993, The anaerobic degredation of organic matter in Danish coastal sediments: Iron reduction, manganese reduction, and sulfate reduction: Geochimica et Cosmochimica Acta, v. 57, p. 3867-3883.

Chan, M., 1992, Oolitic ironstone in the Cretaceous Western Interior Seaway, East-central Utah: Journal of Sedimentary Petrology, v. 62, p. 693-705.

Chapelle, F. H., and Lovley, D. R., 1992, Competitive exclusion of sulfate reduction by Fe(III)-reducing bacteria: A mechanism for producing discrete zones of high-iron ground water: Ground Water, v. 30, p 29-36.

Claypool, G. E., and Kaplan, I. R., 1974, The origin and distribution of methane in marine sediments. *in* Kaplan, I. R., ed., Natural Gases in Marine Sediments: New York, Plenum Press, p. 99-139.

Coleman, M. L., 1985, Geochemistry of diagenetic non-silicate minerals: kinetic considerations: Philosophical Transactions of the Royal Society of London, v. A 315, p. 39-56.

Coleman, M. L. and Raiswell, R., 1995, Source of carbonate and origin of zonation in pyritiferous concretions: Evaluation of a dynamic model: American Journal of Science, v. 295, p. 282-308.

Curtis, C. D., 1987a, Donnees recentes sur le reactions entre matieres organiques et substances minerales dans les sediments et sur leurs consequences mineralogiques: Societe Geologique de France, Memoires, v. 151, p. 127-141.

Curtis, C. D., 1987b, Inorganic chemistry and petroleum exploration. *in* Brooks, J., and Welte, D., eds., Advances in Petroleum Geochemistry II: Academic Press, p. 91-140.

Curtis, C. D., Coleman, M. L., and Love, L. G., 1986, Pore water evolution during sediment burial from isotope and mineral chemistry of calcite, dolomite and siderite concretions: Geochimica et Cosmochimica Acta, v. 50, p. 2321-2334.

Curtis, C. D., and Spears, D. A., 1968, The formation of sedimentary iron minerals: Economic Geology, v. 63, p. 257-270.

Froelich, P. N., Klinkhammer, G. P., Bender, M. L., Luedke, N. A., Heath, G. R., Cullen, D., Dauphin, P., Hammond, D., Hartmann, B., and Maynard, V., 1979, Early oxidation of organic matter in pelagic sediments of the eastern equatorial Atlantic: suboxic diagenesis: Geochimica et Cosmochimica Acta, v. 43, p. 1057-1090.

Gautier, D. L., 1982, Siderite concretions: indicators of early diagenesis in the Gammon Shale (Cretaceous): Journal of Sedimentary Petrology, v. 52, p. 859-871.

Hallam, A., 1963, Observations on the palaeoecology and ammonite sequence of the Frodingham Ironstone (Lower Jurassic): Palaeontology, v. 6, p. 554-574.

Heggie, D. T., Dyring, G. W., O'Brien, G. W., Reimers, C., Herczeg, A., Moriarty, D. J. W., Burnett, W. C., and Milnes, A. R. 1990, Organic carbon cycling and modern phosphorite formation on the East Australian continental margin: an overview, *in* Notholt, A. J. G., and Jarvis, I., eds., Phosphorite Research and Development: London, Geological Society of London Special Publication 52, p. 87-117.

Henrichs, S.M. and Reeburgh, W.S., 1987, Anaerobic mineralization of marine sediment organic matter: Rates and the role anaerobic processes in the ocean carbon economy: Geomicrbiological Journal, v. 5, p. 191-237.

Hines, M. E., Bazylinski, D. A., Tugel, J. B., and Lyons, W. B., 1991, Anaerobic microbial biogeochemistry in sediments from two basins in the Gulf of Maine — evidence for iron and manganese reduction: Estuarine and Coastal Shelf Science, v. 32, p. 313-324.

Howard, A. S., 1985, Lithostratigraphy of the Staithes Sandstone and Cleveland Ironstone formations (Lower Jurassic) of north-east Yorkshire: Proceedings of the Yorkshire Geological Society, v. 45, p. 261-275

Hunt, D., and Tucker, M. E., 1992. Stranded parsequences and the forced regressive wedge systems tract: Deposition during base level fall: Sedimentary Geology, v. 81, p. 1-9.

Irwin, H, Curtis, C. D., and Coleman, M. L., 1977, Isotopic evidence for source of diagenetic carbonates formed during burial of organic-rich sediments: Nature, v. 269, p. 209-213.

Jervey, M. T., 1988, Quantitative geological modeling of siliciclastic rock sequences and their seismic expression, *in* Wilgus, C. K., Hastings, B. S., Kendall, C. G. St. C., Posamentier, H. G., Ross, C. A., and Van Wagoner, J. C., eds., Sea-level Change: An Integrated Approach: Tulsa, Society for Sedimentary Geology (SEPM) Special Publication 42, p. 47-69.

Leithold, E. L., 1994, Stratigraphical architecture at the muddy margin of the Cretaceous Western Interior Seaway, southern Utah: Sedimentology, v. 41, p. 521-542.

Lesuer, P., and Tastet, J. P., 1994, Facies, internal structures and sequence of modern Gironde derived muds on the Aquitaine inner shelf, France: Marine Geology, v. 120, p. 267-290

Loutit, T. S., Hardenbol, J.,Vail, P. R., and Baum, G. R., 1988, Condensed sections: The key to age dating and correlation of continental margin sequences, *in* Wilgus, C. K., Hastings, B. S., Kendall, C. G. St. C., Posamentier, H. G., Ross, C. A., and Van Wagoner, J. C., eds., Sea-level Change: An Integrated Approach: Tulsa, Society for Sedimentary Geology (SEPM) Special Publication 42, p. 183-216.

Lovley, D. R., and Phillips, E. J. P., 1987. Competitive mechanisms for inhibition of sulfate reduction and methane production in the zone of ferric iron reduction in sediments: Applied and Environmental Microbiology, v. 53, p. 2636-2641.

Lovley, D. R., and Phillips, E. J. P., 1988, Novel mode of microbial metabolism: organic carbon oxidation coupled to dissimilatory reduction of iron and manganese: Applied and Environmental Microbiology, v. 54, p. 1472-1480.

Lovley, D. R., and Lonergan, D. J., 1990, Anaerobic oxidation of toluene, phenol, and p-cresol by the dissimilatory iron-reducing organism, GS-15: Applied and Environmental Microbiology., v. 56, p. 1858-1864.

Macquaker, J. H. S., 1994a, A lithofacies study of the Lower Oxford Clay, an example of sediment bypass in mudstone successions: Journal of the Geological Society of London, v. 151, p. 161-172.

Macquaker, J. H. S., 1994b, Palaeoenvironmental significance of 'bone-beds' in organic-rich mudstone successions: an example from the Upper Triassic of south west Britain: Zoological Journal of the Linnean Society, v. 112, p. 285-308.

Macquaker, J. H. S., and Gawthorpe, R. L., 1993, Mudstone lithofacies in the Kimmeridge Clay Formation, Wessex Basin, Southern England: Implications for the origin and controls of the distribution of mudstones: Journal of Sedimentary Petrology, v. 63, p. 1129-1143.

Macquaker, J. H. S., Gawthorpe, R. L., Taylor, K. G. and Oates, M. J.,1998, Heterogeneity, stacking patterns and sequence stratigraphic interpretation in distal mudstone successions: examples from the Kimmeridge Clay Formation, UK., *in* Schieber, J., ed., Shales and Mudstones: Tulsa, Society for Seidimentary Geology (SEPM) Special Publication, in press, 1999.

Macquaker, J. H. S., and Taylor, K. G., 1996, A sequence stratigraphic interpretation of a mudstone-dominated succession: The Lower Jurassic Cleveland Ironstone Formation, UK.: Journal of the Geological Society of London, v. 153, p. 759-770.

Macquaker, J. H. S., and Taylor, K. G., 1997, Discussion on a sequence stratigraphic interpretation of a mudstone-dominated succession: The Lower Jurassic Cleveland Ironstone Formation, UK.: Journal of the Geological Society of London, v. 154, p. 913-916.

MINOURA, K., NAKAYA, S., AND TAKEMURA, A., 1991, Origin of manganese carbonates in Jurassic red shale, central Japan: Sedimentology, v. 38, p. 137-152.

MOZLEY, P. S. 1996, The internal structure of carbonate concretions in mudsrocks: A critical evaluation of the conventional concentric model of concretion growth: Sedimentary Geology, v. 103, p. 85-91.

MYERS, C. R., AND NEALSON, K. H., 1988, Microbial reduction of manganese oxides: interactions with iron and sulfur: Geochimica et Cosmochimica Acta, v. 52, p. 2727-2732.

O'BRIEN, G. W., MILNES, A. R., VEEH, H. H., HEGGIE, D. T., RIGGS, S. R., CULLEN, D. J., MARSHALL, J. F., AND COOK, P. J. 1990, Sedimentation dynamics and redox iron-cycling: Controlling factors for the apatite-glauconite association on the East Australian continental margin, *in* Notholt, A. J. G., and Jarvis, I., eds., Phosphorite Research and Development: London, Geological Society of London Special Publication 52, p. 61-86.

O'BYRNE, C. J., AND FLINT, S., 1995, Sequence, parasequence and intraparasequence architecture of the Grassy Member, Blackhawk Formation, Book Cliffs, Utah, U.S.A., *in* Van Wagoner, J. C., and Bertram, G. T., eds, Sequence Stratigraphy of Foreland Basin Deposits: Tulsa, American Association of Petroleum Geologists Memoir 64, p. 225-255.

ODIN, G. S., 1988, Green Marine Clays, Oolitic Ironstone Facies, Verdine Facies, Glaucony Facies and Celadonite-bearing facies — A comparative study: Developments in Sedimentology 45: Amsterdam, Elsevier, 445 p.

OKITA, P. M., MAYNARD, J. B., SPIKER, E. C., AND FORCE, E. F., 1988, Isotopic evidence for organic matter oxidation by manganese reduction in the formation of stratiform manganese carbonate ore: Geochimica et Cosmochimica Acta, v. 52, p. 2679-2685.

PASSEGA, R., RIZZINI, A. AND BORGHETTI, G., 1967, Transport of sediments by waves, Adriatic coastal shelf, Italy: Bulletin of the American Association of Petroleum Geologists, v. 51, p. 1304-1319.

PATTISON, S. A. J., 1995, Sequence stratigraphic significance of sharp-based lowstand shoreface deposits, Kenilworth Member, Book Cliffs, Utah: American Association of Petroleum Geologists Bulletin, v. 79, p. 444-462.

PORRENGA, D. H., 1965, Chamosite in Recent sediments of the Niger and Orinoco deltas: Geol Mijnbow, v. 44, p. 400-403.

PROSSER, D. J., DAWS, J. A., FALLICK, A. E., AND WILLIAMS, B. P. J., 1993, Geochemistry and diagenesis of stratabound calcite cement layers within the Rannoch Formation of the Brent Group, Murchison Field, North Viking Graben (Northern North Sea): Sedimentary Geology, v. 87, p. 139-164.

RAISWELL, R., 1987, Non-steady state microbiological diagenesis and the origin of concretions and nodular limestones, *in* Marshall, J. D., ed., Diagenesis of Sedimentary Sequences: London, Geological Society of London Special Publication 36, p. 41-54.

RAISWELL, R., 1988, Chemical model for the origin of minor limestone-shale cycles by anaerobic methane oxidation: Geology, v. 16, p. 641-644.

RAISWELL, R., 1997, A geochemical framework for the application of stable sulphur isotopes to fossil pyritisation: Journal of the Geological Society of London, v. 154, p. 343-356.

RAWSON, P. F., GREENSMITH, J. T., AND SHALABY, S. E., 1982, Coarsening-upwards cycles in the uppermost Staithes and Cleveland Ironstone Formations (Lower Jurassic) of the Yorkshire Coast, England: Proceedings of the Geologists' Association, v. 94, p. 91-93.

RUDE, R. P., AND ALLER, R. C., 1989, Early diagenetic alteration of lateritic particle coatings in Amazon Continental Shelf sediment: Journal of Sedimentary Petrology, v. 59, p. 704-716.

RUTTENBERG, K. C., AND BERNER, R. A. 1993, Authigenic apatite formation and burial in sediments from non-upwelling, continental margin environments: Geochimica et Cosmochimica Acta, v. 57, p. 991-1007.

SELLWOOD, B .W., 1972, Regional environmental changes across a Lower Jurassic stage boundary in Britain: Palaeontology, v. 15, p. 125-157.

SELLWOOD, B. W., AND JENKYNS, H. C., 1975, Basins and swells and the evolution of an epeiric sea (Pliensbachian-Bajocian of Great Britain): Journal of the Geological Society of London, v. 131, p. 373-378.

SETHI, P. S., AND LEITHOLD, E. L., 1994, Climatic cyclicity and terrigenous sediment influx to the early Turoniaan Greenhorn Sea, southern Utah: Journal of Sedimentary Research, v. B64, p. 26-39.

SØRENSEN, J., AND JØRGENSEN, B. B., 1987, Early diagenesis from Danish coastal waters: Microbial activity and Mn-Fe-S geochemistry: Geochimica et Cosmochimica Acta, v. 51, p. 1583-1590.

SWIFT, D. J. P., HUDELSON, P. M., BRENNER, R. L., AND THOMPSON, P., 1987, Shelf construction in a foreland basin: Storm beds, shelf sandbodies, and shelf-slope depositional sequences in the Upper Cretaceous Mesaverde Group, Book Cliffs, Utah: Sedimentology, v. 34, p. 423-457.

TAYLOR, K. G., AND CURTIS, C. D., 1995, Stability and facies association of early diagenetic mineral assemblages: An example from a Jurassic ironstone — mudstone succession, U.K.: Journal of Sedimentary Research, v. A65, p. 358-368.

TAYLOR, K. G., GAWTHORPE, R. L., AND VAN WAGONER, J. C., 1995, Stratigraphic control on laterally persistent cementation, Book Cliffs, Utah: Journal of the Geological Society of London, v. 152, p. 225-228.

TAYLOR, K. G., GAWTHORPE, R. L., CURTIS, C. D., FANNON, S., AND VAN WAGONER, J. C., 1997, Predictive diagenesis in a sequence stratigraphic framework (abs.): American Association of Petroleum Geologists Annual Meeting, Dallas, 1997.

TUGEL, J. B., HINES, M. E., AND JONES, G. E., 1986, Microbial iron reduction by enrichment cultures isolated from estuarine sediments: Applied and Environmental Microbiology, v. 52, p. 1167-1172.

VAN BUCHEM, F. S. P., MELNYK, D. H., AND MCCAVE, I. N., 1992, Chemical cyclicity and correlation of Lower Lias mudstones using gamma ray logs, Yorkshire, UK.: Journal of the Geological Society of London., v. 149, p. 991-1002.

VAN WAGONER, J. C., 1995, Sequence stratigraphy and marine to non-marine facies architecture of foreland basin strata, Book Cliffs, Utah, U.S.A., *in* Van Wagoner, J. C., and Bertram, G. T., eds, Sequence Stratigraphy of Foreland Basin Deposits: Tulsa, American Association of Petroleum Geologists Memoir 64, p. 137-223.

VAN WAGONER, J. C., MITCHUM, R. M., CAMPION, K. M., AND RAHMANIAN, V. D., 1990, Siliciclastic sequence stratigraphy in well logs, cores and outcrops: Concepts for high resolution correlation of time and facies: Tulsa, American Association of Petroleum Geologists Methods in Exploration Series 7, 55 p.

WARRINGTON, G., AND IVIMEY-COOK, H. C., 1996, The late Triassic and Early Jurassic of coastal sections in west Somerset and South and Mid-Glamorgan, *in* Taylor, P. D., ed., Field Geology of the British Jurassic: London, Geological Society of London, p. 9- 30

WIGNALL, P. B., 1991, Test of the concepts of sequence stratigraphy in the Kimmeridgian (Late Jurassic) of England and northern France: Marine and Petroleum Geology, v. 8, p. 430-441.

WIGNALL, P. B., 1994, Black Shales: New York, Oxford University Press, 127 p.

YOCUM, D. A., 1996, Iron oolites from the Cretaceous Interior Seaway: Albian Paddy Member of the Peace River Formation, Alberta, Canada, and Campanian Castlegate Formation, Colorado, USA: Unpublished M.S. Thesis, University of Wisconsin, Madison, 98 p.

YOUNG, T. P., PARSONS, D., AND AGGETT, J. R., 1991, The Frodingham Ironstone, *in* Young, T. P., ed., Jurassic and Ordovician Ooidal Ironstones: 13th International Sedimentological Congress, Field Guide 9, p. 32-50.

STRATIGRAPHIC CONDENSATION AND THE REDEPOSITION OF ECONOMIC PHOSPHORITE: ALLOSTRATIGRAPHY OF OLIGO-MIOCENE SHELFAL SEDIMENTS, BAJA CALIFORNIA SUR, MEXICO

KURT A. GRIMM*

Department of Earth and Ocean Sciences, University of British Columbia, Vancouver, British Columbia V6T 1Z4, Canada

ABSTRACT: Understanding the relationship between stratigraphic architecture, sedimentary processes, and the accumulation of economic phosphorites is fundamental to paleogeographic reconstruction and phosphorite mining concerns, and is the primary focus of this paper. Shelfal biosiliceous sediments from the middle Tertiary of Baja California Sur, Mexico, preserve a suite of features recording stratigraphic condensation, a distinctive association of coarse-grained, nearshore-derived sediments atop extensively condensed horizons, and the amalgamation of peloidal phosphorite event strata to form thick economic phosphorite. A disconformity-based, allostratigraphic approach is used to elucidate stratigraphic packaging and origin of these sediments.

Prominent rock types include silt-rich biosiliceous mudrock, phosphatic sandstone, peloidal phosphorite, and coquinoid silty limestone. Sedimentological data suggest mixed hemipelagic (suspension) and tractional event deposition in a muddy marine forearc that was influenced by coastal upwelling and marked shoaling in its uppermost part. Paleontological data and geochemical considerations suggest that concretionary horizons and a diversity of dolomitic hardgrounds are attributable to stratigraphic condensation; these features are described and ranked in relative complexity, a hierarchy interpreted as a proxy for relative intensity of each condensation episode.

Extensively bored dolomitic hardgrounds in the Timbabichi Formation are attributed to maxima in stratigraphic condensation, and are overlain by a very coarse lithofacies (VCL) that includes very coarse lithic arenite, coquinoid limestone, bored dolostone pebble conglomerate, and peloidal phosphorite. Each horizon juxtaposing the VCL atop an extensively condensed marine hardground is interpreted as a Type 1 sequence boundary, attributable to an abrupt decrease in accommodation volume; each of three such episodes resulted in seaward downlapping of coarse-grained, nearshore-derived sediments and redeposition of condensed sediments into more distal environments that were formerly settings of extensive stratigraphic condensation.

The relationship between allostratigraphic packaging and thick phosphorites is illuminating but problematic. The association of some phosphatic grainstones with cross-stratified barnacle coquinas suggests a nearshore origin for phosphorites that was adjacent to hard substrates suitable for barnacle proliferation. Conspicuous burrows produced by event-redeposited callianassid tracemakers in some phosphatic turbidites (Grimm and Föllmi, 1994) suggest that peloidal phosphorites were sourced from shoals that hosted these burrowing crustacea. Depositional and biological amalgamation of phosphatic event strata (shelfal turbidites) formed thick economic phosphorites. These conclusions provide essential insights for improving phosphorite exploration, reserve assessment, and production programs

INTRODUCTION

Much fascination concerning phosphorites derives from their hybrid character, which embraces their genesis as biogeochemical precipitates and the puzzling array of processes that govern their stratigraphic distribution. Phosphatic sediments (sediments that contain phosphatic minerals, generally as apatite) and phosphorite (apatite-enriched sediments; ≥ 18% P_2O_5, Glenn et al., 1994) occur in modern environments; however, these pale before their abundance and the diversity of their associations in ancient settings (Föllmi, 1994; Föllmi and Garrison, 1991; cf. Glenn et al., 1994). Furthermore, the economic and geopolitical implications of rock phosphorites as nonrenewable fertilizer ores highlight practical concerns about reconstructing their stratigraphic architecture, environmental settings, and the sedimentary processes that govern their geometry and distribution (Herring and Fantel, 1993; Grimm, 1997).

Oligo-Miocene sediments of Baja California Sur, Mexico, record the regional heterogeneity of sedimentary deposits influenced by coastal upwelling in the eastern Pacific (Grimm, 1992). In this paper essential features relevant to stratigraphic packaging, sedimentary processes, and paleogeographic settings of condensed beds, phosphorites, and associated sediments are presented. Elucidating the complex relationship between stratigraphic architecture, sedimentary processes and the preservation of economic phosphorite is the primary aim.

Conventional geologic methods and an allostratigraphic approach—one focused on bounding discontinuities and the paleogeographic significance of sediments enclosed by them (Walker, 1992)—were employed. A complete sequence stratigraphic interpretation is outlined in Grimm (1992); a simpler approach is presented here to enhance brevity, clarity, and general applicability of the results. In addition, this paper presents useful criteria for ranking degrees of stratigraphic condensation and summarizes sedimentological features of economic phosphorites.

Overview of Paleoenvironmental Setting

The ubiquitous presence of tuffs and related volcanogenic deposits in Oligo-Miocene sediments in Baja California Sur suggest deposition in an active forearc setting. Hausbach (1984) attributed volcanogenic deposits to eastward-dipping subduction of the Farallon Plate beneath the ancestral arc of the Sierra Madre Occidental on the Mexican mainland, which produced a general north-south strandline and a westward-facing shelf and slope (Grimm, 1992). The deposition of sediments in the study interval (Figs. 1 and 2) was post-dated by the onset of rifting of the proto-Gulf of California at about 12.5 Ma (Hausbach, 1984).

Sediments of the San Gregorio Formation (SG) are restricted to exposures of finely laminated and commonly petroliferous biosiliceous rocks (diatomite, porcelanite, and chert) in the

*email: kgrimm@eos.ubc.ca

Marine Authigenesis: From Global to Microbial, SEPM Special Publication No. 66

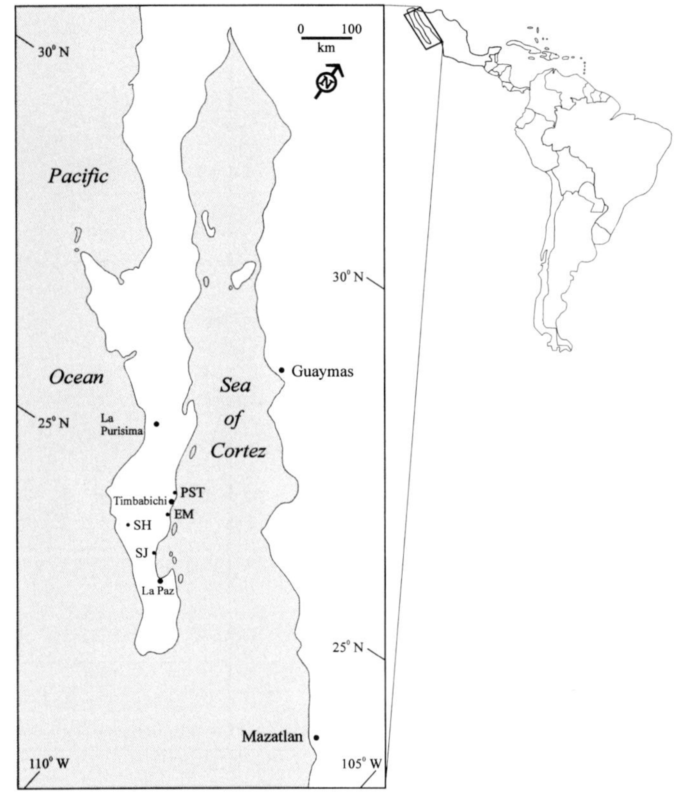

FIG. 1.—Locality Map, Baja California, Mexico (SJ= RoFoMex Mine at San Juan de la Costa; SH = San Hilario (Kilometro Cien); EM = El Mangle; PST = Punta San Telmo. Detailed locality maps and data in Grimm (1992).

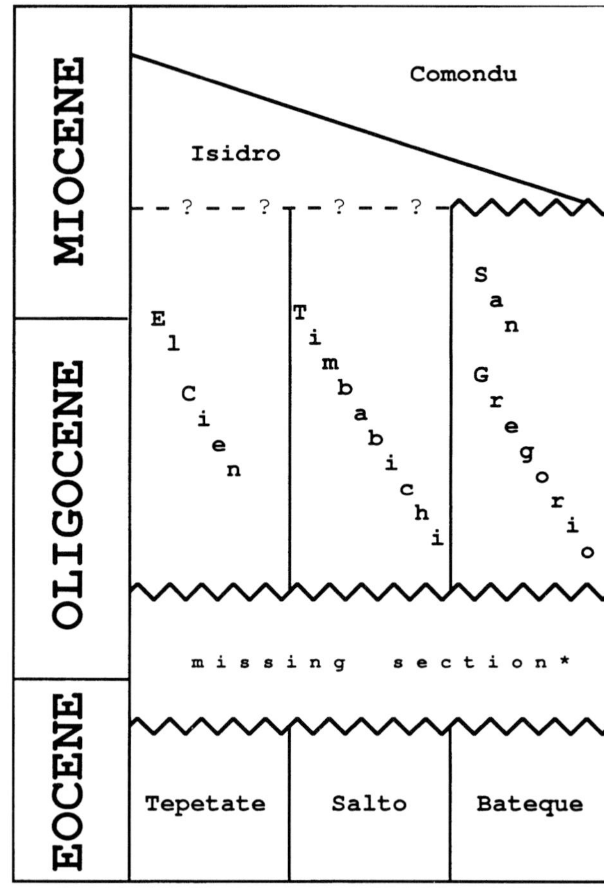

FIG. 2.—General Stratigraphic Diagram, Eocene-Miocene, BCS, Mexico. The Timbabichi Formation crops out between Punta San Telmo and El Mangle (Ensenada La Vibora); it disconformably overlies eolian sands of the Salto Formation, is conformably overlain by neritic sandstones of the Isidro Formation, which are in turn conformably overlain by Miocene volcaniclastic and epiclastic rocks of the Comondu formation (Grimm, 1992). The El Cien Formation crops out between San Juan de la Costa and San Hilario; the San Gregorio Formation crops out in the vicinity of La Purisima. The absolute age of the Timbabichi Formation is very poorly constrained (Grimm, 1992; cf. Schwennicke, 1995b)

vicinity of La Purisima (Applegate, 1986; cf. Fig. 1); these contain abundant tuffaceous strata, minor pristine phosphate, and many sharply based beds of allochthonous granular phosphorite (Grimm and Föllmi, 1994). Sediments in the San Gregorio Formation closely resemble Miocene hemipelagites of the Monterey Formation in California (Garrison et al., 1981), recording upwelling-influenced hemipelagic deposition in basinal or low-gradient basin slope settings within the oxygen minimum zone (Grimm, 1992; Grimm et al., 1990, 1991a,b).

Detrital-rich, biosiliceous sediments in southern Baja California Sur contain thick economic phosphorites and are assignable to the El Cien Formation (Fischer et al., 1995) and the Timbabichi Formation (TF; see Fig. 2, this paper). I concur with the stratigraphic assignments of Fischer et al. (1995); references to the El Cien Formation in this paper correspond to their designation of the San Juan Member of the El Cien (cf. Applegate, 1986). The designation of the Isidro Formation in this paper is equivalent to Fischer et al.'s (1995) Cerro Colorado Member of the El Cien Formation (Gidde, 1992).

Timbabichi Formation

I propose the designation Timbabichi Formation (TF) for a continuous belt of siliceous siltstones, siliceous mudrocks, silty phosphorites, and bored dolostone ledges, with minor very coarse sandstone, dolostone pebble conglomerate, coquinoid limestone, and phosphorite, in the vicinity of Timbabichi, Baja California Sur, Mexico (Figs. 3 and 4; Fig. 13 in Grimm, 1992).

The TF is a silt-rich marine deposit with phosphatic sandstones, phosphorites, and minor coquinoid silty limestone. Poorly preserved diatom frustules are present throughout; many samples contained gypsum from penetrative outcrop weathering, a phenomenon that may account for the absence of calcareous microfossils. Paleontological data—including the abundance of biogenic opal (as minor diatomite and abundant opal CT porcelanite), mixed tropical and cool water (California Current) planktonic forams and diatoms, ubiquitous fish scales, and an abundant vertebrate fauna including cetacean, sea cow, whale, turtle, and shark taxa—are summarized in Grimm (1992), suggesting that the EC and TF were influenced by California Current-dominated, coastal upwelling.

Applegate and Wilson (1976) first noted lithologic and faunal similarities between rocks of the El Cien Formation near San Hilario and coastal outcrops near Timbabichi at Punta San Telmo. Wilson (1979) summarized the history of this important

FIG. 3.—Outcrop photograph of the upper two-thirds of the proposed type section of the Timbabichi Formation at Timbabichi; note trench providing complete exposure. Ledge-forming dolostone beds are the prominent determinant of outcrop geomorphology in the TF; they contain synsedimentary fractures and abundant borings and are interpreted as marine hardgrounds recording stratigraphic condensation. Important stratigraphic surfaces are indicated at arrows; i = Ledge Seven; ii = base of porcelanite cliff; iii = base of the Isidro Formation. The portion of the TF shown in this photo is bracketed on the right margin of Fig. 4.

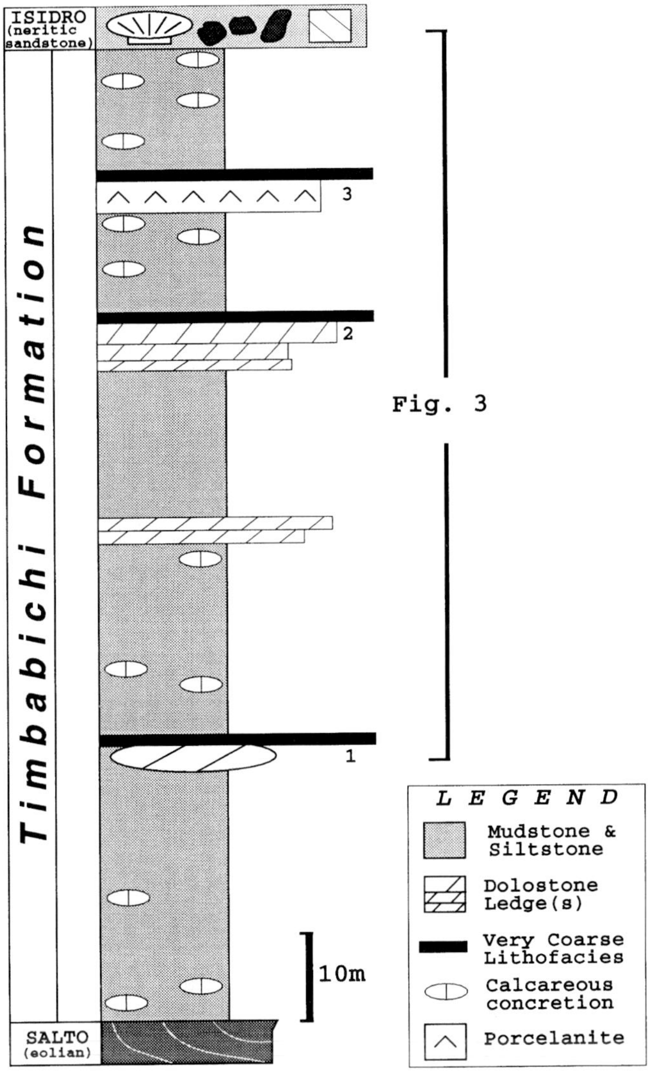

FIG. 4.—Summary stratigraphic diagram of the Timbabichi Formation at Timbabichi. Detailed section is presented in Grimm (1992). Intervals of particular interest are numbered: 1 = Lower Complexly Bored Ledge; 2 = Seven Ledge Cluster [uppermost ledge (Ledge Seven) is extensively bored]; 3 = Porcelanite Cliff (cf. Fig. 3). From left to right, symbols in Isidro Formation refer to molluscan coquinas, gravels and well-developed cross-stratification.

correlation, which led to the discovery of economic phosphorites at San Juan de La Costa.

Thickness of the type section of the TF on the north side of the broad arroyo at Timbabichi spans 120 m, nearly equal to the 115 m thickness reported for the El Cien Formation at San Hilario (Kim and Barron, 1986). I have observed the TF in the central, northernmost and southernmost exposures of an outcrop belt that spans 29 km. Field observations suggest lateral continuity of major bounding discontinuities, and a general uniformity of lithologic succession (Grimm, 1992). Peloidal phosphorite beds are a problematic exception to this generalization of continuity, since substantial changes in thickness and internal structure of phosphorite beds are evident over distances of ≥ 1–2 km.

Detrital sediments in the TF are dominated by recessively weathered siltstones and siliceous silty mudrock, with subordinate fine-sandstone, silty mudrock, and minor medium sandstone (Figs. 3 and 4). Channels are absent. Meter-scale packages of detrital sediment—each containing several to several tens of discrete depositional units—are distinctively tabular and laterally continuous. Massively bedded siltstones and silty mudstones are generally extensively weathered and void of sedimentary structures.

Sharply based, thinly laminated to thinly bedded siliceous siltstones and sandstones commonly occur interbedded with mudstone and silty mudstone (Fig. 5). Siltstones and fine sandstones commonly contain tuffaceous material and preserve tractional sedimentary structures, including sharp erosive bases, irregular plane lamination, and low angle cross-stratification (Fig. 5A and B). Ripple forms are rarely preserved (Fig. 5B). These tractional deposits are commonly overlain by lutites (muds and silty muds) that drape depositional topography with

FIG. 5.—Photographs of acid-etched slabs, Timbabichi Formation. A) Finely interstratified fine sand with light-colored dolomitic partings; note disruption and lateral discontinuity of some dolomitic partings; white arrows indicate probable burrows; stratigraphic thickness of specimens = 4 cm. B) Thickly laminated fine sand and light-colored dolomitic silty mud. Note tractional sedimentary structures [low-angle and irregular plane lamination in sandy laminae; ripple forms with planar tabular foresets (at arrow) are rare in the TF. C) Bold white arrow points in stratigraphic up direction. Note extensively bioturbated dolomitic silty mud; burrows are rare in the TF. Burrows are infilled with sand from the overlying stratum. Vertical length of specimen = 9 cm. D) Banded dolostone bed; note interstratification of sandy laminae with lighter colored dolomitic laminae. Structure at small white arrow is a probable burrow cross-cutting a lamina of dolomitic silty mud. Up direction as in C).

uniform thickness.

The suite of sedimentary structures suggests centimeter-scale interstratification of suspension deposits (muds) with thinly laminated to thinly bedded coarse silt, fine sand, and (rarely) medium sand, that were deposited from tractional, turbulent flows (Grimm, 1992). I interpret coarser interbeds as centimeter-scale tractional—probably density-driven—event strata ("shelf-turbidites"; Leithold, 1989; cf. Dott, 1983), in a distal, muddy-shelf environment; lutite drapes constitute post-event and inter-event hemipelagites. In addition, most bedded associations of mud with tractively interstratified silt and sand record upward-thickening of coarser, tractional laminae and upward increase in the relative proportion of tractively deposited laminae (cf. Fig. 13 in Grimm, 1992). Dolomite, chalcedony, and gypsum cements are present; poorly preserved diatoms in many samples suggest derivation of some chalcedonic cements by diatom dissolution.

Associated deposits include phosphatic siltstones and fine sandstones, sharply based, tabular beds that are massively bedded to crudely normally graded or rarely planar-tabular cross-stratified. Tuffaceous strata occur throughout, ranging in thickness from 1 cm to 1 m, and are commonly weathered. Phenocryst-bearing tuffs are absent. A marked increase in abundance of tuffaceous strata in the uppermost TF is associated with coarsening of the sediment and the appearance of well-developed planar tabular and trough cross-stratification. These changes are attributable to depositional shoaling in the upper TF, which culminated in the conformable transition into coquina-rich and cross-stratified coarse arenites and pebble conglomerates of the Isidro Formation (Grimm, 1992).

INDICATORS OF STRATIGRAPHIC CONDENSATION

The identification of breaks in sedimentary successions facilitates stratigraphic subdivision, correlation, and basin reconstruction. Recent expansion of sequence stratigraphic approaches generated considerable interest in lacunae and condensed sections; however, little attention has focused on the sedimentary processes directing their formation (cf. Clari et al., 1995; Moss and Tucker, 1996). A condensed section is defined as "a local record that is complete but thin relative to the same age strata elsewhere and reflects a period of low net sediment accumulation" (Kidwell, 1991). Condensed sections are related to some subaqueous disconformities that arise from sediment starvation and/or bypassing (Loutit, et al., 1988). Below, a continuum of concretionary horizons and ledge-forming dolomitic beds in the TF are described and attributed to slow or negative rates of sediment accumulation. These distinct horizons are the foundation for stratigraphic subdivision and interpretation of the TF and the phosphorites contained within.

CARBONATE CONCRETIONS

Thirteen concretion-bearing horizons are present in the TF type section (Fig. 4). These concretions are oblate ellipsoids flattened in the plane of bedding and range from pebble to cobble size. Most concretions are calcitic, however, some include minor siderite. Petrographic inspection reveals that concretions precipitated as void-filling cement in silts and silty muds. Some concretions contain conspicuous borings, simple cylindrical tubes ranging from 2–8 mm in width and ≥ 2.5 cm in maximum depth. Most bored concretions are bored only on their upper and lateral extremities; the absence of borings on the underside suggests that concretion formation was postdated by only partial exhumation (cf. Kennedy and Klinger, 1977). A concretion at the 100.0-m level was encrusted with barnacles.

The origin of concretions is a complex topic. Many concretions are postulated to form at very shallow burial depths, from several centimeters to several tens of meters beneath the seafloor (Baker and Burns, 1985; cf. Garrison et al., 1969). Differential compaction around carbonate concretions in many settings testifies to their precipitation prior to significant burial compaction of surrounding strata (Maples, 1986). The presence of carbonate concretions reworked into sedimentary breccias in the Miocene Monterey Formation of California (Garrison and Graham, 1984), as well as isotopic evidence (Kushnir and Kastner, 1984; Baker and Burns, 1985), all suggest precipitation of carbonate concretions at very shallow depths.

Complexly interrelated biogeochemical factors may contribute to enhanced carbonate precipitation during intervals of negative, zero, or very low sedimentation. Although this brief survey is not a comprehensive discussion of this complex topic [See Baker and Burns (1985); Coleman (1985); Raiswell (1987, 1988); and Taylor and Macquaker (this volume) for more complete discussions of geochemical factors], it highlights factors relevant to the formation of concretions and dolomitic hardgrounds.

Compton and Siever (1986) and Compton (1988) considered Mg^{++} flux from seawater to be the rate-limiting step for precipitation of authigenic dolomite concretions and dolomitic beds in the Miocene Monterey Formation. Since the diffusive flux of Mg^{++} (and other cations) decreases exponentially with increased burial depth, a depositional hiatus results in a longer residence time in the shallow burial setting of high Mg^{++} availability .

Middleburg et al. (1990) credited high alkalinity in the zone of anoxic methane oxidation for the initiation of dolomite precipitation in modern anoxic sediments of Indonesia; Raiswell (1987) reached similar conclusions concerning concretionary mudstones from the British Jurassic. Three microbially mediated reactions that occur with increased depth during shallow burial diagenesis are relevant:

1. Sulfate reduction: $2CH_2O + SO_4^{2-} \rightarrow 2HCO_3^- + H_2S$
2. Anoxic methane oxidation:
$$CH_4 + SO_4^{2-} \rightarrow HCO_3^- + HS^- + H_2O$$
3. Methanogenesis: $2CH_2O \rightarrow CH_4 + CO_2$

Reactions 1 and 2 produce bicarbonate ion, and thus increased alkalinity. Under conditions of steady sedimentation, the zone of high alkalinity migrates steadily upward, to maintain a constant depth beneath the seafloor. However, a break in sedimentation can stabilize the zone of excess alkalinity, or cause it to migrate vertically upward, nearer to the sediment/water interface (Raiswell, 1987, 1988). As a result, Middelburg (1990) concluded that anoxic methane oxidation—deriving sulfate from the overlying zone of sulfate reduction and methane from the zone of methanogenesis directly below—propelled elevated alkalinity and the initiation of dolomite precipitation.

In addition, Baker and Kastner (1981) suggested that dissolved sulfate in pore water inhibits dolomite precipitation and credited sulfate reduction (reaction 1) for dolomite precipitation. Furthermore, increased alkalinity also occurs in the zone of sulfate reduction, further favoring precipitation of dolomite and other diagenetic carbonates.

Many diagenetic scenarios for concretion formation are possible (see Taylor and Macquaker, this volume). However, the common factor throughout is a reduction in net sediment accumulation leading to increased alkalinity at shallow burial depth. Thus, fixing reaction 1 and/or 2 at a specific horizon during arrested sedimentation favors carbonate precipitation. Consequently, each concretionary horizon is attributed to carbonate precipitation during a pause or slowdown in sediment accumulation at very shallow depth beneath the seafloor.

Absence of significant change in the character of the sediments above and below each concretionary horizon suggests that no major changes in environment were associated with each hiatus. The formation of carbonate concretions on a generally quiet-water, muddy marine shelf records brief perturbations in sediment supply. Horizons with encrusted and/or bored carbonate concretions suggest shallow burial, concretion formation, partial exhumation by winnowing currents, biological colonization of the hard substrate, and subsequent reburial. These may reflect somewhat longer hiatuses than simple concretions that were unaffected by boring or encrustation (cf. Compton, 1988).

LEDGE-FORMING DOLOSTONE BEDS

Ledge-forming dolostone beds, which are pale yellow to tannish gray, are laterally persistent and occur in stratigraphically clustered intervals (Figs. 3 and 4). Dolostone beds in the TF are subdivided into banded and massive dolostone ledges. Banded

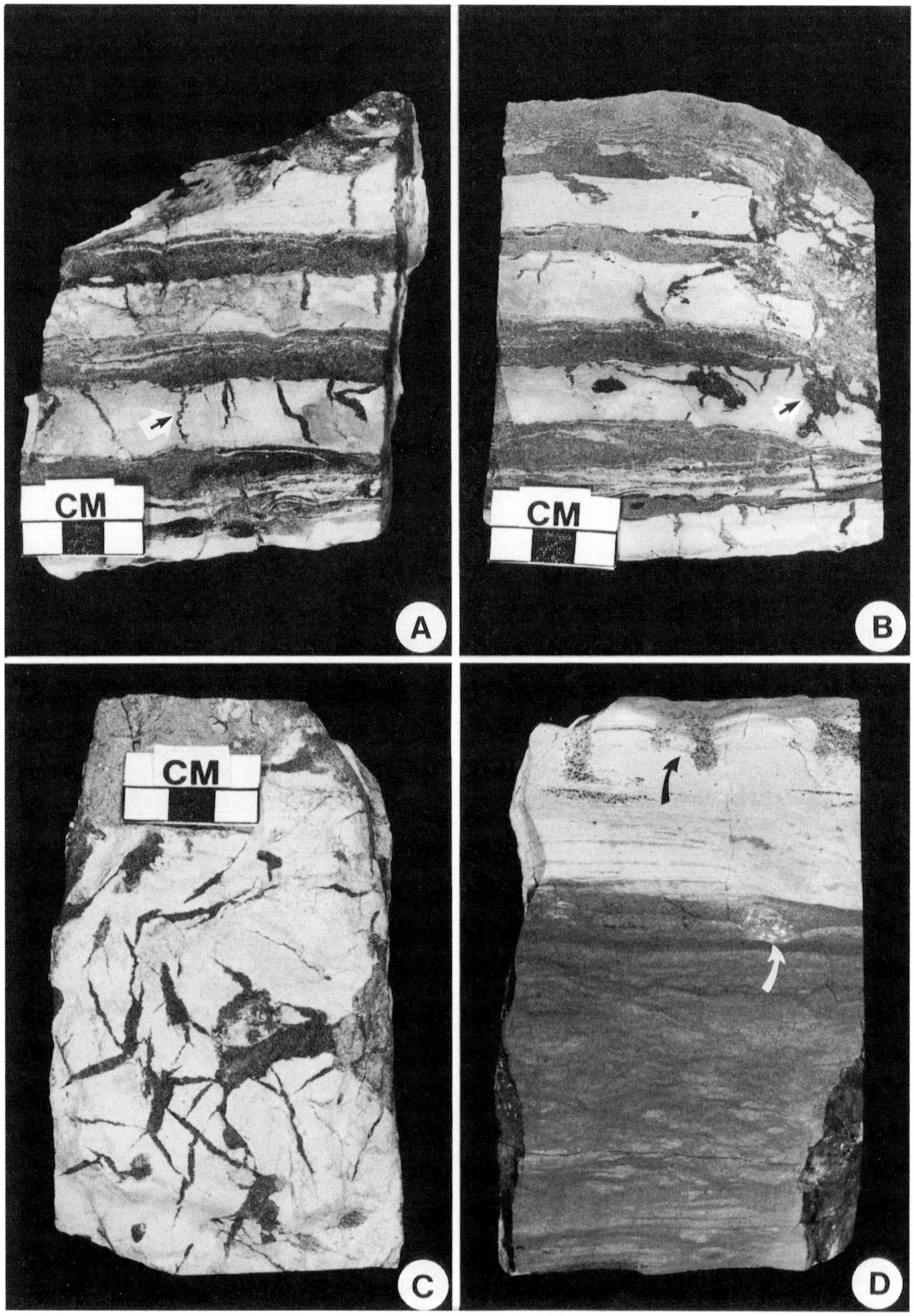

FIG. 6.—Banded dolostone beds in the TF; dolomitic mud and dolomitic silty mud in all specimens is light-colored. A) Thickly laminated dolomitic mud interbedded with fine sandstone; note irregular, wedge-shaped, vertical fractures. Crumpled fracture at arrow is comparable to "molar tooth structure" of Smith (1968). B) Banded dolostone bed with fractures and probable burrows (borings?); upper right corner of specimen shows erosional disruption of partially lithified dolomitic band, thus testifying to synsedimentary lithification. C) Plan view of "syneresis cracks" from the TF. D) Thick dolomitic band above darker sandy siltstone; structure at white arrow is a probable burrow infill; structure at black arrow is a probable boring that penetrated the dolomitic band while it was partially lithified. Mottled depositional fabric in lower two-thirds of specimen is of uncertain origin (dewatering?); vertical length of specimen = 18 cm.

dolostone ledges preserve centimeter-scale alternation of dolomitic silty mud with coarse silt to medium sand and commonly display internal fracturing (Figs. 5 and 6; 7A and B). Massive dolostone ledges lack internal fractures and prominent stratification features, and are characterized by boring and hard-substrate, epifaunal encrustation (Figs. 7C and D; 8-10). The

FIG. 7.—Dolostone beds in the TF. A) Light-colored dolostone bands with simple vertical fractures have minor offsets; middle of specimen shows downward-directed, irregular wedge-shaped fractures. In plan view, many fractures of this sort resemble Fig. 6C. B) Light colored dolostone bands in this specimen contain simple vertical fractures, complex vertical fractures, and probable penetration by burrowing and/or boring organisms. Arrows in upper right corner highlight the erosive basal contact of the overlying sandy stratum, thus demonstrating syndepositional lithification of dolostone bands. C) In some exposures, ledge-forming dolomitic beds contains sharply based stringers and lenses of fine sand (Ledge Seven). Arrows point to simple pholad borings infilled with very coarse lithic arenite; this horizon passes laterally to a complexly bored bed. D) Massive dolostone bed, containing lenses of sharply based sand, a partially crumpled and sand-filled vertical fracture, and a simple boring (b) .

interpretation of dolostone beds underlies genetic subdivision of the stratigraphy and is discussed below.

Banded Dolostone Beds

Banded dolostone beds preserve centimeter-scale depositional alternations of traction-deposited coarse silt to medium sand with suspension interbeds of mud and silty mud (Figs. 5 and 6; 7A and B). Silty lutites within banded intervals were preferentially dolomitized, a common feature in many ancient settings (Hardie, 1987). Most dolomitic bands contain fractures; some are penetrated by shallow burrows and/or shallow borings (Figs. 5C and D; 6B and D).

Fractures cross-cut discrete dolomite bands and are infilled with coarse sediment from the directly overlying layer. Fractured dolostone bands commonly preserve evidence of erosional reworking (Figs. 6B and 7B), thus testifying to syndepositional lithification and fracturing. Vertical fractures fall into three categories, simple, complex, and wedge-shaped.

Simple vertical fractures offset subhorizontal units of thickly laminated to very thinly bedded dolostone. Ductile bending of some dolomitic laminae suggests that some intervals were only partially lithified when fracturing occurred. In addition, the smooth concordance of sandy bands with deformed and fractured dolomitic bands indicates that sandy sediments were unconsolidated when fracturing of dolomitic bands occurred. Furthermore, the presence of fractured dolomitic laminae as clasts demonstrate their syndepositional origin (Figs. 6 and 7; cf. Shinn, 1969; Purser, 1969).

Complex vertical fractures are narrow, accordion-like fractures, generally < 4 mm in diameter (Fig. 6A-C; 7A and B). They resemble "molar tooth" structure described by Smith (1968); however, the irregular fractures he described are filled with calcite spar, whereas complex vertical fractures are infilled with sandy sediment from the overlying stratum.

The development of these accordion-shaped fractures records three stages of deformation: dilational opening of a vertical crack by brittle fracture, partial to complete infilling with

FIG. 8.—Massive dolostone beds with simple borings. A) Simple borings (medium-sized pholad borings) on upper surface of massive dolostone ledge. B) Massive dolostone bed with small- and medium-sized pholad borings; the overlying stratum contains abundant barnacle fragments and peloidal phosphatic grains, prominent components of the VCL (cf. Fig. 13; Table 2). C) Simple, flask-shaped pholad boring in light-colored massive, ledge-forming dolomitic bed; this marine hardground is overlain by sediments of the Very Coarse Lithofacies including bivalve fragments, tuffaceous pebbles, and peloidal phosphorite; Ledge Seven. D) Simple boring in dolostone bed with indistinct horizontal stratification. This hardground is overlain by sediments of the VCL, including abundant mollusc fragments, tuffaceous granules, and phosphorite grains. Note sandstone pebble (p); float specimen, probably from Lower Discontinuous Ledge.

sandy sediment from the overlying stratum, and compactional flattening of vertical fractures that resulted in tight to isoclinal folding. Compactional flattening of fractured dolomitic siltstones indicates that they were weakly lithified when vertical flattening strain was applied during burial compaction, further implying syndepositional lithification.

Irregular wedge-shaped/spindle-shaped fractures comprise a third class of vertical fracture. These downward-directed fractures possess an irregular wedge-shaped profile, that penetrated only partially downward through the dolomudstone stratum in which each occurs (Figs. 6A; 7A and B). In plain view, most of these fractures are discontinuous, linear to gently curved fractures that are tapered at their distal extremities; lengths of these isolated, spindle shaped fractures range from range ≥ 1–3 cm, and ≥ 5 mm in maximum breadth (Fig. 6C). Some fractures intersect. Erosional reworking testifies to their syndepositional origin.

Synsedimentary fractures of this latter sort are commonly referred to as syneresis cracks (cf. Figs. 2 and 3 in Donovan and Foster, 1972). Experimental work suggests that they form by subaqueous dewatering and/or extension of partially consolidated and cohesive, clay-rich sediment (Burst, 1965; Picard, 1966, 1969; Plummer and Gostin, 1981). Contractional shrinkage of semiconsolidated sediments at the seafloor or at shallow burial depth is best accomplished by osmotic dewatering. For example, if a clay-rich colloid is permitted to consolidate under hydrostatic and perhaps very shallow lithostatic overburden, and a higher salinity fluid is introduced directly above it, then the pore fluid of the sediment aggregate will be osmotically leached from above. The result of such osmotically driven fluid escape is sedimentary volume loss, shrinkage-induced contraction, and downward-directed fracturing. As a consequence, the presence of syneresis cracks in dolostone beds requires significantly variable salinity in a subaqueous environment (Plummer and Gostin, 1998; cf. Astin and Rogers, 1991). Consequently, syneresis cracks in the TF are interpreted as a record of synde-

relative duration and/or intensity of each condensation episode. Within the Seven Ledge interval of the TF type section (Figs. 3 and 4), dolomitic beds that lack borings are succeeded in turn by dolostone beds with simple borings and a banded dolostone ledge; the seven ledge interval is capped by an extensively and complexly bored dolomitic ledge (Ledge Seven) that records a maximum in intensity and perhaps duration of stratigraphic condensation. A comparable pattern described in the Cretaceous of southern England was attributed to a probable increase in water depth and associated decrease in environmental energy (Garrison et al., 1987).

Each of the three complexly bored horizons described above record the most extensive stratigraphic condensation in the TF. It is significant that these three horizons are overlain by a very distinctive association of coarse-grained sediments that occur only upon complexly bored horizons; they are termed the Very Coarse Lithofacies and are discussed in the following section.

VERY COARSE LITHOFACIES

Diagnostic components of the Very Coarse Lithofacies (VCL) are very coarse lithic arenite, bored dolostone pebble conglomerate, rounded lithic pebbles, bone fragments, coquina, and phosphorite (Figs. 11–13; Table 2). This association of sediment types occurs only within the VCL, which directly overlies the three most extensively condensed horizons. Several of the components of the VCL also occur only within the VCL; coquinoid limestone is a very prominent component of the VCL; however, it is not exclusively limited to the VCL, since it also occurs in the upper several meters of the TF. In addition, phosphatic grains are common and in some exposures the predominant component of the VCL; however, phosphatic grains are not exclusively limited to the VCL.

The internal stratigraphy of the VCL is complex and laterally variable; for example, (5 distinct and laterally discontinuous depositional units are amalgamated to form the VCL on top of Ledge Seven. Coarse-grained sediments of the VCL commonly infill large borings, yet some components that are preserved within borings may be sparse or locally absent in the directly overlying stratum. These coarse-grained "omission sediments" are included as part of the VCL; their preservation only within some borings suggests that sediment bypassing likely accompanied deposition of some sedimentary units within the VCL (Kidwell, 1991).

TABLE 2.—CONSTITUENTS OF VERY COARSE LITHOFACIES (VCL)

1. Very coarse, angular lithic arenite
2. Coquina (bivalve, barnacle)
3. Rounded lithic pebbles
4. Bored dolostone pebbles and cobbles
5. Marine mammal bone
6. Peloidal phosphorite

Key: The VCL occurs only directly overlying extensively condensed horizons (cf. Table 1); 1–3 records basinward shift of coastal lithofacies, 4–6 records reworking of previously condensed lithofacies.

Very coarse lithic arenite is generally angular to very angular, consistently lacking significant finer detritus. An abundance of strained quartz, polycrystalline quartz, plagioclase feldspar, and metamorphic lithics (Figs. 9D and 13A), as well as minor nucleated and nonnucleated phosphatic grains and sedimentary lithics, are evident in thin sections (Figs. 9D and 13A). These sandstones are compositionally and texturally immature, recording derivation from an unknown, likely proximal metamorphic source area. Very coarse lithic arenite is found only within the VCL.

Lithic pebbles and granules are a volumetrically minor, but significant member of the VCL (Fig. 11D). Most lithic pebbles are well rounded. Lithologies include rhyolite, basalt, chert, and well-indurated tuff. Lithic pebbles are found only in the VCL, and in the upper several meters of the TF. Energetic fluvial and/or nearshore processes are credited for the rounding of clasts.

Coquinoid limestones are the volumetrically dominant component of the VCL. They are not exclusively limited to the VCL, since coquinas also occur in the uppermost several meters of the TF, where it shoals into cross-stratified and fossiliferous sandstones, tuffs, and pebble conglomerates of the Isidro Formation (cf. Figs. 2-4). Molluscan rudstones and grainstones are predominant in coquinoid limestones of the VCL (Figs. 11 and 13D and E); however, barnacle grainstone and barnacle/peloidal phosphorite grainstone facies also occur (Fig. 13F; Grimm et al., 1991a). They are generally highly fragmented and densely cemented by calcite, making precise faunal identifications difficult. Coquinoid limestones in the VCL record turbulent redeposition of coarse-grained, bioclastic sediment from a nearshore setting into a more distal setting that was formerly the site of extensive marine condensation. Fragmented samples of *Diplochaetetes mexicanus* also occur in the VCL.

Bored dolostone pebbles and cobbles occur in clast supported associations and floating in a matrix of coquinoid limestone (Fig. 11 B-D). Many clasts are extensively bored on all sides, perhaps suggesting multiple episodes of transport and boring (Fig. 11C; cf. Kennedy and Klinger, 1972). Some dolomitic clasts contain tabular ("flat pebble") clasts that were only partially lithified when they were deposited (Fig. 11D). Bored dolostone clasts are identical to the complexly bored hardgrounds they overlie. Like other coarse components of the VCL, they testify to high competence of depositional flows.

Large bone fragments are common within the VCL (Figs. 12A and 13C; cf. Fig. 14C). They are generally broken and weathered, however the presence of haversian canals suggests mammalian origin. Abundant bone is common in condensed marine sections in other settings (Macquaker, 1994; Macquaker et al., 1996). Like the bored dolomitic pebbles with which they occur, bone fragments in the VCL may record erosional entrainment and redeposition of sediments derived from a condensed interval. Shark teeth are also present in some exam-

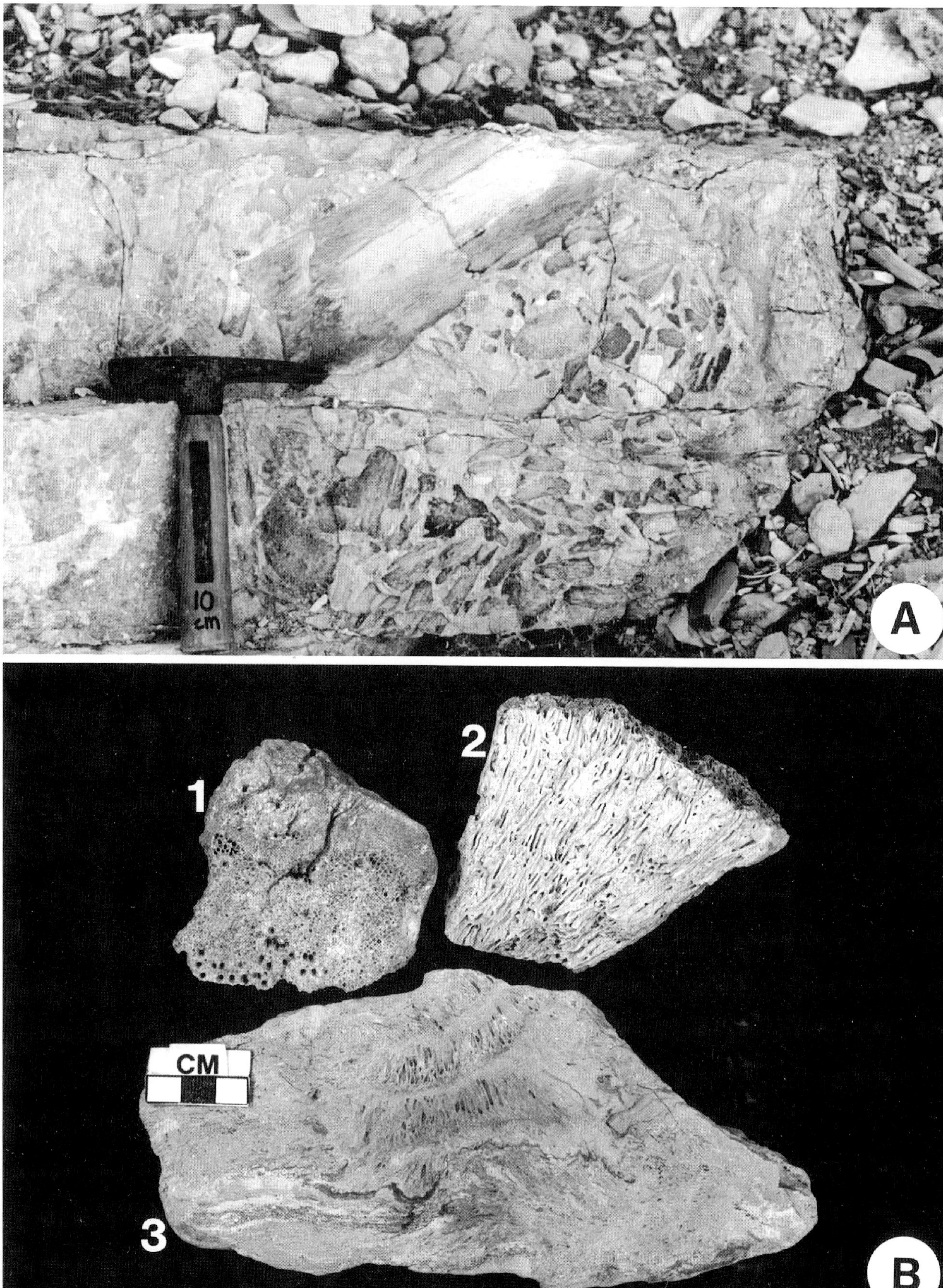

FIG. 12.—A) Bone-rich stratum at sole of thick phosphorite bed possessing coarse-tail grading, San Hilario (Kilometro Cien); redeposited bone beds are characteristic of the VCL (Table 2). B) *Diplochaetetes mexicanus* from conglomerate bed in upper El Cien Formation

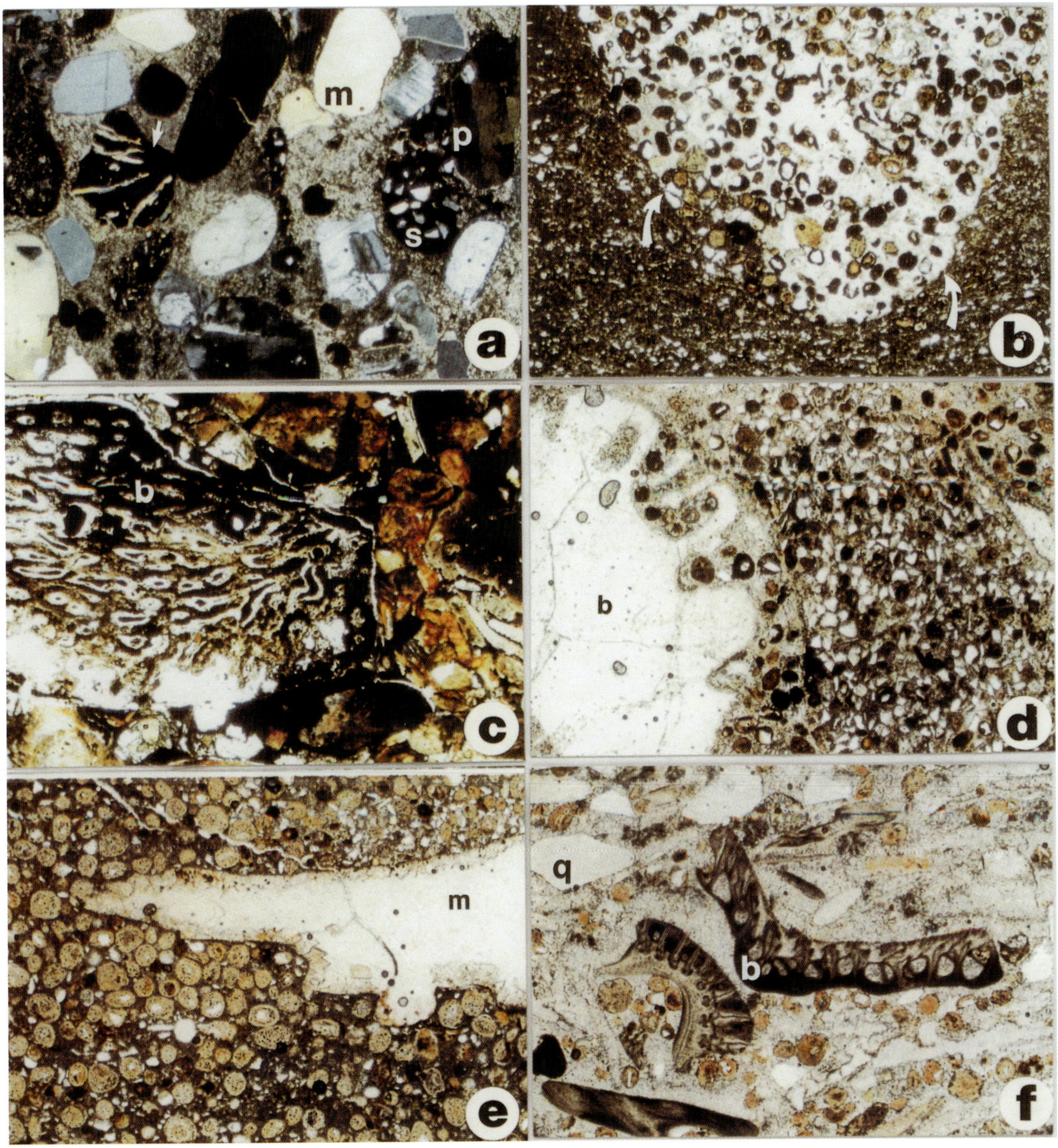

FIG. 13.—Photomicrographs of the Very Coarse Lithofacies (VCL). A) Very coarse lithic arenite: Note monocrystalline quartz (m), polycrystalline quartz (p), and sedimentary lithic fragment (s); white arrow points to sedimentary (mudrock) lithic with calcite (?) filled fractures. Crossed polars, white dot = 500 μm. B) Silty phosphorite infilling bored upper surface of dolomitic silty mud hardground. White arrows point to sharp margins of pholad boring; white dot = 500 μm. C) Bone fragment (b) and peloidal phosphorite grains in the VCL; white dot = 500 μm. D) Bivalve fragments (b) with probable bored margins in a matrix of silty phosphorite are common in the very coarse lithofacies; white dot = 500 μm. E) Coarse mollusc fragment (m) in a matrix enriched in simple nucleated and nonnucleated phosphorite grains (cf. Ledesma, 1992); note dolomite rhombs at the margin of a mollusc fragment; white dot = 500 μm. F) Barnacle fragments (b) with coarse detrital quartz (q) and simple nucleated and nonnucleated phosphorite grains; white dot = 500 μm.

FIG. 14.—Sedimentary features of the upper TF at El Mangle. A) Small "teepee structure" in banded dolostones, El Mangle; note folding of dolomitic lamination (d). Coin diameter = 2.6 cm. B) Dumbbell-shaped, Polydora boring (arrows) and pholad borings (b) on upper surface of thin dolomitic band associated with tepee structure at El Mangle; c.f. Fig. 14A (Scale bar = 1 cm). C) Vertebrae from massive silts at El Mangle, possibly *Cornwallius,* a Tertiary sea cow. Whole vertebral columns and scattered mammal bones are common in massive siltstones above the porcelanite cliff at El Mangle. 1= lateral view of well-preserved vertebra; 2= well-rounded (wave-abraded ?) specimen; 3= apical view of abraded vertebra. D) Internal molds of mussels (F. Mytilaceae) in life position, upper TF at El Mangle. These bivalves probably lived partly buried in a muddy substrate with their plane of commissure resting parallel to the sediment/water interface. Coin diameter = 3.1 cm.

ples of the VCL.

Nucleated and nonnucleated phosphatic grains (Ledesma, 1992; cf. Schwennicke 1995a) are present in all samples of the VCL (Figs. 11 and 13B-F). They are particularly abundant as matrix in some coquinoid limestones within the VCL and within some dolostone pebble rudstones. The common association of peloidal phosphorite with coquinoid limestones and other sediments in the VCL may reflect each's derivation from a shallow water environment.

Components and interpretation of the VCL are listed in Table 2. The VCL occurs only directly overlying the three dolomitic ledges that record extensive condensation. Very coarse lithic arenite, coquinoid limestone, and lithic pebbles in the VCL are attributable to abrupt basinward shift of nearshore depositional facies (Kidwell, 1991; Loutit et al., 1988). The presence of bored dolostone pebbles, phosphatic grains, and bone fragments suggests that reworking of previously deposited sediment accompanied this conspicuous shift. The deposition of the VCL upon bored, dolomitic hardgrounds recording maxima in stratigraphic condensation records a striking environmental change and is the key element for defining stratigraphic architecture of the TF.

ALLOSTRATIGRAPHY

An allostratigraphic unit is defined as a mappable stratiform body of sedimentary rock that is identified on the basis of its bounding discontinuities (Walker, 1992). Allostratigraphy focuses upon significant breaks in sedimentary successions (and their correlative conformities; Walker, 1992; Van Wagoner et al., 1988) and complements lithostratigraphic approaches, since it provides criteria for genetic packaging of sediments. Grimm (1992) presented an interpretation of allostratigraphic packaging in the TF, in the context of the sequence stratigraphy paradigm, that included definition of sequence boundaries, systems tracts, and parasequences. Contacts between complexly bored dolomitic ledges and VCL were defined as Type 1 sequence boundaries; individual concretionary horizons at the base of coarsening and thickening upward packages were defined as flooding surfaces constituting parasequence boundaries (Posamentier et al., 1988a, b; Van Wagoner et al., 1988; Loutit

G). Phosphatic crusts are uncommon but present locally (Fig. 4H). Phosphatic cement is also uncommon, but locally binds phosphatic peloids.

Bioturbation is commonly pervasive in the phosphorite lithofacies. Well preserved, locally silicified *Thalassinoides* burrows are characteristic (Fig. 4I), particularly towards the top of the Meade Peak section, where coarse phosphorite is interbedded on a submeter scale with phosphatic dolowackestone of the basal Franson Member (Figs. 3, 4J, and K). The lack of sedimentary structures in all phosphorite exposed naturally and observed in cut slabs is also attributable to pervasive bioturbation, as has been inferred for structureless phosphorite beds elsewhere (e.g., Glenn and Arthur, 1990).

Organic-rich phosphatic dolomite lithofacies.—

In sections west of and including Mackentire Draw along the southern flank of the Uinta Mountains, and in the Death Valley and Sheep Creek Canyon sections along the northern flank of the range, 4-10 m of organic-rich phosphatic dolomite makes up much of the Meade Peak Member (Fig. 3). Measured TOC values of the organic-rich dolomite range from 0.29% to 2.04%, though these are probably minimum values due to outcrop weathering. Millimeter-scale lamination is the dominant sedimentary structure, though millimeter- to centimeter-scale scours have disrupted this lamination locally (Fig. 4L). Faint, oval-shaped features no larger than a millimeter across are discernible in places. We interpret these features as poorly preserved *Chondrites* burrows. *Chondrites* burrows were also reported by Hendrix (1987) from laminated, organic-rich shale from the Meade Peak in southeastern Idaho. Overall shoaling within organic-rich dolomite of the Uinta study area is suggested by the vertical variation in physical and biogenic structures. Continuous, millimeter-scale laminae are most common toward the base of the lithofacies, and discontinuous laminae, *Chondrites* burrows, and small-scale scours are more typical of

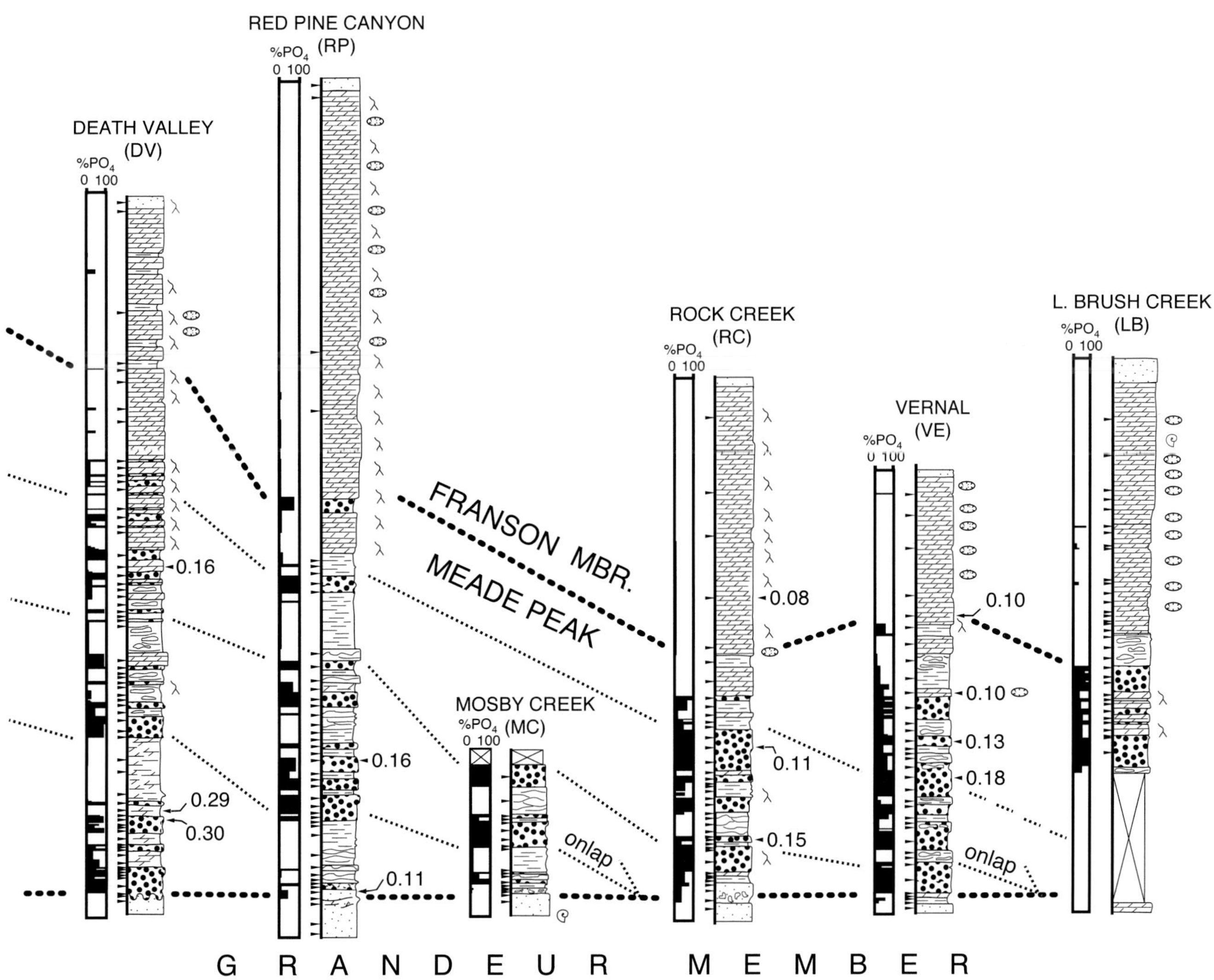

FIG. 3.—Continued.

the upper portion of the lithofacies.

Within the organic-rich dolomite, phosphate occurs most commonly both as irregular blebs and stringers that occupy the interstices between dolomite rhombs and as isolated peloids within the dolomite matrix (Fig. 5). Phosphatic blebs are usually no larger than 100 microns long and are commonly elongate parallel to bedding. To verify their phosphatic composition, we performed wavelength dispersive spectrometric analysis of several samples containing the irregular blebs of phosphate. Fig. 5A is a backscatter image of a polished thin-section of organic-rich dolomite. White areas on the image indicate high backscatter due either to phosphate or diagenetic Fe oxides. To distinguish between the two compositions, the same field of view is presented in the form of an X-ray map in which the microprobe is tuned for P (Fig. 5B). In the X-ray map, phosphatic phases return X-Rays, indicated by the white dots, whereas Fe oxides return no X-Rays. Examination of the X-ray map indicates that irregular blebs and stringers of phosphate are common throughout the sample. As discussed below, we interpret these blebs and stringers to be phosphate that has precipitated *in situ* (that is, "pristine phosphate" of Föllmi et al., 1991).

To identify the specific phosphatic mineral present within the organic-rich dolomite and granular phosphorite lithofacies, we performed XRD on several bulk samples of each lithofacies. In both lithofacies, the main phosphatic mineral phase is carbonate fluorapatite.

Chert facies.—

Well bedded, nonphosphatic chert, representing a distal tongue of the Rex Chert Member of the Phosphoria Formation (McKelvey et al., 1959), is present at the top of the Meade Peak in sections west of and including Mackentire Draw (Fig. 3). We interpret this chert to be biogenic, due to the local presence of recrystallized sponge spicules. Primary sedimentary structures in the chert are minimal and appear to be limited to a local, vague lamination that is best expressed in sections west of Dry Canyon. East of Mackentire Draw, silicified phosphorite has the outcrop appearance of bedded chert but in this case the silica is clearly secondary. Bedded chert is not present in sections along the northeastern flank of the Uintas.

Biogenic chert from sections west of Mackentire Draw appears to grade east into silica-cemented phosphorite containing progressively greater percentages of phosphatic allochems. Chert of the Hanna and Blind Stream sections is devoid of phosphatic grains. Chert of Mackentire Draw contains local floating phosphatic grains. Silicified phosphorite of the Red Pine Canyon, Mosby Creek, and Rock Creek Canyon sections commonly contains phosphate clasts in grain support. Phosphatic peloids comprise 95% of the grains by visual estimate within the silica-cemented phosphorite of these sections, though phosphatic shelly debris is also locally common. A very high degree of sorting is also characteristic of silicified phosphorite from these sections, as is pervasive bioturbation by both *Thalassinoides* and *Chondrites* burrows (Hendrix, 1987).

Abundant secondary chert is present in all measured sections, where it occurs as cement, irregular blebs and stringers, or silicified *Thalassinoides* burrows. East of Rock Creek, chert is present as isolated nodules within beds of phosphorite and gray shale.

Gray shale lithofacies.—

Gray to green shale, which is locally both phosphatic and dolomitic, occurs in every measured section in the study area. In the westernmost sections and along the northeastern flank of the Uintas, gray shale is present above the organic-rich dolomite facies, where it is interbedded at a submeter scale with phosphorite and phosphatic dolowackestone (Fig. 4J and K), though beds of shale locally persist above the last appearance of phosphorite (Fig. 3). East of Dry Canyon, gray shale is interbedded on a submeter scale with phosphorite in the upper part of the Meade Peak Member (Fig. 3). The gray shale facies weathers recessively in all sections.

Primary sedimentary structures in the gray shale facies are limited to several occurrences of millimeter-scale lamination in the upper part of the Sheep Creek Canyon section and in the Red Pine Canyon and Mosby Creek sections. *Chondrites* and silicified *Thalassinoides* burrows are present locally. Displacive chert nodules are common and deform both laminae and burrows. No concretions of dolomite or siderite were observed. Throughout the study area, phosphatic peloids, intraclasts, shelly debris, and steinkerns are common within the gray shale facies. Unlike the organic-rich phosphatic dolomite facies, all phosphatic grains

FIG. 4.—Outcrop, slab, and thin-section characteristics of strata from the study area. A) Slab of the contact between dolomite of the uppermost Grandeur Member and phosphorite of the Meade Peak Member at Farm Creek. Dolomite of the uppermost Grandeur is not phosphatized and is interpreted to be a carbonate hardground at this locality. Note the irregularity of the contact, the presence of borings (b) infilled by phosphorite from the overlying Meade Peak, and the inclusion of clasts of the Grandeur dolomite (d) in phosphorite of the basal Meade Peak. This contact is interpreted to be a transgressive surface. Smallest divisions at the base of the photo are 1 mm. B) Slab of the Grandeur - Meade Peak contact at the type locality of the Phosphoria in southeastern Idaho. The uppermost Grandeur is a phosphatized hardground at this location. Note the reworking of hardground clasts in the basal Meade Peak. Smallest divisions at the base of the photo are 1 mm. C) Thin-section photomicrograph of the Grandeur-Meade Peak contact at the type locality of the Phosphoria in southeastern Idaho. Contact is highlighted by white arrows. The uppermost Grandeur member is a hardground with standard omission fabrics such as boring and truncation of grains by the omission surface. A phosphatized, truncated bryozoan is visible in the center of the photograph just below the contact. Scale bar is 1 mm long. D) Thin-section photomicrograph of quartz-bearing phosphorite collected from the base of the Meade Peak, Hanna. This unit is interpreted to be a transgressive lag and is characterized by phosphatic coated grains (c), abundant phosphatized shelly debris (s), phosphatic peloids (p), and reworked quartz (q) from the uppermost Grandeur member and Pennsylvanian Weber Sandstone. Scale bar is 1 mm long. E) Slab of poorly sorted phosphorite packstone and phosphorite wackestone, Hanna. The interstitial material is dolomite. Note the abundance of coarse bivalve steinkerns (s). Smallest divisions at the base of the photo are 1 mm. F) Thin section photomicrograph of poorly sorted peloidal phosphorite with an abundance of gastropod steinkerns (s), Hanna. Steinkerns are one of the major grain types in Meade Peak phosphorites in the study area. Scale bar is 1 mm long. G) Thin section photomicrograph of peloidal, intraclast-bearing phosphorite packstone, Mackentire Draw. The bimodal grain-sized distribution shown here is characteristic of many of the Meade Peak phosphorites. Scale bar is 1 mm long. H) Slab of phosphatic crust, Mackentire Draw. The crust is the source for abundant intraclasts shown in the slab. Smallest divisions at the base of the photo are 1 mm.

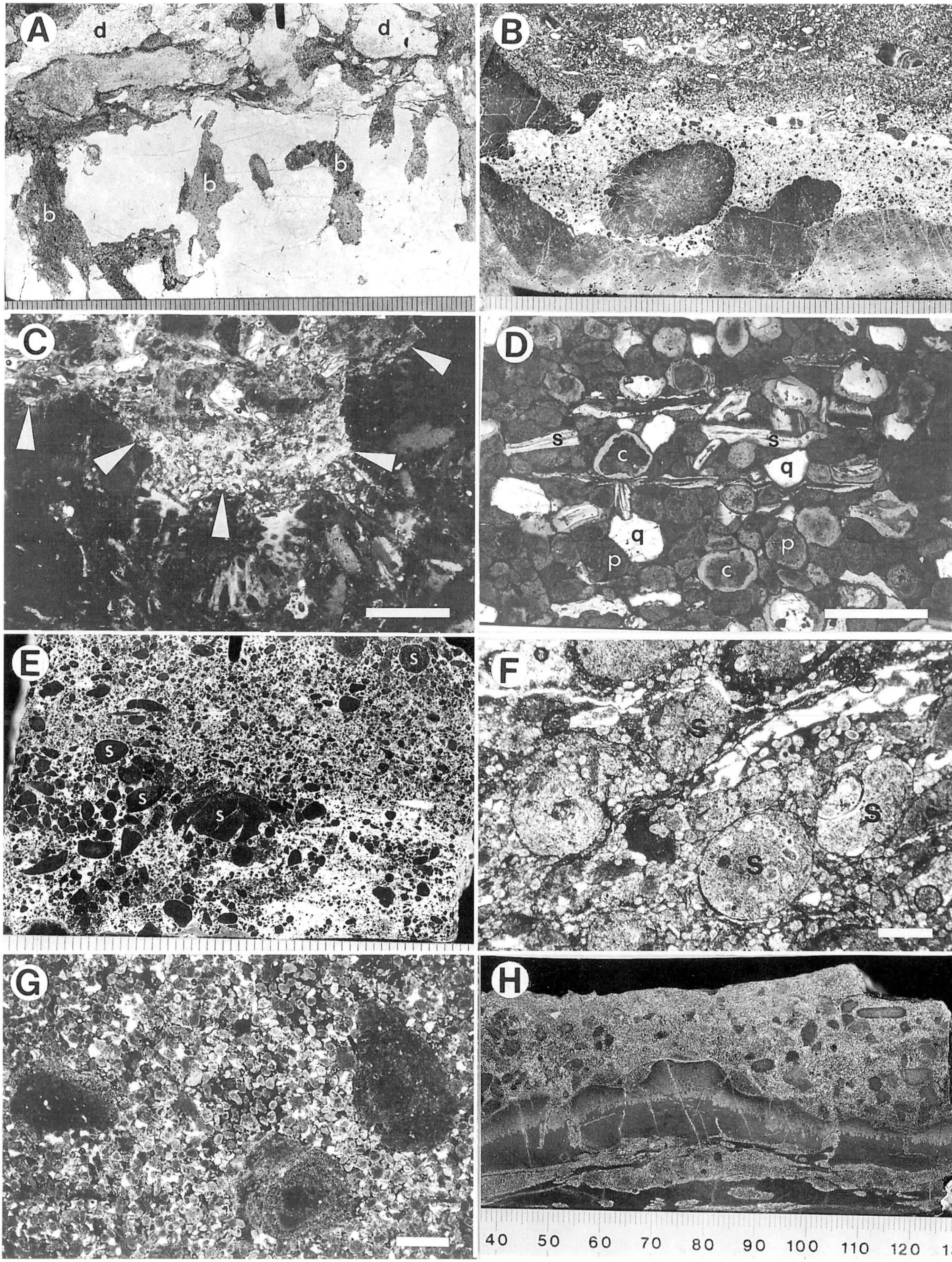
A
d
d
b
b
b
b
B
C
D
s
s
c
q
q
p
p
c
E
s
s
s
s
F
S
S
S
G
H
40
50
60
70
80
90
100
110
120
13

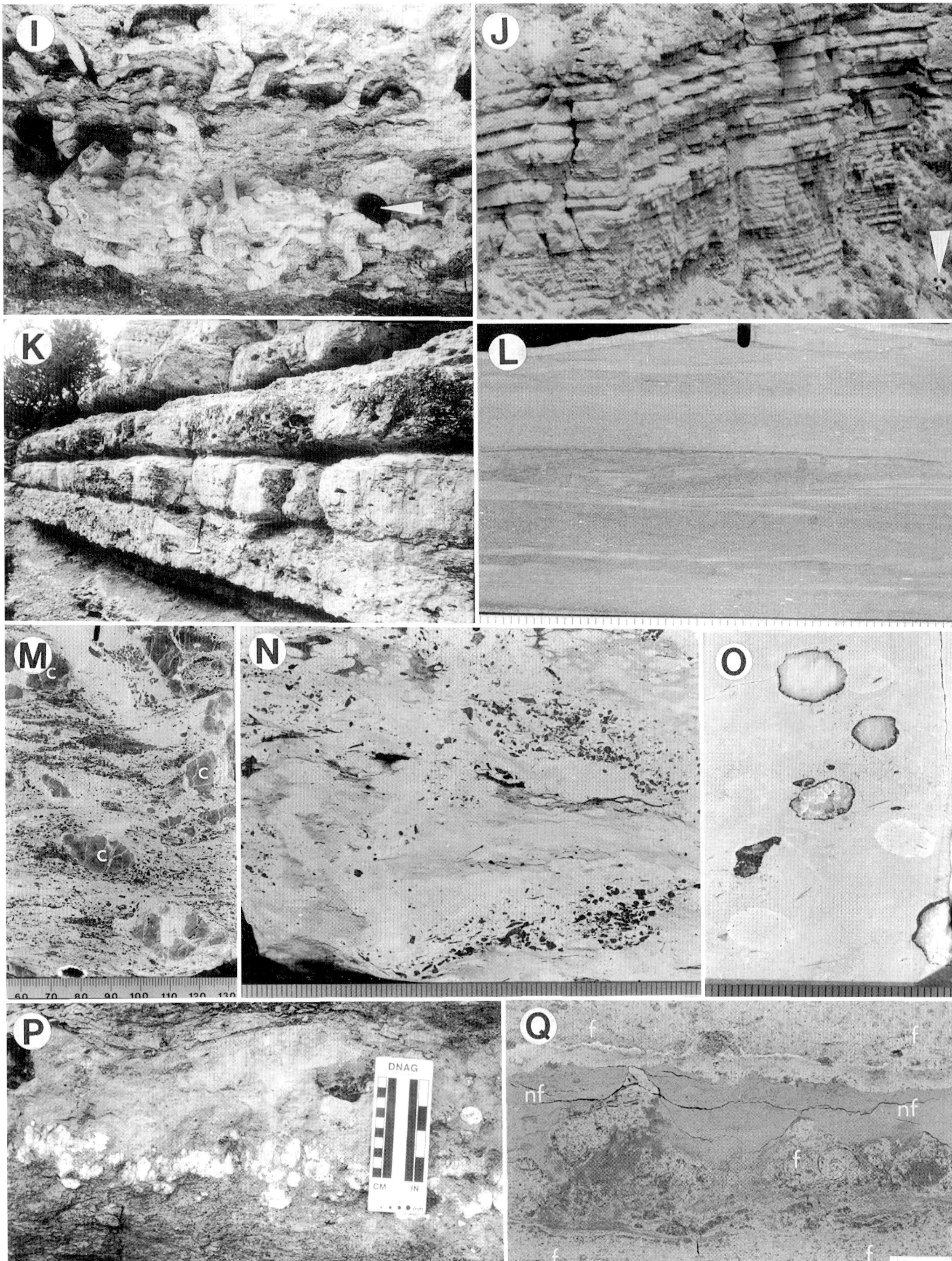
I
J
K
L
M
c
c
c
60 70 80 90 100 110 120 130
N
O
P
DNAG
CM
IN
Q
f
f
nf
nf
f
f
f

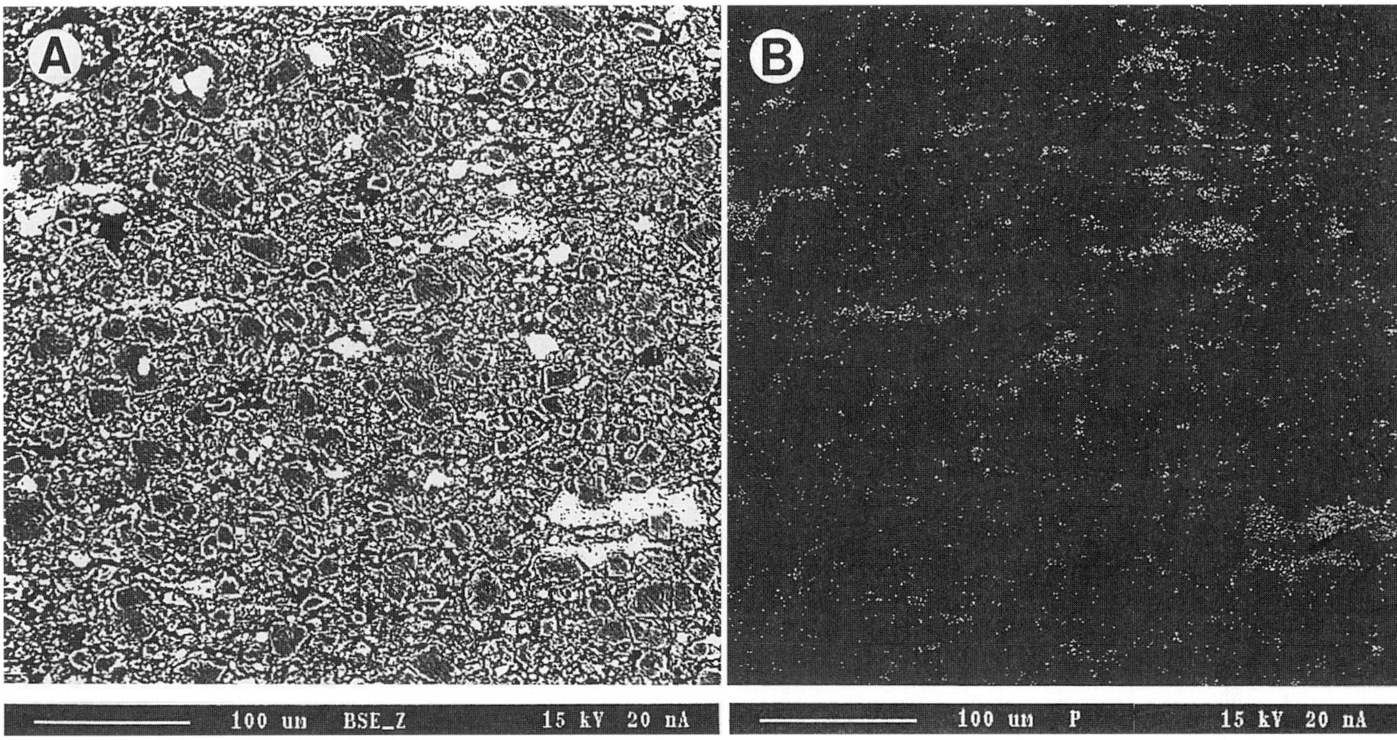

FIG. 5.— (A) Electron backscatter image derived from wavelength dispersive spectrometric analysis of a sample of organic-rich dolomite (Blind Stream). White areas are composed of either phosphate or Fe-oxide. Polygonal gray patches are dolomite rhombs. (B) X-ray image derived from P-tuned wavelength dispersive spectrometry of the same imaged area shown in Fig. 5A. Areas that are P-rich are shown by concentration of white dots; Fe-oxides do not return X-rays; compare to Fig. 5A. Areas of high P are structureless blebs and stringers interpreted to have grown authigenically *in situ*.

FIG. 4 Continued.— I) Outcrop character of phosphorite with abundant Thalassinoides burrows near the Meade Peak-Franson contact, Hanna. Lenscap (highlighted with white arrow) for scale. J) Sub-meter scale interbedded phosphorite, green shale, and phosphatic dolowackestone of the Meade Peak-Franson transition zone, Little Sheep Creek Canyon section. White arrow points to person at base of cliff for scale. The phosphorite and green shale lithofacies weather recessively, whereas the phosphatic dolowackestone weathers resistantly. These submeter-scale interbeds are laterally continuous on a kilometer scale and are interpreted to represent high frequency changes in the position of sea level with possible associated climate change that affected the shelf. K) Decimeter scale interbeds of gray shale and phosphorite (recessive layers) and phosphatic dolowackestone (resistant layers), Hanna section. This style of interbedding is characteristic of the Meade Peak-Franson gradational contact. Hammer (highlighted with white arrow) for scale. Note that the location of this photograph is over 100 km from that of Figure 4J, attesting to the lateral continuity of this bedding style. L) Slab of organic-rich dolomite facies, Blind Stream. Note the discontinuous millimeter-scale lamina and dark color, indicative of abundant preserved organic matter. Smallest divisions at the base of the photo are 1 mm. M) Bioturbated, partially silicified phosphatic dolowackestone, Hanna. The chert (c) is replacing Thalassinoides burrows. Note the presence of a large burrow, unreplaced by chert, at the top of the slab and the presence of abundant sand-sized grains of phosphate, many of which are molluscan steinkerns. Smallest divisions at the base of the photo are 1 mm. N) Thoroughly bioturbated phosphatic dolowackestone, Dry Canyon. Note the presence both of Thallassinoides burrows, partially infilled with sand-sized phosphate grains, and Chondrites burrows. Smallest divisions at the base of the photo are 1 mm. O) Close up of calcite filled molds after anhydrite nodules, Little Brush Creek. Note the pyrite rim around each of the molds. Smallest divisions at the base of the photo are 1 mm. P) Concentration of calcite-filled anhydrite molds along a bedding horizon, Vernal phosphate mine. Q) Cut slab of phosphatic dolomite showing centimeter-scale interbedded fossiliferous dolomite (f) and nonfossiliferous dolomite (nf; Little Brush Creek). Spalling of fossiliferous dolomite probably resulted from desiccation. This fabric is interpreted as representing deposition in a periodically wetted supratidal environment. Scale bar is 1 cm long.

ing, such as abraided margins; phosphate does not appear to have formed in situ. In addition to phosphatic grains, silt-sized grains of detrital quartz and mica are common.

Lithofacies of the Franson Member

Phosphatic dolowackestone facies.—

Massive cliff-forming exposures of phosphatic dolowackestone with little organic matter (TOC = 0.08 – 0.19%; n = 8) comprise most of the Franson Carbonate Member around the Uinta Mountains. This lithofacies also occurs as decimeter-scale interbeds with gray shale, chert, and phosphorite within the uppermost Meade Peak Member (Figs. 3, 4J and K). Few traction transport structures are present in the dolowackestone facies, probably due to abundant *Thalassinoides* and *Chondrites* burrowing (Fig. 4I, M, and N). Phosphate in the dolowackestone facies occurs as reworked granule-, sand-, and silt-sized clasts (for example, Fig. 4M and N). Phosphate is most abundant towards the base of the lithofacies where grain-supported phosphatic peloids and steinkerns locally comprise up to 80% of the rock by visual estimate. The percentage of phosphatic components decreases a few meters upsection, however, and the majority of the Franson Member is best characterized as a phosphatic dolowackestone. In most sections, the top of the Franson is devoid of phosphatic grains (Fig. 3).

Throughout the Franson in the eastern part of the study area and at the top of the Franson in all other sections, calcite-reduced and calcite-filled molds after anhydrite nodules are reduced and calcite-filled molds after anhydrite nodules are common (Figs. 3, 4O and P). The molds, which have subequant dimensions with bedding-parallel long axes, are commonly concentrated along specific horizons within the dolowackestone facies (Fig. 4P). Most molds have a pyrite rim typically no thicker than 0.5 mm (Fig. 4O). Commonly associated with the moldic dolowackestone is a low diversity faunal assemblage apparently consisting entirely of molluscs (Yochelson, 1968), consistent with evaporitic conditions. In the extreme eastern section at Little Brush Creek (Figs. 1 and 3), evidence of periodic wetting and desiccation is present in that finely laminated, locally silicified dolomite containing molluscan bioclasts is interstratified at a centimeter-scale with more massive dolomite horizons devoid of fossil fragments. Commonly, the top of each mollusc-bearing layer is cracked and reworked locally as clasts that are incorporated into the massive dolomite (Fig. 4Q).

Vertical and Lateral Facies Architecture

Several observations with respect to vertical and lateral facies architecture are key to the genetic interpretation of the Meade Peak and Franson Members. These trends in facies architecture are summarized below.

The Meade Peak Member changes laterally across the study area. In sections west of Rock Creek Canyon and in sections on the northeastern flanks of the Uinta Mountains, organic-rich phosphatic dolomite comprises much of the Meade Peak Member (Fig. 3). Between Mackentire Draw and Rock Creek Canyon, this facies pinches out, and the Meade Peak is made up primarily of granular phosphorite with interbedded gray shale. Where exposure permits (for example, artificial exposures in the Vernal phosphate mine), these interbeds are laterally continuous on a kilometer scale.

Perhaps most visually striking in outcrop is a series of submeter to meter scale repetitive interbeds of phosphorite, gray shale, and phosphatic dolowackestone lithofacies that mark the gradational contact between the Meade Peak and Franson throughout the study area (Figs. 3, 4J and K). The exact sequence of lithofacies involved in the interbeds changes vertically across the gradational contact. In most sections, interbedded gray shale and granular phosphorite passes vertically upward into interbeds of gray shale, phosphorite, and phosphatic dolowackestone, which in turn pass into interbeds of phosphorite and phosphatic dolowackestone or gray shale and phosphatic dolowackestone (Fig. 3). Importantly, individual beds are laterally traceable for up to 1 km with no appreciable change in thickness; there appears to be no evidence of thickening or thinning due to scour, channelization, or local aggradation. Extensive *Thalassinoides* burrowing characterizes the granular phosphorite interbeds and commonly projects down into underlying gray shale or phosphatic dolowackestone interbeds that are also extensively bioturbated (Fig. 4I). In many instances, *Thalassinoides* burrows in the phosphatic dolowackestone beds are filled with granular phosphorite from the overlying phosphorite bed.

Strata at the level of the Meade Peak-Franson transition zone thicken upward notably in virtually all sections (for example, Fig. 4J). Within the transition zone itself, individual bed thicknesses are typically at a sub-meter to meter scale, whereas within the Franson Member itself, beds thicken to meter scale.

SEDIMENTOLOGIC INTERPRETATIONS

Evidence for West-facing Ramp Environment

Several independent lines of evidence suggest that the Grandeur-Meade Peak-Franson system developed on a gently sloping ramp system that faced west during Permian time (Fig. 6A). The Grandeur Carbonate Member of the Park City Formation, dated as Roadian (late Leonardian) in age (Wardlaw et al., 1979; Wardlaw and Collinson, 1979a), thickens westward in the field area and is characterized by a westward increase in the proportion of limestone over sandstone (Jado, 1980). Along the southern Uintas, the Grandeur pinches out east of the Mackentire Draw section, and the basal Meade Peak rests unconformably on Pennsylvanian to Wolfcampian eolian sandstone of the upper Weber Formation (Peterson, 1980b). In the vicinity of the Uinta Mountains, it is unclear whether the eastern terminus of the Grandeur Member is a function of erosional truncation or depositional pinch-out. However, a widespread, locally angular intra-Leonardian unconformity has long been recognized across the Rocky Mountains and Midcontinent regions (Thomas, 1940; Merriam, 1963). In their regional synthesis of Permian geology of the Rocky Mountains, Rascoe and Baars (1972) interpreted the unconformity as representing a "marked change in the tectonic framework of the area." Based on these regional data, as well as our own observations, we interpret the basal Meade Peak contact to be an angular unconformity at the scale of the study area (Figs. 3 and 6B).

Within the Meade Peak Member and its equivalents, facies considerations suggest a ramp-style basin that faced the ocean to the west (Fig. 3). These considerations are: 1) westward thickening of the Meade Peak Member; 2) development in the western part of the field area of fine-grained, laminated, organic-rich dolomite with *in situ* phosphate ("pristine phosphatic sediment" of Föllmi et al., 1991); 3) coeval development in the eastern part of the field area of granular phosphorite in which all grains show evidence of reworking ("condensed phosphatic strata" of Föllmi et al., 1991); and 4) deposition of evaporite facies in the equivalent Goose Egg Formation east of the study area in western Colorado (Jado, 1980).

A west-facing ramp environment during Franson time can also be inferred from available facies data. West of the study area, the Franson is equivalent to the Gerster Limestone and the upper part of the more basinal Rex chert; all three have been dated as Wordian (lower Guadalupian; Wardlaw and Collinson, 1979a). In the western part of our study area, the Franson contains a normal marine faunal assemblage that includes brachiopods, echinoderms, sponge spicules, and green algae, attesting to water depths within the photic zone. In the eastern part of the study area, calcite-filled molds after anhydrite nodules suggest evaporitic conditions, which Jado (1980) confirmed for sections of the Franson east of our study area.

Interpreted Depositional Conditions

Grandeur-Meade Peak contact.—

The presence of a normal marine fauna within the upper Grandeur of the Uinta Mountains (Jado, 1980) as well as in the uppermost Grandeur at the Phosphoria type locality (Yochelson, 1968) is consistent with a well oxygenated, warm water carbonate ramp environment. Carbonate hardgrounds of the uppermost Grandeur Member within the Uinta Mountains are interpreted to have been deposited under well-oxygenated conditions in relatively shallow water, well within fair-weather wave base, and likely within the photic zone, as suggested by local stromatolite development (Fig. 3; Jado, 1980; Hendrix, 1987). Occasional subaerial exposure is supported in places by the presence of fenestral fabrics and autoclast development (Hendrix, 1987). Sedimentation rates associated with the hardgrounds are interpreted to be low, characteristic of carbonate omission surfaces (e.g., Glenn and Kronen, 1993). Following hardground development, transgression of the ramp apparently resulted in scouring and local deletion of hardgrounds (Fig. 3); reworking of carbonate hardgrounds from the Grandeur Member and sand grains from the uppermost Weber Formation; deposition of a basal conglomerate; and concentration of coarse, quartz-bearing, phosphatic shelly debris as a transgressive lag.

In contrast to uppermost Grandeur carbonate hardgrounds in the Uinta Mountains study area, uppermost Grandeur hardgrounds of the Phosphoria type section are characterized by wholesale replacement of carbonate grains by phosphate (Fig. 4C). This observation indicates that phosphogenesis occurred very close to the sediment water interface, consistent with depth limitation of phosphate precipitation due to lattice poisoning of phosphatic phases by CO_3^{2-} ions at depths greater than several centimeters (Glenn and Arthur, 1988; Glenn, 1990) and/or the diffusive-access limitations of fluorine required in carbonate fluorapatite (Froelich et al., 1988).

The vertical facies succession at the Grandeur-Meade Peak contact likely represents a temporal change in the abundance of biolimiting nutrients associated with the establishment of an upwelling system. Working in a Late Devonian marine carbonate ramp in Alberta, Caplan et al. (1996) described a very similar vertical facies succession wherein normal marine clastic carbonate strata pass upward into carbonate hardgrounds that

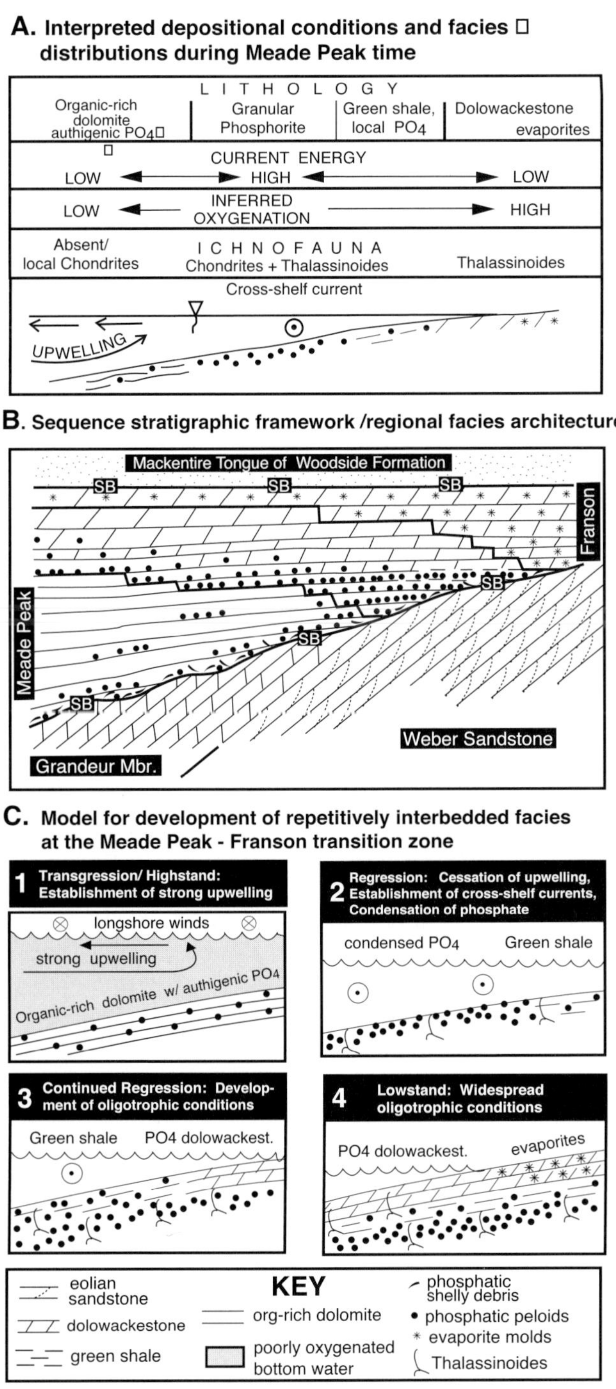

FIG. 6.— A) Overall model for phosphogenesis on a continental ramp as derived from observations and interpretations in the present study. Oceanic upwelling over the outer ramp enriches surface waters in nutrients, fostering high surface water productivity and the development of dysoxic to anoxic bottom waters in which organic-rich dolomite is deposited and local phosphate is precipitated authigenically. Current-induced winnowing of well-oxygenated inner shelf positions results in phosphorite grainstones and packstones that are thoroughly bioturbated. Inner shelf coarse phosphorites pass landward into subtidal green muds, phosphatic dolowackestone of peritidal affinity, and supratidal dolowackestone with evaporites. B) Schematic cross section of stratal architecture in an east-west profile across the study area. Rapid transgression over the Grandeur Member and Pennsylvanian Weber Sandstone resulted in the deposition of a transgressive lag in the basal Meade Peak. Overall high stand conditions were marked by deposition of organic-rich dolomite in more basinal positions and deposition of coarse phosphorite, green shale, phosphatic dolowackestone, and evaporite-bearing dolowackestone in succeedingly more landward positions. As sea level gradually lowered, these facies stepped basinward. SB = sequence boundary. C) Model for the formation of repetitively interbedded lithofacies at the Meade Peak-Franson transition zone. Deposition of interbedded lithofacies was related to high-frequency fluctuations in relative sea level and associated changes in climate that affected the shelf over time frames within the Milankovitch waveband. 1) Transgression and highstand resulted in the establishment of strong upwelling currents, development of eutrophic conditions and poorly oxygenated bottom waters, and deposition of organic-rich dolomite in which phosphate was precipitated authigenically. 2) Regression resulted in cessation of upwelling and condensation of phosphate to form granular phosphorite. 3) Continued regression, development of oligotrophic conditions, and deposition of green shale facies in the central part of the ramp and dolowackestone facies further inboard. 4) Lowstand resulted in widespread oligotrophic conditions and the deposition of dolowackestone across most of the shelf and evaporites in supratidal positions

are in turn overlain by phosphatic and glauconitic allochems. Phosphate and glauconite occur both as homogeneous grains and as grain coatings in the Late Devonian example. As in the Grandeur-Meade Peak contact, allochems are gradationally overlain by organic-rich mudrocks. Caplan et al (1996) interpreted the Devonian succession as representing an increase in the abundance of biolimiting nutrients from oligotrophic conditions (low abundance of nutrients) during deposition of normal marine clastic carbonates, to eutrophic conditions (high abundance of nutrients) during deposition of organic-rich mudrocks. They interpreted the increased abundance of hardground surfaces, mud laminae, phosphate, and glauconite at the top of the carbonate section to reflect significantly diminished rates of carbonate production due to increased flux of organic carbon to the sediment-water interface, attendant lower oxygenation levels, and low rates of net accumulation (see also Glenn and Arthur, 1988; and Föllmi et al., 1994). Clearly many of the attributes of the Grandeur-Meade Peak contact, both in the Uinta Mountains sections as well as in the Phosphoria depocenter in southeastern Idaho, bear similarity to the Devonian section described by Caplan et al. (1996). For this reason, we find this hypothesis attractive and suggest that the vertical facies changes observed at the Grandeur-Meade Peak transition may relate significantly to the development of eutrophic conditions, probably resulting from enhanced upwelling.

The establishment of strong upwelling conditions during deposition of the Meade Peak Member is consistent with a rise in relative sea level. Parrish (1982) pointed out that a minimum depth of 50–100 m is required for Ekman transport to replace turbulence in the water column (her "effective shoreline for upwelling"). In addition, several independent lines of evidence suggest that a rise in relative sea-level rise was an important influence on facies development. First, in the eastern part of the field area, the basal Meade Peak rests directly upon Pennsylvanian eolian sandstone of the Weber Formation. Second, local autobrecciation and the presence of fenestral fabrics and stromatolites in uppermost Grandeur carbonate (Hendrix, 1987) are suggestive of subaerial exposure or very shallow water prior to the commencement of Meade Peak deposition (Fig. 3). Third, a rise in eustatic sea level at the Leonardian-Guadalupian boundary has been interpreted from Permian strata exposed in North America and Russia (Ross and Ross, 1988), in the southwestern United States (Ross and Ross, 1995), and in West Texas (Sarg, 1989).

Meade Peak-Franson depositional system.—

Rapid deepening of the outboard portion of the ramp system, likely coupled with the development of eutrophic conditions, led to the deposition of lime micrite containing pristine phosphate. Blebs of authigenic phosphate in the organic-rich dolomite facies (Fig. 5) are consistent with phosphate supersaturation of pore waters during early diagenesis. Similar blebs of structureless phosphate, interpreted to have precipitated *in situ,* were reported from Jurassic organic-rich siltstone from Peru (Loughman, 1984), Cretaceous organic-rich shale in Egypt (Glenn and Arthur, 1990), and modern organic-rich mud off the coast of Peru (Glenn, 1990). Similarly, the filling of many steinkerns by phosphatic mud (Fig. 4E and F) and the common occurrence of mud peloids in the Meade Peak (Fig. 4D, F, and G) suggest local "micro-environments" supersaturated with carbonate fluorapatite.

The increase in organic matter content, preservation of discontinuous laminae, and only sporadic presence of the low oxygen ichnogenus *Chondrites* (Bromley and Ekdale, 1984) in the organic-rich dolomite suggest deposition in oxygen-deficient waters. Though the formation of early diagenetic dolomite due to sulfate reduction has been postulated for organic-rich deposits under low oxygenated conditions (Baker and Burns, 1985), we interpret Meade Peak dolomitization in this facies to be late diagenetic, because most dolomite crystals are zoned and locally displacive. The presence of small scale scours in places indicates that gentle bottom currents (contour currents?) were operative at times, although nowhere is there unequivocal evidence for vigorous bottom current activity. We interpret these characteristics to reflect deposition in middle to outer shelfal water depths (50–150 m), possibly within the oxygen minimum zone. Alternatively, if the Meade Peak basin were dysoxic or anoxic due to a regional bathymetric sill, as some paleogeographic reconstructions permit (Snyder et al., 1991), it is possible that organic-rich micrite could have been deposited in deeper water below sill depth. Importantly, we observed no evidence of turbidite deposition or siliciclastic input coarser than fine silt, suggesting that this part of the basin was sediment starved.

We interpret that pristine phosphate formed in the organic-rich dolomite facies in relatively deep water was subsequently winnowed by currents, possibly during regression, to form granular condensed phosphorite beds. Granular phosphorite beds contain local phosphatic cement but no phosphatic mud, suggesting that they underwent a significant amount of reworking and likely represent a condensed section ("condensed phosphatic sediment" of Föllmi et al., 1991). The general lack of sedimentary structures associated with the granular phosphorite facies probably reflects pervasive bioturbation, indicating that the rates of biological reworking consistently exceeded the rates of sediment aggradation (Föllmi et al., 1991) and that formation of condensed phosphorite took place under well-oxygenated conditions. Granular phosphorite of the Meade Peak is thickest east of the depositional pinch-out of the organic-rich dolomite facies and reaches economic proportions in the Vernal area, likely marking the easternmost position to which pristine phosphatic sediment accumulated.

We observed no direct evidence that phosphorite beds were sediment gravity flow events (cf. Grimm and Föllmi, 1994). With the exception of several beds in sections west of Mackentire Draw (Fig. 3), most phosphorite in the Uinta Mountains is not interbedded with laminated, fine-grained facies, as reported by Grimm and Föllmi (1994) for allochthonous phosphorites of Baja California. Rather, phosphorite is largely interbedded with green shale or phosphatic dolowackestone, both of which are interpreted to have been deposited under well-oxygenated conditions.

In the extreme eastern part of the field area, condensed phosphorite and interbedded cherty shale pass into phosphatic dolowackestone of the Franson Member. Within 10 km of the easternmost section examined in this study, the Meade Peak pinches out altogether and carbonate of the Franson Member rests directly on the Weber Sandstone (McKelvey et al., 1959). There is considerable evidence that supratidal sedimentation dominated the eastern part of the study area. Much of the Franson Member there contains calcite-filled and calcite-reduced molds after evaporite (anhydrite?) nodules, not surprising given an estimated paleolatitude between 5° and 20° N for the Phosphoria (Scotese, 1992). In addition, supratidal fabrics and a restricted molluscan assemblage are present in the upper part of the Franson Member in the eastern portion of the study area, as discussed above.

Shoaling of the Meade Peak-Franson depositional system.—

In addition to the observed trend of eastward shoaling within the Meade Peak-Franson depositional system, the vertical succession of sedimentary fabrics and the thickening-upward trend across the Meade Peak-Franson contact and throughout the Franson records an overall shoaling through time. Within all of the sections examined in this study, the maximum interpreted water depths occur within the Meade Peak, as suggested by the presence of laminated organic-rich phosphatic dolomite in the western part of the study area and granular phosphorite with associated green shale in the eastern part of the study area. Water depths above this "maximum flooding surface" (Galloway, 1989a, b) decreased, as recorded by the increased volumetric importance of phosphatic dolowackestone containing abundant evaporite molds (Fig. 4O and P) and supratidal fabrics (Fig. 4Q). Throughout the study area, supratidal carbonates in Franson pass vertically upward into subaerial redbeds of the Mackentire Member of the Woodside Formation. Regional erosional truncation of the upper Franson and deletion of several conodont-brachiopod biozones (Wardlaw et al., 1979) is consistent with erosional loss during subaerial exposure. Though karstification of the uppermost Franson was not observed during the course of this study, the replacement of anhydrite nodules by blocky calcite (Fig. 4O and P) suggests dissolution of anhydrite and precipitation of calcite by meteoric water.

Summary of environmental interpretations.—

To summarize our environmental interpretations for the Uinta Mountains study area, the basal Meade Peak records an eastward transgression over a shallow, monotonically shallowing ramp. Evidence consistent with a transgression at the beginning of the Meade Peak time includes a transition from nonmarine or subaerially exposed strata to marine strata of the basal Meade Peak, the development of a hardground surface in uppermost Grandeur carbonate, and the accumulation of a coarse clastic deposit containing a variety of phosphatic allochems at the base of the Meade Peak. We interpret phosphorites of the basal Meade Peak to be a transgressive lag, in addition possibly to marking a change from oligotrophic to eutrophic conditions (Caplan et al., 1996). The Grandeur-Meade Peak contact itself may be an angular unconformity on a regional scale, as suggested by pinch out of the Grandeur Member in the eastern part of the study area and deposition of basal Meade Peak phosphorite directly on the Pennsylvanian eolian sandstone of the Weber Formation.

At the time of maximum water depth during Meade Peak deposition (relative sea-level highstand), a series of distinctive environments characterized the ramp. The most basinal sections in the study area were the sites of pristine phosphatic sedimentation where authigenic phosphate was precipitated in laminated, organic-rich carbonate. Organic-rich carbonate sedimentation passed eastward up-dip into a zone of phosphate condensation that was likely a product of bottom winnowing currents. Further east, the zone of maximum phosphate condensation passed into a subtidal zone of green siliciclastic mud deposition that passed, in turn, into intertidal and supratidal carbonate deposition on an evaporative flat (Fig. 6A). This 'sabkha' environment passed further east into subaerial redbeds (Jado, 1980). Following maximum highstand deposition, the Meade Peak-Franson underwent regression, recorded by an overall shoaling and the progradation of Franson carbonate sedimentation westward across the study area.

INTERPRETED ALLOSTRATIGRAPHIC CONTROLS

Role of Oceanic Upwelling

Many workers have postulated that phosphate concentration in the Phosphoria was in part a product of intense oceanic upwelling along the western margin of Permian North America (McKelvey et al., 1959; Baturin, 1971; Sheldon, 1989). Sheldon (1987) emphasized the association of organic carbon, phosphate, and siliceous sediment in upwelling-related strata. All three lithologies occur within the study area. The presence of a cold water fauna and abundant organic matter in the equatorial Phosphoria (Wardlaw and Collinson, 1986) is consistent with upwelling of cool, nutrient-rich waters, possibly from within the oxygen minimum zone. Oceanic circulation modeling by Baturin and Vitaliano (1981) predicted the presence of upwelling currents off the western coast of Permian North America. Several authors have commented on the likelihood that oceanic circulation would be more vigorous and upwelling more intense during glacial periods with steepened latitudinal oceanic temperature gradients (McKelvey et al., 1959; Sheldon, 1989), although Wardlaw and Collinson (1986) presented an argument against Late Permian glaciation based on the regional distribution of glacial indicators.

Several observations suggest that the Meade Peak Member records the development of eutrophic conditions, consistent with the establishment of significant oceanic upwelling currents in the study area (cf. Caplan et al., 1996). Grandeur carbonates are reported to contain a normal marine (oligotrophic) assemblage (Jado, 1980), whereas the Meade Peak contains a depauperate fauna consisting of conodonts, molluscs, and brachiopods (Yochelson, 1968). The local hardground surface that marks the uppermost Grandeur attests to a reduction in sediment accumulation rates associated with the transition from oligotrophic to eutrophic conditions (e.g., Glenn and Kronen, 1993). Eutrophic conditions during deposition of the Meade Peak are strongly supported by its organic-rich character (Maughan, 1974, 1984) as well as the type and distributions of organic geochemical compounds it contains (Dahl et al., 1993). Once nutrient-rich conditions were established, both pristine and condensed phosphatic sediment accumulated in the Meade Peak. Eventually, eutrophic conditions gave way to more oligotrophic conditions, and the Franson carbonate Member was able to prograde westward across the ramp.

Sequence Stratigraphic Framework: The Role of Relative Sea Level

The Meade Peak-Franson depositional sequence records an environment that was physically, biologically, and chemically dynamic. At a first-order level, the Meade Peak-Franson is interpreted as recording a transgression over a gentle carbonate ramp, establishment of eutrophic conditions in the water column and poorly oxygenated bottom waters conducive to phosphogenesis, and a subsequent return to carbonate production during progradation of the ramp. We interpret strata of the Meade Peak and Franson Members as a single stratigraphic sequence bounded by unconformable sequence boundaries (Mitchum et al., 1977). The lower sequence boundary is the Grandeur-Meade Peak unconformity, which appears to be angular at the scale of the study area. Detailed stratal relationships across the Uinta Mountain study area suggest that the Meade Peak onlaps the top Grandeur unconformity (cf. Mitchum et al., 1977; Figs. 3 and 6B). Phosphatic strata in the Meade Peak are associated with transgressive and highstand systems tracts, consistent with other sequence stratigraphic interpretations of phosphate deposits (Glenn et al., 1993, 1994b). We interpret the surface of maximum flooding to be within the middle Meade Peak Member and the upper sequence boundary to be the subaerial erosion surface marking the contact between the Franson Member and the base of the Mackentire Tongue of the Woodside Formation (Fig. 6B).

The thickness of the Meade Peak-Franson stratigraphic sequence inferred in this study (10–35 m; Fig. 3) is similar to the thickness of parasequences reported from shelfal and peritidal phosphorites from the Cambrian Georgina basin of Australia (Southgate and Shergold, 1991) and Cretaceous phosphatic parasequences in hemipelagic strata of Colombia (Föllmi et al., 1992). Importantly, the down-dip succession of facies reported from the Australian parasequence sets is similar to that which we interpret from the Meade Peak-Franson (Fig. 6A). Namely, the down-dip succession of facies in the Australian parasequences grades from coastal plain carbonates and evaporites to peritidal carbonates, peritidal phosphorites, shelf phosphorites, organic-rich shale with benthic and planktonic fauna, and organic shale with planktonic fauna.

Role of High Frequency Sea Level and Climate Change

Key to understanding the formation and concentration of phosphate in the study area are the sub-meter to meter-scale repetitively interbedded lithofacies that characterize the Meade Peak-Franson transition zone (Fig. 4J and K). Salient observations that require interpretation are kilometer-scale lateral continuity of individual beds, the occurrence of repetitive interbeds across the study area (over 100 km laterally), the demonstrably reworked character of virtually all phosphatic grains associated with the repetitive interbeds, and the ubiquity of bioturbation that characterizes interbedded lithofacies.

The lateral and vertical facies architecture of the rocks, the sedimentary and biogenic fabrics contained in them, and the nature of phosphatic grains generally do not support the idea that the repetitively interbedded lithofacies were "event" beds (cf. Föllmi et al., 1991). Individual beds can be visually traced for over a kilometer with no observable change in thickness; hence, in situ condensation of phosphate by storm events seems unlikely. Storm deposition should have produced beds that thicken and thin on a several meter scale and contain evidence of combined flow, such as hummocky cross-stratification (Dott and Bourgeois, 1982). It is plausible that storms drove reworked phosphatic sediment up-dip to produce the observed granular phosphorites, and that the resulting deposit was subsequently reworked by physical and biogenic processes into a laterally continuous geometry, though we observed no direct sedimentologic evidence to support this interpretation. Similarly, we do not interpret the formation of phosphorite beds as allochthonous sediment gravity flow deposits interbedded with "background" green shale and carbonate deposition (cf. Grimm and Föllmi, 1994). We observed no sedimentary structures consistent with a sediment gravity flow interpretation. Even if such structures were obliterated by the pervasive bioturbation that characterizes the Meade Peak-Franson transition zone, the generation of phosphorite up dip from intertidal to supratidal carbonates containing local evaporite fabrics is problematic.

Based on our field measurements and observations, the local and regional architecture of strata in the field area, and the occurrence and nature of phosphate, we interpret the repetitively interbedded lithofacies that characterize the Meade Peak-Franson gradational contact as resulting from high-order fluctuations in sea level (fourth or fifth order), possibly with accompanying climate change, that affected shelfal circulation on a ramp with low relief and a microtidal range (Fig. 6C). During transgressions and relative high stands, mesotropic to eutrophic conditions overtook much of the shelf, in part because water depths exceeded the minimum value required for Ekman transport and the development of upwelling (Parrish, 1982). Accompanying changes in climate, such as an increase in the intensity of northerly trade winds, may have increased upwelling-related productivity (e.g., Juillet-Leclerc and Schrader, 1987). During relative high stands, deposition of organic-rich sediment containing pristine phosphate took place across much of the ramp (Fig. 6C). During subsequent regression, the effective shoreline for upwelling (Parrish, 1982) moved offshore, and cross-shelf currents reworked pristine phosphate to form condensed granular phosphorite. With continued regression, deposition of the green shale facies became widespread. As the locus of upwelling moved offshore or the intensity of upwelling decreased due to related climate change, nearshore waters became more oligotrophic and conditions favorable to carbonate production overtook much of the study area, resulting in the deposition of phosphatic dolowackestone facies (Fig. 6C).

Episodic reworking of pristine phosphate in the organic-rich dolomite facies likely led to the accumulation of economic quantities of phosphate in and near the Vernal section. Bottom winnowing currents associated with regression were likely very effective at mid-shelf water depths, resulting in the condensation of all pristine phosphate formed in the mid-shelf to granular phosphorite. Thick economic phosphorite accumulations near Vernal likely mark the easternmost position to which pristine phosphatic sediment accumulated during relative high stands of sea level.

The idea that interbedded lithofacies in the Meade Peak-Franson transition zone resulted from high frequency changes in sea level and accompanying changes in climate that affected the shelf is neither unreasonable nor unduly complex. As a benchmark for comparison, it is important to note that both geologic evidence (Fairbanks, 1989) and theoretical modeling studies (Tushingham and Peltier, 1991) indicate that modern sea level has risen well over 100 m within the past 18 ka and that major shifts in worldwide climate and the nature of shelf sedimentation have accompanied this sea-level rise. Though the record of continental glaciation during the middle Permian appears to be substantially less than that of the Carboniferous through Early Permian, this may in part be an artifact of the loss of record due to widespread tectonic extension and corresponding regional transgression (Gustavo and Eyles, 1995). Global climate model results based on the calculated abundance of atmospheric CO_2 predict that the middle Permian was a time of substantially cool global temperatures, consistent with polar glaciation (Berner, 1994). Middle Permian sea level fluctuations could have been of substantially lower magnitude and still produced the observed facies architecture, particularly if sea-level fluctuations were accompanied by changes in climate that affected shelfal circulation and water chemistry through upwelling. Given our interpretation of a broad, shallow ramp on the cratonward side of a flexural basin (see below), small changes in relative sea level would result in significant shifts in shelfal circulation and the position of accompanying facies.

To further test the plausibility of our model, we conducted "back of the envelop" calculations of the frequency of interpreted relative shifts in sea level and possible associated climate change across the study area. Assuming that the Meade Peak and Franson were deposited over 2–6 Ma (Filippelli and Delaney, 1992) and dividing that time interval by the maximum number of observed "cycles" (29 for the Sheep Creek Canyon section; Fig. 3), we calculate a periodicity of between 69 ka and 207 ka, which is within the Milankovitch wave-band (Berger, 1978).

IS THE PHOSPHORIA REALLY ANOMALOUS?

Many workers have commented on the amount of phosphate contained within the Phosphoria and speculated on the conditions needed to accumulate such a giant deposit (Sheldon, 1989, and references therein). Citing the Phosphoria as an example, Bentor (1980) suggested that the Permian may have been an anomalous period in terms of phosphate accumulation. However, calculated rates of phosphate accumulation and burial for the Phosphoria Formation are similar to accumulation and burial rates of modern phosphorites of the Peru-Chile margin, suggesting that the Phosphoria is not anomalous in that respect (Filippelli and Delaney, 1992). In addition, Glenn et al. (1994a) have described a variety of textural, sedimentologic, and geochemical similarities between modern and ancient phosphorites that suggest that phosphorite formation may be a uniformitarian process. Clearly, however, the question remains: was there some peculiarity to the depositional setting of the Phosphoria that resulted in the preservation of such a rich deposit of phosphate? To rephrase: if the Phosphoria is not an anomalous deposit in terms of its rates of phosphate accumulation and burial, why aren't large phosphatic deposits preserved in similar settings (for example, Mississippian of the western continental margin)?

Perhaps part of the answer lies in the paleogeographic and tectonic setting of the Phosphoria basin. The Phosphoria Formation was deposited during a second-order regression of eustatic sea level (Ross and Ross, 1988, 1995). As pointed out by Arthur and Jenkyns (1981), deposition of the Phosphoria during second-order regression stands in contrast to most other phosphorite giants that accumulated in association with transgressions and/or global high stands (for example, Cambrian phosphorites of Asia and Australia; Notholt and Sheldon, 1986; Upper Cretaceous Egyptian phosphorites; Glenn and Arthur, 1990; Paleocene-Eocene phosphorites of northwestern Africa; Cabezon, 1989; Svoboda, 1989; Boujo and Jiddou, 1989; and Oligocene-Miocene phosphorites of the Monterrey Formation; Alatorre, 1988). As a result of the Late Permian regression, many poorly oxygenated shallow seas conducive to phosphate deposition had become subaerially exposed. Because of its position on the western margin of Pangea (Scotese, 1992) and the fact that it remained below sea level for several Ma, Arthur and Jenkyns (1981) suggested that the Phosphoria basin may have been one of the few sites available for fixation of phosphate during that period.

The areal extent of phosphatic strata in Meade Peak Member (estimated at 21,500 km^2; Sheldon, 1989); its accumulation atop continental crust of normal, modern-day thickness (Prodehl and Lipman, 1989; Braile et al., 1989); and its position below sea level during a major eustatic regression (Hallam, 1984; Mitchum et al., 1977) imply the existence of regional tectonic subsidence, which provided accommodation space both for sediment of the Phosphoria basin and sea water of the Phosphoria Sea. The Permian tectonic history of the western North American continental margin is complicated and beyond the scope of this paper. However, in their regional synthesis of the North American Cordillera during the late Paleozoic, Miller et al. (1992) suggested that Early Permian extension in a backarc setting west of the Phosphoria basin (their Havallah basin) gave way by Late Permian time to east-verging contractile deformation, as the Golconda Allochthon was emplaced in response to back-arc collapse. Eastward thrusting within the Golconda Allochthon at the (present day) latitude of the Phosphoria basin involves rocks as young as Leonardian age (Miller et al., 1984), and the allochthon as a whole overthrusts autochthonous rocks as young as Late Permian (Miller et al., 1992). Strata younger than Late Permian are conspicuously absent (Miller et al., 1992), suggesting that east-verging shortening, and possibly emplacement of the entire Golconda allochthon, coincided with deposition of the Phosphoria and that subsidence of the Phosphoria basin was related to tectonic loading by the allochthon. The presence of the west-facing low gradient ramp inferred in the present study is consistent with an eastward decrease in tectonic subsidence expected for a flexural basin that is tectonically loaded on its western margin. The modern southwestern Persian Gulf, characterized by a low-gradient carbonate ramp of ≥.06% slope (United States Defense Mapping Agency, 1991), has analogous bathymetry and a broadly analogous tectonic setting to that which we interpret from the study area. Both the Persian Gulf and the Phosphoria basin are underpinned by continental crust (Ross et al., 1986; Braile et al., 1989), and both appear to be related to crustal flexure, though subsidence in the Persian Gulf is driven by continent-continent collision (Ross et al., 1986) and subsidence in the Phosphoria basin was probably driven by arc-continent collision (Miller et al., 1992).

Importantly, Snyder et al. (1991) suggested that a terrestrial(?) highland and/or the Sonoma Arc partially isolated the Phosphoria basin from the open ocean. This partial isolation does not imply that upwelling currents in the Phosphoria basin were subdued, however, because strong upwelling currents have been well documented or inferred in other restricted basins, such as the Gulf of California (Juillet-Leclerc and Schrader, 1987; Bray and Robles, 1991; Alvarez-Borrego and Rubén Lara-Lara, 1991) and the Sea of Japan (Grimm and Gill, 1994). Furthermore, the existence of this bathymetric sill likely enhanced levels of anoxia within the deeper parts of the basin, leading to conditions still more favorable to phosphate preservation (Arthur and Jenkyns, 1981).

Thus, it appears that, relative to most phosphorite giants, the Phosphoria Formation is anomalous only in terms of the fact that it accumulated during a second order eustatic regression. Neither the accumulation rates nor the modes of occurrence of phosphate appear to be much different than those of modern phosphorites or other ancient phosphorite deposits (Filippelli and Delaney, 1992; Glenn et al., 1994a). Though more information is clearly needed about the paleogeographic and tectonic setting of the Phosphoria basin, it appears that it was the combination of its position on a west-facing continental margin, the presence of a mechanism for subsidence that kept the basin below sea level for several million years, and the likelihood that a eustatic regression rendered the Phosphoria basin one of the few available sites for phosphate accumulation during the Late Permian that led to the accumulation and preservation of this phosphorite giant.

CONCLUSIONS

Based on our observations and interpretations of Permian strata in the study area, we conclude the following:

1. Leonardian-Guadalupian exposures of the uppermost Grandeur Member of the Park City Formation, Meade Peak phosphatic Member of the Phosphoria Formation, and overlying Franson carbonate member of the Park City Formation along the flanks of the Uinta Mountains, Utah, provide the opportunity to study down-dip changes in sedimentary style and phosphate accumulation on an ancient continental ramp.
2. The Meade Peak Shale and Franson Carbonate Members exposed around the Uinta Mountains are described in terms of five major facies that reflect the conditions of sedimentation, the degree of oxygenation, and the nature of phosphate accumulation. 1) Organic-rich dolomite, characterized by discontinuous lamination and local authigenic pristine phosphate, comprises much of the Meade Peak in distal portions of the ramp. 2) Granular condensed phosphorite is concentrated to economic proportions throughout the Meade Peak in the eastern part of the study area and is also common at the base of the Meade Peak and at the Meade Peak-Franson contact throughout the study area. Phosphatic grains include peloids, shelly material, steinkerns, intraclasts, and uncommon coated grains. *Thalassinoides* bioturbation is common. 3) Bedded chert, interpreted to be biogenic, is found at the top of the Meade Peak in the western field area. Further east, chert is interpreted to be secondary, where it fills *Thalassinoides* burrows and is a common cement in granular phosphorite. 4) Green shale is typically associated with granular phosphorite at the Meade Peak-Franson contact. This lithofacies commonly contains *Thalassinoides* burrows and floating grains of reworked phosphate. 5) Locally phosphatic dolowackestone comprises most of the Franson Member. This lithofacies is typically burrowed, particularly at the Meade Peak-Franson transition zone, and contains abundant calcite-filled and calcite-reduced evaporite molds. Supratidal fabrics and a restricted assemblage of molluscs are common at the top of the dolowackestone facies, particularly in the eastern part of the field area.
3. Regional facies architecture, stacking patterns, and preserved sedimentary and biogenic structures suggest that fluctuations in relative sea level, possibly with accompanying changes in climate, were a fundamental control on sedimentation in the study area. At a first order level, the entire Meade Peak-Franson is interpreted as a stratigraphic sequence, recording transgression during early Meade Peak time and regression during deposition of the Franson Member. The Grandeur-Meade Peak contact is interpreted as a transgressive surface, based on the abundance of borings in uppermost Grandeur carbonates and the presence of condensed phosphorite in immediately overlying basal Meade Peak. Field relations suggest that, in the Uinta Mountains region, the Grandeur-Meade Peak contact may be angular at a regional scale. The upper sequence boundary is interpreted to be the subaerial erosion surface that marks the uppermost Franson studied herein. At a more detailed level, small scale (fourth- or fifth-order) fluctuations in sea level, possibly with climate-induced changes in shelf circulation, resulted in the formation of a striking series of submeter- to meter-scale interbedded facies during deposition of the upper Meade Peak and lower Franson.
4. The mechanism and intensity of phosphate accumulation var-

ied with position down-dip on the continental ramp. In most basinward exposures, irregular blebs of phosphate in fine-grained, laminated, organic-rich dolomite suggest that pristine phosphate formed authigenically within the sediment. In contrast, landward sections contain abundant coarse, condensed phosphorite grainstone and packstone. We interpret granular, condensed phosphorites to have resulted from winnowing of pristine phosphatic sediment during episodic sea-level lowering and establishment of cross-shelf currents.

5. Though the Phosphoria often has been cited as an anomalous deposit of phosphate, it appears to differ from most phosphorite giants only in that it was deposited during a second-order eustatic regression, whereas most large phosphorite deposits accumulated during transgression or high stand conditions. Production of accommodation space for Phosphoria sediment likely resulted from tectonic subsidence driven by east-verging contractile deformation within and/or emplacement of the Golconda Allochthon along the flexed western margin of the Phosphoria basin.

ACKNOWLEDGMENTS

This paper stems from a Masters thesis completed by MSH under CWB at the University of Wisconsin-Madison. Thanks are owed to a large number of people, among the most notable of which are the local ranchers, farmers, and other residents of the Uinta Mountains area. Moral and logistical field support provided by the families of J. Briggs and K. Cook following the demise of our field vehicle is especially appreciated. Thanks is also given to the Vernal Phosphate Mine, Monsanto Industrial Chemical Co., the J. R. Simplot Co., and the Stauffer Chemical Co. for access to artificial exposures of the Phosphoria. We are especially grateful for personal interactions by mining geologists M. Weaver (Vernal), R. Bolton and R. Petrun (Monsanto), and W. Johnson (Stauffer). Enthusiastic field assistance by N. Van Wyck was absolutely critical to work completed during the summer of 1986. We thank J. Amato and J. Fournelle at the University of Wisconsin for performing microprobe analyses of the organic-rich dolomite. Very helpful discussions were held with L. C. Pray, T. Simo, G. B. Hieshima, A. R. Carroll, and C. R. Glenn at various points in the evolution of this study. Highly constructive reviews of early versions of this manuscript were provided by A. R. Carroll, S. M. DeBari, K. Föllmi, C. Fonseca, C. R. Glenn, K. Grimm, G. Harwood, A. C. Kendall, J. T. Parrish, and B. Pratt. Partial financial support for this project was provided by the Lewis G. Weeks Fund for Geological Sciences at the University of Wisconsin, the University of Montana Research Office, and the Petroleum Research Fund of the American Chemical Society.

REFERENCES

ALATORRE, A. E., 1988, Stratigraphy and depositional environments of the phosphorite-bearing Monterrey Formation in Baja California Sur, Mexico: Economic Geology, v. 83, p. 1918-1930.

ALVAREZ-BORREGO, S., AND RUBÉN LARA-LARA, J., 1991, The physical environment and primary productivity of the Gulf of California *in* Dauphin, J. P., and Simoneit, B. R. T., eds., The Gulf and Peninsular Province of the Californias: Tulsa, American Association of Petroleum Geologists Memoir 47, p. 555-567.

ARTHUR, M. A., AND JENKYNS, H. C., 1981, Phosphorites and paleoceanography: Oceanologica Acta, v. 4 (supplement), p. 83-96.

BAKER, P. A., AND BURNS, S. J., 1985, Occurrence and formation of dolomite in organic-rich continental margin sediments: American Association of Petroleum Geologists Bulletin, v. 69, p. 1917-1930.

BATURIN, G. N., 1971, States of phosphorite formation on the ocean floor: Nature, v. 232, p. 61-62.

BATURIN, G. N., AND VITALIANO, D. B., 1981, Phosphorites on the sea floor: Origin, composition, and distribution *in* Developments in Sedimentology, v. 33: Amsterdam, Elsevier, 343 p.

BENTOR, Y. K., 1980, Phosphorites—the unsolved problems: Tulsa, Society for Sedimentary Geology (SEPM) Special Publication 29, p. 3-18.

BERGER, A. L., 1978, Long-term variations of caloric insolation resulting from earth's orbital variations: Quaternary Research, v. 9, p. 139-167.

BERNER, R. A., 1994, 3GEOCARB II: A revised model of atmospheric CO_2 over Phanerozoic time: American Journal of Science, v. 294, p. 56-91.

BOUJO, A., AND JIDDOU, E. H. O., 1989, The Eocene phosphorite deposits of Bofal and Loubboira, Mauritania *in* Notholt, A. J. G., Sheldon, R. P., and Davidson, D. F., eds., 1989 Phosphate Deposits of the World, v. 2, Phosphate Rock Resources: Cambridge, Cambridge University Press, p. 207-213.

BRAILE, L. W., HINZE, W. J., VON FRESE, R. R. B., AND KELLER, G. R., 1989, Seismic properties of the crust and uppermost mantle of the conterminous United States and adjacent Canada *in* Pakiser, L. C., and Mooney, W., eds., Geophysical Framework of the Continental United States: Boulder, Geological Society of America Memoir 172, p. 655-680.

BRAY, N. A., AND ROBLES, J. M., 1991, Physical oceanography of the Gulf of California, *in* Dauphin, J. P., and Simoneit, B. R. T., eds., The Gulf and Peninsular Province of the Californias: Tulsa, American Association of Petroleum Geologists Memoir 47, p. 511-553.

BROMLEY, R. G., AND EKDALE, A. A., 1984, *Chondrites;* a trace fossil indicator of anoxia in sediments: Science, v. 224, p. 872-874.

CABEZON, C. M., 1989, The Bu-Craa phosphate deposit, western Sahara *in* Notholt, A. J. G., Sheldon, R. P., and Davidson, D. F., eds., Phosphate Deposits of the World, v. 2, Phosphate Rock Resources: Cambridge, Cambridge University Press, p. 176-183.

CAPLAN, M. L., BUSTIN, R. M., AND GRIMM, K. A., 1996, Demise of a Devonian-Carboniferous carbonate ramp by eutrophication: Geology, v. 24, p. 715-718.

CATHCART, J. B., SHELDON, R. P., AND GULBRANDSEN, R. A., 1984, Phosphate Resources of the United States: Washington, D.C., United States Geological Survey Circular 888, 48 p.

CRESSMAN, E. R., AND SWANSON, R. W., 1964, Stratigraphy and Petrology of the Permian Rocks of Southwestern Montana: Washington, D.C., United States Geological Survey Professional Paper 313-C, p. 275-569.

DAHL, J., MOLDOWAN, J. M., AND SUNDARARAMAN, P., 1993, Relationship of biomarker distribution to depositional environment; Phosphoria Formation, Montana, U. S.A.: Organic Geochemistry, v. 20, p. 1001-1017.

DOTT, R. H., AND BOURGEOIS, J., 1982, Hummocky stratification: significance of its variable bedding sequences: Geological Society of America Bulletin, v. 93, p. 663-680.

EMIGH, G. D., 1967, Petrology and origin of phosphorites in Hale, L. A., ed., Anatomy of the Western Phosphate Field, Fifteenth Annual Field Conference of the Intermountain Association of Geologists, p. 103-114.

FAIRBANKS, R. G., 1989, A 17,000 year glacio-eustatic sea level record: influence of glacial melting rates on the Younger Dryas event and deep-ocean circulation: Nature, v. 342, p. 637-642.

FILIPPELLI, G. M., AND DELANEY, M. L., 1992, Similar phosphorus fluxes in ancient phosphorites and a modern phosphogenic environment: Geology, v. 20, p. 709-712.

FÖLLMI, K. B., 1996, The phosphorus cycle, phosphogenesis and marine phosphate-rich deposits: Earth Science Reviews, v. 40, p. 55-124.

FÖLLMI, K. B., GARRISON, R. E., AND GRIMM, K. A., 1991, Stratification in phosphatic sediments: Illustrations from the Neogene of California *in* Einsele, G., Ricken, W., and Seilacher, A. eds., Cycles and Events in Stratigraphy: New York, Springer-Verlag, p. 492-507.

FÖLLMI, K. B., GARRISON, R. E., RAMIREZ, P. C., ZAMBRANO-ORTIZ, F., KENNEDY,

W. J., AND LEHNER, B. L., 1992, Cyclic phosphate-rich successions in the upper Cretaceous of Colombia: Palaeogeography, Palaeoclimatology, Palaeoecology, v. 93, p. 151-182.

FÖLLMI, K. B., WEISSERT, H., BISPING, M., AND FUNK, H., 1994, Phosphogenesis, carbon-isotope stratigraphy, and carbonate-platform evolution along the Lower Cretaceous northern Tethyan margin: Geological Society of America Bulletin, v. 106, p. 729-746.

FROELICH, P. N., ARTHUR, M. A., BURNETT, W. C., DEAKIN, M., HENSLEY, V. JAHNKE, R., KAUL, L., KIM, K.-H., ROE, K., SOUTAR, A., AND VATHAKANON, C., 1988, Early diagenesis of organic matter in Peru continental margin sediments: phosphorite precipitation: Marine Geology, v. 80, p. 309-343.

GALLOWAY, W. E., 1989a, Genetic stratigraphic sequences in basin analysis I: Architecture and genesis of flooding-surface bounded depositional units: American Association of Petroleum Geologists Bulletin, v. 73, p. 125-142.

GALLOWAY, W. E., 1989b, Genetic stratigraphic sequences in basin analysis II: Application to northwest Gulf of Mexico Cenozoic basin: American Association of Petroleum Geologists Bulletin, v. 73, p. 143-154.

GLENN, C. R., 1990, Pore water, petrologic and stable carbon isotopic data bearing on the origin of Modern Peru margin phosphorites and associated authigenic phases *in* Burnett, W. C., and Riggs, S. R., eds., Phosphate deposits of the world, v. 3, Neogene to modern phosphorites: Cambridge, Cambridge University Press, p. 46-71.

GLENN, C. R., AND ARTHUR, M. A., 1988, Petrology and major element geochemistry of Peru margin phosphorites and associated diagenetic minerals: Authigenesis in modern organic rich sediments: Marine Geology, v. 80, p. 231-268.

GLENN, C. R., AND ARTHUR, M. A., 1990, Anatomy and origin of a Cretaceous phosphorite-greensand giant, Egypt: Sedimentology, v. 37, p. 123-154.

GLENN, C. R., ARTHUR, M. A., RESIG, J. M., BURNETTE, W. C., DEAN, W. E., AND JAHNKE, R. A., 1994a, Are modern and ancient phosphorites really so different? *in* Iijima, A., Abed, A. M., and Garrison, R. E., eds., Siliceous, phosphatic, and glauconitic sediments of the Tertiary and Mesozoic: Utrecht, VSP Publishers, p. 159-188.

GLENN, C. R., FÖLLMI, K. B., RIGGS, S. R., BATURIN, G. N., GRIMM, K. A., TRAPPE, J., ABED, A. M., GALLI-OLIVIER, C., GARRISON, R. E., ILYIN, A. V., JEHL, C., ROHRLICH, V., SADAQAH, R. M. Y., SHIDLOWSKI, M., SHELDON, R. E., AND SIEGMUND, H., 1994b, Phosphorus and phosphorites: Sedimentology and environments of formation: Eclogae geologicae Helvetiae, v. 87, p. 747-788.

GLENN, C. R., AND KRONEN, J. D., 1993, Origin and significance of Late Pliocene phosphatic hardgrounds on the Queensland Plateau, Northeastern Australian Margin *in* McKenzie, J. A., Davies, P. J., Palmer-Julson, A., et al., eds.: Proceedings of the Ocean Drilling Program, Scientific Results, v. 133, p. 525-534.

GLENN, C. R., KRONEN, J. D., SYMONDS, P. A., WEI, W., AND KROON, D., 1993, High-resolution sequence stratigraphy, condensed sections, and flooding events off the Great Barrier Reef: 0–1.5 Ma: Proceedings of the Ocean Drilling Program, Scientific Results, v. 133, p. 353-364.

GRIMM, K. A., AND FÖLLMI, K. B., 1994, Doomed Pioneers: Allochthonous crustacean tracemakers in anaerobic basinal strata, Oligo-Miocene San Gregorio Formation, Baja California Sur, Mexico: Palaios, v. 9, p 313-334.

GRIMM, K. A., AND GILL, A. S., 1994, Fossil phytoplankton blooms and selfish genes: The ecological and evolutionary significance of *Chaetoceros* resting spores in laminated diatomaceous sediments (abs.): Boulder, Geological Society of America Abstracts with Programs, v. 26, no. 7, p. A170.

GULBRANDSEN, R. A., 1966, Chemical composition of phosphorites of the Phosphoria Formation: Geochimica et Cosmochimica Acta, v. 30, p. 769-778.

GUSTAVO, G.-B., AND EYLES, N., 1995, Inverse relation between ice extent and the late Paleozoic glacial record of Gondwana: Geology, v. 23, p. 1015-1018.

HALLAM, A., 1984, Pre-Quaternary sea-level changes: Annual Review of Earth and Planetary Sciences, v. 12, p. 205-243.

HENDRIX, M. S., 1987, Stratigraphy and sedimentology of Permian phosphatic strata, Uinta Mountains, Utah: Unpublished M. S. Thesis, University of Wisconsin, Madison, 135 p.

HENDRIX, M. S., AND BYERS, C. W., 1986, Sedimentology of the Meade Peak shale and Franson carbonate members, Park City Formation (Permian), Uinta Mountains, Utah (abs.): Boulder, Geological Society of America Rocky Mountain Section Abstracts with Programs, v. 19, no. 5, p. 283.

HIATT, E., 1997, A paleoceanographic model for oceanic upwelling in a late Paleozoic epicontinental sea: A chemostratigraphic analysis of the Permian Phosphoria Formation: Unpublished Ph.D. Dissertation, University of Colorado–Boulder, 300 p.

JADO, A. R., 1980, Paleogeographic interpretation of Permian units in northwestern Colorado and northeastern Utah *in* Fouch, T. D., and Magathan, R. R., eds., Paleozoic paleogeography of West-Central United States: West-Central United States Paleogeography Symposium I, p. 293-304.

JARVIS, J., BURNETT, W. C., NATHAN, Y., ALMBAYDIN, F. S. M., ATTIA, A. K. M., CASTRO, L. N., FLICOTEAUX, R., HILMY, M. E., HUSAIN, V., QUTAWNAH, A. A., SERJANI, A., AND ZANIN, Y. N., 1994, Phosphorite geochemistry: State-of-the-art and environmental concerns, *in* Föllmi, K. B., ed., Concepts and controversies in phosphogenesis: Eclogae geologicae Helvetiae, v. 87, p. 643-700.

JUILLET-LECLERC, A., AND SCHRADER, H., 1987, Variations of upwelling intensity recorded in varved sediment from the Gulf of California during the past 3,000 years: Nature, v. 329, p. 146-149.

LOUGHMAN, D. L., 1984, Phosphate authigenesis in the Aramachay Formation (Lower Jurassic) of Peru: Journal of Sedimentary Petrology, v. 54, p. 1147-1156.

MANSFIELD, G. R., 1927, Geography, Geology, and Mineral Resources of Part of Southeastern Idaho: Washington, D.C., United States Geological Survey Professional Paper 152, 453 p.

MAUGHAN, E. K., 1974, Organic Carbon and Selected Element Distribution in the Phosphatic Shale Members of the Permian Phosphoria Formation, Eastern Idaho and Parts of Adjacent States: Washington, D.C., United States Geological Survey Open-File Report OF 76-0577, 92 p.

MAUGHAN, E. K., 1984, Geological setting and some geochemistry of petroleum source in the Permian Phosphoria Formation *in* Woodward, J., Meissner, F. F., and Clayton, J. L., Hydrocarbon source rocks of the Greater Rocky Mountain region: Denver, Rocky Mountain Association of Geologists, p. 281-294.

MCKELVEY, V. E., WILLIAMS, J. S., SHELDON, R. P., CRESSMAN, E. R., CHENEY, T. M., AND SWANSON, R. W., 1959, The Phosphoria, Park City, and Shedhorn Formations in the Western Phosphate Field: Washington, D.C., United States Geological Survey Professional Paper 313-A, 47 p.

MCKELVEY, V.E., WILLIAMS, J. S., SHELDON, R. P., CRESSMAN, E. R., CHENEY, T. M., AND SWANSON, R. W., 1967, The Phosphoria, Park City, and Shedhorn Formations in the Western Phosphate Field *in* Hale, L. A., ed., Anatomy of the Western Phosphate Field: Intermountain Association of Geologists Fifteenth Annual Field Conference, p. 15-33.

MERRIAM, D. F., 1963, The geologic history of Kansas: Kansas Geological Survey Bulletin 162, 317 p.

MILLER, E. L., HOLDSWORTH, B. K., WHITEFORD, W. B., AND ROGERS, D., 1984, Stratigraphy and structure of the Schoonover sequence, northern Nevada; implications for Paleozoic plate margin tectonics: Geological Society of America Bulletin, v. 95, p. 1063-1076.

MILLER, E. L., MILLER, M. M., STEVENS, C. H., WRIGHT, J. E., AND MADRID, R., 1992, Late Paleozoic paleogeographic and tectonic evolution of the western U. S. Cordillera *in* Burchfiel, B. C., Lipman, P. W., and Zoback, M. L., eds., The Cordilleran Orogen: Conterminous U. S.: Boulder, Geological Society of America, The Geology of North America, v. G-3, p. 57-106.

MITCHUM, R. M., VAIL, P. R., AND THOMPSON, S., 1977, Seismic stratigraphy and global changes of sea level, part 2: The depositional sequence as a basic unit for stratigraphic analysis *in* Payton, C., ed., Seismic Stratigraphy—Applications to Hydrocarbon Exploration: Tulsa, American Association of Petroleum Geologists Memoir 26, p. 52-62.

NOTHOLT, A. J. G., AND SHELDON, R. P., 1986, Proterozoic and Cambrian phosphorites-regional review: World resources *in* Cook, P. J., and Shergold, J. H., eds., Phosphate Deposits of the World, v. 1, Proterozoic and Cambrian Phosphorites: Cambridge, Cambridge University Press, p. 9-19.

NOTHOLT, A. J. G., SHELDON, R. P., AND DAVIDSON, D. F., 1989, Phosphate deposits of the world, v. 2, Phosphate Rock Resources, North America and Greenland: Cambridge, Cambridge University Press, p. 1-5.

PARRISH, J. T., 1982, Upwelling and petroleum source beds, with reference to Paleozoic: American Association of Petroleum Geologists Bulletin, v. 66, p. 750-774.

PETERSON, J. A., 1980a, Depositional History and Petroleum Geology of the Permian Phosphoria, Park City, and Shedhorn Formations, Wyoming and Southeastern Idaho: Washington, D.C., United States Geological Survey Open-file Report 80-667, 42 p.

PETERSON, J. A., 1980b, Permian paleogeography and sedimentary provinces, west central United States *in* Fouch, T. D., and Magathan, E. R., eds., Paleozoic paleogeography of west-central United States: Rocky Mountain paleogeography symposium 1, Denver, Colorado, p. 271-292.

PRODEHL, C., AND LIPMAN, P. W., 1989, Crustal structure of the Rocky Mountain region, *in* Pakiser, L. C., and Mooney, W., eds., Geophysical framework of the continental United States: Boulder, Geological Society of America Memoir 172, p. 249-284..

RASCOE, B., AND BAARS, D. L., 1972, Permian System *in* Mallory, W. W., Mudge, M. R., Swanson, V. E., Stone, D. S., and Lumb, W. E., eds., Geologic Atlas of the Rocky Mountain Region: Denver, A. B. Hirschfield Press, p. 143-165.

RICHARDS, R. W., AND MANSFIELD, G. R., 1912, The Bannock overthrust: A major fault in southeastern Idaho and northeastern Utah: Journal of Geology, v. 20, p. 681-709.

ROSS, C. A., UCHUPI, E., AND WHITE, R. S., 1986, The geology of the Persian Gulf—Gulf of Oman region: A synthesis: Reviews of Geophysics, v. 24, p. 537-556.

ROSS, C. A., AND ROSS, J. R. P., 1988, Late Paleozoic transgressive-regressive deposition, *in* Wilgus, C. K., Hastings, B. S., Posamentier, H., Van Wagoner, Ross, C. A., and Kendall, C. G., eds., Sea-level Changes: An Integrated Approach: Tulsa, Society for Sedimentary Geology (SEPM) Special Publication 42, p. 227-247.

ROSS, C. A., AND ROSS, J. R. P., 1995, Foraminiferal zonation of late Paleozoic depositional sequences: Marine Micropaleontology, v. 26, p. 469-478.

SARG, J. F., 1989, Middle-Late Permian depositional sequences, Permian Basin, West and New Mexico *in* Bally, A. W., ed., Atlas of Seismic Stratigraphy: Tulsa, American Association of Petroleum Geologists Studies in Geology, v. 27, p. 140-154.

SCOTESE, C. R., 1992, PALEOMAP Paleogeographic Atlas: Arlington, University of Texas at Arlington, unpaginated.

SHELDON, R. P., 1963, Physical Stratigraphy and Mineral Resources of Permian Rocks in Western Wyoming: Washington, D.C., United States Geological Survey Professional Paper 313-B, p. 49-212.

SHELDON, R. P., 1987, Association of phosphorites, organic-rich shales, chert, and carbonate rocks: Carbonates and Evaporites, v. 2, p. 7-14.

SHELDON, R. P., 1989, Phosphorite deposits of the Phosphoria Formation, western United States, *in* Notholt, A. J. G., Sheldon, R. P., and Davidson, D. F., eds., Phosphate Deposits of the World, v. 2, Phosphate Rock Resources: Cambridge, Cambridge University Press, p. 53-61.

SHELDON, R. P., MAUGHAN, E. K., AND CRESSMAN, E. R., 1967, Sedimentation of rocks of Leonard (Permian) age in Wyoming and adjacent states *in* Hale, L. A., ed., Anatomy of the Western Phosphate Field: Intermountain Association of Geologists Fifteenth Annual Field Conference, p. 1-13.

SNYDER, W. S., SPINOSA, C., AND GALLEGOS, D. M., 1991, Pennsylvanian-Permian tectonism on the western U. S. continental margin *in* Raines, G. L., Lisle, R. E., Schafer, R. W., and Wilkinson, W. H., eds.: Geology and Ore Deposits of the Great Basin; Symposium Proceedings, Reno, Nevada, p. 5-19.

SOUTHGATE, P. N., AND SHERGOLD, J. H., 1991, Application of sequence stratigraphic concepts to Middle Cambrian phosphogenesis, Georgina Basin, Australia: BMR Journal of Australian Geology and Geophysics, v. 12, p. 119-144.

SVOBODA, K., 1989, The Lower Tertiary phosphate deposits of Tunisia, *in* Notholt, A. J. G., Sheldon, R. P., and Davidson, D. F., eds., 1989 Phosphate Deposits of the World, v. 2, Phosphate Rock Resources: Cambridge, Cambridge University Press, p. 284-288.

SWANSON, R. W., 1970, Mineral Resources in Permian Rocks of Southwest Montana: Washington, D.C., United States Geological Survey Professional Paper 313-E, p. 661-777.

SWANSON, R. W., 1973, Geology and Phosphate Deposits of the Permian Rocks in Central-western Montana: Washington, D.C., United States Geological Survey Professional Paper 313-F, 54 p.

THOMAS, H. D., 1940, Pennsylvanian and Permian stratigraphy of central and southeastern Wyoming: Kansas Geological Society Guidebook, 14th Annual Field Conference, western South Dakota and eastern Wyoming, p. 121-126.

THORNBURG, J., 1990, Petrography and sedimentology of a phosphatic shelf deposit: The Permian Shedhorn Sandstone and associated rocks in southwest Montana and northwest Yellowstone National Park: Unpublished Ph.D. Dissertation, University of Colorado–Boulder, 338 p.

TUSHINGHAM, A. M., AND PELTIER, W. R., 1991, Ice-3G: A new global model of Late Pleistocene deglaciation based upon geophysical predictions of post-glacial relative sea level change: Journal of Geophysical Research, v. 96, B3, p. 4497-4523.

UNITED STATES DEFENSE MAPPING AGENCY, 1991, Asia, Persian Gulf Hydrographic Map: Washington, D.C., 1:1,000,000 scale.

WARDLAW, B. R., ed., 1979, Studies of the Permian Phosphoria Formation and Related Rocks, Great Basin-Rocky Mountain Region: Washington, D.C., United States Geological Survey Professional Paper 1163-A-D, 22 p.

WARDLAW, B. R., 1980, Middle-Late Permian paleogeography of Idaho, Montana, Nevada, Utah, and Wyoming *in* Fouch, T. D., and Magathan, E. R., eds., Paleozoic paleogeography of West-Central United States West-Central United States Paleogeography Symposium I, Denver, p. 353-361.

WARDLAW, B. R., AND COLLINSON, J. W., 1979a, Biostratigraphic Zonation of the Park City Group: Washington, D.C., United States Geological Survey Professional Paper 1163-D, p. 17-22.

WARDLAW, B. R., AND COLLINSON, J. W., 1979b; Youngest Permian conodont faunas from the Great Basin and Rocky Mountain regions: Provo, Brigham Young University Geologic Studies, v. 26, part 3, p. 151-164.

WARDLAW, B. R., AND COLLINSON, J. W., 1986, Paleontology and deposition of the Phosphoria Formation: Contributions to Geology, University of Wyoming, v. 24, p. 107-142.

WARDLAW, B. R., COLLINSON, J. W., AND MAUGHAN, E. K., 1979, Stratigraphy of the Park City Group Equivalents (Permian) in Southern Idaho, Northeastern Nevada, and Northwestern Utah: Washington, D.C., United States Geological Survey Professional Paper 1163-C, p. 9-16.

WHALEN, M. T., 1993, Depositional and diagenetic history of the Park City Formation (Permian) carbonate ramp, northwestern USA: Implications for eastern ocean basin carbonate platform development: Unpublished Ph.D. Dissertation, Syracuse University, 269 p.

WHALEN, M. T., 1996, Facies architecture of the Permian Park City Formation, Utah and Wyoming: Implications for the paleogeography and oceanographic setting of Western Pangea, *in* Longman, M. W., and Sonnenfeld, M. D. (eds.), Paleozoic systems of the Rocky Mountain Region: Rocky Mountain Section SEPM (Society for Sedimentary Geology), p. 355-378.

YOCHELSON, E. L., 1968, Paleoecology of the Permian Phosphoria Formation and Related Rocks: Washington, D.C., United States Geological Survey Professional Paper 313-D, p. 571-660.

CHRONOSTRATIGRAPHY OF UPPER CENOZOIC PHOSPHORITES ON THE NORTH CAROLINA CONTINENTAL MARGIN AND THE OCEANOGRAPHIC IMPLICATIONS FOR PHOSPHOGENESIS

STANLEY RIGGS*, SCOTT SNYDER, DOROTHEA AMES

Department of Geology, East Carolina University, Greenville, North Carolina 27858, USA

AND

PETER STILLE

Centre de Géochimie de la Surface (CNRS), 1, rue Blessig, F-67084 Strasbourg, France

ABSTRACT: Integration of biostratigraphic and Sr isotope data constrain the ages of four third-order depositional sequences within the Aurora and Onslow Embayments of the Miocene Pungo River Formation. Within the Aurora Embayment the microsphorite surface capping the Oligocene and marking the base of the Pungo River Formation is ~28 Ma. The bulk of Unit A, which constitutes the Aurora Embayment Sequence (AES), is 23.2–21.6 Ma. Unit B, at 19.1–17.0 Ma, is largely contemporaneous with the Frying Pan Sequence (FPS) in Onslow Embayment. Units C and D, at 16.4–14.8 Ma, are largely contemporaneous with the Onslow Bay Sequence (OBS) in Onslow Embayment. The microsphorite surface at the contact of unit D and the Pliocene Yorktown Formation is 12.7 Ma. Within the Onslow Embayment the microsphorite at the unconformity between the Oligocene and the Miocene Pungo River Formation is 21.0–19.5 Ma. The remainder of the Frying Pan Sequence (FPS) is 20.5–16.7 Ma. Most of the Onslow Bay Sequence (OBS1–OBS3) is 15.4–14.8 Ma, with OBS4 being slightly younger. The lower units of the Bogue Banks Sequence (BBS1–BBS5) are 12.5–8.6 Ma. BBS6-BBS8 have not been constrained biostratigraphically; however, Sr ages suggest that the phosphate was reworked from pre-existing units and deposited between 8.6–7.1 Ma.

Interpretation of depositional patterns in phosphate-rich sediments of North Carolina has led to the development of a sedimentological model for continental margin phosphogenesis. Sedimentation patterns in an idealized phosphogenic episode reflect various stages of sea-level change. Sea-level lowstand produces an erosional unconformity across previously deposited sediment sequences. Marine transgression is characterized by a fining-upward sequence of fine siliciclastic sediments with intermixed phosphate grain types. During early-stage transgression, a basal microsphorite crust forms and erodes to produce rip-up intraclasts that are then reworked into and dominate the lower portion of the subsequent depositional sequence. Mid-stage transgression is dominated by authigenic phosphate, including skeletal grains and fine sand-sized peloids, which form as disseminated grains below the sediment-water interface in response to the degradation of organic matter in nutrient-rich, suboxic shelf environments where there is little or no sediment winnowing or reworking. During late-stage transgression the phosphogenesis is terminated as oxygen-depleted environments migrate landward across the shelf with the formation of organic-rich dolosilts or foraminifer-rich/diatom-rich muds with little to no phosphate. The sea-level highstand is characterized by carbonate-rich sediments deposited under the influence of normal marine, open-shelf conditions. During the subsequent regression, sediments of these shallower depositional environments are often eroded and reworked to form beds of extensively reworked phosphate-rich sediments.

Four episodes of phosphate sedimentation characterize the North Carolina continental margin. Three episodes (I, II, and III) are dominated by primary phosphate formation within the Miocene, whereas one episode (IV) consisting of the Pliocene-Quaternary sediments contains totally reworked phosphates derived from sediments of Episodes II and III. Episode I occurred during the Aquitanian (23.3–21.6 Ma), forms the basal sequence of the Pungo River Formation in the Aurora Embayment, and represents a mid-stage transgression that is only partially preserved. Episode II preserved all sediment facies of an entire sea-level cycle that produced the main Miocene phosphogenic event, which took place during the Burdigalian and Langhian (21.0–14.8 Ma) and extended throughout the Aurora and Onslow Embayments. Episode III was deposited in eastern Onslow Embayment during the uppermost Serravallian and Tortonian (12.7 Ma and likely continuing through 7.1 Ma). Low sea stand during the Messinian led to severe erosion and weathering of Episode III sediments, limiting their distribution and modifying their composition. Episode IV represents many sea-level events that produced Pliocene-Quaternary deposits with local and variable concentrations of reworked phosphate of many types. The four phosphate episodes recognized in North Carolina closely correspond to the Upper Cenozoic phosphorite episodes identified globally by Riggs and Sheldon (1990).

INTRODUCTION

Extensive litho-, chrono-, and biostratigraphic data bases for the Upper Cenozoic section of the North Carolina continental margin have been generated over the past 20 years. The three major objectives of this paper build upon these data bases. 1) Integrate $^{87}Sr/^{86}Sr$ ages from phosphates and associated carbonates with biostratigraphic analyses related to the time scale of Berggren et al. (1995) in order to refine age assignments of complex, cyclical phosphogenic events. 2) Evaluate the oceanographic implications of the phosphogenic events during the Neogene-Quaternary, particularly as they relate to sea-level fluctuations and associated episodes of upwelling. 3) Synthesize this information into a model explaining Neogene-Quaternary phosphogenesis along the North Carolina margin, which will facilitate future comparisons with interpretations of coeval strata throughout the Atlantic continental margin. The brief review of regional geological framework that follows is intended to provide perspective from which to evaluate the aforementioned model, which is the ultimate objective of this study.

GEOLOGIC FRAMEWORK OF THE NORTH CAROLINA PHOSPHORITES

North Carolina phosphorites are part of a series of similar Neogene deposits that formed contemporaneously on many of the world's continental margins equatorward of 45° latitude (Sheldon, 1964, 1980; Cook and McElhinny, 1979). Along the southeastern U.S. continental margin, phosphate and associated authigenic sediments were deposited in the transition zone between dominantly limestone Eocene to Oligocene strata and

*email: riggss@mail.ecu.edu

Marine Authigenesis: From Global to Microbial, SEPM Special Publication No. 66
 ISBN 1-56576-064-6, p. 369-385.

dominantly siliciclastic Pliocene to Quaternary strata (Riggs and Sheldon, 1990). These phosphate deposits extend southward from North Carolina, through South Carolina and Georgia, to form the extensive deposits of the Florida Peninsula (Riggs, 1979b; Riggs and Manheim, 1988; Compton et al., 1993; Mallinson et al., 1994; Mallinson and Compton, 1995, 1997). The vast Carolina and Florida phosphate provinces formed in response to global paleoclimatic and paleoceanographic changes related to development of polar glaciation in the southern hemisphere during the Late Cenozoic (Riggs, 1984; Riggs and Sheldon, 1990).

In North Carolina, specific portions of the Pungo River Formation contain enriched concentrations of phosphorite in three known phosphate districts (Fig. 1). Demonstrated phosphate resources exceed 1 billion metric tons of concentrate for the Aurora Phosphate District, and hypothetical resources approach 21 billion metric tons for the Aurora Embayment (Riggs and Manheim, 1988). Hypothetical phosphate resources in the Frying Pan and Northeast Onslow Bay Phosphate Districts are 10.5 and 10.2 billion metric tons of phosphate concentrate, respectively (Powers et al., 1990). However, much phosphorus is sparsely disseminated and so will never be of economic interest. Riggs and Sheldon (1990) estimated that over 150 billion metric tons of phosphorus were deposited in phosphate-rich beds (>25% phosphate grains) and another 275 billion metric tons of phosphorus were deposited as disseminated grains (<10% phosphate grains) throughout the Neogene section of southeastern U.S.

In North Carolina, economically important phosphorites occur only within the Miocene Pungo River Formation, which on the emergent coastal plain occurs totally in the subsurface (Riggs et al., 1982; Ames, 1994). However, these deposits extend seaward onto the continental shelf (Fig. 1), where they crop out on the sea floor (Riggs, 1984, 1987; Riggs et al., 1985, 1990). On the shelf, the Pungo River Formation forms a truncated sediment sequence (Fig. 2) that strikes north to northeast and dips east to southeast off the flank of the MidCarolina Platform High (Snyder, 1982, 1994; Snyder, 1988; Snyder et al., 1993). The sediments thicken from their western updip limit to 300–500 m at the shelf edge. Miocene phosphorites have been subsequently eroded, reworking the older phosphate grains into younger shelf sediments (Luternauer and Pilkey, 1967; Baturin, 1982; Riggs, 1984; Riggs et al., 1985; Riggs et al., 1990, 1996).

Three third-order depositional sequences within the Miocene section (Fig. 2) were recognized by integrating high-resolution seismic data with lithologic and biostratigraphic data from associated vibracores (Snyder, 1982, 1994; Snyder et al., 1982, 1990, 1993; Snyder, 1988, 1990; Riggs and Mallette, 1990). These three sequences contain at least 18 smaller-scale

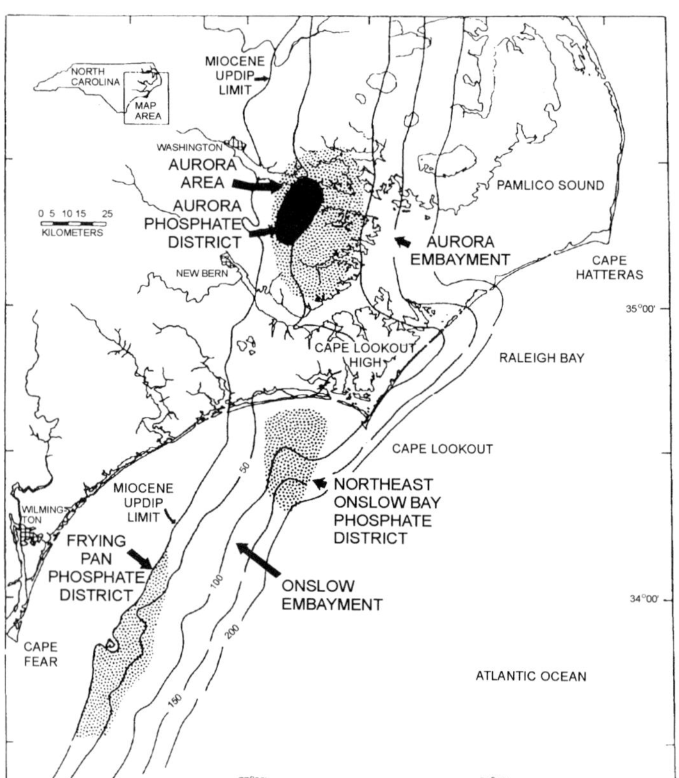

FIG. 1.—Structural contour map (in meters) on the base of the Miocene Pungo River Formation along the North Carolina coastal plain and continental shelf. The map also delineates (1) the Cape Lookout High, a paleotopographic feature on top of Oligocene sediments that defines the Miocene Aurora Embayment on the north and Onslow Embayment on the south; and (2) the three known areas of major phosphate occurrence in the Pungo River Formation. Figure is from Riggs et al. (1985).

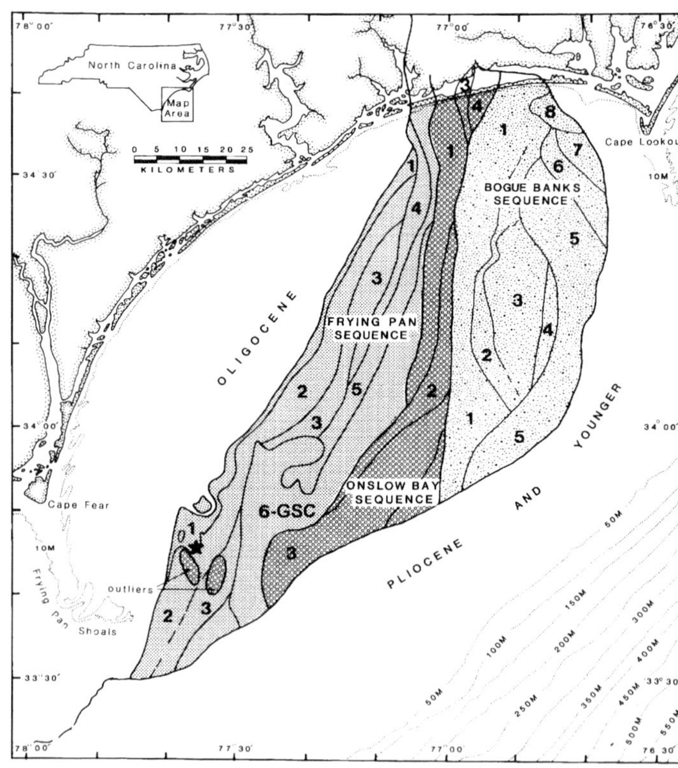

FIG. 2.—Outcrop distribution map of Miocene Pungo River Formation in Onslow Bay, North Carolina. This figure shows (1) the three third-order seismic sequences referred to as the Frying Pan (FPS), Onslow Bay (OBS), and Bogue Banks (BBS) sequences; and (2) 18 fourth-order seismic sequences sequentially numbered from the oldest to the youngest for each third-order sequence, respectively. The uppermost FPS sequence is composed of two parts including the northern shelf sequence labeled 6 and the southern portion labeled 6-GSC (Gulf Stream Channel). The star is the location of phosphate samples utilized in Stille et al. (1994) and Riggs et al. (1997). Figure is modified from Stephen W. Snyder (1982).

seismic sequences (Fig. 2), which were originally interpreted to be distinct depositional episodes produced by fourth-order, glacio-eustatic sea-level fluctuations with durations between 100 ky and 1 my and amplitudes between 50 m and 200 m (Mallette, 1986; Riggs and Mallette, 1990; Riggs et al., 1990; Snyder et al., 1990; Snyder, 1994). However, biostratigraphic, Sr isotopic, and lithologic analyses have only been able to distinguish the larger, third-order sequences.

Recent integration of the lithostratigraphy with detailed Sr isotopic analysis (Stille et al., 1994; Riggs et al., 1997) suggests that individual depositional sequences may encompass multiple fourth-order seismic sequences. Unconformable seismic surfaces previously interpreted to be associated with sea-level events may be produced by periodic sediment starvation or facies boundaries in shallow-water, high-energy shelf environments as outlined by Riggs et al. (1998). Consequently within the Miocene Pungo River Formation, it now appears that the third-order events are significantly more important than the fourth-order events.

Stille et al. (1994) and Riggs et al. (1997) analyzed $^{87}Sr/^{86}Sr$ ratios of four phosphate morphotypes (Table 1) and associated foraminifera in order to test the hypothesis that different authigenic grain types within a given stratum may have distinct ages and processes of formation. These studies found that phosphate grain types yielded different strontium isotopic ages, reflecting different times of formation associated with varying chemical and physical processes during phosphogenesis. Ratios from peloids within the same strata occupied tight clusters and had Sr isotope ages similar to those of associated foraminifera, whereas other grain types showed mixed results that were interpreted to be either reworked or to reflect diagenetic alteration. In addition, peloids had the greatest mineralogical uniformity and chemical stability of all carbonate fluorapatite grain types (Ellington, 1984; Mallette, 1986; Crowson et al., 1990, 1991; Ames et al.,

TABLE 1.—PHOSPHATE GRAIN TYPES IN UPPER CENOZOIC SEDIMENTS OF NORTH CAROLINA (from Riggs, 1979a; Ames et al., 1991; Stille et al., 1994; Riggs et al., 1997).

PHOSPHATE GRAIN TYPES	DESCRIPTION/INTERPRETATION
SKELETAL GRAINS:	Formed directly by organisms
1. Vertebrate vertebrae	Primarily pelagic fish and mammal vertebrae
2. Vertebrate teeth	Primarily pelagic fish and mammal teeth
3. Vertebrate bone frags	Primarily pelagic fish and mammal bones
4. Invertebrate shell	Primarily benthic inarticulate brachiopod shells
PELOIDS AND NODULES:	Formed in the shallow subsurface in response to authigenic and very early-stage diagenetic processes
*5. Peloids	Sand-size, spherical to oval shaped, highly rounded and polished grains with minor inclusions and commonly coated
6. Grapestones	Aggregates of peloidal grains
7. Nodules	Gravel-size, irregular-shaped, highly rounded grains with uniform internal structure and commonly coated
MICROSPHORITE AND INTRACLASTS:	Formed at the sediment-water interface during hiatal periods
*8. Microsphorite	Laminated crusts and bored hardgrounds with common inclusions
9. Intraclasts	Irregular-shaped, rip-up grains from preexisting crusts and nodules
10. Moldic Intraclasts	Benthic invertebrate molds broken from microsphorite crusts, nodules, and intraclasts
SECONDARY PHOSPHATE GRAINS:	Formed in response to middle- to late-stage diagenesis
11. Phosphatized shells	Replaced carbonate shell material of benthic invertebrates and foraminifera
12. Phosphatized diatoms	Replaced siliceous tests of diatoms
13. Complex grains	Primary grains (i.e., skeletal grains) with secondary infilling or overgrowths

* Phosphate grain types utilized in the present study.

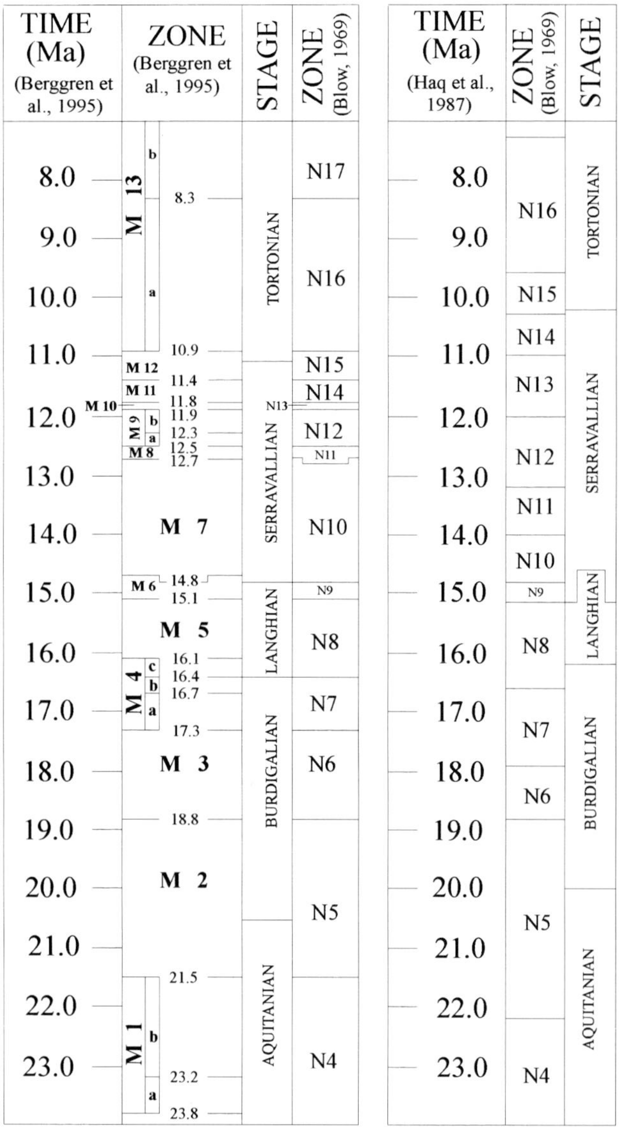

FIG. 3.—Comparison of the time scales from Berggren et al. (1995) and Haq et al. (1987), both shown in relation to the planktic foraminiferal zones of Blow (1969), for the lower and middle upper Miocene.

1991) and were present in sufficient concentrations in all samples to permit analysis. Hence, peloids were interpreted to represent primary sedimentation at the time of phosphogenesis. Consequently, the present chronostratigraphic study utilizes phosphate peloids as the primary grain type for Sr isotope dating. In addition, microsphorite samples at unconformable surfaces were analyzed to help define the timing of early-stage transgression.

Stille et al. (1994) and Riggs et al. (1997) demonstrated that isotopic studies of discrete phosphate grain types can be used to establish high-resolution isotopic stratigraphy that reflects changing paleoceanographic conditions during phosphate formation and deposition. Phosphate peloids and associated grain types date the basal unconformity and four distinct lithofacies in FPS1. Even though the data from all four lithofacies cluster together, when errors from both the analytical techniques and the Hodell et al. (1991) curve are considered, the dates represent a continuum of sedimentation. Based on these Sr age data, the microsphorite at the unconformity formed between 21 and 19.5 Ma. Intraclastic phosphate gravels, skeletal grains, and some peloids were reworked and accumulated to form the basal FPS-1 lithofacies between 19.5 and 19.3 Ma. The primary formation of phosphate peloids and deposition of the associated foraminifera began about 19.0 Ma and continued until about 18.0 Ma.

The biostratigraphy of Miocene sediments from Onslow Bay has been interpreted on the basis of planktic foraminifera and calcareous nannofossils (Snyder et al., 1988), diatoms (Powers, 1988), and radiolaria (Palmer, 1988), and was then constrained by integrating these data (Snyder, 1988). The ages are dependent upon the time scale utilized. In 1988, we opted to relate our biostratigraphic zonal assignments to the time scale and standard stage boundaries presented by Haq et al. (1987) largely because their scheme facilitated comparison with relative changes of coastal onlap (Snyder, 1988). More updated geological time scales (e.g., Berggren et al., 1995) can now be used to interpret the biostratigraphic data from Onslow Bay. Hence, it is appropriate to revisit the biostratigraphic interpretations of the Onslow Bay Miocene section to refine the ages assigned to specific third-order seismic/depositional sequences (Fig. 3).

METHODS

Sr Chronostratigraphy

The Sr chronostratigraphy presented herein is new but is based on analyses done contemporaneously with the data utilized by Stille et al. (1994) and Riggs et al. (1997) in their assessment of phosphate sedimentation and of concurrent phosphate grain types in one, high-frequency phosphate bed in one area of Onslow Bay. Materials and methods for Sr isotopic analysis were detailed in those papers.

Sr ages were calculated using the regression equations for the $^{87}Sr/^{86}Sr$ isotopic composition of sea water as a function of age developed by Hodell et al. (1991). Within the age range of 24–16 Ma (interval N-VI of Hodell et al., 1991), the Sr-derived ages have a stratigraphic resolution of ± 0.48 m.y. for the 80% confidence interval and ± 0.74 m.y. for the 95% confidence interval (Hodell et al., 1991). Within the age range of 16–8 Ma (interval V of Hodell et al., 1991), the Sr-derived ages have a stratigraphic resolution of ± 0.88 m.y. for the 80% confidence interval and ± 1.36 m.y. for the 95% confidence interval (Hodell et al., 1991). In addition, all Sr ages are adjusted to the NBS-987 values of Hodell et al. to make our data directly comparable. In this paper we use the Sr age data rather than the Sr ratios because this study represents a stratigraphic application and age data are more relevant.

Biostratigraphy

The reevaluation of biostratigraphy is based totally upon the data published in Snyder (1988). Materials and methods for planktic foraminifera and nannofossils were detailed in Snyder

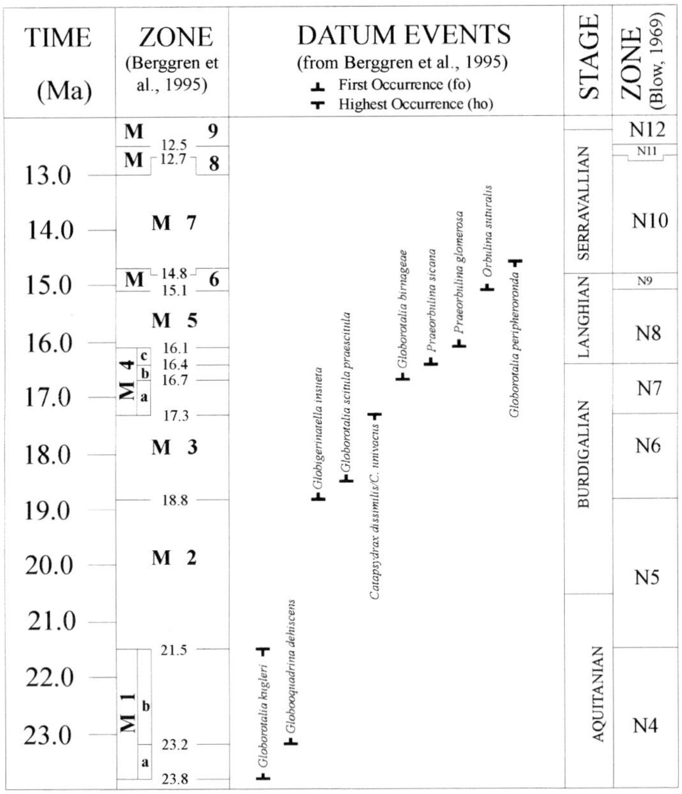

FIG. 4.—Datum events: the first occurrence (fo) and highest occurrence (ho) of selected planktic foraminiferal taxa shown in relation to the time scale of Berggren et al. (1995) and the zonation of Blow (1969). Datum events are from Berggren et al. (1995), as are correlations of the zones of Berggren et al. with those of Blow (1969).

et al. (1988); those for diatoms in Powers (1988); and those for radiolarians in Palmer (1988). Results of this biostratigraphic reassessment of the Miocene Pungo River Formation are presented in the Appendix to the present paper.

Miocene sediments recovered from Onslow Bay often do not contain the species used to define the standard planktic foraminiferal biozones of Blow (1969) or Berggren et al. (1995). In the absence of nominate taxa, secondary species whose relative stratigraphic ranges have been extrapolated from other studies had to be used as proxies to make zonal assignments. Hence, the boundaries presented here often represent chronozones based on species other than those used to define the biozones of standard zonal schemes, and the potential errors for biostratigraphic age assignments are in the vicinity of 1 m.y. Fig. 4 outlines the datum events of selected planktic foraminiferal taxa shown in relation to the time scale of Berggren et al. (1995) and the zonation of Blow (1969). Fig. 5 presents the stratigraphic occurrences of selected planktic foraminiferal taxa through the Miocene sequences of Onslow Bay (data from Snyder et al., 1988).

RESULTS

Sr Stratigraphy of the Upper Cenozoic Phosphorites

Table 2 summarizes 62 Sr age calculations using the regression equations for the $^{87}Sr/^{86}Sr$ isotopic composition of sea water as a function of age developed by Hodell et al. (1991). These data represent 53 samples from the Miocene Pungo River Formation in both the Aurora and Onslow Embayments, and 14 samples from post-Miocene units including the Pliocene Yorktown Formation in Aurora Embayment and Pleistocene strata and Holocene surficial sediments in Onslow Embayment. Many seismic sequences within the Pungo River Formation in Onslow Bay (Fig. 2) were not dated using Sr isotopic analysis, either because they contained too little phosphate or too few foraminifera (FPS3, FPS5, OBS1-OBS4, BBS3-BBS5) or because no vibracores penetrated them (FPS4, BBS6, and BBS7). Phosphate peloids are ubiquitous and abundant enough through the remaining depositional sequences of the Miocene Pungo River Formation to provide stratigraphic comparison.

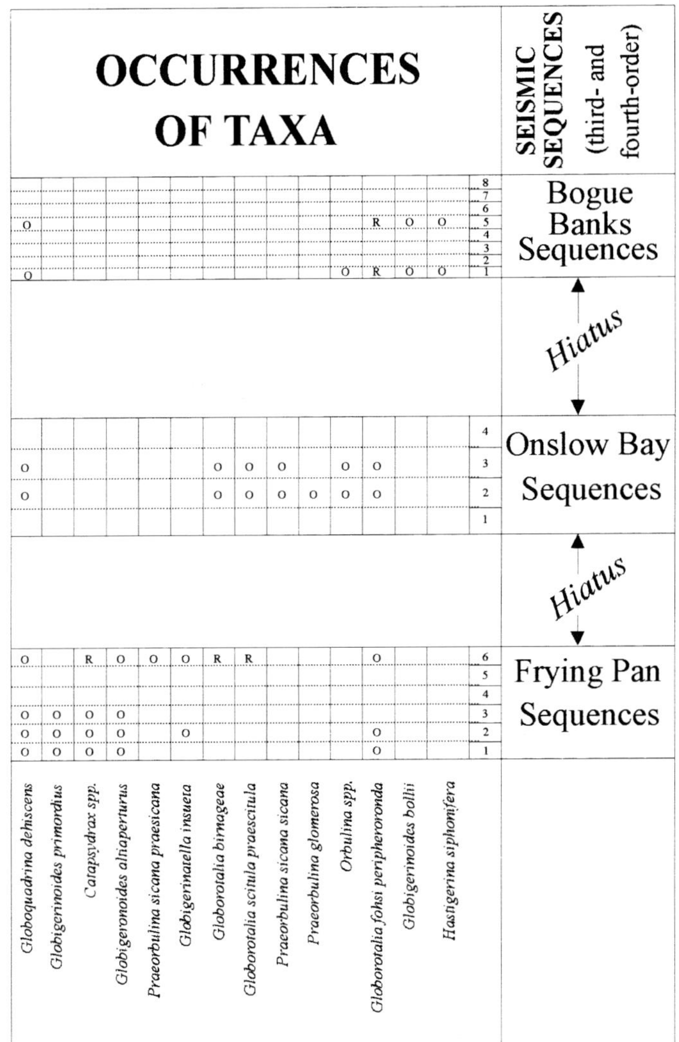

Fig. 5.—Stratigraphic occurrences of selected planktic foraminiferal taxa through the Miocene sequences of Onslow Bay (data from Snyder et al., 1988).

Table 3 presents Sr-age data for paired foraminifera and phosphate peloids from the same samples. Unaltered foraminifera were abundant in only five samples. Strontium isotope analysis of four of these foraminiferal samples produced ratios statistically indistinguishable from those of the associated

TABLE 2.—STRONTIUM ISOTOPIC AGE DATA FOR 62 SAMPLES OF PHOSPHATE PELOIDS AND MICROSPHORITE ON ASSOCIATED UNCONFORMITY SURFACES FOR THE UPPER CENOZOIC SEDIMENT UNITS OF THE NORTH CAROLINA CONTINENTAL MARGIN.

EPOCH	AGE		AURORA EMBAYMENT				ONSLOW EMBAYMENT			
Ma #		$^{87}Sr/^{86}Sr$##	NUMBER*	DEPTH m	UNIT	SR AGE**	NUMBER	DEPTH m	UNIT	SR AGE**
HOLOCENE		0.708854(7)					OB45-1	0.4	HOLO	10.27
0.0-0.01		0.708647(7)					OB24-1	0.0	HOLO	17.34
		0.708628(13)					OB24-2	0.2	HOLO	17.65
		0.708643(7)					OB64-2	0.25	HOLO	17.41
		0.708641(6)					OB64-4	0.75	HOLO	17.44
PLEISTOCENE		0.708685(6)					OB22-3	0.5	PLEIS	16.71
0.01-1.8		0.708638(6)					OB64-6	1.25	PLEIS	17.49
		0.708636(7)					OB64-7	1.50	PLEIS	17.52
		0.708637(6)					OB64-8	1.75	PLEIS	17.51
		0.708639(6)					OB64-10	2.25	PLEIS	17.47
		0.708633(7)					OB64-12	2.75	PLEIS	17.57
PLIOCENE		0.708647(6)	GH96	29.3	U YKT	17.34				
1.8-5.3		0.708615(7)	GH105	32.0	L YKT	17.87				
		0.708608(7)	GH107	32.8	L YKT	17.98				
MIOCENE 5.3-7.1	MESSINIAN									
MIOCENE	TORTONIAN	0.708691(6)					OB1-9	5.0	BBS8	16.61
7.1-11.1		0.708709(7)					OB1-12	5.75	BBS8	16.32
		0.708884(8)					OB53-14	4.0	BBS2	9.00
		0.708893(7)					OB53-24	6.5	BBS2	8.62
		0.708842(7)					OB3-7	2.0	BBS1	10.78
		0.708826(7)					OB3-14	3.75	BBS1	11.46
MIOCENE 11.1-14.8	SERRAVALLIAN	0.708796(8)	GH108	32.9	Microsphorite	12.73				
MIOCENE	LANGHIAN	0.708603(7)	GH109	33.2	PR D	18.07				
14.8-16.4		0.708620(7)	GH119	36.3	PR D	17.79				
MIOCENE	BURDIGALIAN	0.708610(6)	GH126	38.4	PR C	17.95				
16.4-20.6		0.708595(6)	GH133	40.5	PR C	18.20				
		0.708571(7)	GH139	42.4	PR B	18.60				
		0.708561(7)	GH144	43.9	PR B	18.76				
		0.708543(6)	GH149	45.4	PR B	19.06				
		0.708540(5)	GH153	46.6	PR B	19.11				
		0.708608(6)					OB66-2	2.0	FPS6-GSC	17.98
		0.708686(6)					OB45-15	4.0	FPS6	16.70
		0.708602(7)					OB24-4	0.6	FPS1	18.08
		0.708604(10)					OB24-5	0.8	FPS1	18.05
		0.708605(6)					OB24-6	1.0	FPS1	18.03
		0.708610(7)					OB24-7	1.2	FPS1	17.95
		0.708577(6)					OB24-9	1.6	FPS1	18.50
		0.708565(7)					OB24-10	1.8	FPS1	18.69
		0.708519(7)					OB24-11	1.9	Microsphorite	19.45
		0.708574(6)					OB29B-5	1.0	FPS1	18.55
		0.708569(6)					OB29B-9	2.0	FPS1	18.63
		0.708558(7)					OB29B-17	4.0	FPS1	18.81
		0.708557(6)					OB29B-21	5.0	FPS1	18.83
		0.708530(7)					OB29B-25	6.0	FPS1	19.27
		0.708531(7)					OB29B-29	7.0	FPS1	19.26
		0.708606(7)					OB64-14	3.25	FPS1	18.02
		0.708571(8)					OB64-15	3.50	FPS1	18.60
		0.708569(6)					OB64-16	3.75	FPS1	18.63
		0.708572(8)					OB64-18	4.25	FPS1	18.58
		0.708547(6)					OB64-20	4.75	FPS1	18.99
		0.708544(7)					OB64-20	4.75	FPS1	19.04
		0.708541(7)					OB64-21	5.00	FPS1	19.09
		0.708543(7)					OB64-22	5.25	FPS1	19.06
		0.708538(6)					OB64-24	5.75	FPS1	19.14
		0.708512(7)					OB64-25	6.00	FPS1	19.57
		0.708526(7)					OB64-26	6.25	FPS1	19.34
		0.708432(6)					RC3-2	0.35	Microsphorite	20.95
MIOCENE	AQUITANIAN	0.708386(7)	GH163	49.4	PR A2	21.65				
20.6-23.8		0.708380(6)	GH170	51.8	PR A2	21.75				
		0.708301(6)	GH174	53.0	PR A1	23.06				
		0.708295(7)	GH178	54.3	PR A1	23.16				
		0.708298(7)	GH182	55.5	PR A1	23.11				
OLIGOCENE 23.8-28.5	CHATTIAN	0.708110(8)	391-156	47.7	Microsphorite	~28				

\# The stratigraphic age assignments are based upon Berggren et al. (1995).

\## Uncertainties (2σ of the mean) of the $^{87}Sr/^{86}Sr$ ratios are in parentheses. All isotope ratios, except sample 391-156, were corrected to the Hodell et al. (1991) standard by multiplying our original ratios by 0.9999789.

* All samples are from drill core GH8.5, except one sample which is from core 391 (Ames, 1994).

** The Sr age dates were obtained by using the regression analysis of Hodell et al. (1991)

phosphate peloids, suggesting that phosphate peloids are contemporaneous with the foraminifera and supporting the interpretation that phosphogenesis was syndepositional. One peloid sample has a date that is not contemporaneous with the associated foraminifera; this sample is from a large channel structure interpreted to be the product of Gulf Stream scour with subsequent backfill by open marine sediments. Rare phosphate peloids within the channel sediments yield the same Sr age as the underlying FPS units into which the channel structure was eroded. Consequently, this phosphate is interpreted to have been reworked from the pre-existing deposits during the channeling process.

The Sr isotopic ages permit chronostratigraphic assessment of the Miocene third-order sequences in Aurora and Onslow Embayments relative to the time scale of Berggren et al. (1995) (Fig. 6). The Aurora Embayment Sequence (AES), a new third-order depositional sequence that corresponds to units A1 and A2 in the Aurora Embayment, is recognized and was deposited between 23.2 Ma and 21.6 Ma. Based upon stratigraphic evidence from drill cores (Ames, 1994), both older and younger facies of this depositional sequence exist in the down-basin direction. The AES is poorly known because it is the basal sequence in the Aurora Embayment and the phosphates, which are of subeconomic value, occur below the active mining zone in the ongoing mining operation.

Based upon Sr dates of the Frying Pan Sequence (FPS1, FPS2, and FPS6) in Onslow Embayment and unit B1 in Aurora Embayment, deposition occurred between 21.0 and 16.7 Ma (Fig. 6). No strontium dates were obtained from the OBS units in Onslow Embayment, but dates from biostratigraphically equivalent units C and D in Aurora Embayment suggest that these phosphates were totally reworked from underlying unit B and so do not provide chronostratigraphic control. Based upon dates from BBS1 and BBS2 in Onslow Embayment, deposition of the BBS started about 11.5 Ma and continued through at least 8.6 Ma (Fig. 6). BBS3 through BBS7 have no Sr age dates; however, age dates of 16.6–16.3 Ma (Fig. 6) from BBS8 suggest that the phosphate is totally reworked from the uppermost Burdigalian phosphorites. Units BBS3–BBS8 must be stratigraphically younger than 8.6 Ma and probably represent deposition prior to the Messinian lowstand, which began about 7.1 Ma.

Sr age dates of phosphate peloids within some Miocene sequences differ significantly from ages suggested by the biostratigraphy (Fig. 6). Phosphate peloids in the upper portion of unit C and in unit D in the Aurora Embayment, and those from BBS8 in the Onslow Embayment yielded Sr ages significantly older than their biostratigraphic age assignment and stratigraphic position would suggest (Fig. 6). These age differences are interpreted to reflect mixed sedimentation in which older, reworked phosphate is redeposited during the times indicated by the biostratigraphy. All 13 of the phosphate samples from the Pliocene, Pleistocene, and Holocene sediments produced Sr isotopic ages that ranged from 17.7–10.3 Ma (Table 2). This indicates that all Pliocene and Quaternary phosphates have been reworked by erosion of the various Miocene units and that primary phosphogenesis was totally restricted to the Miocene (Stille et al., 1994).

TABLE 3.—STRONTIUM ISOTOPIC AGE DATA ON PAIRED PHOSPHATE PELOID AND FORAMINIFERA SAMPLES FROM THE MIOCENE PUNGO RIVER FORMATION, FRYING PAN SEQUENCE-1 AND -6GSC, ONSLOW BAY, NORTH CAROLINA CONTINENTAL SHELF.

Core-Sample Number	Sample Depth m	Litho-Facies	Strontium Isotopic Age* Phosphate Peloid	Foraminifera
IN SITU PHOSPHATE				
OB29B-17	4	FPS1	18.81 Ma	18.94 Ma
OB29B-21	5	FPS1	18.83 Ma	18.78 Ma
OB64-20	4.75	FPS1	19.04 Ma	19.24 Ma
OB64-24	5.75	FPS1	19.14 Ma	19.19 Ma
REWORKED PHOSPHATE				
OB66	2	FPS6-GSC	17.98 Ma	16.89 Ma

*Strontium ages are based upon and adjusted to Hodell et al. (1991).

Biostratigraphic Reassessment of the Pungo River Formation

The update of planktic foraminiferal datum events by Berggren et al. (1995) included a number of first occurrence (fo) and highest occurrence (ho) datums that are relevant to interpreting the Miocene section of Onslow Bay (Fig. 4). These datums, along with the occurrences of additional taxa whose ranges are well established relative to the planktic foraminiferal zonation of Blow (1969), are central to our chronozonal assignments. The ranges of taxa other than those that have been reassessed by Berggren et el. (1995) are drawn from Bolli and Saunders (1985), Berggren et al. (1985), Kennett and Srinivasan (1983), and Stainforth et al. (1975). Results of the biostratigraphic reassessment of the Pungo River Formation are presented in the Appendix to the present paper.

DISCUSSION

Chronostratigraphic Synthesis

Biostratigraphically reinterpreting Miocene sediments from Onslow Bay and relating revised chronozonal assignments to the time scale of Berggren et al. (1995) results in the following changes for previous age estimates of the three third-order seismic sequences (Fig. 6). The Frying Pan Sequence (FPS) boundaries are changed from 18.4–17.4 Ma (Snyder, 1988) to 20.5–17.2 Ma. The Onslow Bay Sequence (OBS) boundaries are changed from 15.9–14.9 Ma (Snyder, 1988) to 15.4–14.8 Ma, with the caveat that the upper age limit is not well constrained because of the lack of biostratigraphic control in OBS4. The Bogue Banks Sequence (BBS) boundaries are changed

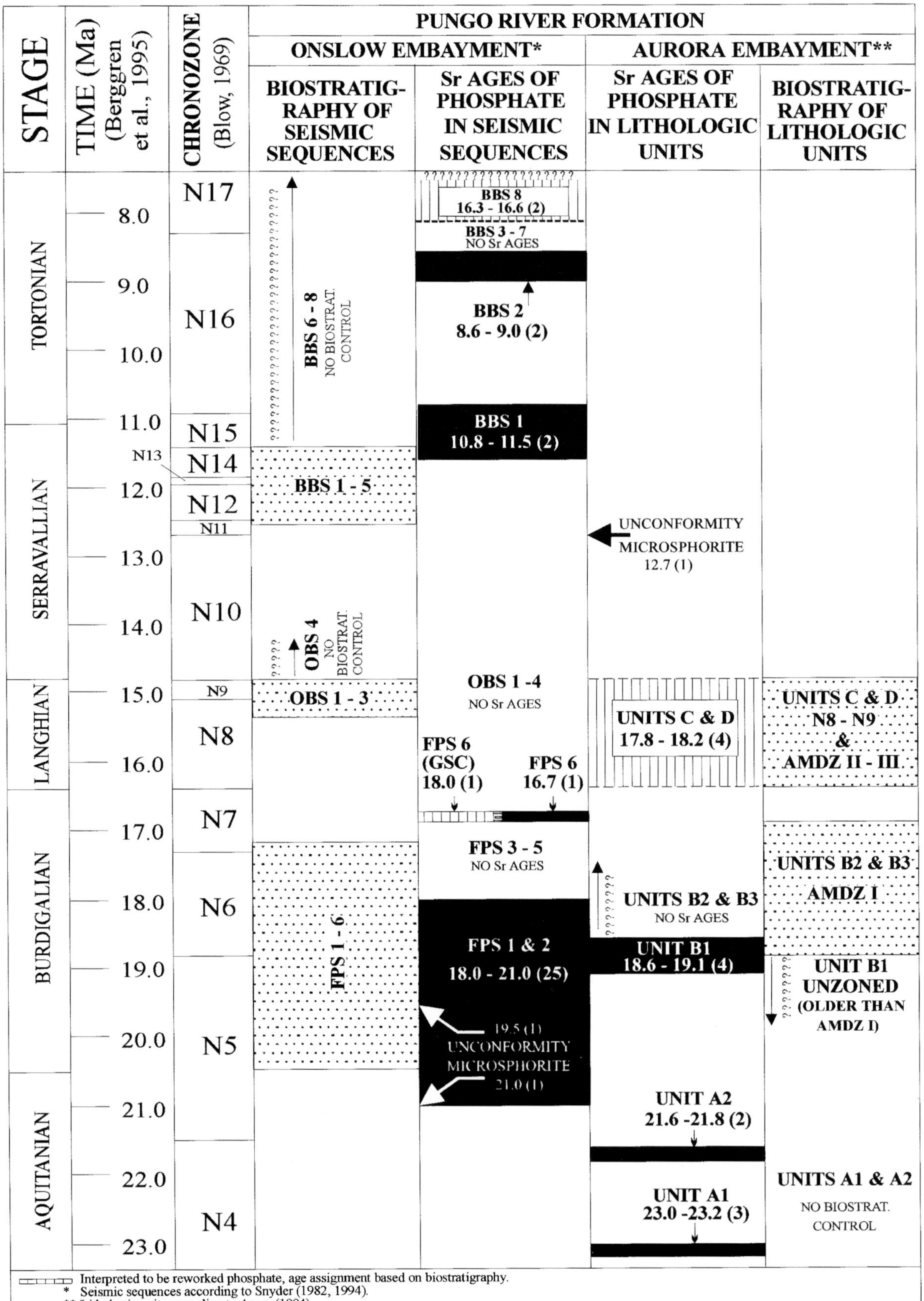

FIG. 6.—Biostratigraphic assignments of fossiliferous Miocene sequences from Onslow and Aurora Embayments. Interpretations are shown in relation to the planktic foraminiferal zones of Blow (1969) and the time scale of Berggren et al. (1995). Relation to geological stages is also from Berggren et al. (1995).

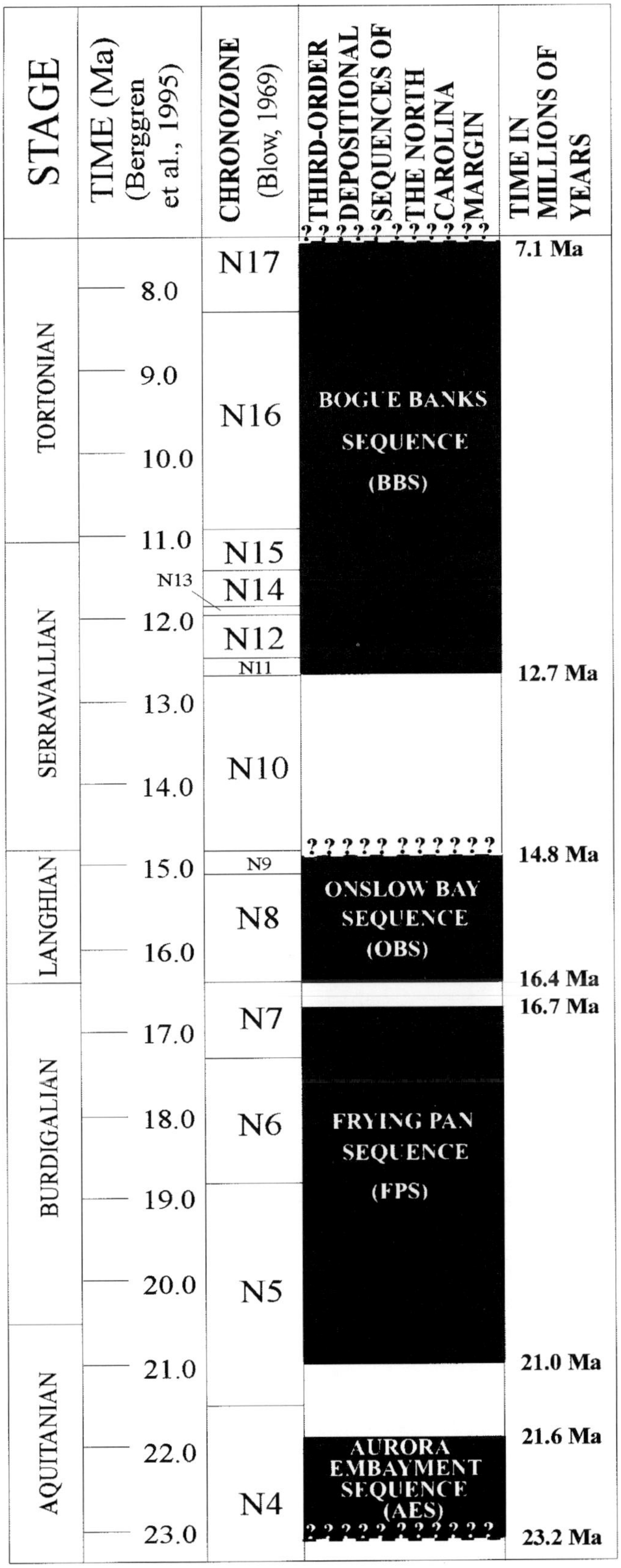

FIG. 7.—Chronostratigraphy of the four third-order Miocene sequences on the North Carolina continental margin. These age assignments represent the integration of the biostratigraphic chronozones and the Sr isotopic age dates and are the earliest and latest dates that occur within these two data bases.

from 13 Ma or younger (Snyder, 1988) to 12.5 Ma or younger, extending to at least 11.3 Ma with the high probability that it could be much younger because there is no biostratigraphic control for BBS6–BBS8.

Based upon the time scale of Berggren et al. (1995), the Onslow Embayment seismic sequences (Snyder, 1982, 1994) correlate biostratigraphically with the Aurora Embayment stratigraphic section (Ames, 1994) as displayed in Fig. 6. Units A1, A2, and B1 have no biostratigraphic control. Units B2 and B3 range from about 18.8 Ma to 16.7 Ma (= AMDZ I) and are generally equivalent to the upper portion of the FPS units. Units C and D range from about 16.4 Ma to 14.8 Ma (= chronozones N8-N9) and are generally equivalent to the OBS1-OBS3 units. No Miocene sediments younger than Langhian (units C and D) have been recognized in the Aurora Embayment; the Langhian sediments are directly overlain by the Pliocene Yorktown Formation.

Chronostratigraphic synthesis can be developed for the North Carolina Upper Cenozoic phosphorites by integrating the $^{87}Sr/^{86}Sr$ age dates on hand-picked phosphate and carbonate grains with the biostratigraphic data base. Resulting age assignments for each of the four third-order depositional sequences of the Pungo River Formation are summarized in Fig. 7. The biostratigraphic and Sr isotopic age data are generally corroborative (Fig. 6); however, there are a few places where minor disagreement exists, such as with units BBS1 and BBS2. Therefore, the integrated age assignments for the deposition of the four third-order depositional sequences (Fig. 7) represent the earliest and latest dates from the combined data bases. Development of the model of phosphogenesis in the next section is based upon the age assignments in Fig. 7.

Oceanographic Implications for Phosphogenesis

Depositional Model of Continental Margin Phosphogenesis.— Interpretation of changing patterns of litho- and biofacies within depositional cycles has led to development of a general depositional model for continental margin phosphogenesis (Riggs, 1984; Riggs and Mallette, 1990). Depositional patterns within a single phosphogenic episode reflect five general stages of sedimentation in response to sea-level fluctuations.

1. Sea-level lowstand: an unconformity forms across previously deposited sediment sequences in response to subaerial exposure and erosion. The upper portions of previously deposited sequences are diagenetically altered to produce distinctive indurated and often moldic surfaces.
2. Early-stage transgression: the indurated surface becomes a marine hardbottom and a basal microsphorite crust forms on the exposed surface. Rip-up intraclasts from the microsphorite surface, along with reworked phosphate skeletal grains, are dominant sediment components in the lower portion of the subsequent depositional sequence. These coarse, reworked phosphorites are concentrated immediately above the microsphorite surface and decrease upsection as a function of increased siliciclastic sedimentation.

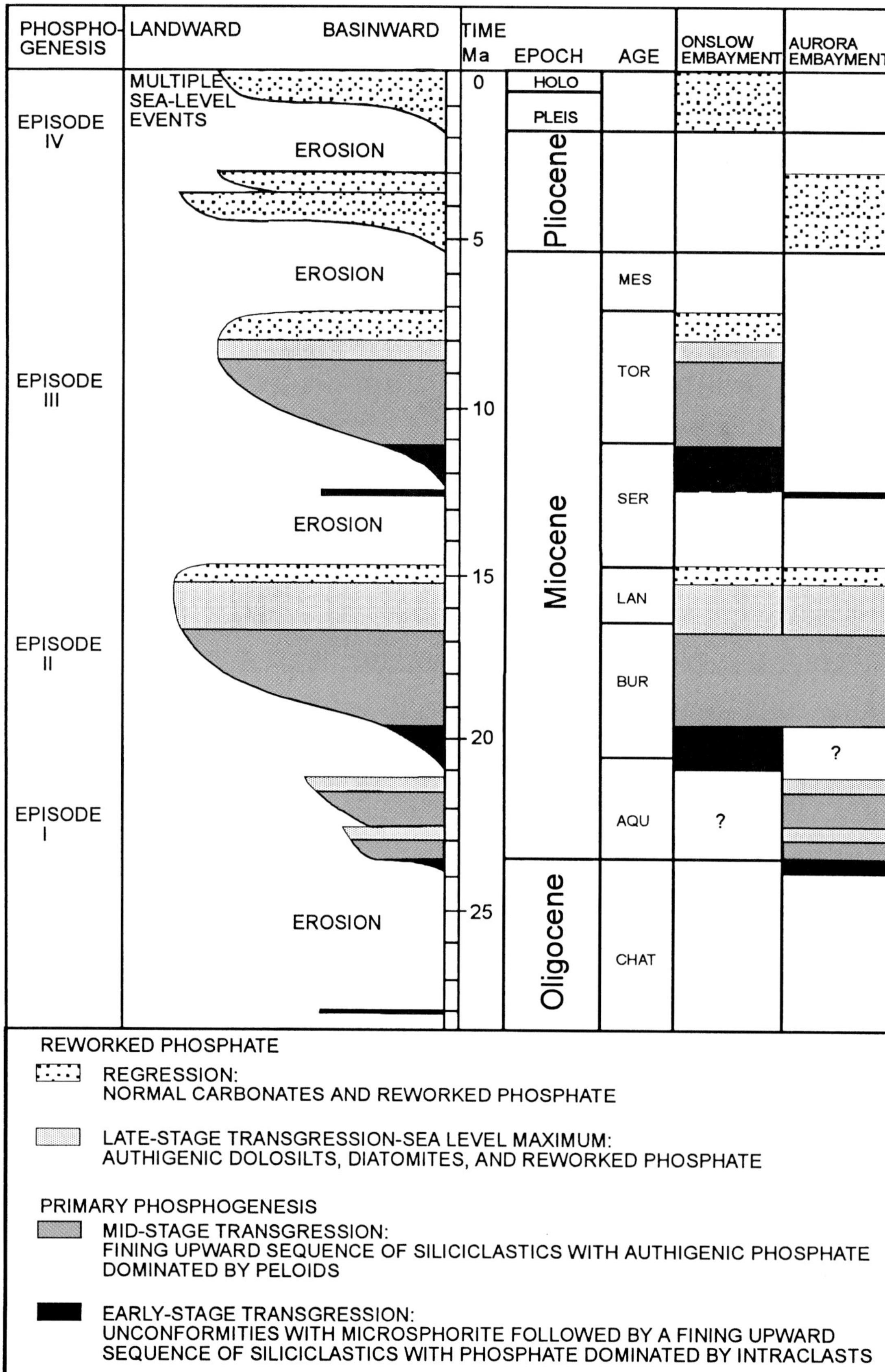

FIG. 8.—Episodes of phosphate formation and deposition for the Upper Cenozoic section of the North Carolina continental margin and their relationship to relative change of coastal onlap. The definition of the phosphate episodes and their relationship to oceanographic sea level cycles is based upon the integration and chronostratigraphic synthesis of biostratigraphic, lithostratigraphic, and Sr isotopic age data.

3. Mid-stage transgression: a fining-upward sequence of siliciclastic muddy sand is deposited concomitantly with primary phosphogenesis of fine sand-sized peloidal and skeletal phosphate grains, which are disseminated through the muddy, fine sand matrix. Phosphogenesis occurs in suboxic environments with high-nutrient levels, high organic production, and low oxygen concentrations (Powers, 1988; Maxwell, 1990; Snyder, 1990; Riggs et al., 1997). Organic matter is degraded in the shallow subbottom zone of sulfate reduction, where authigenic peloidal phosphate forms as disseminated grains in the interstitial environments (Baker and Allen, 1990) with little or no winnowing and reworking of the sediments (Riggs and Mallette, 1990). Conditions for major phosphorite formation are optimal during mid-stage transgression because: a) zonal winds are still strong, causing major upwelling to persist; b) the boundary current has migrated landward, increasing its interaction with the continental margin topography, which intensifies upwelling; and c) dilution from land-derived siliciclastics decreases as the depositional environments deepen. As the transgression proceeds, the continental shelf environments expand, maximizing the distribution of optimum conditions for the formation of phosphate sediments, which results in extensive phosphorite deposition.
4. Maximum transgression and sea-level highstand: deposition of either dolosilts or foraminifer-rich/diatom-rich siliciclastic muds that contain little, if any, phosphate. Phosphogenesis is terminated as oxygen-depleted environments migrate across the continental shelf, and planktic fossils and organic matter become more abundant in the sediments. During maximum transgression, zonal winds are diminished and boundary currents both decrease in intensity and override the continental margin, which terminates upwelling and actively scours and channels the previously deposited continental shelf sediments (Pietrafesa, 1990; Popenoe, 1990; Snyder et al., 1990). Consequently, phosphates occurring within the highstand lithofacies are generally secondary deposits in which phosphate grains have been reworked from the underlying primary phosphorites.
5. Regression: previously deposited sediments are truncated, with phosphates and associated sediments being reworked into a depositional regime characterized by diverse faunas that reflect shallow water, normal marine conditions. The depositional cycle may be capped by thin layers of carbonate sedimentation; however, subsequent low-stand erosion often destroys this caprock. The uppermost, regressive units that survive erosion are often modified by late diagenetic processes during the sea-level lowstand.

Phosphogenic Episodes on the North Carolina Continental Margin.—

Based upon the integrated chronostratigraphy for Aurora and Onslow Embayments (Fig. 7), four episodes of phosphate sedimentation are defined. Fig. 8 relates formation and deposition of different sediment types to stages of transgression and regression associated with proposed third-order sea-level cycles. Three episodes (I, II, and III) are dominated by primary phosphate formation within the Miocene, with some stratigraphic components consisting of reworked phosphate as discussed below. Episodes I and II generally coincide with the sea-level maxima of the TB1 and TB2 second-order cycles of Haq et al. (1987). Episode III partially coincides with the TB3 second-order cycle and coincides with the pre-Messinian highstand. However, there is no evidence of primary phosphogenesis during that portion of the TB3 highstand associated with the Pliocene transgression. The fourth episode (IV) within Pliocene-Quaternary deposits consists totally of reworked phosphate derived from the erosion of sediments formed during primary phosphogenic episodes II and III. These four phosphate episodes of the North Carolina continental margin correspond closely to global episodes identified by Riggs and Sheldon (1990) in their consideration of the temporal distribution of Upper Cenozoic phosphorites. The resulting coastal onlap curve for the North Carolina continental margin is presented in Fig. 8.

Major unconformities occur between each of the three episodes of primary phosphate sedimentation. These unconformable surfaces are characterized by primary phosphogenesis during early-stage transgression resulting in the development of bored microsphorite pavements and associated phosphate rip-up clasts (Riggs, 1979a; Riggs et al., 1997), which form gravelly sands that dominate the basal portion of subsequent transgressive sequences, which in turn grade upsection into *in situ* peloidal phosphorite sand floating within a muddy sand to sandy mud matrix. Unconformities that occur in the post-Miocene section are composed of reworked phosphates but are not characterized by microsphorite pavements or the associated gravelly rip-up intraclasts.

Phosphogenic Episode I occurred during the Aquitanian, forms the basal unit of the Pungo River Formation in the Aurora Embayment, and is poorly defined because of its deep, downbasin occurrence. Units A1 and A2 are interpreted to represent primary phosphogenesis that occurred during the mid-stage transgression between 23.3 Ma and 21.6 Ma. Sediments representing the early and late-stage transgressive, as well as the regressive facies, have not been dated either isotopically or biostratigraphically, but do occur in deeper downbasin drill holes (Ames, 1994). It appears that this episode represents a phosphogenic event of limited extent; however, erosion at the end of the depositional cycle may have destroyed large portions of this sediment sequence.

Phosphogenic Episode II represents the premier phosphogenic event, which took place during the Burdigalian and Langhian (21.0 Ma through 14.8 Ma). Sediments representing this event extend throughout Aurora and Onslow Embayments and include all of units B, C, and D in Aurora Embayment and units FPS1-FPS6 and OBS1-OBS4 in Onslow Embayment. Because all sediment facies of the entire transgression and regression are preserved, this episode serves as the type section for the phosphogenic model.

The basal unconformity at the base of Episode II has a well developed microsphorite deposit yielding Sr dates of 21.0 Ma to 19.5 Ma. FPS1-FPS6 and unit B represent the transgressive phase of Episode II in Onslow and Aurora Embayments, respectively. The transgressive phase consists of a fining upward sequence that begins with an intraclastic phosphorite gravel, which grades sequentially upsection into gravelly intraclastic phosphorite sands, muddy peloidal phosphorite sands, sandy peloidal phosphatic muds, dolomitic muds, and culminates in dolosilts with abundant chert. Minor fluctuations in this idealized succession do occur, suggesting slight oscillations in the depositional environments through the transgression. Intraclastic gravels originated as rip-up clasts from underlying microsphorite surfaces. In the beds of peloidal phosphates, peloids are disseminated within the muddy sands and sandy muds. The peloids have Sr dates equal to associated foraminifera and are interpreted as authigenic, *in situ* deposits. The changing phosphate grain types, sedimentological characteristics, and depositional pattern in this sequence were described in detail by Stille et al. (1994) and Riggs et al. (1997).

Interpretation of the uppermost Burdigalian and Langhian as times of highest sea level during the Upper Cenozoic (Haq et al., 1987) is corroborated by the North Carolina stratigraphic record. Late-stage transgression during the uppermost Burdigalian caused the Gulf Stream to overrun the continental shelf, which terminated upwelling and phosphogenesis and caused severe erosion and reworking of the previously deposited FPS sediments. Vast channel structures were eroded deeply into FPS sediments and extensive erosional scarps were created (the White Oak Scarp of Snyder, 1982; Riggs and Mallette, 1990). FPS6-GSC sediments subsequently filled one of these very large Gulf Stream channels during the uppermost Burdigalian (16.7 Ma Sr date on foraminifera). The FPS6-GSC sediments that backfilled the channel structure consist of marine cut-and-fill deposits containing abundant planktic foraminifera (Snyder et al., 1988). The minor concentrations of phosphate (18.0 Ma Sr age of concurrent phosphate peloids) were reworked from older FPS sediments (Table 2 and Fig. 6).

Maximum transgression and sea-level highstand of Episode II took place during the lowermost Langhian and resulted in extensive reworking of the primary phosphogenic beds (FPS and unit B). The eroded phosphate (18.2–18.0 Ma Sr ages) was redeposited in Aurora Embayment Unit C, with a biostratigraphic age of 16.4–14.8 Ma. Unit C, a clean quartz-phosphorite sand, is the richest bed in the active mine area. Warm Gulf Stream waters extended onto the inner shelf during the maximum transgression and formed the fossiliferous, carbonate interbeds that occur in the upper portion of unit C (Riggs et al., 1982; Scarborough et al., 1982; Riggs, 1984).

The regressive event that ended Episode II formed the sediments of OBS1–OBS4 and unit D in Onslow and Aurora Embayments, respectively. These sediments grade from macrofossiliferous biogenic hash (largely barnacles, bryozoans, and mollusks) in a dolomud matrix in Aurora and northern Onslow Embayments to fine quartz sands interbedded with thin dolomuds in southern Onslow Embayment (Riggs et al., 1982; Scarborough et al., 1982; Riggs and Mallette, 1990). Seismic records demonstrate the regressive nature of the OBS units, which are characterized by clinoform deposits that filled highstand erosional features and led to major shelf progradation (Snyder, 1982, 1994).

The subsequent lowstand, which occurred over some portion of the time frame 14.8–12.7 Ma, eroded and diagenetically altered these regressive units (D and OBS) of Episode II. Diagenesis altered the original micrite matrix of these fossiliferous carbonate sediments to dolosilts. Dolomite rhombs resulted from alternating events of dolomitization and dedolomitization to form complex sequences of highly zoned dolorhombs (Stewart, 1985; Mallette, 1986).

In Aurora Embayment a major unconformity dominated by microsphorite caps Episode II sediments; however, the Episode III depositional sequence is totally missing (Riggs et al., 1982; Ames, 1994). Large phosphatized gravels in the basal portion of the Pliocene Yorktown Formation have ages similar to the underlying microsphorite (12.7 Ma) from which they were derived, whereas associated peloids within the Yorktown Formation have Burdigalian ages (18.0–17.3 Ma).

Phosphogenic Episode III deposited the BBS units in eastern Onslow Embayment (Fig. 1) during the uppermost Serravallian and Tortonian. A discrepancy occurs between the biostratigraphic age assignment and Sr ages of associated phosphate in the BBS (Fig. 6). Sr-derived ages for the 16–8 Ma interval have lower stratigraphic resolution than those ages that are >16 Ma (Hodell et al., 1991); therefore the discrepancy is not as great as it appears in Fig. 6. If the biostratigraphic age assignments are correct, deposition of BBS1–BBS5 occurred in the Serravallian (12.5–11.4 Ma), whereas the phosphate dates are significantly younger (11.5–8.6 Ma or uppermost Serravallian and Tortonian). These younger dates might suggest that the phosphate has undergone subsequent diagenesis or acquired secondary overgrowths. Neither of these possibilities are petrographically obvious (Mallette, 1986), and the former can be rejected for the following reason. All phosphate in the Pliocene and Quaternary has been extensively reworked, with considerable opportunity for diagenesis during sea-level lowstands of the Messinian and Quaternary. However, all reworked post-Messinian phosphates have a tight cluster of dates from the richest and most extensive period of phosphogenesis during the Burdigalian. These data suggest diagenetic alteration has not effected the Sr isotope ratios.

Microsphorite situated upon a major unconformity in the Aurora Embayment has a Sr age date of 12.7 Ma. No other Episode III sediments have been recognized in the Aurora Embayment; the entire record occurs in eastern Onslow Embayment (Figs. 1 and 6) as the BBS depositional sequence. The sea-level cycle began with the deposition of BBS1 through BBS4 between 12.7–8.6 Ma ago. These units consist of a fining-upward sequence that grades from muddy peloidal phosphorite sands, to sandy peloidal phosphatic muds, to dolomuds with abundant diatomite or chert, and are interpreted to represent the transgressive facies and time of phosphogenesis. BBS5 is dominated by biogenic hash and is interpreted to represent the time of maximum transgression. Units BBS6 through BBS8, with ages that are poorly constrained, are interpreted to represent the regressive phases of Episode III. Sr age dates on phosphate peloids in BBS8 (16.6–16.3 Ma) are reworked from an Episode II unit and support this interpretation. BBS6–BBS8 have been diagenetically altered as the micrite muds were severely dolomitized and the microfossil assemblage was totally destroyed in response to weathering during subsequent lowstands.

Significant but highly variable amounts of phosphate occur within the overlying Pliocene, Pleistocene, and Holocene sediments of Phosphogenic Episode IV. All of these phosphates have Sr ages from 18.0 Ma to 10.3 Ma (Table 3), suggesting extensive erosion and reworking of Miocene phosphate-rich sediments. Episode II and III phosphorites, which crop out on the modern seafloor in Onslow Bay, are presently being reworked into the modern continental shelf sediments (Episode IV). Phosphate-rich Miocene units are actively being eroded by both biological and physical processes (Schmid, 1996; Cook, 1997; Riggs et al., 1996, 1998), contributing major amounts of Miocene phosphate to the Holocene sediment regime (Lutenauer and Pilkey, 1967; Riggs et al., 1985; Powers et al., 1990). No evidence supports any post-Miocene primary phosphate formation in North Carolina.

CONCLUSIONS

I. Integrating biostratigraphic and Sr isotope data constrains the ages for third-order depositional sequences and associated unconformities within the Miocene Pungo River Formation in Aurora and Onslow Embayments. The age assignments (Figs. 7 and 8) are based upon the time scale of Berggren et al. (1995).

A. AURORA EMBAYMENT:

1. A microsphorite surface occurs on the basal unconformity between the underlying Oligocene moldic limestone and the Pungo River Formation unit A. The microsphorite has a Sr isotopic age of ~28 Ma.
2. A newly recognized third-order depositional sequence in the Aurora Embayment is called the Aurora Embayment Sequence (AES). Based upon Sr isotopic ages of phosphate in unit A, the AES is 23.2–21.6 Ma. Older and younger facies of this depositional sequence occur in down-basin drill logs.
3. Unit B was deposited 19.1–17.0 Ma and is largely contemporaneous with FPS in Onslow Embayment.
4. Age constraints on units C and D (16.4–14.8 Ma) are based on biostratigraphy because Sr isotopic ages on phosphate suggest that it has been reworked. Units C and D appear to be contemporaneous with OBS in Onslow Embayment.
5. A well developed microsphorite surface marks the contact between unit D and the Pliocene Yorktown Formation. The microsphorite on this unconformity has a Sr isotopic age of 12.7 Ma.

B. ONSLOW EMBAYMENT:

1. A basal unconformity occurs between the underlying Oligocene calcareous quartz sandstone and the Miocene Pungo River Formation. This unconformity contains a microsphorite that has Sr isotopic age dates of 21.0 Ma to 19.5 Ma.
2. The Frying Pan Sequence (FPS) was deposited in the time interval that extends from FPS1 at a maximum age of 21.0 Ma to FPS6 at 16.7 Ma.
3. The Onslow Bay Sequence (OBS) is constrained by biostratigraphic data for OBS1–OBS3, which suggest an age of 15.4 Ma to 14.8 Ma, with OBS4 being slightly younger.
4. The lower units of the Bogue Banks Sequence (BBS1-BBS5) have both biostratigraphic and Sr isotopic age constraints. Deposition occurred 12.5–8.6 Ma. The upper BBS units (BBS6–BBS8) have no biostratigraphic control. Sr ages on phosphate in BBS8 are old, suggesting that the phosphate has been reworked from the FPS units with the deposition of BBS6–BBS8 between 8.6–7.1 Ma.

II. Interpretation of changing patterns of deposition in phosphate-rich sediments of North Carolina leads to development of a sedimentological model for continental margin phosphogenesis. An idealized phosphogenic episode has five stages of sedimentation in response to sea-level change; however, variation in rates, magnitudes, and fluctuations in the cycle may result in varying degrees of development and preservation of each facies.

1. Sea-level lowstand. An erosional unconformity forms across previously deposited sediment sequences in response to subaerial exposure. Dissection by channels and development of hardbottoms on specific lithofacies also occur.
2. Early-stage transgression. Basal microsphorite crusts form on eroded and exposed hardbottom surfaces. Rip-up intraclasts from the microsphorite surface, along with reworked skeletal phosphate grains, dominate the lower portion of subsequent depositional sequence.
3. Mid-stage transgression. A fining-upward sequence of siliciclastic muddy sand to sandy mud is deposited concomitantly with genesis of fine sand-sized peloidal and skeletal phosphate grains. Phosphogenesis occurs in suboxic environments with high nutrient levels, high organic production, and low oxygen concentrations. Organic matter is microbially degraded in the shallow subsurface, where peloidal phosphate forms authigenetically as disseminated grains in interstitial environments with little or no sediment winnowing or reworking.
4. Late-stage transgression. Either dolosilts or foraminifer-rich/diatom-rich muds containing little to no phosphate are deposited. Oxygen-depleted environments migrate landward across the shelf resulting in sediments preserving abundant planktic fossils and associated organic matter.
5. Sea-level highstand and regression. The depositional cycle is capped by a carbonate sediment facies deposited in normal marine conditions under the influence of the Gulf Stream. Previously deposited sediments are severely truncated with extensive reworking of phosphates and associated sediments; fossils consist

of a highly diverse population that reflects normal marine, shallow-water conditions.

III. Based upon the integrated chronostratigraphy for the North Carolina continental margin, four episodes ofphosphate sedimentation have occurred in response to Upper Cenozoic sea-level transgressions and regressions.

1. Three Miocene episodes (I, II, and III) are dominated by sediment facies containing primary phosphate formation with minor facies containing reworked phosphate.
2. Episodes I and II generally coincide with the sea-level maxima of the TB1 and TB2 second-order cycles of Haq et al. (1987). Episode III partially coincides with the TB3 second-order cycle and with the pre-Messinian high. No evidence of primary phosphogenesis exists during the major portion of the TB3 high that is associated with the Pliocene transgression.
3. Episode I occurred during the Aquitanian and forms the basal unit of the Pungo River Formation in the Aurora Embayment. Unit A represents primary phosphogenesis that occurred during mid-stage transgression from 23.3 Ma to 21.6 Ma.
4. Episode II took place throughout Aurora and Onslow Embayments during the Burdigalian and Langhian from 21.0 Ma to 14.8 Ma. This event includes B, C, and D in the Aurora Embayment and FPS1-FPS6 and OBS1–OBS4 in the Onslow Embayment. All sediment facies of Episode II are preserved.
5. Episode III, the final episode of primary phosphogenesis, was deposited as the BBS in eastern Onslow Embayment during the uppermost Serravallian and Tortonian. Episode III began 12.7 Ma and likely continued through the Tortonian. The Messinian low sea level caused severe erosion and weathering of the Episode III sediments, eliminating them from Aurora Embayment.
6. Episode IV involved many rapid sea-level events represented by Pliocene-Quaternary deposits that locally contain variable concentrations of phosphate. The phosphate consists totally of reworked grains of all types derived from the erosion of nearby sediments formed during episodes II and III.

ACKNOWLEDGMENTS

This research was a cooperative project that was partially supported by the *Centre National de la Recherche Scientifique* at the *Centre de Geochimie de la Surface,* Strasbourg, France, and the National Science Foundation grant OCE-8709429 to S. R. Riggs. We thank D. J. Mallinson, Craig R. Glenn, and an anonymous reviewer for critical reviews of the manuscript. This paper is a contribution to International Geological Correlation Program 325 (IGCP 325, "Correlation of Palaeogeography with Phosphorites and Associated Authigenic Minerals").

REFERENCES

AMES, D. V., 1994, Petrology and stratigraphy of the Miocene Pungo River and Pliocene Yorktown Formations in the Aurora Area, North Carolina: Unpublished M.S. Thesis, East Carolina University, Greenville, NC, 126 p.

AMES, D. V., RIGGS, S. R., CROWSON, R. A., AND SHOWERS, W. J., 1991, Geochemical variability of phosphate grain types within the Upper Cenozoic sediments of the Carolina continental margin (abs.): Geological Society of America, Abstracts with Programs, v. 23, n. 1, p. 3.

BAKER, P., AND ALLEN, M., 1990, Occurrence of dolomite in Neogene phosphatic sediments, in Burnett, W. C., and Riggs, S. R., eds., Neogene to Modern Phosphorites: Cambridge, Cambridge University Press, Phosphate Deposits of the World, Volume III, p. 75-86.

BATURIN, G. N., 1982, Phosphorites on the Sea Floor: Origin, Composition, and Distribution: Developments in Sedimentology, v. 33: New York, Elsevier Publishing Company, 343 p.

BERGGREN, W. A., KENT, D. V., AND VAN COUVERING, J. A., 1985, The Neogene, Part 2. Neogene geochronology and chronostratigraphy, in Snelling, N. J., ed., The Chronology of the Geological Record: London, Geological Society of London Memoir 10, p. 141-195.

BERGGREN, W. A., KENT, D. V., SWISHER, C. C., III, AND AUBREY, M. P., 1995, A revised Cenozoic geochronology and chronostratigraphy, in Berggren, W. A., Kent, D. V., Aubrey, M. P., and Hardenbol, J., eds., Geochronology, Time Scales, and Global Stratigraphic Correlations: A Unified Temporal Framework for an Historical Geology: Tulsa, Society for Sedimentary Geology (SEPM) Special Publication 54, p. 129-212.

BLOW, W. H., 1969, Late middle Eocene to Recent planktonic foraminiferal biostratigraphy, in Brönnimann, P., and Renz, H. H., eds., Proceedings of the First International Conference on Planktonic Microfossils, Geneva, 1976: Leiden, E.J. Brill, v. 1, p. 199-422.

BOLLI, H. M., AND SAUNDERS, J. B., 1985, Oligocene to Holocene low latitude planktonic foraminifera, in Bolli, H. M., Saunders, J. B., and Perch-Nielsen, K., eds., Plankton Stratigraphy: Cambridge, Cambridge University Press, p. 155-262.

COMPTON, J. S., HODELL, D. A., GARRIDO, J. R., AND MALLINSON, D. J., 1993, Origin and age of phosphorite from the south-central Florida Platform: Relation of phosphogenesis to sea-level fluctuations and (13C excursions: Geochimica et Cosmochimica Acta, v. 57, p. 131-146.

COOK, J. W., 1997, The benthic infaunal community, bioerosion, and sediment production on Miocene mudstone hardbottoms, Onslow Bay, North Carolina: Unpublished M.S. Thesis, East Carolina University, Greenville, N.C., 224 p.

COOK, P. J., AND MCELHINNY, M. W., 1979, A reevaluation of the spatial and temporal distribution of sedimentary phosphate deposits in the light of plate tectonics: Economic Geology, v. 74, p. 315-330.

CROWSON, R. A., SHOWERS, W. J., AND RIGGS, S. R., 1990, Oxygen isotope geochemistry of Miocene phosphate grains from Onslow Bay, NC: Evidence for Miocene Western Boundary Current induced upwelling (abs.): Geological Society of America, Abstracts with Programs, v. 22, no. 7, p. A354.

CROWSON, R. A., SHOWERS, W. J., AND RIGGS, S. R., 1991, High resolution isotopic stratigraphy of phosphorites from the Miocene section, Onslow Bay, North Carolina (abs.): Geological Society of America, Abstracts with Programs, v. 23, no. 1, p. 19.

ELLINGTON, M. D., 1984, Major and trace element composition of phosphorites of the North Carolina continental margin: Unpublished M.S. Thesis, East Carolina University, Greenville, N.C., 96 p.

HAQ, B. U., HARDENBOL, J., AND VAIL, P. R., 1987, Chronology of fluctuating sea levels since the Triassic: Science, v. 235, p. 1156-1167.

HODELL, D. A., MUELLER, R. A., AND GARRIDO, J. R., 1991, Variations in the strontium isotopic composition of seawater during the Neogene: Geology, v. 19, p. 24-27.

KENNETT, J. P., AND SRINIVASAN, M. S., 1983, Neogene Planktonic Foraminifera: A Phylogenetic Atlas: Stroudsburg, Hutchinson Ross, 265 p.

LUTERNAUER, J. L., AND PILKEY, O. H., 1967, Phosphorite grains: Their application to the interpretation of North Carolina shelf sedimentation: Marine Geology, v. 5, p. 315-320.

MALLETTE, P. M., 1986, Lithostratigraphic analysis of cyclical phosphorite sedimentation within the Miocene Pungo River Formation, North Carolina continental shelf: Unpublished M.S. Thesis, East Carolina University, Greenville, N.C., 186 p.

MALLINSON, D. J., AND COMPTON, J. S., 1995, Mixed carbonate-siliciclastic sequence stratigraphy utilizing Sr-isotopes: Deciphering the Miocene sea-level history on the Florida Platform, in Haq, B., ed., Sequence Stratigraphy and Depositional Response to Eustatic, Tectonic and Climatic Forcing: Provite City, Kluwer Academic Press, p. 25-58.

MALLINSON, D. J., AND COMPTON, J. S., 1997, Linking phosphogenic episodes on the southeast U.S. margin to marine $d^{13}C$ and $d^{18}O$ records: Geology, v. 25, p. 103-106.

MALLINSON, D. J., COMPTON, J. S., SNYDER, STEPHEN W., AND HODELL, D. A., 1994, Strontium isotopes and Miocene sequence stratigraphy across the northeast Florida platform: Journal of Sedimentary Research, v. 64, p. 392-407.

MARTINI, E., 1971, Standard Tertiary and Quaternary calcareous nannoplankton zonation, in Farinacci, A., ed., Proceedings of Second Planktonic Conference, Roma, 1970: Edizioni Tecnoscienza, v. 2, p. 739-785.

MAXWELL, M. E., 1990, Organic carbon and nitrogen concentrations and stable isotopic ratios of organic carbon in the Miocene Pungo River Formation, North Carolina continental shelf: Unpublished M.S. Thesis, University of North Carolina, Chapel Hill, N.C., 84 p.

PALMER, A. A., 1988, Radiolarians from the Miocene Pungo River Formation of Onslow Bay, North Carolina continental shelf, in Snyder, S. W., ed., Micropaleontology of Miocene Sediments in the Shallow Subsurface of Onslow Bay, North Carolina Continental Shelf: Cambridge, Mass., Cushman Foundation for Foraminiferal Research Special Publication 25, p. 163-178.

PIETRAFESA, L., 1990, Upwelling processes associated with Western Boundary Currents, in Burnett, W. C., and Riggs, S. R., eds., Neogene to Modern Phosphorites: Cambridge, Cambridge University Press, Phosphate Deposits of the World, Volume III, p. 3-26.

POPENOE, P., 1990, Paleoceanography and paleogeography of the Miocene of the southeastern United States, in Burnett, W. C., and Riggs, S. R., eds., Neogene to Modern Phosphorites: Cambridge, Cambridge University Press, Phosphate Deposits of the World, Volume III, p. 352-380.

POWERS, E. R., 1988, Diatom biostratigraphy and paleoecology of the Miocene Pungo River Formation, Onslow Bay, North Carolina continental shelf, in Snyder, S. W., ed., Micropaleontology of Miocene Sediments in the Shallow Subsurface of Onslow Bay, North Carolina Continental Shelf: Cambridge, Mass., Cushman Foundation for Foraminiferal Research Special Publication 25, p. 97-161.

POWERS, E. R., CROWSON, R. A., RIGGS, S. R., SNYDER, STEPHEN W., SNYDER, SCOTT W., AND HINE, A. C., 1990, Chemical characteristics and re-evaluation of the phosphate resource potential in Onslow Bay, North Carolina continental shelf: Marine Mining, v. 9, p. 1-41.

RIGGS, S. R., 1979a, Petrology of the Tertiary phosphorite system of Florida: Economic Geology, v. 74, p. 195-220.

RIGGS, S. R., 1979b, Phosphorite sedimentation in Florida—a model phosphogenic system: Economic Geology, v. 74, p. 285-314.

RIGGS, S. R., 1984, Paleoceanographic model of Neogene phosphorite deposition, U.S. Atlantic Continental Margin: Science, v. 223, p. 123-131.

RIGGS, S. R., 1987, Model of Tertiary phosphorites on the world's continental margins, in Teleki, P. G., Dobson, M. R., Moore, J. R., and von Stackelberg, U., eds., Marine Minerals: NATO Advanced Research Workshop, Dordrecht, D. Reidel Publishing Company, p. 99-118.

RIGGS, S. R., AND MALLETTE, P. M., 1990, Patterns of phosphate deposition and lithofacies relationships within the Miocene Pungo River Formation, North Carolina continental margin, in Burnett, W. C., and Riggs, S. R., eds., Neogene to Modern phosphorites: Cambridge, Cambridge University Press, Phosphorite Deposits of the World, v. III, p. 424-443.

RIGGS, S. R., AND MANHEIM, F. T., 1988, Mineral resources of the U.S. Atlantic continental margin, in Sheridan, R. E., and Grow, J. A., eds., The Atlantic Continental Margin, U.S.: Boulder, Geological Society of America, The Geology of North America, v. I-2, p. 501-520.

RIGGS, S. R., AND SHELDON, R. P., 1990, Paleoceanographic and paleoclimatic controls of the temporal and geographic distribution of Upper Cenozoic continental margin phosphorites, in Burnett, W. C., and Riggs, S. R., eds., Neogene to Modern Phosphorites: Cambridge, Cambridge University Press, Phosphate Deposits of the World, Volume III, p. 207-222.

RIGGS, S. R., LEWIS, D. R., SCARBOROUGH, A. K., AND SNYDER, SCOTT W., 1982, Stratigraphy and petrology of the Pungo River Formation, central Coastal Plain, North Carolina: Southeastern Geology, v. 23, p. 205-216.

RIGGS, S. R., SNYDER, STEPHEN W., HINE, A. C., SNYDER, SCOTT W., ELLINGTON, M. D., AND MALLETTE, P. M., 1985, Geologic framework of phosphate resources in Onslow Bay, North Carolina continental shelf: Economic Geology, v. 80, p. 716-738.

RIGGS, S. R., SNYDER, STEPHEN W., SNYDER, SCOTT W., AND HINE, A. C., 1990, Stratigraphic framework for cyclical deposition of Miocene sediments in the Carolina Phosphogenic Province, in Burnett, W. C., and Riggs, S. R., eds., Neogene to Modern Phosphorites: Cambridge, Cambridge University Press, Phosphate Deposits of the World, Volume III, p. 381-395.

RIGGS, S. R., SNYDER, STEPHEN W., HINE, A. C., AND MEARNS, D. L., 1996, Hardbottom morphology and relationship to the geologic framework: Mid-Atlantic continental shelf: Journal of Sedimentary Research, v. 66, p. 830-846.

RIGGS, S. R., STILLE, P., AND AMES, D., 1997, Sr isotopic age analysis of co-occurring Miocene phosphate grain types on the North Carolina continental shelf: Journal of Sedimentary Research, v. 67, p. 65-73.

RIGGS, S. R., AMBROSE, W. G., COOK, J. W., SNYDER, SCOTT W., AND SNYDER, STEPHEN W., 1998, Sediment production on sediment-starved continental margins: The interrelationship between hardbottoms, sedimentological and benthic community processes, and storm dynamics: Journal of Sedimentary Research, v. 68, p. 155-168.

SCARBOROUGH, A. K., RIGGS, S. R., AND SNYDER, SCOTT W., 1982, Stratigraphy and petrology of the Pungo River Formation, central coastal plain, North Carolina: Southeastern Geology, v. 23, p. 205-216.

SCHMID, K. A., 1996, Sedimentation associated with mid-shelf hardbottoms in Onslow Bay, North Carolina: Unpublished M.S. Thesis, East Carolina University, Greenville, N.C., 161 p.

SHELDON, R. P., 1964, Paleolatitudinal and Paleogeographic Distribution of Phosphorites: Washington, D.C., United States Geological Survey Professional Paper 501-C, p. C106-C113.

SHELDON, R. P., 1980, Episodicity of phosphate deposition and deep ocean circulation—an hypothesis, in Bentor, Y. K., ed., Marine Phosphorites—Geochemistry, Occurrence, Genesis: Tulsa, Society for Sedimentary Geology (SEPM) Special Publication 29, p. 239-248.

SNYDER, SCOTT, W., ed., 1988, Micropaleontology of Miocene Sediments in the Shallow Subsurface of Onslow Bay, North Carolina Continental Shelf: Cambridge, Mass., Cushman Foundation Special Publication 25, 189 p.

SNYDER, SCOTT W., 1990, Relationships between benthic foraminiferal assemblages and Neogene phosphatic sediments, North Carolina coastal plain and continental shelf, in Burnett, W. C., and Riggs, S. R., eds., Neogene to Modern Phosphorites: Cambridge, Cambridge University Press, Phosphate Deposits of the World, Volume III, p. 444-464.

SNYDER, SCOTT W., STEINMETZ, J. C., WATERS, V. J., AND MOORE, T. L., 1988, Occurrence and biostratigraphy of planktonic foraminifera and calcareous nannofossils in Pungo River Formation sediments from Onslow Bay, North Carolina continental shelf, in Snyder, Scott W., ed., Micropaleontology of Miocene Sediments in the Shallow Subsurface of Onslow Bay, North Carolina Continental Shelf: Cambridge, Mass., Cushman Foundation for Foraminiferal Research Special Publication 25, p. 15-41.

SNYDER, STEPHEN W., 1982, Seismic stratigraphy within the Miocene Carolina Phosphogenic Province: Chronostratigraphy, paleotopographic controls, sea-level cyclicity, Gulf Stream dynamics, and the resulting depositional framework: Unpublished M.S. Thesis, University of North Carolina, Chapel Hill, 183 p.

SNYDER, STEPHEN W., 1994, Miocene sea-level cyclicity: Frequency and amplitude estimates from the Carolina Platform: Unpublished Ph.D. Dissertation, University of South Florida, St. Petersburg, 688 p.

SNYDER, STEPHEN W., HINE, A. C., AND RIGGS, S. R., 1982, Miocene seismic stratigraphy, structural framework and sealevel cyclicity: North Carolina continental shelf: Southeastern Geology, v. 23, p. 247-266.

SNYDER, STEPHEN W., HINE, A. C., AND RIGGS, S. R., 1990, The seismic stratigraphic record of shifting Gulf Stream flow paths in response to Miocene glacioeustacy: Implications for phosphogenesis along the North Carolina continental margin, in Burnett, W. C., and Riggs, S. R., eds., Neogene to Modern Phosphorites: Cambridge, Cambridge University Press, Phosphate Deposits of the World, Volume III, p. 396-423.

SNYDER, STEPHEN W., HINE, A. C., RIGGS, S .R., AND SNYDER, SCOTT W., 1993, Miocene geology of the continental shelf, Onslow Bay, North Carolina: North Carolina Geological Survey, Map Series No. 3.

Stainforth, R. M., Lamb, J. L., Luterbacher, H., Beard, J. H., and Jeffords, R. M., 1975, Cenozoic planktonic foraminiferal zonation and characteristics of index forms: University of Kansas, Paleontological Contributions, Article 62, p. 1-425.

Stewart, T. L., 1985, Carbonate petrology and sedimentology of the Miocene Pungo River Formation, Onslow Bay, North Carolina continental shelf: Unpublished M.S. Thesis, East Carolina University, Greenville, N.C., 184 p.

Stille, P., Riggs, S., Clauer, N., Ames, D., Crowson, R., and Snyder, Stephen, 1994, Sr and Nd isotopic analysis of phosphorite sedimentation through one Miocene high-frequency depositional cycle on the North Carolina continental shelf: Marine Geology, v. 117, p. 253-273.

APPENDIX: BIOSTRATIGRAPHIC REASSESSMENT OF THE PUNGO RIVER FORMATION

The update of planktic foraminiferal datum events by Berggren et al. (1995) included a number of first occurrence (fo) and highest occurrence (ho) datums that are relevant to interpreting the Miocene section of Onslow Bay (Fig. 4). These datums, along with the occurrences of additional taxa whose ranges are well established relative to the planktic foraminiferal zonation of Blow (1969), are central to our chronozonal assignments. The ranges of taxa other than those that have been reassessed by Berggren et el. (1995) are drawn from Bolli and Saunders (1985), Berggren et al. (1985), Kennett and Srinivasan (1983), and Stainforth et al. (1975). During the following discussions, frequent reference to Figs. 4 and 5 will be helpful. The "M" Zones (e.g., M1, M2, etc.) are from Berggren et al. (1995) and the "N" Zones (e.g., N4, N5, etc.) are from Blow (1969).

Frying Pan Third-Order Sequence (FPS1 - FPS6).—
The fo and ho datums of *Globorotalia kugleri* define Zone M1 (= Zone N4). The absence of *Gr. kugleri* from the oldest Miocene sequences in Onslow Bay is significant because other globorotalids occur commonly in these rich faunas. Hence, the absence of *Gr. kugleri* cannot be attributed to preservational bias, which indicates that sediments equivalent to the lower Aquitanian (23.8 to 23.2 Ma) were not recovered. The fo of *Globoquadrina dehiscens* lies within Zone M1 (= Zone N4) at 23.2 Ma. Its common occurrence upward from the base of the Miocene section, in conjunction with the absence of *Gr. kugleri,* suggests that sediments may be as old as the base of Zone M2 (= Zone N5). *Globigerinoides primordius,* which occurs in FPS1 through FPS3, has a known range from the middle of Zone N4 through Zone N5. Hence, its presence indicates that FPS1 through FPS3 may be assignable to Zone N5. Snyder et al. (1988) did not recognize sediments older than Zone N6, primarily because *Globigerinoides praeimmaturus,* with a fo datum at the base of Zone N6, was recorded in FPS1 (i.e., specimens of *Gs. primordius* were interpreted to be reworked). However, *G. praeimmaturus* is rare, occurs sporadically, and does not characterize lowermost FPS1. Hence, the common occurrence of *Gs. primordius* in FPS1 has been reinterpreted to recognize that sediments assignable to chronozone N5 may be present (Fig. 6). The co-occurrence of *Globigerinoides altiaperturus* (known range of Zone N5 through mid N6) is consistent with this interpretation, as are the biostratigraphic constraints of diatoms, radiolarians, and calcareous nannofossils (Snyder, 1988). Hence, the lower age constraint for the Frying Pan third-order sequence has been modified downward from the base of chronozone N6 to include the upper part of chronozone N5 (Fig. 6). According to the Berggren et al. (1995) time scale, this corresponds to an age that may be as old as 21.5 Ma.

The presence of *Catapsydrax* spp., with ho datums known to approximate the N6-N7 (= M3-M4) zonal boundary, throughout the Frying Pan third-order sequence indicates that the sediments are no younger than chronozone N6. Reassessment of the ho of *Catapsydrax* indicates a ho at the top of Zone M3 (= Zone N6) at 17.3 Ma (Berggren et al., 1995). In addition, *Gs. altiaperturus* occurs throughout the Frying Pan Sequences. Because its ho datum lies in mid N6, the upper boundary of the Frying Pan third-order sequence appears to lie below the N6–N7 chronozonal boundary.

Some minor stratigraphic mixing occurs, perhaps resulting from reworking *Gs. primodius* upward into FPS2 and FPS3. This interpretation is supported by the occurrence of *Globigerinatella insueta* in FPS2. With a fo at the base of Zone M3 (= Zone N6) at 18.8 Ma, its presence in FPS2 through FPS 6 indicates that only the lowest portion of the Frying Pan third-order sequence (i.e., FPS1) lies in chronozone N5. The occurrences of *Globorotalia sictula praescitula* (fo within Zone N6 at 18.5 Ma) and *Praeorbulina sicana praesicana* in FPS6 support this interpretation. Consequently, the upper and lower age limits of the Frying Pan Sequence can be estimated as shown in Fig. 6. The faunas upon which this interpretation is based occur in FPS1, 2, 3, and 6. Hence, the highest and lowest fourth-order sequences are biostratigraphically interpreted, and the age constraints of the encompassing third-order sequence are considered reliable. A conservative estimate for the age of the Frying Pan third-order sequence is somewhere between 21.5 and 17.5 Ma (Fig. 6). Because this estimate merely represents the greatest biostratigraphic resolution based upon our data, the actual age of these sequences may occupy only some portion of this range. In fact, biostratigraphic evidence, although not conclusive, suggests a narrower range of upper chronozone N5 through mid N6 (roughly 19.5 Ma to 18 Ma).

Onslow Bay Third-Order Sequences (OBS1–OBS4).—
Planktic foraminifera were recovered from only two of the Onslow Bay fourth-order sequences (OBS2 and OBS3). These two sequences are characterized by the co-occurrence of a number of taxa that constrain them to upper M5–M6 (approximately equivalent to upper N8-N9). Stratigraphically important taxa include *Globorotalia birnageae* (range of N7–N8), *Globorotalia scitula praescitlua* (ho within Zone N9), *Praeorobulina sicana sicana* (ho within Zone N9), *Praerobulina glomerosa* (range of mid N8–mid N9), and *Orbulina* spp. (fo at the base of N9). Among these taxa, Berggren et al. (1995) placed the fo of *Gr. birnageae* at 16.7 Ma, the first occurrence of *P. sicana sicana* at 16.4 Ma, the fo *P. glomerosa* at 16.1 Ma, and the fo of *Orbulina* at 15.1 Ma.

Again, minor stratigraphic mixing appears to occur, as evidenced by the co-occurrence of *Gr. birnageae* (ho at the top of Zone N8) and *Orbulina* (fo at the base of Zone 9). Co-occurrence of these two taxa would require only the slightest mixing, so the base of the sequences is placed in latest chronozone N8. The upper age limit is conservatively placed at the top of chronozone N9, although it may lie within this zone. The newly interpreted age range of OBS2 and OBS3 is from about 15.4 Ma to 14.8 Ma (Fig. 6). These sequences may occupy only some portion of this range, but these are the most precise constraints possible based on planktic foraminiferal biostratigraphy.

Although the radiolarian and calcareous nannofossil zonations for the Onslow Bay third-order sequence are considerably more inclusive, the diatom evidence corroborates assignments based on planktic foraminifera (Powers, 1988; Snyder, 1988). Diatoms present in OBS1 indicate equivalence to planktic foraminiferal Zone N8, but biostratigraphic data for OBS4 are lacking. Hence, the lower boundary of the Onslow Bay third-order sequence (base of OBS1) may be slightly lower and its upper boundary (top of OBS4) may be somewhat higher than indicated on Fig. 6.

Bogue Banks Third-Order Sequence (BBS1–BBS8).—

Planktic foraminiferal evidence from the Bogue Banks third-order sequence is sparse, as specimens were recovered only from BBS1 and BBS5. Only one taxon that was reassessed by Berggren et al. (1995) occurs in these sediments, and its presence documents stratigraphic mixing. Rare specimens of *Globorotalia peripheroronda* have been recovered from BBS1 and BBS5. According to Berggren et al. (1995), this species' ho datum is in the lower part of Zone M7 (= N10) at 14.8 Ma. However, *Globigeroinoides bolli* and *Hastigerina siphonifera,* both of which first appear at the base of Zone N12 (= M9), are also present in BBS1 and BBS5. Because *Gr. peripheroronda* is rare and poorly preserved, we interpret specimens to be reworked. Hence, planktic foraminifera from the Bogue Banks third-order sequence indicate assignment to chronozone N12 or younger, corresponding to 12.7 Ma or younger. Because BBS1 and BBS6–8 have not yielded planktic foraminifera, the biostratigraphic age constraints for the Bogue Banks third-order sequence remain questionable.

Nannofossils, the only group of microfossils recovered from BBS6-8, indicate only that these sediments are assignable to Zones NN7 (Martini, 1971) or younger (Snyder, 1988). The top of Zone NN7 corresponds to the top of planktic foraminiferal Zone N14, which according to Berggren et al. (1995) equals the top of his Zone M11 at 11.4 Ma. Radiolarians were recovered only from BBS1, and diatoms, which corroborate the chronozonal assignment based on planktic foraminifera, were not recovered from above BBS5. Hence, the upper age limit of the Bogue Banks third-order sequence may extend to as young as 11.4 Ma, or possibly even younger.

PRISTINE TO REWORKED VERDINE: KEYS TO SEQUENCE STRATIGRAPHY IN MIXED CARBONATE-SILICICLASTIC FOREREEF SEDIMENTS (GREAT BARRIER REEF)

JOHN D. KRONEN Jr. AND CRAIG R. GLENN*

Department of Geology and Geophysics, SOEST, University of Hawaii, 1680 East-West Road, Honolulu, HI 96822

ABSTRACT: Thin, well-defined condensed sections containing authigenic green minerals of the verdine facies occur in forereef depositional sequences of the Great Barrier Reef (GBR), Queensland, Australia. Pristine and reworked verdine grains within these sections formed in a mixed carbonate-siliciclastic depositional setting that accumulated during the last 1.5 m.y. in relatively warm waters. Seismic reflection, physical properties, and oxygen isotope data reveal that cyclic deposition occurred at ODP Site 821. Authigenic verdine grains are concentrated and most abundant along maximum flooding surfaces at the top of transgressive systems tracts and are also concentrated along marine flooding surfaces within transgressive systems tracts. These are notably "pristine" and commonly occur as relatively well-preserved green clay minerals that infill primary intraparticle porosity, especially within foraminifer tests. There are also verdine grains that show the effects of reworking and transport. These are usually not contained within foraminifer tests like the pristine verdine, but are abraded to fragmented, less abundant, and typically occur within highstand systems tracts. The reworked nature of these particles is attributed to a higher-energy depositional setting where loss of accommodation space, seaward progradation, and sediment reworking occur. Chemical data from electron microprobe analyses reveal low K_2O concentrations and X-ray diffraction data possibly indicate variations in the degree of verdine mineral maturity.

Verdine is ferric-rich, but because it contains iron in both its oxidized and reduced states, we suggest that it forms in association with mildly reducing, suboxic solutions from which both ferric and ferrous iron may be supplied. Evolution of verdine is terminated shortly after burial. The relatively rapid formation of verdine facies minerals may have implications on reverse weathering and may play an important role in large scale biogeochemical cycling in the oceans. This facies may provide a sink for reverse weathering reactions involving Fe, Mg, and K in clay mineral formation. Although it has been previously suggested that *in situ* formation of verdine minerals is rare in water depths exceeding ca. 60 m and, due to its poor preservation potential, is rare in pre-Holocene sediments, we document here its formation at paleo-water depths of about 200 m within repetitive moderately condensed sections spanning the past ~1.5 m.y.

INTRODUCTION

Condensed sections observed in sedimentary sequences at the Great Barrier Reef (GBR) forereef (see Glenn et al., 1993) contain authigenic marine minerals that belong to the verdine facies. Based on their X-ray diffraction and chemical characteristics (see following discussion), these minerals are identified here as phyllite V, a ferric-rich, dioctahedral-trioctahedral clay containing a mixture of two authigenic mineral phases, 7Å phyllite V and 14Å phyllite V. A pure 7Å phyllite V clay mineral has been formally named "odinite" (Bailey, 1988).

It has been previously suggested that continental margins at low latitudes are ideally suited for formation of the minerals of the verdine facies as long as water temperatures are above 25°C and there is a sufficient supply of originally particulate iron, usually initially delivered to the depositional site by a riverine source or also by a nearby volcanic source (Odin and Sen Gupta, 1988; Odin, 1988, 1990). The sedimentary sequences at the forereef of the GBR margin formed in mixed carbonate/siliciclastic depositional settings and thus received sediments from two different sources and usually at different times (Davies et al., 1991; Glenn et al., 1993). Such mixed settings can display distinct variability between predominantly carbonate-rich and predominantly siliciclastic-rich sedimentary units. These alternations are inferred to result as a response to changes in tectonism, sediment supply, climate, and sea level over time (e.g. Wilgus et al., 1988).

During a three-site transect of Ocean Drilling Program (ODP) Leg 133, coring penetrated holes to depths of 400 m below seafloor (mbsf) at Sites 819, 820, and 821 near Grafton Passage, east-northeast of Cairns (Fig. 1). In this study, we investigate the significance of the verdine contained within these sediments as a proxy for cyclic sedimentation brought about by sea-level change and also their role as a sequence stratigraphic tool. Cyclic changes in sea level and climate, carbonate deposition, and periodic input of fluvially derived continental siliciclastic detritus have influenced the formation of approximately third- (0.5–2 m.y.) to fifth-order (0.01–0.08 m.y.) depositional sequences at the GBR. The timing of verdine formation also provides one of the foundations for a sequence stratigraphic interpretation of the GBR margin (J. D. Kronen and C. R. Glenn, in prep., 1999). We maintain that the condensed sections formed during maximum rates of eustatic rise in sea level, which are conducive to slow net sedimentation that continues over relatively long periods of time, as embraced by Loutit et al. (1988). Among other factors, these are conditions suited for the formation of verdine minerals in low-latitude tropical environments. The relatively rapid formation of verdine facies minerals may also have important implications with regard to providing a global sink for reverse weathering reactions involving Fe, Mg, and K in clay mineral formation (e.g., Aller et al., 1986; Mackenzie and Kump, 1995; Michalopoulos and Aller, 1995).

METHODOLOGY

Lithologic variability in cores recovered at Sites 819, 820, and 821 was examined for % total carbonate and magnetic susceptibility. Smear slides of sample material >63 μm were made from 345 samples for the 400-m interval at Site 821. Specific mineralogies were determined by electron microprobe (EMP) analysis of grain mounts using a Cameca Camebax probe, by X-ray diffraction (XRD) using a Scintag Pad V diffractometer, and

*email: glenn@soest.hawaii.edu

Marine Authigenesis: From Global to Microbial, SEPM Special Publication No. 66

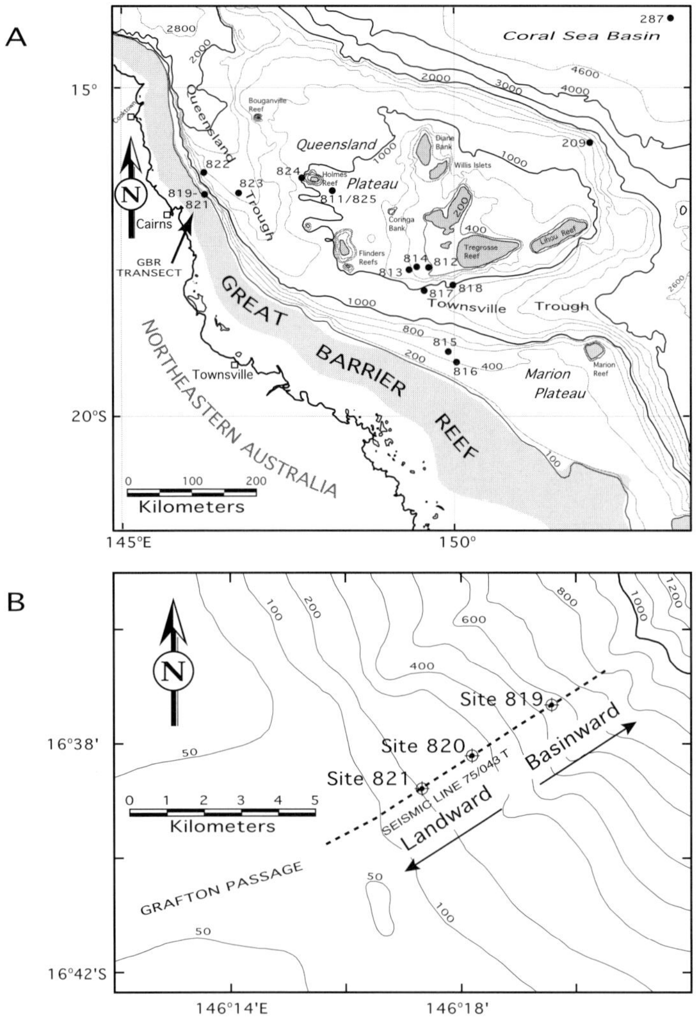

FIG. 1.—(A) Map showing locations of ODP Sites 811-825, DSDP Sites 209 and 287, and the major physiographic features of the northeastern Australian margin (after McKenzie et al., 1993). Bathymetry is in meters. (B) Location of GBR transect ODP Sites 819-821 on the outermost shelf and upper slope (after Feary and Jarrard, 1993). The track of AGSO (formerly BMR) seismic reflection line 75/43T (see Fig. 2) is shown as dashed line. Location of map is shown by arrow in A, bathymetry in meters.

by X-ray fluorescence spectrometry (EDS) using a Princeton Gamma Tech spectrometer.

Twenty-two samples containing green grains (verdine minerals) were selected for further analysis. The >63 μm-size fractions were separated by a Franz Isodynamic separator with a longitudinal slope of 25° and a lateral slope of 17°, settings that have been used in other studies (e.g., Odin et al., 1988a; Purnachandra Rao et al., 1993, 1995). Foraminiferal tests infilled with green material and free green grains were initially attracted at 0.5–0.8 A. Generally, isolation of paramagnetic grains was good, but a few samples were reintroduced for separation at higher currents (0.9–1.2 A). The isolated grains were then powdered and scanned by XRD from 2° to 40° 2θ at 1°2θ/min. Grains were imaged using the scanning electron microscope (SEM) housed in a Cameca SX 50 probe (EMP).

SEQUENCE STRATIGRAPHY

An interpretation of the sequence stratigraphy of the GBR transect (Glenn et al., 1993) as annotated on AGSO seismic line 75/043 (Part T) is shown in Fig. 2. The sequence stratigraphy is based on principles and procedures of sequence stratigraphy as outlined in Van Wagoner et al. (1988; 1990), Posamentier et al. (1988), Posamentier and Vail (1988), Sarg (1988), Vail (1990), Vail et al. (1991), and Haq (1991). Nine sequences, dominated by simple transgressive-highstand systems tract couplets, were identified. Biostratigraphic age data of Wei and Gartner (1993) demonstrates that the durations of the Pleistocene-Holocene sequences at Site 821 range from nearly third- (0.5–2 m.y.) to fifth-order (0.01–0.08 m.y.). A lower, outward building progradational unit (sequences 9 through 6) is distinguishable from an overlying upward building aggradational unit (sequences 5 through 1). Sequence 8 has a thick accumulation of sediments, probably composed of lowstand prograding wedge deposits, that represent a significant episode in slope margin outbuilding.

Sequences at Sites 820 and 821 are interpreted to be separated by reflector truncating sequence boundaries (SB), and are composed of transgressive (TST) and highstand (HST) systems tract deposits (Fig. 2). Approximately meter-thick intervals separate TST and HST deposits and contain concentrations of verdine facies authigenic grains, siliciclastics, and bioclasts (see below), and are interpreted here as maximum flooding surfaces (MFS) that occur along the toes of downlapping reflectors. Authigenic verdine grains are also occasionally concentrated within thin intervals coinciding with marine flooding surfaces within TSTs.

SEDIMENTOLOGY

The sediments recovered at Sites 819, 820, and 821 are predominantly greenish gray, unlithified to partially lithified, calcareous bioclast-rich oozes, chalks, mudstones, wackestones, and packstones. Predominant carbonate components are shallow-water bioclasts, calcareous nannofossils, planktonic and benthic foraminifers, and micrite. Varying percentages of terrigenous clay, quartz, feldspar, with minor amounts of mica, titanium-bearing minerals, and other minor detrital grains are present. The terrigenous component is dominated by fine clay and some silt, and relatively minor percentages are greater than 63 μm; the sand fraction is predominantly very fine-grained quartz (63–125 μm). At Site 821, most proximal to the reef edge, there are meter- to several meter-thick sections of sediments that contain green minerals (verdine) infilling microfossils and as free grains, and detrital quartz. Similarities in log responses and environments of deposition suggest that the sediments at Site 820 also contain sections rich in verdine and quartz. Similar sections at deeper water Site 819 were found to be uncommon and not as well developed.

Magnetically susceptible minerals in the GBR cores are predominantly clays (smectite and illite) and lesser constituents of authigenic minerals (verdine, pyrite, and siderite), biotite, and heavy minerals (Davies et al., 1991; Chamley and Muller, 1993; E. Herrero-Bervera, pers. comm., 1993). Marked increases in

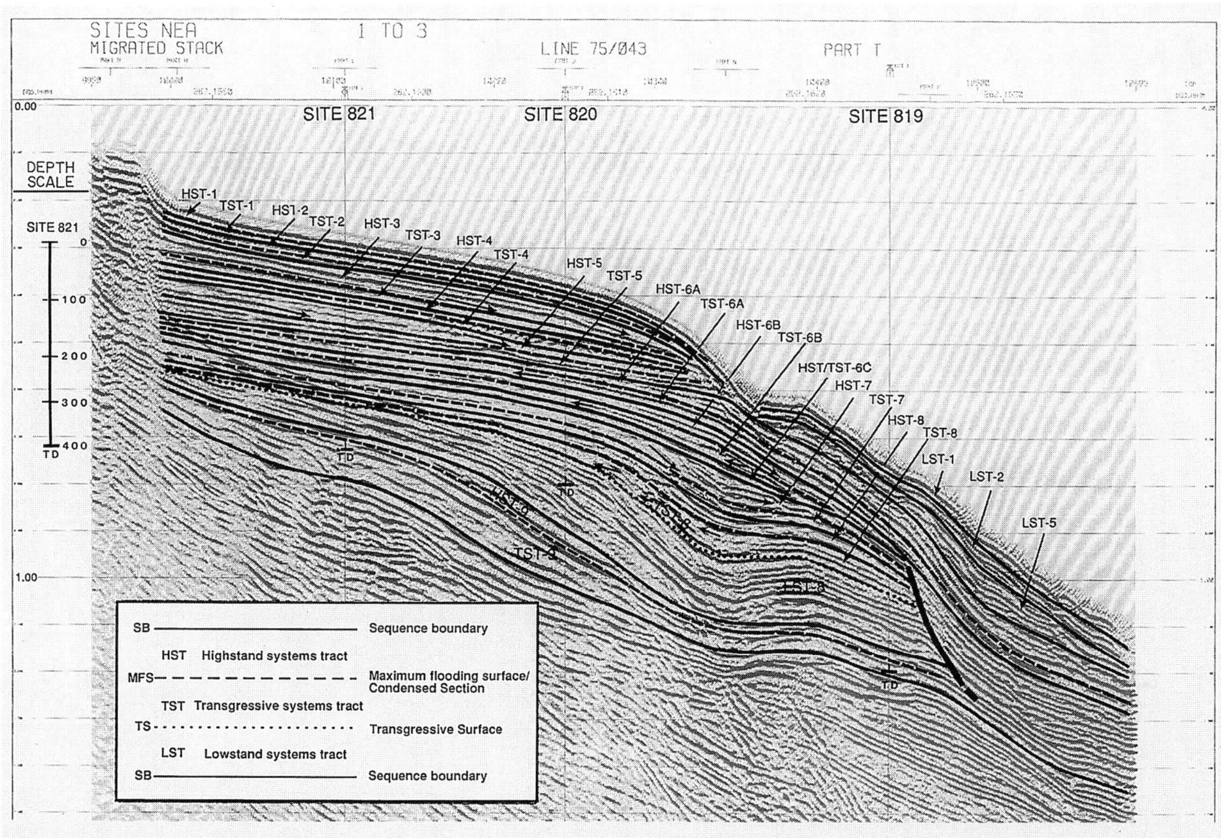

FIG. 2. —Sequence stratigraphic interpretation from Glenn et al. (1993) annotated on seismic reflection profile AGSO Line 75/043, Part T, migrated stack. Site 821 depth scale is shown displaced to left of site. Heavy bold line indicates probable slump surface. TD = total depth. Location of seismic line is shown in Fig. 1B. Site 821: WD = 212 m; Site 820: WD = 278 m; Site 819: WD = 565 m,. Distances: Barrier reef edge to Site 821 = 1.7 km (.91 nmi); Site 821 to Site 820 = 1.854 km (1.0 nmi); Site 820 to Site 819 = 2.78 km (1.5 nmi). Total horizontal dimension of seismic line = 8.78 km (4.73 nmi); vertical exaggeration: ~4.4. Short arrows indicate onlap and downlap reflection terminations.

magnetic susceptibility thus result from recurring fluxes of terrigenous and authigenic minerals on the outer shelf and upper slope. At Site 821, lithologic variability generally correlates with our seismic interpretations by defining surfaces that represent sequence boundaries and condensed sections (Fig. 3).

At Site 821, verdine facies minerals occur within approximately meter-thick intervals that correspond to seismic reflectors interpreted here as condensed sections (CS). The verdine grains at the GBR are abundant and well preserved, or "pristine" (Fig. 4; compare with Fig. 5); they commonly occur as non-abraded authigenic minerals that infill primary intraparticle porosity, especially within foraminifer tests. The tests housing these precipitates show little or no evidence of reworking or abrasion. They are most abundant within meter-thick intervals coinciding with MFSs identified along the tops of the transgressive systems tracts (Figs. 2, 6), and correlate with low-carbonate/high-magnetic susceptibility intervals and rapid increases in the abundance of >63μm siliciclastics/quartz and >63μm bioclasts (Fig. 3).

Verdine grains that show the effects of reworking and possibly transport (free grains) are also present at Site 821. These are usually not contained within foraminifer tests and appear slightly abraded to fragmented (Fig. 5), are less abundant, and typically occur within highstand systems tracts (Figs. 3 and 6). The reworked nature of the particles can be attributed to a higher-energy depositional setting associated with highstand progradation where depositional processes include sedimentary bypassing and starvation (e.g., Kidwell, 1991). It is also presumed that they may have originally formed as authigenic precipitates infilling primary intraparticle pore space. These grains often tend to impart a high luster surface coating and contain many fine cracks (Fig. 5).

Siliciclastics of the GBR forereef are chiefly composed of clay-size minerals, principally smectite, kaolinite, and illite (see Chamley and Muller, 1993), subsidiary amounts of silt- to very fine sand-size detrital quartz, minor amounts of biotite and feldspar, and trace amounts of heavy minerals. X-ray mineralogy of samples from Sites 821 and 820 indicate quartz abundances of 3–40 and 5–26 weight percent (%), respectively (Davies et al., 1991). The terrigenous component probably originates from both fluvial and eolian sources. Surprisingly, the distribution of the >63 μm fraction of siliciclastics, which is mostly eolian quartz (J. D. Kronen and C. R. Glenn, in prep., 1999), generally mimics that of verdine.

Reef-sourced bioclasts are the most abundant carbonate component and are derived from shallow-water inner reef (mid-shelf) and platform reef (outer shelf) environments. Increasing concentrations of bioclasts occur with increases in verdine and

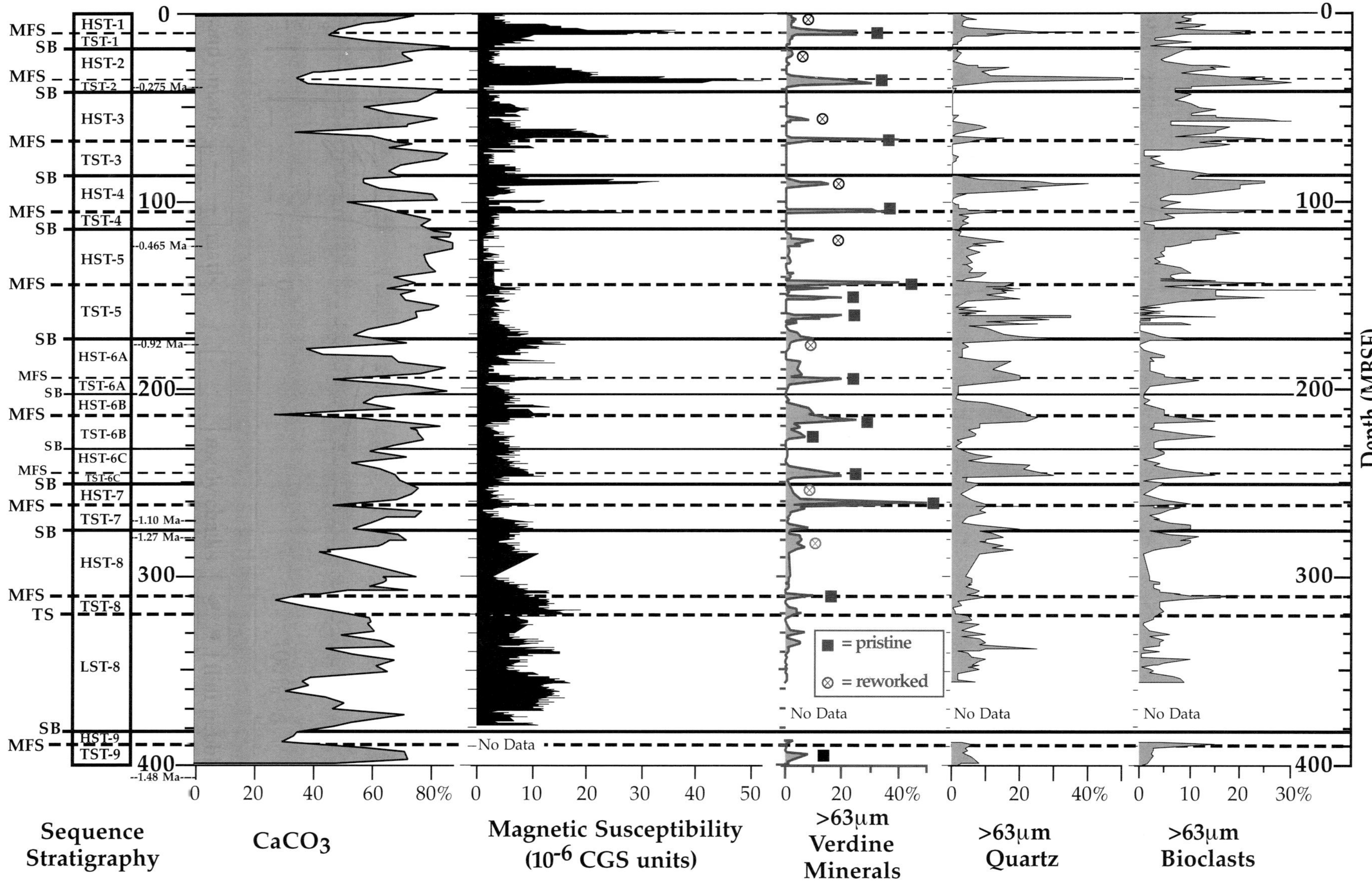

FIG. 3.—Sedimentary, geochemical, gamma-ray, and sequence stratigraphic data for Hole 821A. The sequence boundary at about 170 mbsf (circa 0.92 Ma) separates a largely progradational seismic package below from the largely aggradational seismic package above (Fig. 2); note the inversely correlated relationship between >63 μm quartz and >63 μm bioclast components across this transition. Two kinds of verdine are observed based on degree of preservation: pristine and reworked. See text for discussion.

A. B.

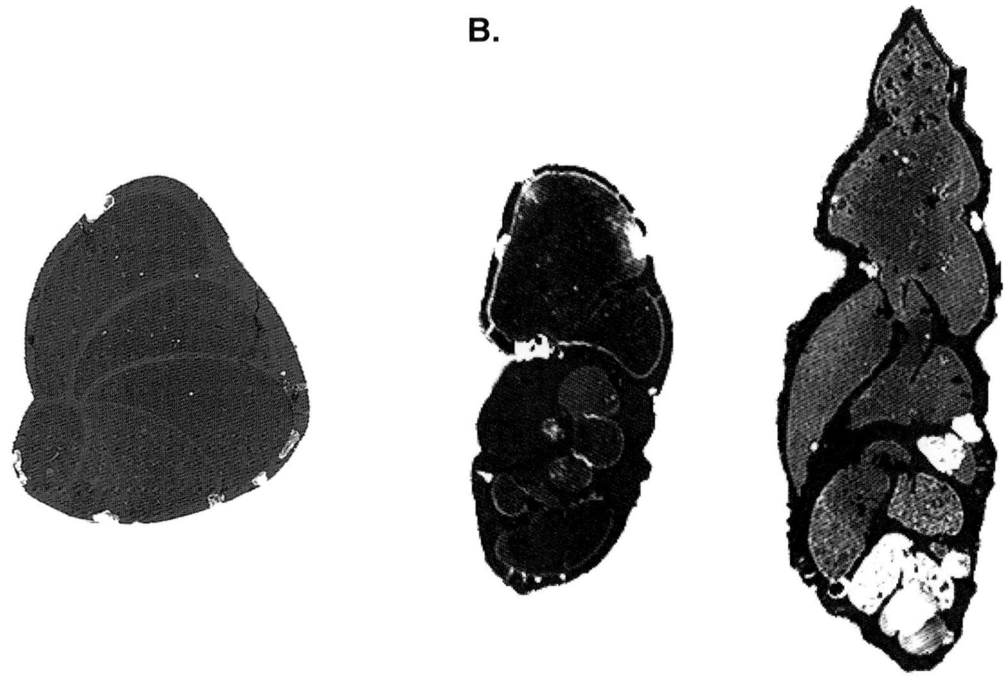

FIG. 4.—Electron microprobe backscatter images of pristine verdine mineralization typical of condensed sections along the MFS at Site 821 (cf. Fig. 3). A) Benthic foraminifer with verdine mineral infill; note undamaged condition of test (Sample 133-821B-2H-1, 6 cm; 9.75 mbsf). B) Benthic foraminifera with verdine mineralization infilling tests still intact (Sample 133-821A-42X-6, 136 cm; 395.26 mbsf). Bright material may be from later pyritization. Grains are approximately 50 x 100 microns.

quartz along surfaces associated with marine flooding and later highstand progradational events (Fig. 3). Inspection of smear slides shows that some bioclasts contained infillings of verdine replacing internal sediment.

DISCUSSION

Condensed Sections

Condensed sections form during reduced to extremely slow net sedimentation that continues over relatively long periods of time depending on the scale of time that applies. These intervals are also frequently marked by the occurrence and, in some instances, a high concentration of authigenic minerals (e.g., Glenn and Arthur, 1988; Loutit et al., 1988; Glenn, 1990; Kidwell, 1991; Southgate and Shergold, 1991; Amorosi, 1995, 1997). Loutit et al. (1988) and their colleagues (Wilgus et al., 1988) define condensed sections as thin marine units composed of hemipelagic to pelagic sediments deposited as the locus of sedimentation along a margin moves landward (retrogradation). They usually occur on the outer shelf and also basinward to the slope over areally expansive regions, and form during maximum rates of eustatic rise on a constantly subsiding margin when accommodation potential is the greatest (Loutit et al., 1988; Posamentier and Vail, 1988). They are associated with limy sediments and marls, a nondepositional environment, apparent marine hiatuses, and may occur as omission surfaces or marine hardgrounds (e.g., see Kennedy and Garrison, 1975).

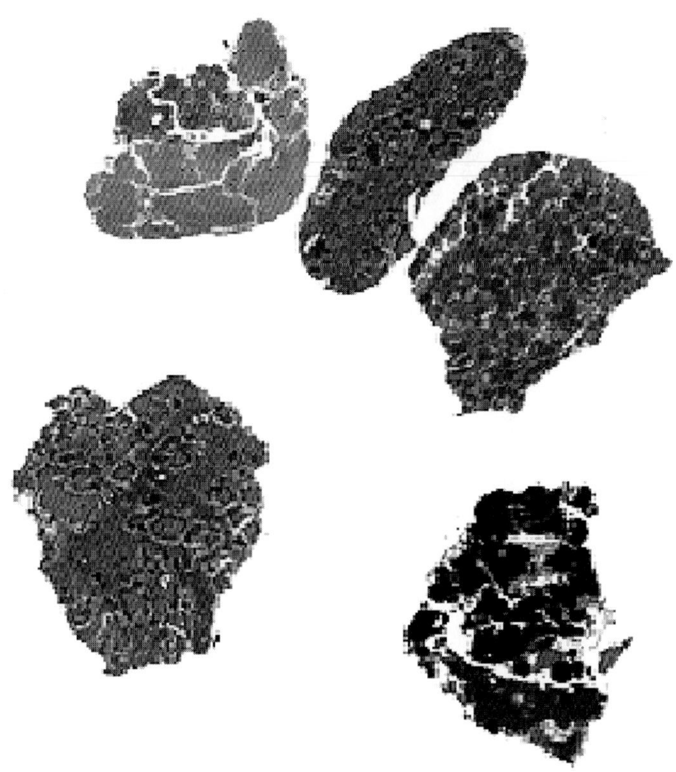

FIG. 5.—Electron microprobe backscatter image of reworked verdine mineral fragments from highstand systems tracts in Hole 821A. The grains represent reworked and fragmented verdine mineral infillings from foraminifer tests, and are typically found in late highstand progradation (cf. Fig. 3) (Sample 133-821A-31X-3, 102 cm; 284.4 mbsf). Each grain is approximately 50 microns across.

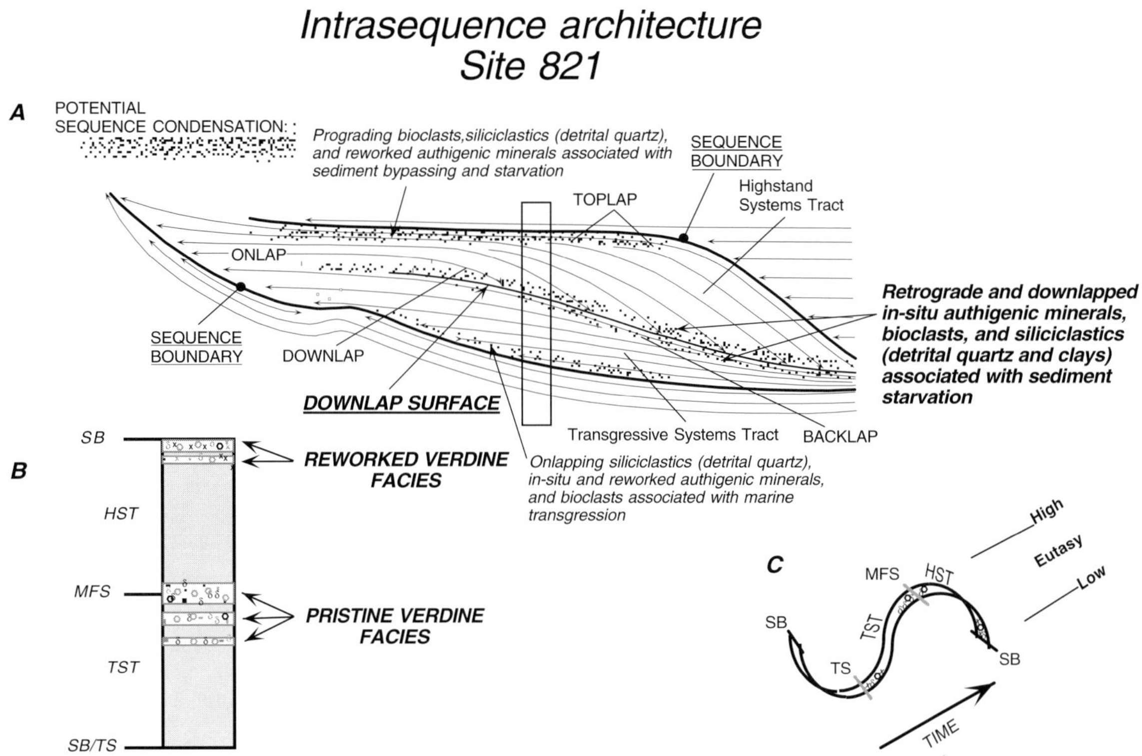

FIG. 6.—Intrasequence architecture at Site 821 (after Glenn et al., 1993). A) Possible positions of sequence condensation within an idealized depositional sequence (stipples) (after Kidwell, 1991). Relative sequence condensation occurs most likely where seismic reflectors converge. B) Schematic illustration of the placement of nonreworked pristine and reworked verdine minerals recovered at Site 821. C) The timing of systems tracts development with respect to one cycle of sea-level change. TST = transgressive systems tract, HST = highstand systems tract, TS = transgressive surface, MFS = maximum flooding surface, SB = sequence boundary. Dots and other graphics on the sea-level curve represent locations of verdine minerals and detrital quartz

Formation of Verdine

The verdine facies is quite similar to and often visually indistinguishable from the glaucony facies. However, the minerals are easily distinguished through chemical and crystallographic analyses. Their formation, as in the manner of glaucony formation, appears to be controlled by a delicate balance between degree of physical confinement and the amount of ionic exchange between the microenvironment, such as within foraminiferal shells, and ambient open seawater (Odin and Matter, 1981; Bailey, 1988; Odin and Fullagar, 1988; Odin and Sen Gupta, 1988). Odin and Matter (1981) originally described a green marine mineral and identified it as berthierine, a 7Å iron layer lattice silicate, as revealed by XRD. On the basis of chemical analyses, four samples from the shelf of Guyana were described as evolved berthierine (not to be confused with the ferrous-rich mineral berthierine of the oolitic ironstone facies), but in later works the same samples have been named phyllite V (Odin et al., 1988a) and subsequently identified as odinite (Bailey, 1988; Odin, 1990). These three names describe the same mineral that belongs to the verdine facies defined by Odin and collaborators (see Odin, 1988). The term "berthierine" of Odin and Matter (1981) has since been replaced by phyllite V or odinite of the verdine facies (Odin, 1990).

The green minerals found at the GBR were originally described as glauconite (Davies et al., 1991; Glenn et al., 1993). Glaucony facies minerals occur mainly in water depths of 100–500 m on modern seafloors and can make up to 90% of the sediments of the outer shelf and upper slope. Their presence can provide a potentially powerful tool for sequence stratigraphic interpretation if their distribution, maturity, and spatial and temporal characteristics are considered (Kidwell, 1991; Amorosi, 1995). They are unique because they contain iron in both its reduced and oxidized states, and are thus characteristic of weakly oxidizing to mildly reducing marine environments (Glenn and Arthur, 1988, 1990; Glenn et al., 1993). It is usually accepted that their formation occurs authigenically and diagenetically in the shallow water marine environment during periods of slow or negative sedimentation (Deer et al., 1992). They are known to be one of the most reliable indicators of sediment starvation by their presence (Amorosi, 1995, 1997).

Minerals of the verdine facies also form authigenically in marine settings near the sediment-water interface where sedimentation rates are low, where there is normal salinity, and often within confined microenvironments. As with glaucony, verdine is ferric-rich, but also contains iron in both its oxidized and reduced states. We suggest, therefore, that both minerals form in association with mildly reducing, suboxic solutions from which both ferric and ferrous iron may be supplied. The environment of verdine formation is different than that for glaucony, however, because verdine tends to form in warmer tropical settings, usually between 16°N and 23°S, with water temperature near 25°C. It forms in shallower shelf settings where water depths are

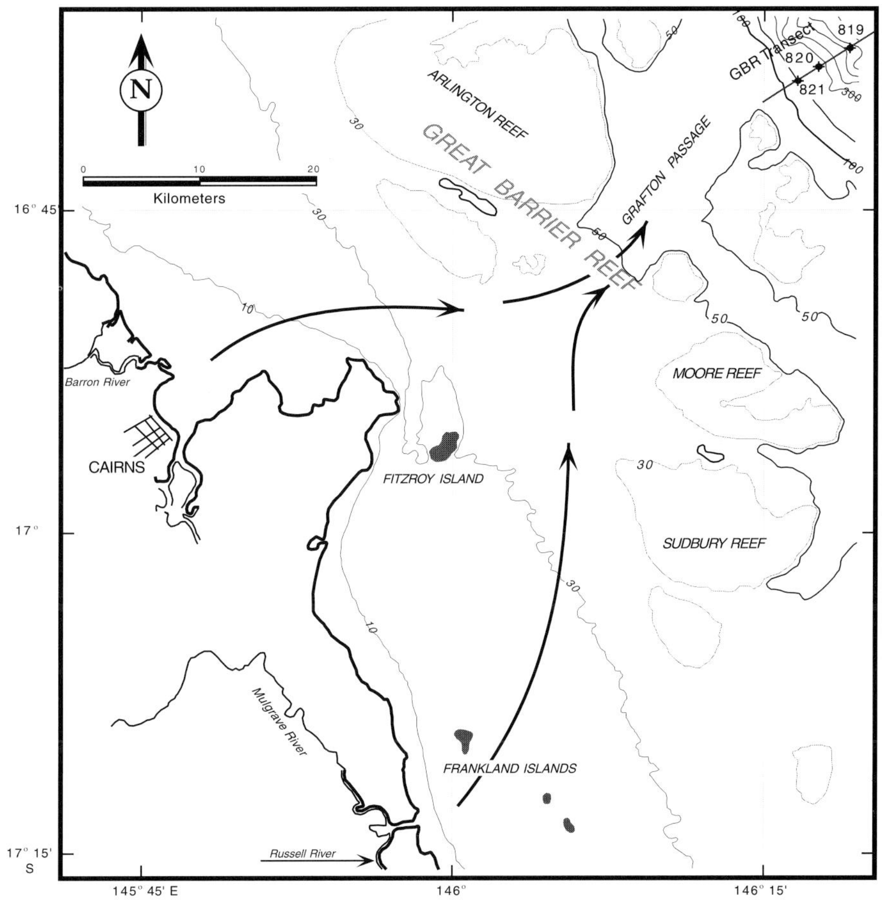

FIG. 7.—Map showing the region of Cairns, Australia and the GBR transect where forereef sediments were drilled at Sites 821, 820, and 819. The Barron, Mulgrave, and Russell Rivers are considered to be significant point sources of particulate iron necessary for authigenesis of verdine. In this case, verdine facies have formed immediately offshore of a barrier reef channel that may have been conducive to transport of iron (arrows) out to the forereef.

most commonly between 15 m and 60 m, although verdine minerals have also been found to form in water depths of 100 m to 200 m (Odin and Sen Gupta, 1988). Regardless of water depth, verdine minerals appear to precipitate as much as an order of magnitude faster (about 1,000 years) than glaucony because of their warmer waters of formation (Bailey, 1988). Verdine mineralization is also known to occur near fluvial influxes. Odin and Sen Gupta (1988) identified deposits associated with rivers along the tropical Atlantic coasts of Africa and found a direct relationship between river length, or size of river drainage basin, and the surficial extent of the associated verdine deposits.

Verdine at the GBR

The region landward of the GBR transect receives a significant continental water supply from two major river systems, the Barron River and the Mulgrave/Russell Rivers, which also discharge terrigenous sediments into the sea close to Grafton Passage (Kershaw et al., 1993; Fig. 7). This region includes some of the highest mountains of Queensland and the extensive basaltic Atherton Tableland, which are ample sources of iron. Rainfall can exceed 400 cm along the coast. The combination of ample rainfall in the region and an extensive inland watershed should produce sufficient supplies of river water and particulate iron to the coast (Fig. 7).

With regard to depth of formation of verdine, it is assumed that tectonic subsidence along the GBR margin was thermal and linear for the time frame considered, based on the work of Katz and Miller (1993). Paleodepth estimates from the benthic foraminiferal studies of Davies et al. (1991) indicate that sediments below 150 mbsf at Site 820 were deposited in an environment fluctuating between upper bathyal (200–600 m) and outer neritic (100–200 m), and sediments above 150 mbsf were all deposited in an upper bathyal environment, suggesting that there may have been increased tectonic subsidence in the upper 150 m of the sediment column.

Chemical analyses (Table 1) clearly show that the green minerals found in Pleistocene-Holocene sediments at the GBR are distinct from Pleistocene-Holocene glauconite of Odin and Matter (1981), but are found within the same fields as odinite of Bailey (1988) and "evolved berthierine" of Odin and Matter (1981) (Fig. 8). What is most noticeable chemically is that the verdine minerals contain significantly less K_2O than glaucony (Fig. 9).

Authigenically formed green minerals of the GBR succes-

TABLE 1.—CHEMICAL ANALYSES OF VERDINE GRAINS OF THE FOREREEF SEDIMENTS OFFSHORE OF THE GREAT BARRIER REEF.

Chemistry obtained by electron microprobe analysis of grain mounts with a Cameca Camebax probe.
Sample numbers correspond to specific core samples. See key at end of table for identification and type.

Sample			1.00					2.00		
subsample	A	B	C	D	E	A	B	C	D	E
SiO_2	23.80	24.59	19.46	13.21	15.16	29.23	31.53	27.34	31.15	29.58
Al_2O3	7.06	7.40	10.50	7.37	8.19	11.49	9.97	13.12	8.05	11.22
Fe2O3 + FeO	40.25	41.10	30.69	20.03	18.82	28.44	30.68	20.67	29.71	28.68
TiO_2	0.04	0.07	0.28	0.20	0.18	0.13	0.10	0.19	0.03	0.13
CaO	1.61	0.85	5.63	19.76	16.17	0.64	0.31	0.63	1.28	0.87
MgO	11.52	11.05	11.95	9.21	10.67	14.94	13.80	10.85	15.65	13.47
MnO	0.09	0.08	0.19	0.14	0.12	0.17	0.05	0.19	0.04	0.14
P_2O_5	1.26	0.91	1.47	1.57	1.28	0.55	0.29	0.32	0.51	0.48
Na_2O	0.19	0.16	0.17	0.15	0.13	0.13	0.13	0.20	0.09	0.09
K_2O	0.22	0.37	0.20	0.16	0.20	0.23	0.37	0.37	0.20	0.35
$H_20 + CO_2$					not determined (n.d.)					
Total	86.04	86.56	80.52	71.79	70.90	85.96	87.23	73.88	86.69	85.02

Sample			3.00					4.00		
subsample	A	B	C	D	E	A	B	C	D	E
SiO_2	31.63	15.45	16.08	19.11	28.38	16.48	26.88	26.80	19.85	29.49
Al_2O_3	11.47	9.41	7.26	8.68	11.82	7.66	11.63	11.63	10.17	12.39
Fe_2O_3 + FeO	20.15	32.07	10.50	12.70	28.29	20.75	26.37	27.29	24.94	26.61
TiO_2	0.10	0.44	0.11	0.01	0.17	0.09	0.09	0.16	0.39	0.26
CaO	3.06	0.90	2.08	0.89	0.91	17.76	2.99	3.04	2.29	1.44
MgO	16.88	8.23	7.95	9.48	14.87	9.68	14.48	13.89	9.93	14.07
MnO	0.01	0.05	0.08	0.03	0.09	0.15	0.14	0.14	0.05	0.09
P_2O_5	0.81	1.20	0.38	0.34	0.68	0.96	0.63	0.61	1.10	0.63
Na_2O	0.08	0.09	0.10	0.06	0.18	0.16	0.10	0.12	0.14	0.09
K_2O	0.40	0.17	0.20	0.25	0.39	0.30	0.22	0.25	0.32	0.31
H20 + CO2					n.d.					
Total	84.58	68.02	44.74	51.53	85.79	74.00	83.53	83.91	69.18	85.38

Sample			5.00					6.00		
subsample	A	B	C	D	E	A	B	C	D	E
SiO_2	27.43	26.97	22.36	26.32	23.77	17.07	24.40	7.35	16.53	2.74
Al_2O_3	10.57	8.95	8.41	8.74	8.19	7.41	11.54	4.32	7.85	1.29
Fe_2O_3 + FeO	20.05	20.58	12.37	24.61	17.48	17.75	23.04	15.83	13.52	3.96
TiO_2	0.16	0.14	0.18	0.08	0.13	0.10	0.16	0.18	0.17	0.07
CaO	0.51	0.88	0.62	0.67	2.45	20.14	4.84	27.65	1.21	44.79
MgO	11.66	12.19	8.63	12.54	11.92	9.42	10.92	6.46	6.32	4.75
MnO	0.01	0.06	0.05	0.07	0.07	0.09	0.10	0.08	0.06	0.02
P_2O_5	0.30	0.22	0.18	0.28	0.39	0.62	0.71	1.01	0.31	1.15
Na_2O	0.08	0.08	0.05	0.13	0.14	0.17	0.10	0.13	0.05	0.17
K_2O	0.41	0.22	0.26	0.22	0.19	0.22	0.42	0.14	0.30	0.09
H20 + CO2					n.d.					
Total	71.18	70.29	53.10	73.66	64.72	72.99	76.24	63.14	46.31	59.03

TABLE 1.—CONTINUED

Sample			7.00					8.00		
subsample	A	B	C	D	E	A	B	C	D	E
SiO_2	27.21	25.55	30.32	23.12	28.18	18.30	8.89	14.12	17.54	21.51
Al_2O_3	12.28	10.38	12.40	10.40	11.17	10.96	5.34	7.30	9.59	10.62
$Fe_2O_3 + FeO$	25.01	24.63	25.21	20.81	27.31	32.50	15.32	9.34	17.58	13.20
TiO_2	0.13	0.25	0.10	0.21	0.13	0.65	0.25	0.24	0.35	0.27
CaO	0.39	1.05	0.57	1.61	1.60	1.69	21.39	15.33	8.92	3.03
MgO	13.62	10.27	16.40	10.30	15.30	8.66	6.38	6.39	9.34	11.58
MnO	0.08	0.05	0.05	0.06	0.09	0.03	0.06	0.04	0.07	0.04
P_2O_5	0.44	0.30	0.50	0.38	0.47	1.04	2.84	0.48	0.72	0.46
Na_2O	0.15	0.18	0.21	0.08	0.13	0.11	0.23	0.08	0.09	0.08
K_2O	0.31	0.39	0.28	0.36	0.19	0.42	0.15	0.31	0.23	0.25
$H_20 + CO_2$					n.d.					
Total	79.62	73.03	86.05	67.32	84.58	74.37	60.84	53.61	64.44	61.05

Sample			9.00					10.00		
subsample	A	B	C	D	E	A	B	C	D	E
SiO_2	9.99	22.35	21.40	14.95	14.58	12.62	30.14	23.96	28.85	26.45
Al_2O_3	4.35	10.51	10.55	7.93	6.98	6.37	16.34	10.86	13.59	13.95
$Fe_2O_3 + FeO$	4.41	11.42	9.62	7.30	7.67	6.82	11.22	11.90	16.64	15.12
TiO_2	0.02	0.09	0.16	0.14	0.08	0.08	0.27	0.10	0.22	0.11
CaO	26.98	2.98	0.99	13.52	14.48	21.44	0.85	0.41	0.53	1.54
MgO	4.73	13.17	9.01	4.04	7.37	8.22	10.03	10.27	14.46	15.56
MnO	0.00	0.00	0.01	0.01	0.01	0.03	0.00	0.04	0.07	0.03
P_2O_5	16.81	1.82	0.57	1.11	8.68	0.53	0.25	0.24	0.33	0.47
Na_2O	0.63	0.12	0.09	0.14	0.34	0.09	0.13	0.14	0.15	0.13
K_2O	0.18	0.16	0.41	0.68	0.17	0.23	0.68	0.50	0.42	0.30
$H_20 + CO_2$					n.d.					
Total	68.10	62.62	52.81	49.82	60.35	56.41	69.90	58.42	75.25	73.65

Sample			11.00					12.00		
subsample	A	B	C	D	E	A	B	C	D	E
SiO_2	30.53	27.82	22.36	31.80	17.70	20.54	26.11	21.49	26.91	21.16
Al_2O_3	12.23	11.75	9.29	11.06	7.32	8.17	9.95	8.64	8.41	7.59
$Fe_2O_3 + FeO$	22.60	17.81	13.34	20.82	10.58	9.76	18.12	10.09	16.11	13.06
TiO_2	0.06	0.18	0.18	0.11	0.10	0.10	0.08	0.11	0.12	0.07
CaO	1.36	0.57	0.95	0.49	11.95	0.70	0.43	1.12	2.98	1.05
MgO	13.93	13.68	10.86	15.31	7.02	6.95	13.98	8.72	9.51	8.82
MnO	0.06	0.00	0.01	0.00	0.03	0.01	0.03	0.04	0.03	0.01
P_2O_5	0.92	0.34	0.64	0.19	4.12	0.12	0.29	0.18	0.42	0.21
Na_2O	0.16	0.11	0.09	0.22	0.20	0.07	0.09	0.33	0.15	0.08
K_2O	0.41	0.52	0.24	0.41	0.32	0.34	0.19	0.33	0.40	0.30
$H_20 + CO_2$					n.d.					
Total	82.25	72.79	57.94	80.40	59.35	46.75	69.26	51.04	65.03	52.35

TABLE 1.—CONTINUED

Sample			13.00					14.00		
subsample	A	B	C	D	E	A	B	C	D	E
SiO_2	23.55	20.41	16.21	31.31	29.94	22.20	31.93	23.42	33.69	29.86
Al_2O_3	7.41	6.94	4.77	7.95	9.76	7.75	9.80	9.37	11.91	10.90
$Fe_2O_3 + FeO$	15.98	13.97	11.64	30.21	23.05	9.54	26.34	17.61	22.75	24.38
TiO_2	0.11	0.09	0.06	0.13	0.17	0.08	0.14	0.08	0.26	0.31
CaO	1.16	1.17	1.82	1.82	0.54	0.92	3.70	1.11	0.71	4.74
MgO	6.03	7.33	5.86	11.98	10.93	6.44	13.49	11.76	13.33	12.90
MnO	0.03	0.07	0.00	0.12	0.12	0.00	0.02	0.03	0.00	0.09
P_2O_5	0.05	0.17	0.10	0.07	0.13	0.25	0.39	0.86	0.37	0.44
Na_2O	0.11	0.08	0.08	0.18	0.15	0.32	0.17	0.08	0.13	0.21
K_2O	0.59	0.36	0.16	0.31	0.58	0.49	0.64	0.18	0.57	0.45
$H_20 + CO_2$					n.d.					
Total	55.01	50.58	40.68	84.08	75.35	48.00	86.61	64.48	83.72	84.28

Sample			15.00					16.00		
subsample	A	B	C	D	E	A	B	C	D	E
SiO_2	22.75	27.68	33.45	32.65	30.23	26.13	28.70	34.58	23.83	32.04
Al_2O_3	8.39	10.65	10.16	12.42	11.16	10.42	12.49	11.34	9.66	11.50
$Fe_2O_3 + FeO$	17.21	20.24	25.02	25.41	23.54	16.47	18.55	23.36	13.87	24.23
TiO_2	0.73	0.11	0.10	0.09	0.10	0.14	0.12	0.33	0.14	0.12
CaO	1.43	1.90	0.65	0.42	0.60	0.81	1.36	2.37	0.55	0.60
MgO	10.57	13.92	17.40	15.64	14.87	12.14	14.26	12.44	9.69	15.36
MnO	0.00	0.03	0.05	0.02	0.06	0.03	0.08	0.06	0.01	0.02
P_2O_5	0.53	0.91	0.43	0.44	0.55	0.69	1.01	0.33	0.33	0.26
Na_2O	0.06	0.14	0.12	0.15	0.10	0.08	0.07	0.35	0.08	0.16
K_2O	0.27	0.31	0.20	0.33	0.28	0.28	0.41	0.63	0.31	0.40
$H_20 + CO_2$					n.d.					
Total	61.93	75.89	87.58	87.56	81.49	67.17	77.05	85.79	58.46	84.67

Sample			17.00					18.00		
subsample	A	B	C	D	E	A	B	C	D	E
SiO_2	31.46	31.20	31.96	32.37	31.44	23.37	34.04	32.20	32.52	25.62
Al_2O_3	8.62	10.11	11.11	9.83	9.61	8.39	11.53	8.86	10.53	10.07
$Fe_2O_3 + FeO$	29.61	28.21	26.65	30.38	30.32	13.56	21.54	30.04	25.75	15.45
TiO_2	0.03	0.06	0.11	0.05	0.05	0.12	0.08	0.04	0.19	0.13
CaO	0.68	1.42	0.77	0.58	0.57	0.39	0.58	0.44	0.65	0.71
MgO	15.99	14.81	13.51	14.12	15.35	12.09	11.76	14.16	15.28	11.67
MnO	0.07	0.07	0.06	0.09	0.08	0.02	0.01	0.12	0.04	0.03
P_2O_5	0.31	0.49	0.22	0.17	0.25	0.29	0.13	0.32	0.38	0.30
Na_2O	0.13	0.18	0.19	0.21	0.14	0.08	0.11	0.10	0.13	0.07
K_2O	0.13	0.31	0.57	0.28	0.18	0.15	0.34	0.24	0.25	0.30
$H_20 + CO_2$					n.d.					
Total	87.01	86.86	85.16	88.08	87.97	58.47	80.11	86.51	85.71	64.33

TABLE 1.—CONTINUED

Sample	19.00					20.00				
subsample	A	B	C	D	E	A	B	C	D	E
SiO_2	26.07	28.19	31.07	33.89	29.28	37.14	43.78	36.42	40.15	31.67
Al_2O_3	9.43	10.73	8.23	11.39	11.44	13.70	15.23	15.15	12.98	15.41
$Fe_2O_3 + FeO$	20.25	20.90	30.43	25.53	21.46	20.72	18.21	16.78	18.42	20.90
TiO_2	0.07	0.04	0.00	0.17	0.06	0.03	0.20	0.05	0.13	0.02
CaO	0.69	0.47	0.50	0.39	0.31	0.24	0.48	0.33	0.25	0.15
MgO	11.77	15.00	15.54	14.96	13.34	16.20	10.98	8.71	10.10	17.77
MnO	0.04	0.04	0.10	0.19	0.02	0.09	0.05	0.04	0.06	0.19
P_2O_5	0.31	0.40	0.11	0.24	0.29	0.06	0.17	0.09	0.01	0.01
Na_2O	0.16	0.14	0.16	0.17	0.16	0.20	0.08	0.17	0.05	0.33
K_2O	0.27	0.18	0.13	0.39	0.35	0.58	0.77	0.73	0.45	0.08
$H_20 + CO_2$					n.d.					
Total	69.08	76.09	86.26	87.28	76.71	88.95	89.96	78.46	82.59	86.53

Sample	21.00					22.00				
subsample	A	B	C	D	E	A	B	C	D	E
SiO_2	37.79	34.79	33.96	25.74	38.76	18.43	28.98	27.98	35.91	25.20
Al_2O_3	12.75	11.60	12.73	10.40	30.34	8.40	9.27	8.13	10.60	7.21
$Fe_2O_3 + FeO$	21.99	27.52	24.19	12.95	11.69	9.40	23.40	20.32	22.54	19.76
TiO_2	0.09	0.10	0.03	0.09	0.02	0.11	0.10	0.11	0.01	0.03
CaO	0.53	0.67	0.31	2.28	0.08	5.27	4.07	0.40	0.36	0.67
MgO	10.45	12.66	14.39	11.44	2.94	12.23	15.10	11.41	18.45	11.24
MnO	0.03	0.14	0.04	0.00	0.17	0.00	0.10	0.00	0.09	0.02
P_2O_5	0.13	0.18	0.23	0.42	0.01	0.32	2.44	0.04	0.01	0.09
Na_2O	0.14	0.15	0.19	0.04	0.20	0.06	0.21	0.11	0.17	0.11
K_2O	2.40	0.69	0.41	0.31	1.41	0.20	0.16	0.30	0.07	0.20
$H_20 + CO_2$					n.d.					
Total	86.28	88.48	86.49	63.67	85.61	54.42	83.82	68.78	88.22	64.52

KEY:

Sample	Hole	Core/Sec/cm	mbsf	Type
1	821B	1H-2-146	2.94	Reworked
2	821B	2H-1-7	9.75	Pristone
3	821B	3H-1-72	19.70	Reworked
4	821B	4H-6-72	36.70	Pristine
5	821B	6H-6-148	56.46	Reworked
6	821B	7H-7-54	66.52	Pristine
7	821B	10H-4-72	90.70	Reworked
8	821B	12H-1-72	105.20	Pristine
9	821B	13H-5-72	120.70	Reworked
10	821B	17V-3-70	143.68	Pristine
11	821-B	19X-4-72	151.80	Pristine
12	821B	20X-4-66	161.34	Pristine

Sample	Hole	Core/Sec/cm	mbsf	Type
13	821A	19X-6-77	173.47	Reworked
14	821A	22X-1-121	195.09	Pristine
15	821A	24X-3-102	216.8	Pristine
16	821A	25X-2-132	225.3	Pristine
17	821A	27X-4-7	246-35	Pristine
18	821A	28X-3-136	255.74	Reworked
19	821A	29X-1-18	261.26	Pristine
20	821A	31X-3-102	284.4	Reworked
21	821A	34X-1-102	310.4	Pristine
22	821A	42X-6-138	395.26	Pristine

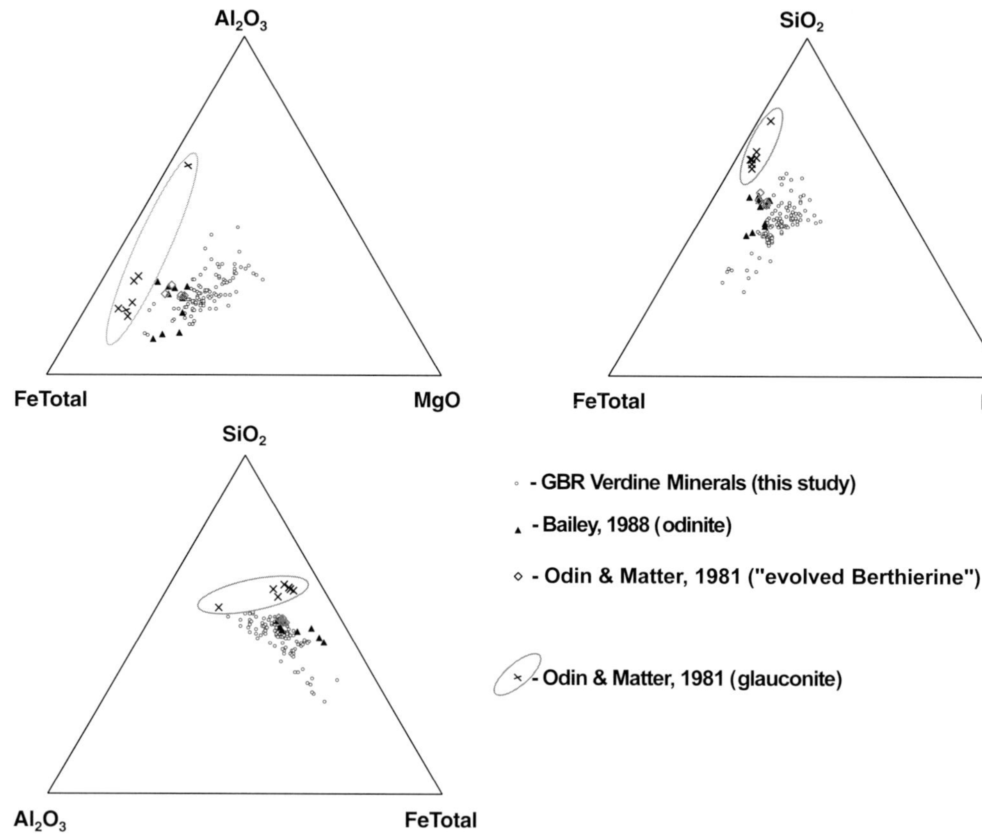

FIG. 8.—Chemical analyses show that green minerals found in sediments at the GBR are distinct from the Pleistocene-Holocene glauconite of Odin and Matter (1981) and are found within the same fields as the odinite of Bailey (1988) and "evolved berthierine" of Odin and Matter (1981).

sion are also chemically distinct from berthierine typically associated with oolitic ironstones that also form in shallow, warm waters (e.g., Odin et al., 1988b; Young and Taylor, 1989). Odin (1990) clearly defined the verdine facies as being distinguishable from the oolitic berthierine and chamosite-bearing ironstone facies because of chemical and genetic attributes; verdine facies can be defined by 1) occurrence: found in young (typically Holocene) sediments from the tropics (oolitic ironstones are from pre-Pleistocene sediments), 2) habit: infilling of tests, and 3) mineralogical composition: most importantly, they are ferric-rich. Ternary plots of chemical analyses clearly show that the green minerals found in sediments at the GBR are distinct from ironstone berthierines and are found within the same fields as odinite of Bailey (1988) (Fig. 10), and we therefore believe that the green minerals found at the GBR are chemically part of the verdine facies and, more specifically, have the same composition as phyllite V clays and likely including odinite.

With regards to sequence stratigraphic interpretation, the presence and nature of verdine minerals can be used as one of the criteria to define moderately condensed intervals. These intervals at the GBR contain verdine as "pristine," based on the degree of physical preservation (Figs. 3, 4, and 5). Chemical analyses comparing pristine and reworked verdine show no distinction between the two (Fig. 11, Table 1). However, XRD analyses published by Odin and Matter (1981) reveal a mineralogical evolution for verdine that suggests they form in a simi-

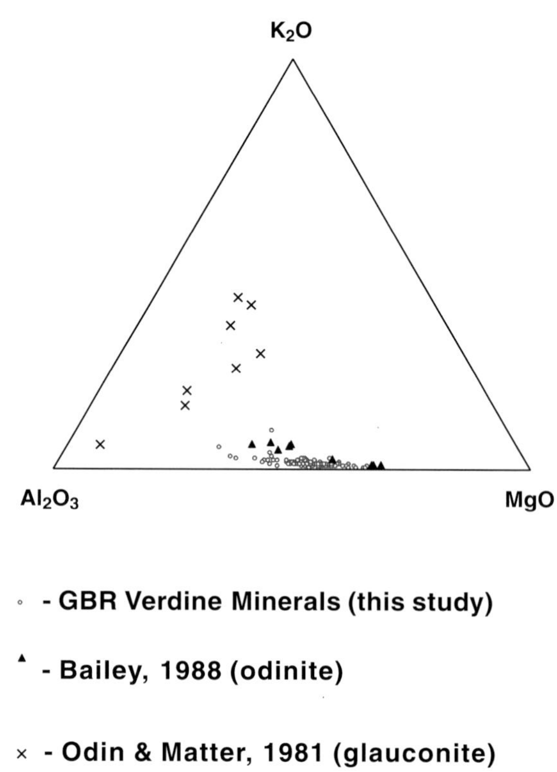

FIG. 9.—Verdine contains significantly less K_2O than glaucony.

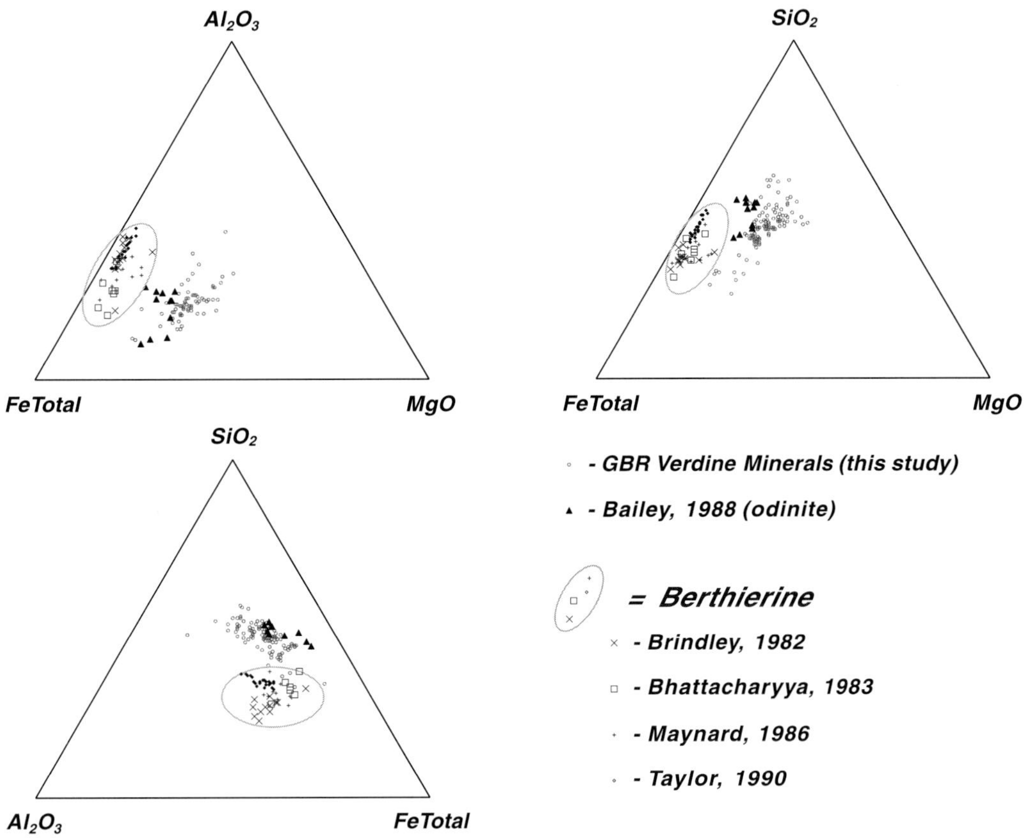

FIG. 10.—Berthierines are typically associated with oolitic ironstones that also form in shallow warm waters (Odin et al., 1988b; Young and Taylor, 1989). Chemical analyses show that green minerals from the GBR are found within the same fields as the odinite of Bailey (1988), are chemically part of the verdine facies, and have the same composition as phyllite V and/or odinite.

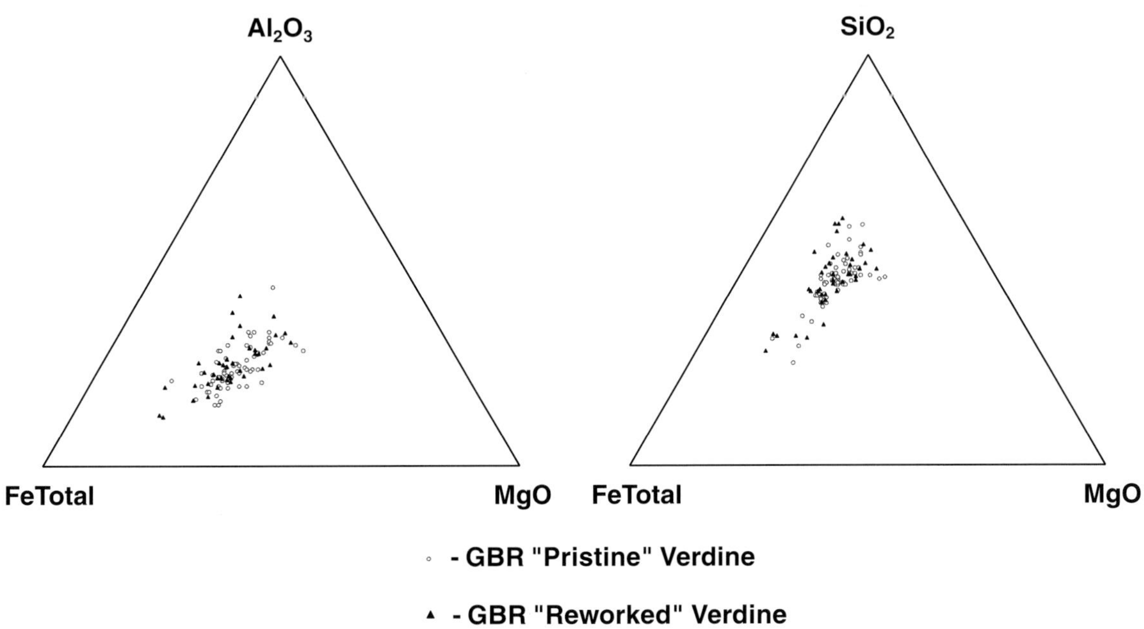

FIG. 11.—Condensed intervals were described containing verdine minerals as either pristine or reworked based on the degree of preservation. Chemical analyses comparing pristine and reworked samples show no distinction between the two.

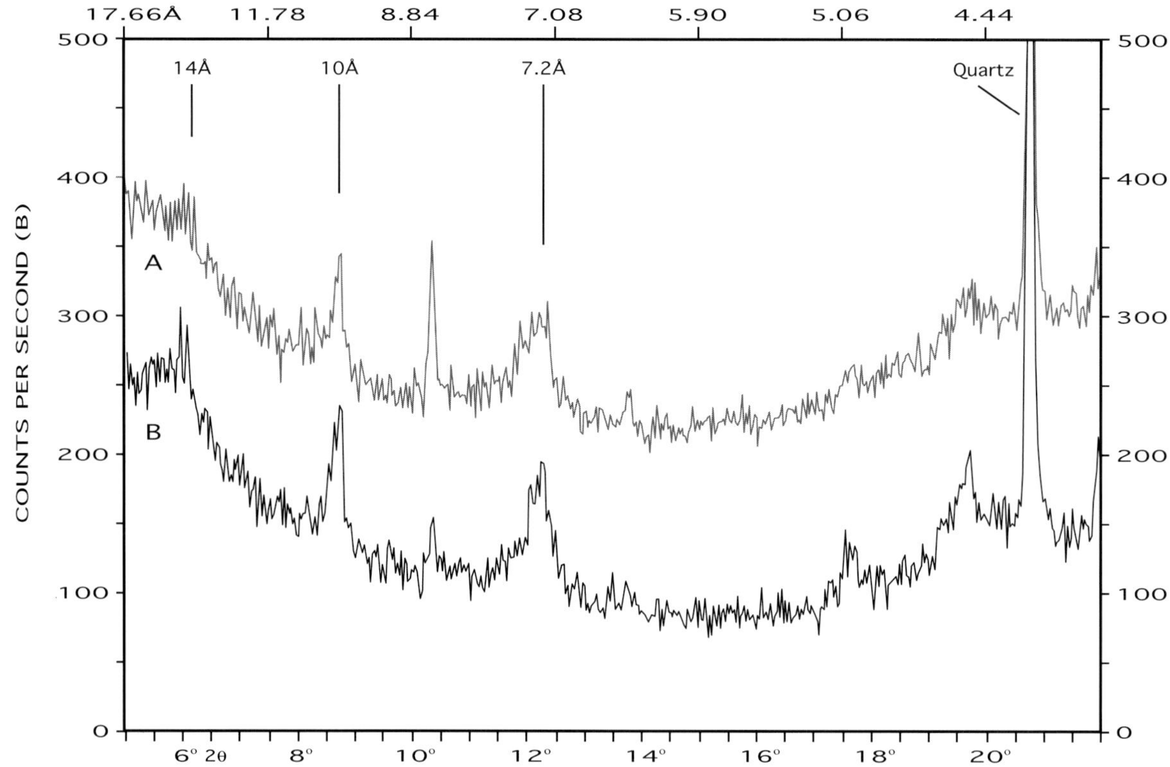

FIG. 12A.—Stacked and offset XRD diffractograms of reworked free verdine found within HST intervals. Relative to "pristine" verdine of maximum flooding surfaces (mfs) (Fig. 12B), these grains are further evolved based on the character of the ~7.2Å and ~14Å diffraction peaks (see Odin and Matter, 1981). A) Sample 133-821A-28X-3, 136 cm; 255.74 mbsf, B) Sample 133-821A-31X-3, 102 cm; 284.4 mbsf. Diffractogram A is offset approximately 125 counts.

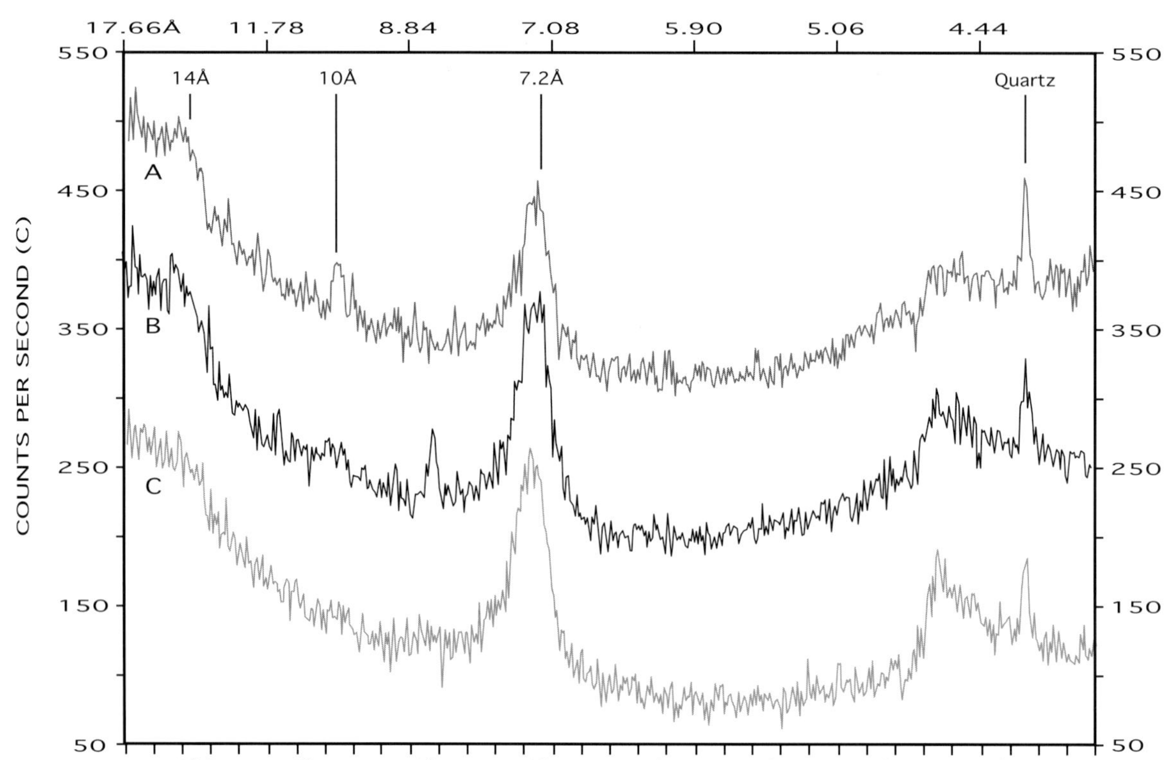

FIG.12B.—Stacked and offset XRD diffractograms show the ~7.2Å peak found in "pristine" verdine is more predominant, suggesting it is less evolved material (Odin et al., 1988a). Owing to the relatively rapid time of formation for verdine (Bailey, 1988), we find pristine verdine are nascent to slightly evolved. A) Sample 133-821B-20X, 66 cm; 161.34 mbsf, B) Sample 133-821A-24X-3, 102 cm; 216.8 mbsf, C) Sample 133-821A-27X-4, 7 cm; 246.35 mbsf. Diffractograms A and B are offset approximately 220 and 110 counts, respectively.

lar mode to glaucony and evolve to more mature forms if they reside for longer periods on the seafloor (e.g., Odin, 1990). Odin et al. (1988a) proposed that less-evolved verdine gives a sharp peak near ~7.2Å, and hypothesized that older, more-evolved verdine shows a main peak at ~7.2Å that is broader and smaller and, in addition, commonly shows the presence of a 14Å peak. In comparison with past studies examining the evolution and maturity of glauconites (see Amorosi, 1995), this observation is the reverse of what one might expect for evolution of an authigenic mineral to a more mature form. However, our results from XRD analyses appear to be consistent with the findings of Odin et al. (1988a) and suggest changes in the crystallography owing to evolution in maturity, and reveal that verdine found within intervals containing reworked free grains (HSTs) may be further evolved, as based on the relatively subdued character of the ~7.2Å diffraction peak (Fig. 12A; also see Odin and Matter, 1981). In comparison, the ~7.2Å peak found in pristine verdines of the condensed sections is more predominant, suggesting it is perhaps younger ("nascent") and less evolved material (Fig. 12B). Although we observe this subtle distinction between the more mature verdine associated with HSTs and the less mature verdine associated with condensed sections, in most cases all of the verdine found at the GBR transect possess ~7.2Å, ~10Å, and ~14Å spacings, with ~7.2Å being most dominant. Considering the relatively rapid time of formation for verdine (Bailey, 1988), we thus find that condensed sections at MFSs (Figs. 3 and 6) contain pristine verdine minerals that are nascent to slightly evolved and believe they have remained so because they were immediately buried by HST progradation and thus preserved from further evolution. The caveat to these interpretations, however, is that the relatively subdued ~7.2Å peaks of the GBR HSTs may also reflect dilution by other components.

HSTs contain reworked verdine grains that appear to be moderately evolved to evolved in the sense of Odin and Matter (1981) and Odin (1988). Relative to the pristine grains of the condensed sections, their higher degree of maturity may result from longer residence time at the sediment/water interface. We suggest that these grains were initially formed authigenically as bioclast-infilling minerals that have been subsequently reworked and abraded into free grains. The reworked nature of the verdine, the relative increase in abundance of bioclasts, and the generally coarser nature of the sediment found in intervals within late highstand systems tract deposits indicate probable decreases in accommodation space that may have promoted both dynamic bypassing of bedload and total passing of suspended load at the time of deposition (e.g., Kidwell, 1991) and in water depths potentially shallower than 100 m. These conditions are conducive to concentrating grains as lag deposits and it is suggested here that verdine grains may have continued to evolve in these conditions until they too were eventually buried. On the southwestern margin of India, verdine grains from the modern shelf at 40 m water depth were found to be reworked and further evolved when compared to the well-preserved grains found on the modern continental slope (see Purnachandra Rao et al., 1993). It is possible that the reworked verdine found within HSTs at the GBR forereef developed in similar conditions that may have promoted further evolution.

We suggest that the reason verdine and glaucony facies minerals typically form within condensed intervals is because of the slow rates of sedimentation that favor their formation on the seafloor and otherwise prevent their burial into the deeper zones of sulfate reduction and methanogenesis. Because verdine typically forms an order of magnitude faster than glaucony, we find verdine formation within intervals interpreted as condensed sections (at 10^3 yr to 10^4 yr scale) in this high resolution stratigraphic setting. The verdine facies minerals develop at and immediately below the sediment/water interface (Odin and Sen Gupta, 1988), their formation is favored by nondeposition, and their evolution is terminated shortly after burial (e.g., Odin et al, 1988a). Furthermore, we suggest that the relatively rapid formation of verdine facies minerals may also have important implications for reverse weathering (authigenic mineral formation) and may play a significant role in large-scale biogeochemical cycling in the oceans. Although we have not quantified the abundance of verdine in the GBR sediments, we believe that this facies has been overlooked in terms of its relative importance with regard to providing a sink for reverse weathering reactions involving Fe, Mg, and K in clay mineral formation (e.g., Aller at al., 1986; Mackenzie and Kump, 1995; Michalopoulos and Aller, 1995). Although *in situ* formation of verdine in waters deeper than about 60 m has been found to be rare and they are rarely found in deposits older than the Holocene (Bailey, 1988; Odin and Sen Gupta, 1988), we document here *in situ* formation of verdine within condensed sections at Site 821, where paleowater depths were probably similar to the water depth of 212 m today, and in sediments spanning the last ~1.5 m.y.

SUMMARY AND CONCLUSIONS

This paper first reveals the presence of the verdine facies off Australia. To date, this makes this the 16th known presence of the facies worldwide. It also explores for the first time the nature of verdine facies emplacement within the context of sea-level change, sequence condensation, and sequence stratigraphic analysis. Approximately meter-thick intervals at maximum flooding surfaces (MFS) are found to be enriched in authigenic verdine minerals and detrital quartz, and are interpreted to represent condensed sections. Concentrations of the authigenic mineral verdine are associated with two different types of depositional processes. Pristine (non-reworked) verdine infilling delicate foraminifer tests are found in relatively condensed sections associated with downlapped MFSs. Lower concentrations of verdine occur within highstand systems tracts and appear abraded and fragmented; the nature of these particles suggests they were moderately reworked from previously deposited nearby sediments and accumulated in a higher-energy depositional setting associated with highstand progradation as sediment accommodation potential diminishes.

We suggest that the reason verdine facies minerals typically form within condensed intervals is because of the relatively slow rates of sedimentation that allow its formation. In the tropical GBR forereef setting verdine minerals apparently developed at and immediately below the sediment/water interface, and their evolution was likely terminated shortly after burial, as is the case for glaucony. Major river systems discharging significant amounts of continental water and terrigenous sediments into the sea close to Grafton Passage most likely provided several point sources of iron needed for the formation of verdine in low latitude depositional settings. As for some glauconites (cf. Odin and Matter, 1981), we believe the foraminiferal and other biogenic tests housing the verdine provided micro-reducing suboxic environments suitable for the incorporation of both Fe^{+++} and Fe^{++} into the verdine minerals. We find that the verdine found in the GBR succession is chemically distinct from chamosite and berthierine minerals, which are typically associated with the oolitic ironstone facies that typically forms in shallower waters. The relatively rapid formation of verdine facies minerals may have important implications for reverse weathering and may play a significant role in oceanic biogeochemical cycling. We stress that this facies may have been long overlooked in terms of its relative importance with regard to providing a sink for reverse weathering reactions in clay mineral formation; as such, it warrants further attention. Although *in situ* formation of verdine in waters deeper than 100 m has been suggested to be uncommon and it has rarely been found in pre-Holocene deposits, we find *in situ* formation of verdine within condensed sections where paleo-water depths may have been approximately 200 m for approximately the last 1.5 m.y. To our knowledge, this is the first described occurrence of verdine found in Pleistocene deposits.

ACKNOWLEDGMENTS

We thank Alessandro Amorosi and an anonymous reviewer for their astute and constructive reviews of this manuscript. Thanks are also extended to Fred Mackenzie, Johanna Resig, Roy Wilkens, Patricia Cooper and Ralph Moberly for their encouragement and for reviewing an earlier draft of this paper. We appreciate the assistance provided by Clifford Todd with EMP analyses and Clark Sherman with XRD analyses. We also thank Jacques Lucas and Liliane Prévôt-Lucas for editorial handling. This is University of Hawaii School of Ocean and Earth Science and Technology contribution number 4832.

REFERENCES

ALLER, R. C., MACKIN, J. E., AND COX, R. T., Jr., 1986, Diagenesis of Fe and S in Amazon inner shelf muds: Apparent dominance of Fe reduction and implication for genesis of ironstones: Continental Shelf Research, v. 6, p. 263-289.

AMOROSI, A., 1995, Glaucony and sequence stratigraphy: A conceptual framework of distribution in siliciclastic sequences: Journal of Sedimentary Research, v. B65, p. 419-425.

AMOROSI, A., 1997, Detecting compositional, spatial, and temporal attributes of glaucony: A tool for provenance research: Sedimentary Geology, v. 109, p. 135-153.

BAILEY, S. W., 1988, Odinite, a new dioctahedral-trioctahedral Fe^{3+}-rich 1:1 clay mineral: Clay Minerals, v. 23, p. 237-247.

BHATTACHARYYA, D. P., 1983, Origin of berthierine in ironstones: Clays and Clay Minerals, v. 31, p. 173-182.

BRINDLEY, G. W., 1982, Chemical compositions of berthierines—A review: Clays and Clay Minerals, v. 30, p. 153-155.

CHAMLEY, H. R., AND MULLER, D. W., 1993, The clay-mineralogical record of the last 10 million years off northeastern Australia, *in* McKenzie, J. A., Davies, P. J., and Palmer-Julson, A., eds, Proceedings of the Ocean Drilling Program, Scientific Results, Volume 133: College Station, Texas, p. 461-470.

DAVIES, P. J., MCKENZIE, J. A., AND PALMER-JULSON, A., et al., 1991, Proceedings of the Ocean Drilling Program, Initial Reports, Volume 133: College Station, Texas, p. 1496.

DEER, W. A., HOWIE, R. A., AND ZUSSMAN, J., 1992, An Introduction to the Rock Forming Minerals: London, Longman, 696 p.

FEARY, D. A., AND JARRARD, R. D., 1993, Sedimentology and downhole log analysis of Site 820, central Great Barrier Reef outer shelf: the factors controlling progradational and aggradational seismic geometry, *in* McKenzie, J. A., Davies, P. J., and Palmer-Julson, A., eds, Proceedings of the Ocean Drilling Program, Scientific Results, Volume 133: College Station, Texas, p. 315-326.

GLENN, C. R., 1990, Depositional sequences of the Duwi, Sibaiya and Phosphate Formations, Egypt: Phosphogenesis and glauconitization in a Late Cretaceous epeiric sea, *in* Notholt, A. J. G., and Jarvis, I., eds., Phosphorite Research and Development, Geological Society Special Publication 52: Oxford, Geological Society, p. 205-222.

GLENN, C. R., AND ARTHUR, M. A., 1988, Petrology and major element geochemistry of Peru margin phosphorites and associated diagenetic minerals: Authigenesis in modern organic-rich sediments: Marine Geology, v. 80, p. 231-267.

GLENN, C. R., AND ARTHUR, M. A., 1990, Anatomy and origin of a Cretaceous phosphorite-greensand giant, Egypt: Sedimentology, v. 37, p. 123-154.

GLENN, C. R., KRONEN, J. D., JR., SYMONDS, P. A., WEI, W., AND KROON, D., 1993, High-resolution sequence stratigraphy, condensed sections, and flooding events off the Great Barrier Reef: 0-1.5 Ma, *in* McKenzie, J. A., Davies, P. J., and Palmer-Julson, A., eds, Proceedings of the Ocean Drilling Program, Scientific Results, Volume 133: College Station, Texas, p. 353-364.

HAQ, B. U., 1991, Sequence stratigraphy, sea level change, and significance for the deep sea: Special Publications of the International Association of Sedimentologists, v. 12, p. 3-39.

KATZ, M. E., AND MILLER, K. E., 1993, Neogene subsidence along the northeastern Australian margin: benthic foraminiferal evidence, *in* McKenzie, J. A., Davies, P. J., and Palmer-Julson, A., eds, Proceedings of the Ocean Drilling Program, Scientific Results, Volume 133: College Station, Texas, p. 75-92.

KENNEDY, W. J., AND GARRISON, R. E., 1975, Morphology and genesis of nodular phosphates in the Cenomanian glauconite marl of South East England: Lethaia, v. 8, p. 339-360.

KERSHAW, A. P., MCKENZIE, G. M., AND MCMINN, A., 1993, A Quaternary vegetation history of northeastern Queensland from pollen analysis of ODP Site 820, *in* McKenzie, J. A., Davies, P. J., and Palmer-Julson, A., eds., Proceedings of the Ocean Drilling Program, Scientific Results, Volume 133: College Station, Texas, p. 107-114.

KIDWELL, S. M., 1991, Condensed deposits in siliciclastic sequences: Expected and observed features, *in* Einsele, G., Ricken, W., and Seilacher, A., eds, Cycles and Events in Stratigraphy: Berlin-Heidelberg-New York, Springer-Verlag, p. 682-695.

LOUTIT, T. S., HARDENBOL, J., VAIL, P. R., AND BAUM, G. R., 1988, Condensed sections: the key to age determination and correlation of continental margin sequences, *in* Wilgus, C. K., Hastings, B. S., Kendall, C. G. St. C., Posamentier, H. W., Ross, C. A., and Van Wagoner, J. C., eds., Sea-Level Changes: An Integrated Approach: Tulsa, Society for Sedimentary Geology (SEPM) Special Publication 42, p. 183-213.

MACKENZIE, F.T., AND KUMP, L.R., 1995, Reverse weathering, clay mineral formation, and oceanic element cycles: Science, v. 270, p. 586-587.

MAYNARD, J. B., 1986, Geochemistry of oolitic iron ores, an electron microprobe study: Economic Geology, v. 81, p. 1473-1483.

MCKENZIE, J. A., DAVIES, P. J., AND PALMER-JULSON, A., 1993, Proceedings of the Ocean Drilling Program, Scientific Results, Volume 133: College Station, Texas, p. 903.

MICHALOPOULOS, P., AND ALLER, R. C., 1995, Rapid clay mineral formation in Amazon Delta sediments: Reverse weathering and oceanic element cycles, Science, v. 270, p. 614-617.

ODIN, G. S., ed., 1988, Green marine clays: Oolitic ironstone facies, verdine facies, glaucony facies, and celadonite-bearing facies—a comparative study, Developments in Sedimentology 45: New York, Elsevier Science Publishing

Company, 445 p.

ODIN, G. S., 1990, Clay mineral formation at the continent-ocean boundary: The verdine facies: Clay Minerals, v. 25, p. 477-483.

ODIN, G. S., AND MATTER, A., 1981. De glauconiarium origine: Sedimentology, v. 28, p. 611-641.

ODIN, G. S., AND FULLAGAR, P. D., 1988, Geological significance of the glaucony facies, *in* Odin, G.S., ed., Green Marine Clays: Amsterdam, Elsevier Science Publishing Company, p. 295-332.

ODIN, G. S., BAILEY, S. W., AMOURIC, M., FRÖHLICH, F., AND WAYCHUNAS, G., 1988a, Mineralogy of the verdine facies, *in* Odin, G. S., ed., Green marine clays: Oolitic ironstone facies, verdine facies, glaucony facies, and celadonite-bearing facies—a comparative study: Developments in Sedimentology 45: New York, Elsevier Science Publishing Company, p. 159-204.

ODIN, G. S., KNOX, R. W. O. B., GYGI, R. A., AND GUERRAK, S., 1988b, Green marine clays from the oolitic ironstone facies: Habit, mineralogy and environment, *in* Odin, G. S., ed., Green marine clays: Oolitic ironstone facies, verdine facies, glaucony facies, and celadonite-bearing facies—a comparative study: Developments in Sedimentology 45: New York, Elsevier Science Publishing Company, p. 29-52.

ODIN, G. S., AND SEN GUPTA, B. K., 1988, Geological significance of the verdine facies, in Odin, G.S., ed., Green marine clays: Oolitic ironstone facies, verdine facies, glaucony facies, and celadonite-bearing facies—a comparative study: Developments in Sedimentology 45: New York, Elsevier Science Publishing Company, p. 205-219.

POSAMENTIER, H. W., JERVEY, M. T., AND VAIL, P. R., 1988, Eustatic controls on clastic deposition I—conceptual framework, *in* Wilgus, C. K., Hastings, B. S., Kendall, C. G. St. C., Posamentier, H. W., Ross, C. A., and Van Wagoner, J. C., eds, Sea-Level Changes: An Integrated Approach.: Tulsa, Society for Sedimentary Geology (SEPM) Special Publication 42, p. 109-124.

POSAMENTIER, H. W., AND VAIL, P. R., 1988. Eustatic controls on clastic deposition II—sequence and systems tract models, *in* Wilgus, C. K., Hastings, B. S., Kendall, C. G. St. C., Posamentier, H. W., Ross, C. A., and Van Wagoner, J. C., eds., Sea-Level Changes: An Integrated Approach: Tulsa, Society for Sedimentary Geology (SEPM) Special Publication 42, p. 125-154.

PURNACHANDRA RAO, V. P., LAMBOY, M., AND DUPEUBLE, P. A., 1993, Verdine and other associated authigenic (glaucony, phosphate) facies from the surficial sediments of the southwestern continental margin of India: Marine Geology, v. 111, p. 133-158.

PURNACHANDRA RAO, V. P., THAMBAN, M., AND LAMBOY, M., 1995, Verdine and glaucony facies from surficial sediments of the eastern continental margin of India: Marine Geology, v. 127, p. 105-113.

SARG, J. F., 1988, Carbonate sequence stratigraphy, *in* Wilgus, C. K., Hastings, B. S., Kendall, C. G.St. C., Posamentier, H. W., Ross, C. A., and Van Wagoner, J. C., eds., Sea-Level Changes: An Integrated Approach: Tulsa, Society for Sedimentary Geology (SEPM) Special Publication 42, p. 155-182.

SOUTHGATE, P. N., AND SHERGOLD, J. H., 1991, Application of sequence stratigraphic concepts to middle Cambrian phosphogenesis: Bureau of Mineral Resources Journal of Australian Geology and Geophysics, v. 12, p. 119-144.

TAYLOR, K. G., 1990, Berthierine from the non-marine Wealden (Early Cretaceous) sediments of south-east England: Clay Minerals, v. 25, p. 391-399.

VAIL, P. R., 1990, Sequence stratigraphic interpretation of seismic, well and outcrop data: American Association of Petroleum Geologists Short Course, Honolulu, 308 p.

VAIL, P. R., AUDEMARD, F., BOWMAN, S. A., EISNER, P. N., AND PEREZ-CRUZ, C., 1991, The stratigraphic signatures of tectonics, eustasy and sedimentology—an overview, *in* Einsele, G., Ricken, W., and Seilacher, A., eds, Cycles and Events in Stratigraphy: Berlin-Heidelberg, Springer-Verlag, p. 617-681.

VAN WAGONER, J. C., POSAMENTIER, H. W., MITCHUM JR., R. M., VAIL, P. R., SARG, J. F., LOUTIT, T. S., AND HARDENBOL, J., 1988. An overview of the fundamentals of sequence stratigraphy and key definitions. *in* Wilgus, C. K., Hastings, B. S., Kendall, C. G. St. C., Posamentier, H. W., Ross, C. A., and Van Wagoner, J. C., eds., Sea-Level Changes: An Integrated Approach.: Tulsa, Society for Seidmentary Geology (SEPM) Special Publication 42, p. 39-45.

VAN WAGONER, J. C., MITCHUM, R. M., CAMPION, K. M., AND RAHMANIAN, V. D., 1990, Siliciclastic Sequence Stratigraphy in Well Logs, Cores, and Outcrops: Concepts for High-Resolution Correlation of Time and Facies: Tulsa, American Association of Petroleum Geologists, Methods in Exploration Series 7, 55 p.

WILGUS, C. K., HASTINGS, B. S., KENDALL, C. G. ST. C., POSAMENTIER, H. W., ROSS, C. A., AND VAN WAGONER, J. C., eds., 1988, Sea-Level Changes: An Integrated Approach: Tulsa, Society for Sedimentary Geology (SEPM) Special Publication 42, p. 407.

WEI, W., AND GARTNER, S., 1993, Neogene calcareous nannofossils from Sites 811 and 819 through 825, offshore northeastern Australia, *in* McKenzie, J. A., Davies, P. J., and Palmer-Julson, A., eds, Proceedings of the Ocean Drilling Program, Scientific Results, Volume 133: College Station, Texas, p. 19-37.

YOUNG, T. P., AND TAYLOR, W. E. G., 1989, Phanerozoic Ironstones: London, Geological Society Special Publication 46, p. 251.

ANATOMY OF A CONDENSED SECTION: THE LOWER CENOMANIAN GLAUCONY-RICH DEPOSITS OF CAP BLANC-NEZ (BOULONNAIS, NORTHERN FRANCE)

ALESSANDRO AMOROSI* AND MARIA CARLA CENTINEO

Dipartimento di Scienze della Terra e Geologico-Ambientali, Via Zamboni 67, 40127 Bologna, Italy

ABSTRACT: Detailed examination of the lowermost Cenomanian, glaucony-rich deposits of Cap Blanc-Nez (Northern France) reveals a hierarchy of depositional units within a stratigraphically condensed succession, providing a fundamental tool for its sequence stratigraphic interpretation.

Condensation is obvious in the highly bioturbated basal glaucony-rich unit (Strouanne Formation), where glaucony is predominantly autochthonous, highly evolved (> 8 % K_2O), and generally constitutes 60–90 % of the sand fraction. Glaucony concentration decreases markedly upwards (to less than 10 %) in the overlying rhythmically bedded marls and chalks (Petit Blanc-Nez Formation), where the green grains are allochthonous and comparatively less evolved (6.0–6.5 % K_2O). Glaucony distribution shows a vertical cyclic pattern within the condensed interval. Maxima of glaucony concentrations are invariably recorded at intensively burrowed omission surfaces, interpreted as bases of either parasequences or parasequence sets.

The characterization of glaucony attributes indicates that a prolonged break in sedimentation, on the order of 0.5–1 Ma, occurred in the study area during the Early Cenomanian. The highly condensed nature of the basal glaucony-rich deposits suggests that either one or two of the third-order Lower Cenomanian depositional sequences recognized in the coeval, less condensed successions of Western Europe are missing at Cap Blanc-Nez. Glaucony maturity in the overlying marl-chalk couplets is consistent with interpretation of these deposits as Milankovitch-scale cycles.

INTRODUCTION

Recent studies show that detailed glaucony characterization may contribute to facies interpretation of glaucony-bearing deposits and provide a means of testing their sequence stratigraphic interpretation (Odin, 1985; Loutit et al., 1988; Amorosi, 1995; Mc Cracken et al., 1996). Previous work, however, has examined the spatial distribution of glaucony predominantly at the scale of third-order depositional sequences (e.g., Amorosi and Centineo, 1997), and models for glaucony distribution within lower-rank units, such as fourth-order depositional cycles, have received little study. Mitchum and Van Wagoner (1991) advanced the hypothesis that glaucony concentrations could mark parasequence boundaries, but no detailed studies have been undertaken to test this hypothesis.

A hierarchy of depositional units, including remarkably different concentrations of glaucony with obvious vertical fluctuations in abundance, characterize the condensed hemipelagic facies of Early Cenomanian age in the Boulonnais (Northern France). These deposits, which have been extensively studied in terms of lithostratigraphy (Robaszynski and Amédro, 1986; Amédro, 1994), biostratigraphy (Amédro, 1993), and sequence stratigraphy (Robaszynski and Amédro, 1993; Robaszynski et al., 1998), are superbly exposed and easily accessible in the white cliff of Cap Blanc-Nez (Fig. 1), approximately 5 km southwest of the French entrance of the Eurotunnel. At low tide the Albian-Cenomanian boundary is exposed on the foreshore and the lowermost Cenomanian glaucony-rich deposits can be studied in detail.

This paper is based on 33 samples of glaucony, collected in the lowermost 4 m of the Cenomanian succession with an average spacing of about 15 cm (Fig. 2). The main focus of this study is to show how the characterization of glaucony attributes can be used for facies analysis and sequence stratigraphic interpretation of a stratigraphically condensed interval. Specific goals are to document how variations in abundance and composition of glaucony can serve to identify lower-rank depositional cycles within otherwise undifferentiated deposits. Refined stratigraphic correlations between the Cenomanian deposits of Cap Blanc-Nez and the coeval units of the Anglo-Paris Basin are ensured by the combination of high-resolution biostratigraphic (ammonite) zonations with lateral tracing of marker beds over tens of kilometers (Amédro, 1994; Gale, 1995; Robaszynski et al., 1998).

FIG. 1.—Location of the study section. Scale bar is 100 km.

THE LOWER CENOMANIAN DEPOSITS OF THE ANGLO-PARIS BASIN

The Albian-Cenomanian boundary is a surface of non-depo-

*email: amorosi@geomin.unibo.it

Marine Authigenesis: From Global to Microbial, SEPM Special Publication No. 66

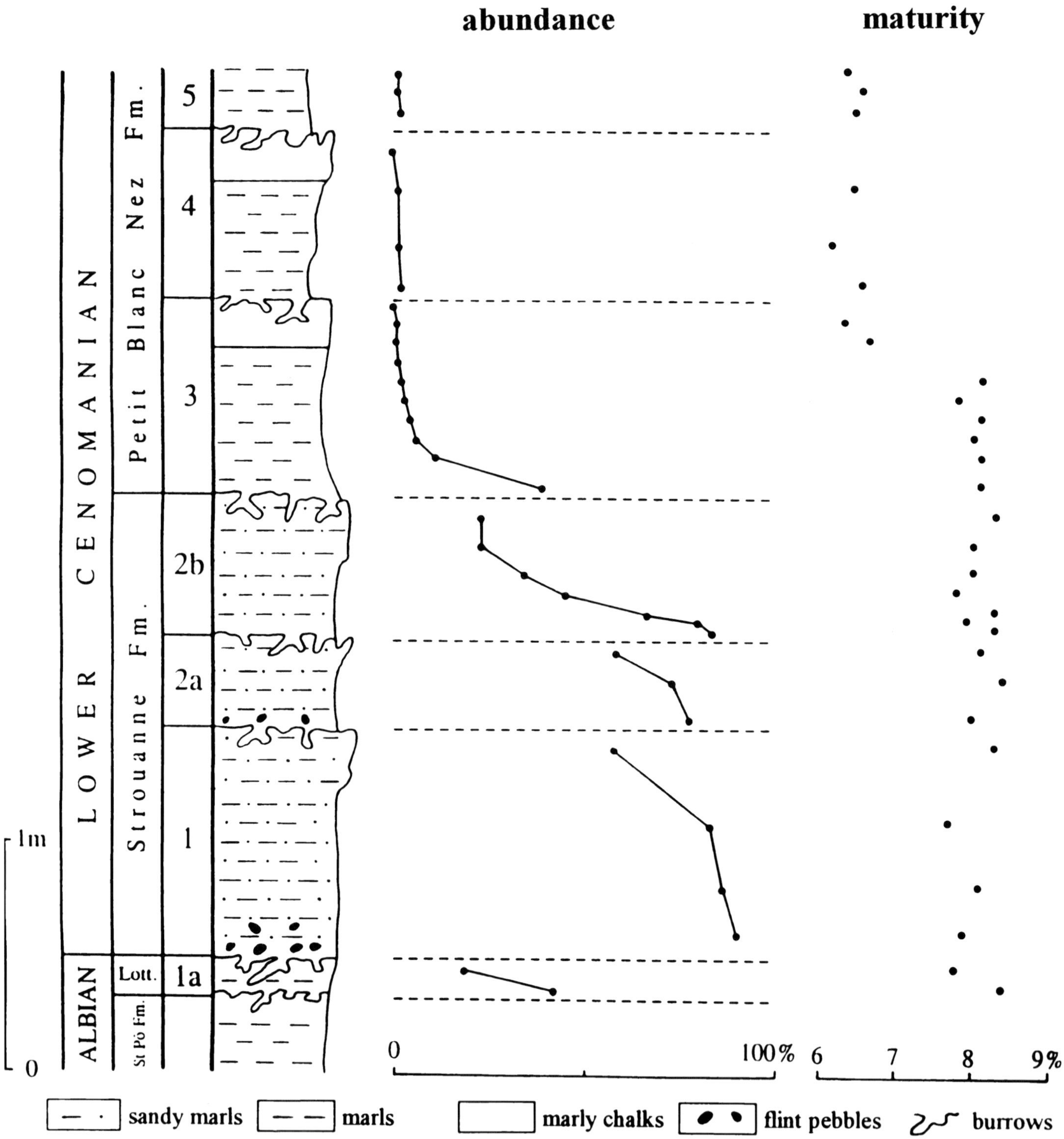

FIG. 2.—Detailed stratigraphic section of the Lower Cenomanian deposits at Cap Blanc-Nez (lithostratigraphy slightly modified after Robaszynski and Amédro, 1993), with indication of glaucony abundance and maturity. Lott. = Lottinghen Formation.

sition or condensation, locally marked by a hardground, across most of the Anglo-Paris Basin. Extremely low sedimentation rates account for the widespread occurrence of glaucony at the base of the overlying transgressive Cenomanian deposits (Gale, 1995). In contrast, the Middle-Upper Cenomanian deposits are generally thicker, due to increased accommodation occurring with rapidly rising sea-level during the Cenomanian (Robaszynski et al., 1998).

The major facies changes recorded within the Cenomanian deposits of Northern France and Southern England reflect the well-known palaeogeography of the Anglo-Paris Basin. During the Cenomanian, clastic sedimentation took place in marginal areas (Devon, southeast England, and most of France), whereas condensed facies (bioturbated hemipelagic chalks and marls) slowly accumulated in the basin center and on structural highs, such as the Boulonnais (Amédro, 1994).

A striking cyclic arrangement characterizes the Cenomanian deposits of the Anglo-Paris Basin; it can be subdivided in a number of "elementary rhythms" (Gale, 1990; Juignet and Breton, 1992). Sedimentary rhythmicity, taking place in decimeter- to meter-thick cycles, is expressed primarily by two types of alternating lithofacies (Destombes and Shephard-Thorn, 1971; Hart, 1987; Robaszynski and Amédro, 1993): marl-chalk couplets, which are well developed in rhythms of Early Cenomanian age, and limestone-flint couplets. Most marl-chalk couplets can be correlated across Western Europe using refined biostratigraphy, marker-bed stratigraphy, and carbon isotope stratigraphy (Gale, 1990, 1995).

Cyclic patterns in sedimentation within the Cenomanian deposits of the Anglo-Paris Basin are interpreted to have been caused by orbital forcing. Four different models (fluctuations in carbonate productivity, terrigenous dilution, redox conditions, and bottom currents) have been advocated as possible forcing mechanisms for the pelagic cyclic sequences (for review, see de Boer and Smith, 1994). Two of these have been suggested to have controlled generation of rhythmic bedding in the Cenomanian deposits of the Anglo-Paris Basin. In particular, marl-chalk cycles have been interpreted to reflect either regular variations of carbonate productivity induced by changing oceanographic conditions (Ditchfield and Marshall, 1989; Paul, 1992) or dilution cycles (Arthur et al., 1986; Bottjer et al., 1986), marking fluctuations in the terrigenous (clay) sediment supply.

Milankovitch forcing has been considered to be the driving mechanism of the Cenomanian microrhythms. A cyclicity presumably forced by precession (21 ka cycles) was suggested by Robinson (1986), Paul (1992), and Gale (1995) for the English Chalk. A possible control by obliquity (41.5 ka cycles) and eccentricity (100 ka cycles) was envisaged by Hart (1987) and Gale (1990), respectively. A Milankovitch-scale cyclicity has been invoked to explain chalk-marl alternations in France (Juignet and Breton, 1992, 1994).

Lower Cenomanian deposits of Cap Blanc-Nez are grouped into two formations (Fig. 2): the Strouanne Formation, approximately 2 m thick, and the overlying Petit Blanc-Nez Formation, about 25 m thick (Robaszynski and Amédro, 1993). Fifteen cm of gray marly clay with abundant glaucony and phosphate pebbles (Lottinghen Formation) separate the Cenomanian deposits from the underlying Albian clays (St. Pô Formation). The Lottinghen Formation, the age of which cannot easily be determined due to a lack of diagnostic ammonite faunas (Gale et al., 1996), consists of a markedly wedge-shaped sedimentary body that thickens northeast, to a maximum thickness of 15 m, in the English Channel area (Amédro, 1994).

Glaucony is present in variable proportions within the lowermost Cenomanian deposits of Cap Blanc-Nez. Glaucony concentration is greatest at the very base of the Cenomanian succession (Strouanne Formation). These glaucony-rich sandy marls, which correspond to the "Tourtia" of French geologists and "Glauconitic Marls" of English literature, can be subdivided into three depositional cycles (units 1 and 2 of Robaszynski and Amédro, 1993, with a further subdivision of unit 2 to units 2a and 2b). These cycles, consisting of lower dark green-black divisions and upper gray portions, are separated by burrowed surfaces that are marked by concentrations of Sponges and *Thalassinoides.* A slight increase in the amount of carbonate is recorded in the upper parts of cycles 2a and 2b. Flint pebbles are common within cycles 1 and 2a.

Rhythmically bedded marls and chalks form the bulk of the overlying Petit Blanc-Nez Formation (Fig. 2). Each couplet includes gray marls overlain by lighter-colored (gray-yellowish) marly chalks, which usually have burrowed upper sections. The upward increase in carbonate content, which parallels a progressive upward decrease in the clay fraction, is reflected in the weathering profile, which shows prominent chalk strata at top of the rhythms. Glaucony is present uniquely in the lower part of the formation (units 3–5 of Robaszynski and Amédro, 1993).

PREVIOUS SEQUENCE STRATIGRAPHIC INTERPRETATIONS

In spite of refined stratigraphic correlations across Western Europe (cf. Gale, 1995; Robaszynski et al., 1998), relationships between sedimentary facies architecture and relative sea-level changes during the Cenomanian have not been fully understood and univocal sequence stratigraphic interpretations are lacking, especially for the Lower Cenomanian deposits (Table 1). The most striking example is that of the boundary between the Lower Cenomanian ammonite subzones *Neostlingoceras carcitanense* and *Mantelliceras saxbii* within the Mantelliceras Mantelli zone (boundary between the units 4 and 5), which has been alternatively interpreted to represent a maximum flooding surface, a sequence boundary, and possibly a transgressive surface (Table 1).

The first sequence stratigraphic interpretation of the Lower Cenomanian units in the Anglo-Paris Basin is that of Haq et al. (1988), who identified two third-order depositional sequences in the type area of the Cenomanian stage, around Le Mans (SE Paris Basin) (Table 1). According to these authors, the lowermost Cenomanian deposits (Glauconitic Chalk) were considered to be an upper transgressive systems tract (TST). A different interpretation of the same units was proposed by Juignet and Breton (1992), who grouped all the lower Cenomanian deposits in a single depositional sequence. In this case, the basal glaucony-rich sands were interpreted as a shelf-margin wedge (SMW).

TABLE 1.—PREVIOUS SEQUENCE STRATIGRAPHIC INTERPRETATIONS OF THE CENOMANIAN DEPOSITS OF THE ANGLO-PARIS BASIN.

Stratigraphy	Ammonite Zones		Formations	Units	Haq et al. 1988 SE Paris Basin	Juignet & Breton 1992 SE Paris Basin	Robaszynski & Amédro, 1993 Boulonnais	Gale 1995 Boulonnais	Owen 1996 Anglo-Paris Basin	Mitchell et al. 1996 Cleveland Basin
LOWER CENOMANIAN	DIXONI		Petit Blanc Nez Fm.	10	SMW	SMW	TST	(TST)	SMW	TST
	MANTELLI	Saxbii		9		HST	SMW		(HST)	SMW
				8	HST		HST	(TST?)		(TST/HST)
				7			TST			
				6						
				5	?				?	?
		Carcitanense		4	TST	TST	TST		TST	(LST)
				3						
			Strouanne Fm.	2						
				1		SMW		?		HST
UPPER ALBIAN	DISPAR		Lottinghen Fm.	1a			SMW		(SMW)	SMW/TST
	INFLATUM		St Pô Fm.							

sequence boundary (thick line) — transgressive surface (thin line) - - - - maximum flooding surface (dashed line)

The shaded units indicate the interval of study. Sequence stratigraphic interpretation by Robaszynski et al. (1998) (see their Fig. 7) is equivalent to the one by Robaszynski and Amédro (1993).
LST: lowstand systems tract; SMW: shelf margin wedge; TST: transgressive systems tract; HST: highstand systems tract.

Controversial sequence stratigraphic interpretations have been also advanced for the Lower Cenomanian deposits of the Boulonnais, based upon the Cap Blanc-Nez section. Robaszynski and Amédro (1993) and Robaszynski et al. (1998) identified three distinct depositional sequences: the glaucony-rich deposits of the Lottinghen Formation were interpreted as the SMW of the lowest sequence, whereas units 1 through 4 were attributed to the TST of the same sequence (cf. Table 1). A different interpretation of the same units, lacking one of the three depositional sequences previously recognized, was provided by Gale (1995; his Fig. 3).

By comparison with the Lower Cenomanian deposits from

various parts of the Anglo-Paris Basin, Owen (1996) reinterpreted the Lower Cenomanian succession as a single depositional sequence, returning to the sequence stratigraphic interpretation of Juignet and Breton (1992). Also, he interpreted the lowermost Cenomanian deposits as a TST. A substantially different interpretation is that of Mitchell et al. (1996), who identified four different depositional sequences in the relatively less condensed Lower Cenomanian deposits of the Cleveland Basin (England).

METHODS

Glaucony separation was performed strictly following the methodology recommended by Odin (1982). Rock samples were dried at 70° C, followed by disaggregation in refined petroleum (2h) and then in water (48h). Wet sieving, with selection of the 50–1000-μm size fraction, was followed by dry sieving, using 160 μm and 500 μm sieves. Pure glaucony was obtained by the following treatments: (1) magnetic separation, using a Frantz isodynamic separator, with longitudinal dip 20° and sideward tilt 15°; (2) magnetic fractionation of the fractions attracted at a current intensity of 0.6 A; and (3) purification by ultrasonic treatment. Chemical analyses were performed by X-ray fluorescence.

The total amount of glaucony was calculated as weight %, and is related here to the > 50 μm size fraction (referred to informally as "sand fraction"). The maturity of glaucony was estimated on the basis of the potassium content, resulting in the traditional distinction among nascent (< 2% K_2O), slightly evolved (4–6% K_2O), evolved (6–8% K_2O), and highly evolved (> 8% K_2O) glaucony (Odin and Matter, 1981).

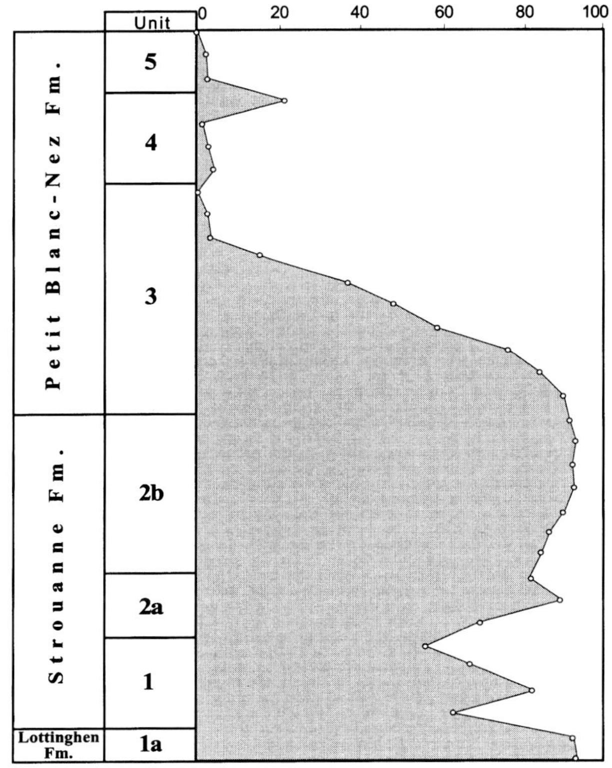

FIG. 3.—Grain-size distribution of glaucony within the 50–500-μm size fraction of the study samples. The gray color indicates the percentage of coarser glaucony (160–500-μm size fractions); the white color indicates the percentage of finer glaucony (50–160-μm size fractions). Dots indicate the sampled interval.

RESULTS

Glaucony is extremely abundant (60-90%) in the lower 2 m (units 1, 2a, and lower 2b), and shows an overall decrease in abundance (less than 40% within upper unit 2b and unit 3, and 0–2 % in units 4 and 5) in the overlying deposits (Fig. 2). An obvious vertical cyclic pattern characterizes glaucony distribution, especially in the basal glaucony-rich deposits. Maximum glaucony abundance is invariably recorded at the very base of the depositional cycles.

Grain-size distribution of glaucony differs markedly from base to top of the study units (Fig. 3). Glauconitization appears to have affected mostly the 160–500-μm size fractions in the lower part of the study succession (units 1, 2a, 2b, and decreasing values in lower unit 3), whereas glaucony from the overlying units is mostly concentrated within the 50–160-μm size fractions.

Two populations of grains are present in the basal glaucony-rich unit. Abundant, dark green lobate grains, often associated with burrows, show surface cracks (Fig. 4A). These grains are interpreted to be autochthonous (that is, grains that have not experienced significant transport from their place of origin) and are associated in variable proportions with dark green grains that lack any cracks and show a remarkably higher abundance of impact marks (Fig. 4B). These latter grains, which are similar in size to the lobes of the lobate grains, have probably experienced transport of considerable duration and are interpreted as parautochthonous (allochthonous intrasequential). SEM of grain surfaces shows the common presence of lamellar nannostructures, which are considered to reflect an advanced stage of maturity of glaucony (Odin and Lamboy, 1988). The lamellar structures are generally well preserved or moderately altered in the autochthonous grains (Fig. 4C), whereas they are commonly strongly modified and locally almost entirely obscured in the allochthonous grains (Fig. 4D). The autochthonous grains are preferentially concentrated in the lower parts of the depositional cycles; the relative proportion of allochthonous grains increases upward.

Glaucony from upper unit 3 and units 4 and 5 is green and generally very small in size (Fig. 3). These grains invariably lack any surface cracks and show typical evidence of transport (abraded and fractured surfaces).

The maturity of glaucony in the units shows substantially different features with respect to the patterns of glaucony distribution (cf. Fig. 2). Neither vertical cyclic variations nor lithologically related trends are recorded, but a major break separates two populations of grains with broadly uniform compositions. In particular, glaucony from the Lottinghen Formation, the Strouanne Formation, and the lower Petit Blanc Nez Formation (unit 3) exhibits very high potassium contents (between 7.7%

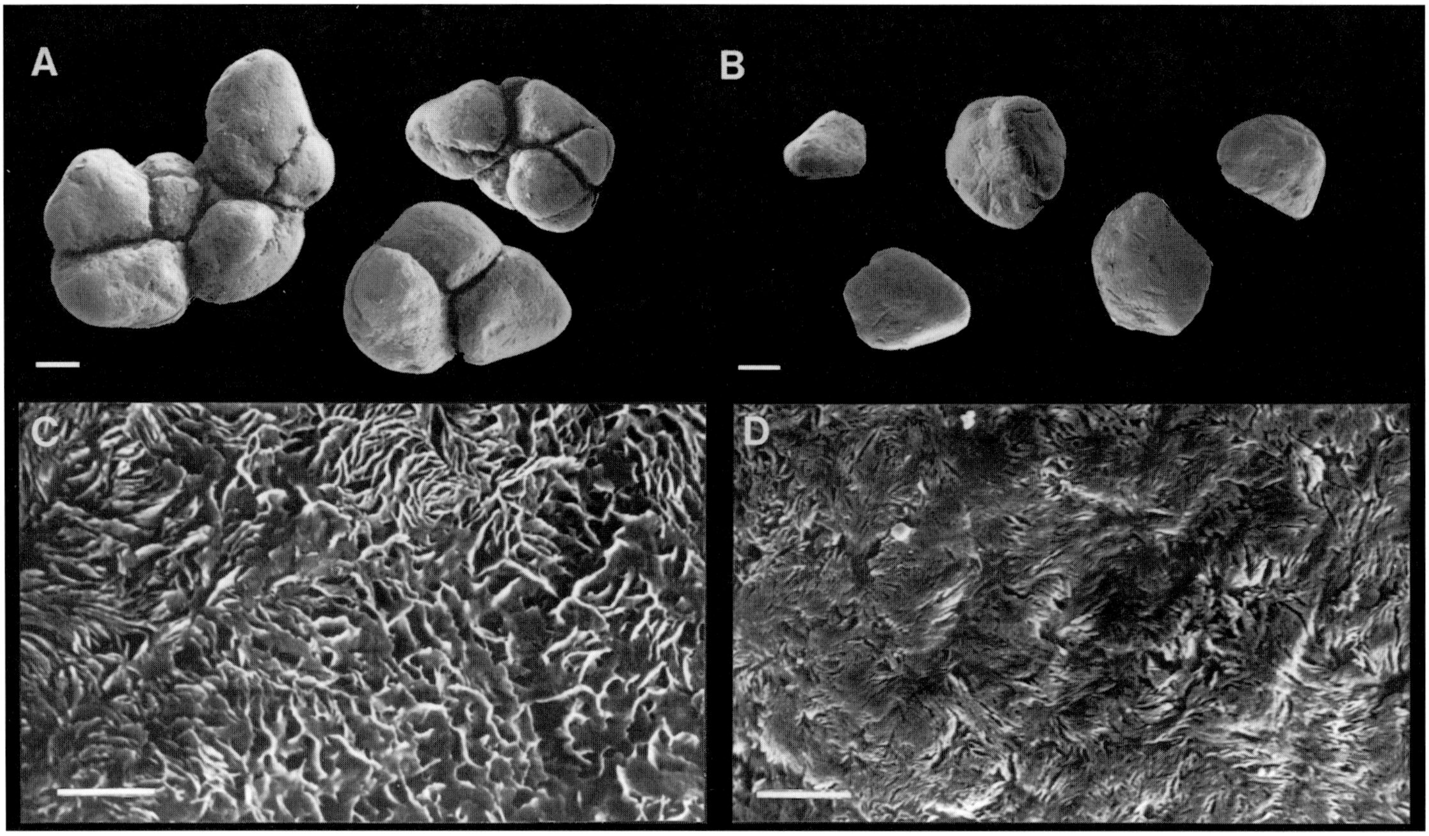

FIG. 4.—SEM microphotographs of glaucony, showing A) autochthonous grains with well developed cracks (scale bar, 100 μm); B) allochthonous grains lacking any crack (scale bar, 100 μm); C) autochthonous glaucony, showing the characteristic lamellar structure of highly evolved glaucony (scale bar, 5 μm); D) modifications of the lamellar structure due to transport in a highly evolved, allochthonous glaucony (scale bar, 5 μm).

and 8.4%) and is classified predominantly as "highly evolved." In contrast, glaucony from the overlying units displays a lower potassium content (between 6.0 % and 6.6 %) and is classified as "evolved." A comparison of the potassium content of glaucony from different size fractions shows that glaucony in the 160–500 μm size fraction is invariably more evolved than glaucony in the 50–160 μm size fraction (Fig. 5), as suggested by Giresse et al. (1980).

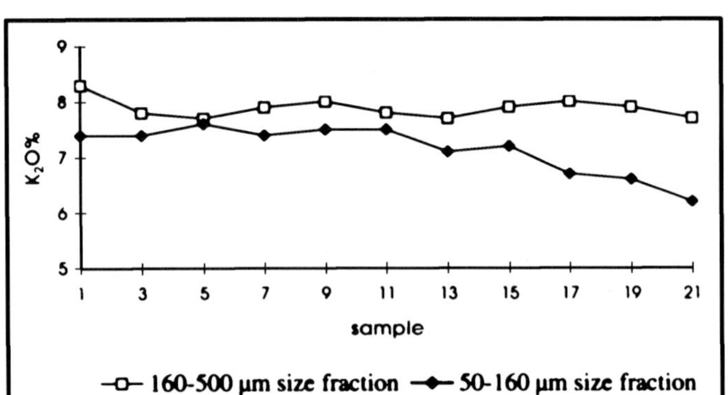

FIG. 5.—Comparison of glaucony maturity from the 160-500-μm and 50-160-μm size fractions of 11 samples from the Lottinghen Formation (sample 1), unit 1 (3, 5), unit 2a (7, 9), unit 2b (11, 13, 15, 17), and unit 3 (19, 21). Note that the 160-500-μm size fraction hosts more evolved glaucony than the 50-160 μm size fraction.

DISCUSSION

The characterization of glaucony attributes allows the distinction of a hierarchy of depositional units within the lowermost Cenomanian deposits of Cap Blanc-Nez. A major separation in terms of abundance and spatial/temporal attributes exists between the Strouanne Formation, where glaucony is abundant, predominantly autochthonous, and highly-evolved, and the overlying Petit Blanc-Nez Formation, where glaucony is scarce, allochthonous, and comparatively less evolved.

The allocthonous origin of glaucony from the Petit Blanc-Nez Formation is strongly suggested by the morphological features discussed in the previous section, combined with the characteristic grain-size distribution of the green grains. Glauconitization is generally favored by the presence of a semiconfined microenvironment at the water-sediment interface, resulting in the preferential accumulation of glaucony within the 160-500-μm size fractions (Giresse et al., 1980; Bornhold and Giresse, 1985). As a consequence, the abundance of glaucony within these size fractions can be regarded as an index of autochthonous origin of the green grains, as in the case of samples from the basal glaucony-rich deposits (Fig. 3). In contrast, when glaucony is submitted to significant transport from its place of formation, fragmentation of the larger grains occurs due to mechanical breakdown, and maxima of glaucony abundance are recorded in comparatively smaller size fractions

(McConchie and Lewis, 1978; Amorosi, 1997), as observed within samples from the Petit Blanc-Nez Formation (Fig. 3).

Further indications of the allochthonous origin of glaucony from the Petit Blanc-Nez Formation are provided by the anomalous combination of very low concentrations (0–2%) with relatively advanced stages of maturity of glaucony (Fig. 2). It has been widely documented that the potassium content of glaucony provides information about the length of time over which glaucony matures before burial (Odin and Matter, 1981; Odin and Fullagar, 1988). According to the diagrams by Odin and coworkers, the evolved glaucony within the marl-chalk couplets in the upper part of the study succession (upper unit 3 and units 4 and 5) should record a break in sedimentation in the order of magnitude of a few tens of thousand years, which is consistent with previous interpretation of the marl-chalk alternations as Milankovitch-scale cycles. This interval of nondeposition generally allows a substantial proportion of substrates to be affected by glauconitization. In this respect, the low concentrations of glaucony very likely reflect an allochthonous origin of these grains.

Whether the allochthonous glaucony was derived from intrasequential or extrasequential sources (parautochthonous versus detrital glaucony, in Amorosi, 1995) cannot be postulated on the sole basis of data from Cap Blanc-Nez, possibly with the exception of samples from the lower part of unit 3. In the lower part of unit 3, the remarkable compositional similarities with glaucony from the units 1 and 2 strongly suggest that these grains were inherited from the underlying deposits, although an autocannibalism from laterally adjacent facies cannot be excluded.

Characterization of glaucony within the basal glaucony-rich unit provides a very different picture. In this case, stratigraphic condensation is suggested by the considerable abundance of glaucony with 8% potassium (Fig. 2): this combination indicates that extremely low sedimentation rates affected the study area in the Early Cenomanian, leading to important accumulations of highly evolved glaucony.

According to diagrams by Odin and Fullagar (1988), the depositional hiatus that is required for the complete maturation of glaucony corresponds to an interval of time of approximately 4–700 ka. By comparison with the coeval deposits of the Vocontian Basin (southern France), Gale (1995) has estimated that 25–30 marl-chalk couplets of Lower Cenomanian age should be missing in the glaucony-rich unit of Cap Blanc-Nez. As these cycles are interpreted by the author to represent depositional cycles, probably controlled by precession, with a mode at 21 ka, it is possible to obtain an independent estimation of approximately 525–630 ka for the time represented by the condensed interval. In a recent paper, Gale et al. (1996) indicate that the Albian-Cenomanian boundary in the eastern Anglo-Paris Basin is marked by a hiatus and condensation of greater magnitude (between 1 and 2 Ma) than previously suspected, comprising a stratigraphic gap in the upper Albian and a condensed interval in the Lower Cenomanian deposits. On the basis of the above considerations, we can assume that the highly condensed glaucony-rich unit at Cap Blanc-Nez should correspond to an interval of time of about 0.5–1 Ma.

The very similar potassium contents of glaucony from the basal glaucony-rich layers, suggesting broadly uniform evolutionary stages, can be only partly ascribed to transport and reworking of grains, as indicated by the abundance of autochthonous grains at any stratigraphic level. Glauconitization is generally assumed to occur either at the sediment-sea water interface or below a moderate thickness of sediments (in the order of magnitude of some dm). It has been suggested, however, that the formation of glaucony and the geochemical closure of the system can occur well beneath this surface (one or a few meters, according to Odin, 1988, and Odin and Fullagar, 1988). This should be the case of an important condensed section such as that documented in this paper. The persistence of a geochemically open system during the slow accumulation of the condensed section may account for the observed uniform composition of the study glaucony throughout the basal glaucony-rich deposits. In this respect, we note that glaucony from the Lottinghen Formation also displays a comparable maturity with respect to glaucony from the overlying Cenomanian deposits.

Despite the remarkably low sedimentation rates suggested by the condensed interval, the rhythmical vertical variations in glaucony distribution indicate that a cyclic alternation of alternately more and less favorable conditions for glauconitization took place in the Early Cenomanian, probably in response to fluctuating sedimentation rates. Fluctuations in sedimentation rates are also suggested by benthic foraminifera distribution in the high-frequency cycles from Cap Blanc-Nez (Paul, 1992). At the base of each cycle, omission surfaces characterized by extensive burrowing are marked by maxima in glaucony abundance. The upward decrease in glaucony concentration from base to top of each high-frequency cycle suggests progressively increasing accumulation rates. The relative abundance of allochthonous glaucony in the upper parts of cycles documents an intensification of transport processes during cycle development that is consistent with this framework.

SEQUENCE STRATIGRAPHIC IMPLICATIONS

Glaucony distribution and maturity at Cap Blanc-Nez provides an input to the evaluation of different sequence stratigraphic models with respect to previous sequence stratigraphic frameworks (Table 1). The stratigraphic condensation of 0.5–1 Ma, documented in this paper, for the lowermost Cenomanian basal glaucony-rich deposits indicates that units 1 and 2 represent a condensed section in the sense of Loutit et al. (1988). Previous sequence stratigraphic studies have recognized two to four distinct third-order depositional sequences in the lowermost Cenomanian successions of Western Europe (*Mantelliceras mantelli* ammonite zone), corresponding to a condensed (less than 10-m thick) stratigraphic unit at Cap Blanc-Nez. According to recent chronostratigraphic studies (Gradstein et al., 1995), the Early Cenomanian would correspond to an interval of approximately 3.1 Ma, the *M. mantelli* zone representing less than 1.5 Ma. Consequently, the depositional hiatus recorded by the basal glaucony-rich deposits at Cap Blanc-Nez has the same order of magnitude as the Lower

Cenomanian third-order depositional sequences recognized across Western Europe.

The data presented in this paper are most consistent with the conclusions of Mitchell et al. (1996) who, on the basis of an independent data set, argued for the condensation in the basal glaucony-rich deposits of Cap Blanc-Nez of one or two of the four sequences recognized in the Cleveland Basin. Our observations are in marked contrast to most previous sequence stratigraphic interpretations, according to which the glaucony-rich deposits at the base of the Cenomanian stage (units 1 and 2 in this paper) simply represent the lower TST of a third-order depositional sequence that includes units 3 and 4 (cf. Robaszynski and Amédro, 1993; Robaszynski et al., 1998, in Table 1). The documentation in this paper of i) an allochthonous origin of glaucony within units 3 and 4, and ii) a completely independent depositional history with respect to glaucony from the underlying deposits, suggests that units 3 and 4 cannot be attributed to the upper TST as proposed by Robaszynski and Amédro (1993).

This paper has shown that high-frequency cycles within the basal glaucony-rich deposits are characterized by diagnostic vertical variations in the glaucony characteristics, with a characteristic maximum concentration of glaucony at the base, followed by a marked upward decline (Fig. 2). Decimeter- to meter-scale depositional cycles in the Cenomanian of the Anglo-Paris Basin have been called parasequences by Juignet and Breton (1992, 1994), who interpreted them as a response to high-frequency, glacio-eustatic oscillations. A primary control by eustatic fluctuations on the stratigraphic architecture of high-frequency cycles has been also proposed by Pomerol and Mortimore (1990), Mitchell et al. (1996), and Robaszynski et al. (1998). Although a link between stratal patterns and sea-level changes has not been clearly established by previous work, and the upward-shoaling trend in the depositional cycles is not obvious (see parasequence definition in Van Wagoner et al., 1988, 1990; Arnott, 1995), the observed patterns of glaucony distribution suggest that the burrowed, glaucony-rich cycle boundaries may correspond to marine flooding surfaces, and the high-frequency cycles reflect short term sea-level fluctuations, at the scale of 4th-order depositional cycles (parasequences or parasequence sets).

CONCLUSIONS

Detailed glaucony characterization within the Lower Cenomanian deposits of Cap Blanc-Nez shows that the stratigraphically condensed interval at the base of the Cenomanian stage includes up to 90% glaucony with approximately 8% K_2O. Elementary cycles within this condensed interval are characterized by rhythmic variations in the abundance of glaucony, with maximum concentrations near the base of each cycle and a systematic upward decrease.

A clear separation in terms of spatial and temporal attributes exists between the predominantly autochthonous, highly evolved glaucony from the basal glaucony-rich unit (Strouanne Formation), and the allochthonous, less evolved glaucony from the overlying Petit Blanc-Nez Formation. No relationships can be envisaged between these two populations of glaucony, which have a completely different origin and document stratigraphic hiatuses differing by at least one order of magnitude.

The condensed nature of the Lower Cenomanian successions in the Boulonnais suggests caution in establishing generalized sequence stratigraphic interpretations based upon the Cap Blanc-Nez section. The characterization of glaucony attributes within the condensed section documents a stratigraphic hiatus of approximately 0.5 to 1 Ma and suggests that one or two of the depositional sequences recognized in the less condensed successions of Western Europe should be missing at Cap Blanc-Nez. The vertical cyclic pattern of glaucony abundance within the condensed section reflects the expected distribution of glaucony within lower-rank parasequences or parasequence sets. This study shows that defining the distribution and characteristics of glaucony may be of value in locating the bounding surfaces of high-frequency cycles within condensed sections when other criteria (lithologic, sedimentologic) are not readily apparent.

ACKNOWLEDGMENTS

We would like to express our gratitude to W. B. Harris and D. McConchie for critical and constructive reviews of the manuscript.

REFERENCES

Amédro, F., 1993, La lithostratigraphie et les biofaciès: des outils de corrélation dans les craies cénomaniennes du Détroit du Pas de Calais: Annales de la Société Géologique du Nord, v. 2, p. 73-80.

Amédro, F., 1994, Correlations dans les craies cénomaniennes traversées par le tunnel sous la Manche (coté France): Bulletin trimestriel de la Société Géologique de Normandie et Amis du Muséum du Havre, v. 81, p. 11-44.

Amorosi, A., 1995, Glaucony and sequence stratigraphy: A conceptual framework of distribution in siliciclastic sequences: Journal of Sedimentary Research, v. B65, p. 419-425.

Amorosi, A., 1997, Detecting compositional, spatial, and temporal attributes of glaucony: A tool for provenance research: Sedimentary Geology, v. 109, 135-153.

Amorosi, A., and Centineo, M. C., 1997, Glaucony from the Eocene of the Isle of Wight (southern UK): Implications for basin analysis and sequence stratigraphic interpretation: Journal of the Geological Society, London, v. 154, p. 887-896.

Arnott, R. W. C., 1995, The parasequence definition—are transgressive deposits inadequately addressed?: Journal of Sedimentary Research, v. B65, p. 1-6.

Arthur, M. A., Bottjer, D. J., Dean, W. E., Fischer, A. G., Hattin, D. E., Kauffman, E. G., Pratt, L. M., and Scholle, P. A., 1986, Rhythmic bedding in Upper Cretaceous pelagic carbonate sequences: Varying sedimentary response to climatic forcing: Geology, v. 14, p. 153-156.

Bornhold, B. D., and Giresse, P., 1985, Glauconitic sediments on the continental shelf off Vancouver Island, British Columbia, Canada: Journal of Sedimentary Petrology, v. 55, p. 653-664.

Bottjer, D. J., Arthur, M. A., Dean, W. E., Hattin, D. E., and Savdra, C. E., 1986, Rhythmic bedding in Cretaceous pelagic carbonate environments—sensitive recorders of climatic cycles: Palaeoceanography, v. 1, p. 467-481.

de Boer, P. L., and Smith, D. G., 1994, Orbital forcing and cyclic sequences, *in* de Boer, P. L., and Smith, D. G., eds., Orbital Forcing and Cyclic Sequences: IAS Special Publication 19, p. 1-14.

Destombes, J. -P., and Shephard-Thorn, E., 1971, Geological results of the

Channel Tunnel site investigation: National Environmental Research Council, Institute of Geological Science, Report n.71/11, 12 p.

DITCHFIELD, P., AND MARSHALL, J., 1989, Isotopic variation within three Cenomanian chalk-marl rhythms: Geology, v. 17, p. 842-845.

GALE, A. S., 1990, A Milankovitch scale for Cenomanian time: Terra Nova, v.1, p. 420-425.

GALE, A. S., 1995, Cyclostratigraphy and correlation of the Cenomanian Stage in Western Europe, in House, M. R., and Gale, A. S., eds., Orbital Forcing Timescales and Cyclostratigraphy: London, Geological Society Special Publication 85, p. 177-197.

GALE, A. S., KENNEDY, W. J., BURNETT, J. A., CARON, M, AND KIDD, B. E., 1996, The Late Albian to Early Cenomanian succession at Mont Risou near Rosans (Drôme, SE France): An integrated study (ammonites, inoceramids, planktonic foraminifera, nannofossils, oxygen and carbon isotopes): Cretaceous Research, v. 17, p. 515-606.

GIRESSE, P., LAMBOY, M., AND ODIN, G. S., 1980, Evolution géométrique des supports de glauconitisation; application à la reconstitution du paléo-environnement: Oceanologica Acta, v. 3, p. 251-260.

GRADSTEIN, F. M., AGTERBERG, F. P., OGG, J. G., HARDENBOL, J., VAN VEEN, P., THIERRY, J., AND HUANG, Z., 1995, A Triassic, Jurassic and Cretaceous Time Scale, *in* Berggren, W. A., Kent, D. E., Aubry, M. -P., and Hardenbol, J., eds., Geocronology Time Scale and Global Stratigraphic Correlation: Tulsa, Society for Sedimentary Geology (SEPM) Special Publication 54, p. 95-126.

HAQ, B. U., HARDENBOL, J., AND VAIL, P. R., 1988, Mesozoic and Cenozoic chronostratigraphy and cycles of sea-level change, *in* Wilgus, C. K., Hastings, B. S., Kendall, C. G. St. C., Posamentier, H. W., Ross, C. A., and Van Wagoner, J. C., eds., Sea Level Changes: An Integrated Approach: Tulsa, Society for Sedimentary Geology (SEPM) Special Publication 42, p. 71-108.

HART, M. B., 1987, Orbitally induced cycles in the Chalk facies of the United Kingdom: Cretaceous Research, v. 8, p. 335-348.

JUIGNET, P., AND BRETON, G., 1992, Mid-Cretaceous sequence stratigraphy and sedimentary cyclicity in the western Paris Basin: Palaeogeography, Palaeoclimatology, Palaeoecology, v. 91, p. 197-218.

JUIGNET, P., AND BRETON, G., 1994, Stratigraphie, rythmes sédimentaires et eustatisme dans les Craies Turoniennes de la region de Fecamp (Seine-Maritime, France). Expression et signification des rythmes de la Craie: Bulletin trimestriel de la Société Géologique de Normandie et Amis du Muséum du Havre, v. 81, p 55-81.

LOUTIT, T. S., HARDENBOL, J., VAIL, P. R., AND BAUM, G. R., 1988, Condensed sections: The key to age determination and correlation of continental margin sequences, *in* Wilgus, C. K., Hastings, B. S., Kendall, C. G. St. C., Posamentier, H. W., Ross, C. A., and Van Wagoner, J. C., eds., Sea Level Changes: An Integrated Approach: Tulsa, Society for Sedimentary Geology (SEPM) Special Publication 42, p. 183-213.

MCCONCHIE, D. M., AND LEWIS, D. W., 1978, Authigenic, perigenic, and allogenic glauconites from the Castle Hill Basin, North Canterbury, New Zealand: New Zealand Journal of Geology and Geophysics, v. 21, p. 199-214.

MCCRACKEN, S. R., COMPTON, J. AND HICKS, K., 1996, Sequence-stratigraphic significance of glaucony-rich lithofacies, *in* Miller, K. G., Mountain, G. S., Blum, P., Poag, C. W., and Twitchell, D. C., eds., Proceedings of the Ocean Drilling Program, Scientific Results 150: College Station, Texas, p. 171-184.

MITCHELL, S. F., PAUL, C. R. C., AND GALE, A. S., 1996, Carbon isotopes and sequence stratigraphy, *in* Howell, J. A., and Aitken, J. F., eds., High Resolution Sequence Stratigraphy: London, Geological Society Special Publication 104, p. 11-24.

MITCHUM, R. M., JR., AND VAN WAGONER, J. C., 1991, High-frequency sequences and their stacking patterns: Sequence-stratigraphic evidence of high-frequency eustatic cycles: Sedimentary Geology, v. 70, p. 131-160.

ODIN, G. S., 1982, How to measure glaucony ages, *in* Odin, G. S., ed., Numerical Dating in Stratigraphy: Chichester, J. Wiley, p. 387-403.

ODIN, G. S., 1985, Significance of green particles (Glaucony, Berthierine, Chlorite) in arenites, in Zuffa, G. G., ed., Provenance of Arenites: Dordrecht, Riedel, p. 279-307.

ODIN, G. S., 1988, Glaucony from the Gulf of Guinea, *in* Odin, G. S., ed., Green Marine Clays: Developments in Sedimentology: Amsterdam, Elsevier, p. 225-247.

ODIN, G. S., AND LAMBOY, M., 1988, Glaucony from the margin off northwestern Spain, *in* Odin, G. S., ed., Green Marine Clays: Developments in Sedimentology: Amsterdam, Elsevier, p. 249-275.

ODIN, G. S., AND FULLAGAR, P. D., 1988, Geological significance of the glaucony facies, *in* Odin, G. S., ed., Green Marine Clays: Developments in Sedimentology: Amsterdam, Elsevier, p. 295-332.

ODIN, G. S., AND MATTER, A., 1981, De glauconiarum origine: Sedimentology, v. 28, p. 611-641.

OWEN, D., 1996, Interbasinal correlation of the Cenomanian Stage; testing the lateral continuity of sequence boundaries, *in* Howell, J. A., and Aitken, J. F., eds., High Resolution Sequence Stratigraphy: London, Geological Society Special Publication 104, p. 269-293.

PAUL, C. R., 1992, Milankovitch cycles and microfossils: Principles and practice of palaeoecological analysis illustrated by Cenomanian chalk-marl rhythms: Journal of Micropaleontology, v. 11, p. 95-105.

POMEROL, B., AND MORTIMORE, R. N., 1990, Lithostratigraphie et cycles sédimentaires dans la Craie du bassin Anglo-Parisien: Comptes Rendus de l'Academie des Sciences de Paris, v. 310, p. 553-558.

ROBASZYNSKI, F., AND AMÉDRO, F., 1986, The Cretaceous of the Boulonnais (France) and a comparison with the Cretaceous of Kent (United Kingdom): Proceedings of the Geological Association, v. 2, p. 171-208.

ROBASZYNSKI, F., AND AMÉDRO, F., 1993, Les falaises crétacées du Boulonnais. La coupe de référence du Cap Blanc-Nez dans un contexte sédimentaire global: Annales de la Société Géologique du Nord, v. 2, p. 31-44.

ROBASZYNSKI, F. GALE, A., JUIGNET, P., AMÉDRO, F., AND HARDENBOL, J., 1998, Sequence stratigraphy in the Upper Cretaceous series of the Anglo-Paris Basin: Exemplified by the Cenomanian Stage, in De Graciansky, P.-C., Hardenbol, J., Jacquin, T., and Vail, P.R., eds., Mesozoic and Cenozoic Sequence Stratigraphy of European Basins: Tulsa, Society for Sedimentary Geology (SEPM) Special Publication 60, p. 363-386.

ROBINSON, N. D., 1986, Fining-upward microrhythms with basal scours in the Chalk of Kent and Surrey, England and their stratigraphic importance: Newsletter on Stratigraphy, v. 7, p. 21-28.

VAN WAGONER, J. C., POSAMENTIER, H. W., MITCHUM, R. M. JR., VAIL, P. R., SARG, J. F., LOUTIT, T. S., AND HARDENBOL, J., 1988, An overview of the fundamentals of sequence stratigraphy and key definitions, *in* Wilgus, C. K., Hastings, B. S., Kendall, C. G. St. C., Posamentier, H. W., Ross, C. A., and Van Wagoner, J. C., eds., Sea Level Changes: An Integrated Approach: Tulsa, Society for Sedimentary Geology (SEPM) Special Publication 42, p. 39-47.

VAN WAGONER, J. C., MITCHUM, R. M., JR., CAMPION, K. M., AND RAHMANIAN, V. D., 1990, Siliciclastic Sequence Stratigraphy in Well Logs, Cores and Outcrops: Tulsa, American Association of Petroleum Geologists, Methods in Exploration Series, v. 7, 55 p.

CARBONATE-PHOSPHATE COMPETITION IN THE NEGEV PHOSPHORITES (SOUTHERN ISRAEL): A MICROSTRUCTURAL STUDY

DAVID SOUDRY*

Geological Survey of Israel, 30 Malkhe Yisrael St., Jerusalem 95501, Israel

ABSTRACT: Study of the carbonate micrite phase of laminated pelletal phosphorites from various phosphate fields of the Negev reveals that this phase was largely formed *in situ*, and that bacterial mats are principally the frame-builders of this phase. These mats are composed of vertically stratified, calcite-mineralized assemblages of bacterial bodies of various shapes, sizes, and organizations. This biostratification is believed to represent the fossilized record of time-successive consumer bacterial populations that degraded planktonic detritus, and mats of filamentous and coccoid microphytes, which colonized the Mishash seafloor during phosphorite formation and were partly preserved in these rocks as apatitic sheaths and envelopes. Carbonate mineralization of the bacterial consumers probably took place during anaerobic degradation of the matted sediments, as indicated by low $\delta^{13}C$ values in many of these carbonates. The diagenetic fate of the filamentous and coccoid Mishash bottom-dwelling microphytes (apatite mineralization versus turnover as bacterial carbonates) was dependent on the balance between the rate of organic matter accumulation and the extent of bottom sediment ventilation, which together controlled the redox conditions of organic matter decomposition and pore-water carbonate ion concentrations. This may explain why primary (pristine) apatite beds are common in condensed proximal Mishash sections around Negev Campanian highs, whereas they are more sporadic in basinal carbonate-rich Mishash sections. It might also explain why the coccoid and filamentous microphyte structures, which matted the Mishash seafloor and are now preserved in many of the Negev phosphorite rocks, were found to be mineralized only by phosphate and never by carbonate.

INTRODUCTION

Carbonate is the most common mineral phase associated with the phosphate fraction in the granular Negev phosphorites. These rocks occur with cherts, porcelanites, and carbonates, and form the Negev Phosphate Member (Soudry et al., 1985), the upper part of the Campanian Mishash Formation (Figs. 1 and 2). As a rule, the carbonate phase in the Negev phosphorites is calcite, mostly micrite, and rarely neospar. Calcite is present in intergranular spaces between phosphate peloids in massive phosphorites, or occurs as carbonate laminae separating phosphate laminae in laminated phosphorites. Dolomite, usually cryptocrystalline, also exists in small amounts with the calcite phase in several carbonate and phosphorite levels of the Mishash section. Many aspects of the Mishash succession have been investigated previously, including its lithostratigraphy (Kolodny, 1967), biostratigraphy and fossil assemblages, (Reiss et al., 1985; Reiss, 1988), and paleogeography (Flexer et al., 1986; Lewy, 1990). These works have together contributed to unravelling the depositional environment of this rock suite and the time and space distribution of its facies. In addition, many studies have been conducted on the petrogenesis of the cherts (Kolodny, 1969) and phosphorites (Soudry, 1987, 1992) in this sequence, and also on the geochemistry of these rocks (Nathan et al., 1979; Kolodny et al., 1980). Yet, in most of these studies, little attention was given to the the micrite fraction of the Mishash section, and its petrofabric has been largely ignored. Most workers have described the Mishash micrites simply as "chalks" presumably because of their common, soft consistencies and massive textures, as well as their whitish colorations in outcrop. However, the extreme scarcity of calcareous nannoplankton in these rocks (Moshkovitz et al., 1983), along with their internal fine lamination and high organic carbon content (up to 30% O.M.) in subsurface, have raised doubts as to the assumed "chalk" composition of the carbonate phase. Moreover, in both the phosphorites and many of the carbonate strata of the Negev phosphate sequence, this phase dis-

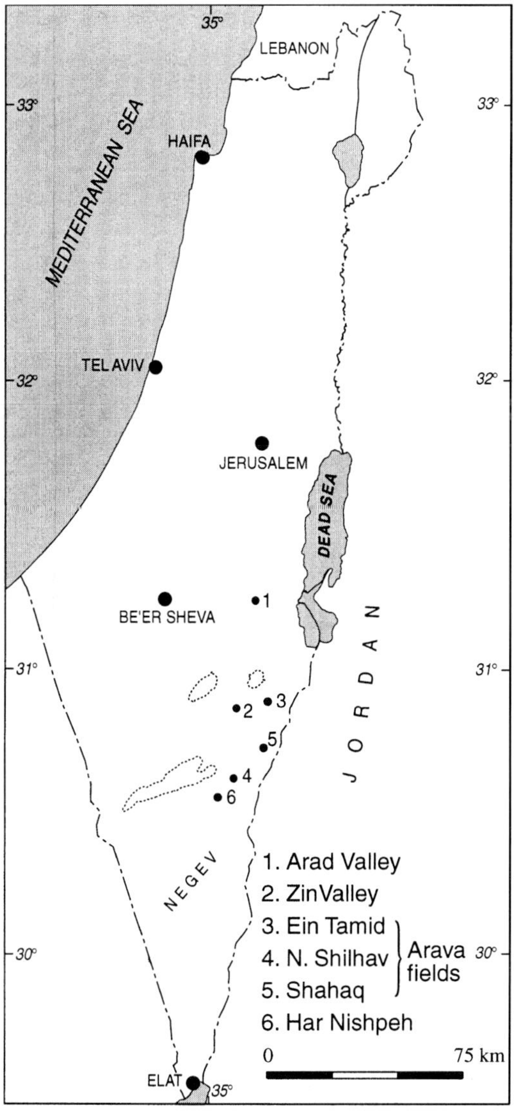

FIG. 1.—Location map showing the various Negev phosphate fields sampled for this study.

*email: david.soudry@mail.gsi.gov.il

Marine Authigenesis: From Global to Microbial, SEPM Special Publication No. 66
 ISBN 1-56576-064-6, p. 415-426.

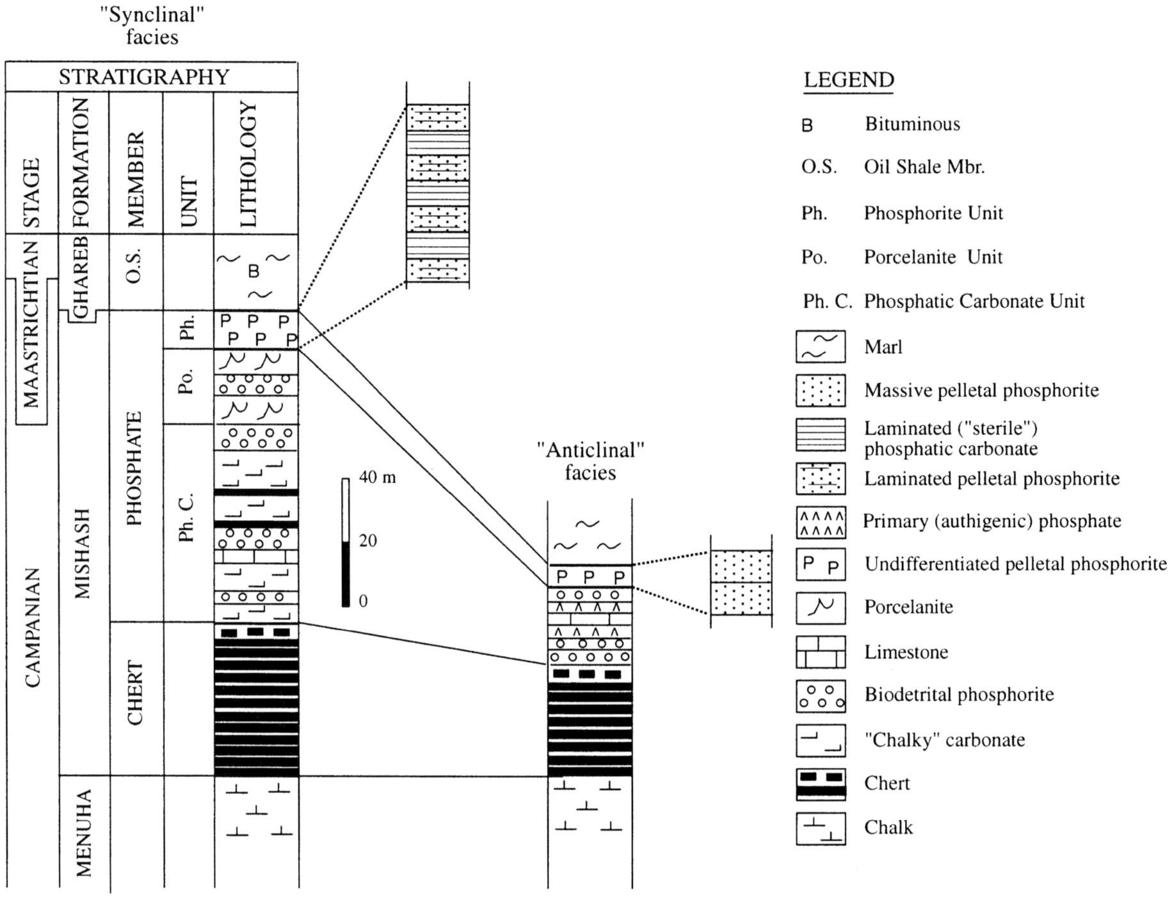

FIG. 2.—Schematic lithostratigraphic columnar sections of the Mishash Formation in the Negev and the facies changes present from "anticlinal" to "synclinal" Campanian positions.

plays curiously low (– 6‰ to -11‰) $\delta^{13}C$ values (Kolodny, 1980; Vengosh et al., 1987), which is quite unusual for "chalks" (Hudson, 1977; Dudley et al., 1986; Ditchfield and Marshall, 1989). Although it was suggested that the low $\delta^{13}C$ of the Mishash micrites may be due to the incorporation of organic-derived HCO_3^- (Kolodny, 1980), the precise origin of this micrite fraction and the mechanism of its accretion have remained obscure and controversial. Given the close association between the carbonate and the phosphate phases in the Mishash phosphorites, these questions need to be clarified if we want to better understand the Negev phosphogenesis and its different facets.

The present study aims to shed light on this problem, focusing on the carbonate matrix of some of the Negev phosphorites (Fig. 1). It presents new evidence from these rocks and examines the significance of these results in the light of geochemical data from previous works and results from modern phosphate-forming environments.

GEOLOGICAL SETTING

Deposition of the Mishash sediments during the Late Cretaceous in the Negev was closely controlled by tectonic, paleogeographic, and paleoceanographic factors prevailing in the area. The Negev was part of a wide and gently folded inner shelf that formed the northern margins of the Arabo-Nubian craton and was flooded by the southern Tethys. Water depths generally did not exceed a few tens of meters in most of the Negev area during Mishash deposition. Folding, which started in the Late Coniacian as a result of compressional forces during the development of the "Syrian Arc" System (Garfunkel, 1981), shaped the Negev into a series of elongated uprising sills and subsiding depressions. The atmospheric circulation pattern during the Late Campanian (Parrish and Curtis, 1982), together with the paleolatitude of the area at that time (Barron et al., 1981), led to an extensive upwelling system in the southern part of the Tethys, which resulted in high photic zone productivity during Mishash deposition. Organic-rich carbonates, phosphorites, and biosiliceous sediments were cyclically deposited in response to time and space changes in depositional and diagenetic conditions and varying oxygen levels in the bottom sediments and the water column. Fossil assemblages in the Mishash strata, including macrofossils (bivalves, gastropods, ammonites) and planktic versus benthic foraminifera (Almogi-Labin et al., 1993), reflect the relative intensity of upwelling across the depositional shelf and the extent of bottom aeration through time and space. Continual growth of the Negev structures during the Campanian, together with variations in deposition rates and redistribution of sediments through downslope transport, resulted in a conspicuous intrashelf differentiation of the Mishash facies (Soudry et al., 1985). Thick, carbonate-rich and phosphate-poor sediment sections (Fig. 2) accumulated in the depressions ("synclinal " facies), while thin (condensed), carbonate-poor and phosphate-rich rock-suites were deposited on the structural highs ("anticlinal" facies).

Composition of the Mishash Formation in the Negev is related to these facies changes and the structural setting. In most of the Negev, the Mishash Formation (Fig. 2) shows two distinct members: (1) the lower Chert Member formed by homogenous and brecciated massive-bedded cherts; (2) the upper Phosphate Member formed by alternations of poorly consolidated carbonates, shelly limestones, cherts, porcelanites, and phosphorite rocks. In structural lows, the Phosphate Member is thick and shows three well-defined units: a lower Phosphatic Carbonate Unit, a middle Porcelanite Unit, and an upper Phosphorite Unit, the latter comprising several phosphate layers of economic value. Toward the paleohighs, this subdivision of the Phosphate Member is less clear due to a decrease in net sedimentation that led to diminution of the carbonate and siliceous lithologies in the sequence. In such settings, the Phosphate Member is strongly condensed and is represented mainly by amalgamated phosphate strata interbedded with thin levels of cherts and carbonates.

PHOSPHORITE LITHOTYPES AND TEXTURAL STATE OF THE CARBONATE PHASE

Textural patterns of the carbonate fraction in the Negev phosphorites are closely related to the fabrics of these rocks and their granular compositions. Three major kinds of phosphorites are present in the economic Negev Phosphorite Unit at the upper part of the Phosphate Member: (1) biodetrital massive phosphorites; (2) massive pelletal phosphorites; and (3) laminated pelletal phosphorites. The relative frequency of these rock types in the Phosphorite Unit is controlled by the paleostructure. Phosphorites with massive fabrics are widespread on the Campanian paleohighs, whereas phosphorites with a laminated arrangement are predominant in the structural depressions.

Biodetrital Massive Phosphorites

These rocks are mostly developed in the two lower units of the Phosphate Member (the Porcelanite Unit and the Phosphatic Carbonate Unit), where they constitute the main form of phosphate rock. In the economic Phosphorite Unit, these rocks are generally displayed as thin layers in the basal part of the pelletal phosphorite beds. As a rule, these phosphorites (Fig. 3A) are coarse-grained, poorly sorted, highly indurated, of low phosphate content, and commonly intensively bioturbated (burrow system mostly of *Thalassinoides* type). Other features commonly displayed by these phosphorites are sharp basal contacts; scour and fill features; crude, graded bedding; and imbricated limestone pebbles and cobbles in the lower parts of the beds. The phosphate fraction of these rocks typically consists of phosphate intraclasts of subjacent derivation, mixed with a few peloids and various amounts of transported bone fragments showing various stages of microphyte infestation at their edges. Carbonates are generally present in these phosphorites as sparry calcite cement with some neomorphic spar and form most of the intergranular fraction. Cherty variants of these rocks, with microquartz (instead of calcite) constituting the intergranular fraction between the phosphate grains and in part replacing them, also exist in the phosphorite section. The overall features of these rocks and their stratal disposition in the sequence has led to their interpretation (Soudry, 1992) as storm-induced lag deposits produced by rapid accumulation of *in situ*-recycled and transported phosphate grains.

Massive Pelletal Phosphorites

These rocks comprise most of the economic phosphate beds of high structural positions in the Negev Phosphorite Unit (Fig. 2). In these settings they form one or a few thick layers commonly interbedded with thin layers of phosphate-poor ("sterile") carbonates. Dominant features of these rocks are poor consolidation, dense packing of the granular fraction, unimodal grain size (generally around 250 μm), and beds with flat soles (Fig. 3B). In addition, discontinuous clay partings locally displayed by these rocks indicate that such phosphorites originated (at least in part) as multi-event (amalgamated) sediment beds. Although typically massive, in places these phosphorites also reveal some faint internal layering reflected by a common orientation of elongated particles. The granular phosphate fraction of these rocks typically consists of sand-sized peloids, intermixed with various amounts of phosphate microoncolites and some bone debris. Previous studies (e.g., Soudry, 1987) have shown that many of these peloids are in fact rounded fragments of early phosphatized microbial mats built by filamentous and coccoid microorganisms, including fungi, cyanobacteria, and bacteria. Carbonates are present in these rocks mostly as micritic calcite filling the intergranular spaces, and more rarely as sparry calcite. Textural and structural evidence (Soudry, 1983) indicates that the intergranular micrite of these rocks is mostly a redeposited, pore-filling, silt-sized mud that infiltrated the void spaces between the accumulated phosphate peloids at times of water quiescence.

Laminated Pelletal Phosphorites

These rocks form part of the "sterile" (phosphate-poor) interbeds between the massive pelletal phosphorite beds in the Phosphorite Unit sections on structural highs. In addition, they build most of the economic phosphorite layers in structural lows (Fig. 2), as well as most of the associated "sterile" intercalated beds. The most distinctive feature of these rocks is an internal flat banding produced by a cyclic alternation of phosphate-rich and phosphate-poor laminae (Fig. 3C). Distribution and relative thicknesses of the different laminae vary according to the strata. The phosphate-rich laminae are predominant and thicker (up to 1 cm) in the economic phosphorite beds, while they are usually thin, discontinuous, and sporadic in the sterile interbeds. The

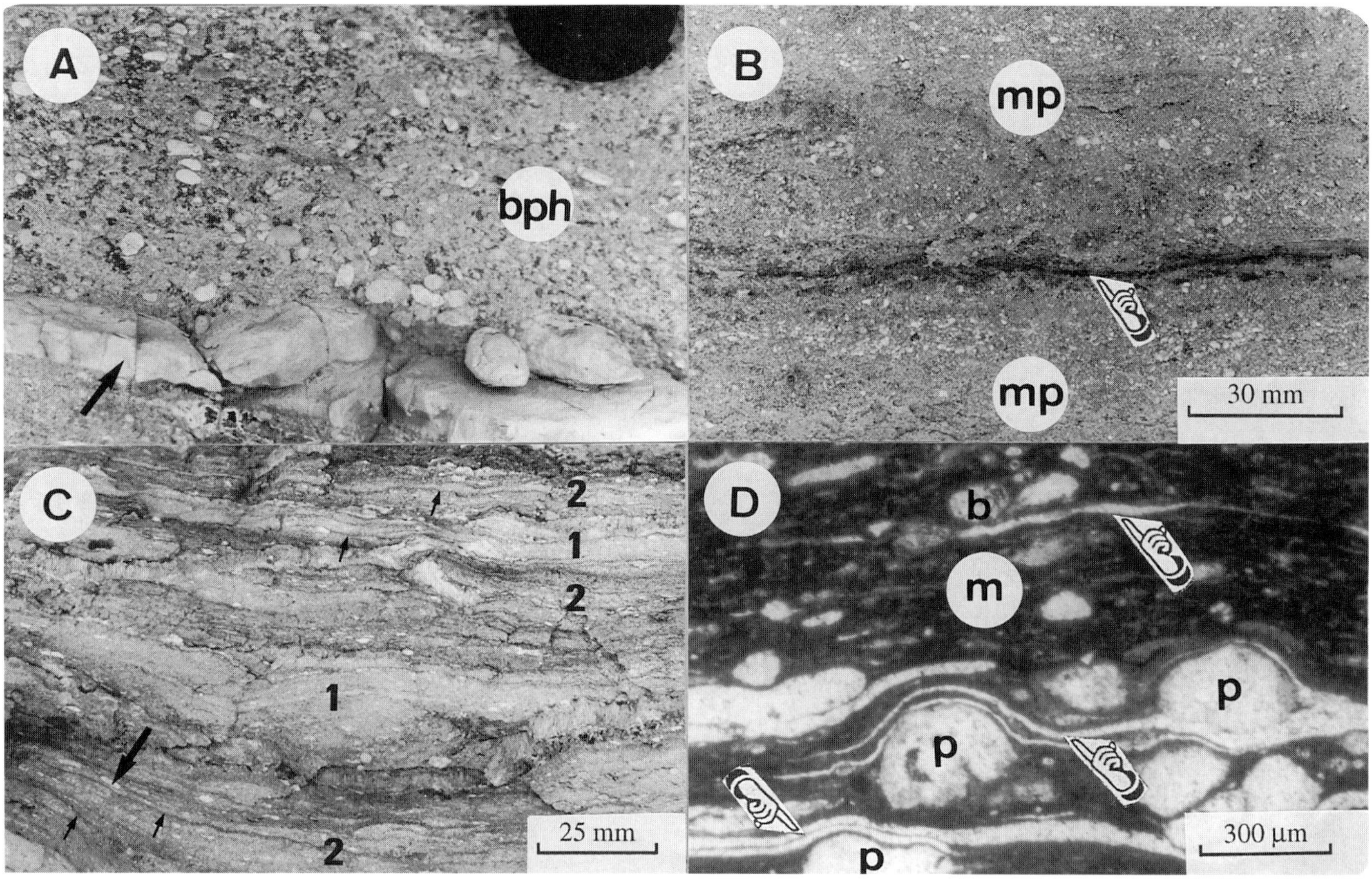

FIG. 3.—Macroscopic and microscopic views of the main phosphorite lithotypes occurring in the Negev phosphate sequence. (A) Field view of biodetrital phosphorites. Coarse-grained and poorly sorted, commonly burrowed phosphorites (bph), composed of redeposited, large-sized phosphate clasts (white components) mixed with various amounts of bone fragments and molluscan shell debris and often imbricated, allogenic limestone pebbles and cobbles (arrow) at the basal part of the phosphorite beds. Carbonates are usually present in these rocks as intergranular calcite spar cement binding together the phosphate components (cover lens for scale; 5 cm diameter). (B) Field view of pelletal massive phosphorites from condensed ("anticlinal") phosphate facies. Two layers of massive pelletal phosphorites (mph) composed of sand-sized phosphate peloids and separated by a partly eroded clay parting (arrowed). Carbonates occur in these rocks mainly as redeposited micrite mud filling the intergranular spaces between phosphate grains. (C) Field view of pelletal laminated phosphorite from basinal Mishash phosphate facies. Alternation of millimeter- to centimeter-thick phosphate-rich (1) and phosphate-poor (2) laminae. The laminated arrangement is also reflected by a common orientation of elongate authigenic phosphate lenses (long arrow) parallel to the bedding. Carbonates are present in these phosphorites as *in situ* micrite making up the intergranular fraction within the phosphate-rich laminae, and as repeated micrite-rich bands (short arrows - dark gray bands) within the carbonate-rich laminae. (D) Photomicrograph (plain light) of a carbonate-rich lamina from a pelletal laminated phosphorite showing a finely laminated micritic groundmass (m) enclosing scattered buliminid shells (b), phosphate peloids (p), and authigenic apatite strips (fingers).

phosphate fraction of these rocks commonly consists of allogenic peloids (60–200 μm) associated with small amounts of phosphate-coated buliminid shells and thin, flexuous strips of *in-situ* formed apatite (Fig. 3D), which bridge part of the phosphate grains. Previous studies (Soudry, 1992) have demonstrated that the *in situ*-formed apatite strips and part of the peloids are composed of early phosphatized extracellular remnants of various microbial mat systems, including coccoid and filamentous forms. Carbonates are present in these rocks mainly as micrite, commonly embedding various amounts of buliminid shells. This micrite fraction composes the major part of the phosphate-poor laminae and forms the intergranular fraction between the phosphate peloids in the phosphate-rich laminae. Minor amounts of iron-rich montmorillonites (Nathan and Soudry, 1982) are commonly associated with the micrite phase of the phosphate-poor laminae. A distinctive feature of this micrite fraction is a delicate, flat microlamination (Fig. 3D), which tends to bend over buliminid shells and phosphate peloids.

ULTRASTRUCTURE OF THE CARBONATE PHASE

Material

Samples from the carbonate fraction were studied under an SEM fitted with an X-ray energy-dispersion spectrometer. To avoid artificial features that might result from dissolution processes (cf., Kirkland et al., 1999), no etching was used and only freshly broken samples were used. All of the samples studied are from the laminated pelletal phosphorites belonging to various Negev phosphate deposits (Arad Valley, Zin Valley, Har Nishpeh, and various Arava fields; Fig. 1). Both the phosphate-rich laminae and the carbonate-rich laminae of these phosphorites were sampled and studied. The samples were oven-dried,

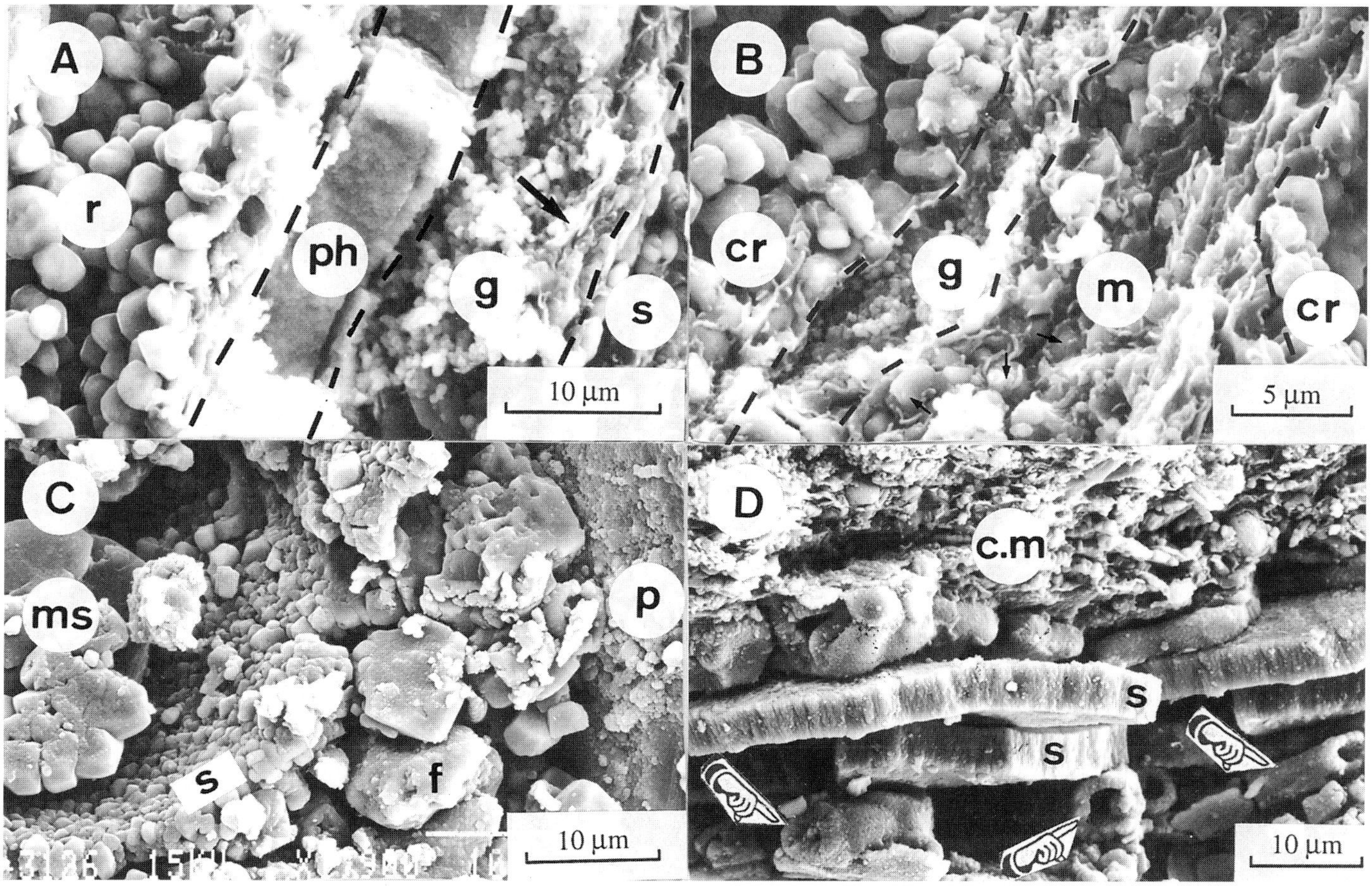

FIG. 4.—Spatial organization of the carbonate fraction in the laminated Negev pelletal phosphorites from basinal (4A, B, D) and proximal (4C) Mishash facies (SEM micrographs). (A) General view of carbonate-rich laminae built by vertically stratified nannolayers composed successively (from right to left) by: (1) sausage-shaped (s) calcite bodies (better view and enlargement in 5B); (2) submicronmeter-sized packed calcite globules (g) (see enlargement in 5K); (3) an authigenic phosphate sheet (ph); (4) rhombohedral calcite microcrystals (r) (crystalline growths from precursor spheroid calcite bodies; see transitional steps in 5I, J). Arrow points to authigenic iron-rich montmorillonites sandwiched between the sausage-shaped and calcite globule layers. Limits between the successive nannolayers are emphasized by dashed lines. (B) Carbonate-rich laminae built by vertically stratified nannolayers successively composed of: (1) erect calcite rods (cr) showing incipient development of crystal face terminations (from precursor spindle-shaped forms; see 5A); (2) montmorillonite-rich layer (m) embedding scattered spheroid calcite structures (arrows); (3) submicrometer-sized massed calcite globules (g). Limits between the successive nannolayers are emphasized by dashed lines. (C) An individual sheet made up of closely packed, sausage-shaped (s) calcite bodies (see enlargement of these structures in 5B) sandwiched between calcite microspar (ms) and strongly deformed phosphate microphyte filaments (f); (p) is a phosphate peloid. (D) Alternating *in-situ* phosphate filaments (fingered) and phosphate sheets (s) from a carbonate-rich laminae, overlain by a calcite and montmorillonite-rich layer (c.m). Similar phosphate filaments form many of the phosphate peloids of the Negev phosphorites and also comprise the intergranular phosphate fraction of some of these rocks.

mounted on stubs, and coated in a first stage with carbon to discriminate between the different mineral phases, and in a second stage they were coated with gold to improve secondary electron imaging resolution. An accelerating voltage of 15–20 kv and a working distance of 20–25 mm were used. Only the laminated pelletal phosphorites were studied, because these rocks were less remobilized and disrupted during their deposition (as compared with other phosphorites), and also because they apparently suffered less diagenetic mineral recrystallization. Minor remobilization and disruption of these rocks during their deposition are suggested by their laminated fabric and also by the presence of *in situ*-formed apatite strips. Minor mineral recrystallization is indicated by the delicate micro-lamination commonly retained by the micrite groundmass of the carbonate-rich laminae.

Nannofabrics and Spatial Arrangement

One of the most impressive features of the carbonate matter of the laminated phosphorites, when observed under SEM, is its remarkable organization in nannolayers. These nannolayers are vertically stacked one upon the other (Fig. 4A, B) or occur as individual sheets bent over the phosphate grains (Fig. 4C). The nannolayered organization is most pronounced in the carbonate-rich laminae of these rocks, whereas it is commonly less conspicuous in the phosphate-rich laminae. The carbonate nannolayers are of a few to tens of μm thick and commonly alternate with iron-rich montmorillonite nannolayers of comparable thickness (Fig. 4B). Also associated with the carbonate nannolayers, and sporadically intercalated between them, are *in situ*-formed strips composed of inter-sandwiched laminae of phosphate tubules and cryptocrystalline phosphate plates

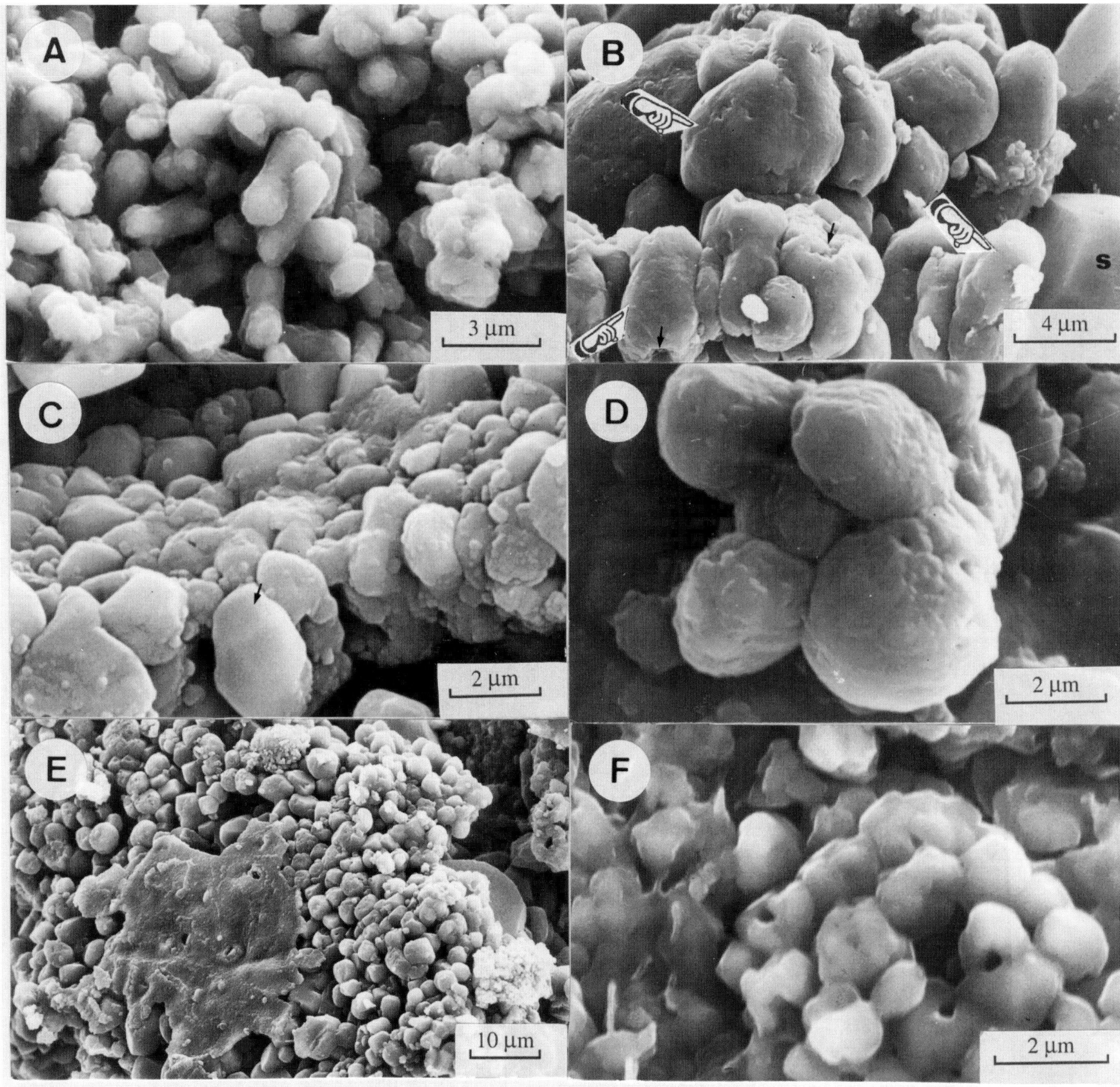

FIG. 5.—Constructional biostructures of the carbonate fraction of the Negev phosphorites (detailed views from the different nannolayers). (A) Spindle-shaped calcite rods. All transitions can be observed in these bodies between fusiform shapes (5A) and crystalline forms (see 4B) developed through increasing mineral accretion. (B) Mutually compressed sausage-shaped calcite bodies showing all transitions between well-preserved morphologies (lower fingers) and random stubby forms (upper finger) produced by enhanced calcite accretion. Note the hollow depressions (arrowed) at the extremities of some of these bodies. (C) Sausage-shaped calcite bodies (enlargement of 4C) showing incipient development of crystal faces (arrowed). (D-J) Spheroidal calcite bodies organized as clusters (5D), sheets (5E), solitary forms (5G), and rows (5J). Some of these structures are hollow (5F, and 5G fingered), while some show incipient binary fission (lower finger in 5G) resulting in paired structures (5H). Fingered in (5I) is a well-preserved spheroidal form; arrows point to incipient development of crystal faces yielding, at a final stage, perfect rhombohedral crystals (see 4A) resembling a void-filling microspar cement. Fig. 5J shows an organization in rows of former spheroidal structures and their transformation into more or less calcite rhombohedrons. (K) Tiny calcite globules, organized in twisted chains, build part of the calcite nannolayers of the Negev carbonate fraction (see 4A, B for an overall view). (L) Stacked calcite rafts (r) composed of tiny calcite rods. Note the clear variations in size of the calcite rods from raft to raft and also within the same individual raft; pg is a phosphate grain.

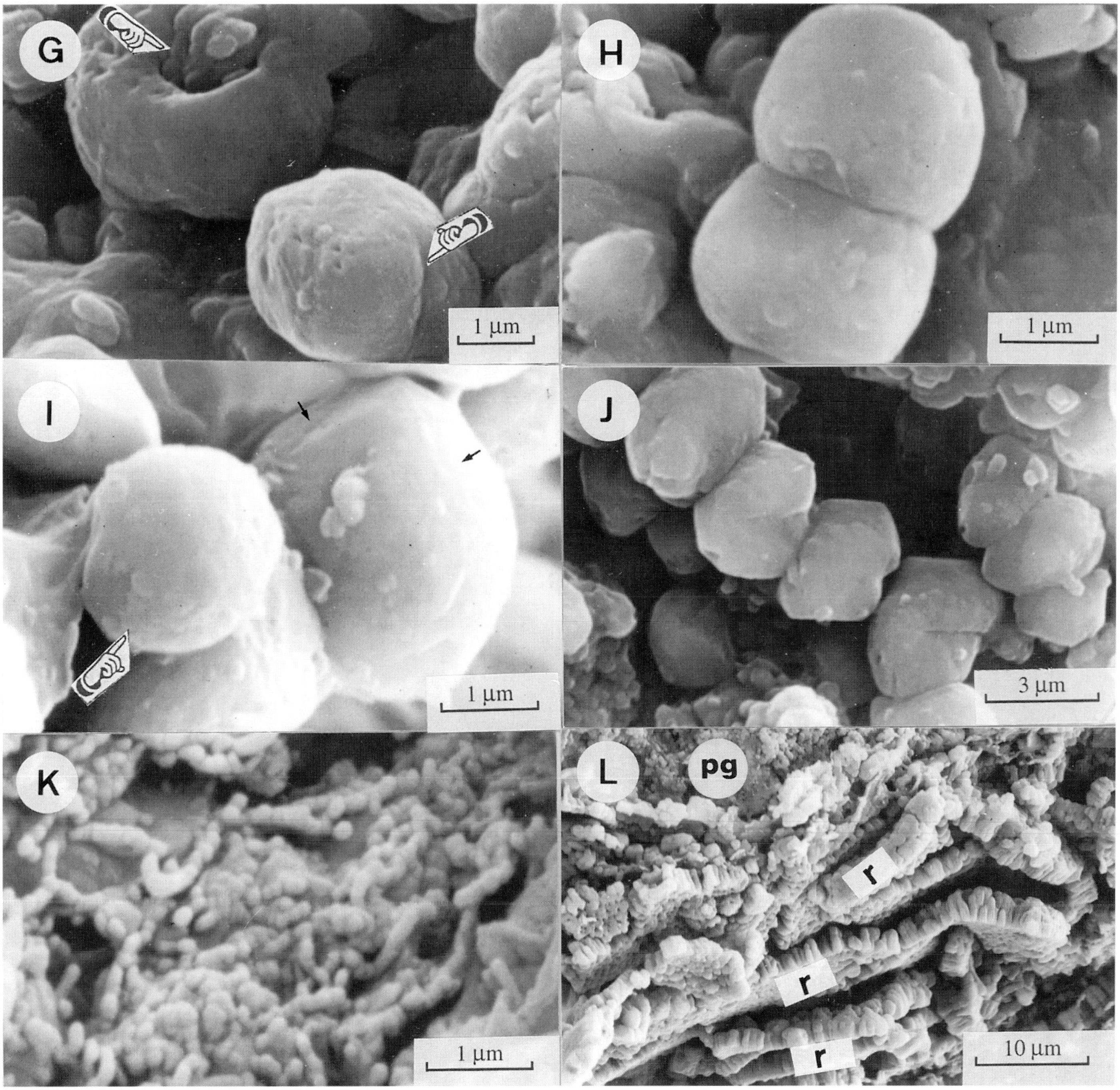

FIG. 5.—Continued.

(Fig. 4D). Each of the carbonate nannolayers is built by a dominant kind of structure, of a particular size, shape, and set-up. The most common appearance consists of straight, vertically oriented calcite rods that are 3 µm in length and 1 µm in width (Fig. 5A). A few of these bodies show incipient developments of crystal face endings (cf., Fig. 4B), whereas the majority occur as spindle-shaped forms (Fig. 5A). All transitions can be observed in these fusiform bodies between terminations that are perfectly convex, and terminations with well-developed crystalline faces. Other kinds of rod-shaped structures consist of sausage-like calcite bodies that are 4–6-µm long and about 2-µm wide and mutually compressed against each other (Fig. 5B). Most of these bodies are erect and show hollow depressions at their extremities, and many of them are partially or strongly deformed by carbonate overgrowths that modify them into bowl-shaped stubby forms. Such bodies alternate with other kinds of carbonate nannolayers or form individual sheets (cf., Fig. 4C) built by vertically oriented, densely packed stubby forms occasionally showing partial development of crystalline faces (Fig. 5C). Additional kinds of structures encountered, and building the carbonate framework of the Negev laminated phosphorites, consist of isolated and clustered spheroidal forms 2–3-μm in diameter (Fig. 5D), in places organized as continuous sheets (Fig. 5E). Part of these globose structures are hollow (Fig. 5F, G), whereas others show incipient (?) binary fission (Fig. 5G) resulting in the production of paired bodies (Fig. 5H). As in the other observed forms, increasing mineral accretion has led to a gradual loss of the spheroidal morphology and to the development of crystalline faces with rhombic junctions (Fig. 5I). Where this process is intensive, calcite rhombohedrons (which could be mistaken for spar-filling cement) are produced from these structures (cf., Fig. 4A). In such cases, only partial deformation or an occasional arrangement in rows of these bodies (Fig. 5J) still testify to their initial spheroidal form. Also comprising the carbonate fraction are tiny (about 0.25-μm diameter.) calcite globules organized as straight and twisted chains up to a few μm in length (Fig. 5K) and distributed in loose meshworks in distinct nannolayers. Another form of carbonate present in the Negev laminated phosphorites are stacked calcite rafts, 1–4-μm thick. These rafts are built by tiny, vertically oriented calcite rods that vary in size from raft to raft and also within the same raft (Fig. 5L).

DISCUSSION

An interesting point that emerges from the above observations is that the major part of the micrite fraction of the Negev phosphorites studied has been generated *in situ*, and is not (as often still thought) a calcitic ooze of partly dissolved skeletal detritus and other reworked carbonate fines. This *in situ*-formed micrite occasionally alternates with phosphate strips, also generated *in situ*, and is composed of sandwiched tubular laminae and cryptocrystalline plates. Similar sandwiched phosphate structures have been observed also to form part of the phosphate matrices of the highly enriched Negev phosphorites (Soudry, 1987). These phosphate structures have been interpreted as permineralized extracellular remnants of filamentous, bottom-dwelling populations that intermittently colonized the Mishash seafloor during phosphorite accumulation. This close interjacence of the phosphate and the carbonate fractions, both generated *in situ*, implies that they are paragenetic sediment-phases alternatively produced by the same phosphogenic environment.

Some hints as to the nature of this *in situ*-formed carbonate micrite fraction are suggested by its constructional fabric and arrangement. The size, shape, and organization in space of most of the structures building the carbonate fraction of the laminated Negev phosphorites are all quite evocative of fabrics of bacterial origin. Many modern bacteria (e.g., Folk, 1993; Jones, 1995) show morphologies and dimensions highly comparable to those of the structures observed in this study. Spindle-shaped bodies, similar to those composing the carbonate fraction of the Negev phosphorites (cf., Fig. 4A), were produced in laboratory experiments as a result of carbonate mineralization of bacterial cells (Krumbein et al., 1977; Castanier et al., 1989). Similar structures also form part of the carbonate precipitates associated with bacterial degradation of dead cyanobacterial filaments (Chafetz and Buczynski, 1992). This was observed (Chafetz and Buczynski, 1992) both in the natural environment beneath a living mat, and in the laboratory using inoculated bacterial cultures collected from microbial mats. Likewise, bacterial cells with sausage-like (bacilli) forms, analogous in size and shape to those building the Negev carbonate matter (cf., Fig. 4B), have been reported from various organic-rich decay environments (Dickman, 1985; Briggs and Kear, 1993; Gerdes et al., 1994). The hollow depressions at the extremities in part of these forms might be due to incomplete carbonate infilling of the inner part of the bacterial structure after cell decay. Likewise, the isolated and clustered spheroidal structures (cf., Fig. 5D, E) are interpreted as carbonate-mineralized solitary and colonial coccoid (cocci) bacteria. Although abiotic minute-sized spheroids have been reported from the fossil record (Cloud and Morrison, 1979) and are also produced artificially in the laboratory (Kirkland et al., 1999; Hirschler et al., 1990), it is believed in the present case that the sum of features exhibited by these Mishash structures, such as their size, shape, hollowness, organization in clusters and chains, and occasional occurrence as paired bodies, all strongly suggest biogenic origin. Many coccoid bacteria show a similar organization in clusters (Gerdes and Krumbein, 1987; Erba 1991) and chains (Adolphe et al., 1989), and paired bodies resulting from binary fission are also common in coccoid microbes (Lanier, 1986; Thompson et al., 1990). Moreover, hollowness is often considered to be a key criterion for biogenicity (Buick, 1991), as purely artificial objects precipitated from mineral phases are expected to be solid bodies and not vacuous forms. In the same way, the 0.25 μm calcite spheroids, organized in chains (Fig. 5K) and building part of the carbonate fraction, are interpreted as nannobacterial cells. Folk (1993) described microfossils with a similar size range from a variety of Paleozoic to Holocene rocks, and modern microbes with sizes as small as 0.2 μm also have been reported from various settings (Sieburth et al., 1978). The stacked calcite rafts (Fig. 5L) are more problematic. At first glance, they vaguely resemble fragments of foraminiferid shells. However, the distinctive calcite rods building these rafts, along with their variations in size, strongly argue against such an origin for these components. Moreover, as the foraminiferal populations in the Negev Mishash rocks are nearly monospecific (almost exclusively buliminids, Reiss, 1988), the likehood that variations in size of the tiny rods are due to different architectures of shells belonging to various taxa is rendered highly improbable. All of this is conducive to envisaging another source, more plausible, for the calcite rafts. A possible explanation, which would fit the other results yielded by this study, is that these rafts represent some sort of calcite-mineralized microbial pellicles, peeled from the bottom sediments and later accumulated as clastic components. Calcitic laminae, also formed by closely packed rods and showing some resemblance to the architecture of these rafts, were recently reported (Szulc and Smyk, 1994) from modern freshwater stromatolites.

The collection of features displayed by the Negev laminated phosphorites indicates that bacterial mats are the frame-builders of a large part of their carbonate micrite fraction. The different microbial morphotypes observed in this fraction probably represent different bacterial populations involved in the building of these mats. Their distribution in distinct layers is similar to present-day (Monty, 1976) and fossil (Adolphe et al., 1989) mat-systems, where a vertical succession of specialized microorganisms also occurs. This biological stratification reflects the vertical changes of the physicochemical gradients across the living mat, and organization of the various taxa according to their ecological optima and their respective metabolic pathways (Bauld, 1981). In modern microbial mats (Golubic, 1976; Cohen et al., 1977), primary-producer microorganisms living at the upper mat surface are replaced at depth by successive zones of consumer and decomposer bacteria. The organic matter initially performed by the primary producers serves as source of carbon and nutrients for the reproduction of the bacterial heterotrophs. The bacterial turnover of the producer biomass is commonly followed by concomitant precipitation of carbonates within the mats (Lyons et al., 1984), and commonly in close relation with bacterial structures (Krumbein et al., 1977; Chafetz and Buczynski, 1992). In Solar Lake (Sinai), Lyons et al. (1984) concluded from pore-water analysis that $CaCO_3$ precipitation took place during anaerobic decay of the organic mat by sulfate-reducing bacteria. Precipitation of $CaCO_3$ started on the bacterial cell surface (Krumbein et al., 1977) and produced carbonate particles in the form of rods, dumbbells, and spheroid-like bodies. Many studies (e.g., Ferris et al., 1987; Castanier et al., 1989) have also stressed the capability of bacterial structures to fix mineral substances including carbonates on their walls. Because of their anionic nature (Beveridge and Fyfe, 1985), the bacterial cell walls will bind metal cations from pore waters, thus providing sites for nucleation and growth of mineral phases. It also has been suggested (Adolphe et al., 1989) that calcitization of bacteria might be enhanced by the ability of some of these micoorganisms to concentrate Ca in their cells. The precipitated carbonate resulting from anaerobic degradation of matted sediments by sulfate-reducing bacteria is strongly depleted in ^{13}C. In Solar Lake, for example, the calcite fraction in the decay zone of mats displays $\delta^{13}C$ values as low as –9.7‰ (Aharon et al., 1977), and similar values are also reported from fossil microbial build-ups (Andrews, 1986; Anadon and Utrilla, 1993).

Comparable conditions also may be envisaged for the formation of the micritic carbonate phase of the Negev laminated pelletal phosphorites. Mishash sedimentation occurred under conditions of high productivity (Reiss, 1988), which resulted in intensive accumulation of plankton-derived organic matter on the seafloor. A highly diversified mat-forming phytobenthos made by filamentous prokaryotic and eukaryotic microorganisms (Soudry, 1987) probably fed on the abundant sedimented organic detritus and colonized the Mishash seafloor during phosphorite formation. Part of this matted biomass was fossilized in a meshwork of apatitic tubules that now build part of the Negev phosphate grains and also forms the intergranular phosphate fraction of some of the pelletal phosphorite rocks (Soudry, 1987). Self-burial of this microbial assemblage by recurrent mat accretion led to its consumption by populations of decomposer bacteria and to an increase in bicarbonate ions in pore waters as a result of suboxic and mainly anoxic decay of organic matter (Berner et al., 1970; Jarvis et al., 1994)). Ca is highly concentrated in marine pore waters, and further Ca supply may also result from breakdown of extracellular polymer substances of the microbial assemblage (Trudinger and Mendelsohn, 1976). High concentrations of both ions led to supersaturation with respect to $CaCO_3$, which started to precipitate on the cell walls of the stratified bacterial decomposers, producing the distinctive arrangement in nannolayers of the micritic fraction of the Negev phosphorites. Precipitation mainly within the sulfate reduction zone was already suggested (Kolodny, 1980) in light of the ^{13}C depletion ($\delta^{13}C$ = –5.9 ‰ to –11.3 ‰) within the micrite phase of many of the Mishash rocks. Sass et al., (1991) similarly attributed the light oxygen compositions of diagenetic spar in Mishash carbonates ($\delta^{18}O$ values as low as –9.5‰ PDB) to decomposition of organic matter by sulfate ions. Anaerobic conditions during the formation of the micrite phase are further confirmed by the common presence of some amounts of dolomite, and also by the iron-rich montmorillonite films that are sandwiched between the calcite nannolayers (cf., Fig. 4B). Laboratory experiments (Harder, 1978) indicate that synthesis of iron-rich smectites is possible only under reducing conditions. Moreover, reducing conditions usually prevail at a few millimeters below the top surface of modern mat systems (Jorgensen and Cohen, 1977), as a result of high O_2 consumption and its inhibited replenishment by diffusion through the dense extracellular slimes. Further carbonate precipitation onto the bacterial structures with increasing burial depth of the Mishash sediments would have led to the gradual obliteration of these bodies and to their conversion into stubby and crystalline forms. Where this process was extensive, perfect rhombohedral crystals were produced, and evidence of the former presence of mat-forming bacteria as frame-builders of carbonate were then totally obliterated in the phosphorites. Although bacterially induced precipitates of rhombohedral-shaped calcite were also produced in the laboratory (Buczynski and Chafetz, 1993), in the present case, however, the rhombohedral crystals are mainly secondary and develop from increasing calcite precipitation onto former bacterial structures.

The recurrent interlayering of authigenic phosphate strips (cf., Fig. 3D) between the calcite nannolayers of the micrite phase calls for some considerations. In Holocene phosphate-forming environments, apatite formation is limited to the upper sediment zone, in association with oxic to suboxic degradation of organic matter. Only there, very close to the sediment-water interface, are the conditions of carbonate alkalinity, fluorine diffusion, and pore water phosphate concentration required for apatite precipitation jointly met (Jahnke et al., 1983; Froelich et al., 1988; Glenn and Arthur, 1988; Jarvis et al., 1994). In addition, production of CO_2 through aerobic oxidation, together with interfacial oxidation of upward-diffusing H_2S produced via microbial sulfate reduction at depth, will result in a lowering of the pH in the topmost sediment column, thus favoring (see Nathan and Sass, 1981) apatite precipitation at this level. Critical in governing the sediment depth of phosphate precipitation is carbonate alkalinity in pore waters (Glenn and Arthur, 1988), since excessive carbonate-ion concentration may prevent francolite crystallization (LeGeros et al., 1967) due to severe lattice distortion. In organic carbon-rich marine settings, the carbonate-ion concentration is directly controlled by the prevalence of the sulfate reduction zone during the decomposition of organic matter, and thus, much of the phosphate phase will precipitate above this zone, via oxic to suboxic diagenesis. Stable carbon isotope values for pore waters and the associated, presently forming phosphate phases in Peru margin phosphorites (Glenn, 1990) clearly reflect the redox interval where apatite precipitation takes place in sediments.

Hence, maximization of apatite-producing events during Mishash deposition requires that much of the organic matter be degraded at shallow burial depths, out of the zone of sulfate reduction where carbonate precipitation prevails. One way to enhance prolonged, near surface oxic to suboxic turnover of the accumulating organic matter, and thus to produce the appropriate chemical conditions for phosphate precipitation, is by balancing the rate of sediment accumulation and the rate of bottom sediment ventilation. This could be eventually promoted by bottom currents sweeping the seafloor, prolonging the residence time of the phosphate-producing zone close to the sediment-water interface (Glenn, 1990). If a suitable balance is sustained between these two parameters, much of the released organic-bound phosphorus will preferentially precipitate within the uppermost sediment zone, producing in this way an authigenic (pristine) phosphate layer. The thickness of this *in situ* phosphate layer will probably depend on the permanence of this balance. The more persistent the sea-bottom ventilation, the longer the residence time of the zone of phosphate precipitation near the sediment surface, and the thicker the authigenic phosphate layer that will be accreted. This may, perhaps, explain why the pristine phosphate layers present in the Mishash Formation (Soudry, 1992) are mainly encountered in proximal areas near the Campanian highs, which are more prone to having been exposed to frequent current activity. Such primary phosphate beds have been found to consist of early phosphatized mat-forming coccoid prokaryotes (Soudry, 1992), and typically occur within condensed, carbonate-poor phosphate sections (cf., Fig. 2). In basinal, carbonate-rich Mishash sections, these authigenic phosphate occurrences are by contrast mostly thin and sporadic. In such confined basinal settings, the accumulation rate and the bottom sediment aeration rate were most likely highly unbalanced. This probably resulted in rapid burial of organic matter, enhanced decomposition via sulfate reduction, and abundant carbonate precipitation in close association with the consumer bacterial populations. Such interplay between the rate of sedimentation and the extent of bottom sediment aeration may perhaps also explain the puzzling selective mineralization of the mat-forming microphytes that colonized the Mishash bottoms. These structures, which now form part of the Negev phosphorite rocks, were found to be mineralized only by phosphate and never by carbonate. The best way for extracellular sheaths to escape total lysis and to be preserved is rapid mineralization (Bartley, 1996). Hence, these structures will undergo rapid fossilization in Mishash sediments (and preserved as phosphate tubules) only where they are durably maintained near the sediment-water interface. Rapidly buried below the sediment-water interface, most of these structures will be anaerobically decomposed by bacterial heterotrophs and concomittantly turned into fine-grained microbial carbonates.

CONCLUSIONS

1. Bacterial mats are the frame-builders of the major part of the carbonate micrite phase of the laminated pelletal phosphorites of the Negev. These mats consist of vertically stratified, calcite-mineralized assemblages of bacterial microorganisms of different sizes, shapes, and organization.
2. These vertically stratified bacterial structures are believed to represent time-successive heterotrophic populations that degraded plankton-derived organic matter, and mats of filamentous and coccoid microphytes that colonized the Mishash sea bottom during phosphorite formation. Carbonate mineralization of these bacterial consumers probably took place during anaerobic degradation of the matted sediments by sulfate-reducing bacteria, as indicated by the low $\delta^{13}C$ values in many of these carbonates.
3. Increasing carbonate mineralization led to obliteration of the bacterial structures and resulted in their conversion into rhombohedral crystals. Where this process was extensive, the former presence of mat-forming bacteria as frame-builders of carbonate in the phosphorites was totally obliterated.
4. The diagenetic fate of the filamentous and coccoid mat-forming phytobenthos, which colonized the Mishash seafloor apparently relied on the balance between the rate of organic matter accumulation and the extent of sediment-bottom ventilation, both of which control the redox interval of organic matter decomposition and pore-water alkalinity. It is suggested that a balance between these two parameters maintained the microphyte structures close to the sediment-water interface where apatite precipitation took place, and resulted in their preservation as phosphate filaments and spheroidal envelopes, now building part of the Negev phosphorite rocks. Imbalance between these two factors led to rapid burial of the microphyte structures below the sediment-water interface, and to their conversion into microbial carbonates by the activity of bacterial heterotrophs.

ACKNOWLEDGMENTS

This work was carried out as part of the Geological Survey of Israel Project 20097. Thanks are due to M. Dvoratchek for SEM services and to B.-S. Cohen for drawing the figures. Y. Nathan critically read an earlier version of the manuscript. Helpful reviews were provided by Henry Chafetz (Houston) and Ian Jarvis (Kingston upon Thames).

REFERENCES

Adolphe, J.-P., Hourimèche, A., Loubière, J.-F., Paradas, J., and Soleilhavoup, F., 1989, Les formations carbonatées d'origine bactérienne. Formations continentales d'Afrique du Nord: Societé Géologique de France, Bulletin, 8ème série, v. 1, p. 55-62.

Aharon, P., Kolodny, Y., and Sass, E., 1977, Recent hot brine dolomitization in the "Solar Lake", Gulf of Elat, isotopic, chemical, and mineralogical study: Journal of Geology, v. 855, p. 27-48.

ALMOGI-LABIN, A., BEIN, A., AND SASS, E., 1993, Late Cretaceous upwelling system along the southern Tethys margin (Israel): Interrelationship between productivity, bottom water environments, and organic matter preservation: Palaeoceanography, v. 8, p. 671-690.

ANADON, P., AND UTRILLA, R., 1993, Sedimentology and isotope geochemistry of lacustrine carbonates of the Oligocene Campins Basin, north-east Spain: Sedimentology, v. 40, p. 699-720.

ANDREWS, J.E., 1986, Microfacies and geochemistry of Middle Jurassic algal limestones from Scotland, Sedimentology, v. 33, p. 499-520.

BARRON, E.J., HARRISON, C.G.A., SLOAN II, J.L., AND HAY, W.W., 1981, Paleogeography, 180 million years ago to the present: Eclogae Geologicae Helvetiae, v. 74, p. 443-470.

BARTLEY, J.K., 1996, Actualistic taphonomy of cyanobacteria: Implications for the Precambrian fossil record: Palaios, v. 11, p. 571-586.

BAULD, J., 1981, Geobiological role of cyanobacterial mats in sedimentary environments: production and preservation of organic matter: BMR Journal of Australian Geology and Geophysics, v. 6, p. 307-317.

BERNER, R.A., SCOTT, M.R., AND THOMLINSON, C., 1970, Carbonate alkalinity in the pore waters of anoxic marine sediments: Limnology and Oceanography, v. 15, p. 544-549.

BEVERIDGE, T.J., AND FYFE, W.S., 1985, Metal fixation by bacterial cell walls: Canadian Journal of Earth Sciences, v. 22, p. 1893-1898.

BRIGGS, D.E.G., AND KEAR, A.J., 1993, Fossilization of soft tissues in the laboratory: Science, v. 259, p. 1439-1442.

BUCZYNSKI, C., AND CHAFETZ, H.S., 1993, Habit of bacterially induced precipitates of calcium carbonate: examples from laboratory experiments and recent sediments, *in* Rezak, R., and Lavoie, D.L., eds., Carbonate Microfabrics: Berlin, Springer-Verlag, p. 105-116.

BUICK, R., 1991, Microfossil recognition in archean rocks: An appraisal of spheroids and filaments from a 3500 M.Y. old Chert-Barite Unit at North Pole, Western Australia: Palaios, v. 5, p. 441-459.

CASTANIER, S., MAURIN, A., AND PERTHUISOT, J.-P., 1989, Production bactérienne expérimentale de corpuscles carbonatés, sphéroidaux à structure-radiaire—Refléxions sur la définition des ooïdes: Societé Géologique de France, Bulletin, 8ème série, no. 3, p. 589-595.

CHAFETZ, H.S., AND BUCZYNSKI, C., 1992, Bacterially induced lithification of microbial mats: Palaios, v. 7, p. 277-293.

CLOUD, P., AND MORRISON, G.R., 1979, On microbial contaminants, micropseudofossils and the oldest records of life: Precambrian Research, v. 9, p. 81-91.

COHEN, Y., KRUMBEIN, W.E., AND SHILO, M., 1977, Solar Lake (Sinai). 3. Bacterial distribution and production: Limnology and Oceanography, v. 22, p. 621-634.

DICKMAN, M., 1985, Seasonal succession and microlamina formation in a meromitic lake displaying varved sediments: Sedimentology, v. 32, p. 109-118.

DITCHFIELD, P., AND MARSHALL, J.D., 1989, Isotopic variation in rhythmically bedded chalks: Paleotemperature variation in the Upper Cretaceous: Geology, v. 17, p. 842-845.

DUDLEY, W.C., BLACKWELDER, P., BRAND, L., AND DUPLESSY, J., 1986, Stable isotopic composition of coccoliths: Marine Micropaleontology, v. 10, p. 1-8.

ERBA, E., 1991, Deep mid-water bacterial mats from anoxic basins of the Eastern Mediterranean: Marine Geology, v. 100, p. 83-101.

FERRIS, F.G., FYFE, W.S., AND BEVERIDGE, T.J., 1987, Bacteria as nucleation sites for authigenic minerals in a metal-contaminated lake sediment: Chemical Geology, v. 63, p. 225-232.

FLEXER, A., ROSENFELD, A., LIPSON-BENITAH, S., AND HONIGSTEIN, A., 1986, Relative sea level changes during the Cretaceous in Israel: American Association of Petroleum Geologists Bulletin, v. 70, p. 1685-1699.

FOLK, R., 1993, SEM imaging of nannobacteria in carbonate sediments and rocks: Journal of Sedimentary Petrology, v. 63, p. 990-999.

FROELICH, P.N., ARTHUR, M., BURNETT, W.C., DEAKIN, M., HENSLEY, V., JAHNKE, R., KAUL, L., KIM, K., ROE, K., SOUTAR, A., AND VATAKANON, C., 1988, Early diagenesis of organic matter in Peru continental margin sediments: phosphorite precipitation: Marine Geology, v. 80, p. 309-346.

GARFUNKEL, Z., 1981, Internal structure of the Dead Sea leaky transform (rift) in relation to plate-kinematics: Tectonophysics, v. 80, p. 1-26.

GERDES, G., AND KRUMBEIN, W.E., 1987, Biolaminated Deposits: Lecture Notes in Earth Sciences, Heidelberg, Springer Verlag, 183 p.

GERDES, G., DUNAJTSCHIK-PIEWAK, K., RIEGE, H., TAHER, A.G., KRUMBEIN, W.E., AND REINECK, H.-E., 1994, Structural diversity of biogenic carbonate particles in microbial mats: Sedimentology, v. 41, p. 1273-1294.

GLENN. C.R., 1990, Pore water, petrologic and stable carbon isotopic data bearing on the origin of Modern Peru margin phosphorites and associated authigenic phases, *in* Burnett, W.C., and Riggs, S.R., eds., Phosphate Deposits of the World, v. 3, Neogene to Modern Phosphorites: Cambridge, Cambridge University Press, p. 46-61.

GLENN, C.R., AND ARTHUR, M.A., 1988, Petrology and major element geochemistry of Peru margin phosphorites and associated diagenetic minerals: authigenesis in modern organic-rich sediments: Marine Geology, v. 80, p. 231-276.

GOLUBIC, S., 1976, Organisms that build stromatolites, in Walter, M.R., ed., Stromatolites: Elsevier, Amsterdam, p. 113-126.

HARDER, H., 1978, Synthesis of iron layer silicate minerals under natural conditions: Clay and Clay Minerals, v. 26, p. 65-72.

HIRSCHLER, A., LUCAS, J., AND HUBERT, J.-C., 1990, Apatite genesis: a biologically induced or biologically controlled mineral formation process?: Geomicrobiology Journal, v. 7, p. 47-57.

HUDSON, J.D., 1977, Stable isotopes and limestone lithification: Journal of the Geological Society of London, v. 133, p. 637-660.

JAHNKE, R.A., EMERSON, S.R., ROE, K.K. AND BURNETT, W.C., 1983, The present day formation of apatite in Mexican continental margin sediments: Geochimica et Cosmochimica Acta, v. 47, p. 259-266.

JARVIS, I., BURNETT, W.C., NATHAN, Y., ALMBAYDIN, F.S.M., ATTIA, A.K.M., CASTRO, L.N., FLICOTEAUX, R., HILMY, M.E., HUSAIN, V., QUATAWNAH, A.A., SERJANI, A. AND ZANIN, Y.N., 1994, Phosphorite geochemistry: State-of-the-art and environmental concerns: Eclogae Geologicae Helvetiae, v. 87, p. 643-700.

JONES, B., 1995, Processes associated with microbial biofilms in the twilight zone of caves: examples from the Cayman Islands: Journal of Sedimentary Research, v. A65, p. 552-560.

JORGENSEN, B.B., AND COHEN, Y., 1977, Solar Lake (Sinai), The sulfur cycle of benthic cyanobacterial mats: Limnology and Oceanography, v. 22, p. 657-666.

KIRKLAND, B.L., LYNCH, F.L., RAHNIS, M.A., FOLK, R.L., MOLINEUX, I.J., AND MCLEAN, R.J.C., 1999, Alternative origins for nannobacteria-like objects in calcite: Geology, v. 27, p. 347-350.

KOLODNY, Y., 1967, Lithostratigraphy of the Mishash Formation, northern Israel: Israel Journal of Earth Sciences, v. 16, p. 57-73.

KOLODNY, Y., 1969, Petrology of siliceous rocks in the Mishash Formation (Negev, Israel): Journal of Sedimentary Petrology, v. 39, p. 166-175.

KOLODNY, Y., 1980, Carbon isotopes and depositional environment of high productivity sedimentary sequence—the case of the Mishash Formation, Israel: Israel Journal of Earth Sciences, v. 29, p. 147-156.

KOLODNY, Y., TARABOULOS, A., AND FRIESLANDER, U., 1980, Participation of fresh water in chert diagenesis: evidence from oxygen isotopes and boron-track mapping: Sedimentology, v. 27, 305-316.

KRUMBEIN, W.E., COHEN, Y., AND SHILO, M., 1977, Solar Lake (Sinai) . 4. Stromatolitic cyanobacterial mats: Limnology and Oceanography, v. 22, p. 635-656.

LANIER, W.P., 1986, Approximate growth rates of Early Proterozoic microstromatolites as deduced by biomass productivity: Palaios, v. 1, p. 525-542.

LEGEROS, R.Z., TRAUTZ, O.R., LEGEROS, J.P., AND KLEIN, E., 1967, Apatite crystallites: Effects of carbonate on morphology: Science, v. 155, p. 1409-1411.

LEWY, Z., 1990, Transgressions, regressions and relative sea level changes on the Cretaceous shelf of Israel and adjacent countries. A critical evaluation of Cretaceous global sea level correlations: Palaeoceanography, v. 5, p. 619-637.

LYONS, W.B., LONG, D.T., HINES, M.E., GAUDETTE, H.E., AND ARMSTRONG, P.B., 1984, Calcification of cyanobacterial mats in Solar Lake, Sinai: Geology, v. 12, p. 623-526.

MONTY, C.L.V., 1976, The origin and development of cryptalgal fabrics, *in* Walter, M.R., ed., Stromatolites: Amsterdam, Elsevier, p. 193-249.

MOSHKOVITZ, S., EHRLICH, A., AND SOUDRY, D., 1983, Siliceous microfossils of the Upper Cretaceous Mishash Formation, central Negev, Israel: Cretaceous Research, v. 4, p. 173-194.

NATHAN, Y., AND SASS, E., 1981, Stability relations of apatites and calcium carbonates: Chemical Geology, v. 34, p. 103-111.

NATHAN, Y., SHILONI, Y., RODED, R., GAL., I., AND DEUTSCH, Y., 1979, The geochemistry of the northern Negev phosphorites: Geological Survey of Israel

Bulletin, No. 73, 41 p.

NATHAN, Y., AND SOUDRY, D., 1982, Authigenic silicate minerals in phosphorites of the Negev, southern Israel: Clay Minerals, v. 17, p. 249-254.

PARRISH, T.J. AND CURTIS, R.L., 1982, Atmospheric circulation, upwelling, and organic rich rocks in the Mesozoic and Cenozoic eras: Palaeogeography, Palaeoclimatology, Palaeoecology, v. 40, p. 31-66.

REISS, Z., 1988, Assemblages from a Senonian high productivity sea: Revue de Paleobiologie, v. 2, p. 323-332.

REISS, Z., ALMOGI-LABIN, A., HONIGSTEIN, A., LEWY, Z., LIPSON-BENITAH, S., MOSHKOVITZ, M., AND ZAKS, Y., 1985, Late Cretaceous multiple framework of Israel: Israel Journal of Earth Sciences, v. 34, p. 147-166.

SASS, E., BEIN, A., AND ALMOGI-LABIN, A., 1991, Oxygen-isotope composition of diagenetic calcite in organic-rich rocks: Evidence for ^{18}O depletion in marine anaerobic pore water: Geology, v. 19, p. 839-842.

SIEBURTH, J.MCN., SMETACEK, V., AND LENZ, J., 1978, Pelagic ecosystem structure: Heterotrophic compartments of the plankton and their relationship to plankton size fractions: Limnology and Oceanography, v. 23, p. 1256-1263.

SZULC, J., AND SMYK, B., 1994, Bacterially controlled calcification of freshwater *Schizotrix*-stromatolites: An example from the Pieniny MTS, southern Poland, *in* Bertrand-Sarfati, J., and Monty, C., eds., Phanerozoic Stromatolites II: Dordrecht, Kluwer Academic Publishers, p. 31-51.

SOUDRY, D., 1983, Étude de la Série Phosphatée de la région d'Ein Yahav (Néguev central, Israel)—Logique séquentielle, pétrologie, approche de la phosphatogénèse: Unpublished Thèse Doctorat d'État: Institut National Polytechnique de Lorraine, Nancy, 177 p.

SOUDRY, D., 1987, Ultra-fine structures and genesis of the Campanian Negev high-grade phosphorites (southern Israel): Sedimentology, v. 34, p. 641-660.

SOUDRY, D., 1992, Primary bedded phosphorites in the Campanian Mishash Formation, Negev, southern Israel: Sedimentary Geology, v. 80, p. 77-88.

SOUDRY, D., NATHAN, Y., AND RODED, R., 1985, The Ashosh-Haroz facies and their significance for the Mishash palaeogeography and phosphorite accumulation in the northern and central Negev: Israel Journal of Earth Sciences, v. 34, p. 211-220.

THOMPSON, J.B., FERRIS, F.G., AND SMITH, D.A., 1990, Geomicrobiology and sedimentology of the mixolimnion and chemocline in Fayetteville Green Lake, New York: Palaios, v. 5, p. 52-75.

TRUDINGER, P.A., AND MENDELSOHN, F., 1976, Biological processes and mineral deposition, *in* Walter, M.R., ed., Stromatolites: Amsterdam, Elsevier, p. 663-672.

VENGOSH, A., KOLODNY, Y., AND TEPPERBERG, M., 1987, Multi-phase oxygen isotopic analysis as a tracer of diagenesis: the example of the Mishash Formation, Cretaceous of Israel: Chemical Geology, v. 65, p. 235-253.

BLACK SHALES AND PHOSPHORITES, RESULTS OF DIFFERENTIAL DIAGENETIC EVOLUTION OF ORGANIC MATTER IN THE OULAD ABDOUN-TIMAHDIT SEDIMENTARY BASIN (MOROCCO)

S. BENALIOULHAJ AND J. TRICHET

Laboratoire de Géochimie Organique, Université d'Orléans, 45067 Orléans cedex 2, France

N. BENALIOULHAJ

Department of Geology, Mohammadia School of Engineering, Agdal, Rabat, Morocco

J. LUCAS

Institut de Géologie, Université Louis Pasteur, 1 Rue Blessig, 67084 Strasbourg cedex, France

B. C. SCHREIBER*

Department of Geology, Appalachian State University, 195 Rankin Science Building, Boone, NC, 28608, USA

AND

R. P. PHILP

School of Geology and Geophysics, University of Oklahoma, 100 East Boyd St., Norman OK, 73019-0628, USA

ABSTRACT: The sedimentary subbasins of Oulad Abdoun and Timahdit (Morocco) were part of a single marine gulf that extended across Morocco for more than 300 km (west to east) from Late Cretaceous through the Eocene. The differential structural development of this basin resulted in the presence of a stable shelf in Oulad Abdoun and a subsident orogenic zone in Timahdit. This structural asymmetry explains the sedimentary asymmetry of the Maastrichtian sequence that consists of a thin, 5-m thick, phosphorite sequence interbedded with dolomitic limestone, marl, and shale near Oulad Abdoun, and an oil shale, 200-m thick, in Timahdit.

Widespread reducing conditions that prevailed in the basin resulted in preservation of large amounts of organic matter. Quantitative and qualitative analysis shows an identical phytoplanktonic origin for all of the organic matter in both of the two subbasins, and the organic matter still remains in an immature state. The subsequent history of this organic matter, however, reflects the specific environments of the two related but disparate depositional settings.

In each part of the basin, the organic material associated with marls, limestones, and shales contains kerogen, whereas in the phosphate-rich strata the organic matter is in two forms: humic compounds inside the phosphatic grains and kerogen in the matrix. Humic compounds from different lithologies have a similar composition, whereas the kerogen in samples from the phosphate-rich strata shows the effect of oxidation (during synsedimentary reworking), increasing with the amounts of P_2O_5 of the total rock. These changes in the phosphate-rich strata have allowed the crystallization of apatite and the genesis and preservation of humic compounds trapped inside these phosphatic grains. In the oil shale, large amounts of associated clay resulted in a rapid and total unsolubilization of the organic matter and its evolution toward kerogen.

The study of the organic matter, which is associated with a wide variety of facies from both parts of the basin, shows the extreme sensitivity of organic compounds to processes that have affected them from their deposition through all phases of diagenesis. The fate of the organic matter during early diagenesis is related to the presence or absence of clays: the presence of clays led to the formation of black shales and the absence of clays led to the formation of phosphorites.

INTRODUCTION

The literature of the two last decades is filled with evidence of the close relationship between phosphorites and organic matter (O.M.) in a variety of deposits covering the whole time scale. As a result, the genesis of phosphatic deposits is now difficult to imagine outside areas of high productivity. Likewise, the frequent occurrence of bituminous shales associated with phosphate deposits supports the hypothesis that there may be a close genetic relationship between both types of sedimentary deposits. One possible cause for this association might be the result of different diagenetic evolution of marine organic matter in two closely spaced but structurally different marine environments.

The Moroccan Late Cretaceous-Tertiary phosphorite deposits, which have a well-defined paleogeographical setting at the edge of the Tethyan Sea, are representative of Tethyan deposits and also of most of the phosphatic facies of all ages, which have many common characteristics. The simple structures and the relationship of the phosphate facies with black shales clearly illustrate the comparison of the organic matter in the two different facies. Because the types of O.M. present in these various facies reflect the environments of sedimentation and early diagenesis, a study of the O.M. may help in the interpretation of the depositional and diagenetic differences and aid in deciphering the main facies that occur in different parts of the basin.

The results of the present study are similar to those of previous studies concerned with the specific characteristics of the O.M. in phosphatic sequences (Baturin et al., 1970; Powell et al., 1975; Sandstrom, 1980, 1986; Waples, 1982; Amit and Bein, 1982; Bein and Amit, 1982; Belayouni, 1983; Belayouni and Trichet, 1983, 1984; Rachidi, 1983; Benalioulhaj, 1989; Krajewski, 1989; Belayouni et al., 1990). However, the specific constraints of the model offered by the synchronous deposition of phosphatic and black shale facies permit new insights into the conditions prevailing during phosphogenesis as well as the role of the sedimentary reworking necessary for the concentration of phosphatic grains within the strata.

*email: geologo@aol.com

Marine Authigenesis: From Global to Microbial, SEPM Special Publication No. 66
 ISBN 1-56576-064-6, p. 427-444.

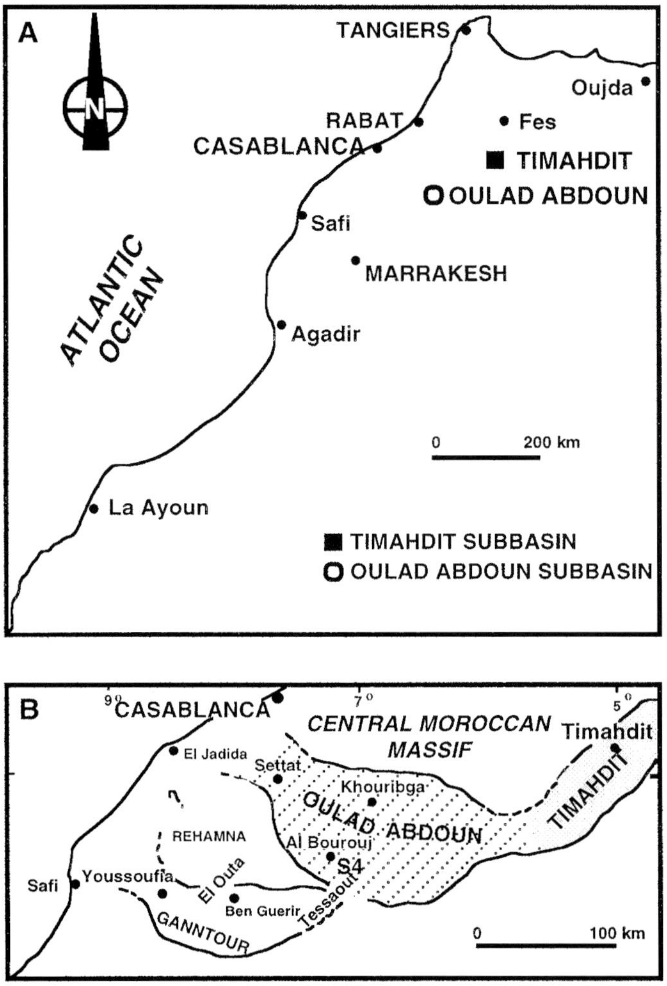

FIG. 1.—Geographic relationships of the Oulad Abdoun and Timahdit subbasins. (A) Geographic locations in Morocco; (B) approximate position and continuity of the two parts of the basin from Late Cretaceous through Middle Eocene.

GEOLOGICAL SETTING

During the Late Cretaceous, mainly at the beginning of Maastrichtian, a transgressive sea invaded the western part of present-day Morocco. Previous orogenies, particularly the Hercynian orogeny, formed a rigid structure that was oriented almost parallel to the present coast and formed several mountain belts (Western High-Atlas, Jebilet, Rehamna, central Morocco). The resulting more or less west-east-oriented marine basins between these mountain belts were narrow paleogulfs, which, depending on different authors, would have been totally independent, locally connected, or even all interconnected, thus forming a continuous epicontinental sea with emerging Hercynian islands (Choubert and Faure-Muret, 1962; Boujo, 1976; Trappe, 1991; Herbig and Trappe, 1994). Maastrichtian to Ypresian phosphate-bearing epicontinental sedimentary sequences were deposited on drowned Hercynian strata between the emerged blocks. These are now the Oulad Abdoun deposit between central Morocco and Rehamna, the Ganntour deposit between Rehamna and Jebilet (Fig.1), and southward, the Meskala deposit between Jebilet and Western High-Atlas, and the Bou Craa deposit that lies south of the anti-Atlas.

Today, since the formation of the Middle-Atlas range, the Timahdit and Oulad Abdoun areas appear to have formed in two separate sedimentary basins. However different, the close correlations between both formations are well supported by palynological results (Rauscher et al., 1986) and by the similarity of the organic matter preserved in both deposits (Meunier-Christmann, 1988; Benalioulhaj, S., 1989; Benalioulhaj, N., 1991). Thus, whatever the actual general paleogeographic situation, the northernmost marine gulf clearly stretched as far as the present-day Middle-Atlas, and the phosphatic deposits of the Oulad Abdoun area and the bituminous shales of the Timahdit area were deposited in one single basin that extended 300-350 km from west to east (Fig. 1).

Tectonically, the huge area of the Oulad Abdoun-Timahdit Basin did not behave symmetrically because of the mobility of the Middle Atlas belt in the eastern part of the basin. In the Timahdit area, the unstable basin is fractured, and thick deposits of sediments accumulated between strongly subsident fault blocks (Rahhali, 1970; Saadi et al., 1981). Maastrichtian sedimentary deposits that are up to 350 m in thickness are present in the Timahdit area, whereas they are a maximum of 5-m thick in the region of Oulad Abdoun (around Khouribga) on the stable Hercynian platform. The result of differential subsidence is clearly displayed at the very edge of this stable margin by the thickening of the Maastrichtian deposits and resulted in a change in sediment thickness from 5 m at Khouribga to 40 m at sites only 40 km distant (Belfkira, 1980) (Fig. 2).

At any given time within the setting just described, subsidence had little influence on the water depth of the basin and the resistant Hercynian platform formed a shoal that was only slightly higher than the surrounding shallow shelf. The very eastern part of the basin, bordering the continent, was supplied with terrigenous material. Due to the dry climate of this period, and also to the upwelling currents and the associated barrenness of the surrounding land area, continental erosion was weak, and the mineral supply consisted mainly of clays. The clays apparently settled close to the shoreline in the subsident but shallow part of the basin. Several times during the Maastrichtian these clays spread, to varying degrees, toward the west. However, these clastic deposits are rare on the stable shoal, where only scarce decimetric deposits are observed.

The western area of the basin, nearest to the Atlantic Ocean, was dominated by a system of active upwelling currents (Tooms et al.,1969; Summerhayes et al., 1971). This led to intense biological activity and high rates of organic sedimentation throughout the entire basin, even in its eastern extremity. Upwelling currents also played an important role in the cycling of the phosphorus and silica that concentrated within the different phosphatic and siliceous strata (Lucas and Lucas-Prévôt, this volume) as well as in the cycling of particular accompanying elements such as Fe, Al, Zn, and Cu (Benalioulhaj, 1991).

This dual sediment source for the basin produced an important variation in facies between the eastern zone (Timahdit), in which black shales are dominant (Fig. 3), and the western part (Oulad Abdoun), where the shoal was covered only by more or less calcareous and siliceous phosphates, with some rare and

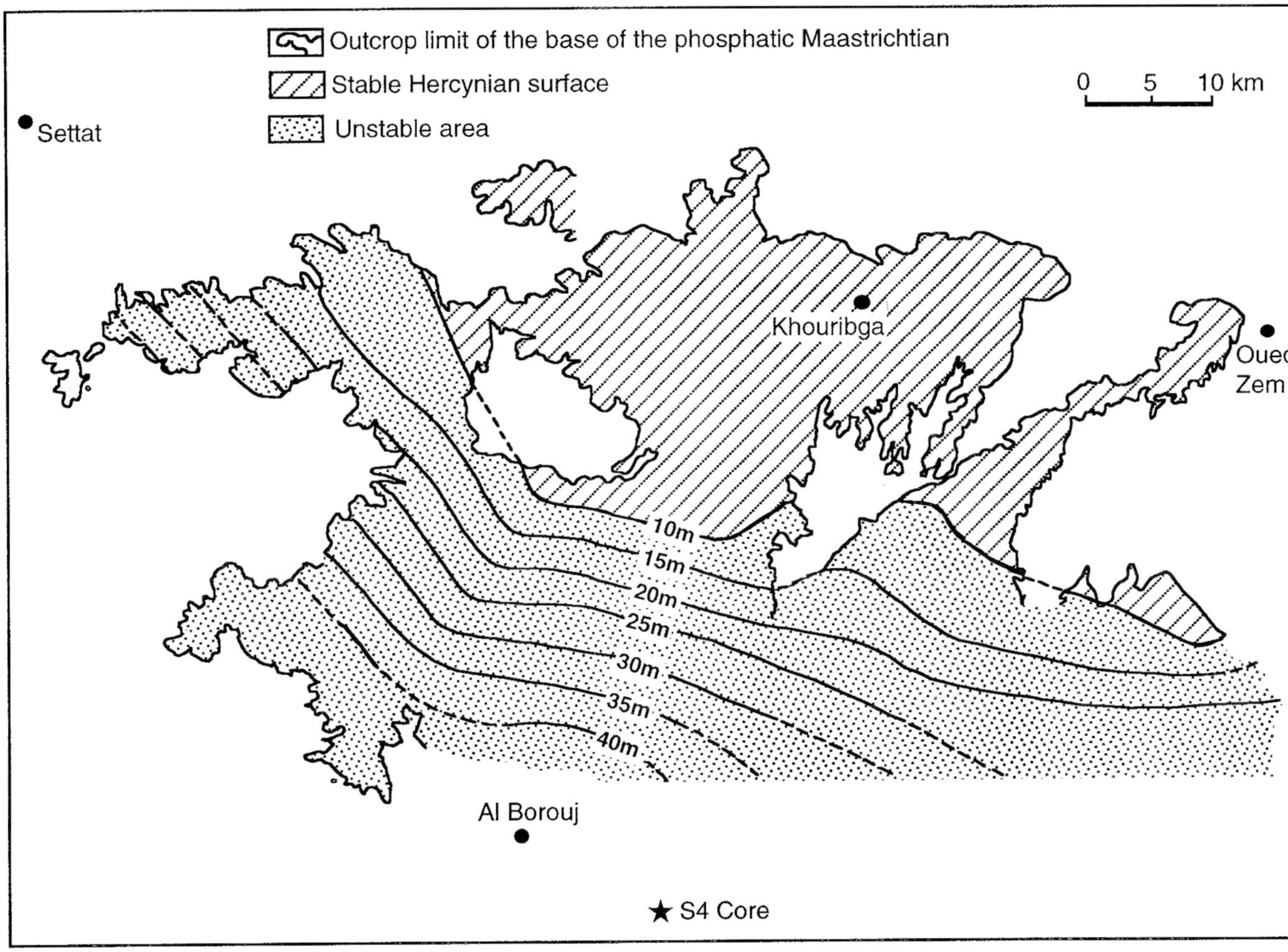

FIG. 2.—Isopachous map of the phosphatic Maastrichtian rocks in the Oulad Abdoun area (modified from Belfkira, 1980).

thin beds of claystone. Between those two extremes, the deposits are mixed with phosphorites and black shales. The thicker the section, the more abundant is the clay fraction. High-grade, phosphate-rich deposits are condensed sequences with very low clay content.

In the upper third of the Maastrichtian of the Timahdit area, the organic carbon-rich black shales are replaced by variably marly sands, in which the organic carbon (C_{org}) content does not exceed 2%. This marked decrease of the C_{org} content and increase in clastic content may be interpreted as the result of a trend toward basinal restriction and possible emersion, which seems confirmed by the thick evaporative succession that was deposited during the ensuing Paleocene. Thus, from Late Maastrichtian, the evolution of the Timahdit area diverged from that of the western part of the basin, where phosphate-carbonate sedimentation continued until Lutetian times.

The present paper compares the organic matter present respectively in the black-shales and in the phosphates, and helps in the understanding of the genesis of both sediments.

ANALYTICAL METHODS

Total Organic Carbon

The total organic carbon of the rock (TOC) was measured by combustion in a carbon analyzer after demineralization of the samples (apatite + calcite) by treatment with HCl 2N at 60°C for 12 hours.

Separation of different fractions of the organic matter.—
Samples of total rock were extracted with chloroform, according to the procedure of Huc et al. (1976). After isolation of the asphaltenes by precipitation in n-pentane at 4°C, the aromatics, saturates, and resins were fractionated on silica gel by thin-layer chromatography. The protocol used for the separation of acid water-soluble compounds, fulvic acids, humic acids, gray humic acids (humic acids closely associated with very fine clay minerals that flocculate with the addition of NaCl), and stable organic residue is as described in Benalioulhaj and Trichet (1990).

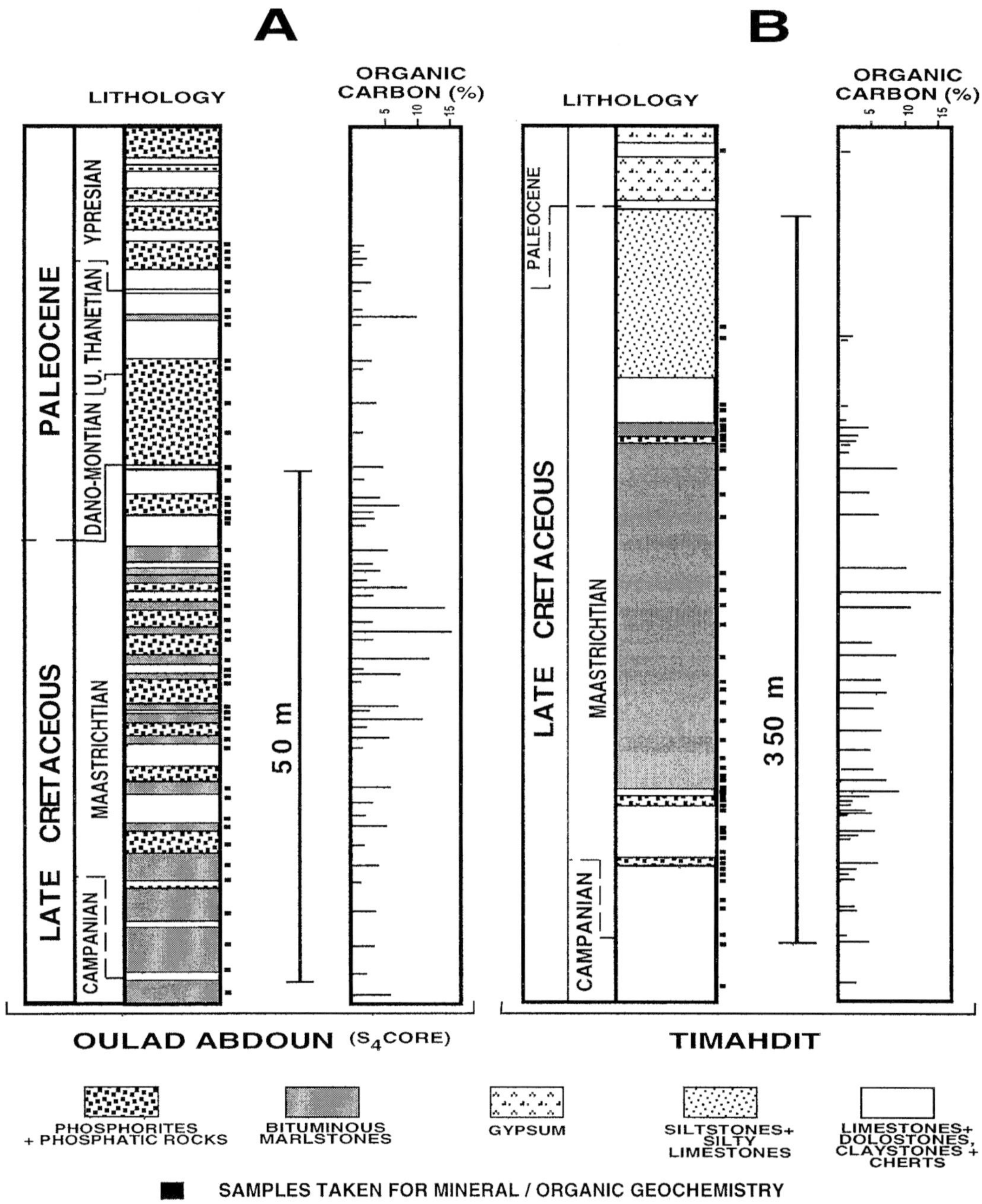

FIG. 3.—Stratigraphic columns together with organic carbon content for the Late Cretaceous through Eocene sections in (A) Oulad Abdoun, and (B) Timahdit.

Gas Chromatography and Gas Chromatography-Mass Spectrometry

Saturate and aromatic fractions were analyzed by gas chromatography using a gas chromatograph (GC) equipped with a 25 m x 0.3 mm i.d. DB-5 capillary column, 110-280° C, at 2° C/mn and by gas chromatography-mass spectrometry (GC-MS).

Elemental Analyses

Elemental analyses were performed in Laboratories WOLFF (formerly ATX). C, H, and N were oxidized under a pure O_2 flux. CO_2, H_2O and N_2 (obtained by reduction of nitrogen oxides) were separated by gas chromatography and their respective amounts measured by variation in thermal conductivity in a catharometric cell. Oxygen is incorporated into CO and then CO_2 by pyrolysis of the organic matter under N_2 flow and then CO_2 is titrated by acidimetry. Sulfur is oxidized into SO_2 under O_2 flow and also titrated by acidimetry.

Rock-Eval Pyrolysis

Rock-Eval analyses were performed according to the procedure of Espitalie et al. (1986) on stable organic residues of samples representative of different lithologies. Hydrogen and oxygen index represent, respectively, the ratio of hydrocarbons (and tars) and CO_2 produced during the pyrolysis of organic samples versus the organic carbon content of the sample.

Organic Petrography

Microscopic examination under reflected light and fluorescent light and reflectance measurements were performed on both polished thin sections of the total rock and polished concentrates of organic particles. Concentration of the organic matter was done by density flotation (mixture of bromoform-ethanol, density = 1.7) and was carried out according to the procedure of Bostick and Alpern (1977).

RESULTS

Organic Petrography

Approximately 100 polished thin sections of rock and about 30 polished sections of organic concentrates were examined in this study. The selected samples represented all of the facies encountered in the basin.

Oil Shale Samples.—
The oil shale samples that were analyzed, regardless of whether they belong to the Oulad Abdoun phosphate or the Timahdit oil-shale sedimentary sequence, have a very similar organic particle facies as determined by reflected and fluorescent light. The organic matter present is associated with fine mineral particles (mostly clay), the latter representing the largest part of the organic content. Observed in transmitted light, these organo-mineral associations display a yellow to brown-red color, as well as a lamellar stratification. They have a yellow to yellow-orange fluorescence. Under reflected light, these matrices have a blackish-red to grayish-purple color, caused by granules of framboidal pyrite.

The most abundant and recognizable organic elements are dinoflagellates. They are distinguishable by their intense green-yellow to yellow fluorescence that also permits the observation of details of their structure. They appear either dispersed in the mass of the rock or aligned along the laminae. Initially subspheric, they are now flattened parallel to stratification. Other microfossils include chlorophyceae, belonging to the Tasmanacae, and diverse acritarch groups such as Pterospermopsis. Continental plant debris includes pollens, spores, cuticles and lignin debris from higher plants. The latter are represented by fusinized particles and some particles of vitrinite and inertinite. Remains of foraminifera are slightly fluorescent in yellow-orange to brown-red colors.

Phosphorite Samples.—
In phosphatic samples, the organic matter is present within both the phosphate grains and their matrix. Within the matrix, the organic matter presents optical characteristics that are similar to those in oil shale samples. Inside the phosphate grains, organic matter shows two different aspects: (1) some is disseminated in the form of particles averaging 10 μm; and (2) some is in the form of an amorphous material, that traces concentric layers responsible for the zoned appearance of some grains. Dinoflagellates are most abundant in both the phosphate grains and in the matrix of phosphate-rich samples, where their number exceeds 10,000 per gram of rock (Rauscher et al., 1986).

Degree of Maturity of the Organic Matter.—
Vitrinite reflectance values measured on rare allochthonous particles transported into the basin range from 0.40 to 0.41 in Oulad Abdoun samples and from 0.33 to 0.38 in Timahdit samples (Table 1), reflecting the immaturity of the organic matter associated with the sediments of the two sedimentary basins. These values are in agreement with the fluorescence colors (green-yellow to yellow) of the phytoplanktonic remains, reflecting the low degree of maturity of the organic matter.

ORGANIC GEOCHEMISTRY

Total Organic Carbon (TOC)

There is a clear evolution of the organic content within the stratigraphic sections of the two subbasins (Fig. 3). The highest amounts of TOC, 14.5% in Oulad Abdoun and 16% in Timahdit, accumulated during the same portion of the Maastrichtian. As will be elaborated below (see *Infrared Spectroscopy*, below), the upper part of the sequence, in the more phosphatic zone (OA), is characterized by reworking processes that resulted in a net depletion of organic content both through the loss of matrix particles containing O.M. and the oxidation of the bulk O.M. within the matrix.

Our petrographic observations permit the establishment of similar discrete lithologic groupings in Oulad Abdoun as well as in Timahdit. These groupings are distinguished in Figure 4, tying together their related P_2O_5 and organic carbon. This permits easy identification of similar lithological facies in each part of the basin.

TABLE 1.—VITRINITE REFLECTANCE MEASURED IN SAMPLES FROM OULAD ABDOUN AND TIMAHDIT

Sample Number	Range	Average Reflectance	σ
TIMAHDIT			
S 4 bis 11	0.25 - 0.45	0.33	0.04
108-23	0.20 - 0.45	0.35	0.07
1-11	0.25 - 0.45	0.38	0.07
1-16	0.25 - 0.50	0.38	0.06
OULAD ABDOUN			
S 4-90	0.30 - 0.50	0.40	0.04
S4-92	0.30 - 0.50	0.41	0.05

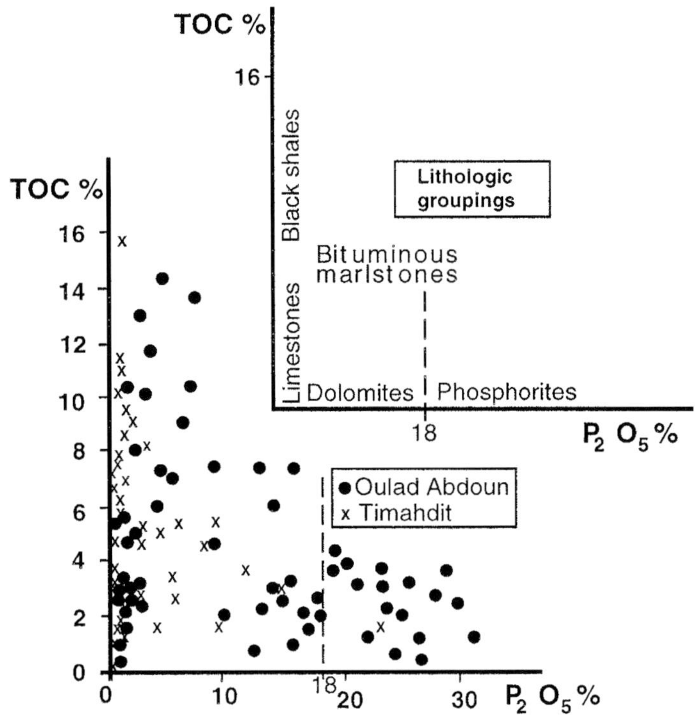

FIG. 4.—Percentages TOC % versus P_2O_5 for samples from Oulad Abdoun and Timahdit. Lithologic groupings indicated (I-VI, and I-V) are clearly tied to percentage composition of carbon versus P_2O_5 (after Benalioulhaj, 1989, modified).

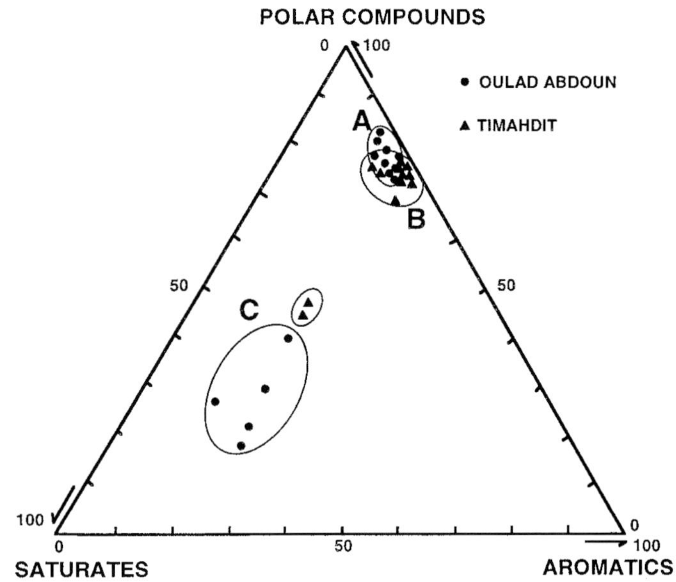

FIG. 5.—Ternary diagram illustrating the lipid extract compositions (polar compounds, saturates, and aromatic hydrocarbons) of the different samples in the two sections (as indicated).

Lipid Extracts.—

Bulk analysis.—The amounts of lipid extract from samples of the Oulad Abdoun area range from 27 mg/g to 516 mg/g TOC, and 0.3 mg/g-28.4 mg/g of the total rock (Table 2). In the samples from Timahdit, the amount of lipid extracts ranges from 47 mg/g-140 mg/g TOC, that is 0.2 mg/g-1.8 mg/g of the total rock. The high quantities of extractable lipids in the Oulad Abdoun samples confirm the high values of extractable lipidic compounds generally reported in phosphatic sediments (Powell et al., 1975; Belayouni and Trichet, 1984; Krajewski, 1989; Belayouni et al., 1990). A good correlation exists between quantities of extract and amounts of TOC (Table 2). Such a correlation is characteristic of sediments containing immature organic matter; most of the lipids are inherited from the original biomass and microorganisms active in its biodegradation (Tissot and Welte, 1984).

The relative amounts of polar compounds, aromatic hydrocarbons, and saturated hydrocarbons proportions are related in a ternary diagram shown in Figure 5. Most of the samples of groups A (Timahdit) and B (Oulad Abdoun) have a similar composition with high amounts of polar compounds, varying from 70% to 80% in keeping with the low degree of maturity of the organic matter. The amounts of saturated hydrocarbons vary from 3% to 10% of the total lipid extract in most samples of both the Timahdit and Oulad Abdoun parts of the basin (Table 2).

However, certain samples from Oulad Abdoun are characterized by very high amounts of saturated hydrocarbons (>40%) in contrast to the low degree of maturity of the organic matter (Group C in Fig. 5). Samples with such a composition are usually found in immature phosphatic deposits (Belayouni, 1983; Belayouni and Trichet, 1984; Belayouni et al., 1990). This high proportion of saturates associated with a regular profile of n-alkanes in a gas-chromatogram is interpreted as the result of bacterial activity (see further) .

Qualitative results.—Gas chromatograms of the saturated fraction of samples from groups A and B (Fig. 5) are characterized by a bimodal distribution of the n-alkanes. The first mode is situated between C_{13} and C_{22} with a predominance of C_{17} and C_{18} and corresponds to an algal planktonic fraction. The second mode is situated between C_{23} and C_{32}, with an odd/even predominance, revealing an input of organic matter derived from higher plants. The pristane to phytane ratio (Pr/Ph) is in all cases smaller than unity, indicating reducing conditions in the environment of deposition. The n-alkanes from samples belonging to group C do not show an odd/even predominance and are characterized by a unimodal distribution between C_{14} and C_{27}, with C_{17} as a mode. Such an n-alkane profile without an odd/even predominance, and generally in the range of over nC_{20}, is attributed to bacterial activity (Parker et al., 1967; Meinschein, 1969; Johnson and Calder, 1973; Baranger, 1989; Simoneit, 1990; Bouloubassi and Saliot, 1991; Bouloubassi et al., 1992).

Polycyclic alkanes are represented by steranes (Fig. 6) and hopanes (Fig. 7), which are characteristic of a primary planktonic contribution and a substantial bacterial influence, respectively. The hopane distribution shows the complete series ranging from C_{27} to C_{35}. The presence of $\beta\beta$ and $\beta\alpha$ isomers among the hopanes confirms the low degree of maturity of the organic matter. Steranes are represented by C_{27}-C_{29} and the methylsteranes C_{28}-C_{30}. Given the abundance of dinocysts revealed by the organic petrography study, the methylsteranes are probably methylated on carbon, position number 4 (Boon et al., 1979; Robinson et al., 1984).

TABLE 2.—QUANTITIES AND COMPOSITIONS OF THE LIPIDS EXTRACTED FROM SAMPLES FROM OULAD ABDOUN AND TIMAHDIT

Sample Number	%TOC	Total Extract in mg/g of rock sample	Total Extract in mg/g of TOC	Saturates % of Total Extract	Aromatics % of Total Extract	Polars % of Total Extract	Resins % of Total Extract	Asphaltenes % of Polars	Asphaltenes / Resins
OULAD ABDOUN									
S 4-68	5.9	5.9	100.0	5.3	14.9	82.2	60.7	39.3	0.7
S4-70	3.0	3.8	126.6	3.3	17.0	79.7	55.0	45.0	0.8
S4-90	14.5	22.0	151.7	5.7	15.5	79.7	49.7	50.3	1.0
S4-92	2.8	3.6	128.5	3.0	16.0	81.3	57.7	42.3	0.7
S4-94	11.9	18.1	152.1	4.8	18.3	80.1	55.3	45.2	0.8
S4-98	1.1	0.3	27.2	4.3	15.9	81.3	58.0	41.8	0.7
S4-100	7.2	9.5	131.9	4.8	15.7	79.2	58.6	41.4	0.7
S4-101	2.2	3.7	168.1	57.1	18.6	24.1	53.3	46.7	0.9
S4-103	10.4	13.6	130.7	5.8	11.7	82.4	60.6	39.4	0.7
S4-104	2.4	4.1	170.8	40.5	17.0	41.1	63.4	36.6	0.6
S4-107	5.5	22.8	414.5	68.6	13.5	17.8	67.0	32.5	0.5
S4-108	1.7	6.1	358.8	60.8	23.8	16.1	49.6	50.3	1.0
TIMAHDIT									
1-11	3.9	3.7	94.8	3.0	18.6	77.9	84.0	16.0	0.2
1-14	8.5	6.6	77.6	4.0	21.0	83.8	85.0	15.0	0.2
1-20	2.9	4.2	144.8	6.6	21.3	72.0	77.2	22.9	0.3
108-20	6.0	2.8	46.6	4.8	20.1	76.5	85.4	14.2	0.2
108-23	15.6	18.0	115.3	4.7	18.0	77.2	82.7	17.3	0.2
108-27	8.5	9.9	116.4	3.7	22.5	73.7	75.7	24.3	0.3
4 bis 26	2.8	2.1	75.0	9.0	20.0	67.4	80.0	20.0	0.3
SAM 3-10	3.3	1.8	54.5	6.3	16.0	79.1	79.0	20.9	0.3
SAM 3-12	2.5	1.7	68.5	7.7	16.8	75.5	78.6	22.3	0.3
SAM 3-17	6.5	4.5	69.2	6.2	12.3	83.1	85.1	14.9	0.2
SAM 3-20	14.2	12.1	85.2	10.0	18.4	70.6	86.0	14.0	0.2
SAM 3-21	4.3	2.9	67.4	6.1	17.2	77.0	82.0	18.1	0.2

Unsaturated hydrocarbons are represented by several families of compounds: diasterenes are represented by C_{27}, C_{28}, and C_{29} R and S (Fig. 8), hop-17(21)-enes are represented by the isomers 22R and 22S ranging from C_{30} to C_{35} (Fig. 9), and dammar-13(17)-enes (20R and 20S) are present only in the oil shale samples from Timahdit. Meunier-Christmann (1988) suggested a planktonic and/or bacterial origin for dammarenes found in rocks from Timahdit. Steradienes and spirosterenes are present in samples from the oil shale series (Timahdit) and only traces are present in samples from the phosphatic series (Oulad Abdoun).

Organic sulfur compounds are represented by alkylthiophenes and isoprenoid thiophenes (Sinninghe-Damste et al., 1986), alkylbenzothiophenes and dibenzothiophenes (Perakis, 1986), thienylhopanes (Valisolalao et al., 1984), and alkylthiolanes (Schmid, 1986). These compounds are particularly abundant in samples from Timahdit and the bituminous marls from Oulad Abdoun (Meunier-Christmann, 1988; Meunier-Christmann et al., 1989; Benalioulhaj, 1989), reflecting anoxic conditions during the deposition of these sediments. In contrast, organic sulfur compounds are rather lean or absent in the phosphatic samples. This

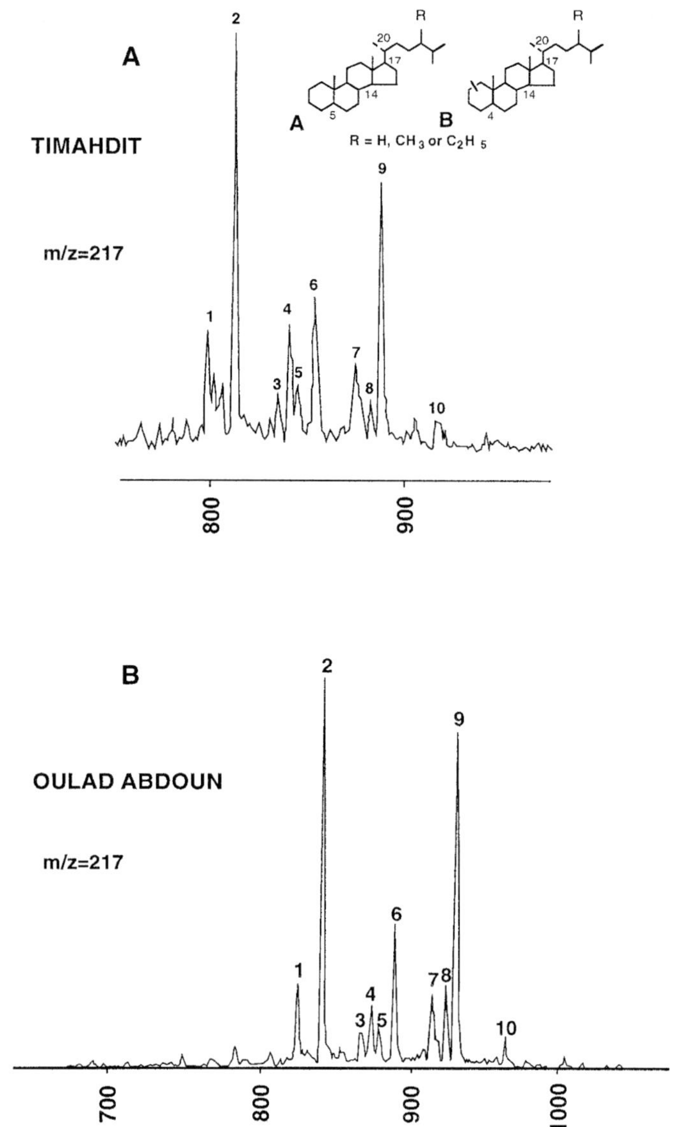

FIG. 6.—Ion chromatogram (m/z = 217) showing the distribution of the steranes in the saturated fraction of the samples 108-23 from Timahdit and S4-90 from Oulad Abdoun. 1= C_{27} 5b 14a 17a type A; 2 = C_{27} 5a 14a 17a type A; 3 = C_{28} 5a 14a 17a type A; 4 = C_{28} 5b 14a 17a type A; 5 = C_{28} type B; 6 = C_{28} 5a 17a type A; 7 = C_{29} 5b 14a 17a type A + C_{29} 5a 14b 17a type A; 8= C_{29} type B; 9 = C_{29} 5a 14a 17a type A; 10 = C_{30} type B.

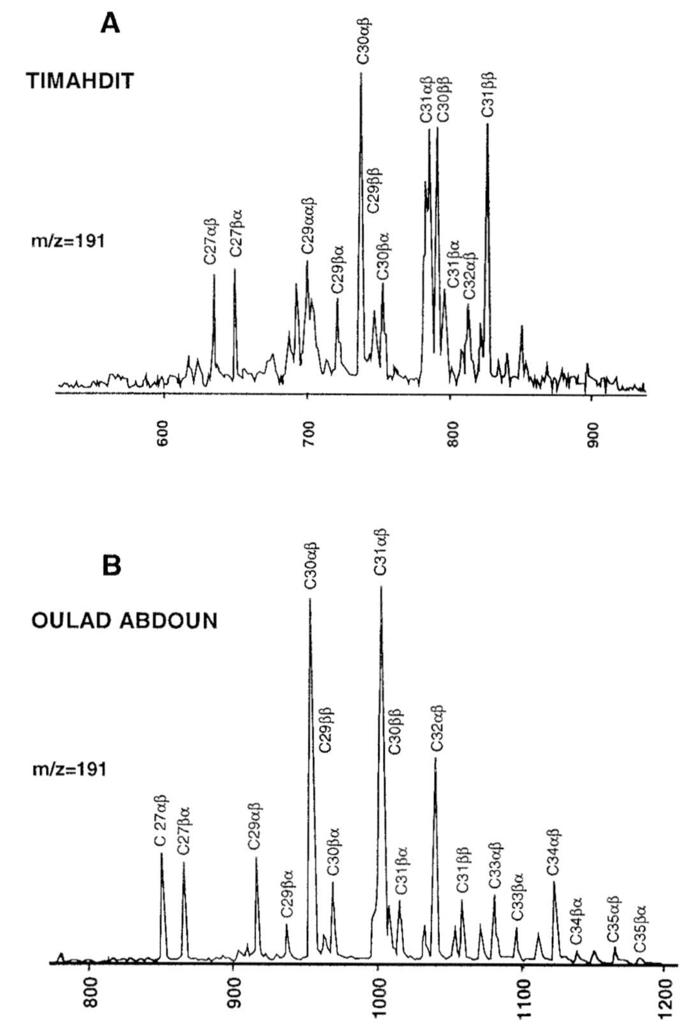

FIG. 7.—Ion chromatogram (m/z 191) showing the distribution of hopanes in the saturate fraction of the samples 108-23 from Timahdit and S4-90 from Oulad Abdoun.

is in agreement with the formation of phosphate-rich sediments in suboxic conditions (Belayouni et al., 1990).

Other aromatic hydrocarbons found in samples from both basins are: benzohopanes in C_{32}, C_{33}, C_{34}, and C_{35}, monoaromatic steroids in C_{27}-C_{29}, methylated chromans, and alkylbenzenes (Meunier-Christmann, 1988). Triaromatic steroids are present exclusively in altered samples in the upper part of the phosphatic series of Oulad Abdoun strata (Benalioulhaj, 1989). The presence of these compounds is unusual in extracts from sediments containing organic matter with a low degree of maturity. Naphthalenes and phenantrenes are present only in traces in samples from each part of the basin. They become more important in samples from the top of the section of Timahdit, where the predominant organic matter is derived from higher plants (Benalioulhaj, 1989).

In conclusion, the composition of the total lipid extract is characterized by a predominance of polar compounds, the distribution of the n-alkanes into four distinct groups, the presence of cyclic hydrocarbons (hopanes and steranes), and the presence of unsaturated compounds (dammarenes, diasterenes, and steradienes). The molecular composition of these fractions confirms the low degree of maturity of the organic matter. The presence of steranes and hopanes shows, respectively, the algal planktonic and bacterial contribution to the original biomass.

Humic Compounds and Stable Residues.—

Bulk Analysis.—Amounts of humic compounds account respectively for 3.4-60% of TOC in the Oulad Abdoun samples and 2.4%-16% of TOC in the Timahdit samples (Table 3). Organic carbon content of the fulvic, humic, and gray humic acids are reported in Table 3 for samples from both sedimentary sequences. The amount of acid-soluble compounds in the sediments of the basin range from 1.1% to 19% of TOC in Oulad Abdoun and 0.6% to 13.5% in Timahdit. Phosphorites and phosphatic rocks (definitions in Table 4) are rich in humic com-

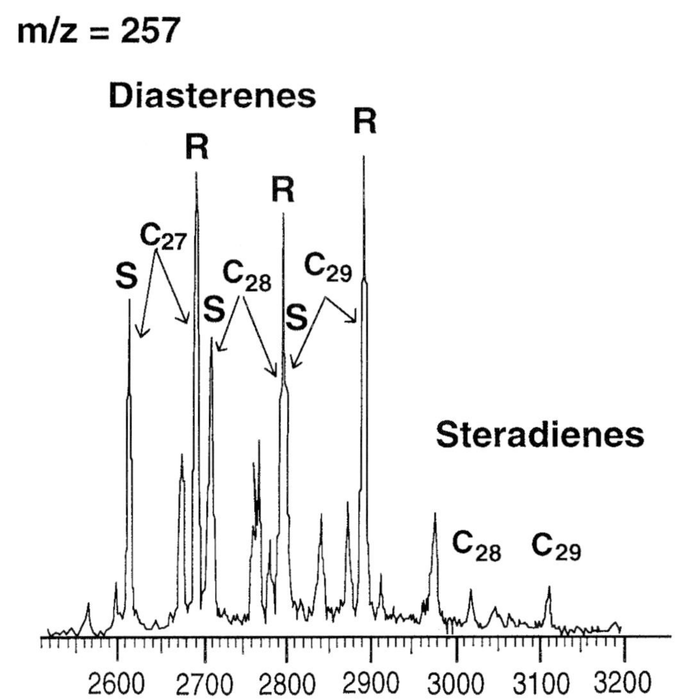

FIG. 8.—Ion chromatogram (m/z =257) showing the distribution of diasterenes and steradienes (sample 108-22 from Timahdit).

pounds (12% > humic carbon > 60% of TOC). The affinity between humic compounds and the apatitic phosphate minerals results from the entrapment and seal of the humic compounds inside the phosphatic grains. This entrapment occurs at an early stage of diagenesis (Belayouni and Trichet, 1984). Similar results were pointed out by numerous authors, for example Amit and Bein (1982), Bein and Amit (1982), Belayouni (1983), Belayouni and Trichet (1983, 1984), Sandstrom (1986, 1990), Trichet et al. (1990), and have been discussed by Nathan (1990). Positive correlation between amounts of P_2O_5 and quantities of

TABLE 3.—PROPORTIONS OF THE ORGANIC CARBON IN THE DIFFERENT HUMIC FRACTIONS AND ACID SOLUBLE COMPOUNDS GIVEN AS A PERCENTAGE OF T.O.C. OF SAMPLES FROM OULAD ABDOUN AND TIMAHDIT

Sample Number	% P_2O_5	% TOC	Acid Soluble Compounds % TOC	Humic Compounds % TOC	Fulvic Acids % TOC	Humic Acids % TOC	Gray Humic Acids % TOC
OULAD ABDOUN							
S4-68	7.5	5.9	1.6	9.4	5.7	1.7	2.1
S4-70	3.9	3.0	7.0	21.0	13.6	6.6	3.2
S4-90	4.4	14.5	6.0	3.4	1.3	0.5	1.1
S4-92	28.0	2.8	2.2	29.3	15.6	7.0	6.5
S4-94	3.5	11.9	1.1	3.8	3.3	0.7	----
S4-98	22.3	1.1	3.1	60.7	48.0	13.0	----
S4-100	4.3	7.2	3.7	12.6	8.4	4.7	----
S4-101	13.0	2.2	12.0	53.0	34.7	5.4	12.2
S4-103	1.4	10.4	2.6	5.9	3.6	0.8	1.3
S4-104	14.5	2.4	1.8	16.0	7.9	4.1	5.0
S4-107	0.8	5.5	1.4	6.9	3.2	0.4	3.1
S4-108	0.8	1.7	8.9	24.6	17.8	2.3	4.0
TIMAHDIT							
1-11	11.4	3.9	3.8	16.1	12.1	3.7	0.1
1-14.	1.3	8.5	0.9	2.6	1.9	0.7	----
1-20	0.3	2.9	8.3	6.1	5.0	1.0	----
1-27	0.8	6.2	2.8	2.9	2.0	0.5	0.3
108-20	0.7	6.0	1.2	4.1	2.2	1.2	0.6
108-23	1.0	15.6	0.6	2.4	1.8	0.3	0.2
108-27	1.0	8.5	1.7	3.2	4.1	0.4	0.2
4 bis 26	11.0	2.8	13.5	11.8	8.8	2.9	0.1
SAM 3-10	11.9	3.3	5.7	13.2	11.1	2.1	----
SAM 3-17	2.6	6.5	1.8	2.8	1.5	0.7	0.4
SAM 3-20	1.0	14.2	1.2	3.1	2.9	0.1	0.1
SAM 3-21	12.9	4.3	6.8	12.1	10.8	0.9	0.3

total humic compounds, and quantities of fulvic acids and humic acids, is illustrated in Figure 10.

An inverse correlation between TOC and the P_2O_5 contents is also evidenced in Figure 10. The relationship between TOC

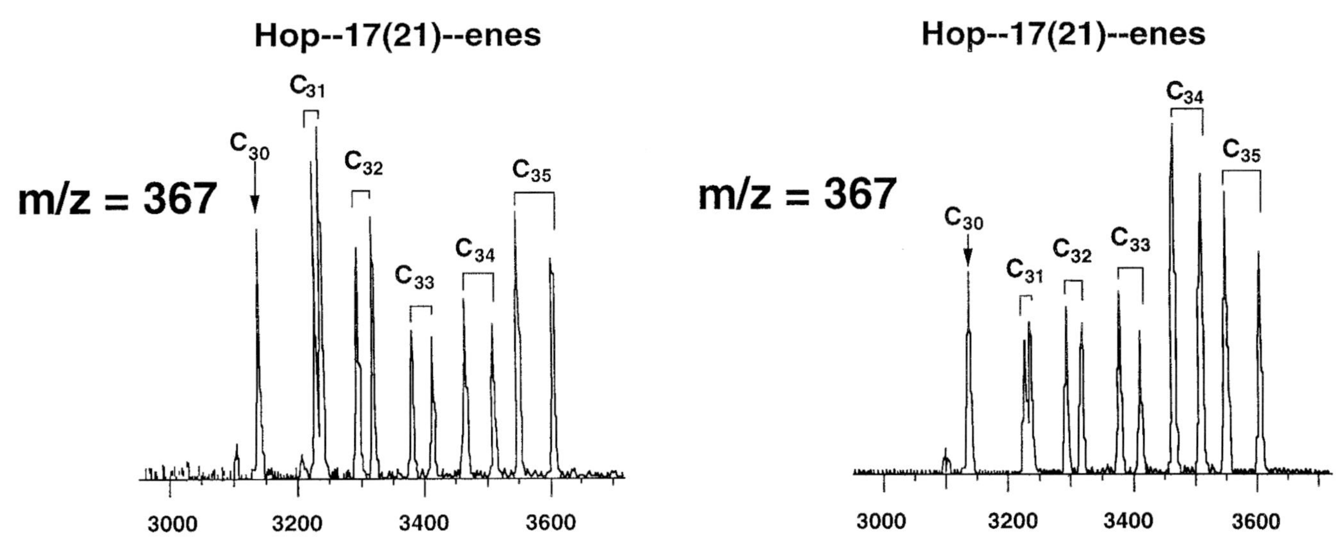

FIG. 9.—Ion chromatogram (m/z = 367) showing the distribution of Hop-17(21)enes in the saturate fraction (a) of sample 108-22 from Timahdit, and sample (b) S4-90 from Oulad Abdoun.

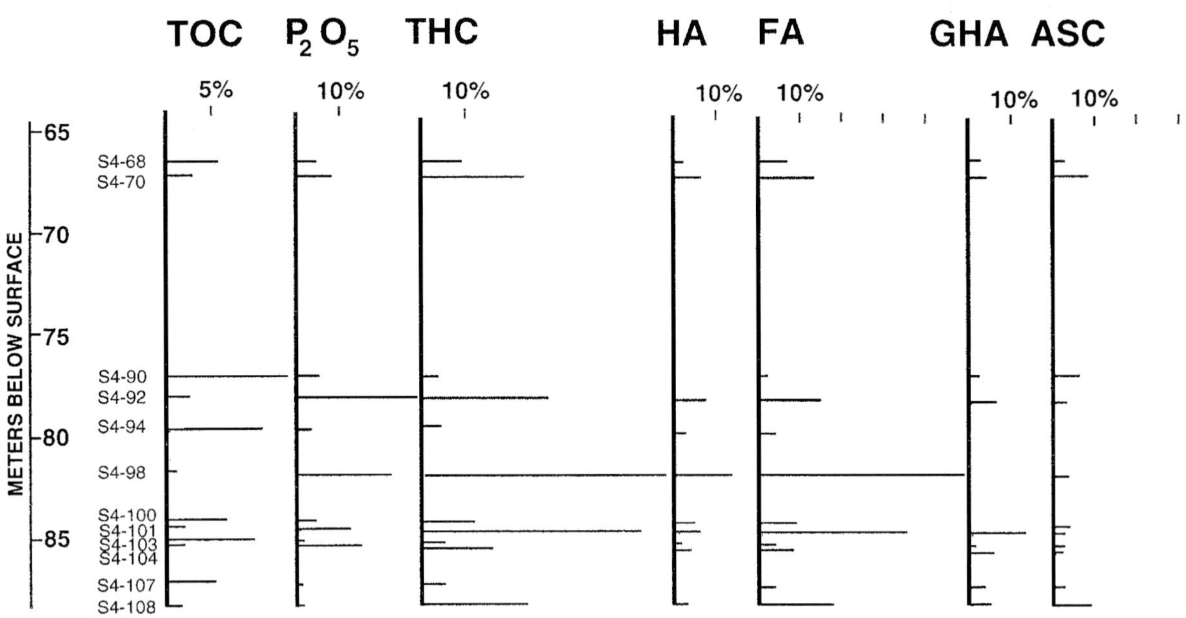

FIG. 10.—Amount of acid-soluble and humic compounds in 12 Oulad Aboun samples: total organic carbon (TOC) and P_2O_5 as a percentage of total rock, carbon in total humic compounds (THC), humic acids (HA), fulvic acids (FA), gray humic acids (GHA), and acid soluble compounds (ASA) versus TOC.

and P_2O_5 has been discussed in the Turonian phosphatic section of Tunisia by Belayouni (1983). A correlation (whether positive or negative) is rarely observed between these two constituents, due to the partition of the organic carbon of the pellets (which is positively correlated with P_2O_5 content) and the organic carbon of the matrix (which is not correlated with P_2O_5 content). In the present case, the observed correlation underlines the relationship that exists globally between the oxidation of the bulk organic matter that has been associated with the liberation of organic phosphorus.

Average amounts of organic carbon contained in the stable residues (i. e., condensed kerogen after extraction of lipids and humic compounds) of samples of different lithologies from the basin are reported in Table 4 as a percentage of organic carbon in the stable residue versus TOC. Generally, samples from Timahdit are richer in stable residues than those from Oulad Abdoun. Phosphate samples and phosphatic rocks contain lesser amounts of stable residue than oil shale samples. The relative quantities of organic carbon contained in the stable organic residue, lipid extract, and the total humic and acid-soluble compounds are shown in the ternary diagram of Figure 11. Relatively low quantities of lipid extract and humic compounds in samples from Timahdit are complementary to their high stable residue content.

Quantitative elemental analysis.—Amounts of carbon in the humic acids from the two sedimentary deposits are reported in

TABLE 4.—PROPORTIONS OF ORGANIC CARBON IN THE STABLE RESIDUES, GIVEN AS A PERCENTAGE OF T.O.C. IN THE DIFFERENT LITHOLOGIC GROUPINGS OF OULAD ABDOUN AND TIMAHDIT.

Lithology	Oulad Abdoun Subbasin	Timahdit Subbasin
Bituminous shale	81	89
Phosphatic rock 10% P_2O_5 18%	60	76
Phosphates P_2O_5 18%	40	--
Average Value for All rocks	60	82

Average values from Benalioulhaj (1989).

TABLE 5.—ELEMENTAL ANALYSIS OF HUMIC ACID OF SAMPLES FROM OULAD ABDDUN AND TIMAHDIT

Sample Number	Calculated Values					Atomic Ratios			
	C %	H %	N %	O %	S %	H/C	N/C	O/C	S/C
Oulad Abdoun									
S4-68	62.47	6.1 4	2.66	20.24	8.47	1.18	0.03	0.24	0.05
S4-70	65.08	5.79	2.35	14.90	11.41	1.08	0.03	0.17	0.07
S4-90	64.57	6.60	2.38	15.56	10.80	1.23	0.03	0.18	0.06
S4-92	60.75	6.46	2.34	10.46	9.97	1.27	0.03	0.25	0.06
S4-101	62.87	6.14	2.88	19.03	9.06	1.17	0.04	0.22	0.05
S4-103	49.27	7.02	3.13	40.57	-----	1.68	0.05	0.61	-----
S4-104	69.11	6.89	2.13	11.51	10.32	1.20	0.03	0.12	0.06
Timahdit									
1-14	62.26	6.64	4.65	19.83	6.59	1.27	0.06	0.23	0.04
1-20	59.95	6.70	3.50	23.06	6.70	1.34	0.05	0.28	0.04
108-20	63.00	6.40	4.70	19.77	5.63	1.21	0.06	0.23	0.03
108-23	62.13	6.70	4.30	20.08	6.68	1.29	0.06	0.02	0.04
108-27	62.68	6.60	4.46	19.22	7.00	1.27	0.05	0.22	0.04
SAM 3-10	62.68	6.90	3.10	16.90	7.30	1.26	0.04	0.19	0.04
SAM 3-17	63.82	6.90	4.20	19.77	5.27	1.29	0.06	0.23	0.03
SAM 3-18	69.43	7.40	2.50	11.85	8.78	1.27	0.03	0.13	0.05
SAM 3-21	66.14	7.30	2.70	16.19	7.16	1.32	0.04	0.18	0.04

(weight percentage values assuming C + H + N + S = 100%, and atomic ratios)

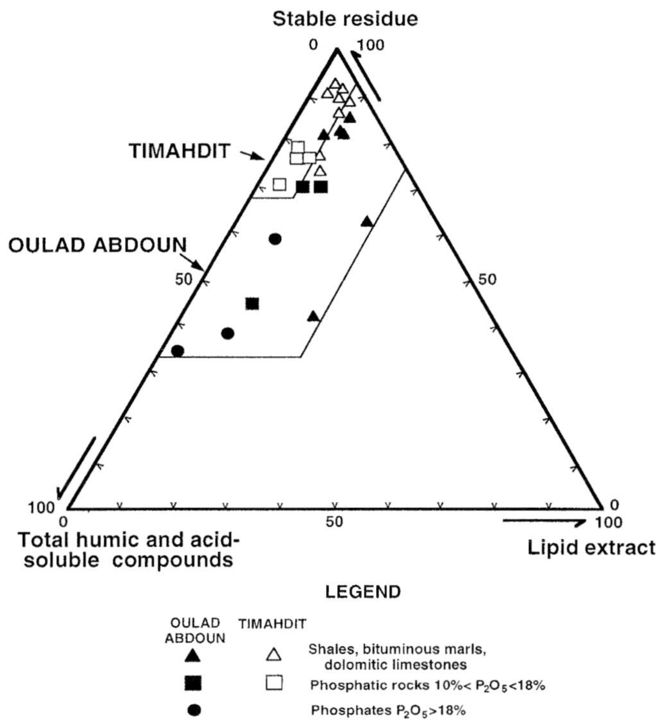

FIG. 11.—Comparative distribution of total organic carbon among the different organic fractions of samples from Timahdit and Oulad Abdoun.

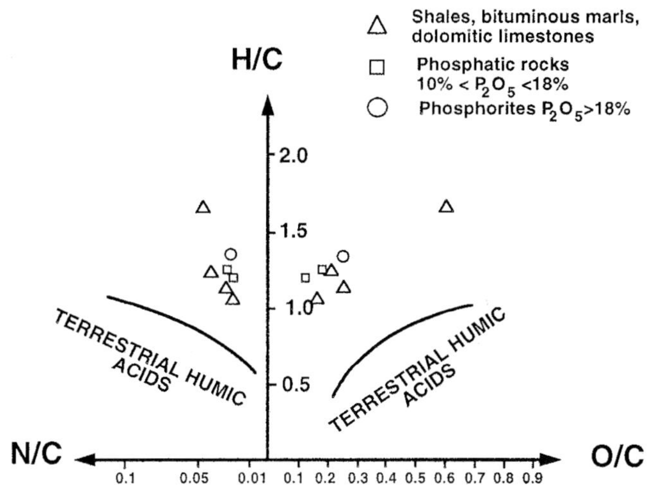

FIG. 12.—Elemental ratios in humic acids from samples taken from Oulad Abdoun. Terrestrial humic acids after Debyser and Gadel (1977).

between 6.14% and 7.40%. Nitrogen and sulfur contents are remarkably high, varying respectively from 2.13% to 3.13% and 8.47% to 11.41% in the Oulad Abdoun samples and from 2.50% to 4.70% and 5.27% to 8.78% in Timadhit. Oxygen content varies over a broad range (10 to 40%). Humic acids with a similar composition have been reported throughout the literature and are considered to derive from planktonic marine biomarkers (Van Krevelen, 1960; Huc and Durand, 1974, 1977; Tissot and Welte, 1984; Debyser et al., 1978; Sandstrom, 1982; Amit and Bein, 1982; Bein and Amit, 1982; Belayouni, 1983; Belayouni et al., 1990)

Ratios for H/C range from 1.08 to 1.68, N/C from 0.03 to 0.06, O/C from 0.02 to 0.6, and S/C from 0.03 to 0.07 for humic acids and are reported in Table 5 and Figure 12. H/C, O/C, and N/C atomic ratios are similar to those reported for humic acids that originate from a marine biomass (Brown et al., 1972; Huc, 1973; Huc and Durand, 1977; Debyser and Gadel, 1977).

The elemental analysis (C, H, O, N, S) and the atomic ratios H/C, O/C, N/C, and S/C for stable organic residues are reported in Table 5. The stable residues of phosphatic samples from the

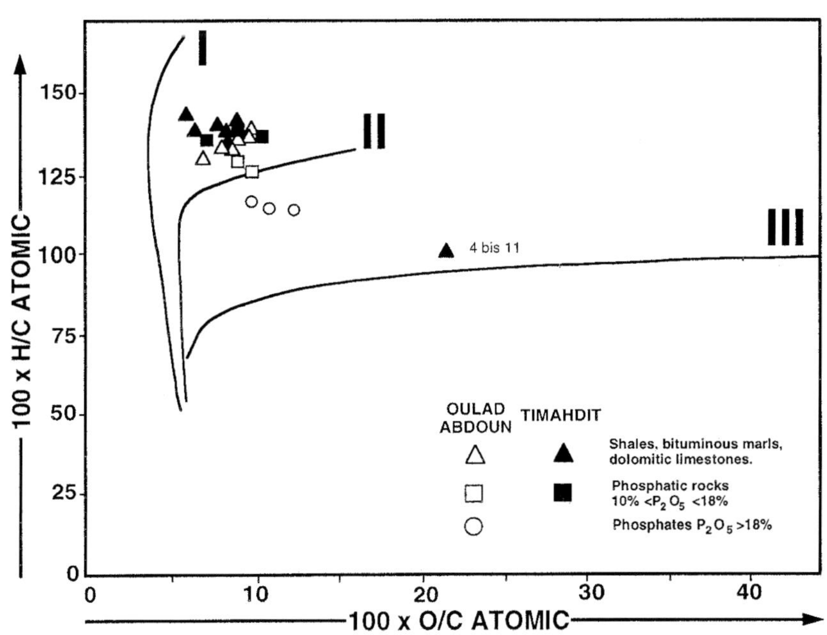

FIG. 13.—Variation of H/C versus O/C (atomic ratios) of the stable residues from samples taken in Timahdit and Oulad Abdoun.

phosphate basin have respectively lower H/C (range 1.02-1.28) and higher O/C atomic ratios (range 0.09-0.12) than stable residues from oil shale samples (H/C range 1.32-1.45, O/C range 0.06-0.11). Atomic ratios S/C are higher in stable residues of samples from the phosphatic rocks (range 0.07-0.8) than those from the oil shales (range 0.03-0.06). Stable residue of the sample "4 bis 11" (from a continental origin) is clearly distinguished by a low H/C and high O/C, N/C atomic ratios. The H/C-O/C diagram (Fig. 13), in which points representing samples from both areas fall close to the evolutionary path of Type II organic matter (i. e., derived from marine planktonic biomasses), illustrates the planktonic marine origin of the organic matter in both sedimentary successions. This suggests more oxidizing conditions (higher H/C, lower O/C) in the environment of deposition in the phosphatic part of the basin, particularly during the deposition and diagenesis of the rich phosphatic strata, than that prevailing in the oil shale part of this basin (Benalioulhaj and Trichet, 1990). The position of "4 bis 11" illustrates the continental origin of the organic matter in this sample, which was derived from higher plants.

Humic acids and stable residues have a remarkably high sulfur content (Tables 5 and 6). In each part of the basin the stable residues are richer in sulfur than the corresponding humic acids. Moreover the humic acids and the stable residues extracted from samples representative of different lithologies in the Oulad Abdoun (phosphate) samples have a higher S/C ratio than the

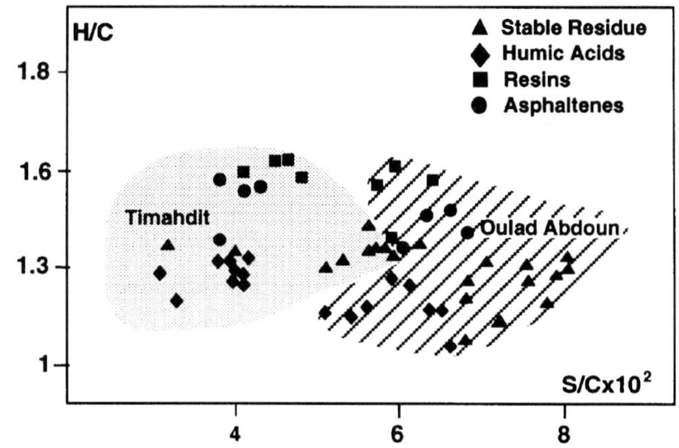

FIG. 14.—Correlation between H/C and S/C (atomic ratios) of the stable residues, humic acids, resins, and asphaltenes extracted from samples taken in Timahdit and Oulad Abdoun.

TABLE 6.—ELEMENTAL ANALYSIS OF STABLE ORGANIC RESIDUES OF SAMPLES FROM OULAD ABDOUN AND TIMAHDIT

Sample Number	% TOC	% P_2O_5	Calculated Values					Atomic ratios			
			C %	H %	N %	O %	S %	H/C	O/C	N/C	S/C
Oulad Abdoun											
S4-56	3.2	25.2	65.43	5.56	10.46	2.94	11.82	1.02	0.12	0.04	0.07
S4-68	5.9	7.5	67.94	7.34	8.57	2.49	13.64	1.29	0.09	0.03	0.08
S4-72	3.0	18.0	68.34	6.36	8.74	2.24	12.90	1.12	0.10	0.03	0.07
S4-90	14.5	4.4	67.91	7.55	7.62	2.25	14.65	1.33	0.08	0.03	0.07
S4-94	11.9	3.5	67.95	7.70	7.48	2.33	14.51	1.36	0.08	0.03	0.08
S4-98	1.1	22.3	67.44	5.95	8.99	2.12	13.10	1.06	0.10	0.03	0.07
S4-100	7.2	4.3	69.27	7.53	6.28	2.24	14.58	1.30	0.07	0.03	0.08
S4-101	2.2	13.0	69.35	7.26	8.98	1.91	12.51	1.25	0.10	0.02	0.07
S4-104	2.4	14.5	68.52	7.33	8.07	2.30	13.76	1.28	0.09	0.02	0.08
S4-107	5.5	0.8	67.16	7.76	9.05	1.75	12.51	1.38	0.10	0.02	0.07
S4-108	1.7	0.8	67.74	7.43	8.66	1.93	13.45	1.31	0.10	0.02	0.07
Timahdit											
4 bis 11	1.4	0.3	70.07	5.89	19.12	3.09	1.87	1.00	0.20	0.04	0.01
4 bis 25	1.2	0.5	70.78	8.31	7.20	2.19	11.49	1.41	0.08	0.08	0.06
4 bis 26	2.8	11.0	69.06	7.96	12.29	4.80	5.86	1.38	0.10	0.06	0.03
4 bis 27	11.7	1.0	73.03	8.02	6.69	2.10	9.91	1.32	0.07	0.02	0.05
108-16	4.9	7.1	69.46	7.75	10.28	2.54	9.94	1.34	0.11	0.03	0.05
108-23	15.5	1.0	71.09	8.04	7.14	2.48	11.22	1.36	0.08	0.03	0.06
1-11	3.9	11.4	74.18	8.42	7.42	1.96	8.00	1.36	0.08	0.02	0.04
1-13	0.9	0.4	73.40	8.66	5.37	1.86	11.07	1.45	0.06	0.02	0.06
SAM 3-10	3.3	11.9	70.80	8.15	7.82	2.14	10.98	1.38	0.08	0.03	0.06
SAM 3-12	2.5	13.2	70.86	8.08	8.06	2.12	10.86	1.37	0.09	0.03	0.06
SAM 3-20	14.2	1.0	70.35	8.16	7.60	2.12	11.57	1.39	0.06	0.03	0.06

[weight percentage values assuming C + H + N + S = 100%, and atomic ratios]

TABLE 7.—ELEMENTAL ANALYSIS OF RESINS AND ASPHALTENE OF SAMPLES FROM OULAD ABDOUN AND TIMAHDIT

Sample Number	Resins									Asphaltenes								
	Calculated Values					Atomic ratio				Calculated Values					Atomic ratio			
	C %	H %	N %	O %	S %	H/C	N/C	O/C	S/C	C %	H %	N %	O %	S %	H/C	N/C	O/C	S/C
Oulad Abdoun																		
S4-68	71.07	9.47	0.99	6.53	12.03	1.59	0.01	0.07	0.06	68.95	8.27	1.02	9.18	12.58	1.43	0.01	0.10	0.07
S4-90	70.17	9.48	0.86	8.51	10.96	1.62	0.01	0.09	0.06	68.18	8.49	0.98	10.27	12.07	1.49	0.01	0.11	0.07
S4-92	70.98	8.35	0.87	8.65	11.15	1.41	0.01	0.09	0.06	68.11	7.83	1.56	11.71	10.78	1.38	0.02	0.13	0.06
S4-94	69.47	9.11	0.80	9.91	10.68	1.57	0.01	0.10	0.06	65.07	8.12	0.88	14.61	10.92	1.49	0.12	0.17	0.06
Timahdit																		
1-27	72.87	9.76	0.96	8.07	8.14	1.60	0.01	0.08	0.04	69.94	8.51	1.63	12.19	7.17	1.40	0.02	0.13	0.04
4 bis 26	71.60	9.42	0.94	8.75	9.27	1.58	0.01	0.09	0.05	67.50	8.71	1.39	12.92	7.49	1.55	0.02	0.14	0.04
SAM 3-17	73.57	9.98	0.94	6.62	8.88	1.63	0.01	0.07	0.05	68.22	8.87	1.41	13.64	7.85	1.56	0.02	0.16	0.04
SAM 3-21	71.32	9.68	0.84	9.05	8.84	1.63	0.01	0.10	0.05	68.69	9.06	1.16	14.01	7.02	1.58	0.02	0.15	0.04

(weight percentage values assuming C + H + N + S = 100%, and atomic ratios)

Abdoun (phosphate) samples have a higher S/C ratio than the corresponding ones in Timahdit (oil shale) (Fig. 14). The high sulfur content in the humic compounds and the stable residues is interpreted to be the result of the reaction of reduced or elemental sulfur with the organic matter contained in these two fractions (termed vulcanization, Lallier-Vergès et al., 1997). Similar data (S/C versus H/C) for asphaltenes and resins in the same samples (Table 7) is included in Figure 14. S/C values are higher in Oulad Abdoun samples than in those from Timahdit. The reaction of reduced or elemental sulfur with the organic matter within all the fractions (polar, humic, and stable residue) of the organics is similar throughout the basin. Reduced sulfur species are considered to result from sulfate reducing processes.

INFRARED SPECTROSCOPY

Results of a detailed study of the humic acids and stable organic residues, using Fourier transform infrared spectroscopy, have been published by Benalioulhaj and Trichet (1990). Here we present only the conclusions of this work that are useful in discussion and interpretation. Benalioulhaj and Trichet (1990) showed from both qualitative and quantitative analyses that both humic and stable residue fractions consist mainly of functional groups (OH, COOH, COO-, C=O), and that their precursors were of a marine origin. The ratio (C=C + C=O + COO^-) /CH, called the oxidation index, is representative of the degree of oxidation of the stable residue.

The correlation of this parameter with the amount of P_2O_5 (Fig. 15) and data concerning the sedimentology and petrography of the studied strata identifies two stages of oxidation of the organic matter in the basin; the first is an early diagenetic oxidation stage during which P_2O_5 and humic compounds accumulate in the pellets. The oxidation index is higher in Oulad Abdoun than in Timahdit. This factor separates the phosphatogenic region from the confined region where shales accumulated. Next is a secondary oxidation stage that was the product of early sediment reworking. For those strata in which the amounts of P_2O_5 exceed 10%, the effect of such oxidation is quite clear in the composition of the organic matter itself. Interpretation of simultaneous increases in the oxidation index of the stable residue and in the P_2O_5 content of the sediment shows that reworking processes lead simultaneously to a progressive oxidation of the organic matter and to a mechanical concentration of phosphatic pellets through removal of the original matrix that laid between pellets.

ROCK-EVAL PYROLYSIS

Rock-Eval pyrolysis data for organic stable residues of samples from the Oulad Abdoun and Timahdit basins are presented

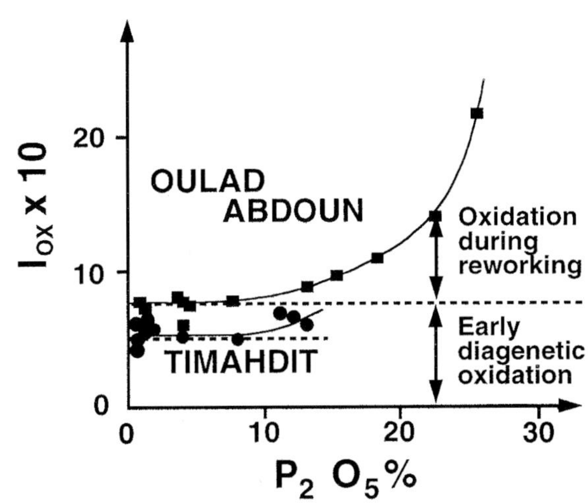

FIG. 15.—Cross-plot showing correlation between percentage P_2O_5 and the degree of oxidation of organic matter (Iox) in the stable residues of Timahdit and Oulad Abdoun. After Benalioulhaj and Trichet (1990).

TABLE 8.—ROCK-EVAL ANALYSIS OF STABLE ORGANIC RESIDUES OF SAMPLES FROM OULAD ABDOUN AND TIMAHDIT

Basin	Sample Number	Tmax °C	S1 mg/g of rock	S2 mg/g of rock	S3 mg/g of rock	T.O.C. %	Petroleum Potential mg/g of TOC (S1 + S2)	H I mg/g of TOC	OI mg/g of TOC
Oulad Abdoun	S4-56 P	404	0.2	230.6	27.7	52.1	443	442	53
	S4-70	411	0.2	311.2	18.9	48.0	648	648	32
	S4-90	405	0.1	389.2	11.6	52.2	745	745	22
	S4-92 P	407	0.3	327.4	18.1	56.0	585	584	32
	S4-94	498	0.1	327.4	13.0	49.6	660	660	26
	S4-98 P	411	0.2	264.3	22.0	52.4	504	504	41
	S4-100	402	0.2	380.6	16.8	56.6	673	672	29
	S4-101	426	0.0	30.0	13.7	50.0	600	600	27
	S4-104	404	1.1	336.2	12.9	54.4	620	618	23
	S4-107	405	1.2	297.8	12.2	45.2	661	658	26
	S4-108	407	1.0	311.8	13.0	50.7	616	614	25
Timahdit	4 bis 11	415	0.2	69.1	18.9	35.0	198	197	54
	4 bis 25	417	1.7	409.4	12.4	54.8	750	747	22
	4 bis 26	417	0.8	308.2	11.0	45.1	685	683	24
	4 bis 27	413	0.2	357.8	15.8	54.9	652	651	28
	108-23	416	0.2	395.8	12.3	53.7	737	737	22
	1-10	415	29.5	455.0	13.2	64.4	752	706	20
	1-11	417	0.1	424.0	12.9	55.0	770	770	23
	SAM 1-13	421	2.9	456.8	10.0	57.0	806	801	17
	SAM 3-17	409	0.1	302.9	17.3	49.9	607	607	35
	SAM 3-20	412	0.0	272.8	13.6	41.5	657	657	32

[S_1= free hydrocarbons and tars; S_2= hydrocarbons and tars produced during hydrolysis; S_3= CO_2 produced during pyrolysis; H_1= Hydrogen Index = S_2/TOC; O_1 = Oxygen Index = S_3/TOC; T_{max}= temperature at the summit of S_2 peak. From Espitalié et al. (1986). P = Phosphate samples]

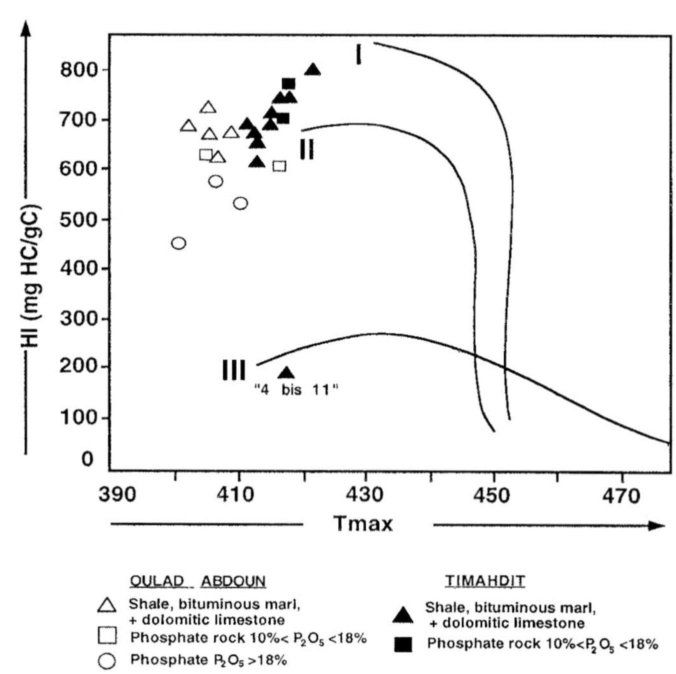

FIG. 16.—Variation in HI (Hydrogen Index) and Tmax in the stable residues of samples from Timahdit and Oulad Abdoun.

in Table 8 and Figure 16. Hydrogen and oxygen indices vary from 440 mg/g to 700 mg/g of TOC and from 22 mg/g to 53 mg/g of TOC, respectively, in the Oulad Abdoun and from 610 mg/g to 800 mg/g of TOC and from 17 to 54 mg CO2/g of organic content, respectively, in the Timahdit. Sample "4 bis 11" (containing organic matter from a continental origin) differs from the other samples by having a low hydrogen index of 197 mg HC/g of TOC and an oxygen index of 54 mg CO_2/g of TOC. The average T_{max} is 410°C in Oulad Abdoun and 415°C in Timahdit, indicating a low thermal maturity in both sedimentary sections (Espitalié et al., 1986; Tissot and Welte, 1984).

Generally, hydrogen indices (HI) are higher and oxygen indices (OI) are lower in Timahdit than in Oulad Abdoun, where phosphate samples are characterized by a relatively lower HI and higher OI, compared to nonphosphatic shaley samples (Table 8). These results confirm those given by elemental analysis and infrared spectroscopy of samples and show up as indicators of oxidation in stable residues from phosphate samples. All these indicators show that the stable residues of phosphatic beds witnessed more intensive oxidation than those of the shale beds, everything being the same for both phosphate and shale strata (depth, temperature of burial, and origin of organic matter).

The position of sample "4 bis 11" in the HI-OI diagram (Fig. 16) confirms the continental origin of its organic matter, as already demonstrated by elemental analysis, in contrast to all the other samples that contain an organic matter derived from marine planktonic biomasses.

DISCUSSION AND CONCLUSION

Though the two basins, Oulad Abdoun and Timahdit, are today clearly separated, their evolution must be considered within the framework of the unique basin in which they formed, which indeed is their only really significant geological aspect. The upwelling current that was active at the western edge of that basin supported the intense productivity that invaded the whole basin and led to the deposition of abundant organic matter. As shown in this paper, the OM had the same composition in both Timahdit and the Oulad Abdoun area. The bulk chemical composition, and the particular presence of hopanes and steranes in the sediments, provide evidence of the planktonic and bacterial contribution to the primary biomass. Some of the organic matter has been shown to originate from the continent, but it represents a minor portion of the bulk of the basins' organic reservoir.

Superimposed on this qualitatively and quantitatively homogeneous marine organic supply, there is a mineral supply, that is mainly clayey and continental. This continental supply was associated with tectonic activity in the Middle Atlas that caused subsidence toward the east, resulting in strain, which decreased toward the west and was totally absent on the stable Hercynian resistant area (Fig. 17). This differential subsidence led to development of increased accommodation space (Lafont, 1994; Homewood, 1996) that varied in different parts of the basin, and which was regularly compensated by sedimentation of clays. Thus, during the Lower and Middle Maastrichtian, tectonic deformation had no or little influence on the water depth and, consequently, on the bioproductivity. Therefore, the only change caused by subsidence seems to be the supply of clay and its dispersal in the basin. This is underlined by the similarity of the argillaceous material over the whole basin, in both Timahdit and Oulad Abdoun. The petrography and chemical composition of the apatite remain invariable, whether from the phosphorites of the condensed section on the stable platform or from those phosphorites in the thicker sections of the nearby subsident zone. In other words, black shales can be considered as the result of diagenesis of mixture made of biogenic material and clays, while phosphorites and associated carbonates would be the result of the same diagenetic transformation of source beds made of the same biogenic material, but without the clay components.

This relationship is shown in Figure 4, which shows the distribution of the various facies in a TOC versus P_2O_5 diagram. (1) The carbonates do not owe anything to O.M. Their accumulation was due to the direct deposition of bioclasts. They contain only a few percent of C_{org} and P_2O_5 (Lucas and Lucas-Prévôt, this volume). In Timahdit, calcite mixed with clays, to form marls, but in Oulad Abdoun, calcite was transformed into dolomite during early diagenesis (Belfkira, 1980; Prévôt, 1990). (2) As the clay content increased, the rocks become richer and richer in residual C_{org}, until they become true bituminous marls and black shales. Their P_2O_5 probably reflects the P_2O_5 content of the primary O.M. (3) The decrease in the amount of clay material settling together with O.M. parallels the relative increase in the amount of residual P_2O_5 and the rocks become more and more enriched in apatite. In Timahdit, where clays were abundant, the rocks are low in phosphate. In Oulad Abdoun, where clays were scarce, there are high-grade phosphorites.

The supply of clays to the O.M.-rich biogenic sediment is highly significant, because it completely modified the diagenetic evolution of the sediment. In all cases, the degree of ther-

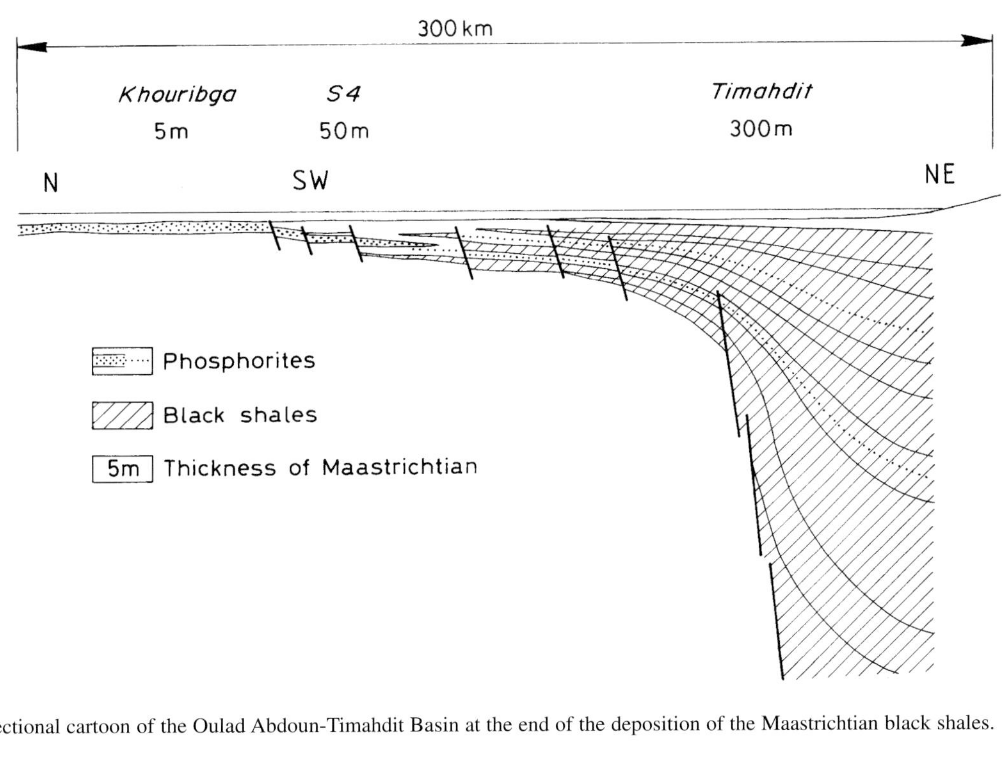

FIG. 17.—Cross-sectional cartoon of the Oulad Abdoun-Timahdit Basin at the end of the deposition of the Maastrichtian black shales.

mal maturity of the O.M. remains low, as shown by the dominance of polar compounds, the distribution of the n-alkanes and cyclic hydrocarbons, the presence of unsaturated compounds, and the Rock-Eval data. The differences observed between Oulad Abdoun and Timahdit samples by Fourier Transform Infrared Spectroscopy, elemental, and Rock-Eval analyses of stable residues show that the O.M. from Oulad Abdoun is more oxidized than that from Timahdit; the more oxidized is associated with the higher apatite content.

Where the supply of clay was low, the sediment was mainly composed of O.M. with a low carbonate content. The chemical environment within this sediment was suboxic, and a slow consumption of O.M. by bacteria led to mineralization of phosphorus and the formation of pellets of apatite containing humic compounds and kerogen within the matrix of the pellets. In the shallow environment, winnowing was common and the pellets were mechanically concentrated, while a part of the O.M. and clays were removed.

When there were clays in the sediment, water exchange between the sediment and the overlying seawater was slowed and the sedimentary environment became more reducing. Decay of O.M. was slower. Moreover, the very fine, sheet-like crystals of clays are believed to play a protective role by insulating O.M. Due to the aligned nature of the clays in the sediment, the pores were so small that bacteria had little possibility of access and thus could not "allow functioning of the hydrolytic enzymes necessary for O.M. decay" (Mayer, 1994). This allowed efficient polymerization and polycondensation of organic molecules (Nissenbaum and Kaplan, 1972; Huc, 1973; Tissot and Welte, 1978; Nissenbaum, 1979). Under these conditions, phosphorus remained scattered and diluted in the sediment in its organic state, being only weakly available for mineralization.

It is interesting to compare the total P_2O_5 content of the bituminous shales of Timahdit with the the total P_2O_5 content of the phosphorite deposits of Oulad Abdoun. After compaction, black shales with a thickness of 200 m were preserved in Timahdit from the primary Maastrichtian source bed. Assuming a density of 2, and an average P_2O_5 content of 3%, approximately 12 t m^{-2} of P_2O_5 was preserved in these black shales. In Oulad Abdoun, the Maastrichtian is represented by 5 m of phosphorites, containing an average P_2O_5 of 30%. Assuming again a density of 2, this yields only 3 t m^{-2} of P_2O_5 Paradoxically, whereas phosphorites, which are high-grade P_2O_5 rocks, are relatively enriched in P_2O_5, the phosphate deposits contain a lower total P_2O_5 content than the black shale. Finally it is necessary to underline that phosphorites are not, among marine sediments, the major stock of phosphorus extracted from seawater.

During early diagenesis, the preservation of O.M. leads to preservation of phosphorus, while the destruction of O.M. leads to the solubilization of phosphorus. Usually, the solubilized phosphorus is then either recycled or mineralized into apatite but dispersed. In some special cases, of which Oulad Abdoun is an example, conditions are met to start a natural "apatite factory" (Lucas and Prèvôt-Lucas, this volume) in which the phosphorus provided by the O.M. is concentrated, but to continue the comparison with a factory, the "output" is low.

ACKNOWLEDGMENTS

We wish to acknowledge the support given to this project by the Commissariat à l'Energie Atomique, France, and the grant given to the Research Foundation of New York by the U.S. Department of Energy Grant #DE-FG02-88ER13961 that has permitted the completion of this study. We also acknowledge the two reviewers, W. E. Dean and Y. Nathan, for their pertinent remarks

REFERENCES

AMIT, O., AND BEIN, A., 1982, Organic matter in Senonian phosphorites from Israel. Origin and diagenesis: Chemical Geology, v. 37, p. 277-287.

BARANGER, P., 1989, Origine et diagenèse des constituants organiques associés aux niveaux salifères tertiaires de Bresse (France). Etude expérimentale de la réduction d'espèces métalliques par la matière organique sédimentaire: Unpublished Ph.D. Dissertation, University of Orléans, France, 325 p.

BATURIN, G. N., KOCHENOV, A. V., AND PETELIN, V. P., 1970, Phosphorite formation on the shelf of South West Africa: Lithology and Mineral Resources, v. 3, p. 266-276.

BEIN, A., AND AMIT, O., 1982, Depositional environments of the Senonian chert, phosphorites and oil shale sequence in Israel as deduced from their organic matter composition: Sedimentology, v. 29, p. 81-90.

BELAYOUNI, H, 1983, Etude de la matière organique dans la série phosphatée du bassin phosphaté de Gafsa-Metlaoui (Tunisie)—Application á la compréhension des mécanismes de la phosphatogenèse: Unpublished Ph.D. Dissertation, University of Orléans, France, 259 p.

BELAYOUNI, H., AND TRICHET, J., 1983, Preliminary data on the origin and diagenesis of the organic matter in the phosphate basin of Gafsa (Tunisia), *in* Bjorøy, M., et al., eds., Advances in Organic Geochemistry 1981: Chichester, Wiley, p. 328-335.

BELAYOUNI, H., AND TRICHET, J., 1984, Hydrocarbons in phosphatized and non phosphatized sediments from the phosphate basin of Gafsa (Tunisia): Organic Geochemistry, v. 6, p. 741-754.

BELAYOUNI, H., SLANSKY, M., AND TRICHET, J., 1990, A study of the organic matter in Tunisian phosphate series: Relevance to phosphorite genesis in the Gafsa basin (Tunisia): Organic Geochemistry, v. 15, p. 47-52.

BELFKIRA, M., 1980, Evolution sédimentologique et géochimique de la série phosphatée du Maastrichtien des Oulad Abdoun (Maroc): Unpublished Ph.D. Dissertation, University of Grenoble, France, 164 p.

BENALIOULHAJ, N., 1991, Les formations à phosphates et à schistes bitumineux du bassin des Oulad Abdoun et du bassin de Timahdit (Maroc): Pétrographie, minéralogie, géochimie et environment de dépôt: Unpublished Ph.D. Dissertation, University Louis Pasteur, Strasbourg, France, 238 p.

BENALIOULHAJ, S., 1989, Géochimie organique de la série phosphatée des Oulad Abdoun et de la série bitumineuse du bassin de Timahdit (Crétacé supérieur-Eocène, Maroc): Unpublished Ph.D. Dissertation, University of Orléans, France, 242 p.

BENALIOULHAJ, S., AND TRICHET, J., 1990, Comparative study by infrared spectroscopy of the organic matter of phosphate-rich (Oulad Abdoun basin) and black shale (Timahdit basin) series (Morocco): Organic Geochemistry, v. 16, p. 649-660.

BOON, J. J., RIJPSTRA, W. I. C., DELANGE, F., DE LEEUW, J. W., YOSHILKA, M., AND SHIMIZU, Y., 1979, Black sea sterol—a molecular fossil for dinoflagellate blooms: Nature, v. 277, p. 125-127.

BOSTICK, N. H., AND ALPERN, B., 1977, Principles of sampling, preparation and constituent selection for microphotometry in measurement of maturation of sedimentary organic matter: Journal of Microscopy, v. 109, p. 411-47.

BOUJO, A., 1976, Contribution à l'étude géologique du gisement de phosphate Cretacé-Eocène des Ganntours (Maroc occidental): Sciences Geologiques, Mem. No. 43, Université de Strasbourg, 227 p.

BOULOUBASSI, I., AND SALIOT, A., 1991, Study of the partioning of hydrocarbons into particulate and dissolved phases in the Rhone delta; consequences on estimations of hydrocarbon fluxes: Water Pollution Research Report, v. 20, p. 231-237.

BOULOUBASSI, I., SALIOT, A., ROUGERIE, F., AND TRICHET, J., 1992, Hydrocarbon geochemistry in coral reef pore waters, French Polynesia, *in* Proceedings 7th International Symposium on Water-Rock Interaction, Park City: Rotterdam, Balkema, p. 271-274.

BROWN, F. S., BAEDECKER, M. J., NISSENBAUM, A., AND KAPLAN, I. R., 1972, Early diagenesis in a reducing jord, Saanich Inlet, British Columbia, III: Changes in organic constituents of sediment: Geochimica et Cosmochimica Acta, v. 36, p. 1185 -1202.

CHOUBERT, G., AND FAURE-MURET, A., 1962, Evolution du Domaine Atlasique marocain depuis les temps paléozoïques. Memoire hors série, Société Géologique de France (Livre mémoire P. Fallot), v. 1, p. 447-527.

DEBYSER, Y., AND GADEL, F., 1977, Etude géochimique des composés humiques et des kérogènes, *in* Géochimie Organique des Sédiments Marins Profonds, ORGON I. Mer de Norvège: Paris, Centre National de la Recherche Scientifique, p. 247-268.

DEBYSER, Y., GADEL, F., LEBLOND, C., AND MARTINEZ, M. J., 1978, Etude des composés humiques, des kérogènes et de la fraction hydrolysable dans les sediments, *in* Géochimie organique des sédiments marins profonds: ORGON II: Paris, Centre National de la Recherche Scientifique, p. 339-356.

ESPITALIÉ, J., DEROO, G., AND MARQUIS, F., 1986, La pyrolyse Rock-Eval et ses applications: Revue de l' Institut Français du Pétrole, v. 12, p. 73-89.

HERBIG, H. G., AND TRAPPE, J., 1994, Stratigraphy of the Subatlas group (Maastrichtian-Middle Eocène, Morocco): Newsletters on Stratigraphy, Berlin, v. 30, p. 125-165.

HOMEWOOD, P. W., 1996, The carbonate feed back system: Interaction between stratigraphic accomodation, ecological succession and the carbonate factory:Bulletin de la Société Géologique de France, v. 167, p. 701-715.

HUC, A. Y., 1973, Contribution à l'étude de l'humus marin et de ses relations avec les kérogènes: Unpublished Ph.D. Dissertation, University of Nancy, France, 59 p.

HUC, A. Y., AND DURAND, B., 1974, Etude des acides humiques et de l'humine de sédiments récents considérés comme précurseurs des Kérogènes, *in* Tissot, B., and Bienner, F., eds., Advances in Organic Geochemistry 1973: Paris, Technip ed., p. 53-72

HUC, A. Y., AND DURAND, B., 1977, Occurrence and significance of humic acids in ancient sediments: Fuel. v. 56, p. 73-80.

HUC, A. Y., ROUCACHE, J., BERNON, M., CAILLET, G., AND DA SILVA, M., 1976, Application de la chromatographie sur couche mince à l'étude quantitative et qualitative des extraits des roches et des huiles: Revue de l'Institut Français du Pétrole, v. 31, p. 67-98.

JOHNSON, C. M., AND CALDER, J. A., 1973, Early diagenesis of fatty acids and hydrocarbons in a salt marsh environment: Geochimica et Cosmochimica Acta, v. 37, p. 1943-1955.

KRAJEWSKI, K. P., 1989, Organic geochemistry of a phosphate to black shale transgressive succession: Wilhelmöya and Janusfjellet formations (Rhaetian-Jurassic) in central Spitzbergen, Arctic Ocean: Chemical Geology, v. 74, p. 249-263.

LAFONT, F., 1994, Influence relative de la subsidence et de l'eustatisme sur la localisation et la géométrie des réservoirs d'un système deltaïque: Mémoire Géosciences, Université de Rennes, 264 p

LALLIER-VERGÈS, E., HAYES, J. M., BOUSSAFIR, M., ZABACK, D. A., TRIBOVILLARD, N. P., CONNAN, J., AND BERTRAND, P., 1997, Productivity-induced sulphur enrichment of hydrocarbon-rich sediments from the Kimmeridge Clay Formation: Chemical Geology, v. 134, p. 277-288.

MAYER, L. M., 1994, Surface area control of organic carbon accumulation in continental shelf sediments: Geochimica et Cosmochimica Acta, v. 58, p. 1271-1284.

MEINSCHEIN, W. G , 1969, Hydrocarbons-saturated, unsaturated and aromatic, *in* Eglinton, G., and Murphy, M. T. J., eds., Organic Geochemistry: New York, Springer-Verlag. p. 330-355.

MEUNIER-CHRISTMANN, C., 1988, Géochimie organique de phosphates et schistes bitumineux marocains: étude du processus de phosphatogenèse: Unpublished Ph.D. Dissertation, University Louis Pasteur, Strasbourg, France, 125 p.

MEUNIER-CHRISTMANN, C., LUCAS, J., AND ALBRECHT, P., 1989, Organic geochemistry of Moroccan phosphates and bituminous shales. A contribution to the problem of phosphatogenesis: Sciences Géologiques Bulletin, Université de Strasbourg, v. 42, p. 205-222.

NATHAN, Y., 1990, Humic substances in phosphorites: occurrences, characterization and significance, *in* Notholt, A. J. G., and Jarvis I., eds., Phosphorites Research and Developement: London, Geological Society Special Publication 52, p. 49-58.

NISSENBAUM, A., 1979, Phosphorus in marine and non marine humic substances: Geochimica et Cosmochimica Acta, v. 43, p. 1973-1978.

NISSENBAUM, A., AND KAPLAN, I. R., 1972, Chemical and isotopic evidence for the in situ origin of marine humic substances: Limnology and Oceanography, v. 17, P. 570-582.

PARKER, P. L., VAN BALLEN, C., AND MAURER, L., 1967, Fatty acids in eleven species of blue-green algae: Science. v. 155, p. 707-708.

PERAKIS, N., 1986, Séparation et détection sélective de composés soufrés dans les fractions lourdes des pétroles. Géochimie des benzo[b]thiophènes: Unpublished Ph.D. Dissertation. University Louis Pasteur, Strasbourg, France, 192 p.

POWELL, T. G., COOK, P. J., AND MCKIRDY, D. M., 1975, Organic geochemistry of phosphorites. Relevance to petroleum genesis: American Association of Petroleum Geology Bulletin, v. 59, p. 618-663.

PRÉVÔT, L., 1990, Geochemistry, petrography, genesis of Cretaceous-Eocene phosphorites. The Ganntour deposit (Morocco), a type example: Société Géologique de France, Paris, Mémoires, v. 158, 232 p.

RACHIDI, M., 1983, Contribution à la connaissance de la matière organique associée à certaines series phosphatées: Unpublished Ph.D. Dissertation, University of Orléans, France, 185 p.

RAHHALI, I., 1970, Foraminifères bentoniques et pélagiques du Crétacé supérieur du synclinal d'Elkoubbat (Moyen-Atlas, Maroc): Notes et Mémoires du Service Géologique du Maroc, v. 30, p. 51-86.

RAUSCHER, R., SCHULER, M., AND BENALIOULHAJ, N., 1986, Les conditions de dépôt des schistes bitumineux et des phosphates Maastrichtiens au Maroc: Une approche palynologique: Documents du Bureau de Recherches Géologiques et Minières, v. 110, p. 127-139.

ROBINSON, N., EGLINGTON, G., BRASSELL, S. C., AND CRANWELL, P. A., 1984, Dinoflagellate origin for sedimentary 4-methyl-steroids and 5(H)-stanols: Nature, v. 308, p. 439-442.

SAADI, M., BOUHAOULI, A., ELARABI, H., IDIR, E., AND RAHHALI, I., 1981, Les roches bitumineuses marocaines: Mines, Géologie et Energie, Rabat (Morocco), v. 50, p. 25-54.

SANDSTROM, M. W., 1980, Organic geochemistry of some Cambrian phosphorites, *in* Douglas, A. G., and Maxwell, J. R., eds., Advances in Organic Geochemistry 1979: Oxford, Pergamon Press, p. 123-131.

SANDSTROM, M. W., 1982, Organic geochemistry of some phosphorites and associated sediments: Unpublished Ph.D. Dissertation, University of Canberra, Australia, 346 p.

SANDSTROM, M. W., 1986, Proterozoic and Cambrian phosphorites—Specialist studies: Geochemistry of organic matter in Middle Cambrian phosphorites from the Georgian basin, north-eastern Australia, *in* Cook, P. J., and Shergold, J. H., eds., Phosphate deposits of the World, v. 1: Cambridge, Cambridge University Press, p. 268-279.

SCHMID, J. C., 1986, Marqueurs biologiques soufrés dans les pétroles: Unpublished Ph.D. Dissertation. University Louis Pasteur, Strasbourg, France, 263 p.

SIMONEIT, B. R. T., 1990, Petroleum generation, an easy and widespread process in hydrothermal systems: An overview: Applied Geochemistry, v. 5, p. 3-15.

SINNINGHE DAMSTÉ, J. S, TEN HAVEN, H. L., DE LEEUW, J. W., AND SCHENCK, P. A., 1986, Organic geochemical studies of a Messinian evaporitic basin, northern Apennines (Italy)—II: Isoprenoid and n-alkyl thiophenes and thiolanes, *in* Leythaeuser, D., and Rullkötter, J., eds., Advances in Organic Geochemistry 1985, Part. 2: Oxford, Pergamon Press, p. 791-805.

SUMMERHAYES, C. P., NUTTER, A. H., AND TOOMS, J. S., 1971, Geology, structure and development of the continental margin of northwest Africa: Marine Geology, v. 11, p. 1-25.

TISSOT, B. P., AND WELTE, D. H., 1978, Petroleum Formation and Occurrence: A New Approach to Oil and Gas Exploration: New York, Springer-Verlag, 538 p.

TISSOT, B. P., AND WELTE, D. H., 1984, Petroleum Formation and Occurrence: 2nd ed.: Berlin, Springer-Verlag, 538 p.

TOOMS, J. S., SUMMERHAYES, C. P., AND CRONAN, D. S., 1969, Geochemistry of marine phosphate and manganese deposits: Oceanography and Marine Biology Review, v. 7, p. 49-100.

TRAPPE, J., 1991, Stratigraphy, facies distribution and paleogeography of marine

Paleogene from the Western High Atlas, Morocco: Neues Jahrbuch für Geologie und Paläontologie: Abhandlungen, v. 180, p. 279-321.

TRICHET, J., RACHIDI, M., AND BELAYOUNI, H., 1990, Organic geochemistry of phosphates: Relative behaviors of phosphorus and nitrogen during the formation of humic compounds in phosphate-bearing sequences, *in* Burnett, W. C., and Riggs, S. R., eds., Phosphate Deposits of the World, v. 3, Neogene to Modern Phosphorites: Cambridge, Cambridge University Press, p. 87-98.

VALISOLALAO, J., PERAKIS, N., CHAPPE, B., AND ALBRECHT, P., 1984, A novel sulfur containing C_{35} hopanoid in sediments: Tetrahedron Letters, v. 25, p. 1183-1186.

VAN KREVELEN, D. W., 1960, Graphical statistical method for the study of structure and reaction processes of coal: Fuel, v. 29, p. 269-284.

WAPLES, D. W., 1982, Phosphorite-rich sedimentary rocks: significance for organic facies and petroleum exploration: Journal of Geochemical Exploration, v. 16, p. 135-160.

RARE-EARTH ELEMENT BEHAVIOR IN PHOSPHATES AND ORGANIC-RICH HOST SHALES: AN EXAMPLE FROM THE UPPER CARBONIFEROUS OF MIDCONTINENT NORTH AMERICA

ANNA M. CRUSE,* TIMOTHY W. LYONS

Department of Geological Sciences, University of Missouri, Columbia, Missouri 65211, USA

AND

DAVID L. KIDDER

Department of Geological Sciences, Ohio University, Athens, Ohio 45701-2979, USA

ABSTRACT: Four distinct shale-normalized rare-earth element (REE) patterns are preserved in the black and gray phosphatic shale facies of the Upper Carboniferous of Midcontinent North America. Two of these patterns [flat and middle REE (MREE) enriched] are observed in both nodular phosphate and the shale host, but MREE-enriched patterns are more common in the nodules. MREE-depleted patterns were observed only in the shales. Subtle negative Ce-anomalies are rare and occur only in samples of nodular phosphate.

MREE-enriched patterns in phosphate are interpreted to have formed during very early diagenesis. Coprecipitation of phosphate and REE may have depleted pore waters in MREE to the extent that host shales, which lithified later, are characterized by MREE depletions. Phosphate diagenesis can thus be a critical factor controlling REE patterns preserved in shales. Flat REE signatures characterize gray shale samples that have low concentrations of phosphate and organic carbon. Such samples were thus not affected by processes of remobilization and scavenging of REE. Black shale samples that have phosphate concentrations intermediate of those in samples characterized by MREE-enrichments and -depletions also display flat patterns. Such patterns likely reflect a composite signal of MREE-enriched phosphate and MREE-depleted shale. Conversely, flat patterns in phosphate nodules could reflect later diagenetic migration of REE from MREE-depleted pore waters into later cement phases in the nodules. Subtle Ce-depleted patterns in phosphate nodules likely reflect the minor exchange of Ce with seawater during episodes of sediment reworking.

INTRODUCTION

The rare-earth elements (REE), La to Lu, are widely utilized as tracers of a range of geological processes because of their strikingly similar electronic configurations that give rise to subtle, predictable differences in chemical behavior along the series (Elderfield, 1988). Recent studies have revealed that REE patterns in sedimentary systems are influenced by both depositional environment (e.g., Murray et al., 1990, 1992) and diagenetic processes (e.g., Milodowski and Zalasiewicz, 1991; Murray et al., 1992). Distributions of REE also vary considerably within modern ocean waters as a function of depth (Murphy and Dymond, 1984; Sholkovitz et al., 1994); latitudinal gradients (Murray and Leinen, 1993); uptake of REE by organic and/or oxyhydroxide grain coatings (Sholkovitz et al., 1994); and variations in water chemistry, in particular redox conditions, between pore water and overlying ocean water (e.g., Elderfield et al., 1990).

Phosphates are often analyzed for REE signatures because phosphate mineralogy commonly stabilizes during early diagenesis with minimal subsequent alteration, and phosphate minerals are naturally enriched in bulk REE concentrations due to substitution for Ca in the apatite lattice (e.g., Wright et al., 1987). However, analysis of the relationships between phosphate minerals and their host sediment is complicated by the common occurrence of phosphate in reworked lag deposits, which allows for the possibility that chemical exchange occurred with overlying seawater at some unknown time after deposition. Winnowing also separates the phosphate mineral from the sediment in which it originally formed. Finally, lag deposits that are more permeable than the surrounding host rock could be preferential conduits for later flow of subsurface fluids. We report here on REE analyses of nonskeletal phosphate (nodules and laminae) and the shale hosts from largely unreworked sections in the Upper Carboniferous of Midcontinent North America in an effort to constrain the sources of and mechanisms for REE concentrations observed in phosphates and the general pathways of REE partitioning in organic-rich shales.

GEOLOGIC SETTING

Phosphatic black shales from Upper Carboniferous (Pennsylvanian) cyclothems in Iowa, Kansas, and Oklahoma are emphasized in the present study (Figs. 1, 2A). Cyclothems in this region of North America consist predominantly of alternating limestone and shale units. Based on extensive study of deposits within the Midcontinent, Heckel (e.g., 1977, 1990) summarized the idealized pattern of cyclicity (Fig. 2B). Within this scheme, the basal unit consists of a nearshore sandy shale that may contain coal, sandy layers, and soil horizons. This shale is overlain by a thin transgressive limestone, which is in turn overlain by the black and gray shales analyzed in the present study. Based on biofacies models of water column oxygenation, the bioturbated gray shales were deposited under an oxic to dysoxic water column, while the black shales were deposited under anoxic waters (Rhoades and Morse, 1971; Savrda and Bottjer, 1991). The black and gray shale horizons are overlain by a thick regressive limestone, and the sequence is typically capped by a second nearshore shaley unit that marks the beginning of the next cycle. The black shale facies has been interpreted as the deepest, most offshore part of the cycle (e.g., Heckel, 1977, 1990), although other workers have favored a shallower depositional environment (Zangerl and Richardson, 1963; Merrill, 1975; Coveney et al., 1991).

Phosphate occurs in the black shale facies as distinct layers

**Present address: Woods Hole Oceanographic Institution, Department of Marine Chemistry and Geochemistry, MS#4, Woods Hole, MA 02543, email:acruse@whoi.edu*

Marine Authigenesis: From Global to Microbial, SEPM Special Publication No. 66

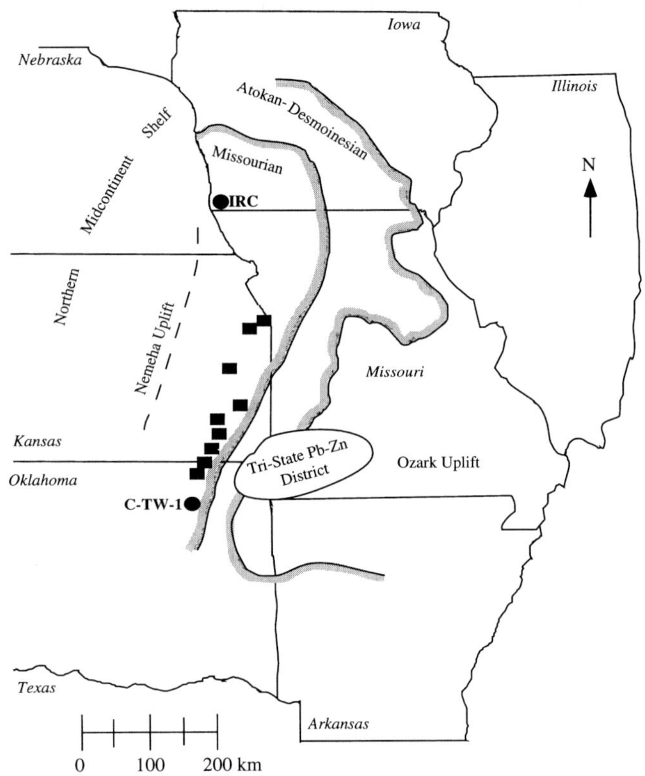

FIG. 1.—Map showing position of northern midcontinent shelf and the limit of Upper Carboniferous (Atokan-Desmoinesian and Missourian) surface exposures (stippled pattern). Filled circles mark positions of drill cores analyzed in the present study; filled squares represent localities from which analyzed phosphate nodules were collected.

of nodules or bedding-parallel microlaminae that are typically 0.5–1 mm thick. The layers either occur in zones proximal to the gray-black transitions or in multiple zones throughout the black, organic-rich horizons. In the latter case, nodules are generally largest near the upper and lower boundaries of the black shale (Kidder, 1985).

METHODS

Documentation of REE variation was conducted on two different scales. Phosphate nodules were examined via a broad reconnaissance of shale outcrops across a region extending from northeastern Oklahoma to northeastern Kansas (Fig. 1). These analyses focused on regional and stratigraphic distributions assessed within the petrographic framework established by Kidder (1985). Analyses of the shales were conducted at a more detailed level. Numerous stratigraphic levels within gray and black shales were analyzed at high-resolution, centimeter-scale intervals across paleoredox gradients to examine local, fine-scale variation within organic-rich shales and between black and gray shale facies. No effort was made to separate phosphatic phases [a nodule found in one sample (289.3-4BR) from core C-TW-1 or laminae in all others] from the shale prior to analysis. For the shale study, samples from two drill cores (IRC, Iowa; C-TW-1, Oklahoma; Fig. 1) were analyzed to avoid alteration related to recent surface weathering effects.

Analytical methods for the nodular phosphate consisted of extracting powders using a carbide-tipped engraver. Centers of

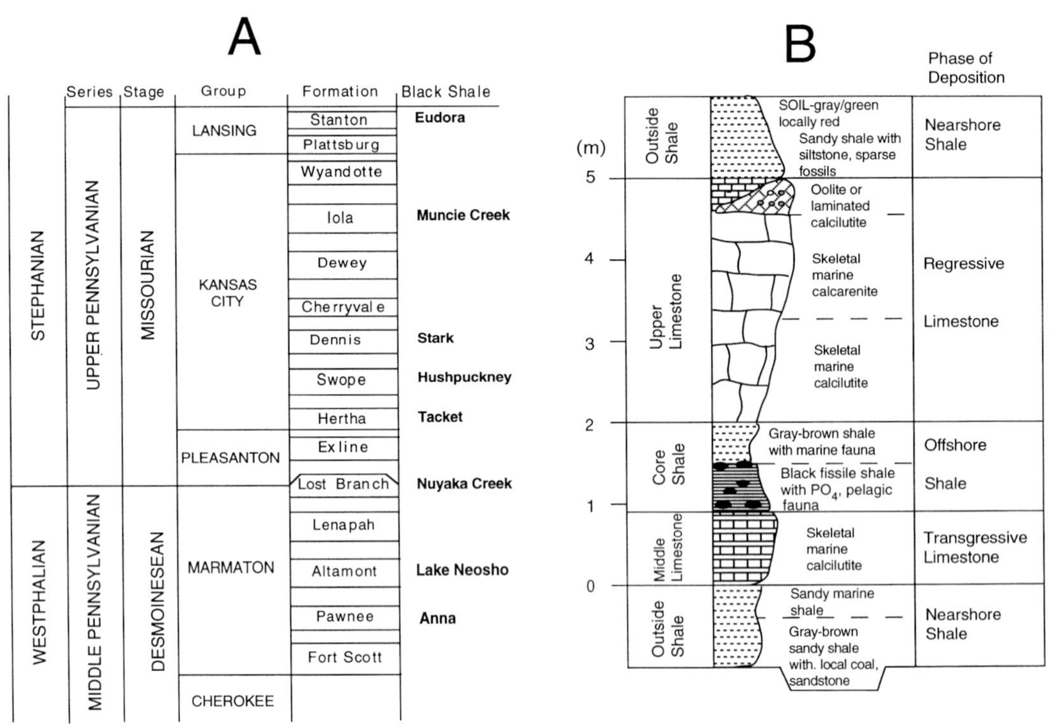

FIG. 2.—(A) Generalized stratigraphy of cyclic units analyzed in this study after Heckel (1990) and Kidder and Eddy-Dilek (1994). The Hushpuckney, Muncie Creek, and Eudora Shales are the units from which analyses are reported, although some of the other units are referred to in the text. (B) Generalized "Kansas" cyclothem modified after Heckel (1994). See text for discussion.

phosphate nodules were preferentially sampled to minimize weathering effects. These samples were analyzed for REE contents using inductively coupled plasma mass spectrometry (ICP-MS). Details of analytical methods can be found in Kidder and Eddy-Dilek (1994).

Bulk shale samples were crushed in a ceramic mill and analyzed by instrumental neutron activation analysis (INAA) with appropriate corrections applied for the effects of U-fission (Glascock et al., 1986). Recognizing that artifacts can arise through the use of INAA on black shales, a selected suite of samples from core IRC (Fig. 1) was also analyzed for REE using ICP-MS to cross-calibrate the two methods. Generally, the two methods gave comparable results except for a spurious enrichment in Sm that was observed in the INAA results, despite the corrections for U-fission (Glascock et al., 1986). Although the specific cause of this artifact is presently unknown, possibilities include previously unidentified interferences from U-fission that are not accounted for in the correction algorithms or ^{187}W (R. Hannigan, pers. commun., 1997).

Concentrations of total and inorganic carbon for the shale samples were determined coulometrically through bulk sample combustion and acidification, respectively. Organic carbon (C_{org}) concentrations were calculated by difference. Inorganic phosphate (P_{inorg}) in shales was extracted by reacting 0.2 g shale with 25 ml of 1 N HCl for 14–16 hours (Aspila et al., 1976; Ruttenberg, 1992; Ingall et al., 1993). P_{inorg} concentrations in the filtered solutions were then measured spectrophotometrically (Murphy and Riley, 1962). Representative results for all analyses are reported in Tables 1 and 2. More complete data can be found in Kidder and Eddy-Dilek (1994) and Cruse (1997).

RARE EARTH ELEMENT PATTERNS AND DISTRIBUTIONS

Four distinct shale-normalized REE patterns in the phosphates and their host shales have been documented (Kidder and Eddy-Dilek, 1994; Cruse, 1997). These are: (1) flat, (2) MREE enriched, (3) MREE-depleted, and (4) patterns characterized by a negative Ce-anomaly (Fig 3). Flat shale-normalized patterns are observed in phosphate nodules, black laminated facies, and, most commonly, in gray bioturbated shales. MREE-enriched patterns are also observed in both phosphate nodules and in black shales with elevated concentrations of phosphate. Patterns with a negative Ce-anomaly are confined to phosphate nodules, while MREE-depleted patterns are observed in both black and gray host shale samples.

Phosphate Nodule Patterns

In nodular phosphate, there appears to be a correlation between geographic position and type of REE pattern observed (Fig. 4). Flat REE patterns are most common in northeastern Kansas, whereas MREE-enriched patterns are most common in southeastern Kansas and northeastern Oklahoma. Kidder and Eddy-Dilek (1994) plotted the slopes of the depletions of LREE (Sm/La) and HREE (Sm/Yb) for nodular phosphates from eight different host shale units against basin position (i.e., distance from an arbitrary baseline that connects the three outcrop locations that were furthest from the shelf edge). Correlation was generally reasonable (r^2 = 0.71–0.77), with MREE-enrichment increasing with distance from the shelf when samples from all eight shales were treated as one data set as well as when samples were grouped according to host shale. One exception was the value of r^2 = 0.35 for the HREE depletion in phosphate from the Muncie Creek Shale. This unit, which displayed the most variability in REE patterns, was affected by local depositional topography and is thus not indicative of regional diagenetic trends. The overall relationship between MREE enrichment and geographic position thus appears to be robust (Fig. 4).

Minor negative Ce-anomalies occur in phosphate nodules collected at three localities in the Stark, Tackett, and Muncie Creek shales. The phosphate from one of these localities is clearly a reworked lag deposit from which the host siliciclastics have largely been winnowed (Kidder, 1985; Kidder and Eddy-Dilek, 1994). The phosphate nodules in the Stark and Tackett shales, however, do not appear to be reworked.

Shale Patterns

Patterns depleted in MREE occur in samples from C-TW-1 with P_{inorg} concentrations generally less than 0.2 wt. % as P_2O_5 (Table 2; Cruse, 1997). The relationship was observed in both black, laminated and gray, bioturbated host shale from core C-TW-1 collected in northeastern Oklahoma (Figs. 3, 4). C-TW-1 was collected in the general region characterized by MREE-enriched phosphate nodules in the study of Kidder and Eddy-Dilek (1994). Black shale samples from C-TW-1 that are enriched in inorganic (authigenic) phosphate (1.2-12.1 wt. % as P_2O_5; Table 2; Cruse, 1997) display an MREE enrichment.

P_{inorg} concentrations in IRC typically range from 0.1 to 0.8 wt. % P_2O_5 in the black shales (an exception is sample 1036.7-4BR with a concentration of 4.4 wt. %) and are less than 0.1 wt. % P_2O_5 in the gray shales (Table 2; Cruse, 1997). These concentrations are less than those characteristic of samples in C-TW-1 that display an MREE-enriched pattern and generally greater than those that display a MREE-depletion. However, all samples from core IRC display flat, shale-normalized patterns (Fig. 3).

DISCUSSION

MREE-enriched and -depleted patterns have not been observed in modern settings, but they have been documented in other ancient apatite samples [nodules and microfossils] (McArthur and Walsh, 1984; Wright et al., 1987). Analyses of foraminifera tests by Palmer (1985) and numeric modeling by Grandjean-Lécuyer et al. (1993) have shown that adsorption of REE by iron oxyhydroxides can result in MREE-enriched patterns. However, petrological and geochemical analyses demonstrate that iron in the nodules of the present study dominantly occurs as pyrite (Kidder, 1985; Ece, 1990). Also, considering the presence of water-column anoxia during black shale deposi-

TABLE 1.—RARE-EARTH ELEMENT ANALYSES FOR PENNSYLVANIAN PHOSPHATE NODULES AND BLACK SHALES FROM IOWA, KANSAS AND OKLAHOMA

Locality		La	Ce	Pr	Nd	Sm	Eu	Gd	Tb	Dy	Ho	Er	Tm	Yb	Lu
Avg. Shale[1]		41	83	10.1	38	7.5	1.61	6.35	1.23	5.49	1.34	3.75	0.63	3.51	0.61
Shale Analyses															
Hushpuckney Shale															
Iowa															
1036.7-4GR		15	29	3.3	13	2.4	0.50	2.13	0.55	1.91	0.41	1.36	0.21	1.69	0.28
1036.7-4BR		49	93	11.5	45	8.3	1.82	8.93	1.25	7.07	1.42	3.79	0.49	2.92	0.46
1036.7-5BR		17	29	3.8	15	2.7	0.62	2.70	0.39	2.26	0.44	1.35	0.22	1.49	0.26
1036.7-8BR		3	11	0.7	3	0.5	0.12	0.43	0.08	0.40	0.10	0.30	0.05	0.47	0.09
1039.2-4BT*		13	23	—	7	9.1	0.44	—	0.40	1.69	—	—	—	1.56	1.11
1036.2-3BT		5	16	1.5	6	1.1	0.26	1.08	0.17	1.01	0.20	0.61	0.10	0.56	0.09
1039.2-1BT		2	14	0.6	2	0.4	0.09	0.33	0.05	0.30	0.06	0.17	0.03	0.22	0.03
1039.2-3GT*		5	8	—	4	0.9	0.10	—	0.10	0.60	—	—	—	0.40	0.10
1039.2-4GT		3	6	0.7	3	0.6	0.15	0.60	0.11	0.51	0.11	0.29	0.05	0.27	0.06
Oklahoma															
289.3-14GR*		42	81	—	45	5.8	1.10	—	1.45	4.13	—	—	—	2.87	0.39
289.3-3GR*		33	50	—	17	2.5	0.50	—	0.50	2.20	—	—	—	1.90	0.30
289.3-4BR*		196	373	—	536	61.6	13.77	—	7.23	50.86	—	—	—	14.44	2.58
289.3-7BR*		26	39	—	11	2.3	0.40	—	0.20	1.60	—	—	—	1.90	0.40
294.0-8BT*		81	182	—	177	23.6	4.60	—	3.30	16.00	—	—	—	5.70	0.70
294.0-1BT*		31	53	—	18	3.7	0.69	—	0.71	3.10	—	—	—	2.47	0.42
294.0-7GT*		34	62	—	26	4.9	0.99	—	1.06	3.17	—	—	—	2.67	0.45
Phosphate Analysis															
Muncie Creek Shale															
IA3	K-32	257	558	91.5	364	76.9	15.90	59.60	10.90	60.60	11.10	29.10	3.30	16.50	2.50
IC3	KTL	144	251	56.4	249	60.7	14.00	50.30	8.80	46.30	7.80	20.00	2.10	10.40	1.40
IC3'	KTL	50	180	32.1	161	46.5	9.30	31.60	5.10	24.40	3.70	8.40	0.80	3.80	0.50
IE4	CAN	84	243	48.0	235	58.9	10.60	37.80	6.50	29.30	4.30	8.90	0.70	2.90	0.40
IF2	NEO	109	250	44.5	217	51.7	10.00	38.70	7.30	39.40	7.00	17.60	1.90	8.90	1.10
IF2'	NEO	120	298	49.7	225	52.7	10.80	40.40	7.40	38.10	6.40	16.60	1.70	8.50	1.00
Eudora Shale															
EA1	CPD	122	385	62.1	314	95.7	19.20	66.00	10.90	48.10	7.30	16.50	1.70	8.00	1.00
EB2	K-10	58	167	25.9	102	21.2	4.70	17.00	3.00	15.80	2.80	7.10	0.80	3.70	0.50
EC1	NQEB	96	269	53.7	237	56.3	12.80	45.20	7.70	38.70	6.30	14.50	1.50	6.30	0.70
ED1	PTC	243	544	79.0	299	59.2	13.70	50.90	9.30	49.20	8.90	24.20	3.00	15.50	1.90
EE1	TYQ	85	405	67.6	320	82.4	18.10	63.70	10.90	55.20	8.60	19.80	1.90	8.40	1.10

[1]Shale values from Piper (1974).

*INAA results.

—: not analyzed.

BR, BT: black shale sample.

GR, GT: gray shale sample.

TABLE 2.—ELEMENTAL ANALYSES (WT.%) FOR PENNSYLVANIAN PHOSPHATE NODULES AND BLACK SHALES IN IOWA, KANSAS AND OKLAHOMA

Locality	SiO_2	Al_2O_3	FeO	CaO	MgO	Na_2O	K_2O	P_2O_5	TiO_2	MnO	F_2O	C_{org}	C_{inorg}
Shale Analyses[1]													
Hushpuckney Shale													
Iowa													
1036.7-4GR	—	14.63	6.23	3.18	3.20	1.29	3.09	0.12	0.83	0.05	—	0.58	0.60
1036.7-4BR	—	12.04	4.54	6.91	3.20	0.94	2.67	4.36	0.48	0.02	—	9.62	0.69
1036.7-5BR	—	9.70	4.71	6.79	3.23	0.70	2.31	0.51	0.46	0.02		14.41	1.46
1036.7-8BR	—	10.62	5.24	4.66	3.10	0.75	2.40	0.14	0.50	0.03	—	15.96	1.03
1039.2-4BT	—	7.77	3.88	3.00	2.64	0.59	1.94	0.28	0.34	0.03	—	33.08	0.62
1039.2-3BT	—	8.11	4.69	6.09	2.91	0.57	2.06	0.73	0.38	0.02		24.34	1.24
1039.2-1BT	—	11.49	4.59	5.08	3.79	0.82	2.71	0.31	0.57	0.05	—	15.25	0.99
1039.2-3GT	—	0.94	0.93	50.84	1.60	0.11	0.17	0.04	0.09*	0.03	—	0.41	10.85
1039.2-4GT	—	1.62	0.81	45.42	2.29	0.18	0.26	0.04	0.09*	0.05	—	0.34	10.29
Oklahoma													
289.3-14GR	—	16.96	4.86	0.65	3.16	0.91	3.47	0.11	0.75	0.08	—	0.81	0.31
289.3-3GR	—	15.57	5.51	0.81	2.57	0.91	3.44	0.05	0.55	0.03	—	1.16	0.07
289.3-4BR	—	7.56	10.97	16.77	1.53	0.56	1.54	12.16	0.17	0.06	—	9.35	0.68
289.3-7BR	—	12.91	8.37	1.27	2.83	0.79	3.08	0.04	0.52	0.05	—	12.25	0.36
294.0-8BT	—	12.13	6.18	7.14	2.67	0.66	2.67	1.99	0.44	0.08	—	6.08	1.39
294.0-1BT	—	13.62	4.18	8.03	2.80	0.83	2.96	0.08	0.57	0.10	—	2.30	1.90
294.0-7GT	—	13.69	4.70	10.41	2.40	0.78	3.27	0.27	0.45	0.10	—	1.16	2.11
Phosphate Analyses[2]													
Muncie Creek Shale													
IC3 KTL	0.03	0.02	0.01	54.54	0.00	0.23	—	37.93	0.01	0.02	7.21	—	—
Eudora Shale													
EA1 CPD	5.44	1.37	1.91	52.91	0.15	0.50	—	37.69	—	0.03	0.00	—	—
ED1 PTC	11.34	6.25	0.90	44.99	0.43	0.29	—	30.72	0.03	0.03	5.03	—	—
EE1 TYQ	6.21	3.30	0.21	48.54	0.00	0.49	—	35.37	0.02	0.01	5.86	—	—

[1]Shale samples analyzed via INAA. Total element reported as oxide.
[2]Phosphate nodules analyzed via electron microprobe. Oxides recalculated to 100% on a volatile-free and reduced iron basis.
*Values represent detection limits.
—: not analyzed.
BR, BT: black shale sample.
GR, GT: gray shale sample.

tion (Heckel 1977, 1990; Cruse, 1997), the presence of appreciable iron oxide phases in the black shales also seems unlikely. Recent experimental results of Byrne et al. (1996) demonstrated that coprecipitation of REE-phosphates with apatite will result in enrichments in MREE in the solid phase, provided there is minimal competition from carbonate ligands in the solution. We suggest that the relationship of MREE-enrichment in phosphate nodules and phosphate-rich black shales and MREE-depletions in overlying and underlying gray and black shales with low P_{inorg} concentrations reflects this mechanism. This idea is strongly supported by the MREE-depleted host shales from Oklahoma (C-TW-1), an area in which phosphate nodules are commonly enriched in MREE. A similar correlation between phosphate-enrichment and MREE-enriched and -depleted host shales was also observed in organic-rich shales of the uppermost Carboniferous (Virgilian) from southeastern Kansas (Baker, 1995).

This remobilization likely proceeded via one of two path-

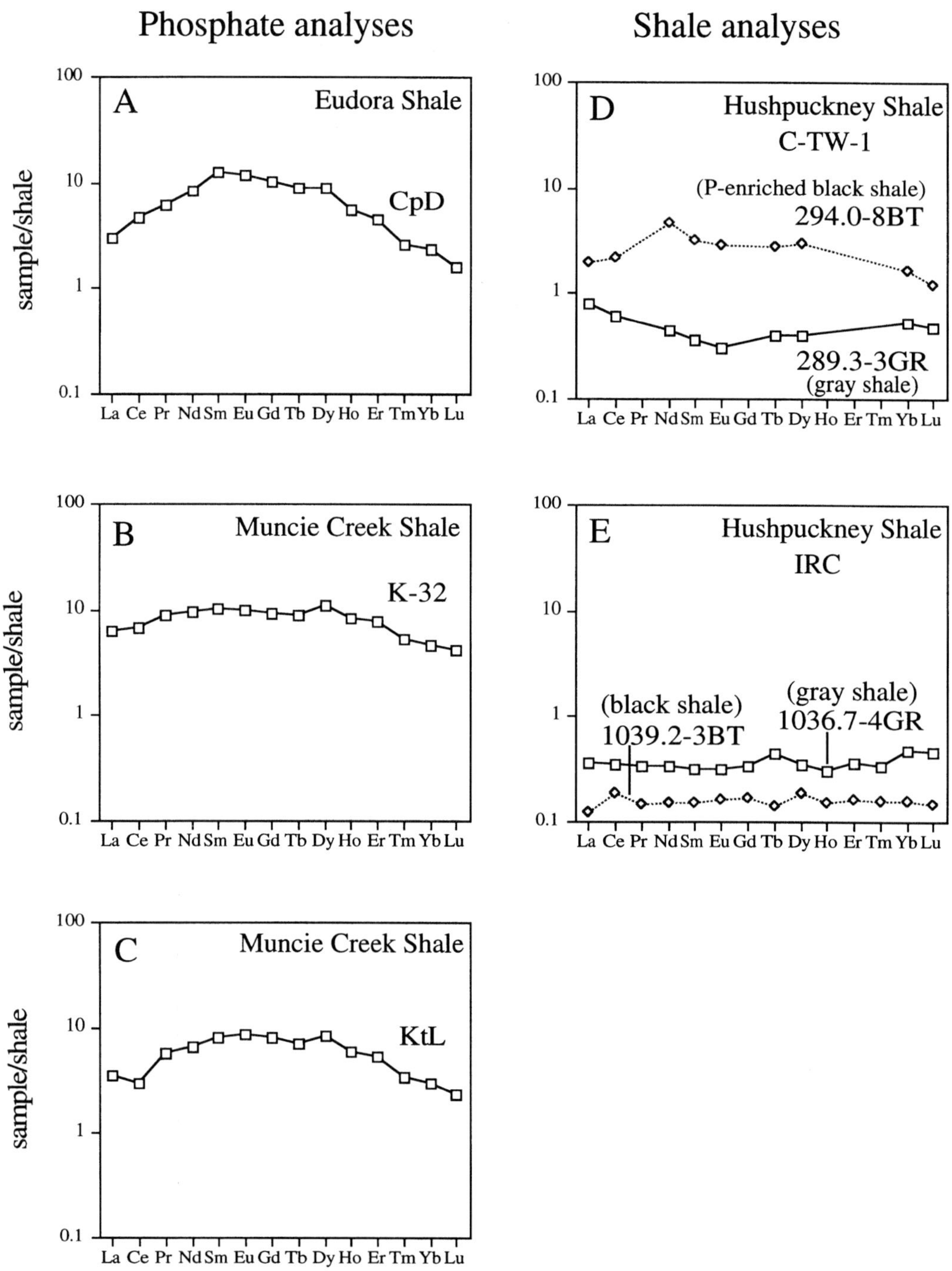

FIG. 3.—Representative examples of REE patterns preserved in phosphate nodules, host black shale, and associated gray shale. Sample numbers refer to localities listed in Tables 1 and 2. (A) MREE-enriched pattern. (B) Flat pattern. (C) Subtle Ce-depleted pattern. (D) MREE-enriched pattern (294.0-8BT) and MREE-depleted pattern (289.3-3GR) observed in core C-TW-1. (E) Flat patterns observed in IRC.

ways that differ only in the relative timing of diffusion of REE through pore waters and adsorption of REE by the sediments. Coprecipitation of REE-phosphates during apatite precipitation could have resulted in the establishment of concentration gradients in the pore waters. As REE diffused to the sites of coprecipitation, REE concentrations may have been sufficiently depleted in the pore-waters of the nonphosphatic sediments to cause desorption of the REE from sediments. Since MREE were preferentially precipitated with the apatite, their concentrations would have been proportionally lower in porewaters. This would result in enhanced remobilization of MREE from the nonphosphatic sediments. However, in Buzzard's Bay, Massachusetts, diagenetic remobilization of REE into pore waters is insufficient to affect sediment concentrations (Elderfield and Sholkovitz,

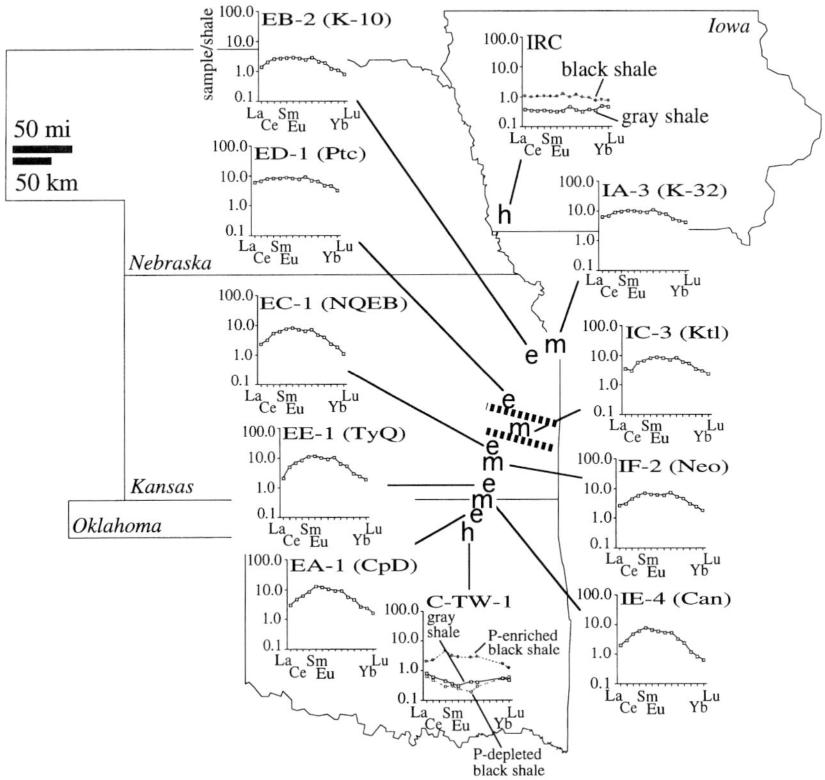

FIG. 4.—Map showing regional variation in shale-normalized REE patterns from phosphate nodules and for two localities of the shale host (IRC and C-TW-1). Plots corresponding to "h" on map are for drill-core samples of the Hushpuckney Shale. Phosphate nodules from the Muncie Creek Shale are represented by "m"; those from the Eudora Shale are indicated by "e". Heavy dashed lines show the position of Bourbon Arch.

1987). A more plausible mechanism is one in which coprecipitation of REE with apatite caused preferential depletion of MREE in pore waters prior to adsorption of the REE by sediments. Our understanding of REE behavior during water-rock interactions is nascent and further work is needed to fully constrain the exact mechanism of this process.

In addition to the inorganic, selective extraction of MREE from solution during coprecipitation of phosphate (Byrne et al., 1996), there may be organic influences on the MREE-enrichment in phosphate. Stanley and Byrne (1990) showed that under certain conditions, algae will preferentially take up MREE. It is plausible that bacteria may be capable of the same process. Thus, we speculate that bacteria present during degradation of phosphate-rich fecal matter may also contribute to REE diagenesis.

Kidder and Eddy-Dilek (1994) interpreted the apparent relationships between REE patterns and geographic distribution as reflecting detrital influence. They suggested that REE from detrital material within the nodules could mask the original MREE-enriched pattern. Phosphate nodules from shales in northeastern Kansas, which tend to display the flat pattern, are closer to potential detrital sources in the Appalachians and the Canadian Shield than equivalent facies in northeastern Oklahoma. Thus, REE in authigenic phosphate nodules from shales closest to the basin margin have greater potential to be affected by incorporated detrital material. However, preliminary geochemical analyses do not support this hypothesis. The major element concentrations of four nodules have been measured using electron microprobe analyses. The Al_2O_3 content of the four samples, used as an indicator of the detrital content, ranges from 0.02 to 6.25 wt. % (Table 2). The concentration of REE associated with the detrital component can be calculated from the following formula:

$$REE_{detrital} = Al_2O_{3sample}*(REE_{shale}/Al_2O_{3shale}),$$

where REE_{shale} and Al_2O_{3shale} refer to the concentrations of a specific REE and Al_2O_3 in a shale standard, and $Al_2O_{3sample}$ is the concentration measured in a given sample. Using a value of 16.90 wt. % Al_2O_3 for average shale (Gromet et al., 1984), a maximum of only 10% of the measured concentrations of each REE can be attributed to incorporated detrital material. Considering that REE concentrations in the nodules are enriched two to three times over average shale concentrations (Table 1), the calculated detrital component is not great enough to affect the observed patterns.

The origin of the flat patterns in phosphates could thus relate to timing of different diagenetic processes. Although the phosphate nodules lithified during early diagenesis as evidenced by shale laminae that commonly compact around them, there are two distinct nodule types. Kidder et al. (1996) illustrated that nodules from Kansas and northeasternmost Oklahoma (including those analyzed for REE in the present study) tend to have a more open structure that is minimally compacted, and they are characterized by micronuclei of radiolaria and peloids on which phosphate and later cements precipitated. These "Kansas-type" nodules may have lithified in the manner suggested by Mozley (1989), in which early lithification consisted of cementation of micronuclei distributed throughout the concretion. Diagenetic fluids generated during compaction and dewatering of host shales may have infiltrated these nodule types and imparted the flat REE pattern to late cement phases. Unfortunately, we are aware of no data for the REE composition of fluids derived from dewatered shales. Microanalysis of the REE composition of late cement phases (apatite, calcite, and silica) in the phosphate nodules could test this hypothesis. Conversely, "Oklahoma-type" nodules may have fully lithified early as indicated by textural relationships within the nodules. This early lithification could have prevented overprinting of the original MREE-enriched pattern during late diagenesis. As the name implies, this nodule type is most common in Oklahoma (Kidder et al., 1996), where MREE-enriched patterns are generally observed in phosphate nodules (Figs. 3, 4).

Overprinting of flat patterns in phosphate nodules during later diagenesis of shales is consistent with the trace element movement from shale to limestone suggested by McHargue and Price (1982). They noted that limestone immediately above and below black/gray shale horizons in the Anna Shale is more heavily dolomitized than the limestone farther above and below the shale, and suggested that this effect was due to release of Mg^{2+} during shale diagenesis. These workers further noted that the dolomite in these limestones was a late diagenetic phase and interpreted it to have formed during generation of fluids from the dewatering of shales during diagenetic compaction.

If phosphate diagenesis exerts a primary control on the REE patterns observed in host shales, MREE-enriched and -depleted patterns should also be observed in IRC. Samples from this drill core contain phosphate concentrations greater than those in samples from C-TW-1 that display an MREE-depleted pattern but less than those from C-TW-1 displaying an MREE-enriched pattern (Table 2; Cruse, 1997). Thus, the observed flat patterns in samples from IRC may reflect a mixed signal of MREE-enrichment in the phosphate and MREE-depletion in the shale. This implies that preservation of MREE-depleted and -enriched patterns indicative of phosphate diagenesis depends on a mass balance between REE concentrations in authigenic phosphate and surrounding shale. Conversely, the apatite present in laminae in IRC could have been reworked from another location rather than having precipitated in situ. In this case, MREE remobilization caused by apatite precipitation occurred in a different location and would not be observed in the sediments that ultimately hosted the phosphate. However, since the apatite grains in the laminae did not appear to have been reworked when observed in thin section, this scenario is highly unlikely.

CONCLUSIONS

The relationships between REE in authigenic phosphate nodules and laminae and host shales are complex. It appears that phosphate diagenesis and the formation of nodules cause the preferential migration of MREE from pore waters in organic-rich shales to authigenic phosphate nodules during early diagenesis. This MREE migration is the result of preferential uptake of MREE from pore waters due to coprecipitation with apatite. Host sediments thus acquire an MREE-depleted signature via an adsorptive process that occurs later. Biogenic factors may also contribute to the MREE-enriched pattern observed in phosphate nodules. Thus, the presence of authigenic phosphate nodules and laminae appears to be a key factor controlling the REE patterns observed in ancient sediments. Organic- and phosphorus-rich black shale samples from southeastern Iowa display flat shale normalized REE patterns, suggesting that the preservation of a phosphate-related signal depends upon the relative abundance of authigenic phosphate. Conversely, flat REE patterns in these shale samples could also reflect the reworked nature of authigenic apatite whereby the signal of the remobilization associated with apatite precipitation was preserved in different sediments.

Flat shale-normalized patterns in phosphate nodules are a late-stage diagenetic phenomenon. We hypothesize that phosphate nodules with open, uncompacted structures may acquire this overprint from detrital clays in the host shale as they undergo diagenesis. Phosphate nodules that formed in areas where high sedimentation rates induced early compaction may have become impermeable before the appropriate shale diagenesis took place.

Further work is required to fully constrain the behavior of REE during diagenesis. These results suggest that shale-normalized REE patterns in authigenic phosphate and host shales can vary temporally during diagenesis. Thus, the patterns observed today may reflect the timing of authigenic nodule precipitation and subsequent lithification in addition to paleoenvironmental changes (e.g., pore water redox conditions). While this has negative implications for the use of REE as simple indicators of paleoenvironmental conditions, these results suggest that the REE could play a critical role in modeling phosphate diagenesis and the systematics of nodule formation.

ACKNOWLEDGMENTS

M. Glascock, S. Herrera, and R. Hannigan provided invaluable assistance with INAA and ICP-MS analyses. We thank D. Piper and R. T. Watkins for thoughtful reviews that helped improve the manuscript. This work was supported in part by a grant from the University of Missouri Research Board.

REFERENCES

Aspila, K. I., Agemain, H., and Chau, A. S. Y., 1976, A semi-automated method for the determination of inorganic, organic and total phosphate in sediments: Analyst, v. 101, p. 187-197.

Byrne, R. H., Liu, X., and Schijf, J., 1996, The influence of phosphate coprecipitation on rare earth distribution in natural waters: Geochimica et Cosmochimica Acta, v. 60, p. 3341-3346.

Coveney, R. M., Jr., Watney, W. L., and Maples, C. G., 1991, Contrasting depositional models for Pennsylvanian black shale discerned from molybdenum abundances: Geology, v. 19, p. 147-150.

Cruse, A. M., 1997, Geochemical trends in the Hushpuckney Shale (Pennsylvanian), Midcontinent U.S.A.: A record of dynamic anoxia [M.S. thesis]: University of Missouri, Columbia, 112 p.

Ece, O. I., 1990, Geochemistry and occurrence of authigenetic phosphate nodules from the Desmoinesian cyclic Excello epeiric sea of the Midcontinent, USA: Marine and Petroleum Geology, v. 7, p. 298-312.

Elderfield, H., 1988, The oceanic chemistry of the rare-earth elements: Philosophical Transactions Royal Society of London, Series A, v. 325, p. 105-126.

Elderfield, H. and Sholkovitz, E. R., 1987, Rare earth elements in the pore waters of reducing nearshore sediments: Earth and Planetary Science Letters, v. 82, p. 280-288.

Elderfield, H., Upstill-Goddard, R., and Sholkovitz, E. R., 1990, The rare earth elements in rivers, estuaries, and coastal seas and their significance to the composition of ocean waters: Geochimica et Cosmochimica Acta, 54, 971-991.

Glascock, M. D., Nabelek, P. I., Weinrich, D. D., and Coveney, R. M., Jr., 1986, Correcting for uranium fission in instrumental neutron activation analysis of high-uranium rocks: Journal of Radioanalytical and Nuclear Chemistry, v. 99, p. 121-131.

Grandjean-Lécuyer, P., Feist, R., and Albarède, F., 1993, Rare earth elements in old biogenic apatites: Geochimica et Cosmochimica Acta, v. 57, p. 2507-2514.

Gromet, L. P., Dymek, R. F., Haskin, L. A., and Korotev, R L., 1984, The "North American shale composite": It's compilation, major and trace element characteristics: Geochimica et Cosmochimica Acta, v. 48, p. 2469-2482.

Heckel, P. H., 1977, Origin of phosphatic black shale facies in Pennsylvanian cyclothems of Midcontinent North America: American Association of Petroleum Geologists Bulletin, v. 61, p. 1045-1068.

Heckel, P. H., 1990, Evidence for global (glacial-eustatic) control over upper Carboniferous (Pennsylvanian) cyclothems in Midcontinent North America, *in* Hardman, R. F. P. and Brooks, J., eds., Tectonic Events Responsible for Britain's Oil and Gas Reserves: London, Geological Society of London, Geological Society Special Publication No. 55, p. 35-47.

Heckel, P. H., 1994, Evaluation of evidence for glacio-eustatic control over marine Pennsylvanian cyclothems in North America and consideration of possible tectonic effects, *in* Dennison, J. M. and Ettensohn, F. R., eds., Tectonic and Eustatic Controls on Sedimentary Cycles: Tulsa, SEPM (Society for Sedimentary Geology), SEPM Concepts in Sedimentology and Paleontology, No. 4, p. 65-87.

Ingall, E. D., Bustin, R. M., and Van Cappellen, P., 1993, Influence of water column anoxia on the burial and preservation of carbon and phosphorus in marine shales: Geochimica et Cosmochimica Acta, v. 57, p. 303-316.

Kidder, D. L., 1985, Petrology and origin of phosphate nodules from the Midcontinent Pennsylvanian epicontinental sea: Journal of Sedimentary Petrology, v. 55, p. 809-816.

Kidder, D. L. and Eddy-Dilek, C.A., 1994, Rare-earth element variation in phosphate nodules from Midcontinent Pennsylvanian cyclothems: Journal of Sedimentary Research, v. A64, p. 584-592.

Kidder, D. L., Hussein, R. A. M., Mapes, R. H., and Eddy-Dilek, C. A., 1996, Phosphates in cyclic cratonic condensed sections, *in* Witzke, B., Ludvigson, G. A., and Day, J., eds., Diagenetic analysis of transgressive-regressive events: Boulder, Geological Society of America, Geological Society of America Special Publication 306, p. 351-358.

McArthur, J. M. and Walsh, J. N., 1984, Rare-earth geochemistry of phosphorites: Chemical Geology, v. 47, p. 191-220.

McHargue, T. R. and Price, R. C., 1982, Dolomite from clay in argillaceous of shale-associated marine carbonates: Journal of Sedimentary Petrology, v. 52, p. 873-886.

Merrill, G. K., 1975, Pennsylvanian conodont biostratigraphy and paleoecology of northwestern Illinois: Geological Society of America Microform Publication, 3.

Milodowski, A. E. and Zalasiewicz, J. A., 1991, Redistribution of rare earth elements during diagenesis of turbidite/hemipelagite mudrock sequences of Llandovery age from central Wales, *in* Morton, A. C., Todd, S. P., and Haughton, P. D. W., eds., Developments in Sedimentary Provenance Studies: London, Geological Society of London, Geological Society Special Publication No. 57, p. 101-124.

Mozley, P. S., 1989, Complex compositional zonation in concretionary siderite: Implications for geochemical studies: Journal of Sedimentary Petrology, v. 59, p. 815-818.

Murphy, J. and Riley, J. P., 1962, A modified single solution method for the determination of phosphate in natural waters: Analytica Chimica Acta, v. 27, p. 31-36.

Murphy, K. and Dymond, J., 1984, Rare earth element fluxes and geochemical budget in the eastern equatorial Pacific: Nature, v. 307, p. 444-447.

Murray, R. W., Buchholtz ten Brink, M. R., Jones, D. L., Gerlach, D. C., and Russ, G. P., III, 1990, Rare earth elements as indicators of different marine depositional environments in chert and shale: Geology, v. 18, p. 268-271.

Murray, R. W., Buchholtz Ten Brink, M. R., Gerlach, D. C., Russ, G. P., III, and Jones, D. L., 1992, Interoceanic variation in the rare earth, major, and trace element depositional chemistry of chert: Perspectives gained from the DSDP and ODP record: Geochimica et Cosmochimica Acta, v. 56, p. 1897-1913.

Murray, R. W. and Leinen, M., 1993, Chemical transport to the seafloor of the equatorial Pacific Ocean across a latitudinal transect at 135°W: Tracking sedimentary major, trace, and rare earth element fluxes at the equator and the intertropical convergence zone: Geochimica et Cosmochimica Acta, v. 57, p. 4141-4163.

Palmer, M. R., 1985, Rare earth elements in foraminifera tests: Earth and Planetary Science Letters, v. 73, p. 285-298.

Piper, D. Z., 1974, Rare earth elements in the sedimentary cycle: A summary: Chemical Geology, v. 14, p. 285-304.

Rhoades, D. C. and Morse, J. W., 1971, Evolutionary and ecologic significance of oxygen-deficient marine basins: Lethia, v. 4, p. 413-428.

Ruttenberg, K. C., 1992, Development of a sequential extraction method for different forms of phosphorus in marine sediments: Limnology and Oceanography, v. 37, p. 1460-1482.

Savrda, C. E. and Bottjer, D. J., 1991, Oxygen-related biofacies in marine strata: An overview and update, *in* Tyson, R. V. and Pearson, T. H., eds., Modern and Ancient Continental Shelf Anoxia: London, Geological Society of London, Geological Society Special Publication No. 58, p. 201-219.

Sholkovitz, E. R., Landing, W. M., and Lewis, B. L., 1994, Ocean particle chemistry: The fractionation of rare earth elements between suspended particles and seawater: Geochimica et Cosmochimica Acta, v. 58, p. 1567-1579.

Stanley, J. K., Jr. and Byrne, R. H., 1990, The influence of solution chemistry on REE uptake by *Ulva lactuca* L. in seawater: Geochimica et Cosmochimica Acta, v. 54, p. 1587-1595.

Wright, J., Schrader, H., and Holser, W. T., 1987, Paleoredox variations in ancient oceans recorded by rare earth elements in fossil apatite: Geochimica et Cosmochimica Acta, v. 51, p. 631-644.

Zangerl, R. and Richardson, E. S., 1963, The paleoecological history of two Pennsylvanian black shales: Chicago, Chicago Natural History Museum, Fieldiana Geology Memoir, v. 4, 352 p.

UPWELLING, SEDIMENTATION AND ANOXIA CONTROL ON DEPOSITION OF PHOSPHATES IN THE LATE CRETACEOUS CALIFORNIA MARGIN

CARLOS FONSECA*

Geological and Environmental Sciences Department, Stanford University, Stanford, CA 94305

ABSTRACT: Sediments of the diatomaceous and laminated Marca Shale provide information on the interaction between anoxia and the occurrence of phosphatic nodules in organic-rich sediments along the Late Cretaceous continental margin of California. Sedimentological observations and geochemical analyses permit differentiation of dysoxic, suboxic, and anoxic facies within the Marca Shale. Characteristics of dysoxic facies include massive mudstones, localized macrobioturbation, occurrence of glauconite, and calcareous concretions. Suboxic facies are predominant in the Marca Shale and characterized by submillimeter laminated sediments, absence of macrobioturbation, occurrence of low-oxygen tolerant benthic foraminifers, episodic occurrence of pristine phosphatic sediments, preferential accumulation of vanadium, and modest content of total organic carbon and hydrogen-rich organic matter. Denitrifying conditions prevailed in the suboxic facies and could be associated with the modest preservation of organic matter in the Marca Shale sediments. Pristine phosphates occur episodically in laminated sequences of the suboxic facies and are locally related to intervals with disrupted laminations, soft-sediment deformation, low siliciclastic content, local abundance of biogenic components, and accumulation of lanthanide elements. Constant rates of pelagic sedimentation, fast burial, and oxygen-deficient environments characterized laminated sequences. In contrast, low sedimentation rates, long-term exposure, and fluid exchange of sediments with ambient sea water are associated with the occurrence of pristine phosphatic nodules and concentration of diagnostic trace metals. Anoxic facies are restricted to laminated sediments and characterized by high values of organic carbon, hydrogen index, and episodic precipitation of sulfides under reducing conditions. The occurrence of late Maestrichtian marine sediments associated with vigorous coastal upwelling along the California margin suggests a change in the paleoceanographic and climatic conditions at the end of the Cretaceous at 65 Ma.

INTRODUCTION

The occurrence of phosphates, laminated diatomaceous shales, and glauconitic mudstones in marine sequences deposited along the upper slope and shelf break of continental margins has long been considered a typical expression of coastal upwelling, high productivity, low sedimentation rates of detrital terrigenous components, and low-oxygen conditions of the bottom water column (Calvert, 1964; Burnett et al., 1983; Baturin, 1983; Ingersoll and Graham, 1983; Reimers and Suess, 1983; Suess et al., 1987; Vercoutere et al., 1987; Pike et al., 1993; Piper, 1991, 1994; Smith, 1992). In the modern ocean, the deposition of sediments associated with coastal upwelling occurs at specific sites of the continental margin and under particular oceanographic conditions. For example, the occurrence of coastal upwelling and deposition of organic-rich laminated sediments has been well documented along the western margin of North and South America, especially in the subtropical arid zones around the 30° latitude along the Peruvian and the California margins (Calvert et al., 1964; Parrish and Curtis, 1982; Dunbar et al., 1988; Froelich et al., 1988; Glenn and Arthur, 1988; Kim and Burnett, 1988; Garrison and Kastner, 1990; Garrison, 1992; Brodie and Kemp, 1994; Christensen et al., 1994; Dean et al., 1994; Thunell et al., 1994; Gardner and Dartnell, 1995). Both of these margins are characterized by the presence of eastern boundary currents and seasonal variations in the strength and direction of coastal winds and resulting Ekman transport and coastal upwelling (Mann and Lazier, 1991). The large amount of nutrient-rich water supplied by the seasonal upwelling fuels high levels of primary productivity in the surface waters of the California margin. Diatoms, dinoflagellates, and coccolithophores are responsible for most of the primary productivity and contribute their skeletons and organic matter to the sediments through processes involving creation of marine snow, self-sedimentation of plankton, flocculation, and production of fecal pellets (Dunbar and Berger, 1981; Grimm et al., 1996).

Much of the organic matter produced during coastal upwelling is consumed by grazing organisms and bacteria as it settles to the sea floor. However, the intermediate water beneath coastal upwelling regimes typically displays an intense oxygen minimum zone at 300–500 m characterized by less than 0.2 ml of dissolved O_2/l. Where this water mass intersects the sea floor it excludes bioturbating macroorganisms and enhances preservation of organic matter and primary sedimentary structures. Dissolved oxygen levels at mid water depths depend primarily on the rate of surface productivity, shape of the basin floor (e.g., silled basin), bottom water circulation, and rate of consumption of oxygen during oxidation of organic matter and respiration. All of these factors affect the intensity of the oxygen minimum zone and yield a distinct series of oxygen-deficient facies marked by various degrees of bioturbation, preservation of organic matter, authigenic mineral assemblages, and type of reducing reactions in bottom water and the sediment-water interface (Demaison and Moore, 1980; Calvert, 1987; Savrda and Bottjer, 1987; Vercoutere et al., 1987; Pedersen and Calvert, 1990; Savrda and Bottjer, 1991; Tyson and Pearson, 1991; Dean et al., 1994).

Late Maestrichtian paleoceanographic and sedimentologic conditions favored deposition and preservation of coastal upwelling deposits along the margins of ancient Tethys Ocean and along the California margin (Parish and Curtis, 1982; Almogi-Labin and Bein, 1993; Ly and Kuhnt, 1994). Deposits

*Present address: Shell International Exploration and Production B.V., Research and Technical Services, P. O. Box 2080 AB Rijswijk, The Netherlands email: c.fonseca@siep.shell.com

Marine Authigenesis: From Global to Microbial, SEPM Special Publication No. 66

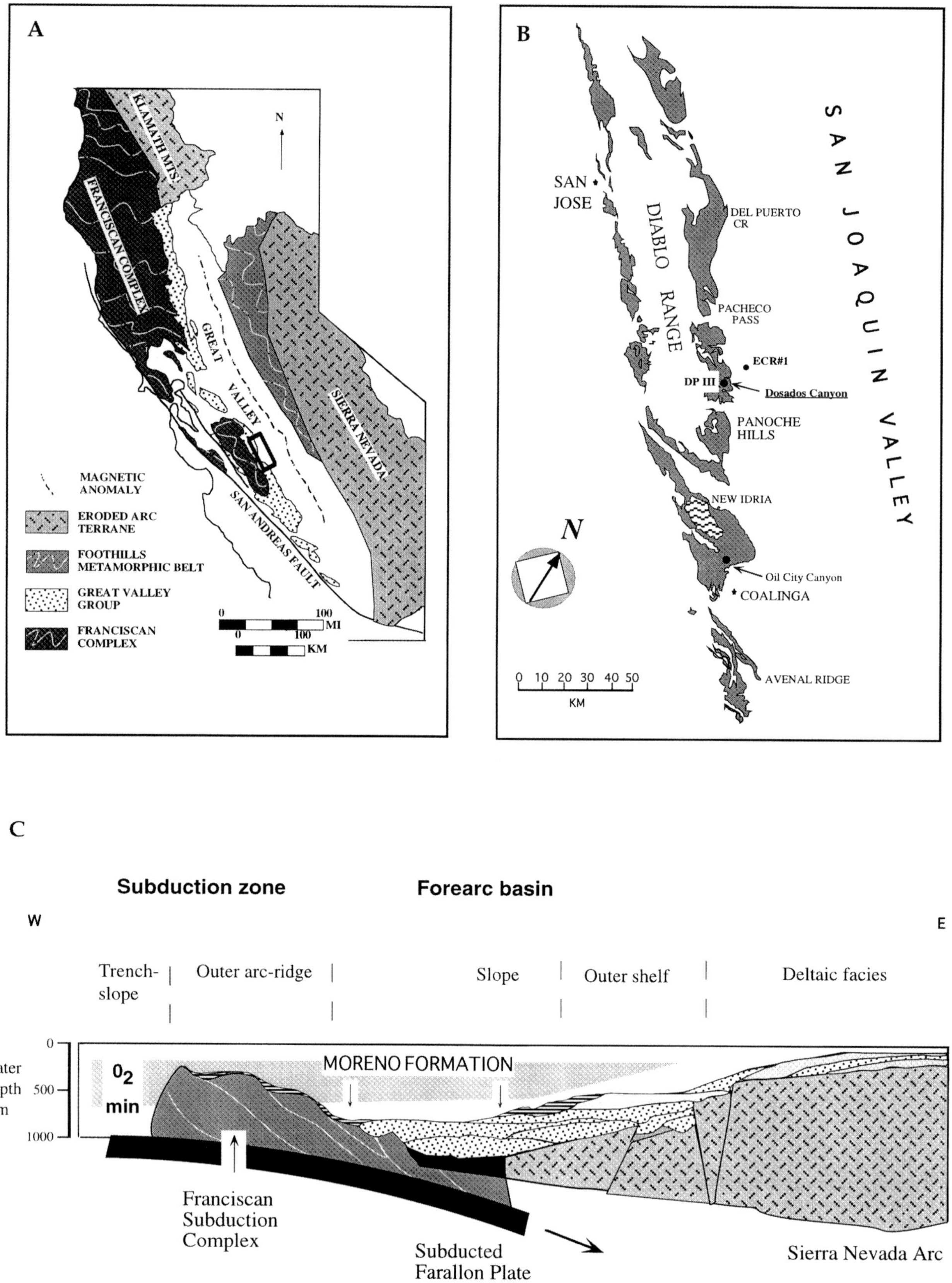

FIG. 1.—(A) General geologic map of California indicating the study area in the San Joaquin Valley. Rectangular box indicates Dosados Canyon area (ct. Fig. 1B). (B) Location map of the Dosados Canyon section in the Panoche Hills, along the San Joaquin Valley, California. ECR #1 and DP III correspond to the location of the cores Exxon Chaney Ranch #1 and Dos Palos Core III. (C) Diagram of the structural elements that constitute the silled-oxygen deficient depositional setting of the Moreno Formation (after Dickinson, 1995).

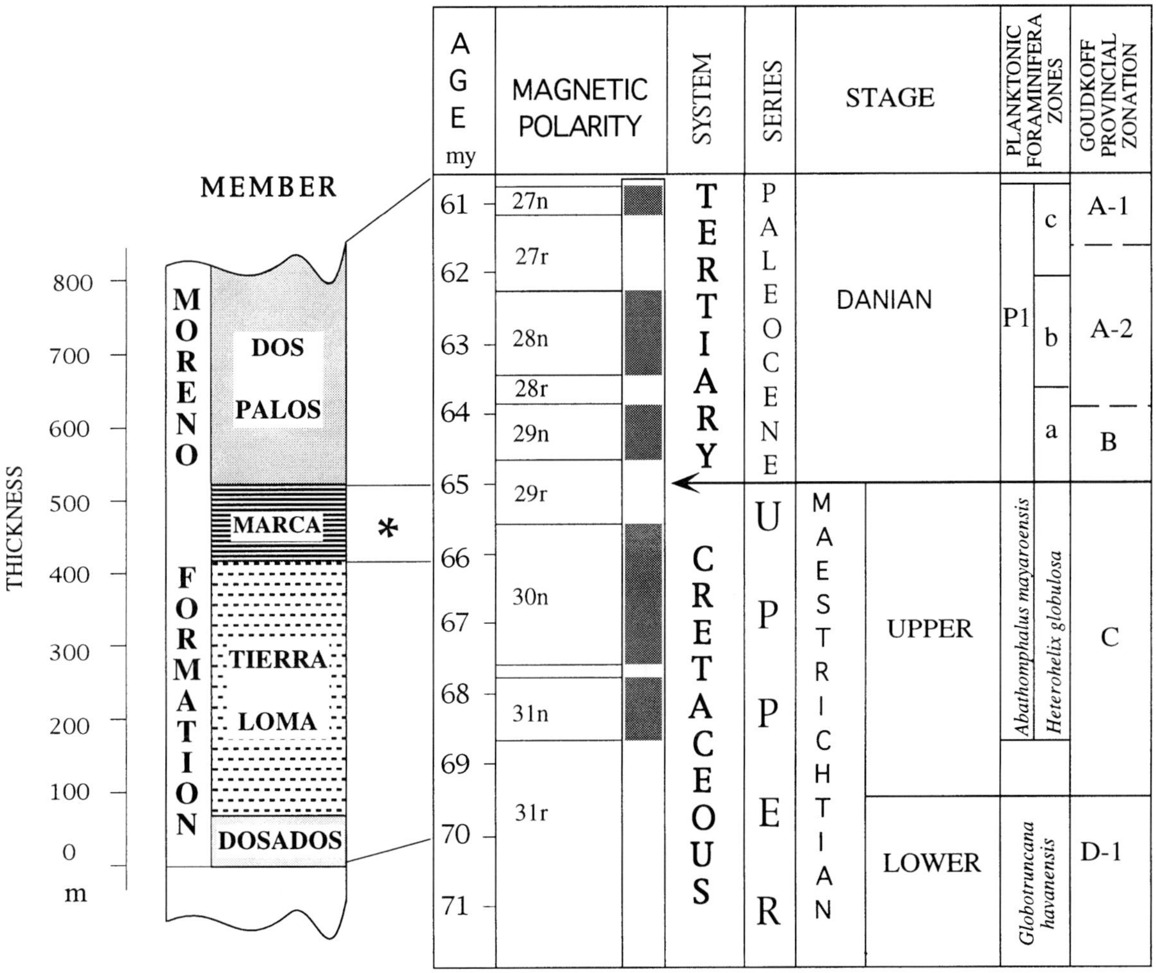

FIG. 2.—Chronostratigraphic chart of the Moreno Formation (Fonseca, 1997), based on the time scale of Gradstein et al., 1994.

in these margin areas also indicate that low-oxygen conditions prevailed during deposition of phosphatic and laminated sediments. In particular, upper Cretaceous sediments of the Marca Shale member of the Moreno Formation contain an exceptional record of high primary productivity, anoxic conditions, and deposition of authigenic phosphate along the upper slope of the Cretaceous California margin. In this paper sedimentological observations and geochemical analyses are integrated to address the occurrence of phosphates and laminated diatomaceous sediments in the Marca Shale and their association with a coastal upwelling environment. Sedimentological observations include outcrop and enlarged scanned images of laminated Marca Shale sediments; geochemical data include bulk elemental XRF analyses, and Rock-Eval pyrolysis data (McGuire, 1988) from surface and subsurface samples.

REGIONAL DEPOSITIONAL SETTING OF THE MORENO FORMATION

The Upper Cretaceous Moreno Formation crops out in the Panoche Hills along the western flank of the San Joaquin Valley of central California (Fig. 1A, B). The Moreno Formation represents the uppermost formation of the Cretaceous Great Valley Sequence deposited in an active ponded forearc basin along the upper Cretaceous margin of California (Ingersoll, 1978; Dickinson and Seely, 1979; Bartow and Nilsen, 1990; Fig 1C). During Maestrichtian time, the forearc basin was bounded by the Sierra Nevada volcanic arc to the east and the Franciscan subduction complex to the west (Ingersoll, 1978; Dickinson and Seely, 1979; Hamilton, 1979; Bartow and Nilsen, 1990; Moxon, 1990; Dickinson, 1995). An outer arc ridge developed by the continuous accretion of the subduction complex partially restricted ocean circulation between the basin and the Pacific Ocean, and apparently acted as an efficient sill to promote anoxia within the basin (Ingersoll, 1978; Dickinson and Seely, 1979; McGuire, 1988; Fig. 1C). Local irregularities that developed on the basin floor due to flexural downwarping or thermal subsidence (Moxon and Graham, 1987; Williams, 1996) created depressions that acted as turbidite depocenters for coarse clastic sediments, while topographic highs collected mainly pelagic sediments (McGuire, 1988). It was along the topographic highs and upper slope that the diatomaceous upwelling facies of the Marca Shale were deposited.

The Moreno Formation averages 800 m in thickness in its type section in Escarpado Canyon (Payne 1951), where it conformably overlies Maestrichtian lower fan sediments of the Panoche Group (McGuire, 1988). An angular unconformity characterizes the contact of the Moreno Formation with the overlying Paleocene Lodo Formation. Payne (1951) originally divided the Moreno Formation into four members, including from oldest to youngest: the Dosados Sand, Tierra Loma Shale, Marca Shale, and Dos Palos Shale (Fig 2). The lowermost unit, the Dosados Sand member, contains arenaceous foraminifera associated with turbiditic sands deposited on the upper fan-lower slope of the Upper Cretaceous forearc basin (Martin, 1964; McGuire, 1988). The Tierra Loma member is characterized by thin basal interlayered mudstones and sandstones that grade into dark brown platy shales and finally into the finely laminated diatomaceous shales of the Marca Shale member (Payne, 1951; McGuire, 1988). Both the Tierra Loma and Marca Shale contain similar benthic foraminiferal assemblages dominated by *Praebulimina prolixa, Siphogenerinoides whitei,* and *Gavelinella turbinata,* species typical of Goudkoff's provincial C zone (Martin, 1964; Almgren, 1986) and representative of an upper-middle slope facies (Sliter and Baker, 1972) and upper middle-bathyal water depths (Williams, 1996).

The uppermost Moreno Formation is characterized by the silty dark brown shales of the Dos Palos member and the fine-grained Cima sandstone. The occurrence of the coral *Flabellum,* the benthic foraminifera *Valvulineria lillisi,* and diverse forms of globigerinoid foraminifera suggest an outer shelf environment for the lower Paleocene Dos Palos Shale member (McGuire, 1988). Planktonic foraminifera within the Dos Palos Shale can be correlated to Goudkoff's provincial substages B and A2 and represent planktonic zone P1a and P1b of the Danian stage (Loeblich, 1958; Martin, 1964). Biostratigraphic analysis of foraminifera, silicoflagellates, dinoflagellates and radiolarians in the Marca Shale and Dos Palos has provided firm evidence that the Moreno Formation spans late Maestrichtian through early Danian time and includes the Cretaceous/Tertiary boundary (Loeblich, 1958; Drugg, 1967; Foreman, 1968; Fonseca, 1997). Extrapolation of the biostratigraphic zones identified within the Marca and Dos Palos shale to the revised Mesozoic magnetic polarity chrons (Gradstein et al., 1994) indicates that the Marca Shale ranges in age from ca 66 to 65 Ma.

METHODS

A 170-m-thick section of the laminated Marca Shale including its lower and upper contacts with the Tierra Loma and Dos Palos Shale members was measured in detail in Dosados Canyon (Fig. 1B). Selected samples of laminated Marca Shale were cut in slabs, scanned, and classified in terms of microfacies according to their texture and the occurrence of synsedimentary structures (i.e., laminated, massive, slumped, gravity flows, etc.) following the scheme of Behl (1995). Special attention was given to documenting occurrences of authigenic phosphates in relation to laminated and massive sediments of the Marca Shale. Textural variations in the sediments appear to be correlated with fluctuations in sedimentation rates, productivity, and anoxic conditions that likely reflect long-term changes in basin shape and paleoceanographic conditions (Fonseca, 1997).

To better understand the variations in texture of laminated Marca Shale, as well as the occurrence of authigenic phosphates, samples from the Dos Palos III C.O.F.R.C. well located in the Escarpado Canyon (1 km south from the Dosados Canyon section) and the Exxon Chaney Ranch #1 well (ECR) (located 13.5 km east from Dosados Canyon; sec. 29, T14S, R13E) were analyzed for organic content, kerogen type, and bulk elemental composition. Due to intense surface weathering of the Marca Shale, subsurface data were applied to better evaluate changes in productivity, anoxia, and preservation of organic matter. Rock-eval pyrolysis data (total organic carbon, TOC and hydrogen index, HI) from Marca Shale core samples from the Exxon Chaney Ranch #1 well (McGuire, 1988) were correlated to the Dosados Canyon section to complement the sedimentological data and provide evidence of organic content in unweathered samples of this unit. Correlation of stratigraphic intervals from the ECR is based on the co-occurrence of a distinct, thin cross-laminated sand at 52 m in the Dosados Canyon section at the 6694-6700 ft depth in the ECR well.

Elemental analyses were performed using X-ray fluorescence (trace elements) on outcrop samples from Dosados Canyon (23 samples) and selected core samples from the DP III well (11 samples). Samples from the Dos Palos Core III correspond to the upper contact of the Marca Shale with the Dos Palos member and have been correlated to the measured Dosados Canyon section based on the co-occurrence of calcareous concretions and the position of the Cretaceous-Tertiary boundary (Drugg, 1967; Fonseca, 1997). X-ray fluorescence analyses of the Dosados Canyon samples were performed at Stanford University on an automated X-ray fluorescence spectrometer, following the methodology of Norrish and Hutton (1969) and Norrish and Chappell (1977). Elemental analyses of the ECR and DP III samples were performed by X-ray Assay Laboratories in Don Mills, Ontario, Canada. Analysis of major element oxides was also performed to calculate the major sedimentary components present in Marca Shale sediments (Fonseca, 1997). Formulas to calculate the sedimentary partitions follow the criteria employed by Isaacs et al. (1990), Piper (1991), Aplin et al. (1992), Medrano and Piper (1992), Piper and Isaacs (1995), and Fonseca (1997). The calculated sedimentary components include biogenic silica, detrital silicates, apatite, and calcium carbonate content. The organic matter content in samples of the Dos Palos Core III was calculated based on the difference between the total amount of carbon and the amount of carbon from carbonates. TOC and HI values of Dos Palos Core III samples are taken from McGuire (1988).

Marca Shale lithologies were classified into oxygen-deficient facies following Tyson and Pearson (1991) and Savrda and Bottjer (1991) based on the preservation of laminated sediments and the occurrence of phosphates, glauconite, organic matter preservation, trace element distribution, and episodic bioturbation. In assigning the different anoxic facies, the recommended terminology of Tyson and Person (1991) has been used; Table 1. Their proposed terms for oxic environmental facies are used when referring to fluctuations in the inferred amounts of dissolved oxygen in the bottom waters and sediment-water interface (e.g., dysoxic, suboxic, anoxic). The term suboxic refers to those sediments where geochemical data suggest that

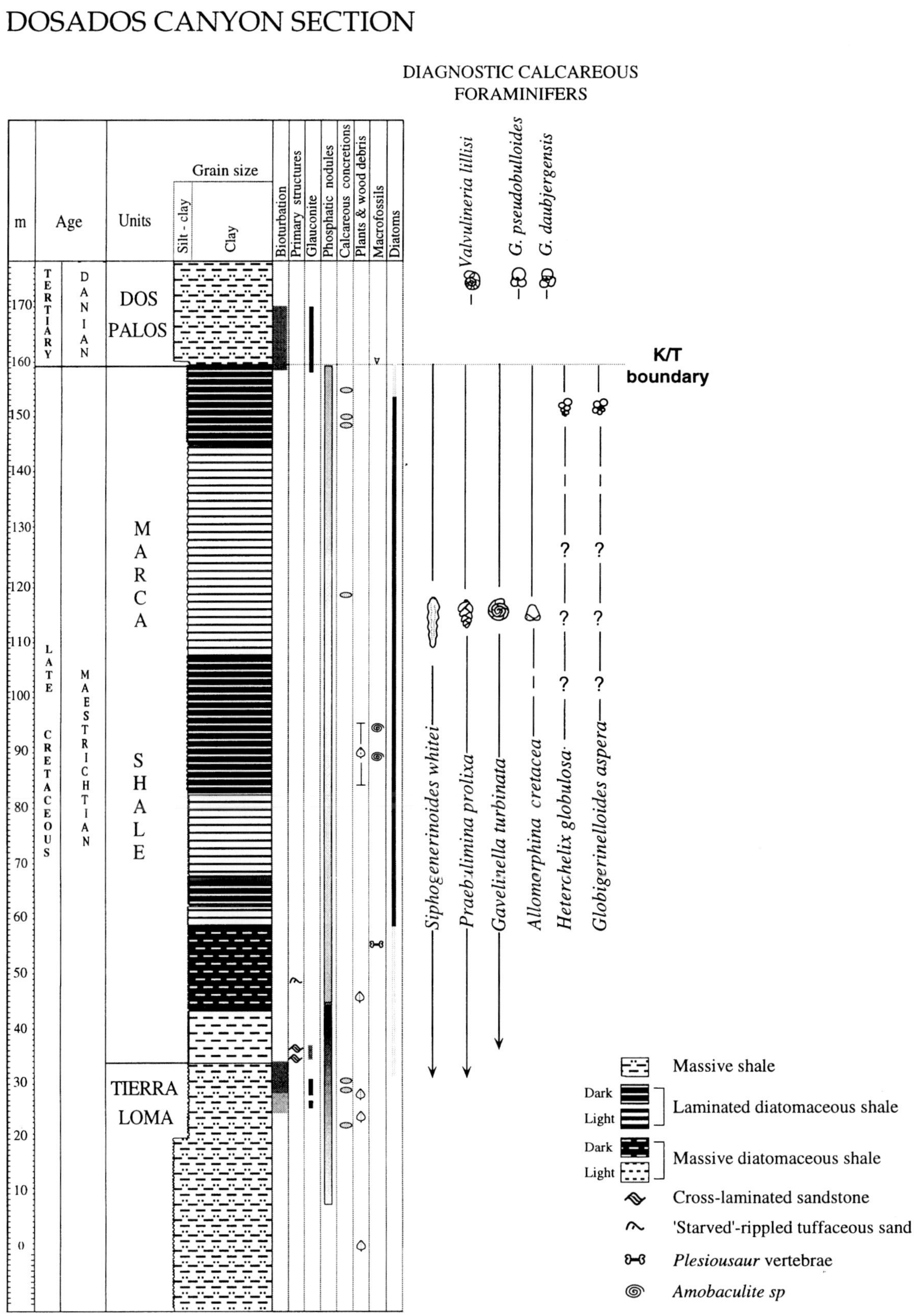

FIG. 3.—General stratigraphic section of the Marca Shale and its lower and upper contacts with the Tierra Loma and the Dos Palos Shale member in the Dosados Canyon.

TABLE 1.—FACIES CLASSIFICATION AND TERMINOLOGY FOR LOW-OXYGEN REGIMES AND RESULTING BIOFACIES

from Tyson and Pearson (1991)

Oxygen ml/l	Oxygenation regime and facies environment	Biofacies
8.0–2.0	Oxic	Aerobic
2.0–0.2	Dysoxic	Dysaerobic
2.0–1.0	moderate	
1.0–0.5	severe	
0.5–0.2	extreme	
0.2–0.0	Suboxic	Quasi-anaerobic
0.0 (H_2S)	Anoxic	Anaerobic

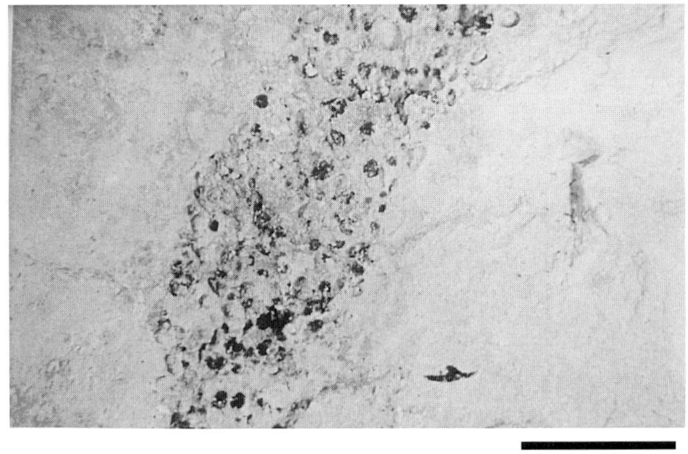

FIG. 4.—Planolites-burrow filled with glauconite grains in the Tierra Loma Shale, Dosados Canyon section. Bar scale is 0.5 cm.

denitrification took place, and anoxic where there is evidence of sulfate reduction. Similarly, the biofacies of Tyson and Pearson (1991) were employed when referring to the presence or absence of benthic fauna in the sediments. For instance, the term dysaerobic refers to the occurrence of macroorganisms or burrows in the sediments; quasi-anaerobic refers to sediments lacking *in situ* macrofauna but having benthic microorganisms, and anaerobic is restricted to the total absence of organisms or burrows. The following section includes a description of the lithology and sedimentology of the Marca Shale and its lower and upper contacts with the Tierra Loma and Dos Palos Shale members. The lowermost interval in the measured section (0–63 m) corresponds to the transition between the Tierra Loma and the overlying Marca Shale member. The intermediate interval (63–155 m) comprises the diatomaceous Marca Shale, and the uppermost interval (155–170 m) marks the contact between the Marca and Dos Palos Shale (Fig. 3).

STRATIGRAPHY AND SEDIMENTOLOGY OF THE MARCA SHALE

Tierra Loma-Marca Shale contact zone (0-63 m)

The transition from the dominantly siliciclastic shales of the Tierra Loma member of the Moreno Formation into the hemipelagic Marca Shale comprises the first 36 m of the measured Dosados Canyon section. Tierra Loma shales are characterized by massive texture and wispy fabric produced by slumping and discontinuous laminations. The lowermost 25 m of the section consists of friable, fissile, purplish brown, silty shale interlayered with millimeter-thick sand layers that include local abundances of terrestrial plant debris. Phosphatized wood has been previously reported within the lower part of the Tierra Loma associated with mudstones and phosphatic nodules (Gulbrandsen et al., 1963).

At the 26-m level in the Dosados section, close to the contact between the Tierra Loma and Marca Shale, a reduction in the amount of terrigenous material and a minor occurrence of carbonate mark a change in the prevailing sedimentary pattern. The mudstones in this interval become finer in grain size, more fissile, and abundant in benthic foraminifers. Discontinuous layers of calcareous concretions averaging 1 m in diameter occur at 25 m and 32 m in the section. The concretions contain foraminiferal casts of primarily benthic species locally concentrated within individual laminae. The most abundant genera include *Praebulimina, Siphogenerinoids,* and various rotaliids. It is likely that planktic foraminifera were originally present in these sediments, as they occur in localized calcareous concretions in the upper part of the section. However, preferential dissolution during deposition and diagenesis likely precluded their preservation in the sediments. Two meters above the first occurrence of calcareous concretions in the section (28-m level in section), non-laminated sediments of the Tierra Loma display an abundance of glauconite-filled *Planolites* burrows (Fig. 4).

Further up-section, calcareous concretions become more common in the Tierra Loma sediments. At 32 m purplish brown silty shales change to light pinkish gray diatomaceous fissile shale. This lithologic transition is marked by the conspicuous occurrence of two continuous thin layers of well-indurated black phosphatic nodules separated by 1.75 m of glauconitic shale. Both layers contain oblate (ca. 7 cm) calcareous sulfide-rich phosphatic nodules that are draped by a 2-cm-thick cross-laminated glauconitic sandstone. The occurrence of indurated black phosphatic nodules and glauconitic sandstones defines the contact between the Tierra Loma and the Marca Shale in the Dosados Canyon section.

Initial deposition of the Marca sediments at 36 m above the base of the Dosados Canyon section is marked by a decrease in the abundance of terrigenous components. Sediments above this level consist of massive siliceous light pinkish gray shale with abundant foraminiferal casts and scattered friable small phosphatic nodules. Six meters above the lower contact, the amount of glauconite decreases whereas foraminifer-rich layers and associated thin phosphatic laminae increase in relative abundance. Authigenic phosphatic and calcareous-rich laminae are common between 44 m and 50 m. The shale in this interval displays fissility and contains scattered, badly preserved fossil wood debris, discontinuous submillimeter laminations, and fish scales. At 61.5 m, the shale is cemented by opal C-T, and con-

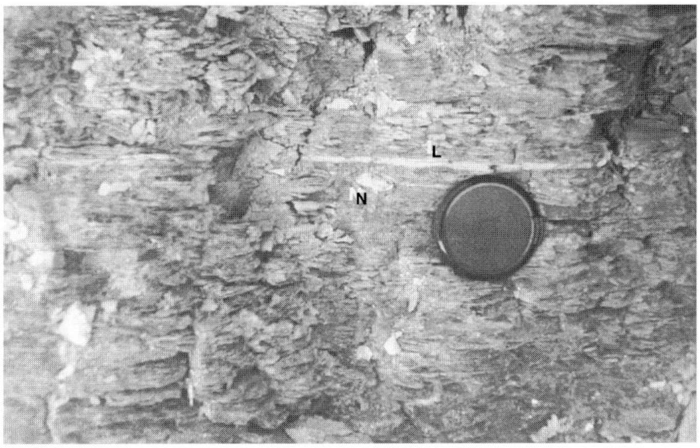

FIG. 5.—Photograph of massive and discontinuously laminated shales of Facies 2, showing the occurrence of phosphatic laminae (L) and friable nodule (N). Lens cover is 5.5 cm diameter.

tains discontinuous submillimeter laminations, common molds of benthic foraminifera, and a singular occurrence of articulated vertebrae bones of a *Plesiosaur* skeleton (J. Gregory, pers. commun., 1996). The skeleton appears to be in the same stratigraphic position from which other vertebrate fossils have been reported in the past (Hesse, 1936; Stock, 1941; Camp, 1942; Campbell, 1943; Wells, 1943; Foster, 1980). A meter above the *Plesiosaur* skeleton, a change in paleoceanographic conditions is signaled by the first appearance of striking light gray and cream colored diatomaceous shales that display well-defined submillimeter laminations, fissility, and excellent preservation of opal A. These textural changes reflect a decrease in terrigenous material and an increase in biogenic components, along with the effects of low dissolved oxygen in the overlying water column.

Laminated Marca Shale (63-155 m)

Common sedimentological characteristics of sediments stratigraphically above the 63-m level include meter-scale dark- and light-colored intervals of finely laminated diatomaceous shales with friable phosphatic nodules and an increased abundance of biosiliceous components. The benthic foraminiferal assemblage in the Marca Shale is dominated by *Siphogenerinoides whitei* and *Praebulimina prolixa*. Sediments of the Marca Shale display continuous submillimeter thick laminations, well-developed fissility, and contain siliceous organisms composed largely of opal-A phase. These latter sediments also display slumping, soft sediment deformation, and the occurrence of low-density, fine-grained gravity flows that produced local scouring of laminae. Macrofauna and burrows are absent in these facies.

Sediments from 66 m to 71 m in the Dosados section contain a higher abundance of massively bedded shale units relative to laminated intervals. Some of the nonlaminated intervals display a wispy fabric composed of millimeter-sized eroded laminae intraclasts. Other massive intervals consist of normally graded turbidites composed of fine, silt-sized material. The graded turbidites usually display erosive lower contacts with silt grains and benthic foraminifers at their base. Most of the gravity flow beds in this unit are thinner than 3 cm and composed of silt and clay. Massive intervals and slumped laminations are also a common feature in this part of the section. Friable autochthonous phosphatic laminae and nodules are present throughout the unit in association with ammonites, fossil wood, and laminae rich in calcareous foraminifers (Fig. 5).

Sediments from 111 m to 155 m reflect the impact of ongoing tectonic uplift of the basin and climate change just prior to the Cretaceous-Tertiary (K-T) boundary. Sediments above 141 m to the upper contact with the Danian Dos Palos Shale member (160 m) display lithologic variations reflecting changes in the abundance of sedimentary components, dissolved oxygen levels, and productivity. Sediments within this interval are characterized by an abundance of slumps, soft sediment deformation, and a siltier texture in response to an increase in terrigenous sedimentation. At the 145.4-m level, three discontinuous layers of calcareous concretions contain abundant benthic foraminifers and mark the end of the diatomaceous facies in the Marca Shale. Siliciclastic sediments overlying the youngest concretionary layer are massive and poorly cemented and display discontinuous laminations and rare calcareous foraminifers.

Marca Shale-Dos Palos Shale Contact Zone (155-165 m)

Between 145 and 155 m, three stacked discontinuous horizons of calcareous concretions containing abundant benthic foraminifera mark the contact of the Marca Shale with the overlying Dos Palos member and mark the top of the coastal upwelling facies within the Moreno Formation. The foraminiferal assemblage in the concretions is similar to that found in the underlying Tierra Loma-Marca Shale transitional interval (32 m) but is marked by higher diversity of benthic species and the rare occurrence of planktonic species. Common benthic foraminifera from this interval include *Allomorphina cretacea, Bulimina prolixa, Dentalina basiplanata, Nonionella cretacea, Praebulimina reussi,* and *Siphogenerinoides withei*. Planktonic foraminifers are represented by the small species *Heterohelix globulosa* and *Globigerinelloides aspera*. At 155 m, massive layers, truncated laminations, and low density turbidites increase. Slumping and soft sediment deformation is also common in this portion of the section, as are minor occurrences of wood debris. The shales become more massive and siltier and lose their fissility beginning at 157 m and lack the benthic foraminifera that characterize the upper Tierra Loma and Marca Shale. At the 160-m level, a 1-cm-thick dark brown bioturbated layer containing goyazite spherules, reworked bone debris, shark teeth, and glauconite marks the K-T boundary. Above the K-T boundary, mudstones of the Dos Palos Shale lack any fissility or laminations, and are heavily bioturbated and rich in glauconite. Sediments from 160 m to 165 m are characterized for their silty texture and lack of calcareous and siliceous microfossils.

FIG. 6.—Microfacies 1 continuous hemipelagic laminations in the Marca Shale. Laminations reflect seasonal sedimentations of light-pelagic components and dark-detrital sediments. Bar scale is 55 mm.

Laminations and microfacies of the Marca Shale.—

It is possible to subdivide the laminated diatomaceous sediments of the Marca Shale into three microfacies following the scheme erected by Behl (1995) for analogous Holocene sediments in the Santa Barbara basin. Microfacies recognized include: (1) well-laminated hemipelagic shale; (2) trace-laminated hemipelagic shale with massive intervals; and (3) well-laminated with graded and massive clayey silt turbidites, slumps, and soft deformation.

Microfacies 1.—In the Dosados Canyon section, microfacies 1 is characterized by cyclic alternation of light and dark laminae or couplets (Fig. 6). The light laminae are composed of up to 45% biogenic opal (Fonseca, 1997) while dark laminae contain mostly clay and silt (up to 80% of detritus).

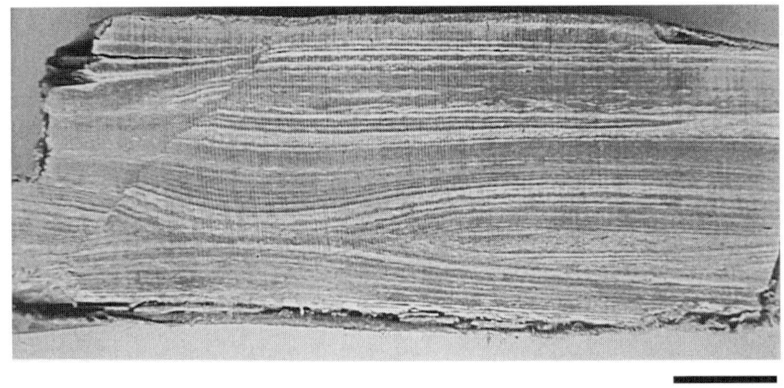

FIG. 7.—Microfacies 2, discontinuous laminations with interbedded channelized gravity flow (lower right) drapped by hemipelagic laminae. The wavy geometry of the laminations could be interpreted as the product of differential compaction of sediments surrounding the channelized flow or the scouring and reworking of older sediments by bottom currents. Bar scale is 1 cm.

Micropaleontological analysis indicates that the light laminae are composed of monoespecific assemblages of diatoms, resting spores and archaemonads that represent bloom conditions during periods of intense upwelling and dry climate. In particular, individual laminae of *Chaetoceros*-type resting spores in the Marca Shale strongly suggest high productivity and depletion of nutrients in surface waters after vigorous seasonal upwelling conditions in the Cretaceous California waters (Fonseca, 1997). In contrast, dark laminae are predominantly composed of clay, fine silt grains, fragmented frustules of transported diatoms, and a robust benthic diatom assemblage dominated by the genus *Paralia*. Dark laminae presumably developed during the rainy season and were marked by weak or no upwelling and relatively low primary productivity (Fonseca, 1996).

Image analysis of the laminated Marca Shale (Fonseca, 1997) indicates that the average thickness of individual laminations is 0.3 mm. The laminations display vertical variations in biogenic and detrital contents throughout the section in response to external forcing climatic mechanisms (Fonseca, 1997). The average calculated sedimentation rate for microfacies 1 is 0.05 cm/yr, based on image analysis of the Marca Shale compacted laminae-couplet thickness. The vertical extent of continuous laminated facies is commonly constrained by the episodic occurrence of very fine-grained, low-density turbidite deposits.

Microfacies 2.—This facies is distinguished by discontinuous laminations displaying scouring and microchannels (Fig. 7). Light lamina within this facies are usually faint and interlayered with massive intervals of clay and disrupted laminae. The non-laminated intervals are commonly bounded by relatively thick (> 0.25 mm) continuous light laminae overlying massive intervals. No unequivocal bioturbation can be distinguished and phosphatic laminae and nodules are common within the massive intervals (Fig. 5). The laminated portions of microfacies 2 contain very thin light biogenic laminae and thicker dark detrital laminae that combine to produce sedimentation rates of 0.07 cm/yr. In this microfacies, light laminae average < 0.14 mm in thickness, which is half the thickness of dark laminae.

Microfacies 3.—Microfacies 3 is marked by the episodic occurrence of massive and graded silty beds. These beds are usually less than 3 cm in thickness, and display scoured bases

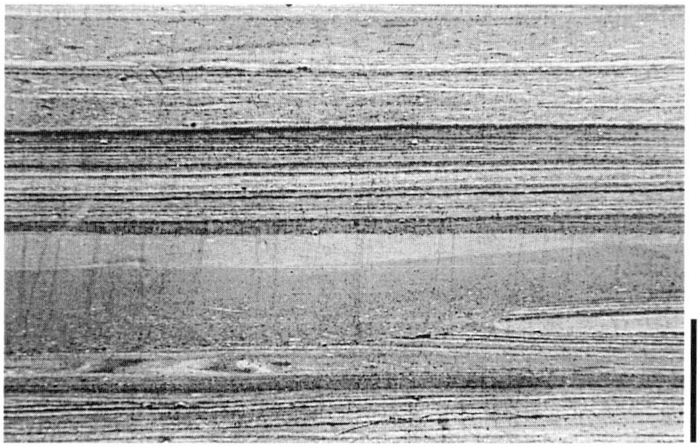

FIG. 8.—Microfacies 3, graded silt-clay turbidite within laminated facies of the Marca Shale. The turbidite displays normal grading with microscopic intraclasts of broken laminae (chips) at the bottom and scouring of underlying laminae (right side). Note the sharp top contact with overlying laminae. Bar scale is 1 cm.

and sharp upper contacts. Some of the beds are normally graded with intraclasts of eroded laminae and silty particles at the base that grade into clay-rich sediments at the top (Fig. 8). Other beds are massive or composed exclusively of microintraclasts of eroded laminae producing a wispy texture. I interpret these beds to be the product of low-density gravity flows within well-laminated sequences (e.g., microfacies 5 of Behl, 1995). Microfacies 3 is also characterized by slumps and soft sediment deformation of preexisting laminae sets. The slumped sets of laminae contain light laminae of unusual thickness (>1 mm). Evidence of soft sediment deformation is present in both laminated and massive intervals. However, a marked increase in massive and slumped laminae in the dark detrital-rich intervals occurs up-section, likely reflecting instability of the upper slope and the uplift of the basin.

The array microfacies follow the cyclic alternation of meter-thick light and dark intervals in the Marca Shale, with oscillations in terrigenous and biogenic opal content (Fonseca, 1997). Light, biogenic-rich intervals are predominantly composed of microfacies 1, followed by facies 3 (F1>>3). In contrast, dark detrital-rich intervals are composed primarily of facies F3>>2. The faint disrupted laminations of microfacies 2 probably reflect suboxic conditions, low sedimentation rates, erosion by bottom currents, and episodic occurrence of low density gravity flows. No macrobioturbation was detected in any of the microfacies.

Elemental Analysis and Sedimentary Components

The co-occurrence of organic-rich diatomaceous sediments and phosphatic particles suggest that the Marca Shale member was deposited beneath an active coastal upwelling system and a well-developed oxygen minimum layer bounded on its lower and upper contacts by more siliciclastic facies that are represented by the Tierra Loma and Dos Palos Shale. Analysis of the minor elements from the detrital fraction in the Marca Shale provides valuable information on the general composition of the sediments, the occurrence of authigenic phosphate, and reducing conditions within the depositional environment. The distribution and concentration of trace metals during sedimentation can be influenced and modified by several factors, including the reducing nature of bottom waters, location of the redox boundary in the sediment water-interface, and changes that take place during early diagenesis (Shaw et al., 1990; Piper, 1991, 1994; Piper and Isaacs, 1994, 1995). In ancient sediments, the distribution of minor elements is also affected by diagenesis and weathering. The majority of the elemental analyses in this study was performed on samples from the Dosados Canyon surface section and a limited number of samples from the Dos Palos III core (Tables 2 and 3).

To assess the mechanisms that fostered the accumulation of minor elements in the Marca Shale sediments it is important to establish their provenance or major primary sources. Two of the major sources of vanadium, copper, and zinc are the detrital fraction that includes terrigenous material, and the biogenic fraction, which includes refractory components (skeletons) and soft organic tissue. A minor source of trace metals is the hydrogenous fraction containing metals associated with hydroxides; distribution depends largely on the nature of the reducing waters. Estimation of the distribution of metals derived from biogenic components versus those derived from terrigenous material can be achieved by normalizing the bulk concentrations of the trace metals to a well-established terrigenous marker. In this case, Al_2O_3 and Ti have been chosen as background terrigenous markers supplied mainly by the clastic fraction (Shaw et al., 1990). Aluminum oxide has been selected because it is a major component in the clay fraction of the Marca Shale, and has been used for the calculation of the sedimentary components (biogenic silica and detritus content, apatite, etc.) (Fonseca, 1997). Ti has been used to calculate the nondetrital fraction that is presumably linked to productivity and associated with the accumulation of trace metals, including V, Cu, and Cr.

Variations in the distribution and concentration of major oxides in outcrop and core samples indicate that sediments of the Marca Shale can be divided into three main groups: authigenic-diagenetic, biogenic, and detrital. The authigenic-diagenetic sediments include phosphatic laminae (2PL/365-75), phosphatic nodules (2PL/365-75 Nodule, NODULES, Nodule-JIC), and calcareous concretions (2PL/335-45). These samples are characterized by their depleted content of terrigenous constituents (<8% Al_2O_3), large content of La, Ce, and nondetrital facies (apatite, calcite). The biogenic fraction is characterized by samples with low amounts of Al_2O_3 (12-9 %), high content of biogenic silica (37-53%), and variable amounts of minor elements (V, Cr, Cu, Zn, etc.). The siliciclastic detrital fraction comprises samples with Al_2O_3 contents higher than 13.5%, small amounts of biogenic silica (< 30%), and minor elements associated with detrital facies.

Comparison of minor element concentrations in the Marca Shale with published values for the World Shale Average (Wedepohl, 1969) indicate that sediments of the Moreno Formation have a marked difference in composition of the detrital fraction. Differences in composition include a higher content

TABLE 2.—SEDIMENTARY COMPONENTS OF THE MARCA SHALE MEASURED VIA XRF. A. DOSADOS CANYON SECTION, B. DOS PALOS CORE III

Stratigraphic level m	Sample number	Total Org. C %	Hydrogen Index	Al203 %	Detritus %	Biogenic Opal %	Calcite %	Apatite %	Org. matter %
A. Dosados Canyon section									
165	DP 1-2 Z0	1.40	80.00	16.11	79.92	18.10	0.54	0.61	-
160	0-1 K/T	1.70	100.00	18.94	82.00	0.00	0.00	11.58	-
148	4-40D	2.00	210.00	19.35	95.87	2.47	2.97	0.00	-
147	4-39.6	2.20	220.00	17.21	85.85	13.55	0.00	0.27	-
145.5	4-37B	1.90	200.00	16.58	82.60	15.90	0.00	0.18	-
119	4-11A	-	-	10.47	52.34	47.10	0.00	0.23	-
115	04/07/09	-	-	9.35	46.87	52.85	0.21	0.04	-
108.5	3B5.5	2.70	210.00	15.76	79.00	22.11	0.00	0.28	-
92	3B	2.50	220.00	19.95	98.23	0.00	0.00	3.38	-
87.5	AMB -B	2.60	220.00	14.1	70.70	10.60	0.00	0.07	-
77	3A	2.70	314.00	10.06	49.81	45.86	0.00	0.32	-
69	2B1	2.50	250.00	13.47	66.91	27.07	0.00	0.73	-
63	2A	2.70	250.00	9.46	46.32	45.46	1.68	0.28	-
58	DS-3	2.50	230.00	11.96	59.93	42.58	0.00	0.10	-
50	DS0	2.80	215.00	18.18	89.89	9.86	0.00	0.00	-
41.5	LENC 7	2.50	160.00	11.97	56.00	37.10	0.00	7.37	-
40	Nodule JCI	-	-	7.28	35.96	8.69	0.00	54.60	-
39	LENC 6	2.60	170.00	14.21	70.54	25.71	0.00	2.90	-
38.2	LENC 5	2.50	174.00	11.36	56.81	30.73	0.00	5.51	-
38	NODULES	-	-	2.08	9.78	1.32	10.98	66.28	-
36.5	LENC 4	3.10	177.00	13.78	68.52	29.31	0.00	1.73	-
25	LENC 3	3.00	180.00	17	83.84	15.68	0.00	0.00	-
1	TLM ENC 1	3.00	300.00	18.66	93.07	6.43	0.00	0.53	-
B. Dos Palos Core III									
170	2PL/263-273 (81)	1.61	102	15.6	94.38	4.30	0.00	0.00	2.97822
166	2PL/288-293 (88)	1.66	69	15.4	93.17	5.00	0.95	0.00	3.06866
163	2PL/297-305 (91)	1.72	77	11.2	67.76	27.50	0.00	0.00	2.78866
160	2PL/305-315 (94)	2.84	115	15.5	93.78	2.45	1.02	0.68	4.3211
156	2PL/315-325 (97)	2.32	81	15	90.75	5.50	0.51	0.02	3.68466
152	2PL/335-345 (103	0.57	153	2.8	16.94	0.00	80.78	0.40	0
146	2PL/355-365 (109	1.84	198	10.2	61.71	31.80	3.50	0.00	2.8203
144	2PL/365-375 (112	2.67	193	13.7	82.89	11.05	3.29	1.07	4.00876
143.8	2PL/365-nodule	1.91	-	4.4	26.62	11.60	5.55	36.34	1.90554
143.6	PL/365-Ph. lamin	1.07	-	7.13	43.14	18.65	22.58	14.23	1.0682
141	2PL/375-385 (115	2.40	196	13.8	83.49	12.70	2.39	0.20	3.24464

*Organic matter values only available for the Dos Palos Core III.

TABLE 3.—CONCENTRATION OF MINOR ELEMENTS IN THE MARCA SHALE. A. DOSADOS CANYON SECTION, B. DOS PALOS CORE III

A. Dosados Canyon section

Stratigraphic level	Sample number	MnO %	V ppm	Cr ppm	Ni ppm	Cu ppm	Zn ppm	Ti ppm	Ba ppm	La ppm	Ce ppm	U ppm	Mo ppm
165.00	DP 1-2 Z0	0.01	137.00	165.00	19.30	36.10	69.10	0.69	383.00	17.20	38.30	9.20	-
160.00	0-1 K/T	0.03	148.00	236.00	101.00	30.00	227.00	0.64	954.00	37.90	131.00	23.90	-
148.00	4-40D	0.03	101.00	139.00	36.30	40.00	146.00	0.72	465.00	17.40	40.60	11.90	-
147.00	4-39.6	0.02	124.00	129.00	43.90	41.70	146.00	0.62	372.00	23.40	32.20	6.40	-
145.50	4-37B	0.03	152.00	116.00	40.10	39.70	145.00	0.58	387.00	6.00	19.40	5.70	-
119.00	4-11A	0.02	143.00	91.70	27.20	27.50	90.20	0.35	223.00	4.90	14.50	5.10	-
115.00	38448	0.03	106.00	80.90	22.70	23.30	69.30	0.34	211.00	5.60	9.50	4.10	-
108.50	3B5.5	0.03	217.00	140.00	44.60	36.70	137.00	0.50	305.00	10.10	17.50	5.50	-
92.00	3B	0.03	195.00	168.00	71.60	52.70	171.00	0.59	427.00	14.70	40.20	11.10	-
87.50	AMB -B	0.04	151.00	147.00	38.80	48.70	152.00	0.59	381.00	12.10	19.60	6.90	-
77.00	3A	0.02	122.00	97.10	57.30	27.10	101.00	0.39	5490.00	23.50	52.40	3.40	-
69.00	2B1	0.04	127.00	109.00	62.10	32.90	113.00	0.45	478.00	38.10	78.40	4.70	-
63.00	2A	0.04	98.90	86.20	40.40	26.90	97.90	0.34	372.00	11.60	17.10	4.20	-
58.00	DS-3	0.01	104.00	87.20	36.40	29.20	83.60	0.42	232.00	21.30	41.40	4.80	-
50.00	DS0	0.02	110.00	157.00	57.40	56.00	143.00	0.64	398.00	3.80	14.30	8.50	-
41.50	LENC 7	0.05	72.80	83.30	140.00	27.00	264.00	0.30	582.00	47.80	63.10	9.80	-
40	Nodule JCI	0.08	54.00	62.60	59.60	26.50	433.00	0.19	1740.00	168.00	333.00	22.00	-
39.00	LENC 6	0.04	30.90	41.60	41.20	19.80	266.00	0.08	1460.00	53.20	60.90	14.40	-
38.20	LENC 5	0.03	218.00	136.00	80.20	12.30	157.00	0.51	547.00	26.10	41.20	8.20	-
38.00	NODULES	0.09	107.00	98.60	81.20	29.40	153.00	0.49	415.00	19.10	44.20	9.60	-
36.50	LENC 4	0.03	102.00	113.00	68.80	35.30	93.00	0.50	372.00	11.90	29.10	6.30	-
1.00	TLM ENC 1	0.05	128.00	134.00	94.30	47.10	202.00	0.66	479.00	14.50	22.70	8.20	-

B. Dos Palos Core III

Stratigraphic level	Sample number	MnO %	V ppm	Cr ppm	Ni ppm	Cu ppm	Zn ppm	Ti ppm	Ba ppm	La ppm	Ce ppm	U ppm	Mo ppm
	2PL/263-273	0.04	110	130	9	33	32.3	133	586	22.3	43	7.7	15
	2PL/288-293	0.04	106	140	10	43	30.9	320	611	23	45	5.7	5
	2PL/297-305	0.03	88	97	7.8	32	25.8	246	399	14.9	30	4.9	11
	2PL/305-315	0.04	105	140	9.7	44	36.2	159	589	27.3	52	9	8
	2PL/315-325	0.04	104	120	9.7	43	34.5	141	595	23.6	45	5.6	7
	2PL/335-345	0.06	29	29	2.6	6	5.2	25.5	0	5.2	9	1	0
	2PL/355-365	0.03	84	94	7.9	29	25	74.2	383	13	26	3.1	0
	2PL/365-375	0.04	158	120	9	39	33	130	545	21	40	6	3
	2PL/365-nodule	0.07	66	71	10	84	12.6	39.1	1690	85.2	168	21.7	0
	2PL/365-Ph. lamin	0.06	119	100	4.2	15	14.9	47.6	665	83.9	155	9.3	2
	2PL/375-385	0.03	157	96	8.7	35	31.4	103	482	17	34	4.1	3

*Molybdene values only available on samples of the Dos Palos Core III of Al_2O_3 (16.7%), lower ratios of La/Al_2O_3 (1.2), Ce/Al_2O_3 (2.5), and lower average content of Cu (22.6 ppm) and V (36 ppm).

Based on concentrations of siliciclastic detritus (Al_2O_3 content) in the samples from the Dosados Canyon section and Dos Palos III core, the distribution of individual minor elements was partitioned into detrital and marine or nondetrital (excess) fractions. The nondetrital element fractions were calculated by subtracting the background terrigenous contribution from the bulk sample content (Medrano and Piper, 1992):

Marine fraction of X =
X bulk content of sample – [kX • sample detritus content]

The minimum detrital content-element coefficient (kX) represents the value of the sample with the lowest ratio of X/Al_2O_3 in the section. The product of kX by the detrital content of a given sample yields an estimate of the total detrital fraction of X in the sample. The nondetrital-marine fraction is obtained by subtracting the detrital fraction from the bulk content of X. This reasoning is only valid if the mineral composition of the detrital fraction remains constant throughout the section and if there has not been large remobilization of element X during diagenesis. In general, there is a modest correlation of a given minor element to total detritus content, which indicates the existence of other nondetrital sources. The marine fraction of individual trace metals shows distinct distributional trends that correspond to their relative concentration in the sediments, enrichment in organic matter, and preferential precipitation or solubility under different reducing conditions.

The analysis of minor elements in the Dos Palos core #III shows a fair correlation between the concentrations of Cu and organic matter, which suggests the incorporation of this element into the soft tissues of marine primary producers during settling of the water column (Fig. 9A). It is also important to note that the occurrence of apatite in sediments of the Marca Shale is associated with low concentrations of siliciclastic detritus and a higher concentration of trace metals (Fig. 9B). In general, sediments containing more than 15% of marine apatite have concentrations of MnO, U , Ba , and lanthanides (Ce, La) above the mean average of the detrital fraction (Fig. 9C, D; Tables 2 and 3). In contrast, samples with low V and Cr (e.g., 2PL/263-273 and 2PL/288-293) apparently reflect the cessation of coastal upwelling, higher detrital content, and very low primary productivity, typical of the massive facies in the Marca Shale and the lower Dos Palos member (Fig. 9E, F).

The distribution of V and Cu in the Marca Shale indicates that they are also adsorbed into the organic matter of primary producers. Thus, their relative concentrations over detrital Ti can provide additional information on variations in primary productivity and reducing conditions through time. Deep weathering in the sediments of the Marca Shale in the Dosados Canyon exposures precludes any direct correlation of trace metals to organic matter variations. However, it is observed that larger concentrations of V/Ti and Cu/Ti coincide with deposition of laminated diatomaceous facies of the Marca Shale and with larger hydrogen index values (HI) in correlative samples from the Exxon Chaney Ranch (Fig. 10). Correlated HI values in the laminated Marca Shale are 200–350 mg hydrocarbons/gm of organic carbon corresponding to a II marine-type kerogen of moderately oil-prone marine facies B (Tissot and Welte, 1984; Jones, 1987; McGuire, 1988). The ratios V/Ti and Cu/Ti follow the general distribution and variations in abundance of biogenic opal throughout the section. The latter relationship suggests that mild reducing conditions were established during deposition of the laminated sediments of the Marca Shale and that primary productivity was closely related to the enhancement of anoxic conditions. Similar distribution patterns of Cu/Ti and biogenic opal also indicate that siliceous plankton were likely the main source of this metal.

The co-occurrence of high hydrogen index values and high biosiliceous productivity reflects the coupled system between productivity and preservation of organic carbon under anoxic conditions (Dean et al., 1994; Gardner and Dartnell, 1995). Exceptions to this assumption occur at 90 m in the Dosados Canyon section where low biogenic-silica content contrasts with high values of hydrogen index and V/Ti, which suggests other nonbiosilica sources of organic matter and the importance of bottom anoxia for preservation of organic matter.

The average enrichment of nondetrital minor elements is small in the Marca Shale and may be a reflection of the modest average total organic carbon content (ca. 3%) of these sediments. In particular, metals such as Mo, V, Cr, and Cu are known to be good indicators of reducing conditions in bottom waters and at the sediment water-interface, as their insoluble reduced species precipitate at low Eh values (Lewan, 1984; Shaw et al., 1990; Wanty and Goldhaber, 1992; Piper, 1994; Piper and Isaacs, 1995). In general, Marca Shale sediments are enriched in V, Cr, and Zn, followed by Ce and La. These elements also display a similar stratigraphic distribution to that of biogenic silica and nondetrital components. According to their redox characteristics, denitrifying and mild reducing conditions would be necessary to sequester these elements from sea water via adsorption to organic matter (Piper and Isaacs, 1994). Under these conditions, organic matter would be oxidized by bacteria via nitrate reduction, after dissolved oxygen falls to ca. 5% of normal oxygenated waters (Froelich et al., 1979; Berner, 1980; Stumm and Morgan, 1981). Low-oxygen conditions and denitrification could have been accompanied by marine precipitation of V_2O_4 and $Cr(OH)_3$. In contrast, suboxic conditions and precipitation of phosphate could be associated with the relatively small abundance of MnO, La, and Ce in the Marca Shale (Murray et al., 1983; Piper and Isaacs, 1994).

According to Piper and Isaacs (1994), Cr and Ce would have precipitated before V under less reducing conditions (0.54 and 0.320 Eh, respectively) immediately after nitrification was initiated as a major oxidizing process in oxygen-depleted waters. Therefore, if remobilization did not occur during burial under reducing conditions, the distribution of Cr and Ce can be used to delimit the initiation of denitrification in sediments of the Marca Shale. The distribution of Cr/Ti and Ce in the Marca Shale (Figs. 9C and 10, Tables 3A, B) indicates that these metals were only concentrated during periods with relatively larger content of dissolved oxygen and phosphate precipitation. The small but rela-

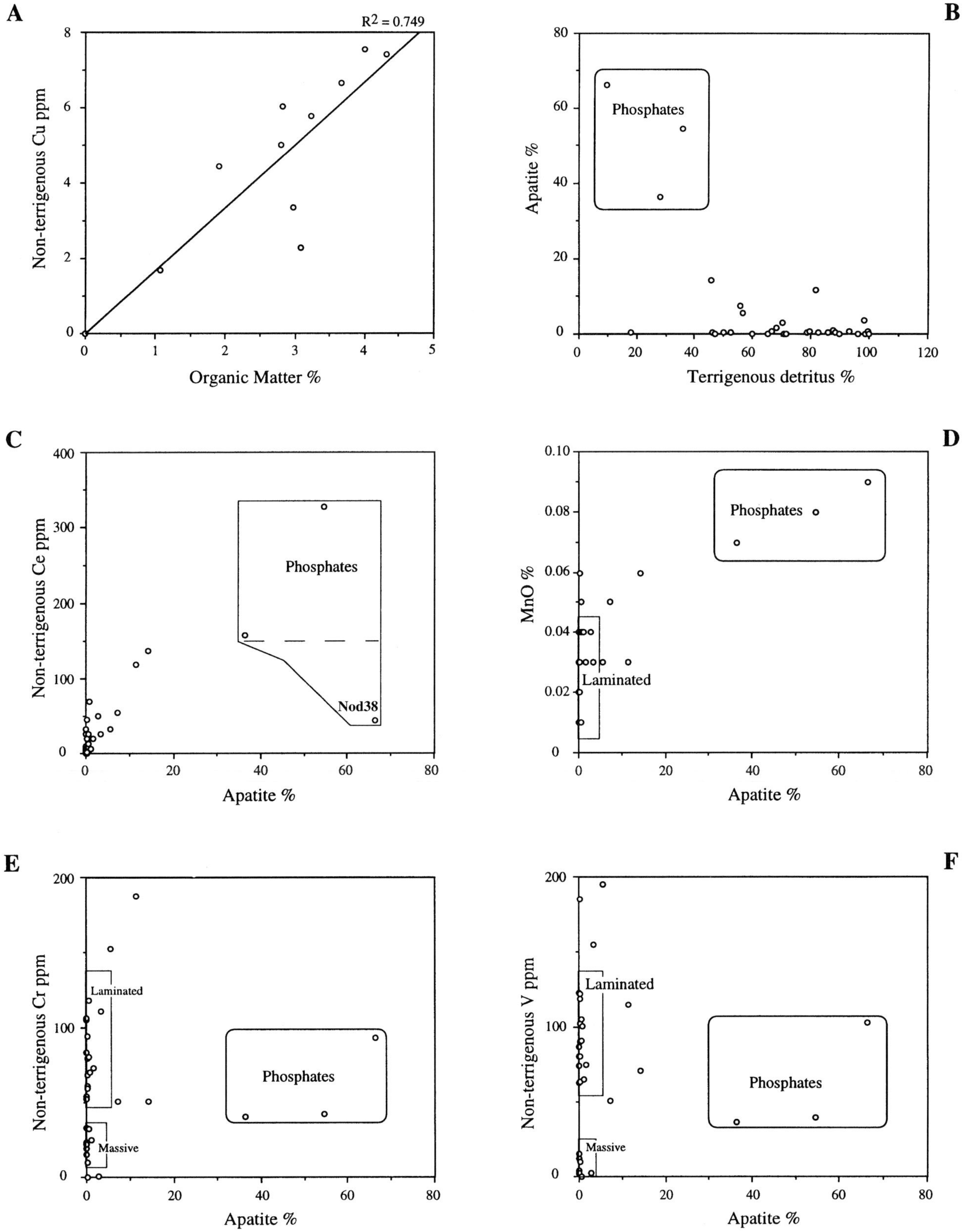

FIG. 9.—Correlation plots of diagnostic trace metals, organic matter and marine apatite. A. Correlation (R2 = 0.75) of non-terrigenous Cu/Organic matter content from sediments of the Dos Palos Core III, suggesting adsorption of Cu to the soft organic matter as it settles through the water column. B-F Display the abundance of non-terrigenous elements and oxides respect to the content of authigenic apatite. Low content of terrigenous components and relatively high contents of Ce and MnO suggest phosphate precipitation occurred under non-reducing conditions and low sedimentation rates. Laminated intervals display high content of V and Cr and low contents of MnO.

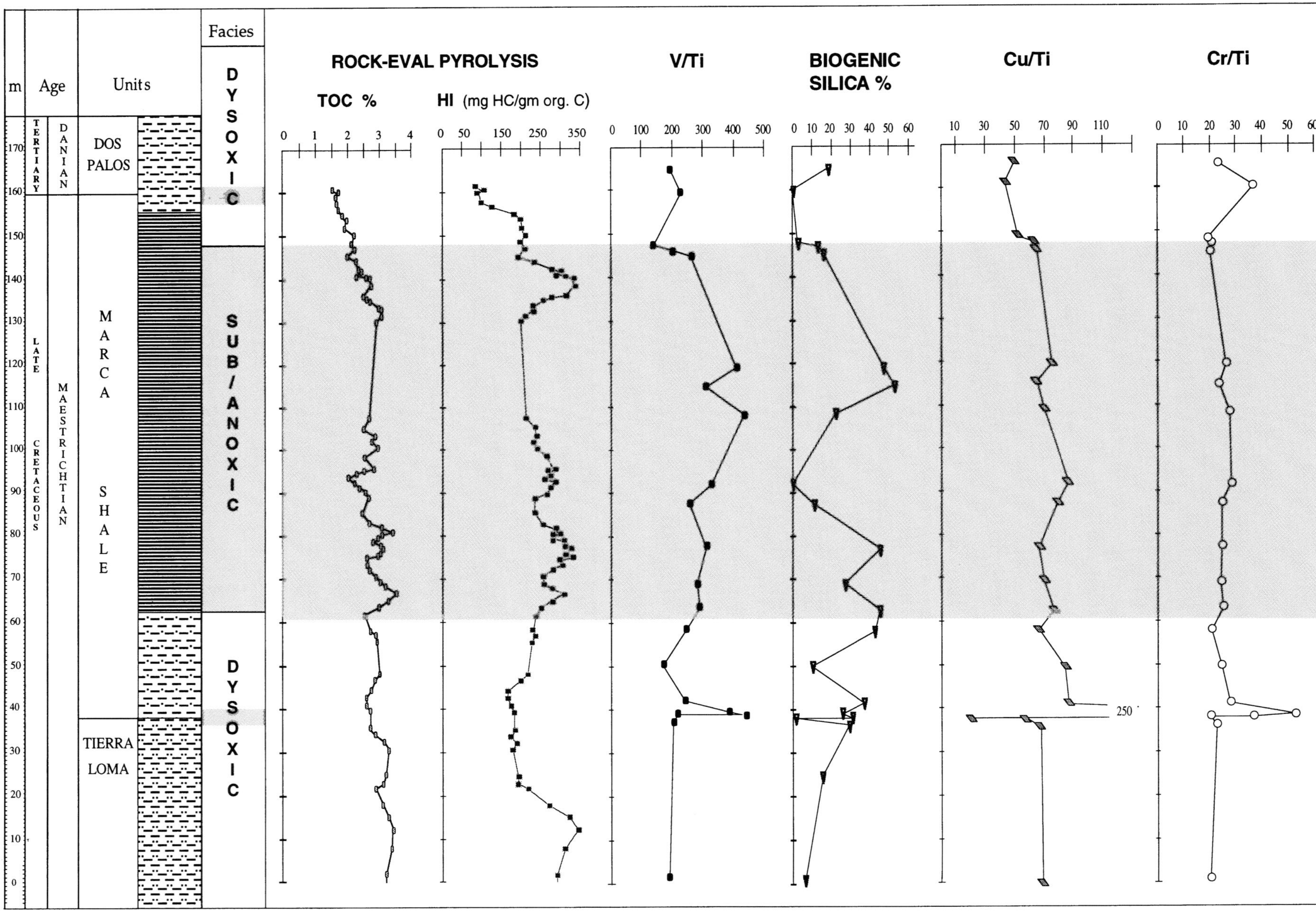

FIG. 10.—Distribution of minor elements in the Dosados Canyon section. Large shaded area corresponds to relatively high concentrations of V/Ti, biogenic silica and Cu/Ti in the anoxic laminated facies of the Marca Shale. The small shaded anoxic events (facies column) at the lower and upper contacts of the Marca Shale are related to the occurrence of phosphatic-rich condensed intervals. Rock-eval pyrolysis data taken from McGuire (1988).

tively higher concentrations of MnO in the associated phosphatic nodules and laminae also support the existence of mildly or perennial oxidizing conditions during phosphate deposition. In contrast, the precipitation of V would have taken place under more reducing conditions (–0.05 Eh) when nitrate depletion was reached, and just prior to the beginning of sulfate reduction and precipitation of sulfides. If the environmental conditions became more reducing rather than denitrifying, precipitation of Mo and Cu sulfides would have occurred (Fig. 11). The low concentrations of Mo in samples from the Dos Palos III core indicate that strong reducing conditions were not reached, or at least not permanently, suggesting that the system oscillated between the nitrate-sulfate (NO_3^-/SO_4^{2-}) reduction boundary (Piper and Isaacs, 1994). This would have allowed intermittent precipitation of phosphates and incorporation of lanthanide elements (Ce, La) during less-reducing and suboxic conditions.

Although it is possible that strong reducing conditions were established during deposition of some portions of the laminated facies of the Marca Shale, the sediments (core and outcrop) do not indicate such an environment. This may reflect sampling bias or may reflect the diluted biogenic content of the Marca Shale (~30%). However, it should also be noted that reducing conditions may not be necessary to preserve laminations and prevent bioturbation by macroorganisms (Dean et al., 1994). The already low oxygen content of denitrifying waters (<0.2 ml/l) could have precluded major bioturbation and enhanced preservation of organic matter. Examples of laminated sediments associated with coastal upwelling and denitrifying, nonsulfidic conditions have also been reported for ancient and Cenozoic sediments offshore California (Jahnke et al., 1990; Dean et al., 1994), including the much-studied Monterey Formation (Piper and Isaacs, 1994). These coastal upwelling sediments tend to have a modest organic matter content and hydrogen index values that are consistent with deposition under anoxic-suboxic conditions and a lack of sulfidic-reducing bottom waters.

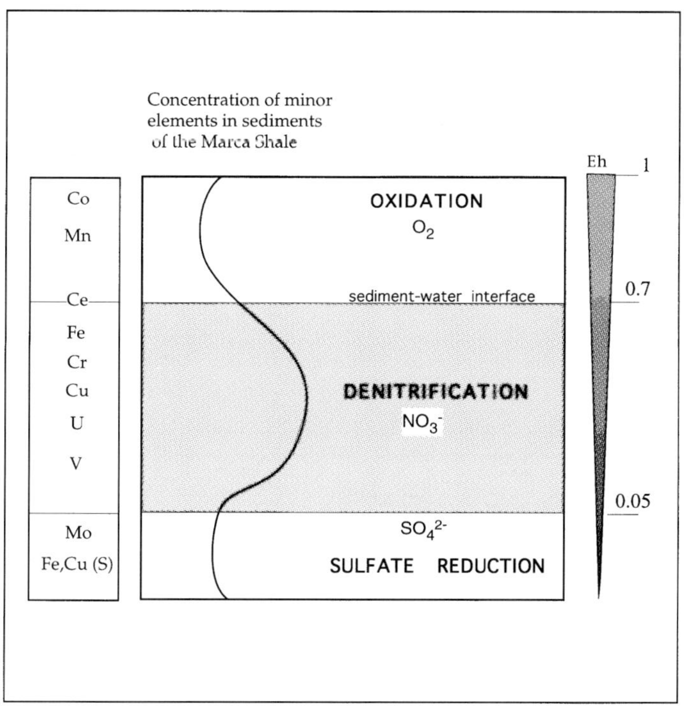

FIG. 11.—Concentration profile of trace metals in sediments of the Marca Shale according to their redox-solubilities (modified from Shaw et al., 1990 and Piper and Isaacs, 1995). The shaded interval highlights denitrification as the more active process that operated during deposition of the Marca Shale.

Fluctuations in rates of sedimentation, primary productivity, bottom water anoxia, and reducing conditions in the sediments are reflected in the distribution of diagnostic trace metals and the occurrence of laminated and phosphatic sediments. The relevance of understanding the occurrence of phosphatic sediments under suboxic conditions of the Marca Shale lies in the observation that factors promoting phosphogenesis beneath high-productivity upwelling systems do not always favor burial and preservation of hydrogen-rich organic matter (Ingall and Van Cappellen, 1990; Ingall et al., 1993) and represent potential discontinuities in the sedimentary record.

PHOSPHATIC SEDIMENTS

Phosphatic-rich sediments in the Marca Shale occur as phosphatized discontinuous millimeter thick-laminations, friable phosphatic nodules with diverse nuclei (wood, coprolites, ammonites), and well-indurated phosphatic nodules. The formation of friable light brown phosphatic nodules in the Marca Shale appears to be related to the occurrence of centimeter-thick massive layers. These layers are usually characterized by their homogenous texture, disrupted laminations, and sharp upper contacts with overlying continuous laminations (Fig. 12A, B). Friable phosphatic nodules occurring within these layers are up to 2 cm in diameter, and have diffuse boundaries and an oblate shape. The nodules are invariably restricted to nonlaminated sequences and appear to have grown during early diagenesis before major compaction took place, as evidenced by bent overlying laminations. Massive layers containing the friable phosphatic nodules are also often related to slumped laminations and to local concentrations of calcareous foraminifera. It is unclear if slumping and associated gravity flows provided a mechanism for episodic burial of phosphate-bearing microbial communities (e.g. mats; Föllmi and Garrison, 1991) or overturned pelagic sediments introducing oxygen and increasing homogeneity and fluid exchange. The local concentration of foraminifera in some of the massive layers appears to be the product of winnowing of clays and preferential concentration of heavier grains by the action of bottom currents. Macrobioturbation is absent from massive layers and associated phosphatic sediments.

Unlike nodules, phosphatic laminae are laterally continuous, usually 0.5 cm in thickness, and occur in thin layers rich in carbonate particles associated with downslope transport by low density fine-grained turbidites. The phosphatic laminae often have a lower content of apatite and higher proportion of detritus than phosphatic nodules. In addition, the laminae associated with phosphatic laminations often contain small, displaced laminae intraclasts and silt-sized clastic grains.

The friable phosphatic nodules and laminae can be regarded as pristine phosphates or autochthonous deposits in the sense

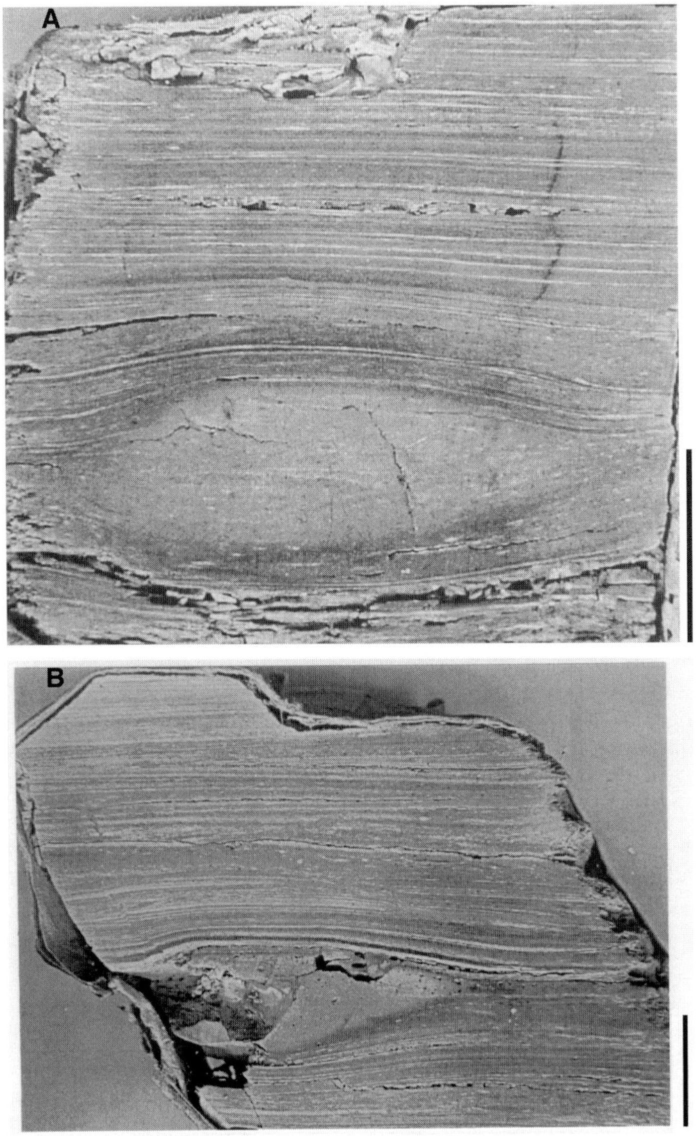

FIG. 12.—Friable phosphatic nodules in microfacies 2 of the Marca Shale. A. Friable phosphatic nodule (bottom) developed within a massive interval and overlied by continuous thin light laminae. Scale bar is 1 cm long. B. Friable nodule in a massive interval and interlayered with continuous fine laminations. Note the distinctive thick light laminae draping the nodule and marking the re-establishment of continuous sedimentation and the end of phosphogenesis. Scale bar is 1 cm long.

that they were formed *in situ* during early diagenesis of the sediments and lack any characteristics indicative of reworking or transport (Föllmi and Garrison, 1991; Föllmi, 1996). The occurrence of pristine phosphates is generally controlled by a complex set of factors involving the intensity and location of the midwater oxygen minimum, sedimentation rates, primary productivity, bottom currents, water chemistry variation, the concentration of biogenic calcareous and phosphatic debris, and sediment dislocation (Burnett et al., 1983; Froelich et al., 1988; Glenn and Arthur, 1988; Piper et al., 1988; Van Cappellen and Berner, 1988; Garrison and Kastner, 1990; Ingall and Van Cappellen, 1990; Föllmi et al., 1991b; Ingall and Jahnke, 1994; Föllmi, 1996). In a simple way, the formation of pristine phosphates involves two mechanisms, one that supplies phosphorus to the sediments and another that fosters precipitation of dissolved phosphate. Phosphorus can be supplied to the sediments from continental detritus, soft tissue of primary producers (diatoms, dinoflagellates, etc.), fecal pellets, fish debris, (Slansky, 1986; Froelich et al., 1988), bacterial mats lying at the sediment water interface (Gallardo, 1977; Froelich et al., 1988; Reimers et al., 1990), sorbed phosphate in particulate iron hydroxides (Froelich et al., 1979; De Lange, 1986; Freoelich et al., 1988; Glenn and Arthur, 1988), and from phosphorus regeneration in reducing sediments (Ingall and Jahnke, 1994).

Current models of phosphate precipitation indicate that particulate phosphate reaching the sea floor is incorporated into the sediments, where it can dissociate and mobilize whenever reducing conditions prevail. Free phosphate in the reducing sediment-water interface will diffuse upwards to more oxygenated conditions where it will associate with carbonate and fluorine to form fluorapatite (Froelich et al., 1988; Glenn and Arthur, 1988; Jarvis et al., 1994). Under reducing conditions, microbes within the sediment-water interface may also liberate phosphate as a result of anaerobic respiration, while in oxic environments they will store phosphate (Gätcher and Meyer, 1993). Ingall and Jahnke (1994) suggest that some of the phosphate stored by microbes under oxygenated conditions may be in a refractory form that would prevent its recycling by other organisms. Föllmi and Garrison (1991) have also suggested that microbial mats can be suddenly buried by sedimentary events or laterally dislocated, preventing any recycling of microbial phosphorus-generation. In general, it can be said that in organic-rich marine sediments, precipitation of phosphate (phosphogenesis) usually takes place along the redox boundary and with microbial participation under suboxic environments that prevail over highly reducing conditions (Lucas and Prévôt, 1985; Froelich et al., 1988; Reimers et al., 1990; Gätcher and Meyer, 1993; Ingall and Jahnke, 1994; Föllmi, 1996).

The organic-rich sediments of the upper Moreno Formation contain much sedimentological and geochemical evidence, including biomarker data, that indicates that microbial activity occurred during deposition of the laminated Marca Shale (Fonseca, 1997). The precipitation of pristine phosphates in the Marca Shale sediments would have occurred during early diagenesis at shallow depths (<10 cm) in the unconsolidated sedimentary column (Sholkovitz, 1973; Froelich et al, 1988; Glenn and Arthur, 1988; Van Cappellen and Berner, 1988; Glenn et al., 1994). Relevant to the formation of phosphatic nodules in the Marca Shale are the highly variable sedimentation and burial rates and the occurrence of nonlaminated intervals characterizing this unit. Depositional patterns indicate that periods of constant sedimentation were typified by the continuous deposition of light and dark laminae and alternated with periods of nondeposition, erosion, mass wasting, and event-deposition of fine grained gravity flows. During periods of continuous pelagic sedimentation and anoxia, laminated sediments are preserved and phosphate is regenerated within the sediments and released to the bottom waters (Ingall and Van Capellen, 1990). Periods of low sedimentation rates, strong bottom currents, slumping of sediments, and erosion by event-gravity flows enhance the precipitation of phosphate if suboxic conditions exist. The occurrence of authigenic phosphate in the Marca Shale is closely related to fluctuations in sedimentation rates and the occurrence of depositional events that supplied carbonate, promoted fluid exchange, and enhanced suboxic conditions through winnowing (concentration of foraminifers), turbulent transport (gravity flows, bottom currents), and mass wasting (slumping of sediments). Upon reestablishment of high background sedimentation rates and deposition of seasonal laminae, phosphogenesis was interrupted and phosphatic lamina and friable nodules were subsequently buried and draped by laminated hemipelagic sediments that restricted fluid exchange and could have promoted reducing conditions.

Oxygen-deficient Sedimentary Facies

Sediments of the Marca Shale, including the lower and upper contacts with the Tierra Loma and Dos Palos Shale, represent an excellent example of a sedimentary facies developed beneath an ancient oxygen-minimum layer associated with an intense coastal upwelling system. Fluctuations in bioturbation intensity, variations in the content of organic carbon, the occurrence of marine authigenic phases, and geochemical signatures preserved within the organic matter and sediments represent sedimentary and biological responses to paleoceanographic changes and fluctuations in the dissolved oxygen content of intermediate water and its impingement within the basin slope.

Because the Marca Shale was deposited on the upper slope of an active forearc basin during late Maestrichtian time, its sedimentary record reveals the interaction of global eustatic changes and local tectonic uplift of the California margin. In the following section, the Dosados Canyon section is divided into dysoxic, suboxic, and anoxic facies based on lithology, bioturbation, organic carbon content, occurrence of laminations, and relative abundance of minor elements.

Lower Dysoxic Facies (0.2-0.1 ml/l O_2).—The dysoxic facies include the Tierra Loma-Marca Shale contact zone and the lower 63 m of the section. The general sedimentological interpretation based on facies distribution and benthic foraminifera from this portion of the section indicates that the Tierra Loma Member was deposited on the upper slope of the forearc basin (Sliter and Baker, 1972; McGuire, 1988). In the basal part of the section, plant material and medium-fine grained sand indicates detrital sedimentation by distal turbidity currents. Massive shales of the Tierra Loma and the occurrence of glauconite-filled Planolites-type burrows at 28 m suggest dysoxic bottom water conditions during the bioturbation of organic-rich sediments (Kauffman and Sageman, 1990). The absence of particles hydraulically equivalent to glauconite indicates that the latter grains were not transported, but rather represent autochthonous particles that were formed under mild reducing conditions during early diagenesis at the sediment-water interface (Odin, 1988; Froelich et al., 1988). Total organic carbon (3%) and hydrogen index values (150 mg HC/gm org. C) from a correlative interval in the Exxon Chaney Ranch well indicate modest concentrations and low preservation of the organic matter in this part of the section. The occurrence of calcareous concretions, bioturbation, glauconite, and low hydrogen index values are interpreted to be a response to low sedimentation rates and dysoxic conditions localized at the boundary between reducing and oxic environments, similar to the modern upperslope environment off Central California (Mullins et al., 1985). Low rates of detrital sedimentation may have been related to a relative rise in sea level that inundated the shoreline, producing sediment starvation in the adjacent upper slope and basin floor as terrigenous sediments were being trapped in the inner shelf (Donovan et al., 1988; Van Wagoner et al., 1988; Amorosi, 1995). In addition, relative rise in sea level could have expanded the oxygen-minimum zone by allowing the incursion of oxygen-deficient intermediate waters into the basin (Fig. 13A). As noted by Vercoutere et al. (1987), Glenn and Arthur (1988), Odin (1988), and Glenn et al. (1994), autochthonous glauconite develops in oxygen-poor marine environments that usually coincide with the upper and lower boundaries of the oxygen minimum zone where it impinges at the upper slope or in bathymetric highs during periods of low detrital sedimentation rates and winnowing of fine particles by strong bottom currents.

The occurrence of black, well-lithified sulfide-bearing phosphatic nodules and cross-laminated glauconitic sand at 36 m and 38 m in the Dosados Canyon indicates an important change in the oceanographic regime of the basin. The oblate shape, poor sorting, random arrangement of the nodules, and the glauconitic sand suggest local reworking of sediments by strong bottom currents on submerged topographic highs of the basin. Like the phosphatic nodules reported from the modern Peruvian continental margin and offshore Baja California, the phosphatic nodules of this part of the Dosados Canyon section represent condensed intervals that were likely formed through cycles of shallow burial, cementation, and winnowing/erosion of overlying sediments (Burnett et al., 1983; Garrison and Kastner, 1990; Glenn et al., 1994). These condensed intervals reflect much reduced net sedimentation rates that could be related to maximum flooding events at the beginning of maximum eustatic highstands and followed by reworking of authigenic grains in the shelf break and topographic highs as marine facies prograded landwards and bottom currents impinge on the slope and basin's margin. It is proposed here that the nodular layers at the 36-m level in the Dosados section might represent flooding surfaces that mark the beginning of deposition of the Marca Shale and could correspond to the onlap cycle UZA4.5 of Haq et al. (1987) and UZA4.6 of Donovan et al. (1988) for the upper Maestrichtian ca. 66 Ma according to the Mesozoic time scale of Gradstein et al. (1994) (Fig. 14). This onlap cycle has also been reported for the glauconite-bearing transgressive cycles in the Upper Cretaceous of New Jersey and Alabama (Donovan et al., 1988; Sugarman et al., 1995).

At 61.5 m, the occurrence of a marine reptile fossil in sediments of the Cretaceous upwelling facies resembles the relatively common occurrence of marine mammal skeletons in Tertiary and modern upwelling sediments (Fischer and Arthur, 1977; R. Garrison, pers. commun., 1996), and denote a transition to a quasi-anaerobic facies (Savrda and Bottjer, 1991). The

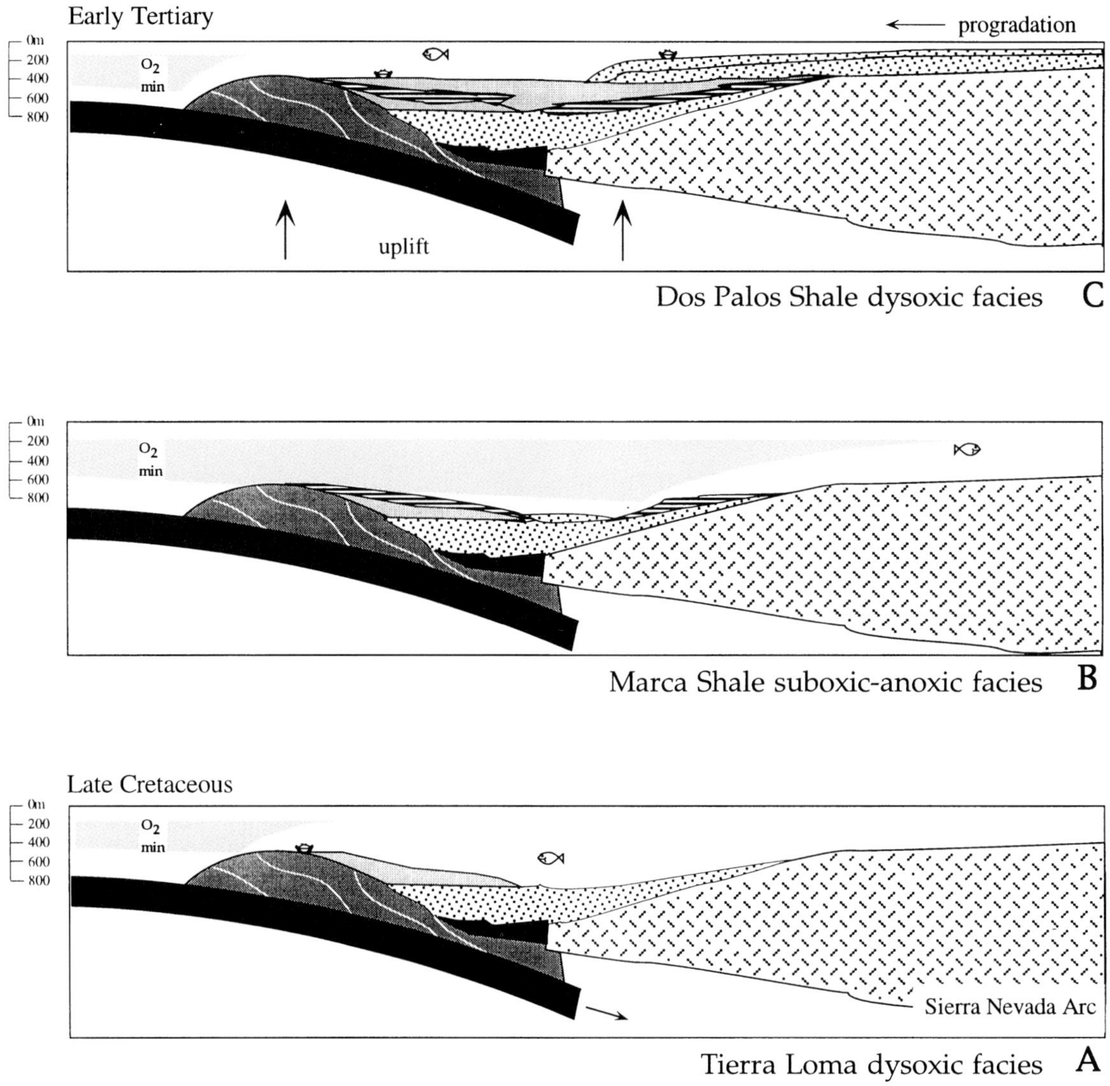

FIG. 13.—Bathymetric fluctuations related to onset of anoxic conditions during deposition of the Marca Shale in the Dosados Canyon. A. Initial stage during deposition of the Tierra Loma and occurrence of dysoxic conditions in the upper slope. B. Onset of anoxia as water depth increases and the oxygen minimum layer expands into the basin. This stage corresponds to deposition of the Marca Shale under sub-oxic/anoxic conditions. C. Final stage of the basin showing progradation of deltaic sediments, encroaching of the basin and regression of the ocean during the early Tertiary.

absence of hardground, reworked clasts or concentration of authigenic grains associated with the *Plesiosaur* skeleton in the Marca Shale does not support the occurrence of a condensed section or flooding surface at this horizon in the Moreno Formation.

The basal sediments of the Marca Shale record the transition between dysoxic into more restrictive suboxic environments that precluded active infaunal bioturbation and formation of glauconite. Higher abundances of biogenic components and lower detrital content are reflected by the color and composition of the Marca shale. Darker colored shales, abundant fish scales, opal C-T cementation, and well-preserved laminations indicate the onset of suboxic conditions and an increase in biosiliceous productivity beginning at 60 m in the section.

Suboxic-anoxic Facies (0-0.1 ml/l O_2).—
Sediments between 63 m and 155 m in the Dosados section were deposited beneath the core of the oxygen-minimum zone and represent a marked sedimentological change produced by the increase in primary productivity, lower detrital sedimentation rates, and enhanced preservation of organic matter in oxygen-deficient bottom waters. It is suggested that this section of the Marca Shale was deposited during a maximum highstand of sea level that was responsible for the apparent low flux of terrigenous sediments and intrusion of oxygen-deficient intermediate water over the basin's sill, creating anoxic bottom conditions on the upper slope and outer shelf (Fig. 13B). The upper slope was especially sensitive to climatic and oceanographic changes during the highstand interval, which is clearly recorded in the conspicuous light- and dark-colored, meter-scale cyclic intervals of laminated diatomite within the Marca Shale (Fonseca, 1996). High productivity and preservation of organic matter during deposition of the laminated Marca Shale is reflected by the concentration of total organic carbon (2-3%), hydrogen

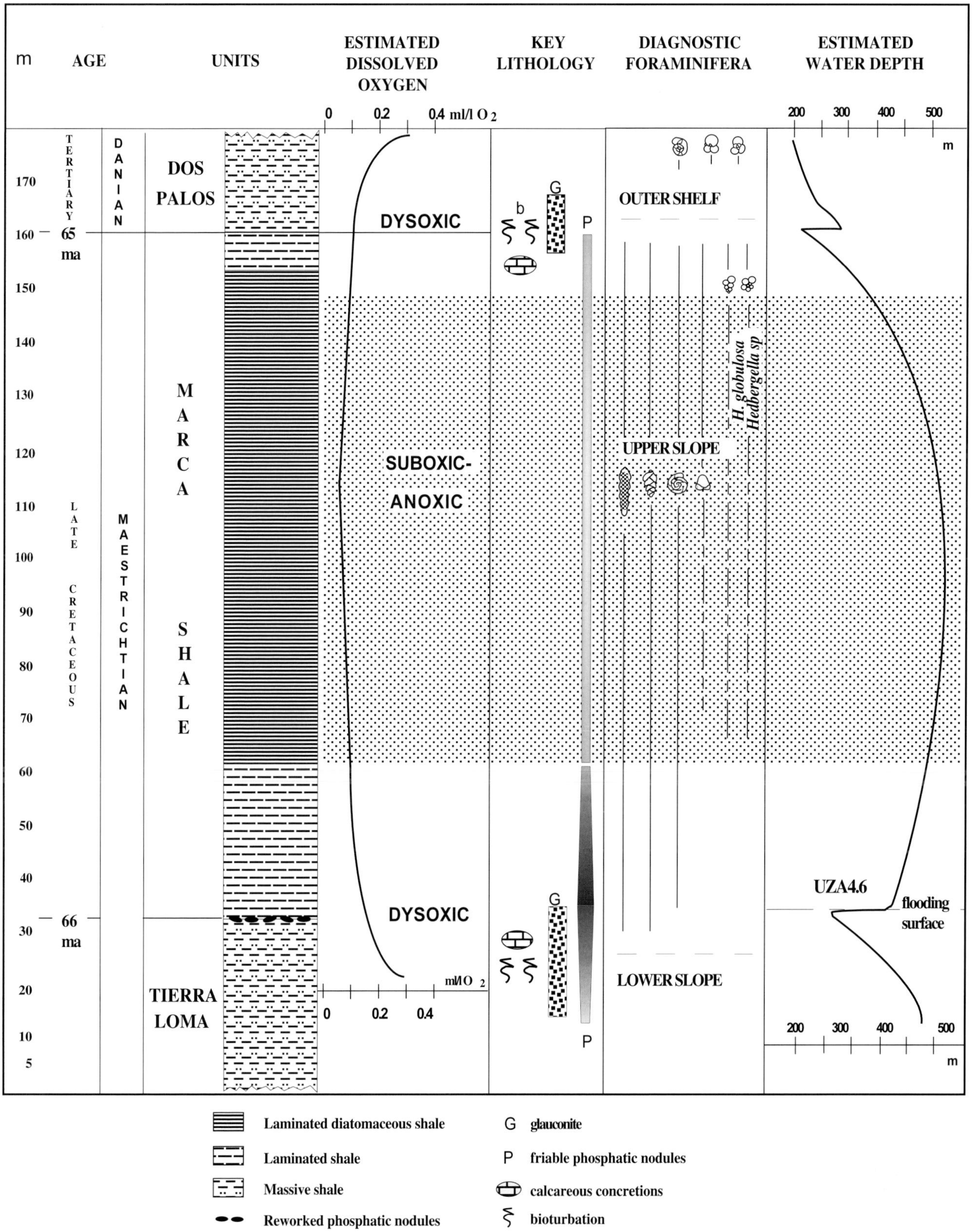

FIG. 14.—Diagram of the vertical distribution of oxygen-deficient facies in the Dosados Canyon section. b = bioturbation; G = glauconite; P = phosphates. Minor occurrence of the small planktonic foraminifers *Heterohelix globulosa* and *Hedbergella sp.* at the top of the Marca Shale, suggests upwelling conditions and shelfal depths. The Maestrichtian diagnostic benthic foraminifers are, from left to right: *Siphogenerinoides whitei, Praebulimina prolixa, Gavelinella turbinata* and *Allomorphina sp.* The Tertiary foraminifera are, from left to right: *Valvulineria lillisi, Globigerina pseudobulloides, Globigerina daubjergensis.* Water depth is estimated after Lagoe and McDougall (1986) and Williams (1996).

index values (250-330), and biogenic silica (up to 50%).

Although the contact between the laminated diatomaceous Marca Shale and lower dysoxic facies of the Tierra Loma is conformable, the laminated texture and diatom content of the anoxic facies presents a major contrast that is reinforced by the higher hydrogen index values and biogenic silica at the 63-m level. Unique characteristics of the quasi-anaerobic facies in the Marca Shale (Tyson and Pearson, 1991) include excellent preservation of continuous light and dark submillimeter laminations (e.g., "varves"); the occurrence of friable phosphatic laminae and nodules; thin interlayered, very fine-grained, low density turbidites; occasional centimeter-scale slumps; and superb preservation of siliceous microfossils composed predominantly of opal A. Although the laminated intervals lack any visible bioturbation and suggest deposition under completely anoxic conditions, close examination reveals the presence of poorly preserved but locally abundant benthic foraminifera in both light and dark laminae. The foraminiferal assemblages are dominated by juvenile and mature specimens of *Praebulimina prolixa* and *Siphogenerinoides whitei*, indicators of oxygen-deficient benthic environments associated with the development of intense oxygen minima beneath the Cretaceous coastal upwelling system (Martin, 1964; Macellari and DeVries, 1987; Almogi-Labin and Bein, 1993; Ly and Kuhnt, 1994). In particular, *Siphogenerinoides* has been reported to occur elsewhere in sediments thought to have been deposited beneath the core of the oxygen-minimum zone at water depths of about 500 m (Ly and Kuhnt, 1994). A similar association between the Tertiary homeomorph *Siphogenerina* and a low-oxygen slope environment was also reported by Lagoe and McDougall (1986). The presence of infaunal benthic foraminifera is interpreted as evidence of the lack of strong reducing conditions in the water column and interstitial sediments that may have allowed microbioturbation. Concentration of trace elements in the sediments also indicates that denitrification was the dominant process in degrading organic matter and points to suboxic conditions associated with the precipitation of pristine phosphates. It is possible that fully anoxic and reducing conditions were reached in the bottom waters and sediments for short periods or in localized areas, allowing the formation of iron sulfides and the total exclusion of microbenthos.

The organic-rich laminated sequence within the suboxic-anoxic facies represents the onset of upwelling in surface waters overlying the upper slope. Variations in character of the laminae reflect the susceptibility of the upwelling system to short-term seasonal variations in terrigenous material and primary productivity. Accelerated coastal upwelling and primary productivity presumably intensified the anoxic-suboxic conditions in the underlying intermediate water column.

Upper Dysoxic Facies (0.2-0.1 ml/l O_2).—
The contact between the anoxic and upper dysoxic facies is transitional and marks the upper Marca Shale and the lower Dos Palos Shale boundary interval. Similar to the lower dysoxic facies, the upper dysoxic facies are characterized by the occurrence of bioturbation, glauconite, and calcareous concretions, which suggest nonreducing conditions in the sediments and increasing content of dissolved oxygen in bottom waters. Low diversity of small planktonic foraminiferal taxa preserved in the calcareous concretions at the 145-m interval indicate shelfal water depths and the ongoing occurrence of coastal upwelling (Leckie, 1987; Keller, 1993) during deposition of the upper Marca Shale. Benthic foraminiferal assemblages in this interval continue to be dominated by *Praebuliminoids* and *Siphogenerinoides*. However, *Nonionella cretacea* and *Allomorphina cretacea* become important species and suggest shoaler water depths for this interval. The increase in the occurrence of massive layers, low-density turbidites, and truncated laminations at 155 m also reflect a higher rate of sedimentation. Common slumping and soft sediment deformation in this portion of the section also indicate instability of slope probably related to the ongoing uplift of the basin and Laramide orogeny (Moxon and Graham, 1987; McGuire, 1988; Moxon, 1988). At the 160-m level the occurrence of concentrated fish bones, shark teeth, goyazite spherules, glauconite, and bioturbation of the K-T boundary layer is interpreted as a condensed interval representing a short-term eustatic highstand, low net sedimentation rates, and reworking of authigenic and impact-ejecta related grains (Fonseca, 1997). Immediately above the K-T boundary, sediments of the Dos Palos Shale are barren of calcareous and siliceous microfossils, possibly representing the collapse of primary productivity and a dissolution interval (Huber, 1988). I consider the latter to be the combined product of the stressful environmental and oceanographic conditions caused by the impact of a bolide at the K/T boundary (e.g., acid rain: Alvarez et al., 1980; Hsü and McKenzie, 1985; Huber, 1988; Sigurdsson et al., 1992; D'Hondt and Lindinger, 1994), and subsequent increase in fine-grained terrigenous material as the shoreline prograded and the basin shoaled up (Fig. 13C). Tertiary calcareous foraminifera are reported to occur at 34 m and 106 m above the K-T boundary in the Dos Palos Shale (Martin, 1964; Drugg, 1967) and indicate a return to normal oceanographic conditions in the basin. At 132 m above the K-T boundary, the occurrence of the coral *Flabellum* in the Cima Sandstone (Goudkoff, 1945; Payne, 1951) and foraminifera indicate that water depths had shallowed to inner shelf environments by middle Danian time (McGuire, 1988).

DISCUSSION

Sediments of the Marca Shale member of the Moreno Formation strongly indicate the occurrence of an active coastal upwelling system along the Maestrichtian margin of California, ca. 66 Ma. Surface waters were characterized by high primary productivity and an associated midwater oxygen minimum layer. Sediments deposited beneath this system were organic-rich with preservation of laminations, authigenic phosphates, and glauconite. Tertiary and Recent upwelling deposits along the western margin of the Americas (i.e., California, Baja California Sur, Peru, and Chile) display lithologic components that represent sedimentary processes similar to those responsible for deposition of the Late Cretaceous Moreno Formation in California (Calvert, 1964; Burnett, 1977; Reimers and Suess,

1983; Kim, 1987; Dunbar et al., 1988; Froelich et al., 1988; Kim and Burnett, 1988; Piper et al., 1988; Suess et al., 1988; Galli-Olivier et al., 1990; Garrison and Kastner, 1990; Fonseca, 1991; Piper, 1991; Schwennike, 1991; Garrison, 1992; Grimm, 1992; Dean et al., 1994; Behl, 1995). The Late Cretaceous global ocean was much different in configuration and circulation than the modern ocean (Parrish and Curtis, 1982; Haq, 1984; Barron and Peterson, 1993) and marked by circum-global flow along the equator and the absence of circum-global flow in the Antarctic/Southern Hemisphere region. Despite these differences, the physics controlling basic oceanographic processes such as coastal upwelling were the same for the Cretaceous ocean as for the modern ocean. Hence, patterns of Cretaceous coastal upwelling can be confidently hindcast based on principles derived from the study of modern systems (e.g., Parrish and Curtis, 1982).

Of great oceanographic significance, the global surface temperature gradient in the Late Cretaceous ocean was relatively smaller than in the modern ocean, based upon both isotopic data and model reconstructions (Barrera, et al., 1987; Barrera and Huber, 1990; Huber et al., 1995; D'Hondt and Arthur, 1996; Bush and Philander, 1997). This pattern was due in part to the lack of permanent polar ice caps during this period, although polar temperatures likely dropped below freezing on a seasonal basis (Bush and Philander, 1997). Despite the relatively shallow global thermal gradient, the climate model reconstruction of Bush and Philander (1997) clearly shows that the Late Cretaceous sea-surface temperature gradient off the midlatitude coast of California was as steep, if not steeper, than the modern gradient in this same area. However, Late Cretaceous sea-surface temperatures off California differed from the modern and have been estimated to have averaged about 5°C higher than today.

An early study of Late Cretaceous upwelling by Parrish and Curtis (1982) positions a large subtropical high pressure system in the midlatitude eastern Pacific Ocean off California which, in turn, is thought to have generated vigorous coastal upwelling along the adjacent continental margin. The diatomaceous and phosphatic deposits of the Marca shale confirm this prediction. Other Maestrichtian biosiliceous shale deposits reported from the Arctic region (Kitchell and Clark, 1982; Dell'Agnese and Clark, 1994) and Antarctica (Harwood, 1988) also confirm the presence of predicted upwelling systems in these regions (Parrish and Curtis, 1982). Maestrichtian upwelling deposits have also been identified in the Venezuela Basin (Sigurdsson et al., 1995) and the Shefela and Zin Basins in Israel (Almogi-Labin and Bein, 1993), however, these latter facies are characterized by a mix of carbonates, biosiliceous shales, and phosphorites that reflect their origin within Tethyan equatorial upwelling systems.

The lower pole-to-equator thermal gradient of the Late Cretaceous is commonly thought to have resulted in slower rates of atmospheric and oceanic circulation, but this idea has been recently challenged by the model results of Bush and Philander (1997). One likely effect of the Cretaceous circum-global Tethys equatorial current system was the forcing of large subtropical high pressure systems slightly poleward relative to their modern positions (Savin, 1977; Barron and Washington, 1982; Saltzman and Barron, 1982). These high pressure systems could have promoted strong wind circulation that resulted in an effective Ekman transport that allowed upwelling of cooler waters along the margin, and the development of semiarid conditions and climate seasonality.

Estimates of past ocean temperatures derived from oxygen isotope data have revealed that a cooling trend took place during the Late Cretaceous (Maestrichtian) as recorded in both surface and deep water data sets (Margolis et al., 1977; Douglas and Woodruff, 1981; Boersma, 1984; Frakes, 1986; Barrera and Huber, 1990; Huber et al., 1995). In summary, the most recent isotopic and model studies of latest Cretaceous climate indicate that the Maestrichtian was characterized by (1) a well-stratified oceanic water column, (2) equatorial surface waters that were about 6°C cooler than those of the modern ocean, (3) strong heat transfer from the tropics to the polar regions, (4) an atmospheric carbon dioxide content about two to four times higher than that of today, (5) a lower pole-to-equator thermal gradient than that of the modern world, and (6) polar temperatures of 0°C with sea-ice formation in the Antarctic region and permanent ice at elevations above 1000 m (Spicer and Parrish, 1990; Barron and Peterson, 1993; Barron, 1994; Sellwood et al., 1994; D'Hondt and Arthur, 1996). Thus, the Maestrichtian ocean-atmosphere system was likely far more dynamic than that of the fully developed "greenhouse" conditions of earlier Cretaceous time and quite capable of sustaining vigorous coastal upwelling wherever conditions permitted this process to take place. Coeval upper Cretaceous upwelling deposits around the world suggest that equatorial circulation of the ancient Tethys Ocean likely forced different types of upwelling, including the coastal upwelling prevailing along the California margin. The coeval onset in upwelling in many areas of the Maestrichtian ocean resulted in large amounts of organic matter successfully trapped in upwelling facies with potential repercussions for the global carbon cycle and climate cooling between 70–65 Ma.

Micropaleontological and biomarker analyses of the Marca Shale indicate that diatoms and dinoflagellates were the main precursors for the accumulated organic matter (Fonseca, 1997), and could have been the major source of phosphorus to the sediments. Paleogeographic reconstructions of the Late Cretaceous California margin (Ingersoll, 1978; Bartow and Nilsen, 1990; Trumbley, 1990) suggest the occurrence of a proximal deltaic system during deposition of the Moreno Formation in the ancient San Joaquin Valley. This fluvial system transported arkosic sediments derived from the Sierra Nevada batholith and could also have provided dissolved phosphorous to the coastal waters, similar to other Cretaceous phosphatic deposits (Glenn et al., 1994). High sedimentation rates associated with the progradation of the shelf, the burial history, and paleogeography of the basin could have precluded the formation and concentration of phosphatic sands in the Moreno Formation.

The upwelling deposits of the Moreno Formation thus represent a combined record of regional and global paleoceanographic changes that took place during late Maestrichtian time. These changes included stratification of the ocean, high primary productivity, and establishment of a well-developed midwater oxygen minima. The lithostratigraphic record of the Moreno

Formation strongly suggests that a coastal upwelling system was already active on the California margin ca. 66 Ma under different climatic and oceanographic conditions than those of Tertiary and modern deposits of the western margin of North and South America.

CONCLUSIONS

The laminated biogenic sediments in the California margin during the Late Cretaceous are the combined product of coastal upwelling, high primary productivity, the occurrence of an oxygen-minimum zone, and the existence of a structural sill outboard. Similar to Tertiary and Quaternary upwelling sediments in California, the sediments of the Marca Shale record the occurrence of oxygen-deficient facies. These facies are characterized by the occurrence and absence of bioturbation, laminated sediments, autochthonous glauconite, pristine phosphates, well-preserved organic matter, and denitrifying conditions in the bottom waters. In the Dosados Canyon, oxygen-deficient facies extend from the upper slope to the shelf edge similar to analogous modern slope-to-shelf edge oxygen minima gradients off California and other areas of the eastern Pacific. Variations in water depth, primary productivity, and basin circulation controlled the stratigraphic distribution of the facies. Dysoxic facies deposited at the margins of the oxygen-minimum zone are characterized by the occurrence of massive glauconitic mudstones, planolite-type bioturbation, calcareous concretions, and phosphatic nodules. The lower and upper contact of the Marca Shale with the Tierra Loma and Dos Palos Shale correspond to the dysoxic facies in the Dosados Canyon section (Fig. 14).

Laminated diatomaceous sediments of the Marca Shale correspond to the upwelling-anoxic facies deposited on the upper slope at the core of the oxygen minimum zone, with the episodic occurrence of phosphatic nodules and absence of macrobenthos. Sediments of the Marca Shale in the Exxon Chaney Ranch Core #1 and Dos Palos Core # III contain modest values of hydrogen index (HI) and total organic carbon (TOC) that suggest a suboxic depositional environment for most of the section. The relatively large values of HI in the kerogen of these sediments correlate with biosiliceous-rich productivity cycles in the Marca Shale and indicate good preservation of the organic matter under anoxic conditions (Peters, 1986; McGuire, 1988). High periods of primary productivity are reflected by large values of biogenic silica, total organic carbon, and well-preserved organic matter. Sediments containing large amounts of authigenic apatite, MnO, and lanthanide correlate with low values of clastic detritus, biogenic opal, organic matter, and hydrogen index, indicating a decrease in net sedimentation rates and probable recycling of organic matter under suboxic conditions (Froelich et al., 1988; Ingall et al., 1993). Deposition of pristine phosphates in the laminated Marca Shale mark probable discontinuities and the occurrence of disrupting-depositional events during otherwise continuous pelagic sedimentation. Slumping of unstable laminated sediments in the upperslope, calcareous fine-grained turbidites and reworking of sediments by bottom currents are among the disruptive-depositional events associated with the occurrence of phosphatic laminae and nodules in the Moreno Formation.

ACKNOWLEDGMENTS

Caroline Isaacs generously assisted with the elemental analysis of the Dos Palos Core III samples. Dave Piper and Margie Medrano provided guidance on the normative calculation of the sedimentary components of the Marca Shale. I thank Jim Ingle for reviewing the early drafts of this manuscript. Thanks are gratefully extended to Kurt Grimm and an anonymous reviewer for carefully reviewing and improving this manuscript. This study was partially supported by a Geological Society of America student research grant, the Shell fund at the GES department, Stanford University, and the Consejo Nacional de Ciencia y Tecnología, México.

REFERENCES

Almgren, A. A., 1986, Benthic foraminiferal zonation and correlations of upper Cretaceous strata of the Great Valley of California—a modification, *in* Abbot, P. L., ed., Cretaceous Stratigraphy Western North America: Los Angeles, Society for Sedimentary Geology (SEPM), Pacific Section, p. 137-152.

Almogi-Labin, A., and Bein, A., 1993, Late Cretaceous upwelling system along the southern Tethys Margin (Israel): Interrelationship between productivity, bottom water environments, and organic matter preservation: Paleoceanography, v. 8, p. 671-690.

Alvarez, L. W., Alvarez, W., Asaro, F., and Michel, H. V., 1980, Extraterrestrial cause for the Cretaceous-Tertiary extinction: Science, v. 208, p. 1095-1108.

Amorosi, A., 1995, Glaucony and sequence stratigraphy: a conceptual framework of distribution in siliciclastic sequences: Journal of Sedimentary Research, v. B65, p. 419-425.

Aplin, A. C., Bishop, A. N., Clayton, C. J., Kearsley, A. T. , Mossmann, J.-R., Patience, R. L., Rees, W. G. A., and Rowland, S. J., 1992, A lamina-scale geochemical and sedimentological study of sediments from the Peru Margin (Site 680, ODP Leg 112), *in* Summerhayes, C. P., Prell, W. L., and Emeis, K. C., eds., Upwelling Systems: Evolution since the Early Miocene: London, Geological Society Special Publication 64, p. 131-149.

Barrera, E., and Huber, B. T., 1990, Evolution of Antarctic waters during the Maestrichtian: Foraminifer oxygen and carbon isotope ratios, ODP Leg 113, *in* Baker, P. F., Kennett, J. P., et al., eds., Proceedings of the Ocean Drilling Program, Scientific Results: College Station, Texas, Ocean Drilling Program, v. 113, p. 813-823.

Barrera, E., Huber, B. T., Savin, S. M., and Webb, P., 1987, Antarctic marine temperatures: Late Campanian through early Paleocene: Paleoceanography, v. 2., p. 21-47.

Barron, E. J., 1994, Chill over the Cretaceous: Nature, v. 370, p. 415.

Barron, E. J., and Washington, W. M., 1982, Cretaceous climate: a comparison of atmospheric simulations with the geologic record: Palaeogeography, Palaeoclimatology, Palaeoecology, v. 40, p. 103-133.

Barron, E. J., and Peterson, W. H., 1993, Past climate and the role of ocean heat transport: model simulations for the Cretaceous: Paleoceanography, v. 8, p. 785-798.

Bartow, A. J., and Nilsen, T. H., 1990, Review of the Great Valley Sequence, Eastern Diablo Range and Northern San Joaquin Valley, Central California, *in* Kuespert, J. G., and Reid, S. A., eds., Structure, Stratigraphy and Hydrocarbon Occurrences of the San Joaquin Basin, California: Tulsa, Society for Sedimentary Geology (SEPM), Pacific Section, p. 253-265.

Baturin, G. N., 1983, Some unique sedimentological and geochemical features of deposits in coastal upwelling regions, *in* Thiede, J., and Suess, E., eds., Coastal Upwelling, Its sediment record, Part B: Sedimentary Records of Ancient Coastal Upwelling: New York, Plenum Press, p. 11-27.

Behl, R. J., 1995, Sedimentary facies and sedimentology of the Late Quaternary Santa Barbara Basin, Site 893, *in* Kennett, J. P., Baldauf, J. G., and Lyle, M., eds., Proceedings of the Ocean Drilling Program, Scientific Results: College Station, Texas, Ocean Drilling Program, v. 146, part 2, p. 295, 305.

BERNER, R. A., 1980, Early diagenesis: a theoretical approach: New Jersey, Princeton University Press, 241 p.

BOERSMA, A., 1984, Campanian through Paleocene paleotemperature and carbon isotope sequence and the Cretaceous-Tertiary boundary, *in* Berggren, W. A., and Van Couvering, J. A., eds., Catastrophes and Earth History: Princeton, New Jersey, Princeton University Press, p. 247-277.

BRODIE, I., AND KEPM, A. E. S., 1994, Variation in biogenic and detrital fluxes and formation of laminae in late Quaternary sediments from the Peruvian coastal upwelling zone: Marine Geology, v.116, p. 385-398.

BURNETT, W.C., 1977, Geochemistry and origin of phosphorite deposits from off Peru and Chile: Geological Society of America Bulletin, v. 8, p. 813-823.

BURNETT, W. C., ROE, K. K., AND PIPER, D. Z., 1983, Upwelling and phosphorite formation in the ocean, *in* Suess, E., and Thiede, J., eds., Coastal Upwelling, its sediment record, Part A: Responses of the Sedimentary Regime to Present Coastal Upwelling: New York, Plenum Press, p. 377-398.

BUSH, A. B. G., AND PHILANDER, S. G. H., 1997, The late Cretaceous: simulation with a coupled atmosphere-ocean general circulation model: Paleoceanography, v. 12, p. 495-516.

CALVERT, S. E., 1964, Origin of diatom-rich, varved sediments from the Gulf of California, *in* Andel, Tj. H. Van, and Shor, G. G., eds., Marine Geology of the Gulf of California: Tulsa, American Association of Petroleum Geologists Memoir 3, p. 546-564.

CALVERT, S. E., 1987, Oceanographic controls on the accumulation of organic matter in marine sediments, *in* Brooks, J., and Fleet, J. A., eds., Marine Petroleum Source Rocks: London, Geological Society Special Publication, v. 26, p. 137-151.

CAMP, C. L., 1942, California mosasaurs (abs.): University of California Memoirs, v. 13, p. 10.

CAMPBELL, A. S., 1943, List of Foraminifera and notes in Elasmosaurid Plesiosaurs, etc. (abs.): University of California Memoirs , v. 13, p. 128-129.

CHRISTENSEN, C. J., GORSLINE, D. S., HAMMOND, D. E., AND LUEND, S. P., 1994, Non-annual laminations and expansion of anoxic basin-floor conditions in Santa Monica Basin, California Borderland, over the past four centuries: Marine Geology, v. 116, p. 399-418.

D'HONDT, S., AND LINDINGER, M., 1994, A stable isotopic record of the Maastrichtian ocean-climate system: South Atlantic DSDP site 528: Palaeogeography, Palaeoclimatology, Palaeoecology, v. 112, p. 363-378.

D'HONDT, S., AND ARTHUR, A. M., 1996, Late Cretaceous Oceans and the Cool Tropic Paradox: Science, v. 271, p.1838-1841.

DEAN, W. E., GARDNER, J. V., AND ANDERSON, R. Y., 1994, Geochemical evidence for enhanced preservation of organic matter in the oxygen minimum zone of the continental margin of northern California during the late Pleistocene: Paleoceanography, v. 9, p. 47-61.

DE LANGE, G. J., 1986, Early diagenetic reactions in interbedded pelagic and turbiditic sediments in the Nares abyssal plain (western North Atlantic), consequences for the composition of sediment and interstitial water: Geochimica et Cosmochimica Acta, v. 50, p. 2543-2561.

DELL'AGNESE, D. J., AND CLARK, D. L., 1994, Siliceous microfossils from the warm Cretaceous and early Cenozoic Arctic Ocean: Journal of Paleontology, v. 68, No. 1, p. 31-47.

DEMAISON, G. J., AND MOORE, G. T., 1980, Anoxic environments and oil source bed genesis: American Association of Petroleum Geologists Bulletin 64, p. 1179-1209.

DICKINSON, W. R., AND SEELY, D. R., 1979, Structure and Stratigraphy of forearc regions: American Association of Petroleum Geologists Bulletin 63, p. 2-31.

DICKINSON, W. R., 1995, Forearc basins, *in* Busby, C. J., and Ingersoll, R. V., eds., Tectonics of Sedimentary Basins: Cambridge, Blackwell Science, p. 221-261.

DONOVAN, A. D., BAUM, G. R., BLECHSCHMIDT, G. L., LOUTIT, T. S., PFLUM, C. E., AND VAIL, P. R., 1988, Sequence stratigraphic setting of the Cretaceous-Tertiary boundary in Central Alabama, *in* Wilgus, C. K., Hastings, B. S., Kendall, C. G. St. C., Posamentier, H. W., Ross, C. A., and VAN WAGONER, J. C., eds., Sea-Level Changes: An Integrated Approach: Tulsa, Society for Sedimentary Geology (SEPM) Special Publication 42, p. 299-308.

DOUGLAS, R. G., AND WOODRUFF, F., 1981, Deep-sea benthic foraminifera, *in* Emiliani, C., ed., In the Sea, v. 7: New York, John Wiley & Sons, Inc., p. 1233-1327.

DRUGG, W. S., 1967, Palynology of the Upper Moreno Formation (Late Cretaceous-Paleogene) Escarpado Canyon, California: Paleontographica Abt. B, v. 120, p. 1-71.

DUNBAR, R. B., AND BERGER, W. H., 1981, Fecal pellet flux to modern bottom sediment of Santa Barbara Basin (California) based on sediment trapping: Geological Society of America Bulletin, v. 92, p. 212-218.

DUNBAR, R. B., MARTY, R. C., AND BAKER, P. A., 1988, Cenozoic marine sedimentation in the Sechura and Pisco basins, Peru: Palaeogeography, Palaeoclimatology, Palaeoecology, vol. 77, p. 235-261.

FISCHER, A., AND ARTHUR, M. A., 1977, Secular variations in the pelagic realm, *in* Cook, H. E., and Enos, P., eds., Deep-water carbonate environments: Tulsa, Society for Sedimentary Geology (SEPM) Special Publication 25, p. 19-50.

FÖLLMI, K. B., 1996, The phosphorus cycle, phosphogenesis and marine phosphate-rich deposits: Earth-Science Reviews, v. 40, p. 55-124.

FÖLLMI, K. B., AND GARRISON, R. E., 1991, Phosphatic sediments, ordinary or extraordinary deposits? The example of the Miocene Monterey Formation (California), *in* Müller, D. W., McKenzie, J. A., and Weissert, H., eds., Controversies in Modern Geology: New York, Academic Press, p. 55-84.

FÖLLMI, K. B., GARRISON, R. E., AND GRIMM, K. A., 1991b, Stratification in phosphatic sediments: Illustrations from the Neogene of California, *in* Einsele et al., eds., Cycles and Events in Stratigraphy: Berlin, Springer-Verlag, p. 493-507.

FONSECA, C., 1991, Variaciones petrológicas en la fosforita de la Capa Humboldt con respecto a su espesor, Formación San Gregorio, San Juan de la Costa, Baja California Sur México: Unpublished B.Sc. Dissertation, Universidad Autonoma de Baja California Sur, 126 pp.

FONSECA, C., 1996, Cyclic organic rich sedimentation in the Marca Shale, Moreno Formation and the onset of Late Cretaceous Upwelling along the California Margin (abs.): American Association of Petroleum Geologists Bulletin, v. 5, p. 47.

FONSECA, C., 1997, Late Cretaceous-Early Tertiary paleoceanography and cyclic sedimentation along the California Margin: Evidence from the Moreno Formation: Unpublished Ph.D. Dissertation, Stanford University, California, 478 p.

FOREMAN, H. P., 1968, Upper Maestrichtian radiolaria of California: London, The Paleontological Association, Special Papers in Palaeontology 3, 82 pp.

FOSTER, D. E., 1980, *Osteopygis* sp., a Marine Turtle from the Late Cretaceous Moreno Formation of California, PaleoBios: Berkeley, University of California, Museum of Paleontology Publication 34, p 33-34.

FRAKES, L. A., 1986, Mesozoic-Cenozoic climate history and causes of the glaciation, *in* Hsü, K. J., ed., Mesozoic and Cenozoic oceans: Washington, D.C., American Geophysical Union Geodynamics Series, v. 15, p. 33-48.

FROELICH, P. N., ARTHUR, M. A., BURNETT, W. C., DEAKING, M., HENSLYE, V, JAHNKE, R., KAUL, L., KIM, K-H., ROE, K., SOUTAR, A., AND VATHAKANON C., 1988, Early diagenesis of organic matter in Peru continental margin sediments: Phosphorite precipitation, *in* Burnett, W. C., and Froelich, P. N., eds., The Origin of Marine Phosphorite. The Results of the R. V. *Robert D. Conrad* Cruise 23-06 to the Peru shelf: Marine Geology, v.80, p. 309-343.

FROELICH, P. N., KLINKHAMMER, G. P., BENDER, M. L., LUEDTKE, N. A., HEATH, G. R., CULLEN, D., AND DAUPHIN, P., 1979, Early oxidation of organic matter in pelagic sediments of the eastern equatorial Atlantic, suboxic diagenesis: Geochimica et Cosmochimica Acta, v. 43, p. 1075-1090.

GALLARDO, V. A., 1977, Large benthic microbial communities in sulfide biota under Peru-Chile subsurface countercurrent: Nature, v. 268, p. 331-332.

GALLI-OLIVIER, C., GARDUÑO, G., AND GAMIÑO, J., 1990, Phosphorite deposits in the upper Oligocene, San Gregorio Formation at San Juan de la Costa, Baja California Sur, Mexico, *in* Burnett, W. C., and Riggs, W. R., eds., Neogene to Modern Phosphorites: Cambridge, Cambridge University Press, v. 3, p. 122-126.

GARDNER, J. V,. AND DARTNELL, P., 1995, Centennial-scale late Quaternary stratigraphies of carbonate and organic carbon from Santa Barbara Basin, Hole 893A, and their paleoceanographic significance, *in* Kennett, J. P., Baldauf, J. G., and Lyle, M., eds., Proceedings of the Ocean Drilling Program, Scientific Results: College Station, Texas, Ocean Drilling Program, v. 146, part 2, p. 103-124.

GARRISON, R. E., AND KASTNER, M., 1990, Phosphatic sediments and rocks recovered from the Peru Margin during ODP Leg 112, *in* Suess, E., Von Huene, R., eds., Proceedings of the Ocean Drilling Program, Scientific Results: College Station, Texas, Ocean Drilling Program, v. 112, p. 111-134.

GARRISON, R. E., 1992, Neogene phosphogenesis along the Eastern Margin of the Pacific Ocean: Revista Geológica de Chile, v. 19, p. 91-111.

GÄTCHER, R., AND MEYER, J. S., 1993, The role of microorganisms in mobilization and fixation of phosphorus in sediments: Hydrobiologia, v. 253, p. 103-121.

GLENN, C. R., AND ARTHUR, M. A., 1988, Petrology and major element geochemistry of Peru margin phosphorites and associated diagenetic minerals: Authigenesis in modern organic-rich sediments, *in* Burnett, W. C., and Froelich, P. N., eds., The Origin of Marine Phosphorite. The Results of the R. V. *Robert D. Conrad* Cruise 23-06 to the Peru shelf: Marine Geology, v. 80, p. 231-267.

GLENN, C. R., ARTHUR, M. A., RESIG, J. M., BURNETT, W. C., DEAN, W. E. AND JAHNKE, R. A., 1994, Are modern and ancient phosphorites really so different?, *in* Iijima, A., Abed, A. M., and Garrison, R. E., 1994, Proceedings of the 29TH International Geological Congress Part C. Siliceous, Phosphatic and Glauconitic Sediments of the Tertiary and Mesozoic, p. 159-188.

GOUDKOFF, P. P., 1945, Stratigraphic relations of Upper Cretaceous in Great Valley, California: American Association of Petroleum Geologists Bulletin, v. 29, p. 956-1007.

GRADSTEIN, F. M., AGTERBERG, F. P., OGG, J. O., HARDENBOL, J., VEEN, P., THIERRY, J., AND HUANG, Z., 1994, A Mesozoic time scale: Journal of Geophysical Research, v. 99, B12, p. 24051-24074.

GRIMM, K. A., 1992, The Sedimentology of Coastal Upwelling Systems: Unpublished Ph.D. Dissertation, University of California, Santa Cruz, 431 p.

GRIMM, K. A., LANGE, C. B., AND GILL, A. S., 1996, Biological forcing of hemipelagic sedimentary laminae: Evidence from ODP site 893, Santa Barbara Basin, California: Journal of Sedimentary Research, Section A: Sedimentary Petrology and Processes, v. 66, p. 613-624.

GULBRANDSEN, R. A., JONES, D. L., TAGG, K. M., AND REESER, D. W., 1963, Apatitized wood and leucophosphite in nodules in the Moreno Formation, California: Washington, D.C., United States Geological Survey Professional Paper 475-C, p 100-104.

HAMILTON, W., 1979, Tectonics of the Indonesian region: United States Geological Survey Professional Paper 1078, 345 p.

HAQ, B. U., 1984, Paleoceanography: a synoptic overview of 200 million years of ocean history, *in* Haq, B.U., and Milliman, J. D., eds., Marine geology and oceanography of Arabian Sea and caostal Pakistan: New York, Van Nordstrand Reinhold Co., p. 201-232.

HAQ, B. U., HARDENBOL, J., AND VAIL, P. R., 1987, Chronology of fluctuating sea levels since the Triassic: Science, v. 235, p. 1156-1167.

HARWOOD, D. M., 1988, Upper Cretaceous and lower Paleocene diatom and silicoflagellate biostratigraphy of Seymour Island, eastern Antarctic Peninsula, *in* Feldmann, M. R., and Woodburne, M. E., eds., Geology and Paleontology of Seymour Island, Antarctic Peninsula: Boulder, Geological Society of America Memoir 169, p. 55-129.

HESSE, C. J., AND WELLS, S. P., 1936, The first record of a dinosaur from the west coast: Science, v. 84, p. 157-158.

HSÜ, K. J., AND MCKENZIE, J. A., 1985, A "strangelove" ocean in the earliest Tertiary, *in* Sundquist, E. T., and Broecker, W. S., eds., The Carbon Cycle and Atmospheric CO_2: Natural Variations Archean to Present: Washington D.C., American Geophysical Union, Geophysical Monograph 32, p. 487-492.

HUBER, B. T., 1988, Upper Campanian-Paleocene foraminifera from the James Ross Island region, Antarctic Peninsula, *in* Feldmann, M. R., and Woodburne, M. E., eds., Geology and Paleontology of Seymour Island, Antarctic Peninsula: Boulder, Geological Society of America Memoir 169, p. 55-129.

HUBER, B. T., HODELL, D. A., AND HAMILTON, C. P., 1995, Middle-Late Cretaceous climate of the southern high latitudes: Stable isotopic evidence for minimal equator-to-pole thermal gradients: Geological Society of America Bulletin, v. 107, No. 10, p. 1164-1191.

INGALL, E., AND VAN CAPELLEN, P., 1990, Relation between sedimentation rate and burial of organic phosphorus and organic carbon in marine sediments: Geochimica et Cosmochimica Acta, v. 54, p. 373-386.

INGALL, E., AND JAHNKE, R., 1994, Evidence for enhanced phosphorus regeneration from marine sediments overlain by oxygen depleted waters: Geochimica et Cosmochimica Acta, v. 58, p. 2571-2575.

INGALL, E., BUSTIN, R. M., AND VAN CAPELLEN, P., 1993, Influence of water column anoxia on the burial and preservation of carbon and phosphorus in marine shales: Geochimica et Cosmochimica Acta, v. 57, p. 303-316.

INGERSOLL, R. V., 1978, Paleogeography and paleotectonics of the Late Mesozoic forearc basin of northern and central California, *in* Howell, D. G., and McDougall, K. A., eds., Mesozoic Paleogeography of the Western United States: Tulsa, Society for Sedimentary Geology (SEPM)-Pacific Coast Paleogeography Symposium 2, p. 471-482.

INGERSOLL, R. V,. AND GRAHAM, S. A., 1983, Recognition of the Shelf-Slope Break along ancient, tectonically active continental margins: Tulsa, Society for Sedimentary Geology (SEPM) Special Publication 33, p. 107-117.

ISAACS, C. M., TOMSON, J. H., TAGGART, J. E. JR., AND JACKSON, L. L., 1990, Abundances of major elements and sedimentary components in Miocene and Early Pliocene cuttings from a well in the South Elwood Oil Field Area, offshore Santa Barbara-Ventura Basin, Southern California: Washington, D. C., United States Geological Survey Open-file report 89-79, 27 p.

JAHNKE, A. R., REIMERS, C. E., AND CRAVEN, D. B., 1990, Intensification of recycling of organic matter at the sea floor near ocean margins: Nature, v. 348, p. 50-54.

JARVIS, I., BURNETT, W. C., NATHAN, Y., ALMBAYDIN, F., ATTIA, K. M., CASTRO, L. N., FLICOTEAUX, R., HILMY, M. E., HUSAIN, V., QUTAWNA, A. A., SERJANI, A., AND ZANIN, Y. N., 1994, Phosphorite geochemistry: State-of-the-art and environmental concerns, *in* Föllmi, K. B., ed., Concepts and Controversies in Phosphogenesis: Eclogae Geologicae Helvetiae, v. 87, p. 643-700.

JONES, R. W., 1987, Organic facies, *in* Brooks, J., and Welte, D., eds, Advances in Petroleum Geochemistry 2: London, Academic Press, p. 1-90.

KAUFFMAN, E. G., AND SAGEMAN, B. B., 1990, Biological sensing of benthic environments in dark shales and related oxygen-restricted facies, *in* Ginsburg, R. N., and Beaudoin, B., eds, Cretaceous Resources, Events and Rhythms: Dordrecht, Netherlands, Kluwer Academic Publishers, p. 121-138.

KELLER, G., 1993, The Cretaceous-Tertiary boundary transition in the Antarctic Ocean and its global implications: Marine Micropaleontology, v. 21, p. 1-45.

KIM, W. H., 1987, Biostratigraphy and Depositional History of the San Gregorio and San Isidro Formations, Baja California Sur, Mexico: Unpublished Ph.D. Dissertation, Stanford University, California, 206 pp.

KIM, K. H., AND BURNETT, W. C., 1988, Accumulation and biological mixing of Peru margin sediments, *in* Burnett, W. C., and Froelich, P. N., eds., The Origin of Marine Phosphorite. The Results of the R. V. *Robert D. Conrad* Cruise 23-06 to the Peru shelf: Marine Geology, v. 80, p. 181-194.

KITCHELL, J. A., AND CLARK, D. L., 1982, Late Cretaceous-Paleogene paleogeography and paleocirculation: Evidence of North Polar Upwelling: Palaeogeography, Palaeoclimatology, Palaeoecology, v. 40, p. 135-165.

LAGOE, M. B., AND MCDOUGALL, K., 1986, Paleoenvironmental control of benthic foraminiferal ranges across a middle Miocene basin-margin, Central California: Journal of Foraminiferal Research, v. 16, p. 232-243.

LECKIE, M. R., 1987, Paleoecology of mid-Cretaceous planktonic foraminifera: A comparison of open ocean and Epicontinental Sea assemblages: Micropaleontology, v. 33, p. 164-176.

LEWAN, M. D., 1984, Factors controlling the proportionality of vanadium to nickel in crude oils: Geochimica et Cosmochimica Acta, v. 48, p. 2231-2238.

LOEBLICH, A. R., Jr., 1958, Danian stage of Paleocene in California: American Association of Petroleum Geologists Bulletin, v. 42, p. 2260-2261.

LUCAS, J., AND PRÉVÔT, L., 1985, The synthesis of apatite by bacterial activity: Mechanism, Mémoires des Sciences Géologiques, v. 77, p. 83-92.

LY, A., AND KUHNT, W., 1994, Late Cretaceous benthic foraminiferal assemblages of the Casamance Shelf (Senegal, NW Africa) indication of a Late Cretaceous Oxygen Minimum Zone: Revue de Micropaléontologie, v. 37, p. 49-74.

MACELLARI, C. E., AND DE VRIES, T. J., 1987, Late Cretaceous upwelling and anoxic sedimentation in northwestern South America: Palaeogeography, Palaeoclimatology, Palaeoecology, v. 59, p. 279-292.

MANN, K. H., AND LAZIER, J. R. N., 1991, Dynamics of marine ecosystems: Cambridge, Blackwell Scientific Publications, p. 161-212.

MARGOLIS, S. V., KROOPNICK, P. M., AND GOODNEY, D. E., 1977, Cenozoic and late Mesozoic paleoceanographic and paleoglacial history recorded in circum-Antarctic deep-sea sediments: Marine Geology, v. 25, p. 131-147.

MARTIN, L., 1964, Upper Cretaceous and Lower Tertiary foraminifera from Fresno County, California: Jahrbuch Der Geologischen Bundesanstalt, Sonderband 9, 128 p.

MCCAFFREY, M. A., FARRINGTON, J. W., AND REPETA, D. J., 1989, Geochemical implications of the lipid composition of Thioploca spp. from the Peru upwelling region-15°S: Organic Geochemistry, v. 14, p. 61-68.

MCGUIRE, D. J., 1988, Stratigraphy, depositional history, and hydrocarbon source-rock potential of the Upper Cretaceous-Lower Tertiary Moreno Formation, Central San Joaquin Basin, California: Unpublished Ph.D. Dissertation, Stanford University, California, 309 p.

MEDRANO, M. D., AND PIPER, D. Z., 1992, Partitioning of minor elements and major element oxides between rock components and calculation of the marine derived fraction of the minor elements in rocks of the Phosphoria Formation, Idaho and Wyoming: United States Geological Survey Open-file report 95-270, 79 p.

MOXON, I. W., AND GRAHAM, S. A., 1987, History and controls of subsidence in the Late Cretaceous-Tertiary Great Valley forearc basin, California: Geology, v. 15, p. 626-629.

MOXON, I. W., 1988, Sequence stratigraphy of the Great Valley Basin in the context of onvergent margin tectonics, *in* Graham, S. A., ed., Studies of the Geology of the San Joaquin Basin, v. 60: Tulsa, Society for Sedimentary Geology (SEPM) Pacific Section, p. 3-28.

MOXON, I. W., 1990, Stratigraphic and Structural architecture of the San Joaquin-Sacramento Basin: Unpublished Ph.D. Dissertation, Stanford University, 371 p.

MULLINS, H. T., THOMPSON, J. B., MCDOUGALL, K., AND VERCOUTERE, T. L., 1985, Oxygen-minimum zone edge effects; evidence from the Central California coastal upwelling system: Geology, v. 13, p. 491-494.

MURRAY, J. W., SPELL, B., AND PAUL, B., 1983, The contrasting geochemistry of manganese and chromium in the eastern tropical Pacific Ocean, *in* Wong, C. S., Boyle, Edward A., Bruland, K. W., Burton, J. D., Goldberg, E. D., eds., NATO Conference Series. IV: Marine Sciences, v. 9, p. 643-669.

NORRISH, K., AND CHAPPELL, B. W., 1977, X-ray fluorescence spectrometry, *in* Zussman, J., ed., Physical Methods in Determinative Mineralogy (2nd ed.): New York, Academic Press, p. 210-272.

NORRISH, K., AND HUTTON, J. T., 1969, An accurate x-ray spectrographic met for the analysis of a wide range of geological samples: Geochimica et Cosmochimica Acta, v. 33, p. 431-453.

ODIN, G. S., 1988, Green Marine Clays, Developments in Sedimentology 45: Netherlands, Elsevier, p. 295-332.

PARRISH, J. T., AND CURTIS, R. L., 1982, Atmospheric circulation, upwelling and organic-rich rocks in the Mesozoic and Cenozoic Eras: Palaeogeography, Palaeoclimatology, Palaeoecology, 40, p. 31-66.

PAYNE, M. B., 1951, Type Moreno Formation and overlying Eocene strata on the west side of the San Joaquin Valley, Fresno and Merced Counties, California: California Division of Mines and Geology Special Report 9, 29 p.

PEDERSEN, T. F., AND CALVERT, S. E., 1990, Anoxia vs. productivity: What controls the formation of organic-carbon-rich sediments and sedimentary rocks?: American Association of Petroleum Geologists Bulletin, v. 74, p. 454-466.

PETERS, K. E., 1986, Guidelines for evaluating petroleum source rock using programmed pyrolysis: American Association of Petroleum Geologists Bulletin, v. 70, p. 318-329.

PIKE, J., KEMP, A. E. S., AND SANCETTA, C., 1993, Complex signals from laminated sediments, Gulf of California (abs.): EOS Transactions, American Geophysical Union, v. 74, Supplement, p. 363.

PIPER, D. Z., 1991, Geochemistry of a Tertiary sedimentary phosphate deposit: Baja California Sur, Mexico: Chemical Geology, v. 92, p. 283-316.

PIPER, D. Z., 1994, Seawater as the source of minor elements in black shales, phosphorites, and sedimentary deposits: Chemical Geology, v. 114, p. 95-114.

PIPER, D. Z., AND ISAACS, C. M., 1994, Geochemistry of minor elements in the Monterey Formation, CA: Seawater chemistry of deposition: Washington, D.C., United States Geological Survey Professional Paper 1566, 41 p.

PIPER, D. Z., AND ISAACS, C. M., 1995, Minor elements in Quaternary sediment from the Sea of Japan: a record of surface-water productivity and intermediate-water redox conditions: Geological Society of America Bulletin, v. 107, p. 54-67.

PIPER, D. Z., BAEDECKER, P. A., CROCK, J. G., BURNETT, W. C., AND LOEBNER, B. J., 1988, Rare Earth elements in the phosphatic-enriched sediment of the Peru Shelf, *in* Burnett, W. C., and Froelich, P. N., eds., The Origin of Marine Phosphorite. The Results of the R. V. *Robert D. Conrad* Cruise 23-06 to the Peru shelf: Marine Geology, v.80, p. 269-285.

REIMERS, C. E., KASTNER, M., AND GARRISON, R. E., 1990, The role of bacterial mats in phosphate mineralization with particular reference to the Monterey Formation, *in* Riggs, S. R., and Burnett, W. C., eds., Phosphate deposits of the world, v. 3, Genesis of Neogene to Modern Phosphorites: Cambridge, Cambridge University Press, p. 301-311.

REIMERS, C. E., AND SUESS, E., 1983, Spatial and temporal patterns or organic matter accumulation on the Peru continental margin, *in* Suess, E., and Theide, J., eds., Coastal Upwelling—its sedimentary record, Part B: Sedimentary Records of Ancient Coastal Upwelling: New York, Plenum Press, p. 311-332.

SALTZMAN, E. S., AND BARRON, E. J., 1982, Deep circulation in the Late Cretaceous: oxygen isotope paleotemperatures from Inoceramus remains in D.S.D.P. cores: Palaeogeography, Palaeoclimatology, Palaeoecology, 40, p. 167-181.

SAVIN, S. M., 1977, The history of the Earth's surface temperature during the past 100 million years: Annual Reviews of the Earth and Planetary Sciences, v. 5, p. 319-355.

SAVRDA, C., AND BOTTJER, D. J., 1987, The exaerobic zone: a new oxygen-deficient marine biofacies: Nature, No. 327, p. 54-56.

SAVRDA, C., AND BOTTJER, D. J., 1991, Oxygen-related biofacies in marine strata: an overview and update, *in* Tyson, R. V., and Pearson, T. H., eds., Modern and Ancient Continental Shelf Anoxia: Geological Society Special Publication 58, p. 201-219.

SCHWENNICKE, T., 1991, On some aspects of the depositional history of the phosphorite-bearing lower member of the El Cien Formation in Baja California Sur, Mexico—Preliminary results (abs): First Annual Conference on the Geology of Baja California Sur, La Paz, Mexico, v. 1, p. 67.

SELLWOOD, B. W., PRICE, G. D., AND VALDES, P. J., 1994, Cooler estimates of Cretaceous temperatures: Nature, v. 370, p. 453-455.

SHAW, T. J., GIESKES, J. M., AND JAHNKE, R. A., 1990, Early diagenesis in differing depositional environment: The response of transition metals in pore water: Geochimica et Cosmochimica Acta, v. 54, p. 1233-1246.

SHOLKOVITZ, E., 1973, Interstitial water chemistry of the Santa Barbara Basin sediments: Geochimica et Cosmochimica Acta, v. 37, p. 2043-2073.

SIGURDSSON, H. D., D'HONDT, S., AND CAREY, S., 1992, The impact of the Cretaceous/Tertiary bolide on evaporite terrane and generation of a major sulfuric acid aerosol: Earth and Planetary Science Letters, v. 109, p. 543-559.

SIGURDSSON, H. D., LECKIE, M. R., AND ACTON, G., 1995, Prospectus Leg 165 Caribbean Ocean History and the Cretaceous/Tertiary Boundary Event: JOIDES Journal, v. 21, p. 26-33.

SLANSKY, M., 1986, Geology of sedimentary phosphates: New York, Elsevier, 210 p.

SLITER, W. V., AND BAKER, R. A., 1972, Cretaceous bathymetric distribution of benthic foraminifers: Journal of Foraminiferal Research, v. 2, p. 167-183.

SMITH, R. L., 1992, Coastal upwelling in the modern ocean, *in* Summerhayes, C. P., Prell, W. L., and Emeis, K. C., eds., Upwelling Systems: Evolution Since the Early Miocene: Geological Society Special Publication 64, p. 9-28.

SPICER, R. A., AND PARRISH, J. T., 1990, Late Cretaceous-early Tertiary palaeoclimates of northern high latitudes: a quantitative view: Journal of the Geological Society, London, v. 147, p. 329-341.

STOCK, C., 1941, Duckbill dinosaur from the Moreno Cretaceous California (abs.): Geological Society of America Bulletin, v. 52, p. 1956.

STUMM, W., AND MORGAN, J., 1981, Aquatic Chemistry: New York, John Wiley & Sons, p. 418-475.

SUESS, E., KULM, L. D., AND KILLINGLEY, J. S., 1987, Coastal upwelling and a history of organic-rich mudstone deposition off Peru, *in* Brooks, J., and Fleet, J. A., eds., Marine Petroleum Source Rocks: London, Geological Society Special Publication, v. 26, p. 181.

SUESS, E., ET AL., 1988, Proceedings of the Ocean Drilling Project Initial Reports, v. 112: College Station, Texas, Ocean Drilling Program, p. 5-23.

SUGARMAN, P. J., MILLER, K. G., BUKRY, D., AND FEIGENSON, M. D., 1995, Uppermost Campanian-Maestrichtian strontium isotopic, biostratigraphic, and sequence stratigraphic framework of the New Jersey Coastal Plain: Geological Society of America Bulletin, v. 107, p. 19-37.

THUNELL, R. C., PRIDE, C., J., AND TAPPA, E., 1994, Biogenic silica fluxes and accumulation rates in the Gulf of California: Geology, v. 22, p. 303-306.

TISSOT, B. P., AND WELTE, D. H., 1984, Petroleum Formation and Occurrence: New York, Springer-Verlag, p. 583-608.

TRUMBLY, P. N., 1990, Review of the Great Valley Sequence, Eastern Diablo Range and Northern San Joaquin Valley, Central California, *in* Kuespert, J. G., and Reid, S. A., eds., Structure, Stratigraphy and Hydrocarbon Occurrences of the San Joaquin Basin, California: Tulsa, Society for Sedimentary Geology (SEPM) Pacific Section, p. 267-276.

TYSON, R. V,. AND PEARSON, T. H., 1991, Modern and ancient continental shelf anoxia: an overview, *in* Tyson, R. V., and Pearson, T. H., eds., Modern and Ancient Continental Shelf Anoxia: London, Geological Society Special Publication 58, p. 1-24.

VAN CAPPELLEN, P., AND BERNER, R. A., 1988, A mathematical model for the early diagenesis of phosphorus and fluorine in marine sediments: Apatite precipitation: American Journal of Science, v. 288, p. 289-333.

VAN WAGONER, J. C., POSAMENTIER, W. H., MITCHUM, R. M., VAIL, P. R., SARG, J. F., LOUTIT, T. S., AND HARDENBOL, J., 1988, An overview of the fundamentals of sequence stratigraphy and key definitions, *in* Wilgus, C. K., Hastings, B. S., Kendall, C. G. St. C,, Posamentier, H. W., Ross, C. A., and Van Wagoner, J. C., eds., Sea-Level Changes: An Integrated Approach: Tulsa, Society for Sedimentary Geologists (SEPM) Special Publication 42, p. 39-45.

VERCOUTERE, T. L., MULLINS, H. T., MCDOUGALL, K., AND THOMPSON, J. B., 1987, Sedimentation across the Central California oxygen minimum zone; an alternative coastal upwelling sequence: Journal of Sedimentary Petrology, v. 57, p. 709-722.

WANTY, R. B., AND GOLDHABER, M. B., 1992, Thermodynamics and kinetics of reactions involving vanadium in natural systems: Accumulation of vanadium in sedimentary rocks: Geochimica et Cosmochimica Acta, v. 56, p. 1471-1483.

WEDEPOHL, K. H., 1969, Handbook of geochemistry: New York, Springer-Verlag, 231 p.

WELLS, S. P., 1943, Elasmosaurid plesiosaurs with description of new material from California and Colorado: University of California Memoirs, v. 13, p. 125-124.

WILLIAMS, T., 1996, History and Controls of Subsidence, Cretaceous Great Valley Group, Sacramento Basin, California: Unpublished Ph.D. dissertation, Stanford University, California, 412 p.

MARINE PHOSPHOGENESIS IN SHALLOW-WATER ENVIRONMENTS: CAMBRIAN, TERTIARY, AND RECENT EXAMPLES

TOBIAS SCHWENNICKE

Departamento de Geología, Universidad Autónoma de Baja California Sur, Apdo Postal 19-B, 23080 La Paz, B.C.S., Mexico

HENDRIK SIEGMUND*

Institut und Museum für Geologie und Paläontologie, Universität Tübingen, Herrenberger Str. 51, 72070 Tübingen, Germany

AND

CAROLINE JEHL

Laboratoire H.E.A., OSTROM, 911 Avenue Agropolis, BP 5045, 34032 Montpellier cedex 1, France

ABSTRACT: In an attempt to summarize the characteristics and importance of shallow marine phosphogenetic settings we describe three examples from fossil and recent environments. These are the lowermost Cambrian Zhongyicun Formation phosphorites from China; the Tertiary Salada Formation phosphorites from Baja California, Mexico; and recent atoll phosphorites from French Polynesia. Common features of these settings, which range from clastic-dominated to pure carbonate, are a lagoonal or inner shelf to nearshore environment with somewhat restricted water circulation and the influence of an increased supply of nutrients for higher bioproductivity. Organic matter is considered as the major source of phosphorus, but in settings with clastic influence other sources like continental input are feasible. The study of modern atoll phosphorites enabled the introduction of a new phosphogenetic model involving cyanobacterial mats, which provide an environment favorable for phosphogenesis and serve as nucleation sites for apatite crystals. This model is also valid for the Cambrian phosphatized microbial mats and stromatolites. However, the Salada Formation phosphorites lack discernible microbial mat structures. Here the generation of a phosphogenetic environment may be due to fine-grained sediment layers, which prevent free access of oxygenated water into the sediment. Although only one of the above shallow marine phosphogenetic settings leads to the formation of economic phosphorites, we argue that more time for phosphate accumulation and concentration may have produced such deposits in the other localities as well.

INTRODUCTION

Marine authigenic phosphogenesis is well known from ancient and recent outer shelf/upper slope environments that are influenced by upwelling (for example, Peru, Baja California). In these settings, upwelling supplies nutrients that trigger a high rate of bioproductivity in the photic zone. The resulting deposition and fixation of organic carbon and phosphate in the sediment may ultimately facilitate authigenic apatite precipitation, as outlined by Baturin (1982). For some time this type of environment generally has been considered the most important setting for marine phosphogenesis throughout the Mesozoic and Cenozoic. Shallow water phosphogenetic systems appear to have been more widespread only in the Neoproterozoic and Early Paleozoic (e.g., Cook and Shergold, 1979). However, during the last 2 decades, numerous studies of ancient phosphorite-bearing sequences and modern phosphorites revealed that many deposits do not fit into the classical upwelling model; in these studies, phosphogenesis has been related to inner shelf environments such as estuaries, lagoons, tidal flats, and insular atolls (Dahanayake and Krumbein, 1985; Lucas et al., 1986; Southgate, 1986; Soudry and Lewy, 1988; Abed and Fakhouri, 1990; Birch, 1990; Bremner and Rogers, 1990; Jehl and Rougerie, 1995). New models involving endo-upwelling (Rougerie and Wauthy, 1986) have replaced the early and often challenged guano model (Dixon 1877) and volcanic sources of phosphate for insular phosphorites (Bourrouilh-Le Jan, 1980, 1989). Some phosphorite accumulation, or at least phosphogenetic processes, have also been described from regions that seem to almost entirely lack upwelling (O'Brien et al., 1990; Heggie et al., 1990; Ruttenberg and Berner, 1993).

This leaves us with a wide variety of settings with different environmental conditions that seem to promote phosphogenesis. Recently, Trappe (1996) proposed a general four-stage phosphogenetic box model that explains sedimentary phosphogenesis as a sequence of favorable environmental conditions and processes. These conditions, and the different phosphogenetic stages of this model, may be accomplished by a variety of processes that are not necessarily restricted to specific environments. Phosphogenesis may, therefore, occur in a variety of marine and nonmarine settings, provided the required conditions are met in the sometimes very small actual phosphogenetic microenvironments.

The formation of major phosphorite deposits commonly requires reworking of primary phosphate-bearing sediments, which may also include transportation, and its subsequent accumulation (Glenn et al., 1994b; Föllmi, 1996). In this paper we review Cambrian, Tertiary, and recent examples of phosphogenesis in different shallow marine environments. We will then summarize the environmental conditions and processes of phosphogenesis, evaluate their meaning for the interpretation of phosphogenesis in general, and discuss pathways for how shallow-water phosphogenetic processes may eventually lead to the formation of phosphorite deposits.

EXAMPLES OF SHALLOW MARINE PHOSPHOGENESIS

The Cambrian Zhongyicun Formation Phosphorites at Meishucun, Yunnan, China

Location.—

The sections investigated are located in the Kunyang phosphate mine at Meishucun some 30 km southwest of Kunming, the capital of Yunnan province (Fig. 1). Additional data were collected from reference sections near Haikou, Wangjiawan, and

*email: h.siegmund@em.uni-frankfurt.de

Marine Authigenesis: From Global to Microbial, SEPM Special Publication No. 66

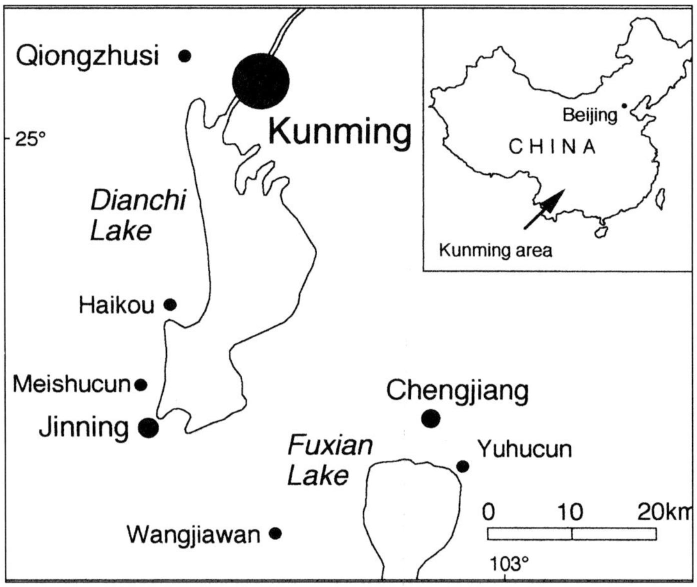

FIG. 1.—Location of the Meishucun phosphate mine near Kunming, Yunnan province, China.

Chengjiang. However, most descriptions refer to the section "excavation site 1" near Meishucun.

Lithology and facies.—

With a thickness of 10–30 m, the phosphorites of the Zhongyicun Formation constitute a considerable part of the lowermost Cambrian sedimentary sequence of Yunnan. The Meishucun section can be subdivided into a sequence of six beds (Fig. 2) that correspond to the standard beds 3–8 of the Precambrian/Cambrian type section of the area (Siegmund, 1995). Beds 3–4 are granular phosphorites, with interbedded dolomite and phosphate-rich layers. A gray, phosphate-bearing shale in bed 5 is succeeded by silica and dolomite-cemented granular phosphorite in bed 6. Coarser, dolomite hosted phosphorite constitutes bed 7, while bed 8 is a pure dolomite above an erosional base with some phosphatic rip-up clasts. A phosphate- and glauconite-bearing basal conglomerate separates this dolomite from the overlying black shales (Siegmund, 1995).

Each of these units possess a unique inventory of sedimentary structures. Planar bedding and cross-bedding is found in units 3–4 and 6. Absence of bedding or thin parallel to wavy bedding marks unit 5. Only in the Meishucun excavation site 1 section this unit is terminated by an inversely graded, fluidized flow deposit. Unit 7 is characterized by somewhat coarser, normally graded beds with erosional bed truncation and amalgamation such as is often found in proximal tempestites (Dott and Burgeois, 1982). Regular, symmetrical undulations of bedding planes with 10–25 cm wavelength are interpreted as wave ripples (Fig. 3). In the Haikou reference section lateral meter-scale variations in thickness and bedding plane inclination of individual beds may represent hummocky cross-stratification. Bedding surfaces occasionally show patterns interpreted as mud cracks (Fig. 4). Up-section, bed 7 comprises several phosphatic condensation horizons and laminated microbial mat/stromatolite phosphorite beds near its top.

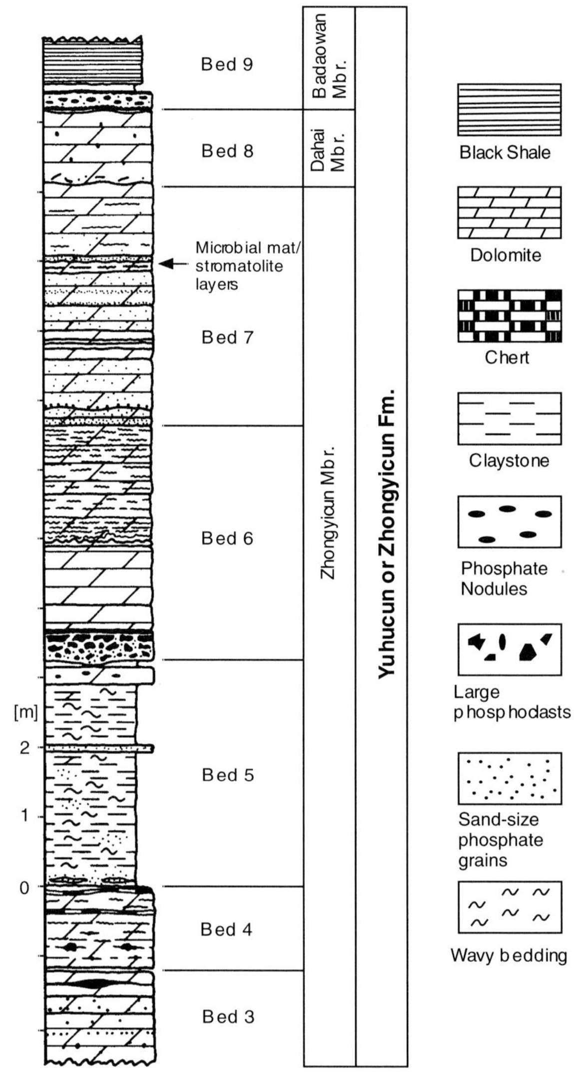

FIG. 2.—Lithology of the Meishucum section at excavation site 1, with standard subdivision adopted from the nearby type section.

Microfacies.—In the units 3–4 and 6–7, phosphoclastic pack or grainstone facies are the predominant phosphorite microfacies. A positive identification of packstone or grainstone is often hampered by widespread diagenetic recrystallization of the matrix or cement. A phosphatic bindstone facies is also present. It constitutes the darker, phosphate-rich layers in bed 3–4 and the microbial/stromatolite phosphorite of unit 7, where sediment fixation occurred due to microbial activity.

Phosphatic grains.—Phosphate particles are dominated by rounded and internally structureless grains ("peloids") of 20 µm to ca. 1 mm in size. Phosphate-coated grains (ooids, oncoids), phosphobioclasts, and rare large phosphatic intraclasts are also present. The grains usually have a complex genetic history (Ye and Chen, 1989; Ye et al., 1989). Comparison of the internally structureless phosphoclasts with *in situ* phosphorites of other facies belts served to identify them as products of erosional reworking, concentration, and transport of predecessor sediments, for example, phosphatized lime mudstone and microconcretions (Siegmund, 1995). Phosphatic intraclasts are readily

FIG. 3.—Undulation patterns (arrows) on a bedding surface in perpendicular section, interpreted as wave ripples. Bed 7, Meishucum section. Lens cap is 52 mm.

FIG. 4.—Pattern interpreted as mud cracks on a bedding surface within bed 7 of the Meishucun section. Lens cap is 52 mm.

identified as eroded fragments of granular phosphorites lithified during previous depositional phases. Small, shelly fossils or fragmented algae make up the observed phosphobioclasts.

Non-phosphatic grains.—This group includes carbonate and siliciclastic grains. Carbonate grains are abundant and comprise mostly recrystallized micritic peloids or larger reworked lithoclasts and some rare bioclasts. Like most nuclei of phosphate-coated grains they have retained a carbonate composition, even if cemented by apatite. Their origin is considered to be related to fecal pellets and eroded, previously semilithified lithoclasts. Sand-sized, mostly subangular siliciclastic grains may constitute up to 30 % of all particles. The siliciclastic fraction is dominated by quartz, with a minor portion of feldspar, heavy minerals, and Fe-rich glauconite in individual beds.

Grain size.—Grain-size distribution analysis from thin sections revealed a mean grain size of 170 μm for beds 3–4, 90–120 μm in bed 5 and 6, and 200 μm in bed 7. Similar grain sizes and variation trends were observed in the reference sections Chengjiang and Wangjiawan (Siegmund, 1995).

Facies interpretation.—

Grain types, grain size, and sedimentary structures present at Meishucun and the reference sections suggest a central to outer shelf environment for beds 3–6, probably with some episodic storm influence. This interpretation is consistent with the dominating phosphoclastic packstone and grainstone microfacies, similar to the one found in Moroccan shelf phosphorites (Trappe, 1992).

Bed 7 shows an increase in depositional energy at its base, reflecting a progradational shift to inner shelf conditions with strong influence of episodic storms. Repeated phosphatic condensation horizons in its upper part serve to indicate intervals of restricted water circulation and sediment supply and a return to more persistently low depositional energy. Locally, supratidal exposition is suggested by mud cracks. Phosphatized microbial mat/stromatolite associations are found in this upper part of bed 7.

Phosphatized microbial mats and stromatolites.—

Throughout the entire Meishucun section, phosphatized microbial structures are preserved as thin (< 10 μm) phosphatic laminae that cover and penetrate individual sediment layers and occlude the available pore space (Fig. 5). In the upper part of bed 7, microbial mats may laterally grade into centimeter-sized stromatolite domes locally with entrapped sediment. They are associated with coarse sediment, sponges, and algae comparable to *Thallophyca ramosa* from Late Sinian (Neoproterozoic) phosphorites in Hubei province (Zhang, 1989). Although usually found as fragmented and reworked bioclasts in the sediment, one stromatolite dome appears to be preserved *in situ*, indicating a photic zone environment (Siegmund, 1995). The stromatolites consist of a succession of laminated phosphatic layers similar to those in microbial mats. These are interbedded with an irregular meshwork of cyanobacterial sheaths, preserved as hollow phosphatized tubes (Fig. 6). The tubes may occasion ally contain well-preserved remnants of the actual organisms (Fig. 7). Phosphate mineralization is visible as palisades of small francolite crystals with their c-axes oriented perpendicularly to the substrate. When phosphate-coated grains are incorporated into a microbial mat, laminae from the grain cortex and adjacent mats may merge, thus perhaps indicating an *in situ* grain origin, as suggested by Dahanayake et al. (1985) and Dahanayake and Krumbein (1985).

Often, especially along the northern margin of the Mediterranean Tethys, phosphatic stromatolites and microbial mats are found in condensed sections of open marine facies (Föllmi, 1990; Martín-Algarra and Sánchez-Navas, 1995). The association of the Meishucun phosphatized microbial mats and stromatolites with sediments of inner shelf facies, probably with photic influence, however, tends to suggest a somewhat restricted shallow water, perhaps lagoonal, environment. These conditions are known as well from other Neoproterozoic to Cambrian phosphogenic settings (Cook and Shergold, 1979).

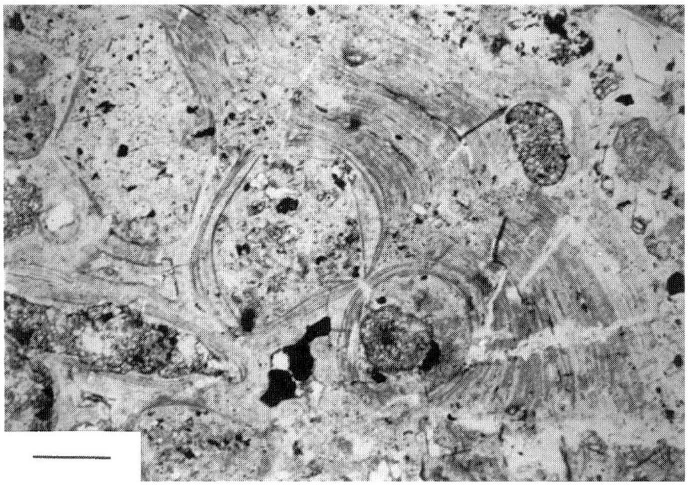

FIG. 5.—Microbial mats covering sediment layers. They also grow in the pore space. Scale bar is 25 μm.

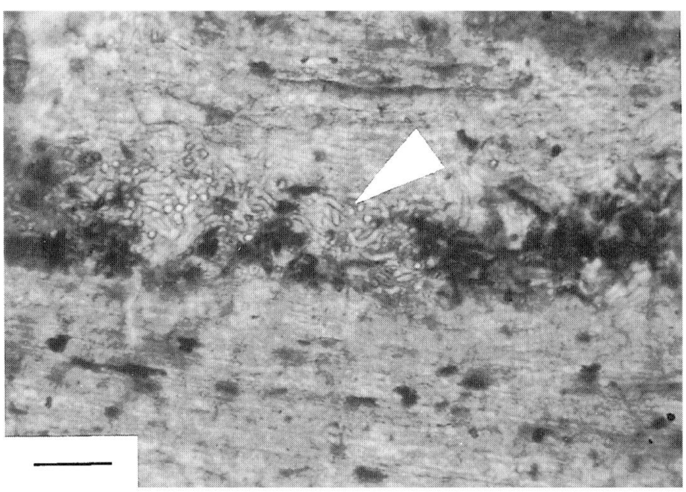

FIG. 6.—Microbial structures preserved in the mats. Layers of hollow tubes (arrow) interbedded with laminated layers. Scale bar is 100 μm.

FIG. 7.—SEM image of tubes of Fig. 6. Arrow points to preserved microbial remains.

Phosphogenetic environment, phosphate sources, and phosphatization process.—

Good preservation of delicate cyanobacterial remains requires mineralization shortly after the death of the organisms. The carbonate preserved in apatite-cemented grains also contradicts a possible late diagenetic bulk apatite replacement of a predecessor mineral. The observed apatite crystals may have evolved diagenetically and do not necessarily represent the initial mineralization, but radiaxial and palisade crystals are identical to those found in undoubtably early diagenetic phosphate concretions (Siegmund, 1995). Therefore, phosphatization is thought to have occurred during earliest diagenesis in the original shallow marine sedimentary environment. In this environment, microbial mats on the sediment surface separated the sediment from a probably oxygenated water column. Similar conditions are observed in modern atolls (see below) and probably facilitated the environmental setting for apatite precipitation in the sediment.

Several sources of phosphorus for apatite precipitation are feasable in marine shallow water settings. Organic matter with incorporated P is generally considered a major source of P in many phosphogenetic environments (Föllmi, 1996). Another source, particularly in the absence of considerable amounts of organic matter, is the input from continental weathering (e.g., Glenn, 1990; Glenn et al., 1994a). A factor analysis revealed the strong clastic influence on the overall geochemical composition of the Meishucun phosphorites (Siegmund, 1995), rendering a P contribution into the Meishucun phosphogenetic system highly probable. Extraction of P from the sea water via Fe- and Mn-oxyhydroxy shuttling, as described by Heggie et al. (1990), may be suggested as well as partial leaching and reprecipitation of apatite grains in the sediment. Phosphorus, however, is a biolimiting nutrient, and any available P in the productive zone of the ocean tends to be immediately incorporated into biomass, leaving only trace concentrations in shallow ocean water and sediment (Bentor, 1980). A direct extraction of P from sea water near the sediment surface would require persistent convection and diffusion of water through the pore space in the sediments during periods of low sedimentation rate (e.g., Föllmi, 1990). Since the pore space and sediment surface in the Meishucun section are effectively closed by microbial mats (Fig. 5), diffusion and convection were probably highly reduced, if not completely prevented. Therefore, the major immediate source of P for apatite cementation in the Meishucun phosphatized microbial mats and stromatolites is considered to have been organic matter from algae, sponges, and organisms, which are preserved as small shelly fossils. The wide spatial distribution of Neoproterozoic and Lower Cambrian phosphorites and contemporaneous black shales along the Palaeotethyan belt may also be interpreted to suggest a raised bioproductivity due to upwelling conditions in the productive epeiric seas (Parrish et al., 1986; Brasier, 1991; Siegmund, 1995). Additional input from other P sources, for example, the clastic input, probably also contributed to the amount of P bound in the mineralized microbial structures of the Meishucun phosphorites. The phosphogenetic process itself then probably proceeded as suggested in the general box model of Trappe (1996), involving microbial decomposition of

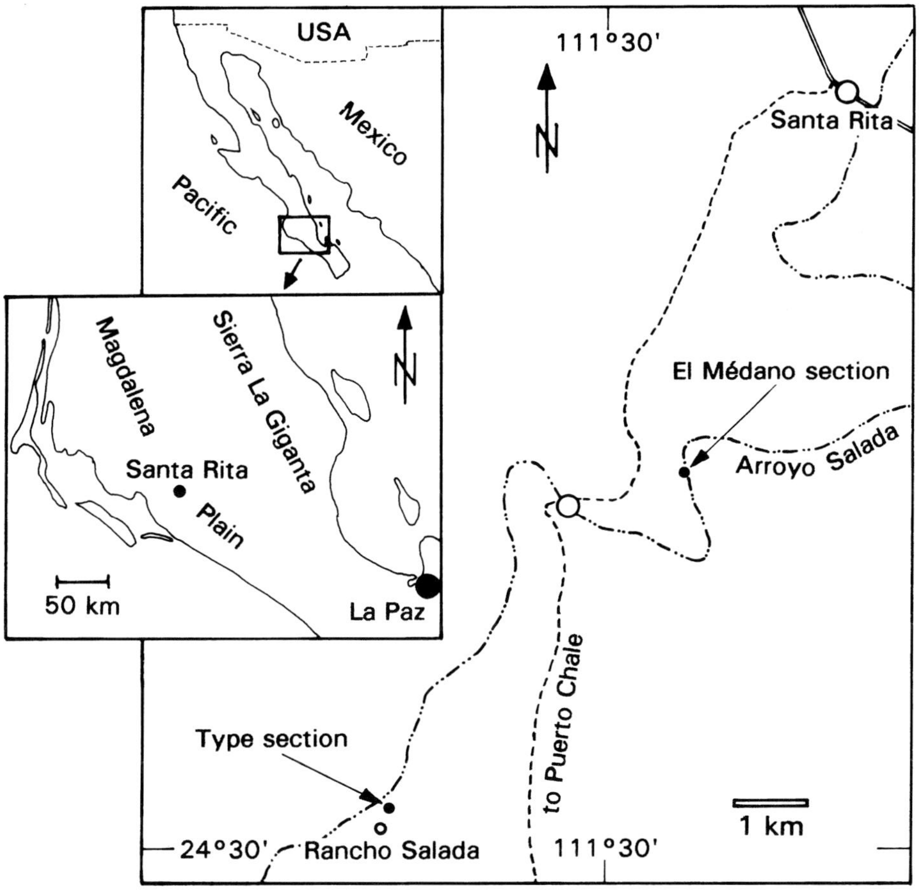

FIG. 8.—Location map of arroyo Salada in southern Baja California.

organic matter, liberalization of phosphate into at least partially closed system pore space, and its subsequent precipitation upon a suitable substrate, such as microbial remains (Briggs et al., 1993).

The Pliocene Salada Formation of Baja California, Mexico

Location.—

The Salada Formation of southern Baja California was named by Heim (1922) for sedimentary rocks exposed at a 15-m-high and about 600-m-long cliff in Arroyo (creek) Salada, about 150 km northwest of La Paz (Fig. 8). The locality is referred to as the type section of the formation (Figs. 8 and 9; Smith, 1992; Schwennicke, 1996).

General sedimentary facies and interpretation.—

About 12 m of Salada beds are exposed at the type locality. Shark teeth suggest a late Pliocene age (Ashby, 1987, Schwennicke and González-Barba, 1995). The basis of the formation is covered and the top is unconformably overlain by marine sandstone of the youngest Pliocene or Pleistocene. About 6 km northeast of the type section, near Rancho El Médano (Fig. 8), the formation unconformably overlies sandstone beds of the Uppermost Cretaceous to Eocene Tepetate Formation (Fulwider, 1991). There, the total thickness of the Salada beds is about 14 m; the upper contact is erosional and is covered by young fluvial deposits (Figs. 8 and 9). Tepetate sandstone is also exposed 150 m upstream from the type locality and correlates to a stratigraphic position right below the base of the cliff section.

The lowermost bed exposed at the type locality is a 4-m-thick silty sandstone that is rich in clayey matrix; the latter is widely replaced by calcite. The rock contains phosphatic uncoated and coated grains of fine sand size (about 10%) in addition to some small phosphatic intraclasts, small fragments of phosphatized coquina and bones, and a few phosphatized molds of small gastropods (see discussion of phosphate petrography below). The sediment is commonly completely mottled, but some relics of wave ripple bedding and small-scale trough cross-bedding are preserved. An outcrop of the same bed halfway between the type locality and El Médano exhibits fine laminations and some wave ripple bedding. The bioturbation structures include burrows of *Thalassinoides* and deposit feeder traces, indicating a well-oxygenated environment. Overall the bed reflects low sedimentation rates, some influence of waves and currents (possibly tidal currents), and favorable conditions for bottom fauna. We suggest a lagoonal depositional environment for this deposit (Schwennicke, 1996). The bed does not exist at El Médano, which remained within a terrestrial setting at that time.

At the type locality, a sandy, pebble- and fossil-rich lag deposit (bed 2, Figs. 9 and 10) overlies the lagoonal sediments,

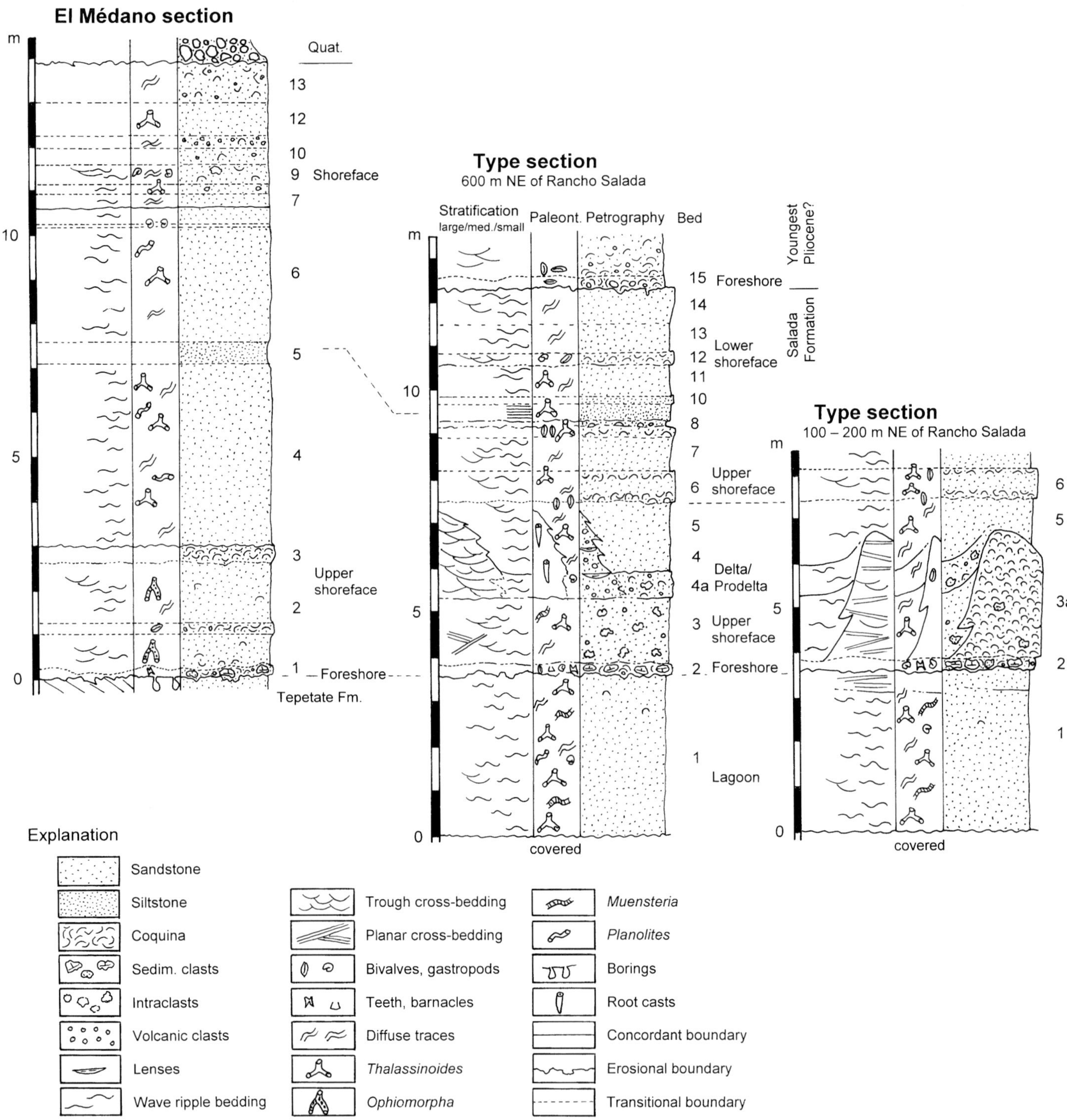

FIG. 9.—Sections of the Salada Formation in Arroyo Salada; type locality sections, El Médano section.

with an erosional contact between both beds. The thickness varies between 10 cm and 30 cm and it grades upwards into sandstone. The lag deposit correlates to the basal bed of the El Médano section (see below). The sandy portion of the bed contains about 20% of sand-sized phosphatic grains. The rock is rich in micritic matrix, but the latter has largely been replaced by dolomite (probably during early diagenesis; Baker and Allen, 1990; Mazzullo et al., 1995). Dolomite also replaced some phosphate. The bed contains small volcanic pebbles and large sedimentary clasts. The latter are irregularly shaped, superficially phosphatized, and bored (see section on phosphate petrography). In addition, the bed contains many poorly preserved mollusks and other fossils (for example, barnacles, shark teeth, and bone fragments). Superficially phosphatized molds of gastropods and bivalves and phosphatized coquina are common. The shark fauna of bed 2 includes open marine to nearshore and

FIG. 10.—Bed 2 of the type section, mainly composed of large superficially phosphatized clasts. The base is erosive and the sediment of bed 2 filled in burrows of Thalassinoides (arrows) below the bed.

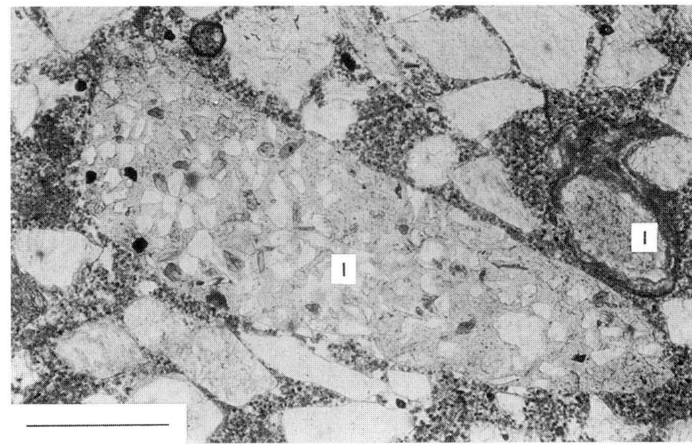

FIG. 11.—Small phosphatic intraclasts (I, center and right). Bed 3 of type section. Scale bar is 500 μm.

lagoonal species, but specimens indicative of nearshore and lagoonal, subtropical, and tropical waters are the most abundant (Ashby, 1987; Schwennicke and González-Barba, 1995).

At El Médano, the basal sediment correlates to bed 2 of the type section (Fig. 9). It is a clast and marine fossil-bearing layer, unconformably superimposed on rocks of the Tepetate Formation. Here, some pebbles are composed of reworked Tepetate sandstone. Many of the clasts and parts of the underlying Tepetate beds are superficially phosphatized. In addition, there are some large, bulbous pholad borings filled with basal Salada sediment in Tepetate rocks. Some phosphate precipitated in these borings (Del Rosal-Pardo et al., 1997).

Bed 2 of the type section, corresponding to the basal layer at El Médano, is a transgressive lag deposit. The bed marks the change from lagoonal low-energy to nearshore moderate- to high-energy conditions at the type locality and represents the basal marine sediment at El Médano. The pholad borings found at El Médano are typical for rocky shorelines. At all localities, the bed is overlain by sandstone that reflects a shoreface environment (see below).

Bed 3 of the type section is composed of a fine- to medium-grained sandstone rich in micritic matrix that is partially replaced by dolomite. Bed 3 contains about 5% phosphatic grains (uncoated angular to ovular grains, a few coated grains, some fine phosphatic coquina, and a few small phosphatic intraclasts) and some unphosphatized mudstone intraclasts. The bed is intensely bioturbated (*Thalassinoides*), but still presents some relics of medium-scale cross-bedding, small-scale wave ripple bedding and locally large-scale planar cross-bedding. The latter was probably produced by migrating sandwaves, which suggests the presence of strong tidal currents. Overall, bed 3 reflects the moderate- to high-energy conditions typical of an upper shoreface environment.

In the southwestern part of the type section cliff, bed 2 is overlain by a lens-shaped, sandy coquina deposit (Fig. 9, bed 3a) of about 150 m in length and 2 m thick at most. The deposit thins to the northeast and the southwest and interfingers with bed 3. Its base is nearly flat and the upper surface is slightly convex. Phosphatic coquina is nearly absent, but the sandy matrix contains many phosphatic grains. Crude cross-bedding and the lens shape of bed 3 suggest a submerged or semiemerged bar origin within an upper shoreface environment.

In the northeastern portion of the type section cliff, bed 3 is overlain by a nearly 4-m-thick package of poorly sorted sandstone rich in mudstone intraclasts that presents medium- to large-scale trough cross-stratification (bed 4). The deposit lacks bioturbation but shows numerous large mangrove root casts (Schwennicke, 1996). Bed 4 thins to the southwest and turns into bed 4a (Fig. 9), which is a poorly sorted, cross-bedded and bioturbated sandstone containing some phosphatic coquina, mudstone intraclasts, and small volcanic pebbles.

Beds 4 and 4a are interpreted as fan delta deposits; bed 4 is related to a delta front environment, while bed 4a represents prodelta sediments. The delta surface was colonized by mangrove.

The upper section of the type locality is composed of fine-grained sandstone, silty sandstone, and siltstone with a carbonatic or clayey matrix. Their bedding structures (lamination, small-scale wave ripple bedding, and medium-scale trough cross-bedding) are commonly disturbed by intense bioturbation (*Thalassinoides*). Several unstratified or crudely cross-stratified beds or horizons rich in coarse and unphosphatized coquina are intercalated; commonly, these strata are less bioturbated. They partially contain some fine phosphatized coquina, small phosphatic molds, and a few percent of sand-sized phosphatic grains. Nevertheless, phosphatic particles are rather rare in this part of the Salada Formation. In conclusion, the beds represent shoreface deposits (Fig. 9). The coquina-rich horizons are proximal tempestites.

A bed-by-bed correlation of beds 5 to 14 to the El Médano section is nearly impossible due to lateral facies changes. At El Médano, the upper beds are composed of well to poorly sorted sandstone that is partly rich in small volcanic pebbles and phosphatic particles (fossil molds, coquina; bed 9). The strata represent shoreface deposits (Fig. 9).

Phosphate petrography.—

Different types of phosphatic material occur in the Salada Formation, including sand-sized phosphatic grains, intraclasts, large superficially phosphatized clasts, phosphatic molds, phosphatized shells, bone fragments, and teeth. To understand the phosphogenetic processes it is neccessary to describe the petrographic features of that material.

Uncoated angular to ovular grains.—Uncoated grains range in size from fine to medium sand; the particles are rounded, subrounded, or angular. Some grains are barely phosphatized (terrigenous silt found dispersed inside the grains is still unphosphatized); more intense phosphatized grains often exhibit fissures described as desiccation cracks by some authors (e.g., Abed and Fakhouri, 1996). The angular grains clearly represent fragments of larger intraclasts and suffered only little reworking or transport. Most ovular grains were probably produced by the reworking of angular grains; some could have also been authigenically formed within muddy sediment. A minor portion (about 20–30%) of the grains was evidently produced by microbial boring activity in bone debris and subsequent phosphate precipitation; all stages ranging from unaltered bone fragments to particles presenting nearly complete obliteration of the original bone structure occur in the Salada beds. The end products are structureless phosphatic grains (Soudry and Nathan, 1980; Soudry, 1994). Overall the uncoated grains make up about 80% of the sand-sized phosphatic grains.

Nucleated grains.—These fine to medium sand-sized, angular to rounded grains are characterized by a detrital grain as a nucleus surrounded by a mass of collophane. Mechanical processes might have played an essential role in the formation of these grains because all transitions—from intraclasts containing detritus to nucleated rounded grains—occur (Quintus-Bosz, 1980; Soudry and Nathan, 1980). Commonly, nucleated grains constitute less than 1% of the sand-sized phosphatic grains.

Coated grains.—Most grains are fine sand exhibiting only a few crude coatings. Some grains are only barely phosphatized, presenting fine silt and clay particles in the coatings, which indicates that the precursor was fine-grained (muddy) siliciclastic material. The grains lack evidence for microbial coating (e.g., Soudry and Champetier, 1983). However, how the coated grains formed remains somewhat unclear. We suggest an origin within the same coastal environment in which the other phosphatic material was produced. Coated grains were either formed in muddy sediment by the phosphatization of mud around sand grains (Birch, 1990; Schwennicke et al., 1996) or they were redeposited from areas with intermittent subaerial exposure, where mudstone phosphorite was transformed into vadoids (Soudry and Southgate, 1989). However, coated grains are rare (commonly less than 1%).

Small intraclasts.—These particles are up to a few millimeters in size; most are subrounded, subangular, or angular. They are usually not completely phosphatized and terrigenous silt is still visible within their matrices (Fig. 11). The petrographic features indicate formation by the reworking of preexisting mudstone or wackestone phosphorite (Riggs, 1980). Several intraclasts composed of mudstone, silt, and sandstone rich in clayey or carbonatic matrix, or micritic limestone, are only slightly phosphatized, suggesting that more intensely phosphatized intraclasts were formed by phosphatization of these precursors. Small phosphatic intraclasts are common, forming 5–10 % of the sand-sized phosphatic material.

Large sedimentary clasts.—Large sedimentary clasts are only found in bed 2 of the type section and the basal deposit at El Médano (Figs. 9 and 10). They are composed of sandstone rich in clayey or micritic matrix, silty or sandy limestone, and dense micritic limestone. The sediment of sandy clasts contains sand-sized phosphatic grains.

The size of the clasts ranges from a few centimeters to more than 30 cm. They exhibit an irregular shape and internal bioturbation structures. Some large clasts are burrow molds of *Thalassinoides* with scratch marks (indicating their formation in consolidated sediment). Various pebbles represent relics of internal bedding structures, for example, small-scale wave ripple bedding. Many clasts, including the burrow molds, have been bored by unidentified small gastropods (commonly no larger than 1 cm) that are found as phosphatized molds inside the borings. Boring took place before and after clast formation. Some borings are cut by the actual clast surface, indicating boring of a hardened surface before reworking. More commonly, clasts have been bored from different sides, which is evidence for boring after clast formation. Most sedimentary clasts exhibit partial or complete peripheral phosphatization and some borings are completely infilled by phosphate. Nevertheless, many clasts remain unphosphatized. Initial dolomitization took place in most clasts after their superficial phosphatization. Dolomite partially replaces not only the matrix of the clasts but also other phosphatic grains and the phosphatic rim of the clasts.

Within the large sedimentary clasts, spot-like phosphatization unrelated to the superficial phosphatization occurs. These pristine phosphatizations are no larger than a few millimeters and frequently the boundary to the surrounding sediment is not well defined. They were produced by precipitation of phosphate transported by interstitial water, probably before significant consolidation of the sediment had occured. The reworking of unconsolidated sediment containing these nodules could have produced small phosphatic intraclasts.

In summary, the large sedimentary clasts suggest that they were formed by (1) deposition of calcareous and generally silty or sandy mud and bioturbation; (2) subsequent hardground formation, phosphatization and boring (hardgrounds, however, are not preserved); and (3) reworking and clast formation, and a second phase of boring and phosphatization. In addition, the basal layer at El Médano contains some partly phosphatized sandstone clasts produced by the reworking of Tepetate rocks.

Molds.—Superficially phosphatized molds (up to a few centimeters in length) are of gastropod and bivalve origin. Some molds are only slightly phosphatized; their composition resembles that of the large sedimentary clasts. The molds were probably formed by rapid lithification of the infilling sediment by apatite in a favorable microenvironment before the dissolution of the calcareous skeletons (Soudry and Lewy, 1988).

Phosphatized coquina.—The coquina stems from mollusks, mainly from bivalves. Intensely phosphatized fragments (up to 3 cm in size) are predominant, produced by the replacement of carbonate by apatite (Lamboy, 1993).

Bone fragments and teeth.—Bone fragments are up to a few centimeters in size. Many fragments are bored or rephosphatized, resulting in partial obliteration of the original bone structure. Teeth are mainly from sharks (Ashby, 1987; Schwennicke and González-Barba, 1995).

Distribution.—Most phosphatic grains are found in the lower beds of the sequence, especially in beds 1 and 2 of the type section (Fig. 9). The phosphatic portion is about 10% in bed 1 and about 20% in the sandy matrix of bed 2. The occurrence of superficially phosphatized, large sedimentary clasts is restricted to bed 2. The upper beds mostly lack phosphatic grains, with the exception of sand-sized phosphatic grains, phosphatic coquina, and some molds found in minor amounts (up to 10%) in all tempestites.

Depositional and phosphogenetic model of the Salada Formation at Arroyo Salada.—

Deposition of the Salada Formation took place in a shallow marine, large bay-like environment, as suggested by sedimentological, paleontological, and paleogeographical data. The waters were warm and subtropical (Quintus-Bosz, 1980; Ashby, 1987; Schwennicke and González-Barba, 1995). Delta deposits indicate temporal brackish water influence. The terrigenous sediments were shed from the Sierra La Giganta, which formed a major zone of erosion east and northeast of the depositional area (Fig. 8). The petrology of the phosphatic particles suggests that most or all grains were not redeposited from older phosphorite-bearing formations exposed in southern Baja California (Schwennicke, 1995), but were formed during deposition of the Salada Formation.

Deposition of the Salada Formation started with an east or northeast-directed transgression, which did not reach El Médano. Subsequently a lagoonal environment developed, as suggested by bed 1 of the type section. The bed was not deposited at El Médano, which indicates that this area still remained terrestrial. At the type locality, deposition of mainly fine-grained and partially calcareous sediment under shallow and calm conditions took place (Fig. 12A). The bioturbation indicates favorable conditions for bottom fauna. Locally the mud was phosphatized; the reworking of the sediment produced small phosphatic intraclasts and sand-sized angular grains. Other angular grains were formed by the bio-alteration of bone debris. Phosphatic ovular grains were mainly produced by the reworking of angular grains or bio-corrosion of bone fragments. Coated grains were either formed below the sediment surface or possibly in areas with intermittent subaerial exposure. The phosphate also precipitated in mollusks and replaced shells (Fig. 12B). Fine-grained (muddy) carbonate sedimentation occasionally dominated in the lagoon, possibly caused by intense insulation, low rates of terrigenous sedimentation (Nordlund, 1989), or via degradation of accumulated organic matter. Finally, carbonate hardgrounds started to form, which were later bored and superficially phosphatized (Fig. 12C). In summary, phosphogenesis occured in an oxygenated macroenvironment that was probably influenced by coastal upwelling (Garrison, 1992). The upwelled water was supposedly brought into the lagoons by tidal currents, enhancing the bioproductivity of the lagoon (Acosta-Ruiz and Lara-Lara, 1978; Millan-Nuñez et al., 1982).

In a second stage of transgression, the shoreline moved to the east or northeast of El Médano. Because of the transgression, the presumed lagoonal barrier disappeared or moved to the east or northeast of El Médano and an environment with higher energy conditions developed in the study area. The transgression led to the reworking of the lagoonal hardgrounds embedded in unconsolidated sediments; loose mud, silt, and sand was transported away, causing the concentration of clasts, molds, and phosphatized coquina in the transgressive lag deposit. During that transgression, a second event of phosphogenesis occurred in very shallow, nearshore water. The newly formed clasts, burrow molds, and some coquina continued to be phosphatized and bored (Fig. 12D). In addition, rocks of the Tepetate Formation exposed at the sea-bottom near El Médano were slightly phosphatized and bored by pholads. Some phosphate was formed inside the borings.

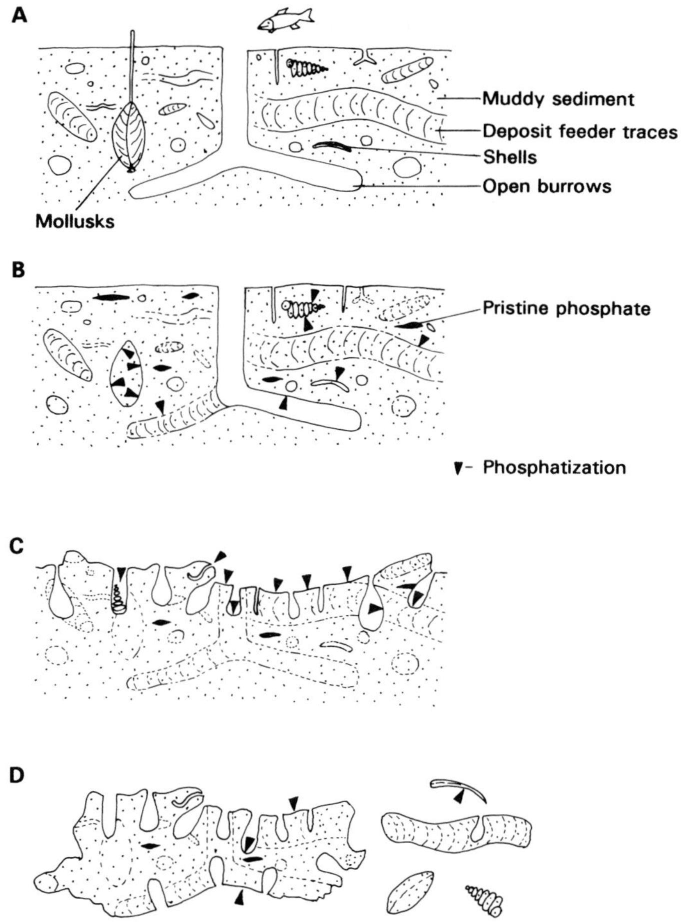

FIG 12.—Suggested scheme for the formation of phosphatic material in the Salada Formation. (A) Deposition of mainly calcareous, sandy mud in a lagoonal setting. Colonization and bioturbation. (B) Internal phosphatization of mollusks, phosphatization of shells, formation of small phosphatic grains and some pristine phosphate. (C) Hardground formation, boring, phosphatization. (D) Reworking and clast formation, further phosphatization of clasts, boring and phosphatization of coquina in a nearshore environment.

As the transgression continued, mostly sandy and silty sediments were deposited, probably in a large, bay-like environment. A fan delta temporarily developed but disappeared with the progressive deepening of the depositional area. Sporadic storms produced coquina-rich tempestites. Phosphogenetic processes continued in the northern or northeastern marginal areas, as indicated by phosphatic grains and coquina found in tempestites, but terminated in the region between the type locality and El Médano.

Modern Atoll Phosphogenesis

Geomorphology, hydrology, and ecology of atolls.—
Atolls such as those forming the Tuamotu Archipelago (French Polynesia, Fig. 13) are also called low-islands, in contrast to volcanic high islands: their rim is just exposed at the ocean's surface and extends a few meters above sea-level. The reef rim constitutes the top of the barrier reef. Behind it, a succession of detrital accumulations (motu or islets) intersected by channels (hoa) encircle the central lagoon. The islets constitute the habitable and cultivatable part of the atolls.

The hydrologic system of a motu consists of a groundwater lens (Ghyben-Herzberg lens) with fresh to brackish water floating on the haline interstitial water that impregnates the carbonate substrate (Fig. 14). The volume of this phreatic lens depends on the height and surface area of the motu. The waters are composed of mixed meteoric water (major part) and marine water (minor part). The marine component is subdivided into oceanic surface water (infiltrations) and water originating from the Antarctic Intermediate Water (AIW), deep oceanic water that is rich in nutrients, brought to the top of the atoll structure by geothermal endo-upwelling (Rougerie and Wauthy, 1993).

In most atolls, the ocean is in direct connection with the lagoon via channels, but some atolls are almost or completely closed by detrital accumulations in the channel system. In closed atolls, the characteristics of the lagoonal water quickly differentiate from those of the oceanic surface water:

1. When evaporation exceeds the phreatic discharge into the lagoon, hypersaline conditions develop.
2. When the phreatic discharge exceeds evaporation, the lagoons become more or less brackish.

Biologic productivity strongly increases in either case, as evident through the development of green algae and eutrophication at Taiaro atoll (hypersaline), or the invasion of thick cyanobacterial mats at Niau atoll (brackish). This shows that the phreatic water is the source of nutrients that in open lagoons are lost to the ocean.

The initial step of phosphogenesis is the closing of large, high-rimmed atolls, which have an important phreatic volume with regard to the lagoonal volume. As stated by Rodgers (1994), the groundwater lens has a central role in redistributing phosphorus in an atoll ecosystem. Colonization by cyanobacterial mats occurs with the desalting in ponds on beaches (Mataiva: Delesalle et al., 1985), at the bottom of lagoons (Niau: Landret, 1978; Aldabra, Seychelles: Braithwaite et al., 1989), and brackish lakes (Palau, West Pacific: Burnett et al., 1993). Endo-upwelling fluxes provide a constant input of nutrients from the deep ocean (Jehl and Barsczus, 1996), which promote growth of autotrophic cyanobacteria. Cyanobacterial mats also

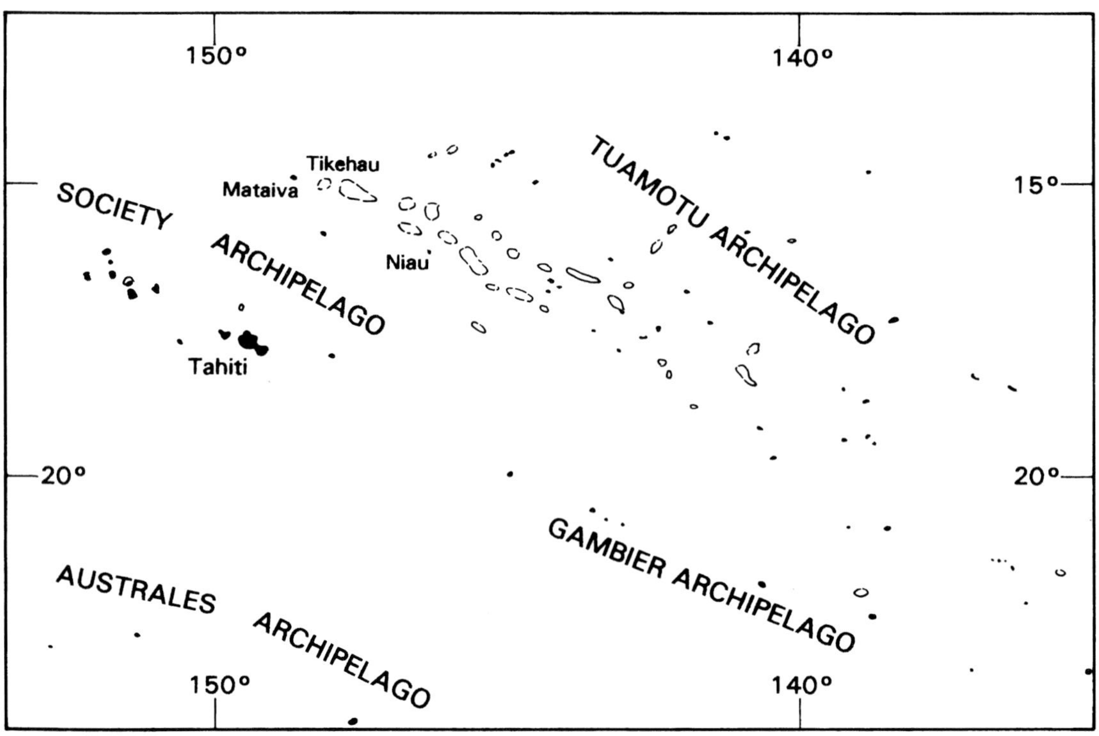

FIG. 13.—Tuamotu Archipelago in French Polynesia.

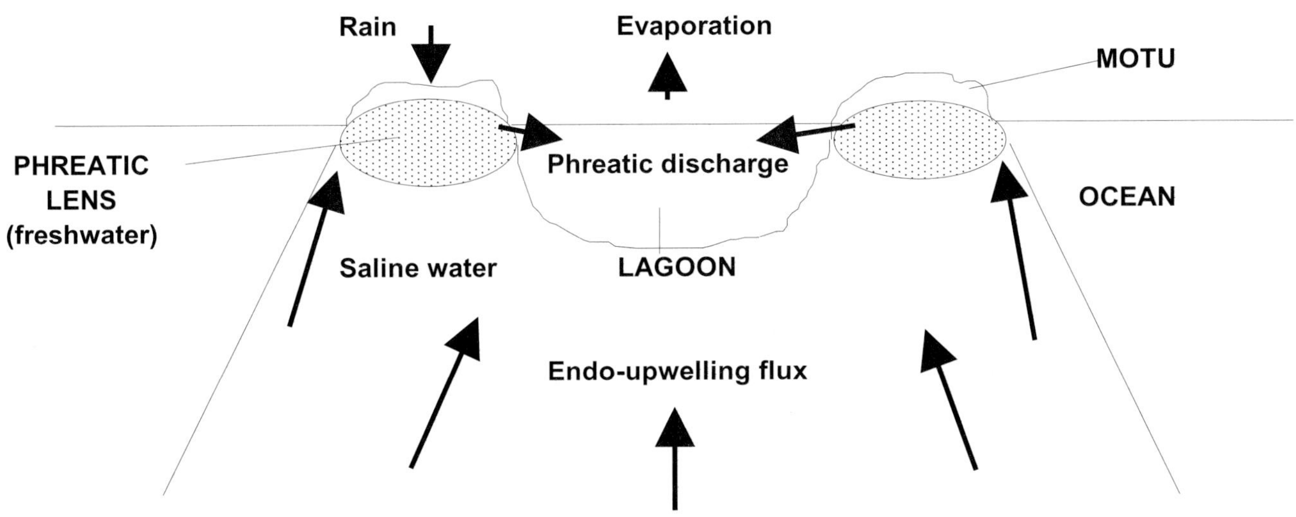

FIG. 14.—Geomorphology of an atoll and its hydrologic system.

thrive in ponds (Fig. 15) and lakes, where the brackish phreatic water emerges on the exposed parts of atolls.

The salinities in the waters overlying the mats range from 5 g to 20 g NaCl/l (Jehl and Rougerie, 1995). The pH values vary from 9 in surface water to 6 in cyanobacterial mat interstitial water due to bacterial metabolism. This acidity of the pore water slowly dissolves the underlying carbonate basement, so the presence of cyanobacterial mats promotes deepening of the depression.

Cyanobacterial mats and accumulation of organic matter.—
The cyanobacterial mats (called kopara by the Tuamotu natives in French Polynesia) have a green-orange surface color, and the organic matter accumulated is red, due to carotenoid pigments. The uppermost centimeters contain 20–30% of organic carbon in weight. Cores show that the mats have stromatolitic properties that exhibit an alternation of red layers and

FIG. 15.—Kopara pond and its desiccation polygons in surface: the water level limits the productivity. Thickness of several tens of centimeters of organic matter lays out on several hundred square meters.

FIG. 16.—Kopara core, showing the green autotrophic layer (G) in surface, the red organic matter (R) below, and the white stromatolitic laminations.

white laminations of carbonate precipitated *in situ* (Trichet, 1967). Old kopara has a pink or yellow color. The lowermost layers of kopara are almost entirely composed of magnesian calcite. In ponds, cyanobacterial mats may reach several tens of centimeters in thickness, thanks to the richness of nutrients in the brackish phreatic water (Fig. 16). The age of the bottom of a 40-cm-long mat core may reach 1–2 Ka. The productivity of the mat is only limited by water level. The Palau lake (Caroline Islands) described by Orem et al. (1991) contains kopara that is 15 m in thickness.

Kopara mats are composed of an association of autotrophic cyanobacteria and heterotrophic bacteria (micro-organisms feeding on cyanobacterial remains). Four dominant groups of prokaryotes are present in kopara: cyanobacteria, sulfo-oxidizing bacteria, sulfate-reducing bacteria, and methanogens. The growth of these different groups along oxygen and light gradients (Winogradsky succession) has led to the formation of microzones with concentric stratification of communities and steep gradients of oxygen within the first centimeter of the mats. Although sulfate-reducing bacteria dominate throughout the first centimeters of kopara by far, the quantities of methanogens are also considerable.

The cyanobacterial covers are characterized by rapid formation of reducing conditions below the uppermost layers. Bacterial activity consumes most of the oxygen, and highly negative redox potentials develop in microniches. The mucilageous mats constitute an impermeable cover, limiting gas and fluid exchange between water and sediment. This reducing environment is very favorable to the preservation of organic matter (Bauld, 1981). Organic matter diagenesis then is primarily due to bacterial degradation, which occurs mainly at the top of the mat. Microbial denitrification and methanogenesis lead to a loss of nitrogen and carbon, so a relative enrichment in P develops with increasing decay of organic matter.

The distribution of humic substances in kopara samples reveals the immaturity of the organic material, with fulvic acids and acido-soluble compounds dominating, despite their polycondensation to humin (Defarge, 1983; Rougerie et al., 1997). Decreasing (C/P) ratios in all humic substances indicate an enrichment of P in kopara with increasing depth and time, except for humic acids, which show a loss of P to the interstitial medium during formation.

After its fixation in the organic matter of kopara during autotrophic growth of cyanobacteria, P is first indiscriminately associated with various humic fractions and later with fulvic acids (FA). The highly reducing conditions, which prevail below the first centimeter of the mat, allow long-term preservation of the organic material with its associated P.

The hydrocarbon composition of the lipidic fraction of kopara organic matter and its vertical profile are consistent with the community structure and stratification of the mats, the influence of the surrounding area, and environmental changes that have occurred in the past. They reflect contributions from various mat organisms as well as rapid post-depositional changes.

The diagenetic evolution of the organic matter along each generation of kopara seems to show an increase in hydrocarbons with age. Early diagenesis already occurs at the surface of the mat, as shown by mature products synthetized in the green layer (Jehl, 1995). The diagenetic evolution tends to concentrate some compounds with depth: odd n-alkanes from C25 to C31, heavy n-alkenes (C29:3, C30:1, C31:3), C20:5, C30 hopenes, hopanes (normoretane, norneohopane, C31 and C35 homohopanes), some tetracyclic terpenoids, euphadiene, and fernene. Some of these compounds are unstable. Their presence in the oldest sections implies that the structure and development of kopara is characterized by highly reducing conditions that substantially moderate diagenesis and allow preservation of unstable molecules.

Oxidation.—

Among the hydrocarbons identified, many redox indicators reveal less reducing conditions at the kopara-carbonate interface. At the interface with the rock foundation, the presence and circulation of suboxic interstitial water may induce a slight oxidation and degradation of the remaining organic matter (already P-enriched) in the deepest and oldest layers of kopara. The P remineralized and liberated by this oxidation at the bottom of the cyanobacterial sediment can be efficiently trapped by FA and humin, which possess a high affinity for phosphorus. This diagenetic process is illustrated by the decrease of C/P ratios with age in the raw material and humic compounds, except in the humic acids (HA). In the confined systems of the kopara ponds, the transformations of FA to primary HA and humin to secondary HA lead to a loss of P into the interstitial water. Such a release of P promotes a gradual enrichment of the interstitial waters both in ponds and large lagoons occupied by microbial mats (Rougerie et al., 1997).

A core from the Niau lagoon showed strata with 1.24 % P_2O_5 at 8 m, and 1.33 % P_2O_5 at 15 m depth (Fikri, 1991). Similar results were found in an anoxic lake of Palau (Caroline Islands) in the West Pacific (Burnett et al., 1993). This mechanism may lead to enrichment and saturation of interstitial waters, favoring authigenic apatite precipitation or *in situ* carbonate substitution. However, the kinetics of the oxidation stage leading to the liberation of the phosphorus entrapped in the lagoonal biologic sediment are still unknown.

Apatite precipitation.—

Rare earth elements (Jehl and Barsczus, 1996), hydrocarbon composition and distribution (Fikri, 1991; Jehl, 1995), and humic substance repartition (Rougerie et al., 1997) show an identical signature in kopara and phosphate rock found in modern atolls. They thus confirm that the cyanobacterial accumulations constitute the phosphogenic sediment preceeding atoll phosphorites (Jehl and Rougerie, 1995). Trichet (1967) and Trichet and Defarge (1993) found that the stromatolitic properties of kopara are highly favorable for the nucleation of carbonate minerals. After the death of cyanobacteria, the polysaccharidic membranes of cyanobacteria reorganize in a tridimensional network, inducing the formation of pores where the nucleation occurs (Defarge et al., 1994). This mechanism has been observed *in situ* during the formation of aragonite and magnesium calcite. The carbonate fluorapatite (CFA), which is the most widely occurring mineral in phosphorites, could also be synthetized by kopara in the presence of high concentrations of phosphorus in the pore waters, provided sufficient amounts of F are supplied by the endo-upwelling system. The formation of pores is particularly important for providing a microenvironment where the chemical conditions may be different from the global interstitial water (redox conditions, alkalinity, presence/absence of magnesium, calcium, fluorine, phosphate, etc.). Related chemical data are discussed elsewhere (Fikri, 1991; Jehl, 1995; Jehl and Rougerie, 1995; Rougerie et al., 1997).

Microscopic observation of phosphate samples from the Mataiva deposit emphasizes the strong implication of cyanobacteria in apatite crystal nucleation. Some cyanobacteria colonized the cavities of the phosphorite after its formation. The biconcave aspect of coccoid-shaped structures (Fig. 17) and the collapse of filaments (Fig. 18A) are artifacts linked to the vacuum created by the observational technique. Only the soft structures (nonmineralized) have been affected by this phenomenon. The spheric microbial bodies (Fig. 18A) associated with the collapsed filaments seem to be consolidated by a precipitation of apatite needles (Fig. 18B). This observation implies that the cyanobacterial bodies provided suitable surfaces for apatite crystallization, which is afterwards extended to the filaments (Fig. 19). In vitro experiments also show that cyanobacteria are able to promote the precipitation of apatite (Gerasimenko et al., 1993).

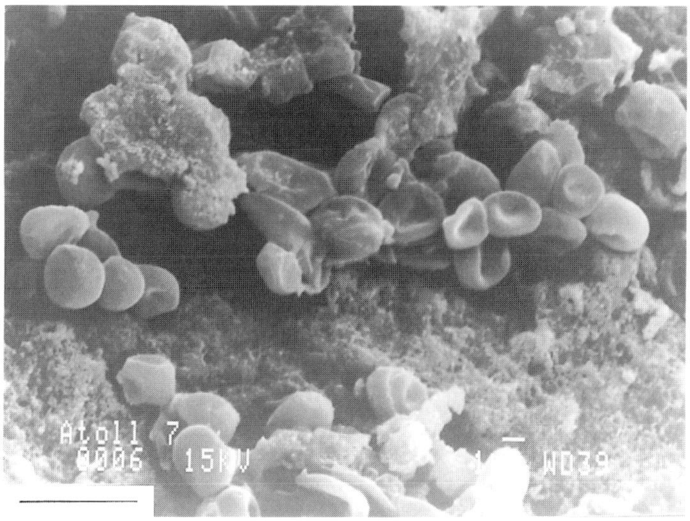

FIG. 17.—Electronic microscopy observation of cyanobacteria included in pores of a phosphate sample from Mataiva atoll: unphosphatized microbial bodies collapsed by the vacuum producing biconcave structures (arrow). Scale bar is 5 μm.

Microscopic observations and analyses have shown the importance of dissolution-reprecipitation of phosphatic material, with a P-enrichment of the secondary apatites in deeper zones (Jehl, 1995). This implicates a continuity of P enrichment in the deposit through time, which can be due to many stages of kopara growth and degradation (linked to isostatic or eustatic variations) and a permanent supply of P from the interstitial reservoir. Where the cyanobacterial mats are large and thick, maturation and oxidation of kopara may finally lead to the formation of a major phosphate deposit (Rougerie et al., 1997).

The new model for insular phosphogenesis (Fig. 20), following previous work by Rougerie and Wauthy (1989) and Fikri (1991), may be summarized as follows:

1. Closing of the lagoonal ecosystem.
2. Trapping of the P permanently supplied by endo-upwelling.
3. Growth of cyanobacterial mats.
4. Accumulation of organic matter in anoxic conditions below the cover of cyanobacterial mats.
5. P-enrichment of the deepest strata (decrease of C/P ratio).
6. Slow oxidative degradation of organic matter by suboxic groundwater or by other means linked to sea-level changes.
7. Mineralization and release of P into the pore water, saturation of the pore water.
8. Apatite formation.
9. Secondary enrichment of the primary deposit.

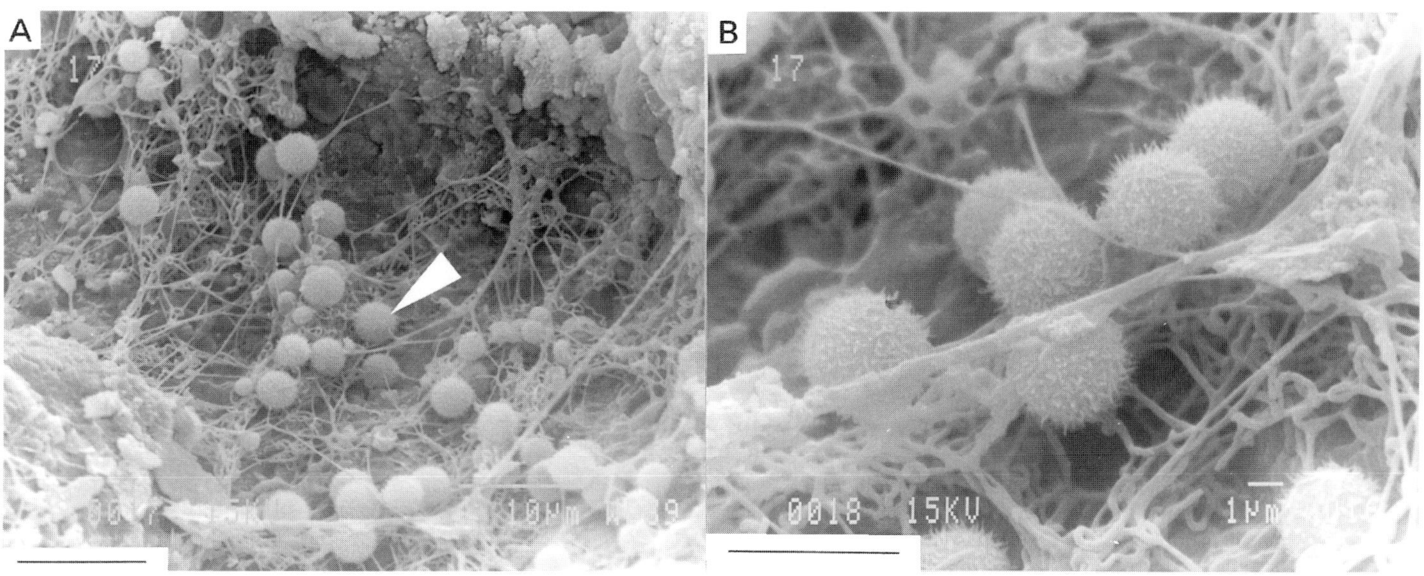

FIG. 18.—Electronic microscopy observation of cyanobacteria preserved in pores of a phosphate sample from Mataiva atoll. (A) Individual coccoid-shaped cyanobacteria (arrow) in a collapsed filamentous network. Scale bar is 10 μm. (B) Detail of the microbial bodies covered by fine apatite crystals. Scale bar is 5 μm.

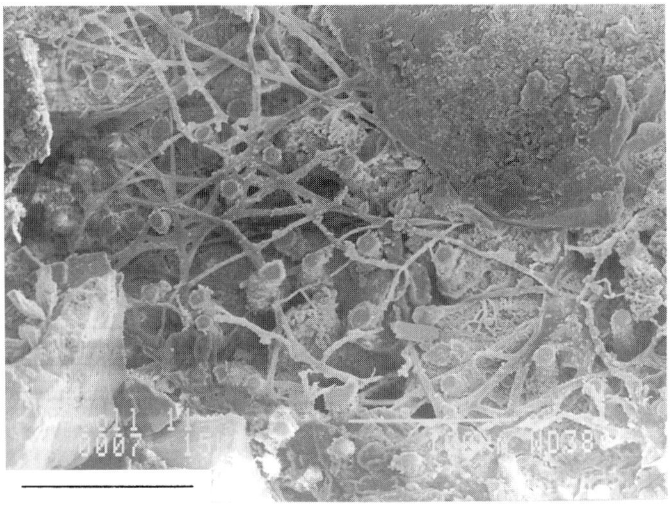

FIG. 19.—Mineralized filamentous cyanobacteria (arrow) observed in electronic microscopy in a phosphate sample from Mataiva atoll. Scale bar is 10 μm.

Shallow marine phosphogenesis

The case studies documented above comprise both modern and ancient phosphorites. There are, however some problems that tend to render any comparison a difficult task. The modern atoll example is definitely of shallow marine origin and serves to observe biological and chemical phosphogenetic processes *in situ*. The fossil examples of the Cambrian and Tertiary have to rely on geologic data but still provide ample sedimentary evidence for a similar environmental interpretation. Unfortunately they lack most data on the conditions of the actual phosphogenetic environment. However, we still consider it useful to include all examples into the following discussion, because they demonstrate the variety of shallow water phosphogenetic processes.

The phosphogenetic settings described above comprise pure carbonate (atolls), mixed carbonate-siliciclastic (Zhongyicun Formation), and mixed carbonate-siliciclastic to siliciclastic-dominated (Salada Formation) marine shallow water environments. The Cambrian fossilized stromatolitic inner shelf and perhaps lagoonal phosphorites supply data on phosphate mineralization and preservation and demonstrate that processes comparable to modern shallow water phosphogenesis probably already existed some 530 Ma ago. The Tertiary Salada Formation phosphorites show that extensive phosphogenesis is possible even without evident cyanobacterial activity other than the boring of grains. Further studies will shed more light on this problem.

Reviewing the data provided by the different shallow-water settings, we conclude that there is a number of general factors that may facilitate phosphogenesis in these environments:

1. A considerable supply of nutrients that enable rich bioproductivity. This may be accomplished by some form of upwelling of nutrient-enriched water, either as coastal upwelling or as endo-upwelling in atolls, or by additional P influx from the hinterland in nearshore settings. In the latter case, a hinterland with a tropical climate and dominated by intense chemical weathering seems to be favorable (Glenn et al., 1994a).
2. Trapping and storage of organic matter and phosphate in an environment protected from recycling into open ocean water. In pure carbonate and mixed carbonate-siliciclastic settings, cyanobacterial mats serve to concentrate organic matter and

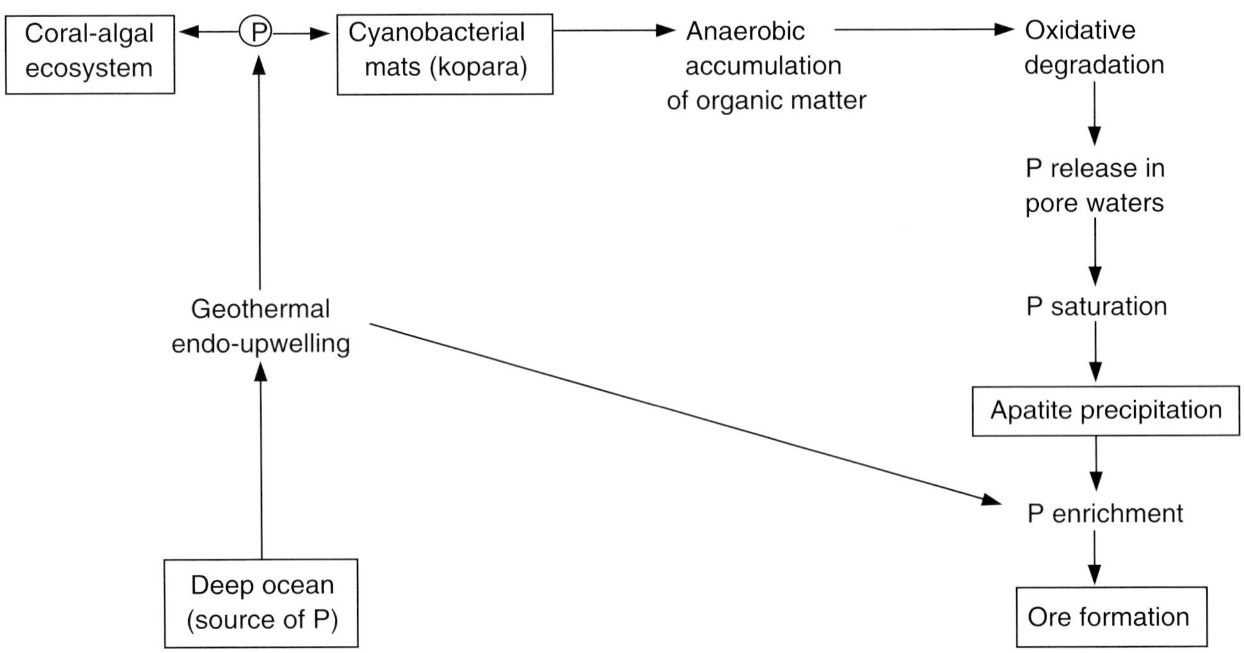

FIG. 20.—Insular phosphogenesis model showing the deep oceanic and interstitial phosphorus reservoirs to cyanobacterial mats accumulation and apatite precipitation and accumulation.

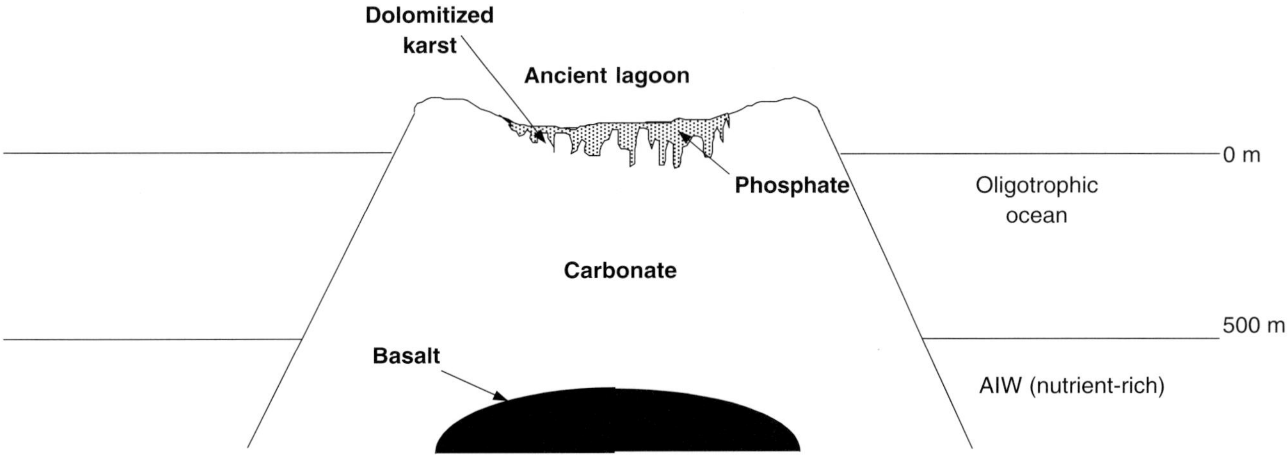

FIG. 21.—Location of phosphorite in karstic depressions, formed in the ancient lagoon of an uplifted atoll (e.g., Makatea, Nauru, Ocean Island).

entrapped phosphate and separate the interstitial phosphogenetic microenvironment from the oxygenated surface waters. In settings with some fine clastic input, impermeable mud layers may help to separate sedimentary pore fluids from the open water above and prevent recycling of organic matter and phosphate. Low sedimentation rates seem to be favorable.

3. Liberation and concentration of phosphate in a favorable microenvironment, probably accomplished by the microbial degradation of organic matter.
4. Apatite precipitation. In settings with cyanobacterial mats, these provide a suitable substrate for apatite crystal nucleation (atolls, Zhongyicun phosphorites). Otherwise, replacement of carbonate or siliciclastic predecessors or precipitation from pore fluids may occur (Salada Formation).

These factors are very similar to the stages of phosphogenesis proposed in the general model by Trappe (1996). The sequence of processes encountered in the presented shallow marine phosphogenetic settings actually constitute some combinations of the wide variety of processes discussed there. It is, therefore, not surprising that shallow marine phosphogenesis is not only widespread, but also is found in very different sedimentary macroenvironments.

Enrichment of Primary Phosphate Deposits

Although each of the examples described makes a strong case for shallow marine phosphogenesis, only the Cambrian deposit has evolved into what could be considered an economic phosphorite (Siegmund, 1995). In the other examples some processes and conditions required for the transformation of shallow water primary phosphate deposits into rich phosphorites were not met. Some factors considered important for this transformation process are discussed below.

1. Time.—

Time is considered a major factor because it strongly controls the amount of phosphate available from a particular phosphogenetic setting. In atolls, microscopic observations and analysis have shown the importance of the dissolution-reprecipitation of phosphatic material, leading to a P-enrichment in the secondary apatites (Jehl, 1995). This implies a continuation of P-enrichment in the phosphate deposit through time, which can be due to many stages of cyanobacterial growth and degradation and a permanent supply of P from the interstitial reservoir. Also, other restricted, but coastal and more clastic-influenced phosphogenetic environments (for example, the Salada Formation deposit), will benefit from more time since the phosphate factory will continue to accumulate apatite as long as the conditions are favorable. Time is a critical factor since many shallow-water environments are very unstable.

2. Sedimentary dynamics.—

A low sedimentation rate is another critical factor for phosphogenesis (Föllmi, 1996). As long as environmental conditions with low clastic input exist, phosphogenesis may persist. Sedimentary dynamics may modify the depositional conditions of the primary phosphates. They also have a profound influence on the grade of a phosphorite. Sedimentary dynamics provide the physical means of reworking and concentration and are generally responsible for the creation of phosphorite deposits (Glenn et al., 1994b; Föllmi, 1996). Commonly, the absence of mechanical concentration will prevent formation of high-grade phosphorites. In the environments of the examples discussed, storms or tsunamis, waves, or tidal currents may have played an important role in the reworking and concentration of parts of the deposits.

3. Sea level.—

Sea level has a major influence on phosphogenesis and phosphorite genesis. It governs the phosphogenetic processes in all of the investigated settings. Cambrian phosphogenesis in China was terminated after a profound sea-level rise, which was documented by the deposition of black shales over the phosphorites. The Mexican Salada Formation phosphogenetic environment was strongly influenced by sea-level-induced facies changes. Due to a transgression the phosphate-producing environment shifted laterally and one secondary bed enriched in phosphatic material was formed.

Isostatic and eustatic variations are responsible for important fluctuations of the groundwater level on atolls. These water movements lead to the remobilization of P, induce a circulation of fluids more or less enriched in P, and finally result in the reprecipitation of apatite in deeper zones (Nathan, 1990). Successive phases of dissolution and reprecipitation will affect the phosphates located below the level of phreatic lenses (Rodgers, 1994). Dissolution affects primarily the carbonated phosphates (Nathan et al., 1990), which are less stable. During times of low sea level, the meteoric discharge helps the dissolution of carbonated minerals (foraminifera, coral debris, and CFA). Not infrequently, intense karstification results in the formation of a karrenfeld and the redeposition of high-grade phosphatic ore into deep depressions (Fig. 21).

CONCLUSIONS

The Lowermost Cambrian Zhongyicun Formation phosphorites from China, the phosphorites of the Tertiary Salada Formation of Baja California, Mexico, and recent atoll phosphorites from French Polynesia are all examples of extensive shallow marine phosphogenesis. Common features of these more clastic-dominated to pure carbonate settings are a lagoonal or nearshore environment with restrictions in water circulation and the influence of some form of upwelling to supply the nutrients for a high bioproductivity. Organic matter is considered the major source of P, but in the settings with terrigenous influence, sources such as input from continental weathering are probable. The study of modern atoll phosphorites enabled the introduction of a refined phosphogenetic model. It involves cyanobacterial mats that provide the microenvironment favorable for phosphogenesis and serve as nucleation sites for apatite crystals. In light of environmental and microbial similarities, we consider this model also valid for the Cambrian phosphatized microbial mats and stromatolites, but the Salada Formation phosphorites lack discernible microbial mat structures. Here the generation of a phosphogenetic environment may be due to fine-grained sediment layers preventing free access of oxygenated water into the sediment. The characteristics found in shallow marine phosphogenetic settings, especially the modern atoll example, provide vital clues for the understanding of phosphogenesis in general. Although only one of these shallow marine phosphogenetic settings lead to the formation of an economic phosphorite, we argue that more time for phosphate accumulation and concentration might have generated such deposits in the other cases as well.

ACKNOWLEDGMENTS

We thank K. B. Föllmi and D. Soudry for their critical reviews of the manuscript. C. R. Glenn encouraged us to write the present paper and provided useful information. We are grateful to Jochen Halfar, who helped to improve the English.

REFERENCES

Abed, A. M., and Fakhouri, K., 1990, Role of microbial processes in the genesis of Jordanian Upper Cretaceous phosphorites, *in* Notholt, A. J. G., and Jarvis, I, eds., Phosphorite Research and Development: London, Geological Society of London Special Publication 52, p. 193-203.

Abed, A. M., and Fakhouri, K., 1996, On the chemical variability of phosphatic particles from Jordanian phosphorite deposits: Chemical Geology, v. 131, p. 1-13.

Acosta-Ruiz, M. J, and Lara-Lara, J. R., 1978, Resultados físico-químicos en un estudio de variación diurna en el área central de Bahía Magdalena, B.C.S.: Ciencias Marinas, v. 5, p. 37-46.

Ashby, J. R., 1987, Vertebrate paleoecology of the late Pliocene Arroyo Salada local fauna, Baja California Sur, Mexico: Ciencias Marinas, v. 13, p. 23-30.

Baker, P., and Allen, M., 1990, Occurrence of dolomite in Neogene phosphatic sediments, *in* Burnett, W. C., and Riggs, S. R., eds., Phosphate Deposits of the World, Volume 3—Neogene to Modern Phosphorites: Cambridge, Cambridge University Press, pp 75-86.

Baturin, G. N., 1982, Phosphorites on the sea floor. Origin, composition and distribution: Developments in Sedimentology, v. 33, 343 p.

Bauld, J., 1981, Geobiological role of cyanobacterial mats in sedimentary environments: Production and preservation of organic matter: Journal of Australian Geology and Geophysics, v. 6, p. 307-317.

Bentor, Y. K., 1980, Phosphorites—the unsolved problems: Tulsa, Society for Sedimentary Geology (SEPM) Special Publication 29, p. 3-18.

Birch, G., 1990, Phosphorite deposits on the Namibian continental shelf, *in* Burnett, W. C., and Riggs, S. R., eds., Phosphate Deposits of the World, Volume 3—Neogene to Modern Phosphorites: Cambridge, Cambridge University Press, p. 153-158.

Bourrouilh-Le Jan, F. G., 1980, Phosphates, sols bauxitiques et karsts dolomitiques du Centre et SW Pacifique. Comparaisons sédimentologiques et géochimiques: Géologie Comparée des Gisements de Phosphate et de Pétrole, Colloque International, Orléans, 6-7 novembre 1979, Documents du B.R.G.M., v. 24, p. 113-128.

Bourrouilh-Le Jan, F. G., 1989, Phosphate, bauxite et diagenèse carbonatée dans les récifs miocènes du Sud Pacifique: Géologie Méditerranéenne, v. 16, p. 79-96.

Braithwaite, C. R. J., Casanova, J., Frever, t. T., and Whitton, B. A., 1989, Recent stromatolites in landlocked pools on Aldabra, Western Indian Ocean: Palaeogeography, Palaeoclimatology, Palaeoecology, v. 69, p. 145-165.

Brasier, M. D., 1991, Nutrient flux and the evolutionary explosion across the Precambrian/Cambrian boundary interval: Historical Biology, v. 5, p. 85-93.

Bremner , J. M., and Rogers, J., 1990, Phosphorite deposits on the Namibian continental shelf, *in* Burnett, W. C., and Riggs, S. R., eds., Phosphate Deposits of the World, Volume 3—Neogene to Modern Phosphorites: Cambridge, Cambridge University Press, p. 143-152.

Briggs, D. E. G., Kear, A. J., Martill, D. M., and Wilby, P. R., 1993, Phosphatization of soft-tissues in experiments and fossils: Journal of the Geolocical Society of London, v. 150, p. 1035-1038.

Burnett, W. C., Riggs, S. R., Orem, W., Stanton, G., Cable, P., and Gilliam, B., 1993, The origin of insular phosphate deposit of Palau (abs.): The Pamphlet, v. 3, p. 49.

Cook, P. J., and Shergold, J. H., 1979, Proterozoic and Cambrian phosphorites: Report on the First International Meeting of IGCP Project 156: Canberra, Australia National University Press, 106 p.

Dahanayake, K., Gerdes, G., and Krumbein, W. E., 1985, Stromatolites, oncolites and oolites biogenically formed in situ: Naturwissenschaften, v. 72, p. 513-518.

Dahanayake, K., and Krumbein, W. E., 1985, Ultrastructure of a microbial mat-generated phosphorite: Mineralium Deposita, v. 20, p. 260-265.

DEFARGE, C., 1983, Contribution à l'étude géochimique et pétrologique des formations protostromatolitiques de Polynésie. Application à la connaissance des mécanismes de la précipitation des carbonates de calcium au sein de matières organiques sédimentaires: Unpublished Doctoral Dissertation, University of Orléans, Orléans, France, 167 p.

DEFARGE, C., TRICHET, J., MAURIN, A., AND HUCHER, M., 1994, Kopara in polynesian atolls: Early stages of formation of calcareous stromatolites: Sedimentary Geology, v. 89, p. 9-23.

DEL ROSAL-PARDO, A., SANCHEZ-DORAME, F., AND SCHWENNICKE, T., 1997, Origen de perforaciones pliocénicas fosfatizadas (Formación Salada) en rocas cretácicas (Formación Tepetate), arroyo Salada, Baja California Sur, México (abs.): 4 Reunión Internacional sobre Geología de la Península de Baja California, abril 1997, Ensenada.

DELESALLE, B., BELL, J., DE VAUGELAS, J., GALZIN, R., AND RARO MOANA, G. I. E., 1985, Atoll de Mataiva, Archipel des Tuamotu (abs.): Proceedings Fifth International Coral reef Congress, Tahiti, v. 3, p. 365-370.

DIXON, W. A., 1877, The guano and other phosphatic deposits occurring on Malden Island: Journal of the Royal Society N.S.W., v. XI, 176 p.

DOTT, R. H., AND BOURGEOISE, J., 1982, Hummocky cross-stratification: Significance of its variable bedding sequences: Geological Society of America Bulletin, v. 93, p. 663-680.

FIKRI, A., 1991, Contribution à l'étude de la géochimie organique des phosphates insulaires. Implications dans la phosphatogenèse: Unpublished Doctoral Disssertation, University of Orléans, Orléans, France, 200 p.

FÖLLMI, K. B., 1990, Condensation and phosphogenesis: Example of the Helvetic mid-Cretaceous (northern Tethyan margin), in Notholt, A. J. G., and Jarvis, I., eds., Phosphorite Research and Development: London, Geological Society of London Special Publication 52, p. 237-252.

FÖLLMI, K. B., 1996, The phosphorus cycle, phosphogenesis and marine phosphate-rich deposits: Earth-Science Reviews, v. 40, p. 55-124.

FULWIDER, R., 1991, Biostratigraphy of the Tepetate Formation, Baja California Sur (abs.): Primera Reunión Internacional sobre Geología de la Peninsula de Baja California, La Paz, p. 28.

GARRISON, R. E., 1992, Neogene phosphogenesis along the eastern margin of the Pacific Ocean: Revista Geológica de Chile, v. 19, p. 91-111.

GERASIMENKO, L. M., GONCHAROVA, I. V., POCHTAREVA, I. V., ROZANOV, A. J., USHATINSKAYA, G. T., AND ZHEGALLO, G. T., 1993, Laboratory experiments on the deposition of calcium phosphate with the help of cyanobacteria (abs.): The Pamphlet, v. 3, p. 51.

GLENN, C. R., 1990, Depositional sequences of the Duwi, Sibaiya and Phosphate Formations, Egypt: Phosphogenesis and glauconitization in a Late Cretaceous epieric sea, *in* Notholt, A. J. G., and Jarvis, I., eds., Phosphorite Research and Development: London, Geological Society of London Special Publication 52, p. 205-222.

GLENN, C. R., ARTHUR, M. A., RESIG, J. M., BURNETT, W. C., DEAN, W. E., AND JAHNKE, R. A., 1994a, Are modern and ancient phosphorites really so different?, *in* Iijima, A., Abed, A. M., and Garrison, R. E., eds., Siliceous, Phosphatic and Glauconitic Sediments of the Tertiary and Mesozoic: Proceedings of the 29th International Geological Congress Part C, Utrecht, VSP, p. 159-188.

GLENN, C. R., et al., 1994b, Phosphorus and phosphorites: Sedimentology and environments of formation: Eclogae Geologicae Helveticae, v. 87, p. 747-788.

HEGGIE, D. T., SKYRING, G. W., O'BRIEN, G. W., REIMERS, C., HERCZEG, A., MORIARTY, D. J. W., BURNETT, W. C., AND MILNES, A. R., 1990, Organic carbon cycling and modern phosphorite formation of the East Australian continental margin: An overview, *in* Notholt, A. J. G., and Jarvis, I., eds., Phosphorite Research and Development: London, Geological Society of London Special Publication 52, p. 87-117.

HEIM, A., 1922, Notes on the Tertiary of southern lower California (Mexico): Geological Magazine, v. 59, p. 529-547.

JEHL, C., 1995, Les mattes algaires cyanobactériennes (kopara) des atolls des Tuamotu: Biochimie, productivité et signification écologique. Insertion dans un modèle de phosphatogenèse: Unpublished Doctoral Dissertation, Université Française Pacifique, 318 p.

JEHL, C., AND BARSCZUS, H. G., 1996, Origin of rare earth elements in cyanobacterial mats (kopara) from Tikehau atoll (Tuamotu, French Polynesia): C. R. Académie de Sciences Paris, v. 322, serie IIa, p. 205-212.

JEHL, C., AND ROUGERIE, F., 1995, Phosphatogenèse en atolls polynésiens: La filiation mattes cyanobactériennes-phosphorites: Oceanologica Acta, v. 18, p. 79-93.

LANDRET, J. P., 1978, Rapport de la mission effectuée sur l'atoll fermé de Niau: Rapport Interne, COP-CNEXO, Tahiti, 10 p.

LAMBOY, M., 1993, Phosphatization of calcium carbonate in phosphorites: Microstructure and importance: Sedimentology, v. 40, p. 53-62.

LUCAS, J., ILYIN, A. V., AND KÜHN, A., 1986, Proterozoic and Cambrian phosphorites—deposits: Volta Basin, West Africa, *in* Cook, P. J., and Shergold, J. H., eds., Phosphate Deposits of the World, Volume 1—Proterozoic and Cambrian Phosphorites: Cambridge, Cambridge University Press, p. 235-243.

MARTÍN-ALGARRA, A., AND SÁNCHEZ-NAVAS, A., 1995, Phosphate stromatolites from condensed cephalopod limestones, Upper Jurassic, southern Spain: Sedimentology, v. 42, p. 893-919.

MAZZULLO, S. J., BISCHOFF, W. D., AND TEAL, T. S., 1995, Holocene shallow-subtidal dolomitization by near-normal seawater, northern Belize: Geology, v. 23, p. 341-344.

MILLAN-NUÑEZ, R., ALVAREZ-BORREGO, S., AND NELSON, D. M., 1982, Effects of physical phenomena on the distribution of nutrients and phytoplankton productivity in a coastal lagoon: Estuarine, Coastal, and Shelf Science, v. 15, p. 317-335.

NATHAN, Y., 1990, Humic substances in phosphorites: Occurrence, characterization and significance, in Notholt, A. J. G., and Jarvis, I., eds., Phosphorite Research and Development: London, Geological Society of London Special Publication 52, p. 49-58.

NATHAN, Y., SOUDRY, D., AND AVIGOUR, A., 1990, Geological significance of carbonate substitution in apatites: Israeli phosphorites as an example, *in* Notholt, A. J. G., and Jarvis, I., eds., Phosphorite Research and Development: London, Geological Society of London Special Publication 52, p. 179-191.

NORDLUND, U., 1989, Genesis of phosphatic hardgrounds in the Lower Ordovician of northern Öland, Sweden: Geologiska Föreningens i Stockholm Förhandlingar, v. 111, p. 161-170.

O'BRIEN, G. W., MILNES, A. R., VEEH, H. H., HEGGIE, D. T., RIGGS, S. R., CULLEN, D. J., MARSHALL, J. F., AND COOK, P. J., 1990, Sedimentation dynamics and redox iron-cycling: Controlling factors for the apatite-glauconite association on the east Australian continental margin, *in* Notholt, A. J. G., and Jarvis, I., eds., Phosphorite Research and Development: London, Geological Society of London Special Publication 52, p. 61-86.

OREM, W. H., BURNETT, W. C., LANDING, W. M., LYONS, W. B., AND SHOWERS, W., 1991, Jellyfish Lake, Palau: Early diagenesis of organic matter in sediments of an anoxic marine lake: Limnology and Oceanography, v. 36, p. 526-543.

PARRRISH, J. T., ZIEGLER, A. M., SCOTESE, C. R., HUMPHREVILLE, R. G., AND KIRSCHVINK, J. L., 1986, Proterozoic and Cambrian phosphorites—specialist studies: Early Cambrian palaeogeography, palaeooceanology and phosphorites, *in* Cook, P. J., and Shergold, J. H., eds, Phosphorite Deposits of the World Volume 1—Proterozoic and Cambrian Phosphorites: Cambridge, Cambridge University Press, p. 280-294.

QUINTUS-BOSZ, R. L., 1980, Petrology and distribution of phosphate in the lower Salada Formation, Santa Rita, Baja California Sur, Mexico: Unpublished M. S. Thesis, Colorado School of Mines, Golden, 126 p.

RIGGS, S. R., 1980, Intraclast and pellet phosphorite sedimentation in the Miocene of Florida: Journal of the Geological Society of London, v. 137, p. 741-748.

RODGERS, K. A., 1994, The cycle of phosphorus in an atoll ecosystem: Applied Geochemistry, v. 9, p. 311-322.

ROUGERIE, F., JEHL, C., AND TRICHET, J., 1997, Phosphorus pathways in atolls: Interstitial nutrient pool, cyanobacterial accumulation and carbonate-fluoroapatite (CFA) precipitation: Marine Geology, v. 139, p. 201-217.

ROUGERIE, F., AND WAUTHY, B., 1986, Le concept d'endo-upwelling dans le fonctionnement des atolls-oasis: Oceanologica Acta, v. 9, p. 133-148.

ROUGERIE, F., AND WAUTHY, B., 1989, Une nouvelle hypothèse sur la genèse des phosphates d'atolls: Le rôle du processus d'endo-upwelling: C. R. Académie de Sciences Paris, v. 308, series II, p. 1043-1047.

ROUGERIE, F., AND WAUTHY, B., 1993, The endo-upwelling concept: From geothermal convection to reef construction: Coral Reefs, v. 12, p. 19-30.

RUTTENBERG, K. C., AND BERNER, R. A., 1993, Authigenic apatite formation and burial in sediments from non-upwelling, continental margin environments: Geochimica et Cosmochimica Acta, v. 57, p. 991-1007.

SCHWENNICKE, T., 1995, Phosphatic grains of the Upper Oligocene San Juan Member (El Cien Formation) of Baja California Sur, Mexico: Boletín del Departamento de Geología de la Universidad de Sonora, v. 12, 1, p. 41-64.

SCHWENNICKE, T., 1996, Sedimentology and depositional environment of the Pliocene Salada Formation at Arroyo Salada, Baja California Sur, Mexico: Boletín del Departamento de Geología de la Universidad de Sonora, v. 12, 2, p. 19-34.

SCHWENNICKE, T., NATHAN, Y., VÁZQUEZ-GARCÍA, A., AND DE ANDA-FRANCO, N., 1996: Petrographic evidences for modern phosphogenetic processes in some coastal lagoons of Baja California Sur, Mexico (abs.): XIII Convención Geológica Nacional, La Paz, p. 37-38.

SCHWENNICKE, T., AND GONZÁLEZ-BARBA, G., 1995, Lithology, sedimentology and depositional environment of the Salada Formation of Baja California Sur, Mexico, at its type locality (abs.): Tercera Reunión Internacional sobre Geología de la Península de Baja California, La Paz.

SIEGMUND, H., 1995, Fazies und Genese unterkambrischer Phosphorite und mariner Sedimente der Yangtze-Plattform, Südchina: Berliner geowissenschaftliche Abhandlungen, v. 173, A, 114 p.

SMITH, J. T., 1992, The Salada Formation of Baja California Sur, Mexico, *in* Carrillo-Chavez, A., and Alvarez-Arellano, A., eds., Primera Reunión Internacional sobre Geología de la Península de Baja California, Memorias, La Paz, p. 23-32.

SOUDRY, D., 1994, Bottom-dwelling microphytes and phosphate sediment accretion: The case of the Campanian phosphorites, Negev, southern Israel: Kaupia, v. 4, p. 61-70.

SOUDRY, D., AND CHAMPETIER, Y., 1983, Microbial processes in the Negev phosphorites (southern Israel): Sedimentology, v. 30, p. 411-423.

SOUDRY, D., AND LEWY, Z., 1988, Microbially influenced formation of phosphate nodules and megafossil moulds (Negev, southern Israel): Palaeogeography, Palaeoclimate, and Palaeoecology, v. 64, p. 15-34.

SOUDRY, D., AND NATHAN, Y., 1980, Phosphate peloids from the Negev phosphorites: Journal of the Geological Society of London, v. 137, p. 749-755.

SOUDRY, D., AND SOUTHGATE, P. N., 1989, Ultrastructure of a Middle Cambrian primary nonpelletal phosphorite and its early transformation into phosphate vadoids: Georgina Basin, Australia: Journal of sedimentary Petrology, v. 59, p. 53-64.

SOUTHGATE, P. N., 1986, Middle Cambrian phosphatic hardgrounds, phoscrete profiles and stromatolites and their implications for phosphogenesis, *in* Cook, P. J., and Shergold, J. H., eds., Phosphate Deposits of the World, Volume 1—Proterozoic and Cambrian Phosphorites: Cambridge, Cambridge University Press, p. 327-351.

TRAPPE, J., 1992, Microfacies zonation and spatial evolution of a carbonate ramp: Marginal Moroccan phosphate sea during the Paleogene: Geologische Rundschau, v. 81, p. 105-126.

TRAPPE, J., 1996 Phanerozoic Phosphorite Depositional Systems: Habilitationsschrift, Universität Bonn, Bonn, 237 p.

TRICHET, J., 1967, Essai d'explication du dépôt d'aragonite sur des substrats organiques: C.R. Académie de Sciences Paris, v. 265, p. 1464-1467.

TRICHET, J., AND DEFARGE, C., 1993, From biominerals to "organominerals": The example of the modern lacustrine calcareous stromatolites from Polynesian atolls (abs.): Proceedings of the 7th International Symposium on Biomineralization, Monaco, p. 80.

YE, L. J., AND CHEN, Q. Y., 1989, Composite-process and poly-episodic aspect of the Chinese sedimentary mineral deposits (abs.): Science Geologicae Sinae, v. 1989, p. 126-127.

YE, L. J., CHEN, Q. Y., ZHAO, D. X., CHEN, Z. M., CHEN, Y. M., AND LIU, K. W., 1989, The Phosphorites of China: Beijing, Science Press, 339 p.

ZHANG, Y., 1989, Multicellular thallophytes with differentiated tissues from Late

BACTERIALLY MEDIATED AUTHIGENESIS IN MESOZOIC STROMATOLITES FROM CONDENSED *PELAGIC* SEDIMENTS (BETIC CORDILLERA, SOUTHERN SPAIN)

AGUSTÍN MARTÍN-ALGARRA*

Departamento de Estratigrafía y Paleontología and IACT, Facultad de Ciencias, Universidad de Granada, 18071-Granada, Spain

AND

ANTONIO SÁNCHEZ-NAVAS

Departamento de Mineralogía y Petrología and IACT, Facultad de Ciencias, Universidad de Granada, 18071-Granada, Spain

ABSTRACT: Facies, fossils, morphology, lamination features, microstratigraphy, and composition of phosphate and iron-manganese nodules and crusts found in condensed limestones associated with stratigraphic discontinuity surfaces in Mesozoic successions of the Betic Cordillera of southern Spain, demonstrate that these structures are not diagenetic concretions, but biosedimentary microbial accretions (stromatolites) that grew rhythmically at the sediment-water interface in pelagic environments. These structures accreted by bacterially mediated precipitation of authigenic minerals, by trapping and binding of fine-grained, pelagic, siliciclastic, and carbonate particles, and by encrustation of benthic foraminifera. SEM examination of stromatolite laminae reveals the presence of dense accumulation of spheroidal, ovoidal, and sausage-shaped strings of micrometer-sized bodies with a bacteria-like size and shape. XRD and microprobe (BSE and X-ray images) studies show the presence of authigenic minerals such as francolite, goethite, hematite, barite, pyrite, and glauconite, among others. Authigenic clays and complex poorly crystalline Fe-Mn-Al- (Si)-oxyhydroxides are always closely associated with these minerals. Marine authigenesis in stromatolites of the pelagic realm was achieved through microbially mediated synsedimentary precipitation of poorly crystalline or amorphous phases that preceded crystallization of the observed minerals. Deposition of fine-grained metals and silicate precipitates on bacterial surfaces favored preservation of microbial remains as external moulds. After degradation of bacterial organic matter, other components, such as francolite, nucleated within empty bacterial cells and other voids from bacterially precipitated amorphous precursors. Texture, mineralogy, and chemistry indicate the vertical and temporal evolution of the physicochemical conditions during the synsedimentary bacterial precipitation of amorphous precursors and the oscillations of the oxic-anoxic boundary (usually from oxic to postoxic conditions, although sometimes sulfidic conditions have been registered, allowing pyrite precipitation) during which the maturation of authigenic mineral phases took place. In conclusion, periods of very low sedimentation rate, now seen as sequence boundaries and/or transgressive surfaces in Mesozoic Alpine-Mediterranean paleomargins, favored accretion of stromatolites in the pelagic realm and bacterially mediated precipitation of a wide spectrum of authigenic minerals that provide abundant information about paleoceanography, sedimentation, and early diagenesis in open marine mesozoic environments.

INTRODUCTION

The presence of phosphate-rich and iron-manganese-rich nodules and crusts is a common feature of condensed pelagic facies in the Mesozoic record of the Betic Cordillera (González-Donoso et al., 1983; García-Cervigón et al., 1986-1987; Martín-Algarra, 1987; Castro and Ruiz-Ortiz, 1991; Martín-Algarra and Vera, 1994; Vera and Martín-Algarra, 1994; Martín-Algarra and Sánchez-Navas, 1995; Nieto-Albert, 1997) and many other Alpine-mediterranean belts, such as the Alps (Bourbon, 1980; Föllmi, 1989; Delamette, 1990), Jura (Burkhalter, 1995), Apennines (Clari et al., 1995) and Sicily (Jenkyns, 1970, 1971), Carpathians (Krajewski, 1981, 1983, 1984), Dinarides (Wendt, 1973) and Hellenides (Pomoni-Papaioannou and Solakius, 1991; Pomoni-Papaioannou, 1994), as well as some condensed Meso-Cenozoic successions of Israel (Lewy, 1990; Soudry, 1994) and Oman (Blendinger, 1991), and many Paleozoic sites of the Variscan belt (Tucker, 1973, 1974; Wendt et al., 1984; Huckriede, 1996). These structures formed in moderately deep submarine plateaus and seamounts of tectonic and/or volcanic origin (pelagic swells) formed in distal regions of Paleozoic and especially Mesozoic tethysian paleomargins (Jenkyns, 1978, 1986). They systematically concentrate in strongly condensed facies and stratigraphic discontinuity surfaces, which can be recognized in large areas.

These P-rich and Fe-Mn-rich structures sometimes have been considered ancient equivalents of phosphoritic crusts and hardgrounds (Soudry, 1994; Föllmi, 1989, 1996, and references therein) and iron-manganese nodules of the modern deep ocean (Jenkyns, 1970, 1978). In spite of their systematic association with pelagic carbonate facies rich in planktonic-nektonic organisms (neritic, shallow water fossils always being absent), a detailed examination of the internal architecture of these crusts and nodules reveals that many are stromatolites because microbes actively participate in their accretion (see references above). These peculiar microbial accretions, formed in the pelagic realm, are mainly composed of typical marine authigenic minerals, such as francolite and Fe-Mn oxyhydroxides, with variable amounts of glauconite, pyrite, and other minerals. When they are studied in detail, it is possible to detect complex interactions between microbially controlled mechanical trapping and binding of pelagic, mainly fine-grained, bio- and siliciclastic particles, synsedimentary biomineralization, and early to late diagenetic mineralization. Some common controls in the genesis of these structures can be recognized through study of P-rich and Fe-Mn-rich Jurassic-Cretaceous examples of different ages and sites in the Betic Cordillera. The aims of this paper are: (1) to illustrate, from selected examples, the main morphological, textural, and mineralogical features of such peculiar primary biosedimentary structures; (2) to discuss the role of microbes in the precipitation of authigenic mineral phases; (3) to present a general model for their accretion; and (4) to comment on some aspects of their paleoegeographical significance.

*email: agustin@goliat.ugr.es

Marine Authigenesis: From Global to Microbial, SEPM Special Publication No. 66

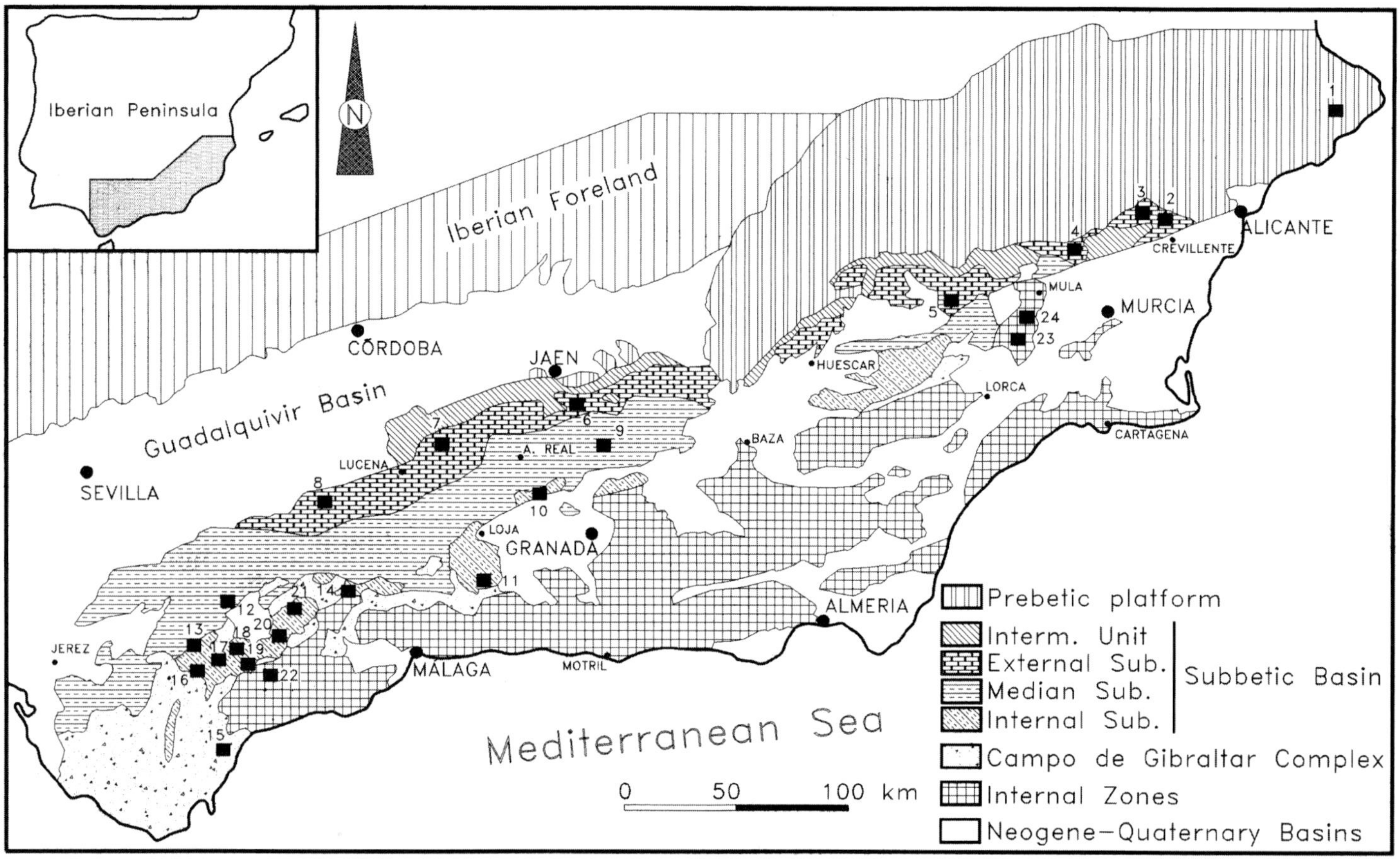

FIG. 1.—Map of the Betic Cordillera, with locations of sites bearing stromatolites associated with condensed pelagic sediments (see Table 1 for details). (1) Aixorta. (2) Reclot. (3) Asiento. (4) Lúgar. (5) Quípar. (6) Pandera. (7) Cabra. (8) Estepa. (9) Campotéjar. (10) Illora. (11) Sierra Gorda. (12) Montecorto. (13) Grazalema. (14) Chorro. (15) Canutos. (16) Berrueco. (17) Benaoján. (18) Montejaque. (19) Hacho. (20) Boquerón. (21) Ortejícar. (22) Almola. (23) Espuña. (24) Espuña-N.

GEOLOGICAL AND PALEOGEOGRAPHICAL SETTING

The Betic Cordillera of Southern Spain, the westernmost Alpine-Mediterranean chain, is subdivided in three main geological ensembles (Fig. 1): the External Zones, the Internal Zones, and the Campo de Gibraltar Complex, which is located between them.

Paleogeographically (Fig. 2), the External Zones constituted the continental paleomargin of Southern Iberia. The proximal area of this margin, or Prebetic, was a pericontinental platform with shallow marine carbonate sedimentation that received some siliciclastic contribution from the Iberian continent. Toward the southeast, the Prebetic evolved into an open platform and slope setting that distally was connected to a complex basin called the Subbetic. The Subbetic basin originated after a Liassic phase of rifting that affected a wide shallow carbonate platform that covered the External Zones in pre-Middle Liassic times. The rifting phase differentiated the Subbetic realm from the Prebetic platform, and transformed the formerly unique shallow water area into a complex and topographically irregular basin where two swells (External Subbetic and Internal Subbetic or Penibetic), separated by one deeper area (Median Subbetic trough) have been distinguished (Azéma et al., 1979; García-Hernández et al., 1980; Sanz de Galdeano et al., 1993). These realms were bordered by faults that were tectonically active during the Middle-Late Jurassic and the Lower Cretaceous. At the same time submarine volcanism and hydrothermalism locally played an important role in parts of the Median Subbetic trough.

The Southern Iberian paleomargin distally changed into a narrow oceanic basin where sediments of very deep facies were deposited. These sediments now constitute the Campo de Gibraltar Complex (Martín-Algarra, 1987; López-Galindo and Martín-Algarra, 1992; Martín-Algarra et al., 1992; Guerrera et al., 1993). This basin also includes ophiolite-bearing successions that belong to the lower tectonic units of the Internal Zones; they are composed of high-pressure metamorphic rocks (metasediments associated with eclogites and serpentinites: Puga et al., 1995). This narrow and deep basin connected the central Atlantic with the Piemont-Ligurian Alpino-Apenninic Ocean (Fig. 2), and was later strongly deformed and subjected to alpine metamorphism as a consequence of compressional tectonics and subduction that probably started during the latest part of the Cretaceous.

Towards the SE of these deep oceanic areas, another faulted, topographically complex Mesozoic continental margin existed, from which the highest tectonic units of the Internal Zones are derived. This margin, similar to the Austroalpine zone of the Alps, constituted the northern part of a continental crust block (Mesomediterranean Terrane: Fig. 2) located between Iberia and Africa, which was squeezed during the alpine orogenesis (Guerrera et al., 1993).

Jurassic-Cretaceous condensed pelagic limestones, bearing

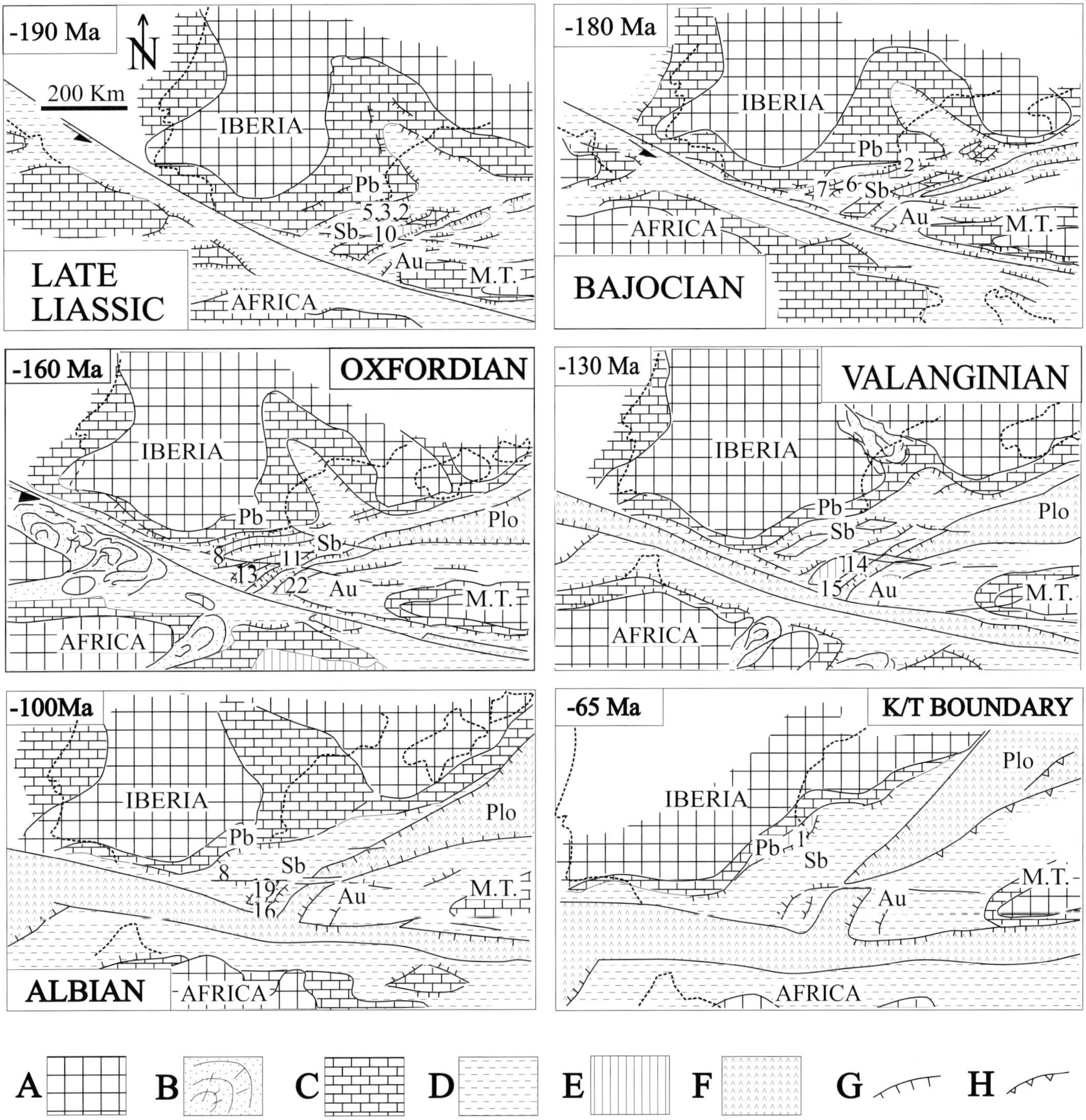

Fig. 2.—Jurassic-Cretaceous paleogeographic evolution of the Betic Cordillera. (Pb) Prebetic. (Sb) Subbetic. (Au) Australpine-type margin of the M.T. (M.T.) Mesomediterranean Terrane. (Plo) Piemont-Ligurian ocean. (A) Land. (B) Terrigenous clastics. (C) Carbonate platforms. (D) Basinal (deep-sea) pelagic sediments. (E) Condensed facies on pelagic swells. (F) Oceanic crust and associated deep sediments. (G) Normal faults. (HG) Subduction. Numbers indicate the paleogeographical locations of the sites indicated in Figure 1.

P-rich and Fe-Mn-rich nodules and crusts, are frequently found close to sediment-starved areas of pelagic swells located both in the Austroalpine-type margin of the Internal Zones and, especially, in the Southern Iberian Paleomargin, particularly in the External Subbetic and in the Western Internal Subbetic (Penibetic), although some examples have been found in some Median Subbetic and Internal Prebetic outcrops (Table 1). In all cases, stromatolite-bearing condensed successions are rich in open marine macro- and microfossils. Stromatolites associated with pelagic sediments concentrate in the chronostratigraphic intervals and settings summarized in Table 1.

MATERIALS AND METHODS

Data presented below come from some selected examples of phosphate-rich and Fe-Mn oxyhydroxide-rich stromatolites of

TABLE 1.—LOCATION, AGE, DESCRIPTION, FOSSILS, AND AUTHIGENIC MINERALS FOUND IN P-FE-MN-STROMATOLITES OF THE BETIC CORDILLERA, AND FACIES OF ASSOCIATED SEDIMENTS

SITE (Fig. 1)	AGE	DESCRIPTION	FOSSILS	AUTHIGENIC MINERALS
PREBETIC				
1.- Aixorta	K-T boundary	Yellowish, phosphatic, dm-thick laterally continuous stromatolite pavement on, and endostromatolites within, strongly burrowed Upper Maastrichtian gray-yellowish, biomicritic limestones, buried under Lower Paleocene light gray-white pelagic marls and marlstones	PF, CN	CFA, gl
EXTERNAL SUBBETIC				
2.- Reclot	Middle Liassic	Fe-rich pavements on dm-cm-thick bed of red condensed fossiliferous limestones including Fe-Mn-(P) mm to cm crusts and oncoids lying on Carixian crinoidal limestones and covered by Domerian yellowish marly limestones and marls	A, Br, Cr, B, BF (Bv), EF	goeth., Fe-oxyhydr. (CFA)
	Toarc.-Aalen.	Complex succession of cm-mm-thick Fe-(Mn)-rich crusts alternating with Toarcian-Aelenian red biomicrites including encrusted macrofossils and oncoids	A, B, EF, (F)	Fe-(Mn) oxyhydr., phosphates, Gl
	Up. Bath.-Call.	Fe-Mn-rich pavement made by a dm-thick bed of bored and encrusted red condensed cephalopod limestones, including centimetric Fe-Mn-oncoids encrusting ammonites	A, B, F, EF, BF, (PF, CN)	Fe-Mn oxyhydr., CFA
3.- Asiento	Up. Bathonian	Fe-Mn-P-pavement on red condensed cephalopod limestones including cm-dm oncoids encrusting macrofossils	A, B, F, EF, BF, (PF, CN)	goeth, Fe-Mn-Al oxyhydr., CFA
4.- Lúgar	Up. Bath.-Oxf.	Fe-Mn-P-pavement on a dm-thick bed of red condensed cephalopod limestones including oncoids encrusting macrofossils	A, B, F, EF, CN, BF	goeth, Fe-Mn oxyhydr., CFA, gl
5.- Quípar	Middle Liassic	Thin Fe-(Mn) crusts on a centimetric bed of red-brown fossiliferous limestones covered by Domerian gray marly-limy rhythmites	Br, B, Cr, BF, (A, Bv, G)	Fe-Mn oxyhydr.
	Low. Bajocian	Erosion surface on Aalenian-Lower Bajocian limestones encrusted by cm-thick, Fe-(Mn, P)-crusts covered by red pelagic limestones with Fe-(Mn) oncoids	A, B, F, EF, PF, (BF)	Fe-Mn oxyhydr., CFA, gl
	Up. Bathonian	Fe-Mn-P pavement formed by a cm-thick crust on red nodular limestones including big Bathonian ammonites covered by a dm-thick bed of red limestone including numerous oncoids and heavily encrusted macrofossils of Callovian-Oxfordian age, endostromatolites within macrofossils and stromatolite breccias	A, B, Cr, EF, PF, BF, CN, (Br, SCo)	Fe-Mn oxyhydr., CFA, gl
6.- Pandera	Upper Jurassic	Phosphatic oncoids and mm crusts on macrofossils, endostromatolites encrusting the walls of neptunian dykes opened in pink-red rosso ammonitico nodular limestones	A, B, E, Bv, Cr, PF, CN	CFA, (gl, Fe-Mn oxyhydr.)
7.- Cabra	Up. Bathonian	P-Fe-rich crusts (cm-thick) on irregular dissolution surface (palaeokarst) on Bathonian, shallow marine (bahamian-type), cross-bedded oolitic-crinoidal limestones, associated with a decimetric bed of brown-red nodular limestones with abundant cephalopods and centimetric Fe-P oncoids	A, B, PF, EF, CN, Se	Fe-Mn oxyhydr., CFA, gl
8.- Estepa	Upper Aptian-Upper Albian	Bed of limestone breccia rich in Aptian-Albian planktonic foraminifera with centimetric oncoids and fragments of P-stromatolite crusts, lying on Lower Valanginian rosso ammonitico limestones, and covered by Turonian red scaglia rossa-like marls	PF, CN, EF, BF, Cr	CFA, goeth. Gl
MEDIAN SUBBETIC				
9.- Campotéjar	Kimmeridgian	P-Fe-Mn microstromatolite crusts on macrofossils within red-pink condensed cephalopod limestones lying on basaltic pillow lavas (volcanic seamount)	A, B, Ap, (Cr, EF, BF, S)	Fe-Mn oxyhydr., gl, (CFA)
10.- Illora	Middle Liassic	dm-thick bed of red cephalopod limestones on Carixian crinoidal limestones with abundant Fe-rich oncoids, buried by Domerian, gray marly-limy rhythmites	A, B, Cr	Fe-Mn oxyhydr.
INTERNAL SUBBETIC / PENIBETIC				
11.- S. Gorda	Up. Oxfordian-Low. Kimmer.	Phosphate crusts (mm-thick) and isolated microstromatolites on and within framework pores of macrofossils, in brown condensed fossiliferous limestones lying on thick bedded, filament-bearing pelagic limestones and buried under Kimmeridgian-Tithonian red marly nodular limestones including calcareous turbidites	A, B, Ap, PF, (EF, BF, CN	CFA, gl, (Fe oxyhydr.)
12.- Montecorto	Kimmer.-Tith.-Berriasian	Fe-Mn-crusts (mm-cm-thick) and centimetric oncoids made of arborescent Fe-Mn-rich microstromatolites on clasts and within macrofossils, associated with red condensed fossiliferous limestones filling neptunian dykes opened within Triassic continental red pelites and cross-bedded sandstones	A, B, EF, BF, CN, (C)	Fe-Mn oxyhydr., (CFA)
13.- Grazalema	Up. Oxfordian	Phosphatic oncoids and mm-thick crusts within brown condensed fossiliferous limestones lying on a palaeokarst surface carved on Bathonian crinoidal-oolitic, cross bedded limestones of bahamian type, and buried by brown-yellowish pelagic limestones	A, B, PF, EF, Cr, (BF)	CFA, (Fe-oxyhydr.)

TABLE 1.—CONTINUED

SITE (Fig. 1)	AGE	DESCRIPTION	FOSSILS	AUTHIGENIC MINERALS
INTERNAL SUBBETIC / PENIBETIC (cont.)				
14.- Chorro	Up. Valangin.- Low. Hauteriv.	Bed (dm-thick) of light brown-greenish limestone with abundant phosphatic, centimetric oncoids and cm-thick stromatolite crusts and breccias, lying on Upper Berriasian-Lower Valanginian white/red nodular and pelagic oolitic limestones and buried by mid-Albian green marls	C, CN, EF, (Cr, BF, PF), T	CFA, AC, Fe-oxyhydr., Gl
15.- Canutos	Up. Valangin.- Low. Hauteriv.	Same as in the Chorro outcrop, but the oncoid-rich limestone bed is white-yellowish and is buried by Upper Hauterivian cherty marlstones with ammonites in one outcrop but, in other three ones, the same bed is fossilized by Upper Albian scaglia bianca type white marls	C, CN, EF, A (PF, Cr, BF)	CFA, AC, Gl, (Fe-oxyhydr.)
	Upper Albian	Breccia of phosphatic stromatolite crusts with a marly-calcarenitic matrix rich in planktonic foraminifera	PF, CN, B	CFA, Gl, AC (Fe-oxyhydr.)
16.- Berrueco	Upper Albian	Red-orange, dm-thick, P-(Fe)-rich pavement, encrusting a very irregular karstic palaeorelief carved on white Berriasian pelagic oolitic limestones, that includes numerous oncoids and stromatolite breccias and red biomicritic limestones rich in planktonic foraminifera. It is fossilized by Upper Albian scaglia bianca-like marls	PF, CN, EF (BF, E)	CFA, goeth, hm, (Fe-Mn-oxyhydr.), AC, Gl
17.-Benaoján	Upper Albian	Phosphatic pavement made of dense stromatolite crusts associated with white-yellowish pelagic limestones with some oncoids and breccias of stromatolites, encrusting a very irregular karstic palaeorelief carved on white Berriasian pelagic oolitic limestone and fossilized by Upper Albian, scaglia bianca-like marls	PF, CN, EF (B, BF)	CFA, goeth, hm, (Fe-Mn oxyhydr.), AC, Gl,
18.- Montejaque	Upper Albian	P-(Fe)-rich, red-orange pavement with some oncoids and abundant stromatolite breccias associated with Upper Albian red limestones, that encrusts a very irregular and heavily bored karstic palaeorelief carved on white Berriasian pelagic (oolitic) limestones. It is fossilized by Upper Albian-Cenomanian scaglia bianca-like marls.	PF, CN, EF (BF, B)	CFA, goeth, (Fe-Mn-oxyhydr.), AC, Gl
19.- Hacho	Upper Albian	Phosphatic pavement made of dense stromatolite crusts associated with white -yellowish pelagic limestones with numerous oncoids and breccias of stromatolites, encrusting a karstic palaeorelief carved on Berriasian white pelagic oolitic limestones and fossilized by Upper Albian scaglia bianca-like marls	PF, CN, EF (B, BF)	CFA, goeth, (Fe-Mn oxyhydr.), AC, Gl, ba, py
20.- Boquerón	Upper Albian	Phosphatic pavement made of dense stromatolite crusts, encrusting a very irregular karstic palaeorelief carved on white Berriasian pelagic oolitic limestone and fossilized by ammonite-bearing Upper Albian dark gray marls and light gray, scaglia bianca-like marls	PF, CN, EF (BF, B, A)	CFA, goeth, (Fe-Mn oxyhydr.), AC, Gl, (py)
21.- Ortegícar	Upper Albian	P-(Fe)-rich, red-orange pavement with some oncoids and abundant stromatolite breccias associated with red limestones rich in Upper Albian planktonic foraminifera, encrusting a karstic palaeorelief on light brown-white Berriasian pelagic oolitic limestones and buried by Upper Albian scaglia bianca-like marls	PF, CN, EF (BF, B)	CFA, goeth, (Fe-Mn oxyhydr.), AC, Gl
INTERNAL ZONES / AUSTROALPINE-TYPE MARGIN AND MALAGUIDES				
22.- Almola	Up. Oxfordian- Low. Kimmer.	Bed (dm-thick) of brown condensed cephalopod limestone with dark brown-black phosphatic oncoids and mm-cm-thick stromatolite crusts, breccias of phosphate stromatolites and isolated microstromatolites on and within macrofossils	A, B, Ap, EF, BF, Se, PF, CN	CFA, hm, AC, Fe-Si-Al oxyhydr.
23.- Espuña	Middle Liassic	Bed (dm-thick) of red-yellow pelagic limestone rich in ammonites, with abundant Fe-rich centimetric oncoids and iron ooids and pisolites	A, B, EF, BF	Fe-Mn oxyhydr.
24.- Espuña-N	Upper Albian	Phosphoritic and glauconitic greensand and phosphatic-glauconitic microstromatolitic crusts on an irregular (probably palaeokarstic) dissolution surface, associated with greenish-light gray pelagic limestones and marls	B, E, PF, (A, BF)	CFA, Gl

FOSSILS: A: ammonites; B: belemnites; Ap: aptychi; Cr: crinoids; E: echinoids; Se: serpulids; EF, BF and PF: encrusting, benthic (non-encrusting) and planktonic foraminifera; F: (Bositra) filaments; C: Calpionellids; T: fish teeth; CN: calcareous nannoplankton; SCo: solitary corals; Br: brachiopods; Bv: bivalves; G: gastropods. MINERALS: CFA: Carbonate Fluorapatite; goeth: goethite; hm: hematite; Fe-Mn-(Al, Si)-oxyhydr.: oxyhydroxides of Fe-Mn-(Al, Si); AC: authigenic clays (mainly Fe-rich smectites); Gl: glauconite; gl: glaucony; ba: barite; py: pyrite. Detailed studies on most of these outcrops have been published by García-Cervigón et al., 1986-87, Martín-Algarra (1987), Castro and Ruiz-Ortiz (1991), Martín-Algarra and Vera (1994), Vera and Martín-Algarra (1994), Martín-Algarra and Sánchez-Navas, (1995), Nieto-Albert (1997) and Sánchez-Navas et al. (1998).

the sites and ages indicated in Figure 1 and Table 1. Most of them were previously catalogued and studied by González-Donoso et al. (1983), Martín-Algarra (1987), Martín-Algarra and Vera (1994), and Vera and Martín-Algarra (1994). Polished slabs and thin sections were studied by binocular magnifying glass and optical microscopy to identify architectural elements and mineral phases of the stromatolites, as well as microfacies and microfossils of the associated sediments. More detailed textural and compositional laboratory studies were performed by X-ray diffraction (XRD), scanning electron microscopy (SEM), and electron probe microscopy (EPM). Mineralogical studies were performed in a first phase by XRD of whole rock samples, although fractions purified after acid attack with HCl were also studied. A diffractometer with graphite monocromator, automatic slit, CuKα radiation, and on-line connection with a microcomputer, was used to identify the mineral phases. Secondary electron micrographs of thin sections and small stromatolite pieces and chips, some of them observed after etching with diluted HCl to emphasize micromorphology, were obtained with a scanning electron microscope (SEM) equipped with an energy dispersion X-ray (EDX) microanalytical system. An automated four-spectrometer electron microprobe was used to determine

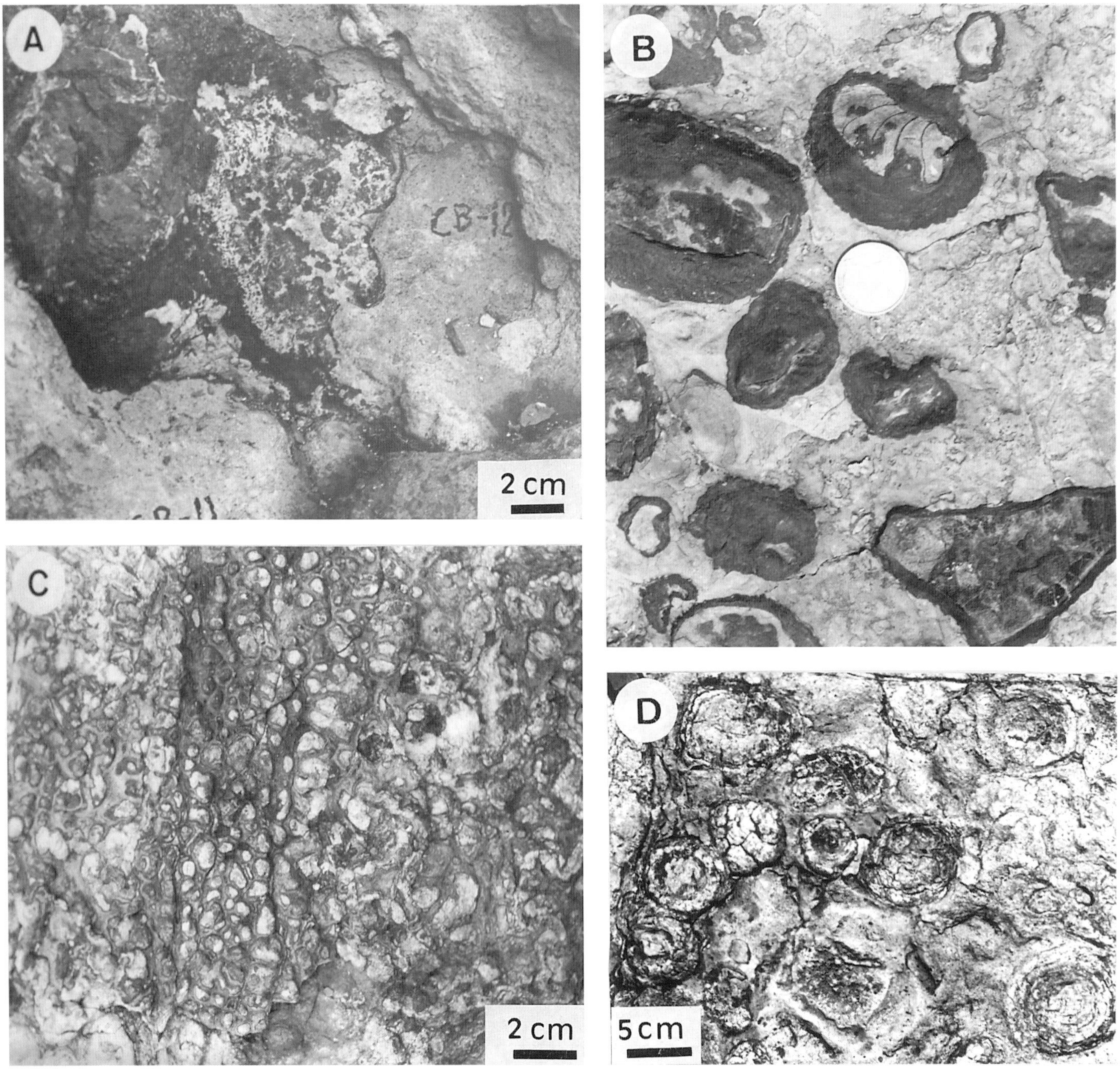

Fig. 3.—Field view of Jurassic-Cretaceous P-Fe-Mn stromatolites of the Betic Cordillera (present-day and paleogeographical locations of each site in Figures 1 and 2, respectively). (A) Upper Albian phosphatic crust (black zone on the upper left) and microstromatolites (site 20). (B) Upper Bathonian-Callovian Fe-Mn-P oncoids encrusting ammonites (site 5). Coin is 3 cm in diameter. (C) Top of a phosphate stromatolite crust. Cretaceous-Tertiary boundary (site 1). (D) Late Valanginian phosphate oncoids (site 15).

the composition of mineral phases of phosphate and Fe-Mn stromatolites and of glaucony pellets. Data were reduced using a ϕ-ρ-z procedure (see Pouchou and Pichoir, 1985, for details), using simple oxides (Al_2O_3, Fe_2O_3, $MnTiO_4$, MgO) and silicates (albite, orthoclase, wollastonite) as calibration standards. Operating conditions were: 20 kV accelerating voltage, 20 nA beam current, and 10 μm beam spot. No alkali loss was observed during electron radiation, as confirmed by repeated analyses on the same spot (see García-Casco et al., 1993, for details). Scanning back-scattered electron (BSE) and X-ray images of the textures studied were performed on areas smaller than approximately 100x100 μm.

ARCHITECTURE OF THE MICROBIAL ACCRETIONS

Macroscopic Features

Field and laboratory examination of stromatolite-bearing intervals and hand specimens of several ages and sites permit establishment of their stratigraphic relationships with under- and overlying formations, recognition of their basic composition and architecture, and identification of the facies of the associated pelagic sediments (Table 1). The color of both stromatolites and related pelagic sediments changes between red, brown, greenish, or yellowish, and sometimes also from gray to black. Microfacies of pelagic limestones are always micritic or biomicritic (mudstones to wackestones-packstones). They are usually rich in bioclasts of pelagic macrofossils (ammonites, belemnites, aptychi, etc.) and microfossils, mainly benthic and planktonic foraminifera, coccoliths, and other pelagic microfossils that allow accurate biostratigraphic dating (see Table 1 for details).

Stromatolites are dominantly phosphatic, although samples very rich in Fe-Mn minerals, sometimes without phosphate, are also common, especially during the Jurassic. Nevertheless, significant differences in macro- or microscopic features do not exist, and two basic morphologies can be recognized by the naked eye in most stratigraphic intervals: (1) flat to undulose stromatolitic crusts a few millimeters to few decimeters-thick (Figs. 3A,C); and (2) centimeter- to decimeter-sized, spheroidal, ovoidal, or ellipsoidal, sometimes flattened nodules (oncoids: Figs. 3B, D, and 4). Oncoids and crusts are commonly found associated with each other, forming decimeter-thick, complex stromatolitic pavements with aggregates made of pelagic sediments and microstromatolites (Fig. 3A and 4). Some building units of stromatolitic crusts and oncoids are constituted by microstromatolite-rich laminae (Figs. 5A and 6B), but these microstructures are also commonly found either isolated or forming 1–2-mm-thick irregular and laterally discontinuous crusts (Fig. 6A, C, D). Stromatolite crusts and, especially, microstromatolites sometimes constitute cavity-dwelling stromatolites (endostromatolites) that indistinctly encrust roofs, walls, or floors of interstices, fissures, and cavities carved in fragmented and corroded macrofossil shells or in other substrates.

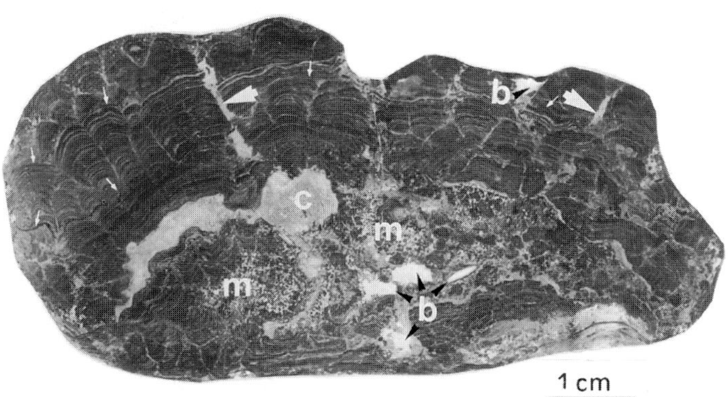

Fig. 4.—Macroscopic aspect of one flattened phosphate oncoid from site 18, Upper Albian, with core made of pelagic limestone mud clasts (c) and soft sediment containing abundant microstromatolites (m). Note asymmetric growth, with thicker stromatolite laminae in the upper part, erosion surfaces or unconformities (small arrows) and retraction fissures (arrowheads) filled with pelagic sediment (large arrows). Later borings (b) are filled with lighter (and younger) fossiliferous pelagic sediment.

Stromatolite nodules, crusts, and pavements are always associated with important stratigraphic gaps and lacunae, as evidenced by macroscopic erosion features and especially by biostratigraphy (see Vera and Martín-Algarra, 1994, for details). Reencrustation events and the presence of micro-unconformities cross-cutting the stromatolite lamination are common at all scales (Fig. 4). The surface aspect of both nodules and crusts changes from smooth, sometimes polished, to clotted and earthy, due to the presence of a last microstromatolitic crust made of numerous microcolumnar or dendritic microstromatolites that are buried by pelagic sediment. The surface is also frequently mammilated (Fig. 3C), as this is the expression of an internal architecture constituted by laminated columns or pseudocolumns several millimeters wide consisting of stacked hemispheroids (Fig. 4). Nodules, crusts, and pavements frequently appear cracked, fragmented, and/or bored (Fig. 4), and stromatolite breccias are common.

Nodules appear either isolated, or, more frequently, concentrated (Fig. 3B, D) within limestone beds up to few decimeters thick that lie on the underlying sediment on a sharp, planar to moderately irregular surface showing clear evidence of erosion and/or of stratigraphic gap. This evidence includes abrasion surfaces on macrofossils or on previous sedimentary structures, lag deposits (concentration of macrofossil shells, breccias), or marked facies changes relative to the underlying formation that, in some cases, is made of shallow platform carbonates and, in other cases, is made of condensed pelagic facies deposited in moderately deep pelagic swells (Table 1). Nodules have a core constituted by one or several bioclasts (Fig. 3B), pieces of former stromatolite crusts, chips of pelagic lime mud with evidence of soft deformation, and/or by fragments of the underlying rocks. Adjacent nodules can coalesce and be enveloped by outer and later stromatolite crusts and form greater nodules with several cores, or be heavily encrusted by irregular to planar crusts forming complex pavements up to few decimeters thick, where macrofossils, bioclasts, and pieces of former stromatolitic crusts of the associated soft pelagic sediment or of the underlying formation are always present. These thick pavements always intercalate more or less laterally continuous millimeter- to

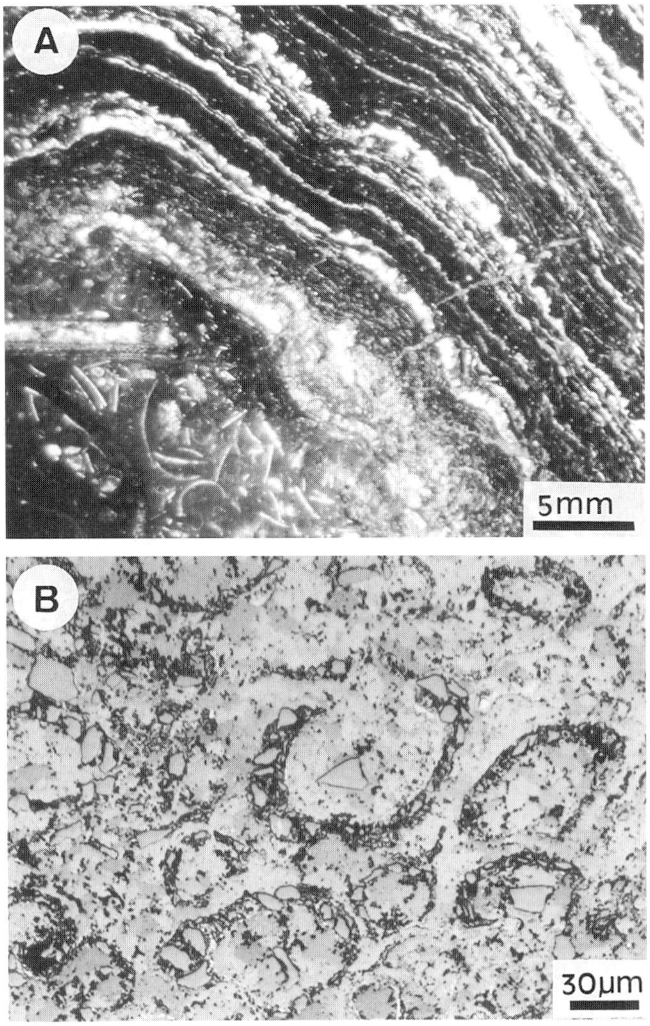

Fig. 5.—Microfacies of Jurassic stromatolites from site 5 (Upper Bathonian) under optical microscope. (A) Section through one P-Fe(-Mn) oncoid with a core made of a chip of pelagic filament (Bositra sp.) limestone. Note P-rich (light) and Fe(-Mn)-rich (finest black laminae) microbial laminations alternating with sediment-rich laminae (dark gray). Transmitted light, parallel nicols. (B) Agglutinating foraminifera within the same sample as (A). Agglutinated quartz grains in the foraminifera walls are clearly visible. Pelagic sediments cemented by phosphate fill the chambers. Reflected light, parallel nicols.

centimeter-thick veneers and thin laminae of pelagic sediment that commonly include green peloids, clasts, and crusts of glaucony.

Lamination Features and Rhythmicity

Sectioned and polished hand specimens of stromatolitic nodules and crusts show a well-defined laminated architecture that is mainly due to changes in color from greenish or yellowish to red, brown, or even black. This is due to: (a) variations in the mineralogical composition and texture of successive laminae, which are alternatively dominated by authigenic minerals or by pelagic limy sediments; (b) changes in the relative abundance of Fe-Mn mineral phases versus phosphatic and/or glaucony phases; or (c) changes in the oxidation state of Fe. Lamination is usually consistent, with small lateral changes in thickness. If diagenetic overprinting, fragmentation and remineralization episodes have not destroyed primary features, a cyclic arrangement is clearly visible at the naked eye in most crusts and nodules of all ages and sites (Fig. 4) as well as under the microscope (Figs. 5A and 6B). Depending on the scale of observation, lamination can change slightly, but some general patterns are recognized at diverse scales. Basal laminae of each cycle encrusting the substratum show a dense packing and form laterally continuous planar sets that tend to eliminate underlying irregularities. Upwards, laminations tend to crenulate and define domes and hemispheroids that are either laterally connected or isolated. At the same time, laminae of pelagic sediment alternate with laminae dominated by authigenic components, producing a rhythmic arrangement (Figs. 4, 5A, and 6A-D). Successive laminae are regularly piled up, giving rise to typical stromatolite pseudocolumns (the most common case) when adjacent hemispheroids are attached to each other (Fig. 6A), or to true columns when they are isolated, but with a moderate to low synoptic relief, so that a gentle irregularity characterizes the stromatolite surface instead of a marked unevenness. Encrusting foraminifera are commonly present in these zones (Fig. 5B) and, where abundant, they confer a spongy aspect to the lamination. Finally, millimeter-thick microstromatolitic veneers (Fig. 6C) topped by sediment-rich laminae (Fig. 6D, see also Fig. 9A) mark the end of the stromatolite-growth cycle, which may be repeated several times within the same nodule or crust without significant variation.

Microstromatolites are finely laminated structures that are small in size (tens of micrometers to few millimeters) and range from isolated to grouped. They show delicate and morphologically very complex architectures: cupolar (Fig. 6A,B), pseudocolumnar to columnar (Fig. 6D), club-shaped, or dendritic (Figs. 6D and 9A). Microstromatolites may be found isolated within condensed limestone facies without direct relationship to any important stratigraphic discontinuity surface; for example, they may encrust more or less corroded macrofossils. They are, however, more frequently related to erosion features on the top of the underlying formation, which show evidence of long episodes of sedimentary omission: sharp, planar to irregular, abrasion, corrosion, or dissolution surfaces, fissures of mechanical origin, borings, and burrows. They also commonly encrust chips of lime mud eroded from contemporaneous pelagic sediments and clasts of the substrate (siliciclasts or carbonate lithoclasts). Microstromatolite crusts frequently occur within thicker stromatolites and oncoids, forming thin veneers that are systematically associated with partially phosphatized pelagic sediments (Figs. 4, 5A, and 6B,D). They also frequently mark the outer border of nodules and crusts in the transition to the surrounding or overlying pelagic sediment (Fig. 6D).

Architectural Elements

Three main basic elements define the stromatolite architecture: (1) authigenic minerals that define stromatolite lamination and are constituted by isolated to grouped structures that evoke mineralized microbial remains; (2) pelagic sediments; and (3)

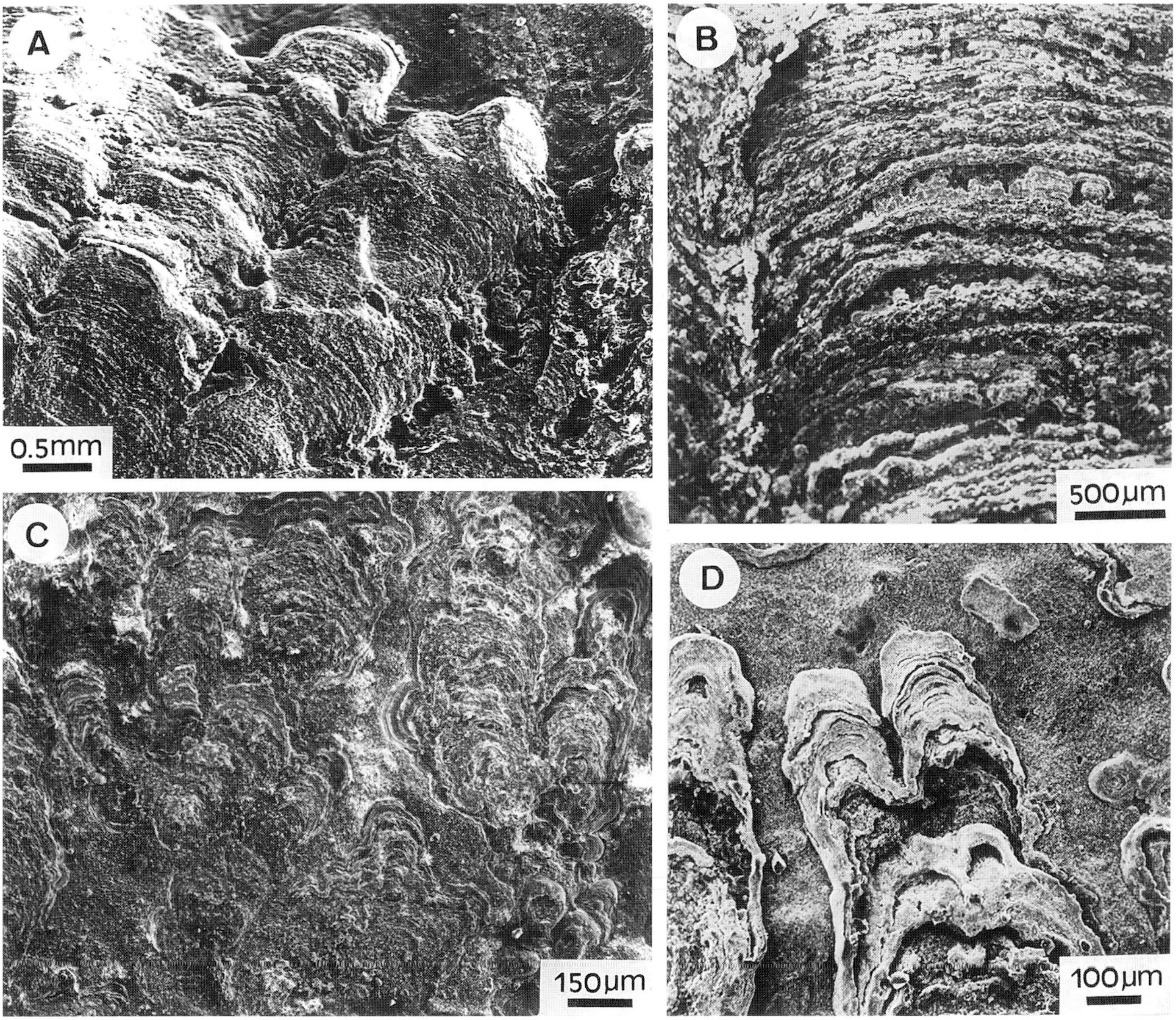

Fig. 6.—Lamination in phosphate stromatolites (SEM views, etched samples). (A) Dense, laminated stromatolite microcrusts, site 22 (Upper Jurassic). (B) Alternating lamination within a stromatolite pseudocolumn within one Upper Valanginian oncoid, site 15. (C) Arborescent microstromatolites, Cretaceous-Tertiary boundary, site 1. (D) Albian columnar microstromatolites from site 18.

encrusters dominated by benthic agglutinating foraminifera. Replacement and void filling textures are also present.

Microbial remains are represented by authigenic minerals and residual organic matter. They are better preserved in P-rich than in Fe-Mn-rich stromatolites. Optical (Fig. 7) and SEM images (Fig. 8) of both etched and unattacked samples of stromatolitic laminae show globular and flocculent ultrastructures and dense agglomerates of spheroidal, ovoidal, or sausage-shaped, micrometer-sized bodies that are the size and shape of bacteria. In many cases, they form a few μm-long strings that lie either parallel or oblique to the main lamination (Fig. 7). Qualitative EDX analyses of these bodies indicate that they are mainly composed of calcium phosphate, with minor amounts of Si, Fe, Al, and K systematically present. The existence of numerous pores and voids within authigenic laminae, parallel to the stromatolite lamination, is another striking feature that can be especially seen in BSE images (Fig. 9B). This was confirmed by observation under TEM, which further reveals that the bacteria-like objects are constituted by cores made of hexagonal francolite crystallites with sizes ranging between 0.1 μm and 1 μm, surrounded by a cortex formed by poorly crystalline clays associated with Fe-Si-Al-rich (and sometimes Mn-rich, when Mn is present) phases (see Martín-Algarra and Sánchez-Navas, 1995; and Sánchez-Navas et al., 1996, 1998). These poorly crystalline Fe-Si-Al envelopes confer the rounded morphology to the phosphate grains observed under SEM.

Pelagic sediments are trapped and bound between successive stromatolite laminae either as isolated particles or forming distinct sediment layers. They are silt- to clay-sized siliciclastics and often biogenic carbonates mainly consisting of coccolith

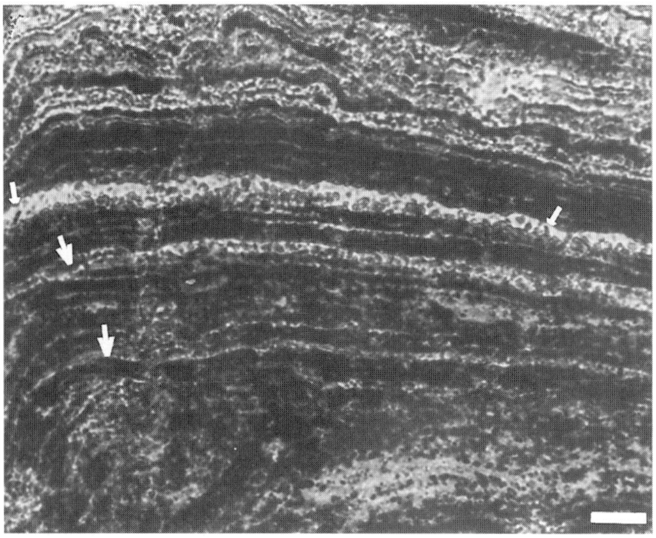

FIG. 7.—Stromatolite lamination and bacterial remains under the optical microscope (transmitted light, Upper Albian, site 18). Repetitive lamination in a domal stromatolite formed by dense successions of microbial mats crowded with spheroidal to ellipsoidal and sausage-shaped structures, some of which form prostrate and erect filaments interpreted as coccoidal and bacillary bacteria. Microrhythms formed by microbial laminae with changing texture and color, sometimes limited by microunconformities (arrows) are clearly visible. Bar = 25 μm.

debris (Fig. 8A). In addition, planktonic foraminifera are particularly abundant in some Cretaceous stromatolites (Figs. 9A). Authigenic minerals such as glaucony and Fe-Mn oxyhydroxides are commonly present as sand- to silt-sized intraclasts and peloids.

The common presence, and sometimes abundance, of encrusting benthic foraminifera is an important feature of stromatolite crusts and oncoids of all ages, sites, and compositions (Figs. 5B, 10A, 11A, and 12-14). Some of them, up to 1-2 mm long, with chambers reaching up to 0.8 mm in diameter, have agglutinating walls where silt- to clay-sized clastic particles of any composition have been cemented together (Fig. 5B). These encrusters are ammodiscids belonging to the Tolypamminninae subfamily; they concentrate in the upper part of the largest rhythmic cycles mentioned above and especially between stromatolite columns or pseudocolumns. Other smaller encrusting foraminifera (tens to a few hundreds of micrometers in size, with chambers rarely larger than 0.1 mm in diameter) have micritic walls where no individual clastic particles can be differentiated (Fig. 10A). These appear everywhere within the stromatolite architecture. Either by agglutinating sediment particles in their walls or by precipitating calcite, these encrusters can be significant contributors to the total volume of stromatolites of all ages, sites, and compositions. Besides, they produce a high primary porosity that has been filled with fine-grained pelagic sediments, microbial remains, and authigenic minerals, mainly phosphates (Fig. 5B). Phosphates also cement together and frequently replace sediment particles (Fig. 8A).

Mineral Phases and Their Relationships to the Stromatolite Texture

The stromatolites studied are mostly phosphatic and, in a much lesser extent, ferromanganesiferous. Most frequently, Jurassic examples are Fe-Mn-rich, while those of the Cretaceous are P-rich, but mixed situations commonly are found. Mineralogy is variable in detail, but two main groups of minerals can be clearly differentiated: detrital and authigenic. Detrital minerals are fine-grained siliciclastics and, dominantly, bioclastic and lithoclastic carbonates. Siliciclasts consist mainly of quartz, with minor amounts of feldspars and muscovite (detrital

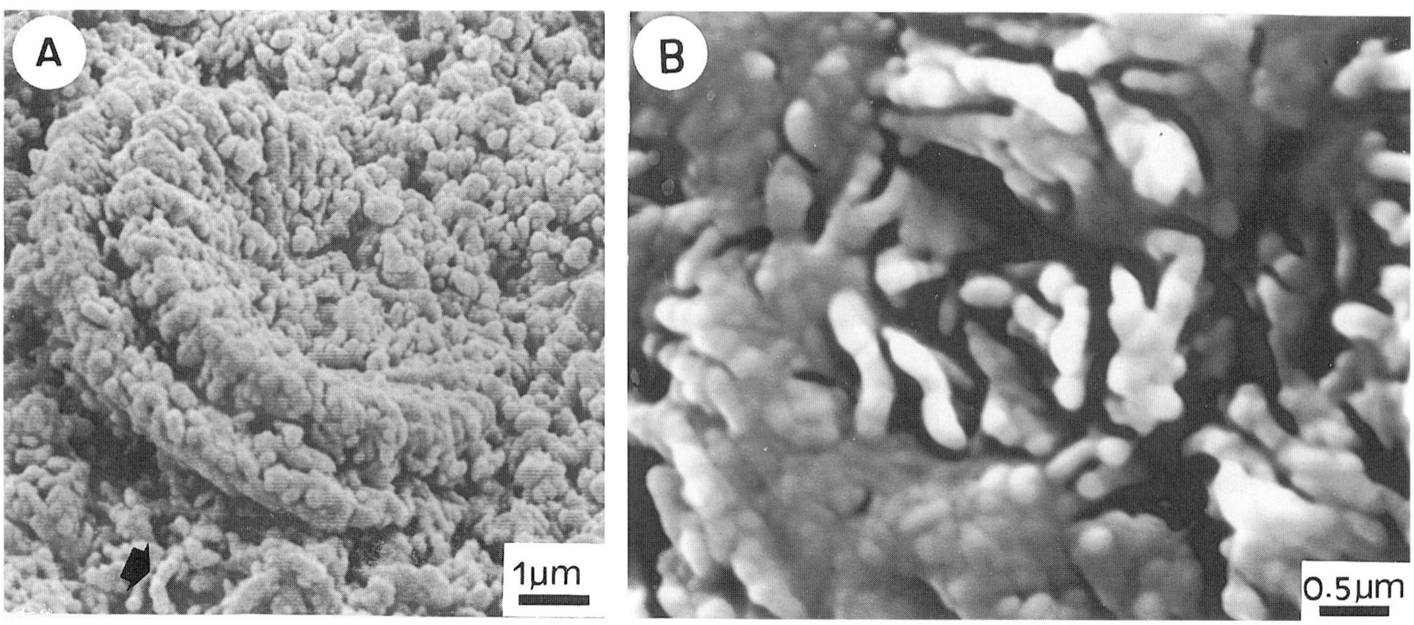

FIG. 8.—(A) Secondary electron image (etched sample of a phosphatized coccolith replaced by a crowded mass of bacteria-like objects. It is lying on a microbial mat that is also made up of similar phosphatic objects. Arrow points to a filament made of coccoids. Upper Jurassic, site 22. (B) Secondary electron micrograph (nonetched sample) of the top of a stromatolite mat constituted by a dense mass of globular and sausage-shaped phosphatic objects. Upper Albian, site 16.

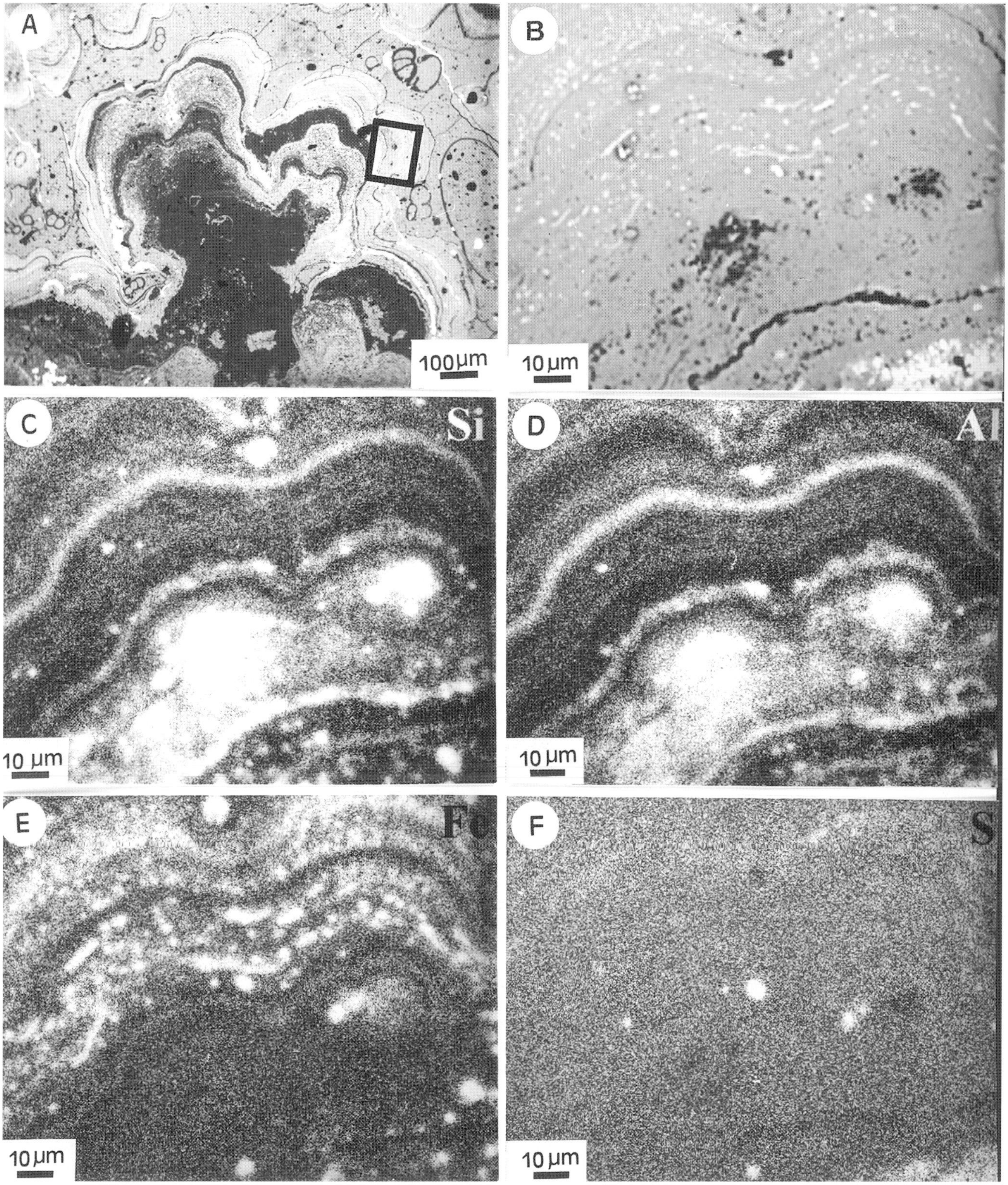

Fig. 9.—BSE (A, B) and XR (C-F) images of an Upper Albian stromatolite from site 18. (A) General view of a finely laminated, mixed phosphatic and carbonatic microstromatolite surrounded by pelagic sediment. Black areas correspond to calcite, siliciclastic grains, and/or organic matter; white zones to barite, light gray to phosphate, and darker gray colors to more or less phosphatic-carbonatic sediment. Note the numerous (nonphosphatized) planktonic foraminifera and the section of one ammonite chamber (right-hand side) included within the sediment, which is made of phosphatic micrite (fossilized bacteria?) and partially phosphatized coccoliths. (B) Close-up of the area selected in (A). (C), (D), (E), and (F) XR images of, respectively, Si, Al, Fe, and S, in the area shown in (B). Images (C) and (D) show the existence of authigenic clay-rich bands intergrown with phosphate laminae and of grains of clastic clays, while images (E) and (F) demonstrate the existence of Fe-oxyhydroxide bands and isolated pyrite grains.

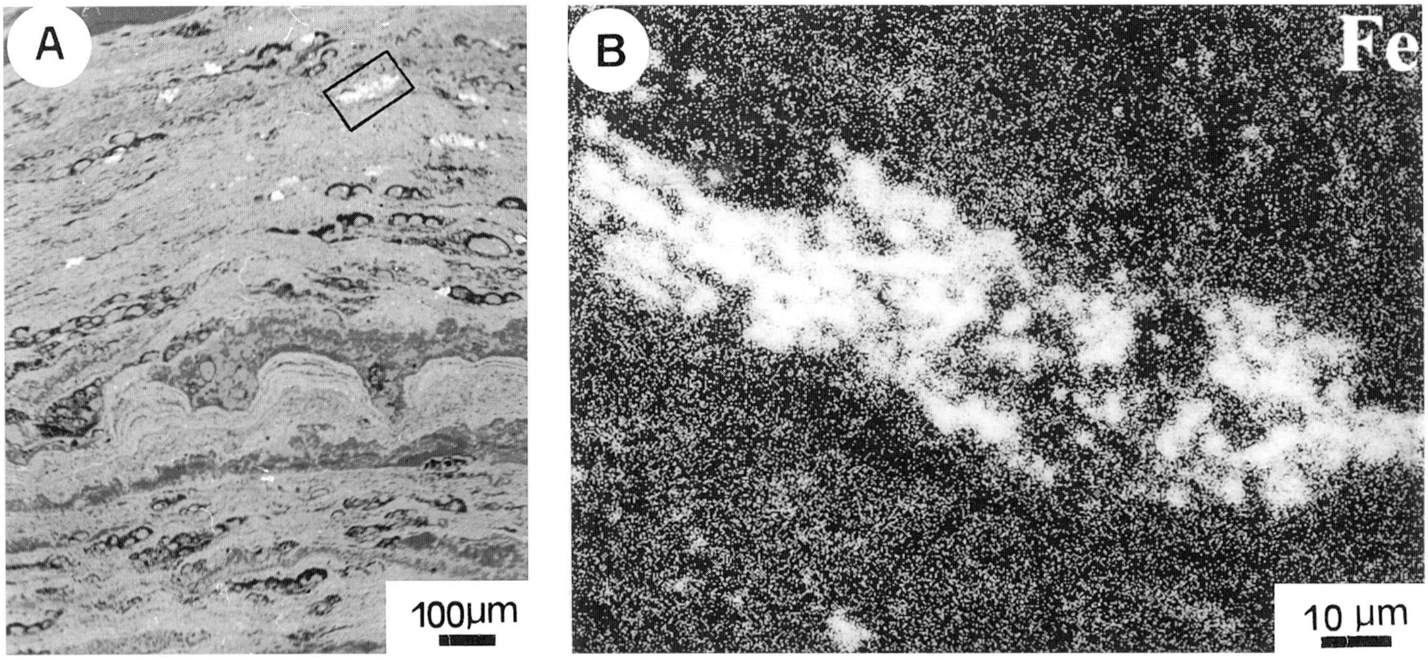

Fig. 10.—BSE (A) and XR image (B) of another site in the same sample as that in Figure 9, but within the central part of one stromatolite rhythm. At the naked eye and under optical microscope, stromatolitic phosphate in this area is very dark due to relatively high content of organic matter and pyrite. (A) General view of microbial laminae made of authigenic minerals (light gray is phosphate; white is pyrite), sediment-rich laminae (dark gray) and encrusting foraminifera. (B) XR image of Fe in the area selected in (A).

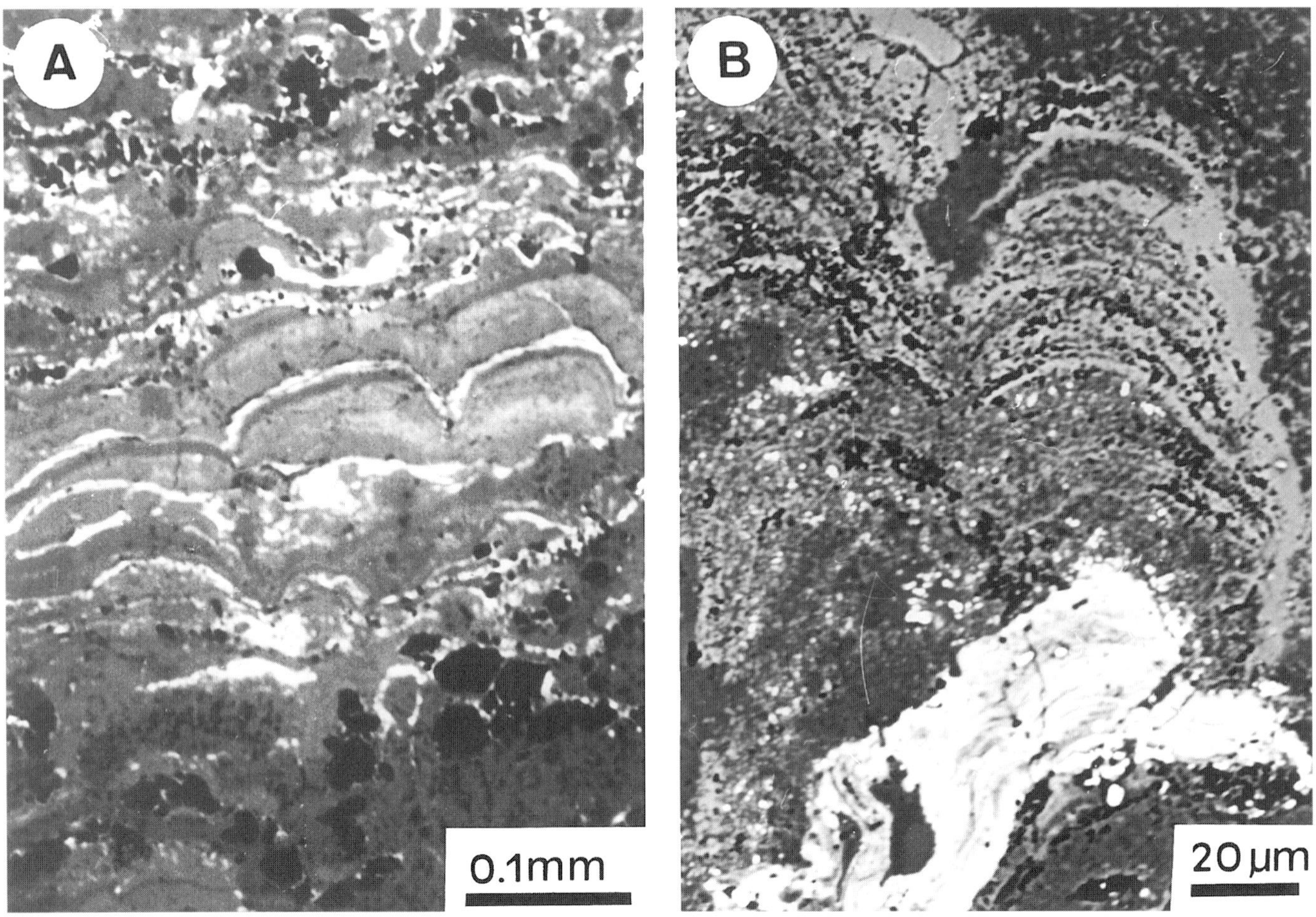

Fig. 11.—BSE images of P-Fe (Mn) stromatolites. Black spots are quartz grains, dark gray is carbonate plus clays; light gray is phosphate; white corresponds to Fe-Mn phases. (A) Lower Jurassic, site 3. Note that in this case each authigenic phase constitutes an independent lamina. (B) Upper Albian, site 16.

illite) among other minerals. Bioclasts and lithoclasts are mainly calcite and less frequently dolomitic, although phosphatic bioclasts such as fish remains (teeth) are at times locally found in the sediments associated with the stromatolites. Authigenic minerals appear either *in situ* within the stromatolite structure or as reworked sand- to silt-sized allochems (intraclasts or peloids) trapped within the stromatolite laminae or included within associated sediments.

Authigenic phases consist of phosphates (carbonate fluorapatite), iron minerals (goethite, hematite, and pyrite), glaucony (both glauconite and Fe-rich authigenic clays) and sometimes barite, (Fe)-Mn oxyhydroxides, and calcite. Carbonate-fluorapatite, goethite, calcite, and quartz have been identified by X-ray powder diagrams of whole-rock samples of phosphate stromatolites (see also García-Cervigón et al., 1986-1987). X-ray diffractograms of samples purified by acid attack with diluted HCl, show peaks with spacings and relative intensities that closely match those of marine francolite studied by Jarvis (1992). Besides these main mineral components, minor amounts of barite, glauconite, and pyrite have been detected by BSE and X-ray images. These images also allow determination of the distribution of mineral phases and their textural relationship within the stromatolite architecture.

Thus, most of these authigenic phases (phosphate, Fe-Mn-oxyhydroxides, barite, glaucony) may form laminated (stromatolitic) textures, isolated grains (intraclasts or peloids, sometimes coated), void-fillings and replacement textures after sediment particles of different previous composition. Mineral associations found in the authigenic stromatolitic laminae vary in relation to the predominance of P or Fe. Typical associations are:

- *Type a*, found in P-rich stromatolites: *a.1*) francolite + Fe-rich smectite + non-crystalline Fe-Si-Al oxyhydroxides + goethite/hematite; and *a.2*) francolite + glaucony + goethite/hematite + barite + pyrite + (non-crystalline substances).
- *Type b*, typical of Fe-P-rich stromatolites: francolite + goethite/hematite + (clay minerals), with increasing proportion of Fe (and Mn) minerals.
- *Type c*, found in Fe-Mn-rich stromatolites: goethite/hematite + pyrolusite + Fe-Mn-Al noncrystalline substances + (francolite).

The features of the type a.1 mineral association are those of the Jurassic phosphate stromatolites discussed by Martín-Algarra and Sánchez-Navas (1995) and Sánchez-Navas et al. (1998). Some comments about the other types will be made now. BSE images in Figures 9A and 10A correspond to general views of one Upper Albian phosphate stromatolites of type a.2. Lamination is defined by francolite, barite, goethite, pyrite, and calcite, alternating with silt- to clay-sized clastic components, basically calcite (dark areas), although quartz, feldspars, muscovite, and grains of goethite are commonly detected. Sediments that bury microstromatolitic laminae are made of partially phosphatized and/or phosphate-cemented micritic sediments that are very rich in coccoliths (frequently phosphatized) and planktonic foraminifera (non-phosphatized; Fig. 9A). A detailed analysis of microstromatolitic, phosphate-rich laminae indicates that lamination is defined by the alternation of Fe-oxyhydroxide-rich and clay-rich fine bands with phosphate laminae (Figs. 9B-F), which also include barite that follows stromatolite lamination (Figs. 9A-B) and minor amounts of pyrite (Figs. 9E-F, and 10B). Some of the observed clays in these figures correspond to clastic components as well as quartz and calcite (black areas of BSE image of Figs. 9A-B, 10A, and 11A; see also Fig. 7B in Martín-Algarra and Sánchez-Navas, 1995, for P-rich, type a.1 stromatolites).

In type b Fe-P-rich stromatolites, Fe-P-(Mn)-rich laminae appear either intergrown with phosphate laminae (Fig. 11A; Jurassic) and/or forming isolated microstromatolites (Fig. 11B, Cretaceous). Fe-rich laminae are mainly made of goethite with minor amounts of hematite, as indicated by XRD analyses (García-Cervigón et al., 1986-1987). Fe-Mn-rich oxyhydroxides appear in the form of grains and fine stromatolite laminae within Fe-microstromatolites (Fig. 11A). Clay minerals closely associate with authigenic Fe-oxyhydroxides in Fe-rich (type c) stromatolites (Figs. 12-14). Coated grains with a thin, laminated cortex of authigenic minerals, mainly of phosphate but also of glaucony and of Fe-Mn phases, are also common in association with the three types of stromatolites.

Fe-Mn stromatolites without associated phosphate (type c) have been found in some Jurassic samples (Figs. 12-15). In these examples, the primary biosedimentary structure is partially obliterated due to strong diagenetic overprinting (Fig. 13A). Preserved lamination features (Fig. 13B) and the existence of encrusting foraminifera (Fig. 13A) and trapped and bound pelagic sediments (Fig. 14D) permit confirmation of a stromatolitic origin. Lamination is defined by the alternation between Fe-oxyhydroxides, Mn-oxyhydroxides, undetermined Al-Mn-rich phases, and carbonate laminae (Figs. 12, 14, and 15). Clay minerals concentrate in Fe-rich laminae (Fig. 14C, E, F). Particles formed by pyrolusite crystals (Fig. 13A, B), quartz, and clays (Fig. 14E, F) have been found cemented together with Fe-Mn-oxyhydroxides (Figs. 13C and 14B, C) in the walls of agglutinating encrusting foraminifera of stromatolites of Jurassic age. Mn-Al oxyhydroxides are also found filling voids within foraminifera chambers (Fig. 13B, D).

With the exception of barite, which sometimes appears in the form of crystal aggregates in type a.2 stromatolites, it is noteworthy to mention that the crystal size of all described authigenic minerals forming the stromatolite laminae is extremely small, as visible in BSE and secondary electron images. TEM studies of authigenic laminae of phosphate stromatolites by Martín-Algarra and Sánchez-Navas (1995, 1996, and in prep., 1999) and by Sánchez-Navas et al. (1996, 1998) have shown the general low crystallinity of francolite, clays, and Fe-Mn oxyhydroxides, and the presence of noncrystalline phases tightly intergrown with them. In particular, amorphous phosphate phases and noncrystalline Fe-Si-Al-phases appear closely intergrown with extremely small (5-15 nm in diameter and up to 70-80 nm in length), ribbon-plate-like crystals of francolite, and fusiform smectite crystals, respectively.

Microprobe analyses (Table 2, Fig. 16) corresponding to one

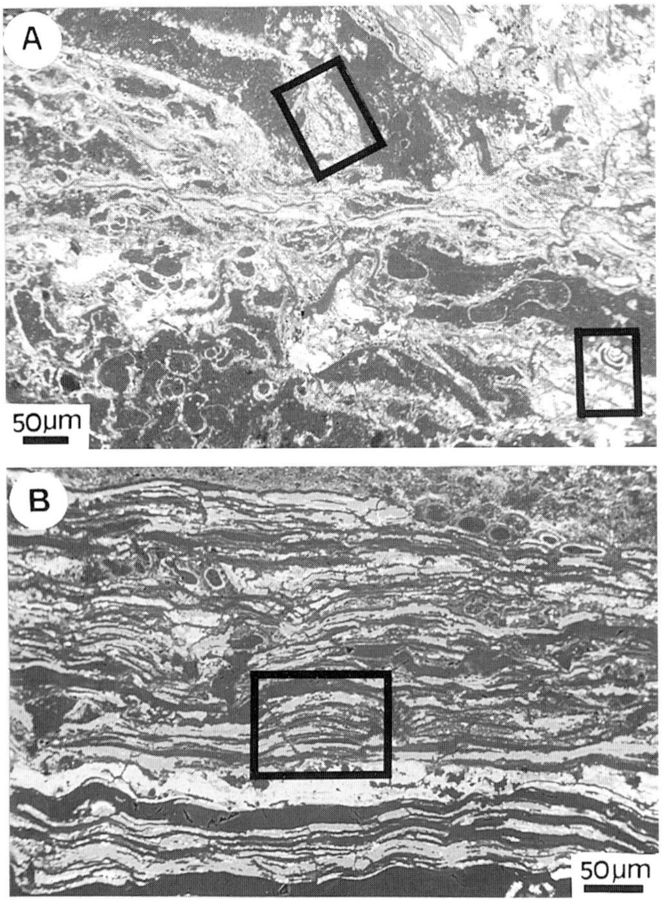

Fig. 12.—(A) and (B) BSE images of Mn-Fe stromatolites of Early Jurassic age from site 2. Both samples are strongly affected by strong diagenetic overprinting that, however, did not completely destroy either the laminated architecture or the encrusting foraminifera. Photo (A) is stratigraphically upside down. White colors correspond to Mn-rich phases, light gray to Fe-Al-rich phases, and darker zones to carbonatic and siliciclastic sediments. Close ups in Figures 13, 14, and 15.

homogeneous glauconite grain and to one glauconitized planktonic foraminifera of the same sample (type a.2. stromatolite) show a wide compositional range and a positive correlation between K_2O and Fe_2O_3. These analyses have been selected according to the following crystal chemistry constraints:

Si: 3.5-4
$\Sigma^{(VI)}$: 1.95-2.2
K+Na+2Ca: < 1

Chemical analyses (Table 2, Fig. 16) show that glauconitic minerals have a variable composition, ranging from glauconitic smectite (K_2O contents lesser than 6%) to glauconitic mica (Odin and Matter, 1981). Bivariate plots (Fig. 17) indicate the existence of important beidellitic ($Si + \square^{XII} = Al^{VI} + [K+Na+Ca]$, where $\square$ means the vacuum sites in the clay structure) and nontronitic ($Al^{VI} = Fe^{3+}$) substitutions for the glauconitic minerals (diamonds in Fig. 17). Both substitutions have been also described by Sánchez-Navas et al. (1998) for authigenic smectites, which are intergrown with francolite crystals of Jurassic phosphate stromatolites (circles in Fig. 17). In brief, the plot of Figure 17 shows that Fe-rich smectites and glauconitic mica constitute end-members of the observed chemical trends.

Replacement textures of authigenic minerals after carbonate particles are also frequent within the stromatolite structure. Bioclasts and fossils, in particular coccoliths, may be found partially or totally phosphatized (Fig. 8A), glauconitized, baritized, or replaced by Fe-Mn oxyhydroxides with variable degrees of preservation. Where phosphatized or glauconitized, these pseudomorphs are typically constituted by bacteria-like objects similar to those found within the stromatolite laminae (Martín-Algarra and Sánchez-Navas, 1995). Micritic sediments, intercalated between stromatolitic laminae forming distinct layers or internal sediments within framework voids such as microfossil chambers, also appear partially cemented and/or replaced by phosphate and other minerals (Fig. 9A). Drusiform aggregates of barite crystals, up to 2-3-mm long, are a peculiar type of texture that has only be found in type a.2 mineral associations within Late Albian phosphate stromatolites. These crystal aggregates form discrete layers on top of the underlying mat and are encrusted by overlying stromatolite laminae. Crystals in these aggregates sometimes show pointed apices that in other cases appear cross-cut and eroded by encrusting overlying laminae (see Fig. 18 in Martín-Algarra and Vera, 1994). Barite crystals have also been found filling fractures that cross-cut stromatolite lamination. Aggregates of subidiomorphic, probably pyrolusite crystals between Mn-rich primary stromatolite laminae are also present in Jurassic Fe-Mn-stromatolites with type c mineral associations (Fig. 14A, B).

DISCUSSION

Mechanisms of Accretion and Differences with Shallow-water Stromatolites

Phosphate and iron-manganese nodules and crusts of the Alpine-Mediterranean regions are frequently seen as fossil counterparts (Jenkyns, 1970) of their modern compositional equivalents of the deep ocean, which are usually interpreted as inorganic concretions (Usui, 1983/1984, 1985; von Stackelberg, 1987; Mangini et al., 1987; Stow et al., 1996) or diagenetic replacements after carbonate particles (although other authors have demonstrated the microbial mediation in their genesis: Monty, 1973; Greenslate, 1974; Ehrlich 1975, 1990; Giresse et al., 1998). Data presented above demonstrate that the structures studied did not form during diagenesis after burial under a more or less thick cover of sediments. On the contrary, they are primary sedimentary structures of microbial origin that must be designated as pelagic stromatolites (see Jenkyns, 1971, 1978, 1986; Bernoulli and Jenkyns, 1974; Massari and Dieni, 1983; Krajewsky, 1981, 1983, 1984; Martín-Algarra and Vera, 1994; Vera and Martín-Algarra, 1994, and references therein). Actually, they accreted rhythmically at the sediment-water interface due to the activity of benthic microbial communities that were able to trap and bind pelagic sediment particles and, especially, to *in situ* precipitate marine authigenic minerals. In structures of different age, this origin is demonstrated by: (a) their accretionary, cauliflower-like, finely laminated architecture,

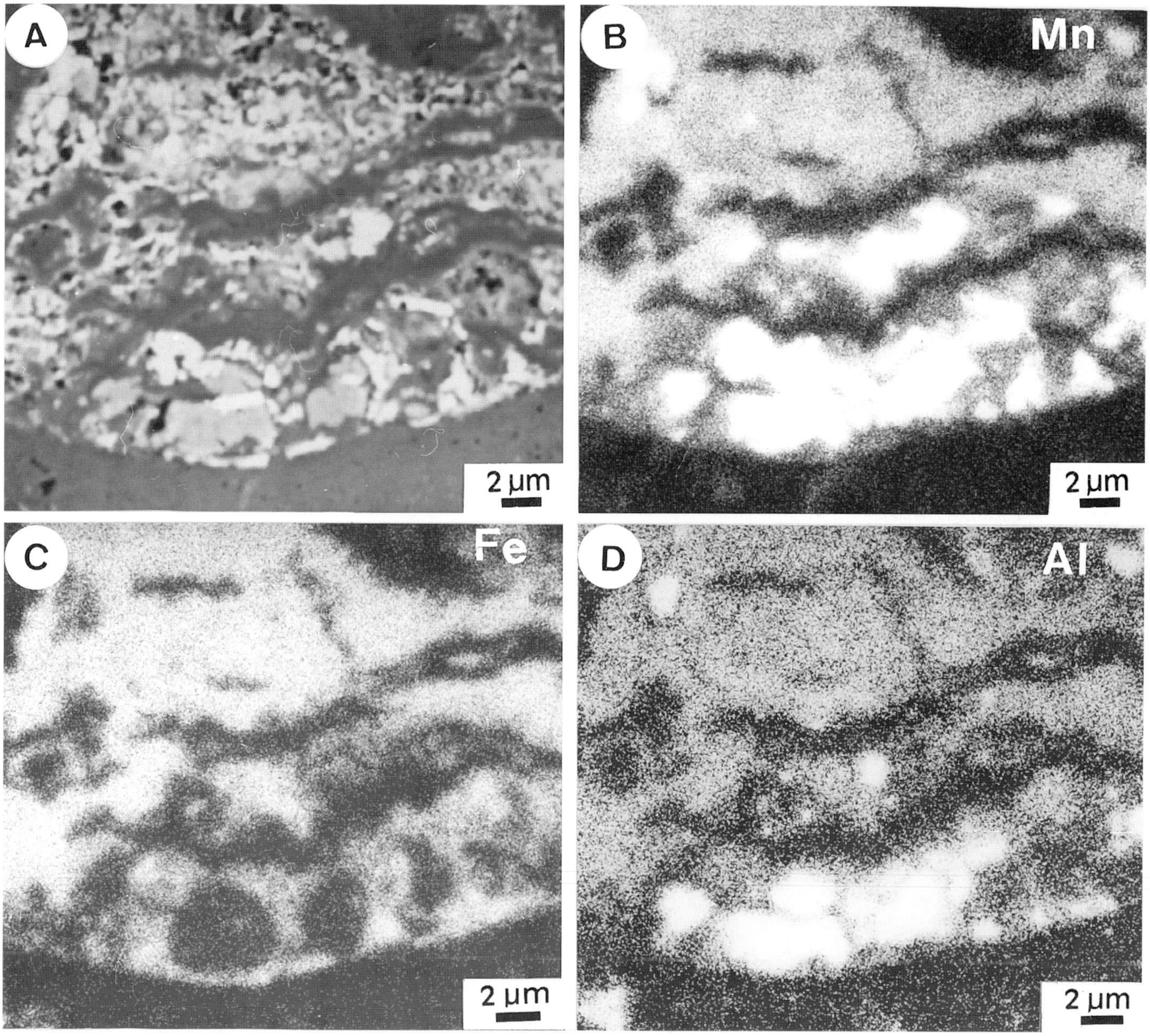

Fig. 13.—Close-up of Figure 12A. (A) BSE image showing a benthic foraminifera encrusted by microbial mats. Carbonate sediment is dark gray, white corresponds to Mn-phases (pyrolusite), and intermediate gray tones correspond to Mn-Fe-Al (-Si) phases. (B), (C), and (D) are XR images of Mn, Fe, and Al, respectively, of the same area as (A). Note that Mn-Al-rich crystals fill the chambers of the foraminifer, while agglutinating pyrolusite grains cemented together with very small Fe-rich and other particles form its walls.

which consists of alternations between laminae composed of siliciclastic and bioclastic sediments rich in pelagic fossils and laminae made of authigenic minerals; (b) the existence of microunconformities and reencrustations after fragmentation of the structure; (c) the normal presence, in the pelagic sediments associated with nodules and crusts, of sand- to silt-sized grains (forming intraclasts or peloids) constituted by the same authigenic minerals as those formed within the nodules and crusts; (d) the repeated encrustations of authigenic laminae by benthic agglutinating foraminifera and vice-versa; and (e) the existence of objects that evoke microbial remains, preserved as authigenic minerals. The common presence of microunconformities cross-cutting the internal lamination, of intraclasts of authigenic minerals (such as hematite, francolite, glauconite and Fe-Mn oxy-hydroxides), and of reencrustations of stromatolite clasts, demonstrate that early submarine lithification and mineralization of these structures occurred in or near the sediment-water interface without any significant burial under thick drapes of pelagic sediment. So, the seafloor was subjected to repeated events of shrinkage plus reworking and fragmentation of the bottom sediment due to early mineralization, boring and/or (probably weak) bottom current action, and to periods of microbial re-colonization and reencrustation.

The main differences between the stromatolites studied and classical shallow water stromatolites are: (1) in the studied stromatolites the mechanical trapping and binding of allochtonous

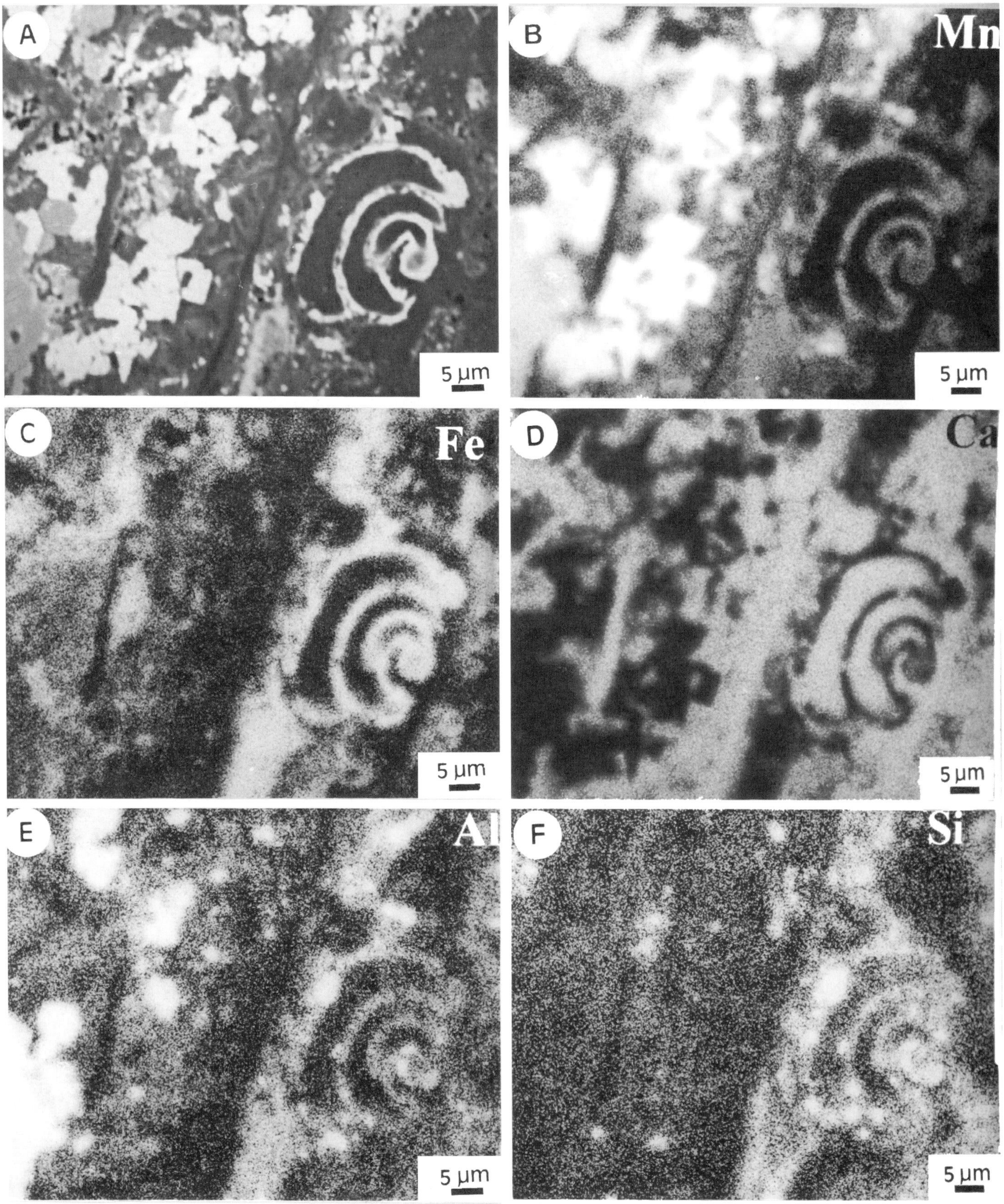

Fig. 14.—(A) Close-up of Figure 12A. BSE image. Top towards the left. (B), (C), (D), (E), (F) XR images of Mn, Fe, Ca, Al, and Si of the area shown in (A). Note that crystals are Mn-phases (white is probably pyrolusite, while light gray zones correspond to an undetermined Mn-Al-rich phase). The foraminifer made its walls with quartz and clay particles as well as with extremely small Fe(-Mn)-rich particles that were probably cemented together with organic substances. Small Fe (-Mn) particles would probably be precipitated by Fe-Mn-oxidizing bacteria suspended in the water column.

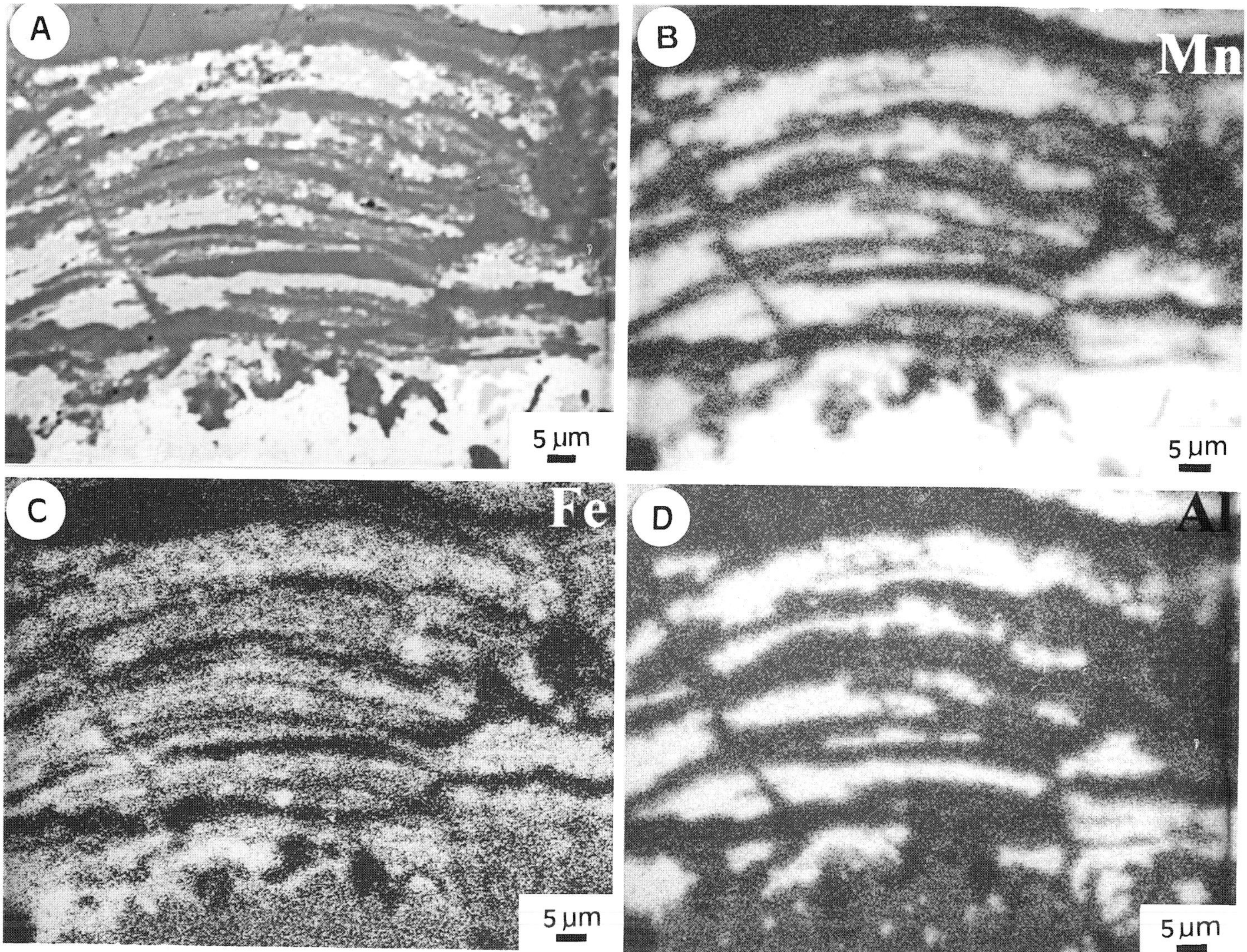

Fig. 15.—Close-up of Figure 12B. (A) BSE image. Lighter colors correspond to Mn-rich phases. Light gray colors are Fe-Al-Mn oxyhydroxides plus Fe-rich authigenic clays. Dark color corresponds to carbonates. (B), (C), and (D) are XR images of Mn, Fe, and Al, respectively.

sediments involve systematically fine-grained pelagic sediments and never those nearshore; (2) they precipitated preferentially marine authigenic minerals other than carbonates; and (3) they accreted during long intervals of geological time, associated with episodes of extreme sediment starvation. In shallow marine and continental stromatolites, the trapping and binding mechanism of accretion is mainly due to the light-induced motility of photosynthetic filamentous microbes, generally cyanobacteria. It is also favored by the presence, around microbial cells, of thick sticky sheaths formed by extracellular jelly-like polysaccharides, which promote the adherence of particles, confer stability to the microbial accretion, and provide favorable templates for precipitation of many different minerals (see many papers in Walter, 1976; Défarge et al., 1996; Jones and Renaut, 1997; Jones et al., 1997). The activity of heterotrophic bacteria that degrade the organic matter of the sheath also creates local supersaturation conditions that allow *in situ* carbonate precipitation (Folk et al., 1985; Chafetz et al., 1991; Guo and Riding, 1992a, b; Vasconcelos and McKenzie, 1997). In addition, micritization of trapped particles is not rare in shallow marine stromatolites due to boring by endolithic bacteria.

In the nodules and crusts studied, the former existence of such mucilaginous substances around microbes and their role as templates for precipitation of different minerals can be reasonably inferred by the presence of isolated and filamentous structures that evoke stromatolite-forming microbes (compare with Folk, 1993; Folk and Lynch, 1997). They are preserved within colloform textures made by extremely small crystals of authigenic minerals closely associated with noncrystalline phases, for which a precursory noncrystalline jelly-like stage is proposed (see below). The former existence of endolithic bacteria can also be inferred by the presence of degraded (corroded) carbonate particles that now appear partially or totally replaced and pseudomorphosed by authigenic minerals. Nevertheless, several facts point toward an origin in the aphotic zone for these open marine stromatolites, for example, in relatively deep-water and not in neritic, shallow-water settings: (a) the pelagic nature of the trapped and associated sediments that precludes a shallow-water origin; (b) the strong morphological similarities between the structures studied and present-day deep sea nodules and

TABLE 2.— MICROPROBE ANALYSES OF GLAUCONIES.

No.*	1	2	3	4	5	6	7	8	9	10	11	12	13	14	15
SiO_2	53.535	55.786	56.482	54.789	53.310	53.326	54.122	53.878	53.395	53.779	53.749	54.053	53.977	53.360	54.387
TiO_2	0.046	0.072	0.061	0.070	0.054	0.087	0.059	0.070	0.074	0.055	0.047	0.061	0.046	0.074	0.071
Al_2O_3	8.739	11.828	12.169	11.448	11.249	9.178	9.058	8.840	8.390	8.413	8.715	8.839	8.786	9.962	9.674
Cr_2O_3	0.170	0.462	0.479	0.464	0.444	0.172	0.172	0.173	0.162	0.148	0.166	0.144	0.147	0.202	0.203
Fe_2O_3**	20.768	15.139	14.883	15.530	15.322	19.800	20.405	20.809	21.120	20.967	20.283	20.436	20.868	18.523	19.500
MnO	0.000	0.024	0.000	0.000	0.024	0.006	0.001	0.000	0.000	0.001	0.013	0.002	0.000	0.000	0.020
MgO	4.591	4.751	4.938	4.962	4.911	4.703	4.555	4.461	4.551	4.558	4.633	4.685	4.698	4.649	4.635
CaO	0.526	0.773	0.648	0.720	1.578	0.740	0.605	0.540	0.494	0.474	0.545	0.567	0.560	1.844	0.714
Na_2O	0.225	0.568	0.555	0.529	0.534	0.247	0.194	0.210	0.198	0.201	0.245	0.241	0.221	0.255	0.243
K_2O	7.597	5.973	5.884	6.000	6.010	7.456	7.596	7.610	7.733	7.768	7.600	7.437	7.685	6.617	6.970
Total	94.125	93.869	94.608	92.959	91.901	93.785	94.723	94.506	94.001	94.319	93.974	94.418	94.913	93.646	94.463
Cations***															
Si	3.711	3.776	3.782	3.755	3.716	3.704	3.720	3.716	3.713	3.725	3.727	3.725	3.711	3.692	3.725
Al^{IV}	0.289	0.224	0.218	0.245	0.284	0.296	0.280	0.284	0.287	0.275	0.273	0.275	0.289	0.308	0.275
Al^{VI}	0.425	0.720	0.742	0.680	0.641	0.455	0.453	0.435	0.400	0.412	0.439	0.443	0.423	0.505	0.506
Cr	0.009	0.025	0.025	0.025	0.024	0.009	0.009	0.009	0.009	0.008	0.009	0.008	0.008	0.011	0.011
Ti	0.002	0.004	0.003	0.004	0.003	0.005	0.003	0.004	0.004	0.003	0.002	0.003	0.002	0.004	0.004
Fe	1.083	0.771	0.750	0.801	0.804	1.035	1.055	1.080	1.105	1.093	1.058	1.060	1.080	0.965	1.005
Mn	0.000	0.001	0.000	0.000	0.001	0.000	0.000	0.000	0.000	0.000	0.001	0.000	0.000	0.000	0.001
Mg	0.474	0.479	0.493	0.507	0.510	0.487	0.467	0.459	0.472	0.471	0.479	0.481	0.481	0.479	0.473
Σ^{VI}	1.982	1.970	1.985	1.987	1.955	1.977	1.975	1.974	1.977	1.975	1.976	1.983	1.984	1.949	1.984
Ca	0.039	0.056	0.046	0.053	0.118	0.055	0.045	0.040	0.037	0.035	0.040	0.042	0.041	0.137	0.052
Na	0.030	0.075	0.072	0.070	0.072	0.033	0.026	0.028	0.027	0.027	0.033	0.032	0.029	0.034	0.032
K	0.672	0.516	0.503	0.525	0.535	0.661	0.666	0.670	0.686	0.686	0.672	0.654	0.674	0.584	0.609
Σ^{XII}	0.741	0.646	0.621	0.648	0.725	0.749	0.736	0.738	0.749	0.749	0.746	0.728	0.745	0.755	0.694

* Number of the analysis
** Total Fe expressed as Fe_2O_3
*** Oxygen normalization (10 O and 2 OH)

TABLE 2.— (Continued.)

No.*	16	17	18	19	20	21	22	23	24	25	26	27	28	29
SiO_2	53.935	54.468	53.969	54.294	53.586	54.069	53.435	53.542	53.707	53.609	53.284	55.527	56.266	55.009
TiO_2	0.046	0.048	0.053	0.088	0.044	0.051	0.064	0.069	0.134	0.048	0.068	0.110	0.072	0.071
Al_2O_3	9.735	9.744	9.897	10.086	9.513	9.704	9.353	9.713	9.628	9.617	9.203	12.150	12.315	11.711
Cr_2O_3	0.247	0.239	0.167	0.212	0.247	0.222	0.236	0.232	0.232	0.231	0.234	0.479	0.461	0.483
Fe_2O_3**	19.276	19.170	18.990	19.183	19.536	19.594	20.014	19.405	19.697	19.641	20.398	14.547	14.636	14.445
MnO	0.042	0.021	0.000	0.017	0.006	0.000	0.000	0.000	0.000	0.000	0.006	0.011	0.023	0.005
MgO	4.779	4.628	4.635	4.753	4.786	4.701	4.501	4.514	4.579	4.466	4.620	4.934	4.959	5.143
CaO	0.660	0.728	1.160	0.741	0.672	0.699	0.498	0.515	0.533	0.535	0.550	0.797	1.193	0.974
Na_2O	0.238	0.244	0.209	0.219	0.210	0.198	0.288	0.336	0.336	0.346	0.345	0.634	0.637	0.576
K_2O	7.000	6.996	6.959	7.028	7.025	6.840	7.044	6.996	7.001	6.979	6.830	5.702	5.701	5.719
Total	94.044	94.390	94.137	94.699	93.668	94.116	93.456	93.378	93.888	93.524	93.509	93.434	94.849	92.711
Cations***														
Si	3.713	3.732	3.712	3.708	3.708	3.716	3.711	3.713	3.708	3.714	3.700	3.767	3.764	3.767
Al^{IV}	0.287	0.268	0.288	0.292	0.292	0.284	0.289	0.287	0.292	0.286	0.300	0.233	0.236	0.233
Al^{VI}	0.504	0.519	0.514	0.521	0.485	0.502	0.477	0.507	0.491	0.500	0.453	0.738	0.735	0.712
Cr	0.013	0.013	0.009	0.011	0.014	0.012	0.013	0.013	0.013	0.013	0.013	0.026	0.024	0.026
Ti	0.002	0.002	0.003	0.005	0.002	0.003	0.003	0.004	0.007	0.003	0.004	0.006	0.004	0.004
Fe	0.999	0.988	0.983	0.986	1.017	1.013	1.046	1.013	1.023	1.024	1.066	0.743	0.737	0.744
Mn	0.002	0.001	0.000	0.001	0.000	0.000	0.000	0.000	0.000	0.000	0.000	0.001	0.001	0.000
Mg	0.490	0.473	0.475	0.484	0.494	0.481	0.466	0.467	0.471	0.461	0.478	0.499	0.494	0.525
Σ^{VI}	1.993	1.980	1.972	1.990	1.996	1.997	1.989	1.987	1.986	1.985	1.997	1.980	1.967	1.982
Ca	0.049	0.053	0.085	0.054	0.050	0.051	0.037	0.038	0.039	0.040	0.041	0.058	0.086	0.071
Na	0.032	0.032	0.028	0.029	0.028	0.026	0.039	0.045	0.045	0.046	0.046	0.083	0.083	0.076
K	0.615	0.612	0.611	0.612	0.620	0.600	0.624	0.619	0.617	0.617	0.605	0.493	0.487	0.500
Σ^{XII}	0.695	0.697	0.724	0.696	0.698	0.678	0.700	0.702	0.701	0.703	0.692	0.635	0.655	0.648

* Number of the analysis
** Total Fe expressed as Fe_2O_3
*** Oxygen normalization (10 O and 2 OH)

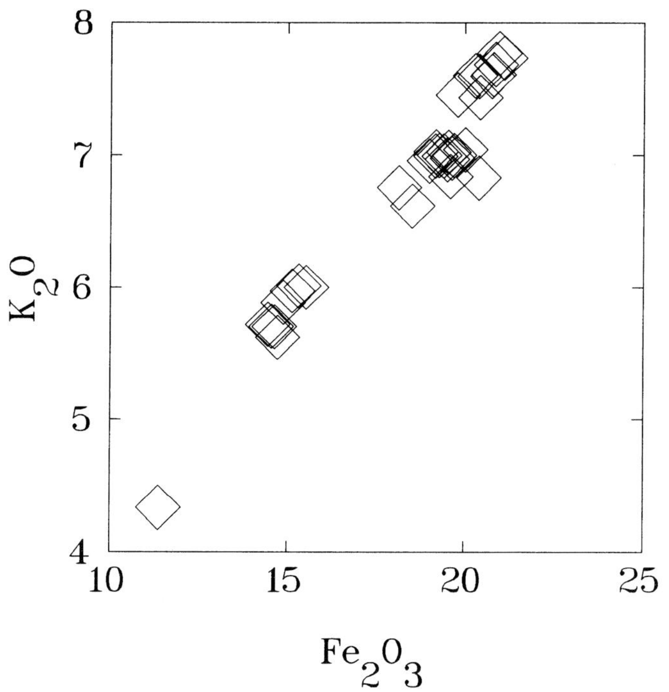

Fig. 16.—K_2O versus Fe_2O_3 plot for three different glauconies. The diamond in the lower left correspond to the average of analyses of Fe-rich authigenic smectites coming from one Jurassic phosphate stromatolite published by Sánchez-Navas et al. (1998). The other two groups of analyses respectively correspond to EPMA data obtained for two glauconite grains trapped between phosphate stromatolite laminae in the same sample as that reproduced in Figure 9. One is a typical glauconite peloid (>15% Fe_2O_3) and the other is an internal mould of one planktonic foraminifer (*circa* 15% Fe_2O_3) located just a few hundred micrometers of the former in the same sediment layer.

crusts; (c) the features of the encrusting benthic foraminifera, which are very similar to those commonly found in modern deep-sea nodules (Wendt, 1974; Resig and Glenn, 1997); and (d) the common presence of endostromatolites in the accretions studied of any age. This is a particularly interesting observation, because it demonstrates that no light is necessary for the microbial communities to accrete this type of stromatolites, which should form certainly below a minimum depth of several tens of meters, probably around few hundred of meters. By consequence, the stromatolite-forming microbes certainly were not autotrophic photosynthetic cyanobacteria or bacteria living in photic environments, but probably were complex chemosynthetic bacterial communities (see Jannasch, 1984a). They probably were formed by: (a) hetero-organotrophic bacteria living on degraded organic matter fallen down to seabottom; (b) heterolithotrophic bacteria able to digest organic matter plus inorganic sediment particles and use inorganic compounds as a source of electrons; and/or (c) chemoautotrophic bacteria capable of obtaining energy for biosynthesis by precipitating authigenic minerals after ion concentration either directly from sea water (chemoautolithotrophic) or indirectly from degrading organic matter (chemoautoorganotrophic).

Bacterial Authigenesis of Marine Minerals

In the stromatolites studied and in the associated sediments, marine authigenic minerals such as francolite, glauconitic minerals, pyrite, iron, and manganese oxyhydroxides frequently occur together, both forming isolated grains (peloids, coated grains) and autochthonous deposits. The same minerals appear as isolated grains in winnowed and reworked greensands and granular recent and ancient phosphorites (Glenn, 1990; Glenn

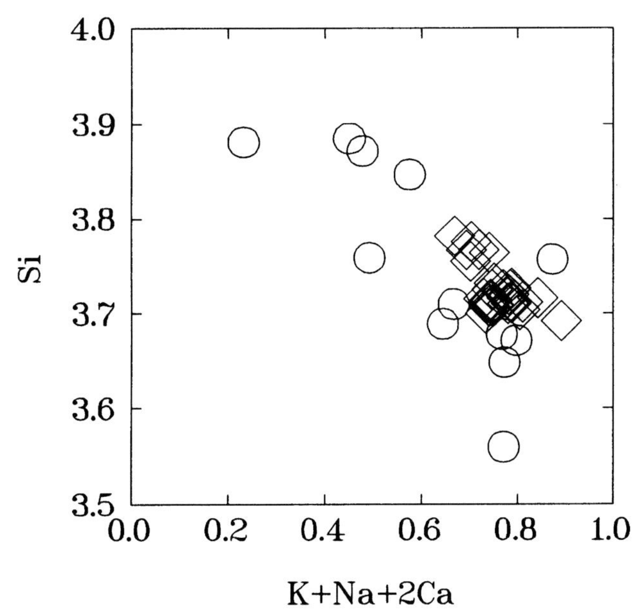

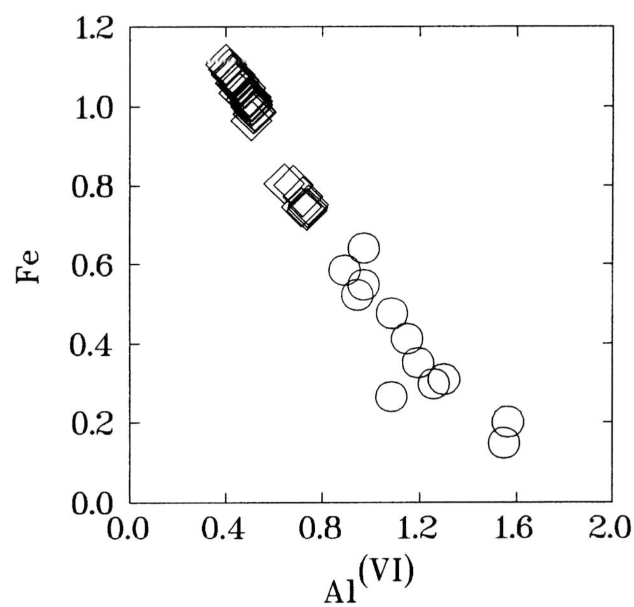

Fig. 17.—X-Y plots for montmorillonitic, beidellitic, and nontronitic substitutions in smectites and glauconites. Diamonds correspond to microprobe data obtained from glauconite analyses reproduced in Table 2, circles are HRTEM analyses for authigenic Fe-rich smectites published by Sánchez-Navas et al. (1998).

and Arthur, 1990; Zanin et al., 1990; Jarvis, 1992). They also constitute autochthonous deposits in crusts, nodules, and hardgrounds either in the crests of seamounts (Burnett et al., 1987) or in the deep sea (Usui, 1985; Glenn and Arthur, 1988; Lamboy, 1990a and b; Glenn and Kronen, 1993; Glenn et al., 1994). Authigenic clays are common components of phosphorites (Nathan and Soudry, 1982, Soudry, 1992; Martín-Algarra and Sanchez-Navas, 1995; Sánchez-Navas et al., 1998) as well as Fe-Mn oxyhydroxides (Glenn et al., 1994). Francolite, authigenic clays and iron oxyhydroxides (goethite, limonite) also commonly form discrete laminae in marine ooidal ironstones and ferruginous microbialites of the fossil record (Sturesson and Bauert, 1994; Burkhalter, 1995), and in phosphate and iron-manganese nodules and crusts of the recent deep ocean (Usui, 1979a, b, 1983/1984, 1985; Burnett et al., 1987).

The participation of bacteria in the precipitation of francolite in sedimentary phosphorites, some of them stromatolitic, has been proposed by many authors (O'Brien et al., 1981; Soudry and Champetier, 1983; Soudry and Lewy, 1988; Soudry, 1987, 1994; Abed and Fakhouri, 1990; Lamboy, 1990 a, b, 1993; Sisodia and Chauhan, 1990; Lewy, 1990; Föllmi and Garrison, 1991; Föllmi et al., 1992; Martín-Algarra and Vera, 1994; Krajewski et al., 1994; Glenn et al., 1994; Föllmi, 1996). In recent ovoidal and rod-like phosphorites that sometimes form filaments, apatite particles from phosphatic crusts similar to those found in the studied stromatolites have been interpreted as phosphatized bacteria (Lamboy, 1990a, 1993). Multilayered mats of chemosynthetic bacteria are found encased in heavy Mn-Fe deposits around hydrothermal deep-sea vents (Jannasch, 1984b, Jannasch and Mottl, 1985). Monty (1973), Greenslate (1974), Ehrlich (1975), and Krumbein (1983a), among many others, have also suggested the influence of bacteria in the reactions of Mn in deep-sea manganese nodules. Framboidal pyrite intimately mixed with francolite in phosphorites has also been interpreted as microbially influenced (Lamboy, 1990a). Pyrite concretions are usually interpreted as precipitated diagenetically by sulphur- and sulphate-reducing bacteria living buried within organic-rich sediments (Berner, 1970). Degens and Ittekkot (1982) have shown structures resembling bacteria surrounded by idiomorphic sulfide crystals.

The participation of Fe-oxidizing bacteria in the precipitation of Fe-oxide-hydroxides from marine hydrothermal solutions is well known, and a biogenic origin from hydrothermal solutions has also been proposed for certain recent ferruginous sediments (Holm, 1987a) for stromatolite-bearing Precambrian banded iron formations (Holm, 1987b) and for mesozoic ironstones (Burkhalter, 1995). The participation of hydrothermal solutions in the origin of some of the Jurassic Fe-Mn-rich stromatolites studied can also be reasonably invoked. Actually, some time intervals characterized by a common and basin wide occurrence of Jurassic Fe-Mn-rich and P-rich stromatolites in the Betic Cordillera (such as those of the Upper Bathonian and of the Upper Albian) coincide with periods of basin revolution associated with active faulting, platform drowning, changes in subsidence, and reactivation of the submarine volcanism (and hydrothermalism?) in wide areas of the Subbetic realm and in other alpine-Mediterranean and Atlantic regions (Bernoulli and Jenkyns, 1974; Azéma et al., 1979; García-Hernández et al., 1980; Martín-Algarra and Vera, 1994).

In the studied stromatolites, the trapping and binding mechanism is definitely subordinated to the authigenesis of mainly noncarbonate minerals and subordinately to that of calcite. Such intimate textural relationships exist among them that some common biogeochemical controls for their precipitation must be invoked. These relationships include alternations of thin Fe-rich and P-rich bands in type b stromatolites and the presence of minor amounts of phases other than francolite within pristine P-rich laminae, such as clays, pyrite, and Fe oxyhydroxides in those of type a.1. These relationships—and the common occurrence of amorphous substances, especially when they present a very small crystal size, which is the most common case—clearly indicate a common origin for all of these phases (Martín-Algarra and Sánchez-Navas, 1995; Sánchez-Navas et al., 1998). Moreover, the existence of envelopes made of amorphous Fe-Si-Al-(Mn) oxyhydroxides and Fe-rich clays surrounding francolite crystals and forming bacteria-like structures in all types described, especially in types a and b of different ages, indicates that authigenesis took place around bacteria by biomineralization processes of the microbes that formed (or degraded) the microbial mats, probably via precipitation of amorphous precursor phases on or near the bacterial wall. Thus, the stromatolites studied must be seen as natural laboratories for the authigenesis of many marine minerals via bacterial chemosynthesis.

Precursor Phases

The participation of bacteria in the precipitation of crystalline mineral phases has been shown by laboratory studies of microbial cultures (Lowenstam, 1972; 1981; Krumbein, 1983b; Lucas and Prévôt, 1984, 1985; Hirschler et al., 1990 a, b; and many others), and after analysis of natural samples obtained from deep oceans to soils (Greenslate, 1974; Ehrlich, 1975, 1990; Holms, 1987 a, b; Jannasch, 1984 a, b; Ferris et al., 1986, 1987, 1988, 1989 a, b; Urrutia and Beveridge, 1993, 1994, 1995; Dietrich and Jacob, 1996). Nevertheless, many bacterially precipitated minerals may be preceded by precursor noncrystalline or poorly crystalline phases. Amorphous phosphate minerals have been described both in synthetic and natural phosphates (McConnell and Ward, 1978; Posner et al., 1984; LeGeros and LeGeros, 1984). From a kinetic point of view, amorphous calcium phosphate constitutes a precursor of the apatite (Eanes et al., 1965; Termine and Posner, 1970, Termine et al., 1970; Nylen et al., 1972). See also Van Capellen and Berner, 1991, and Krajewski et al., 1994, for a recent review about kinetics of phosphate precipitation from marine waters in simulated natural environments. Extracellular formation of fine-grained, compositionally heterogeneous, noncrystalline and poorly crystalline Fe-Si-Al, phosphate, and metal precipitates around bacteria extracted from soils recently has been demonstrated in the laboratory (Urrutia and Beveridge, 1993, 1994). These precipitates slowly evolve to crystalline phases, among which short, range-ordered aluminosilicates have been identified (Urrutia and

Beveridge, 1995). Holm (1987b) described how Fe-oxidizing bacteria produce amorphous Fe-oxide-hydroxide, and explained the presence of apatite-rich bands in some banded iron-formations as a result of primary precipitation of calcium phosphate by scavenging of FeOOH, from which apatite separates in the diagenesis when goethite and other iron oxyhydroxides transform into hematite. Non-crystalline β-FeOOH.Cl_n (akaganeite) constitutes the primary precipitate that later transforms to goethite or hematite depending on the temperature (Holm, 1985). A colloidal form also precedes the two line-form δ-MnO_2 in manganese nodules (Usui, 1979a).

The existence of amorphous calcium phosphate and Fe-Si-Al amorphous substances closely associated with crystalline phases of very small crystal size (francolite, Fe-rich montmorillonites, and goethites) observed by the authors using TEM in phosphate stromatolites of Upper Jurassic age (Martín-Algarra and Sánchez-Navas, 1995, 1996; Sánchez-Navas et al., 1996, 1998) demonstrate that the precipitation of most authigenic minerals observed in the stromatolites studied was preceded by the formation of amorphous precursors. These were precipitated extracellularly on bacterial surfaces living on the stromatolite mat. Sánchez-Navas et al. (1998) concluded that after dehydration and recrystallization, P-Fe-Si-Al-rich jelly-like substances concentrated and evolved to poorly crystalline clays and amorphous Fe-Si-Al oxyhydroxides. The latter fossilize bacterial cell walls and probably represent a residue of the primary gel, in the same way that Eggleton (1987) has demonstrated how amorphous Fe-Si-Al oxyhydroxides, formed by weathering of silicate minerals, evolve to clay minerals (smectite or kaolinite, according to the gel composition).

Similar conclusions are reached for phosphate and iron-manganese stromatolites of other ages and sites studied here. In all cases the fixation of chemical components by organisms within microbial mats led to the formation of primary (stromatolitic) authigenic phases: francolite, iron and manganese oxy-hydroxides, glaucony minerals (Fe-rich smectites in particular), pyrite, and barite. In some cases, bacteria seem to have precipitated crystalline phases directly, such as barite, or have provided favorable sites for nucleation of barite crystals. Kinetic constraints favor, however, the precipitation of amorphous precursors instead of direct crystallization of the other minerals from a solution. Besides, metastable phases such as noncrystalline Fe-Si-Al oxyhydroxides (Eggleton, 1987) can persist for a long time far from equilibrium due to the slow chemical reaction rates at low temperatures (Berner, 1984). Then, the sheaths of stromatolite-forming bacteria were reactive substrates able to adsorb and concentrate phosphate and metals from degraded organic matter and seawater, providing supersaturation levels in PO_4^{3-} and other ions required to precipitate an amorphous precursor for francolite, possibly amorphous calcium phosphate (ACP: Martín-Algarra and Sánchez-Navas, 1995) and other precursor phases for other authigenic minerals. As well as for amorphous phosphate precipitates studied in the laboratory (Termine et al., 1970), the high content of organic substances in bacterial sheaths, especially polysaccharides, would probably favor stabilization of these metastable phases.

Physicochemical Evolution During Mineral Authigenesis

According to most authors, marine authigenesis in phosphorites results from early diagenetic processes that involve the precipitation, along with francolite, of other minerals such as glauconite, chamosite, pyrite, goethite, calcite, dolomite, siderite, and sulphates. Their mineralogenetic significance led to establishment of the physicochemical evolution of the sedimentary environment (Berner, 1981; Glenn and Arthur, 1988; Zanin et al., 1990; Sturensson and Bauer, 1994; etc.). The authigenesis of these minerals seems to be preceded by the formation, under oxic conditions, of precursor phases that later evolve through increasingly reduced conditions during burial, favored by degradation of organic matter and microbial sulphate reduction (Odin and Matter, 1981; Nathan and Soudry, 1982; Krajewski, 1984; Soudry, 1992; Glenn and Arthur, 1988; Sturesson and Bauert, 1994; Martín-Algarra and Sánchez-Navas, 1995; Sánchez-Navas et al., 1998). The scavenging of phosphate and manganese ions by Fe-oxyhydroxides under oxic conditions (Berner, 1981) explains the occurrence of Fe-Mn minerals in phosphorites (Cook and McElhinny, 1979; Holm, 1987b; O'Brien et al., 1990; Heggie et al., 1990; Burkhalter, 1995; Maynard, 1983).

The mineral associations described above of type a.1 (francolite + Fe-rich smectite + non-crystalline Fe-Si-Al oxy-hydroxides + goethite/hematite), of type b (francolite + goethite/hematite + clay minerals), and especially that of type c (goethite/hematite + pyrolusite + Fe-Mn-Al noncrystalline substances + francolite) formed mainly under oxic conditions that, in the case of associations type a.1 and type b, evolved to postoxic during early diagenesis. Association type a.2 (francolite + glaucony + goethite/hematite + barite + pyrite + noncrystalline substances) is, however, indicative of poorly oxic to anoxic conditions.

In the stromatolites studied, the organic matter contribution (directly from the primary production of the microbial mat, or indirectly, by settling of particulate organic matter derived from the pelagic rain from surface waters) and the degree of oxygenation of bottom waters are important factors for mineral authigenesis. Chemical elements (P, Ca, Fe, Si, Al, Mn, S, Ba, and others) were concentrated within the mat either from degrading organic matter, trapped particulate sediments (both bioclasts and siliciclasts), or directly from seawater. Organic gels secreted by the stromatolite-forming microbes played an essential role in their concentration. Although P may be derived from seawater and from submarine volcanic sources, an important cellular origin may be invoked for it, because bacterial cytoplasm and cell walls are especially rich in P (Urrutia and Beveridge, 1994). Enzymes from heterotrophic bacteria would rapidly release phosphate ions from organic matter (Hirschler et al., 1990b), producing high supersaturation levels necessary for amorphous calcium phosphate precipitation (Krajewski et al., 1994) within the microbial mat. Colloidal wall-Fe complexes, forming a FeOOH-coated surface surrounded by organic compounds rich in mucopolysaccharides, would play an important role in the retention of elements that are present in very low amounts in seawater (such as Al, Mn, and Si) through a chemisorption process. The biocorrosion of detrital clay particles is an additional source of Fe, Mn, Si, and especially Al, that is necessary to form Fe-Al-Si gels prior to authigenic smectites.

The precipitation of noncrystalline Fe-Si-Al-oxyhydroxides, amorphous calcium phosphate, and Mn-oxyhydroxides took place under oxidizing conditions at the mat-water interface. These were especially persistent in stromatolites characterized by the mineral association of type c. Oxygen consumption as a consequence of decomposition of bacterial organic matter led to post-oxic conditions within the fossilized mat. The transformation of amorphous precursors into crystalline phases took place under such reducing conditions. Thus, amorphous calcium phosphate transformed to francolite, and part of the noncrystalline Fe-Si-Al-oxyhydroxides transformed to Fe-rich smectite (types a.1 and b). This agrees with experimental work on the low-temperature synthesis of Fe-rich smectites from silica and Fe-hydroxides, which is only possible under reducing conditions (Harder, 1978), and with the common occurrence of authigenic Fe-rich smectites in close association with pristine phosphate of Neguev phosphorites (Nathan and Soudry, 1982; Soudry, 1992). The release of iron from oxyhydroxides, combined with the available sulphur associated with sulphate-reducing bacteria, formed pyrite in few sites of the structure of stromatolites with a type a.2 mineral association where more reducing (sulphidic) conditions were reached.

Fe-rich smectite constitutes the chemical precursor of glauconite and is the first step in the glauconitization process. Such close genetic relationships between authigenic Fe-rich smectite formed around bacteria, and glauconitic mica occurring as isolated grains within the sediment associated with the phospate stromatolites is supported by the chemical trend reproduced in Figure 17. An intense reworking, which is commonly detected in glauconite-bearing sediments (Odin and Matter, 1981; Glenn and Arthur, 1988), and repeated exhumation of precursor authigenic clays with change from post-oxic to oxic environments and exposure to bottom waters rich in K, would favor the maturation of Fe-rich smectites to glauconites. Submarine erosion and winnowing due to current action and burrowing recycle authigenic minerals. Thus, glaucony pellets and intraclasts of other authigenic minerals (phosphate, hematite, and Mn-oxyhydroxides) were incorporated into stromatolite structure.

Although Fe-Mn-Al oxyhydroxides are synsedimentary, textures observed in Figures 12-15 indicate oxidation, cation-leaching, and recrystallization processes during early diagenesis. Oxidation of iron led to the formation of goethite-limonite. Leaching of Si, Mg, and K principally affected authigenic clay minerals that formed as part of the primary phases of the stromatolites, along with amorphous Fe-Mn-Si-Al oxyhydroxides. Recrystallization processes gave rise to the increase in crystal size observed in these Jurassic samples. Oxidizing conditions prevalent during reworking and erosion would have renewed precipitation of Mn, determined late Mn enrichments, and favored the occurrence of well-crystallized Mn-oxyhydroxides. Anyway, this does not contradict the hypothesis of a bacterially mediated primary precipitation of Mn associated with iron oxyhydroxides. Although the bacteriology of Mn precipitates is not exactly known, a certain bacterial catalytic effect is known for the precipitation of Mn oxides (Maynard, 1983).

Sedimentological and Paleogeographical Significance

The physicochemical significance of the mineral associations found in the stromatolites discussed above is also supported by the nature of the sediments vertically and laterally associated with the stromatolite-bearing surfaces in outcrop, and by that of the successions laterally related to and coeval with the stromatolite-bearing intervals in the same basin in the zones where pelagic sedimentation was continuous and no discontinuity surfaces formed. In the case of stromatolites showing the mineral associations type b and especially c, such sediments are systematically red, sometimes nodular, and always strongly condensed limestones. This observation supports an essentially oxic sedimentary (and postsedimentary) environment. In the case of stromatolites with the association type a.1, the related facies are brown condensed limestones, that formed under oxic to suboxic conditions, and iron minerals were not completely oxidized. In the case of stromatolites with the mineral association type a.2, the only one where pyrite has been systematically found, the facies associated with the stromatolites are usually light grey to green marls and limestones. These facies laterally change to darker, organic matter-rich sediments, as in the case of the Albian stromatolites of the Penibetic, which were coeval with widespread oceanic anoxic events (Schlanger and Jenkyns, 1976; Jenkyns, 1980).

Open marine, relatively deep aphotic settings are usually relatively stable and quiet water environments with rather uniform sedimentary conditions. These are to be expected in areas of the sedimentary basin laterally related to zones where these stromatolites formed. There, relatively poorly diversified pelagic facies associations are dominated by fine-grained sediments with only limited compositional and textural facies changes that produce, at most, monotonous rhythmically bedded successions, formed by marly-limy and/or clayey-radiolaritic alternations (in subsiding troughs) or by cephalopod nodular limestones with *Rosso Ammonitico* facies (in pelagic swells). In the stromatolites studied, however, many different textural elements and minerals are involved. This reveals that many different mechanical and, especially, chemical processes acted successively and overlapped each other with some chronological order during geological time, in the same sedimentary microenvironment. These alternating processes caused the fine, rhythmically laminated architecture of the stromatolites and cycles of accretion of different types. In conclusion, environments where these stromatolites formed changed in time, and stromatolite lamination actually represents a high-resolution microstratigraphic record of these changes during a period of extreme basin starvation and condensation.

One possible explanation for accretion rhythms of the smallest size is that microbially mediated precipitation of authigenic minerals, especially of phosphate, was governed by autocyclic processes related to ecological succession of benthic microbial communities that progressively modified the physicochemistry of the microbial mat microenvironment, thus creating conditions alternatively favorable and unfavorable for the fixation of different chemical elements to the mat and/or for the authigenesis of diverse minerals. Another more probable hypothesis that does not preclude the first one is that stromatolite rhythms of different scales are due to allocyclic processes as a response to cyclic variations in the sedimentary environment related to changes in sediment contribution from surface waters and in paleoceanography that alternatively favored or inhibited microbial accretion. Variations in sediment supply from surface waters due to changes in productivity and bottom current activity, probably climatically controlled, seem to have played a major role in the accretion of the stromatolites. Relatively high inputs of particulate sediments would have hindered stromatolite accretion, forcing the development of dendritic microstromatolitic edifices, which are systematically associated with sediment-rich intervals. Besides, the contribution of nutrients via particulate organic matter probably played a main role in controlling the nature of the microbial community and in triggering stromatolite accretion due to heterotrophic bacteria. Another essential environmental parameter was certainly the oxygenation of bottom waters, which should be related to bottom current action and to the total amount of organic matter to be oxidized at the sediment-water interface either if this organic matter is derived from increased productivity of surface waters or from the microbial mat itself.

To avoid burial under sediment particles and microbial precipitates, stromatolite accretion requires a certain motility of the microbes that build it. The motility of nonphotosynthetic bacteria such as those that probably accreted the stromatolites is, however, rather limited. It could be due to chemotactic responses or, more probably, it is the natural passive result of the accretion of the whole microbial community. If enough nutrients were not supplied to microbial mats, which is probably the case for stromatolites accreted in starved pelagic environments, this low motility would result in strong competition for nutrients. This in turn would have favored concentration around bacteria of the chemical elements necessary to precipitate the observed authigenic minerals. A very slow sedimentary rate is essential for the structure to be formed and preserved, because high inputs in biogenic or terrigenous sediments to the seafloor would bury microbial communities and would hinder stromatolite accretion. This interpretation is fully supported by geological evidence, because these stromatolites are systematically associated with important stratigraphic lacunae. Moreover, trapped coccoliths allow for some calculations of average accretion rate of these structures, which is on the order of 0.5-1 mm per thousand years (González-Donoso et al., 1983; Martín-Algarra, 1987; Martín-Algarra and Vera, 1994; Vera and Martín-Algarra, 1994; Martín-Algarra and Sánchez-Navas, 1995).

Finally, some hypothesis can be made relative to bottom current action and its incidence in this oscillating microevironment. Most morphological, architectural, and textural features of the stromatolites and associated sediments suggest an essentially quiet-water environment of formation (fine-grained sediments, delicate microstromatolite edifices, asymmetrical development of laminae in oncoids, etc.). Nevertheless, the existence of periods of stronger and more persistent current action is supported by many features: the common existence of microunconformities, brecciation episodes, clasts formed by stromatolite fragments and allochems (peloids and intraclasts) made of authigenic minerals that were primarily formed within the stromatolite structure and/or that needs reworking processes to mature (such as glauconite) as well as changing styles of stromatolite lamination. Upward evolution of laminae macroscopically visible in stromatolite crusts, and especially in oncoids, can be interpreted as a consequence of an upward-decreasing energy in bottom currents (Martín-Algarra and Vera, 1994): flat, planar or undulose laminae were probably formed under relatively important bottom currents that favored repeated rolling and winnowing of pelagic sediments, and that hindered upward accretion of microbes that, by consequence, were forced to produce laterally continuous mats with low synoptic relief. Decreasing energy of bottom currents favored outward accretion of microbial mats, allowing the development of columns and pseudocolumns and finally giving rise to microstromatolitic structures during periods of very low current action or quiet waters.

CONCLUSIONS

(1) Sediment-starved zones of pelagic swells in the Mesozoic of Betic Cordillera were favored sites for microbially mediated marine authigenesis. This phenomenon was systematically related to the accretion of biosedimentary structures at the sediment-water interface. Bacterial communities accreted finely laminated stromatolites of different sizes and morphologies (crusts and pavements, oncoids and microstromatolites) at the sediment-water interface and/or within interstices in the topmost sediment including previous erosion cavities, the interior of macrofossil shells, syneresis cracks, burrows, and features due to biocorrosion or dissolution.

(2) Stromatolite accretion was essentially due to the precipitation of authigenic minerals, but trapping and binding of fine-grained siliciclastics and pelagic biogenic carbonates and encrustation by benthic foraminifera are systematically detected in stromatolites of all ages and sites. Taking into account their morphology, architectural features, and microbially mediated genesis by trapping and binding of pelagic sediments and biogeochemical precipitation of authigenic minerals, as well as their systematic association with fine-grained, pelagic macro- and microfossils and fine-grained sediments, these peculiar biosedimentary structures accreted by bacteria at the sediment-water interface in the open ocean must be called pelagic stromatolites.

(3) Phosphates (francolite) and Fe-Mn oxyhydroxides (goethite, hematite, and pyrolusite) are the most common authigenic minerals found in these stromatolites, but authigenic clays precursory of glauconite (Fe-rich smectites), glauconite, pyrite, barite, carbonates and other authigenic minerals are always present. Their precipitation, which was bacterially mediated, was related to chemosynthetic bacteria.

(4) Some minerals directly precipitated on the microbial mat but more frequently precipitation of noncrystalline precursor phases on bacterial surfaces took place in oxic environments. These phases changed later into crystalline phases during early diagenesis under postoxic conditions, that

favored the formation of francolite, Fe-rich smectites and probably hematite and barite. When sulphidic conditions were reached within the buried microbial mat the pyrite also formed. The persistence of oxic conditions in the sedimentary environment determined the formation of mineral associations rich in Fe-Mn- oxyhydroxides. The repeated reworking, winnowing and re-exposure of the stromatolites and of the associated sediments to oxic, marine bottom waters rich in K, preceded by periods of burial with generation of reducing pore waters, favored the maturation of Fe-rich smectites to glauconite.

(5) The source of the necessary elements may have been seawater, and their concentration may have occurred directly on polysaccharide-rich bacterial surfaces by autholithotrophic bacteria. But more probably the most important source of elements was organic matter that fell to the seafloor from productive surface waters, along with detrital clay and carbonate particles. This organic matter served as food for the stromatolite-building benthic microbial communities that, by consequence, may have been dominated by heterotrophic bacteria. The influence of hydrothermal solutions certainly was remarkable in some cases of basin-wide periods of formation of P-Fe-Mn stromatolites, which coincide with periods of active faulting, sudden changes in subsidence, and submarine volcanism in large areas of the Western Tethys.

(6) Mineral authigenesis related to open marine stromatolites is independent of age: similar associations of authigenic minerals are found in different sites and ages, but they always formed in starved depositional environments where similar microbial chemosynthesis processes occurred. During periods of basin starvation, these processes were especially favored in aphotic, moderately deep pelagic swells (probably some hundred meters deep at most) that received hardly any biogenic or siliciclastic sediment and that probably were swept away by bottom currents. This paleogeographic frame allowed hardground development and colonization of the sea-bottom by microbial communities dominated by chemosynthetic bacteria and encrusting foraminifera.

(7) Stromatolite lamination reveals that sediment-starved pelagic environments were not stable, but instead changing sedimentary environments that were affected by the subtle changes that controlled sediment fluxes and the evolution of physicochemical parameters both in the productive surface waters and at the sediment-water interface. Questions remaining unanswered are those of the duration of stromatolite rhythms, the periodicity of the successions of events that determined each type of rhythm, and their ultimate causes. Resolving these problems will require detailed micropaleontological studies of coccoliths trapped and bound within stromatolite laminae, along with accurate identification and quantification of the complete succession of rhythms through optical image analysis and semiquantitative chemical methods. This study must be followed by detailed microstratigraphic correlation first between rhythmic successions recognized in different parts of the same stromatolite sample and later between different samples of the same and of progressively more distant outcrops.

(8) The opening of alpine-Mediterranean paleomargins determined a generalization of pelagic conditions to wide areas, and a general starvation of sedimentary environments. Under such conditions, microbially generated Fe-Mn-rich and P-rich nodules and crusts were formed in Mesozoic pelagic condensed successions of the Betic Cordillera and other Alpine-Mediterranean domains. Their genesis was chronologically related to periods of major paleogeographical revolution associated with episodes of plate reorganization and tectonoeustatic change. These revolutions produced strong modifications in wide areas of the Western Tethys, where narrow seaways opened throughout the Jurassic-Cretaceous period connected this ocean and the newly formed central Atlantic. During such periods, sediment supply and distribution, subsidence rates and general paleoceanographic controls (current system, submarine hydrothermalism, and bottom oxygenation) were seriously modified.

ACKNOWLEDGMENTS

The authors thank Craig Glenn for his understanding and patience throughout the preparation of this paper. M. A. Hidalgo, M. A. González and I. Guerra (Centro de Instrumentación Científica, University of Granada) are also acknowledged for their help during EPM and SEM study, as well as M. Martín-Martín for his help with computer drawings. The critical and helpful revisions of Karl Föllmi and one anonymous reviewer are gratefully acknowledged. This is a contribution of the DGICYT Research Projects n°1 PB96-1430 and n1 PB96-1383 and the Research Group n°1 4089 of the Junta de Andalucía.

REFERENCES

ABED, A. M., AND FAKHOURI, K., 1990, Role of microbial processes in the genesis of Jordanian Upper Cretaceous phosphorites, *in* Notholt, A. J. G., and Jarvis, I., eds., Phosphorite Research and Development: Oxford, Geological Society Special Publication 52, p. 193-203.

AZÉMA, J., FOUCAULT, A., FOURCADE, E., GARCÍA-HERNÁNDEZ, M., LINARES, A., LINARES, D., LÓPEZ-GARRIDO, A. C., RIVAS, P., AND VERA, J. A., 1979, Las Microfacies del Jurásico y Cretácico de las Zonas Externas de las Cordilleras Béticas: Granada, Spain, Secretariado de Publicaciones, Universidad de Granada, 83 p.

BERNER, R. A., 1970, Sedimentary pyrite formation: American Journal of Science, v. 268, p. 1-23.

BERNER, R. A., 1981, A new geochemical classification of sedimentary environments: Sedimentary Petrology, v. 51, p. 359-365.

BERNER, R. A., 1984, Kinetics of weathering and diagenesis, *in* Lasaga, A. C., and Kirkpatrick, R. J., eds., Kinetics of geochemical processes: Mineralogical Society of America Reviews in Mineralogy, v. 8, p. 111-134.

BERNOULLI, D., AND JENKYNS, H. C., 1974, Alpine Mediterranean and central Atlantic Mesozoic facies in relation to the early evolution of the Tethys, *in* Dott, R. H., and Shaver, R. H., eds., Modern and Ancient Geosynclinal Sedimentation: Tulsa, Society for Sedimentary Geology (SEPM) Special Publication 19, p.129-160.

BLENDINGER, W., 1991, Upper Triassic (Norian) cephalopod limestone of the Hallstatt-type, Oman: Sedimentology, v. 38, p. 223-242.

BOURBON, M., 1980, Evolution d'un secteur de la marge nordTethysienne en milieu pélagique: la Zone Briançonnaise près de Briançon entre le début du Malm et l'Eocène inférieur: Master's Thesis, University Louis Pasteur,

Strasbourg, 335 p.

BURKHALTER, R. M., 1995, Ooidal ironstones and ferruginous microbialites: Origin and relation to sequence stratigraphy (Aalenian and Bajocian, Swiss Jura mountains): Sedimentology, v. 42, p. 57-74.

BURNETT, W. C., CULLEN, D. J. AND MCMURTRY, G. M., 1987, Open-ocean phosphorites: In a class by themselves?, *in* Teleki, P. G., Dobson, M. R., Moore, J. R., and von Stackelberg, U., eds., Marine Minerals, Advances in Research and Source Assessment, NATO ASI Series C, v. 194: Dordrecht, Reidel, p. 119-134.

CASTRO, J. M., AND RUIZ-ORTIZ, P. A., 1991, Nivel condensado con estromatolitos pelágicos en el Cretácico de la Sierra de Estepa (Subbético Externo, Prov. Sevilla): Revista Sociedad Geológica de España, v. 4, p. 305-319.

CLARI, P. A., DELLA PIERRE, F., AND MARTIRE, L., 1995, Discontinuities in carbonate successions: Identification, interpretation and classification of some Italian examples: Sedimentary Geology, v. 100, p. 97-127.

COOK, P. J., AND MCELHINNY, M. W., 1979, A reevaluation of the spatial and temporal distribution of sedimentary phosphate deposits in the light of plate tectonics: Economic Geology, v. 74, p. 315-330.

CHAFETZ, H. S., PATRICK F. R., AND UTECH, N. M., 1991, Microenvironmental controls on mineralogy and habit of $CaCO_3$ precipitates: An example from an active travertine system: Sedimentology, v. 38, p. 107-126.

DÉFARGE, C., TRICHET, J., JAUNET, A. M., ROBERT, M., TRIBBLE, J., AND SANSONE, F., 1996, Texture of microbial sediments revealed by crioscanning electron microscopy: Journal of Sedimentary Research A, v. 66, p. 935-947.

DEGENS, E. T., AND ITTEKKOT, V., 1982, In situ metal staining of biological membranes in sediments: Nature, v. 298, p. 262-264.

DELAMETTE, M., 1990, Aptian, Albian and Cenomanian microbialites from the condensed phosphatic deposits of the Helvetic shelf, Western Alps: Eclogae Geologicae Helvetiae, v. 83/1, p. 99-121.

DIETRICH, S., AND JACOB, K. H., 1996, Self-organization of recent rhythmic iron-manganese precipitation in underground mines in Harz mountains: Geologische Rundschau, v. 85, p. 29-37.

EANES, E. D., GILLESEN, I. H., AND POSNER, A. S., 1965, Intermediate states in the precipitation of hydroxyapatite: Nature, v. 208, p. 365-367.

EHRLICH, H. L., 1975, The formation of ores in the sedimentary environment of the deep sea with microbial participation: The case for ferromanganese concretions: Soil Science, v. 119, p. 36-41.

EHRLICH, H. L., 1990, Geomicrobiology: New York, Marcel Decker, 646 p.

EGGLEton, R. A., 1987, Noncrystalline Fe-Si-Al-oxyhydroxides: Clays and Clay Minerals, v. 35, p. 29-37.

FERRIS, F. G., BEVERIDGE, T. J., AND FYFE, W. S., 1986, Iron-silica crystallite nucleation by bacteria in a geothermal sediment: Nature, v. 320, p. 609-611.

FERRIS, F. G., FYFE, W. S., AND BEVERIDGE, T. J., 1987, Bacteria as nucleation sites for authigenic minerals in a metal-contaminated sediment: Chemical Geology, v. 63, p. 225-232.

FERRIS, F. G., FYFE, W. S., AND BEVERIDGE, T. J., 1988, Metallic ion binding by *Bacillus subtilis*. Implications for the fossilization of microorganisms: Geology, v. 16, p. 149-152.

FERRIS, F. G., SCHULTZE, S., WHITTEN, T. C., FYFE, W. S., AND BEVERIDGE, T. J., 1989a, Metal interactions with microbial biofilms in acidic and neutral pH environments: Applied Environmental Microbiology, v. 55, p. 1249-1257.

FERRIS, F. G., SHOTYK, W., AND FYFE, W. S., 1989b, Mineral formation and decomposition by microorganisms, *in* Beveridge, T. J., and Doyle, R. J., eds., Metal Ions and Bacteria: New York, Wiley, p. 413-441.

FOLK, R. L., 1993, SEM imaging of bacteria and nannobacteria in carbonate sediments and rocks: Journal of Sedimentary Petrology, v. 63, p. 990-1000.

FOLK, R. L., CHAFETS, H. S., AND TIEZZI, P. A., 1985, Bizarre forms of depositional and diagenetic calcite in hot spring travertines, central Italy, *in* Schneidermann, N., and Harris, P. M., Carbonate Cements: Tulsa, Society for Sedimentary Geology (SEPM) Special Publication 36, p. 349-369.

FOLK, R. L., AND LYNCH, F. L., 1997, The possible role of nannobacteria (dwarf bacteria) in clay-mineral diagenesis and the importance of careful preparation in high-magnification SEM study: Journal of Sedimentary Research, v. 67A, p. 583-589.

FÖLLMI, K. B., 1989, Evolution of Mid-Cretaceous Triad (Platform Carbonates, phosphatic sediments and pelagic carbonates) along the Northern Tethys margin, *in* Bhattacharji, S., Friedman, G. M., Neugebauer, H. J., and Seilacher, A., eds., Lecture Notes in Earth Sciences, v. 23: Berlin, Springer-Verlag, p. 1-153.

FÖLLMI, K. B., 1996, The phosphorous cycle, phosphogenesis and marine phosphate-rich deposits: Earth Science Review, v. 40, p. 55-124.

FÖLLMI, K. B., AND GARRISON, R. E., 1991, Phosphatic sediments, ordinary or extraordinary deposits? The example of the Miocene Monterey Formation (California), *in* Müller, D., Weissert, H., and McKenzie, J. A., eds., Controversies in Modern Geology: New York, Academic Press, p. 55-84.

FÖLLMI, K. B.; GARRISON, R. E.; RAMÍREZ, P. C.; ZAMBRANO-ORTIZ, F.; KENNEDY, W. J. AND LEHNER, B. L., 1992, Cyclic phosphate-rich successions in the upper Cretaceous of Colombia: Palaeogeography, Palaeoclimatology, and Palaeoecology, v. 93, p. 151-182.

GARCÍA-CASCO, A., SÁNCHEZ-NAVAS, A., AND TORRES-ROLDÁN, R. L., 1993, Disequilibrium decomposition and breakdown of muscovite in high P.T gneisses, Betic alpine belt (southern Spain): American Mineralogist, v. 78, p. 158-177.

GARCÍA-CERVIGÓN, A., MARTÍN-ALGARRA, A., MONTEALEGRE, L., AND VERA, J. A., 1986-1987, Estromatolitos pelágicos fosfatados relacionados con discontinuidades estratigráficas en el Cretácico del Penibético (provincia de Málaga): Acta Geológica Hispánica, v. 21-22, p. 361-371.

GARCÍA-HERNÁNDEZ, M., LÓPEZ-GARRIDO, A. C., RIVAS, P., SANZ DE GALDEANO, C., AND VERA, J. A., 1980, Mesozoic palaeogeographic evolution of the external zones of the Betic Cordillera: Geologie en Mijnbouw, v. 59, p. 155-168.

GIRESSE, P., WIEWIÓRA, A., AND LACKA, B., 1998, Processes of Holocene ferro-manganese-coated grains (oncolites) in the nearshore shelf of Cameroon: Journal of Sedimentary Research, v. 68A, p. 20-36.

GLENN, C. R., 1990, Depositional sequences of Duwi, Sibâîya and Phosphate Formations, Egypt: Phosphogenesis and glauconitization in a Late Cretaceous epeiric sea, *in* Notholt, A. J. G., and Jarvis, I., eds., Phosphorite Research and Development: Oxford, Geological Society Special Publication 52, p. 205-222.

GLENN, C. R., AND ARTHUR, M. A., 1988, Petrology and major element geochemistry of Peru margin phosphorites and associated diagenetic minerals: Authigenesis in modern organic-rich sediments: Marine Geology, v. 80, p. 231-267.

GLENN, C. R., FÖLMI, K. B., RIGGS, S. R., BATURIN, G. N., GRIMM, K. A., TRAPPE, J., ABED, A. M., GALLI-OLIVER, C., GARRISON, R. E., ILYAN, A., JEHL, C., ROHRLICH, V., SADAQAH, R. M., SCHIDLOWSKI, M., SHELDON, R. E., AND SIEGMUND, H., 1994, Phosphorous and phosphorites: Sedimentology and environments of formation: Eclogae Geologicae Helvetiae, v. 87, p. 747-788.

GLENN, C. R., AND KRONEN, Jr., J. D., 1993, Origin and significance of late Pliocene phosphatic hardgrounds on the Queensland Plateau, Northeastern Australian Margin, *in* McKenzie, J. A., et al., eds., Proceedings of the Ocean Drilling Program, Scientific Results: v. 133, p. 525-534.

GONZÁLEZ-DONOSO, J. M., LINARES, D., MARTÍN-ALGARRA, A., REBOLLO, M., SERRANO, F., AND VERA, J. A., 1983, Discontinuidades estratigráficas durante el Cretácico en el Penibético (Cordillera Bética): Estudios Geológicos, v. 39, p. 71-116.

GREENSLATE, J., 1974, Microorganisms participate in the construction of manganese nodules: Nature, v. 249, p.181-183.

GUERRERA, F., MARTÍN-ALGARRA, A., AND PERRONE, V., 1993, Late Oligocene-Miocene syn-/ late-orogenic successions in western and central Mediterranean chains from the Betic Cordillera to the southern Apennine: Terra Nova, v. 5, p. 525-544

GUO, L., AND RIDING, R., 1992a, Microbial carbonates in uppermost Permian reefs, Sichuan Basin, southern China: Some similarities with recent travertines: Sedimentology, v. 39, p. 37-53.

GUO, L., AND RIDING, R., 1992b, Aragonite laminae in hot water travertine crusts, Rapolano Terme, Italy: Sedimentology, v. 39, p. 1067-1079.

HARDER, H., 1978, Synthesis of iron layer silicate minerals under natural conditions. Clays and Clay Minerals, v. 26, p. 65-72.

HEGGIE, D. T., SKYRING, G. W., O'BRIEN, G. W., REIMERS, C., HERCZEG, A., MORIARTY, D. J. W., BURNETT, W. C., AND MILNES, A. R., 1990, Organic carbon cycling and modern phosphorite formation on the East Australian continental margin: An overview, *in* Notholt, A. J. G., and Jarvis, I., eds., Phosphorite Research and Development: Oxford, Geological Society Special Publication 52, p. 87-107.

HIRSCHLER, A.; LUCAS, J., AND HUBERT, J. C., 1990a, Bacterial involvement in apatite genesis: FEMS Microbiology Geology, v. 73, p. 211-220.

HIRSCHLER, A., LUCAS, J., AND HUBERT, J. C., 1990b, Apatite genesis: A biologically induced or biologically controlled mineral formation process?:

Geomicrobiological Journal, v. 7, p. 47-57.

HOLM, N. G., 1985, New evidence for a tubular structure of β-iron (III) oxide hydroxide–akaganéite: Origin of Life, v. 15, p. 131-139.

HOLM, N. G., 1987a, Biogenic influences on the geochemistry of certain ferruginous sediments of hydrothermal origin: Chemical Geology, v. 63, p. 45-57.

HOLM, N. G., 1987b, Possible biological origin of banded iron-formations from hydrothermal solutions: Origin of Life, v. 17, p. 229-250.

HUCKRIEDE, H., 1996, Lower Carboniferous stratiform iron-manganese mineralizations (Rheinisches Schiefergebirge, Germany): Products of submarine hydrothermal activity and diagenetic manganese redistribution: Geologie en Mijnbouw, v. 75, p. 57-68.

JANNASCH, H. W., 1984a, Chemosynthesis: The Nutritional Basis for Life at Deep-Sea Vents: Oceanus, v. 27, p. 73-77.

JANNASCH, H. W., 1984b, Microbial processes at deep sea hydrothermal vents, *in* Rona, P. A., Bostrom, K., Laubier, L., and Smith, K. L., Jr., eds., Hydrothermal Processes at Seafloor Spreading Centers: New York, Plenum Press, p. 677-709.

JANNASCH, H. W., AND MOTTL, M. J., 1985, Geomicrobiology of Deep-Sea Vents: Science, v. 229, p. 717-725.

JARVIS, I., 1992, Sedimentology, geochemistry and origin of phosphatic chalks: The Upper Cretaceous deposits of NW Europe: Sedimentology, v. 39, p. 55-97.

JENKYNS, H. C., 1970, Fossil manganese nodules from the west Sicilian Jurassic: Eclogae Geologicae Helvetiae, v. 63, p. 741-774.

JENKYNS, H. C., 1971, The genesis of condensed sequences in the Tethyan Jurassic: Lethaia, v. 4, p. 327-352.

JENKYNS, H. C., 1978, Pelagic Environments, *in* Reading, H. G., ed., Sedimentary Environments and Facies, 1st. Edition: Oxford, Blackwell Scientific Publishers, p. 314-371.

JENKYNS, H. C., 1980, Cretaceous anoxic events: From continents to oceans: Journal of the Geological Society of London, v. 137, p. 171-188.

JENKYNS, H. C., 1986, Pelagic Environments, *in* Reading, H. G., ed., Sedimentary Environments and Facies, 2nd. Edition: Oxford, Blackwell Scientific Publishers, p. 343-397.

JONES, B., AND RENAUT, R. W., 1997, Formation of silica oncoids around geysers and hot springs at El Tatio, northern Chile: Sedimentology, v. 44, p. 287-304.

JONES, B., ROSEN, M. R., AND RENAUT, R. W., 1997, Silica-cemented beachrock from Lake Taupo, North Island, New Zealand: Journal of Sedimentary Research, v. 67A, p. 805-814.

KRAJEWSKI, K. P., 1981, Phosphate microstromatolites in the High-Tatric Albian Limestone in the Polish Tatra Mountains: Bulletin Académie Polonaise des Sciences, v. 39, p. 175-183.

KRAJEWSKI, K. P., 1983, Albian pelagic phosphate-rich macrooncoids from the Tatra Mts. (Poland), *in* Peryt, T. M., ed., Coated Grains: Berlin, Springer, p. 344-357.

KRAJEWSKI, K. P., 1984, Early diagenetic phosphate cements in the Albian condensed glauconitic limestones of the Tatra Mountains, Western Carpathians: Sedimentology, v. 31, p. 443-470.

KRAJEWSKI, K. P., VAN CAPELLEN, P., TRICHET, J., KUHN, O., LUCAS, J., MARTÍN-ALGARRA, A., PRÉVÔT, L., TEWARY, V. C., GASPAR, L., KNIGHT, R., AND LAMBOY, M., 1994, Biological controls and apatite formation in sedimentary environments: Eclogae Geologicae Helvetiae, v. 87/3, p. 701-745.

KRUMBEIN, W. E., ed., 1983a, Microbial Geochemistry: London, Blackwell Scientific Publishers, 330 p.

KRUMBEIN, W. E., 1983b, Stromatolites—the challenge of a term in space and time: Precambrian Research, v. 20, p. 493-531.

LAMBOY, M., 1990a, Microstructures of a phosphatic crust from the Peruvian continental margin: Phosphatized bacteria and associated phenomena: Oceanologica Acta, v. 13/4, p. 439-451.

LAMBOY, M., 1990b, Microbial mediation in phosphatogenesis: New data from the Cretaceous phosphatic chalks of northern France, *in* Notholt, A. J. G., and Jarvis, I., eds., Phosphorite Research and Development: Oxford, Geological Society Special Publication 52, p. 157-167.

LAMBOY, M., 1993, Phosphatization of calcium carbonate in phosphorites: Microstructure and importance: Sedimentology, v. 40, p. 53-62.

LEGEROS, R. Z., AND LEGEROS, J. P., 1984, Phosphate minerals in human tissues, *in* Nriagu, J. O., and Moore, P. B., eds., Phosphate Minerals: New York, Springer, p. 351-385.

LEWY, Z., 1990, Pebbly phosphate and granular phosphorite (Late Cretaceous, southern Israel) and their bearing on phosphatization processes, *in* Notholt, A. J. G., and Jarvis, I., eds., Phosphorite Research and Development: Oxford, Geological Society Special Publication 52, p. 169-178.

LÓPEZ-GALINDO, A., AND MARTÍN-ALGARRA, A., 1992, Palaeogeography and clay mineralogy of mid-Cretaceous flysches in the Gibraltar Arc area: Cretaceous Research, v. 13, p. 421-423.

LOWENSTAM, H. A., 1972, Phosphate hard tissue of marine invertebrates: Their nature and mechanical function and some fossil implications: Chemical Geology, v. 9, p. 153-166.

LOWENSTAM, H. A., 1981, Minerals formed by organisms: Science, v. 211, p. 1126-1131.

LUCAS, J., AND PRÉVÔT, L., 1984, Synthèse de l'apatite par voie bacterienne à partir de matière organique phosphatée et de divers carbonates de calcium dans les eaux douce et marine naturelles: Chemical Geology, v. 42, p. 101-118.

LUCAS, J., AND PRÉVÔT, L., 1985, The synthesis of apatite by bacterial activity: Mechanism, *in* Lucas, J., and Prévôt, L., eds., Phosphorites: Strasbourg, Mémoires Sciences Géologiques, v. 77, p. 243-251.

MANGINI, A., HALBACH, P., PUTEANUS, D., AND SEGL, M., 1987, Chemistry and growth history of Central Pacific Mn-crusts and their economic importance, *in* Teleki, P. G., Dobson, M. R., Moore, J. R., and von Stackelberg, U., eds., Marine Minerals, Advances in Research and Source Assessment, NATO ASI Series C, v. 194: Dordrecht, Reidel, p. 205-220.

MARTÍN-ALGARRA, A., 1987, Evolución geológica alpina del contacto entre las Zonas Internas y las Zonas Externas de la Cordillera Bética (sector central y occidental): Ph.D. Thesis, University of Granada, Publ. Departamento Estratigrafía Paleontología, 1171 p.

MARTÍN-ALGARRA A., RUIZ-ORTIZ, P. A., AND VERA, J. A., 1992, Factors controlling Cretaceous turbidite deposition in the Betic Cordillera: Revista Sociedad Geológica España, v. 5, p. 53-80.

MARTÍN-ALGARRA, A., AND SÁNCHEZ-NAVAS, A., 1995, Phosphate stromatolites from condensed cephalopod limestones, Upper Jurassic, Southern Spain: Sedimentology, v. 42, p. 893-919.

MARTÍN-ALGARRA, A., AND SÁNCHEZ-NAVAS, A., 1996, Stromatolitic phosphates: microstructural evidences of the existence of an amorphous precursor for phosphogenesis, *in* Lucas, J., and Prévôt-Lucas, L., eds., Deposystems of phosphorites and related authigenic minerals: processes, pathways and products. An International Symposium and Workshop—I.U.G.S.-I.G.C.P., Project 325: Correlation of palaeogeography with phosphorites and associated authigenic minerals. Strasbourg, France, Abstracts, p. 28.

MARTÍN-ALGARRA, A., AND VERA, J. A., 1994, Mesozoic pelagic phosphate stromatolites from the Penibetic (Betic Cordillera Southern Spain), *in* Bertrand-Sarfati, J., and Monty, C., Phanerozoic Stromatolites II: Dordrecht, Kluwer Academic Publishers, p. 345-391.

MASSARI, F., AND DIENI, I., 1983, Pelagic oncoids and ooids in the Middle-Upper Jurassic of Eastern Sardinia, *in* Peryt, T. M., ed., Coated Grains: Berlin, Springer, p. 367-376.

MAYNARD, J. B., 1983, Geochemistry of sedimentary ore deposits: New York, Springer, 305 p.

MCCONNELL, D., AND WARD, P., 1978, Nautical uroliths composed of phosphatic hydrogels: Science, v. 199, p. 208-209.

MONTY, C. L. V., 1973, Les nodules de manganèse sont des stromatolites océaniques: Comptes Rendus Académie Sciences Paris, Series D, v. 276, p. 3285-3288.

NATHAN, Y., AND SOUDRY, D., 1982, Authigenic silicate minerals in phosphorites of the Negev, southern Israel: Clay Minerals, v. 17, p. 249-254.

NIETO-ALBERT, L. M., 1997, La cuenca subbética mesozoica en el sector oriental de las Cordilleras Béticas: Ph.D. Thesis, University of Jaen, University of Murcia Press, 562 p.

NYLEN, M. U., EANES, E. D., AND TERMINE, J. D., 1972, Molecular and ultrastructural studies of non-crystalline calcium phosphates: Calcium Tissue Research, v. 9, p. 95-108.

O'BRIEN, G. W., HARRIS, J. R., MILNES, A. R., AND VEEH, H. H., 1981, Bacterial origin of East Australian continental margin phosphorites: Nature, v. 294, p. 442-444.

O'BRIEN, G. W., MILNES, A. R., VEEH, H. H., HEGGIE, D. T., RIGGS, S. R., CULLEN, D. J., MARSHALL, J. F., AND COOK, P. J., 1990, Sedimentation dynamics and redox iron-cycling: controlling factors for the apatite-glauconite association on the East Australian continental margin, *in* Notholt, A. J. G., and Jarvis, I., eds., Phosphorite Research and Development: Oxford, Geological Society Special Publication 52, p. 61-86.

ODIN, G. S., AND MATTER, A., 1981, De glauconiarum origine: Sedimentology, v. 28, p. 611-641.

POMONI-PAPAIOANNOU, F., 1994, Palaeoenvironmental reconstruction of a condensed hardground-type depositional sequence at the Cretaceous-Tertiary contact in the Parnassus-Ghiona zone, central Greece: Sedimentary Geology, v. 93, p. 7-24.

POMONI-PAPAIOANNOU, F., AND SOLAKIUS, N., 1991, Phosphatic hardgrounds and stromatolites from the limestone/shale boundary section at Prossilion (Maastrictian-Paleocene) in the Parnassus-Ghiona Zone, Central Greece: Palaeogeography, Palaeoclimatology, and Palaeoecology, v. 86, p. 243-254.

POSNER, A. S., BLUMENTHAL, N. C., AND BETTS, F., 1984, Chemistry and structure of precipitated hydroxyapatites, *in* Nriagu, J. O., and Moore, P. B., Phosphate Minerals: New York, Springer, p. 330-350.

POUCHOU, J. L., AND PICHOIR, F., 1985, "PAP" $\varphi(\rho z)$ procedure for improved quantitative microanalysis, *in* Armstrong, J. T., ed., Microbeam Analysis: San Francisco, San Francisco Press, p. 104.

PUGA, E., DÍAZ DE FEDERICO, A., AND DEMANT, A., 1995, The eclogitized pillows of the Betic ophiolitic Association: Relics of the Tethys Ocean floor incorporated in the Alpine chain after subduction: Terra Nova, v. 7, p. 31-43.

RESIG J. M., AND GLENN, C. R., 1997, Foraminifera encrusting phosphoritic hardgrounds of the Peruvian upwelling zone. Taxonomy, geochemistry and distribution: Journal of Foraminiferal Research, v. 27/2, p.133-150.

SÁNCHEZ-NAVAS, A., MARTÍN-ALGARRA, A., AND NIETO, F., 1996, Authigenic clays from phosphate stromatolites. Genetic implications, *in* Lucas, J., and Prévôt-Lucas, L., eds., Deposystems of phosphorites and related authigenic minerals: processes, pathways and products. An International Symposium and Workshop—I.U.G.S.-I.G.C.P., Project 325: Correlation of palaeogeography with phosphorites and associated authigenic minerals: Strasbourg, France, Abstracts, p. 31.

SÁNCHEZ-NAVAS, A., MARTÍN-ALGARRA, A., AND NIETO, F., 1998, Bacterially mediated authigenesis of clays in phosphate stromatolites: Sedimentology, v. 45, 519-533.

SANZ DE GALDEANO, C., MARTÍN-ALGARRA, A., VERA, J. A., AND RIVAS, P., 1993, Comments on "Structure and palaeogeography of the External Betic Cordillera, southern Spain" by C. L. Blankenship: Marine and Petroleum Geology, v. 10, p. 518-521.

SCHLANGER, S. O., AND JENKYNS, H. C., 1976, Cretaceous anoxic events: Causes and consequences: Geologie en Mijnbouw, v. 55, p. 179-184.

SISODIA, M. S., AND CHAUHAN, D. S., 1990, The influence of magnesium ions during the formation of stromatolitic phosphorites of Udaipur, Rajasthan, India, *in* Notholt, A. J. G., and Jarvis, I., eds., Phosphorite Research and Development: Oxford, Geological Society Special Publication 52, p. 313-320.

SOUDRY, D., 1987, Ultra-fine structures and genesis of the Campanian Negev high phosphorites (southern Israel): Sedimentology, v. 34, p. 641-660.

SOUDRY, D., 1992, Primary bedded phosphorites in the Campanian Mishash Formation, Negev, southern Israel: Sedimentary Geology, v. 80, p. 77-88.

SOUDRY D., 1994, Stromatolitic phosphorites in the Eocene of the Negev (Southern Israel), *in* Bertrand-Sarfati, J., and Monty, C., Phanerozoic Stromatolites II: Dordrecht, Kluwer Academic Publishers, p. 255-276.

SOUDRY, D., AND CHAMPETIER, Y., 1983, Microbial processes in the Negev phosphorites (southern Israel): Sedimentology, v. 30, p. 411-423.

SOUDRY, D., AND LEWY, Z., 1988, Microbially influenced formation of phosphate nodules and megafossil moulds (Negev, southern Israel): Palaeogeography, Palaeoclimatology, and Palaeoecology, v. 64, p.15-34.

STOW, D. A. V., READING, H. G., AND COLLINSON, J. D., 1996, Deep seas, *in* Reading, H. G., ed., Sedimentary Environments: Processes, Facies and Stratigraphy, 3rd. Edition: Oxford, Blackwell Scientific Publishers, p. 395-453.

STURENSSON, U., AND BAUERT, H., 1994, Origin and palaeogeographical distribution of the Viruan iron and phosphate ooids in Estonia: Evidence from mineralogical and chemical compositions: Sedimentary Geology, v. 93, p. 51-72.

TERMINE, J. D., PECKAUSKAS, R. A., AND POSNER, A. S., 1970, Calcium phosphate formation in vitro. II. Effects of Environment on Amorphous-Crystalline Transformation: Archives of Biochemistry, v. 140, p. 318-325.

TERMINE, J. D., AND POSNER, A. S., 1970, Calcium phosphate formation in vitro. I. Factors affecting initial phase separation: Archives of Biochemistry, v. 140, p. 307-317.

TUCKER, M. E., 1973, Ferromanganese nodules from the Devonian of the Montagne Noire (S France and West Germany): Geologische Rundschau, v. 62, p. 137-153.

TUCKER, M. E., 1974, Sedimentology of Paleozoic pelagic limestones: The Devonian Griotte (Southern France) and Cephalopodenkalk (Germany), *in* Hsü, K. J., and Jenkyns, H. C., eds., Pelagic Sediments: On Land and Under the Sea: Oxford, International Association of Sedimentologists Special Publication 1, p. 71-92.

URRUTIA, M. M., AND BEVERIDGE, T. J., 1993, Mechanisms of silicate binding to the bacterial cell wall in *Bacillus subtilis*: Journal of Bacteriology, v. 175, p. 1936-1945.

URRUTIA, M. M., AND BEVERIDGE, T. J., 1994, Formation of fine-grained metal and silicate precipitates on a bacterial surface (*Bacillus subtilis*): Chemical Geology, v. 116, p. 261-280.

URRUTIA, M. M., AND BEVERIDGE, T. J., 1995, Formation of short range ordered aluminosilicates in the presence of bacterial surface (*Bacillus subtilis*) and organic ligands: Geoderma, v. 65, p. 149-165.

USUI, A., 1979a, Nickel and copper accumulations as essential elements in 10Å manganite of deep-sea manganese nodules: Nature, v. 279, p. 411-413.

USUI, A., 1979b, Minerals, metal contents and mechanism of formation of manganese nodules from the central Pacific basin, *in* Bischoff, J. L., and Piper, D. Z., eds., Marine Geology and Oceanography of the Pacific Manganese nodule Province: New York, Plenum Press, p. 651-679.

USUI, A., 1983/1984, Regional variation of Manganese nodule facies on the Wake-Tahiti transect: Morphological, chemical and mineralogical study: Marine Geology, v. 54, p. 27-51.

USUI, A., 1984, Mineralogy and internal structure of manganese nodules of the GH80-5 area: Geological Survey of Japan, v. 20, p. 227-241.

USUI, A., 1985, Mineralogical and geochemical characteristics of manganese nodules and crusts collected during GDP-5 and 8 cruises in the western Pacific, *in* Shiki, T., ed., Geology of the Northern Philippine Sea: Tokay University Press, p. 179-201.

VAN CAPELLEN, P., AND BERNER, R. A., 1991, Fluorapatite crystal growth from modified seawater solutions: Geochimica et Cosmochimica Acta, v. 55, p. 1219-1234.

VASCONCELOS, C., AND MCKENZIE, J. A., 1997, Microbial mediation of modern dolomite precipitation and diagenesis under anoxic conditions (Lagoa Bermelha, Rio de Janeiro, Brazil): Journal of Sedimentary Research, v. 67A, p. 378-390.

VERA, J. A., AND MARTÍN-ALGARRA, A., 1994, Mesozoic stratigraphic breaks and pelagic stromatolites in the Betic Cordillera (Southern Spain), *in* Bertrand-Sarfati, J., and Monty, C., eds., Phanerozoic Stromatolites II: Dordrecht, Kluwer, p. 319-344.

VON STACKELBERG, U., 1987, Growth history and variability of manganese nodules of equatorial north Pacific, *in* Teleki, P. G., Dobson, M. R., Moore, J. R., and von Stackelberg, U., eds., Marine Minerals, Advances in Research and Source Assessment, NATO ASI Series C, v. 194: Dordrecht, Reidel, p. 189-204.

WALTER, M. R., ed., 1976, Stromatolites: Developments in Sedimentology, v. 20, Elsevier, 790 p.

WENDT, J., 1973, Cephalopod accumulations in the Middle Triassic Hallstatt-Limestone of Jugoslavia and Greece: Neues Jahrbuch Geologie Paläontologie Mh., p. 624-640.

WENDT, J., 1974, Encrusting organism in deep-sea manganese nodules, *in* Hsü, K. J., and Jenkyns, H. C., eds., Pelagic Sediments; On Land and Under the Sea: Oxford, International Association of Sedimentologists Special Publication 1, p. 437-447.

WENDT, J., AIGNER, T., AND NEUGEBAUER, J., 1984, Cephalopod limestone deposition on a shallow pelagic ridge: The Tafilalt Platform (upper Devonian, eastern Anti-Atlas, Morocco): Sedimentology, v. 31, p. 601-745.

ZANIN, Y. N., ZVEREV, K. V., AND SOLOTCHINA, E. P., 1990, Clay minerals and phosphorite genesis in the Upper Cretaceous of the northern Siberian Platform, *in* Notholt, A. J. G., and Jarvis, I., eds., Phosphorite Research and Development: Oxford, Geological Society Special Publication 52, p. 223-235.

INDEX

J

K

L

M

N

O

P

Q

R

S